Lecture Notes in Computer Science 14224

The series Lecture Notes in Computer Science (LNCS), including its subseries Lecture Notes in Artificial Intelligence (LNAI) and Lecture Notes in Bioinformatics (LNBI), has established itself as a medium for the publication of new developments in computer science and information technology research, teaching, and education.

LNCS enjoys close cooperation with the computer science R & D community, the series counts many renowned academics among its volume editors and paper authors, and collaborates with prestigious societies. Its mission is to serve this international community by providing an invaluable service, mainly focused on the publication of conference and workshop proceedings and postproceedings. LNCS commenced publication in 1973.

Hayit Greenspan · Anant Madabhushi ·
Parvin Mousavi · Septimiu Salcudean ·
James Duncan · Tanveer Syeda-Mahmood ·
Russell Taylor
Editors

Medical Image Computing and Computer Assisted Intervention – MICCAI 2023

26th International Conference
Vancouver, BC, Canada, October 8–12, 2023
Proceedings, Part V

Editors
Hayit Greenspan
Icahn School of Medicine, Mount Sinai,
NYC, NY, USA

Tel Aviv University
Tel Aviv, Israel

Parvin Mousavi
Queen's University
Kingston, ON, Canada

James Duncan
Yale University
New Haven, CT, USA

Russell Taylor
Johns Hopkins University
Baltimore, MD, USA

Anant Madabhushi
Emory University
Atlanta, GA, USA

Septimiu Salcudean
The University of British Columbia
Vancouver, BC, Canada

Tanveer Syeda-Mahmood
IBM Research
San Jose, CA, USA

ISSN 0302-9743 ISSN 1611-3349 (electronic)
Lecture Notes in Computer Science
ISBN 978-3-031-43903-2 ISBN 978-3-031-43904-9 (eBook)
https://doi.org/10.1007/978-3-031-43904-9

This Springer imprint is published by the registered company Springer Nature Switzerland AG
The registered company address is: Gewerbestrasse 11, 6330 Cham, Switzerland

Paper in this product is recyclable.

Preface

We are pleased to present the proceedings for the 26th International Conference on Medical Image Computing and Computer-Assisted Intervention (MICCAI). After several difficult years of virtual conferences, this edition was held in a mainly in-person format with a hybrid component at the Vancouver Convention Centre, in Vancouver, BC, Canada October 8–12, 2023. The conference featured 33 physical workshops, 15 online workshops, 15 tutorials, and 29 challenges held on October 8 and October 12. Co-located with the conference was also the 3rd Conference on Clinical Translation on Medical Image Computing and Computer-Assisted Intervention (CLINICCAI) on October 10.

MICCAI 2023 received the largest number of submissions so far, with an approximately 30% increase compared to 2022. We received 2365 full submissions of which 2250 were subjected to full review. To keep the acceptance ratios around 32% as in previous years, there was a corresponding increase in accepted papers leading to 730 papers accepted, with 68 orals and the remaining presented in poster form. These papers comprise ten volumes of Lecture Notes in Computer Science (LNCS) proceedings as follows:

- Part I, LNCS Volume 14220: Machine Learning with Limited Supervision and Machine Learning – Transfer Learning
- Part II, LNCS Volume 14221: Machine Learning – Learning Strategies and Machine Learning – Explainability, Bias, and Uncertainty I
- Part III, LNCS Volume 14222: Machine Learning – Explainability, Bias, and Uncertainty II and Image Segmentation I
- Part IV, LNCS Volume 14223: Image Segmentation II
- Part V, LNCS Volume 14224: Computer-Aided Diagnosis I
- Part VI, LNCS Volume 14225: Computer-Aided Diagnosis II and Computational Pathology
- Part VII, LNCS Volume 14226: Clinical Applications – Abdomen, Clinical Applications – Breast, Clinical Applications – Cardiac, Clinical Applications – Dermatology, Clinical Applications – Fetal Imaging, Clinical Applications – Lung, Clinical Applications – Musculoskeletal, Clinical Applications – Oncology, Clinical Applications – Ophthalmology, and Clinical Applications – Vascular
- Part VIII, LNCS Volume 14227: Clinical Applications – Neuroimaging and Microscopy
- Part IX, LNCS Volume 14228: Image-Guided Intervention, Surgical Planning, and Data Science
- Part X, LNCS Volume 14229: Image Reconstruction and Image Registration

The papers for the proceedings were selected after a rigorous double-blind peer-review process. The MICCAI 2023 Program Committee consisted of 133 area chairs and over 1600 reviewers, with representation from several countries across all major continents. It also maintained a gender balance with 31% of scientists who self-identified as women. With an increase in the number of area chairs and reviewers, the reviewer load on the experts was reduced this year, keeping to 16–18 papers per area chair and about 4–6 papers per reviewer. Based on the double-blinded reviews, area chairs' recommendations, and program chairs' global adjustments, 308 papers (14%) were provisionally accepted, 1196 papers (53%) were provisionally rejected, and 746 papers (33%) proceeded to the rebuttal stage. As in previous years, Microsoft's Conference Management Toolkit (CMT) was used for paper management and organizing the overall review process. Similarly, the Toronto paper matching system (TPMS) was employed to ensure knowledgeable experts were assigned to review appropriate papers. Area chairs and reviewers were selected following public calls to the community, and were vetted by the program chairs.

Among the new features this year was the emphasis on clinical translation, moving Medical Image Computing (MIC) and Computer-Assisted Interventions (CAI) research from theory to practice by featuring two clinical translational sessions reflecting the real-world impact of the field in the clinical workflows and clinical evaluations. For the first time, clinicians were appointed as Clinical Chairs to select papers for the clinical translational sessions. The philosophy behind the dedicated clinical translational sessions was to maintain the high scientific and technical standard of MICCAI papers in terms of methodology development, while at the same time showcasing the strong focus on clinical applications. This was an opportunity to expose the MICCAI community to the clinical challenges and for ideation of novel solutions to address these unmet needs. Consequently, during paper submission, in addition to MIC and CAI a new category of "Clinical Applications" was introduced for authors to self-declare.

MICCAI 2023 for the first time in its history also featured dual parallel tracks that allowed the conference to keep the same proportion of oral presentations as in previous years, despite the 30% increase in submitted and accepted papers.

We also introduced two new sessions this year focusing on young and emerging scientists through their Ph.D. thesis presentations, and another with experienced researchers commenting on the state of the field through a fireside chat format.

The organization of the final program by grouping the papers into topics and sessions was aided by the latest advancements in generative AI models. Specifically, Open AI's GPT-4 large language model was used to group the papers into initial topics which were then manually curated and organized. This resulted in fresh titles for sessions that are more reflective of the technical advancements of our field.

Although not reflected in the proceedings, the conference also benefited from keynote talks from experts in their respective fields including Turing Award winner Yann LeCun and leading experts Jocelyne Troccaz and Mihaela van der Schaar.

We extend our sincere gratitude to everyone who contributed to the success of MICCAI 2023 and the quality of its proceedings. In particular, we would like to express our profound thanks to the MICCAI Submission System Manager Kitty Wong whose meticulous support throughout the paper submission, review, program planning, and proceeding preparation process was invaluable. We are especially appreciative of the effort and dedication of our Satellite Events Chair, Bennett Landman, who tirelessly coordinated the organization of over 90 satellite events consisting of workshops, challenges and tutorials. Our workshop chairs Hongzhi Wang, Alistair Young, tutorial chairs Islem

Rekik, Guoyan Zheng, and challenge chairs, Lena Maier-Hein, Jayashree Kalpathy-Kramer, Alexander Seitel, worked hard to assemble a strong program for the satellite events. Special mention this year also goes to our first-time Clinical Chairs, Drs. Curtis Langlotz, Charles Kahn, and Masaru Ishii who helped us select papers for the clinical sessions and organized the clinical sessions.

We acknowledge the contributions of our Keynote Chairs, William Wells and Alejandro Frangi, who secured our keynote speakers. Our publication chairs, Kevin Zhou and Ron Summers, helped in our efforts to get the MICCAI papers indexed in PubMed. It was a challenging year for fundraising for the conference due to the recovery of the economy after the COVID pandemic. Despite this situation, our industrial sponsorship chairs, Mohammad Yaqub, Le Lu and Yanwu Xu, along with Dekon's Mehmet Eldegez, worked tirelessly to secure sponsors in innovative ways, for which we are grateful.

An active body of the MICCAI Student Board led by Camila Gonzalez and our 2023 student representatives Nathaniel Braman and Vaishnavi Subramanian helped put together student-run networking and social events including a novel Ph.D. thesis 3-minute madness event to spotlight new graduates for their careers. Similarly, Women in MICCAI chairs Xiaoxiao Li and Jayanthi Sivaswamy and RISE chairs, Islem Rekik, Pingkun Yan, and Andrea Lara further strengthened the quality of our technical program through their organized events. Local arrangements logistics including the recruiting of University of British Columbia students and invitation letters to attendees, was ably looked after by our local arrangement chairs Purang Abolmaesumi and Mehdi Moradi. They also helped coordinate the visits to the local sites in Vancouver both during the selection of the site and organization of our local activities during the conference. Our Young Investigator chairs Marius Linguraru, Archana Venkataraman, Antonio Porras Perez put forward the startup village and helped secure funding from NIH for early career scientist participation in the conference. Our communications chair, Ehsan Adeli, and Diana Cunningham were active in making the conference visible on social media platforms and circulating the newsletters. Niharika D'Souza was our cross-committee liaison providing note-taking support for all our meetings. We are grateful to all these organization committee members for their active contributions that made the conference successful.

We would like to thank the MICCAI society chair, Caroline Essert, and the MICCAI board for their approvals, support and feedback, which provided clarity on various aspects of running the conference. Behind the scenes, we acknowledge the contributions of the MICCAI secretariat personnel, Janette Wallace, and Johanne Langford, who kept a close eye on logistics and budgets, and Diana Cunningham and Anna Van Vliet for including our conference announcements in a timely manner in the MICCAI society newsletters. This year, when the existing virtual platform provider indicated that they would discontinue their service, a new virtual platform provider Conference Catalysts was chosen after due diligence by John Baxter. John also handled the setup and coordination with CMT and consultation with program chairs on features, for which we are very grateful. The physical organization of the conference at the site, budget financials, fund-raising, and the smooth running of events would not have been possible without our Professional Conference Organization team from Dekon Congress & Tourism led by Mehmet Eldegez. The model of having a PCO run the conference, which we used at

MICCAI, significantly reduces the work of general chairs for which we are particularly grateful.

Finally, we are especially grateful to all members of the Program Committee for their diligent work in the reviewer assignments and final paper selection, as well as the reviewers for their support during the entire process. Lastly, and most importantly, we thank all authors, co-authors, students/postdocs, and supervisors for submitting and presenting their high-quality work, which played a pivotal role in making MICCAI 2023 a resounding success.

With a successful MICCAI 2023, we now look forward to seeing you next year in Marrakesh, Morocco when MICCAI 2024 goes to the African continent for the first time.

October 2023

Tanveer Syeda-Mahmood
James Duncan
Russ Taylor
General Chairs

Hayit Greenspan
Anant Madabhushi
Parvin Mousavi
Septimiu Salcudean
Program Chairs

Organization

General Chairs

Tanveer Syeda-Mahmood	IBM Research, USA
James Duncan	Yale University, USA
Russ Taylor	Johns Hopkins University, USA

Program Committee Chairs

Hayit Greenspan	Tel-Aviv University, Israel and Icahn School of Medicine at Mount Sinai, USA
Anant Madabhushi	Emory University, USA
Parvin Mousavi	Queen's University, Canada
Septimiu Salcudean	University of British Columbia, Canada

Satellite Events Chair

Bennett Landman	Vanderbilt University, USA

Workshop Chairs

Hongzhi Wang	IBM Research, USA
Alistair Young	King's College, London, UK

Challenges Chairs

Jayashree Kalpathy-Kramer	Harvard University, USA
Alexander Seitel	German Cancer Research Center, Germany
Lena Maier-Hein	German Cancer Research Center, Germany

Tutorial Chairs

Islem Rekik	Imperial College London, UK
Guoyan Zheng	Shanghai Jiao Tong University, China

Clinical Chairs

Curtis Langlotz	Stanford University, USA
Charles Kahn	University of Pennsylvania, USA
Masaru Ishii	Johns Hopkins University, USA

Local Arrangements Chairs

Purang Abolmaesumi	University of British Columbia, Canada
Mehdi Moradi	McMaster University, Canada

Keynote Chairs

William Wells	Harvard University, USA
Alejandro Frangi	University of Manchester, UK

Industrial Sponsorship Chairs

Mohammad Yaqub	MBZ University of Artificial Intelligence, Abu Dhabi
Le Lu	DAMO Academy, Alibaba Group, USA
Yanwu Xu	Baidu, China

Communication Chair

Ehsan Adeli	Stanford University, USA

Publication Chairs

Ron Summers	National Institutes of Health, USA
Kevin Zhou	University of Science and Technology of China, China

Young Investigator Chairs

Marius Linguraru	Children's National Institute, USA
Archana Venkataraman	Boston University, USA
Antonio Porras	University of Colorado Anschutz Medical Campus, USA

Student Activities Chairs

Nathaniel Braman	Picture Health, USA
Vaishnavi Subramanian	EPFL, France

Women in MICCAI Chairs

Jayanthi Sivaswamy	IIIT, Hyderabad, India
Xiaoxiao Li	University of British Columbia, Canada

RISE Committee Chairs

Islem Rekik	Imperial College London, UK
Pingkun Yan	Rensselaer Polytechnic Institute, USA
Andrea Lara	Universidad Galileo, Guatemala

Submission Platform Manager

Kitty Wong	The MICCAI Society, Canada

Virtual Platform Manager

John Baxter	INSERM, Université de Rennes 1, France

Cross-Committee Liaison

Niharika D'Souza	IBM Research, USA

Program Committee

Sahar Ahmad	University of North Carolina at Chapel Hill, USA
Shadi Albarqouni	University of Bonn and Helmholtz Munich, Germany
Angelica Aviles-Rivero	University of Cambridge, UK
Shekoofeh Azizi	Google, Google Brain, USA
Ulas Bagci	Northwestern University, USA
Wenjia Bai	Imperial College London, UK
Sophia Bano	University College London, UK
Kayhan Batmanghelich	University of Pittsburgh and Boston University, USA
Ismail Ben Ayed	ETS Montreal, Canada
Katharina Breininger	Friedrich-Alexander-Universität Erlangen-Nürnberg, Germany
Weidong Cai	University of Sydney, Australia
Geng Chen	Northwestern Polytechnical University, China
Hao Chen	Hong Kong University of Science and Technology, China
Jun Cheng	Institute for Infocomm Research, A*STAR, Singapore
Li Cheng	University of Alberta, Canada
Albert C. S. Chung	University of Exeter, UK
Toby Collins	Ircad, France
Adrian Dalca	Massachusetts Institute of Technology and Harvard Medical School, USA
Jose Dolz	ETS Montreal, Canada
Qi Dou	Chinese University of Hong Kong, China
Nicha Dvornek	Yale University, USA
Shireen Elhabian	University of Utah, USA
Sandy Engelhardt	Heidelberg University Hospital, Germany
Ruogu Fang	University of Florida, USA

Reviewers

Alaa Eldin Abdelaal
John Abel
Kumar Abhishek
Shahira Abousamra
Mazdak Abulnaga
Burak Acar
Abdoljalil Addeh
Ehsan Adeli
Sukesh Adiga Vasudeva
Seyed-Ahmad Ahmadi
Euijoon Ahn
Faranak Akbarifar
Alireza Akhondi-asl
Saad Ullah Akram
Daniel Alexander
Hanan Alghamdi
Hassan Alhajj
Omar Al-Kadi
Max Allan
Andre Altmann
Pablo Alvarez
Charlems Alvarez-Jimenez
Jennifer Alvén
Lidia Al-Zogbi
Kimberly Amador
Tamaz Amiranashvili
Amine Amyar
Wangpeng An
Vincent Andrearczyk
Manon Ansart
Sameer Antani
Jacob Antunes
Michel Antunes
Guilherme Aresta
Mohammad Ali Armin
Kasra Arnavaz
Corey Arnold
Janan Arslan
Marius Arvinte
Muhammad Asad
John Ashburner
Md Ashikuzzaman
Shahab Aslani
Mehdi Astaraki
Angélica Atehortúa
Benjamin Aubert
Marc Aubreville
Paolo Avesani
Sana Ayromlou
Reza Azad
Mohammad Farid Azampour
Qinle Ba
Meritxell Bach Cuadra
Hyeon-Min Bae
Matheus Baffa
Cagla Bahadir
Fan Bai
Jun Bai
Long Bai
Pradeep Bajracharya
Shafa Balaram
Yaël Balbastre
Yutong Ban
Abhirup Banerjee
Soumyanil Banerjee
Sreya Banerjee
Shunxing Bao
Omri Bar
Adrian Barbu
Joao Barreto
Adrian Basarab
Berke Basaran
Michael Baumgartner
Siming Bayer
Roza Bayrak
Aicha BenTaieb
Guy Ben-Yosef
Sutanu Bera
Cosmin Bercea
Jorge Bernal
Jose Bernal
Gabriel Bernardino
Riddhish Bhalodia
Jignesh Bhatt
Indrani Bhattacharya
Binod Bhattarai
Lei Bi
Qi Bi
Cheng Bian
Gui-Bin Bian
Carlo Biffi
Alexander Bigalke
Benjamin Billot
Manuel Birlo
Ryoma Bise
Daniel Blezek
Stefano Blumberg
Sebastian Bodenstedt
Federico Bolelli
Bhushan Borotikar
Ilaria Boscolo Galazzo
Alexandre Bousse
Nicolas Boutry
Joseph Boyd
Behzad Bozorgtabar
Nadia Brancati
Clara Brémond Martin
Stéphanie Bricq
Christopher Bridge
Coleman Broaddus
Rupert Brooks
Tom Brosch
Mikael Brudfors
Ninon Burgos
Nikolay Burlutskiy
Michal Byra
Ryan Cabeen
Mariano Cabezas
Hongmin Cai
Tongan Cai
Zongyou Cai
Liane Canas
Bing Cao
Guogang Cao
Weiguo Cao
Xu Cao
Yankun Cao
Zhenjie Cao

Jaime Cardoso
M. Jorge Cardoso
Owen Carmichael
Jacob Carse
Adrià Casamitjana
Alessandro Casella
Angela Castillo
Kate Cevora
Krishna Chaitanya
Satrajit Chakrabarty
Yi Hao Chan
Shekhar Chandra
Ming-Ching Chang
Peng Chang
Qi Chang
Yuchou Chang
Hanqing Chao
Simon Chatelin
Soumick Chatterjee
Sudhanya Chatterjee
Muhammad Faizyab Ali Chaudhary
Antong Chen
Bingzhi Chen
Chen Chen
Cheng Chen
Chengkuan Chen
Eric Chen
Fang Chen
Haomin Chen
Jianan Chen
Jianxu Chen
Jiazhou Chen
Jie Chen
Jintai Chen
Jun Chen
Junxiang Chen
Junyu Chen
Li Chen
Liyun Chen
Nenglun Chen
Pingjun Chen
Pingyi Chen
Qi Chen
Qiang Chen
Runnan Chen
Shengcong Chen
Sihao Chen
Tingting Chen
Wenting Chen
Xi Chen
Xiang Chen
Xiaoran Chen
Xin Chen
Xiongchao Chen
Yanxi Chen
Yixiong Chen
Yixuan Chen
Yuanyuan Chen
Yuqian Chen
Zhaolin Chen
Zhen Chen
Zhenghao Chen
Zhennong Chen
Zhihao Chen
Zhineng Chen
Zhixiang Chen
Chang-Chieh Cheng
Jiale Cheng
Jianhong Cheng
Jun Cheng
Xuelian Cheng
Yupeng Cheng
Mark Chiew
Philip Chikontwe
Eleni Chiou
Jungchan Cho
Jang-Hwan Choi
Min-Kook Choi
Wookjin Choi
Jaegul Choo
Yu-Cheng Chou
Daan Christiaens
Argyrios Christodoulidis
Stergios Christodoulidis
Kai-Cheng Chuang
Hyungjin Chung
Matthew Clarkson
Michaël Clément
Dana Cobzas
Jaume Coll-Font
Olivier Colliot
Runmin Cong
Yulai Cong
Laura Connolly
William Consagra
Pierre-Henri Conze
Tim Cootes
Teresa Correia
Baris Coskunuzer
Alex Crimi
Can Cui
Hejie Cui
Hui Cui
Lei Cui
Wenhui Cui
Tolga Cukur
Tobias Czempiel
Javid Dadashkarimi
Haixing Dai
Tingting Dan
Kang Dang
Salman Ul Hassan Dar
Eleonora D'Arnese
Dhritiman Das
Neda Davoudi
Tareen Dawood
Sandro De Zanet
Farah Deeba
Charles Delahunt
Herve Delingette
Ugur Demir
Liang-Jian Deng
Ruining Deng
Wenlong Deng
Felix Denzinger
Adrien Depeursinge
Mohammad Mahdi Derakhshani
Hrishikesh Deshpande
Adrien Desjardins
Christian Desrosiers
Blake Dewey
Neel Dey
Rohan Dhamdhere

Maged Goubran
Sobhan Goudarzi
Mark Graham
Alejandro Granados
Mara Graziani
Thomas Grenier
Radu Grosu
Michal Grzeszczyk
Feng Gu
Pengfei Gu
Qiangqiang Gu
Ran Gu
Shi Gu
Wenhao Gu
Xianfeng Gu
Yiwen Gu
Zaiwang Gu
Hao Guan
Jayavardhana Gubbi
Houssem-Eddine Gueziri
Dazhou Guo
Hengtao Guo
Jixiang Guo
Jun Guo
Pengfei Guo
Wenzhangzhi Guo
Xiaoqing Guo
Xueqi Guo
Yi Guo
Vikash Gupta
Praveen Gurunath Bharathi
Prashnna Gyawali
Sung Min Ha
Mohamad Habes
Ilker Hacihaliloglu
Stathis Hadjidemetriou
Fatemeh Haghighi
Justin Haldar
Noura Hamze
Liang Han
Luyi Han
Seungjae Han
Tianyu Han
Zhongyi Han
Jonny Hancox
Lasse Hansen
Degan Hao
Huaying Hao
Jinkui Hao
Nazim Haouchine
Michael Hardisty
Stefan Harrer
Jeffry Hartanto
Charles Hatt
Huiguang He
Kelei He
Qi He
Shenghua He
Xinwei He
Stefan Heldmann
Nicholas Heller
Edward Henderson
Alessa Hering
Monica Hernandez
Kilian Hett
Amogh Hiremath
David Ho
Malte Hoffmann
Matthew Holden
Qingqi Hong
Yoonmi Hong
Mohammad Reza Hosseinzadeh Taher
William Hsu
Chuanfei Hu
Dan Hu
Kai Hu
Rongyao Hu
Shishuai Hu
Xiaoling Hu
Xinrong Hu
Yan Hu
Yang Hu
Chaoqin Huang
Junzhou Huang
Ling Huang
Luojie Huang
Qinwen Huang
Sharon Xiaolei Huang
Weijian Huang
Xiaoyang Huang
Yi-Jie Huang
Yongsong Huang
Yongxiang Huang
Yuhao Huang
Zhe Huang
Zhi-An Huang
Ziyi Huang
Arnaud Huaulmé
Henkjan Huisman
Alex Hung
Jiayu Huo
Andreas Husch
Mohammad Arafat Hussain
Sarfaraz Hussein
Jana Hutter
Khoi Huynh
Ilknur Icke
Kay Igwe
Abdullah Al Zubaer Imran
Muhammad Imran
Samra Irshad
Nahid Ul Islam
Koichi Ito
Hayato Itoh
Yuji Iwahori
Krithika Iyer
Mohammad Jafari
Srikrishna Jaganathan
Hassan Jahanandish
Andras Jakab
Amir Jamaludin
Amoon Jamzad
Ananya Jana
Se-In Jang
Pierre Jannin
Vincent Jaouen
Uditha Jarayathne
Ronnachai Jaroensri
Guillaume Jaume
Syed Ashar Javed
Rachid Jennane
Debesh Jha
Ge-Peng Ji

Luping Ji
Zexuan Ji
Zhanghexuan Ji
Haozhe Jia
Hongchao Jiang
Jue Jiang
Meirui Jiang
Tingting Jiang
Xiajun Jiang
Zekun Jiang
Zhifan Jiang
Ziyu Jiang
Jianbo Jiao
Zhicheng Jiao
Chen Jin
Dakai Jin
Qiangguo Jin
Qiuye Jin
Weina Jin
Baoyu Jing
Bin Jing
Yaqub Jonmohamadi
Lie Ju
Yohan Jun
Dinkar Juyal
Manjunath K N
Ali Kafaei Zad Tehrani
John Kalafut
Niveditha Kalavakonda
Megha Kalia
Anil Kamat
Qingbo Kang
Po-Yu Kao
Anuradha Kar
Neerav Karani
Turkay Kart
Satyananda Kashyap
Alexander Katzmann
Lisa Kausch
Maxime Kayser
Salome Kazeminia
Wenchi Ke
Youngwook Kee
Matthias Keicher
Erwan Kerrien
Afifa Khaled
Nadieh Khalili
Farzad Khalvati
Bidur Khanal
Bishesh Khanal
Pulkit Khandelwal
Maksim Kholiavchenko
Ron Kikinis
Benjamin Killeen
Daeseung Kim
Heejong Kim
Jaeil Kim
Jinhee Kim
Jinman Kim
Junsik Kim
Minkyung Kim
Namkug Kim
Sangwook Kim
Tae Soo Kim
Younghoon Kim
Young-Min Kim
Andrew King
Miranda Kirby
Gabriel Kiss
Andreas Kist
Yoshiro Kitamura
Stefan Klein
Tobias Klinder
Kazuma Kobayashi
Lisa Koch
Satoshi Kondo
Fanwei Kong
Tomasz Konopczynski
Ender Konukoglu
Aishik Konwer
Thijs Kooi
Ivica Kopriva
Avinash Kori
Kivanc Kose
Suraj Kothawade
Anna Kreshuk
AnithaPriya Krishnan
Florian Kromp
Frithjof Kruggel
Thomas Kuestner
Levin Kuhlmann
Abhay Kumar
Kuldeep Kumar
Sayantan Kumar
Manuela Kunz
Holger Kunze
Tahsin Kurc
Anvar Kurmukov
Yoshihiro Kuroda
Yusuke Kurose
Hyuksool Kwon
Aymen Laadhari
Jorma Laaksonen
Dmitrii Lachinov
Alain Lalande
Rodney LaLonde
Bennett Landman
Daniel Lang
Carole Lartizien
Shlomi Laufer
Max-Heinrich Laves
William Le
Loic Le Folgoc
Christian Ledig
Eung-Joo Lee
Ho Hin Lee
Hyekyoung Lee
John Lee
Kisuk Lee
Kyungsu Lee
Soochahn Lee
Woonghee Lee
Étienne Léger
Wen Hui Lei
Yiming Lei
George Leifman
Rogers Jeffrey Leo John
Juan Leon
Bo Li
Caizi Li
Chao Li
Chen Li
Cheng Li
Chenxin Li
Chnegyin Li

Dawei Li
Fuhai Li
Gang Li
Guang Li
Hao Li
Haofeng Li
Haojia Li
Heng Li
Hongming Li
Hongwei Li
Huiqi Li
Jian Li
Jieyu Li
Kang Li
Lin Li
Mengzhang Li
Ming Li
Qing Li
Quanzheng Li
Shaohua Li
Shulong Li
Tengfei Li
Weijian Li
Wen Li
Xiaomeng Li
Xingyu Li
Xinhui Li
Xuelu Li
Xueshen Li
Yamin Li
Yang Li
Yi Li
Yuemeng Li
Yunxiang Li
Zeju Li
Zhaoshuo Li
Zhe Li
Zhen Li
Zhenqiang Li
Zhiyuan Li
Zhjin Li
Zi Li
Hao Liang
Libin Liang
Peixian Liang
Yuan Liang
Yudong Liang
Haofu Liao
Hongen Liao
Wei Liao
Zehui Liao
Gilbert Lim
Hongxiang Lin
Li Lin
Manxi Lin
Mingquan Lin
Tiancheng Lin
Yi Lin
Zudi Lin
Claudia Lindner
Simone Lionetti
Chi Liu
Chuanbin Liu
Daochang Liu
Dongnan Liu
Feihong Liu
Fenglin Liu
Han Liu
Huiye Liu
Jiang Liu
Jie Liu
Jinduo Liu
Jing Liu
Jingya Liu
Jundong Liu
Lihao Liu
Mengting Liu
Mingyuan Liu
Peirong Liu
Peng Liu
Qin Liu
Quan Liu
Rui Liu
Shengfeng Liu
Shuangjun Liu
Sidong Liu
Siyuan Liu
Weide Liu
Xiao Liu
Xiaoyu Liu
Xingtong Liu
Xinwen Liu
Xinyang Liu
Xinyu Liu
Yan Liu
Yi Liu
Yihao Liu
Yikang Liu
Yilin Liu
Yilong Liu
Yiqiao Liu
Yong Liu
Yuhang Liu
Zelong Liu
Zhe Liu
Zhiyuan Liu
Zuozhu Liu
Lisette Lockhart
Andrea Loddo
Nicolas Loménie
Yonghao Long
Daniel Lopes
Ange Lou
Brian Lovell
Nicolas Loy Rodas
Charles Lu
Chun-Shien Lu
Donghuan Lu
Guangming Lu
Huanxiang Lu
Jingpei Lu
Yao Lu
Oeslle Lucena
Jie Luo
Luyang Luo
Ma Luo
Mingyuan Luo
Wenhan Luo
Xiangde Luo
Xinzhe Luo
Jinxin Lv
Tianxu Lv
Fei Lyu
Ilwoo Lyu
Mengye Lyu

Qing Lyu
Yanjun Lyu
Yuanyuan Lyu
Benteng Ma
Chunwei Ma
Hehuan Ma
Jun Ma
Junbo Ma
Wenao Ma
Yuhui Ma
Pedro Macias Gordaliza
Anant Madabhushi
Derek Magee
S. Sara Mahdavi
Andreas Maier
Klaus H. Maier-Hein
Sokratis Makrogiannis
Danial Maleki
Michail Mamalakis
Zhehua Mao
Jan Margeta
Brett Marinelli
Zdravko Marinov
Viktoria Markova
Carsten Marr
Yassine Marrakchi
Anne Martel
Martin Maška
Tejas Sudharshan Mathai
Petr Matula
Dimitrios Mavroeidis
Evangelos Mazomenos
Amarachi Mbakwe
Adam McCarthy
Stephen McKenna
Raghav Mehta
Xueyan Mei
Felix Meissen
Felix Meister
Afaque Memon
Mingyuan Meng
Qingjie Meng
Xiangzhu Meng
Yanda Meng
Zhu Meng
Martin Menten
Odyssée Merveille
Mikhail Milchenko
Leo Milecki
Fausto Milletari
Hyun-Seok Min
Zhe Min
Song Ming
Duy Minh Ho Nguyen
Deepak Mishra
Suraj Mishra
Virendra Mishra
Tadashi Miyamoto
Sara Moccia
Marc Modat
Omid Mohareri
Tony C. W. Mok
Javier Montoya
Rodrigo Moreno
Stefano Moriconi
Lia Morra
Ana Mota
Lei Mou
Dana Moukheiber
Lama Moukheiber
Daniel Moyer
Pritam Mukherjee
Anirban Mukhopadhyay
Henning Müller
Ana Murillo
Gowtham Krishnan Murugesan
Ahmed Naglah
Karthik Nandakumar
Venkatesh Narasimhamurthy
Raja Narayan
Dominik Narnhofer
Vishwesh Nath
Rodrigo Nava
Abdullah Nazib
Ahmed Nebli
Peter Neher
Amin Nejatbakhsh
Trong-Thuan Nguyen
Truong Nguyen
Dong Ni
Haomiao Ni
Xiuyan Ni
Hannes Nickisch
Weizhi Nie
Aditya Nigam
Lipeng Ning
Xia Ning
Kazuya Nishimura
Chuang Niu
Sijie Niu
Vincent Noblet
Narges Norouzi
Alexey Novikov
Jorge Novo
Gilberto Ochoa-Ruiz
Masahiro Oda
Benjamin Odry
Hugo Oliveira
Sara Oliveira
Arnau Oliver
Jimena Olveres
John Onofrey
Marcos Ortega
Mauricio Alberto Ortega-Ruíz
Yusuf Osmanlioglu
Chubin Ou
Cheng Ouyang
Jiahong Ouyang
Xi Ouyang
Cristina Oyarzun Laura
Utku Ozbulak
Ece Ozkan
Ege Özsoy
Batu Ozturkler
Harshith Padigela
Johannes Paetzold
José Blas Pagador Carrasco
Daniel Pak
Sourabh Palande
Chengwei Pan
Jiazhen Pan

Jin Pan
Yongsheng Pan
Egor Panfilov
Jiaxuan Pang
Joao Papa
Constantin Pape
Bartlomiej Papiez
Nripesh Parajuli
Hyunjin Park
Akash Parvatikar
Tiziano Passerini
Diego Patiño Cortés
Mayank Patwari
Angshuman Paul
Rasmus Paulsen
Yuchen Pei
Yuru Pei
Tao Peng
Wei Peng
Yige Peng
Yunsong Peng
Matteo Pennisi
Antonio Pepe
Oscar Perdomo
Sérgio Pereira
Jose-Antonio Pérez-Carrasco
Mehran Pesteie
Terry Peters
Eike Petersen
Jens Petersen
Micha Pfeiffer
Dzung Pham
Hieu Pham
Ashish Phophalia
Tomasz Pieciak
Antonio Pinheiro
Pramod Pisharady
Theodoros Pissas
Szymon Płotka
Kilian Pohl
Sebastian Pölsterl
Alison Pouch
Tim Prangemeier
Prateek Prasanna
Raphael Prevost
Juan Prieto
Federica Proietto Salanitri
Sergi Pujades
Elodie Puybareau
Talha Qaiser
Buyue Qian
Mengyun Qiao
Yuchuan Qiao
Zhi Qiao
Chenchen Qin
Fangbo Qin
Wenjian Qin
Yulei Qin
Jie Qiu
Jielin Qiu
Peijie Qiu
Shi Qiu
Wu Qiu
Liangqiong Qu
Linhao Qu
Quan Quan
Tran Minh Quan
Sandro Queirós
Prashanth R
Febrian Rachmadi
Daniel Racoceanu
Mehdi Rahim
Jagath Rajapakse
Kashif Rajpoot
Keerthi Ram
Dhanesh Ramachandram
João Ramalhinho
Xuming Ran
Aneesh Rangnekar
Hatem Rashwan
Keerthi Sravan Ravi
Daniele Ravì
Sadhana Ravikumar
Harish Raviprakash
Surreerat Reaungamornrat
Samuel Remedios
Mengwei Ren
Sucheng Ren
Elton Rexhepaj
Mauricio Reyes
Constantino Reyes-Aldasoro
Abel Reyes-Angulo
Hadrien Reynaud
Razieh Rezaei
Anne-Marie Rickmann
Laurent Risser
Dominik Rivoir
Emma Robinson
Robert Robinson
Jessica Rodgers
Ranga Rodrigo
Rafael Rodrigues
Robert Rohling
Margherita Rosnati
Łukasz Roszkowiak
Holger Roth
José Rouco
Dan Ruan
Jiacheng Ruan
Daniel Rueckert
Danny Ruijters
Kanghyun Ryu
Ario Sadafi
Numan Saeed
Monjoy Saha
Pramit Saha
Farhang Sahba
Pranjal Sahu
Simone Saitta
Md Sirajus Salekin
Abbas Samani
Pedro Sanchez
Luis Sanchez Giraldo
Yudi Sang
Gerard Sanroma-Guell
Rodrigo Santa Cruz
Alice Santilli
Rachana Sathish
Olivier Saut
Mattia Savardi
Nico Scherf
Alexander Schlaefer
Jerome Schmid

Adam Schmidt
Julia Schnabel
Lawrence Schobs
Julian Schön
Peter Schueffler
Andreas Schuh
Christina Schwarz-Gsaxner
Michaël Sdika
Suman Sedai
Lalithkumar Seenivasan
Matthias Seibold
Sourya Sengupta
Lama Seoud
Ana Sequeira
Sharmishtaa Seshamani
Ahmed Shaffie
Jay Shah
Keyur Shah
Ahmed Shahin
Mohammad Abuzar Shaikh
S. Shailja
Hongming Shan
Wei Shao
Mostafa Sharifzadeh
Anuja Sharma
Gregory Sharp
Hailan Shen
Li Shen
Linlin Shen
Mali Shen
Mingren Shen
Yiqing Shen
Zhengyang Shen
Jun Shi
Xiaoshuang Shi
Yiyu Shi
Yonggang Shi
Hoo-Chang Shin
Jitae Shin
Keewon Shin
Boris Shirokikh
Suzanne Shontz
Yucheng Shu
Hanna Siebert
Alberto Signoroni
Wilson Silva
Julio Silva-Rodríguez
Margarida Silveira
Walter Simson
Praveer Singh
Vivek Singh
Nitin Singhal
Elena Sizikova
Gregory Slabaugh
Dane Smith
Kevin Smith
Tiffany So
Rajath Soans
Roger Soberanis-Mukul
Hessam Sokooti
Jingwei Song
Weinan Song
Xinhang Song
Xinrui Song
Mazen Soufi
Georgia Sovatzidi
Bella Specktor Fadida
William Speier
Ziga Spiclin
Dominik Spinczyk
Jon Sporring
Pradeeba Sridar
Chetan L. Srinidhi
Abhishek Srivastava
Lawrence Staib
Marc Stamminger
Justin Strait
Hai Su
Ruisheng Su
Zhe Su
Vaishnavi Subramanian
Gérard Subsol
Carole Sudre
Dong Sui
Heung-Il Suk
Shipra Suman
He Sun
Hongfu Sun
Jian Sun
Li Sun
Liyan Sun
Shanlin Sun
Kyung Sung
Yannick Suter
Swapna T. R.
Amir Tahmasebi
Pablo Tahoces
Sirine Taleb
Bingyao Tan
Chaowei Tan
Wenjun Tan
Hao Tang
Siyi Tang
Xiaoying Tang
Yucheng Tang
Zihao Tang
Michael Tanzer
Austin Tapp
Elias Tappeiner
Mickael Tardy
Giacomo Tarroni
Athena Taymourtash
Kaveri Thakoor
Elina Thibeau-Sutre
Paul Thienphrapa
Sarina Thomas
Stephen Thompson
Karl Thurnhofer-Hemsi
Cristiana Tiago
Lin Tian
Lixia Tian
Yapeng Tian
Yu Tian
Yun Tian
Aleksei Tiulpin
Hamid Tizhoosh
Minh Nguyen Nhat To
Matthew Toews
Maryam Toloubidokhti
Minh Tran
Quoc-Huy Trinh
Jocelyne Troccaz
Roger Trullo

Chialing Tsai
Apostolia Tsirikoglou
Puxun Tu
Samyakh Tukra
Sudhakar Tummala
Georgios Tziritas
Vladimír Ulman
Tamas Ungi
Régis Vaillant
Jeya Maria Jose Valanarasu
Vanya Valindria
Juan Miguel Valverde
Fons van der Sommen
Maureen van Eijnatten
Tom van Sonsbeek
Gijs van Tulder
Yogatheesan Varatharajah
Madhurima Vardhan
Thomas Varsavsky
Hooman Vaseli
Serge Vasylechko
S. Swaroop Vedula
Sanketh Vedula
Gonzalo Vegas Sanchez-Ferrero
Matthew Velazquez
Archana Venkataraman
Sulaiman Vesal
Mitko Veta
Barbara Villarini
Athanasios Vlontzos
Wolf-Dieter Vogl
Ingmar Voigt
Sandrine Voros
Vibashan VS
Trinh Thi Le Vuong
An Wang
Bo Wang
Ce Wang
Changmiao Wang
Ching-Wei Wang
Dadong Wang
Dong Wang
Fakai Wang
Guotai Wang
Haifeng Wang
Haoran Wang
Hong Wang
Hongxiao Wang
Hongyu Wang
Jiacheng Wang
Jing Wang
Jue Wang
Kang Wang
Ke Wang
Lei Wang
Li Wang
Liansheng Wang
Lin Wang
Ling Wang
Linwei Wang
Manning Wang
Mingliang Wang
Puyang Wang
Qiuli Wang
Renzhen Wang
Ruixuan Wang
Shaoyu Wang
Sheng Wang
Shujun Wang
Shuo Wang
Shuqiang Wang
Tao Wang
Tianchen Wang
Tianyu Wang
Wenzhe Wang
Xi Wang
Xiangdong Wang
Xiaoqing Wang
Xiaosong Wang
Yan Wang
Yangang Wang
Yaping Wang
Yi Wang
Yirui Wang
Yixin Wang
Zeyi Wang
Zhao Wang
Zichen Wang
Ziqin Wang
Ziyi Wang
Zuhui Wang
Dong Wei
Donglai Wei
Hao Wei
Jia Wei
Leihao Wei
Ruofeng Wei
Shuwen Wei
Martin Weigert
Wolfgang Wein
Michael Wels
Cédric Wemmert
Thomas Wendler
Markus Wenzel
Rhydian Windsor
Adam Wittek
Marek Wodzinski
Ivo Wolf
Julia Wolleb
Ka-Chun Wong
Jonghye Woo
Chongruo Wu
Chunpeng Wu
Fuping Wu
Huaqian Wu
Ji Wu
Jiangjie Wu
Jiong Wu
Junde Wu
Linshan Wu
Qing Wu
Weiwen Wu
Wenjun Wu
Xiyin Wu
Yawen Wu
Ye Wu
Yicheng Wu
Yongfei Wu
Zhengwang Wu
Pengcheng Xi
Chao Xia
Siyu Xia
Wenjun Xia
Lei Xiang

Tiange Xiang
Deqiang Xiao
Li Xiao
Xiaojiao Xiao
Yiming Xiao
Zeyu Xiao
Hongtao Xie
Huidong Xie
Jianyang Xie
Long Xie
Weidi Xie
Fangxu Xing
Shuwei Xing
Xiaodan Xing
Xiaohan Xing
Haoyi Xiong
Yujian Xiong
Di Xu
Feng Xu
Haozheng Xu
Hongming Xu
Jiangchang Xu
Jiaqi Xu
Junshen Xu
Kele Xu
Lijian Xu
Min Xu
Moucheng Xu
Rui Xu
Xiaowei Xu
Xuanang Xu
Yanwu Xu
Yanyu Xu
Yongchao Xu
Yunqiu Xu
Zhe Xu
Zhoubing Xu
Ziyue Xu
Kai Xuan
Cheng Xue
Jie Xue
Tengfei Xue
Wufeng Xue
Yuan Xue
Zhong Xue
Ts Faridah Yahya
Chaochao Yan
Jiangpeng Yan
Ming Yan
Qingsen Yan
Xiangyi Yan
Yuguang Yan
Zengqiang Yan
Baoyao Yang
Carl Yang
Changchun Yang
Chen Yang
Feng Yang
Fengting Yang
Ge Yang
Guanyu Yang
Heran Yang
Huijuan Yang
Jiancheng Yang
Jiewen Yang
Peng Yang
Qi Yang
Qiushi Yang
Wei Yang
Xin Yang
Xuan Yang
Yan Yang
Yanwu Yang
Yifan Yang
Yingyu Yang
Zhicheng Yang
Zhijian Yang
Jiangchao Yao
Jiawen Yao
Lanhong Yao
Linlin Yao
Qingsong Yao
Tianyuan Yao
Xiaohui Yao
Zhao Yao
Dong Hye Ye
Menglong Ye
Yousef Yeganeh
Jirong Yi
Xin Yi
Chong Yin
Pengshuai Yin
Yi Yin
Zhaozheng Yin
Chunwei Ying
Youngjin Yoo
Jihun Yoon
Chenyu You
Hanchao Yu
Heng Yu
Jinhua Yu
Jinze Yu
Ke Yu
Qi Yu
Qian Yu
Thomas Yu
Weimin Yu
Yang Yu
Chenxi Yuan
Kun Yuan
Wu Yuan
Yixuan Yuan
Paul Yushkevich
Fatemeh Zabihollahy
Samira Zare
Ramy Zeineldin
Dong Zeng
Qi Zeng
Tianyi Zeng
Wei Zeng
Kilian Zepf
Kun Zhan
Bokai Zhang
Daoqiang Zhang
Dong Zhang
Fa Zhang
Hang Zhang
Hanxiao Zhang
Hao Zhang
Haopeng Zhang
Haoyue Zhang
Hongrun Zhang
Jiadong Zhang
Jiajin Zhang
Jianpeng Zhang

Jiawei Zhang
Jingqing Zhang
Jingyang Zhang
Jinwei Zhang
Jiong Zhang
Jiping Zhang
Ke Zhang
Lefei Zhang
Lei Zhang
Li Zhang
Lichi Zhang
Lu Zhang
Minghui Zhang
Molin Zhang
Ning Zhang
Rongzhao Zhang
Ruipeng Zhang
Ruisi Zhang
Shichuan Zhang
Shihao Zhang
Shuai Zhang
Tuo Zhang
Wei Zhang
Weihang Zhang
Wen Zhang
Wenhua Zhang
Wenqiang Zhang
Xiaodan Zhang
Xiaoran Zhang
Xin Zhang
Xukun Zhang
Xuzhe Zhang
Ya Zhang
Yanbo Zhang
Yanfu Zhang
Yao Zhang
Yi Zhang
Yifan Zhang
Yixiao Zhang
Yongqin Zhang
You Zhang
Youshan Zhang
Yu Zhang
Yubo Zhang
Yue Zhang
Yuhan Zhang
Yulun Zhang
Yundong Zhang
Yunlong Zhang
Yuyao Zhang
Zheng Zhang
Zhenxi Zhang
Ziqi Zhang
Can Zhao
Chongyue Zhao
Fenqiang Zhao
Gangming Zhao
He Zhao
Jianfeng Zhao
Jun Zhao
Li Zhao
Liang Zhao
Lin Zhao
Mengliu Zhao
Mingbo Zhao
Qingyu Zhao
Shang Zhao
Shijie Zhao
Tengda Zhao
Tianyi Zhao
Wei Zhao
Yidong Zhao
Yiyuan Zhao
Yu Zhao
Zhihe Zhao
Ziyuan Zhao
Haiyong Zheng
Hao Zheng
Jiannan Zheng
Kang Zheng
Meng Zheng
Sisi Zheng
Tianshu Zheng
Yalin Zheng
Yefeng Zheng
Yinqiang Zheng
Yushan Zheng
Aoxiao Zhong
Jia-Xing Zhong
Tao Zhong
Zichun Zhong
Hong-Yu Zhou
Houliang Zhou
Huiyu Zhou
Kang Zhou
Qin Zhou
Ran Zhou
S. Kevin Zhou
Tianfei Zhou
Wei Zhou
Xiao-Hu Zhou
Xiao-Yun Zhou
Yi Zhou
Youjia Zhou
Yukun Zhou
Zongwei Zhou
Chenglu Zhu
Dongxiao Zhu
Heqin Zhu
Jiayi Zhu
Meilu Zhu
Wei Zhu
Wenhui Zhu
Xiaofeng Zhu
Xin Zhu
Yonghua Zhu
Yongpei Zhu
Yuemin Zhu
Yan Zhuang
David Zimmerer
Yongshuo Zong
Ke Zou
Yukai Zou
Lianrui Zuo
Gerald Zwettler

Outstanding Area Chairs

Mingxia Liu	University of North Carolina at Chapel Hill, USA
Matthias Wilms	University of Calgary, Canada
Veronika Zimmer	Technical University Munich, Germany

Outstanding Reviewers

Kimberly Amador	University of Calgary, Canada
Angela Castillo	Universidad de los Andes, Colombia
Chen Chen	Imperial College London, UK
Laura Connolly	Queen's University, Canada
Pierre-Henri Conze	IMT Atlantique, France
Niharika D'Souza	IBM Research, USA
Michael Götz	University Hospital Ulm, Germany
Meirui Jiang	Chinese University of Hong Kong, China
Manuela Kunz	National Research Council Canada, Canada
Zdravko Marinov	Karlsruhe Institute of Technology, Germany
Sérgio Pereira	Lunit, South Korea
Lalithkumar Seenivasan	National University of Singapore, Singapore

Honorable Mentions (Reviewers)

Kumar Abhishek	Simon Fraser University, Canada
Guilherme Aresta	Medical University of Vienna, Austria
Shahab Aslani	University College London, UK
Marc Aubreville	Technische Hochschule Ingolstadt, Germany
Yaël Balbastre	Massachusetts General Hospital, USA
Omri Bar	Theator, Israel
Aicha Ben Taieb	Simon Fraser University, Canada
Cosmin Bercea	Technical University Munich and Helmholtz AI and Helmholtz Center Munich, Germany
Benjamin Billot	Massachusetts Institute of Technology, USA
Michal Byra	RIKEN Center for Brain Science, Japan
Mariano Cabezas	University of Sydney, Australia
Alessandro Casella	Italian Institute of Technology and Politecnico di Milano, Italy
Junyu Chen	Johns Hopkins University, USA
Argyrios Christodoulidis	Pfizer, Greece
Olivier Colliot	CNRS, France

Lei Cui	Northwest University, China
Neel Dey	Massachusetts Institute of Technology, USA
Alessio Fagioli	Sapienza University, Italy
Yannik Glaser	University of Hawaii at Manoa, USA
Haifan Gong	Chinese University of Hong Kong, Shenzhen, China
Ricardo Gonzales	University of Oxford, UK
Sobhan Goudarzi	Sunnybrook Research Institute, Canada
Michal Grzeszczyk	Sano Centre for Computational Medicine, Poland
Fatemeh Haghighi	Arizona State University, USA
Edward Henderson	University of Manchester, UK
Qingqi Hong	Xiamen University, China
Mohammad R. H. Taher	Arizona State University, USA
Henkjan Huisman	Radboud University Medical Center, the Netherlands
Ronnachai Jaroensri	Google, USA
Qiangguo Jin	Northwestern Polytechnical University, China
Neerav Karani	Massachusetts Institute of Technology, USA
Benjamin Killeen	Johns Hopkins University, USA
Daniel Lang	Helmholtz Center Munich, Germany
Max-Heinrich Laves	Philips Research and ImFusion GmbH, Germany
Gilbert Lim	SingHealth, Singapore
Mingquan Lin	Weill Cornell Medicine, USA
Charles Lu	Massachusetts Institute of Technology, USA
Yuhui Ma	Chinese Academy of Sciences, China
Tejas Sudharshan Mathai	National Institutes of Health, USA
Felix Meissen	Technische Universität München, Germany
Mingyuan Meng	University of Sydney, Australia
Leo Milecki	CentraleSupelec, France
Marc Modat	King's College London, UK
Tiziano Passerini	Siemens Healthineers, USA
Tomasz Pieciak	Universidad de Valladolid, Spain
Daniel Rueckert	Imperial College London, UK
Julio Silva-Rodríguez	ETS Montreal, Canada
Bingyao Tan	Nanyang Technological University, Singapore
Elias Tappeiner	UMIT - Private University for Health Sciences, Medical Informatics and Technology, Austria
Jocelyne Troccaz	TIMC Lab, Grenoble Alpes University-CNRS, France
Chialing Tsai	Queens College, City University New York, USA
Juan Miguel Valverde	University of Eastern Finland, Finland
Sulaiman Vesal	Stanford University, USA

Wolf-Dieter Vogl	RetInSight GmbH, Austria
Vibashan VS	Johns Hopkins University, USA
Lin Wang	Harbin Engineering University, China
Yan Wang	Sichuan University, China
Rhydian Windsor	University of Oxford, UK
Ivo Wolf	University of Applied Sciences Mannheim, Germany
Linshan Wu	Hunan University, China
Xin Yang	Chinese University of Hong Kong, China

Contents – Part V

Computer-Aided Diagnosis I

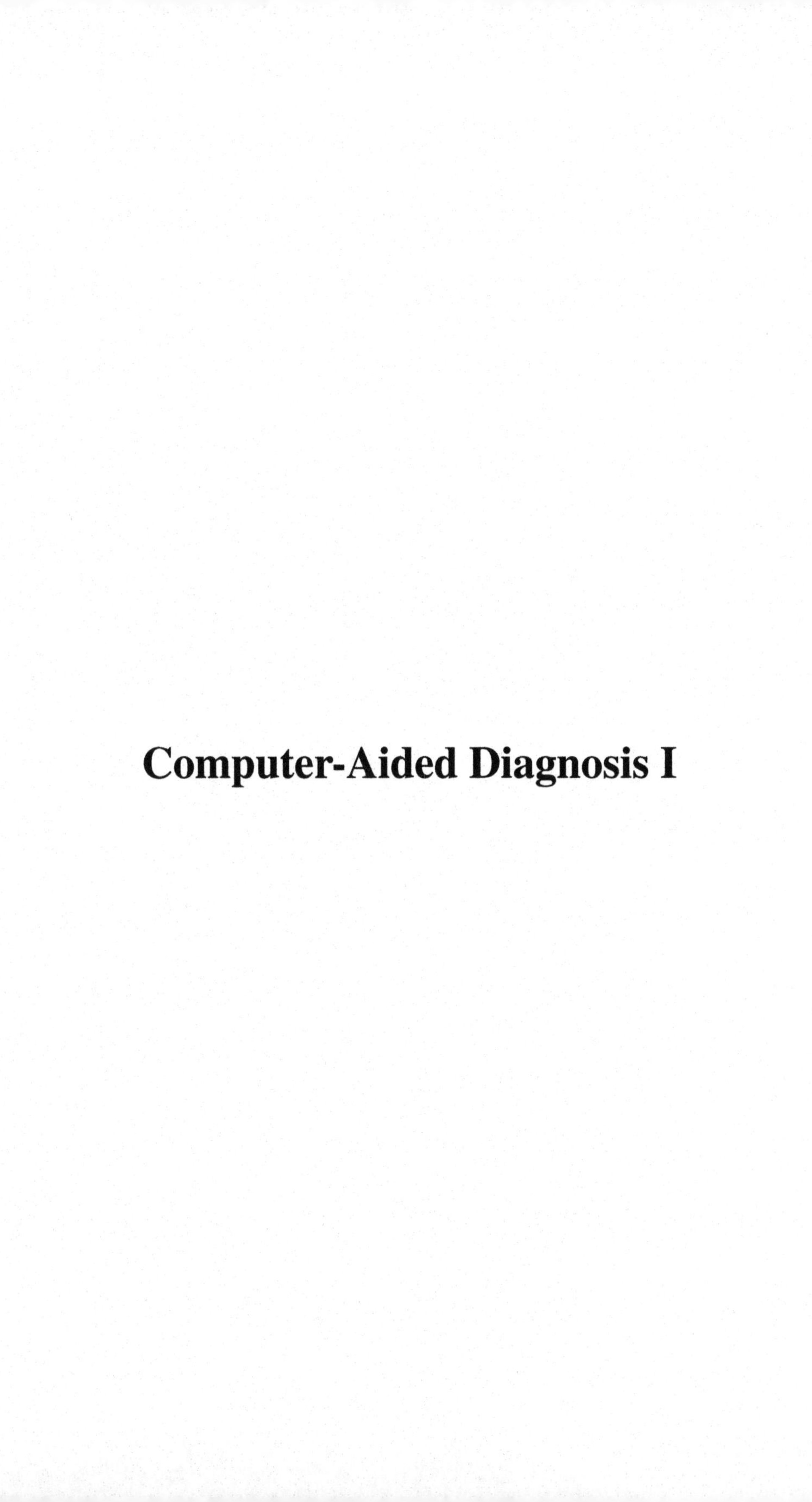

Computer-Aided Diagnosis I

Automatic Bleeding Risk Rating System of Gastric Varices

Yicheng Jiang[1], Luyue Shi[1], Wei Qi[2], Lei Chen[2], Guanbin Li[5], Xiaoguang Han[3,4], Xiang Wan[1], and Siqi Liu[1](✉)

[1] Shenzhen Research Institute of Big Data, Shenzhen, China
siqiliu@sribd.cn

[2] Department of Gastroenterology, The Second Hospital of Hebei Medical University, Hebei Key Laboratory of Gastroenterology, Hebei Institute of Gastroenterology, Hebei Clinical Research Center for Digestive Diseases, Shijiazhuang, China

[3] FNii, The Chinese University of Hong Kong, Shenzhen, Shenzhen, China

[4] School of Science and Engineering, The Chinese University of Hong Kong, Shenzhen, Shenzhen, China

[5] School of Computer Science and Engineering, Sun Yat-sen University, Guangzhou, China

Abstract. An automated bleeding risk rating system of gastric varices (GV) aims to predict the bleeding risk and severity of GV, in order to assist endoscopists in diagnosis and decrease the mortality rate of patients with liver cirrhosis and portal hypertension. However, since the lack of commonly accepted quantification standards, the risk rating highly relies on the endoscopists' experience and may vary a lot in different application scenarios. In this work, we aim to build an automatic GV bleeding risk rating method that can learn from experienced endoscopists and provide stable and accurate predictions. Due to the complexity of GV structures with large intra-class variation and small inter-class variations, we found that existing models perform poorly on this task and tend to lose focus on the important varices regions. To solve this issue, we constructively introduce the segmentation of GV into the classification framework and propose the region-constraint module and cross-region attention module for better feature localization and to learn the correlation of context information. We also collect a GV bleeding risks rating dataset (**GVbleed**) with 1678 gastroscopy images from 411 patients that are jointly annotated in three levels of risks by senior clinical endoscopists. The experiments on our collected dataset show that our method can improve the rating accuracy by nearly 5% compared to the baseline. Codes and dataset will be available at https://github.com/LuyueShi/gastric-varices.

Y. Jiang, L. Shi and W. Qi—Contributed equally to this work.

Supplementary Information The online version contains supplementary material available at https://doi.org/10.1007/978-3-031-43904-9_1.

H. Greenspan et al. (Eds.): MICCAI 2023, LNCS 14224, pp. 3–12, 2023.
https://doi.org/10.1007/978-3-031-43904-9_1

Keywords: Gastric Varices · Bleeding Risk Rating · Cross-region Attention

1 Introduction

Esophagogastric varices are one of the common manifestations in patients with liver cirrhosis and portal hypertension and occur in about 50 percent of patients with liver cirrhosis [3,6]. The occurrence of esophagogastric variceal bleeding is the most serious adverse event in patients with cirrhosis, with a 6-week acute bleeding mortality rate as high as 15%–20% percent [14]. It is crucial to identify high-risk patients and offer prophylactic treatment at the appropriate time. Regular endoscopy examinations have been proven an effective clinical approach to promptly detect esophagogastric varices with a high risk of bleeding [7]. Different from the grading of esophageal varices (EV) that is relatively complete [1], the bleeding risk grading of gastric varices (GV) involves complex variables including the diameter, shapes, colors, and locations. Several rating systems have been proposed to describe GV based on the anatomical area. Sarin et al. [16] described and divided GV into 2 groups according to their locations and extensions. Hashizume et al. [10] published a more detailed examination describing the form, location, and color. Although the existing rating systems tried to identify the risk from different perspectives, they still lack clear quantification standard and heavily rely on the endoscopists' subjective judgment. This may cause inconsistency or even misdiagnosis due to the variant experience of endoscopists in different hospitals. Therefore, we aim to build an automatic GV bleeding risk rating method that can learn a stable and robust standard from multiple experienced endoscopists.

Recent works have proven the effectiveness and superiority of deep learning (DL) technologies in handling esophagogastroduodenoscopy (EGD) tasks, such as the detection of gastric cancer and neoplasia [4]. It is even demonstrated that AI can detect neoplasia in Barrett's esophagus at a higher accuracy than endoscopists [8]. Intuitively we may regard the GV bleeding risk rating as an image classification task and apply typical classification architectures (e.g., ResNet [12]) or state-of-the-art gastric lesion classification methods to it. However, they may raise poor performance due to the large intra-class variation between GV with the same bleeding risk and small inter-class variation between GV and normal tissue or GV with different bleeding risks. First, the GV area may look like regular stomach rugae as it is caused by the blood vessels bulging and crumpling up the stomach (see Fig. 1). Also, since the GV images are taken from different distances and angles, the number of pixels of the GV area may not reflect its actual size. Consequently, the model may fail to focus on the important GV areas for prediction as shown in Fig. 3. To encourage the model to learn more robust representations, we constructively introduce segmentation into the classification framework. With the segmentation information, we further propose a region-constraint module (RCM) and a cross-region attention module (CRAM) for better feature localization and utilization. Specifically, in RCM, we utilize

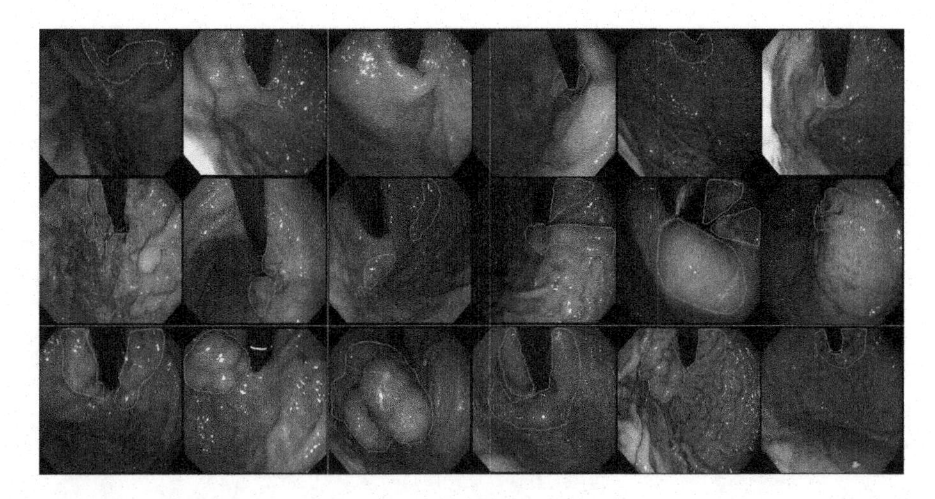

Fig. 1. Visualization of our GVbleed dataset. The three rows from up to down show the typical samples of gastric varices with mild, moderate, and severe bleeding risk, respectively. The green line represents the contour of the target gastric varices regions, from which we can sense the complex patterns involving the diameter, shapes, and colors.

the segmentation results to constrain the CAM heatmaps of the feature maps extracted by the classification backbone, avoiding the model making predictions based on incorrect areas. In CRAM, the varices features are extracted using the segmentation results and combined with an attention mechanism to learn the intra-class correlation and cross-region correlation between the target area and the context.

To learn from experienced endoscopists, GV datasets with bleeding risks annotation is needed. While most works and public datasets focus on colonoscopy [13,15] and esophagus [5,9], with a lack of study on gastroscopy images. In the public dataset of EndoCV challenge [2], the majority are colonoscopies while only few are gastroscopy images. In this work, we collect a GV bleeding risks rating dataset (**GVbleed**) that contains 1678 gastroscopy images from 411 patients with different levels of GV bleeding risks. Three senior clinical endoscopists are invited to grade the bleeding risk of the retrospective data in three levels and annotated the corresponding segmentation masks of GV areas.

In sum, the contributions of this paper are: 1) a novel GV bleeding risk rating framework that constructively introduces segmentation to enhance the robustness of representation learning; 2) a region-constraint module for better feature localization and a cross-region attention module to learn the correlation of target GV with its context; 3) a GV bleeding risk rating dataset (**GVbleed**) with high-quality annotation from multiple experienced endoscopists. Baseline methods have been evaluated on the newly collected GVbleed dataset. Experimental results demonstrate the effectiveness of our proposed framework and modules, where we improve the accuracy by nearly 5% compared to the baseline model.

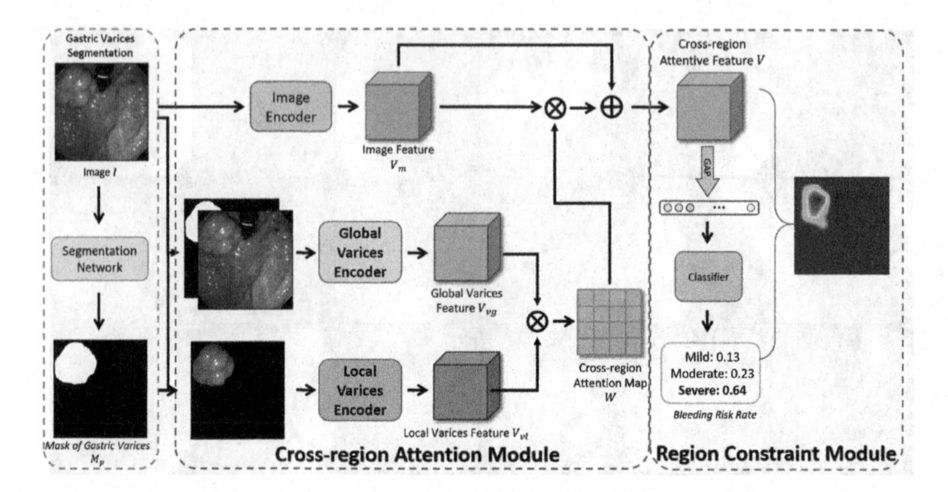

Fig. 2. Pipeline of the proposed automatic bleeding risk rating framework. The framework consists of a segmentation module, a cross-region attention module, and a region constraint module.

2 Methodology

The architecture of the proposed framework is depicted in Fig. 2, which consists of a segmentation module (SM), a region constraint module (RCM), and a cross-region attention module (CRAM). Given a gastroscopy image, the SM is first applied to generate the varices mask of the image. Then, the image together with the mask are fed into the CRAM to extract the cross-region attentive feature map, and a class activation map (CAM) is calculated to represent the concentrated regions through RCM. Finally, a simple classifier is used to predict the bleeding risk using the extracted feature map.

2.1 Segmentation Module

Due to the large intra-class variation between GV with the same bleeding risk and small inter-class variation between GV and normal tissue or GV with different bleeding risks, existing classification models exhibit poor perform and tend to lose focus on the GV areas. To solve this issue, we first embed a segmentation network into the classification framework. The predict the varices mask is then used to assist the GV feature to obtain the final bleeding risk rate. Specifically, we use SwinUNet [11] as the segmentation network, considering its great performance, and calculate the DiceLoss between the segmentaion result M_p and ground truth mask of vaices region M_{gt} for optimizing the network:

$$l_{se} = 1 - \frac{2\Sigma M_p * M_{gt}}{\Sigma M_p^2 + \Sigma M_{gt}^2 + \epsilon}, \tag{1}$$

where ϵ is a smooth constant equals to 10^{-5}.

A straightforward strategy to utilize the segmentation mask is directly using it as an input of the classification model, such as concatenating the image with the mask as the input. Although such strategy can improve the classification performance, it may still lose focus in some hard cases where the GV area can hardly be distinguished. To further regularize the attention and fully utilize the context information around the GV area, on top of the segmentation framework we proposed the cross-region attention module and the region-constraint module.

2.2 Cross-Region Attention Module

Inspired by the self-attention mechanism [17,18], we propose a cross-region attention module (CRAM) to learn the correlation of context information. The CRAM consists of an image encoder f_{im}, a varices local encoder f_{vl} and a varices global encoder f_{ve}. Given the image I and the predicted varices mask M_p, a feature extraction step is first performed to generate the image feature V_m, the local varices feature V_{vl} and global varices feature V_{vg}:

$$V_m = f_{im}(I), \quad V_{vl} = f_{vl}(I * M_p), \quad V_{vg} = f_{vg}(concat[I, M_p]), \tag{2}$$

Then, through similarity measuring, we can compute the attention with

$$A = (V_{vl})^T V_{vg}, \quad W_{ij} = \frac{exp(A_{ij})}{\Sigma_p(exp(A_{pj}))}, \tag{3}$$

which composes of two correlations: self-attention over varices regions and cross-region attention between varices and background regions. Finally, the output feature is calculated as:

$$V = \gamma V_m W + V_m, \tag{4}$$

where γ is a learnable parameter. Then the cross-region attentive feature V is fed into a classifier to predict the bleeding risk.

2.3 Region Constraint Module

To improve the focus ability of the model, we propose the region constraint module (RCM) to add a constraint on the class activation map (CAM) of the classification model. Specifically, we use the feature map after the last convolutional layer to calculate the CAM [19], which computes the weighted sum of feature maps from the convolutional layer using the weights of the FC layer. After getting the CAM, we regularize CAM by calculating the dice loss between the CAM and ground truth mask of varices region l_{co}.

2.4 Network Training

In our framework, we use the cross entropy loss as the classification loss:

$$l_{cl} = -\sum_{c=1}^{C} log \frac{exp(p_c)}{\Sigma_{i=1}^{C} exp(p_i)} y_c \tag{5}$$

Table 1. Distribution of the three-level GV bleeding risks in the GVBleed dataset.

	Mild	Moderate	Severe	Total
Train	398	484	455	1337
Test	94	145	102	341
Total	462	629	557	1678

where p is the prediction of the classifier and y is the ground-truth label. And the total loss of our framework can be summarized as:

$$L_{total} = \frac{1}{N} \sum(\omega_s l_{se} + \omega_{co} l_{co} + \omega_{cl} l_{cl}), \tag{6}$$

where N is the total number of samples, ω_s, ω_{co} and ω_{cl} are weights of the three losses, respectively.

The training process of the proposed network consists of three steps: 1) The segmentation network is trained first; 2) The ground-truth segmentation masks and images are used as the inputs of the CRAM, the classification network, including CRAM and RCM, are jointly trained; 3) The whole framework is jointly fine-tuned.

3 GVBleed Dataset

Data Collection and Annotation. The GVBleed dataset contains 1678 endoscopic images with gastric varices from 527 cases. All of these cases are collected from 411 patients in a Grade-III Class-A hospital during the period from 2017 to 2022. In the current version, images from patients with ages elder than 18 are retained[1]. The images are selected from the raw endoscopic videos and frames. To maximize the variations, non-consecutive frames with larger angle differences are selected. To ensure the quality of our dataset, senior endoscopists are invited to remove duplicates, blurs, active bleeding, chromoendoscopy, and NBI pictures.

Criterion of GV Bleeding Risk Level Rating. Based on the clinical experience in practice, the GV bleeding risks in our dataset are rated into three levels, i.e., mild, moderate, and severe. The detailed rating standard is as follows: 1) Mild: low risk of bleeding, and regular follow-up is sufficient (usually with a diameter less than or equal to 5 mm). 2) Moderate: moderate risk of bleeding, and endoscopic treatment is necessary, with relatively low endoscopic treatment difficulty (usually with a diameter between 5 mm and 10 mm). 3) Severe: high risk of bleeding and endoscopic treatment is necessary, with high endoscopic treatment difficulty. The varices are thicker (usually with a diameter greater than 10 mm) or less than 10mm but with positive red signs. Note that the diameter is only one reference for the final risk rating since the GV is with

[1] Please refer to the supplementary material for more detailed information about our dataset.

Table 2. Quantitative results of the proposed method and modules.

Models	Accuracy(%)	F1-Score(%)		
		Mild	Moderate	Severe
ResNet18	65.40	60.17	60.38	**73.25**
ResNet18 + SegMask (Concat)	66.86	69.30	66.46	65.98
ResNet18 + SegMask + Attention	67.16	69.21	65.50	66.62
ResNet18 + SegMask + Attention + Constraint	**70.97**	**73.33**	**68.32**	72.88

various 3D shapes and locations. The other facts are more subjectively evaluated based on the experience of endoscopists. To ensure the accuracy of our annotation, three senior endoscopists with more than 10 years of clinical experience are invited to jointly label each sample in our dataset. If three endoscopists have inconsistent ratings for a sample, the final decision is judged by voting. A sample is selected and labeled with a specific bleeding risk level only when two or more endoscopists reach a consensus on it.

The GVBleed dataset is partitioned into training and testing sets for evaluation, where the training set contains 1337 images and the testing set has 341 images. The detailed statistics of the three levels of GV bleeding risk in each set are shown in Table 1. The dataset is planned to be released in the future.

4 Experiments

4.1 Implementation Details

In experiments, the weights ω_s, ω_{co}, and ω_{cl} of the segmentation loss, region constraint loss, and classification loss are set to 0.2, 1, and 1, respectively. The details of the three-step training are as follows: 1) **Segmentation module**: We trained the segmentation network for 600 epochs, using Adam as the optimizer, and the learning rate is initialized as 1e−3 and drops to 1e−4 after 300 epochs. 2) **Cross-region attention module and region constraint module**: We used the ground-truth varices masks and images as the inputs of the CRAM, and jointly trained the CRAM and RCM for 100 epochs. Adam is used as the optimizer, the learning rate is set to 1e−3; 3) **Jointly fine-tuning**: The whole framework is jointly fine-tuned for 100 epochs with Adam as optimizer and the learning rate set to 1e−3. In addition, common data augmentation techniques such as rotation and flipping were adopted here.

4.2 Results Analysis

Table 2 reports the quantitative results of different models and Fig. 3 shows the CAM visualizations. We tested several baseline models, including both simple CNN models and state-of-the-art transformer-based models. However, the transformer-based models achieves much worse performances since they always require more training data, which is not available in our task. Thus, we selected the simple CNN models as baselines since they achieve better performances. As shown in the figure, the baseline model cannot focus on the varices regions due to the complexity of GV structures with large intra-class variations and small inter-class variations. By introducing the segmentation of GV into the framework, concatenating the image with its segmentation mask as the inputs of the classifier can improve the classification accuracy by 1.2%. And the focus ability of the classifier is stronger than the baseline model. With the help of CRAM, the performance of the model can be further improved. Although the model can extract more important context information at the varices regions, the performance improvement is not very large since the focus ability is not the best and the model may still make predictions based on the incorrect regions for some hard images. By adding the RCM to the CRAM, the focus ability of the model can be further improved, and thus the model has a significant improvement in performance by 5% compared to the baseline model, this proves the effectiveness of our proposed modules. Note that, the baseline model tends to predict the images as severe, thus the f1-score of severe is high but the f1-scores of mild and moderate are significantly lower than other models. More quantitative and visualization results are shown in supplementary material. In addition, given the input image with resolution 512×512, the parameters and computational cost of our framework are 40.2M, and 52.4G MACs, and 29 ms inference time for a single image on GPU RTX2080. For comparison, a single ResNet152 model has 60.19M parameters with 60.62 G MACs.

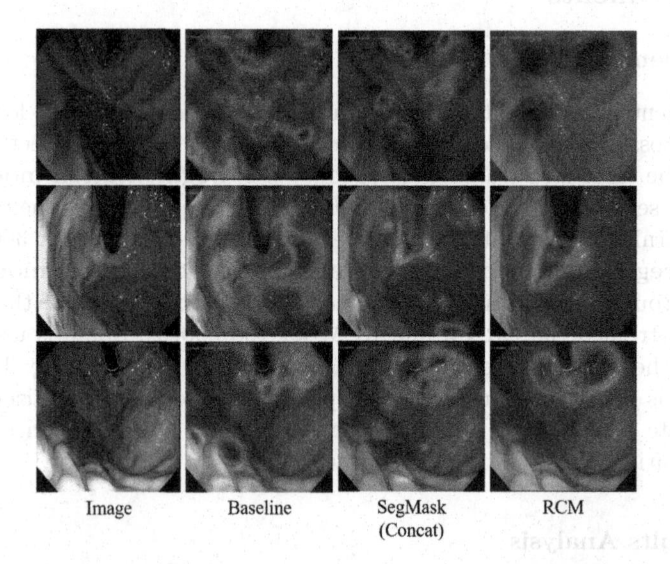

Fig. 3. Visualization of CAM. Green lines represent the contours of varices regions. The blue areas represent the primary areas of concern. With the proposed SegMask and RCM module, the network tends to focus on the important gastric varices areas as endoscopists do.

5 Conclusions

In this paper, we propose a novel bleeding risk rating framework for gastric varices. Due to the large intra-class variation between GV with the same bleeding risk and small inter-class variation between GV and normal tissue or GV with different bleeding risks, existing classification models cannot correctly focus on the varices regions and always raise poor performance. To solve this issue, we constructively introduce segmentation to enhance the robustness of representation learning. Besides, we further design a region-constraint module for better feature localization and a cross-region attention module to learn the correlation of target GV with its context. In addition, we collected the GVBleed dataset with high-quality annotation of three-level of GV bleeding risks. The experiments on our dataset demonstrated the effectiveness and superiority of our framework.

Acknowledgements. This work is supported by Chinese Key-Area Research and Development Program of Guangdong Province (2020B0101350001), and the Shenzhen Science and Technology Program (JCYJ20220818103001002), and the Guangdong Provincial Key Laboratory of Big Data Computing, The Chinese University of Hong Kong, Shenzhen. It was also supported by the Shenzhen Sustainable Development Project (KCXFZ20201221173008022), the National Natural Science Foundation of China (NO. 61976250), the Guangdong Basic and Applied Basic Research Foundation (NO. 2020B1515020048).

References

1. Abby Philips, C., Sahney, A.: Oesophageal and gastric varices: historical aspects, classification and grading: everything in one place. Gastroenterol. Rep. **4**(3), 186–195 (2016)
2. Ali, S., Ghatwary, N., Jha, D., Halvorsen, P.: EndoCV 2021 3rd International Workshop and Chal-lenge on Computer Vision in En-doscopy
3. Brocchi, E., et al.: Prediction of the first variceal hemorrhage in patients with cirrhosis of the liver and esophageal varices: a prospective multicenter study. New England J. Med. **319**(15), 983–989 (1988)
4. Chen, H., Sung, J.J.: Potentials of AI in medical image analysis in gastroenterology and hepatology. J. Gastroenterol. Hepatol. **36**(1), 31–38 (2021)
5. Du, W., et al.: Improving the classification performance of esophageal disease on small dataset by semi-supervised efficient contrastive learning. J. Med. Syst. **46**, 1–13 (2022)
6. Garcia-Tsao, G., Abraldes, J.G., Berzigotti, A., Bosch, J.: Portal hypertensive bleeding in cirrhosis: risk stratification, diagnosis, and management: 2016 practice guidance by the American association for the study of liver diseases. Hepatology **65**(1), 310–335 (2017)
7. Gralnek, I.M., et al.: Endoscopic diagnosis and management of esophagogastric variceal hemorrhage: European society of gastrointestinal endoscopy (ESGE) guideline. Endoscopy **54**(11), 1094–1120 (2022)
8. de Groof, A.J., et al.: Deep-learning system detects neoplasia in patients with Barrett's esophagus with higher accuracy than endoscopists in a multistep training and validation study with benchmarking. Gastroenterology **158**(4), 915–929 (2020)

9. Guo, L., et al.: Real-time automated diagnosis of precancerous lesions and early esophageal squamous cell carcinoma using a deep learning model (with videos). Gastrointest. Endosc. **91**(1), 41–51 (2020)
10. Hashizume, M., Kitano, S., Yamaga, H., Koyanagi, N., Sugimachi, K.: Endoscopic classification of gastric varices. Gastrointest. Endosc. **36**(3), 276–280 (1990)
11. Hatamizadeh, A., Nath, V., Tang, Y., Yang, D., Roth, H.R., Xu, D.: Swin UNETR: swin transformers for semantic segmentation of brain tumors in MRI images. In: Crimi, A., Bakas, S. (eds.) BrainLes 2021. LNCS, vol. 12962, pp. 272–284. Springer, Cham (2022). https://doi.org/10.1007/978-3-031-08999-2_22
12. He, K., Zhang, X., Ren, S., Sun, J.: Deep residual learning for image recognition. In: Proceedings of the IEEE Conference on Computer Vision and Pattern Recognition, pp. 770–778 (2016)
13. Kang, J., Gwak, J.: Ensemble of instance segmentation models for polyp segmentation in colonoscopy images. IEEE Access **7**, 26440–26447 (2019)
14. Lv, Y., et al.: Covered tips versus endoscopic band ligation plus propranolol for the prevention of variceal rebleeding in cirrhotic patients with portal vein thrombosis: a randomised controlled trial. Gut **67**(12), 2156–2168 (2018)
15. Ma, Y., Chen, X., Cheng, K., Li, Y., Sun, B.: LDPolypVideo benchmark: a large-scale colonoscopy video dataset of diverse polyps. In: de Bruijne, M., et al. (eds.) MICCAI 2021. LNCS, vol. 12905, pp. 387–396. Springer, Cham (2021). https://doi.org/10.1007/978-3-030-87240-3_37
16. Sarin, S., Kumar, A.: Gastric varices: profile, classification, and management. Am. J. Gastroenterol. (Springer Nature) **84**(10) (1989)
17. Wang, X., Girshick, R., Gupta, A., He, K.: Non-local neural networks. In: Proceedings of the IEEE Conference on Computer Vision and Pattern Recognition, pp. 7794–7803 (2018)
18. Ye, L., Rochan, M., Liu, Z., Wang, Y.: Cross-modal self-attention network for referring image segmentation. In: Proceedings of the IEEE/CVF Conference on Computer Vision and Pattern Recognition, pp. 10502–10511 (2019)
19. Zhou, B., Khosla, A., Lapedriza, A., Oliva, A., Torralba, A.: Learning deep features for discriminative localization. In: Proceedings of the IEEE Conference on Computer Vision and Pattern Recognition, pp. 2921–2929 (2016)

You Don't Have to Be Perfect to Be Amazing: Unveil the Utility of Synthetic Images

Xiaodan Xing[1], Federico Felder[1,2], Yang Nan[1], Giorgos Papanastasiou[4], Simon Walsh[1,2], and Guang Yang[3,5,6,7](✉)

[1] National Heart and Lung Institute, Imperial College London, London, UK
[2] Royal Brompton Hospital, London, UK
[3] Bioengineering Department and Imperial-X, Imperial College London, London W12 7SL, UK
g.yang@imperial.ac.uk
[4] University of Essex, Essex, UK
[5] National Heart and Lung Institute, Imperial College London, London SW7 2AZ, UK
[6] Cardiovascular Research Centre, Royal Brompton Hospital, London SW3 6NP, UK
[7] School of Biomedical Engineering & Imaging Sciences, King's College London, London WC2R 2LS, UK

Abstract. Synthetic images generated from deep generative models have the potential to address data scarcity and data privacy issues. The selection of synthesis models is mostly based on image quality measurements, and most researchers favor synthetic images that produce realistic images, i.e., images with good fidelity scores, such as low Fréchet Inception Distance (FID) and high Peak Signal-To-Noise Ratio (PSNR). However, the quality of synthetic images is not limited to fidelity, and a wide spectrum of metrics should be evaluated to comprehensively measure the quality of synthetic images. In addition, quality metrics are not truthful predictors of the utility of synthetic images, and the relations between these evaluation metrics are not yet clear. In this work, we have established a comprehensive set of evaluators for synthetic images, including fidelity, variety, privacy, and utility. By analyzing more than 100k chest X-ray images and their synthetic copies, we have demonstrated that there is an inevitable trade-off between synthetic image fidelity, variety, and privacy. In addition, we have empirically demonstrated that the utility score does not require images with both high fidelity and high variety. For intra- and cross-task data augmentation, mode-collapsed images and low-fidelity images can still demonstrate high utility. Finally, our experiments have also showed that it is possible to produce images with both high utility and privacy, which can provide a strong rationale for the use of deep generative models in privacy-preserving applications. Our study can shore up comprehensive guidance for the evaluation of synthetic

Supplementary Information The online version contains supplementary material available at https://doi.org/10.1007/978-3-031-43904-9_2.

H. Greenspan et al. (Eds.): MICCAI 2023, LNCS 14224, pp. 13–22, 2023.
https://doi.org/10.1007/978-3-031-43904-9_2

images and elicit further developments for utility-aware deep generative models in medical image synthesis.

Keywords: Synthetic Data Augmentation · Medical Image Synthesis

1 Introduction

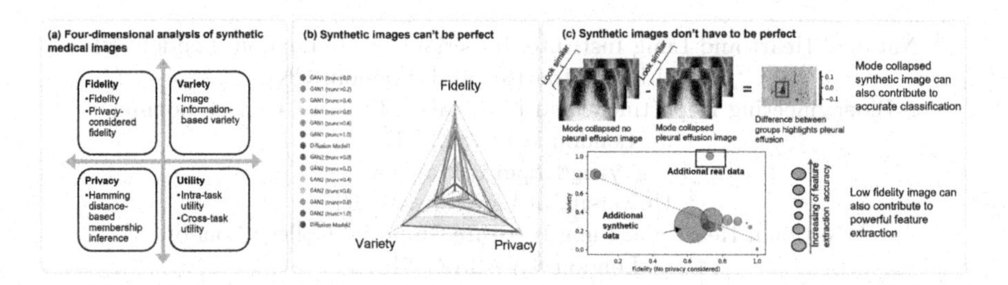

Fig. 1. The major contribution (a) and conclusions (b,c) of this work. Our study proposes a comprehensive and disentangled four-dimensional framework for synthetic medical images, incorporating fidelity, variety, privacy, and utility. We conduct extensive experiments to investigate the interplay between these dimensions and identify a set of best practices (c) for selecting synthetic models for downstream tasks.

In 2002, SMOTE [1] was proposed to generate synthetic samples to increase the accuracy of classification tasks. Since then, synthetic data has emerged as a promising solution for addressing the data scarcity problem by generating additional training data to supplement the limited real-world datasets. In addition, the potential of synthetic data for privacy preservation has led to the development of generative deep learning models that have shown promise in producing high-quality synthetic data which maintain the statistical properties of the original data while preserving the privacy of individuals in the training data.

Deep learning practitioners have been using various metrics to evaluate synthetic images, including Fréchet Inception Distance (FID) [4], Inception Score (IS) [10], precision, and recall [7]. However, these measurements are entangled, i.e., they are only able to measure image quality holistically, not a specific aspect. For example, FID is defined by

$$\mathrm{FID} = |(|\mu_1 - \mu_2|)|^2 + \mathrm{Tr}(\Sigma_1 + \Sigma_2 - 2\sqrt{(\Sigma_1 \Sigma_2)}). \tag{1}$$

Here, where μ_1, μ_2 Σ_1, and Σ_2 are the mean vectors and covariance matrices of the feature representations of two sets of images. A high difference between the image diversity (Σ_1 and Σ_2) also leads to a high FID score, which further complicates fidelity evaluation. Another entangled fidelity evaluation is the precision. As is shown in Fig. 2 (a), a high-precision matrix cannot identify non-authentic synthetic images which are copies of real data. Thus, a high-precision matrix can either be caused by high fidelity, or a high privacy breach.

When evaluating these entangled metrics, it is difficult to find the true weakness and strengths of synthetic models. In addition, large-scale experiments are currently the only way to measure the utility of synthetic data. The confusion of evaluation metrics and this time and resource-consuming evaluation of synthetic data utility increase the expenses of synthetic model selection and hinder the real-world application of synthetic data.

In this study, we aim to provide a set of evaluation metrics that are mathematically disentangled and measure the potential correlation between different aspects of the synthetic image as in Fig. 1 (a). Then, we aim to analyze the predictive ability of these proposed metrics to image utilities for different downstream tasks and provide a set of best practices for selecting synthetic models in various clinical scenarios. We compare two state-of-the-art deep generative models with different parameters using a large open-access X-ray dataset that contains more than 100k data [5]. Through our experiments, we empirically show the negative correlations among synthetic image fidelity, variety, and privacy (Fig. 1 (b)). After analyzing their impacts on downstream tasks, we discovered that the common problems in data synthesis, i.e., mode collapse and low fidelity, can sometimes be a merit according to the various motivations of different downstream tasks.

Overall, our study contributes new insights into the use of synthetic data for medical image analysis and provides a more objective and reproducible approach to evaluating synthetic data quality. By addressing these fundamental questions, our work provides a valuable foundation for future research and practical applications of synthetic data in medical imaging.

2 Deep Generative Models and Evaluation Metrics

In this study, we conducted an empirical evaluation using two state-of-the-art deep generative models: StyleGAN2, which has brought new standards for generative modeling regarding image quality [11] and Latent Diffusion Models (LDM) [9]. We proposed an analysis framework for synthetic images based on four key dimensions: fidelity, variety, utility, and privacy. Manual evaluation of synthetic image fidelity typically involves human experts assessing whether synthetic images appear realistic. However, this evaluation can be subjective and have high intra-observer variance. While many algorithms can be used to measure synthetic image quality, most of them are designed to capture more than one of the four key dimensions.

With this motivation, we aimed to redefine the conventional evaluation metrics of synthetic images and disentangle them into four independent properties. In this study, we employed a metric-based membership inference attack in an unsupervised manner to evaluate the privacy of our model. We presume that synthetic records should have some similarity with the records used to generate them [2]. Since differential privacy settings could significantly reduce image fidelity, we chose not to perform differential privacy analysis in this study.

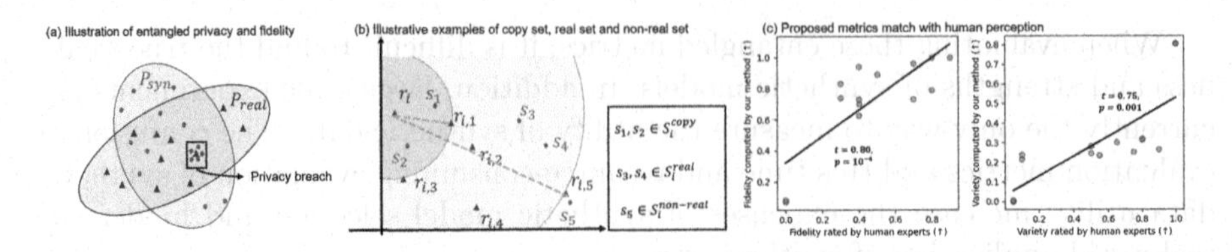

Fig. 2. Illustrative examples of entangled privacy and fidelity (a), our proposed algorithms to adjust the fidelity computation (b) and their validity measured by Pearson's correlation analysis (c). In (b), we selected the 5th-nearest neighbor to compute the fidelity, i.e., $k = 5$. The human experts were two radiologists with one and two years of experience, respectively.

Fidelity and Privacy. Considering we have M real image points and N synthetic image points. For each real image $\{r_i|i \in [0, M)\}$, we select its k nearest neighbors $\{r_{i,1}, r_{i,2}, \dots, r_{i,k}\}$ according to the similarity $D(r_i, r_j)$ of these images in the image space. Then, as shown in Fig. 2 (b), we split the synthetic dataset into three sets according to these k nearest neighbors:

- Copy set S_i^{copy}: Synthetic sets that are realistic but are also copies of this real image r_i. Synthetic images will belong to this set if this synthetic image is closer to r_i than any other real images, i.e., $D(s_j, r_i) < D(r_i, r_{i,1})$.
- Real set S_i^{real}: Synthetic sets that are realistic and are not copies of this real image r_i. Synthetic images will belong to this set if this synthetic image is in the kth-nearest neighbor of r_i, i.e., $D(s_j, r_i) \in [(D(r_i, r_{i,1}), D(r_i, r_{i,k})]$
- Non-real set $S_i^{\text{non-real}}$: Synthetic sets that are not realistic compared to r_i. Synthetic images will belong to this set if this synthetic image is not the kth-nearest neighbor of r_i, i.e., $D(s_j, r_i) > D(r_i, r_{i,k})$

We can compute the privacy preservation score of a single synthetic image $s_j|j \in [0, N)]$ with the definition of three sets. If the synthetic data p_j is in the copy set of any real data, the privacy protection ability of this p_j is 0, i.e.,

$$p_j = \begin{cases} 0 & \text{if } \exists i \in [0, M) \rightarrow s_j \in S_i^{\text{copy}} \\ 1 & \text{otherwise} \end{cases}. \tag{2}$$

For the synthetic data, the overall privacy protection ability $P \in [0, 1]$ was then defined by

$$P = \frac{\sum_0^N p_j}{N}. \tag{3}$$

With this privacy definition, we have adjusted the original fidelity evaluation [7] shown in Eq. (4) to a privacy violation considered formula f^p in Eq. 5,

$$f_j = \begin{cases} 0 & \text{if } \exists i \in [0, M) \rightarrow s_j \in S_i^{\text{copy}} \cup S_i^{\text{real}} \\ 1 & \text{otherwise} \end{cases} \text{ and } F = \frac{\sum_0^N f_j}{N}; \tag{4}$$

$$f_j^p = \begin{cases} 0 & \text{if } \exists i \in [0, M) \rightarrow s_j \in S_i^{\text{real}} \\ 1 & \text{otherwise} \end{cases} \text{ and } F^p = \frac{\sum_0^N f_j^p}{N}. \tag{5}$$

The measurements of image distance can be tricky due to the high resolution of the original images (512×512). Thus, we first used VQ-VAE [8,13] to quantize all images to integral latent feature maps, i.e., each pixel in the latent feature maps is a Q-way categorical variable, sampling from 0 to $Q-1$, and then compute the Hamming distance between these images. In our experiments, we used $Q = 256$.

Variety. To measure the variety, we introduced the JPEG file size of the mean image. The lossless JPEG file size of the group average image was used to measure the inner class variety in the ImageNet dataset [3]. This approach was justified by the authors who presumed that a dataset containing diverse images would result in a blurrier average image, thus reducing the lossless JPEG file size of the mean image. To ensure that the variety score is positively correlated with the true variety, we normalized it to $[0, 1]$ across all groups of synthetic images, and then subtracted it from 1 to obtain the final variety score. It is worth noting that variety can also be quantified by the standard deviation of the discrete latent features. However, in our study, we chose to measure variety in the original image space to better align with human perception.

Utility. We divided our X-ray dataset into four groups: training datasets A1 and A2, validation set B, and testing set C. We also included an additional open-access pediatric X-ray dataset, D. For our simulation, we treated A1 as a local dataset and A2 as a remote dataset that cannot be accessed by A1. We evaluated the utility of synthetic data in two conditions:

1. A1 vs. adding synthetic data generated from A1. In this condition, no privacy issue is considered.
2. A1 vs. adding synthetic data generated from A2. In this condition, synthetic data will be evaluated using privacy protection skills.

Under both conditions, we evaluated the intra-task augmentation utility and cross-task augmentation utility to simulate real-world use cases for synthetic data. Intra-task augmentation utility is measured by the percentage improvement in classification accuracy of C when adding synthetic data to the training dataset. We used a paired Wilcoxon signed-rank test to assess the significance of the accuracy improvement. If the improvement is significant, it indicates that the synthetic images are useful. We compared the augmentation utility with simple augmentations, such as random flipping, rotating, and contrasting.

The cross-task augmentation utility is determined by the power of features extracted from the models trained with synthetic data. We used the models to extract features from D and trained a Support Vector Machine classifier on these features to measure accuracy. This allowed us to evaluate whether synthetic images can provide powerful features that facilitate downstream tasks. Similarly, the cross-task augmentation utility is also the percentage improvement in classification accuracy compared to the model trained only on A1.

3 Experimental Settings and Parameters

We primarily evaluated the performance of synthetic data on the CheXpert dataset, with a focus on identifying the presence of pleural effusion (PE). To perform our evaluation, we split the large dataset into four subsets: A1 (15004 with PE and 5127 without PE), A2 (30692 with PE and 10118 without PE), B (3738 with PE and 1160 PE), and C (12456 with PE and 3863 without PE). To evaluate the cross-task utility of synthetic models, we used an X-ray dataset D[1] consisting of 5863 images of pediatric patients with pneumonia and normal controls. We resized all X-ray images to a resolution of 512×512 before evaluation.

For the StyleGAN2 method, we utilized six truncation parameters during sampling to generate six sets of synthetic images ($\phi \in [0.0, 0.2, 0.4, 0.6, 0.8, 1.0]$). In total, we trained 16 classification models on different combinations of datasets, including A1, A1+A2, A1+StyleGAN2-synthesized A1 (6 models), A1+LDM-synthesized A1, A1+StyleGAN2-synthesized A2 (6 models), and A1+LDM- synthesized A2. For further information on implementation details, hyperparameters and table values, please refer to our supplementary file and our publicized codes https://github.com/ayanglab/MedSynAnalyzer.

4 Experimental Results

4.1 The Proposed Metrics Match with Human Perception

In our work, we proposed to use VQ-VAE to extract discrete features from original high-resolution X-ray images. To prove the validity of our method, we selected twenty images from each synthetic dataset and dataset A1 and invited two clinicians to rate the fidelity and variety manually. The human perceptual fidelity is rated from 0 to 1; and the human perceptual variety is computed by the percentage of different scans identified from the selected twenty synthetic images, i.e., if they thought all twenty patients were derived from the same scan, the human perceptual variety score is $1/20 = 0.05$. To assure a fair comparison, we allow discussion between them. The result is shown in Fig. 2 (c). The fidelity and variety score calculated with our method matched perfectly with human perception ($p < 0.05$), and FID, which was highly influenced by mode collapse and increased over the diversity, failed to provide a valid analysis of image fidelity.

4.2 The Trade-Off Between Fidelity and Variety

All of our experiments showed a strong negative correlation between variety and fidelity, with a Pearson's correlation coefficient of -0.92 ($p < 0.01$). As it is widely known in GAN-based models, fidelity and variety are in conflict [6]. In this study, we further validated this by introducing the LDM model and demonstrating empirically that deep generative models inevitably face the trade-off between variety and fidelity (as shown by the grey lines in Fig. 3 (a-b)).

[1] https://www.kaggle.com/datasets/paultimothymooney/chest-xray-pneumonia?resource=download.

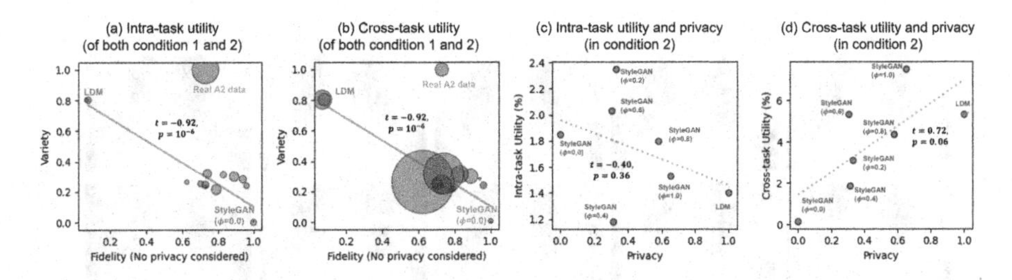

Fig. 3. The relationships between fidelity, variety, and utility of adding additional synthetic data. The sizes of points in (a) and (b) are the improvement brought by adding synthetic data compared to no synthetic data being added. It should be noted that we visualized the point size in a power of 7 to better compare the improvements. Green points indicate significance using Signed Ranked T-test, and red indicates no significance. (Color figure online)

4.3 What Kind of Synthetic Data is Desired by Downstream Tasks When Privacy is Not an Issue?

In this study, we cannot find a significant correlation of utility between neither dimension (fidelity, variety nor privacy), indicating that there is currently no way to measure the utility except for large-scale experiments. However, we did observe a similar pattern of utility in our experiments shown in Fig. 3 (a-b).

First, the intra-task augmentation utility favors synthetic data with higher fidelity (i.e., high F), even under mode collapses. For instance, when $\phi = 0.2$ for StyleGAN, a mode collapse was observed. The model is producing images that look similar to each other, resulting in a mean image with a sharp contrast (Fig. 4 (a4)). However, this mode collapse seems to highlight the difference between the presence and non-presence of PE (Fig. 4 (c)), leading to a performance improvement in intra-task augmentation. The PE in X-ray images is fluid in the lowest part of the chest, forming a concave line obscuring the costophrenic angle and part or all of the hemidiaphragm. These opacity differences were highlighted in the mode-collapsed images, which, on the other hand, improved the classification accuracy of PE identification.

For cross-task augmentation, synthetic data with a higher variety is favored for its utility as mode collapse can limit the focus of classification networks and lead to poor generalization performance on other tasks. For instance, a network trained to focus on lung opacity differences near the hemidiaphragm may not help in the accurate diagnosis of pediatric pneumonia, which is the motivation behind dataset D. As shown in Fig. 3 (b), synthetic data with low variety but high fidelity is unable to contribute to powerful feature extraction. Therefore, variety in synthetic data is crucial for effective cross-task augmentation.

As mentioned, we invited two radiologists to visually assess synthetic images. The visual inspection showed that all twenty LDM-synthesized images were easily recognized as fake due to their inability to capture the texture of X-ray images, as shown in the supplementary file. However, the shape and boundaries

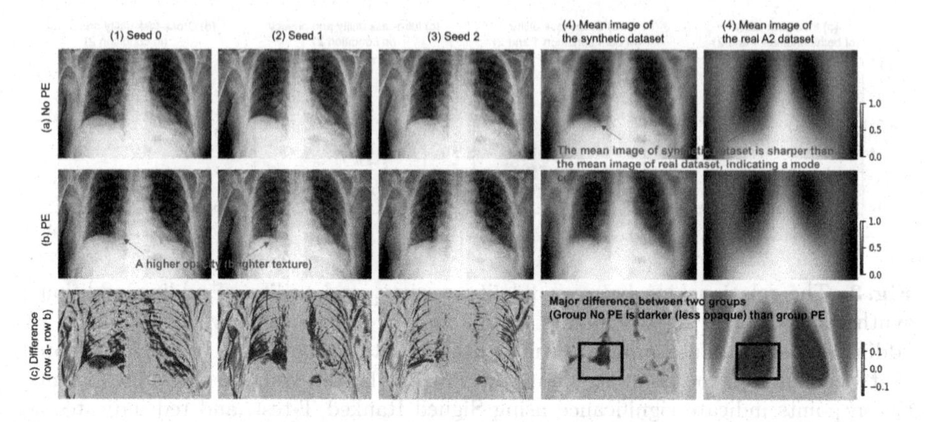

Fig. 4. Example synthetic images sampled from the StyleGAN2 model with a truncation $\phi = 0.2$ trained on A2 with different sampling seeds (1–3), as well as the mean image of the two groups calculated on the entire synthetic dataset (4). All the images in (a-b) were normalized to [0,1], and the difference between row (a) and row (b) is shown in row (c). The presence of PE in X-ray images is characterized by increased opacity of the lung near the hemidiaphragm, which matches the group difference of mode-collapsed images. Therefore, even in the case of severe mode collapse during synthesis, these synthetic images can still contribute to the improvement of intra-task classification accuracy.

of the lungs were accurately captured by LDM. Despite their low visual fidelity, we demonstrated that these synthetic images still contribute to powerful feature extraction, which is crucial for cross-task utility.

4.4 What Kind of Synthetic Data is Desired by Downstream Tasks When Privacy is an Issue?

It is also discussed in the literature about the dilemma between utility and privacy [12]. We reached a similar conclusion, i.e., there is a trade-off between utility and privacy in intra-class classification tasks, for which fidelity is considered to be crucial. Thus, a shift in the image domain could lead to a decrease in intra-task utility.

However, as shown in Fig. 4 (d), we observed that privacy and utility are not always conflicting. As discussed earlier, for cross-task augmentation, the utility favors synthetic images with high variety rather than high fidelity. Therefore, we demonstrated that it is possible to achieve both privacy and utility in cross-task augmentation scenarios. Figure 4 (d) shows an interesting positive correlation between privacy and cross-task utility. However, it is important to note that this does not imply a causal relationship between privacy and cross-task utility. Rather, the positive correlation is caused by the mode collapse during synthesis. Mode collapse can lead to a lack of diversity in generated data, which in turn can make it easier to identify individuals or sensitive information in the generated data, i.e., mode collapses are more likely to result in a high possibility of privacy breach as well as low cross-task utility.

5 Conclusion

In this work, we proposed a four-dimensional evaluation metric for synthetic images, including a novel privacy evaluation score and utility evaluation score. Through intensive experiments in over 100k chest X-ray images, we drew three major conclusions which we can envision that have broad applicability in medical image synthesis and analysis.

Firstly, there is an inevitable trade-off among different aspects of synthetic images, especially between fidelity and variety. Secondly, different downstream tasks require different properties of synthetic images, and synthetic images do not necessarily have to reach high metric scores across all aspects to be useful. Traditionally, low fidelity and mode collapses have been treated as disadvantages in data synthesis, and numerous algorithms have been proposed to fix these issues. However, our work demonstrates that these failures of synthetic data do not always sabotage their utility as expected. Lastly, we have showed that it is possible to achieve both privacy and utility in transfer learning problems.

In conclusion, our work contributes to the development of synthetic data as a valuable solution to enrich real-world datasets, to evaluate thoroughly medical image synthesis as a pathway to overall enhance medical image analysis tasks.

Acknowledgement. This study was supported in part by the ERC IMI (101005122), the H2020 (952172), the MRC (MC/PC/21013), the Royal Society (IEC\NSFC\211235), the NVIDIA Academic Hardware Grant Program, the SABER project supported by Boehringer Ingelheim Ltd, Wellcome Leap Dynamic Resilience, and the UKRI Future Leaders Fellowship (MR/V023799/1).

References

1. Chawla, N.V., Bowyer, K.W., Hall, L.O., Kegelmeyer, W.P.: Smote: synthetic minority over-sampling technique. J. Artif. Intell. Res. **16**, 321–357 (2002)
2. Chen, D., Yu, N., Zhang, Y., Fritz, M.: Gan-leaks: A taxonomy of membership inference attacks against generative models. In: Proceedings of the 2020 ACM SIGSAC Conference on Computer and Communications Security, pp. 343–362. CCS '20, Association for Computing Machinery, New York, NY, USA (2020). https://doi.org/10.1145/3372297.3417238
3. Deng, J., Dong, W., Socher, R., Li, L.J., Li, K., Fei-Fei, L.: Imagenet: a large-scale hierarchical image database. In: 2009 IEEE Conference on Computer Vision and Pattern Recognition, pp. 248–255. IEEE (2009)
4. Heusel, M., Ramsauer, H., Unterthiner, T., Nessler, B., Hochreiter, S.: GANs trained by a two time-scale update rule converge to a local nash equilibrium. Adv. Neural Inform. Process. Syst. **30** (2017)
5. Irvin, J., et al.: Chexpert: A large chest radiograph dataset with uncertainty labels and expert comparison. In: Proceedings of the AAAI Conference on Artificial Intelligence, vol. 33, pp. 590–597 (2019)

6. Kynkäänniemi, T., Karras, T., Laine, S., Lehtinen, J., Aila, T.: Improved precision and recall metric for assessing generative models. Adv. Neural Inform. Process. Syst. **32** (2019)
7. Lucic, M., Kurach, K., Michalski, M., Gelly, S., Bousquet, O.: Are GANs created equal? a large-scale study. Adv. Neural Inform. Process. Syst. **31** (2018)
8. Razavi, A., Van den Oord, A., Vinyals, O.: Generating diverse high-fidelity images with vq-vae-2. Adv. Neural Inform. Process. Syst. **32** (2019)
9. Rombach, R., Blattmann, A., Lorenz, D., Esser, P., Ommer, B.: High-resolution image synthesis with latent diffusion models. In: Proceedings of the IEEE/CVF Conference on Computer Vision and Pattern Recognition, pp. 10684–10695 (2022)
10. Salimans, T., Goodfellow, I., Zaremba, W., Cheung, V., Radford, A., Chen, X.: Improved techniques for training GANs. Adv. Neural Inform. Process. Syst. **29** (2016)
11. Sauer, A., Schwarz, K., Geiger, A.: Stylegan-xl: Scaling stylegan to large diverse datasets. In: ACM SIGGRAPH 2022 Conference Proceedings, pp. 1–10 (2022)
12. Stadler, T., Oprisanu, B., Troncoso, C.: Synthetic data-anonymisation groundhog day. In: 31st USENIX Security Symposium (USENIX Security 22), pp. 1451–1468 (2022)
13. Van Den Oord, A., Vinyals, O., et al.: Neural discrete representation learning. Adv. Neural Inform. Process. Syst. **30** (2017)

SHISRCNet: Super-Resolution and Classification Network for Low-Resolution Breast Cancer Histopathology Image

Luyuan Xie[1,3], Cong Li[1,3], Zirui Wang[2,3], Xin Zhang[1,3], Boyan Chen[1,3], Qingni Shen[1,3(✉)], and Zhonghai Wu[1,3(✉)]

[1] School of Software and Microelectronics, Peking University, Beijing, China
qingnishen@ss.pku.edu.cn, wuzh@pku.edu.cn

[2] National Engineering Research Center for Software Engineering, Peking University, Beijing 100871, China

[3] Tencent Cloud Media, Shenzhen, China

https://github.com/xiely-123/SHISRCNet

Abstract. The rapid identification and accurate diagnosis of breast cancer, known as the killer of women, have become greatly significant for those patients. Numerous breast cancer histopathological image classification methods have been proposed. But they still suffer from two problems. (1) These methods can only hand high-resolution (HR) images. However, the low-resolution (LR) images are often collected by the digital slide scanner with limited hardware conditions. Compared with HR images, LR images often lose some key features like texture, which deeply affects the accuracy of diagnosis. (2) The existing methods have fixed receptive fields, so they can not extract and fuse multi-scale features well for images with different magnification factors. To fill these gaps, we present a **S**ingle **H**istopathological **I**mage **S**uper-**R**esolution **C**lassification network (SHISRCNet), which consists of two modules: Super-Resolution (SR) and Classification (CF) modules. SR module reconstructs LR images into SR ones. CF module extracts and fuses the multi-scale features of SR images for classification. In the training stage, we introduce HR images into the CF module to enhance SHISRCNet's performance. Finally, through the joint training of these two modules, super-resolution and classified of LR images are integrated into our model. The experimental results demonstrate that the effects of our method are close to the SOTA methods with taking HR images as inputs.

Keywords: breast cancer · histopathological image · super-resolution · classification · joint training

1 Introduction

Breast cancer is one of the high-mortality cancers among women in the 21st century. Every year, 1.2 million women around the world suffer from breast cancer and about 0.5 million die of it [3]. Accurate identification of cancer types

H. Greenspan et al. (Eds.): MICCAI 2023, LNCS 14224, pp. 23–32, 2023.
https://doi.org/10.1007/978-3-031-43904-9_3

will make a correct assessment of the patient's risk and improve the chances of survival. However, the traditional analysis method is time-consuming, as it mainly depends on the experience and skills of the doctors. Therefore, it is essential to develop computer-aided diagnosis (CADx) for assisting doctors to realize the rapid detection and classification.

Due to being collected by various devices, the resolution of histopathological images extracted may not always be high. Low-resolution (LR) images lack of lots of details, which will have an important impact on doctors' diagnosis. Considering the improvement of histopathological images' acquisition equipment will cost lots of money while significantly increasing patients' expense of detection. The super-resolution (SR) algorithms that improve the resolution of LR images at a small cost can be a practical solution to assist doctors in diagnosis. At present, most single super-resolution methods only have fixed receptive fields [7,10,11,18]. These models cannot capture multi-scale features and do not solve the problems caused by LR in various magnification factors well. MRC-Net [6] adopted LSTM [9] and Multi-scale Refined Context to improve the effect of reconstructing histopathological images. It considered the problem of multi-scale, but only fused two scales features. This limits its performance in the scenarios with various magnification factors. Therefore, designing an appropriate feature extraction block for SR of the histopathological images is still a challenging task.

In recent years, a series of deep learning methods have been proposed to solve the breast cancer histopathological image classification issue by the high-resolution (HR) histopathological images. [12,21,22] improved the specific model structure to classify breast histopathology images, which showed a significant improvement in recognition accuracy compared with the previous works [1,20]. SSCA [24] considered the problem of multi-scale feature extraction which utilized feature pyramid network (FPN) [15] and attention mechanism to extract discriminative features from complex backgrounds. However, it only concatenates multi-scale features and does not consider the problem of feature fusion. So it is still worth to explore the potential of extraction and fusion of multi-scale features for breast images classification.

To tackle the problem of LR breast cancer histopathological images reconstruction and diagnosis, we propose the **S**ingle **H**istopathological **I**mage **S**uper-**R**esolution **C**lassification network (SHISRCNet) integrating Super-Resolution (SR) and Classification (CF) modules. The main contributions of this paper can be described as follows:

(1) In the SR module, we design a new block called Multi-Features Extraction block (MFEblock) as the backbone. MFEblock adopts multi-scale receptive fields to obtain multi-scale features. In order to better fuse multi-scale features, a new fusion method named multi-scale selective fusion (MSF) is used for multi-scale features. These make MFEblock reconstruct LR images into SR images well.

(2) The CF module completes the task of image classification by utilizing the SR images. Like SR module, it also needs to extract multi-scale features. The difference is that the CF module can use the method of downsampling to capture multi-scale features. So we combine the multi-scale receptive fields (SKNet) [13]

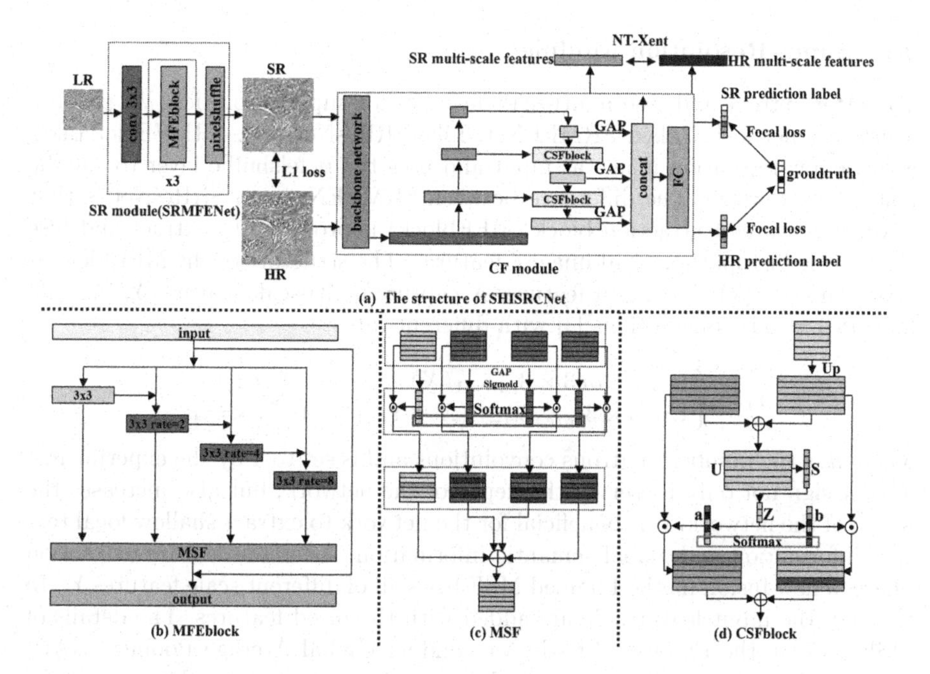

Fig. 1. The structure of the SHISRCNet.

with the feature pyramid network (FPN) to achieve the feature extraction of this module. In FPN, we design a cross-scale selective fusion block (CSFblock) to fuse features of different scales.

(3) Through the joint training of these two designed modules, the super-resolution and classification of low-resolution histopathological images are integrated into our model. For improving the performance of CF module and reducing the error caused by the reconstructed SR images, we introduce HR images to CF module in the training stage. The experimental results demonstrate that the effects of our method are close to those of SOTA methods that take HR breast cancer histopathological images as inputs.

2 Methods

This section describes the proposed SHISRCNet. The overall pipeline of the proposed network is shown in Fig. 1(a). It composes two modules: SR module and CF module. The SR module reconstructs the LR image into the SR image. The CF module utilize the reconstructed SR images to diagnose histopathological images. In the training stage, we introduce HR images to improve the performance of CF module and alleviate the error caused by SR images.

2.1 Super-Resolution Module

To better extract and fuse multi-scale features for super-resolution, we propose a new SR network, called SRMFENet. Like SRResNet [11], SRMFENet takes a single low-resolution image as input and uses the pixelshuffle layer to get the restructured image. The difference between SRMFENet and SRResNet is that a Multi-Features Extraction block (MFEblock) is proposed to extract and fuse multi-scale histopathological images' features. The structure of the MFEblock is shown in Fig. 1(b). The input features X capture multi-scale features Y_i through four 3×3 atrous convolutions [4] with different rates:

$$Y_i = \begin{cases} Cov3 \times 3_{rate=1}(X) & i = 1 \\ Cov3 \times 3_{rate=2(i-1)}(X + Y_{i-1}) & 1 < i <= n \end{cases}$$

where n is the number of atrous convolutions and is set to 4 by the experiments. This design not only preserves the depth of the network, but also increases the width of the network. It is beneficial for the network to extract shallow local texture information and global semantic information. After the feature extraction phase, a new fusion method named MSF fuses all of different scale features Y_i. In the end, the input features X are added with the fused features. The details of MSF show in the Fig. 1(c). Firstly, we conduct Global Average Pooling (GAP) [14] on the multi-scale features to obtain their average channel-wise weights. Then using Sigmoid activation function to map weight to 0 to 1. Next softmax operation normalizes the same position of the obtained multi-scale average channel-wise weights. Finally, the features are multiplied by the corresponding normalized weights and the processed features are added together to generate the new multi-scale features. MFEblock is very applicable to process histopathological images of different magnification factors, as it employs convolution and attention operations to capture local and global image context information and fuse them well.

2.2 Classification Module

The task of the CF module is to classify the reconstructed SR images. It can use the method of downsampling to capture multi-scale features. So we combine the multi-scale receptive fields (SKNet as backbone network) with the FPN (a downsampling method) to achieve the feature extraction of this module. In Fig. 1(a), the multi-sacle features extracted by SKNet are the input of FPN. We propose a new fusion method, called cross-scale selective fusion block (CSFblock) to effectively fuse high-resolution and low-resolution features in FPN. After the fused features are processed by GAP, they are aggregated into a new multi-scale feature by concatenate operation. Finally, the aggregated multi-scale features are classified through the fully connected (FC) layer. The structure of CSFblock is shown in Fig. 1(d). The inputs of CSFblock are two-way inputs which are the high-resolution features $X_h \in W \times H \times C$ and the low-resolution features $X_l \in W_1 \times H_1 \times C$. In CSFblock, the upsampling operations are performed on the low-resolution features X_l to realize consistency with X_h dimension. X_h and restructured X_l are fused via an element-wise summation:

$$U = X_h + Up(X_l)$$

Then, using GAP along the channel dimension to get the global information S. A FC layer generates a compact feature vector Z which guides the feature selection procedure. And Z is reconstructed into two weight vectors a, b of the same dimension as S through two FC layers, which can be defined as:

$$Z = \delta(W_c S), \quad a = W_a Z, \quad b = W_b Z$$

where δ denotes ReLU and W_a, W_b, W_c, means the weight of the FC layers. Specifically, a softmax operator is applied on a and b 's channel-wise digits:

$$a[i] = \frac{e^{a[i]}}{e^{a[i]} + e^{b[i]}}, \quad b[i] = \frac{e^{b[i]}}{e^{a[i]} + e^{b[i]}}, \quad i \in C$$

The fused feature map F is obtained through the attention weights on multi-scale features:

$$F[i] = a[i] \times X_h[i] + b[i] \times Up(X_l)[i], \quad i \in C$$

2.3 Loss Function

The SR module and the CF module exploit different loss functions for training. In the SR module, $L1$ Loss is used for super-resolution. In the CF module, we introduce HR images to CF module in the training stage for improving the performance of CF module and reducing the error caused by the reconstructed SR images. We use $Focal$ Loss [16] to alleviate the class imbalanced data problem of the HR and SR images' classification. Inspired by the contrastive learning algorithm SimCLR [5], the HR and SR of the same images are similar to two different views. So the $NT - Xent$ loss [19] is adopted to calculate similarity between SR multi-scale features and HR multi-scale features for CF module's robustness. The total loss function can be expressed as:

$$L_{total} = \lambda_1 L_{L1} + \lambda_2 L_{FL} + \lambda_3 L_{NT-Xent}, \quad \lambda_1 + \lambda_2 + \lambda_3 = 1$$

where λ_1, λ_2 and λ_3 are the weights of $L1$ Loss, $Focal$ Loss and $NT - Xent$ Loss, respectively. In the inference stage, only SR images are taken as inputs by CF module. In our experiment, λ_1, λ_2 and λ_3 are set to 0.6, 0.3 and 0.1, respectively. And the temperature parameter in $NT - Xent$ Loss is set to 0.5.

3 Experiment

Dataset: This work uses the breast cancer histopathological image database (BreaKHis)[1] [20]. The images in the dataset have four magnification factors

[1] https://web.inf.ufpr.br/vri/databases/breast-cancer-histopathological-database-breakhis/.

(40x, 100x, 200x, 400x) and eight breast cancer classes. This dataset includes four distinct histological types of benign breast tumors: adenosis (A), fibroadenoma (F), phyllodes tumor (PT), and tubular adenona (TA); and four malignant tumors (breast cancer): carcinoma (DC), lobular carcinoma (LC), mucinous carcinoma (MC) and papillary carcinoma (PC). The original dataset is randomly divided into training set and testing set for each magnification at a ratio of 7:3 following previous work.

Implementation Details: For all experiments, we conduct 5-fold cross validation, and report the mean. We use LR histopathological images with size 48×48, 96×96, 192×192 as input for different single image SR tasks (x8, x4, x2) and set batch size to 8. For the corresponding LR and HR images in the training dataset, the same data augmentation is adopted, such as rotation, color jitter. The model is trained using the ADAM optimizer [25] with the learning rate set to $1\text{x}10^{-3}$. The learning rate is multiplied by 0.9 for every two epochs. We use SKNet-26 [13] as the backbone network in the CF module. The total training epochs are 100.

4 Results and Discussion

4.1 The Results of Super-Resolution and Classification

Table 1 shows the results of the super-resolution phase. We adopt Peak Signal to Noise Ratio (PSNR) and structural similarity index (SSIM) [6] to evaluate the performance of the SR model. MRC-Net and our proposed SRMFENet (SR module) achieves better metrics than the other algorithms. This proves the effectiveness of multi-scale features extraction. Compared with MRC-Net, our MFEblocks can extract and fuse multi-scale features well. And the joint training of SRMFENet and CF module improves the performance of super-resolution. Figure 2 demonstrates that our model can recover more details with less blurring.

We compare our introduced CF module with five state-of-the-art breast cancer histopathological image models and Diagnosis Network with MRC-Net [6], as shown in Table 2. The results illustrate that the CF module reaches the best

Table 1. Average PSNR/SSIM for x8, x4, x2 SR.

Methods	x8		x4		x2	
	PSNR(db)	SSIM	PSNR(db)	SSIM	PSNR(db)	SSIM
Bicubic	20.75	0.4394	23.21	0.6305	26.60	0.9151
SRCNN	21.12	0.4872	24.04	0.6634	28.36	0.8631
WA-SRGAN	21.76	0.5141	26.20	0.7930	30.93	0.9351
EDSR	22.13	0.6063	26.22	0.8005	30.79	0.9325
MRC-Net	22.72	0.6213	26.86	0.8222	31.73	0.9433
SRMFENet (ours)	23.44	0.6325	27.21	0.8370	33.96	0.9566
SHISRCNet (ours)	**24.21**	**0.6814**	**27.97**	**0.8413**	**35.61**	**0.9680**

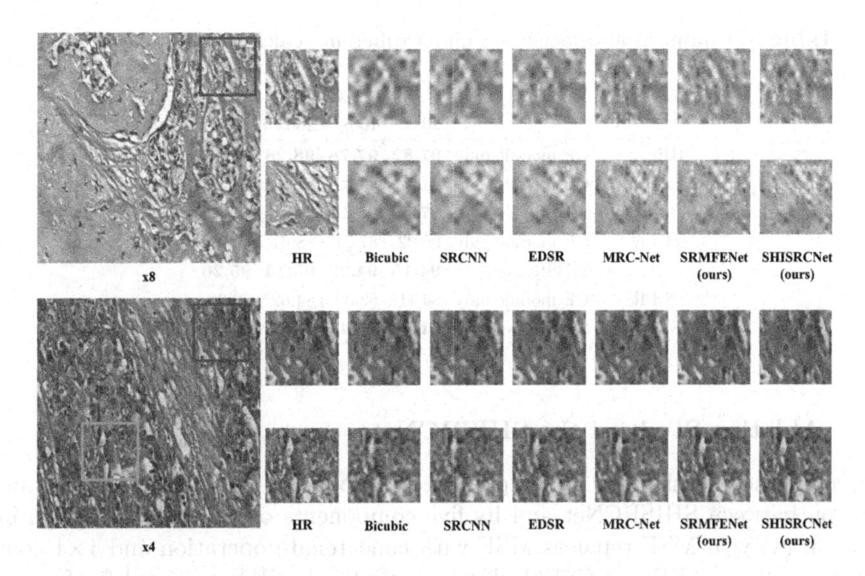

Fig. 2. Qualitative Comparison with SR methods on breast cancer histopathological images x8 and x4.

Table 2. Compare results with state-of-the-art on image level (* means that inputs are HR images, # means that inputs are down sample to half resolution from HR images.).

Methods	Years	ACC(%)			
		40x	100x	200x	400x
AlexNet variant* [21]	2016	85.6	83.5	82.7	80.7
Inception V3* [2]	2018	90.2	85.6	86.1	82.5
DSoPN* [12]	2020	96	96.16	98.01	95.97
FE-BkCapsNet* [22]	2021	92.71	94.52	94.03	93.54
SSCA* [24]	2022	96.93	97.32	95.31	96.24
CF module only* (ours)	-	**97.82**	**97.78**	**98.28**	**98.15**
Diagnosis Network with MRC-Net# [6]	2021	94.43	94.45	94.73	93.92
SHISRCNet (ours)#	-	**97.49**	**96.19**	**97.60**	**97.04**

performance in four different magnification factors. This indicates the effectiveness of our proposed combination of two multi-scale feature extraction methods. SHISRCNet, which uses down sample to half resolution (x2) from HR images, outperforms the SSCA at 40x, 200x and 400x. And it gets results close to the CF module at all magnification factors. Meanwhile, compared with the Diagnosis Network that also uses LR images as input, SHISCNet has remarked performance advantages. Table 3 compares our results with the CF module using different resolution images. The performance of the CF module decreases significantly with the reduction of resolution. In contrast, SHISRCNet greatly improves the CF module performance of different scale low-resolution images.

Table 3. Comparison of accuracy under different scales on the image level.

resolution	Model	ACC(%)			
		40x	100x	200x	400x
HR	CF module only	**97.82**	**97.78**	**98.28**	**98.15**
x2 LR	CF module only	94.47	89.92	92.64	91.30
	SHISRCNet	**97.49**	**96.19**	**97.60**	**97.04**
x4 LR	CF module only	90.32	87.71	88.61	86.89
	SHISRCNet	**94.15**	**94.22**	**95.14**	**95.26**
x8 LR	CF module only	84.11	82.32	84.62	82.23
	SHISRCNet	**91.47**	**92.43**	**92.98**	**92.78**

4.2 Ablation Study of the SHISRCNet

To verify the effectiveness of the proposed components in SHISRCNet, a comparison between SHISRCNet and its five components on x2 images is given in Table 3. (1) w/o MSF repalces MSF with concatenate operation and 1×1 convolution. (2) w/o FPN + CSFblock means that only SKNet is used for feature extraction in the CF module. (3) w/o CSFblock, w/o HR images and w/o NT-Xent loss remove the corresponding operation, respectively. As shown in table 3, firstly, the performance of super-resolution in the SHISRCNet is significantly reduced when we remove MSF. It indicates the importance of MSF for multi-scale feature fusion in the SR module. Secondly, only SKNet is used to extract multi-scale features in the CF module, and the accuracy decreased significantly. This again proves the effectiveness of our proposed combination of two multi-scale feature extraction methods. Thirdly, compared with using FPN alone, the performance of SHISRCNet is further improved by adding CSFblock to FPN. Finally, the introduction of HR images further promotes the performance of SHISRCNet. Because the training method of HR and SR images proposed by us helps to improve the generalization of the SHISRCNet (Table 4).

Table 4. Ablation study of SHISRCNet on x2 images.

	PSNR(db)	SSIM	ACC(%)			
			40x	100x	200x	400x
SHISRCNet	**35.61**	**0.9680**	**97.49**	**96.19**	**97.60**	**97.04**
w/o MSF	33.13	0.9554	96.02	95.73	95.98	95.78
w/o FPN+CSFblock	34.41	0.9609	93.98	92.11	91.35	92.15
w/o CSFblock	34.57	0.9619	94.37	93.21	93.99	94.71
w/o HR images	34.43	0.9611	93.13	94.28	93.86	94.17
w/o NT-Xent loss	34.54	0.9623	95.53	95.20	95.01	95.36

5 Conclusion

This paper proposes SHISRCNet for the low-resolution breast cancer histopathological images' super-resolution and classification problem. The SR module employs MFEblock to extract and fuse multi-scale features for reconstructing low-resolution histopathological images into high-resolution ones. The CF module adopts two different multi-scale features extraction methods to capture features for the breast cancer diagnosis. We introduce high-resolution images into the CF module in the training stage to improve SHISRCNet's robustness. Through the joint training of the two modules, the super-resolution and classification of the low-resolution histopathological images are integrated in one model. Our method's results are close to the SOTA methods, which require using high-resolution breast cancer histopathological images instead of low-resolution ones.

Acknowledgements. This work was supported by the National Key R&D Program of China under Grant No.2022YFB2703301.

References

1. Akay, M.F.: Support vector machines combined with feature selection for breast cancer diagnosis. Expert Syst. Appl. **36**, 3240–3247 (2009)
2. Benhammou, Y., Tabik, S., Achchab, B., Herrera, F.: A first study exploring the performance of the state-of-the art CNN model in the problem of breast cancer. In: proceedings of the International Conference on Learning and Optimization Algorithms: Theory and Applications, pp. 1–6 (2018)
3. Bray, F., Ferlay, J., Soerjomataram, I., Siegel, R.L., Torre, L.A., Jemal, A.: Global cancer statistics 2018: Globocan estimates of incidence and mortality worldwide for 36 cancers in 185 countries. CA: Cancer J. Clin. **68**(6), 394–424 (2018)
4. Chen, L.C., Papandreou, G., Kokkinos, I., Murphy, K., Yuille, A.L.: Deeplab: semantic image segmentation with deep convolutional nets, atrous convolution, and fully connected CRFs. IEEE Trans. Pattern Anal. Mach. Intell. **40**(4), 834–848 (2017)
5. Chen, T., Kornblith, S., Mohammad, N., Hinton, G.: A simple framework for contrastive learning of visual representations. arXiv preprint arXiv:2002.05709 (2020)
6. Chen, Z., Guo, X., Woo, P.Y., Yuan, Y.: Super-resolution enhanced medical image diagnosis with sample affinity interaction. IEEE Trans. Med. Imaging **40**(5), 1377–1389 (2021)
7. Dong, C., Loy, C.C., He, K., Tang, X.: Learning a deep convolutional network for image super-resolution. In: Fleet, D., Pajdla, T., Schiele, B., Tuytelaars, T. (eds.) Computer Vision – ECCV 2014: 13th European Conference, Zurich, Switzerland, September 6-12, 2014, Proceedings, Part IV, pp. 184–199. Springer, Cham (2014). https://doi.org/10.1007/978-3-319-10593-2_13
8. Gandomkar, Z., Brennan, P.C., Mello-Thoms, C.: Mudern: multi-category classification of breast histopathological image using deep residual networks. Artif. Intell. Med. **88**, 14–24 (2018)
9. Hochreiter, S., Schmidhuber, J.: Long short-term memory. Neural Comput. **9**(8), 1735–1780 (1997)

10. Lai, W.S., Huang, J.B., Ahuja, N., Yang, M.H.: Deep Laplacian pyramid networks for fast and accurate super-resolution. In: Proceedings of the IEEE Conference on Computer Vision and Pattern Recognition, pp. 624–632 (2017)
11. Ledig, C., et al.: Photo-realistic single image super-resolution using a generative adversarial network. In: Proceedings of the IEEE Conference on Computer Vision and Pattern Recognition, pp. 4681–4690 (2017)
12. Li, J., et al.: Breast cancer histopathological image classification based on deep second-order pooling network. In: 2020 International Joint Conference on Neural Networks (IJCNN), pp. 1–7. IEEE (2020)
13. Li, X., Wang, W., Hu, X., Yang, J.: Selective kernel networks. In: Proceedings of the IEEE/CVF Conference on Computer Vision and Pattern Recognition, pp. 510–519 (2019)
14. Lin, M., Chen, Q., Yan, S.: Network in network. arXiv preprint arXiv:1312.4400 (2013)
15. Lin, T.Y., Dollár, P., Girshick, R., He, K., Hariharan, B., Belongie, S.: Feature pyramid networks for object detection. In: Proceedings of the IEEE Conference on Computer Vision and Pattern Recognition, pp. 2117–2125 (2017)
16. Lin, T.Y., Goyal, P., Girshick, R., He, K., Dollár, P.: Focal loss for dense object detection. In: Proceedings of the IEEE International Conference on Computer Vision, pp. 2980–2988 (2017)
17. Lim, B., Son, S., Kim, H., Nah, S., Lee, K.: Enhanced Deep Residual Networks for Single Image Super-Resolution. In: 2017 IEEE Conference on Computer Vision and Pattern Recognition Workshops, CVPR Workshops 2017, pp. 1132–1140. IEEE (2017)
18. Shahidi, F.: Breast cancer histopathology image super-resolution using wide-attention GAN with improved Wasserstein gradient penalty and perceptual loss. IEEE Access **9**, 32795–32809 (2021)
19. Sohn, K.: Improved deep metric learning with multi-class n-pair loss objective. Adv. Neural Inform. Process. Syst. **29** 1857–1865 (2016)
20. Spanhol, F.A., Oliveira, L.S., Petitjean, C., Heutte, L.: A dataset for breast cancer histopathological image classification. IEEE Trans. Biomed. Eng. **63**(7), 1455–1462 (2016). https://doi.org/10.1109/TBME.2015.2496264
21. Spanhol, F.A., Oliveira, L.S., Petitjean, C., Heutte, L.: Breast cancer histopathological image classification using convolutional neural networks. In: 2016 International Joint Conference on Neural Networks (IJCNN), pp. 2560–2567 (2016). https://doi.org/10.1109/IJCNN.2016.7727519
22. Wang, P., Wang, J., Li, Y., Li, P., Li, L., Jiang, M.: Automatic classification of breast cancer histopathological images based on deep feature fusion and enhanced routing. Biomed. Signal Process. Control **65**, 102341 (2021)
23. Woo, S., Park, J., Lee, J.Y., Kweon, I.S.: Cbam: Convolutional block attention module. In: Proceedings of the European Conference on Computer Vision (ECCV), pp. 3–19 (2018)
24. Xu, B., Zhang, W.: Selective scale cascade attention network for breast cancer histopathology image classification. In: ICASSP 2022–2022 IEEE International Conference on Acoustics, Speech and Signal Processing (ICASSP), pp. 1396–1400. IEEE (2022)
25. Zhang, Z.: Improved adam optimizer for deep neural networks. In: 2018 IEEE/ACM 26th International Symposium on Quality of Service (IWQoS), pp. 1–2. IEEE (2018)

cOOpD: Reformulating COPD Classification on Chest CT Scans as Anomaly Detection Using Contrastive Representations

Silvia D. Almeida[1,2,3](✉), Carsten T. Lüth[4,5], Tobias Norajitra[1,3], Tassilo Wald[1,5], Marco Nolden[1], Paul F. Jäger[4,5], Claus P. Heussel[3,6], Jürgen Biederer[3,7], Oliver Weinheimer[3,7], and Klaus H. Maier-Hein[1,3,5]

[1] Division of Medical Image Computing, German Cancer Research Center, Heidelberg, Germany
silvia.diasalmeida@dkfz-heidelberg.de
[2] Medical Faculty, Heidelberg University, Heidelberg, Germany
[3] Translational Lung Research Center Heidelberg (TLRC), Member of the German Center for Lung Research (DZL), Heidelberg, Germany
[4] Interactive Machine Learning Group, German Cancer Research Center, Heidelberg, Germany
[5] Helmholtz Imaging, German Cancer Research Center, Heidelberg, Germany
[6] Diagnostic and Interventional Radiology with Nuclear Medicine, Thoraxklinik at University Hospital, Heidelberg, Germany
[7] Diagnostic and Interventional Radiology, University Hospital, Heidelberg, Germany

Abstract. Classification of heterogeneous diseases is challenging due to their complexity, variability of symptoms and imaging findings. Chronic Obstructive Pulmonary Disease (COPD) is a prime example, being underdiagnosed despite being the third leading cause of death. Its sparse, diffuse and heterogeneous appearance on computed tomography challenges supervised binary classification. We reformulate COPD binary classification as an anomaly detection task, proposing cOOpD: heterogeneous pathological regions are detected as Out-of-Distribution (OOD) from normal homogeneous lung regions. To this end, we learn representations of unlabeled lung regions employing a self-supervised contrastive pretext model, potentially capturing specific characteristics of diseased and healthy unlabeled regions. A generative model then learns the distribution of healthy representations and identifies abnormalities (stemming from COPD) as deviations. Patient-level scores are obtained by aggregating region OOD scores. We show that cOOpD achieves the best performance on two public datasets, with an increase of 8.2% and 7.7% in terms of AUROC compared to the previous supervised state-of-the-art. Additionally, cOOpD yields well-interpretable spatial anomaly maps and

S. D. Almeida and C. T. Lüth—These authors contributed equally to this work.

Supplementary Information The online version contains supplementary material available at https://doi.org/10.1007/978-3-031-43904-9_4.

H. Greenspan et al. (Eds.): MICCAI 2023, LNCS 14224, pp. 33–43, 2023.
https://doi.org/10.1007/978-3-031-43904-9_4

patient-level scores which we show to be of additional value in identifying individuals in the early stage of progression. Experiments in artificially designed real-world prevalence settings further support that anomaly detection is a powerful way of tackling COPD classification. Code is at https://github.com/MIC-DKFZ/cOOpD.

Keywords: COPD classification · Anomaly detection · Contrastive learning

1 Introduction

By virtue of the human body's complexity, most diseases present phenotypic variability in terms of symptoms, rate of progression and imaging findings, which challenges diagnostic criteria. Among many hard to diagnose diseases, Chronic Obstructive Pulmonary Disease (COPD) stands out, as it is extensively under- and misdiagnosed [20], despite being the 3rd leading cause of death worldwide, with an estimated global prevalence of 10.3% [3]. Its pathological manifestations in the lung range from emphysema to airway disease, leading to a sparse, diffuse, and heterogeneous appearance, as shown in Fig. 1a. Appropriate and earlier diagnosis that accounts for all of its manifestations is therefore of paramount importance for public health [1].

Considering the limitations of spirometry as the standard diagnostic method [1], computed tomography (CT) has emerged as a complementary tool for COPD characterization. Initial efforts focused on identifying typical intensity and texture-level imaging features from either inspiration or expiration CT scans [4]. With the advent of deep learning (DL), more complex supervised approaches have been proposed to tackle binary classification of COPD. In this context, due to GPU memory constraints and large size of the images, different strategies to parcel a single 3D image as 2D slices [9,23] or 3D patches [19] have been pursued by supervised DL methods. Significant emphasis has been put on multiple instance learning (MIL) approaches [8,22,25], considering the spatial heterogeneity of COPD and that only a binary label is needed in case-finding scenarios. Typically, for a supervised model to learn good decision boundaries, the labeled training dataset needs good coverage of the appearances of all classes. However, good coverage of the diseased class can be difficult for low prevalence and heterogeneous diseases, making supervised models susceptible to fail on novel data points [13] (Fig. 1b). COPD fits exactly in this scenario, as its manifestations in the lung are diverse, in contrast to healthy individuals whose lungs are generally more uniform in appearance. This raises questions about the suitability of supervised models for COPD classification.

Instead of attempting to learn all possible complex manifestations of the disease, we ask: *Could COPD be more accurately detected if considered as an anomaly from the distribution of healthy lungs?*

As previously reported for anomaly detection [17,26], modeling the distribution of normal samples in the latent space, instead of in the voxel space, has shown to be both feasible and desirable. With this in mind, our contribution is two-fold:

1. We show the benefit of reformulating COPD prediction as an anomaly detection task. Inspired by [16], we develop a generative model operating on the self-supervised representation space (Fig. 1c), learning the distribution of labeled healthy features and identifying unknown abnormal ones (stemming from **COPD**) as **O**ut-of-**D**istribution (**cOOpD**). cOOpD outperforms all compared DL-based supervised methods on two distinct public datasets, whilst maintaining performance in a scenario using a simulated real-world prevalence training dataset.
2. We highlight the benefit of moving from voxels to representation space through a supervised method that leverages contrastive representations of lower dimensionality than voxels and outperforms voxel-based classifiers.

To the best of our knowledge, this work is the first to investigate anomaly detection in the context of a heterogeneous lung disease classification and has the potential to be applied to a wide range of diffuse diseases affecting large body areas.

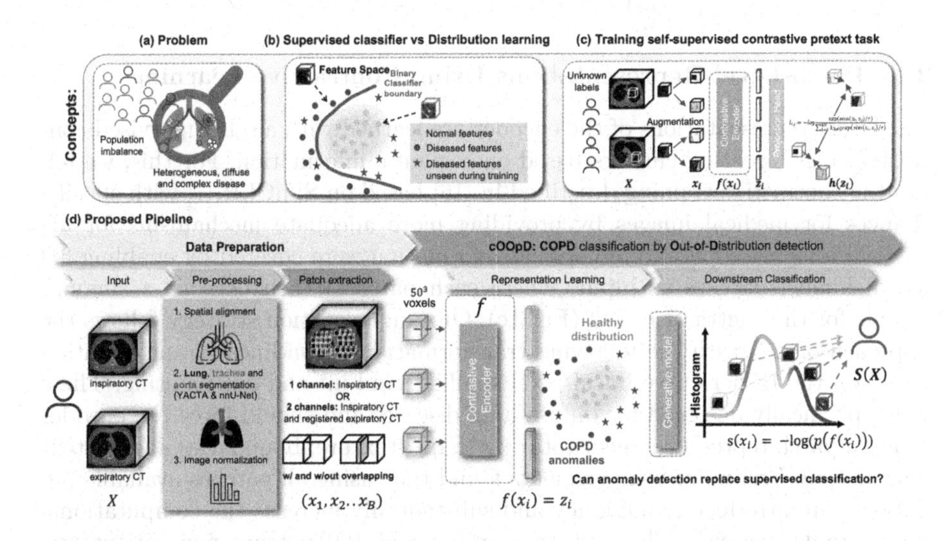

Fig. 1. (a) Unbalanced prevalence of heterogeneous diseases in the population. (b) In the feature space, traditional classification methods may struggle with rare cases, while anomaly detection might lead to improved decision boundaries. (c) Self-supervised contrastive learning extracts meaningful representations from unlabeled lung patches, transitioning from voxels to features. (d) cOOpD during inference: paired CT scans are pre-processed, then patches are extracted and representations are obtained. Anomaly scores are assigned per patch, based on the distribution of healthy representations, which are then aggregated by patient.

2 Method

Our proposed method cOOpD aims at reformulating COPD classification as anomaly detection. It is a self-supervised anomaly detection framework, inspired by the strategy of [16], optimized for diffuse lung diseases covering 3D multi-channel inputs, suitable augmentation strategies and patient-level aggregation of patch-level scores. During inference (Fig. 1d), a sequence of B 3D lung patches $\{x_i\}_{i=1}^{B}$ from a single or paired CT scan X is extracted. Then, for each patch, a representation is obtained using a trained self-supervised contrastive encoder $z_i = f(x_i)$ (Sect. 2.1). Having learned the distribution of healthy patch-representations (Sect. 2.2), the patch-level latent representation is given to a generative model $p(z)$, being attributed an anomaly score defined as the negative log-likelihood: $s(x_i) = -log(p(f(x_i)))$. Several aggregation strategies $S(x)$ of these scores to patient-level were tested (supplementary) including the mean, which was found to be the best performing and most conceptually meaningful strategy, as outlined in Eq. 1.

$$S(X) = \sum_{i=1}^{B} s(x_i)/B = -\log\left(\prod_{i=1}^{B} p\left(f(x_i)\right)^{1/B}\right) \tag{1}$$

2.1 Patch-Level Representations Using Contrastive Learning

The latent representations of the encoder are learned with a self-supervised contrastive task, creating clusters based on semantic information. For this, we follow the contrastive training described in [16] based on SimCLR [7] with specific changes for medical images by providing more adequate mechanisms for 3D medical imaging. Context and spatial information were covered by enabling 3D multi-channel patches as input, where each patch is then used as a singular sample for the contrastive task (Fig. 1c). Our augmentation strategy follows the approach of [27] with the following transformations: Non-linear transformation based on the Bézier curve, local-pixel shuffling and in- and out-painting. These were specifically designed for diffuse lung diseases and should force the encoder to learn patch representations capturing shape, texture, boundaries and context information. Preliminary experiments found that using all patches available per patient can introduce redundancy and substantially increase the computational cost (supplementary). Therefore, a maximum of 100 patches per patient was set for training the self-supervised contrastive task. As an encoder, different 3D ResNet configurations (18 and 34) were tested.

2.2 Generative Models Operating on Representation Space

Once having extracted the latent representations, the distribution of normal representations is modeled, by fitting a generative model $p(z)$ on the representations of purely normal patches. Patch normality is defined by $\%_{\text{emphysema}} < 1\%$ strictly applied to normal individuals, a very restrictive bound to guarantee that no intensity alterations could be present in the definition of normality. $\%_{\text{emphysema}}$ is defined as the percentage of low attenuation areas less than a threshold of −950 Hounsfield units [4]. As generative models, Gaussian Mixture Model (GMM) and Normalizing Flow (NF) are employed. While both are density estimation methods used to model $p(z)$, GMMs model the probability density function of the data as a weighted sum of Gaussian distributions, whereas our NF model uses the change of variable formula with a Gaussian prior. The implementation of the NF is identical to [16] consisting of fully connected affine coupling blocks and permutations based on the RealNVP architecture. We fit several GMM with $\kappa \in 1, 2, 4, 8$ and a NF on representations of the encoder from the purely normal patches of healthy patients from the training dataset without any transformations. The best performing generative model is selected based on the validation set performance.

3 Experiment Setup

Dataset & Preprocessing: Paired inspiratory and expiratory volumetric CT images were used from two nationwide multi-center studies (COPDGene [18][1] and COSYCONET [12]), from which 5244 and 484 unique individuals were randomly selected, respectively (supplementary). Binary classes were defined based on the Global Initiative for Chronic Obstructive Lung Disease (GOLD), a discrete score between 0–4. The negative class (healthy) included never-smokers and individuals with a GOLD score of 0, while the positive class (diseased) included those with a GOLD score of 1 or higher. This resulted in the prevalence of the positive class being 57% for COPDGene and 85% for COSYCONET. All trainings were performed on COPDGene which was split randomly into training (50%), validation (25%) and test (25%) sets on the patient-level. COSYCONET was entirely used as an external test dataset. The data preparation process is illustrated in Fig. 1d and comprises the following sequential steps:

Spatial Alignment of Paired Inspiratory and Expiratory CT Images: Considering the potential of adding the expiratory scan as an extra channel as an indirect measure of gas trapping [4], the paired images were geometrically aligned. Having the inspiratory image as the fixed image, an adaptation of [21] was performed.

Lung Parenchyma Segmentation for Patch Extraction: Lung masks were generated on the inspiratory image space using a nnU-Net model [11] on YACTA [2] segmentation masks, a validated intensity-based method.

Intensity Normalization: Inter-scanner variability was addressed by normalizing the intensity values to a scale between 0 (air) and 1 (tissue) [14]. Mean intensity values for air and tissue were derived from segmented tracheal and aortic regions,

[1] The COPDGene study (dbGaP #28978) was funded by NIH grants U01HL089856 and U01 HL089897 and also supported by the COPD Foundation through contributions made by an Industry Advisory Board comprised of Pfizer, AstraZeneca, Boehringer Ingelheim, Novartis, and Sunovion.

respectively, obtained using a pre-trained nnU-Net model (Task_055_SegTHOR). Additionally, all images were resampled to an isotropic resolution of 0.5 mm.

Patch Extraction: Volumetric patches (50^3 voxels) containing $> 70\%$ of the lung were extracted from the lung parenchyma of aligned inspiratory and expiratory CT images. The chosen size covered the secondary pulmonary lobule, the basic unit of lung structure [24]. Two different patch overlapping strategies were implemented (0% and 20%) on inspiratory (1-channel) and inspiratory + registered expiratory (2-channels) images. Thus, four different configurations of input patches were tested.

Baselines: State-of-the-art (SotA) baselines were applied to 2D slices and 3D patches. A *2D-CNN* [9] was employed at the patient-level. An end-to-end 3D patch classifier with score aggregation (*PatClass*), an MIL approach with a Recurrent Neural Network as aggregation (*MIL+RNN*) [5] and an Attention-based MIL (*MIL+Att*) (similar to [22], adapted from [10]) were employed at the patch-level. Implementation was performed as described in the original works, with adaptations to 3D, when required (supplementary).

Contrastive Representations Ablation: The contrastive latent representations' usefulness was evaluated with a supervised method (ReContrastive) that maps the latent representations back to their position in the original image, producing a 4D image, where the 4th dimension is the length of the latent representation vector (Suppl). The produced image is then used as input for a CNN classifier. Training was performed for 500 epochs using the SGD Optimizer, a learning rate of 1e−2, Cosine Annealing [15] and a weight decay of 3e−5. A combination of random cropping, random scaling, random mirroring, rotations, and Gaussian blurring was employed as transformations.

Evaluation Metrics: We used Area Under Receiver Operator Curve (AUROC) and Area Under Precision Recall Curve (AUPRC) as the default multi-threshold metric for classification. AUROC is used as the main evaluation metric since it is less sensitive to class balance changes.

Final Method Configurations: These were chosen based on the highest AUROC on three experiment runs on the validation set. The best patch extraction configuration for all tested 3D methods was two-channel (inspiratory and registered expiratory) with 20% patch overlap. The best performance was always achieved with a ResNet34. For our proposed cOOpD method, GMM with $\kappa = 4$ was found to be the best performing generative model.

Real-World Prevalence Ablation: Given the global prevalence of COPD at 10.3% [3], we further evaluated the top two performing approaches in scenarios designed to approximate this real world prevalence. To better reflect these conditions, the diseased class in the COPDGene training set was undersampled to 5%, 10.3% and 15% while keeping all samples from the normal class, limiting the diversity of the diseased class in the training set (instead of oversampling the normal class).

4 Results

As shown in Table 1, cOOpD outperforms all SotA supervised methods, achieving statistically significant improvements in terms of AUROC of 8.2% compared to the best method on COPDGene (PatClass+RNN), and 7.7% on COSYCONET (MIL+RNN). ReContrastive, as a supervised ablation for assessing the advantage of using representations, also outperformed all the other voxel-based supervised strategies on the internal test set, by an AUROC difference of 3.8%

Table 1. Mean ± standard deviation in % of 3 independent runs on the internal (COPDGene) and external (COSYCONET) test sets. Levels of statistical significance are denoted by ($p<0.05^{*}/0.01^{**}$) in comparison to the proposed method cOOpD (paired samples t-test).

Input	Methods	COPDGene		COSYCONET	
		AUROC	AUPRC	AUROC	AUPRC
2D image	*2D-CNN* [9]	$55.6 \pm 2.5^{**}$	$72.0 \pm 1.5^{**}$	$57.0 \pm 8.0^{**}$	$84.6 \pm 1.4^{**}$
3D patch	*PatClass + RNN*	$76.1 \pm 0.2^{**}$	$86.3 \pm 0.1^{**}$	$56.2 \pm 0.7^{**}$	$95.3 \pm 0.1^{*}$
	MIL + RNN [5]	$73.0 \pm 0.6^{**}$	$84.5 \pm 0.5^{**}$	$60.2 \pm 4.2^{*}$	$95.7 \pm 0.4^{*}$
	MIL + Att [10,22]	$65.8 \pm 1.2^{**}$	$80.9 \pm 0.8^{**}$	$57.7 \pm 1.3^{**}$	$95.1 \pm 0.2^{*}$
	ReContrastive (ours)	$79.9 \pm 0.3^{**}$	$88.5 \pm 0.2^{*}$	$53.3 \pm 0.1^{**}$	$95.0 \pm 0.1^{**}$
	cOOpD (ours)	$\mathbf{84.3 \pm 0.3}$	$\mathbf{89.7 \pm 0.2}$	$\mathbf{67.9 \pm 0.7}$	$\mathbf{96.5 \pm 0.4}$

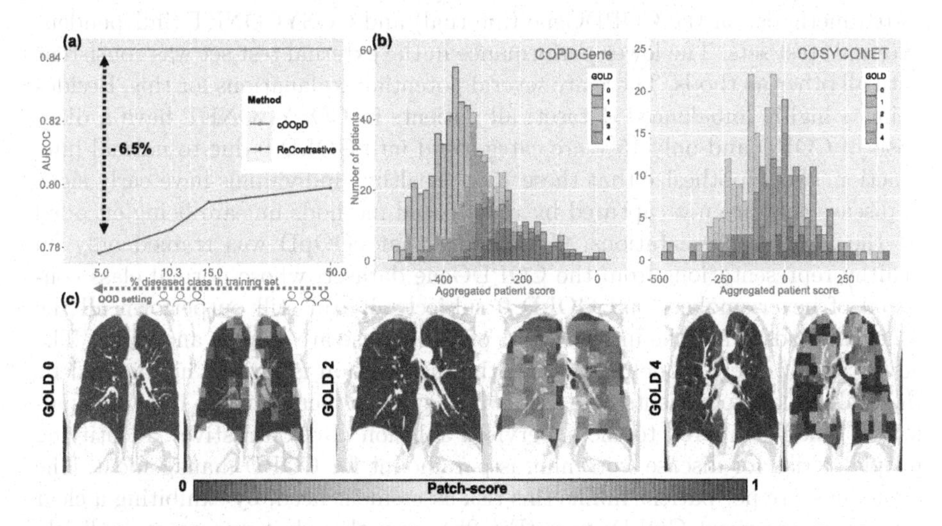

Fig. 2. (a) Real world AUROC (5%, 10.3% and 15% of diseased class prevalence) for the top two performing methods compared to the baseline (50%). (b) Patient distribution by mean aggregated scores from cOOpD, colored by the function risk score (GOLD), for COPDGene and COSYCONET. (c) Coronal view of the cOOpD score map on three subjects with different degrees of severity. Min-max normalization was applied to patch scores, corresponding to the 5th–95th percentiles of the internal testset.

but shows a large performance drop leading to the worst AUROC on the external test set. In the real-world ablation, as seen in Fig. 2a, the best performing supervised method (ReContrastive) performance decreases with the diseased class prevalence, reaching a drop of 6.5% compared to cOOpD.

5 Discussion

Final Method Configurations: The best working method configurations reflect the following properties of the task: Using both inspiratory and expiratory images provides information about pulmonary vascular alterations and airway wall thickness not visible on the inspiratory scan alone, as in line with [6]. Finer grained information is captured using overlapping patches, which tended to work better in conjunction with deeper encoders. Regarding our proposed method cOOpD, we note the following: We hypothesize that the latent space's complexity level is low, being easily covered with a simple generative model. As for the aggregation strategy, considering the spatial distribution of COPD, it can happen that only a small part of the lung is diseased. As the negative-log-likelihood has a lower bound but not an upper bound, a single patch having a high score leads to a high overall score when using mean aggregation, which is the desired behavior.

Should COPD Binary Classification be Formulated as Anomaly Detection? cOOpD performance shows to be significantly superior compared to all tested methods, on the COPDGene (internal) and COSYCONET (independent external) test sets. The lower performance in the external test set was consistent with all other methods. There are several potential explanations for this. Besides being a highly imbalanced dataset, all patients in COSYCONET have a diagnosis of COPD and only 15% are categorized into GOLD 0 due to normal lung function. We hypothesize that these 15% "healthy" individuals have early signs of disease that are not captured by voxel-based methods but are being encoded by the latent representations. Considering that cOOpD was trained only on healthy representations from the COPDGene dataset, whose normal class consisted of never-smokers and GOLD 0 subjects, it can still outperform all the other methods, since the unseen traits of the disease are seen as anomalies. The advantage of solely modeling the healthy distribution is further highlighted by the real-world experiments (Fig. 2a), where cOOpD performance remains unaffected, when compared to the supervised ablation (ReContrastive). Identifying people at risk for disease worsening is paramount for COPD management. The anomaly score per patient fulfills this risk assessment need, by exhibiting a clear relation to the exact GOLD stage (Fig. 2b), even though it was never explicitly given the GOLD stage as a multi-class label. Further, the lung region scores enable spatial localization of anomalies, giving interpretability to the method (Fig. 2c). These findings support our approach of reformulating COPD binary classification as an anomaly detection task.

Are Self-supervised Patch-Level Latent Representations Advantageous over Voxels? Both methods working on the representation space (cOOpD and ReContrastive) outperform all voxel-based baselines on the internal test set. Although for ReContrastive this improvement is no longer seen for the external test set, being the worst performing method, the early signs of disease for the healthy class of COSYCONET are likely being encoded by the latent representations, as mentioned earlier. We hypothesize that this performance drop stems from the problem of supervised models depicted in Fig. 1b. Combined with the cOOpD findings, this still supports the hypothesis that patch-level latent representations provide meaningful information and reduce the complexity of the problem.

6 Conclusion

Our proposed reformulation of COPD binary classification into an anomaly detection task (cOOpD) demonstrated superior performance compared to SotA methods. Additionally, the advantage of using latent representations was demonstrated. The cOOpD approach also demonstrated stability in performance when trained on datasets with simulated real-world class imbalance. Future work should focus on further validation on larger and more diverse datasets, longitudinal evaluation, and exploring its application to other heterogeneous diseases where annotated diseased data is scarce and access to healthy data is abundant.

Acknowledgments. This research was funded by the State Ministry of Baden-Wuerttemberg for Sciences, Research and Arts, Germany grant number 32-5400/58/3 and Helmholtz Imaging (HI), a platform of the Helmholtz Incubator on Information and Data Science.

References

1. Global strategy for the diagnosis, management, and prevention of chronic obstructive pulmonary disease (2023). https://goldcopd.org/wp-content/uploads/2023/03/GOLD-2023-ver-1.3-17Feb2023_WMV.pdf. Accessed 08 Mar 2023
2. Achenbach, T., Weinheimer, O., Buschsieweke, C., Heussel, C., Thelen, M., Kauczor, H.: Vollautomatische detektion und quantifizierung des lungenemphysems in dünnschicht-MD-CT des thorax durch eine neue, speziell entwickelte software. Fortschr Röntgenstr **176**(10), 1409–1415 (2004). https://doi.org/10.1055/s-2004-813530
3. Adeloye, D., Song, P., Zhu, Y., Campbell, H., Sheikh, A., Rudan, I.: Global, regional, and national prevalence of, and risk factors for, chronic obstructive pulmonary disease (COPD) in 2019: a systematic review and modelling analysis. Lancet Respir. Med. **10**(5), 447–458 (2022). https://doi.org/10.1016/S2213-2600(21)00511-7
4. Bhatt, S.P., et al.: Imaging advances in chronic obstructive pulmonary disease. insights from the genetic epidemiology of chronic obstructive pulmonary disease (COPDGene) study. Am. J. Respir. Crit. Care Med. **199**(3), 286–301 (2019). https://doi.org/10.1164/rccm.201807-1351SO

5. Campanella, G., Hanna, M., Geneslaw, L., et al.: Clinical-grade computational pathology using weakly supervised deep learning on whole slide images. Nat. Med. **25**(8), 1301–1309 (2019). https://doi.org/10.1038/s41591-019-0508-1
6. Cao, X., Gao, X., Yu, N., et al.: Potential value of expiratory CT in quantitative assessment of pulmonary vessels in COPD. Front. Med. **8**, 761804 (2021). https://doi.org/10.3389/fmed.2021.761804
7. Chen, T., Kornblith, S., Norouzi, M., Hinton, G.: A Simple Framework for Contrastive Learning of Visual Representations. arXiv.cs.CV (2020). https://doi.org/10.48550/ARXIV.2002.05709
8. Cheplygina, V., Sorensen, L., Tax, D.M., Pedersen, J.H., Loog, M., Bruijne, M.D.: Classification of COPD with multiple instance learning. In: International Conference on Pattern Recognition, pp. 1508–1513. IEEE (2014). https://doi.org/10.1109/ICPR.2014.268
9. González, G., Ash, S.Y., Vegas-Sánchez-Ferrero, et al.: Disease staging and prognosis in smokers using deep learning in chest computed tomography. Am. J. Respir. Crit. Care Med. **197**(2), 193–203 (2018). https://doi.org/10.1164/rccm.201705-0860OC
10. Ilse, M., Tomczak, J.M., Welling, M.: Attention-based deep multiple instance learning. arXiv.stat.ML (2018). https://doi.org/10.48550/ARXIV.1802.04712
11. Isensee, F., Jaeger, P.F., Kohl, S.A.A., Petersen, J., Maier-Hein, K.H.: nnU-Net: a self-configuring method for deep learning-based biomedical image segmentation. Nat. Methods **18**(2), 203–211 (2021). https://doi.org/10.1038/s41592-020-01008-z
12. Karch, A., Vogelmeier, C., Welte, T., et al.: The German COPD cohort COSYCONET: aims, methods and descriptive analysis of the study population at baseline. Respir. Med. **114**, 27–37 (2016). https://doi.org/10.1016/j.rmed.2016.03.008
13. Kim, C.: Multicentre external validation of a commercial artificial intelligence software to analyse chest radiographs in health screening environments with low disease prevalence. Eur. Radiol. (2023). https://doi.org/10.1007/s00330-022-09315-z
14. Kim, S.S., et al.: Improved correlation between CT emphysema quantification and pulmonary function test by density correction of volumetric CT data based on air and aortic density. Eur. J. Radiol. **83**(1), 57–63 (2014). https://doi.org/10.1016/j.ejrad.2012.02.021
15. Loshchilov, I., Hutter, F.: SGDR: Stochastic gradient descent with warm restarts. arXiv.cs.LG (2016). https://doi.org/10.48550/ARXIV.1608.03983
16. Lüth, C.T., et al.: CRADL: contrastive representations for unsupervised anomaly detection and localization. arXiv.cs.CV (2023). https://doi.org/10.48550/ARXIV.2301.02126
17. Marimont, S.N., Tarroni, G.: Anomaly detection through latent space restoration using vector quantized variational autoencoders. In: 2021 IEEE 18th International Symposium on Biomedical Imaging (ISBI), pp. 1764–1767. IEEE (2021)
18. Regan, E.A., et al.: Genetic epidemiology of COPD (COPDGene) study design. COPD: J. Chronic Obstructive Pulm. Disease **7**(1), 32–43 (2011). https://doi.org/10.3109/15412550903499522
19. Singla, S., Gong, M., Ravanbakhsh, S., Sciurba, F., Poczos, B., Batmanghelich, K.N.: Subject2Vec: generative-discriminative approach from a set of image patches to a vector. In: Frangi, A.F., Schnabel, J.A., Davatzikos, C., Alberola-López, C., Fichtinger, G. (eds.) MICCAI 2018. LNCS, vol. 11070, pp. 502–510. Springer, Cham (2018). https://doi.org/10.1007/978-3-030-00928-1_57

20. Soriano, J.B., Zielinski, J., Price, D.: Screening for and early detection of chronic obstructive pulmonary disease. Lancet **374**(9691), 721–732 (2009). https://doi.org/10.1016/S0140-6736(09)61290-3
21. Staring, M., Klein, S., Reiber, J.H., Niessen, W.J., Stoel, B.C.: Pulmonary image registration with elastix using a standard intensity-based algorithm. Med. Image Anal. Clin.: Grand Challenge, 73–79 (2010)
22. Sun, J., Liao, X., Yan, Y., et al.: Detection and staging of chronic obstructive pulmonary disease using a computed tomography-based weakly supervised deep learning approach. Eur. Radiol. **32**(8), 5319–5329 (2022). https://doi.org/10.1007/s00330-022-08632-7
23. Tang, L.Y.W., Coxson, H.O., Lam, S., Leipsic, J., Tam, R.C., Sin, D.D.: Towards large-scale case-finding: training and validation of residual networks for detection of chronic obstructive pulmonary disease using low-dose CT. Lancet Digit. Health **2**(5), e259–e267 (2020). https://doi.org/10.1016/S2589-7500(20)30064-9
24. Webb, W.R.: Thin-section CT of the secondary pulmonary lobule: Anatomy and the image-the 2004 Fleischner lecture. Radiology **239**(2), 322–338 (2006). https://doi.org/10.1148/radiol.2392041968
25. Xu, C., et al.: DCT-MIL: deep CNN transferred multiple instance learning for COPD identification using CT images. Phys. Med. Biol. **65**(14), 145011 (2020). https://doi.org/10.1088/1361-6560/ab857d
26. Zhang, H., Li, A., Guo, J., Guo, Y.: Hybrid models for open set recognition. In: Vedaldi, A., Bischof, H., Brox, T., Frahm, J.-M. (eds.) ECCV 2020. LNCS, vol. 12348, pp. 102–117. Springer, Cham (2020). https://doi.org/10.1007/978-3-030-58580-8_7
27. Zhou, Z., et al.: Models genesis: generic autodidactic models for 3D medical image analysis. In: Shen, D., et al. (eds.) MICCAI 2019. LNCS, vol. 11767, pp. 384–393. Springer, Cham (2019). https://doi.org/10.1007/978-3-030-32251-9_42

YONA: You Only Need One Adjacent Reference-Frame for Accurate and Fast Video Polyp Detection

Yuncheng Jiang[1,2,4], Zixun Zhang[1,2,4], Ruimao Zhang[3], Guanbin Li[5], Shuguang Cui[1,2], and Zhen Li[1,2,4](✉)

[1] SSE, The Chinese University of Hong Kong, Shenzhen, China
lizhen@cuhk.edu.cn
[2] FNii, The Chinese University of Hong Kong, Shenzhen, China
[3] SDS, The Chinese University of Hong Kong, Shenzhen, China
[4] Shenzhen Research Insititute of Big Data, Shenzhen, China
[5] School of Computer Science and Engineering, Sun Yat-sen University, Guangzhou, China

Abstract. Accurate polyp detection is essential for assisting clinical rectal cancer diagnoses. Colonoscopy videos contain richer information than still images, making them a valuable resource for deep learning methods. However, unlike common fixed-camera video, the camera-moving scene in colonoscopy videos can cause rapid video jitters, leading to unstable training for existing video detection models. In this paper, we propose the **YONA** (**Y**ou **O**nly **N**eed one **A**djacent Reference-frame) method, an efficient end-to-end training framework for video polyp detection. YONA fully exploits the information of one previous adjacent frame and conducts polyp detection on the current frame without multi-frame collaborations. Specifically, for the foreground, YONA adaptively aligns the current frame's channel activation patterns with its adjacent reference frames according to their foreground similarity. For the background, YONA conducts background dynamic alignment guided by inter-frame difference to eliminate the invalid features produced by drastic spatial jitters. Moreover, YONA applies cross-frame contrastive learning during training, leveraging the ground truth bounding box to improve the model's perception of polyp and background. Quantitative and qualitative experiments on three public challenging benchmarks demonstrate that our proposed YONA outperforms previous state-of-the-art competitors by a large margin in both accuracy and speed.

Keywords: Video Polyp Detection · Colonoscopy · Feature Alignment · Contrastive Learning

Y. Jiang and Z. Zhang—Equal contribution.

Supplementary Information The online version contains supplementary material available at https://doi.org/10.1007/978-3-031-43904-9_5.

H. Greenspan et al. (Eds.): MICCAI 2023, LNCS 14224, pp. 44–54, 2023.
https://doi.org/10.1007/978-3-031-43904-9_5

1 Introduction

Colonoscopy plays a crucial role in identifying and removing early polyps and reducing mortality rates associated with rectal cancer. Over the past few years, the research community has devoted great effort to understanding colonoscopy videos using either optical flow [22,23] or temporal information aggregation [5, 12,16,19] between multiple frames.

However, those works are mainly designed based on the experience of previous natural video object detection studies, ignoring the inherent uniqueness of the colonoscopy motion patterns. Thus, we rethink the video polyp detection task and conclude three core challenges in colonoscopy videos. **1) Fast motion speed**. In Fig. 1(a), we show the target motion speed [26][1] on ImageNetVID [14] (natural) and LDPolypVideo [9] (colonoscopy) dataset. The motion speed in ImageNetVID evenly distributes in three intervals. In contrast, most targets in LDPolypVideo fall in the fast speed zone, leading to a large variance in the adjacent foreground features, like motion blur or occlusion, as shown in Fig. 1(c). Thus we conjecture that collaborating too many frames for polyp video detection will increase the misalignment between adjacent frames and leads to poor detection performance. Figure 1(b) shows the performance of FGFA [26] on two datasets with increasing reference frames. The different trends of the two lines confirm our hypothesis. **2) Complex background**. Different from the common camera-fixed videos, the camera-moving of colonoscopy video will introduce large disturbances between adjacent frames (e.g., specular reflection, bubbles, water,

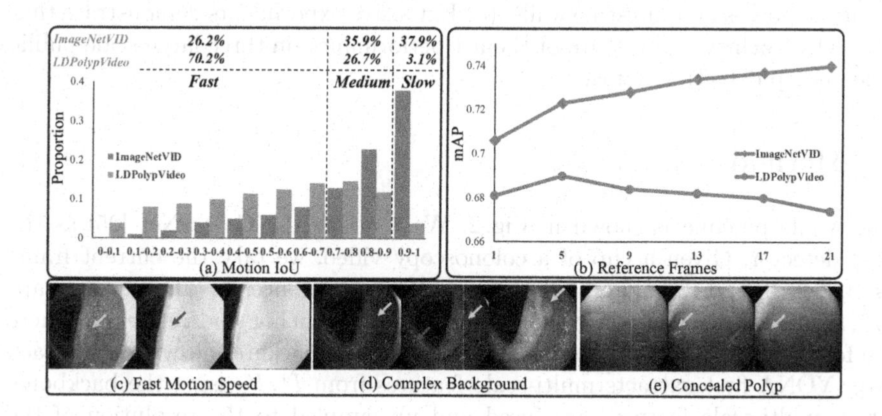

Fig. 1. (a) The histogram of the motion IoUs distribution on two datasets. Lower motion IoU denotes a faster target moving speed. The proportion of slow, medium and fast-moving targets is listed at the top of the figure. (b) The performance of FGFA [26] using multiple reference frames increases on ImageNetVID while decreasing on LDPolypVideo. (c) The typical challenges in colonoscopy videos. Yellow arrows point to the polyp, and red arrows point to distraction that causes false detection. (Color figure online)

[1] averaged intersection-over-union scores of target in the nearby frames (±10 frames).

etc.), as shown in Fig. 1(d). Those abnormalities disrupt the integrity of background structures and thus affect the effect of multi-frame fusion. **3) Concealed polyps.** As shown in Fig. 1(e), we noticed that some polyps could be seen as concealed objects in the colonoscopy video since such polyps have a very similar appearance to the intestine wall. The model will be confused by such frames in inference and result in high false-positive or false-negative predictions.

To address the above issues, we propose the **YONA** framework, which fully exploits the reference frame information and only needs one adjacent reference frame for accurate video polyp detection. Specifically, we propose the Foreground Temporal Alignment (FTA) module to explicitly align the foreground channel activation patterns between adjacent features according to their foreground similarity. In addition, we design the Background Dynamic Alignment (BDA) module after FTA that further learns the inter-frame background spatial dynamics to better eliminate the influence of motion speed and increase the training robustness. Finally, parallel to FTA and BDA, we introduce the Cross-frame Box-assisted Contrastive Learning (CBCL) that fully utilizes the box annotations to enlarge polyp and background discrimination in embedding space.

In summary, our contributions are in three-folds: (1) To the best of our knowledge, we are the first to investigate the obstacles to the development of existing video polyp detectors and conclude that two-frame collaboration is enough for video polyp detection. (2) We propose the YONA, a novel framework for video polyp detection. It composes the foreground and background alignment modules to align the features under the fast-moving condition. It further introduces the cross-frame contrastive learning module to enhance the model's discrimination ability of polyps and intestine walls. (3) Extensive experiments demonstrate that our YONA achieves new state-of-the-art performance on three large-scale public video polyp detection datasets.

2 Method

The whole pipeline is shown in Fig. 2. We leverage the CenterNet [25] as the base detector. Given a clip of a colonoscopy video, we take the current frame as anchor I^a and its adjacent previous frame as reference I^r. The binary maps M^a, M^r are generated using the bounding box of anchor and reference, where the foreground pixels are assigned with 1 while the background with 0. At each step, YONA first extracts multi-scale features from I^a, I^r using the backbone. Then, multi-scale features are fused and up-sampled to the resolution of the first stage as the intermediate features F^a, F^r. Then, we conduct foreground temporal alignment (Fig. 2(a)) on intermediate features to align their channel activation pattern. Next, the enhanced anchor feature $\tilde{F}$ is further refined by the background dynamic alignment module (Fig. 2(b)) to mitigate the rapid dynamic changes in the spatial field. The BDA's output F^* is used to compute the detection loss. Meanwhile, the intermediate features and binary maps are used to calculate the contrastive loss during training to improve the model's perception of polyp and background (Fig. 2(c)).

Overall, the whole network is optimized with the combination loss function in an end-to-end manner. The final loss is composed of the same detection loss with CenterNet and our proposed contrastive loss, formulated as $\mathcal{L} = \mathcal{L}_{\text{detection}} + \lambda_{contrast}\mathcal{L}_{\text{contrast}}$.

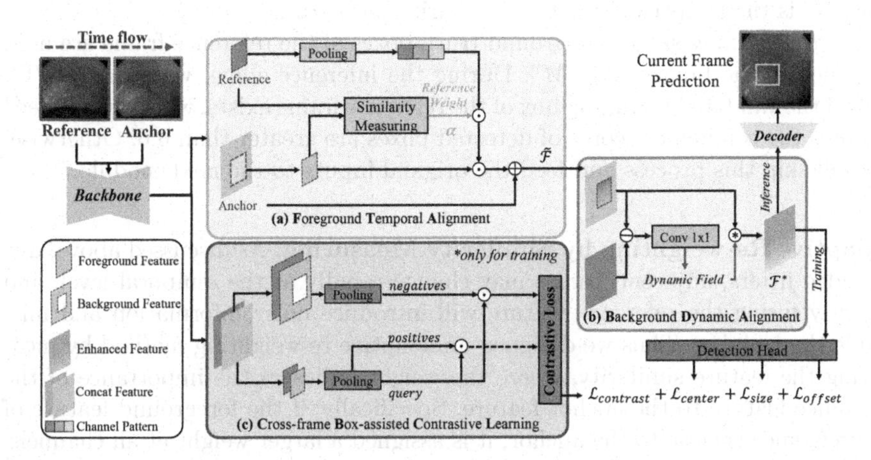

Fig. 2. Illustration of our proposed video polyp detection framework, YONA. It first aligns the foreground channel patterns between the anchor and reference frame in (a). Then it extracts polyp context guided by dynamic field in (b). Meanwhile, YONA enhances the discrimination ability via contrastive learning in (c) during training. The final output of (b) is used to predict the bounding box of the current frame.

2.1 Foreground Temporal Alignment

Since the camera moves at a high speed, the changes in the frame are very drastic for both foreground and background targets. As a result, multi-frame (reference>3) fusion may easily incorporate more noise features into the aggregation features. On the other hand, the occluded or distorted foreground context may also influence the quality of aggregation. Thus we propose to conduct temporal alignment between adjacent features by leveraging the foreground context of only **one** adjacent reference frame. It is designed to align the certain channel's activation pattern of anchor feature to its preceding reference feature. Specifically, given the intermediate features F^a, F^r and reference binary map M^r, we first pooling F^r to 1D channel pattern f^r by the binary map on the spatial dimension ($\mathbb{R}^{N\times C\times H\times W} \rightarrow \mathbb{R}^{N\times C\times 1}$) and normalize it to $[0, 1]$:

$$\begin{aligned} f^r &= \text{norm}\left[\text{Pooling}(F^r)\right] \\ \text{Pooling}(F^r) &= \text{sum}_{HW}\left[F^r(x,y)\right]/\text{sum}[M^r(x,y)] \quad \text{if } M^r(x,y)=1 \end{aligned} \tag{1}$$

Then, the foreground temporal alignment is implemented by channel attention mechanism, where the attention maps are computed by weighted dot-product.

We obtain the enhanced anchor feature by adding the attention maps with the original anchor feature through skip connection to keep the gradient flow.

$$\tilde{\mathcal{F}} = [\alpha f^r \odot F^a(x,y)] \oplus F^a \quad \text{if } M^r(x,y) = 1 \tag{2}$$

where α is the adaptive weight by similarity measuring.

At the training stage, the ground truth boxes of the reference frame are used to generate the binary map M^r. During the inference stage, we conduct FTA only if the validated bounding box of the reference frame exists, where "validated" denotes the confidence scores of detected boxes are greater than 0.6. Otherwise, we will skip this process and feed the original inputs to the next module.

Adaptive Re-weighting by Similarity Measuring. As discussed above, due to video jitters, adjacent frames may change rapidly at the temporal level, and directly fusing the reference feature will introduce noisy information and misguide the training. Thus we designed an adaptive re-weighting method by measuring the feature similarity, where the weight indicates the importance of the reference feature to the anchor feature. Specifically, if the foreground feature of the reference is close to the anchor, it is assigned a larger weight at all channels. Otherwise, a smaller weight is assigned. For efficiency, we use the cosine similarity metric [8] to measure the similarity, where f^a is the 1D channel pattern of F^a computed with Eq. 1:

$$\alpha = exp\left(\frac{f^r \cdot f^a}{|f^r||f^a|}\right) \tag{3}$$

2.2 Background Dynamic Alignment

The traditional convolutional-based object detector can detect objects well when the background is stable. However, once it receives obvious interference, such as light or shadow, the background changes may cause the degradation of spatial correlation and lead to many false-positive predictions. Motivated by the inter-frame difference method [20], we first mine the dynamic field of adjacent background contents, then consult to deformable convolution [3] to learn the inherent geometric transformations according to the intensity of the dynamic field. In practice, given the enhanced anchor feature $\tilde{\mathcal{F}}$ from FTA and reference feature F^r, the inter-frame difference is defined as the element-wise subtraction of enhanced anchor and reference feature. Then a 1×1 convolution is applied on the difference to generate dynamic field $\mathcal{D}$, which encodes all spatial dynamic changes between adjacent frames.

$$\mathcal{D} = \text{Conv}_{1\times1}(\tilde{\mathcal{F}} - F^r) \tag{4}$$

Finally, a 3×3 deformable convolution embeds the spatial dynamic changes of $\mathcal{D}$ on the enhanced anchor feature $\tilde{F}$.

$$\mathcal{F}^* = \text{DeConv}_{3\times3}(\tilde{\mathcal{F}}, \mathcal{D}) \tag{5}$$

where D works as the deformable offset and F^* is the final aligned anchor feature. Then the enhanced anchor feature is fed into three detection heads composed of a 3×3 Conv and a 1×1 Conv to produce center, size, and offset features for detection loss:

$$\mathcal{L}_{\text{detection}} = \mathcal{L}_{\text{focal}}^{\text{center}} + \lambda_{size}\mathcal{L}_{\text{L1}}^{\text{size}} + \lambda_{off}\mathcal{L}_{\text{L1}}^{\text{offset}} \tag{6}$$

where $\mathcal{L}_{\text{focal}}$ is focal loss and $\mathcal{L}_{\text{L1}}$ is L1 loss.

2.3 Cross-Frame Box-Assisted Contrastive Learning

Typically, in colonoscopy videos, some concealed polyps appear very similar to the intestine wall in color and texture. Thus, an advanced training strategy is required to distinguish such homogeneity. Inspired by recent studies on supervised contrastive learning [18], we select the foreground and background region on both two frames guided by ground truth boxes to conduct contrastive learning. In practice, Given a batch of intermediate feature maps $F^a, F^r \in \mathbb{R}^{N\times T\times C\times H\times W}$ and corresponding binary maps $M^a, M^r \in \mathbb{R}^{N\times T\times H\times W}$, we first concatenate the anchor and reference at the batch-wise level as $\hat{F} \in \mathbb{R}^{NT\times C\times H\times W}$ and $\hat{M} \in \mathbb{R}^{NT\times H\times W}$ to exploit the cross-frame information. Then we extract the foreground and background channel patterns of cross-frame feature $\hat{F}$ using the Eq. 1 base on $\hat{M}(x,y) = 1$ and $\hat{M}(x,y) = 0$, respectively. After that, for each foreground channel pattern, which is the "query", we randomly select another different foreground feature as the "positive", while all the background features in the same batch are taken as the "negatives". Finally, we calculate the one-step contrastive loss by InfoNCE [18]:

$$\mathcal{L}_j^{\text{NCE}} = -\log\frac{\exp(q_j \cdot i^+/\tau)}{\exp(q_j \cdot i^+/\tau) + \sum_{i^- \in \mathcal{N}_j}\exp(q_j \cdot i^-/\tau)} \tag{7}$$

where $q_j \in \mathbb{R}^C, j = 0, ..., NT$ is the query feature, $i^+ \in \mathbb{R}^C$ and $i^- \in \mathbb{R}^{NT\times C}$ are positives and negatives. $\mathcal{N}_j$ denote embedding collections of the negatives. We repeat this process until every foreground channel pattern is selected and sum all steps as the final contrastive loss:

$$\mathcal{L}_{\text{contrast}} = \frac{1}{NT}\sum_{j=1}^{NT}\mathcal{L}_j^{NCE} \tag{8}$$

3 Experiments

We evaluate the proposed method on three public video polyp detection benchmarks: SUN Colonoscopy Video Database [7,10] (train set: 19,544 frames, test set: 12,522 frames), LDPolypVideo [9] (train set: 20,942 frames, test set: 12,933 frames), and CVC-VideoClinicDB [1] (train set: 7995 frames, test set: 2030 frames). For the fairness of the experiments, we keep the same dataset settings for YONA and all other methods.

Table 1. Performance comparison with other image/video-based detection models. P, R, and F1 denote the precision, recall, and F1-score. †: results from the original paper with the same data division. The best score is marked as red, while the second best score is marked as blue.

Methods	SUN Database			LDPolypVideo			CVC-VideoClinic			FPS
	P	R	F1	P	R	F1	P	R	F1	
Faster-RCNN [13]	77.2	69.6	73.2	68.8	46.7	55.6	84.6	98.2	90.9	44.7
FCOS [17]	75.7	64.1	69.4	65.1	46.0	53.9	92.1	74.1	82.1	42.0
CenterNet [25]	74.6	65.4	69.7	70.6	43.8	54.0	92.0	80.5	85.9	51.5
Sparse-RCNN [15]	75.5	73.7	74.6	71.6	47.9	57.4	85.1	96.4	90.4	40.0
DINO [21]	81.5	72.3	76.6	68.3	51.1	58.4	93.1	89.3	91.2	23.0
FGFA [26]	78.9	70.4	74.4	68.8	48.9	57.2	94.5	89.2	91.7	1.8
OptCNN† [23]	–	–	–	–	–	–	84.6	97.3	90.5	–
AIDPT† [22]	–	–	–	–	–	–	90.6	84.5	87.5	–
MEGA [2]	80.4	71.6	75.7	69.2	50.1	58.1	91.6	87.7	89.6	8.1
TransVOD [24]	79.3	69.6	74.1	69.2	49.2	57.5	92.1	91.4	91.7	8.4
STFT [19]	81.5	72.4	76.7	72.1	50.4	59.3	91.9	92.0	92.0	12.5
Ours-YONA	83.3	74.9	78.9	75.4	53.1	62.3	92.8	93.8	93.3	46.3

We use ResNet-50 [6] as our backbone and CenterNet [25] as our base detector. Following the same setting in CenterNet, we set $\lambda_{size} = 0.1$ and $\lambda_{off} = 1$. We set $\lambda_{contrast} = 0.3$ by ablation study. Detailed results are listed in the supplement. We randomly crop and resize the images to 512×512 and normalize them using ImageNet settings. Random rotation and flip with probability $p = 0.5$ are used for data augmentation. We set the batch size $N = 32$. Our model is trained using the Adam optimizer with a weight decay of 5×10^{-4} for 64 epochs. The initial learning rate is set to 10^{-4} and gradually decays to 10^{-5} with cosine annealing. All models are trained with PyTorch [11] framework. The training setting of other competitors follows the best settings given in their paper.

3.1 Quantitative and Qualitative Comparison

Quantitative Comparison. The comparison results are shown in Table 1. Following the standard of [1], the Precision, Recall, and F1-scores are used for evaluation. Firstly, compared with the CenterNet baseline, our YONA with three novel designs significantly improved the F1 score by 9.2%, 8.3%, and 7.4% on three benchmarks, demonstrating the effectiveness of the model design. Besides, YONA achieves the best trade-off between accuracy and speed compared with all other image-based SOTAs across all datasets. Second, for video-based competitors, previous video object detectors with multiple frame collaborations lack the ability for accurate detection on challenging datasets. Specifically, YONA surpasses the second-best STFT [19] by 2.2%, 3.0%, and 1.3% on F1 score on three datasets and 33.8 on FPS. All the results confirm the superiority of our proposed framework for accurate and fast video polyp detection.

Qualitative Comparison. Figure 3 visualizes the qualitative results of YONA with other competitors [19,25]. Thanks to this one-adjacent-frame framework, our YONA can not only prevent the false positive caused by part occlusion (1st and 2nd clips) but also capture useful information under severe image quality (2nd clip). Moreover, our YONA shows robust performance even for challenging scenarios like concealed polyps (3rd clip).

3.2 Ablation Study

We investigated the effectiveness of each component in YONA on the SUN database, as shown in Table 2. It can be observed that all the modules are necessary for precise detection compared with the baseline results. Due to the large variance of colonoscopy image content, the F1 score slightly decreases if directly adding FTA without the adaptive re-weighting strategy. Adding the adaptive weight greatly improves the F1 score by 5.4. Moreover, we use other two mainstream channel attention mechanisms to replace our proposed FTA for comparison. Compared with them, our FTA with adaptive weighting achieves the largest gain over the baseline and higher FPS. Overall, by combining all the proposed methods, our model can achieve new state-of-the-art performance.

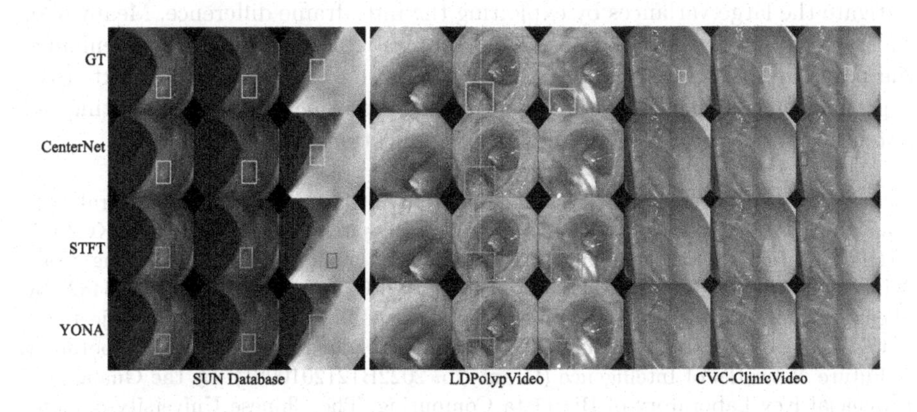

Fig. 3. Qualitative results of polyp detection on some video clips. The yellow, green, and red denote the ground truth, true positive, and false positive, respectively (Color figure online)

Table 2. Ablation studies of YONA under different settings. Ada means the adaptive re-weighting by similarity measuring; CW denotes the channel-wise attention [4]; CA denotes the channel-aware attention [19].

FTA	CW [4]	CA [19]	Ada	BDA	CBCL	Precision	Recall	F1	FPS
						74.6	65.4	69.7	**51.5**
✓						74.0	63.9	$68.6_{\downarrow 1.1}$	49.7
✓			✓			80.9	70.1	$75.1_{\uparrow 5.4}$	48.5
	✓		✓			78.0	65.2	$71.1_{\uparrow 1.4}$	48.3
		✓	✓			80.4	68.4	$73.9_{\uparrow 4.2}$	45.2
✓			✓	✓		82.0	72.2	$76.8_{\uparrow 7.1}$	46.3
✓			✓	✓	✓	**83.3**	**74.9**	$\mathbf{78.9}_{\uparrow 9.2}$	46.3

4 Conclusion

Video polyp detection is a currently challenging task due to the fast-moving property of colonoscopy video. In this paper, We proposed the YONA framework that requires only one adjacent reference frame for accurate and fast video polyp detection. To address the problem of fast-moving polyps, we introduced the foreground temporal alignment module, which explicitly aligns the channel patterns of two frames according to their foreground similarity. For the complex background content, we designed the background dynamic alignment module to mitigate the large variances by exploiting the inter-frame difference. Meanwhile, we employed a cross-frame box-assisted contrastive learning module to enhance the polyp and background discrimination based on box annotations. Extensive experiment results confirmed the effectiveness of our method, demonstrating the potential for practical use in real clinical applications.

Acknowledgements. This work was supported in part by Shenzhen General Program No. JCYJ20220530143600001, by the Basic Research Project No. HZQB-KCZYZ-2021067 of Hetao Shenzhen HK S&T Cooperation Zone, by Shenzhen-Hong Kong Joint Funding No. SGDX20211123112401002, NSFC with Grant No. 62293482, by Shenzhen Outstanding Talents Training Fund, by Guangdong Research Project No. 2017ZT07X152 and No. 2019CX01X104, by the Guangdong Provincial Key Laboratory of Future Networks of Intelligence (Grant No. 2022B1212010001), by the Guangdong Provincial Key Laboratory of Big Data Computing, The Chinese University of Hong Kong, Shenzhen, by the NSFC 61931024&81922046, by the Shenzhen Key Laboratory of Big Data and Artificial Intelligence (Grant No. ZDSYS201707251409055), and the Key Area R&D Program of Guangdong Province with grant No. 2018B030338001, by zelixir biotechnology company Fund, by Tencent Open Fund.

References

1. Bernal, J.J., et al.: Polyp detection benchmark in colonoscopy videos using GTCreator: a novel fully configurable tool for easy and fast annotation of image databases. In: Proceedings of 32nd CARS Conference (2018)

2. Chen, Y., Cao, Y., Hu, H., Wang, L.: Memory enhanced global-local aggregation for video object detection. In: Proceedings of the IEEE/CVF Conference on Computer Vision and Pattern Recognition, pp. 10337–10346 (2020)
3. Dai, J., et al.: Deformable convolutional networks. In: Proceedings of the IEEE International Conference on Computer Vision, pp. 764–773 (2017)
4. Fu, J., et al.: Dual attention network for scene segmentation. In: Proceedings of the IEEE/CVF Conference on Computer Vision and Pattern Recognition, pp. 3146–3154 (2019)
5. González-Bueno Puyal, J., et al.: Polyp detection on video colonoscopy using a hybrid 2d/3d CNN. Med. Image Anal. **82**, 102625 (2022)
6. He, K., Zhang, X., Ren, S., Sun, J.: Deep residual learning for image recognition. In: Proceedings of the IEEE Conference on Computer Vision and Pattern Recognition, pp. 770–778 (2016)
7. Itoh, H., Misawa, M., Mori, Y., Oda, M., Kudo, S.E., Mori, K.: Sun colonoscopy video database (2020). https://amed8k.sundatabase.org/
8. Luo, C., Zhan, J., Xue, X., Wang, L., Ren, R., Yang, Q.: Cosine normalization: using cosine similarity instead of dot product in neural networks. In: Kůrková, V., Manolopoulos, Y., Hammer, B., Iliadis, L., Maglogiannis, I. (eds.) ICANN 2018. LNCS, vol. 11139, pp. 382–391. Springer, Cham (2018). https://doi.org/10.1007/978-3-030-01418-6_38
9. Ma, Y., Chen, X., Cheng, K., Li, Y., Sun, B.: LDPolypVideo benchmark: a large-scale colonoscopy video dataset of diverse polyps. In: de Bruijne, M., et al. (eds.) MICCAI 2021. LNCS, vol. 12905, pp. 387–396. Springer, Cham (2021). https://doi.org/10.1007/978-3-030-87240-3_37
10. Misawa, M., et al.: Development of a computer-aided detection system for colonoscopy and a publicly accessible large colonoscopy video database (with video). Gastrointest. Endosc. **93**(4), 960–967 (2021)
11. Paszke, A., et al.: Pytorch: an imperative style, high-performance deep learning library. Neural Inf. Process. Syst. (2019)
12. Qadir, H.A., Balasingham, I., Solhusvik, J., Bergsland, J., Aabakken, L., Shin, Y.: Improving automatic polyp detection using CNN by exploiting temporal dependency in colonoscopy video. IEEE J. Biomed. Health Inf. **24**(1), 180–193 (2019)
13. Ren, S., He, K., Girshick, R., Sun, J.: Faster R-CNN: towards real-time object detection with region proposal networks. In: Advances in Neural Information Processing Systems, vol. 28 (2015)
14. Russakovsky, O., et al.: ImageNet large scale visual recognition challenge. Int. J. Comput. Vis. (IJCV) **115**(3), 211–252 (2015)
15. Sun, P., et al.: Sparse R-CNN: end-to-end object detection with learnable proposals. In: Proceedings of the IEEE/CVF Conference on Computer Vision and Pattern Recognition, pp. 14454–14463 (2021)
16. Tajbakhsh, N., Gurudu, S.R., Liang, J.: Automatic polyp detection in colonoscopy videos using an ensemble of convolutional neural networks. In: 2015 IEEE 12th International Symposium on Biomedical Imaging (ISBI), pp. 79–83. IEEE (2015)
17. Tian, Z., Shen, C., Chen, H., He, T.: FCOS: fully convolutional one-stage object detection. In: Proceedings of the IEEE/CVF International Conference on Computer Vision, pp. 9627–9636 (2019)
18. Wang, W., Zhou, T., Yu, F., Dai, J., Konukoglu, E., Van Gool, L.: Exploring cross-image pixel contrast for semantic segmentation. In: Proceedings of the IEEE/CVF International Conference on Computer Vision, pp. 7303–7313 (2021)

19. Wu, L., Hu, Z., Ji, Y., Luo, P., Zhang, S.: Multi-frame collaboration for effective endoscopic video polyp detection via spatial-temporal feature transformation. In: de Bruijne, M., et al. (eds.) MICCAI 2021. LNCS, vol. 12905, pp. 302–312. Springer, Cham (2021). https://doi.org/10.1007/978-3-030-87240-3_29
20. Zhan, C., Duan, X., Xu, S., Song, Z., Luo, M.: An improved moving object detection algorithm based on frame difference and edge detection. In: Fourth International Conference on Image and Graphics (ICIG 2007), pp. 519–523 (2007)
21. Zhang, H., et al.: DINO: DETR with improved denoising anchor boxes for end-to-end object detection. In: The Eleventh International Conference on Learning Representations (2022)
22. Zhang, Z., et al.: Asynchronous in parallel detection and tracking (AIPDT): real-time robust polyp detection. In: Martel, A.L., et al. (eds.) MICCAI 2020. LNCS, vol. 12263, pp. 722–731. Springer, Cham (2020). https://doi.org/10.1007/978-3-030-59716-0_69
23. Zheng, H., Chen, H., Huang, J., Li, X., Han, X., Yao, J.: Polyp tracking in video colonoscopy using optical flow with an on-the-fly trained CNN. In: 2019 IEEE 16th International Symposium on Biomedical Imaging (ISBI 2019), pp. 79–82. IEEE (2019)
24. Zhou, Q., et al.: Transvod: end-to-end video object detection with spatial-temporal transformers. IEEE Trans. Pattern Anal. Mach. Intell. (2022)
25. Zhou, X., Wang, D., Krähenbühl, P.: Objects as points. arXiv preprint arXiv:1904.07850 (2019)
26. Zhu, X., Wang, Y., Dai, J., Yuan, L., Wei, Y.: Flow-guided feature aggregation for video object detection. In: Proceedings of the IEEE International Conference on Computer Vision, pp. 408–417 (2017)

Personalized Patch-Based Normality Assessment of Brain Atrophy in Alzheimer's Disease

Jianwei Zhang[1,2] and Yonggang Shi[1,2(✉)]

[1] Stevens Neuroimaging and Informatics Institute, Keck School of Medicine, University of Southern California (USC), Los Angeles, CA 90033, USA
yonggans@usc.edu

[2] Ming Hsieh Department of Electrical and Computer Engineering, Viterbi School of Engineering, University of Southern California (USC), Los Angeles, CA 90089, USA

Abstract. Cortical thickness is an important biomarker associated with gray matter atrophy in neurodegenerative diseases. In order to conduct meaningful comparisons of cortical thickness between different subjects, it is imperative to establish correspondence among surface meshes. Conventional methods achieve this by projecting surface onto canonical domains such as the unit sphere or averaging feature values in anatomical regions of interest (ROIs). However, due to the natural variability in cortical topography, perfect anatomically meaningful one-to-one mapping can be hardly achieved and the practice of averaging leads to the loss of detailed information. For example, two subjects may have different number of gyral structures in the same region, and thus mapping can result in gyral/sulcal mismatch which introduces noise and averaging in detailed local information loss. Therefore, it is necessary to develop new method that can overcome these intrinsic problems to construct more meaningful comparison for atrophy detection. To address these limitations, we propose a novel personalized patch-based method to improve cortical thickness comparison across subjects. Our model segments the brain surface into patches based on gyral and sulcal structures to reduce mismatches in mapping method while still preserving detailed topological information which is potentially discarded in averaging. Moreover, the personalized templates for each patch account for the variability of folding patterns, as not all subjects are comparable. Finally, through normality assessment experiments, we demonstrate that our model performs better than standard spherical registration in detecting atrophy in patients with mild cognitive impairment (MCI) and Alzheimer's disease (AD).

Keywords: Personalized Atlas · Alzheimer's disease · Shape Analysis

1 Introduction

Neurodegenerative diseases like Alzheimer's disease (AD) are the main causes of cognitive impairment and earlier neuropathological alterations [7]. Cortical

Y. Shi—This work is supported by the National Institute of Health (NIH) under grants R01EB022744, RF1AG077578, RF1AG056573, RF1AG064584, R21AG064776, P41EB015922, U19AG078109.

H. Greenspan et al. (Eds.): MICCAI 2023, LNCS 14224, pp. 55–62, 2023.
https://doi.org/10.1007/978-3-031-43904-9_6

thickness change is an essential feature that can quantify the potential brain atrophy in these diseases [9,12]. To analyze cortical thickness across different subjects, one essential step is to find the correspondence between surface points for thickness analysis [14,15]. This is traditionally achieved through 1-to-1 mapping or averaging feature in the region of interest (ROI). Mapping methods typically rely on surface registration techniques on a canonical space such as the unit sphere or high-dimensional embedding space [5,6,10].However,surface registration methods have limitations in accounting for the intrinsic variations in cortical topography among different individuals. For instance, a one-to-one mapping using surface registration could wrongly map a gyral region into a sulcal region due to distinguished local topography between subjects,as example shown in Fig. 1(B), thereby introducing noise into cortical thickness analysis since gyral and sulcal regions have distinct thickness profiles,as example shown in Fig. 1(A). Alternatively, ROI-based methodologies rely on the correspondence of anatomically segmented ROIs and establish comparisons through mean feature extraction. This approach disregards intra-ROI variability and the averaging process obscures detailed folding patterns. To mitigate the inherent challenges of surface mapping and mean ROI feature extraction, novel methodologies that consider the variability in folding patterns for identifying correspondences across distinct topographies are imperative.

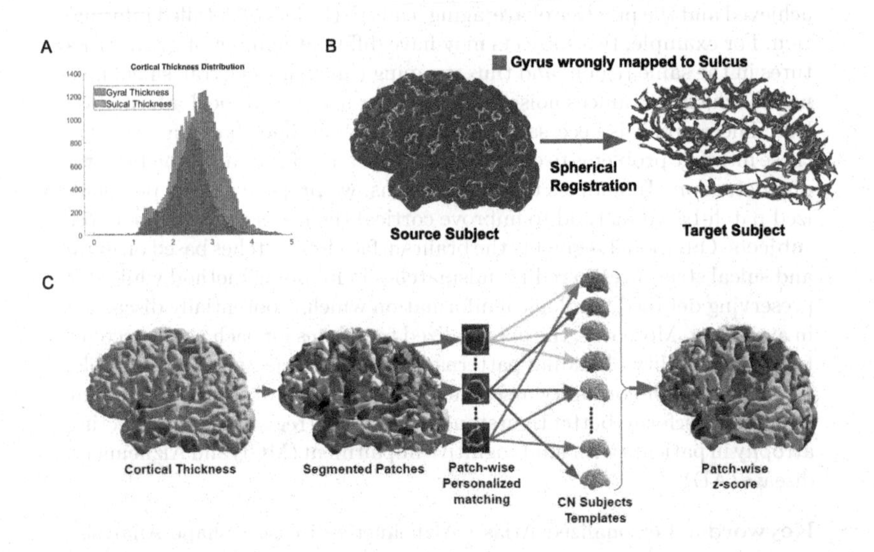

Fig. 1. (A) The plot shows the distinct cortical thickness distributions of gyral/sulcal regions. (B) From gyral/sulca segmentation [13], the red parts of the left plot highlight the mismatched gyral region to sulcal region. The right plot displays mismatched sulcal region. (C) The diagram covers the process of personalized matching, where cortical surface is segmented into small patches and matched onto its own personalized set of templates for normality assessment using z-score.

In this paper, we propose a novel personalized patch-based folding analysis method that finds correspondence through segmented patch similarity and personalized template set. Comparing to conventional methods, our method accounts for the gyral/sulcal mismatch problem by explicitly matching gyral and sulcal patches to their respective regions and selecting a personalized set of templates for each patch, as shown in Fig. 1(C), such that only comparable features are measured together to increase sensitivity in brain atrophy detection and reduce noise introduced by mismatching.

2 Methods

In this section, we will present the technical details of our method, which involve three main parts: surface segmentation, similarity metric and personalized templates. By integrating these 3 parts, we demonstrate the effectiveness of our method through normality assessment experiments on patients with mild cognitive impairment (MCI) and Alzheimer's disease (AD) on data from Alzheimer Disease Neuroimaging Initiative(ADNI) [11].

2.1 Brain Surface Segmentation

We employ FreeSurfer to extract 3D surface mesh from T1-weighted Volumetric MRI data. [2]. We extracted both pial and white surfaces, which have vertex correspondence by reconstruction method. The meshes are decimated to 50000 vertices for computational cost. The brain mesh is then subdivided into gyral and sulcal sub meshes following the methods in [13] using shape index and graph cut. The intricate folding pattern of the brain's cortical surface can make it difficult to accurately classify certain gyral structures that extend deeper from the outer surface. However, these structures can be better detected on the white surface, which has more prominent gyrus. Therefore, the segmentations from pial and white surfaces are combined through vertex correspondence. Specifically, given two array $mask_p$ and $mask_w$ as segmentation mask from pial and white surfaces, the final mask is $mask_p|mask_w$, where entry 1 is gyrus and 0 is sulcus. The resulting gyral/sulcal surfaces are 3D meshes with disconnected components and boundaries due to noise in reconstruction and the intrinsic geometry of the folding pattern. Therefore, filtering is applied to eliminate components that have less vertices than a threshold, which is empirically set at 50.

The segmented patches are generated for gyral and sulcal sub meshes separately, which is shown in Fig. 2(B). Given a 3d mesh $M(V, F)$, V as set of vertices and F as set of faces, the boundary of M is defined as $B(M)$, the set of vertices that has at least one neighboring edge that is contained in only one triangular face. Next, the distance transform of vertex $v \in V$ is defined as:

$$DT(v) = \min_{b \in B(M)} Geo(v, b) \tag{1}$$

The Geo denotes the geodesic distance between v and b. The distance transform computation is implemented using fast marching algorithm. [8] From the distance

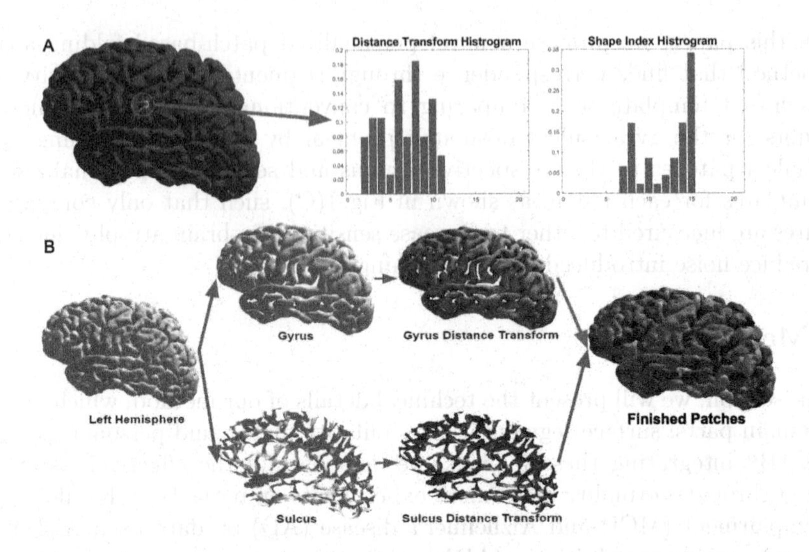

Fig. 2. (A) Example of segmented patch and its histogram shape descriptor (B) Process of the segmentation, from gyral/sulcal segmentation, distance transform computation to patch generation

transform, we can obtain a set L of local maximum vertices, defined as $\{v \in V : DT(v) >= DT(w), \forall w \in Neighbor(w)\}$. The neighbors of a vertex v is the set of vertices that connect to v by one edge. With L and Geo, we can compute the Voronoi diagram that consists of Voronoi cells defined as:

$$C_i = \{x \in V : Geo(x, c_i) <= Geo(x, c_j); \forall c_j \in L\} \tag{2}$$

Each Voronoi cell defines a segmented patch. Since the surface is not perfectly smooth, the mesh is usually over segmented. To address this issue, we employ a region grow strategy to merge patches. Starting from one patch, we compute the geodesic distances between its center vertex and those of all neighboring patches. If the computed distance falls below a specified threshold, the two patches are merged into a single patch. In the event that no neighboring patch meets this criterion, we proceed to an unvisited patch. The iteration repeats until all patches are visited. The threshold is empirically set as twice the mean edge length on the mesh. Patches are labeled as gyral or sulcal based on the sub-mesh they're generated from.

2.2 Patch Similarity Metric

We employ shape similarity metrics to identify comparable patches. Histogram-based shape descriptors are used for topological similarity. Specifically, a histogram H is constructed for a function F defined on all vertices in a patch. The input values for histogram generation are normalized by the root mean square (RMS) of the function. The maximum bin value of the histogram is set as the maximum feature value, and the number of bins is fixed at 20 to ensure comparability between histograms. The values in each bin are normalized with respect to the total number of vertices in the patch, ensuring that the histogram values add up to 1. This normalization is done to avoid the scaling problem, as patches with similar topography can have different sizes. The distance transform, described in previous section, and shape index are used for constructing histogram, H_{dt} and H_{si}. Distance transform is used since the distance to boundary encodes information about the gyral/sulcal patches,such as width of the gyrus. For example, a crest gyral structure and a plateau one have different distance to boundary distribution. For topological comparison, chi square distance is computed as similarity. Given histograms H_i and H_j, the chi square distance is defined below:

$$D_{\chi 2}(H_i, H_j) = \Sigma_k \frac{(H_i[k] - H_j[k])^2}{H_i[k] + H_j[k]} \tag{3}$$

$H_i[k]$ denotes the value in the kth bin of H_i. For any two patches,i and j, we can define a distance vector $v_{dist}(i,j) = [D_{\chi 2}(H_i^{dt}, H_j^{dt}), D_{\chi 2}(H_i^{si}, H_j^{si}), D(C_i, C_j)]$. The final similarity score $S(i,j)$ is defined as $\frac{1}{||w \times v_{dist}(i,j)||}$, where w is a weighting vector.

2.3 Personalized Template Set

To mitigate the impact of inter-subject variability in brain structure folding patterns, a personalized set of templates is chosen for each patch. This involves selecting a cohort of N cognitively normal (CN) subjects as the templates for comparison, while a separate group of M mild cognitive impairment (MCI) and M Alzheimer's disease (AD) subjects is selected for testing purposes.

For all subjects in both the template and test groups, we first obtain the sets T_{temp} and T_{test} of segmented patches as described in the previous section. From the output of Freesurfer, we compute the vertex-wise spherical registered coordinates (stored in the sphere.reg file) and cortical thickness (stored in the .thickness file) as $SR(v)$ and $CT(v)$, respectively, where v denotes a vertex [4]. For a query patch P_i, which is a set of vertices and has a gyral/sulcal label from the sub mesh they are generated from, we first compute its patch spherical registered coordinates $PSR_i = \frac{1}{|P_i|}\sum_{v \in P_i} SR(v)$ and patch cortical thickness $PCT_i = \frac{1}{|P_i|}\sum_{v \in P_i} CT(v)$, where $|P_i|$ is the cardinality of P_i. Then, for each patch $P_i \in T_{test}$, we compute the nearest 200 patches $P^{1...200} \subset T_{temp}$ that has the same gyral/sulcal label as query patch in terms of $||PSR_i - PSR_j||$, where $P_j \in T_{test}$. This is to leverage the location information of the patch with respect to the whole hemisphere, as matched patches should be from the neighborhood region of the query patch and label restriction is to reduce mismatch of gyral/sulcal regions. From $P^{1...200}$, we select the 50 most similar patches based on their similarity scores. Specifically, we compute the similarity score $S(i,j)$ for patch P_i and patch $P_j \in P^{1...200}$, and choose the top 50 patches in $P^{1...200}$ with the largest similarity scores to form a personalized template set T_{P_i} for patch P_i. We repeat this process for each patch in the test set.

In the next step, we define the normality metric used to measure the normality of a patch. For each patch P_i and its personalized template set T_{P_i}, we first compute the mean (μ) and standard deviation (std) of the patch cortical thickness for all patches in T_{P_i}. We then define the z-score of P_i as $Z(P_i) = |\frac{PCT_{P_i} - \mu}{std}|$. This z-score is used as a normality measure because if P_i is abnormal, it will have a larger z-score and vice versa. Finally, we use this z-score for each patch in the test subject set to conduct normality assessment experiments.

3 Results

3.1 Normality Assessment Experiments

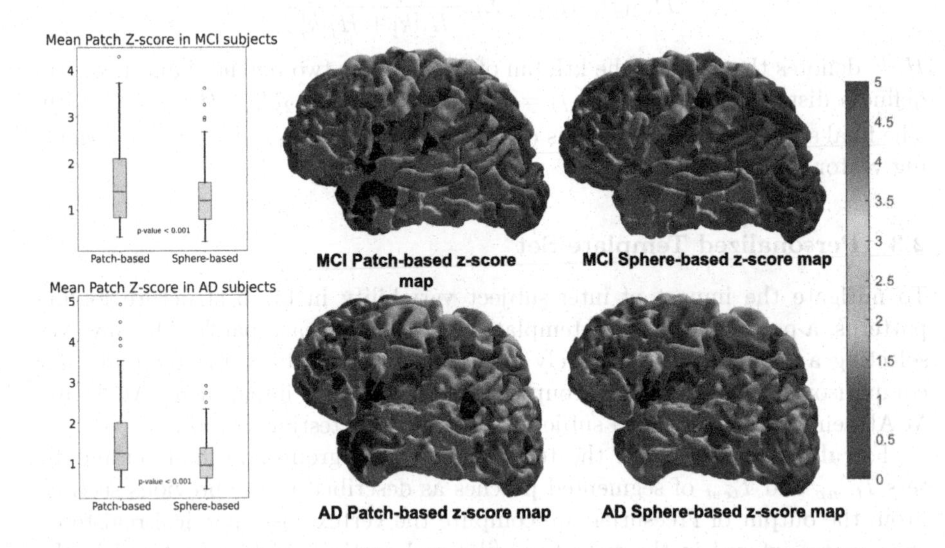

Fig. 3. The Left 2 plots show the results of normality assessment experiment. The right 4 plots show the visualization of patch-based and sphere-based z-score map on MCI and AD subjects

To investigate the discriminative power of our patch-based personalized matching approach, we randomly selected 200 cognitive normal (CN) subjects as templates, 100 mild cognitive impairment (MCI) and 100 Alzheimer's disease (AD) subjects from the ADNI dataset to compare our method to Freesurfer's Spherical Registration. The experiments are limited to the left hemisphere for computing cost. Specifically, we used the sphere.reg file output from Freesurfer, which was decimated to 50000 vertices. For each query subject's surface, we computed the spherical registration of each vertex to a template surface by finding the closest vertex in the template's vertices based on Euclidean distance on the unit sphere space, as described by the sphere.reg file. We then computed a patch-based z-score for each patch P_i using the thickness of P_i and the spherically matched vertices from each template subject, which we denote as the sphere-based z-score. Alternatively, we matched patches as previously described in the Method section, and computed a patch-based z-score for each patch in the MCI and AD subject sets.

For each MCI and AD subject, the mean sphere-based z-score and mean patch-based z-score are computed by averaging the respective scores of patches for 1 subject. These mean z-scores represent how abnormal a subject is. The resulting distributions are shown in Fig. 3. The results show that our method yields a higher z-score for MCI and AD subjects compared to the sphere-based z-score, indicating that our model is more sensitive in detecting brain atrophy. As a qualitative analysis, the patch-based z-score maps and sphere-based z-score maps for selected MCI and AD subjects are also shown in Fig. 3. The patch-based z-score map shows more brain atrophy than the sphere-based z-score map, which further verifies our model's advantage in brain atrophy detection.

3.2 CN vs MCI, AD Prediction Experiment

In this experiment, we conduct CN vs MCI and CN vs AD prediction accuracy test. Similar to previous experiment, we choose 200 CN subjects as templates and another 100 CN subjects, 100 MCI subjects and 100 AD subjects for testing. Afterwards, sphere-based z-scores and patch-based z-scores are computed for all patches in CN, MCI and AD subjects from 200 template subjects. For training feature, we use the mean z-score of patches in each ROI in one subject, which is a length-34 vector for 34 ROIs from Freesurfer(*.aparc.annot*) [3]. Each subject has one feature vector generated from patch-based z-score and one from sphere-based z-score. The CN, MCI and AD are used as labels. The feature vectors and labels are randomly split into train and test set at 8:2 ratio. The classifier we use is a generic support vector machine from sklearn package [1]. The test is conducted for CN vs AD and CN vs MCI binary classifications. We used 5-fold cross validation for evaluation on stability. The resulting test accuracy is shown in Table 1. The results show that our method performs better in accuracy than sphere-based method.

Table 1. test accuracy for CN vs AD and CN vs MCI

Feature Type	CN vs. AD Accuracy(%)	CN vs. MCI Accuracy(%)
Sphere-based	68.33	50.84
Patch-based	**71.66**	**58.50**

4 Conclusion

In this paper, we proposed a novel personalized patch-based method for brain atrophy detection by matching segmented patches based on gyral/sulcal label, location, shape similarity and constructing personalized template set for abnormal detection. Through normality assessment and MCI AD prediction experiments, the method is shown to be more effective at detecting brain atrophy.

References

1. Buitinck, L., et al.: API design for machine learning software: experiences from the scikit-learn project. In: ECML PKDD Workshop: Languages for Data Mining and Machine Learning, pp. 108–122 (2013)
2. Dale, A.M., Fischl, B., Sereno, M.I.: Cortical surface-based analysis: I. segmentation and surface reconstruction. NeuroImage **9**(2), 179–194 (1999)
3. Fischl, B., et al.: Automatically parcellating the human cerebral cortex. Cereb. Cortex **14**(1), 11–22 (2004)
4. Fischl, B., Sereno, M.I., Tootell, R.B., Dale, A.M.: High-resolution intersubject averaging and a coordinate system for the cortical surface. Hum. Brain Mapp. **8**(4), 272–284 (1999)
5. Gahm, J.K., Shi, Y.: Riemannian metric optimization on surfaces (RMOS) for intrinsic brain mapping in the Laplace-Beltrami embedding space. Med. Image Anal. **46**, 189–201 (2018)
6. Gu, X., Wang, Y., Chan, T., Thompson, P., Yau, S.T.: Genus zero surface conformal mapping and its application to brain surface mapping. IEEE Trans. Med. Imaging **23**(8), 949–958 (2004)
7. Jack, C.R., Jr., et al.: NIA-AA research framework: toward a biological definition of Alzheimer's disease. Alzheimer's Dement. **14**(4), 535–562 (2018)
8. Kimmel, R., Sethian, J.A.: Computing geodesic paths on manifolds. Proc. Natl. Acad. Sci. **95**(15), 8431–8435 (1998)
9. Lerch, J.P., Pruessner, J.C., Zijdenbos, A., Hampel, H., Teipel, S.J., Evans, A.C.: Focal decline of cortical thickness in Alzheimer's disease identified by computational neuroanatomy. Cereb. Cortex **15**(7), 995–1001 (2004)
10. Lyu, I., Kang, H., Woodward, N., Styner, M., Landman, B.: Hierarchical spherical deformation for cortical surface registration. Med. Image Anal. **57** (2019)
11. Mueller, S.G., et al.: The Alzheimer's disease neuroimaging initiative. Neuroimaging Clin. North Am. **15**(4), 869–877 (2005). Alzheimer's Disease: 100 Years of Progress
12. Querbes, O., et al.: Early diagnosis of Alzheimers disease using cortical thickness: impact of cognitive reserve. Brain: J. Neurol. **132**, 2036–47 (2009)
13. Shi, Y., Thompson, P.M., Dinov, I., Toga, A.W.: Hamilton-Jacobi skeleton on cortical surfaces. IEEE Trans. Med. Imaging **27**(5), 664–673 (2008)
14. Thompson, P.M., et al.: Mapping hippocampal and ventricular change in Alzheimer disease. Neuroimage **22**(4), 1754–1766 (2004)
15. Thompson, P.M., et al.: Mapping cortical change in Alzheimer's disease, brain development, and schizophrenia. Neuroimage **23**, S2–S18 (2004)

Patients and Slides are Equal: A Multi-level Multi-instance Learning Framework for Pathological Image Analysis

Fei Li[1,2], Mingyu Wang[1,2], Bin Huang[1,2], Xiaoyu Duan[4], Zhuya Zhang[1,2], Ziyin Ye[3(✉)], and Bingsheng Huang[1,2(✉)]

[1] Medical AI Lab, School of Biomedical Engineering, Shenzhen University Medical School, Shenzhen University, Shenzhen, China
huangb@szu.edu.cn

[2] Marshall Laboratory of Biomedical Engineering, Shenzhen University, Shenzhen, China

[3] Department of Pathology, The First Affiliated Hospital, Sun Yat-sen University, Guangzhou, China
yeziyin@mail.sysu.edu.cn

[4] Department of Pathology, The Seventh Affiliated Hospital, Sun Yat-sen University, Shenzhen, China

Abstract. In current pathology image classification, methods mostly rely on patch-based multi-instance learning (MIL), which only considers the relationship between patches and slides. However, in clinical medicine, doctors use slide-level labels to summarize patient-level labels as a diagnostic result, indicating the involvement of three levels of patch, slide, and patient in actual pathology image analysis, which we refer to as the multi-level multi-instance learning (ML-MIL) problem. To address this issue, we propose a novel and general framework called Patients and Slides are Equal (P&SrE), inspired by the doctor's diagnostic process of repeatedly confirming labels at the patient and slide level. In this framework, we treat patients and slides as instances at the same level and use transformers and attention mechanisms to build connections between them. This allows for interaction between patient-level and slide-level information and the correction of their respective features to achieve better classification performance. We evaluate our method on two datasets using two state-of-the-art MIL methods as baselines. The results show that our method improves the performance of the baselines on both slide and patient levels. Our method provides a simple and effective solution to the common problem of ML-MIL in medical clinical scenarios and has broad potential applications.

Keywords: Multiple instance learning · Multi-level Labels · Pathology Images · Transformer

F. Li, M. Wang and B. Huang—Contribute equally to this work.

H. Greenspan et al. (Eds.): MICCAI 2023, LNCS 14224, pp. 63–71, 2023.
https://doi.org/10.1007/978-3-031-43904-9_7

1 Introduction

Pathological image analysis is a vital area of research within medical image analysis, focused on utilizing computer technology to aid doctors in diagnosing and treating diseases by analyzing pathological tissue slide images [5]. Advancements in pathological image analysis have been made in early cancer diagnosis, tumor localization, and grading, and treatment planning [3,10]. Multi-instance learning [2] is the primary analysis method used, which involves analyzing tasks based on slide labels and patches. Despite this, the clinical pathological analysis presents certain challenges and complexities, with the ultimate diagnosis relying on patients rather than slides.

Specifically, in clinical problems of pathological image analysis, doctors usually summarize patient-level labels based on slide labels as the diagnostic results [1,6]. For example, for the pathological discrimination diagnosis task of intestinal tuberculosis(ITB) and Crohn's desease(CD), the categories of postoperative slides are divided into three types (normal, CD, ITB), and doctors will summarize the binary results of patients (ITB or CD) based on slide-level labels [6]. Similar situations exist in other tasks, such as the classification of breast cancer metastases in lymph nodes, where slide categories may have different classifications, and the corresponding diagnosis of the same patient is whether the cancer has spread to the regional lymph nodes (N-stage) [1]. Therefore, as shown in Fig. 1, actual pathological image analysis involves the relationships of patches, slides, and patients, which is called a multi-level multi-instance learning (ML-MIL) problem. Among them, for patients and slides, patients are bags while slides are instances, and for slides and patches, slides are bags while patches are instances.

There are generally two methods to solve the ML-MIL problem. The first method is to directly average the prediction values of slides or take the maximum prediction value [9]. This method is relatively simple, but the information exchange between slides is not fully utilized, which may lead to errors in the summary result. The second method is to treat slide-patient as a new MIL problem according to the traditional MIL thinking, where slides are regarded as instances and patient labels as bags. Although this method seems reasonable, the number of patients is usually relatively small, and deep learning models usually require a large amount of data for training. Therefore, the insufficient number of samples at the slide-patient level may make it difficult for the model to learn enough information.

To address the multi-level multi-instance learning (ML-MIL) problem in medical field, we propose a novel framework called Patients and Slides are Equal (P&SrE). Inspired by the iterative labeling process in medical diagnosis, this framework treats patients and slides as instances at the same level and uses transformers and attention mechanisms to build connections between them. This simple yet effective method allows for interaction between patient-level and slide-level information to correct their respective features and improve classification performance. Our framework consists of two steps: first, at the patch-slide level, a common MIL framework is used to train a MIL neural network and obtain

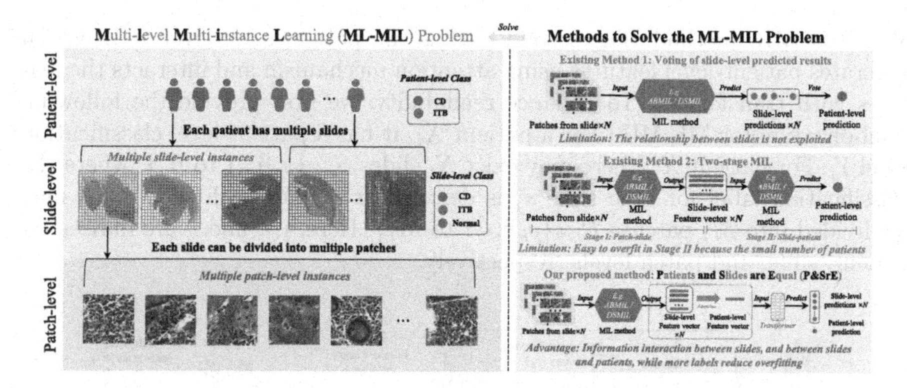

Fig. 1. Description and solutions for the ML-MIL problem.

slide-level feature vectors; then, at the slide-patient level, we use self-attention mechanisms to combine the slides of the same patient into patient-level feature vectors, and treat these patient-level feature vectors together with all slide-level feature vectors of the same patient as instances at the same level, which are inputted into transformers for feature interaction and prediction of patient- and slide-level labels. Our method can effectively solve the problem of difficult training due to the scarcity of samples at the highest level in ML-MIL, and can be integrated into two state-of-the-art methods to further improve performance. We conducted rigorous experiments on two datasets and demonstrated the effectiveness of our method. Our contributions include:

1) Proposing a novel general framework to address the unique "patch-slide-patient" ML-MIL problem in the medical field. Before this, no other framework had directly tackled this specific problem, making our proposal a ground-breaking step in the application of ML-MIL in healthcare;
2) Proposing a simple yet highly effective method that leverages self-attention mechanisms and transformer models to enhance the interaction between slide and patient information. This innovative approach not only improves the classification performance at the patient level but also at the slide level, showcasing its effectiveness and versatility;
3) Conducting extensive experiments on two separate datasets. Our method was seamlessly integrated with two prior state-of-the-art methods, demonstrating its compatibility and adaptability. The experiments resulted in improved performance, indicating that our method enhances the efficacy of these existing approaches.

2 Method

2.1 Overview

Our proposed method P&SrE is illustrated in Fig. 2. Specifically, the framework consists of two parts. The first part is the slide-patch level MIL based on a state-of-the-art MIL method. The second part is the patient-slide level MIL, which generates patient-level features using attention mechanism and interacts the features with transformer. To enhance readability, we first provide the following symbolization for ML-MIL: For a patient X_i, it has a patient-level classification label Y_i. For patient X_i, there may exist N_i slides $S_i=\{s_j|\text{j}=1 \text{ to } N_i\}$, where the classification label for each slide s_j is denoted as z_j. For each slide s_j, it may be divided into M_j patches $P_j=\{p_k|\text{k}=1 \text{ to } M_j\}$. Here, i,j, and k are indices for patient, slide, and patch levels, respectively.

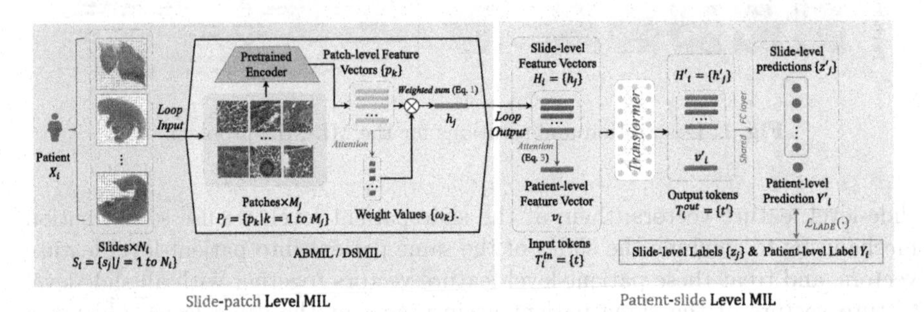

Fig. 2. Overview of the proposed framework P&SrE. This framework consists of two-level MIL parts: Slide-patch level MIL and patient-slide level MIL

2.2 Slide-Patch Level MIL

Our proposed framework has strong scalability as it can be based on any attention-based MIL method. Therefore, we directly use the state-of-the-art (SOTA) MIL methods, ABMIL [8] and DSMIL [9] for the slide-patch stage. These two methods differ in their attention computing approach for each patch.

For ABMIL, the attention of each patch is computed by an MLP. Specifically, for M_j patches p_k, an encoder is applied to obtain the patch feature matrix F_i, where,$F_i \in R^{M_j \times 1024}$. Then, F_i is passed through an fc layer followed by a Tanh activation and another fc layer followed by a sigmoid activation to obtain two feature matrices, F'_i and F''_i, both $\in R^{M_j \times 128}$. These matrices are element-wise multiplied and then passed through an fc layer to obtain the weight of each patch, ω_k.

For DSMIL, the attention of each patch is based on the cosine distance between instances and key instances. First, an fc layer is applied to the patch feature matrix F_i to obtain the importance score θ_k for each patch. The patch with the highest score is selected as the key instance. Then, the feature matrix F_i is mapped to a matrix $Q_i \in R^{M_j \times 128}$ and the cosine similarity between all instances and the key instance is computed as the weight of each patch, ω_k.

Although ABMIL and DSMIL compute attention differently, both methods compute the attention-weighted sum of patch instances features as the bag representation of the slide. Therefore, the slide feature output by both methods can be generalized as:

$$h_j = \sum_{k=1}^{M_j} \omega_k * p_k / \sum_{k=1}^{M_j} \omega_k \tag{1}$$

Finally, we obtain the feature vector set $H_i=\{h_j|\text{j}=1 \text{ to } N_i\}$ for all slides $\{s_j\}$ of patientX_i through patch-slide MIL.

2.3 Patient-Slide Level MIL

After performing patch-slide level MIL, we move on to patient-slide level MIL. In general MIL algorithms, the patient is regarded as the bag and the slide as the instance. However, considering the diagnostic process in clinical practice, we propose to treat both patients and slides as instances at the same level. Specifically, our P&SrE framework for patient-slide level consists of two parts: patient-level feature generation based on self-attention and patient-slide feature interaction based on Transformer [11].

Patient-Level Feature Generation Based on Self-attention. Doctors usually select certain key slides for careful observation and information aggregation during diagnosis, similar to the self-attention mechanism. Therefore, we directly use a fully connected (FC) layer to integrate the feature-level features into patient-level features v_i through attention mechanism, serving as patient instances. Specifically, given the feature vector collection $\{h_j\}$ from multiple slides in the previous step, we input it to the FC layer and apply the sigmoid activation function to output the weight α_j for each h_j. Then, we perform a weighted average of the vectors based on this weight to obtain the patient feature v_i:

$$\alpha_j = FC(\{h_j|j = 1 \ to \ N_i\}) \tag{2}$$

$$v_j = \sum_{j=1}^{N_i} \alpha_j * h_j / \sum_{j=1}^{N_i} \alpha_j \tag{3}$$

Patient-Slide Feature Interaction Based on Transformer. This process is where our framework shines. After doctors summarize the patient-level results, they typically review the slides to double-check the diagnosis results. This patient-slide feature interaction (PSFI) naturally lends itself to the construction of a Transformer, and information exchange and integration between slides and patient level are bidirectional. Thus, self-attention is more ideal for this purpose than other kinds of attention (such as cross-attention or doctors' attention). By using the self-attention-based transformer structure, each input token is treated equally (i.e., viewed as the same instance level), and tokens can interact extensively with each other, enabling mutual correction between patients and slides and even between slides. Specifically, we merge the slide feature set $\{h_j\}$ and the patient feature v_i into the input tokens $T_i^{in} = \{h_1, h_2, ..., h_{N_i}, v_i\} = \{t\}$, and then input them into a multi-layer transformer through self-attention and feed-forward neural network layers to obtain the interaction information between slides and output tokens T_i^{out}:

$$\beta_{k,l} = softmax(W^Q t_k{}^T (W^K t_l)/\sqrt{d}) \tag{4}$$

$$\bar{t}_k = \sum_{l=1}^{N_i+1} \beta_{k,l} W^V t_l \tag{5}$$

$$t'_k = RELU(\bar{t}_k W^R + b_1) W^O + b_2 \tag{6}$$

where d is the dimension of the token, and t_k and t_l come from T_i^{in}. $\beta_{k,l}$ is multi-head attention matrix, and W^Q, W^K, and W^V are weight matrices of query, key, and value, respectively. W^R and W^O are transformation matrices. b_1 and b_2 are bias vectors. This update procedure is repeated for L layers, where the t'_k are fed to the successive transformer layer. Finally, we obtain the output tokens T_i^{out}=$\{h'_1, h'_2, ..., h'_{N_i}, v'_i\}$. Then, all output tokens are input into a shared FC layer, and the patient's predicted logits Y'_i and the predicted classification logits $\{z'_j | j = 1\ to\ N_i\}$ for each slide are output.

Training Progress and Loss Function. During training, we sampled one patient at a time and pre-extracted their batch-level features for all slides, in order to save GPU memory. Due to the issue of class imbalance in both slide level and patient level, we use the LADE [7] loss function.

3 Experiments and Results

3.1 Dataset and Evaluation

CD-ITB Dataset. CD-ITB is a private dataset consisting of 853 slides from 163 patients, with binary patient-level labels of CD or ITB in a ratio of 103:60 and tri-class slide-level labels of CD, ITB, and normal slides in a ratio of 436:121:296, respectively. On average, there were 5 slides per patient. The slides were scanned at a magnification of 40× (0.25 µm/px), and annotations were curated by experienced pathologists. We adopted a patient-level stratification approach for 5-fold cross-validation, with 20% of the training set randomly assigned as the validation set for each fold. The dataset comprises an average of 2.3k instances per bag, with the largest bag containing over 16k instances.

Camelyon17 Dataset. Camelyon17 [1] is a publicly dataset, and its training set comprises 500 slides from 100 breast cancer patients with lymph node metastases. The slides are classified into four distinct categories, namely negative, ITC, micro, and macro, in proportions of 318:36:59:87, respectively. There were 5 slides per patient on average. The patients are divided into two groups based on their pN stage, namely lymph node positive and lymph node negative, in proportions of 24:76, respectively. The data folding method is the same as the CD-ITB dataset. The average number of instances per bag is approximately 6.1k, and the largest bag contains over 23k instances.

Metrics. We report class-wise weighted accuracy (Acc), precision(Pre), Recall, and F1-score (F1). To avoid randomness, we run all experiments five times and report the averaged metrics.

3.2 Implementation Details

We utilized ResNet50, which was pre-trained on ImageNet1K, to extract features from patches. Each patch was of size 512×512 pixels. For both ABMIL and DSMIL networks, we kept the original parameters for the number of channels at each layer. Following the reference [4], we employed a transformer with 8 heads and 8 layers in the patient-slide feature interactions. All networks are implemented using PyTorch and trained on a NVIDIA RTX TITAN GPU with 24 GB memory. We employed two Adam optimizers with a maximum learning rate of 1e−4 and a cosine annealing update strategy that gradually decreased the learning rate to 1e−12 over 300 epochs.

3.3 Comparisons and Results

We compared our strategy with two state-of-the-art MIL methods to evaluate its performance. To investigate the impact of self-attention and transformers on slide-level and case-level results, we conducted ablation experiments: "ABMIL + P&SrE (with/without PSFI)" and "DSMIL + P&SrE (with/without PSFI)", respectively. For slide-level classification, we used mean pooling and max pooling to pool feature vectors of patches into a representative vector for the slide, which was then fed into a fully connected layer for classification. At the patient level, we used two approaches for prediction: MaxS, where the feature of the instance that achieves the maximum positive probability from the slide-level MIL model is selected to patient-level model, and MaxMinS, where the mean value of features of the maximum and minimum positive probability from the slide-level MIL model is selected to patient-level model.

The results of 5-fold CV at the slide and patient levels are reported in Table 1 and Table 2, respectively. Our P&SrE framework improves both ABMIL and DSMIL methods at both levels. ABMIL with P&SrE improves the F1 score from 0.565 to 0.579 for the CD-ITB dataset and from 0.529 to 0.571 for the Camelyon17 dataset at the slide-level, and improves the F1 score from 0.522 to 0.599 for the CD-ITB dataset and from 0.842 to 0.861 for the Camelyon17 dataset at the patient-level. Therefore, the ablation experiments demonstrate the effectiveness of P&SrE in enhancing the classification performance at both the slide and patient levels.

Table 1. Slide-level 5-fold CV results (%)

Method	CD-ITB dataset				Camelyon17 dataset			
	Pre	Recall	Acc	F1	Pre	Recall	Acc	F1
Mean pooling (on patch-level feature vectors)	45.0	40.2	40.2	41.6	47.4	31.6	31.6	35.5
Max pooling (on patch-level feature vectors)	39.6	33.1	33.1	34.9	46.9	25.4	25.4	29.9
ABMIL [8]	57.6	57.2	57.2	55.4	55.3	63.7	63.7	50.9
(ours) ABMIL + P&SrE (w/o PSFI)	57.0	57.3	57.3	56.5	58.6	65.2	65.2	52.9
(ours) ABMIL + P&SrE	**59.0**	**59.7**	**59.7**	**57.9**	**61.0**	**66.9**	**66.9**	**57.1**
DSMIL [9]	57.0	57.3	57.3	56.9	55.4	63.8	63.8	50.5
(ours) DSMIL + P&SrE (w/o PSFI)	56.7	57.4	57.4	56.6	55.5	64.6	64.6	51.7
(ours) DSMIL + P&SrE	**58.5**	**59.1**	**59.1**	**57.3**	**55.8**	**66.4**	**66.4**	**52.3**

Table 2. Patient-level 5-fold CV results (%)

Method	CD-ITB dataset				Camelyon17 dataset			
	Pre	Recall	Acc	F1	Pre	Recall	Acc	F1
ABMIL (MaxS)	36.9	63.3	46.6	46.6	78.0	94.4	75.0	85.0
ABMIL+ (MaxMinS)	38.2	65.0	48.5	48.2	78.3	94.7	76.0	85.7
ABMIL(baseline+Mean) (on probabilities of slides)	38.2	43.3	53.4	40.6	**98.3**	75.0	**80.0**	85.1
(ours) ABMIL + P&SrE (w/o PSFI)	48.3	60.0	59.5	52.2	81.5	87.1	75.2	84.2
(ours) ABMIL + P&SrE	**56.8**	**71.0**	**66.3**	**59.9**	78.0	**96.1**	76.4	**86.1**
DSMIL + (MaxS)	50.8	56.7	63.8	53.5	78.7	97.4	78	87.1
DSMIL + (MaxMinS)	50.0	60.0	63.2	54.6	79.3	85.5	72	82.3
DSMIL(baseline)+Mean (on probabilities of slides)	43.8	57.7	53.3	48.1	**100**	73.7	**80.0**	84.9
(ours) DSMIL + P&SrE (w/o PSFI)	49.7	**61.7**	62.8	54.9	78.0	**97.9**	77.4	86.8
(ours) DSMIL + P&SrE	**60.0**	56.7	**69.7**	**57.7**	80.2	96.8	79.4	**87.7**

4 Limitations

Our study has some limitations that should be addressed. For instance, we did not explore the possibility of treating patches as an equivalent level to slides and patients. The primary reason is that the vast number of patches required for analysis is significantly larger than that of slides and patients, which presents a computational challenge for training. As a result, we have not yet explored this avenue. In the future, we plan to leverage clustering and active learning methods to reduce the number of patches and enable the interaction of all three levels with the Transformer, which would further enhance the accuracy and efficiency of our proposed method.

5 Conclusion

This study proposes a highly scalable and versatile framework to address M-MIL problems. We first classify the process from patch to slide to the patient in medical pathology diagnosis as a multi-level MIL problem. Based on existing state-of-the-art MIL methods, we then extend the framework to P&SrE, which conducts feature extraction and interaction at the slide-patient level. By introducing a transformer, the framework enables iterative interaction and correction of information between patients and slides, resulting in better performance at both the patient level and slide level compared to existing state-of-the-art algorithms on two validation datasets.

Acknowledgements. This study was supported by Guangdong Basic and Applied Basic Research Foundation (No. 2020A1515010571), National Natural Science Foundation of China (No. 82271958, 81971684, 81801761), Shenzhen-Hong Kong Institute of Brain Science-Shenzhen Fundamental Research Institutions (No. 2022SHIBS0003), and Guangdong Provincial Clinical Research Center for Digestive Diseases (No. 2020B1111170004).

References

1. Bandi, P., et al.: From detection of individual metastases to classification of lymph node status at the patient level: the camelyon17 challenge. IEEE Trans. Med. Imaging **38**(2), 550–560 (2018)
2. Carbonneau, M.A., Cheplygina, V., Granger, E., Gagnon, G.: Multiple instance learning: a survey of problem characteristics and applications. Pattern Recognit. **77**, 329–353 (2018)
3. Cui, M., Zhang, D.Y.: Artificial intelligence and computational pathology. Lab. Invest. **101**(4), 412–422 (2021)
4. Dosovitskiy, A., et al.: An image is worth 16x16 words: Transformers for image recognition at scale. arXiv preprint arXiv:2010.11929 (2020)
5. Fuchs, T.J., Buhmann, J.M.: Computational pathology: challenges and promises for tissue analysis. Comput. Med. Imaging Graph. **35**(7–8), 515–530 (2011)
6. Gecse, K.B., Vermeire, S.: Differential diagnosis of inflammatory bowel disease: imitations and complications. Lancet Gastroenterol. Hepatol. **3**(9), 644–653 (2018)
7. Hong, Y., Han, S., Choi, K., Seo, S., Kim, B., Chang, B.: Disentangling label distribution for long-tailed visual recognition. In: Proceedings of the IEEE/CVF Conference on Computer Vision and Pattern Recognition, pp. 6626–6636 (2021)
8. Ilse, M., Tomczak, J., Welling, M.: Attention-based deep multiple instance learning. In: International Conference on Machine Learning, pp. 2127–2136. PMLR (2018)
9. Li, B., Li, Y., Eliceiri, K.W.: Dual-stream multiple instance learning network for whole slide image classification with self-supervised contrastive learning. In: Proceedings of the IEEE/CVF Conference on Computer Vision and Pattern Recognition, pp. 14318–14328 (2021)
10. Serag, A., et al.: Translational AI and deep learning in diagnostic pathology. Front. Med. **6**, 185 (2019)
11. Vaswani, A., et al.: Attention is all you need. In: Advances in Neural Information Processing Systems, vol. 30 (2017)

Liver Tumor Screening and Diagnosis in CT with Pixel-Lesion-Patient Network

Ke Yan[1,2(✉)], Xiaoli Yin[3], Yingda Xia[1], Fakai Wang[1], Shu Wang[3], Yuan Gao[1,2], Jiawen Yao[1,2], Chunli Li[1,2,3], Xiaoyu Bai[1,2], Jingren Zhou[1,2], Ling Zhang[1], Le Lu[1], and Yu Shi[3]

[1] DAMO Academy, Alibaba Group, Hangzhou, China
yanke.yan@alibaba-inc.com
[2] Hupan Lab, 310023 Hangzhou, China
[3] Department of Radiology, Shengjing Hospital of China Medical University, Shenyang 110004, China

Abstract. Liver tumor segmentation and classification are important tasks in computer aided diagnosis. We aim to address three problems: liver tumor screening and preliminary diagnosis in non-contrast computed tomography (CT), and differential diagnosis in dynamic contrast-enhanced CT. A novel framework named Pixel-Lesion-pAtient Network (PLAN) is proposed. It uses a mask transformer to jointly segment and classify each lesion with improved anchor queries and a foreground-enhanced sampling loss. It also has an image-wise classifier to effectively aggregate global information and predict patient-level diagnosis. A large-scale multi-phase dataset is collected containing 939 tumor patients and 810 normal subjects. 4010 tumor instances of eight types are extensively annotated. On the non-contrast tumor screening task, PLAN achieves 95% and 96% in patient-level sensitivity and specificity. On contrast-enhanced CT, our lesion-level detection precision, recall, and classification accuracy are 92%, 89%, and 86%, outperforming widely used CNN and transformers for lesion segmentation. We also conduct a reader study on a holdout set of 250 cases. PLAN is on par with a senior human radiologist, showing the clinical significance of our results.

Keywords: Liver tumor · Lesion segmentation and classification · CT

1 Introduction

Liver cancer is the third leading cause of cancer death world-wide in 2020 [14]. Early detection and accurate diagnosis of liver tumors may improve overall

Partially supported by the National Natural Science Foundation of China (grant 82071885), Basic Research Projects of Liaoning Provincial Department of Education (LJKMZ20221160), the National Youth Talent Support Program of China, and Science and Technology Innovation Talent Project in Shenyang (RC210265).

Supplementary Information The online version contains supplementary material available at https://doi.org/10.1007/978-3-031-43904-9_8.

H. Greenspan et al. (Eds.): MICCAI 2023, LNCS 14224, pp. 72–82, 2023.
https://doi.org/10.1007/978-3-031-43904-9_8

patient outcomes, in which imaging plays a key role [11]. Computed tomography (CT) is one of the most important imaging modalities for liver tumors. Dynamic contrast-enhanced (DCE) CT is widely used for diagnostics, but it requires iodine contrast injection which can cause reaction and potential risks in patients. Recently, non-contrast (NC) CT scans are gaining attention as they are cheaper and safer to acquire, thus can be potential tools for opportunistic tumor screening [18,20]. Meanwhile, finding and diagnosing tumors in NC CTs is also extremely challenging because of the poor contrast between tumors and normal tissues compared to those in DCE CTs. Prior works on pancreas [18] and esophagus [20] have shown that latest deep learning techniques can detect subtle texture and shape changes in NC CT that even human eyes may miss. Thus, we aim to investigate the performance of liver tumor segmentation and classification in NC CTs. Such an approach will be helpful to discover asymptomatic incidental tumors [12] from routine NC CT scans indicated for general diagnostic purposes at no additional cost and radiation exposure. After an incidental tumor is found, the patient may undergo further imaging examination such as a multi-phase DCE CT for differential diagnosis [11], which can provide useful discriminative information such as the vascularity of lesions and the pattern of contrast agent enhancement [19]. Liver is largest solid organ in body and is the site of many tumor types [11]. Therefore, accurate tumor type classification is important for the decision of treatment plans and prognosis.

Many researchers have developed algorithms to automatically segment [1,9, 13,15,23] or classify [19,21,25] liver tumors in CT to help radiologists improve their accuracy and efficiency. For example, public datasets such as the Liver Tumor Segmentation Benchmark (LiTS) [1] fostered a series of works aiming to segment liver tumors with improved convolutional neural network (CNN) backbones [9,13] and lesion edge information [15]. LiTS only has single-phase CTs (venous phase). Several studies investigated methods to exploit multi-phase CT by methods such as hetero-phase fusion [5] and modality-aware mutual learning [23]. There are few work discussing liver tumor analysis in NC CT [5]. Besides lesion segmentation, CNN-based lesion classification algorithms have been studied to distinguish common lesion types [19,21,25].

In this paper, we build a comprehensive framework to address both tumor screening and diagnosis. (1) Tumor screening involves finding tumor patients in a large pool of healthy subjects and patients. Most existing works in tumor segmentation and detection did not explicitly consider it since their training and testing images are all tumor patients. Such models may generate false positives in real-world screening scenario when facing diverse tumor-free images. We collect a large-scale dataset with both tumor and non-tumor subjects, where the non-tumor subjects includes not only healthy ones, but also patients with various diffuse liver diseases such as steatosis and hepatitis to improve the robustness of the algorithm. (2) Most works studied liver tumor segmentation alone without differentiating tumor types, while a few works classify liver tumors on cropped tumor patches [19,21,25]. Meanwhile, we learn tumor segmentation and classification with one network using an instance segmentation framework [3]. We train two networks for NC and multi-phase DCE CTs, respectively. (3) For evaluation, previous segmentation works typically use *pixel-level* metrics such as Dice coefficient. Such metrics cannot reflect the *lesion-level* accuracy (how many lesion instances are correctly detected and classified) and may bias to large lesions when a patient has multiple tumors. *Patient-level* metrics (e.g. classifying whether a subject has malignant tumors) are also useful for treatment recommendation in clinical practice [18,20]. Therefore, we assess our algorithm thoroughly with pixel, lesion, and patient-level metrics.

Algorithms for liver tumor segmentation have focused on improving the feature extraction backbone of a fully-convolutional CNN [9,13,15,23]. The pixel-wise segmentation architectures may not be optimal for lesion and patient-level evaluation metrics since they cannot consider a lesion or an image holistically. Recently, a series of mask transformer algorithms [3,4,17] have emerged in the computer vision community and achieved the state-of-the-art performance in instance segmentation tasks. In brief, they use object queries to interact with image feature maps and with each other to produce mask and class predictions for each instance. Inspired by them, we propose a novel end-to-end framework named Pixel-Lesion-pAtient Network (PLAN) for lesion segmentation and classification, as well as patient classification. It contains three branches with bottom-up cooperation: The segmentation map from the *pixel branch* helps to initialize the *lesion branch*, which is an improved mask transformer aiming to segment and classify each lesion; The *patient branch* aggregates information from the whole image and predicts image-level labels of each lesion type, with regularization terms to encourage consistency with the lesion branch.

We collected a large-scale multi-phase dataset containing 810 non-tumor subjects and 939 tumor patients. 4010 tumor instances of eight types are extensively annotated based on pathological reports. On the non-contrast tumor screening and diagnosis task, PLAN achieves 95.0%, 96.4%, and 0.965 in patient-level sensitivity, specificity, and average AUC for malignant and benign patients, in contrast to 94.4%, 93.7%, and 0.889 for the widely-used nnU-Net [8]. On multi-phase DCE CT, our lesion-level detection precision, recall, and classification accuracy are 92.2%, 89.0%, 85.9%, outperforming nnU-Net [8] and Mask2Former [3]. We further conduct a reader study on a holdout set of 250 cases. Our algorithm is on par with a senior radiologist (16 yrs experience), showing the clinical significance of our results. Our codes will be made public upon institutional approval.

2 Method

2.1 Preliminary on Mask Transformer

Mask transformers are a series of latest works achieving superior accuracy on various segmentation tasks [3,4,17,22]. Different from traditional fully-convolutional segmentators [8] that predict a class label for each pixel, mask transformers predict a class label and a binary mask for each object. Take Mask2Former [3] as an example. It includes a pixel encoder and a pixel decoder that extract a high-resolution pixel embedding tensor $\mathbf{P} \in \mathbb{R}^{M \times D \times H \times W}$ from

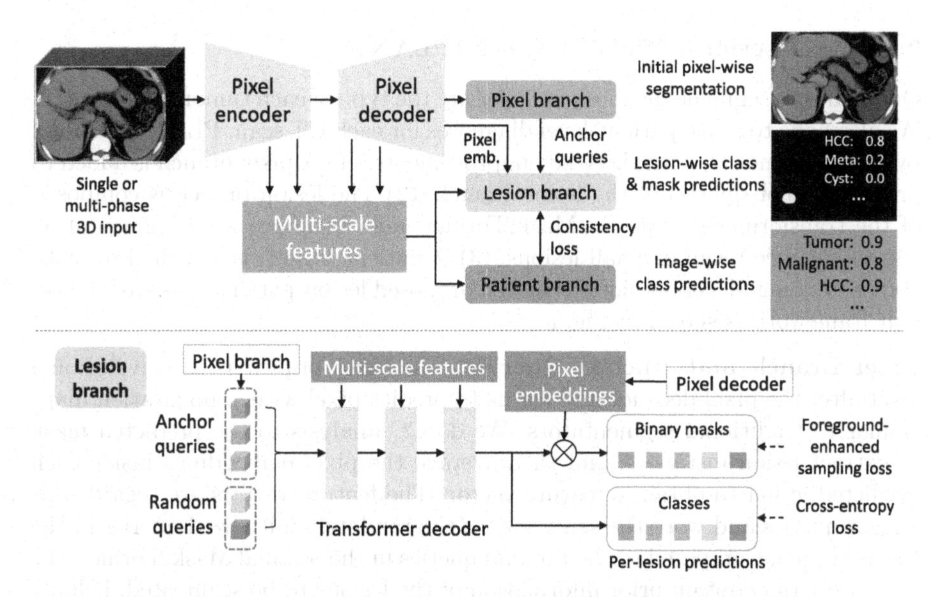

Fig. 1. Framework of the Pixel-Lesion-pAtient Network (PLAN).

the image, where M is the embedding dimension, $D \times H \times W$ is the shape of the 3D image. A group of Q learnable feature vectors $\{\mathbf{q}_i \in \mathbb{R}^M\}_{i=1}^Q$ are randomly initialized as object queries. They are processed by a transformer decoder to interact with multi-scale image features and each other using cross and self-attention operations. After processing, each query is supposed to contain information of one object, which can be used to predict the class probability $\mathbf{c} \in \mathbb{R}^{C+1}$ of the object. Here C is the number of object classes, and we add 1 to indicate an additional "no-object" class if the query does not match with any object. In training, Mask2Former uses bipartite matching [2] to assign each query to a ground-truth object (or "no-object"). Multiplying $\mathbf{q}_i$ with $\mathbf{P}$ gives the binary mask $\mathbf{m}_i \in \mathbb{R}^{D \times H \times W}$ of object i. During inference, the class and mask predictions of all queries can be merged by matrix multiplication to obtain the final semantic segmentation result $\hat{\mathbf{Y}} \in \mathbb{R}^{C \times D \times H \times W}$. We refer readers to [3] for more details.

Mask transformers have various advantages when applied to our task. They can classify a lesion as a whole instead of classifying each pixel, thus can view each lesion holistically. Cross-attention is used to aggregate global features for each lesion. Inter-lesion relation can also be exploited by self-attention operations. In liver CT, inter-lesion relation is diagnostically useful, e.g., metastases and cysts are often multiple. Therefore, We pioneer mask transformers' adaptation for lesion segmentation and classification in 3D medical images. Given a ground-truth or a predicted lesion mask image, we perform connected component (CC) analysis and treat each CC as a lesion instance for training and evaluation.

2.2 Pixel-Lesion-Patient Network (PLAN)

Our goal is to segment the mask and classify the type of each tumor in a liver CT. We also hope to make patient-level diagnoses for each CT scan. PLAN is inspired by Mask2Former [3] with three key improvements: (1) A pixel branch is added to provide anchor queries to the lesion branch. (2) The lesion branch is composed of the transformer decoder in Mask2Former, and we improve its segmentation loss to enhance recall of small lesions. (3) A patient branch is attached to make dedicated image-level predictions with a proposed lesion-patient consistency loss. Our framework is shown in Fig. 1.

Pixel Branch and Anchor Queries. The pixel branch is a convolutional layer after the pixel decoder and learns to predict pixel-wise segmentation maps similar to traditional segmentators. We do CC analysis to the predicted mask to extract lesion instances, and then average the pixel embeddings inside each predicted lesion to obtain a feature vector. The feature vectors are regarded as anchor queries and work the same way as the randomly initialized queries in the lesion branch. Compared to the random queries in the original Mask2Former, the anchor queries contain prior information of the lesions to be segmented, helping the lesion branch to match with the lesion targets more easily [10].

Lesion Branch and Foreground-Enhanced Sampling Loss. Similar to Mask2Former, the lesion branch predicts a binary mask and a class label for each query, see Fig. 1. Mask2Former calculates its segmentation loss on K sampled pixels instead of on the whole image, which is shown to both improve accuracy and reduce GPU memory usage [3]. However, in lesion segmentation, some tumors are very small compared to the whole 3D image. The importance sampling strategy [3] can hardly select any foreground pixels in such cases, so the loss only contains background pixels, degrading the segmentation recall of small lesions. We propose a simple approach to remedy this issue by sampling an extra n foreground pixels for each lesion.

Patient Branch. A patient-level diagnosis is useful for triage. For example, diagnosing the subject as normal, benign, or malignant will result in completely different treatments [24]. Intuitively, we can also infer patient-level labels from segmentation results by checking if there is any lesion in the predicted mask. However, certain tumors are often related to signs outside the tumor, e.g. hepatocellular carcinoma and cirrhosis, cholangiocarcinoma and bile duct dilatation, etc. We equip PLAN with a dedicated patient branch to aggregate such global information to make better patient-level prediction. Since one patient can have multiple liver tumors of different types, in our problem, we give each image several hierarchical binary labels. The first label classifies normal and tumor subjects (whether the image contains any tumor); The second and third labels indicate the existence of respectively benign and malignant tumors; The rest C labels suggest the existence of C fine-grained types of tumors. We employ the dual-path transformer block [17] to fuse multi-scale features from the pixel encoder and decoder to generate a feature map, followed by global average pooling and a linear classification layer to predict the $C + 3$ labels.

A **lesion-patient consistency loss** is further proposed to encourage coherence of the lesion and patient-level predictions. Inspired by multi-instance learning [6], we compute a pseudo patient-level prediction $\tilde{\mathbf{c}} \in \mathbb{R}^C$ from the lesion-level predictions by max-pooling the class probability of each class across all lesion queries (discarding the no-object class). We also have the probability vector from the patient branch $\tilde{\mathbf{p}} \in \mathbb{R}^C$ corresponding to the C fine-grained classes. Then, we compute the L2 loss between them: $\mathcal{L}_{\text{consist}} = \|\tilde{\mathbf{p}} - \tilde{\mathbf{c}}\|^2$.

The overall loss of PLAN is listed in Eq. 1, where $\mathcal{L}_{\text{pixel}}$ is the combined cross-entropy (CE) and Dice loss for the pixel branch as in nnU-Net [8]; $\mathcal{L}_{\text{lesion-class}}$ is the CE loss [3] for lesion classification in the lesion branch; $\mathcal{L}_{\text{lesion-mask}}$ is the combined CE and Dice loss [3] for binary lesion segmentation in the lesion branch with the foreground-enhanced sampling strategy; $\mathcal{L}_{\text{patient}}$ is the binary CE loss for the multi-label classification task in the patient branch.

$$\mathcal{L} = \lambda_1 \mathcal{L}_{\text{pixel}} + \lambda_{2c} \mathcal{L}_{\text{lesion-class}} + \lambda_{2m} \mathcal{L}_{\text{lesion-mask}} + \lambda_3 \mathcal{L}_{\text{patient}} + \lambda_4 \mathcal{L}_{\text{consist}}. \quad (1)$$

3 Experiments

Data. Our dataset contains 810 normal subjects and 939 patients with liver tumors. Each normal subject has a non-contrast (NC) CT, while each patient has a dynamic contrast-enhanced (DCE) CT scan with NC, arterial, and venous phases. We use DEEDS [7] to register NC and arterial phases to the venous phase, and then invite a senior radiologist with 10 years of experience to annotate on the multi-phase CTs using CT Labeler [16]. The 3D mask and the type of all liver tumors are annotated based on pathological reports and magnetic resonance scans if necessary. Eight tumor types are considered in our study: hepatocellular carcinoma (HCC), intrahepatic cholangiocarcinoma (ICC), metastasis (meta), hepatoblastoma (hepato), hemangioma (heman), focal nodular hyperplasia (FNH), cyst, and others (all other tumor types). If a lesion's type cannot be determined according to image signs [11] and pathology, it will be marked as "unknown" and ignored in training and evaluation. In total, 4010 tumor instances are annotated, whose volumes range from 11 to 3.7×10^6 mm^3. Detailed statistics and examples of the lesions are shown in the supplementary material. We train two separate networks for NC and DCE CTs. In the former setting, both normal and patient data are used and randomly split into 1149 training, 100 validation, and 500 testing. In the latter one, only patient data are used with 641 training, 100 validation, and 200 testing. Another hold-out set of 150 patients and 100 normal CTs are used for reader study to compare our accuracy with two radiologists.

Implementation Details. Each CT is resampled to $0.7 \times 0.7 \times 5$mm in spacing. We first train an nnU-Net on public datasets to segment liver and surrounding organs (gallbladder, hepatic vein, spleen, stomach, and pancreas), and then crop the liver region to train PLAN. To help PLAN differentiate liver tumors and other organs, we train the network to segment both tumors and organs

Table 1. Patient-level performance on the test set of 500 cases. Spec. 1: specificity on the 202 completely normal cases; Spec. 2: specificity on the 100 hard non-tumor cases.

	NC tumor screening (%)			NC diagnosis AUC		DCE diagnosis AUC
	Sens.	Spec. 1	Spec. 2	Malignant	Benign	8-class Average
nnU-Net [8]	94.4	95.1	91.0	0.948	0.829	0.863
Mask2Former [3]	93.9	97.0	**94.0**	0.924	0.828	0.873
PLAN (ours)	**95.0**	**97.5**	**94.0**	**0.961**	**0.968**	**0.898**

using the predicted organ labels. PLAN is built on top of the nnU-Net framework [8]. Its pixel encoder is a U-Net encoder, whereas its pixel decoder is a light-weight feature pyramid network [3]. The lesion branch incorporates three transformer decoder blocks with masked attention [3] which use feature maps of strides 16, 8, 4 from the pixel decoder. The number of random queries is $Q = 20$; the embedding dimension is $M = 64$; the number of sampled pixels is $K = 12544$ [3], foreground pixels $n = 3$; the loss weight is 0.1 for the no-object class while 1 for other classes in the lesion branch [3]. The weights in Eq. 1 are $\lambda_1 = \lambda_{2c} = 2, \lambda_{2m} = 5, \lambda_3 = 1, \lambda_4 = 0.1$. We use the RAdam optimizer with an initial learning rate of 0.0001. Each training batch contains two patches of size $256 \times 256 \times 24$. For DCE CT, the three phases form a 3-channel image as the network input. Extensive data augmentation is applied including random cropping, scaling, flipping, elastic deformation, and brightness adjustment [8]. During training, we first pretrain the backbone and the pixel branch for 500 epochs, and then train the whole network for another 500 epochs.

Patient-Level Results. This paper has three major goals: tumor screening in NC CT (classifying a subject as normal or tumor), preliminary diagnosis in NC CT (predicting the existence of malignant and benign tumors), and fine-grained diagnosis in DCE CT (predicting the existence of 8 tumor types). Among the 8 tumor types, HCC, ICC, meta, and hepato are malignant; heman, FNH, and cyst are benign. "Others" can be either malignant or benign, thus are excluded in the preliminary diagnosis task. The NC test set contains 198 tumor cases, 202 completely normal cases, and 100 "hard" non-tumor cases which may have larger image noise, artifact, ascites, diffuse liver diseases such as hepatitis and steatosis. These cases are used to test the robustness of the model in real-world screening scenario with diverse tumor-free images. We compare PLAN with a widely-used strong baseline, nnU-Net [8]. The recent mask transformer, Mask2Former [3], is also adapted to 3D for comparison. For the baselines, patient-level labels are inferred from their predicted masks by counting lesion pixels. As displayed in Table 1, PLAN achieves the best accuracy on all tasks, especially in NC preliminary diagnosis tasks, which demonstrates the effectiveness of its dedicated patient branch that can explicitly aggregate features from the whole image.

Lesion and Pixel-Level Results. In lesion-level evaluation, we treat a prediction as a true positive if its overlap with a ground-truth lesion is >0.2 in Dice.

Table 2. Lesion-level performance (precision, recall, recall of lesions with different radius, classification accuracy of 8 tumor types), and pixel-level performance (Dice per case). Precision, recall, and Dice are computed without considering the tumor types.

		Prec.	Recall	R<5 mm	5∼10	10∼20	>20 mm	Acc.	Dice
NC	nnU-Net	78.8	77.3	19.7	63.6	90.1	96.5	75.7	**78.3**
	Mask2Former	**85.7**	74.0	10.0	60.5	**91.9**	97.4	77.9	76.4
	PLAN	80.1	**81.9**	**21.9**	**64.6**	90.1	**98.3**	**78.5**	77.2
DCE	nnU-Net	88.1	88.3	22.5	**76.4**	93.7	**98.3**	83.1	**84.2**
	Mask2Former	90.3	83.5	11.7	74.4	**94.6**	97.4	84.8	82.9
	PLAN	**92.2**	**89.0**	**25.6**	74.9	**94.6**	**98.3**	**85.9**	**84.2**

Table 3. Reader study results on 150 tumor cases and 100 normal cases. 3-class acc. means classification accuracy of normal vs. benign vs. malignant.

	NC			DCE
	Sens.	Spec.	3-class Acc.	8-class Acc.
Radiologist 1	94.1	**99.0**	90.8	**75.6**
Radiologist 2	85.5	**99.0**	72.0	40.5
PLAN	**96.7**	98.0	**91.3**	**75.6**

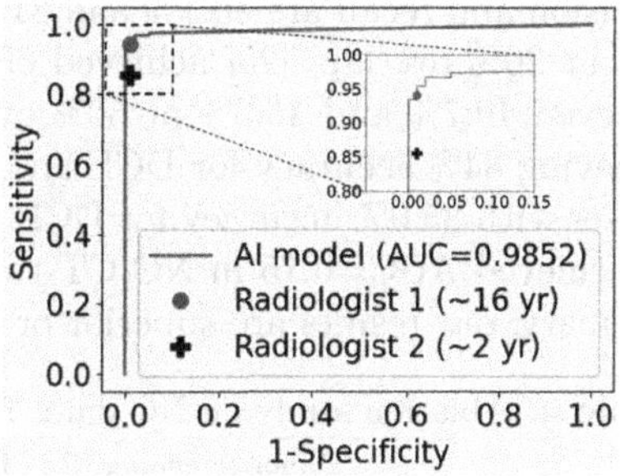

Fig. 2. ROC curve of our method versus 2 radiologists' performance.

Lesions smaller than 3 mm in radius are ignored. As shown in Table 2, the pixel-level accuracy of nnU-Net and PLAN are comparable, but PLAN's lesion-level accuracy is consistently higher than nnU-Net. In this work, we focus more on patient and lesion-level metrics. Although NC images have low contrast, they can still be used to segment and classify lesions with $\sim 80\%$ precision, recall, and classification accuracy. It implies the potential of NC CT, which has been under-studied in previous works. Mask2Former has higher precision but lower recall in NC CT, especially for small lesions, while PLAN achieves the best recall using the foreground-enhanced sampling loss. Both PLAN and Mask2Former achieve better classification accuracy, which illustrates the mask transformer architecture is good at lesion-level classification.

Comparison with Radiologists. In the reader study, we invited a senior radiologist with 16 years of experience in liver imaging, and a junior radiologist with 2 years of experience. They first read the NC CT of all subjects and provided a diagnosis of normal, benign, or malignant. Then, they read the DCE scans and provided a diagnosis of the 8 tumor types. We consider patients with only one tumor type in this study. Their reading process is without time constraint. In Table 3 and Fig. 2, all methods get good specificity probably because the normal subjects are completely healthy. Our model achieves comparable accuracy with the senior radiologist but outperforms the junior one by a large margin in sensitivity and classification accuracy.

An ablation study for our method is shown in Table 4. It can be seen that our proposed anchor queries produced by the pixel branch, FES loss, and lesion-patient consistency loss are useful for the final performance. The efficacy of the lesion and patient branches has been analyzed above based on the lesion and patient-level results. Due to space limit, we will show the accuracy for each tumor type and more qualitative examples in the supplementary material.

Comparison with Literature. In the pixel level, we obtain Dice scores of 77.2% and 84.2% using NC and DCE CTs, respectively. The current state of the art (SOTA) of LiTS [1] achieved 82.2% in Dice using CTs in venous phase; [23] achieved 81.3% in Dice using DCE CT of two phases. In the lesion level, our precision and recall are 80.1% and 81.9% for NC CT, 92.2% and 89.0% for DCE CT, at 20% overlap. [25] achieved 83% and 93% for DCE CT. SOTA of LiTS achieved 49.7% and 46.3% at 50% overlap. [21] classified lesions into 5 classes, achieving 84% accuracy for DCE and 49% for NC CT. We classify lesions into 8 classes with 85.9% accuracy for DCE and 78.5% for NC CT. In the patient level, [5] achieved AUC=0.75 in NC CT tumor screening, while our AUC is 0.985. In summary, our results are superior or comparable to existing works.

Table 4. Ablation study on NC data. FES loss: foreground enhanced sampling loss.

	Tumor screening (%)		Prelim. diagnosis AUC		Lesion and pixel-level (%)			
	Sens.	Spec.	Malignant	Benign	Precision	Recall	Acc.	Dice
PLAN (proposed)	**95.0**	96.4	**96.1**	**96.8**	80.1	**81.9**	**78.5**	**77.2**
w/o anchor queries	94.4	95.4	94.9	93.5	78.9	78.1	77.1	75.0
w/o FES loss	93.4	96.0	94.0	96.4	**86.6**	75.1	77.7	**77.2**
w/o consistency loss	93.9	**96.7**	95.4	96.3	79.1	80.7	78.2	76.6

4 Conclusion

Three tasks are investigated in this paper: liver tumor screening and preliminary diagnosis in NC CT, and the diagnosis of 8 tumor types in DCE CT. The pixel-lesion-patient network is proposed that can accomplish lesion-level segmentation and classification, and patient-level classification. Comprehensive evaluation on a large-scale dataset confirms the effectiveness and clinical significance of our method. It can serve as a powerful tool for automated screening and diagnosis of various liver tumors. Our future work includes further improving the specificity of hard non-tumor cases and sensitivity of small lesions.

References

1. The Liver Tumor Segmentation Benchmark (LiTS). Med. Image Anal. **84** (2023)
2. Carion, N., Massa, F., Synnaeve, G., Usunier, N., Kirillov, A., Zagoruyko, S.: End-to-end object detection with transformers. In: Vedaldi, A., Bischof, H., Brox, T., Frahm, J.-M. (eds.) ECCV 2020. LNCS, vol. 12346, pp. 213–229. Springer, Cham (2020). https://doi.org/10.1007/978-3-030-58452-8_13

3. Cheng, B., Misra, I., Schwing, A.G., Kirillov, A., Girdhar, R.: Masked-attention mask transformer for universal image segmentation. In: CVPR, pp. 1280–1289 (2022)
4. Cheng, B., Schwing, A.G., Kirillov, A.: Per-Pixel classification is not all you need for semantic segmentation. In: NeurIPS, vol. 22, pp. 17864–17875 (2021)
5. Cheng, C.T., Cai, J., Teng, W., Zheng, Y., Huang, Y.T.: A flexible three-dimensional hetero-phase computed tomography hepatocellular carcinoma (HCC) detection algorithm for generalizable and practical HCC screening. Hepatol. Commun. (2022)
6. Cheplygina, V., de Bruijne, M., Pluim, J.P.: Not-so-supervised: a survey of semi-supervised, multi-instance, and transfer learning in medical image analysis. Med. Image Anal. **54**, 280–296 (2019)
7. Heinrich, M.P., Jenkinson, M., Brady, M., Schnabel, J.A.: MRF-based deformable registration and ventilation estimation of lung CT. IEEE Trans. Med. Imaging **32**(7), 1239–1248 (2013)
8. Isensee, F., Jaeger, P.F., Kohl, S.A., Petersen, J., Maier-Hein, K.H.: nnU-Net: a self-configuring method for deep learning-based biomedical image segmentation. Nat. Methods **18**(2), 203–211 (2021)
9. Li, X., Chen, H., Qi, X., Dou, Q., Fu, C.W., Heng, P.A.: H-DenseUNet: hybrid densely connected UNet for liver and tumor segmentation from CT volumes. IEEE Trans. Med. Imaging **37**(12), 2663–2674 (2018)
10. Liu, S.: DAB-DETR : dynamic anchor boxes are better queries for DETR. In: ICLR, pp. 1–19 (2022)
11. Marrero, J.A., Ahn, J., Rajender Reddy, K.: Americal college of gastroenterology: ACG clinical guideline: the diagnosis and management of focal liver lesions. Am. J. Gastroenterol. **109**(9), 1328–1347 (2014)
12. Semaan, A., et al.: Incidentally detected focal liver lesions-a common clinical management dilemma revisited. Anticancer Res. **36**(6), 2923–2932 (2016)
13. Seo, H., Huang, C., Bassenne, M., Xiao, R., Xing, L.: Modified U-Net (mU-Net) with incorporation of object-dependent high level features for improved liver and liver-tumor segmentation in CT images. IEEE Trans. Med. Imaging **39**(5), 1316–1325 (2020)
14. Sung, H., et al.: Global cancer statistics 2020 : GLOBOCAN estimates of incidence and mortality worldwide for 36 cancers in 185 countries. CA: Cancer J. Clin. **71**(3), 209–249 (2021)
15. Tang, Y., Tang, Y., Zhu, Y., Xiao, J., Summers, R.M.: E^{2}Net: an edge enhanced network for accurate liver and tumor segmentation on CT scans. In: Martel, A.L., et al. (eds.) MICCAI 2020. LNCS, vol. 12264, pp. 512–522. Springer, Cham (2020). https://doi.org/10.1007/978-3-030-59719-1_50
16. Wang, F., et al.: A Cascaded Approach for Ultraly High Performance Lesion Detection and False Positive Removal in Liver CT Scans (2023). http://arxiv.org/abs/2306.16036
17. Wang, H., Adam, H., Yuille, A., Chen, L.c.: MaX-DeepLab : end-to-end panoptic segmentation with mask transformers. In: CVPR, pp. 5463–5474 (2021)
18. Xia, Y., et al.: Effective pancreatic cancer screening on non-contrast CT scans via anatomy-aware transformers. In: de Bruijne, M., et al. (eds.) MICCAI 2021. LNCS, vol. 12905, pp. 259–269. Springer, Cham (2021). https://doi.org/10.1007/978-3-030-87240-3_25
19. Xu, X., Zhu, Q., Ying, H., Li, J., Cai, X., Li, S.: A knowledge-guided framework for fine-grained classification of liver lesions based on multi-phase CT images. IEEE J. Biomed. Health Inf. **27**(1), 386–396 (2023)

20. Yao, J., et al.: Effective opportunistic esophageal cancer screening using noncontrast CT imaging. In: Wang, L., Dou, Q., Fletcher, P.T., Speidel, S., Li, S. (eds.) MICCAI 2022. LNCS, vol. 13433, pp. 344–354. Springer, Cham (2022). https://doi.org/10.1007/978-3-031-16437-8_33
21. Yasaka, K., Akai, H., Abe, O., Kiryu, S.: Deep learning with convolutional neural network for differentiation of liver masses at dynamic contrast-enhanced CT: a preliminary study. Radiology **286**(3), 887–896 (2018)
22. Yu, Q., et al.: K-means mask transformer. In: Avidan, S., Brostow, G., Cissé, M., Farinella, G.M., Hassner, T. (eds.) ECCV 2022. LNCS, vol. 13689, pp. 288–307. Springer, Cham (2022). https://doi.org/10.1007/978-3-031-19818-2_17
23. Zhang, Y., Yang, J., Tian, J., Shi, Z., Zhong, C., He, Z.: Modality-aware mutual learning for multi-modal medical image segmentation. In: de Bruijne, M., Cattin, P.C., Cotin, S., Padoy, N., Speidel, S., Zheng, Y., Essert, C. (eds.) MICCAI 2021. LNCS, vol. 12901, pp. 589–599. Springer, Cham (2021). https://doi.org/10.1007/978-3-030-87193-2_56
24. Zhao, T., et al.: 3D graph anatomy geometry-integrated network for pancreatic mass segmentation, diagnosis, and quantitative patient management. In: CVPR, pp. 13738–13747 (2021)
25. Zhou, J., et al.: Automatic detection and classification of focal liver lesions based on deep convolutional neural networks: a preliminary study. Front. Oncol. **10**, 1 (2021)

Self- and Semi-supervised Learning for Gastroscopic Lesion Detection

Xuanye Zhang[1], Kaige Yin[2], Siqi Liu[1], Zhijie Feng[2], Xiaoguang Han[3,4], Guanbin Li[5](✉), and Xiang Wan[1](✉)

[1] Shenzhen Research Institute of Big Data, CUHK-Shenzhen, Shenzhen, China
wanxiang@sribd.cn
[2] Department of Gastroenterology, Hebei Key Laboratory of Gastroenterology, Hebei Institute of Gastroenterology, Hebei Clinical Research Center for Digestive Diseases, The Second Hospital of Hebei Medical University, Shijiazhuang, Hebei, China
[3] FNii, CUHK-Shenzhen, Shenzhen, China
[4] School of Science and Engineering, CUHK-Shenzhen, Shenzhen, China
[5] School of Computer Science and Engineering, Research Institute of Sun Yat-sen University in Shenzhen, Sun Yat-sen University, Guangzhou, China
liguanbin@mail.sysu.edu.cn

Abstract. Gastroscopic Lesion Detection (GLD) plays a key role in computer-assisted diagnostic procedures. However, this task is not well studied in the literature due to the lack of labeled data and the applicable methods. Generic detectors perform below expectations on GLD tasks for 2 reasons: 1) The scale of labeled data of GLD datasets is far smaller than that of natural-image object detection datasets. 2) Gastroscopic images exhibit distinct differences from natural images, which are usually of high similarity in global but high diversity in local. Such characteristic of gastroscopic images also degrades the performance of using generic self-supervised or semi-supervised methods to solve the labeled data shortage problem using massive unlabeled data. In this paper, we propose Self- and Semi-Supervised Learning (SSL) for GLD tailored for using massive unlabeled gastroscopic images to enhance GLD tasks performance, which consists of a Hybrid Self-Supervised Learning (HSL) method for backbone pre-training and a Prototype-based Pseudo-label Generation (PPG) method for semi-supervised detector training. The HSL combines patch reconstruction with dense contrastive learning to boost their advantages in feature learning from massive unlabeled data. The PGG generates pseudo-labels for unlabeled data based on similarity to the prototype feature vector to discover potential lesions and avoid introducing much noise. Moreover, we contribute the first Large-scale GLD Datasets (LGLDD), which contains 10,083 gastroscopic images with 12,292 well-annotated boxes for four-category lesions. Experiments on LGLDD demonstrate that SSL can bring significant improvement.

X. Zhang and K. Yin—Contribute equally to this paper.

Supplementary Information The online version contains supplementary material available at https://doi.org/10.1007/978-3-031-43904-9_9.

H. Greenspan et al. (Eds.): MICCAI 2023, LNCS 14224, pp. 83–93, 2023.
https://doi.org/10.1007/978-3-031-43904-9_9

Keywords: Gastroscopic Lesion Detection · Self-Supervised Backbone Pre-training · Semi-supervised Detector Training

1 Introduction

Gastroscopic Lesion Detection (GLD) plays a key role in computer-assisted diagnostic procedures. Although deep neural network-based object detectors achieve tremendous success within the domain of natural images, directly training generic object detectors on GLD datasets performs below expectations for two reasons: 1) The scale of labeled data in GLD datasets is limited in comparison to natural images due to the annotation costs. Though gastroscopic images are abundant, those containing lesions are rare, which necessitates extensive image review for lesion annotation. 2) The characteristic of gastroscopic images exhibits distinct differences from the natural images [18,19,21] and is often of high similarity in global but high diversity in local. Specifically, each type of lesion may have diverse appearances though gastroscopic images look quite similar. Some appearances of lesions are quite rare and can only be observed in a few patients. Generic self-supervised backbone pre-training or semi-supervised detector training methods can solve the first challenge for natural images but its effectiveness is undermined for gastroscopic images due to the second challenge.

Self-Supervised Backbone Pre-training methods enhance object detection performance by learning high-quality feature representations from massive unlabelled data for the backbone. The mainstream self-supervised backbone pre-training methods adopt self-supervised contrast learning [3,4,7,9,10] or masked

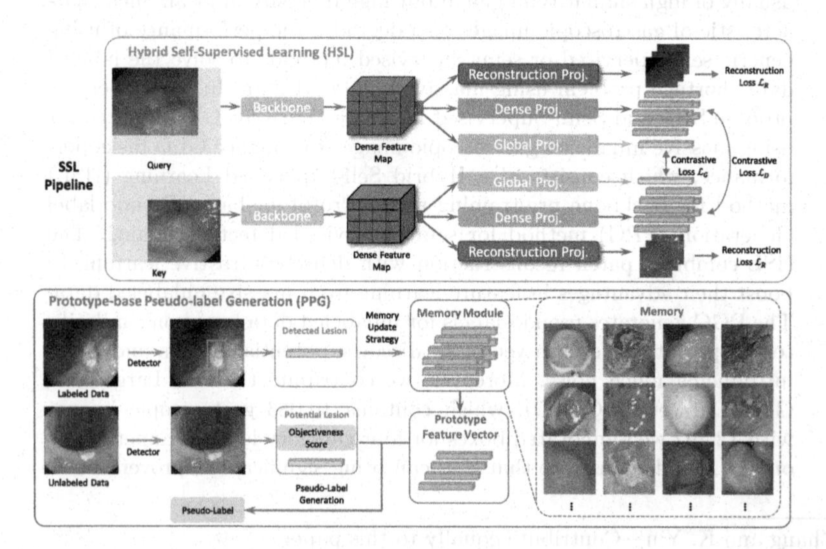

Fig. 1. Pipeline of Self- and Semi-Supervised Learning (SSL) for GLD. SSL consists of a Hybrid Self-Supervised Learning (HSL) method and a Prototype-based Pseudo-label Generation (PPG) method. HSL combines patch reconstruction with dense contrastive learning. PPG generates pseudo-labels for potential lesions based on the similarity to the prototype feature vectors.

image modeling [8,15]. Self-supervised contrastive learning methods [3,4,7,9] can learn discriminative global feature representations, and [10] can further learn discriminative local feature representations by extending contrastive learning to dense paradigm. However, these methods usually cannot grasp enough local detailed information. On the other hand, masked image modeling is expert in extracting local detailed information but is weak in preserving the discriminability of feature representation. Therefore, both types of methods have their own weakness for GLD tasks.

Semi-Supervised object detection methods [12,14,16,17,20,22,23] first use detectors trained with labeled data to generate pseudo-labels for unlabeled data and then enhance object detection performance by regarding these unlabeled data with pseudo-labels as labeled data to train the detector. Current pseudo-label generation methods rely on the objectiveness score threshold to generate pseudo-labels, which makes them perform below expectations on GLD, because the characteristic of gastroscopic lesions makes it difficult to set a suitable threshold to discover potential lesions meanwhile avoiding introducing much noise.

The motivation of this paper is to explore how to enhance GLD performance using massive unlabeled gastroscopic images to overcome the labeled data shortage problem. The main challenge for this goal is the characteristic of gastroscopic lesions. Intuitively, such a challenge requires local feature representations to contain enough detailed information, meanwhile preserving discriminability. Enlightened by this, we propose the **S**elf- and **S**emi-Supervised **L**earning (**SSL**) framework tailored to address challenges in daily clinical practice and use massive unlabeled data to enhance GLD performance. SSL overcomes the challenges of GLD by leveraging a large volume of unlabeled gastroscopic images using self-supervised learning for improved feature representations and semi-supervised learning to discover and utilize potential lesions to enhance performance. Specifically, it consists of a **H**ybrid **S**elf-Supervised **L**earning (**HSL**) method for self-supervised backbone pre-training and a **P**rototype-based **P**seudo-label **G**eneration (**PPG**) method for semi-supervised detector training. The HSL combines the dense contrastive learning [10] with the patch reconstruction to inherit the advantages of discriminative feature learning and grasp the detailed information that is important for GLD tasks. The PPG generates pseudo-labels based on the similarity to the prototype feature vectors (formulated from the feature vectors in its Memory Module) to discover potential lesions from unlabeled data, and avoid introducing much noise at the same time. Moreover, we propose the first Large-scale GLD Datasets (LGLDD), which contains 10,083 gastroscopic images with 12,292 well-annotated lesion bounding boxes of four categories of lesions (polyp, ulcer, cancer, and sub-mucosal tumor). We evaluate SSL with multiple detectors on LGLDD and SSL brings significant improvement compared with baseline methods (CenterNet [6]: +2.7AP, Faster RCNN [13]: +2.0AP). In summary, our contributions include:

- A Self- and Semi-supervise Learning (SSL) framework to leverage massive unlabeled data to enhance GLD performance.
- A Large-scale Gastroscopic Lesion Detection datasets (LGLDD)

– Experiments on LGLDD demonstrate that SSL can bring significant enhancement compared with baseline methods.

2 Methodology

In this section, we introduce the main ideas of the proposed **SSL** for GLD. The proposed approach includes 2 main components and is illustrated in Fig. 1.

2.1 Hybrid Self-supervised Learning

The motivation of Hybrid Self-Supervised Learning (HSL) is to learn the local feature representations of high discriminability meanwhile contain detailed information for the backbone from massive unlabeled gastroscopic images. Among existing backbone pre-training methods, dense contrastive learning can preserve local discriminability and masked image modeling can grasp local detailed information. Therefore, to leverage the advantages of both types of methods, we propose Hybrid Self-Supervised Learning (HSL), which combines patch reconstruction with dense contrastive learning to achieve the goal.

Structure. HSL heritages the structure of the DenseCL [10] but adds an extra reconstruction projection head to reconstruct patches. Specifically, HSL consists of a backbone network and 3 parallel sub-heads. The global projection head and the dense projection head heritages from the DenseCL [10], and the proposed reconstruction projection head is inspired by the Masked Image Modeling. Enlightened by the SimMIM [15], we adopt a lightweight design for the reconstruction projection head, which only contains 2 convolution layers.

Learning Pipeline. Like other self-supervised contrastive learning methods, HSL randomly generates 2 different "views" of the input image, uses the backbone to extract the dense feature maps $\mathbf{F}_1, \mathbf{F}_2 \in \mathbb{R}^{H\times W\times C}$, and then feeds them to the following projection heads. The global projection head of HSL uses $\mathbf{F}_1$, $\mathbf{F}_2$ to obtain the global feature vector $\mathbf{f}_{g1}$, $\mathbf{f}_{g2}$ like MoCo [9]. The dense projection head and the reconstruction projection head crop the dense feature maps $\mathbf{F}_1$, $\mathbf{F}_2$ into $\mathbf{S}\times\mathbf{S}$ patches and obtain the local feature vector sets $\mathbb{F}_1$ and $\mathbb{F}_2$ of each view ($\mathbb{F} = \{f_1, f_2, ..., f_{\mathbf{S}^2}\}$). The dense projection head use $\mathbb{F}_1$ and $\mathbb{F}_2$ to obtain local feature vector sets $\mathbb{F}_{l1}$ and $\mathbb{F}_{l2}$ ($\mathbb{F}_l = \{f_{l1}, f_{l2}, ..., f_{l\mathbf{S}^2}\}$) like DenseCL [10]. The reconstruction projection head uses each feature vector in $\mathbb{F}_1, \mathbb{F}_2$ to reconstruct corresponding patches and obtains the patch set $\mathbb{P}_1, \mathbb{P}_2(\mathbb{P} = \{p_{i1}, p_{i2}, ..., p_{i\mathbf{S}^2}\}$.

Training Objective. The HSL formulates the two contrastive learning as dictionary look-up tasks like DenseCL [10] while the reconstruction learning as a regression task. The global contrastive learning uses the global feature vector $\mathbf{f}_g$ of an image as query $\mathbf{q}$ and feature vectors from the alternate view of the query image and the other images within the batch as keys $\mathbb{K} = \{k_1, k_2, ...,\}$. For each query $\mathbf{q}$, the only positive key $\mathbf{k}_+$ is the different views of the same images and the others are all negative keys ($\mathbf{k}_-$) like MoCo [9]. We adopt the InfoNCE loss function for it:

$$\mathcal{L}_G = -log \frac{exp(q \cdot k_+/\tau)}{exp(q \cdot k_+) + \sum_{k_-} exp(q \cdot k_-/\tau)}$$

The dense contrastive learning uses the local feature vector in $\mathbb{F}_{li}$ as query r and keys $\mathbb{T}_l = \{t_1, t_2, ..., \}$. The negative keys t_- here are the feature vectors of different images while the positive key t_+ is the correspondence feature vector of r in another view of the images. Specifically, we adopt the correspondence methods in DenseCL [10] to obtain the positive key t_+, which first conducts the matching process based on vector-wise cosine similarity between r and feature vectors in $\mathbb{T}$ and then selects the t_j of highest similarity as the t_+. The loss function is also the InfoNCE loss but in a dense paradigm:

$$\mathcal{L}_L = \frac{1}{S^2} \sum -log \frac{exp(r^s \cdot t_+^s/\tau)}{exp(r^s \cdot t_+^s) + \sum_{t_-^s} exp(r^s \cdot t_-^s/\tau)}$$

The reconstruction task uses the feature vector in $\mathbb{F}$ to reconstruct each patch and obtain $\mathbb{P}_i$. The ground truth is the corresponding patches $\mathbb{V}_i = \{v_{i1}, v_{i2}, ..., v_{i\mathbf{S}^2}\}$ of the input view. We adopt the MSE loss function for it:

$$\mathcal{L}_R = \frac{1}{2S^2} \sum E(v_i - p_i)^2$$

The overall loss function is the weighted sum of these losses:

$$\mathcal{L}_H = \mathcal{L}_G + \lambda_D \mathcal{L}_D + \lambda_R \mathcal{L}_R$$

where λ_D and λ_R are the weights of $\mathcal{L}_D$ and $\mathcal{L}_R$ and are set to 1 and 2.

2.2 Prototype-Based Pseudo-label Generation Method

We propose the Prototype-based Pseudo-label Generation method (PPG) to discover potential lesions from unlabeled gastroscopic data meanwhile avoid introducing much noise to further enhance GLD performance. Specifically, PPG adopts a Memory Module to remember feature vectors of the representative lesions as memory and generates prototype feature vectors for each class based on the memories stored. To preserve the representativeness of the memory and further the prototype feature vectors, PPG designs a novel Memory Update Strategy. In semi-supervised learning, PPG generates pseudo-labels for unlabeled data relying on the similarity to the prototype feature vectors, which achieves a better balance between lesion discovery and noise avoidance.

Memory Module. Memory Module stores a set of lesion feature vectors as memory. For a C-class GLD task, the Memory Module stores $C \times N$ feature vectors as memory. Specifically, for each lesion, we denote the feature vector used to classify the lesion in the detector as f_c. PPG stores N feature vectors for each class c to formulate the class memory $m_c = \{f_{c1}, f_{c2}, ..., f_{cN}\}$, and the memory $\mathbb{M}$ of PPG can be expressed as $\mathbb{M} = \{m_1, m_2, ..., m_C\}$. Then, PPG obtains the prototype feature vector p_c by calculating the center of each class memory m_c, and the prototype feature vector set can be expressed as $\mathbb{P}_t = \{p_1, p_2, ..., p_C\}$. Moreover, the prototype feature vectors further serve as supervision for detector training under a contrastive clustering formulation and adopt a contrastive loss:

$$\mathcal{L}_{cc} = \|f_c, p_c\| + \sum_{j \neq c}^{C} max(0, 1 - \|f_c, p_j\|)$$

If the detector training loss is $\mathcal{L}_{Det}$, the overall loss $\mathcal{L}$ can be expressed as:

$$\mathcal{L} = \mathcal{L}_{Det} + \lambda_{cc}\mathcal{L}_{cc}$$

where the λ_{cc} is the weight of the contrastive learning loss and is set to 0.5.

Memory Update Strategy. Memory Update Strategy directly influences the representativeness of the class memory m_c and further the prototype feature vector p_c. Therefore, PPG adopts a novel Memory Update Strategy, which follows the idea that "The Memory Module should preserve the more representative feature vector among similar feature vectors". The pipeline of the strategy is as follows: 1) Acquisition the lesion feature vector f'_c. 2) Identification of the most similar f_s to f_c from corresponding class memory m_c based on similarity:

$$f'_s = \max_j sim(f_{cj}, f'_c)$$

3) Updating the memory by selecting more unique features f_s of $F' = \{f'_s, f'_c\}$ compared to the class prototype feature vector p_c based upon similarity:

$$f_s = \underset{f' \in F'}{argmin}\, sim(f', p_c)$$

The similarity function $sim(u, v)$ can be expressed as $sim(u, v) = u^T v / \|u\| \|v\|$. To initialize the memories, we empirically select 50 lesions randomly for each class. To maintain stability, we start updating the memory and calculating its loss after fixed epochs, and only the positive sample feature vector can be selected to update the memory.

Pseudo-label Generation. PPG proposes to generate pseudo-labels based on the similarity between the prototype feature vectors and the feature vector of potential lesions. To be specific, PPG first detects a large number of potential lesions with a low objectiveness score threshold τ_u and then matches them with all the prototype feature vectors $\mathbb{P}$ to find the most similar one:

$$c = \underset{p_c \in \mathbb{P}}{argmax}\, sim(p_c, f_u)$$

PPG assigns the pseudo-label c for similarity value $sim(p_c, f_u)$ greater than the similarity threshold τ_s otherwise omits it. We set $\tau_u = 0.5$ and $\tau_s = 0.5$

3 Datasets

We contribute the first **L**arge-scale **G**stroscopic **L**esion **D**etection **D**atasets (LGLDD) in the literature.

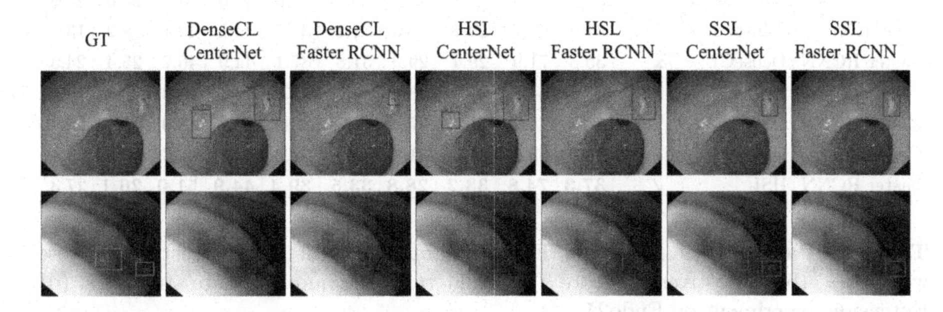

Fig. 2. Qualitative Results of SSL on LGLDD. SSL can actually enhance the GLD performance for some challenging cases.

Collection : LGMDD collects about 1M+ gastroscopic images from 2 hospitals of about 500 patients and their diagnosis reports. After consulting some senior doctors and surveying gastroscopic diagnosis papers [1], we select to annotate 4-category lesions: polyp(pol), ulcer(ulc), cancer(can) and sub-mucosal tumor(smt). We invite 10 senior doctors to annotate them from the unlabeled endoscopic images. To preserve the annotation quality, doctors can refer to the diagnosis reports, and each lesion is annotated by a doctor and checked by another. Finally, they annotates 12,292 lesion boxes in 10,083 images after going through about 120,000 images. The polyp, ulcer, cancer, and sub-mucosal tumor numbers are 7,779, 2,171, 1,164and 1,178, respectively. The train/val split of LGMDD is 8,076/2,007. The other data serves as unlabeled data.

Evaluation Metrics : We use standard object detection metrics to evaluate the GLD performance, which computes the average precision (AP) under multiple intersection-of-union (IoU) thresholds and then evaluate the performance using the mean of APs (mAP) and the AP of some specific IoU threshold. For mAP, we follow the popular object detection datasets COCO [11] and calculate the mean of 11 APs of IoU from 0.5 to 0.95 with stepsize 0.05 (mAP @[.5:.05:.95]). We also report AP under some specific IoU threshold (AP_{50} for .5, AP_{75} for .75) and AP of different scale lesions (AP_S, AP_M, AP_L) like COCO [11].

4 Experiments

Please kindly refer to the **Supplemental Materials** for implementation details and training setups.

Table 1. Quantitative Results of SSL on LGLDD. Both components of SSL (HSL & PPG) can bring significant performance enhancement for GLD tasks.

Detector	Pre-training	PPG	AP	AP_{50}	AP_{75}	AP_S	AP_M	AP_L	pol	stm	uls	can
CenterNet	Supervised	x	29.3	57.2	25.4	22.0	31.6	31.3	41.6	36.0	27.3	12.1
Faster RCNN	Supervised	x	34.1	70.6	28.1	28.2	29.2	35.6	44.0	44.4	24.0	24.2
CenterNet	DenseCL	x	31.9	60.7	29.5	22.6	32.1	34.7	43.2	43.1	28.2	13.2
Faster RCNN	DenseCL	x	35.3	71.9	29.4	29.9	31.8	37.1	44.9	46.7	25.4	24.3
CenterNet	HSL	x	33.7	64.2	30.6	23.1	33.7	35.9	42.5	45.3	28.8	18.0
Faster RCNN	HSL	x	36.4	74.0	31.4	27.9	31.8	38.3	43.7	48.0	26.1	**27.6**
CenterNet	HSL	✓	34.6	65.6	31.6	21.7	32.9	37.3	43.1	46.3	29.3	19.6
Faster RCNN	HSL	✓	**37.3**	**74.8**	**33.2**	**28.8**	**33.5**	**39.4**	**44.9**	**51.0**	**26.1**	27.3

Table 2. Parameters Analysis Experiment Results. (a) Reconstruction loss weight λ_R. (b) Objectiveness Score Threshold τ_u. (c) Memory update strategies. (d) Extension experiment on Endo21.

(a)

λ_R	AP	AP_{50}	AP_{75}
0.5	35.8	73.1	30.7
1	**36.4**	**74.0**	**31.4**
2	36.3	73.4	31.8
5	35.5	71.6	29.5

(b)

τ_u	AP	AP_{50}	AP_{75}
w/o	36.4	74.0	31.4
0.7	36.7	74.0	32.1
0.6	36.2	73.7	31.8
0.5	35.8	72.4	30.9
PPG	**37.3**	**74.8**	**33.2**

(c)

	AP	AP_{50}	AP_{75}
Q-like	37.0	74.2	31.4
PPG	**37.3**	**74.8**	**33.2**

(d)

	AP	AP_{50}	AP_{75}
YOLO v5	60.5	81.0	66.4
Faster RCNN	57.8	79.1	68.1
+DenseCL	59.0	80.9	66.0
+HSL	61.4	83.0	67.3
+PPG	**61.9**	**83.0**	**69.2**

Main Results. Table 1 shows the quantitative results of SSL on LGLDD. As is illustrated, when compared with the DenseCL [10] baseline, SSL can enhance 2.0AP and 2.7AP for Faster RCNN and CenterNet respectively. When compared with the supervised pre-training (ImageNet [5] weights) baseline, SSL can boost more AP enhancement (CenterNet: +5.3AP, FasterRCNN: +3.2AP). Qualitative Results are shown in Fig. 2. It can be noticed, SSL can actually enhance the GLD performance for both types of detectors, especially for some challenging cases.

Ablation Studies. We further analyze each component of SSL (HSL & PPG). HSL can bring 1.8 AP and 1.1 AP enhancement for CenterNet and FasterRCNN respectively compared with DenseCL. PPG can bring extra 0.9AP and 0.9AP enhancement for CenterNet and FasterRCNN respectively.

Parameter Analysis. We conduct extra experiments based on Faster RCNN to further analyze the effect of different parameter settings on LGLDD.
1) **Reconstruction Loss Weight** λ_R is designed to balance the losses of contrastive learning and the reconstruction, which is to balance the discriminability and the detailed information volume of local feature representations. As illustrated in Table 2.a, only suitable λ_R can fully boost the detection performance.
2) **Objectiveness score threshold** τ_u: We compare **PPG** with objectiveness score-based pseudo-label generation methods with different τ_u (Table 2.b). The Objectiveness score threshold controls the quality of pseudo-labels. a) A

low threshold generates noisy pseudo-labels, leading to reduced performance (-0.6/-0.2 AP at thresholds 0.5/0.6). b) A high threshold produces high-quality pseudo-labels but may miss potential lesions, resulting in only slight performance improvement (+0.3 AP at threshold 0.7). c) PPG approach uses a low threshold (0.5) to identify potential lesions, which are then filtered using prototype feature vectors, resulting in the most significant performance enhancement (+0.9 AP).
3) **Memory Update Strategy** influences the representativeness of memory and the prototype feature vectors. We compare our Memory Update Strategy with a queue-like ('Q-like') memory update strategy (first in & first out). Experiment results (Table 2.c) show our Memory Update Strategy performs better.
4) **Endo21**: To further evaluate the effectiveness of SSL, we conduct experiments on Endo21 [2] Sub-task 2 (Endo21 challenge consists of 4 sub-tasks and only the Sub-task 2 train/test split is available according to the [2]). Experimental results in Table 2.d show that SSL can bring significant improvements to publicly available datasets. Moreover, SSL overperforms current SOTA (YOLO v5 [2]).

5 Conclusion

In this work, we propose Self- and Semi-Supervised Learning (SSL) for GLD tailored for using massive unlabeled gastroscopic to enhance GLD performance. The key novelties of the proposed method include a Hybrid Contrastive Learning method for backbone pre-training and a Prototype-based Pseudo-Label Generation method for semi-supervised learning. Moreover, we contribute the first Large-scale GLD Datasets (LGLDD). Experiments on LGLDD prove that SSL can bring significant improvements to GLD performance. Since annotation cost always limits of datasets scale of such tasks, we hope SSL and LGLDD could fully realize its potential, as well as kindle further research in this direction.

Acknowledgement. This work is supported by Chinese Key-Area Research and Development Program of Guangdong Province (2020B0101350001), the National Natural Science Foundation of China (NO. 61976250), the Guangdong Basic and Applied Basic Research Foundation (NO. 2020B1515020048), the Shenzhen Science and Technology Program (JCYJ20220818103001002, JCYJ20220530141211024), the Shenzhen Sustainable Development Project (KCXFZ20201221173008022), the Guangdong Provincial Key Laboratory of Big Data Computing, and the Chinese University of Hong Kong, Shenzhen.

References

1. Ali, S., et al.: Endoscopy disease detection challenge 2020. arXiv preprint arXiv:2003.03376 (2020)
2. Ali, S., et al.: Assessing generalisability of deep learning-based polyp detection and segmentation methods through a computer vision challenge. arXiv preprint arXiv:2202.12031 (2022)

3. Chen, T., Kornblith, S., Norouzi, M., Hinton, G.: A simple framework for contrastive learning of visual representations. In: International Conference on Machine Learning, pp. 1597–1607. PMLR (2020)
4. Chen, X., He, K.: Exploring simple Siamese representation learning. In: Proceedings of the IEEE/CVF Conference on Computer Vision and Pattern Recognition, pp. 15750–15758 (2021)
5. Deng, J., Dong, W., Socher, R., Li, L.J., Li, K., Fei-Fei, L.: ImageNet: a large-scale hierarchical image database. In: 2009 IEEE Conference on Computer Vision and Pattern Recognition, pp. 248–255. IEEE (2009)
6. Duan, K., Bai, S., Xie, L., Qi, H., Huang, Q., Tian, Q.: CenterNet: keypoint triplets for object detection. In: Proceedings of the IEEE/CVF International Conference on Computer Vision, pp. 6569–6578 (2019)
7. Grill, J.B., et al.: Bootstrap your own latent-a new approach to self-supervised learning. In: Advances in Neural Information Processing Systems, vol. 33, pp. 21271–21284 (2020)
8. He, K., Chen, X., Xie, S., Li, Y., Dollár, P., Girshick, R.: Masked autoencoders are scalable vision learners. In: Proceedings of the IEEE/CVF Conference on Computer Vision and Pattern Recognition, pp. 16000–16009 (2022)
9. He, K., Fan, H., Wu, Y., Xie, S., Girshick, R.: Momentum contrast for unsupervised visual representation learning. In: Proceedings of the IEEE/CVF Conference on Computer Vision and Pattern Recognition, pp. 9729–9738 (2020)
10. Li, X., et al.: Dense semantic contrast for self-supervised visual representation learning. In: Proceedings of the 29th ACM International Conference on Multimedia, pp. 1368–1376 (2021)
11. Lin, T.-Y., et al.: Microsoft COCO: common objects in context. In: Fleet, D., Pajdla, T., Schiele, B., Tuytelaars, T. (eds.) ECCV 2014. LNCS, vol. 8693, pp. 740–755. Springer, Cham (2014). https://doi.org/10.1007/978-3-319-10602-1_48
12. Liu, Y.C., et al.: Unbiased teacher for semi-supervised object detection. arXiv preprint arXiv:2102.09480 (2021)
13. Ren, S., He, K., Girshick, R., Sun, J.: Faster R-CNN: towards real-time object detection with region proposal networks. In: Advances in Neural Information Processing Systems, vol. 28 (2015)
14. Sohn, K., Zhang, Z., Li, C.L., Zhang, H., Lee, C.Y., Pfister, T.: A simple semi-supervised learning framework for object detection. arXiv preprint arXiv:2005.04757 (2020)
15. Xie, Z., et al.: SimMIM: a simple framework for masked image modeling. In: Proceedings of the IEEE/CVF Conference on Computer Vision and Pattern Recognition, pp. 9653–9663 (2022)
16. Xu, M., et al.: End-to-end semi-supervised object detection with soft teacher. In: Proceedings of the IEEE/CVF International Conference on Computer Vision, pp. 3060–3069 (2021)
17. Yan, P., et al.: Semi-supervised video salient object detection using pseudo-labels. In: Proceedings of the IEEE/CVF International Conference on Computer Vision, pp. 7284–7293 (2019)
18. Zhang, R., et al.: Lesion-aware dynamic kernel for polyp segmentation. In: Wang, L., Dou, Q., Fletcher, P.T., Speidel, S., Li, S. (eds.) International Conference on Medical Image Computing and Computer-Assisted Intervention, pp. 99–109. Springer, Cham (2022). https://doi.org/10.1007/978-3-031-16437-8_10
19. Zhang, R., Li, G., Li, Z., Cui, S., Qian, D., Yu, Y.: Adaptive context selection for polyp segmentation. In: Martel, A.L., et al. (eds.) MICCAI 2020. LNCS, vol.

12266, pp. 253–262. Springer, Cham (2020). https://doi.org/10.1007/978-3-030-59725-2_25
20. Zhang, R., Liu, S., Yu, Y., Li, G.: Self-supervised correction learning for semi-supervised biomedical image segmentation. In: de Bruijne, M., et al. (eds.) MICCAI 2021. LNCS, vol. 12902, pp. 134–144. Springer, Cham (2021). https://doi.org/10.1007/978-3-030-87196-3_13
21. Zhao, X., Fang, C., Fan, D.J., Lin, X., Gao, F., Li, G.: Cross-level contrastive learning and consistency constraint for semi-supervised medical image segmentation. In: 2022 IEEE 19th International Symposium on Biomedical Imaging (ISBI), pp. 1–5 (2022). https://doi.org/10.1109/ISBI52829.2022.9761710
22. Zhao, X., et al.: Semi-supervised spatial temporal attention network for video polyp segmentation. In: Wang, L., Dou, Q., Fletcher, P.T., Speidel, S., Li, S. (eds.) International Conference on Medical Image Computing and Computer-Assisted Intervention, pp. 456–466. Springer, Cham (2022). https://doi.org/10.1007/978-3-031-16440-8_44
23. Zhou, H., et al.: Dense teacher: dense pseudo-labels for semi-supervised object detection. In: Avidan, S., Brostow, G., Cissé, M., Farinella, G.M., Hassner, T. (eds.) Computer Vision-ECCV 2022: 17th European Conference, Tel Aviv, Israel, October 23–27, 2022, Proceedings, Part IX, pp. 35–50. Springer, Cham (2022). https://doi.org/10.1007/978-3-031-20077-9_3

DiffULD: Diffusive Universal Lesion Detection

Peiang Zhao[1,2], Han Li[1,2], Ruiyang Jin[1,2], and S. Kevin Zhou[1,2,3](✉)

[1] School of Biomedical Engineering, Division of Life Sciences and Medicine, University of Science and Technology of China, Hefei, Anhui 230026, People's Republic of China
{pazhao,hanli21,ryjin}@mail.ustc.edu.cn, skevinzhou@ustc.edu.cn
[2] Center for Medical Imaging, Robotics, Analytic Computing & Learning (MIRACLE), Suzhou Institute for Advanced Research, University of Science and Technology of China, Suzhou, Jiangsu 215123, People's Republic of China
[3] Key Lab of Intelligent Information Processing of Chinese Academy of Sciences (CAS), Institute of Computing Technology, CAS, Beijing 100190, People's Republic of China

Abstract. Universal Lesion Detection (ULD) in computed tomography (CT) plays an essential role in computer-aided diagnosis. Promising ULD results have been reported by anchor-based detection designs, but they have inherent drawbacks due to the use of anchors: **(i) Insufficient training target** and **(ii) Difficulties in anchor design.** Diffusion probability models (DPM) have demonstrated outstanding capabilities in many vision tasks. Many DPM-based approaches achieve great success in natural image object detection without using anchors. But they are still ineffective for ULD due to the insufficient training targets. In this paper, we propose a novel ULD method, DiffULD, which utilizes DPM for lesion detection. To tackle the negative effect triggered by insufficient targets, we introduce a novel Center-aligned bounding box (BBox) padding strategy that provides additional high-quality training targets yet avoids significant performance deterioration. DiffULD is inherently advanced in locating lesions with diverse sizes and shapes since it can predict with arbitrary boxes. Experiments on the benchmark dataset DeepLesion [32] show the superiority of DiffULD when compared to state-of-the-art ULD approaches.

Keywords: Universal Lesion Detection · Diffusion Model

1 Introduction

Universal Lesion Detection (ULD) in computed tomography (CT) [3,13,14, 16,20,28–30,33–37,40,44] plays an important role in computer-aided diagnosis (CAD) [42,43]. The design of detection-only instead of identifying the lesion

Supplementary Information The online version contains supplementary material available at https://doi.org/10.1007/978-3-031-43904-9_10.

H. Greenspan et al. (Eds.): MICCAI 2023, LNCS 14224, pp. 94–105, 2023.
https://doi.org/10.1007/978-3-031-43904-9_10

types in ULD [1,6,17,19,22,24,39,41] prominently decreases the difficulty of this task for a specific organ (e.g., lung, liver), but it is still challenging for lesions vary in shapes and sizes among whole human body.

Previous arts in ULD are mainly motivated by the anchor-based detection framework, *e.g.*, Faster-RCNN [21]. These studies focus on adapting the detection backbone to universally locate lesions in CT scans. For instance, Li *et al.* [16] propose the so-called MVP-Net, a multi-view FPN with a position-aware attention mechanism to assist ULD training. Yang *et al.* [36–38] propose a series of 3D feature fusion operators to incorporate context information from several adjacent CT slices for better performance. Li *et al.* [15] introduce a plug-and-play transformer block to form hybrid backbones which can better model long-distance feature dependency.

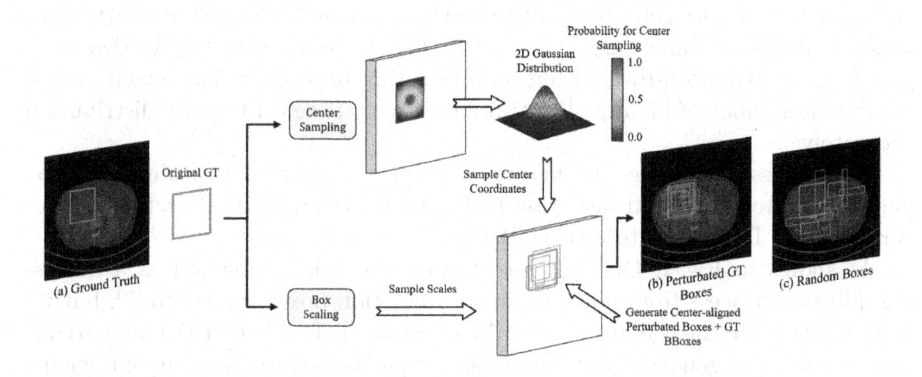

Fig. 1. Overview of the proposed center-aligned BBox padding strategy.

While achieving success, these anchor-based methods have inherent drawbacks: **(i) Insufficient training target.** In stage-1, anchor-based methods identify the positive (lesion) anchors and label them as the region of interest (RoI) based on the IoU between anchors and ground-truth (GT) bounding boxes (BBoxes). An anchor is considered positive if its IoU with any GT BBox is greater than the IoU threshold and negative otherwise [21]. The positive anchors are sufficient in natural images as they usually have many targets per image [13]. However, the number of lesions per CT scan is limited, most CT slices only contain one or two lesions (*i.e.*, detection targets in ULD) per CT slice [14]. Still applying the IoU-based anchor matching mechanism with such limited targets can lead to severe data imbalance and further hinders network convergence. Simply lowering the positive IoU threshold in the anchor-selecting mechanism can alleviate the shortage of positive anchors to some degree, but it leads to a higher false positive (FP) rate by labeling more low-IoU anchors as positive. **(ii) Difficulties in anchor design.** In anchor-based methods, the size, ratio and number of anchors are pre-defined hyper-parameters that significantly influence the detection performance [26]. Thus a proper design of anchor hyper-parameters is of great importance. However, tuning anchor hyper-parameters is a challenging task in ULD because of the variety of lesions (target) with diverse diameters (from 0.21 to 342.5 mm). Even with a careful design, the fixed rectangle anchor boxes can be a kind of obstruction in capturing heterogeneous lesions, which have irregular shapes and vary in size.

To get rid of the drawbacks caused by the anchor mechanism, researchers resort to anchor-free detection, *e.g.*, FCOS [31] and DETR [5]. But these methods experience difficulties in achieving state-of-the-art results in ULD, as they lack the initialization of position prior provided by anchors.

Recently, the diffusion probabilistic model (DPM) [2,8,10,11,23] has demonstrated its outstanding capabilities in various vision tasks. Chen *et al.* follow the key idea of DPM and propose a noise-to-box pipeline, DiffusionDet [7], for natural image object detection. They achieved success on natural images with sufficient training targets, but still experience difficulties in dealing with tasks with insufficient training targets like ULD. This is because the DPM's denoising is a dense distribution-to-distribution forecasting procedure that heavily relied on a large number of high-quality training targets to learn targets' distribution accurately.

To address this issue, we hereby introduce a novel center-aligned BBox padding strategy in DPM detection to form a diffusion-based detector for Universal Lesion Detection, termed DiffULD.

As shown in Fig. 1, DiffULD also formulates lesion detection as a denoising diffusion process from noisy boxes to prediction boxes similar to [7], but we further introduce the center-aligned BBox padding before DiffULD's forward diffusion process to generate perturbated GT. Specifically, we add random perturbations to both scales and center coordinates of the original GT BBox, resulting in *perturbated boxes* whose centers remain aligned with the corresponding original GT BBox. Next, original GT boxes paired with these perturbated boxes are called *perturbated GT boxes* for simplicity. Finally, we feed the perturbated GT boxes to the model as the training objective during training. Compared with other training target padding methods (*e.g.*, padding with random boxes), our strategy can provide additional targets of higher quality, *i.e.*, center aligned with the original GT BBox. This approach effectively expands the insufficient training targets on CT scans, enhancing DPM's detection performance and avoiding deterioration triggered by adding random targets.

The following DPM training procedure contains two diffusion processes. i) In the forward training process, DiffULD corrupts the perturbated GT with Gaussian noise gradually to generate noisy boxes step by step. Then the model is trained to remove the noise and reconstruct the original perturbated GT boxes. ii) In the reverse inference process, the trained DiffULD can refine a set of randomly generated boxes iteratively to obtain the final detect predictions.

Our method gets rid of the drawbacks of pre-defined anchors and the deterioration of training DPM with insufficient GT targets. Besides, DiffULD is inherently advanced in locating targets with diverse sizes since it can predict with arbitrary boxes, which is an advantageous feature for detecting lesions of irregular shapes and various sizes. To validate the effectiveness of our method, we conduct experiments against seven state-of-the-art ULD methods on the public dataset DeepLesion [32]. The results demonstrate that our method achieves competitive performance compared to state-of-the-art ULD approaches.

2 Method

In this section, we first formulate our overall detection process for DiffULD and then specify the training manner, inference process and backbone design.

2.1 Diffusion-Based Detector for Lesion Detection

Universal Lesion Detection can be formulated as locating lesions in input CT scan I_{ct} with a set of boxes predictions $z_0^{'}$. For a particular box $z^{'}$, it can be denoted as $z^{'} = [x_1, y_1, x_2, y_2]$, where x_1, y_1 and x_2, y_2 are the coordinates of the top-left and bottom-right corners, respectively.

We design our model based on a diffusion model mentioned in [7]. As shown in Fig. 2, our method consists of two stages, a forward diffusion (or training) process and a reverse refinement (or inference) process. In the forward process, We denote GT BBoxes as z_0 and generate corrupted training samples $z_1, z_2, ..., z_T$ for latter DiffULD training by adding Gaussian noise iteratively, which can be defined as:

$$(z_t \mid z_0) = \mathcal{N}\left(z_t \mid \sqrt{\bar{\alpha}_t} z_0, (1 - \bar{\alpha}_t) I\right) \tag{1}$$

where $\bar{\alpha}_t$ represents the noise variance schedule and $t \in \{0, 1, ..., T\}$. Subsequently, a neural network $f_\theta(z_t, t, I_{ct})$ conditioned on the corresponding CT scan I_{ct} is trained to predict z_0 from a noisy box z_T by reversing the noising process step by step. During inference, for an input CT scan I_{ct} with a set of random boxes, the model is able to refine the random boxes to get a lesion detection prediction box $z_0^{'}$, iteratively.

2.2 Training

In this section, we specify the training process with our novelty introduced 'Center-aligned BBox padding'.

Center-Aligned BBox Padding. As shown in Fig. 1, we utilize Center-aligned BBox padding to generate *perturbated boxes.* Then, the perturbated boxes are paired with original GT BBoxes, forming *perturbated GT boxes* which are used to generate corrupted training samples $z_1, z_2, ..., z_T$ for the latter DiffULD training by adding Gaussian noise iteratively.

For clarity, we denote $[x_c^i, y_c^i]$ as center coordinates of original GT BBox, where z^i, $[w^i, h^i]$ are the width and height of z^i.

We consider the generation in two parts: box scaling and center sampling. (i) *Box scaling:* We set a hyper-parameter $\lambda_{scale} \in (0, 1)$ for scaling. For z^i,

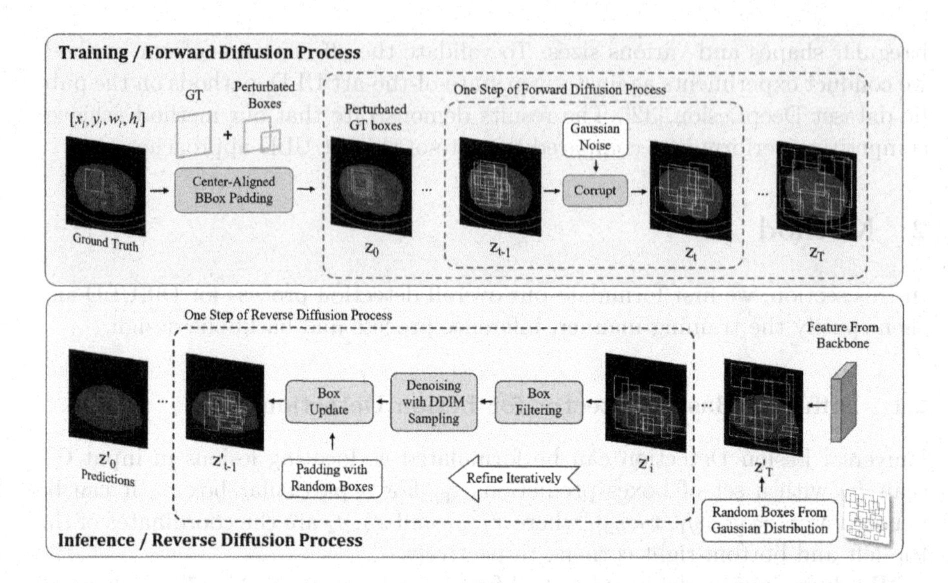

Fig. 2. Overview of DiffULD. The backbone extracts feature representation from an input CT scan. Then, the model takes the feature map and a set of noisy boxes as input and makes predictions.

The width and height of the corresponding perturbated boxes are randomly sampled in uniform distributions on $U_w \sim [(1-\lambda_{scale})w^i, (1+\lambda_{scale})w^i]$ and $U_h \sim [(1-\lambda_{scale})h^i, (1+\lambda_{scale})h^i]$. (ii) *Center sampling:* We sample the center coordinates $[x_c^{new}, y_c^{new}]$ of perturbated boxes from a 2D Gaussian distribution $\mathcal{N}$ whose probability density function can be denoted as:

$$f(x, y) = \exp\left(-\frac{\left(x - x_c^i\right)^2 + \left(y - y_c^i\right)^2}{2\sigma^2}\right), \qquad (x, y) \in z_i \tag{2}$$

where σ is a size-adaptive parameter, which can be calculated according to the z^i's width and height:

$$\sigma = \frac{1}{6}(w^i + h^i). \tag{3}$$

Besides, for each input CT scan I_{ct}, we collect all GT BBoxes in z and add random perturbations to them and generate multiple *perturbated boxes* for each of them. Thus the number of perturbated boxes in an image varies with its number of GT BBoxes. For better training, we fix the number of perturbated boxes as N for all training images.

As shown in Fig. 1., the perturbated boxes cluster together and their centers are still aligned with the corresponding original GT BBox. Subsequently, perturbated GT boxes z_0 are sent for corruption as the training objective.

Box Corruption. As shown in Fig. 1, we corrupt the parameters of z_0 with Gaussian noises. The noise scale is controlled by $\bar{\alpha}_t$ (in Eq. 1), which adopts the decreasing cosine scheduler in the different time step t.

Loss function. As the model generates the same number of (N) predictions for the input image, termed as a prediction set, the loss function should be set-wised [5]. Specifically, each GT is matched with the prediction by the least matching cost, and the overall training loss [5] can be represented as:

$$\mathcal{L} = \lambda_{L1box} \cdot \mathcal{L}_{L1box} + \lambda_{giou} \cdot \mathcal{L}_{giou} \tag{4}$$

where $\mathcal{L}_{L1box}$ and $\mathcal{L}_{giou}$ are the pairwise box loss. We adopt $\lambda_{L1box} = 2.0$ and $\lambda_{giou} = 5.0$.

2.3 Inference

At the inference stage, with a set of random boxes sampled from Gaussian distribution, the model does refinement step by step to obtain the final predictions $z_0^{'}$. For better performance, two key components are used:

Box Filtering. In each refinement step, the model receives a set of box proposals from the last step. As the prediction starts from arbitrary boxes and the lack of GT (lesion), most of these proposals are very far from lesions. Keeping refining them in the following steps will hinder network training. Toward efficient detection, we send the proposals to the detection head and remove the boxes whose confidential scores are lower than a particular threshold λ_{conf}. The remaining high-quality proposals are sent for followed DDIM sampling.

Box Update with DDIM Sampling. DDIM [27] is utilized to further refine the received box proposals by denoising. Next, these refined boxes are sent to the next step and start a new round of refinement. After multiple steps, final predictions are obtained.

However, we observe that if we just filter out boxes with low scores during iterative refinement, the model runs out of usable box proposals rapidly, which also leads to a deterioration in performance. Therefore, after the box updating, we add new boxes sampled from a Gaussian Distribution to the set of remaining boxes with. The number of box proposals per image is padded to the fixed number N before they are sent to the next refinement step.

2.4 Backbone Design

Our overall backbone design is identical to [16]. The input CT scan is rendered with different window widths and window levels. Then, multi-window features are extracted with a ConvNeXt-T [18] shared network and sent to 3D context feature fusion module. Subsequently, the fused feature is sent to the detector.

Multi-window Input. Most prior arts in ULD use a single and fixed window (*e.g.*, a wide window of [1024, 4096]) to render the input CT scan, which suppresses organ-specific information and makes it hard for the network to focus on the various lesions. Therefore, taking cues from [16], we introduce 3 organ-specific HU windows to highlight multiple organs of interest. Their window widths and window levels are: $W_1 = [1200, -600]$ for chest organs, $W_2 = [400, 50]$ for soft tissues and $W_3 = [200, 30]$ for abdomenal organs.

3D Context Feature Fusion. We modify the original A3D [37] DenseNet backbone for context fusion. We remove the first Conv3D Block and use the truncated network as our 3D context fusion module, which fuses the multi-window features from the last module.

3 Experiments

3.1 Settings

Our experiments are conducted on the standard ULD dataset DeepLesion [32]. The dataset contains 32,735 lesions on 32,120 axial slices from 10,594 CT studies of 4,427 unique patients. We use the official data split of DeepLesion which consists of 70%, 15%, 15% for training, validation, and test, respectively. Besides, we also evaluate the performance of 3 methods based on a revised test set from [4].

Training Details. DiffULD is trained on CT scans of size 512 × 512 with a batch size of 4 on 4 NVIDIA RTX Titan GPUs with 24GB memory. For hyper-parameters, the threshold N for box padding is set to 300. λ_{scale} for box scaling is set to 0.4. λ_{conf} for box filtering is set to 0.5. We use the Adam optimizer with an initial learning rate of $2e-4$ and the weight decay as $1e-4$. The default training schedule is 120K iterations, with the learning rate divided by 10 at 60K and 100K iterations. Data augmentation strategies contain random horizontal flipping, rotation, and random brightness adjustment.

Evaluation Metrics. The lesion detection is classified as true positive (TP) when the IoU between the predicted and the GT BBox is larger than 0.5. Average sensitivities computed at 0.5, 1, 2, and 4 false-positives (FP) per image are reported as the evaluation metrics on the test set for a fair comparison (Table 2).

Table 1. Sensitivity (%) at various FPPI on the standard test set of DeepLesion. DKA-ULD [25] and SATr [15] are up-to-date SOTA ULD methods under the settings of 3 slices and 7 slices, respectively. The numbers in brackets indicate the performance gains of our method comparing to the previous SOTA methods under the same settings.

Methods	Slices	@0.5	@1	@2	@4	Avg.[0.5,1,2,4]
3DCE [33]	27	62.48	73.37	80.70	85.65	75.55
RetinaNet [44]	3	72.18	80.07	86.40	90.77	82.36
MVP-Net [16]	9	73.83	81.82	87.60	91.30	83.64
MULAN [34]	27	76.10	82.50	87.50	90.90	84.33
AlignShift [36]	3	73.00	81.17	87.05	91.78	83.25
A3D [37]	3	74.10	81.81	87.87	92.13	83.98
DiffusionDet [7]	3	76.71	83.49	88.21	91.70	85.03
DKA-ULD [25]	3	77.38	84.06	89.28	92.04	85.79
SATr [15]	7	81.02	86.64	90.69	93.30	87.91
DiffULD (Ours)	3	**77.84** (0.46↑)	**84.57** (0.51↑)	**89.41** (0.13↑)	**92.31** (0.40↑)	**86.03** (0.58↑)
DiffULD (Ours)	7	80.43	**87.16** (0.52↑)	**91.20** (0.51↑)	93.21	**88.00** (0.08↑)
FCOS [31]	3	56.12	67.31	73.75	81.44	69.66
CenterNet++ [9]	3	67.34	75.95	82.73	87.72	78.43
DN-DETR [12]	3	69.27	77.90	83.97	88.59	81.02

Table 2. Sensitivity (%) at various FPPI on the revised test set introduced by Lesion-Harvester [4].

Methods	Slices	@0.5	@1	@2	@4	Avg.[0.5,1,2,4]
A3D [37]	7	88.73	91.23	93.89	95.91	92.44
SATr [15]	7	91.04	93.75	95.58	96.73	94.28
DiffULD (Ours)	7	**91.65** (0.61↑)	**94.33** (0.58↑)	**95.97** (0.39↑)	**96.86** (0.13↑)	**94.70** (0.42↑)

Table 3. Ablation study of padding strategies at various FPs per image (FPPI).

Baseline	Duplicate	Gaussian	Uniform	Center-Aligned	FPPI = 0.5	FPPI = 1
✓					76.71	83.49
✓	✓				77.01	83.61
✓		✓			77.68	83.98
✓			✓		77.22	83.75
✓				✓	**77.84** (1.13↑)	**84.57** (1.08↑)

3.2 Lesion Detection Performance

We evaluate the effectiveness of DiffULD against anchor-based ULD approaches such as 3DCE [33], MVP-Net [16], A3D [37] and SATr [15] on DeepLesion. Several anchor-free natural image detection methods such as FCOS [31] and DN-DETR [12] are also introduced for comparison. We report the performance of DiffusionDet [7] trained with our proposed backbone in 2.4 as well. In addition, we conduct an extensive experiment to explore DiffULD's potential on an revised test set of completely annotated DeepLesion volumes, introduced by Lesion-Harvester [4].

Table 1 demonstrates that our proposed DiffULD achieves favorable performances when compared to recent state-of-the-art anchor-based ULD approaches such as SATr on both 3 slices and 7 slices. It outperforms prior well-established methods such as A3D and MULAN by a non-trivial margin. This validates that with our padding strategy, the concise DPM can be utilized in general medical object detection tasks such as ULD and attain impressive performance.

3.3 Ablation Study

We provide an ablation study about our proposed approach: center-aligned BBox padding. As shown in Table 3., we compared it with various other padding strategies, including: (i) duplicating original GT boxes; (ii) padding random boxes sampled from a uniform distribution; (iii) padding random boxes sampled from a Gaussian distribution; (iv) padding with center-aligned strategy.

Our baseline method is training the diffusion model [7] directly with no box padding, using our proposed backbone in 2.4. The performance is increased by 0.30% by simply duplicating the original GT boxes. Padding random boxes following uniform and Gaussian distributions brings 0.51% and 0.91% improvement respectively. Our center-aligned padding strategy accounts for 1.13% improvement from the baseline. We attribute this performance boost to center-aligned padding's ability to provide high-quality additional training targets. It effectively expands the insufficient training targets on CT scans and enhances DPM's detection performance while avoiding deterioration triggered by adding random targets. This property is favorable for utilizing DPMs on a limited amount of GT like ULD.

4 Conclusion

In this paper, we propose a novel ULD method termed DiffULD by introducing the diffusion probability model (DPM) to Universal Lesion Detection. We present a novel center-aligned BBox padding strategy to tackle the performance deterioration caused by directly utilizing DPM on CT scans with sparse lesion BBoxes. Compared with other training target padding methods (e.g., padding with random boxes), our strategy can provide additional training targets of higher quality and boost detection performance while avoiding significant deterioration. DiffULD is inherently advanced in locating lesions with diverse sizes and shapes since it can predict with arbitrary boxes, making it a promising method for ULD. Experiments on both standard and revised DeepLesion datasets show that our proposed method can achieve competitive performance compared to state-of-the-art ULD approaches.

Acknowledgement. Supported by Natural Science Foundation of China under Grant 62271465 and Open Fund Project of Guangdong Academy of Medical Sciences, China (No. YKY-KF202206).

References

1. Baumgartner, M., Jäger, P.F., Isensee, F., Maier-Hein, K.H.: nnDetection: a self-configuring method for medical object detection. In: de Bruijne, M., et al. (eds.) MICCAI 2021. LNCS, vol. 12905, pp. 530–539. Springer, Cham (2021). https://doi.org/10.1007/978-3-030-87240-3_51
2. Boah, K., et al.: DiffuseMorph: unsupervised deformable image registration using diffusion model. In: Avidan, S., Brostow, G., Cissé, M., Farinella, G.M., Hassner, T. (eds.) ECCV, pp. 347–364. Springer, Cham (2022). https://doi.org/10.1007/978-3-031-19821-2_20
3. Cai, J., et al.: Deep volumetric universal lesion detection using light-weight pseudo 3D convolution and surface point regression. In: Martel, A.L., et al. (eds.) MICCAI 2020. LNCS, vol. 12264, pp. 3–13. Springer, Cham (2020). https://doi.org/10.1007/978-3-030-59719-1_1
4. Cai, J., et al.: Lesion-harvester: iteratively mining unlabeled lesions and hard-negative examples at scale. IEEE Trans. Med. Imaging **40**(1), 59–70 (2020)
5. Carion, N., Massa, F., Synnaeve, G., Usunier, N., Kirillov, A., Zagoruyko, S.: End-to-end object detection with transformers. In: Vedaldi, A., Bischof, H., Brox, T., Frahm, J.-M. (eds.) ECCV 2020. LNCS, vol. 12346, pp. 213–229. Springer, Cham (2020). https://doi.org/10.1007/978-3-030-58452-8_13
6. Chen, J., Zhang, Y., Wang, J., Zhou, X., He, Y., Zhang, T.: EllipseNet: anchor-free ellipse detection for automatic cardiac biometrics in fetal echocardiography. In: de Bruijne, M., et al. (eds.) MICCAI 2021. LNCS, vol. 12907, pp. 218–227. Springer, Cham (2021). https://doi.org/10.1007/978-3-030-87234-2_21
7. Chen, S., et al.: DiffusionDet: diffusion model for object detection. arXiv preprint arXiv:2211.09788 (2022)
8. Chen, T., et al.: A generalist framework for panoptic segmentation of images and videos. arXiv preprint arXiv:2210.06366 (2022)
9. Duan, K., et al.: CenterNet++ for object detection. arXiv preprint arXiv:2204.08394 (2022)
10. Ho, J., et al.: Denoising diffusion probabilistic models. In: Advances in Neural Information Processing Systems, vol. 33, pp. 6840–6851 (2020)
11. Holmquist, K., et al.: DiffPose: multi-hypothesis human pose estimation using diffusion models. arXiv preprint arXiv:2211.16487 (2022)
12. Li, F., et al.: DN-DETR: accelerate DETR training by introducing query DeNoising. In: IEEE CVPR, pp. 13619–13627 (2022)
13. Li, H., Han, H., Zhou, S.K.: Bounding maps for universal lesion detection. In: Martel, A.L., et al. (eds.) MICCAI 2020. LNCS, vol. 12264, pp. 417–428. Springer, Cham (2020). https://doi.org/10.1007/978-3-030-59719-1_41
14. Li, H., Chen, L., Han, H., Chi, Y., Zhou, S.K.: Conditional training with bounding map for universal lesion detection. In: de Bruijne, M., et al. (eds.) MICCAI 2021. LNCS, vol. 12905, pp. 141–152. Springer, Cham (2021). https://doi.org/10.1007/978-3-030-87240-3_14
15. Li, H., et al.: SATr: slice attention with transformer for universal lesion detection. In: Wang, L., Dou, Q., Fletcher, P.T., Speidel, S., Li, S. (eds.) MICCAI, pp. 163–174. Springer, Cham (2022). https://doi.org/10.1007/978-3-031-16437-8_16

16. Li, Z., Zhang, S., Zhang, J., Huang, K., Wang, Y., Yu, Y.: MVP-net: multi-view FPN with position-aware attention for deep universal lesion detection. In: Shen, D., et al. (eds.) MICCAI 2019. LNCS, vol. 11769, pp. 13–21. Springer, Cham (2019). https://doi.org/10.1007/978-3-030-32226-7_2
17. Lin, C., Wu, H., Wen, Z., Qin, J.: Automated malaria cells detection from blood smears under severe class imbalance via importance-aware balanced group softmax. In: de Bruijne, M., et al. (eds.) MICCAI 2021. LNCS, vol. 12908, pp. 455–465. Springer, Cham (2021). https://doi.org/10.1007/978-3-030-87237-3_44
18. Liu, Z., et al.: A ConvNet for the 2020s. In: IEEE CVPR, pp. 11976–11986 (2022)
19. Luo, L., Chen, H., Zhou, Y., Lin, H., Heng, P.-A.: OXnet: deep omni-supervised thoracic disease detection from chest X-Rays. In: de Bruijne, M., et al. (eds.) MICCAI 2021. LNCS, vol. 12902, pp. 537–548. Springer, Cham (2021). https://doi.org/10.1007/978-3-030-87196-3_50
20. Lyu, F., Yang, B., Ma, A.J., Yuen, P.C.: A segmentation-assisted model for universal lesion detection with partial labels. In: de Bruijne, M., et al. (eds.) MICCAI 2021. LNCS, vol. 12905, pp. 117–127. Springer, Cham (2021). https://doi.org/10.1007/978-3-030-87240-3_12
21. Ren, S., et al.: Faster R-CNN: towards real-time object detection with region proposal networks. In: Advances in Neural Information Processing Systems, vol. 28 (2015)
22. Ren, Y., et al.: Retina-match: ipsilateral mammography lesion matching in a single shot detection pipeline. In: de Bruijne, M., et al. (eds.) MICCAI 2021. LNCS, vol. 12905, pp. 345–354. Springer, Cham (2021). https://doi.org/10.1007/978-3-030-87240-3_33
23. Saharia, C., et al.: Photorealistic text-to-image diffusion models with deep language understanding. In: Advances in Neural Information Processing Systems, vol. 35, pp. 36479–36494 (2022)
24. Shahroudnejad, A., et al.: TUN-Det: a novel network for thyroid ultrasound nodule detection. In: de Bruijne, M., et al. (eds.) MICCAI 2021. LNCS, vol. 12901, pp. 656–667. Springer, Cham (2021). https://doi.org/10.1007/978-3-030-87193-2_62
25. Sheoran, M., et al.: An efficient anchor-free universal lesion detection in CT-scans. In: 2022 IEEE 19th International Symposium on Biomedical Imaging (ISBI), pp. 1–4. IEEE (2022)
26. Sheoran, M., et al.: DKMA-ULD: domain knowledge augmented multi-head attention based robust universal lesion detection. arXiv preprint arXiv:2203.06886 (2022)
27. Song, J., et al.: Denoising diffusion implicit models. arXiv preprint arXiv:2010.02502 (2020)
28. Tang, Y., et al.: ULDor: a universal lesion detector for CT scans with pseudo masks and hard negative example mining. In: 2019 IEEE 16th International Symposium on Biomedical Imaging (ISBI 2019), pp. 833–836. IEEE (2019)
29. Tang, Y., et al.: Weakly-supervised universal lesion segmentation with regional level set loss. In: de Bruijne, M., et al. (eds.) MICCAI 2021. LNCS, vol. 12902, pp. 515–525. Springer, Cham (2021). https://doi.org/10.1007/978-3-030-87196-3_48
30. Tao, Q., Ge, Z., Cai, J., Yin, J., See, S.: Improving deep lesion detection using 3D contextual and spatial attention. In: Shen, D., et al. (eds.) MICCAI 2019. LNCS, vol. 11769, pp. 185–193. Springer, Cham (2019). https://doi.org/10.1007/978-3-030-32226-7_21
31. Tian, Z., et al.: FCOS: fully convolutional one-stage object detection. In: IEEE ICCV, pp. 9627–9636 (2019)

32. Yan, K., Wang, X., Lu, L., Summers, R.M.: DeepLesion: automated mining of large-scale lesion annotations and universal lesion detection with deep learning. J. Med. Imaging **5**(3), 036501–036501 (2018)
33. Yan, K., Bagheri, M., Summers, R.M.: 3D context enhanced region-based convolutional neural network for end-to-end lesion detection. In: Frangi, A.F., Schnabel, J.A., Davatzikos, C., Alberola-López, C., Fichtinger, G. (eds.) MICCAI 2018. LNCS, vol. 11070, pp. 511–519. Springer, Cham (2018). https://doi.org/10.1007/978-3-030-00928-1_58
34. Yan, K., et al.: MULAN: multitask universal lesion analysis network for joint lesion detection, tagging, and segmentation. In: Shen, D., et al. (eds.) MICCAI 2019. LNCS, vol. 11769, pp. 194–202. Springer, Cham (2019). https://doi.org/10.1007/978-3-030-32226-7_22
35. Yan, K., et al.: Learning from multiple datasets with heterogeneous and partial labels for universal lesion detection in CT. IEEE Trans. Med. Imaging **40**(10), 2759–2770 (2020)
36. Yang, J., et al.: *AlignShift*: bridging the gap of imaging thickness in 3D anisotropic volumes. In: Martel, A.L., et al. (eds.) MICCAI 2020. LNCS, vol. 12264, pp. 562–572. Springer, Cham (2020). https://doi.org/10.1007/978-3-030-59719-1_55
37. Yang, J., He, Y., Kuang, K., Lin, Z., Pfister, H., Ni, B.: Asymmetric 3D context fusion for universal lesion detection. In: de Bruijne, M., et al. (eds.) MICCAI 2021. LNCS, vol. 12905, pp. 571–580. Springer, Cham (2021). https://doi.org/10.1007/978-3-030-87240-3_55
38. Yang, J., et al.: Reinventing 2D convolutions for 3D images. IEEE J. Biomed. Health Inform. **25**(8), 3009–3018 (2021)
39. Yu, X., et al.: Deep attentive panoptic model for prostate cancer detection using biparametric MRI scans. In: Martel, A.L., et al. (eds.) MICCAI 2020. LNCS, vol. 12264, pp. 594–604. Springer, Cham (2020). https://doi.org/10.1007/978-3-030-59719-1_58
40. Zhang, S., et al.: Revisiting 3D context modeling with supervised pre-training for universal lesion detection in CT slices. In: Martel, A.L., et al. (eds.) MICCAI 2020. LNCS, vol. 12264, pp. 542–551. Springer, Cham (2020). https://doi.org/10.1007/978-3-030-59719-1_53
41. Zhao, Z., Pang, F., Liu, Z., Ye, C.: Positive-unlabeled learning for cell detection in histopathology images with incomplete annotations. In: de Bruijne, M., et al. (eds.) MICCAI 2021. LNCS, vol. 12908, pp. 509–518. Springer, Cham (2021). https://doi.org/10.1007/978-3-030-87237-3_49
42. Zhou, S.K., et al.: Handbook of Medical Image Computing and Computer Assisted Intervention. Academic Press (2019)
43. Zhou, S.K., et al.: A review of deep learning in medical imaging: imaging traits, technology trends, case studies with progress highlights, and future promises. Proc. IEEE **109**(5), 820–838 (2021)
44. Zlocha, M., Dou, Q., Glocker, B.: Improving RetinaNet for CT lesion detection with dense masks from weak RECIST labels. In: Shen, D., et al. (eds.) MICCAI 2019. LNCS, vol. 11769, pp. 402–410. Springer, Cham (2019). https://doi.org/10.1007/978-3-030-32226-7_45

Graph-Theoretic Automatic Lesion Tracking and Detection of Patterns of Lesion Changes in Longitudinal CT Studies

Beniamin Di Veroli[1], Richard Lederman[2], Jacob Sosna[2], and Leo Joskowicz[1](✉)

[1] School of Computer Science and Engineering, The Hebrew University of Jerusalem, Jerusalem, Israel
beniamin.diveroli@mail.huji.ac.il, josko@cs.huji.ac.il

[2] Department of Radiology, Hadassah University Medical Center, Jerusalem, Israel

Abstract. Radiological follow-up of oncological patients requires the analysis and comparison of multiple unregistered scans acquired every few months. This process is currently partial, time-consuming and subject to variability. We present a new, generic, graph-based method for tracking individual lesion changes and detecting patterns in the evolution of lesions over time. The tasks are formalized as graph-theoretic problems in which lesions are vertices and edges are lesion pairings computed by overlap-based lesion matching. We define seven individual lesion change classes and five lesion change patterns that fully summarize the evolution of lesions over time. They are directly computed from the graph properties and its connected components with graph-based methods. Experimental results on lung (83 CTs from 19 patients) and liver (77 CECTs from 18 patients) datasets with more than two scans per patient yielded an individual lesion change class accuracy of 98% and 85%, and identification of patterns of lesion change with an accuracy of 96% and 76%, respectively. Highlighting unusual lesion labels and lesion change patterns in the graph helps radiologists identify overlooked or faintly visible lesions. Automatic lesion change classification and pattern detection in longitudinal studies may improve the accuracy and efficiency of radiological interpretation and disease status evaluation.

Keywords: longitudinal follow-up · lesion matching · lesion change analysis

1 Introduction

The periodic acquisition and analysis of volumetric CT and MRI scans of oncology patients is essential for the evaluation of the disease status, the selection of the treatment, and the response to treatment. Currently, scans are acquired every 2–12 months according to the patient's characteristics, disease stage, and treatment regime. The scan interpretation consists of identifying lesions (primary tumors, metastases) in the affected

Supplementary Information The online version contains supplementary material available at https://doi.org/10.1007/978-3-031-43904-9_11.

H. Greenspan et al. (Eds.): MICCAI 2023, LNCS 14224, pp. 106–115, 2023.
https://doi.org/10.1007/978-3-031-43904-9_11

organs and characterizing their changes over time. Lesion changes include changes in the size of existing lesions, the appearance of new lesions, the disappearance of existing lesions, and complex lesion changes, e.g., the formation of conglomerate lesions. As treatments improve and patients live longer, the number of scans in longitudinal studies increases and their interpretation is more challenging and time-consuming.

Radiological follow-up requires the quantitative analysis of lesions and patterns of lesion changes in subsequent scans. It differs from diagnostic reading since the goal is to find and quantify the differences between the scans, rather than to find abnormalities in a single scan. In current practice, quantification of lesion changes is partial and approximate. The RECIST 1.1 guidelines call for finding new lesions (if any), identifying up to the five largest lesions in each scan in the CT slice where they appear largest, manually measuring their diameters, and comparing their difference [1]. While volumetric measures of individual lesions and of all lesions (tumor burden) have long been established as more accurate and reliable than partial linear measurements, they are not used clinically because they require manual lesion delineation and lesion matching in unregistered scans, which is usually time-consuming and subject to variability [2].

In a previous paper, we presented an automatic pipeline for the detection and quantification of lesion changes in pairs of CT liver scans [3]. This paper describes a graph-based lesion tracking method for the comprehensive analysis of lesion changes and their patterns at the lesion level. The tasks are formalized as graph-theoretic problems (Fig. 1). Complex lesion changes include merged lesions, which occurs when at least two lesions grow and merge into one (possible disease progression), split lesions, which occurs when a lesion shrinks and cleaves into several parts (possible response to treatment) and conglomeration of lesions, which occurs when clusters of lesions coalesce. While some of these lesion changes have been observed [4], they have been poorly studied. Comprehensive quantitative analysis of lesion changes and patterns is of clinical importance, since response to treatment may vary among lesions, so the analysis of a few lesions may not be representative.

The novelties of this paper are: 1) identification and formalization of longitudinal lesion matching and patterns of lesion changes in CT in a graph-theoretic framework; 2) new classification and detection of changes of individual lesions and lesion patterns based on the properties of the lesion changes graph and its connected components; 3) a simultaneous lesion matching method with more than two scans; 4) graph-based methods for the detection of changes in individual lesions and patterns of lesion changes. Experimental results on lung (83 CTs, 19 patients) and liver (77 CECTs, 18 patients) datasets show that our method yields high classification accuracy.

To the best of our knowledge, ours is the first method to perform longitudinal lesion matching and lesion changes pattern detection. Only a few papers address lesion matching in pairs of CT/MRI scans [5–13] – none performs simultaneous matching of all lesions in more than two scans. Also, very few methods [3, 14] handle matching of split/merged lesions. Although many methods exist for object tracking in optical images and videos [15–17], they are unsuited for analyzing lesion changes since they assume many consecutive 2D images where objects have very similar appearance and undergo small changes between images. Overlap-based methods pair two lesions in registered scans when their segmentations overlap, with a reported accuracy of 66–98% [3, 5–11,

18]. These methods assume that organs and lesions undergo minor changes, are very sensitive to registration errors, and cannot handle complex lesion changes. Similarity-based methods pair two lesions with similar features, e.g., intensity, shape, location [13–16] with an 84–96% accuracy on the DeepLesion dataset [14]. They are susceptible to major changes in the lesion appearance and do not handle complex lesion changes. Split-and-merge matching methods are used for cell tracking in fluorescence microscopy [19]. They are limited to 2D images, assume registration between images, and do not handle conglomerate changes.

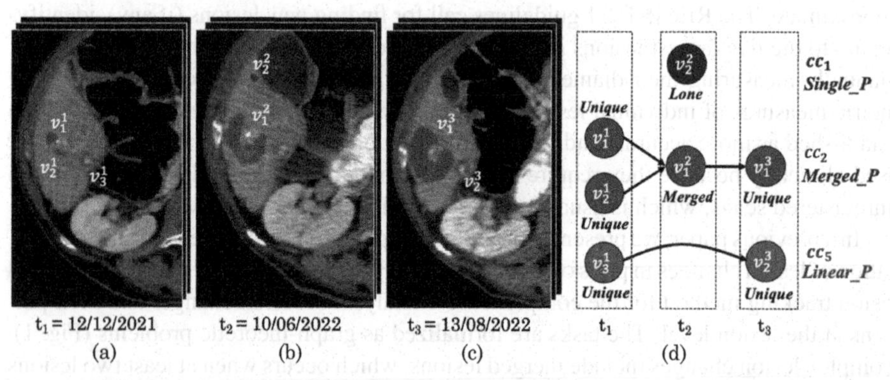

Fig. 1. Longitudinal study of a patient with liver metastases (color overlays): three consecutive (a–c) illustrative slices of unregistered CECT scans acquired at times t_i; matching colors correspond to matching lesions; (d) lesion changes graph: nodes correspond to lesions v_j^i where j is the lesion number at time t_i (dotted rectangle); edges correspond to lesion matches (consecutive straight, non-consecutive curved). The individual lesion changes labels are shown below each node. The graph has three undirected connected components cc_m (red, brown, green), corresponding to three lesion changes patterns, ***Single_P***, ***Merged_P***, ***Linear_P***. Note that lesion v_2^2 labeled as ***Lone*** in t_2 is likely a false positive (not a lesion); lesion v_3^2 (not shown, light green) is likely a false negative (missed lesion) since there is a non-consecutive edge between v_3^1 and v_2^3. (Color figure online)

2 Method

We present a new generic model-based method for the automatic detection and classification of changes in individual lesions and patterns of lesion changes in consecutive CT scans. The tasks are formalized in a graph-theoretic framework in which nodes represent lesions, edges represent lesion matchings, and paths and connected components represent patterns of lesion changes. Lesion matchings are computed with an overlap-based lesion pairing method after establishing a common reference frame by deformable registration of the scans and organ segmentations. Changes in individual lesions and patterns of lesion changes are computed from the graph's properties and its connected components. We define seven individual lesion change classes and five lesion change patterns that fully describe the evolution of lesions over time.

The method inputs the scans and the organ and lesion segmentations in each scan. Its outputs are the lesion matchings, the labels of the changes in individual lesions, and the patterns of the lesion changes. The method is a pipeline of four steps: 1) pairwise deformable registration of each prior scan, organ and lesion segmentations, with the most recent (current) scan as in [3]; 2) overlap-based lesion matching; 3) construction of the lesion change graph from the individual lesion segmentations and lesion matches; 4) detection of changes in individual lesions and patterns of lesion changes from the graph properties and from analysis of its connected components.

2.1 Problem Formalization

Let $S = \langle S^1, \ldots, S^N \rangle$ be a series of $N \geq 2$ consecutive patient scans acquired at times $t_i, 1 \leq i \leq N$. Let $G = (V, E)$ be a directed acyclic graph where $V = \{V^i\}, 1 \leq i \leq N$ and $V^i = \{v_1^i, v_2^i, \ldots, v_{n_i}^i\}$ is a set of vertices v_j^i corresponding to the lesions associated with the lesion segmentation masks $L^i = \{l_1^i, l_2^i, \ldots, l_{n_i}^i\}$, where $n_i \geq 0$ is the number of lesions in scan S^i at time t_i. By definition, any two lesions $v_j^i, v_l^i, j \neq l$ in L^i are disjoint in their voxels. Let $E = \left\{e_{j,l}^{i,k} = \left(v_j^i, v_l^k\right) | v_j^i \in V^i, v_l^k \in V^k, 1 \leq i < k \leq N\right\}$ be a set of forward-directed edges connecting vertices in V^i to vertices in V^k. Edge $e_{j,l}^{i,k}$ indicates that the lesions corresponding to vertices v_j^i, v_l^k are the same lesion, i.e., that the lesion appears in scans S^i, S^k in the same location. Edges of consecutive scans S^i, S^{i+1} are called consecutive edges; edges of non-consecutive scans, $S^i, S^k, i < k - 1$, are called non-consecutive edges. The in- and out-degree of a vertex v_j^i, $d_{in}(v_j^i)$ and $d_{out}(v_j^i)$, are the number of incoming and outcoming edges, respectively.

Let $CC = \{cc_m\}_{m=1}^M$ be the set of connected components of the undirected graph version of G, where M is the number of connected components and $cc_m = (V_m, E_m)$ is a sub-graph of G, $V_m \subseteq V, E_m \subseteq E$. By definition, for each $1 \leq m \leq M$, the sets V_m, E_m are mutually disjoint and their unions are V, E, respectively. In a connected component cc_m, there is an undirected path between any two vertices v_j^i, v_l^k consisting of a sequence of undirected edges in E_m. . In this setup, connected components correspond to matched lesions and their pattern of evolution over time (Fig. 1d).

We define **seven** mutually exclusive individual lesion change labels for lesion v_j^i in scan S^i based on the vertex in- and out-degrees (Fig. 2). In the following definitions we refer to the indices: $1 \leq k < i < l \leq N$; 1) ***Lone***: a lesion present in scan S^i and absent in all previous scans S^k and subsequent scans S^l; 2) ***New***: a lesion present in scan S^i and absent in all previous scans S^k; 3) ***Disappeared***: a lesion present in scan S^i and absent in all subsequent scans S^l; 4) ***Unique***: a lesion present in scan S^i and present as a single lesion in a previous scan S^k and/or in a subsequent scan S^l; 5) ***Merged***: a lesion present in scan S^i and present as two or more lesions in a previous scan S^k; 6) ***Split***: a lesion present in scan S^i and present as two or more lesions in a subsequent scan S^l; 7) ***Complex***: a lesion present as two or more lesions in at least one previous scan S^k and at least one subsequent scan S^l. We also define as ***Existing*** a lesion present in scan S^i and present in at least one previous scan S^k and one subsequent scan S^l, $(d_{in}(v_j^i) \geq 1, d_{out}(v_j^i) \geq 1)$. For the first and current scans S^1 and S^N, we set $d_{in}(v_j^1) = 1$, $d_{out}(v_j^N) = 1$, i.e., the lesion existed before the first scan or remains after the last scan. Thus, lesions in the first (last) scan can only be ***Unique***, ***Disappeared*** or ***Split*** (***Unique***, ***New*** or ***Merged***). Finally, when lesion v_j^i is ***Merged*** and $d_{out}(v_j^i) = 0$, $i < N$, it is also labeled ***Disappeared***; when it is ***Split*** and $d_{in}(v_j^i) = 0$, $i > 1$, it is also labeled ***New***.

We define five patterns of lesion changes based on the properties of the connected components cc_m of G and on the labels of lesion changes: 1) ***Single_P***: a connected component $cc_m = \{v_j^i\}$ consisting of a single lesion labeled as ***Lone***, ***New***, ***Disappeared***; 2) ***Linear_P***: a connected component consisting of a single earliest vertex v_j^{first} (can be ***New***), a single latest vertex v_l^{last} (can be ***Disappeared***) connected by a sequence. (possibly empty) of ***Unique*** vertices v_j^k, $1 \le first < k < last \le N$; ; 3) ***Merged_P***: a connected component whose undirected graph is a tree rooted at a single latest vertex v_l^{last} connected to earlier vertices v_j^k, $1 \le k < last \le N$, one or more labeled ***Merged***; 4) ***Split_P***: a connected component whose undirected graph is a tree rooted at a single earliest vertex v_l^{first} connected to later vertices v_j^k, $1 \le first < k \le N$, one or more labeled ***Split***; 5) ***Complex_P***: all other connected components. Note that ***Merged_P***, ***Split_P*** and ***Complex_P*** connected components can themselves be subdivided into ***Linear_P***, ***Merge_P***, ***Split_P***, and ***Complex_P*** subcomponents, correspondingly.

(a) CHANGES IN INDIVIDUAL LESION							
Vertex degree				***Existing***			
	Lone	***New***	***Disapp-eared***	***Unique***	***Merged***	***Split***	***Complex***
$d_{in}(v_j^i)$	0	0	1	1 (0 first)	≥ 2	0,1	≥ 2
$d_{out}(v_j^i)$	0	1	0	1 (0 last)	0,1	≥ 2	≥ 2
Node pattern							

(b) PATTERNS OF LESION CHANGES					
	Single_P	***Linear_P***	***Merged_P***	***Split_P***	***Complex_P***
Conditions, node labels	*Lone, New, Disappear*	*Single path, at least two nodes*	*Tree, at least one Merge node*	*Tree, at least one Split node*	*All others*
Connected component pattern					

Fig. 2. (a) Individual lesion change classes of a vertex v_j^i defined by its in- and out-degrees $d_{in}(v_j^i)$, $d_{out}(v_j^i)$. Illustrative node pattern: node (circle), edges (arrows); (b) Patterns of lesion changes defined by node labels and connected component properties and illustrative patterns.

The changes in individual lesions and the detection and classification of patterns of lesion changes consist of constructing a graph whose vertices are the corresponding lesion in the scans, computing the graph consecutive and non-consecutive edges that correspond to lesion matchings, computing the connected components of the resulting graph, and assigning an individual lesion change label to each vertex and a lesion change pattern label to each connected component according to the categories above.

2.2 Lesion Matching Computation

Lesion matchings are determined by the location and relative proximity of the lesions in two or more registered scans. The lesion matching rule is ***lesion voxel overlap***: when the lesion segmentation voxels l_j^i, l_l^k of vertices v_j^i, v_l^k overlap, $1 \leq i < k \leq N$, they are matched and the edge $e_{j,l}^{i,k} = \left(v_j^i, v_l^k\right)$ is added to E. Lesion matchings are computed first on consecutive pairs and then on non-consecutive pairs of scans.

Consecutive lesion matching on scans (S^i, S^{i+1}) is performed with an iterative greedy strategy whose aim is to compensate for registration errors: 1) the lesion segmentations in L^i and L^{i+1} are isotropically dilated in 3D by d millimeters; 2) for all pairs of lesions $\left(v_j^i, v_l^{i+1}\right)$, compute the intersection % of their corresponding lesion a segmentations (l_j^i, l_l^{i+1}) as $\max(\left|l_j^i \cap l_l^{i+1}\right| / \left|l_j^i\right|, \left|l_j^i \cap l_l^{i+1}\right| / \left|l_l^{i+1}\right|)$; ; 3) if the % intersection is $\geq p$, then edge $e_{j,l}^{i,i+1} = \left(v_j^i, v_l^{i+1}\right)$ is added to E_c; ; 4) remove the lesion segmentations l_j^i, l_l^{i+1} from L^i, L^{i+1}, respectively. Steps 1–4 are repeated r times. This yields the consecutive edges graph $G_C = (V, E_c)$. The values of d, r are pre-defined empirically.

Lesion matching on non-consecutive scans searches for lesion matchings that were not found previously due to missing lesions (unseen or undetected). It is performed by examining the pairs of connected components of G_C and finding possible edges (lesion pairings) between them. Formally, let $CC = \{cc_m\}_{m=1}^M$ be the set of undirected connected components of G_C. Let $\tau_m = \left[t_m^{first}, t_m^{last}\right]$ be the time interval between the first and last scans of cc_m, and let $centroid(cc_m)$ be the center of mass of all lesions in cc_m at all times. Let $G_{CC} = (V_{cc}, E_{cc})$ be the graph of connected components of G_C such that each vertex cc_m of V_{cc} corresponds to a connected component cc_m and edges $e_{i,j}^{CC} = (cc_i, cc_j)$ of E_{CC} satisfy three conditions: 1) the time interval of cc_i is disjoint and precedes by at least one time point that of cc_j, i.e., $t_i^{first} > t_i^{last} + 1$; 2) the connected components are not too far from each other, i.e., the distance between the connected components centroids is smaller than a fixed distance δ, $\left\|centroid(cc_i) - centroid(cc_j)\right\| \leq \delta$; 3) there is at least one pairwise matching between a lesion in t_i^{last} and a lesion in t_j^{first} computed with the consecutive lesion matching method described above. When the consecutive lesion matching between the lesions in t_i^{last} in cc_i and the lesions in t_j^{first} in cc_j yields a non-empty set of edges, these edges are added as non-consecutive edges to E_c. . Iterating over all non-ordered pairs (cc_i, cc_j) yields the set of consecutive and non-consecutive edges of E.

We illustrate this process with the graph of Fig. 1. First, the consecutive edges (straight lines) of the graph are computed by consecutive lesion matching. This yields a graph with four connected components: $cc_1 = \{v_2^2\}, cc_2 = \{v_1^1, v_2^1, v_1^2, v_1^3\}, cc_3 = \{v_3^1\}, cc_4 = \{v_2^3\}$ corresponding to one ***Merged_P*** and three ***Single_P*** patterns of lesion changes. Connected component cc_2 has no edges to the other connected components since it spans the entire time interval $\tau_1 = [t_1, t_3]$. Connected component cc_1 has no edges to cc_3 and cc_4 since its centroid is too far from them. The connected component pair (cc_3, cc_4) fulfills all three conditions and thus the non-consecutive edge (curved line) $e_{3,2}^{1,3} = (v_3^1, v_2^3)$ is added to E. This results in a new connected component $cc_5 = \{v_3^1, v_2^3\}$ that replaces cc_3 and cc_4 and corresponds to a ***Linear_P*** pattern.

2.3 Classification of Changes in Lesions and in Patterns of Lesion Changes

The changes in individual lesions are directly computed for each lesion from the resulting graph with the in- and out-degree of each vertex (Fig. 2a). The connected components $CC = \{cc_m\}$ of G are computed by graph Depth First Search (DFS). The patterns of lesion changes (Fig. 2b) are computed with path and tree graph algorithms.

The changes in individual lesions, patterns of lesion changes, and lesion changes graph serve as the basis for individual lesion tracking, which consists of following the path from the lesion in the most recent scan backwards to its origins in earlier scans and recording the merged, split and complex lesion changes labels. Summarizing longitudinal studies and queries can also be performed with graph-based algorithms.

3 Experimental Results

We evaluated our method with two studies on retrospectively collected patient datasets that were manually annotated by an expert radiologist.

Dataset: Lung and liver CT studies were retrospectively obtained from two medical centers (Hadassah Univ Hosp Jerusalem Israel) during the routine clinical examination of patients with metastatic disease. Each patient study consists of at least 3 scans.

DLUNG consists of 83 chest CT scans from 19 patients with a mean 4.4 ± 2.0 scans/patient, a mean time interval between consecutive scans of 125.9 ± 81.3 days, and voxel sizes of 0.6–1.0 × 0.6–1.0 × 1.0–3.0 mm^3. DLIVER consists of 77 abdominal CECT scans from 18 patients with a mean 4.3 ± 2.0 scans/patient, a mean time interval between consecutive scans of 109.7 ± 93.5 days, and voxel sizes of 0.6–1.0 × 0.6–1.0 × 0.8–5.0 mm^3.

Lesions in both datasets were annotated by an expert radiologist, yielding a total of 1,178 lung and 800 liver lesions, with a mean of 14.2 ± 19.1 and 10.4 ± 7.9 lesions/scan (lesions with <20 voxels were excluded). Ground-truth lesion matching graphs and lesion changes labeling were produced by running the method on the datasets and then having the radiologist review and correct the resulting node labels and edges.

Study 1: *Lesion changes labeling, lesion matching, evaluation of patterns of lesion changes.* We ran our method on the DLUNGS and DLIVER lesion segmentations. The settings of the parameters were: dilation distance $d = 1$ mm, overlap percentage $p =$ 10%, number of iterations $r = 5$ and 7, and centroid maximum distance $\delta = 17$ and 23 mm for the lungs and liver lesions, respectively.

We compared the computed and ground-truth lesion changes graphs with two metrics: 1) lesion changes classification accuracy, which is the % of correct computed labels from the ground truth labels; 2) lesion matching precision and recall based on the presence/absence of computed vs. ground truth edges. The precision and recall definitions were adapted so that wrong or missed non-consecutive edges are counted as True Positive when there is a path between their vertices in either the ground-truth or the computed graph. Table 1 summarizes the results. The distribution of lesion changes labels for DLUNGS (1,178 lesions) is ***Unique*** 785 (67%), ***New*** 215 (18%), ***Lone*** 109 (9%), ***Disappeared*** 51 (4%), ***Merged*** 12 (1%), ***Split*** 6 (1%), ***Complex*** 0 (0%) with class accuracy ≥ 96% for all except ***Split*** (66%). For DLIVER (800 lesions) it is ***Unique*** 450 (56%),

New 185 (23%), ***Lone*** 45 (6%), ***Disappeared*** 77 (10%), ***Merged*** 27 (3%), ***Split*** 18 (2%), ***Complex*** 1 (0.05%) with class accuracy $\geq$ 81% for all except ***Disappeared*** (71%) and ***Split*** (67%).

For the patterns of lesion changes, we compared the computed and ground truth patterns of lesion changes. The accuracy is the % of identical connected components in each category. Table 1 summarizes the results. Note that the ***Split_P***, ***Merged_P*** and ***Complex_P*** patterns jointly account for 3% and 8% of the cases. These patterns are hard to detect manually but their correct classification and tracking are crucial for the proper application of the RECIST 1.1 follow-up protocol [1].

Table 1. (a) Individual lesion change classification (Correctly computed, Ground Truth, Accuracy) and (b) lesion matching (Consecutive, Nonconsecutive, All and Ground Truth, True Positive, False Positive, False Negative, Precision, Recall) results; (c) patterns of lesion changes (Computed, Ground Truth, Accuracy).

Dataset	**(a) Individual lesion change classification**		**(b) Lesions matching Edge detection**						
	Accuracy			*GT*	*TP*	*FP*	*FN*	*Precision*	*Recall*
DLUNGS	*Correct*	1,151	*Cons*	636	628	10	8	0.98	0.99
	GT	1,178	*Noncons*	27	24	2	3	0.96	0.89
	Accuracy	98%	*All*	663	652	12	11	0.98	0.98
DLIVER	*Correct*	679	*Cons*	458	397	39	61	0.91	0.87
	GT	800	*Noncons*	25	17	11	8	0.68	0.80
	Accuracy	85%	*All*	483	414	50	69	0.90	0.86

Dataset	**(c) Patterns of lesion changes**						
		Total	*Single_P*	*Linear_P*	*Merged_P*	*Split_P*	*Complex_P*
DLUNGS	*Computed*	518	210 (41%)	293 (57%)	9 (2%)	3 (1%)	3 (1%)
	GT	516	207 (40%)	295 (57%)	9 (2%)	3 (1%)	2 (0%)
	Accuracy	96%	97%	89%	67%	50%	96%
DLIVER	*Computed*	346	169 (49%)	150 (43%)	11 (3%)	7 (2%)	9 (3%)
	GT	325	157 (48%)	139 (43%)	13 (4%)	8 (2%)	8 (2%)
	Accuracy	76%	85%	76%	54%	0%	25%

Study 2: *Detection of missed lesions in the ground truth.* The expert radiologist was asked to examine non-consecutive edges and lesions labeled as ***Lone*** in the lesion changes graph and determine if lesions were unseen or undetected (actual or presumed false negative) in the skipped or contiguous scans (Fig. 1d). For each non-consecutive edge connecting lesions v_j^i, v_l^k, he analyzed the corresponding region in the skipped scans S^j at $t_j \in]t_i, t_k[$ for possible missed lesions. For the DLUNGS dataset, 25 visible and 5 faintly visible or surmised to be present unmarked lesions were found for 27 non-consecutive edges. For the DLIVER dataset, 20 visible and 21 faintly visible or surmised to be present unmarked lesions were found for 25 non-consecutive edges.

After reviewing the 42 and 37 lesions labeled as ***Lone*** in DLUNGS and DLIVER with > 5mm diameter, the radiologist determined that 1 and 8 of them had been wrongly identified as a cancerous lesion. Moreover, he found that 14 and 16 lesions initially labeled as ***Lone***, had been wrongly classified: for these lesions he found 15 and 21 previously unmarked matching lesions in the next or previous scans. In total, 45 and 62 missing lesions were added to the ground truth DLUNGS and DLIVER datasets, respectively. These hard-to-find ground-truth False Negatives (3.7%, 7.2% of all lesions) may change the radiological interpretation and the disease status. See the Supplemental Material for examples of these scenarios.

4 Conclusion

The use of graph-based methods for lesion tracking and detection of patterns of lesion changes was shown to achieve high accuracy in classifying changes in individual lesion and identifying patterns of lesion changes in liver and lung longitudinal CT studies of patients with metastatic disease. This approach has proven to be useful in detecting missed, faint, and surmised to be present lesions, otherwise hardly detectable by examining the scans separately or in pairs, leveraging the added information provided by evaluating all patient's scans simultaneously using the labels from the lesion changes graph and non-consecutive edges.

References

1. Eisenhauer, E.A., Therasse, P., Bogaerts, J.: New response evaluation criteria in solid tumors: revised RECIST guideline (version 1.1). Eur. J. Cancer **45**(2), 228–247 (2009)
2. Joskowicz, L., Cohen, D., Caplan, N., Sosna, J.: Inter-observer variability of manual contour delineation of structures in CT. Eur. Radiol. **29**(3), 1391–1399 (2019)
3. Szeskin, A., Rochman, S., Weis, S., Lederman, R., Sosna, J., Joskowicz, L.: Liver lesion changes analysis in longitudinal CECT scans by simultaneous deep learning voxel classification with SimU-Net. Med. Image Anal. **83**(1) (2023)
4. Shafiei, A., et al.: CT evaluation of lymph nodes that merge or split during the course of a clinical trial: limitations of RECIST 1.1. Radiol. Imaging Cancer **3**(3) (2021)
5. Beyer, F., et al.: Clinical evaluation of a software for automated localization of lung nodules at follow-up CT examinations. RoFo: Fortschritte auf dem Gebiete der Rontgenstrahlen und Nuklearmedizin **176**(6), 829–836 (2004)
6. Lee, K.W., Kim, M., Gierada, D.S., Bae, K.T.: Performance of a computer-aided program for automated matching of metastatic pulmonary nodules detected on follow-up chest CT. Am. J. Roentgenol. **189**(5), 1077–1081 (2007)
7. Koo, C.W., et al.: Improved efficiency of CT interpretation using an automated lung nodule matching program. Am. J. Roentgenol. **199**(1), 91–95 (2012)
8. Tao, C., Gierada, D.S., Zhu, F., Pilgram, T.K., Wang, J.H., Bae, K.T.: Automated matching of pulmonary nodules: evaluation in serial screening chest CT. Am. J. Roentgen. **192**(3), 624–628 (2009)
9. Beigelman-Aubry, C., Raffy, P., Yang, W., Castellino, R.A., Grenier, P.A.: Computer-aided detection of solid lung nodules on follow-up MDCT screening: evaluation of detection, tracking, and reading time. Am. J. Roentgenol. **189**(4), 948–955 (2007)
10. Moltz, J.H., Schwier, M., Peitgen, H.O.: A general framework for automatic detection of matching lesions in follow-up CT. In: Proceedings IEEE International Symposium on Biomedical Imaging, pp. 843–846 (2009)
11. Rafael-Palou, X., et al.: Re-identification and growth detection of pulmonary nodules without image registration using 3D Siamese neural networks. Med. Image Anal. **67**, 101823 (2021)

12. Cai, J., et al.: Deep lesion tracker: monitoring lesions in 4D longitudinal imaging studies. In: Proceedings IEEE Conference on Computer Vision and Pattern Recognition, pp. 15159–15169 (2021)
13. Tang, W., Kang, H., Zhang, H., Yu, P., Arnold, C.W., Zhang, R.: Transformer lesion tracker. arXiv preprint arXiv:2206.06252 (2022)
14. Yan, K., Wang, X., Lu, L., Summers, R.M.: DeepLesion: automated mining of large-scale lesion annotations and universal lesion detection with deep learning. J. Med. Imaging **5**(3), 036501 (2018)
15. Bolme, D.S., Beveridge, J.R., Draper, B.A., Lui, Y.M.: Visual object tracking using adaptive correlation filters. In: Proceedings IEEE Conference Computer Vision & Pattern Recognition, pp. 2544–2550 (2010)
16. Li, B., Wu, W., Wang, Q., Zhang, F., Xing, J., Yan, J.S.: Evolution of Siamese visual tracking with very deep networks. In: Proceedings IEEE Conference Computer Vision & Pattern Recognition, pp. 16–20 (2019)
17. Teed, Z., Deng, J.: RAFT: recurrent all-pairs field transforms for optical flow. In: Proceedings European Conference on Computer Vision, pp. 402–419 (2020)
18. Santoro-Fernandes, V., et al.: Development and validation of a longitudinal soft-tissue metastatic lesion matching algorithm. Phys. Med. Biol. **66**(15), 155017 (2021)
19. Padfield, D., Rittscher, J., Roysam, B.: Coupled minimum-cost flow cell tracking for high-throughput quantitative analysis. Med. Image Anal. **15**(4), 650–668 (2011)

Learning with Synthesized Data for Generalizable Lesion Detection in Real PET Images

Xinyi Yang[1], Bennett Chin[1], Michael Silosky[1], Daniel Litwiller[2], Debashis Ghosh[1], and Fuyong Xing[1](✉)

[1] University of Colorado Anschutz Medical Campus, Aurora, USA
fuyong.xing@cuanschutz.edu
[2] GE Healthcare, Denver, USA

Abstract. Deep neural networks have recently achieved impressive performance of automated tumor/lesion quantification with positron emission tomography (PET) imaging. However, deep learning usually requires a large amount of diverse training data, which is difficult for some applications such as neuroendocrine tumor (NET) image quantification, because of low incidence of the disease and expensive annotation of PET data. In addition, current deep lesion detection models often suffer from performance degradation when applied to PET images acquired with different scanners or protocols. In this paper, we propose a novel single-source domain generalization method, which learns with human annotation-free, list mode-synthesized PET images, for hepatic lesion identification in real-world clinical PET data. We first design a specific data augmentation module to generate out-of-domain images from the synthesized data, and incorporate it into a deep neural network for cross domain-consistent feature encoding. Then, we introduce a novel patch-based gradient reversal mechanism and explicitly encourage the network to learn domain-invariant features. We evaluate the proposed method on multiple cross-scanner ^{68}Ga-DOTATATE PET liver NET image datasets. The experiments show that our method significantly improves lesion detection performance compared with the baseline and outperforms recent state-of-the-art domain generalization approaches.

Keywords: Lesion detection · PET images · domain generalization

1 Introduction

Deep neural networks have recently shown impressive performance on lesion quantification in positron emission tomography (PET) images [6]; however, they usually rely on a large amount of well-annotated, diverse data for model training.

Supplementary Information The online version contains supplementary material available at https://doi.org/10.1007/978-3-031-43904-9_12.

H. Greenspan et al. (Eds.): MICCAI 2023, LNCS 14224, pp. 116–126, 2023.
https://doi.org/10.1007/978-3-031-43904-9_12

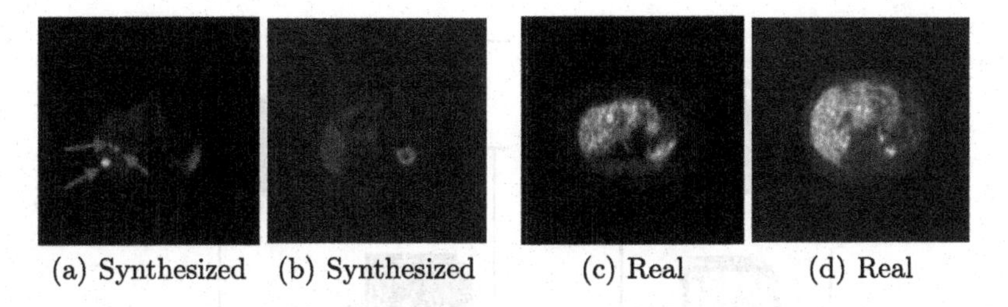

Fig. 1. Example PET images. Diseased subjects with (a) simulated and (c) real lesions (red arrows), and normal (b) synthesized and (d) real subjects without lesions. (Color figure online)

This is difficult or even infeasible for some applications such as lesion identification in neuroendocrine tumor (NET) images, because NETs are rare tumors and lesion annotation in low-resolution, noisy PET images is expensive. To address the data shortage issue, we propose to train a deep model for lesion detection with synthesized PET images generated from list mode PET data, which is low-cost and does not require human effort for manual data annotation.

Synthesized PET images may exhibit a different data distribution from real clinical images (see Fig. 1), i.e., a domain shift, which can pose significant challenges to model generalization. To address domain shifts, domain adaptation requires access to target data for model training [5,29], while domain generalization (DG) trains a model with only source data [39] and has recently attracted increasing attention in medical imaging [1,13,15,18]. Most of current DG methods rely on multiple sources of data to learn a generalizable model, i.e., multi-source DG (MDG); however, multi-source data collection is often difficult in real practice due to privacy concerns or budget deficits. Although single-source DG (SDG) using only one source dataset has been applied to medical images [12,14,32], very few studies focus on SDG with PET imaging and the current SDG methods may not be suitable for lesion identification on PET data. For instance, many existing methods use a complicated, multi-stage model design pipeline [10,23,30], which introduces an additional layer of algorithm variability. This situation will become worse for PET images, which typically have a poor signal-to-noise ratio and low spatial resolution. Several other SDG approaches [26,31,34] leverage unique characteristics of the imaging modalities, e.g., color spectrum of histological stained images, which are not applicable to PET data.

In this paper, we propose a novel single-stage SDG framework, which learns with human annotation-free, list mode-synthesized PET images for generalizable lesion detection in real clinical data. Compared with domain adaptation and MDG, the proposed method, while more challenging, is quite practical for real applications due to the relatively cheaper NET data collection and annotation. Specifically, we design a new data augmentation module, which generates out-of-domain samples from single-source data with multi-scale random convolutions. We integrate this module into a deep lesion detection neural network and introduce a cross-domain consistency constraint for feature encoding between

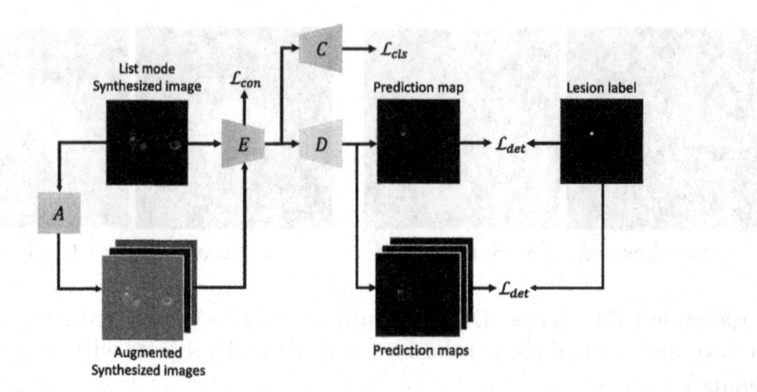

Fig. 2. The proposed SDG framework for generalizable lesion detection. The A, E, D and C represents the data augmentation module, feature encoder, decoder and domain classifier, respectively. The $\mathcal{L}_{det}, \mathcal{L}_{cls}$ and $\mathcal{L}_{con}$ denote the losses for lesion detection, domain classification and cross-domain consistency, respectively.

original synthesized and augmented images. Furthermore, we incorporate a novel patch-based gradient reversal mechanism into the network and accomplish a pretext task of domain classification, which explicitly promotes domain-invariant, generalizable representation learning. Trained with a single-source synthesized dataset, the proposed method provides superior performance of hepatic lesion detection in multiple cross-scanner real clinical PET image datasets, compared with the reference baseline and recent state-of-the-art SDG methods.

2 Methodology

Figure 2 presents the proposed SDG framework. Given a source-domain dataset of list mode-synthesized 3D PET images and corresponding lesion labels ($\boldsymbol{X}_S$, $\boldsymbol{Y}_S$), the goal of the framework is to learn a lesion detection model H, composed of E and D, which generalizes to real clinical PET image data. The framework first feeds synthesized images $\boldsymbol{X}_S$ into a random-convolution data augmentation module A and generates out-of-domain samples $\boldsymbol{X}_A = A(\boldsymbol{X}_S)$. Then, it provides both original and augmented images, $\boldsymbol{X}_S$ and $\boldsymbol{X}_A$, to a feature encoder E, which is followed by a decoder D for lesion detection. The framework imposes a cross-domain consistency constraint on the encoder E to promote consistent feature learning between $\boldsymbol{X}_S$ and $\boldsymbol{X}_A$. Meanwhile, it uses a patch gradient reversal-based domain classifier C to differentiate $\boldsymbol{X}_A$ from $\boldsymbol{X}_S$ and further encourages the encoder E to learn domain-agnostic representations for H.

2.1 Synthesized Data Augmentation

In the synthesized PET image dataset, each subject have multiple simulated lesions of varying size with known boundaries [11], and thus no human annotation is required. However, this synthesized dataset presents a significant domain shift from real clinical data, as they have markedly different image textures and voxel intensity values (see Fig. 1). Inspired by previous domain generalization work [39], we introduce a specific data augmentation module to generate out-of-domain samples from this single-source synthesized dataset for generalizable model learning (see Fig. 2). Specifically, we tailor a random convolution technique [33] for synthesized PET image augmentation with the following substantial improvement: 1) extend it from single value-prediction image classification to a more challenging dense prediction task of lesion detection; 2) refine it to produce realistic augmented images where the organ regions are brighter than image background, instead of randomly switching the foreground and background intensity values; 3) place a cross-domain consistency constraint on the encoding features, rather than output predictions, of original synthesized and augmented images, so as to directly encourage consistent representation learning between the source and other domains. This module can preserve global shapes or the structure of objects (e.g., lesions and livers) in images but distorts local textures, so that the lesion detection model learned with these augmented images can generalize to unseen real-world PET image data, which typically have high lesion heterogeneity and divergent texture styles.

Given a synthesized input image $\boldsymbol{x}_S \in \boldsymbol{X}_S$, our data augmentation module A first performs a *random* convolution operation $R(\boldsymbol{x}_S)$ with a $k \times k$ kernel R, where the kernel size k and the convolutional weights are randomly sampled from a multi-scale set $\mathcal{K} = \{1, 3, 5, 7\}$ and a normal distribution $\mathcal{N}(0, 1/k^2)$, respectively. Then, inspired by [7,35,36], we mix $R(\boldsymbol{x}_S)$ and $\boldsymbol{x}_S$ to generate a new mixed image $\boldsymbol{x}_M$ via a convex combination, $\boldsymbol{x}_M = \alpha \boldsymbol{x}_S + (1 - \alpha) R(\boldsymbol{x}_S)$, where $\alpha \in [0, 1]$ is randomly sampled from a uniform distribution $\mathcal{U}(0, 1)$. This data mixing strategy allows continuous interpolation between the source domain and a randomly generated out-of-distribution domain to improve model generalizability. Finally, if the foreground (i.e., lesion region) of $\boldsymbol{x}_M$ has a higher mean intensity value than the background (non-lesion region), we use $\boldsymbol{x}_M$ as the final augmented image, $\boldsymbol{x}_A = \boldsymbol{x}_M$. Otherwise, we invert the image intensity of $\boldsymbol{x}_M$ to obtain $\boldsymbol{x}_A = x_M^{max}\mathbf{1} + x_M^{min}\mathbf{1} - \boldsymbol{x}_M$, where x_M^{max}/x_M^{min} is the maximum/minimum intensity of $\boldsymbol{x}_M$ and $\mathbf{1}$ is a matrix with all elements being one and the same dimension as $\boldsymbol{x}_M$. This intensity inversion operation is to ensure the lesion region has higher intensity values than other regions, mimicking the image characteristics of real-world PET data in our experiments. Here we calculate the mean intensity value of the background from the regions that have a distance greater than half of the image width from the closest lesion center.

In our modeling, for each synthesized training image $\boldsymbol{x}_S$, we generate multiple augmented images (i.e., 3), $\{\boldsymbol{x}_A^i\}_{i=1}^3$, and feed them into the encoder E for feature learning. Due to the distance preservation property of random convolutions [33], the module A changes local textures but preserves object shapes at different scales, and thus $\boldsymbol{x}_S$ and $\{\boldsymbol{x}_A^i\}_{i=1}^3$ should have identical semantic content, such as lesion presence, quantity and positions. Therefore, they should have consistent representations in the feature space, i.e., $E(\boldsymbol{x}_S) \approx E(\boldsymbol{x}_A^i)$, $i = 1, 2, 3$. To this end, we place a cross-domain consistency loss $\mathcal{L}_{con}$ on top of the encoder E as

$$\mathcal{L}_{con} = \frac{1}{3}\sum_{i=1}^{3}\mathcal{L}_{con}^{i}, \tag{1}$$

$$\mathcal{L}_{con}^{i} = \mathbb{E}_{\boldsymbol{x}_S \sim \boldsymbol{X}_S}[\frac{1}{|E(\boldsymbol{x}_S)|}||E(\boldsymbol{x}_S) - E(\boldsymbol{x}_A^i)||_F^2],\ \forall i \in \{1,2,3\}, \tag{2}$$

where $\mathbb{E}$ is an expectation operator, $|E(\boldsymbol{x}_S)|$ is the number of elements in $E(\boldsymbol{x}_S)$, and $||\cdot||_F$ denotes the Frobenius norm. Unlike the previously reported work [33] promotes consistent output-layer predictions, the loss $\mathcal{L}_{con}$ in Eq. (1) directly encourages the encoder E to extract cross-domain consistent representations, which improves model generalization more effectively for dense prediction tasks [8], such as lesion detection. We hypothesize that forcing similar feature encoding between $\boldsymbol{x}_S$ and $\boldsymbol{x}_A^i$ can facilitate image content preservation for lesion detection. In addition, we adopt a mean squared error (MSE) to measure the consistency, different from [33] using the Kullback-Leibler divergence for image classification, which is not suitable for our application. Note that the convolution weights in module A are randomly sampled within each iteration and are not updated during model training.

2.2 Patch Gradient Reversal

Because of random convolution weights, the original synthesized $\boldsymbol{X}_S$ and augmented $\boldsymbol{X}_A$ data can have substantially different image appearances. Consequently, the use of the loss $\mathcal{L}_{con}$ in Eq. (1) may not be sufficient to enforce consistent feature encoding. To address this issue, we propose to use a pretext task as an additional information resource for the encoder E and to further promote domain-agnostic representation learning. Specifically, we incorporate a domain classifier C on top of the encoder E to perform a pretext task of domain discrimination, i.e., predict whether each input image is from the original synthesized data $\boldsymbol{X}_S$ or augmented data $\boldsymbol{X}_A$. This domain classification accompanies the main task of lesion detection (see Fig. 2) to assist with feature learning. In this way, the encoder E improves feature invariance to domain changes by penalizing domain classification accuracy, while retaining feature discriminativeness to lesion prediction via the decoder D. This is different from other methods [2,25] that use intrinsic supervision signals within a single image to perform an auxiliary task, e.g., solving jigsaw puzzles, for model generalization enhancement.

In general, the classifier C will encourage the encoder E to learn discriminative features for accurate domain classification. In order to make features invariant to different domains, we reverse the gradient propagated from the domain classifier C with a multiplication of -1 [3] and send this reversed gradient to the encoder E, while keeping all the other gradient flows unchanged during the backpropagation for model training. Note that the computation in forward propagation of our network is the same as that in a standard feed-forward neural network. Compared with [3], we make the following significant improvements: 1) Instead of back propagating the reversed gradient from a single-valued prediction of the domain label of the entire input image, we introduce a patch-based gradient reversal to enhance feature representation invariance to local texture or style changes. Inspired by [9], we design the domain classifier C with a fully convolutional network and produce a prediction map, where each element corresponds to a local patch of input image, i.e., conducting small patch categorization. We then apply the reversal operation to the gradient propagated from the prediction map and feed it into the encoder E for feature learning. 2) Motivated by [17], we remove the sigmoid layer in [3] and replace the cross-entropy loss by an MSE loss, which can facilitate the adversarial training caused by the gradient reversal. With the MSE loss, the patch-based gradient reversal penalizes image structures and enhances feature robustness and invariance to style shifts at the local-patch level, so that the lesion detection model H (i.e., E followed by D) learned with source data annotations is directly applicable to unseen domains [4,24], based on the covariate shift assumption [20].

Formally, let $\boldsymbol{X} = \{\boldsymbol{X}_S, \boldsymbol{X}_A\}$ denote the input data for the encoder E and $\boldsymbol{Z} = \{\boldsymbol{Z}_S, \boldsymbol{Z}_A\}$ represent the corresponding domain category labels, with $\boldsymbol{Z}_S$ and $\boldsymbol{Z}_A$ for the original source images $\boldsymbol{X}_S$ and corresponding random convolution-augmented image $\boldsymbol{X}_A$, respectively. Each label $\boldsymbol{z} \in \boldsymbol{Z}$ is a 3D image with all voxel intensity being 0's for $\boldsymbol{z} \in \boldsymbol{Z}_S$ or 1's for $\boldsymbol{z} \in \boldsymbol{Z}_A$. We define the domain classification objective $\mathcal{L}_{cls}$ as follows

$$\mathcal{L}_{cls} = \mathbb{E}_{(\boldsymbol{x},\boldsymbol{z})\sim(\boldsymbol{X},\boldsymbol{Z})}[\frac{1}{|\boldsymbol{z}|}||\boldsymbol{z} - \hat{\boldsymbol{z}}||_F^2], \tag{3}$$

where $\hat{\boldsymbol{z}} = C(E(\boldsymbol{x}))$ is the prediction of $\boldsymbol{x}$.

For source-domain data $(\boldsymbol{X}_S, \boldsymbol{Y}_S)$, the augmented images $\boldsymbol{X}_A$ have the same gold-standard lesion labels $\boldsymbol{Y}_A = \boldsymbol{Y}_S$, each of which is a 3D binary image with $1's$ for lesion voxels and $0's$ for non-lesion regions. Let $\boldsymbol{Y} = \{\boldsymbol{Y}_S, \boldsymbol{Y}_A\}$. We formulate the lesion detection objective $\mathcal{L}_{det}$ as

$$\begin{aligned}\mathcal{L}_{det} = \beta\mathbb{E}_{(\boldsymbol{x},\boldsymbol{y})\sim(\boldsymbol{X},\boldsymbol{Y})}[\frac{-1}{|\boldsymbol{y}|}\sum_{j=1}^{|\boldsymbol{y}|}(\gamma y^j \log \hat{y}^j + (1-y^j)\log(1-\hat{y}^j))] \\ +\mathbb{E}_{(\boldsymbol{x},\boldsymbol{y})\sim(\boldsymbol{X},\boldsymbol{Y})}[1 - \frac{2\sum_{j=1}^{|\boldsymbol{y}|} y^j \hat{y}^j + \epsilon}{\sum_{j=1}^{|\boldsymbol{y}|} y^j + \sum_{j=1}^{|\hat{\boldsymbol{y}}|} \hat{y}^j + \epsilon}],\end{aligned} \tag{4}$$

where the first and second terms in Eq. (4) are a weighted binary cross-entropy loss and a Dice loss, respectively. We add a smooth term, $\epsilon = 10^{-6}$, to the Dice loss to avoid division by zero. The y^j and $\hat{y}^j$ are the j-th values of $\boldsymbol{y}$ and corresponding prediction $\hat{\boldsymbol{y}}$, respectively. The β controls the relative importance between the two losses, and γ emphasizes the lesions in each image. The combo loss $\mathcal{L}_{det}$ can further help address the data imbalance issue [22], i.e., lesions have significantly fewer voxels than the non-lesion regions including the background.

With the losses in Eqs. (1)–(4), we define the following full objective as

$$\mathcal{L} = \mathcal{L}_{det} + \lambda_{con}\mathcal{L}_{con} + \lambda_{cls}\mathcal{L}_{cls}, \tag{5}$$

where λ_{con} and λ_{cls} are weighting parameters. Note that while we minimize $\mathcal{L}$ for model training, we reverse the gradient propagated from the domain classifier C before sending it to the encoder E during the backpropagation.

3 Experiments

Datasets. We evaluate the proposed method with multiple ^{68}Ga-DOTATATE PET liver NET image datasets that are acquired using different PET/CT scanners and/or imaging protocols. The synthesized source-domain dataset contains 103 simulated subjects, with an average of 5 lesions and 153 transverse slices per subject. This dataset is synthesized using list mode data from a single real, healthy subject acquired on a GE Discovery MI PET/CT scanner with list mode reconstruction [11,37]. We collect two additional real ^{68}Ga-DOTATATE PET liver NET image datasets that serve as unseen domains. The first dataset (*Real*1) has 123 real subjects with about 230 hepatic lesions in total and is acquired using clinical reconstructions with a photomultiplier tube-based PET/CT scanner (GE Discovery STE). The second real-world dataset (*Real*2) consists of 65 cases with around 113 lesions and is acquired from clinical reconstructions using a digital PET/CT scanner (GE Discovery MI). Following [28,38], we randomly split the synthesized dataset and the *Real*1 dataset into 60%, 20% and 20% for training, validation and testing, respectively. Due to the relatively small size of *Real*2, we use a two-fold cross-validation for model evaluation on this dataset. Here we split the real datasets to learn fully supervised models for a comparison with the proposed method.

Table 1. Domain generalization evaluation on different datasets. Each method is run 5 times, and the mean and standard deviation (std) of each metric (%) are reported: *mean* (*std*). The $*$ means a statistically significant difference (p-value < 0.05) between our method and others. The highest F_1 score is highlighted with **bold** font.

	Real1			Real2		
	F_1	Precision	Recall	F_1	Precision	Recall
CMSDG [18]	57.7* (3.2)	64.2 (13.0)	54.6 (7.9)	49.8 (6.5)	45.9 (8.3)	56.5 (9.8)
RandConv [33]	58.1* (1.3)	72.8 (10.4)	49.4 (5.9)	48.1* (2.3)	81.2 (4.4)	34.8 (2.1)
L2D [27]	46.4* (1.6)	41.1 (4.8)	54.7 (7.6)	41.9* (4.4)	30.1 (4.5)	71.0 (3.6)
Baseline	39.4* (1.9)	30.9 (2.4)	54.6 (2.2)	45.8* (5.0)	66.7 (10.0)	37.4 (4.2)
Aug.	51.5* (5.0)	45.4 (7.4)	60.3 (1.1)	44.3* (3.7)	53.5 (12.6)	41.6 (8.3)
Aug.+$\mathcal{L}_{con}$	55.1* (3.3)	60.0 (14.5)	52.3 (7.5)	44.5* (2.2)	67.0 (12.3)	35.0 (5.5)
Aug.+$\mathcal{L}_{con}$+*gGR*	58.7* (1.8)	62.5 (4.7)	55.5 (3.2)	48.2* (1.7)	69.6 (6.5)	37.8 (3.5)
Ours	**63.1** (0.5)	74.2 (9.2)	55.6 (5.4)	**53.8** (2.1)	58.0 (4.3)	50.9 (6.1)
Upper-bound	75.5 (4.3)	81.7 (3.6)	70.6 (7.0)	63.5 (7.6)	57.5 (7.3)	75.7 (5.1)

Implementation Details and Evaluation Metrics. We implement the encoder E and the decoder D with a U-Net architecture [19], with four downsampling and upsampling layers in the encoder and decoder, respectively. We build the domain classifier C using three stacked stride-1 convolutional layers of kernel size of 4, and each convolution is followed by a batch normalization and a leaky ReLU activation [16]. We set $\beta = 6, \gamma = 5$ in Eq. (4) and $\lambda_{con} = 1, \lambda_{cls} = 1$ in Eq. (5). We train the model using stochastic gradient descent with Nesterov momentum with learning rate $= 5 \times 10^{-4}$, momentum $= 0.99$ and batch size $= 1$. We perform standard image augmentation including random scaling, noise adding and image contrast adjustment before applying random convolutions in the module A. In the testing stage, we adopt the model H to produce a prediction map for each input image, and identify lesions with a threshold (i.e., 0.1) to binarize the map followed by a connected component analysis, which helps detect individual lesions by identifying connected regions from the binarized map. We use precision, recall and F_1 score as model evaluation metrics [21,28,38].

Comparison with State of the Art. We compare our method with several recent state-of-the-art SDG approaches, including causality-inspired SDG (*CISDG*) [18], *RandConv* [33], and learning to diversify (*L2D*) [27]. We run each model 5 times with different random seeds and report the mean and standard deviation. Table 1 presents the comparison results on the two unseen-domain datasets. Our method significantly outperforms the state-of-the-art approaches in terms of F_1 score, with p-value < 0.05 in Student's t-test for almost all cases on both datasets. In addition, our method gives lower standard deviation of F_1 than others. This indicates that compared with the competitor approaches, our method is relatively more effective and stable in learning generalizable representations for lesion detection in a very challenging situation, i.e., learning with a single-source synthesized PET image dataset to generalize to real clinical data. The qualitative results are provided in the Supplementary Material.

Ablation Study. In Table 1, the *Baseline* represents a lesion detection model trained with the source data but without the data augmentation module A, $\mathcal{L}_{con}$ or $\mathcal{L}_{cls}$. We then evaluate different variants of our method by sequentially adding one component to the *Baseline* model: 1) using only the module A for model training (*Aug.*), 2) using module A and $\mathcal{L}_{con}$ (*Aug.*+$\mathcal{L}_{con}$), and 3) using module A, $\mathcal{L}_{con}$ and $\mathcal{L}_{cls}$ (*Ours*). We also report the performance of the model, *Aug.*+$\mathcal{L}_{con}$+*gGR*, which does not use the proposed patch-based gradient reversal but outputs a single-value prediction for the entire input image, i.e., global gradient reversal (gGR). The *Upper-bound* means training with real-world images and gold-standard labels from the testing datasets. We note that using the data augmentation module A can significantly improve the lesion detection performance compared with the *Baseline* on the *Real*1 dataset, and combining data augmentation and patch gradient reversal can further close the gap to the *Upper-bound* model. Our method also outperforms the *Baseline* model by a large margin on the *Real*2 dataset, suggesting the effectiveness of our method.

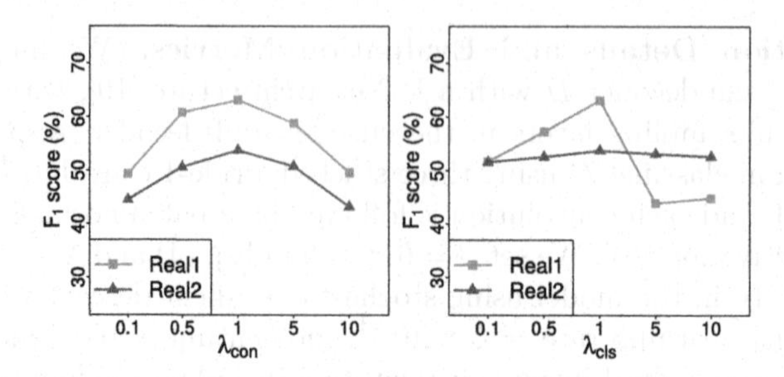

Fig. 3. The F_1 score of our method with different values of λ_{con} (left) and λ_{cls} (right).

Effects of Parameters. We evaluate the effects of λ_{con} and λ_{cls} of our method on lesion detection in Fig. 3. The lesion detection performance improves when increasing λ_{con} from 0.1 to 1. However, a further emphasis on consistent feature encoding, e.g., $\lambda_{con} \geq 5$, decreases the F_1 score. This suggests the importance of an appropriate λ_{con} value. In addition, we observe a similar trend of the F_1 curve for the λ_{cls}, especially for the *Real*1 dataset, and this indicates the necessity of the domain classification pretext task.

4 Conclusion

We propose a novel SDG framework that uses only a single dataset for hepatic lesion detection in real clinical PET images, without any human data annotations. With a specific data augmentation module and a new patch-based gradient reversal, the framework can learn domain-invariant representations and generalize to unseen domains. The experiments show that our method outperforms the reference baseline and recent state-of-the-art SDG approaches on cross-scanner or -protocol real PET image datasets. A potential limitation may be the need of a proper selection of weights for different tasks during model training.

References

1. Cai, J., et al.: Generalizing nucleus recognition model in multi-source Ki67 immunohistochemistry stained images via domain-specific pruning. In: de Bruijne, M., et al. (eds.) MICCAI 2021. LNCS, vol. 12908, pp. 277–287. Springer, Cham (2021). https://doi.org/10.1007/978-3-030-87237-3_27
2. Carlucci, F.M., D'Innocente, A., Bucci, S., Caputo, B., Tommasi, T.: Domain generalization by solving jigsaw puzzles. In: CVPR, pp. 2229–2238 (2019)
3. Ganin, Y., Lempitsky, V.: Unsupervised domain adaptation by backpropagation. In: ICML, pp. 1180–1189 (2015)
4. Geirhos, R., et al.: ImageNet-trained CNNs are biased towards texture; increasing shape bias improves accuracy and robustness. In: ICLR, pp. 1–12 (2019)
5. Guan, H., Liu, M.: Domain adaptation for medical image analysis: a survey. IEEE TBME **69**(3), 1173–1185 (2021)

6. Hatt, M., Laurent, B., Ouahabi, A., Fayad, H., Tan, S.: The first MICCAI challenge on pet tumor segmentation. MedIA **44**, 177–195 (2018)
7. Hendrycks, D., et al.: AugMix: a simple data processing method to improve robustness and uncertainty. In: ICLR, pp. 1–11 (2020)
8. Hong, W., Wang, Z., Yang, M., Yuan, J.: Conditional generative adversarial network for structured domain adaptation. In: CVPR, pp. 1335–1344 (2018)
9. Isola, P., Zhu, J.Y., Zhou, T., Efros, A.A.: Image-to-image translation with conditional adversarial networks. In: CVPR, pp. 1125–1134 (2017)
10. Kamraoui, R.A., et al.: DeepLesionBrain: towards a broader deep-learning generalization for multiple sclerosis lesion segmentation. MedIA **76**, 102312 (2022)
11. Leung, K.H., et al.: A physics-guided modular deep-learning based automated framework for tumor segmentation in pet. Phys. Med. Biol. **65**(24), 245032 (2020)
12. Li, H., et al.: Domain generalization for medical imaging classification with linear-dependency regularization. In: NeurIPS, pp. 3118–3129 (2020)
13. Li, Z., et al.: Domain generalization for mammography detection via multi-style and multi-view contrastive learning. In: de Bruijne, M., et al. (eds.) MICCAI 2021. LNCS, vol. 12907, pp. 98–108. Springer, Cham (2021). https://doi.org/10.1007/978-3-030-87234-2_10
14. Liu, Q., Chen, C., Dou, Q., Heng, P.A.: Single-domain generalization in medical image segmentation via test-time adaptation from shape dictionary. In: AAAI, pp. 1756–1764 (2022)
15. Liu, Q., Dou, Q., Heng, P.-A.: Shape-aware meta-learning for generalizing prostate MRI segmentation to unseen domains. In: Martel, A.L., et al. (eds.) MICCAI 2020. LNCS, vol. 12262, pp. 475–485. Springer, Cham (2020). https://doi.org/10.1007/978-3-030-59713-9_46
16. Maas, A., Hannun, A., Ng, A.: Rectifier nonlinearities improve neural network acoustic models. In: ICML, pp. 1–6 (2013)
17. Mao, X., et al.: Least squares generative adversarial networks. In: ICCV, pp. 2813–2821 (2017)
18. Ouyang, C., Chen, C., Li, S., Li, Z., Qin, C.: Causality-inspired single-source domain generalization for medical image segmentation. IEEE TMI, 1–12 (2022)
19. Ronneberger, O., Fischer, P., Brox, T.: U-Net: convolutional networks for biomedical image segmentation. In: Navab, N., Hornegger, J., Wells, W.M., Frangi, A.F. (eds.) MICCAI 2015. LNCS, vol. 9351, pp. 234–241. Springer, Cham (2015). https://doi.org/10.1007/978-3-319-24574-4_28
20. Shimodaira, H.: Improving predictive inference under covariate shift by weighting the log-likelihood function. J. Stat. Plan. Inference **90**(2), 227–244 (2000)
21. Song, Y., et al.: Lesion detection and characterization with context driven approximation in thoracic FDG PET-CT images of NSCLC studies. IEEE TMI **33**(2), 408–421 (2014)
22. Taghanaki, S.A., et al.: Combo loss: handling input and output imbalance in multi-organ segmentation. CMIG **75**, 24–33 (2019)
23. Vesal, S., et al.: Domain generalization for prostate segmentation in transrectal ultrasound images: a multi-center study. MedIA **82**, 102620 (2022)
24. Wang, H., Ge, S., Lipton, Z., Xing, E.P.: Learning robust global representations by penalizing local predictive power. In: NeurIPS, pp. 10506–10518 (2019)
25. Wang, S., Yu, L., Li, C., Fu, C.-W., Heng, P.-A.: Learning from extrinsic and intrinsic supervisions for domain generalization. In: Vedaldi, A., Bischof, H., Brox, T., Frahm, J.-M. (eds.) ECCV 2020. LNCS, vol. 12354, pp. 159–176. Springer, Cham (2020). https://doi.org/10.1007/978-3-030-58545-7_10

26. Wang, X., et al.: A generalizable and robust deep learning algorithm for mitosis detection in multicenter breast histopathological images. MedIA **84**, 102703 (2023)
27. Wang, Z., Luo, Y., Qiu, R., Huang, Z., Baktashmotlagh, M.: Learning to diversify for single domain generalization. In: ICCV, pp. 834–843 (2021)
28. Wehrend, J., et al.: Automated liver lesion detection in 68Ga DOTATATE PET/CT using a deep fully convolutional neural network. EJNMMI Res. **11**(1), 1–11 (2021)
29. Wilson, G., Cook, D.J.: A survey of unsupervised deep domain adaptation. ACM TIST **11**(5), 1–46 (2020)
30. Xie, L., Wisse, L.E., Wang, J., Ravikumar, S., Khandelwal, P.: Deep label fusion: a generalizable hybrid multi-atlas and deep convolutional neural network for medical image segmentation. MedIA **83**, 102683 (2023)
31. Xu, C., Wen, Z., Liu, Z., Ye, C.: Improved domain generalization for cell detection in histopathology images via test-time stain augmentation. In: Wang, L., et al. (eds.) MICCAI 2022, pp. 150–159. Springer, Cham (2022). https://doi.org/10.1007/978-3-031-16434-7_15
32. Xu, Y., et al.: Adversarial consistency for single domain generalization in medical image segmentation. In: Wang, L., et al. (eds.) MICCAI 2022, pp. 671–681. Springer, Cham (2022). https://doi.org/10.1007/978-3-031-16449-1_64
33. Xu, Z., Liu, D., Yang, J., Raffel, C., Niethammer, M.: Robust and generalizable visual representation learning via random convolutions. In: ICLR, pp. 1–12 (2021)
34. Yamashita, R., Long, J., Banda, S., Shen, J., Rubin, D.L.: Learning domain-agnostic visual representation for computational pathology using medically-irrelevant style transfer augmentation. IEEE TMI **40**(12), 3945–3954 (2021)
35. Yun, S., et al.: CutMix: regularization strategy to train strong classifiers with localizable features. In: ICCV, pp. 6022–6031 (2019)
36. Zhang, H., Cisse, M., Dauphin, Y.N., Lopez-Paz, D.: mixup: beyond empirical risk minimization. In: ICLR, pp. 1–13 (2018)
37. Zhang, Z., et al.: Optimization-based image reconstruction from low-count, list-mode TOF-pet data. IEEE TBME **65**(4), 936–946 (2018)
38. Zhao, Y., et al.: Deep neural network for automatic characterization of lesions on 68Ga-PSMA-11 PET/CT. EJNMMI **47**, 603–613 (2020)
39. Zhou, K., Liu, Z., Qiao, Y., Xiang, T., Loy, C.C.: Domain generalization: a survey. IEEE TPAMI **45**(4), 4396–4415 (2022)

Robust Exclusive Adaptive Sparse Feature Selection for Biomarker Discovery and Early Diagnosis of Neuropsychiatric Systemic Lupus Erythematosus

Tianhong Quan[1], Ye Yuan[2], Yu Luo[3], Teng Zhou[4,5](✉), and Jing Qin[4]

[1] School of Computer Science and Technology, Beijing Institute of Technology, Beijing, China
[2] College of Engineering, Shantou University, Shantou, China
[3] School of Computer Science and Technology, Guangdong University of Technology, Guangzhou, China
[4] Centre for Smart Health, The Hong Kong Polytechnic University, Hung Hom, Hong Kong
teng.zhou@polyu.edu.hk
[5] School of Cyberspace Security, Hainan University, Haikou, China

Abstract. The symptoms of neuropsychiatric systemic lupus erythematosus (NPSLE) are subtle and elusive at the early stages. ^{1}H-MRS (proton magnetic resonance spectrum) imaging technology can detect more detailed early appearances of NPSLE compared with conventional ones. However, the noises in ^{1}H-MRS data often bring bias in the diagnostic process. Moreover, the features of specific brain regions are positively correlated with a certain category but may be redundant for other categories. To overcome these issues, we propose a robust exclusive adaptive sparse feature selection (REASFS) algorithm for early diagnosis and biomarker discovery of NPSLE. Specifically, we employ generalized correntropic loss to address non-Gaussian noise and outliers. Then, we develop a generalized correntropy-induced exclusive $\ell_{2,1}$ regularization to adaptively accommodate various sparsity levels and preserve informative features. We conduct sufficient experiments on a benchmark NPSLE dataset, and the experimental results demonstrate the superiority of our proposed method compared with state-of-the-art ones.

Keywords: NPSLE · Feature Selection · non-Gaussian Noise

1 Introduction

Neuropsychiatric systemic lupus erythematosus (NPSLE) refers to a complex autoimmune disease that damages the brain nervous system of patients. The clinical symptoms of NPSLE include cognitive disorder, epilepsy, mental illness, etc., and patients with NPSLE have a nine-fold increased mortality compared to the general population [11]. Since the pathogenesis and mature treatment of NPSLE have not yet been found, it is extremely important to detect NPSLE at

H. Greenspan et al. (Eds.): MICCAI 2023, LNCS 14224, pp. 127–135, 2023.
https://doi.org/10.1007/978-3-031-43904-9_13

its early stage and put better clinical interventions and treatments to prevent its progression. However, the high overlap of clinical symptoms with other psychiatric disorders and the absence of early non-invasive biomarkers make accurate diagnosis difficult and time-consuming [3].

Although conventional magnetic resonance imaging (MRI) tools are widely used to detect brain injuries and neuronal lesions, around 50% of patients with NPSLE present no brain abnormalities in structural MRI [17]. In fact, metabolic changes in many brain diseases precede pathomorphological changes, which indicates proton magnetic resonance spectroscopy (^{1}H-MRS) to be a more effective way to reflect the early appearance of NPSLE. ^{1}H-MRS is a non-invasive neuroimaging technology that can quantitatively analyze the concentration of metabolites and detect abnormal metabolism of the nervous system to reveal brain lesions. However, the complex noise caused by overlapping metabolite peaks, incomplete information on background components, and low signal-to-noise ratio (SNR) disturb the analysis results of this spectroscopic method [15]. Meanwhile, the individual differences in metabolism and the interaction between metabolites under low sample size make it difficult for traditional learning methods to distinguish NPSLE. Figure 1 shows spectra images of four participants including healthy controls (HC) and patients with NPSLE. It can be seen that the visual differences between patients with NPSLE and HCs in the spectra of the volumes are subtle. Therefore, it is crucial to develop effective learning algorithms to discover metabolic biomarkers and accurately diagnose NPSLE.

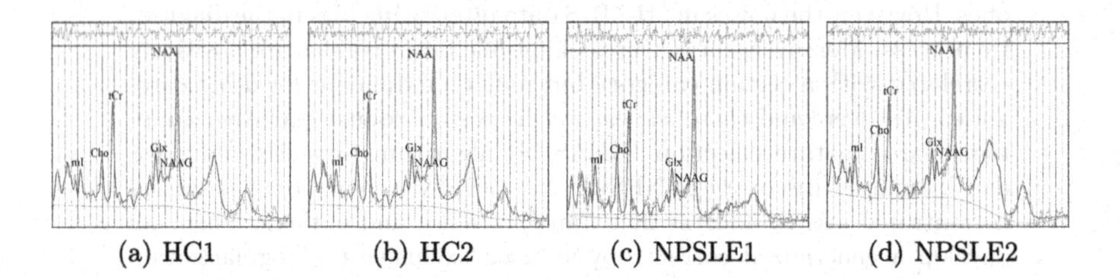

(a) HC1 (b) HC2 (c) NPSLE1 (d) NPSLE2

Fig. 1. Images (a) to (d) represent the spectra of the volumes obtained by LCModel for four individual participants. Abbreviation: N-acetylaspartate (NAA), N-acetylaspartylglutamate (NAAG), choline (Cho), total creatine (tCr), myo-inositol (mI), glutamate+glutamine (Glu + Gln = Glx).

The machine learning application for biomarker analysis and early diagnosis of NPSLE is at a nascent stage [4]. Most studies focus on the analysis of MR images using statistical or machine learning algorithms, such as Mann-Whitney U test [8], support vector machine (SVM) [7,24], ensemble model [16,22], etc. Generally, Machine learning algorithms based on the minimum mean square error (MMSE) criterion heavily rely on the assumption that noise is of Gaussian distribution. However, measurement-induced non-Gaussian noise in ^{1}H-MRS data undoubtedly limits the performance of MMSE-based machine learning methods.

On the other hand, for the discovery task of potential biomarkers, sparse coding-based methods (e.g., $\ell_{2,1}$ norm, $\ell_{2,0}$ norm, etc.) force row elements to zero that remove some valuable features [12,21]. More importantly, different brain regions have different functions and metabolite concentrations, which implies that the metabolic features for each brain region have different sparsity levels. Therefore, applying the same sparsity constraint to the metabolic features of all brain regions may not contribute to the improvement of the diagnostic performance of NPSLE.

In light of this, we propose a robust exclusive adaptive sparse feature selection (REASFS) algorithm to jointly address the aforementioned problems in biomarker discovery and early diagnosis of NPSLE. Specifically, we first extend our feature learning through generalized correntropic loss to handle data with complex non-Gaussian noise and outliers. We also present the mathematical analysis of the adaptive weighting mechanism of generalized correntropy. Then, we propose a novel regularization called generalized correntropy-induced exclusive $\ell_{2,1}$ to adaptively accommodate various sparsity levels and preserve informative features. The experimental results on a benchmark NPSLE dataset demonstrate the proposed method outperforms comparing methods in terms of early non-invasive biomarker discovery and early diagnosis.

2 Method

Dataset and Preprocessing: The T2-weighted MR images of 39 participants including 23 patients with NPSLE and 16 HCs were gathered from our affiliated hospital. All images were acquired at an average age of 30.6 years on a SIGNA 3.0T scanner with an eight-channel standard head coil. Then, the MR images were transformed into spectroscopy by multi-voxel ^{1}H-MRS based on a point-resolved spectral sequence (PRESS) with a two-dimensional multi-voxel technique. The collected spectroscopy data were preprocessed by a SAGE software package to correct the phase and frequency. An LCModel software was used to fit the spectra, correct the baseline, relaxation, and partial-volume effects, and quantify the concentration of metabolites. Finally, we used the absolute NAA concentration in single-voxel MRS as the standard to gain the absolute concentration of metabolites, and the NAA concentration of the corresponding voxel of multi-voxel ^{1}H-MRS was collected consistently. The spectra would be accepted if the SNR is greater than or equal to 10 and the metabolite concentration with standard deviations (SD) is less than or equal to 20%. The absolute metabolic concentrations, the corresponding ratio, and the linear combination of the spectra were extracted from different brain regions: RPCG, LPCG, RDT, LDT, RLN, LLN, RI, RPWM, and LPWM. A total of 117 metabolic features were extracted, and each brain region contained 13 metabolic features: Cr, phosphocreatine (PCr), Cr+PCr, NAA, NAAG, NAA+NAAG, NAA+NAAG/Cr+PCr, mI, mI/Cr+PCr, Cho+phosphocholine (PCh), Cho+PCh/Cr+PCr, Glu+Gln, and Glu+Gln/Cr+PCr.

2.1 Sparse Coding Framework

Given a data matrix $\mathbf{X} = [\mathbf{x}^1; \cdots; \mathbf{x}^n] \in \mathbb{R}^{n \times d}$ with n sample, the i-th row is represented by $\mathbf{x}^i$, and the corresponding label matrix is denoted as $\mathbf{Y} = [\mathbf{y}^1; \cdots; \mathbf{y}^n] \in \mathbb{R}^{n \times k}$, where $\mathbf{y}_i$ is one-hot vector. The Frobenius norm of projection matrix $\mathbf{W} \in \mathbb{R}^{d \times k}$ is $\|\mathbf{W}\|_F = \sqrt{\sum_{i=1}^{d} \sum_{j=1}^{k} W_{ij}^2}$. For sparse coding-based methods, the general problem can be formulated as

$$\min_{\mathbf{W}} J(\mathbf{W}) = L(\mathbf{W}) + \lambda R(\mathbf{W}), \tag{1}$$

where $L(\mathbf{W})$, $R(\mathbf{W})$, and λ are the loss function, the regularization term, and the hyperparameter, respectively. For least absolute shrinkage and selection operator (LASSO) [20], $L(\mathbf{W}) = \|\mathbf{Y} - \mathbf{X}\mathbf{W}\|_F^2$ and $R(\mathbf{W}) = \|\mathbf{W}\|_1$. For multi-task feature learning [6], $\ell_{2,1}$ norm is the most widely used regularization to select class-shared features via row sparsity, which is defined as $\|\mathbf{W}\|_{2,1} = \sum_{i=1}^{d} \|\mathbf{w}^i\|_2 = \sum_{i=1}^{d} \left(\sum_{j=1}^{k} W_{ij}^2 \right)^{1/2}$. Due to the row sparsity of $\ell_{2,1}$ norm, the features selected for all different classes are enforced to be exactly the same. Thus, the inflexibility of $\ell_{2,1}$ norm may lead to the deletion of meaningful features.

2.2 Generalized Correntropic Loss

Originating from information theoretic learning (ITL), the correntropy [2] is a local similarity measure between two random variables A and B, given by

$$V(A, B) = \mathbf{E}[k_\sigma(A, B)] = \int k_\sigma(a, b) dF_{AB}(a, b), \tag{2}$$

where $\mathbf{E}$, $k_\sigma(\cdot, \cdot)$, and $F_{AB}(a, b)$ denote the mathematical expectation, the Gaussian kernel, and the joint probability density function of (A, B), respectively.

When applying correntropy to the error criterion, the boundedness of the Gaussian kernel limits the disturbance of large errors caused by outliers on estimated parameters. However, the kernelized second-order statistic of correntropy is not suitable for all situations. Therefore, the generalized correntropy [1], a more flexible and powerful form of correntropy, was proposed by substituting the generalized Gaussian density (GGD) function for the Gaussian kernel in correntropy, and the GGD function is defined as

$$G_{\alpha,\beta}(e) = \frac{\alpha}{2\beta\Gamma\left(\frac{1}{\alpha}\right)} \exp\left(-|\frac{e}{\beta}|^\alpha\right) = \gamma \exp\left(-s|e|^\alpha\right), \tag{3}$$

where $\Gamma(\cdot)$ is the gamma function, α, $\beta > 0$ are the shape and bandwidth parameters, respectively, $s = 1/\beta^\alpha$ is the kernel parameter, $\gamma = \alpha/(2\beta\Gamma(1/\alpha))$ is the normalization factor. Specifically, when $\alpha = 2$ and $\alpha = 1$, GGD degenerates to Gaussian distribution and Laplacian distribution, respectively. As an adaptive similarity measure, generalized correntropy can be applied to machine learning and adaptive systems [14]. The generalized correntropic loss function between A and B can be defined as

$$L_{GC}(A,B) = G_{\alpha,\beta}(0) - V_{\alpha,\beta}(A,B). \tag{4}$$

To analyze the adaptive weighting mechanism of generalized correntropy, we consider an alternative problem of (1), where $L(\mathbf{W}) = \hat{V}_{\alpha,\beta}(\mathbf{Y},\mathbf{XW}) = \sum_{i=1}^{n} \exp(-s\|\mathbf{y}^i - \mathbf{x}^i\mathbf{W}\|_2^\alpha)$ and $R(\mathbf{W}) = \|\mathbf{W}\|^2 = \sum_{i=1}^{d} \|\mathbf{w}^i\|_2^2$. The optimal projection matrix $\mathbf{W}$ should satisfy $(\partial J(\mathbf{W})/\partial \mathbf{W}) = 0$, and we have

$$\mathbf{W} = \left(\mathbf{X}^T \Lambda \mathbf{X} + \lambda \mathbf{I}\right)^{-1} \mathbf{X}^T \Lambda \mathbf{Y}, \tag{5}$$

where Λ is a diagonal matrix with error-based diagonal elements $\Lambda_{ii} = \exp(-s\|\mathbf{y}^i - \mathbf{x}^i\mathbf{W}\|_2^\alpha)\|\mathbf{y}^i - \mathbf{x}^i\mathbf{W}\|_2^{\alpha-2}$ for adaptive sample weight.

2.3 Generalized Correntropy-Induced Exclusive $\ell_{2,1}$

To overcome the drawback of $\ell_{2,1}$ norm and achieve adaptive sparsity regularization on metabolic features of different brain regions, we propose a novel GCIE $\ell_{2,1}$. A flexible feature learning algorithm exclusive $\ell_{2,1}$ [9] is defined as

$$\min_{\mathbf{W}} J(\mathbf{W}) = L(\mathbf{W}) + \lambda_1 \|\mathbf{W}\|_{2,1} + \lambda_2 \|\mathbf{W}\|_{1,2}^2, \tag{6}$$

where $\|\mathbf{W}\|_{1,2}^2 = \sum_{i=1}^{d} \|\mathbf{w}^i\|_1^2 = \sum_{i=1}^{d} \left(\sum_{j=1}^{k} |W_{ij}|\right)^2$. Based on the exclusive $\ell_{2,1}$, we can not only removes the redundant features shared by all classes through row sparsity of $\ell_{2,1}$ norm but also selects different discriminative features for each class through exclusive sparsity of $\ell_{1,2}$ norm. Then we propose to introduce generalized correntropy to measure the sparsity penalty in the feature learning algorithm. We apply generalized correntropy to the row vector $\mathbf{w}^i$ of $\mathbf{W}$ to achieve the adaptive weighted sparse constraint, and the problem (6) can be rewritten as

$$\min_{\mathbf{W}} J(\mathbf{W}) = \hat{L}_{GC}(\mathbf{Y},\mathbf{XW}) + \lambda_1 \sum_{i=1}^{d} (1 - \exp(-s\|\mathbf{w}^i\|_2)) + \lambda_2 \|\mathbf{W}\|_{1,2}^2, \tag{7}$$

where $\hat{L}_{GC}(\mathbf{Y},\mathbf{XW}) = \sum_{i=1}^{n}(1 - exp(-s\|\mathbf{y}^i - \mathbf{x}^i\mathbf{W}\|_2^\alpha))$. Since minimizing $\|\mathbf{w}^i\|_2$ is equivalent to maximizing $\exp(-s\|\mathbf{w}^i\|_2)$, we add a negative sign to this term. Through the GCIE $\ell_{2,1}$ regularization term in (7), each feature is expected to be enforced with a sparsity constraint of a different weight according to its sparsity level of metabolic information in different brain regions.

Optimization: Since the GCIE $\ell_{2,1}$ is a non-smooth regularization term, the final problem (7) can be optimized by the stochastic gradient method with appropriate initialization of $\mathbf{W}$. To this end, We use the closed-form solution (5) to initialize $\mathbf{W}$ reasonably, and the error-driven sample weight has reached the optimum based on half-quadratic analysis [13]. Once we obtained the solution to the problem (7), the importance of feature i is proportional to $\|\mathbf{w}_i\|_2$. We then rank the importance of features according to $(\|\mathbf{w}_i\|_2, \cdots, \|\mathbf{w}_d\|_2)$, and select the top m ranked features from the sorted order for further classification.

3 Experimental Results and Conclusion

Experimental Settings: The parameters α and λ_1 are are set to 1, while β and λ_2 are searched form $\{0.5, 1, 2, 5\}$ and $\{0.1, 0.5, 1\}$, respectively. We use Adam optimizer and the learning rate is 0.001. To evaluate the performance of classification, we employ a support vector machine as the basic classifier, where the kernel is set as the radial basis function (RBF) and parameter C is set to 1. We average the 3-fold cross-validation results.

Results and Discussion: We compare the classification accuracy of the proposed REASFS with several SOTA baselines, including two filter methods: maximal information coefficient (MIC) [5], Gini [23], and four sparse coding-based methods: multi-task feature learning via $\ell_{2,1}$ norm [6,12], discriminative feature selection via $\ell_{2,0}$ norm [21], feature selection via $\ell_{1,2}$ norm [10] and exclusive $\ell_{2,1}$ [9]. The proposed REASFS is expected to have better robustness and flexibility. It can be seen from Fig. 2 that the sparse coding-based methods achieve better performance than filter methods under most conditions, where "0%" represents no noise contamination. The highest accuracy of our REASFS demonstrates the effectiveness and flexibility of the proposed GCIE $\ell_{2,1}$.

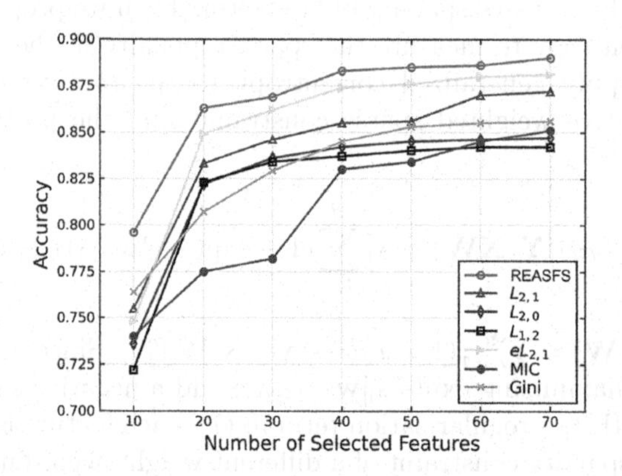

Fig. 2. REASFS versus SOTA baselines on NPSLE dataset (0% noise).

Generally speaking, the probability of samples being contaminated by random noise is equal. Therefore, we randomly select features from the training set and replace the selected features with pulse noise. The number of noisy attributes is denoted by the ratio between the numbers of selected features and total features, such as 15% and 30%. The classification performances of the NPSLE dataset contaminated by attribute noise are shown in Fig. 3(a) and Fig. 3(b), where one clearly perceives that our REASFS achieves the highest accuracy under all conditions. Besides, it is unreasonable to apply the same level of sparse regularization to noise features and uncontaminated features, and our GCIE $\ell_{2,1}$ can adaptively increase the sparse level of noise features to remove redundant information, and vice versa. For label noise, we randomly select samples from the training set and replace classification labels of the selected samples with opposite values, *i.e.*, $0 \to 1$ and $1 \to 0$. The results are shown in Fig. 3(c) and Fig. 3(d), where the proposed REASFS is superior to other baselines. It can be seen from Fig. 3 that our REASFS achieves the highest accuracy in different noisy environments, which demonstrates the robustness of generalized correntropic loss.

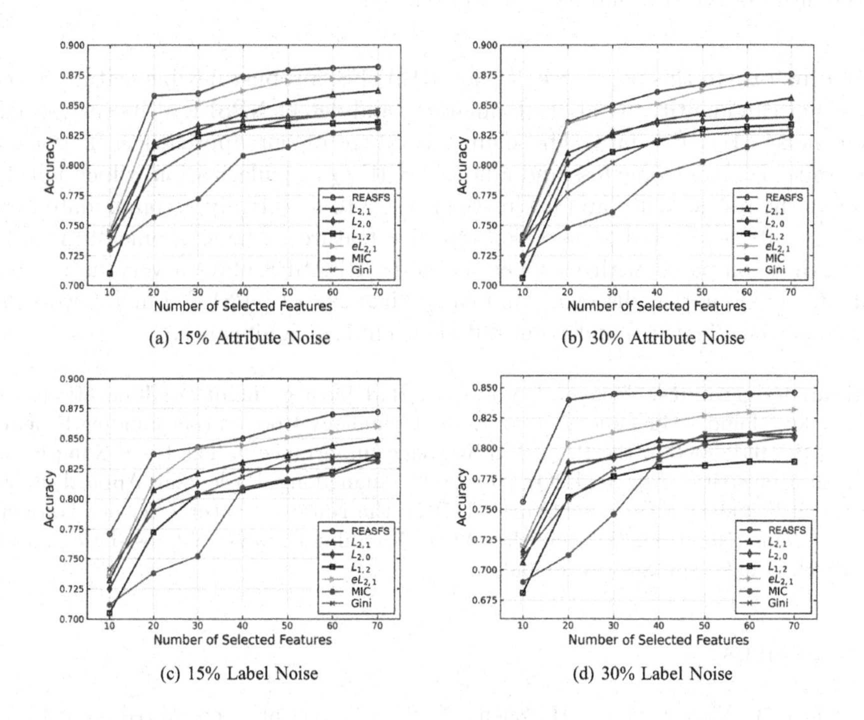

Fig. 3. REASFS versus SOTA baselines on NPSLE dataset with different noise.

For non-invasive biomarkers, our method shows that some metabolic features contribute greatly to the early diagnosis of NPSLE, i.e., NAAG, mI/Cr+PCr, and Glu+Gln/Cr+PCr in RPCG; Cr+PCr, NAA+NAAG, NAA+NAAG/Cr+PCr, mI/Cr+PCr and Glu+Gln in LPCG; NAA, NAAG, and Cho+PCh in LDT; PCr, Cr+PCr, Cho+PCh, Cho+PCh/Cr+PCr and Glu+Gln/Cr+PCr in RLN; MI/Cr+PCr, Cho+PCh and Cho+PCh/Cr+PCr in LLN; NAA+NAAG/Cr+PCr and Cho+PCh in RI; Cho+PCh/Cr+PCr and

Glu+Gln/Cr+PCr in RPWM; And PCr, NAAG and NAA+NAAG/Cr+PCr in LPWM. Moreover, we use isometric feature mapping (ISOMAP) [19] to analyze these metabolic features and find that this feature subset is essentially a low-dimensional manifold. Meanwhile, by combining the proposed REASFS and ISOMAP, we can achieve 99% accuracy in the early diagnosis of NPSLE. In metabolite analysis, some studies have shown that the decrease in NAA concentration is related to chronic inflammation, damage, and tumors in the brain [18]. In the normal white matter area, different degrees of NPSLE disease is accompanied by different degrees of NAA decline, but structural MRI is not abnormal, suggesting that NAA may indicate the progress of NPSLE. We also found that Glu+Gln/Cr+PCr in RI decreased, which indicates that the excitatory neurotransmitter Glu in the brain of patients with NPSLE may have lower activity. To sum up, the proposed method provides a shortcut for revealing the pathological mechanism of NPSLE and early detection.

Conclusion: In this paper, we develop REASFS, a robust flexible feature selection that can identify metabolic biomarkers and detect NPSLE at its early stage from noisy ^{1}H-MRS data. The main advantage of our approach is its robust generalized correntropic loss and a novel GCIE $\ell_{2,1}$ regularization, which jointly utilizes the row sparsity and exclusive sparsity to adaptively accommodate various sparsity levels and preserve informative features. The experimental results show that compared with previous methods, REASFS plays a very important role in the biomarker discovery and early diagnosis of NPSLE. Finally, we analyze metabolic features and point out their clinical significance.

Acknowledgements. This work was supported by a grant of the Innovation and Technology Fund - Guangdong-Hong Kong Technology Cooperation Funding Scheme (No. GHP/051/20GD), the Project of Strategic Importance in The Hong Kong Polytechnic University (No. 1-ZE2Q), the 2022 Guangdong Basic and Applied Basic Research Foundation (No. 2022A1515011590), the National Natural Science Foundation of China (No. 61902232), and the 2020 Li Ka Shing Foundation Cross-Disciplinary Research Grant (No. 2020LKSFG05D).

References

1. Chen, B., Xing, L., Zhao, H., Zheng, N., Prı, J.C., et al.: Generalized correntropy for robust adaptive filtering. IEEE Trans. Signal Process. **64**(13), 3376–3387 (2016)
2. He, R., Zheng, W.S., Hu, B.G.: Maximum correntropy criterion for robust face recognition. IEEE Trans. Pattern Anal. Mach. Intell. **33**(8), 1561–1576 (2010)
3. Jeltsch-David, H., Muller, S.: Neuropsychiatric systemic lupus erythematosus: pathogenesis and biomarkers. Nat. Rev. Neurol. **10**(10), 579–596 (2014)
4. Kingsmore, K.M., Lipsky, P.E.: Recent advances in the use of machine learning and artificial intelligence to improve diagnosis, predict flares, and enrich clinical trials in lupus. Curr. Opin. Rheumatol. **34**(6), 374–381 (2022)
5. Kinney, J.B., Atwal, G.S.: Equitability, mutual information, and the maximal information coefficient. Proc. Natl. Acad. Sci. **111**(9), 3354–3359 (2014)

6. Liu, J., Ji, S., Ye, J.: Multi-task feature learning via efficient l2, 1-norm minimization. In: Proceedings of the Twenty-Fifth Conference on Uncertainty in Artificial Intelligence, pp. 339–348 (2009)
7. Luo, X., et al.: Multi-lesion radiomics model for discrimination of relapsing-remitting multiple sclerosis and neuropsychiatric systemic lupus erythematosus. Eur. Radiol. **32**(8), 5700–5710 (2022)
8. Mackay, M., Tang, C.C., Vo, A.: Advanced neuroimaging in neuropsychiatric systemic lupus erythematosus. Curr. Opin. Neurol. **33**(3), 353 (2020)
9. Ming, D., Ding, C.: Robust flexible feature selection via exclusive l21 regularization. In: Proceedings of the 28th International Joint Conference on Artificial Intelligence, pp. 3158–3164 (2019)
10. Ming, D., Ding, C., Nie, F.: A probabilistic derivation of LASSO and L12-norm feature selections. In: Proceedings of the AAAI Conference on Artificial Intelligence, vol. 33, pp. 4586–4593 (2019)
11. Monahan, R.C., et al.: Mortality in patients with systemic lupus erythematosus and neuropsychiatric involvement: a retrospective analysis from a tertiary referral center in the Netherlands. Lupus **29**(14), 1892–1901 (2020)
12. Nie, F., Huang, H., Cai, X., Ding, C.: Efficient and robust feature selection via joint $\ell 2$, 1-norms minimization. In: Advances in Neural Information Processing Systems, vol. 23 (2010)
13. Nikolova, M., Ng, M.K.: Analysis of half-quadratic minimization methods for signal and image recovery. SIAM J. Sci. Comput. **27**(3), 937–966 (2005)
14. Quan, T., Yuan, Y., Song, Y., Zhou, T., Qin, J.: Fuzzy structural broad learning for breast cancer classification. In: 2022 IEEE 19th International Symposium on Biomedical Imaging (ISBI), pp. 1–4. IEEE (2022)
15. Ruiz-Rodado, V., Brender, J.R., Cherukuri, M.K., Gilbert, M.R., Larion, M.: Magnetic resonance spectroscopy for the study of CNS malignancies. Prog. Nucl. Magn. Reson. Spectrosc. **122**, 23–41 (2021)
16. Simos, N.J., et al.: Quantitative identification of functional connectivity disturbances in neuropsychiatric lupus based on resting-state fMRI: a robust machine learning approach. Brain Sci. **10**(11), 777 (2020)
17. Tamires Lapa, A., et al.: Reduction of cerebral and corpus callosum volumes in childhood-onset systemic lupus erythematosus: a volumetric magnetic resonance imaging analysis. Arthritis Rheumatol. **68**(9), 2193–2199 (2016)
18. Tannous, J., et al.: Altered neurochemistry in the anterior white matter of bipolar children and adolescents: a multivoxel 1h MRS study. Mol. Psychiatry **26**(8), 4117–4126 (2021)
19. Tenenbaum, J.B., Silva, V.D., Langford, J.C.: A global geometric framework for nonlinear dimensionality reduction. Science **290**(5500), 2319–2323 (2000)
20. Tibshirani, R.: Regression shrinkage and selection via the LASSO. J. Roy. Stat. Soc.: Ser. B (Methodol.) **58**(1), 267–288 (1996)
21. Wang, Z., Nie, F., Tian, L., Wang, R., Li, X.: Discriminative feature selection via a structured sparse subspace learning module. In: IJCAI, pp. 3009–3015 (2020)
22. Yuan, Y., Quan, T., Song, Y., Guan, J., Zhou, T., Wu, R.: Noise-immune extreme ensemble learning for early diagnosis of neuropsychiatric systemic lupus erythematosus. IEEE J. Biomed. Health Inform. **26**(7), 3495–3506 (2022)
23. Zhang, S., Dang, X., Nguyen, D., Wilkins, D., Chen, Y.: Estimating feature-label dependence using Gini distance statistics. IEEE Trans. Pattern Anal. Mach. Intell. **43**(6), 1947–1963 (2019)
24. Zhuo, Z., et al.: Different patterns of cerebral perfusion in SLE patients with and without neuropsychiatric manifestations. Hum. Brain Mapp. **41**(3), 755–766 (2020)

Multi-view Vertebra Localization and Identification from CT Images

Han Wu[1], Jiadong Zhang[1], Yu Fang[1], Zhentao Liu[1], Nizhuan Wang[1], Zhiming Cui[1(✉)], and Dinggang Shen[1,2,3(✉)]

[1] School of Biomedical Engineering, ShanghaiTech University, Shanghai, China
cuizm.neu.edu@gmail.com, dgshen@shanghaitech.edu.cn

[2] Shanghai United Imaging Intelligence Co. Ltd., Shanghai, China

[3] Shanghai Clinical Research and Trial Center, Shanghai, China

Abstract. Accurately localizing and identifying vertebra from CT images is crucial for various clinical applications. However, most existing efforts are performed on 3D with cropping patch operation, suffering from the large computation costs and limited global information. In this paper, we propose a multi-view vertebra localization and identification from CT images, converting the 3D problem into a 2D localization and identification task on different views. Without the limitation of the 3D cropped patch, our method can learn the multi-view global information naturally. Moreover, to better capture the anatomical structure information from different view perspectives, a multi-view contrastive learning strategy is developed to pre-train the backbone. Additionally, we further propose a Sequence Loss to maintain the sequential structure embedded along the vertebrae. Evaluation results demonstrate that, with only two 2D networks, our method can localize and identify vertebrae in CT images accurately, and outperforms the state-of-the-art methods consistently. Our code is available at https://github.com/ShanghaiTech-IMPACT/Multi-View-Vertebra-Localization-and-Identification-from-CT-Images.

Keywords: Vertebra localization and identification · Contrastive learning · Sequence Loss

1 Introduction

Automatic Localization and identification of vertebra from CT images are crucial in clinical practice, particularly for surgical planning, pathological diagnosis, and post-operative evaluation [1,9,10]. However, the process is challenging due to the significant shape variations of vertebrae with different categories, such as lumbar and thoracic, and also the close shape resemblance of neighboring vertebrae. Apart from these intrinsic challenges, the arbitrary field-of-view (FOV) of different CT scans and the presence of metal implant artifacts also introduce additional difficulties to this task.

Supplementary Information The online version contains supplementary material available at https://doi.org/10.1007/978-3-031-43904-9_14.

H. Greenspan et al. (Eds.): MICCAI 2023, LNCS 14224, pp. 136–145, 2023.
https://doi.org/10.1007/978-3-031-43904-9_14

With the advance of deep learning, many methods are devoted to tackling these challenges. For example, Lessmann et al. [11] employed a one-stage segmentation method to segment vertebrae with different labels for localization and identification. It is intuitive but usually involves many segmentation artifacts. Building upon this method, Masuzawa et al. [12] proposed an instance memory module to capture the neighboring information, but the long-term sequential information is not well studied. Recently, two or multi-stage methods [4,5,13–15], that first localize the vertebra and further classify the detected vertebra patches, are proposed to achieve the state-of-the-art performance. And some additional modules, such as attention mechanism [5], graph optimization [13], and LSTM [15], are integrated to capture the sequential information of adjacent vertebrae. However, all these methods are performed on 3D patches, where the global information of the CT scan is destroyed and cannot be well-captured. Moreover, due to the lack of pre-trained models in 3D medical imaging, networks trained from scratch using a small dataset often lead to severe overfitting problems with inferior performance.

In this paper, to tackle the aforementioned challenges, we present a novel framework that converts the 3D vertebra labeling problem into a multi-view 2D vertebra localization and identification task. Without the 3D patch limitation, our network can learn 2D global information naturally from different view perspectives, as well as leverage the pre-trained models from ImageNet [6]. Specifically, given a 3D CT image, we first generate multi-view 2D Digitally Reconstructed Radiograph (DRR) projection images. Then, a multi-view contrastive learning strategy is designed to further pre-train the network on this specific task. For vertebra localization, we predict the centroid of each vertebra in all DRR images and map the 2D detected centroids of different views back into the 3D CT scan using a least-squares algorithm. As for vertebra identification, we formulate it as a 2D segmentation task that generates vertebra labels around vertebra centroids. Particularly, a Sequence Loss, based on dynamic programming, is introduced to maintain the sequential information along the spine vertebrae in the training stage, which also serves as a weight to vote the multi-view 2D identification results into the 3D CT image for more reliable results. Our proposed method is validated on a public challenging dataset [17] and achieved the state-of-the-art performance both in vertebra localization and identification. Moreover, more evaluation results on a large-scale in-house dataset collected in real-world clinics (with 500 CT images) are provided in the supplementary materials, further demonstrating the effectiveness and robustness of our framework.

2 Methodology

An overview of our proposed method for vertebra localization and identification using multi-view DRR from CT scans is shown in Fig. 1, which mainly consists of three steps. Step 1 is to generate DRR images, followed by a multi-view contrastive learning strategy to pre-train the backbone. Step 2 aims to finish 2D single-view vertebra localization and identification, and step 3 is to map the

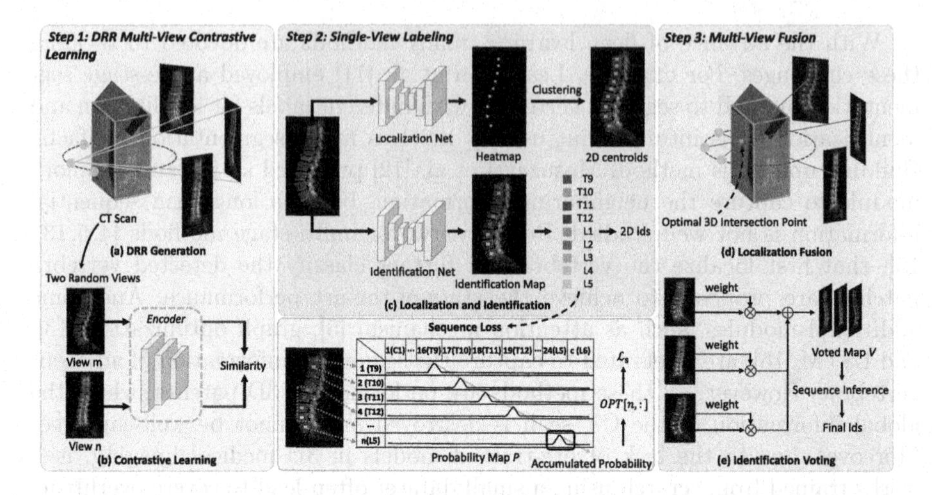

Fig. 1. An overview of the proposed method, including (a) DRR Generation, (b) DRR multi-view contrastive learning, (c) single-view vertebra localization and identification, (d) multi-view localization fusion, and (e) multi-view identification voting. The implementation of Sequence Loss is also illustrated. (Color figure online)

2D results back to 3D with a multi-view fusion strategy. We will elaborate our framework in this section.

2.1 DRR Multi-view Contrastive Learning

DRR Generation. To accurately localize and identify each vertebra in CT images, we convert the 3D task into 2D, where global information can be naturally captured from different views, avoiding the large computation of 3D models. To achieve this, DRR (Digitally Reconstructed Radiograph) technique, a simulation procedure for generating a radiograph similar to conventional X-ray image, is performed by projecting a CT image onto a virtual detector plane with a virtual X-ray source. In this way, we can generate K DRR projection images of a CT image for every $360/K$ degree. The 3D labeling problem can then be formulated as a multi-view localization and identification task in a 2D manner. Specifically, the 2D ground-truth are generated by projecting the 3D centroids and labels onto the 2D image following the DRR projection settings.

DRR Multi-view Contrastive Learning. After DRR generation, our goal is to localize and identify the vertebra on DRR images. However, as the dataset for the vertebra task is relatively small due to time-consuming manual annotation, we design a new multi-view contrastive learning strategy to better learn the vertebrae representation from various views. Unlike previous contrastive learning methods, where the pretext is learned from numerous augmented negative and positive samples [2,3,7,8], e.g., random crop, image flip, rotation and resize, the multi-view DRR images generated from the same CT image share consistent anatomical information, which are natural positive samples. Based on this insight, we pre-train our network backbone using the Simsiam [3] approach to encode two random views from the same CT image as a key and query, as shown in Fig. 1 (b), in the aims of learning the invariant vertebrae representation from different views.

2.2 Single-view Vertebra Localization

With multi-view DRR images, the 3D vertebra localization problem is converted into a 2D vertebra centroid detection task, followed by a multi-view fusion strategy (as introduced in Sect. 2.4) that transforms the 2D results to 3D. To achieve this, we utilize the commonly-used heatmap regression strategy for 2D vertebra centroid detection. Specifically, for each vertebra in a DRR image, our model is trained to learn the contextual heatmap defined on the ground-truth 2D centroid using a Gaussian kernel. During inference, we apply a fast peak search clustering method [16] to localize the density peaks on the regressed heatmap as the predicted centroid. Benefiting from the pre-trained models from multi-view contrastive learning, our method can capture more representative features from different views. Further, compared to existing 3D methods, our approach performs vertebra localization on several DRR images with a fusion strategy, making it more robust to the situation of missing detection in certain views.

2.3 Single-View Vertebra Identification

After the vertebrae localization, we further predict the label of each vertebra using an identification network on multi-view DRR images. Unlike other 3D methods that require cropping vertebra patches for classification, our identification network performs on 2D, allowing us to feed the entire DRR image into the network, which can naturally capture the global information. Specifically, we use a segmentation model to predict the vertebra labels around the detected vertebra centroids, i.e., a 22 mm × 22 mm square centered at the centroid. During the inference of single-view, we analyze the pixel-wise labels in each square and identify the corresponding vertebra with the majority number of labels.

Sequence Loss. In the identification task, we observe that the vertebra labels are always in a monotonically increasing order along the spine, which implies the presence of sequential information. To better exploit this property and enhance our model to capture such sequential information, we propose a Sequence Loss as an additional network supervision, ensuring the probability distribution along the spine follows a good sequential order. Specifically, as shown in Fig. 1, we compute a probability map $P \in \mathbb{R}^{n \times c}$ for each DRR image by averaging the predicted pixel-wise possibilities in each square around the vertebra centroid from the identification network. Here, n is the number of vertebrae contained in this DRR image, and c indicates the number of vertebra categories (i.e., from C1 to L6). Due to the sequential nature of the vertebra identification problem, the optimal distribution of P is that the index of the largest probability in each row is in ascending order (green line in Fig. 1). To formalize this notion, we compute the largest accumulated probability in ascending order, starting from each category in the first row and ending at the last row, using dynamic programming. The higher accumulated probability, the better sequential structure presented by this distribution. We set this accumulated probability as target profit, and aim to maximize it to enable our model to better capture the sequential structure in this DRR image. The optimal solution (OPT) based on the dynamic programming algorithm is as:

$$OPT[i,j] = \begin{cases} P[i,j] & if\, j = 1\, or\, i = 1 \\ OPT[i-1, j-1] + D & otherwise \end{cases} \tag{1}$$
$$D = \max(\alpha P[i,j-1], \beta P[i,j], \alpha P[i,j+1]),$$

where $i \in [1,n]$ and $j \in [1,c]$. Here, α and β are two parameters that are designed to alleviate the influence of wrong-identified vertebra. Sequence Loss ($\mathcal{L}_s$) is then defined as:

$$\mathcal{L}_s = 1 - \frac{max(OPT[n,:])}{\beta n}. \tag{2}$$

The overall loss function $\mathcal{L}_{id}$ for our identification network is:

$$\mathcal{L}_{id} = \mathcal{L}_{ce} + \gamma \mathcal{L}_s, \tag{3}$$

where $\mathcal{L}_{ce}$ and $\mathcal{L}_s$ refer to the Cross-Entropy loss and Sequence Loss, respectively. γ is a parameter to control the relative weights of the two losses.

2.4 Multi-view Fusion

Localization Multi-view Fusion. After locating all the vertebrae in each DRR image, we fuse and map the 2D centroids back to 3D space by a least-squares algorithm, as illustrated in Fig. 1 (d). For a vertebra located in K views, we can track K target lines from the source points in DRR technique to the detected centroid on the DRR images. Ideally, the K lines should intersect at a unique point in the 3D space, but due to localization errors, this is always unachievable in practice. Hence, instead of finding a unique intersection point, we employ the least-squares algorithm to minimize the sum of perpendicular distances from the optimal intersection point to all the K lines, given by:

$$D(\boldsymbol{p}; \boldsymbol{A}, \boldsymbol{N}) = \sum_{k=1}^{K} D(\boldsymbol{p}; \boldsymbol{a}_k, \boldsymbol{n}_k) = \sum_{k=1}^{K} (\boldsymbol{a}_k - \boldsymbol{p})^T (\boldsymbol{I} - \boldsymbol{n}_k \boldsymbol{n}_k^T)(\boldsymbol{a}_k - \boldsymbol{p}), \tag{4}$$

where $\boldsymbol{p}$ denotes the 3D coordinate of the optimal intersection point, $\boldsymbol{a}_k$ and $\boldsymbol{n}_k$ represent the point on the k_{th} target line and the corresponding direction vector. By taking derivatives with respect to $\boldsymbol{p}$, we get a linear equation of $\boldsymbol{p}$ as shown in Eq. (5), where the optimal intersection point can be obtained by achieving the minimum distance to the K lines.

$$\begin{aligned} \frac{\partial D}{\partial \boldsymbol{p}} &= \sum_{k=1}^{K} -2(\boldsymbol{I} - \boldsymbol{n}_k \boldsymbol{n}_k^T)(\boldsymbol{a}_k - \boldsymbol{p}) = 0 \Rightarrow \boldsymbol{S}\boldsymbol{p} = \boldsymbol{q}, \\ \boldsymbol{S} &= \sum_{k=1}^{K} (\boldsymbol{I} - \boldsymbol{n}_k \boldsymbol{n}_k^T),\ \boldsymbol{q} = \sum_{k=1}^{K} (\boldsymbol{I} - \boldsymbol{n}_k \boldsymbol{n}_k^T)\boldsymbol{a}_k. \end{aligned} \tag{5}$$

Identification Multi-view Voting. The Sequence Loss evaluates the quality of the predicted vertebra labels in terms of their sequential property. During inference, we further use this Sequence Loss of each view as weights to fuse the probability maps obtained from different views. We obtain the final voted identification map V of K views as:

$$V = \sum_{k=1}^{K} W_k P_k,\ W_k = \frac{(1 - \mathcal{L}_s^k)}{\sum_{a=1}^{K} (1 - \mathcal{L}_s^a)}. \tag{6}$$

For each vertebra, the naive solution for obtaining vertebra labels is to extract the largest probability from each row in voted identification map V. Despite the promising performance of the identification network, we still find some erroneous predictions. To address this issue, we leverage the dynamic programming (described in Eq. (1)) again to correct the predicted vertebra labels in this voted identification map V. Specifically, we identify the index of the largest accumulated probability in the last row as the last vertebra category and utilize it as a reference to correct any inconsistencies in the prediction.

3 Experiments and Results

3.1 Dataset and Evaluation Metric

We extensively evaluate our method on the publicly available MICCAI VerSe19 Challenge dataset [17], which consists of 160 spinal CT with ground truth annotations. Specifically, following the public challenge settings, we utilize 80 scans for training, 40 scans for testing, and 40 scans as hidden data. To evaluate the performance of our method, we use the mean localization error (L-Error) and identification rate (Id-Rate) as the evaluation metrics, which are also adopted in the challenge. The L-Error is calculated as the average Euclidean distance between the ground-truth and predicted vertebral centers. The Id-Rate is defined as the ratio of correctly identified vertebrae to the total number of vertebrae.

3.2 Implementation Details

All CT scans are resampled to an isotropic resolution of 1 mm. For DRR Multi-View Contrastive Learning, we use ResNet50 as encoder and apply the SGD

Table 1. Results on the VerSe19 challenge dataset.

Method	Test Dataset		Hidden Dataset	
	Id-Rate(%)	L-Error(mm)	Id-Rate(%)	L-Error(mm)
Payer C. [17]	95.65	4.27	94.25	4.80
Lessmann N. [17]	89.86	14.12	90.42	7.04
Chen M. [17]	96.94	4.43	86.73	7.13
Sekuboyina A. [18]	89.97	5.17	87.66	6.56
Ours	**98.12**	**1.79**	**96.45**	**2.17**

optimizer with an initial learning rate of 0.0125, which follows the cosine decay schedule. The weight decay, SGD momentum, batch size and loss function are set to 0.0001, 0.9, 64, and cosine similarity respectively. We employ U-Net for both the localization and identification networks, using the pre-trained ResNet50 from our contrastive learning as backbone. Adam optimizer is set with an initial learning rate of 0.001, which is divided by 10 every 4000 iterations. Both networks are trained for 15k iterations. We empirically set $\alpha = 0.1$, $\beta = 0.8$, $\gamma = 1$. All methods were implemented in Python using PyTorch framework and trained on an Nvidia Tesla A100 GPU with 40 GB memory.

3.3 Comparison with SOTA Methods

We train our method on 70 CT images and tune the hyperparameter on the rest 10 CT images from the training data. We then evaluate it on both testing and hidden datasets, following the same setting as the challenge. In the comparison, our method is compared with four methods which are the first four positions in the benchmark of this challenge [17]. The experimental results are presented in Table 1. Our method achieves Id-Rate of 98.12% and L-Error of 1.79 mm on the test dataset, and Id-Rate of 96.45% and L-Error of 2.17 mm on the hidden dataset, which achieves the leading performance both in localization and identification tasks with just two 2D networks. Compared to these methods performed on 3D with random cropping or patch-wise method (Payer C. [17], Lessmann N. [17] and Chen M. [17]), our 2D strategy can capture more reliable global and sequential information in all 2D projection images which can improve the labeling performance, especially the localization error. Compared to those using 2D MIP (Sekuboyina A. [18]), our DRR multi-view projection and fusion strategy can provide superior performance by analyzing more views and introducing the geometry information carried by varied DRR projections naturally.

3.4 Ablation Study

Ablation Study of Key Components. We conduct an ablation study on the VerSe19 dataset to demonstrate the effectiveness of each component. As presented in Table 2, we build the basic network for the vertebra localization and

Table 2. Ablation study results of key components.

Baseline	Pre-train	Sequence Loss	Voting	Id-Rate(%)	
				Test Dataset	Hidden dataset
✓				84.00	83.45
✓	✓			85.58	86.52
✓	✓	✓		89.41	90.54
✓	✓	✓	✓	**98.12**	**96.45**

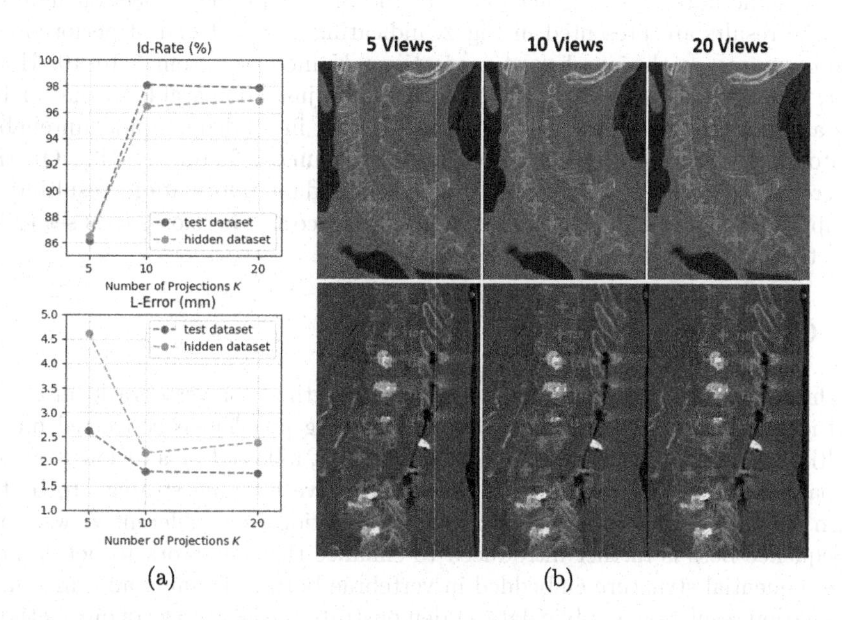

Fig. 2. (a) The Id-Rate and L-Error of different K. (b) Comparison between different K from the final predicted CT scan on limited FOV and metal artifacts cases (red for ground truth and green for predictions). (Color figure online)

identification with a bagging strategy, where for each vertebra, we opt for the ID that is predicted by the majority of views, when not using weighted voting, and $K = 10$, denoted as Baseline. Pre-train, Sequence Loss, and voting in Table 2 represent the addition of the multi-view contrastive learning, Sequence Loss, and multi-view voting one by one. Pre-trained from ImageNet is used when not utilizing our contrastive learning pre-trained parameters. Specifically, the Baseline achieves Id-Rate of 84.00% and 83.54% on two datasets. With the contrastive learning pre-trained parameters, we achieve 1.88% and 2.98% improvements over the ImageNet pre-trained, respectively. This shows the pre-trained parameters of the backbone obtained from our contrastive learning can effectively facilitate the network to learn more discriminative features for identification than the model learning from scratch. Sequence Loss provides extra supervision for sequential information, and results in 3.53% and 4.02% increase, illustrating the significance of capturing the sequential information in the identification task. Finally, multi-view weighted voting yields the best results with 98.12% and 96.45% on the two datasets, indicating the robustness of our multi-view voting when the identification errors occurred in a small number of DRR images can be corrected by other DRR prediction results.

Ablation Study of Projection Number. We also conduct an ablation study on the same dataset to further evaluate the impact of the projection number K. The results are presented in Fig. 2, indicating a clear trend of performance improvements as the number of projections K increases from 5 to 10. However, when K increases to 20, the performance is just comparable to that of 10. We analyze that using too few views may result in inadequate and unreliable anatomical structure representation, leading to unsatisfactory results. On the other hand, too many views may provide redundant information, resulting in comparable results but with higher computation cost. Therefore, K is set to 10 as a trade-off between accuracy and efficiency.

4 Conclusion

In this paper, we propose a novel multi-view method for vertebra localization and identification in CT images. The 3D labeling problem is converted into a multi-view 2D localization and identification task, followed by a fusion strategy. In particular, we propose a multi-view contrastive learning strategy to better learn the invariant anatomical structure information from different views. And a Sequence Loss is further introduced to enhance the framework to better capture sequential structure embedded in vertebrae both in training and inference. Evaluation results on a public dataset demonstrate the advantage of our method.

References

1. Burns, J.E., Yao, J., Muñoz, H., Summers, R.M.: Automated detection, localization, and classification of traumatic vertebral body fractures in the thoracic and lumbar spine at CT. Radiology **278**(1), 64–73 (2016)
2. Chen, T., Kornblith, S., Norouzi, M., Hinton, G.: A simple framework for contrastive learning of visual representations. In: International Conference on Machine Learning, pp. 1597–1607. PMLR (2020)
3. Chen, X., He, K.: Exploring simple Siamese representation learning. In: Proceedings of the IEEE/CVF Conference on Computer Vision and Pattern Recognition, pp. 15750–15758 (2021)
4. Cheng, P., Yang, Y., Yu, H., He, Y.: Automatic vertebrae localization and segmentation in CT with a two-stage Dense-U-Net. Sci. Rep. **11**(1), 1–13 (2021)
5. Cui, Z., et al.: VertNet: accurate vertebra localization and identification network from CT images. In: de Bruijne, M., et al. (eds.) MICCAI 2021. LNCS, vol. 12905, pp. 281–290. Springer, Cham (2021). https://doi.org/10.1007/978-3-030-87240-3_27

6. Deng, J., Dong, W., Socher, R., Li, L.J., Li, K., Fei-Fei, L.: ImageNet: a large-scale hierarchical image database. In: 2009 IEEE Conference on Computer Vision and Pattern Recognition, pp. 248–255. IEEE (2009)
7. Grill, J.B., et al.: Bootstrap your own latent-a new approach to self-supervised learning. Adv. Neural. Inf. Process. Syst. **33**, 21271–21284 (2020)
8. He, K., Fan, H., Wu, Y., Xie, S., Girshick, R.: Momentum contrast for unsupervised visual representation learning. In: Proceedings of the IEEE/CVF Conference on Computer Vision and Pattern Recognition, pp. 9729–9738 (2020)
9. Knez, D., Likar, B., Pernuš, F., Vrtovec, T.: Computer-assisted screw size and insertion trajectory planning for pedicle screw placement surgery. IEEE Trans. Med. Imaging **35**(6), 1420–1430 (2016)
10. Kumar, R.: Robotic assistance and intervention in spine surgery. In: Li, S., Yao, J. (eds.) Spinal Imaging and Image Analysis. LNCVB, vol. 18, pp. 495–506. Springer, Cham (2015). https://doi.org/10.1007/978-3-319-12508-4_16
11. Lessmann, N., Van Ginneken, B., De Jong, P.A., Išgum, I.: Iterative fully convolutional neural networks for automatic vertebra segmentation and identification. Med. Image Anal. **53**, 142–155 (2019)
12. Masuzawa, N., Kitamura, Y., Nakamura, K., Iizuka, S., Simo-Serra, E.: Automatic segmentation, localization, and identification of vertebrae in 3D CT images using cascaded convolutional neural networks. In: Martel, A.L., et al. (eds.) MICCAI 2020. LNCS, vol. 12266, pp. 681–690. Springer, Cham (2020). https://doi.org/10.1007/978-3-030-59725-2_66
13. Meng, D., Mohammed, E., Boyer, E., Pujades, S.: Vertebrae localization, segmentation and identification using a graph optimization and an anatomic consistency cycle. In: Lian, C., Cao, X., Rekik, I., Xu, X., Cui, Z. (eds.) Machine Learning in Medical Imaging, pp. 307–317. Springer Nature Switzerland, Cham (2022). https://doi.org/10.1007/978-3-031-21014-3_32
14. Payer, C., Stern, D., Bischof, H., Urschler, M.: Coarse to fine vertebrae localization and segmentation with SpatialConfiguration-Net and U-Net. In: VISIGRAPP (5: VISAPP), pp. 124–133 (2020)
15. Qin, C., et al.: Vertebrae labeling via end-to-end integral regression localization and multi-label classification network. IEEE Trans. Neural Netw. Learn. Syst. **33**(6), 2726–2736 (2021)
16. Rodriguez, A., Laio, A.: Clustering by fast search and find of density peaks. Science **344**(6191), 1492–1496 (2014)
17. Sekuboyina, A., et al.: VerSe: a vertebrae labelling and segmentation benchmark for multi-detector CT images. Med. Image Anal. **73**, 102166 (2021)
18. Sekuboyina, A., et al.: Btrfly net: vertebrae labelling with energy-based adversarial learning of local spine prior. In: Frangi, A.F., Schnabel, J.A., Davatzikos, C., Alberola-López, C., Fichtinger, G. (eds.) MICCAI 2018. LNCS, vol. 11073, pp. 649–657. Springer, Cham (2018). https://doi.org/10.1007/978-3-030-00937-3_74

Cluster-Induced Mask Transformers for Effective Opportunistic Gastric Cancer Screening on Non-contrast CT Scans

Mingze Yuan[1,2,3], Yingda Xia[1(✉)], Xin Chen[4(✉)], Jiawen Yao[1,3], Junli Wang[5], Mingyan Qiu[1,3], Hexin Dong[1,2,3], Jingren Zhou[1], Bin Dong[2,6], Le Lu[1], Li Zhang[2], Zaiyi Liu[4(✉)], and Ling Zhang[1]

[1] DAMO Academy, Alibaba Group, Hangzhou, China
yingda.xia@alibaba-inc.com
[2] Peking University, Beijing, China
[3] Hupan Lab, 310023 Hangzhou, China
[4] Guangdong Province People's Hospital, Guangzhou, China
wolfchenxin@163.com, zyliu@163.com
[5] The First Affiliated Hospital of Zhejiang University, Hangzhou, China
[6] Peking University Changsha Institute for Computing and Digital Economy, Changsha, China

Abstract. Gastric cancer is the third leading cause of cancer-related mortality worldwide, but no guideline-recommended screening test exists. Existing methods can be invasive, expensive, and lack sensitivity to identify early-stage gastric cancer. In this study, we explore the feasibility of using a deep learning approach on non-contrast CT scans for gastric cancer detection. We propose a novel cluster-induced Mask Transformer that jointly segments the tumor and classifies abnormality in a multi-task manner. Our model incorporates learnable clusters that encode the texture and shape prototypes of gastric cancer, utilizing self- and cross-attention to interact with convolutional features. In our experiments, the proposed method achieves a sensitivity of 85.0% and specificity of 92.6% for detecting gastric tumors on a hold-out test set consisting of 100 patients with cancer and 148 normal. In comparison, two radiologists have an average sensitivity of 73.5% and specificity of 84.3%. We also obtain a specificity of 97.7% on an external test set with 903 normal cases. Our approach performs comparably to established state-of-the-art gastric cancer screening tools like blood testing and endoscopy, while also being more sensitive in detecting early-stage cancer. This demonstrates the potential of our approach as a novel, non-invasive, low-cost, and accurate method for opportunistic gastric cancer screening.

Keywords: Gastric cancer · Large-scale cancer screening · Mask Transformers · Non-contrast CT

M.Yuan—Work was done during an internship at DAMO Academy, Alibaba Group.

H. Greenspan et al. (Eds.): MICCAI 2023, LNCS 14224, pp. 146–156, 2023.
https://doi.org/10.1007/978-3-031-43904-9_15

1 Introduction

Gastric cancer (GC) is the third leading cause of cancer-related deaths worldwide [19]. The five-year survival rate for GC is approximately 33% [16], which is mainly attributed to patients being diagnosed with advanced-stage disease harboring unresectable tumors. This is often due to the latent and nonspecific signs and symptoms of early-stage GC. However, patients with early-stage disease have a substantially higher five-year survival rate of around 72% [16]. Therefore, early detection of resectable/curable gastric cancers, preferably before the onset of symptoms, presents a promising strategy to reduce associated mortality. Unfortunately, current guidelines do not recommend any screening tests for GC [22]. While several screening tools have been developed, such as Barium-meal gastric photofluorography [5], upper endoscopy [4,7,9], and serum pepsinogen levels [15], they are challenging to apply to the general population due to their invasiveness, moderate sensitivity/specificity, high cost, or side effects. Therefore, there is an urgent need for novel screening methods that are noninvasive, highly accurate, low-cost, and ready to distribute.

Non-contrast CT is a commonly used imaging protocol for various clinical purposes. It is a non-invasive, relatively low-cost, and safe procedure that exposes patients to less radiation dose and does not require the use of contrast injection that may cause serious side effects (compared to multi-phase contrast-enhanced CT). With recent advances in AI, opportunistic screening of diseases using non-contrast CT during routine clinical care performed for other clinical indications, such as lung and colorectal cancer screening, presents an attractive approach to early detect treatable and preventable diseases [17]. However, whether early detection of gastric cancer using non-contrast CT scans is possible remains unknown. This is because early-stage gastric tumors may only invade the mucosal and muscularis layers, which are difficult to identify without the help of stomach preparation and contrast injection. Additionally, the poor contrast between the tumor and normal stomach wall/tissues on non-contrast CT scans and various shape alterations of gastric cancer, further exacerbates this challenge.

In this paper, we propose a novel approach for detecting gastric cancer on non-contrast CT scans. Unlike the conventional "segmentation for classification" methods that directly employ segmentation networks, we developed a cluster-induced Mask Transformer that performs segmentation and global classification simultaneously. Given the high variability in shape and texture of gastric cancer, we encode these features into learnable clusters and utilize cluster analysis during inference. By incorporating self-attention layers for global context modeling, our model can leverage both local and global cues for accurate detection. In our experiments, the proposed approach outperforms nnUNet [8] by 0.032 in AUC, 5.0% in sensitivity, and 4.1% in specificity. These results demonstrate the potential of our approach for opportunistic screening of gastric cancer in asymptomatic patients using non-contrast CT scans.

2 Related Work

Automated Cancer Detection. Researchers have explored automated tumor detection techniques on endoscopic [13,14], pathological images [20], and the prediction of cancer prognosis [12]. Recent developments in deep learning have significantly improved the segmentation of gastric tumors [11], which is critical for their detection. However, our framework is specifically designed for non-contrast CT scans, which is beneficial for asymptomatic patients. While previous studies have successfully detected pancreatic [25] and esophageal [26] cancers on non-contrast CT, identifying gastric cancer presents a unique challenge due to its subtle texture changes, various shape alterations, and complex background, *e.g.*, irregular gastric wall; liquid and contents in the stomach.

Mask Transformers. Recent studies have used Transformers for natural and medical image segmentation [21]. Mask Transformers [3,24,29] further enhance CNN-based backbones by incorporating stand-alone Transformer blocks, treating object queries in DETR [1] as memory-encoded queries for segmentation. CMT-Deeplab [27] and KMaX-Deeplab [28] have recently proposed interpreting the queries as clustering centers and adding regulatory constraints for learning the cluster representations of the queries. Mask Transformers are locally sensitive to image textures for precise segmentation and globally aware of organ-tumor morphology for recognition. Their cluster representations demonstrate a remarkable balance of intra-cluster similarity and inter-class discrepancy. Therefore, Mask Transformers are an ideal choice for an end-to-end joint segmentation and classification system for detecting gastric cancer.

3 Methods

Problem Formulation. Given a non-contrast CT scan, cancer screening is a binary classification with two classes as $\mathcal{L} = \{0, 1\}$, where 0 stands for"normal" and 1 for"GC" (gastric cancer). The entire dataset is denoted by $\mathcal{S} = \{(\mathbf{X}_i, \mathbf{Y}_i, \mathbf{P}_i)|i = 1, 2, \cdots, N\}$, where $\mathbf{X}_i$ is the i-th non-contrast CT volume, with $\mathbf{Y}_i$ being the voxel-wise label map of the same size as $\mathbf{X}_i$ and K channels. Here, $K = 3$ represents the background, stomach, and GC tumor. $\mathbf{P}_i \in \mathcal{L}$ is the class label of the image, confirmed by pathology, radiology, or clinical records. In the testing phase, only $\mathbf{X}_i$ is given, and our goal is to predict a class label for $\mathbf{X}_i$.

Knowledge Transfer from Contrast-Enhanced to Non-contrast CT. To address difficulties with tumor annotation on non-contrast CTs, the radiologists start by annotating a voxel-wise tumor mask on the contrast-enhanced CT, referring to clinical and endoscopy reports as needed. DEEDs [6] registration is then performed to align the contrast-enhanced CT with the non-contrast CT and the resulting deformation field is applied to the annotated mask. Any misaligned ones are revised manually. In this manner (Fig. 1d), a relatively coarse yet highly reliable tumor mask can be obtained for the non-contrast CT image.

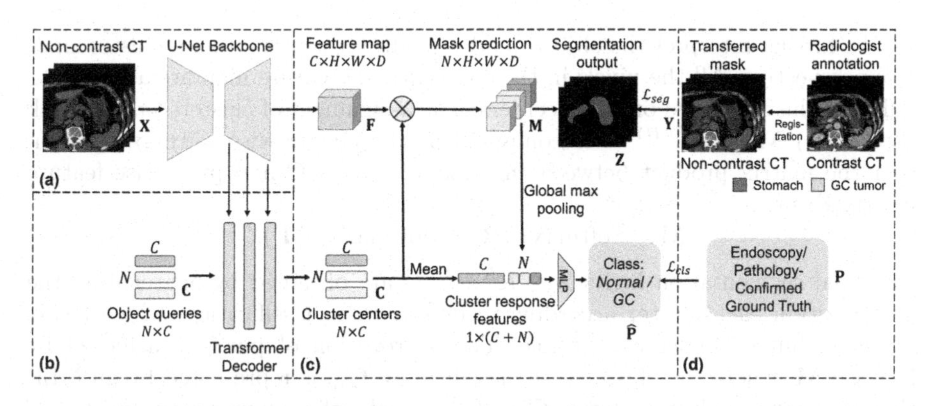

Fig. 1. Method overview. (a) The non-contrast CT image is first forwarded into a U-Net [8,18] to extract a feature map. (b) Learnable object queries interact with the multi-level U-Net features through a Transformer Decoder and produce learned cluster centers. (c) All the pixels are assigned to cluster centers by matrix multiplication. The cluster assignment (*a.k.a.* mask prediction) is further used to generate the final segmentation output and the classification probability. (d) The entire network is supervised by transferred masks from radiologists' annotation on contrast-enhanced CT and endoscopy or pathology-confirmed ground truth.

Cluster-Induced Classification with Mask Transformers. Segmentation for classification is widely used in tumor detection [25,26,32]. We first train a UNet [8,18] to segment the stomach and tumor regions using the masks from the previous step. This UNet considers local information and can only extract stomach ROIs well during testing. However, local textures are inadequate for accurate gastric tumor detection on non-contrast CTs, so we need a network of both local sensitivity to textures and global awareness of the organ-tumor morphology. Mask transformer [3,24] is a well-suited approach to boost the CNN backbone with stand-alone transformer blocks. Recent studies [27,28] suggest interpreting object queries as cluster centers, which naturally exhibit intra-cluster similarity and inter-class discrepancy. Inspired by this, we further develop a deep classification model on top of learnable cluster representations.

Specifically, given image $\mathbf{X} \in \mathbb{R}^{H\times W\times D}$, annotation $\mathbf{Y} \in \mathbb{R}^{K\times HWD}$, and patient class $\mathbf{P} \in \mathcal{L}$, our model consists of three components: 1) a CNN backbone to extract its pixel-wise features $\mathbf{F} \in \mathbb{R}^{C\times HWD}$ (Fig. 1a), 2) a transformer module (Fig. 1b), and 3) a multi-task cluster inference module (Fig. 1c). The transformer module gradually updates a set of randomly initialized object queries $\mathbf{C} \in \mathbb{R}^{N\times C}$, *i.e.*, to meaningful mask embedding vectors through cross-attention between object queries and multi-scale pixel features,

$$\mathbf{C} \leftarrow \mathbf{C} + \arg\max_{N}(\mathbf{Q}^c(\mathbf{K}^p)^{\mathrm{T}})\mathbf{V}^p, \tag{1}$$

where c and p stand for query and pixel features, $\mathbf{Q}^c, \mathbf{K}^p, \mathbf{V}^p$ represent linearly projected query, key, and value. We adopt cluster-wise argmax from KMax-DeepLab [28] to substitute spatial-wise softmax in the original settings.

We further interpret the object queries as cluster centers from a cluster analysis perspective. All the pixels in the convolutional feature map are assigned to different clusters based on these centers. The assignment of clusters (*a.k.a.* mask prediction) $\mathbf{M} \in \mathbb{R}^{N \times HWD}$ is computed as the cluster-wise softmax function over the matrix product between the cluster centers $\mathbf{C}$ and pixel-wise feature matrix $\mathbf{F}$, *i.e.*,

$$\mathbf{M} = \mathrm{Softmax}_N(\mathbf{R}) = \mathrm{Softmax}_N(\mathbf{CF}). \tag{2}$$

The final segmentation logits $\mathbf{Z} \in \mathbb{R}^{K \times HWD}$ are obtained by aggregating the pixels within each cluster according to cluster-wise classification, which treats pixels within a cluster as a whole. The aggregation of pixels is achieved by $\mathbf{Z} = \mathbf{C}_K\mathbf{M}$, where the cluster-wise classification $\mathbf{C}_K$ is represented by an MLP that projects the cluster centers $\mathbf{C}$ to K channels (the number of segmentation classes).

The learned cluster centers possess high-level semantics with both inter-cluster discrepancy and intra-cluster similarity for effective classification. Rather than directly classifying the final feature map, we first generate the cluster-path feature vector by taking the channel-wise average of cluster centers $\bar{\mathbf{C}} = \frac{1}{N}\sum_{i=1} \mathbf{C}_i \in \mathbb{R}^C$. Additionally, to enhance the consistency between the segmentation and classification outputs, we apply global max pooling to cluster assignments $\mathbf{R}$ to obtain the pixel-path feature vector $\bar{\mathbf{R}} \in \mathbb{R}^N$. This establishes a direct connection between classification features and segmentation predictions. Finally, we concatenate these two feature vectors to obtain the final feature and project it onto the classification prediction $\hat{\mathbf{P}} \in \mathbb{R}^2$ via a two-layer MLP.

The overall training objective is formulated as,

$$\mathcal{L} = \mathcal{L}_{seg}(\mathbf{Z}, \mathbf{Y}) + \mathcal{L}_{cls}(\hat{\mathbf{P}}, \mathbf{P}), \tag{3}$$

where the segmentation loss $\mathcal{L}_{seg}(\cdot, \cdot)$ is a combination of Dice and cross entropy losses, and the classification loss $\mathcal{L}_{cls}(\cdot, \cdot)$ is cross entropy loss.

4 Experiments

4.1 Experimental Setup

Dataset and Ground Truth. Our study analyzed a dataset of CT scans collected from Guangdong Province People's Hospital between years 2018 and 2020, with 2,139 patients consisting of 787 gastric cancer and 1,352 normal cases. We used the latest patients in the second half of 2020 as a hold-out test set, resulting in a training set of 687 gastric cancer and 1,204 normal cases, and a test set of 100 gastric cancer and 148 normal cases. We randomly selected 20% of the training data as an internal validation set. To further evaluate specificity in a larger population, we collected an external test set of 903 normal cases from Shengjing Hospital. Cancer cases were confirmed through endoscopy (and pathology) reports, while normal cases were confirmed by radiology reports and a two-year follow-up. All patients underwent multi-phase CTs with a median spacing of $0.75 \times 0.75 \times 5.0$ mm and an average size of (512, 512, 108) voxel. Tumors

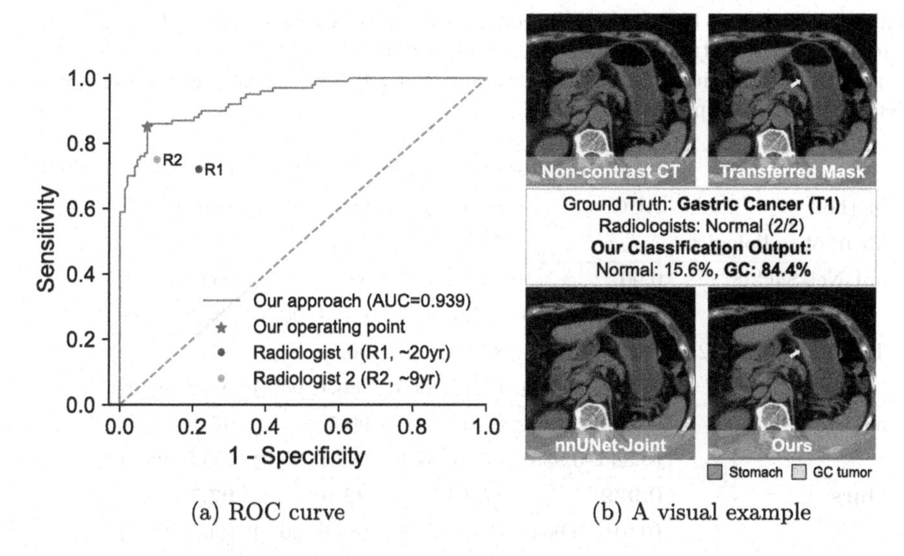

(a) ROC curve (b) A visual example

Fig. 2. (a) ROC curve for our model versus two experts on the hold-out test set of $n = 248$ patients for binary classification. (b) A visual example in the test set. This early-stage GC case is miss-detected by both radiologists and nnUNet [8] but our model succeeds to locate the tumor.

were annotated on the venous phase by an experienced radiologist specializing in gastric imaging using CTLabeler [23], while the stomach was automatically annotated using a self-learning model [31].

Implementation Details. We resampled each CT volume to the median spacing while normalizing it to have zero mean and unit variance. During training, we cropped the 3D bounding box of the stomach and added a small margin of (32, 32, 4). We used nnUNet [8] as the backbone, with four transformer decoders, each taking pixel features with output strides of 32, 16, 8, and 4. We set the number of object queries N to 8, with each having a dimension of 128, and included an eight-head self-attention layer in each block. The patch size used during training and inference is (192, 224, 40) voxel. We followed [8] to augment data. We trained the model with RAdam using a learning rate of 10^{-4} and a (backbone) learning rate multiplier of 0.1 for 1000 epochs, with a frozen backbone of the pre-trained nnUNet [8] for the first 50 epochs. To enhance performance, we added deep supervision by aligning the cross-attention map with the final segmentation map, as per KMax-Deeplab [27]. The hidden layer dimension in the two-layer MLP is 128. We also trained a standard UNet [8,18] to localize the stomach region in the entire image in the testing phase.

Evaluation Metrics and Reader Study. For the binary classification, model performance is evaluated using area under ROC curve (AUC), sensitivity (Sens.), and specificity (Spec.). And successful localization of the tumors is considered when the overlap between the segmentation mask generated by the model and

Table 1. Results on binary classification: gastric cancer vs. normal. The 95% confidence interval of each metric is listed. †: $p < 0.05$ for DeLong test (ours vs. nnUNet-S4C). *: $p < 0.05$ for permutation test (ours vs. nnUNet-S4C and radiologist experts). Sens.: Sensitivity. Spec.: Specificity.

	Internal Hold-out ($n = 248$)			External ($n = 903$)
Method	AUC	Sens.(%)	Spec.(%)	Spec.(%)
Mean of radiologists	-	73.5	84.1	-
nnUNet-S4C [8]	0.907 (0.862, 0.942)	80.0 (72.0, 87.5)	88.5 (83.3, 93.5)	96.6 (95.2, 97.8)
TransUNet-S4C [2]	0.916 (0.876, 0.952)	82.0 (74.7, 89.5)	90.5 (86.1, 94.8)	96.0 (94.8, 97.2)
nnUNet-Joint [8]	0.924 (0.885, 0.959)	81.0 (73.0, 87.9)	90.5 (85.1, 95.0)	97.6 (96.5, 98.6)
Ours	**0.939**† (0.910, 0.964)	**85.0*** (78.1, 91.1)	**92.6*** (88.0, 96.5)	**97.7** (96.7, 98.7)

Table 2. Patient-level detection and tumor-level localization results (%) over gastric cancer across different T-stages. Tumor-level localization evaluates how segmented masks overlap with the ground-truth cancer (Dice > 0.01 for correct detection). Miss-T: Missing of T stage information.

Method	Criteria	T1	T2	T3	T4	Miss-T
nnUNet-Joint [8]	Patient	30.0(3/10)	66.7(6/9)	94.1(32/34)	100.0(9/9)	86.1(31/36)
	Tumor	20.0(2/10)	55.6(5/9)	94.1(32/34)	100.0(9/9)	80.6(29/36)
Ours	Patient	60.0(6/10)	77.8(7/9)	94.1(32/34)	100.0(9/9)	86.1(31/36)
	Tumor	30.0(3/10)	66.7(6/9)	94.1(32/34)	100.0(9/9)	80.6(30/36)
Radiologist 1	Patient	50.0(5/10)	55.6(5/9)	76.5(26/34)	88.9(8/9)	77.8(28/36)
Radiologist 2	Patient	30.0(3/10)	55.6(5/9)	85.3(29/34)	100.0(9/9)	80.6(29/36)

the ground truth is greater than 0.01, measured by the Dice score. A reader study was conducted with two experienced radiologists, one from Guangdong Province People's Hospital with 20 years of experience and the other from The First Affiliated Hospital of Zhejiang University with 9 years of experience in gastric imaging. The readers were given 248 non-contrast CT scans from the test set and asked to provide a binary decision for each scan, indicating whether the scan showed gastric cancer. No patient information or records were provided to the readers. Readers were informed that the dataset might contain more tumor cases than the standard prevalence observed in screening, but the proportion of case types was not disclosed. Readers used ITK-SNAP [30] to interpret the CT scans without any time constraints.

Table 3. Comparison with a state-of-the-art blood test on gastric cancer detection [10], UGIS and endoscopy screening performance in large population [4], and early stage gastric cancer detection rate of senior radiologists on narrow-band imaging with magnifying endoscopy (ME-NBI) [7]. (∗: We leave out two tumors *in situ* within the test set in accordance with the setting in [10]. †: We also merely consider early-stage gastric cancer cases, including Tumor *in situ*, T1, and T2 stages, among whom we successfully detect 17 of 19 cases.)

Method	Spec.(%)	Sens.(%)	Our sensitivity(%) at the same specificity
Blood Test [10]	99.5	66.7	69.4*
Upper-gastrointestinal series [4]	96.1	36.7	85.0
Endoscopy screening [4]	96.0	69.0	85.0
ME-NBI (early-stage) [7]	74.2	76.7	89.5†

Compared Baselines. Table 1 presents a comparative analysis of our proposed method with three baselines. The first two approaches belong to "Segmentation for classification" (S4C) [26,32], using nnUNet [8] and TransUNet [2]. A case is classified as positive if the segmented tumor volume exceeds a threshold that maximizes the sum of sensitivity and specificity on the validation set. The third baseline (denoted as "nnUNet-Joint") integrates a CNN classification head into UNet [8] and trained end-to-end. We obtain the 95% confidence interval of AUC, sensitivity, and specificity values from 1000 bootstrap replicas of the test dataset for statistical analysis. For statistical significance, we conduct a DeLong test between two AUCs (ours vs. compared method) and a permutation test between two sensitivities or specificities (ours vs. compared method and radiologists).

4.2 Results

Our method Outperforms Baselines. Our method outperforms three baselines (Table 1) in all metrics, particularly in AUC and sensitivity. The advantage of our approach is that it captures the local and global information simultaneously in virtue of the unique architecture of mask transformer. It also extracts high-level semantics from cluster representations, making it suitable for classification and facilitating a holistic decision-making process. Moreover, our method reaches a considerable specificity of 97.7% on the external test set, which is crucial in opportunistic screening for less false positives and unnecessary human workload.

AI Models Surpass Experienced Radiologists on Non-contrast CT Scans. As shown in Fig. 2a, our AI model's ROC curve is superior to that of two experienced radiologists. The model achieves a sensitivity of 85.0% in detecting gastric cancer, which significantly exceeds the mean performance of doctors (73.5%) and also surpasses the best performing doctor (R2: 75.0%), while maintaining a high specificity. A visual example is presented in Fig. 2b. This early-stage cancer (T1) is miss-detected by both radiologists, whereas classified and localized precisely by our model.

Subgroup Analysis. In Table 2, we report the performance of patient-level detection and tumor-level localization stratified by tumor (T) stage. We compare our model's performance with that of both radiologists. The results show that our model performs better in detecting early stage tumors (T1, T2) and provides more precise tumor localization. Specifically, our model detects 60.0% (6/10) T1 cancers, and 77.8% (7/9) T2 cancers, surpassing the best performing expert (50% T1, 55.6% T2). Meanwhile, our model maintains a reliable detection rate and credible localization accuracy for T3 and T4 tumors (2 of 34 T3 tumors missed).

Comparison with Established Screening Tools. Our method surpasses or performs on par with established screening tools [4,7,10] in terms of sensitivity for gastric cancer detection at a similar specificity level with a relatively large testing patient size ($n = 1151$ by integrating the internal and external test sets), as shown in Table 3. This finding sheds light on the opportunity to employ automated AI systems to screen gastric cancer using non-contrast CT scans.

5 Conclusion

We propose a novel Cluster-induced Mask Transformer for gastric cancer detection on non-contrast CT scans. Our approach outperforms strong baselines and experienced radiologists. Compared to other screening methods, such as blood tests, endoscopy, upper-gastrointestinal series, and ME-NBI, our approach is non-invasive, cost-effective, safe, and more accurate for detecting early-stage tumors. The robust performance of our approach demonstrates its potential for opportunistic screening of gastric cancer in the general population.

Acknowledgement. This work was supported by Alibaba Group through Alibaba Research Intern Program. Bin Dong and Li Zhang was partly supported by NSFC 12090022 and 11831002, and Clinical Medicine Plus X-Young Scholars Project of Peking University PKU2023LCXQ041.

References

1. Carion, N., Massa, F., Synnaeve, G., Usunier, N., Kirillov, A., Zagoruyko, S.: End-to-end object detection with transformers. In: Vedaldi, A., Bischof, H., Brox, T., Frahm, J.-M. (eds.) ECCV 2020. LNCS, vol. 12346, pp. 213–229. Springer, Cham (2020). https://doi.org/10.1007/978-3-030-58452-8_13
2. Chen, J., et al.: TransuNet: transformers make strong encoders for medical image segmentation. arXiv preprint arXiv:2102.04306 (2021)
3. Cheng, B., Schwing, A., Kirillov, A.: Per-pixel classification is not all you need for semantic segmentation. In: NeurIPS, vol. 34, pp. 17864–17875 (2021)
4. Choi, K.S., et al.: Performance of different gastric cancer screening methods in Korea: a population-based study. PLoS One **7**(11), e50041 (2012)
5. Hamashima, C., et al.: The Japanese guidelines for gastric cancer screening. Jpn. J. Clin. Oncol. **38**(4), 259–267 (2008)

6. Heinrich, M.P., Jenkinson, M., Brady, M., Schnabel, J.A.: MRF-based deformable registration and ventilation estimation of lung CT. IEEE Trans. Med. Imaging **32**(7), 1239–1248 (2013)
7. Hu, H., et al.: Identifying early gastric cancer under magnifying narrow-band images with deep learning: a multicenter study. Gastrointest. Endosc. **93**(6), 1333–1341 (2021)
8. Isensee, F., Jaeger, P.F., Kohl, S.A., Petersen, J., Maier-Hein, K.H.: NNU-net: a self-configuring method for deep learning-based biomedical image segmentation. Nat. Methods **18**(2), 203–211 (2021)
9. Jun, J.K., et al.: Effectiveness of the Korean national cancer screening program in reducing gastric cancer mortality. Gastroenterology **152**(6), 1319–1328 (2017)
10. Klein, E., et al.: Clinical validation of a targeted methylation-based multi-cancer early detection test using an independent validation set. Ann. Oncol. **32**(9), 1167–1177 (2021)
11. Li, H., et al.: 3d IFPN: improved feature pyramid network for automatic segmentation of gastric tumor. Front. Oncol. **11**, 618496 (2021)
12. Li, J., et al.: CT-based delta radiomics in predicting the prognosis of stage iv gastric cancer to immune checkpoint inhibitors. Front. Oncol. **12**, 1059874 (2022)
13. Li, L., et al.: Convolutional neural network for the diagnosis of early gastric cancer based on magnifying narrow band imaging. Gastric Cancer **23**, 126–132 (2020)
14. Luo, H., et al.: Real-time artificial intelligence for detection of upper gastrointestinal cancer by endoscopy: a multicentre, case-control, diagnostic study. The Lancet Oncol. **20**(12), 1645–1654 (2019)
15. Miki, K.: Gastric cancer screening using the serum pepsinogen test method. Gastric Cancer **9**, 245–253 (2006)
16. National Cancer Institute, S.P.: Cancer stat facts: Stomach cancer. https://seer.cancer.gov/statfacts/html/stomach.html (2023)
17. Pickhardt, P.J.: Value-added opportunistic CT screening: state of the art. Radiology **303**(2), 241–254 (2022)
18. Ronneberger, O., Fischer, P., Brox, T.: U-net: convolutional networks for biomedical image segmentation. In: Navab, N., Hornegger, J., Wells, W.M., Frangi, A.F. (eds.) MICCAI 2015. LNCS, vol. 9351, pp. 234–241. Springer, Cham (2015). https://doi.org/10.1007/978-3-319-24574-4_28
19. Smyth, E.C., Nilsson, M., Grabsch, H.I., van Grieken, N.C., Lordick, F.: Gastric cancer. The Lancet **396**(10251), 635–648 (2020)
20. Song, Z., et al.: Clinically applicable histopathological diagnosis system for gastric cancer detection using deep learning. Nat. Commun. **11**(1), 4294 (2020)
21. Tang, Y., et al.: Self-supervised pre-training of Swin transformers for 3d medical image analysis. In: CVPR, pp. 20730–20740 (2022)
22. USPSTF: U.S. Preventive Services Task Force, Recommendations. https://www.uspreventiveservicestaskforce.org/uspstf/topic_search_results?topic_status=P (2023)
23. Wang, F., et al.: A cascaded approach for ultraly high performance lesion detection and false positive removal in liver CT scans. arXiv preprint arXiv:2306.16036 (2023)
24. Wang, H., Zhu, Y., Adam, H., Yuille, A., Chen, L.C.: Max-deeplab: end-to-end panoptic segmentation with mask transformers. In: CVPR, pp. 5463–5474 (2021)
25. Xia, Y., et al.: Effective pancreatic cancer screening on non-contrast CT scans via anatomy-aware transformers. In: de Bruijne, M., et al. (eds.) MICCAI 2021. LNCS, vol. 12905, pp. 259–269. Springer, Cham (2021). https://doi.org/10.1007/978-3-030-87240-3_25

26. Yao, J., et al.: Effective opportunistic esophageal cancer screening using noncontrast CT imaging. In: Wang, L., Dou, Q., Fletcher, P.T., Speidel, S., Li, S. (eds.) Medical Image Computing and Computer Assisted Intervention. MICCAI 2022. LNCS, vol. 13433, pp. 344–354. Springer, Cham (2022). https://doi.org/10.1007/978-3-031-16437-8_33
27. Yu, Q., et al.: CMT-DeepLab: clustering mask transformers for panoptic segmentation. In: CVPR, pp. 2560–2570 (2022)
28. Yu, Q., et al.: k-means mask transformer. In: Avidan, S., Brostow, G., Cissé, M., Farinella, G.M., Hassner, T. (eds.) Computer Vision. ECCV 2022. LNCS, vol. 13689, pp. 288–307. Springer, Cham (2022). https://doi.org/10.1007/978-3-031-19818-2_17
29. Yuan, M., et al.: Devil is in the queries: advancing mask transformers for real-world medical image segmentation and out-of-distribution localization. In: CVPR, pp. 23879–23889 (2023)
30. Yushkevich, P.A., et al.: User-guided 3d active contour segmentation of anatomical structures: significantly improved efficiency and reliability. Neuroimage **31**(3), 1116–1128 (2006)
31. Zhang, L., Gopalakrishnan, V., Lu, L., Summers, R.M., Moss, J., Yao, J.: Self-learning to detect and segment cysts in lung CT images without manual annotation. In: ISBI, pp. 1100–1103 (2018)
32. Zhu, Z., Xia, Y., Xie, L., Fishman, E.K., Yuille, A.L.: Multi-scale coarse-to-fine segmentation for screening pancreatic ductal adenocarcinoma. In: Shen, D., et al. (eds.) MICCAI 2019. LNCS, vol. 11769, pp. 3–12. Springer, Cham (2019). https://doi.org/10.1007/978-3-030-32226-7_1

Detecting Domain Shift in Multiple Instance Learning for Digital Pathology Using Fréchet Domain Distance

Milda Pocevičiūtė[1,2(✉)], Gabriel Eilertsen[1,2], Stina Garvin[3], and Claes Lundström[1,2,4]

[1] Center for Medical Imaging Science and Visualization, Linköping University, University Hospital, Linköping, Sweden
{milda.poceviciute,gabriel.eilertsen,claes.lundstrom}@liu.se
[2] Department of Science and Technology, Linköping University, Linköping, Sweden
[3] Department of Clinical Pathology, Department of Biomedical and Clinical Sciences, Linköping University, Linköping, Sweden
garvin.stina@regionostergotland.se
[4] Sectra AB, Linköping, Sweden

Abstract. Multiple-instance learning (MIL) is an attractive approach for digital pathology applications as it reduces the costs related to data collection and labelling. However, it is not clear how sensitive MIL is to clinically realistic domain shifts, i.e., differences in data distribution that could negatively affect performance, and if already existing metrics for detecting domain shifts work well with these algorithms. We trained an attention-based MIL algorithm to classify whether a whole-slide image of a lymph node contains breast tumour metastases. The algorithm was evaluated on data from a hospital in a different country and various subsets of this data that correspond to different levels of domain shift. Our contributions include showing that MIL for digital pathology is affected by clinically realistic differences in data, evaluating which features from a MIL model are most suitable for detecting changes in performance, and proposing an unsupervised metric named Fréchet Domain Distance (FDD) for quantification of domain shifts. Shift measure performance was evaluated through the mean Pearson correlation to change in classification performance, where FDD achieved 0.70 on 10-fold cross-validation models. The baselines included Deep ensemble, Difference of Confidence, and Representation shift which resulted in 0.45, −0.29, and 0.56 mean Pearson correlation, respectively. FDD could be a valuable tool for care providers and vendors who need to verify if a MIL system is likely to perform reliably when implemented at a new site, without requiring any additional annotations from pathologists.

Supported by Swedish e-Science Research Center, VINNOVA, The CENIIT career development program at Linköping University, and Wallenberg AI, WASP funded by the Knut and Alice Wallenberg Foundation.

Supplementary Information The online version contains supplementary material available at https://doi.org/10.1007/978-3-031-43904-9_16.

H. Greenspan et al. (Eds.): MICCAI 2023, LNCS 14224, pp. 157–167, 2023.
https://doi.org/10.1007/978-3-031-43904-9_16

Keywords: Domain shift detection · attention-based MIL · Digital pathology

1 Introduction

The spreading digitalisation of pathology labs has enabled the development of deep learning (DL) tools that can assist pathologists in their daily tasks. However, supervised DL methods require detailed annotations in whole-slide images (WSIs) which is time-consuming, expensive and prone to inter-observer disagreements [6]. Multiple instance learning (MIL) alleviates the need for detailed annotations and has seen increased adoption in recent years. MIL approaches have proven to work well in academic research on histopathology data [1,17,29] as well as in commercial applications [26]. Most MIL methods for digital pathology employ an attention mechanism as it increases the reliability of the algorithms, which is essential for successful clinical adoption [14].

Domain shift in DL occurs when the data distributions of testing and training differs [20,34]. This remains a significant obstacle to the deployment of DL applications in clinical practice [7]. To address this problem previous work either use domain adaptation when data from the target domain is available [32], or domain generalisation when the target data is unavailable [34]. Domain adaptation has been explored in the MIL setting too [22,23,27]. However, it may not be feasible to perform an explicit domain adaptation, and an already adapted model could still experience problems with domain shifts. Hence, it is important to provide indications of the expected performance on a target dataset without requiring annotations [5,25]. Another related topic is out-of-distribution (OOD) detection [33] which aims to detect individual samples that are OOD, in contrast to our objective of estimating a difference of expected performances between some datasets. For supervised algorithms, techniques of uncertainty estimation have been used to measure the effect of domain shift [4,15,18] and to improve the robustness of predictions [19,21,30]. However, the reliability of uncertainty estimates can also be negatively affected by domain shifts [11,31]. Alternatively, a drop in performance can be estimated by comparing the model's softmax outputs [8] or some hidden features [24,28] acquired on in-domain and domain shift datasets. Although such methods have been demonstrated for supervised algorithms, as far as we know no previous work has explored domain shift in the specific context of MIL algorithms. Hence, it is not clear how well they will work in such a scenario.

In this work, we evaluate an attention-based MIL model on unseen data from a new hospital and propose a way to quantify the domain shift severity. The model is trained to perform binary classification of WSIs from lymph nodes of breast cancer patients. We split the data from the new hospital into several subsets to investigate clinically realistic scenarios triggering different levels of domain shift. We show that our proposed unsupervised metric for quantifying domain shift correlates best with the changes in performance, in comparison to multiple baselines. The approach of validating a MIL algorithm in a new site without collecting new labels can greatly reduce the cost and time of quality assurance efforts and ensure that the models perform as expected in a variety of settings. The novel contributions of our work can be summarised as:

1. Proposing an unsupervised metric named Fréchet Domain Distance (FDD) for quantifying the effects of domain shift in attention-based MIL;
2. Showing how FDD can help to identify subsets of patient cases for which MIL performance is worse than reported on the in-domain test data;
3. Comparing the effectiveness of using uncertainty estimation versus learnt representations for domain shift detection in MIL.

2 Methods

Our experiments center on an MIL algorithm with attention developed for classification in digital pathology. The two main components of our domain shift quantification approach are the selection of MIL model features to include and the similarity metric to use, described below.

2.1 MIL Method

As the MIL method for our investigation, we chose the clustering-constrained-attention MIL (CLAM) [17] because it well represents an architecture of MIL with attention, meaning that our approach can equally be applied to many other such methods.

2.2 MIL Features

We explored several different feature sets that can be extracted from the attention-based MIL framework: learnt embedding of the instances (referred to as *patch features*), and penultimate layer features (*penultimate features*). A study is conducted to determine the best choices for type and amount of patch features. As a baseline, we take a mean over all patches ignoring their attention scores (*mean patch features*). Alternatively, the patch features can be selected based on the attention score assigned to them. *Positive evidence* or *Negative evidence* are defined as the K patch features that have the K highest or lowest attention scores, respectively. *Combined evidence* is a combination of an equal number of patch features with the highest and lowest attention scores. To test if the reduction of the number of features in itself has a positive effect on domain shift quantification, we also compare with K randomly selected patch features.

2.3 Fréchet Domain Distance

Fréchet Inception Distance (FID) [10] is commonly used to measure similarity between real and synthetically generated data. Inspired by FID, we propose a metric named Fréchet Domain Distance (FDD) for evaluating if a model is experiencing a drop in performance on some new dataset. The Fréchet distance (FD) between two multivariate Gaussian variables with means μ_1, μ_2 and covariance matrices $\mathbf{C}_1$, $\mathbf{C}_2$ is defined as [3]:

$$FD((\mu_1, \mathbf{C}_1), (\mu_2, \mathbf{C}_2)) = \|\mu_1 - \mu_2\|^2 + Tr(\mathbf{C}_1 + \mathbf{C}_2 - 2(\mathbf{C}_1\mathbf{C}_2)^{\frac{1}{2}}). \quad (1)$$

We are interested in using the FD for measuring the domain shift between different WSI datasets $\mathcal{X}_d$. To this end, we extract features from the MIL model applied to all the WSIs in $\mathcal{X}_d$, and arrange these in a feature matrix $\mathbf{M}_d \in \mathbb{R}^{W_d \times J}$, where W_d is the number of WSIs in $\mathcal{X}_d$ and J is the number of features extracted by the MIL model for one WSI $x \in \mathcal{X}_d$. For penultimate layer and mean patch features, we can apply this strategy directly. However, for positive, negative, and combined evidence we have K feature matrices $\mathbf{M}_{d,k}$, with K feature descriptions for each WSI. To aggregate evidence features over a WSI, we take the mean $\mathbf{M}_d^K = \frac{1}{K}\sum_{k=1}^{K}\mathbf{M}_{d,k}$. Now, we can compute means μ_d^K and covariance matrices $\mathbf{C}_d^K$ from $\mathbf{M}_d^K$ extracted from two datasets $d = 1$ and $d = 2$. By measuring the FD, we arrive at our proposed FDD:

$$FDD_K(\mathcal{X}_1, \mathcal{X}_2) = FD((\mu_1^K, \mathbf{C}_1^K), (\mu_2^K, \mathbf{C}_2^K)). \quad (2)$$

FDD_K uses the K aggregated positive evidence features, but in the results we also compare to $\mathbf{M}_d$ described from penultimate features, mean patch features, and the other evidence feature selection strategies.

3 Datasets

Grand Challenge Camelyon data [16] is used for model training (770 WSIs of which 293 WSIs contain metastases) and in-domain testing (629 WSIs of which 289 WSIs contain metastases). For domain shift test data, we extracted 302 WSIs (of which 111 contain metastases) from AIDA BRLN dataset [12]. To evaluate clinically realistic domain shifts, we divided the dataset in two different ways, creating four subsets:

1a. 161 WSIs from sentinel node biopsy cases (54 WSIs with metastases): a small shift as it is the same type of lymph nodes as in Grand Challenge Camelyon data [16].
1b. 141 WSIs from axillary nodes dissection cases (57 WSIs with metastases): potentially large shift as some patients have already started neoadjuvant treatment as well as the tissue may be affected from the procedure of sentinel lymph node removal.
2a. 207 WSIs with ductal carcinoma (83 WSIs with metastases): a small shift as it is the most common type of carcinoma and relatively easy to diagnose.
2b. 68 WSIs with lobular carcinoma (28 WSIs with metastases): potentially large shift as it is a rare type of carcinoma and relatively difficult to diagnose.

The datasets of lobular and ductal carcinomas each contain 50 % of WSIs from sentinel and axillary lymph node procedures. The sentinel/axillary division is motivated by the differing DL prediction performance on such subsets, as observed by Jarkman et al. [13]. Moreover, discussions with pathologists led to the conclusion that it is clinically relevant to evaluate the performance difference between ductal and lobular carcinoma. Our method is intended to avoid requiring dedicated WSI labelling efforts. We deem that the information needed to do this type of subset divisions would be available without labelling since the patient cases in a clinical setting would already contain such information. All datasets are publicly available to be used in legal and ethical medical diagnostics research.

4 Experiments

The goal of the study is to evaluate how well FDD_K and the baseline methods correlate with the drop in classification performance of attention-based MIL caused by several potential sources of domain shifts. In this section, we describe the experiments we conducted.

4.1 MIL Training

We trained, with default settings, 10 CLAM models to classify WSIs of breast cancer metastases using a 10-fold cross-validation (CV) on the training data. The test data was kept the same for all 10 models. The classification performance is evaluated using the area under receiver operating characteristic curve (ROC-AUC) and Matthews correlation coefficient (MCC) [2]. Following the conclusions of [2] that MCC well represents the full confusion matrix and the fact that in clinical practice a threshold needs to be set for a classification decision, MCC is used as a primary metric of performance for domain shift analysis while ROC-AUC is reported for completeness. Whereas extremely large variations in label prevalence could reduce the reliability of the MCC metric, this is not the case here as label prevalence is similar (35–45%) in our test datasets. For Deep ensemble [15] we trained 4 additional CLAM models for each of the 10 CV folds.

4.2 Domain Shift Quantification

As there is no related work on domain shift detection in the MIL setting, we selected methods developed for supervised algorithms as baselines:

- The model's accumulated uncertainty between two datasets. Deep ensemble [15] (DE) and Difference in Confidence with entropy [8] (DoC) compare the mean entropy over all data points. DE uses an ensemble to estimate better-calibrated uncertainty than the single model in DoC.
- The accumulated confidence of a model across two datasets. DoC [8] can be measured on the mean softmax scores of two datasets. A large difference indicates a potential drop in performance.
- The hidden features produced by an algorithm. Representation Shift [28] (RS) has shown promising results in detecting domain shift in convolutional neural networks and it is the method most similar to FDD. However, it is not trivial which hidden features of MIL that are most suitable for this task, and we evaluate several options (see Sect. 2.2) with both methods.

For all possible pairs of Camelyon and the other test datasets, and for the 10 CV models, we compute the domain shift measures and compare them to the observed drop in performance. The effectiveness is evaluated by Pearson correlation and visual investigation of corresponding scatter plots. All results are reported as mean and standard deviation over the 10-fold CV.

5 Results

The first part of our results is the performance of the WSI classification task across the subsets, summarized in Table 1. While showing similar trends, there is some discrepancy in the level of domain shift represented by the datasets due to the differences between the MCC and ROC-AUC measures.

As we deemed MCC to better represent the clinical use situation (see Sect. 4.1), it was used for our further evaluations. Overall, the performance is in line with previously published work [17,29].

Table 1. Classification performance reported in mean (standard deviation) of MCC and ROC-AUC metrics, computed over the 10-fold CV models. A threshold for MCC is determined on validation data.

Dataset	MCC	ROC-AUC
Camelyon	0.67 (0.02)	0.87 (0.01)
BRLN	0.61 (0.06)	0.89 (0.01)
Sentinel	0.67 (0.05)	0.90 (0.01)
Axillary	0.56 (0.08)	0.86 (0.01)
Ductal	0.72 (0.04)	0.92 (0.01)
Lobular	0.61 (0.04)	0.85 (0.03)

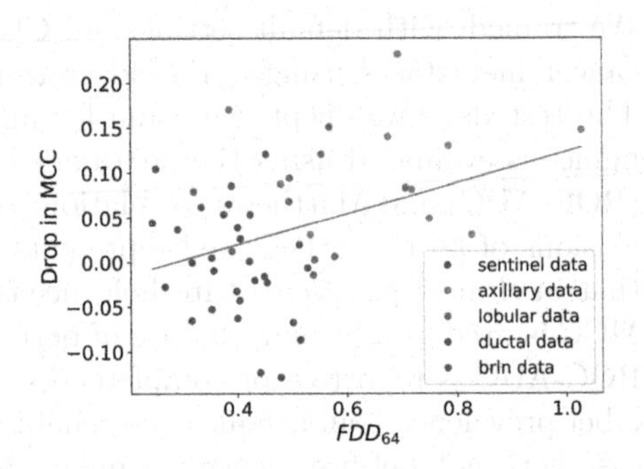

Fig. 1. Scatter plot of FDD_{64} against drop in MCC for all model-dataset combinations. A fitted line is included for better interpretability.

We observe the largest domain shift in terms of MCC on axillary nodes followed by lobular carcinoma and full BRLN datasets. There seems to be no negative effect from processing the sentinel nodes data. CLAM models achieved better performance on ductal carcinoma compared to the in-domain Camelyon test data.

Table 2 summarises the Pearson correlation between the change in performance, i.e., the MCC difference between Camelyon and other test datasets, and the domain shift measures for the same pairs. FDD_{64} outperforms the baselines substantially, and has the smallest standard deviation. Figure 1 shows how individual drop in performance of model-dataset combinations are related to the FDD_{64} metric. For most models detecting larger drop in performance (> 0.05) is easier on axillary lymph nodes data than on any other analysed dataset.

Table 2. Pearson correlations between domain shift measure and difference in performance (MCC metric) of Camelyon test set and the other datasets. The mean and standard deviation values are computed over the 10 CV folds.

Measure	Features used	Pearson correlation
DoC	Softmax score	−0.29 (0.46)
DoC	Entropy score	−0.32 (0.47)
DE	Entropy score	0.45 (0.38)
RS	Penultimate feat.	0.48 (0.25)
RS	Mean patch feat.	0.41 (0.32)
RS_{64}	Positive evidence	0.56 (0.19)
FD	Mean patch feat.	0.49 (0.28)
FD	Penultimate feat.	0.61 (0.19)
FDD_{64} (our)	Positive evidence	0.70 (0.15)

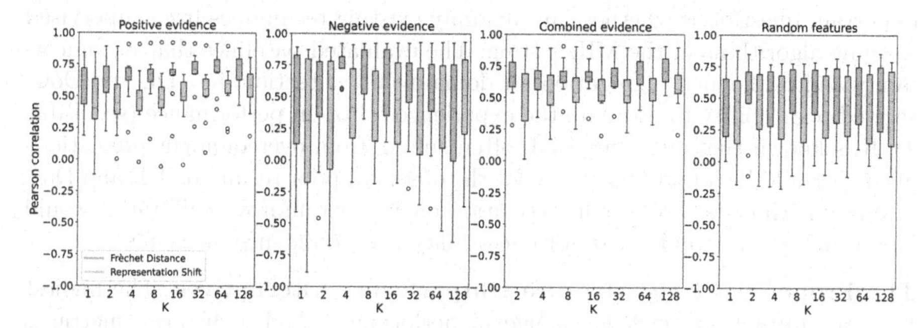

Fig. 2. Box plots of Pearson correlations achieved by Fréchet distance and Representation shift metric using attention-based features, i.e., positive, negative, combined evidence, and randomly selected features. Varying number of extracted patch representations K is considered: from 1 to 128. The reported results are over the 10 cross validation models.

A study of the number and type of MIL attention-based features and FD and RS metrics is presented in Fig. 2. The baseline of randomly selecting patch features resulted in the worst outcome on domain shift detection. Negative evidence with FD achieved high Pearson correlation when $K = 4$. However, the results were among the worst with any other number of K. Both combined and positive evidence achieved peak performance of 0.68 (0.17) and 0.70 (0.13), respectively, when FD and $K = 64$ were used. We conclude that in our setup the best and most reliable performance of domain shift quantification is achieved by positive evidence with FD and $K = 64$, i.e. FDD_{64}.

6 Discussion and Conclusion

MIL is Affected by Domain Shift. Some previous work claim that MIL is more robust to domain shift as it is trained on more data due to the reduced costs of data annotation [1,17]. We argue that domain shift will still be a factor to consider as an algorithm deployed in clinical practice is likely to encounter unseen varieties of data. However, it may require more effort to determine what type of changes in data distribution are critical. Our results show that domain shift is present between the WSIs from the same hospital (Camelyon data) and another medical centre (BRLN data). However, as clinically relevant subsets of BRLN data are analysed, stark differences in performance and reliability (indicated by the standard deviation) are revealed. Therefore, having a reliable metric for unsupervised domain shift quantification could bring value for evaluating an algorithm at a new site.

Domain Shift Detection for Supervised DL Struggles for MIL. An important question is whether can we apply existing techniques from supervised learning algorithms in the MIL setting. The evaluated baselines that use uncertainty and confidence aggregation for domain shift detection, i.e., DE and DoC, showed poor ability to estimate the experienced drop in performance (see Table 2). It is known that supervised DL often suffers from overconfident predictions [9]. This could be a potential cause for the observed poor results by DE and DoC in our experiments. Further investigation on how to improve calibration could help to boost the applicability of uncertainty and confidence measures.

The Proposed FDD_K Measure Outperforms Alternatives. The highest Pearson correlation between change in performance and a distance metric is achieved by Fréchet distance with 64 positive evidence features, FDD_{64} (see Table 2). RS_{64} approach performed better than uncertainty/confidence-based methods but still was substantially worse than FDD_{64}. Furthermore, FDD_{64} resulted in the smallest standard deviation which is an important indicator of the reliability of the metric. Interestingly, using penultimate layer features, which combine all patch features and attention scores, resulted in much worse outcome than FDD_{64}, 0.61 versus 0.70. Thus, it seems a critical component in domain shift measurement in attention-based MIL is to correctly make use of the attention scores. From Fig. 1 we can see that if we further investigated all model-dataset combinations that resulted in FDD_{64} above 0.5, we would detect many cases with a drop in performance larger than 0.05. However, the drop is easier to detect on axillary and lobular datasets compared to others. An interesting aspect is that the performance was better for the out-of-domain ductal subset compared to in-domain Camelyon WSIs. In practical applications, it may be a problem when the domain shift quantification cannot separate between shifts having positive or negative effect on performance. Such differentiation could be the topic of future work.

Conclusion. We carried out a study on how clinically realistic domain shifts affect attention-based MIL for digital pathology. The results show that domain shift may raise challenges in MIL algorithms. Furthermore, there is a clear benefit of using attention for feature selection and our proposed FDD_K metric for quantification of expected performance drop. Hence, FDD_K could aid care providers and vendors in ensuring safe deployment and operation of attention-based MIL in pathology laboratories.

References

1. Campanella, G., et al.: Clinical-grade computational pathology using weakly supervised deep learning on whole slide images. Nat. Med. **25**(8), 1301–1309 (2019)
2. Chicco, D., Jurman, G.: The advantages of the Matthews correlation coefficient (MCC) over f1 score and accuracy in binary classification evaluation. BMC Genom. **21**(1), 1–13 (2020)
3. Dowson, D., Landau, B.: The fréchet distance between multivariate normal distributions. J. Multivar. Anal. **12**(3), 450–455 (1982)
4. Elder, B., Arnold, M., Murthi, A., Navratil, J.: Learning prediction intervals for model performance. In: Proceedings of the AAAI Conference on Artificial Intelligence (2020)
5. Elsahar, H., Gallé, M.: To annotate or not? Predicting performance drop under domain shift. In: Proceedings of the 2019 Conference on Empirical Methods in Natural Language Processing and the 9th International Joint Conference on Natural Language Processing (EMNLP-IJCNLP), pp. 2163–2173 (2019)
6. Gomes, D.S., Porto, S.S., Balabram, D., Gobbi, H.: Inter-observer variability between general pathologists and a specialist in breast pathology in the diagnosis of lobular neoplasia, columnar cell lesions, atypical ductal hyperplasia and ductal carcinoma in situ of the breast. Diagn. Pathol. **9**(1), 121 (2014)
7. Guan, H., Liu, M.: Domain adaptation for medical image analysis: a survey. IEEE Trans. Biomed. Eng. **69**(3), 1173–1185 (2021)
8. Guillory, D., Shankar, V., Ebrahimi, S., Darrell, T., Schmidt, L.: Predicting with confidence on unseen distributions. In: 2021 IEEE/CVF International Conference on Computer Vision (ICCV), pp. 1114–1124 (2021)
9. Guo, C., Pleiss, G., Sun, Y., Weinberger, K.Q.: On calibration of modern neural networks. In: Proceedings of the 34th International Conference on Machine Learning. Proceedings of Machine Learning Research, vol. 70, pp. 1321–1330. PMLR (2017)
10. Heusel, M., Ramsauer, H., Unterthiner, T., Nessler, B., Hochreiter, S.: GANs trained by a two time-scale update rule converge to a local Nash equilibrium (2017)
11. Hoebel, K., et al.: Do i know this? Segmentation uncertainty under domain shift. In: Proceedings of SPIE, the International Society for Optical Engineering, vol. 12032, pp. 1203211–1203216 (2022)
12. Jarkman, S., Lindvall, M., Hedlund, J., Treanor, D., Lundström, C., van der Laak, J.: Axillary lymph nodes in breast cancer cases (2019)
13. Jarkman, S., et al.: Generalization of deep learning in digital pathology: experience in breast cancer metastasis detection. Cancers **14**(21), 5424 (2022)
14. Javed, S.A., Juyal, D., Padigela, H., Taylor-Weiner, A., Yu, L., Prakash, A.: Additive MIL: Intrinsically interpretable multiple instance learning for pathology. In: Advances in Neural Information Processing Systems (2022)

15. Lakshminarayanan, B., Pritzel, A., Blundell, C.: Simple and scalable predictive uncertainty estimation using deep ensembles. In: Advances in Neural Information Processing Systems, vol. 30. Curran Associates, Inc. (2017)
16. Litjens, G., et al.: 1399 H&E-stained sentinel lymph node sections of breast cancer patients: the CAMELYON dataset. GigaScience **7**(6), giy065 (2018)
17. Lu, M.Y., et al.: Data-efficient and weakly supervised computational pathology on whole-slide images. Nat. Biomed. Eng. **5**(6), 555–570 (2021)
18. Maggio, S., Bouvier, V., Dreyfus-Schmidt, L.: Performance prediction under dataset shift. In: 2022 26th International Conference on Pattern Recognition (ICPR), pp. 2466–2474 (2022)
19. Martinez, C., et al.: Segmentation certainty through uncertainty: uncertainty-refined binary volumetric segmentation under multifactor domain shift. In: Proceedings of the IEEE/CVF Conference on Computer Vision and Pattern Recognition (CVPR) Workshops (2019)
20. Moreno-Torres, J.G., et al.: A unifying view on dataset shift in classification. Pattern Recogn. **45**(1), 521–530 (2012)
21. Pocevičiūtė, M., Eilertsen, G., Jarkman, S., Lundström, C.: Generalisation effects of predictive uncertainty estimation in deep learning for digital pathology. Sci. Rep. **12**(1), 1–15 (2022)
22. Prabono, A.G., Yahya, B.N., Lee, S.L.: Multiple-instance domain adaptation for cost-effective sensor-based human activity recognition. Futur. Gener. Comput. Syst. **133**, 114–123 (2022)
23. Praveen, R.G., Granger, E., Cardinal, P.: Deep weakly supervised domain adaptation for pain localization in videos. In: 2020 15th IEEE International Conference on Automatic Face and Gesture Recognition (FG 2020), pp. 473–480. IEEE (2020)
24. Rabanser, S., Günnemann, S., Lipton, Z.: Failing loudly: an empirical study of methods for detecting dataset shift. In: NeurIPS 2019 (2019)
25. Schelter, S., Rukat, T., Bießmann, F.: Learning to validate the predictions of black box classifiers on unseen data. In: Proceedings of the 2020 ACM SIGMOD International Conference on Management of Data, pp. 1289–1299 (2020)
26. Silva, L.M., et al.: Independent real-world application of a clinical-grade automated prostate cancer detection system. J. Pathol. **254**, 147–158 (2021)
27. Song, R., Cao, P., Yang, J., Zhao, D., Zaiane, O.R.: A domain adaptation multi-instance learning for diabetic retinopathy grading on retinal images. In: 2020 IEEE International Conference on Bioinformatics and Biomedicine (BIBM), pp. 743–750. IEEE (2020)
28. Stacke, K., Eilertsen, G., Unger, J., Lundstrom, C.: Measuring domain shift for deep learning in histopathology. IEEE J. Biomed. Health Inform. **25**, 325–326 (2021)
29. Su, Z., et al.: Attention2majority: weak multiple instance learning for regenerative kidney grading on whole slide images. Med. Image Anal. **79**, 102462 (2022)
30. Thagaard, J., Hauberg, S., van der Vegt, B., Ebstrup, T., Hansen, J.D., Dahl, A.B.: Can you trust predictive uncertainty under real dataset shifts in digital pathology? In: Martel, A.L., Abolmaesumi, P., Stoyanov, D., Mateus, D., Zuluaga, M.A., Zhou, S.K., Racoceanu, D., Joskowicz, L. (eds.) MICCAI 2020. LNCS, vol. 12261, pp. 824–833. Springer, Cham (2020). https://doi.org/10.1007/978-3-030-59710-8_80
31. Tomani, C., Gruber, S., Erdem, M.E., Cremers, D., Buettner, F.: Post-hoc uncertainty calibration for domain drift scenarios. In: Proceedings of the IEEE/CVF Conference on Computer Vision and Pattern Recognition (CVPR), pp. 10124–10132 (2021)

32. Xiaofeng, L., et al.: Deep unsupervised domain adaptation: a review of recent advances and perspectives. APSIPA Trans. Signal Inf. Process. **11**(1), 1–10 (2022)
33. Yang, J., Zhou, K., Li, Y., Liu, Z.: Generalized out-of-distribution detection: a survey. arXiv preprint arXiv:2110.11334 (2021)
34. Zhou, K., et al.: Domain generalization: a survey. IEEE Trans. Pattern Anal. Mach. Intell. (2022)

Positive Definite Wasserstein Graph Kernel for Brain Disease Diagnosis

Kai Ma, Xuyun Wen, Qi Zhu, and Daoqiang Zhang(✉)

College of Computer Science and Technology, Nanjing University of Aeronautics and Astronautics, MIIT Key Laboratory of Pattern Analysis and Machine Intelligence, Key Laboratory of Brain-Machine Intelligence Technology, Ministry of Education, Nanjing 211106, China

dqzhang@nuaa.edu.cn

Abstract. In brain functional networks, nodes represent brain regions while edges symbolize the functional connections that enable the transfer of information between brain regions. However, measuring the transportation cost of information transfer between brain regions is a challenge for most existing methods in brain network analysis. To address this problem, we propose a graph sliced Wasserstein distance to measure the cost of transporting information between brain regions in a brain functional network. Building upon the graph sliced Wasserstein distance, we propose a new graph kernel called sliced Wasserstein graph kernel to measure the similarity of brain functional networks. Compared to existing graph methods, including graph kernels and graph neural networks, our proposed sliced Wasserstein graph kernel is positive definite and a faster method for comparing brain functional networks. To evaluate the effectiveness of our proposed method, we conducted classification experiments on functional magnetic resonance imaging data of brain diseases. Our experimental results demonstrate that our method can significantly improve classification accuracy and computational speed compared to state-of-the-art graph methods for classifying brain diseases.

Keywords: Graph kernel · Brain functional network · Brain diseases · Classification · Wasserstein distance

1 Introduction

Brain functional networks characterize functional interactions of human brain, where brain regions correspond to nodes and functional interactions between brain regions are considered as edges. Brain functional networks are widely utilized to classify brain diseases, including Alzheimer's disease [13], attention deficit hyperactivity disorder (ADHD) [7], major depressive disorder [4] and schizophrenia [26]. In these studies, various network characteristics, e.g., degree, clustering coefficient [17], ordinal pattern [16,27] are utilized to represent brain functional network and then used to calculate network measurements for classify brain diseases. However, these network characteristics are built on local connections (e.g., node degree) and ignore the transfer of information between brain regions and the global geometric information in brain functional networks.

H. Greenspan et al. (Eds.): MICCAI 2023, LNCS 14224, pp. 168–177, 2023.
https://doi.org/10.1007/978-3-031-43904-9_17

Recently, Wasserstein distance has attracted broad attention in image processing [6], computer vision [18] and neural network [1]. The Wasserstein distance, also known as the optimal transport distance or Earth Mover's distance [28], was originally proposed to investigate the translocation of masses [8]. In mathematics, the Wasserstein distance is used to quantify the dissimilarity between two probability distributions based on a given ground metric [11,28]. Graph data can be represented as probability distributions using graph or network embedding methods [20]. Graph kernels based on the Wasserstein distance can capture the global geometric information of graphs, making them useful for measuring the similarities between graphs, including brain networks. For example, in the graph kernel based on Wasserstein distance [19], Wasserstein distance is used to measure the similarity of bag-of-vectors between two graphs where the eigen-decompositions of adjacency matrix are used to represent the graphs as bag-of-vectors. In Wasserstein Weisfeiler-Lehman graph kernels [22], the Weisfeiler-Lehman scheme is used to generate node label sequences, and the Wasserstein distance is utilized to measure the similarity of label sequences between two graphs. In the optimal transport based ordinal pattern tree kernel [16], optimal transport distance (i.e., Wasserstein distance) is used to measure the similarity of ordinal pattern tree in brain functional networks and is then applied to classify brain diseases. However, these graph kernels based on Wasserstein distance ignore the transfer of information between brain regions, and their positive definition is hardly guaranteed.

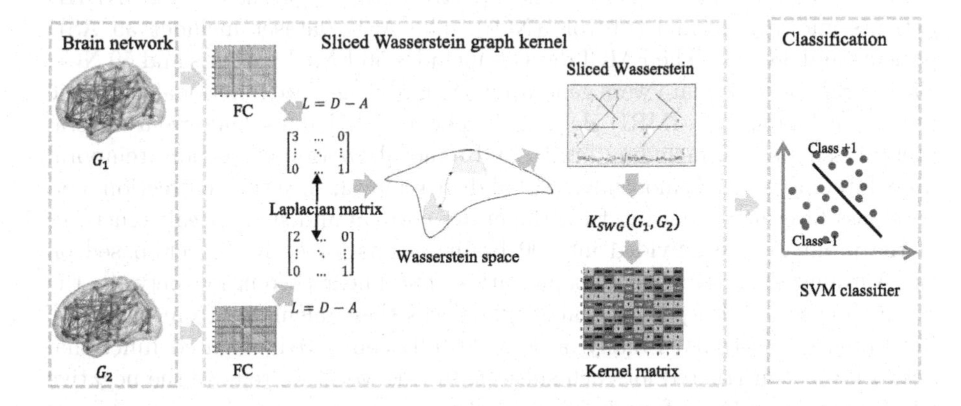

Fig. 1. Illustration of the sliced Wasserstein graph kernel. FC is functional connection matrix.

To tackle these problems, we develop a sliced Wasserstein graph kernel to measure the similarity of brain functional networks. Firstly, we use Laplacian embedding as a feature projection function to project each brain functional network into a set of points in Euclidean space. Then, we calculate the eigen-decomposition on these embeddings and acquire the Laplacian eigenvalues and eigenvectors for a brain functional network. On Laplacian eigenvectors, each row of Laplacian eigenvectors is the representation of a node in brain functional network. We utilize the Wasserstein distance to measure the cost of transporting information between nodes based on these Laplacian eigenvectors. At last, we calculate one-dimensional empirical measures for the points in Euclidean space and calculate sliced Wasserstein graph kernel. We apply the proposed sliced Wasserstein graph kernel to support vector machine (SVM) classifier [14] for brain disease diagnosis. Figure 1 presents the schematic diagram of the proposed framework with each network representing a specific subject. Specifically, our work has following advantages:

- We provide a new approach for investigating the transfer of information between brain regions with sliced Wasserstein distance.
- The proposed sliced Wasserstein graph kernel is positive definite and a faster method for comparing brain functional networks.
- The proposed sliced Wasserstein graph kernel can improve classification accuracy compared to state-of-the-art graph methods for classifying brain diseases.

2 Methods

2.1 Data and Preprocessing

The functional network data used in the experiments are based on three datasets of brain diseases: ADHD[1], autistic spectrum disorder (ASD)[2], and early mild cognitive impairment (EMCI)[3]. The ADHD dataset consists of 121 ADHD patients and 101 normal controls (NCs). The ASD dataset includes 36 ASD patients and 38 NCs. The EMCI dataset includes 56 EMCI patients and 50 NCs. These brain network data were generated using resting-state functional magnetic resonance imaging (rs-fMRI) data [24]. The rs-fMRI data underwent several preprocessing steps, including brain skull removal, motion correction, temporal pre-whitening, spatial smoothing, global drift removal, slice time correction, and bandpass filtering. Following this, the entire cortical and subcortical structures of the brain were subdivided into 90 brain regions for each subject, based on the Automated Anatomical Labeling atlas. The linear correlation between the mean time series of a pair of brain regions was then calculated to measure the functional connectivity. Finally, a 90×90 fully-connected weighted functional network was constructed for each subject. In this work, we remove the negative connections from brain functional networks.

2.2 Sliced Wasserstein Graph Kernel

Sliced Wasserstein graph kernel is used to measure the similarity of paired brain functional networks. In this subsection, we firstly introduce graph sliced Wasserstein distance and then use it to calculate sliced Wasserstein graph kernel.

[1] http://www.nitrc.org/plugins/mwiki/index.php/neurobureau:AthenaPipeline.
[2] http://fcon-1000.projects.nitrc.org/indi/abide/.
[3] http://adni.loni.usc.edu/.

Throughout the paper, we will refer to brain functional network when mentioning the graph, unless noted otherwise.

In image processing, computer vision, and graph comparison, many efficient algorithms of machine learning are available in Euclidean space [10]. We define graph sliced Wasserstein distance in Euclidean space. In other words, Euclidean space is as metric space, i.e., $M \subseteq \mathbb{R}^d$, and we use Euclidean distance as the ground distance, i.e., $d(x, y) = |x - y|$, when defining Wasserstein and graph sliced Wasserstein distances. Here, we provide the definition of the Wasserstein distance in Euclidean space.

Let r and c be two probability measures on $\mathbb{R}^d$. The Wasserstein distance between r and c is defined as

$$W_p(r, c) := (inf_{\gamma \in \Gamma(r,c)} \int_{\mathbb{R}^d \times \mathbb{R}^d} (x - y)^p d\gamma(x, y))^{\frac{1}{p}} \tag{1}$$

where $p \in [1, \infty]$ and $\Gamma(r, c)$ denotes the set of all transportation plans of r and c.

The sliced Wasserstein kernel on Wasserstein distance has been studied in [2,10]. In this paper, we extend sliced Wasserstein distance to graph domain and design graph sliced Wasserstein distance and sliced Wasserstein graph kernel.

Graph Sliced Wasserstein Distance: Given two graphs $G_1 = (V_1, E_1)$ and $G_2 = (V_2, E_2)$, the feature projection function $\Phi : \mathcal{G} \rightarrow \mathbb{R}^{n \times d}$ projects graph G_1 or G_2, $G_1, G_2 \in \mathcal{G}$ into a set of n points (i.e., $\Phi(G_1), \Phi(G_2) \in \mathbb{R}^{n \times d}$) in d-dimensional Euclidean space. Let $r_{\Phi(G_1)}$ and $c_{\Phi(G_2)}$ denote probability measures on d-dimensional feature representation of graph G_1 and G_2. The graph sliced Wasserstein distance between G_1 and G_2 is defined as

$$D^{\Phi}_{GSW}(G_1, G_2) = GSW_2^2(r_{\Phi(G_1)}, c_{\Phi(G_2)}) \tag{2}$$

where Φ denotes the d-dimensional feature projection. $r_{\Phi(G_1)}$ and $c_{\Phi(G_2)}$ are the probability measures of graph G_1 and G_2, respectively, in d-dimensional Euclidean space. The 2-sliced Wasserstein distance is defined as

$$GSW_2(r_{\Phi(G_1)}, c_{\Phi(G_2)}) := (\int_{\mathbb{R}^d \times \mathbb{R}^d} W_2^2(g_{\theta\#} r_{\Phi(G_1)}, g_{\theta\#} c_{\Phi(G_2)}) d\theta)^{\frac{1}{2}} \tag{3}$$

where $g_{\theta\#} r_{\Phi(G_1)}$ and $g_{\theta\#} c_{\Phi(G_2)}$ are the one-dimensional projections of the measure $r_{\Phi(G_1)}$ and $c_{\Phi(G_2)}$. θ is a one dimensional absolutely continuous positive probability density function.

Theorem 1. *Graph sliced Wasserstein distance* $D^{\Phi}_{GSW} : \mathcal{G} \times \mathcal{G} \rightarrow \mathbb{R}^+$ *is distance metric.*

Proof. According to [10], W_2^2 is distance metric which is symmetric, nonnegativity, identity of indiscernibles, and triangle inequality. D^{Φ}_{GSW} is an integral of W_2^2 terms. Hence, D^{Φ}_{GSW} is also a distance metric and satisfies nonnegativity, symmetry, identity of indiscernibles, and triangle inequality.

Algorithm 1. Compute sliced Wasserstein graph kernel

Input: Two graphs G_1, G_2, parameter λ
Output: kernel value $K_{SWG}(G_1, G_2)$
$X_{G_1} \leftarrow \Phi(G_1)$; %Compute feature representations of graph G_1 and G_2%
$X_{G_2} \leftarrow \Phi(G_2)$;
$r \leftarrow r(X_{G_1})$; %Compute probability measure on feature representations%
$c \leftarrow c(X_{G_2})$
$D^{\Phi}_{GSW}(G_1, G_2) \leftarrow GSW_2^2(r, c)$ %Compute graph sliced Wasserstein distance
$K_{SWG}(G_1, G_2) \leftarrow e^{-\lambda D^{\Phi}_{GSW}(G_1,G_2)}$;

According to the definition of graph sliced Wasserstein distance in Eq.(3), we can find the thought behind graph sliced Wasserstein distance is to achieve the one-dimensional representations for the probability measures on d-dimensional feature representation of graph. Then, the distance between two input probability measures is computed as a function on Wasserstein distance of their corresponding one-dimensional representations.

Feature Projection: We use Laplacian embedding as a feature projection function to project a graph into a set of points in Euclidean space. Then, we calculate the eigen-decomposition on these embeddings and acquire the Laplacian eigenvalues and eigenvectors. On Laplacian eigenvectors, we construct the points in Euclidean space where each row of Laplacian eigenvectors is a node representation.

In Eq.(3), $r_{\Phi(G_1)}$ and $c_{\Phi(G_2)}$ are the feature projection of graph G_1 and G_2 by using Laplacian embedding. Assume that $X_v = \{x_i^v\}_{i=1}^n$ and $Y_v = \{y_i^v\}_{j=1}^n$ is the Laplacian eigenvectors of graph G_1 and G_2. We reformulate the graph sliced Wasserstein distance as a sum rather than an integral. Following the work of [9], we calculate the graph sliced Wasserstein distance by sorting the samples and calculate the L_2 distance between the sorted samples. The graph sliced Wasserstein distance between r and c can be approximated from their node representation which is defined as

$$GSW_2(r_{\Phi(G_1)}, c_{\Phi(G_2)}) \approx (\frac{1}{L}\sum_{l=1}^{L}\sum_{n=1}^{N} |g_{\theta_l}(x^v_{i[n]}) - g_{\theta_l}(y^v_{j[n]})|^2)^{\frac{1}{2}} \quad (4)$$

where $i[n]$ and $j[n]$ are the indices of sorted $\{g_{\theta_l}(x_i^v)\}_{i=1}^N$ and $\{g_{\theta_l}(y_i^v)\}_{j=1}^N$, $g_{\theta_l}(x^v_{i[n]}) =< \theta_l, x^v_{i[n]} >$.

By combining graph sliced Wasserstein distance and feature projection on graphs, we can construct a new graph kernel called sliced Wasserstein graph kernel which can be used to measure the similarity between the paired graphs.

Sliced Wasserstein Graph Kernel: Given two graphs $G_1 = (V_1, E_1)$, $G_2 = (V_2, E_2)$ and graph sliced Wasserstein distance on them (i.e., $D^{\Phi}_{GSW}(G_1, G_2)$). We define the sliced Wasserstein graph (SWG) kernel as

$$K_{SWG}(G_1, G_2) = e^{-\lambda D^{\Phi}_{GSW}(G_1,G_2)} \quad (5)$$

Obviously, sliced Wasserstein graph kernel is a case of Laplacian kernel. The procedure of calculating SWG kernel is described in Algorithm 1. According the theorems in [10], one-dimensional Wasserstein space is a flat space. The graph sliced Wasserstein distance is induced from one-dimensional Wasserstein distance. Hence, the graph sliced Wasserstein distance is isometric. SWG kernel based on graph sliced Wasserstein distance is positive definite.

Theorem 2. *Sliced Wasserstein graph kernel is positive definite and differentiable for all* $\lambda > 0$.

Proof. The sliced Wasserstein graph kernel is the extension of sliced Wasserstein kernel on graphs. According to [2,10], sliced Wasserstein kernel is positive definite. Hence, sliced Wasserstein graph kernel is also positive definite.

2.3 Sliced Wasserstein Graph Kernel Based Learning

We use the image processing method described in the data preprocessing to analyze the rs-fMRI data for all subjects and create a brain functional network for each subject. In these brain networks, brain regions are represented as nodes, while the functional connections between paired brain regions are represented as edges. After constructing the brain functional networks for all subjects, we compute the sliced Wasserstein graph kernel using Eq.(5) and apply SVM for disease classification.

3 Experiments

3.1 Experimental Setup

In the experiments, we compare our proposed method with the state-of-the-art graph methods including graph kernels and graph neural networks. Graph kernels include Weisfeiler-Lehman subtree (WL-ST) kernel [21], Weisfeiler-Lehman shortest path (WL-SP) kernel [21], random walk (RW) kernel [23], Wasserstein Weisfeiler-Lehman (WWL) kernel [22], GraphHopper (GH) kernel [5], depth-first-based ordinal pattern (DOP) kernel [15], optimal transport based ordinal pattern tree (OT-OPT) kernel [16]. Graph neural networks include DIFFPOOL [25] and BrainGNN [12]. SVM [3] as the final classifier is exploited to conduct the classification experiment. We perform the leave-one-out cross-validation for all the classification experiments. In the experiments, uniform weight λ is chosen from $\{10^{-2}, 10^{-1}, \cdots, 10^{2}\}$ and the tradeoff parameter C in the SVM is selected from $\{10^{-3}, 10^{-2}, \cdots, 10^{3}\}$.

Table 1. Comparison of different methods on three classification tasks

Method	ADHD vs. NCs		ASD vs. NCs		EMCI vs. NCs	
	ACC(%)	AUC	ACC(%)	AUC	ACC(%)	AUC
WL-ST kernel [21]	70.13	0.682	72.35	0.697	69.41	0.653
WL-SP kernel [21]	68.25	0.664	67.55	0.643	63.18	0.607
RW kernel [23]	69.78	0.643	70.11	0.691	66.87	0.641
WWL kernel [22]	73.63	0.688	72.26	0.715	71.13	0.695
GH kernel [5]	67.56	0.639	71.36	0.674	65.56	0.649
DOP kernel [15]	73.28	0.715	78.68	0.774	75.47	0.737
OT-OPT kernel [16]	76.42	0.738	78.38	0.778	80.36	0.784
BrainGNN [12]	76.46	0.741	82.14	0.801	81.71	0.809
DIFFPOOL [25]	74.63	0.727	80.50	0.798	82.18	0.812
SWG kernel (Proposed)	**78.83**	**0.762**	**90.54**	**0.863**	**85.44**	**0.823**

3.2 Classification Results

We compare the proposed SWG kernel with the state-of-the-art graph kernels on three classification tasks, i.e., ADHD vs. NCs, ASD vs. NCs and EMCI vs. NCs classification. Classification performance is evaluated by accuracy (ACC) and area under receiver operating characteristic curve (AUC). The classification results are shown in Table 1. From Table 1, we can find that our proposed method achieves the best performance on three tasks. For instance, the accuracy achieved by our method is respectively 78.83%, 90.54%, and 85.44% in ADHD vs. NCs, ASD vs. NCs and EMCI vs. NCs classification, which is better than the second best result obtained by BrainGNN and DIFFPOOL. This demonstrates that the proposed SWG kernel is good at distinguishing the patients with brain diseases (i.e., ADHD, ASD, and EMCI) from NCs, compared with the state-of-the-art graph kernels and graph neural networks.

We select four graph methods, including DOP kernel, OT-OPT kernel, BrainGNN and DIFFPOOL, whose classification accuracies are superior to those of other graph methods, except for the SWG kernel. We have recorded the computational time required by these methods, as shown in Table 2. The results in

Table 2. Computational time (in seconds) of different methods.

Method	ADHD vs. NCs	ASD vs. NCs	EMCI vs. NCs
DOP kernel [15]	≥100	≥100	≥100
OT-OPT kernel [16]	≥100	≥100	≥100
BrainGNN [12]	≥100	≥100	≥100
DIFFPOOL [25]	≥100	≥100	≥100
SWG kernel (Proposed)	**83.27**	**8.17**	**15.29**

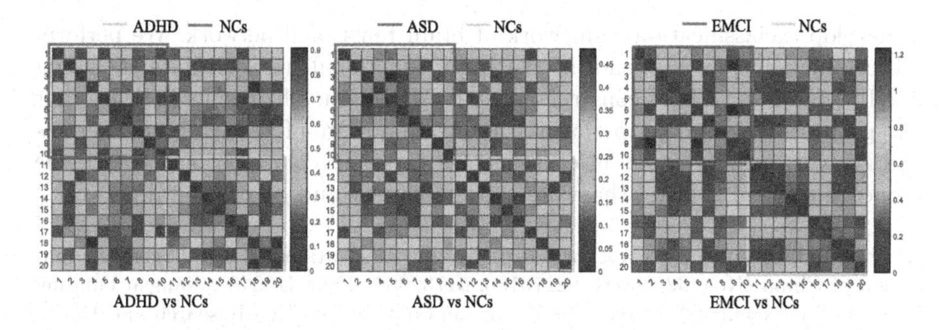

Fig. 2. Distance matrix heatmap for ADHD, ASD and EMCI.

Table 2 demonstrate that our proposed method can enhance computational efficiency as compared to other graph kernels and graph neural networks for the classification of brain diseases.

3.3 Analysis on Wasserstein Distance

The Wasserstein distance was initially proposed to examine mass translocation [28]. In this subsection, we utilize sliced Wasserstein distance to investigate the transportation cost of brain functional networks and brain regions (i.e., nodes). We respectively selected 10 patients and 10 NCs from each dataset, and calculated sliced Wasserstein distances for all selected subjects. We generated a matrix heatmap of the sliced Wasserstein distances for ADHD, ASD, and EMCI, as shown in Fig. 2. The results in Fig. 2 indicate that the sliced Wasserstein distances of patients with brain diseases (e.g., ADHD, ASD, and EMCI) are greater than those of NCs. These results suggest that the transportation cost of information transfer in the brain for patients is higher than that of NCs.

We use Eq.(4) to calculate the sliced Wasserstein distances between brain regions for ADHD, ASD and EMCI and then conduct statistical analysis on these distances. We identify three important brain regions where the sliced Wasserstein distances of patients with brain diseases significantly differed from those of NCs. In ADHD, important brain regions involve right paracentral lobule (PCL.R), left parahippocampal gyrus (PHG.L), and right angular gyrus (ANG.R). In ASD, important brain regions involve left superior frontal gyrus-medial orbital (ORB-supmed.L), right posterior cingulate gyrus (PCG.R), and right superior temporal gyrus (STG.R). In EMCI, important brain regions include right supplementary motor area (SMA.R), right superior occipital gyrus (SOG.R), and right inferior temporal gyrus (ITG.R).

4 Conclusion

In this paper, we propose a sliced Wasserstein graph kernel to measure the similarities between a pair of brain functional networks. We use this graph kernel to develop a classification framework of brain functional network. We perform the classification experiments in the brain functional network data including ADHD, ASD, and EMCI constructed from fMRI data. The results indicate that our proposed method outperforms the existing state-of-the-art graph kernels and graph neural networks in classification tasks. In computational speed, the proposed method is faster than latest graph kernels and graph neural networks.

Acknowledgement. This work was supported by the National Natural Science Foundation of China (Nos. 62136004, 62276130, 61732006, 62050410348), Jiangsu Funding Program for Excellent Postdoctoral Talent, and also by the Key Research and Development Plan of Jiangsu Province (No. BE2022842).

References

1. Akbari, A., Awais, M., Fatemifar, S., Kittler, J.: Deep order-preserving learning with adaptive optimal transport distance. IEEE Trans. Pattern Anal. Mach. Intell. **45**, 313–328 (2022)
2. Carriere, M., Cuturi, M., Oudot, S.: Sliced Wasserstein kernel for persistence diagrams. In: International Conference on Machine Learning, pp. 664–673. PMLR (2017)
3. Chang, C.C., Lin, C.J.: LIBSVM: a library for support vector machines. ACM Trans. Intell. Syst. Technol. **2**, 1–27 (2011)
4. Fee, C., Banasr, M., Sibille, E.: Somatostatin-positive gamma-aminobutyric acid interneuron deficits in depression: cortical microcircuit and therapeutic perspectives. Biol. Psychiat. **82**(8), 549–559 (2017)
5. Feragen, A., Kasenburg, N., Petersen, J., De Bruijne, M., Borgwardt, K.M.: Scalable kernels for graphs with continuous attributes. In: Advances in Neural Information Processing Systems, pp. 216–224 (2013)
6. Ge, Z., Liu, S., Li, Z., Yoshie, O., Sun, J.: Ota: optimal transport assignment for object detection. In: Proceedings of the IEEE Conference on Computer Vision and Pattern Recognition, pp. 303–312 (2021)
7. Hartmut, H., Thomas, H., Moll, G.H., Oliver, K.: A bimodal neurophysiological study of motor control in attention-deficit hyperactivity disorder: a step towards core mechanisms? Brain **4**, 1156–1166 (2014)
8. Kantorovitch, L.: On the translocation of masses. Manage. Sci. **5**(1), 1–4 (1958)
9. Kolouri, S., Nadjahi, K., Simsekli, U., Badeau, R., Rohde, G.: Generalized sliced Wasserstein distances. In: Advances in Neural Information Processing Systems, vol. 32 (2019)
10. Kolouri, S., Zou, Y., Rohde, G.K.: Sliced Wasserstein kernels for probability distributions. In: Proceedings of the IEEE Conference on Computer Vision and Pattern Recognition, pp. 5258–5267 (2016)
11. Le, T., Yamada, M., Fukumizu, K., Cuturi, M.: Tree-sliced variants of Wasserstein distances. In: Advances in Neural Information Processing Systems (2019)
12. Li, X., et al.: BrainGNN: interpretable brain graph neural network for fMRI analysis. Med. Image Anal. **74**, 102233 (2021)
13. Liu, M., Zhang, J., Adeli, E., Shen, D.: Joint classification and regression via deep multi-task multi-channel learning for Alzheimer's disease diagnosis. IEEE Trans. Biomed. Eng. **66**(5), 1195–1206 (2019)

14. Ma, K., Huang, S., Wan, P., Zhang, D.: Optimal transport based pyramid graph kernel for autism spectrum disorder diagnosis. Pattern Recogn. 109716 (2023)
15. Ma, K., Huang, S., Zhang, D.: Diagnosis of mild cognitive impairment with ordinal pattern kernel. IEEE Trans. Neural Syst. Rehabil. Eng. **30**, 1030–1040 (2022)
16. Ma, K., Wen, X., Zhu, Q., Zhang, D.: Optimal transport based ordinal pattern tree kernel for brain disease diagnosis. In: Wang, L., Dou, Q., Fletcher, P.T., Speidel, S., Li, S. (eds.) Medical Image Computing and Computer Assisted Intervention. MICCAI 2022. LNCS, vol. 13433, pp. 186–195. Springer, Cham (2022). https://doi.org/10.1007/978-3-031-16437-8_18
17. Ma, K., Yu, J., Shao, W., Xu, X., Zhang, Z., Zhang, D.: Functional overlaps exist in neurological and psychiatric disorders: a proof from brain network analysis. Neuroscience **425**, 39–48 (2020)
18. Ma, Z., Wei, X., Hong, X., Lin, H., Qiu, Y., Gong, Y.: Learning to count via unbalanced optimal transport. In: Association for the Advancement of Artificial Intelligence, vol. 35, pp. 2319–2327 (2021)
19. Nikolentzos, G., Meladianos, P., Vazirgiannis, M.: Matching node embeddings for graph similarity. In: Association for the Advancement of Artificial Intelligence (2017)
20. Peng, C., Wang, X., Pei, J., Zhu, W.: A survey on network embedding. IEEE Trans. Knowl. Data Eng. **31**(5), 833–852 (2019)
21. Shervashidze, N., Schweitzer, P., Jan, E., Leeuwen, V., Borgwardt, K.M.: Weisfeiler-Tehman graph kernels. J. Mach. Learn. Res. **12**(3), 2539–2561 (2011)
22. Togninalli, M., Ghisu, E., Llinares-López, F., Rieck, B., Borgwardt, K.: Wasserstein weisfeiler-lehman graph kernels. In: Advances in Neural Information Processing Systems, pp. 6439–6449 (2019)
23. Vishwanathan, S.V.N., Schraudolph, N.N., Kondor, R., Borgwardt, K.M.: Graph kernels. J. Mach. Learn. Res. **11**(2), 1201–1242 (2010)
24. Wang, M.L., Shao, W., Hao, X.K., Zhang, D.Q.: Machine learning for brain imaging genomics methods: a review. Mach. Intell. Res. **20**(1), 57–78 (2023)
25. Ying, Z., You, J., Morris, C., Ren, X., Hamilton, W., Leskovec, J.: Hierarchical graph representation learning with differentiable pooling. In: Advances in Neural Information Processing Systems, vol. 31 (2018)
26. Yu, Q., Sui, J., Kiehl, K.A., Pearlson, G.D., Calhoun, V.D.: State-related functional integration and functional segregation brain networks in schizophrenia. Schizophr. Res. **150**(2), 450–458 (2013)
27. Zhang, D., Huang, J., Jie, B., Du, J., Tu, L., Liu, M.: Ordinal pattern: a new descriptor for brain connectivity networks. IEEE Trans. Med. Imaging **37**(7), 1711–1722 (2018)
28. Zhao, P., Zhou, Z.: Label distribution learning by optimal transport. In: Association for the Advancement of Artificial Intelligence (2018)

A Reliable and Interpretable Framework of Multi-view Learning for Liver Fibrosis Staging

Zheyao Gao[1], Yuanye Liu[1], Fuping Wu[2], Nannan Shi[3], Yuxin Shi[3], and Xiahai Zhuang[1(✉)]

[1] School of Data Science, Fudan University, Shanghai, China
zxh@fudan.edu.cn
[2] Nuffield Department of Population Health, University of Oxford, Oxford, UK
[3] Department of Radiology, Shanghai Public Health Clinical Center, Shanghai, China
http://www.sdspeople.fudan.edu.cn/zhuangxiahai/

Abstract. Staging of liver fibrosis is important in the diagnosis and treatment planning of patients suffering from liver diseases. Current deep learning-based methods using abdominal magnetic resonance imaging (MRI) usually take a sub-region of the liver as an input, which nevertheless could miss critical information. To explore richer representations, we formulate this task as a multi-view learning problem and employ multiple sub-regions of the liver. Previously, features or predictions are usually combined in an implicit manner, and uncertainty-aware methods have been proposed. However, these methods could be challenged to capture cross-view representations, which can be important in the accurate prediction of staging. Therefore, we propose a reliable multi-view learning method with interpretable combination rules, which can model global representations to improve the accuracy of predictions. Specifically, the proposed method estimates uncertainties based on subjective logic to improve reliability, and an explicit combination rule is applied based on Dempster-Shafer's evidence theory with good power of interpretability. Moreover, a data-efficient transformer is introduced to capture representations in the global view. Results evaluated on enhanced MRI data show that our method delivers superior performance over existing multi-view learning methods.

Keywords: Liver fibrosis · Multi-view learning · Uncertainty

Z. Gao and Y. Liu—These two authors contribute equally.

X. Zhuang—This work was funded by the National Natural Science Foundation of China (grant No. 61971142 and 62111530195).

Supplementary Information The online version contains supplementary material available at https://doi.org/10.1007/978-3-031-43904-9_18.

H. Greenspan et al. (Eds.): MICCAI 2023, LNCS 14224, pp. 178–188, 2023.
https://doi.org/10.1007/978-3-031-43904-9_18

1 Introduction

Viral or metabolic chronic liver diseases that cause liver fibrosis impose great challenges on global health. Accurate staging for the severity of liver fibrosis is essential in the diagnosis of various liver diseases. Current deep learning-based methods [24,25] mainly use abdominal MRI and computed tomography (CT) data for liver fibrosis staging. Usually, a square sub-region of the liver instead of the whole image is cropped as input features, since the shape of the liver is irregular and unrelated anatomies in the abdominal image could disturb the training of deep learning models. To automatically extract the region of interest (ROI), a recent work [8] proposes to use slide windows to crop multiple image patches around the centroid of the liver for data augmentation. However, it only uses one patch as input at each time, which only captures a sub-view of the liver. To exploit informative features across the whole liver, we formulate this task as a multi-view learning problem and consider each patch as a view.

The aim of multi-view learning is to exploit complementary information from multiple features [23]. The central problem is how to integrate features from multiple views properly. In addition to the naive method that concatenates features at the input level [5], feature-level fusion strategies seek a common representation between different views through canonical correlation analysis [12,22] or maximizing the mutual information between different views using contrastive learning [1,21]. In terms of decision-level fusion, the widely used methods are decision averaging [18], decision voting [14], and attention-based decision fusion [9]. However, in the methods above, the weighting of multi-view features is either equal or learned implicitly through model training, which undermines the interpretability of the decision-making process. Besides, they are not capable of quantifying uncertainties, which could be non-trustworthy in healthcare applications.

To enhance the interpretability and reliability of multi-view learning methods, recent works have proposed uncertainty-aware decision-level fusion strategies. Typically, they first estimate uncertainties through Bayesian methods such

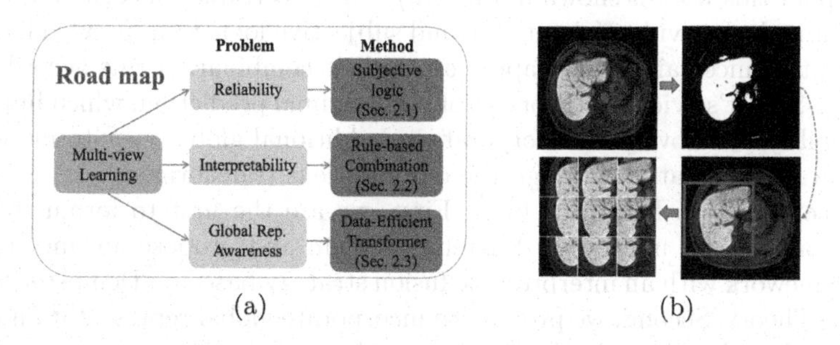

Fig. 1. (a) The road map of this work. (b) The pipeline to extract sub-views of the liver. First, the foreground is extracted using intensity-based segmentation. Based on the segmentation, a square region of interest (ROI) centered at the centroid of the liver is cropped. Then overlapped sliding windows are used in the ROI to obtain nine sub-views of the liver.

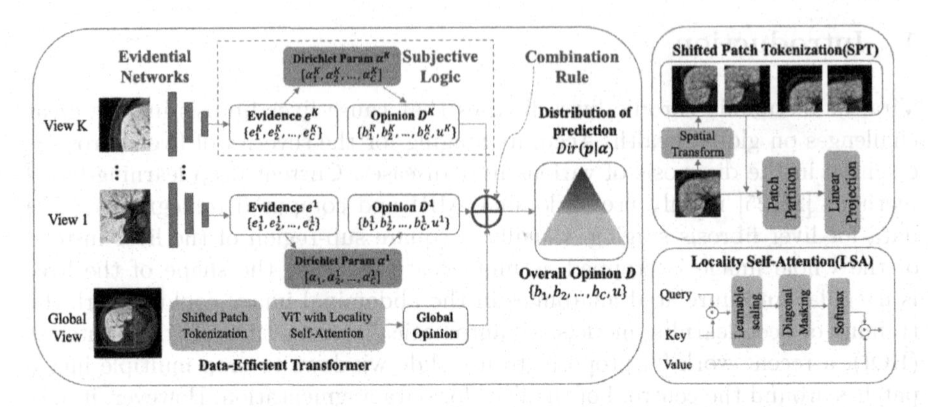

Fig. 2. The left side shows the main framework. Multi-view images are first encoded as evidence vectors by evidential networks. For each view, an opinion with uncertainty u is derived from evidence, under the guidance of subjective logic. Finally, the opinions are combined based on an explicit rule to derive the overall opinion, which can be converted to the distribution of classification probabilities. The right side illustrates the SPT and LSA modules in the data-efficient transformer that serves as the evidential network for the global view.

as Monte-Carlo dropout [20], variational inference [19], ensemble methods [4], and evidential learning [17]. Then, the predictions from each view are aggregated through explicit uncertainty-aware combination rules [7,20], as logic rules are commonly acknowledged to be interpretable in a complex model [26]. However, the predictions before the combination are made based on each independent view. Cross-view features are not captured to support the final prediction. In our task, global features could also be informative in the staging of liver fibrosis.

In this work, we propose an uncertainty-aware multi-view learning method with an interpretable fusion strategy of liver fibrosis staging, which captures both global features across views and local features in each independent view. The road map for this work is shown in Fig. 1(a). The uncertainty of each view is estimated through the evidential network and subjective logic to improve reliability. Based on the uncertainties, we apply an explicit combination rule according to Dempster-Shafer's evidence theory to obtain the final prediction, which improves explainability. Moreover, we incorporate an additional global view to model the cross-view representation through the data-efficient transformer.

Our contribution has three folds. First, we are the first to formulate liver fibrosis staging as a multi-view learning problem and propose an uncertainty-aware framework with an interpretable fusion strategy based on Dempster-Shafer Evidence Theory. Second, we propose to incorporate global representation in the multi-view learning framework through the data-efficient transformer network. Third, we evaluate the proposed framework on enhanced liver MRI data. The results show that our method outperforms existing multi-view learning methods and yields lower calibration errors than other uncertainty estimation methods.

2 Methods

The aim of our method is to derive a distribution of class probabilities with uncertainty based on multiple views of a liver image. The pipeline for view extraction is shown in Fig. 1(b). A square region of interest (ROI) is cropped based on the segmentation of the foreground. Then nine sub-views of the liver are extracted in the ROI through overlapped sliding windows. The multi-view learning framework is shown in Fig. 2. Our framework mainly consists of three parts, *i.e.*, evidential network, subjective logic, and combination rule. The evidential networks encode local views and the whole ROI as global view to evidence vectors $\boldsymbol{e}$. For local views, the networks are implemented with the convolutional structure. While for the global view, a data-efficient vision transformer with shifted patch tokenization (SPT) and locality self-attention (LSA) strategy is applied. Subjective logic serves as a principle that transforms the vector $\boldsymbol{e}$ into the parameter $\boldsymbol{\alpha}$ of the Dirichlet distribution of classification predictions, and the opinion $\boldsymbol{D}$ with uncertainty u. Then, Dempster's combination rule is applied to form the final opinion with overall uncertainty, which can be transformed into the final prediction. The details of subjective logic, Dempster's combination rule, the data-efficient transformer, and the training paradigm are discussed in the following sections.

2.1 Subjective Logic for Uncertainty Estimation

Subjective logic, as a generalization of the Bayesian theory, is a principled method of probabilistic reasoning under uncertainty [10]. It serves as the guideline of the estimation of both uncertainty and distribution of predicted probabilities in our framework. Given an image x_k from view k, $k \in \{1, 2, \cdots, K\}$, the evidence vector $\boldsymbol{e}^k = [e_1^k, e_2^k, ..., e_C^k]$ with non-negative elements for C classes is estimated through the evidential network, which is implemented using a classification network with softplus activation for the output.

According to subjective logic, the Dirichlet distribution of class probabilities $Dir(\boldsymbol{p}^k|\boldsymbol{\alpha}^k)$ is determined by the evidence. For simplicity, we follow [17] and derive the parameter of the distribution by $\boldsymbol{\alpha}^k = \boldsymbol{e}^k + 1$. Then the Dirichlet distribution is mapped to an opinion $\boldsymbol{D}^k = \{\{b_c^k\}_{c=1}^C, u^k\}$, subject to

$$u^k + \sum_{c=1}^{C} b_c^k = 1, \tag{1}$$

where $b_c^k = \frac{\alpha_c^k - 1}{S^k}$ is the belief mass for class c, $S^k = \sum_{c=1}^C \alpha_c^k$ is the Dirichlet strength, and $u^k = \frac{C}{S^k}$ indicates the uncertainty.

The predicted probabilities $\tilde{\boldsymbol{p}}^k \in \mathbb{R}^C$ of all classes are the expectation of Dirichlet distribution, *i.e.*, $\tilde{\boldsymbol{p}}^k = \mathbb{E}_{Dir(\boldsymbol{p}^k|\boldsymbol{\alpha}^k)}[\boldsymbol{p}^k]$. Therefore, the uncertainty u^k and predicted probabilities $\tilde{\boldsymbol{p}}^k$ can be derived in an end-to-end manner.

2.2 Combination Rule

Based on opinions derived from each view, Dempster's combination rule [11] is applied to obtain the overall opinion with uncertainty, which could be converted to the distribution of the final prediction. Specifically, given opinions

$\boldsymbol{D}^1 = \{\{b_c^1\}_{c=1}^C, u^1\}$ and $\boldsymbol{D}^2 = \{\{b_c^2\}_{c=1}^C, u^2\}$, the combined opinion $\boldsymbol{D} = \{\{b_c\}_{c=1}^C, u\} = \boldsymbol{D}^1 \oplus \boldsymbol{D}^2$ is derived by the following rule,

$$b_c = \frac{1}{N}(b_c^1 b_c^2 + b_c^1 u^2 + b_c^2 u^1), u = \frac{1}{N} u^1 u^2, \tag{2}$$

where $N = 1 - \sum_{i \neq j} b_i^1 b_j^2$ is the normalization factor. According to Eq. (2), the combination rule indicates that the combined belief b_c depends more on the opinion which is confident (with small u). In terms of uncertainty, the combined u is small when at least one opinion is confident.

For opinions from K local views and one global view, the combined opinion could be derived by applying the above rule for K times, *i.e.*, $\boldsymbol{D} = \boldsymbol{D}^1 \oplus \cdots \oplus \boldsymbol{D}^K \oplus \boldsymbol{D}^{Global}$.

2.3 Global Representation Modeling

To capture the global representation, we apply a data-efficient transformer as the evidential network for the global view. We follow [13] and improve the performance of the transformer on small datasets by increasing locality inductive bias, *i.e.*, the assumption about relations between adjacent pixels. The standard vision transformer (ViT) [3] without such assumptions typically require more training data than convolutional networks [15]. Therefore, we adopt the SPT and LSA strategy to improve the locality inductive bias.

As shown in Fig. 2, SPT is different from the standard tokenization in that the input image is shifted in four diagonal directions by half the patch size, and the shifted images are concatenated with the original images in the channel dimension to further utilize spatial relations between neighboring pixels. Then, the concatenated images are partitioned into patches and linearly projected as visual tokens in the same way as ViT.

LSA modifies self-attention in ViT by sharpening the distribution of the attention map to pay more attention to important visual tokens. As shown in Fig. 2, diagonal masking and temperature scaling are performed before applying softmax to the attention map. Given the input feature $\boldsymbol{X}$, The LSA module is formalized as,

$$L(\boldsymbol{X}) = \text{softmax}(\mathcal{M}(\boldsymbol{q}\boldsymbol{k}^T)/\tau)\boldsymbol{v}, \tag{3}$$

where $\boldsymbol{q}, \boldsymbol{k}, \boldsymbol{v}$ are the query, key, and value vectors obtained by linear projections of $\boldsymbol{X}$. $\mathcal{M}$ is the diagonal masking operator that sets the diagonal elements of $\boldsymbol{q}\boldsymbol{k}^T$ to a small number (*e.g.*,$-\infty$). $\tau \in \mathbb{R}$ is the learnable scaling factor.

2.4 Training Paradigm

Theoretically, the proposed framework could be trained in an end-to-end manner. For each view k, we use the integrated cross-entropy loss as in [17],

$$\mathcal{L}_{ice}^k = \mathbb{E}_{\boldsymbol{p}^k \sim Dir(\boldsymbol{p}^k|\boldsymbol{\alpha}^k)}[\mathcal{L}_{CE}(\boldsymbol{p}^k, \boldsymbol{y}^k)] = \sum_{c=1}^{C} y_c^k(\psi(S^k) - \psi(\alpha_c^k)), \tag{4}$$

where ψ is the digamma function and $\boldsymbol{y}^k$ is the one-hot label. We also apply a regularization term to increase the uncertainty of misclassified samples,

$$\mathcal{L}^k = \mathcal{L}^k_{ice} + \lambda KL[Dir(\boldsymbol{p}^k|\tilde{\boldsymbol{\alpha}}^k)||Dir(\boldsymbol{p}^k|\mathbf{1})], \tag{5}$$

where λ is the balance factor which gradually increases during training and $\tilde{\boldsymbol{\alpha}}^k = \boldsymbol{y}^k + (1 - \boldsymbol{y}^k) \odot \boldsymbol{\alpha}^k$. The overall loss is the summation of losses from all views and the loss for the combined opinion,

$$\mathcal{L}_{Overall} = \mathcal{L}_{Combined} + \mathcal{L}_{Global} + \sum_{k=1}^{K} \mathcal{L}^k, \tag{6}$$

where $\mathcal{L}_{Combined}$ and $\mathcal{L}_{Global}$ are losses of the combined and global opinions, implemented in the same way as $\mathcal{L}^k$. In practice, we pre-train the evidential networks before training with Eq. (6). For local views, we use the model weights pre-trained on ImageNet, and the transformer is pre-trained on the global view images.

3 Experiments

3.1 Dataset

The proposed method was evaluated on Gd-EOB-DTPA-enhanced [25] hepatobiliary phase MRI data, including 342 patients acquired from two scanners, *i.e.*, Siemens 1.5T and Siemens 3.0T. The gold standard was obtained through the pathological analysis of the liver biopsy or liver resection within 3 months before and after MRI scans. Please refer to supplementary materials for more data acquisition details. Among all patients, 88 individuals were identified with fibrosis stage S1, 41 with S2, 40 with S3, and 174 with the most advanced stage S4. Following [25], the slices with the largest liver area in images were selected. The data were then preprocessed with z-score normalization, resampled to a resolution of $1.5 \times 1.5\,\text{mm}^2$, and cropped to 256×256 pixel. For multi-view extraction, the size of the ROI, window, and stride were 160, 96, 32, respectively.

For all experiments, a four-fold cross-validation strategy was employed, and results of two tasks with clinical significance [25] were evaluated, *i.e.*, staging cirrhosis (S4 vs S1-3) and identifying substantial fibrosis (S1 vs S2-4). To keep a balanced number of samples for each class, we over-sampled the S1 data and under-sampled S4 data in the experiments of staging substantial fibrosis.

3.2 Implementation Details

Augmentations such as random rescale, flip, and cutout [2] were applied during training. We chose ResNet34 as the evidential network for local views. For configurations of the transformer, please refer to supplementary materials. The framework was trained using Adam optimizer with an initial learning rate of $1e-4$ for 500 epochs, which was decreased by using the polynomial scheduler.

Table 1. Comparison with multi-view learning methods. Results are evaluated in accuracy (ACC) and area under the receiver operating characteristic curve (AUC) for both tasks.

Method	Cirrhosis(S4 vs S1-3)		Substantial Fibrosis(S1 vs S2-4)	
	ACC	AUC	ACC	AUC
SingleView [8]	77.1 ± 3.17	78.7 ± 4.17	78.2 ± 7.18	75.0 ± 11.5
Concat [5]	80.0 ± 2.49	81.8 ± 3.17	80.5 ± 2.52	83.3 ± 3.65
DCCAE [22]	80.6 ± 3.17	82.7 ± 4.03	83.1 ± 5.30	84.5 ± 4.77
CMC [21]	80.6 ± 1.95	83.5 ± 3.67	83.4 ± 3.22	85.3 ± 4.06
PredSum [18]	78.8 ± 4.16	78.2 ± 4.94	81.1 ± 2.65	84.9 ± 3.21
Attention [9]	76.2 ± 0.98	78.9 ± 3.72	81.4 ± 4.27	84.4 ± 5.34
Ours	$\mathbf{84.4 \pm 1.74}$	$\mathbf{89.0 \pm 0.03}$	$\mathbf{85.5 \pm 1.91}$	$\mathbf{88.4 \pm 1.84}$

The balance factor λ was set to increase linearly from 0 to 1 during training. The transformer network was pre-trained for 200 epochs using the same setting. The framework was implemented using Pytorch and was run on one Nvidia RTX 3090 GPU.

3.3 Results

Comparison with Multi-view Learning Methods. To assess the effectiveness of the proposed multi-view learning framework for liver fibrosis staging, we compared it with five multi-view learning methods, including Concat [5], DCCAE [22], CMC [21], PredSum [18], and Attention [9]. Concat is a commonly used method that concatenates multi-view images at the input level. DCCAE and CMC are feature-level strategies. PredSum and Attention are based on decision-level fusion. Additionally, SingleView [8] was adopted as the baseline method for liver fibrosis staging, which uses a single patch as input.

As shown in Table 1, our method outperformed the SingleView method by 10.3% and 12% in AUC on the two tasks, respectively, indicating that the proposed method could exploit more informative features than the method using single view. Our method also set the new state of the art, when compared with other multi-view learning methods. This could be due to the fact that our method was able to capture both the global and local features, and the uncertainty-aware fusion strategy could be more robust than the methods with implicit fusion strategies.

Comparison with Uncertainty-Aware Methods. To demonstrate reliability, we compared the proposed method with other methods. Specifically, these methods estimate uncertainty using Monte-Carlo dropout (Dropout) [20], variational inference (VI) [19], ensemble [4], and softmax entropy [16], respectively. Following [6], we evaluated the expected calibration error (ECE), which measures the gap between model confidence and expected accuracy.

Table 2. Comparison with uncertainty-aware methods. The expected calibration error (ECE) is evaluated in addition to ACC and AUC. Methods with lower ECE are more reliable.

Method	Cirrhosis(S4 vs S1-3)			Substantial Fibrosis(S1 vs S2-4)		
	ACC	AUC	ECE	ACC	AUC	ECE
Softmax	77.1 ± 3.17	78.7 ± 4.17	0.256 ± 0.040	78.2 ± 7.18	83.3 ± 3.65	0.237 ± 0.065
Dropout [20]	77.1 ± 4.89	79.8 ± 4.50	0.183 ± 0.063	80.2 ± 5.00	83.8 ± 6.12	0.171 ± 0.067
VI [19]	77.6 ± 2.20	79.5 ± 4.50	0.229 ± 0.020	81.1 ± 2.08	82.2 ± 6.12	0.191 ± 0.023
Ensemble [4]	78.1 ± 1.91	80.8 ± 3.13	0.181 ± 0.040	79.3 ± 5.11	80.4 ± 3.90	0.193 ± 0.031
Ours	$\mathbf{84.4 \pm 1.74}$	$\mathbf{89.0 \pm 0.03}$	$\mathbf{0.154 \pm 0.028}$	$\mathbf{85.5 \pm 1.91}$	$\mathbf{88.4 \pm 1.84}$	$\mathbf{0.156 \pm 0.019}$

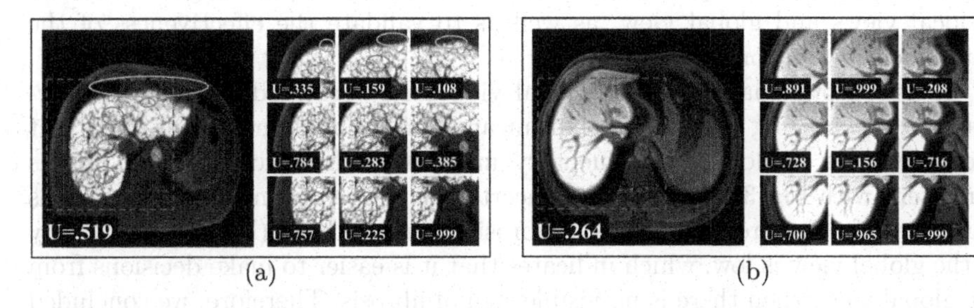

Fig. 3. Typical samples of stage 4 (a) and stage 1 (b). Visible signs of liver fibrosis are highlighted by circles. Yellow circles indicate the nodular surface contour and green circles denote numerous regenerative nodules. Uncertainties (U) of local and global views estimated by our model were demonstrated. Notably, local views of lower uncertainty contain more signs of fibrosis. Please refer to supplementary materials for more high-resolute images (Color figure online)

Table 2 shows that our method achieved better results in ACC and AUC for both tasks than the other uncertainty-ware multi-view learning methods. It indicates that the uncertainty in our framework could paint a clearer picture of the reliability of each view, and thus the final prediction was more accurate based on the proposed scheme of rule-based combination. Our method also achieved the lowest ECE, indicating that the correspondence between the model confidence and overall results was more accurate.

Interpretability. The proposed framework could explain which view of the input image contains more decisive information for liver fibrosis staging through uncertainties. To evaluate the quality of explanations, we compared the estimated uncertainties with annotations from experienced physicians. Views that contain more signs of fibrosis are supposed to have lower uncertainties. According to Fig. 3, the predicted uncertainties are consistent with annotations in local views of the S4 sample (a). In the S1 sample (b), the uncertainty of global view is low. It is reasonable since there are no visible signs of fibrosis in this stage. The model needs to capture the entire view to discriminate the S1 sample.

Table 3. Ablation study for the roles of local and global views, and effectiveness of the data-efficient transformer.

Method	Cirrhosis(S4 vs S1-3)			Substantial Fibrosis(S1 vs S2-4)		
	ACC	AUC	ECE	ACC	AUC	ECE
Global View solely	76.8 ± 2.81	79.4 ± 4.76	0.192 ± 0.071	82.4 ± 3.45	84.9 ± 5.42	0.192 ± 0.071
Local Views solely	84.1 ± 6.47	88.0 ± 8.39	$\mathbf{0.148 \pm 0.086}$	82.0 ± 6.07	86.9 ± 6.68	0.180 ± 0.060
Both views by CNN	82.9 ± 3.17	87.8 ± 3.09	0.171 ± 0.029	82.0 ± 3.54	87.1 ± 3.47	0.174 ± 0.039
Ours	$\mathbf{84.4 \pm 1.74}$	$\mathbf{89.0 \pm 0.03}$	0.154 ± 0.028	$\mathbf{85.5 \pm 1.91}$	$\mathbf{88.4 \pm 1.84}$	$\mathbf{0.156 \pm 0.019}$

Ablation Study. We performed this ablation study to investigate the roles of local views and global view, as well as to validate the effectiveness of the data-efficient transformer.

Table 3 shows that using the global view solely achieved the worst performance in the staging of cirrhosis. This means that it could be difficult to extract useful features without complementary information from local views. This is consistent with Fig. 3(a), where the uncertainty derived from the global view is high, even if there are many signs of fibrosis. While in Fig. 3(b), the uncertainty of the global view is low, which indicates that it is easier to make decisions from the global view when there is no visible sign of fibrosis. Therefore, we concluded that the global view was more valuable in identifying substantial fibrosis. Compared with the method that only used local views, our method gained more improvement in the substantial fibrosis identification task, which further confirms the aforementioned conclusion. Our method also performed better than the method that applied a convolution neural network (CNN) for the global view. This demonstrates that the proposed data-efficient transformer was more suitable for the modeling of global representation than CNN.

4 Conclusion

In this work, we have proposed a reliable and interpretable multi-view learning framework for liver fibrosis staging. Specifically, uncertainty is estimated through subjective logic to improve reliability, and an explicit fusion strategy is applied which promotes interpretability. Furthermore, we use a data-efficient transformer to model the global representation, which improves the performance.

References

1. Chen, T., Kornblith, S., Norouzi, M., Hinton, G.: A simple framework for contrastive learning of visual representations. In: International Conference on Machine Learning, pp. 1597–1607. PMLR (2020)
2. DeVries, T., Taylor, G.W.: Improved regularization of convolutional neural networks with cutout. arXiv preprint arXiv:1708.04552 (2017)
3. Dosovitskiy, A., et al.: An image is worth 16x16 words: Transformers for image recognition at scale. arXiv preprint arXiv:2010.11929 (2020)

4. Durasov, N., Bagautdinov, T., Baque, P., Fua, P.: Masksembles for uncertainty estimation. In: Proceedings of the IEEE/CVF Conference on Computer Vision and Pattern Recognition, pp. 13539–13548 (2021)
5. Feichtenhofer, C., Pinz, A., Zisserman, A.: Convolutional two-stream network fusion for video action recognition. In: Proceedings of the IEEE Conference on Computer Vision and Pattern Recognition, pp. 1933–1941 (2016)
6. Guo, C., Pleiss, G., Sun, Y., Weinberger, K.Q.: On calibration of modern neural networks. In: International Conference on Machine Learning, pp. 1321–1330. PMLR (2017)
7. Han, Z., Zhang, C., Fu, H., Zhou, J.T.: Trusted multi-view classification with dynamic evidential fusion. IEEE Trans. Pattern Anal. Mach. Intell. **45**, 2551–2566 (2022)
8. Hectors, S., et al.: Fully automated prediction of liver fibrosis using deep learning analysis of gadoxetic acid-enhanced MRI. Eur. Radiol. **31**, 3805–3814 (2021)
9. Ilse, M., Tomczak, J., Welling, M.: Attention-based deep multiple instance learning. In: International Conference on Machine Learning, pp. 2127–2136. PMLR (2018)
10. Jøsang, A.: Subjective Logic, vol. 4. Springer, Cham (2016). https://doi.org/10.1007/978-3-319-42337-1
11. Jøsang, A., Hankin, R.: Interpretation and fusion of hyper opinions in subjective logic. In: 2012 15th International Conference on Information Fusion, pp. 1225–1232. IEEE (2012)
12. Karami, M., Schuurmans, D.: Deep probabilistic canonical correlation analysis. In: Proceedings of the AAAI Conference on Artificial Intelligence, vol. 35, pp. 8055–8063 (2021)
13. Lee, S.H., Lee, S., Song, B.C.: Vision transformer for small-size datasets. arXiv preprint arXiv:2112.13492 (2021)
14. Liu, X., et al.: Late fusion incomplete multi-view clustering. IEEE Trans. Pattern Anal. Mach. Intell. **41**(10), 2410–2423 (2018)
15. Neyshabur, B.: Towards learning convolutions from scratch. Adv. Neural. Inf. Process. Syst. **33**, 8078–8088 (2020)
16. Pearce, T., Brintrup, A., Zhu, J.: Understanding softmax confidence and uncertainty. arXiv preprint arXiv:2106.04972 (2021)
17. Sensoy, M., Kaplan, L., Kandemir, M.: Evidential deep learning to quantify classification uncertainty. In: Advances in Neural Information Processing Systems, vol. 31 (2018)
18. Simonyan, K., Zisserman, A.: Two-stream convolutional networks for action recognition in videos. In: Advances in Neural Information Processing Systems, vol. 27 (2014)
19. Subedar, M., Krishnan, R., Meyer, P.L., Tickoo, O., Huang, J.: Uncertainty-aware audiovisual activity recognition using deep Bayesian variational inference. In: Proceedings of the IEEE/CVF International Conference on Computer Vision, pp. 6301–6310 (2019)
20. Tian, J., Cheung, W., Glaser, N., Liu, Y.C., Kira, Z.: Uno: uncertainty-aware noisy-or multimodal fusion for unanticipated input degradation. In: 2020 IEEE International Conference on Robotics and Automation (ICRA), pp. 5716–5723. IEEE (2020)
21. Tian, Y., Krishnan, D., Isola, P.: Contrastive multiview coding. In: Vedaldi, A., Bischof, H., Brox, T., Frahm, J.-M. (eds.) ECCV 2020. LNCS, vol. 12356, pp. 776–794. Springer, Cham (2020). https://doi.org/10.1007/978-3-030-58621-8_45

22. Wang, W., Arora, R., Livescu, K., Bilmes, J.: On deep multi-view representation learning. In: International Conference on Machine Learning, pp. 1083–1092. PMLR (2015)
23. Yan, X., Hu, S., Mao, Y., Ye, Y., Yu, H.: Deep multi-view learning methods: a review. Neurocomputing **448**, 106–129 (2021)
24. Yasaka, K., Akai, H., Kunimatsu, A., Abe, O., Kiryu, S.: Deep learning for staging liver fibrosis on CT: a pilot study. Eur. Radiol. **28**, 4578–4585 (2018)
25. Yasaka, K., Akai, H., Kunimatsu, A., Abe, O., Kiryu, S.: Liver fibrosis: deep convolutional neural network for staging by using gadoxetic acid-enhanced hepatobiliary phase mr images. Radiology **287**(1), 146–155 (2018)
26. Zhang, Y., Tiňo, P., Leonardis, A., Tang, K.: A survey on neural network interpretability. IEEE Trans. Emerg. Top. Comput. Intell. **5**(5), 726–742 (2021)

Utilizing Longitudinal Chest X-Rays and Reports to Pre-fill Radiology Reports

Qingqing Zhu[1], Tejas Sudharshan Mathai[2], Pritam Mukherjee[2], Yifan Peng[3], Ronald M. Summers[2], and Zhiyong Lu[1](✉)

[1] National Center for Biotechnology Information, National Library of Medicine, National Institutes of Health, Bethesda, MD, USA
zhiyong.lu@nih.gov

[2] Imaging Biomarkers and Computer-Aided Diagnosis Laboratory, Department of Radiology and Imaging Sciences, National Institutes of Health Clinical Center, Bethesda, MD, USA

[3] Department of Population Health Sciences, Weill Cornell Medicine, New York, NY, USA

Abstract. Despite the reduction in turn-around times in radiology reporting with the use of speech recognition software, persistent communication errors can significantly impact the interpretation of radiology reports. Pre-filling a radiology report holds promise in mitigating reporting errors, and despite multiple efforts in literature to generate comprehensive medical reports, there lacks approaches that exploit the longitudinal nature of patient visit records in the MIMIC-CXR dataset. To address this gap, we propose to use longitudinal multi-modal data, i.e., previous patient visit CXR, current visit CXR, and the previous visit report, to pre-fill the "findings" section of the patient's current visit. We first gathered the longitudinal visit information for 26,625 patients from the MIMIC-CXR dataset, and created a new dataset called *Longitudinal-MIMIC*. With this new dataset, a transformer-based model was trained to capture the multi-modal longitudinal information from patient visit records (CXR images + reports) via a cross-attention-based multi-modal fusion module and a hierarchical memory-driven decoder. In contrast to previous works that only uses current visit data as input to train a model, our work exploits the longitudinal information available to pre-fill the "findings" section of radiology reports. Experiments show that our approach outperforms several recent approaches by ≥3% on F1 score, and ≥2% for BLEU-4, METEOR and ROUGE-L respectively. Code will be published at https://github.com/CelestialShine/Longitudinal-Chest-X-Ray.

Keywords: Chest X-Rays · Radiology reports · Longitudinal data · Report Pre-Filling · Report Generation

Supplementary Information The online version contains supplementary material available at https://doi.org/10.1007/978-3-031-43904-9_19.

H. Greenspan et al. (Eds.): MICCAI 2023, LNCS 14224, pp. 189–198, 2023.
https://doi.org/10.1007/978-3-031-43904-9_19

1 Introduction

In current radiology practice, a signed report is often the primary form of communication, to communicate results of a radiological imaging exam between radiologist. Speech recognition software (SRS), which converts dictated words or sentences into text in a report, is widely used by radiologists. Despite SRS reducing the turn-around times for radiology reports, correcting any transcription errors in the report has been assumed by the radiologists themselves. But, persistent report communication errors due to SRS can significantly impact report interpretation, and also have dire consequences for radiologists in terms of medical malpractice [1]. These errors are most common for cross-sectional imaging exams (e.g., CT, MR) and chest radiography [2]. Problems also arise when re-examining the results from external examinations and in interventional radiology procedural reports. Such errors are due to many factors, including SRS finding a nearest match for a dictated word, the lack of natural language processing (NLP) for real-time recognition and dictation conversion [2], and unnoticed typographical mistakes. To mitigate these errors, a promising alternative is to automate the pre-filling of a radiology report with salient information for a radiologist to review. This enables standardized reporting via structured reporting.

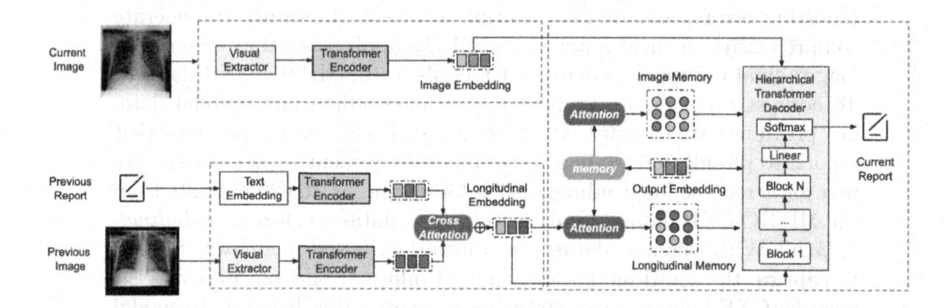

Fig. 1. Our proposed approach uses the CXR image and report from a previous patient visit and the current visit CXR image to pre-fill the "findings" section of the current visit report. The transformer-based model uses a cross-attention-based multi-modal fusion module and a hierarchical memory-driven decoder to generate the required text.

A number of methods to generate radiology reports have been proposed previously, with significant focus on CXR images [3–11]. Various attention mechanisms were proposed [4,6,12] to drive the encoder and the decoder to emphasize more informative words in the report, or visual regions in the CXR, and improve generation accuracy. Other approaches [8–10] effectively used Transformer-based models with memory matricies to store salient information for enhanced report generation quality. Despite these advances, there has been scarce research into harnessing *the potential of longitudinal patient visits* for improved patient care.

In practice, CXR images from multiple patient visits are usually examined simultaneously to find interval changes; e.g., a radiologist may compare a patient's current CXR to a previous CXR, and identify deterioration or improvement in the lungs for pneumonia. Reports from longitudinal visits contain valuable information regarding the patient's history, and harnessing the longitudinal multi-modal data is vital for the automated pre-filling of a comprehensive "findings" section in the report.

In this work, we propose to use longitudinal multi-modal data, i.e., previous visit CXR, current visit CXR, and previous visit report, to pre-fill the "findings" section of the patient's current visit report. To do so, we first gathered the longitudinal visit information for 26,625 patients from the MIMIC-CXR dataset[1] and created a new dataset called *Longitudinal-MIMIC*. Using this new dataset, we trained a transformer-based model containing a cross-attention-based multi-modal fusion module and a hierarchical memory-driven decoder to capture the features of longitudinal multi-modal data (CXR images + reports). In contrast to current approaches that only use the current visit data as input, our model exploits the longitudinal information available to pre-fill the "findings" section of reports with accurate content. Experiments conducted with the proposed dataset and model validate the utility of our proposed approach. Our main contribution in this work is training a transformer-based model that fully tackles the longitudinal multi-modal patient visit data to pre-fill the "findings" section of reports.

2 Methods

Dataset. The construction of the Longitudinal-MIMIC dataset involved several steps, starting with the MIMIC-CXR dataset, which is a large publicly available dataset of 377,110 chest X-ray images corresponding to 227,835 radiographic reports from 65,379 patients [13]. The first step in creating the Longitudinal-MIMIC dataset was to pre-process MIMIC-CXR to ensure consistency with prior works [8,9]. Specifically, patient visits where the report did not contain a "findings" section were excluded. For each patient visit, there was at least one chest X-ray image (frontal, lateral or other view) and a corresponding medical report. In our work, we only generated pre-filled reports with the "findings" section.

Table 1. A breakdown of the MIMIC-CXR dataset to show the number of patients with a specific number of visit records.

# visit records	1	2	3	4	5	>5
# patients	33,922	10,490	5,079	3,021	1,968	6,067

Next, the pre-processed dataset was partitioned into training, validation, and test sets using the official split provided with the MIMIC-CXR dataset. Table 1

[1] https://physionet.org/content/mimic-cxr-jpg/2.0.0/.

shows that 26,625 patients in MIMIC-CXR had $\geq$ 2 visit records, providing a large cohort of patients with longitudinal study data that could be used for our goal of pre-filling radiology reports. For patients with $\geq$2 visits, consecutive pairs of visits were used to capture richer longitudinal information. The dataset was then arranged chronologically based on the "StudyTime" attribute present in the MIMIC-CXR dataset. "StudyTime" represents the exact time at which a particular chest X-ray image and its corresponding medical report were acquired.

Following this, patients with $\geq$2 visit records were selected, resulting in 26,625 patients in the final *Longitudinal-MIMIC* dataset with a total of 94,169 samples. Each sample used during model training consisted of the current visit CXR, current visit report, previous visit CXR, and the previous visit report. The final dataset was divided into training (26,156 patients and 92,374 samples), validation (203 patients and 737 samples), and test (266 patients and 2,058 samples) splits. We aimed to create the *Longitudinal-MIMIC* dataset to enable the development and evaluation of models leveraging multi-modal data (CXR + reports) from longitudinal patient visits.

Model Architecture. Figure 1 shows the pipeline to generate a pre-filled "findings" section in the current visit report R_C, given the current visit CXR image I_C, previous visit CXR image I_P, and the previous visit report R_P. Mathematically, we can write: $p(R_C \mid I_C, I_P, R_P) = \prod_{t=1} p\left(w_t \mid w_1, \ldots, w_{t-1}, I_C, I_P, R_P\right)$, where w_i is the i-th word in the current report.

Encoder. Our model uses an *Image Encoder* and a *Text Encoder* to process the CXR images and text input separately. Both encoders were based on transformers. First, a pre-trained ResNet-101 [14] extracted image features $F = [f_1, \ldots, f_S]$ from the CXR images, where S is the number of patch features. They were then passed to the *Image Encoder*, which consisted of a stack of blocks. The encoded output was a list of encoded hidden states $H = [h_1, \ldots, h_S]$. The CXR images from the previous and the current visits were encoded in the same manner, and denoted by H^{I_P} and H^{I_C} respectively.

The *Text Encoder* encoded text information for language feature embedding using a previously published method [15]. First, the radiology report R_P was tokenized into a sequence of M tokens, and then transformed into vector representations $V = [v_1, \ldots, v_M]$ using a lookup table [16]. They were then fed to the *text encoder*, which had the same architecture as the *image encoder*, but with distinct network parameters. The final text feature embedding H^{R_P} was defined as: $H^{R_P} = \theta_R^E(V)$, where θ_R^E refers to the parameters of the report text encoder.

Cross-Attention Fusion Module. A multi-modal fusion module integrated longitudinal representations of images and texts using a cross-attention mechanism [17], which was defined as: $H^{I_P^*} = \operatorname{softmax}\left(\frac{q(H^{I_P})k(H^{R_P})^\top}{\sqrt{d_k}}\right) v\left(H^{R_P}\right)$ and $H^{R_P^*} = softmax\left(\frac{q(H^{R_P})k(H^{I_P})^\top}{\sqrt{d_k}}\right) v\left(H^{I_P}\right)$, where $q(\cdot)$, $k(\cdot)$, and $v(\cdot)$ are linear transformation layers applied to features of proposals. d_k is the number of attention heads for normalization. Finally, $H^{I_P^*}$ and $H^{R_P^*}$ were concatenated to obtain the multi-modal longitudinal representations H^L.

Hierarchical Decoder with Memory. Our model's backbone decoder is a Transformer decoder with multiple blocks (The architecture of an example block is shown in the supplementary material). The first block takes partial output embedding H^O as input during training and a pre-determined starting symbol during testing. Subsequent blocks use the output from the previous block as input. To incorporate the encoded H^L and H^{I_C}, we use a hierarchical structure for each block that divides it into two sub-blocks: D^I and D^L.

Sub-block-1 uses H^{I_C} and consists of a self-attention layer, an encoder-decoder attention layer, and feed-forward layers. It also employs residual connections and conditional layer normalization [8]. The encoder-decoder attention layer performs multi-head attention over H^{I_C}. It also uses a memory matrix M to store output and important pattern information. The memory representations not only store the information of generated current reports over time in the decoder, but also the information across different encoders. Following [8], we adopted a matrix M to store the output over multiple generation steps and record important pattern information. Then we enhance M by aligning it with H^{I_C} to create an attention-aligned memory M^{I_C} matrix. Different from [8], we use M^{I_C} while transforming the normalized data instead of M. The decoding process of sub-block-1 D^I is formalized as: $H^{dec,b,I} = D^I(H^O, H^{I_C}, M^{I_C})$, where b stands for the block index. The output of sub-block 1 is combined with H^O through a fusion layer: $H^{dec,b} = (1-\beta)H^O + \beta H^{dec,b,I}$. β is a hyper-parameter to balance H^O and $H^{dec,b,I}$. In our experiment, we set it to 0.2.

The input to sub-block-2 D^L is $H^{dec,b}$. This structure is similar to sub-block-1, but interacts with H^L instead of H^{I_C}. The output of this block is $H^{dec,b,L}$ and combined with $H^{dec,b,I}$ by adding them together. After fusing these embeddings and doing traditional layer normalization for them, we use these embeddings as the output of a block. The output of the previous block is used as the input of the next block. After N blocks, the final hidden states are obtained and used with a Linear and Softmax layer to get the target report probability distributions.

3 Experiments and Results

Baseline Comparisons. We compared our proposed method against prior image captioning and medical report generation works respectively. The same *Longitudinal-MIMIC* dataset was used to train all baseline models, such as AoANet [18], CNNTrans [16], Transformer [15], R2gen [8], and R2CMN [9]. Implementation of these methods is detailed in the supplementary material.

Evaluation Metrics. Conventional natural language generation (NLG) metrics, such as $BLEU$ [19], $METEOR$ [20], and $Rouge_L$ [21] were used to evaluate the utility of our approach against other baseline methods. Similar to prior work [8,16], the CheXpert labeler [22] classified the predicted report for the presence

Table 2. Results of the NLG metrics (BLEU (BL), Meteor (M), Rouge R_L) and clinical efficacy (CE) metrics (Accuracy, Precision, Recall and F-1 score) on the *Longitudinal-MIMIC* dataset. Best results are highlighted in bold.

Method	NLG metrics						CE metrics			
	BL-1	BL-2	BL-3	BL-4	M	R_L	A	P	R	F-1
AoANet	0.272	0.168	0.112	0.080	0.115	0.249	0.798	0.437	0.249	0.317
CNN+Trans	0.299	0.186	0.124	0.088	0.120	0.263	0.799	0.445	0.258	0.326
Transformer	0.294	0.178	0.119	0.085	0.123	0.256	0.811	0.500	0.320	0.390
R2gen	0.302	0.183	0.122	0.087	0.124	0.259	0.812	0.500	0.305	0.379
R2CMN	0.305	0.184	0.122	0.085	0.126	0.265	0.817	0.521	0.396	0.449
Ours	**0.343**	**0.210**	**0.140**	**0.099**	**0.137**	**0.271**	**0.823**	**0.538**	**0.434**	**0.480**
Baseline	0.294	0.178	0.119	0.085	0.123	0.256	0.811	0.500	0.320	0.390
+ report	0.333	0.201	0.133	0.094	0.135	0.268	0.823	0.539	0.411	0.466
+ image	0.320	0.195	0.130	0.092	0.130	0.268	0.817	0.522	0.34	0.412
simple fusion	0.317	0.193	0.128	0.090	0.130	0.266	0.818	0.521	0.396	0.450

of 14 disease conditions[2] and compared them against the labels of the ground-truth report. Clinical Efficacy (CE) metrics, such as; accuracy, precision, recall, and F-1 score, were used to evaluate model performance.

Results. Table 2 shows the summary of the NLG metrics and CE metrics for the 14 disease observations for our proposed approach when compared against prior baseline approaches. In particular, our model achieves the best performance over previous baselines across all NLG and CE metrics.

Generic image captioning approaches like AoANet resulted in unsatisfactory performance on the *Longitudinal-MIMIC* dataset as they failed to capture specific disease observations. Moreover, our approach outperforms previous report generation methods, R2Gen and R2CMN that also use memory-based models, due to the added longitudinal context arising from the use of longitudinal multi-modal study data (CXR images + reports). In our results, the BLEU scores show a substantial improvement, particularly in BLEU-4, where we achieve a 1.4% increase compared to the previous method R2CMN. BLEU scores measure how many continuous sequences of words appear in predicted reports, while $Rouge_L$ evaluates the fluency and sufficiency of predicted reports. The highest $Rouge_L$ score demonstrates the ability of our approach to generate accurate reports, rather than meaningless word combinations. We also use METEOR for evaluation, taking into account the precision, recall, and alignment of words and phrases in generated reports and the ground truth. Our METEOR score shows a 1.1% improvement over the previous outstanding method, which further solidifies the effectiveness of our approach. Meanwhile, our model exhibits a significant

[2] No Finding, Enlarged Cardiomediastinum, Cardiomegaly, Lung Lesion, Airspace Opacity, Edema, Consolidation, Pneumonia, Atelectasis, Pneumothorax, Pleural Effusion, Pleural Other, Fracture and Support Devices.

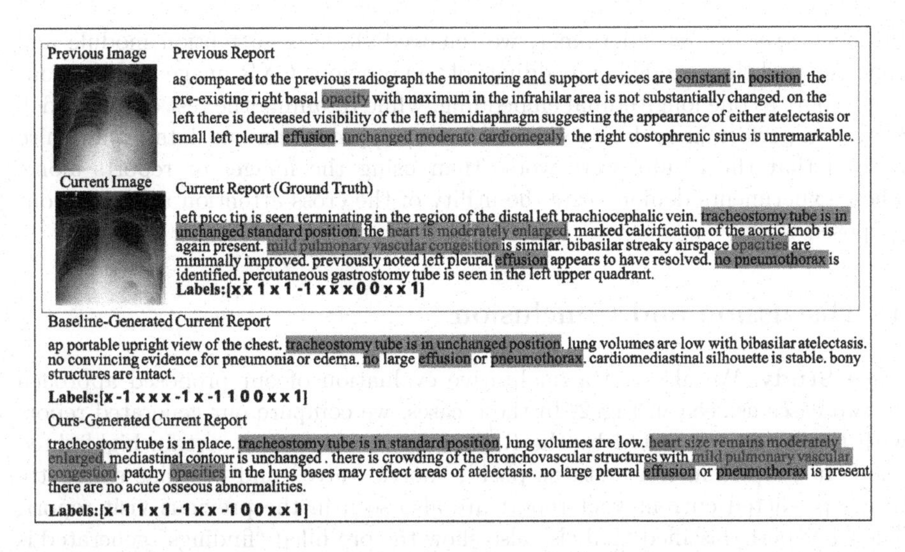

Fig. 2. Two examples of pre-filled "findings" sections of reports. Gray highlighted text indicates the same words or words with similar meaning that appear in the current reports and other reports. Purple highlighted text represents similar words in the current visit report generated by our approach, previous visit reports, and groundtruth current visit report. The red highlighted text indicates similar words that only exist in the report generated by our approach and the current ground truth report. R2Gen was the baseline method that generated the report. The "Labels" array shows the CheXpert classification of 14 disease observations (see text for details) as positive (1), negative (-1), uncertain (0) or unmentioned ($\times$). (Color figure online)

improvement in clinical efficacy metrics compared to other baselines. Notably, F1 is the most important metric, as it provides a balanced measure of both precision and recall. Our approach outperforms the best-performing method by 3.1% in terms of F1 score. These observations are particularly significant, as higher NLG scores do not necessarily correspond to higher clinical scores [8], confirming the effectiveness of our proposed method.

Effect of Model Components. We also studied the contribution of different model components and detail results in Table 2. The *Baseline* experiment refers to a basic Transformer model trained to generate a pre-filled report given a chest CXR image without any additional longitudinal information. The NLG and CE metrics are poor for the vanilla transformer compared to our proposed approach. We also analyze the contributions of the previous chest CXR image *+ image* and previous visit report *+ report* when added to the model separately. These two experiments included memory-enhanced conditional normalization. We observed that with each added feature enhanced the pre-filled report quality compared to the baseline, but the previous visit report had a higher impact than the previous CXR image. We hypothesize that the previous visit reports contain more text that can be directly transferred to the current visit reports.

In our *simple fusion* experiment, we removed the cross-attention module and concatenated the encoded embeddings of the previous CXR image and previous visit report as one longitudinal embedding, while retaining the rest of the model. We saw a performance drop compared to our approach on our dataset, and also noticed that the results were worse than using the images or reports alone. These experiments demonstrate the utility of the cross-attention module in our proposed work.

4 Discussion and Conclusion

Case Study. We also ran a qualitative evaluation of our proposed approach on two cases as seen in Fig. 2. In these cases, we compare our generated report with the report generated by the R2Gen. In the first case, certain highlighted words in purple, such as "status post", "aortic valve" and "cardiac silhouette in the predicted current visit report are also seen in the previous visit report. The CheXpert classified "Labels" also show the pre-filled "findings" generated is highly consistent with the ground truth report in contrast to the R2Gen model. For example, the "cardiac silhouette enlarged" was not generated by the R2Gen model, but our prediction contains them and is consistent with the word "cardiomegaly" in the ground truth report. In the second case, our generated report is also superior. Not only does our report generate more of the same content as the ground truth, but the positive diagnosis labels classified by CheXpert in our report are completely consistent with those in the ground truth. We also provide more cases in the supplementary material.

Error Analysis. To analyze errors from our model, we examine generated reports alongside ground truths and longitudinal information. It is found that the label accuracy of the observations in the generated reports is greatly affected by the previous information. For example, as time changes, for the same observation "pneumothorax", the label can change from "positive" to "negative". And such changing examples are more difficult to generate accurately. According to our statistics, on the one hand, when the label results of current and previous report are the same, 88.96% percent of the generated results match them. On the other hand, despite mentioning the same observations, when the labels of current and previous report are different, there is an 84.42% probability of generated results being incorrect. Thus how to track and generate the label accurately of these examples is a possible future work to improve the generated radiology reports. One possible way to address this issue is to use active learning [23] or curriculum learning [24] methods to differentiate different types of samples and better train the machine learning models.

Conclusion. In this paper, we propose to pre-fill the "findings" section of chest X-Ray radiology reports by considering the longitudinal multi-modal (CXR images + reports) information available in the MIMIC-CXR dataset. We gathered 26,625 patients with multiple visits to constitute the new *Longitudinal-MIMIC* dataset, and proposed a model to fuse encoded embeddings of multimodal data along with a hierarchical memory-driven decoder. The model generated a pre-filled "findings" section of the report, and we evaluated the generated results against prior image captioning and medical report generation works. Our model yielded a $\geq 3\%$ improvement in terms of the clinical efficacy F-1 score on the *Longitudinal-MIMIC* dataset. Moreover, experiments that evaluated the utility of different components of our model proved its effectiveness for the task of pre-filling the "findings" section of the report.

Acknowledgements. This research was supported by the Intramural Research Program of the National Library of Medicine and Clinical Center at the NIH. The authors thank to Qingyu Chen and Xiuying Chen for their time and effort in providing thoughtful comments and suggestions to revise this paper. This work was also supported by the National Institutes of Health under Award No. 4R00LM013001 (Peng), NSF CAREER Award No. 2145640 (Peng), and Amazon Research Award (Peng).

References

1. Smith, J.J., Berlin, L.: Signing a colleague's radiology report. Am. J. Roentgenol. **176**(1), 27–30 (2001). PMID: 11133532
2. Ringler, M.D., Goss, B.C., Bartholmai, B.J.: Syntactic and semantic errors in radiology reports associated with speech recognition software. Health Inform. J. **23**(1), 3–13 (2017)
3. Shin, H.-C., Roberts, K., Lu, L., Demner-Fushman, D., Yao, J., Summers, R.M.: Learning to read chest x-rays: Recurrent neural cascade model for automated image annotation. In Proceedings of the IEEE Conference on Computer Vision and Pattern Recognition, pp. 2497–2506 (2016)
4. Jing, B., Xie, P., Xing, E.: On the automatic generation of medical imaging reports. In: Proceedings of the 56th Annual Meeting of the Association for Computational Linguistics (Volume 1: Long Papers), pp. 2577–2586 (2018)
5. Li, Y., Liang, X., Hu, Z., Xing, E.P.: Hybrid retrieval-generation reinforced agent for medical image report generation. In: Advances in Neural Information Processing Systems, vol. 31 (2018)
6. Wang, X., Peng, Y., Lu, L., Lu, Z., Summers, R.M.: Tienet: text-image embedding network for common thorax disease classification and reporting in chest x-rays. In Proceedings of the IEEE Conference on Computer Vision and Pattern Recognition, pp. 9049–9058 (2018)
7. Jing, B., Wang, Z., Xing, E.: Show, describe and conclude: on exploiting the structure information of chest x-ray reports. arXiv preprint arXiv:2004.12274 (2020)
8. Chen, Z., Song, Y., Chang, T.-H., Wan, X.: Generating radiology reports via memory-driven transformer. In: Proceedings of the 2020 Conference on Empirical Methods in Natural Language Processing (EMNLP), pp. 1439–1449 (2020)
9. Teo, T.W., Choy, B.H.: STEM education in Singapore. In: Tan, O.S., Low, E.L., Tay, E.G., Yan, Y.K. (eds.) Singapore Math and Science Education Innovation. ETLPPSIP, vol. 1, pp. 43–59. Springer, Singapore (2021). https://doi.org/10.1007/978-981-16-1357-9_3
10. Wang, J., Bhalerao, A., He, Y.: Cross-modal prototype driven network for radiology report generation. In: Avidan, S., Brostow, G., Cissé, M., Farinella, G.M., Hassner, T. (eds.) Computer Vision. ECCV 2022. LNCS, vol. 13695, pp. 563–579. Springer, Cham (2022). https://doi.org/10.1007/978-3-031-19833-5_33

11. Liu, F., Wu, X., Ge, S., Fan, W., Zou, Y.: Exploring and distilling posterior and prior knowledge for radiology report generation. In: Proceedings of the IEEE/CVF Conference on Computer Vision and Pattern Recognition, pp. 13753–13762 (2021)
12. Xue, Y., et al.: Multimodal recurrent model with attention for automated radiology report generation. In: International Conference on Medical Image Computing and Computer-Assisted Intervention (2018)
13. Johnson, A.E.W., et al.: MIMIC-CXR-JPG, a large publicly available database of labeled chest radiographs. arXiv preprint arXiv:1901.07042 (2019)
14. Simonyan, K., Zisserman, A.: Very deep convolutional networks for large-scale image recognition. arXiv preprint arXiv:1409.1556 (2014)
15. Vaswani, A., et al.: Attention is all you need. In: Advances in Neural Information Processing Systems, vol. 30 (2017)
16. Moon, J.H., Lee, H., Shin, W., Choi, E.: Multi-modal understanding and generation for medical images and text via vision-language pre-training. arXiv preprint arXiv:2105.11333 (2021)
17. Nagrani, A., Yang, S., Arnab, A., Jansen, A., Schmid, C., Sun, C.: Attention bottlenecks for multimodal fusion. Adv. Neural. Inf. Process. Syst. **34**, 14200–14213 (2021)
18. Huang, L., Wang, W., Chen, J., Wei, X.-Y.: Attention on attention for image captioning. In: Proceedings of the IEEE/CVF International Conference on Computer Vision, pp. 4634–4643 (2019)
19. Papineni, K., Roukos, S., Ward, T., Zhu, W.-J.: Bleu: a method for automatic evaluation of machine translation. In: ACL 2002, pp. 311–318 (2002)
20. Denkowski, M., Lavie, A.: Meteor universal: language specific translation evasukhbaatar2015endluation for any target language. In: Proceedings of the Ninth Workshop on Statistical Machine Translation, pp. 376–380 (2014)
21. Chin-Yew Lin. Rouge: A package for automatic evaluation of summaries. In: Text Summarization Branches Out, pp. 74–81 (2004)
22. Irvin, J., et al.: Chexpert: a large chest radiograph dataset with uncertainty labels and expert comparison. Proc. AAAI Conf. Artif. Intell. **33**, 590–597 (2019)
23. Settles, B.: Active Learning. Synthesis Lectures on Artificial Intelligence and Machine Learning, vol. 6, pp. 1–114. Springer, Cham (2012). https://doi.org/10.1007/978-3-031-01560-1
24. Bengio, Y., Louradour, J., Collobert, R., Weston, J.: Curriculum learning. In: Proceedings of the 26th Annual International Conference on Machine Learning, pp. 41–48 (2009)

Parse and Recall: Towards Accurate Lung Nodule Malignancy Prediction Like Radiologists

Jianpeng Zhang[1,4(✉)], Xianghua Ye[2(✉)], Jianfeng Zhang[1,4], Yuxing Tang[5], Minfeng Xu[1,4], Jianfei Guo[1,4], Xin Chen[3], Zaiyi Liu[3], Jingren Zhou[1,4], Le Lu[5], and Ling Zhang[5]

[1] DAMO Academy, Alibaba Group, Hangzhou, China
jianpeng.zhang0@gmail.com
[2] The First Affiliated Hospital of College of Medicine, Zhejiang University, Hangzhou, China
hye1982@zju.edu.cn
[3] Guangdong Provincial People's Hospital, Guangzhou, China
[4] Hupan Lab, Hangzhou 310023, China
[5] DAMO Academy, Alibaba Group, New York, USA

Abstract. Lung cancer is a leading cause of death worldwide and early screening is critical for improving survival outcomes. In clinical practice, the contextual structure of nodules and the accumulated experience of radiologists are the two core elements related to the accuracy of identification of benign and malignant nodules. Contextual information provides comprehensive information about nodules such as location, shape, and peripheral vessels, and experienced radiologists can search for clues from previous cases as a reference to enrich the basis of decision-making. In this paper, we propose a radiologist-inspired method to simulate the diagnostic process of radiologists, which is composed of context parsing and prototype recalling modules. The context parsing module first segments the context structure of nodules and then aggregates contextual information for a more comprehensive understanding of the nodule. The prototype recalling module utilizes prototype-based learning to condense previously learned cases as prototypes for comparative analysis, which is updated online in a momentum way during training. Building on the two modules, our method leverages both the intrinsic characteristics of the nodules and the external knowledge accumulated from other nodules to achieve a sound diagnosis. To meet the needs of both low-dose and noncontrast screening, we collect a large-scale dataset of 12,852 and 4,029 nodules from low-dose and noncontrast CTs respectively, each with pathology- or follow-up-confirmed labels. Experiments on several datasets demonstrate that our method achieves advanced screening performance on both low-dose and noncontrast scenarios.

Supplementary Information The online version contains supplementary material available at https://doi.org/10.1007/978-3-031-43904-9_20.

H. Greenspan et al. (Eds.): MICCAI 2023, LNCS 14224, pp. 199–209, 2023.
https://doi.org/10.1007/978-3-031-43904-9_20

1 Introduction

Lung cancer screening has a significant impact on the rate of mortality associated with lung cancer. Studies have proven that regular lung cancer screening with low-dose computed tomography (LDCT) can lessen the rate of lung cancer mortality by up to 20% [1,14]. As most (e.g., 95% [13]) of the detected nodules are benign, it is critical to accurately assess their malignancy on CT to achieve a timely diagnosis of malignant nodules and avoid unnecessary procedures such as biopsy for benign ones. Particularly, the evaluation of nodule (i.e., 8–30mm) malignancy is recommended in the guidelines [13].

One of the major challenges of lung nodule malignancy prediction is the quality of datasets [6]. It is characterized by a lack of standard-of-truth of labels for malignancy [16, 27], and due to this limitation, many studies use radiologists' subjective judgment on CT as labels, such as LIDC-IDRI [3]. Recent works have focused on collecting pathologically labeled data to develop reliable malignancy prediction models [16,17,19]. For example, Shao et al. [16] collated a pathological gold standard dataset of 990 CT scans. Another issue is most of the studies focus on LDCT for malignancy prediction [10]. However, the majority of lung nodules are incidentally detected by routine imaging other than LDCT [4,15], such as noncontrast chest CT (NCCT, the most frequently performed CT exam, nearly 40% [18]).

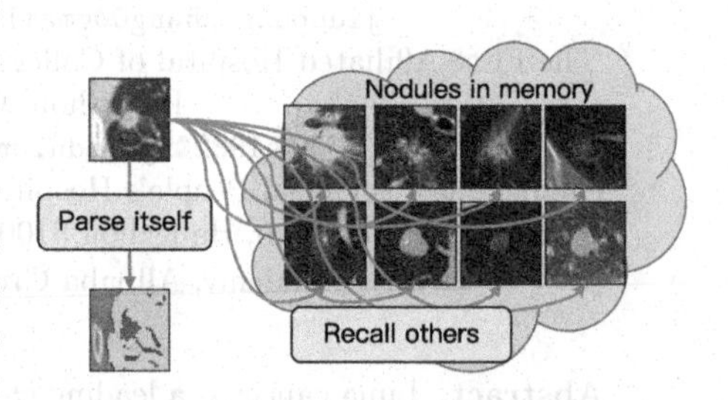

Fig. 1. In PARE, a nodule is diagnosed from two levels: first **parsing** the contextual information contained in the nodule itself, and then **recalling** the previously learned nodules to look for related clues.

Technically, current studies on lung nodule malignancy prediction mainly focus on deep learning-based techniques [10,12,17,23,24]. Liao et al. [10] trained a 3D region proposal network to detect suspicious nodules and then selected the top five to predict the probability of lung cancer for the whole CT scan, instead of each nodule. To achieve the nodule-level prediction, Xie et al. [24] introduced a knowledge-based collaborative model that hierarchically ensembles multi-view predictions at the decision level for each nodule. Liu et al. [12] extracted both nodules' and contextual features and fused them for malignancy prediction. Shi et al. [17] effectively improved the malignancy prediction accuracy by using a transfer learning and semi-supervised strategy. Despite their advantages in representation learning, these methods do not take into account expert diagnostic knowledge and experience, which may lead to a bad consequence of poor generalization. We believe a robust algorithm should be closely related to the diagnosis experience of professionals, working like a radiologist rather than a black box.

In this paper, we suggest mimicking radiologists' diagnostic procedures from intra-context **par**sing and inter-nodule **re**calling (see illustrations in Fig. 1),

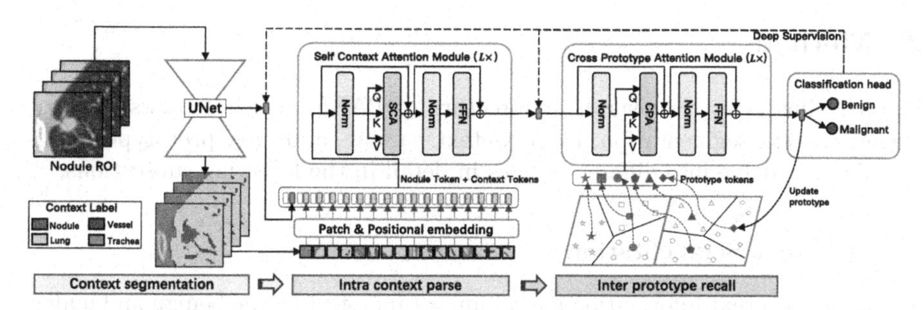

Fig. 2. Overview architecture of our proposed PARE model.

abbreviated as **PARE**. At the intra-level, the contextual information of the nodules provides clues about their shape, size, and surroundings, and the integration of this information can facilitate a more reliable diagnosis of whether they are benign or malignant. Motivated by this, we first segment the context structure, *i.e.*, nodule and its surroundings, and then aggregate the context information to the nodule representation via the attention-based dependency modeling, allowing for a more comprehensive understanding of the nodule itself. At the inter-level, we hypothesize that the diagnosis process does not have to rely solely on the current nodule itself, but can also find clues from past learned cases. This is similar to how radiologists rely on their accumulated experience in clinical practice. Thus, the model is expected to have the ability to store and recall knowledge, *i.e.*, the knowledge learned can be recorded in time and then recalled as a reference for comparative analysis. To achieve this, we condense the learned nodule knowledge in the form of prototypes, and recall them to explore potential inter-level clues as an additional discriminant criterion for the new case. To fulfill both LDCT and NCCT screening needs, we curate a large-scale lung nodule dataset with pathology- or follow-up-confirmed benign/malignant labels. For the LDCT, we annotate more than 12,852 nodules from 8,271 patients from the NLST dataset [14]. For the NCCT, we annotate over 4,029 nodules from over 2,565 patients from our collaborating hospital. Experimental results on several datasets demonstrate that our method achieves outstanding performance on both LDCT and NCCT screening scenarios.

Our contributions are summarized as follows: (1) We propose context parsing to extract and aggregate rich contextual information for each nodule. (2) We condense the diagnostic knowledge from the learned nodules into the prototypes and use them as a reference to assist in diagnosing new nodules. (3) We curate the largest-scale lung nodule dataset with high-quality benign/malignant labels to fulfill both LDCT and NCCT screening needs. (4) Our method achieves advanced malignancy prediction performance in both screening scenarios (0.931 AUC), and exhibits strong generalization in external validation, setting a new state of the art on LUNGx (0.801 AUC).

2 Method

Figure 2 illustrates the overall architecture of PARE, which consists of three stages: context segmentation, intra context parsing, and inter prototype recalling. We now delve into different stages in detail in the following subsections.

2.1 Context Segmentation

The nodule context information has an important effect on the benign and malignant diagnosis. For example, a nodule associated with vessel feeding is more likely to be malignant than a solitary one [22]. Therefore, we use a U-like network (UNet) to parse the semantic mask m for the input image patch x, thus allowing subsequent context modeling of both the nodule and its surrounding structures. Specifically, each voxel of m belongs to $\{0 : \text{background}, 1 : \text{lung}, 2 : \text{nodule}, 3 : \text{vessel}, 4 : \text{trachea}\}$. This segmentation process allows PARE to gather comprehensive context information that is crucial for an accurate diagnosis. For the diagnosis purpose, we extract the global feature from the bottleneck of UNet as the nodule embedding q, which will be used in later diagnostic stages.

2.2 Intra Context Parse

In this stage, we attempt to enhance the discriminative representations of nodules by aggregating contextual information produced by the segmentation model. Specifically, the context mask is tokenized into a set of sequences via the overlapped patch embedding. The input image is also split into patches and then embedded into the context tokens to keep the original image information. Besides, positional encoding is added in a learnable manner to retain location information. Similar to the class token in ViT [7], we prepend the nodule embedding token to the context sequences, denoted by $[q; t_1, ..., t_g] \in \mathbb{R}^{(g+1)\times D}$. Here g is the number of context tokens, and D represents the embedding dimension. Then we perform the self-attention modeling on these tokens simultaneously, called Self Context Attention (SCA), to aggregate context information into the nodule embedding. The nodule embedding token at the output of the last SCA block serves as the updated nodule representation. We believe that explicitly modeling the dependency between nodule embedding and its contextual structure can lead to the evolution of more discriminative representations, thereby improving discrimination between benign and malignant nodules.

2.3 Inter Prototype Recall

Definition of the Prototype: To retain previously acquired knowledge, a more efficient approach is needed instead of storing all learned nodules in memory, which leads to a waste of storage and computing resources. To simplify this process, we suggest condensing these pertinent nodules into a form of prototypes. As for a group of nodules, we cluster them into N groups $\{C_1, ..., C_N\}$ by minimizing the objective function $\sum_{i=1}^{N}\sum_{p\in C_i} d(p, \boldsymbol{P}_i)$ where d is the Euclidean distance function and p represents the nodule embedding, and refer the center of each cluster, $\boldsymbol{P}_i = \frac{1}{|C_i|}\sum_{p\in C_i} p$, as its prototype. Considering the differences between benign and malignant nodules, we deliberately divide the prototypes into benign and malignant groups, denoted by $\boldsymbol{P}^B \in \mathbb{R}^{N/2\times D}$ and $\boldsymbol{P}^M \in \mathbb{R}^{N/2\times D}$.

Cross Prototype Attention: In addition to parsing intra context, we also encourage the model to capture inter-level dependencies between nodules and external prototypes. This enables PARE to explore relevant identification basis beyond individual nodules. To accomplish this, we develop a Cross-Prototype Attention (CPA) module that utilizes nodule embedding as the query and the prototypes as the key and value. It allows the nodule embedding to selectively attend to the most relevant parts of prototype sequences. The state of query at the output of the last CPA module servers as the final nodule representation to predict its malignancy label, either "benign" ($y = 0$) or "malignant" ($y = 1$).

Updating Prototype Online: The prototypes are updated in an online manner, thereby allowing them to adjust quickly to changes in the nodule representations. As for the nodule embedding q of the data (x, y), its nearest prototype is singled out and then updated by the following momentum rules,

$$\begin{cases} \boldsymbol{P}^B_{\arg\min_j d(q,\boldsymbol{P}^B_j)} = \lambda \cdot \boldsymbol{P}^B_{\arg\min_j d(q,\boldsymbol{P}^B_j)} + (1-\lambda)\cdot q & if\ y=0 \\ \boldsymbol{P}^M_{\arg\min_j d(q,\boldsymbol{P}^M_j)} = \lambda \cdot \boldsymbol{P}^M_{\arg\min_j d(q,\boldsymbol{P}^M_j)} + (1-\lambda)\cdot q & otherwise \end{cases} \tag{1}$$

where λ is the momentum factor, set to 0.95 by default. The momentum updating can help accelerate the convergence and improve the generalization ability.

2.4 Training Process of PARE

The Algorithm 1 outlines the training process of our PARE model which is based on two objectives: segmentation and classification. The Dice and cross-entropy loss are combined for segmentation, while cross-entropy loss is used for classification. Additionally, deep classification supervision is utilized to enhance the representation of nodule embedding in shallow layers like the output of the UNet and SCA modules.

3 Experiment

3.1 Datasets and Implementation Details

Data Collection and Curation: NLST is the first large-scale LDCT dataset for low-dose CT lung cancer screening purpose [14]. There are 8,271 patients enrolled in this study. An experienced radiologist chose the last CT scan of each

Algorithm 1. Training process of PARE model.

1: **for** *iteration* = 1, 2, ... **do**
2: $\{x, m, y\} \leftarrow$ Sample(D) ▷ Sample a data
3: $\{q, s\} \leftarrow$ UNet(x) ▷ Infer UNet backbone
4: $z_0 \leftarrow [q; t_1, ..., t_g]$ ▷ Patch embedding and positional encoding
5: $p_3 \leftarrow$ MLP(q) ▷ Deep classification supervision
6: **for** l = 1, ..., L **do**
7: $z_l' \leftarrow \mathrm{SCA}(\mathrm{LN}(z_{l-1})) + z_{l-1}$ ▷ self context attention
8: $z_l \leftarrow \mathrm{MLP}(\mathrm{LN}(z_l')) + z_l'$
9: **end for**
10: $p_2 \leftarrow \mathrm{MLP}(z_L^0)$ ▷ Deep classification supervision
11: **for** l = 1, ..., L **do**
12: $z_l' \leftarrow \mathrm{CPA}(\mathrm{LN}(z_{l-1})) + z_{l-1}$ ▷ Cross prototype attention
13: $z_l \leftarrow \mathrm{MLP}(\mathrm{LN}(z_l')) + z_l'$
14: **end for**
15: $p_1 \leftarrow \mathrm{MLP}(z_L^0)$ ▷ Classification head
16: Update prototype according to Eq. 1
17: $J \leftarrow \mathrm{seg\_loss}(m, s) + \sum_{i=1}^{3} \mathrm{cls\_loss}(y, p_i)$ ▷ Update loss
18: **end for**

patient, and localized and labeled the nodules in the scan as benign or malignant based on the rough candidate nodule location and whether the patient develops lung cancer provided by NLST metadata. The nodules with a diameter smaller than 4mm were excluded. **The in-house cohort** was retrospectively collected from 2,565 patients at our collaborating hospital between 2019 and 2022. Unlike NLST, this dataset is noncontrast chest CT, which is used for routine clinical care. **Segmentation annotation:** We provide the segmentation mask for our in-house data, but not for the NLST data considering its high cost of pixel-level labeling. The nodule mask of each in-house data was manually annotated with the assistance of CT labeler [20] by our radiologists, while other contextual masks such as lung, vessel, and trachea were generated using the TotalSegmentator [21].

Train-Val-Test: The training set contains 9,910 (9,413 benign and 497 malignant) nodules from 6,366 patients at NLST, and 2,592 (843 benign and 1,749 malignant) nodules from 2,113 patients at the in-house cohort. The validation set contains 1,499 (1,426 benign and 73 malignant) nodules from 964 patients at NLST. The NLST test set has 1,443 (1,370 benign and 73 malignant) nodules from 941 patients. The in-house test set has 1,437 (1,298 benign and 139 malignant) nodules from 452 patients. We additionally evaluate our method on the LUNGx [2] challenge dataset, which is usually used for external validation in previous work [6,11,24]. LUNGx contains 83 (42 benign and 41 malignant) nodules, part of which (13 scans) were contrast-enhanced. **Segmentation**: We also evaluate the segmentation performance of our method on the public nodule segmentation dataset LIDC-IDRI [3], which has 2,630 nodules with nodule segmentation mask. **Evaluation metrics**: The area under the receiver operating characteristic curve (AUC) is used to evaluate the malignancy prediction performance.

Implementation: All experiments in this work were implemented based on the nnUnet framework [8], with the input size of $32 \times 48 \times 48$, batch size of 64, and total training iterations of $10K$. In the context patch embedding, each patch token is generated from a window of $8 \times 8 \times 8$. The hyper-parameters of PARE are empirically set based on the ablation experiments on the validation set. For example, the Transformer layer is set to 4 in both SCA and CPA modules, and the number of prototypes is fixed to 40 by default. More details can be found in the ablation. Due to the lack of manual annotation of nodule masks for the NLST dataset, we can only optimize the segmentation task using our in-house dataset, which has manual nodule masks.

3.2 Experiment Results

Ablation Study: In Table 1, we investigate the impact of different configurations on the performance of PARE on the validation set, including Transformer layers, number of prototypes, embedding dimension, and deep supervision. We observe that a higher AUC score can be obtained by increasing the number of Transformer layers, increasing the number of prototypes, doubling the channel dimension of token embeddings, or using deep classification supervision. Based on the highest AUC score of 0.931, we empirically set L=4, N=40, D=256, and DS=True in the following experiments. In Table 2, we investigate the ablation study of different methods/modules on the validation set and observe the following results: (1) The pure segmentation method performs better than the pure classification method, primarily because it enables greater supervision at the pixel level, (2) the joint segmentation and classification is superior to any single method, indicating the complementary effect of both tasks, (3) both context parsing and prototype comparing contribute to improved performance on the strong baseline, demonstrating the effectiveness of both modules, and (4) segmenting more contextual structures such as vessels, lungs, and trachea provide a slight improvement, compared to solely segmenting nodules.

Table 1. Ablation comparison of hyper-parameters (Transformer layers (L), number of prototypes (N), embedding dimension (D), and deep supervision (DS))

L	N	D	DS	AUC
1	20	128	✓	0.912
2	20	128	✓	0.918
4	20	128	✓	0.924
4	10	128	✓	0.920
4	40	128	✓	0.924
4	40	256	✓	**0.931**
4	40	256	×	0.926

Table 2. Effectiveness of different modules. MT: multi-task learning. Context: intra context parsing. Prototype: inter prototype recalling. *: only nodule mask was used in the segmentation task.

Method	AUC
Pure classification	0.907
Pure segmentation	0.915
MT	0.916
MT+Context*	0.921
MT+Context	0.924
MT+Context+Prototype	**0.931**

Table 3. Comparison of different methods on both NLST and in-house test sets. †: pure classification; ‡: pure segmentation; ◇: multi-task learning; *: ensemble of deep supervision heads. Note that we add the segmentation task in CA-Net.

Method	NLST test set				In-house test set			
	<10 mm	10∼20 mm	>20 mm	All	<10 mm	10∼20 mm	>20 mm	All
CNN†	0.742	0.706	0.780	0.894	0.851	0.797	0.744	0.901
ASPP [5]†	0.798	0.716	0.801	0.902	0.854	0.788	0.743	0.901
MiT [25]†	0.821	0.755	0.810	0.908	0.858	0.784	0.751	0.904
nnUnet [8]‡	0.815	0.736	0.815	0.910	0.863	0.804	0.750	0.911
CA-Net [12]◇	0.833	0.759	0.807	0.916	0.878	0.786	0.779	0.918
PARE◇	0.882	0.770	0.826	0.928	0.892	0.817	**0.783**	0.927
PARE◇*	**0.890**	**0.781**	**0.827**	**0.931**	**0.899**	**0.821**	0.780	**0.931**

Comparison to Other Methods on Both Screening Scenarios: Table 3 presents a comparison of PARE with other advanced methods, including pure classification-based, pure segmentation-based, and multi-task-based methods. Stratification assessments were made in both test sets based on the nodule size distribution. The results indicate that the segmentation-based method outperforms pure classification methods, mainly due to its superior ability to segment contextual structures. Additionally, the multi-task-based CA-Net outperforms any single-task method. Our PARE method surpasses all other methods on both NLST and In-house test sets. Moreover, by utilizing the ensemble of multiple deep supervision heads, the overall AUC is further improved to 0.931 on both datasets.

External Evaluation on LUNGx: We used LUNGx [2] as an external test to evaluate the generalization of PARE. It is worth noting that these compared methods have never been trained on LUNGx. Table 4 shows that our PARE model achieves the highest AUC of 0.801, which is 2% higher than the best method DAR [11]. We also conducted a reader study to compare PARE with two experienced radiologists, who have 8 and 13 years of lung nodule diagnosis experience respectively. Results in Fig. 3 reveal that our method achieves performance comparable to that of radiologists.

Generalization on LDCT and NCCT: Our model is trained on a mix of LDCT and NCCT datasets, which can perform robustly across low-dose and regular-dose applications. We compare the generalization performance of the models obtained under three training data configurations (LDCT, NCCT, and a combination of them). We find that the models trained on either LDCT or NCCT dataset alone cannot generalize well to other modalities, with at least a 6% AUC drop. However, our mixed training approach performs best on both LDCT and NCCT with almost no performance degradation.

Table 4. Comparison to other competitive methods on LUNGx [2].

Method	AUC
NLNL [9]	0.683
D2CNN [26]	0.746
KBC [24]	0.768
DAR [11]	0.781
PARE (Ours)	**0.801**

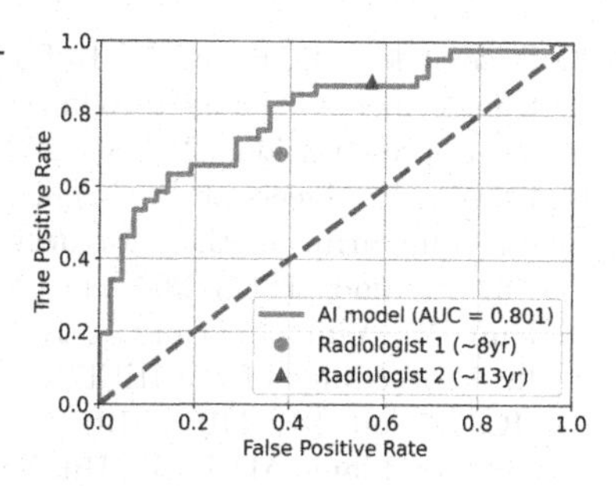

Fig. 3. Reader study compared with AI.

4 Conclusion

In this work, we propose the PARE model to mimic the radiologists' diagnostic procedures for accurate lung nodule malignancy prediction. Concretely, we achieve this purpose by parsing the contextual information from the nodule itself and recalling the previous diagnostic knowledge to explore related benign or malignancy clues. Besides, we curate a large-scale pathological-confirmed dataset with up to 13,000 nodules to fulfill the needs of both LDCT and NCCT screening scenarios. With the support of a high-quality dataset, our PARE achieves outstanding malignancy prediction performance in both scenarios and demonstrates a strong generalization ability on the external validation.

References

1. Ardila, D., et al.: End-to-end lung cancer screening with three-dimensional deep learning on low-dose chest computed tomography. Nat. Med. **25**(6), 954–961 (2019)
2. Armato, S.G., III., Drukker, K., Li, F., et al.: LUNGx challenge for computerized lung nodule classification. J. Med. Imaging **3**(4), 044506 (2016)
3. Armato, S.G., III., McLennan, G., Bidaut, L., et al.: The lung image database consortium (LIDC) and image database resource initiative (IDRI): a completed reference database of lung nodules on CT scans. Med. Phys. **38**(2), 915–931 (2011)
4. Bi, W.L., et al.: Artificial intelligence in cancer imaging: clinical challenges and applications. CA Cancer J. Clin. **69**(2), 127–157 (2019)
5. Chen, L.C., Papandreou, G., Kokkinos, I., Murphy, K., Yuille, A.L.: DeepLab: semantic image segmentation with deep convolutional nets, atrous convolution, and fully connected CRFs. IEEE Trans. Pattern Anal. Mach. Intell. **40**(4), 834–848 (2017)
6. Choi, W., Dahiya, N., Nadeem, S.: CIRDataset: a large-scale dataset for clinically-interpretable lung nodule radiomics and malignancy prediction. In: Wang, L., Dou, Q., Fletcher, P.T., Speidel, S., Li, S. (eds.) Medical Image Computing and Computer Assisted Intervention-MICCAI 2022. MICCAI 2022. Lecture Notes in Computer Science. vol. 13435. Springer, Cham (2022). https://doi.org/10.1007/978-3-031-16443-9_2

7. Dosovitskiy, A., Beyer, L., Kolesnikov, A., et al.: An image is worth 16x16 words: transformers for image recognition at scale. In: International Conference on Learning Representations (ICLR) (2021)
8. Isensee, F., Jaeger, P.F., Kohl, S.A., Petersen, J., Maier-Hein, K.H.: nnU-Net: a self-configuring method for deep learning-based biomedical image segmentation. Nat. Methods **18**(2), 203–211 (2021)
9. Kim, Y., Yim, J., Yun, J., Kim, J.: NLNL: negative learning for noisy labels. In: Proceedings of the IEEE/CVF International Conference on Computer Vision (ICCV), pp. 101–110 (2019)
10. Liao, F., Liang, M., Li, Z., Hu, X., Song, S.: Evaluate the malignancy of pulmonary nodules using the 3D deep leaky noisy-or network. IEEE Trans. Neural Netw. Learn. Syst. **30**(11), 3484–3495 (2019)
11. Liao, Z., Xie, Y., Hu, S., Xia, Y.: Learning from ambiguous labels for lung nodule malignancy prediction. IEEE Trans. Med. Imaging **41**(7), 1874–1884 (2022)
12. Liu, M., Zhang, F., Sun, X., Yu, Y., Wang, Y.: CA-Net: leveraging contextual features for lung cancer prediction. In: de Bruijne, M., et al. (eds.) MICCAI 2021. LNCS, vol. 12905, pp. 23–32. Springer, Cham (2021). https://doi.org/10.1007/978-3-030-87240-3_3
13. Mazzone, P.J., Lam, L.: Evaluating the patient with a pulmonary nodule: a review. JAMA **327**(3), 264–273 (2022)
14. National Lung Screening Trial Research Team: Reduced lung-cancer mortality with low-dose computed tomographic screening. N. Engl. J. Med. **365**(5), 395–409 (2011)
15. Osarogiagbon, R.U., et al.: Lung cancer diagnosed through screening, lung nodule, and neither program: a prospective observational study of the detecting early lung cancer (DELUGE) in the mississippi delta cohort. J. Clin. Oncol. **40**(19), 2094 (2022)
16. Shao, Y., et al.: LIDP: a lung image dataset with pathological information for lung cancer screening. In: Wang, L., Dou, Q., Fletcher, P.T., Speidel, S., Li, S. (eds.) Medical Image Computing and Computer Assisted Intervention-MICCAI 2022. MICCAI 2022. Lecture Notes in Computer Science. vol. 13433. Springer, Cham (2022). https://doi.org/10.1007/978-3-031-16437-8_74
17. Shi, F., et al.: Semi-supervised deep transfer learning for benign-malignant diagnosis of pulmonary nodules in chest CT images. IEEE Trans. Med. Imaging **41**(4), 771–781 (2021)
18. Sodickson, A., et al.: Recurrent CT, cumulative radiation exposure, and associated radiation-induced cancer risks from CT of adults. Radiology **251**(1), 175–184 (2009)
19. Wang, C., et al.: DeepLN: a multi-task AI tool to predict the imaging characteristics, malignancy and pathological subtypes in CT-detected pulmonary nodules. Front. Oncol. **12**, 683792 (2022)
20. Wang, F., et al.: A cascaded approach for ultraly high performance lesion detection and false positive removal in liver CT scans. arXiv preprint arXiv:2306.16036 (2023)
21. Wasserthal, J., Meyer, M., Breit, H.C., Cyriac, J., Yang, S., Segeroth, M.: Totalsegmentator: robust segmentation of 104 anatomical structures in CT images. arXiv preprint arXiv:2208.05868 (2022)
22. Wu, D., et al.: Stratified learning of local anatomical context for lung nodules in CT images. In: IEEE Computer Society Conference on Computer Vision and Pattern Recognition (CVPR), pp. 2791–2798 IEEE (2010)

23. Xie, Y., Xia, Y., Zhang, J., Feng, D.D., Fulham, M., Cai, W.: Transferable multi-model ensemble for benign-malignant lung nodule classification on chest CT. In: Descoteaux, M., Maier-Hein, L., Franz, A., Jannin, P., Collins, D.L., Duchesne, S. (eds.) MICCAI 2017. LNCS, vol. 10435, pp. 656–664. Springer, Cham (2017). https://doi.org/10.1007/978-3-319-66179-7_75
24. Xie, Y., et al.: Knowledge-based collaborative deep learning for benign-malignant lung nodule classification on chest CT. IEEE Trans. Med. Imaging **38**(4), 991–1004 (2018)
25. Xie, Y., Zhang, J., Xia, Y., Wu, Q.: UniMiSS: universal medical self-supervised learning via breaking dimensionality barrier. In: Avidan, S., Brostow, G., Cissé, M., Farinella, G.M., Hassner, T. (eds.) Computer Vision-ECCV 2022. ECCV 2022. Lecture Notes in Computer Science. vol. 13681. Springer, Cham (2022). https://doi.org/10.1007/978-3-031-19803-8_33
26. Yao, Y., Deng, J., Chen, X., Gong, C., Wu, J., Yang, J.: Deep discriminative CNN with temporal ensembling for ambiguously-labeled image classification. In: Proceedings of the AAAI Conference on Artificial Intelligence (AAAI). vol. 34, pp. 12669–12676 (2020)
27. Zhang, H., Chen, L., Gu, X., et al.: Trustworthy learning with (un)sure annotation for lung nodule diagnosis with CT. Med. Image Anal. **83**, 102627 (2023)

Privacy-Preserving Early Detection of Epileptic Seizures in Videos

Deval Mehta[1,2,3(✉)], Shobi Sivathamboo[4,5,6], Hugh Simpson[4,5], Patrick Kwan[4,5,6], Terence O'Brien[4,5], and Zongyuan Ge[1,2,3,7]

[1] AIM for Health Lab, Faculty of IT, Monash University, Melbourne, Australia
deval.mehta@monash.edu
[2] Monash Medical AI, Monash University, Melbourne, Australia
[3] Faculty of Engineering, Monash University, Melbourne, Australia
[4] Department of Neuroscience, Central Clinical School, Faculty of Medicine Nursing and Health Sciences, Monash University, Melbourne, Australia
[5] Department of Neurology, Alfred Health, Melbourne, Australia
[6] Departments of Medicine and Neurology, The University of Melbourne, Royal Melbourne Hospital, Parkville, VIC, Australia
[7] Airdoc-Monash Research Lab, Monash University, Melbourne, Australia
https://www.monash.edu/it/aimh-lab/home

Abstract. In this work, we contribute towards the development of video-based epileptic seizure classification by introducing a novel framework (SETR-PKD), which could achieve privacy-preserved early detection of seizures in videos. Specifically, our framework has two significant components - (1) It is built upon optical flow features extracted from the video of a seizure, which encodes the seizure motion semiotics while preserving the privacy of the patient; (2) It utilizes a transformer based progressive knowledge distillation, where the knowledge is gradually distilled from networks trained on a longer portion of video samples to the ones which will operate on shorter portions. Thus, our proposed framework addresses the limitations of the current approaches which compromise the privacy of the patients by directly operating on the RGB video of a seizure as well as impede real-time detection of a seizure by utilizing the full video sample to make a prediction. Our SETR-PKD framework could detect tonic-clonic seizures (TCSs) in a privacy-preserving manner with an accuracy of **83.9%** while they are only **half-way** into their progression. Our data and code is available at https://github.com/DevD1092/seizure-detection.

Keywords: epilepsy · early detection · knowledge distillation

1 Introduction

Epilepsy is a chronic neurological condition that affects more than 60 million people worldwide in which patients experience epileptic seizures due to abnormal

---

Supplementary Information The online version contains supplementary material available at https://doi.org/10.1007/978-3-031-43904-9_21.

H. Greenspan et al. (Eds.): MICCAI 2023, LNCS 14224, pp. 210–219, 2023.
https://doi.org/10.1007/978-3-031-43904-9_21

brain activity [17]. Different types of seizures are associated with the specific part of the brain involved in the abnormal activity [8]. Thus, accurate detection of the type of epileptic seizure is essential to epilepsy diagnosis, prognosis, drug selection and treatment. Concurrently, real-time seizure alerts are also essential for caregivers to prevent potential complications, such as related injuries and accidents, that may result from seizures. Particularly, patients suffering from tonic-clonic seizures (TCSs) are at a high risk of sudden unexpected death in epilepsy (SUDEP) [18]. Studies have shown that SUDEP is caused by severe alteration of cardiac activity actuated by TCS, leading to immediate death or cardiac arrest within minutes after the seizure [5]. Therefore, it is critical to accurately and promptly detect and classify epileptic seizures to provide better patient care and prevent any potentially catastrophic events.

The current gold standard practice for detection and classification of epileptic seizures is the hospital-based Video EEG Monitoring (VEM) units [23]. However, this approach is expensive and time consuming which is only available at specialized centers [3]. To address this issue, the research community has developed automated methods to detect and classify seizures based on several modalities - EEG [7,30], accelerometer [16], and even functional neuroimaging modalities such as fMRI [22] and electrocorticography (ECoG) [24]. Although, there have been developments of approaches for the above modalities, seizure detection using videos remains highly desirable as it involves no contact with the patient and is easier to setup and acquire data compared to other modalities. Thus, researchers have also developed automated approaches for the video modality.

Initial works primarily employed hand-crafted features based on patient motion trajectory by attaching infrared reflective markers to specific body key points [4,15]. However, these approaches were limited in performance due to their inability to generalize to changing luminance (night time seizures) or when the patient is occluded (covered by a bed sheet) [14]. Thus, very recently deep learning (DL) models have been explored for this task [1,2,12,21,29]. [29] demonstrated that DL models could detect generalized tonic-clonic seizures (GTCSs) from the RGB video of seizures. Authors in [21] radically used transfer learning (from action recognition task) to train DL networks for distinguishing focal onset seizures (FOSs) from bilateral TCSs using features extracted from the RGB video of seizures. Whereas, the authors in [12] developed a DL model to discriminate dystonia and emotion in videos of Hyperkinetic seizures. However, these developed approaches have two crucial limitations - (1) As these approaches directly operate on RGB videos, there is a possibility of privacy leakage of the sensitive patient data from videos. Moreover, obtaining consent from patients to share their raw RGB video data for building inter-cohort validation studies and generalizing these approaches on a large scale becomes challenging; (2) The current approaches consider the full video of a seizure to make predictions, which makes early detection of seizures impossible. The duration of a seizure varies significantly among patients, with some lasting as short as 30 s while others can take minutes to self-terminate. Thus, it is unrealistic to wait until the completion of a long seizure to make a prediction and alert caregivers.

In this work, we address the above two challenges by building an in-house dataset of privacy-preserved extracted features from a video and propose a framework for early detection of seizures. Specifically, we investigate two aspects - (1) The feasibility of detecting and classifying seizures based only on *optical flow*, a modality that captures temporal differences in a scene while being intrinsically privacy-preserving. (2) The potential of predicting the type of seizure during its progression by analyzing only a fraction of the video sample. Our early detection approach is inspired by recent developments in early action recognition in videos [9,10,19,21,28,31]. We develop a custom feature extractor-transformer framework, named **SE**izure **TR**ansformer (SETR) block for processing a single video sample. To achieve early detection from a fraction of the sample, we propose **P**rogressive **K**nowledge **D**istillation (PKD), where we gradually distill knowledge from SETR blocks trained on longer portions of a video sample to SETR blocks which will operate on shorter portions. We evaluate our proposed SETR-PKD framework on two datasets - an in-house dataset collected from a VEM unit in a hospital and a publicly available dataset of video-extracted features (GESTURES) [21]. Our experiments demonstrate that our proposed SETR-PKD framework can detect TCS seizures with an accuracy of **83.9%** in a privacy-preserving manner when they are only **half-way** into their progression. Furthermore, we comprehensively compare the performance of direct knowledge distillation with our PKD approach on both optical flow features (in-house dataset) and raw video features (public dataset). We firmly believe that our proposed method makes the first step towards developing a privacy-preserving real-time system for seizure detection in clinical practice.

2 Proposed Method

In this section, we first outline the process of extracting privacy-preserving information from RGB video samples to build our in-house dataset. Later, we explain our proposed approach for early detection of seizures in a sample.

2.1 Privacy Preserving Optical Flow Acquisition

Our in-house dataset of RGB videos of patients experiencing seizures resides on hospital premises and is not exportable due to the hospital's ethics agreement[1]. To work around this limitation, we develop a pipeline to extract optical flow information [11] from the videos. This pipeline runs locally within the hospital and preserves the privacy of the patients while providing us with motion semiotics of the seizures. An example of the extracted optical flow video sample can be seen in Fig. 1. We use the TV-L1 algorithm [20] to extract the optical flow features for each video, which we then export out of the hospital for building our proposed approach. We provide more information about our dataset, including the number of patients and seizures, annotation protocol, etc. in Sect. 3.

[1] We have a data ethics agreement approved for collection of data at hospital.

2.2 Early Detection of Seizures in a Sample

Consider an input optical flow video sample V_i as shown in Fig. 1(a) with a time period of T_i, consisting of N frames - $\{f_0, f_1, ...f_{N-1}\}$, and having a ground truth label of $y_i \in \{0, 1, ...C\}$ where is C the total number of categories. Then, the task of early detection is to build a framework that could classify the category of the sample correctly by analyzing the least possible partial segment of the sample. Thus, to define the problem of early detection, we split the sample V_i into k segments -$\{0, 1, ...k-1\}$ starting from the beginning to the end as shown in Fig. 1(b). Here V_i^{k-1} corresponds to the full video sample and the descending segments correspond to the reduced partial video samples. We build these partial segments by equally adding the temporal information throughout the sample i.e. the time period for a partial subset V_i^j of a sample V_i is computed as $(j+1) \times T_i/k$. Thus, the early detection task is to correctly predict the category y_i of the sample V_i from the lowest possible (j) partial segment V_i^j of V_i. In Fig. 1, we illustrate our proposed framework where - (a) First, we build a Seizure Transformer (SETR) block for processing a single optical flow video sample (b) Later, we employ SETR based Progressive Knowledge Distillation (SETR-PKD) to achieve early detection in a sample.

Processing a Single Sample. Since seizure patterns comprise of body movements, we implement transfer learning from a feature extractor pre-trained on action recognition task to extract the spatial features from the optical flow frames. Prior work [21] has shown that Temporal Segment Networks (TSNs) [27] pretrained on RGB videos of various actions are effective at extracting features from videos of seizures. We also utilize TSNs but pretrained on the optical flow modality, since we have privacy-preserved optical flow frames. The TSNs extract a 1D feature sequence for each frame f_j, referred as spatial features in Fig. 1(a). The spatial features are then processed by a linear transformation (1-layer MLP) that maps them into $motion_{tokens} \in \mathbb{R}^{N \times D}$, where each token has D-dimensions.

We leverage transformers to effectively learn temporal relations between the extracted spatial features of the seizure patterns. Following the strategy of ViT [6], after extracting the spatial features, we append a trainable class embedding $class_{embed} \in \mathbb{R}^D$ to the motion tokens. This class embedding serves to represent the temporal relationships between the motion tokens and is later used for classification ($class_{token}$ in Fig. 1(a)). As the order of the $motion_{tokens}$ is not known, we also add a learnable positional encoding $L_{POS} \in \mathbb{R}^{(N+1) \times D}$ to the combined $motion_{tokens}$ and $class_{embed}$. This is achieved using an element-wise addition and we term it as the input X_i for the input sample V_i.

To enable the interaction between tokens and learn temporal relationships for input sample classification, we employ the Vanilla Multi-Head Self Attention (MHSA) mechanism [26]. First, we normalize the input sequence $X_i \in \mathbb{R}^{(N+1) \times D}$ by passing it through a layer normalization, yielding $X_i^{'}$. We then use projection matrices $(Q_i, K_i, V_i) = (X_i^{'} W_i^Q, X_i^{'} W_i^K, X_i^{'} W_i^V)$ to project $X_i^{'}$ into queries (Q), keys (K), and values (V), where $W_i^{Q/K/V} \in \mathbb{R}^{D \times D}$ are the projection matrices

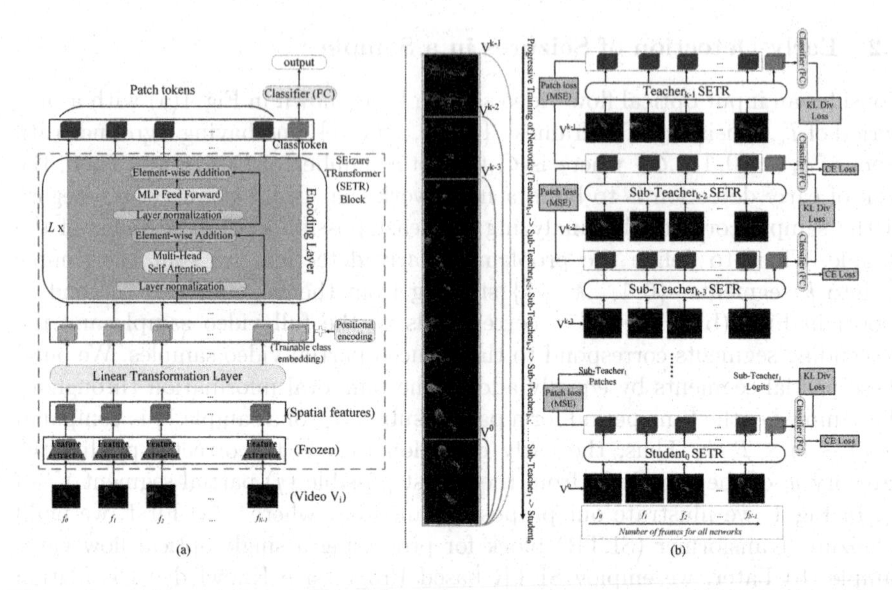

Fig. 1. Our proposed framework - (a) SEizure TRansformer (SETR) block for a single optical flow video sample (b) SETR based Progressive Knowledge Distillation (SETR-PKD) for early detection of seizures in a sample. (Best viewed in zoom and color).

for query, key, and value respectively. Next, we compute a dot product of Q with K and apply a softmax layer to obtain weights on the values. We repeat this self-attention computation N_h times, where N_h is the number of heads, and concatenate their outputs. Eq. 1, 2 depict the MHSA process in general.

$$A_i = Softmax(Q_i K_i) \tag{1}$$

$$MHSA(X_i^{'}) = A_i \times W_i^V, \qquad X_i^{'} = Norm(X_i) \tag{2}$$

Subsequently, the output of MHSA is passed to a two-layered MLP with GELU non-linearity while applying layer normalization and residual connections concurrently. Eq. 3, 4 represent this overall process.

$$m_l^{'} = MHSA(X_{l-1}^{'}) + X_{l-1}, \qquad l = 1...L \tag{3}$$

$$m_l = MLP(Norm(m_l^{'})) + m_l^{'}, \qquad l = 1...L \tag{4}$$

where $m_L \in \mathbb{R}^{(N+1)\times D}$ are the final output feature representations and L is the total number of encoding layers in the Transformer Encoder. Note that the first $\mathbb{R}^{N\times D}$ features correspond to the $patch_{tokens}$, while the final $\mathbb{R}^D$ correspond to the $class_{token}$ of the m_L as shown in Fig. 1(a). As mentioned earlier, we then use a one-layer MLP to predict the class label from the $class_{token}$. We refer to this whole process as a SEizure TRansformer (SETR) block shown in Fig. 1(a).

Progressive Knowledge Distillation. To achieve early detection, we use **K**nowledge **D**istillation in a **P**rogressive manner (PKD), starting from a SETR block trained on a full video sample and gradually moving to a SETR block trained on a partial video sample, as shown in Fig. 1(b). Directly distilling from a SETR block which has seen a significantly longer portion of the video (say V_i^{k-1}) to a SETR block which has only seen a smaller portion of the video sample (say V_i^0) will lead to considerable mismatches between the features extracted from the two SETRs as there is a large portion of the input sample that the $student_0$ SETR has not seen. In contrast, our proposed PKD operates in steps. First we pass the knowledge from teacher ($Teacher_{k-1}$ in Fig. 1(b)) SETR trained on V_i^{k-1} to a student ($Sub-teacher_{k-2}$) SETR that operates on V_i^{k-2}; Later, the $Sub-teacher_{k-2}$ SETR passes its distilled knowledge to its subsequent student ($Sub-teacher_{k-3}$) SETR, and this continues until the final $Sub-teacher_1$ SETR passes its knowledge to the bottom most $Student_0$ SETR. Since the consecutive segments of the videos do not differ significantly, PKD is more effective than direct distillation, which is proven by results in Sect. 3.4.

For distilling knowledge we consider both class token and patch tokens of the teacher and student networks. A standard Kullback-Leibler divergence ($\mathcal{L}_{KL}$) loss is applied between the probabilities generated from class token of the teacher and student SETR, whereas a mean squared error ($\mathcal{L}_{MSE}$) loss is computed between the patch tokens of teacher and student SETR. Overall, a student SETR is trained with three losses - $\mathcal{L}_{KL}$ and $\mathcal{L}_{MSE}$ loss for knowledge distillation, and a cross-entropy ($\mathcal{L}_{CE}$) loss for classification, given by the equations below.

$$\mathcal{L}_{KL} = \tau^2 \sum_j q_j^T (log(q_j^T / q_j^S)) \tag{5}$$

where q_j^S and q_j^T are the soft probabilities (moderated by temperature τ) of the student and teacher SETRs for the jth class, respectively.

$$\mathcal{L}_{mse} = (\sum_{i=0}^{N} |p_i^T - p_i^S\|^2)/N \tag{6}$$

where N is the number of patches and p_i^T and p_i^S are the patches of teacher and student SETRs respectively.

$$\mathcal{L}_{total} = \mathcal{L}_{CE} + \alpha\mathcal{L}_{KL} + \beta\mathcal{L}_{mse} \tag{7}$$

where α and β are the weights for $\mathcal{L}_{KL}$ and $\mathcal{L}_{MSE}$ loss respectively.

3 Datasets and Experimental Results

3.1 In-House and Public Dataset

Our in-house dataset[2] contains optical flow information extracted from high-definition (1920 × 1080 pixels at 30 frames per second) video recordings of TCS

[2] We plan to release the in-house optical flow dataset and corresponding code.

seizures (infrared cameras are used for nighttime seizures) in a VEM unit in hospital. To annotate the dataset, two neurologists examined both the video and corresponding EEG to identify the clinical seizure onset (t_{ON}) and clinical seizure offset (t_{OFF}) times for each seizure sample. We curated a dataset comprising of 40 TCSs from 40 epileptic patients, with one sample per patient. The duration (in seconds) of the 40 TCSs in our dataset ranges from 52 to 367 s, with a median duration of 114 s. We also prepared normal samples (no seizure) for each patient by considering the pre-ictal duration from (t_{ON} - 300) to (t_{ON} - 60) seconds, resulting in dataset of 80 samples (40 normal and 40 TCSs). We refrain from using the 60 s prior to clinical onset as it corresponds to the transition period to the seizure containing preictal activity [13,25]. We use a 5-fold cross validation (split based on patients) for training and testing on our dataset.

We also evaluate the effectiveness of our early detection approach on the GESTURES dataset [21], which contains features extracted from RGB video samples of seizures. The dataset includes two seizure types - 106 focal onset seizures (FOS) and 77 Tonic-Clonic Seizures (TCS). In contrast to our in-house dataset, the features are provided by the authors, and we directly input them into our SETR block without using a feature extractor. To evaluate our method, we adopt the stratified 10-fold cross-validation protocol as used in GESTURES.

3.2 Training Implementation and Evaluation Metrics

We implement all experiments in PyTorch 1.8.1 on a single A100 GPU. The SETR block takes in a total of 64 frames (N) with 512 1-D spatial feature per frame, has 8 MHSA heads (N_h) with a dropout rate of 0.1, 3 encoder layers (L), and 256 hidden dimensions (D). For early detection, we experiment by progressively segmenting a sample into -{4,8,16} parts (k). We employ a grid search to select the weight of 0.2 and 0.5 for KL divergence ($\tau = 10$) and MSE loss respectively. We train all methods with a batch size of 16, a learning rate of 1e-3 and use the AdamW optimizer with a weight decay of 1e-4 for a total 50 epochs. For GESTURES dataset, we implement a weighted BCE loss to deal with the dataset imbalance, whereas for our in-house dataset we implement the standard BCE loss. We use precision, recall and f1-score for benchmarking.

3.3 Performance for Early Detection

Table 1 shows the benchmarking performance of all techniques with varying fractions of input video samples on both datasets. We observed three key findings from the results in Table 1. First, transformer-based methods such as our proposed **SETR-PKD** and OaDTR exhibit better performance retention compared to LSTM-based techniques (RULSTM, Slowfast RULSTM, EgoAKD, GESTURES) with a reduction in the fraction of input sample. Second, **SETR-PKD** performance increases with k=8 from k=4, but saturates at k=16 for in-house dataset, whereas it achieves the best performance for k=4 for GESTURES dataset. The median seizure length for the in-house dataset and GESTURES dataset is 114 s and 71 s, respectively. As a result,

PKD using relatively longer partial segments (k=4) is sufficient for GESTURES, while shorter partial segments (k=8) are required for our dataset. Thus, the optimal value of k for PKD may vary depending on a dataset. Finally, we observed better performance on the GESTURES dataset, which is expected given the more detailed and refined features extracted from RGB video compared to optical flow information.

3.4 Progressive V/s Direct Knowledge Distillation

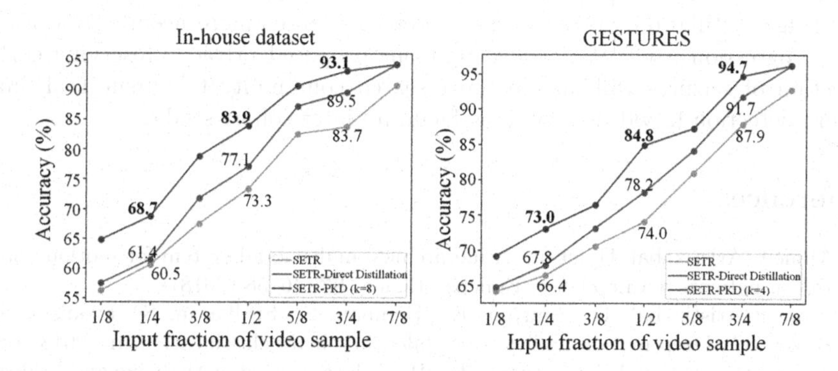

Fig. 2. Performance comparison of direct knowledge distillation and progressive knowledge distillation between SETR blocks for different fractions of input video sample.

To validate our approach of progressive knowledge distillation in a fair manner, we conducted an ablation study to compare it with direct knowledge distillation. Figure 2 shows the comparison of the accuracy of the two approaches for different fractions of the input video sample on both datasets. The results indicate that although direct knowledge distillation can increase performance, it is less effective when the knowledge gap is wide, i.e., from a SETR block trained on a full

Table 1. Benchmarking of different techniques for different fraction **{1/4, 1/2, 3/4, Full}** of input video sample. The performance is presented as mean of - **{Precision/Recall/F1-score}** across the 5-folds & 10-folds for in-house and GESTURES dataset respectively. (Best viewed in zoom).

Method/Dataset	In-house dataset				GESTURES			
	1/4	1/2	3/4	Full	1/4	1/2	3/4	Full
RULSTM [9]	0.57/0.56/0.56	0.72/0.71/0.71	0.79/0.79/0.79	0.95/0.93/0.94	0.65/0.64/0.64	0.71/0.73/0.72	0.84/0.85/0.84	0.93/0.94/0.93
Slowfast RULSTM [19]	0.57/0.56/0.56	0.73/0.72/0.72	0.81/0.80/0.80	0.94/0.94/0.94	0.67/0.65/0.66	0.73/0.72/0.72	0.86/0.84/0.85	0.97/0.95/0.96
EgoAKD [31]	0.64/0.65/0.64	0.79/0.80/0.79	0.89/0.90/0.89	0.95/0.94/0.94	0.70/0.69/0.69	0.80/0.79/0.79	0.93/0.90/91	0.97/0.94/0.95
OaDTR [28]	0.66/0.65/0.65	0.82/0.83/0.82	0.90/0.90/0.90	0.95/0.95/0.95	0.72/0.69/0.70	0.82/0.83/0.82	0.91/0.92/0.91	0.99/0.99/0.99
GESTURES [21]	0.59/0.60/0.59	0.74/0.73/0.73	0.82/0.83/0.82	0.94/0.94/0.94	0.68/0.66/0.66	0.74/0.72/0.73	0.86/0.85/0.85	0.97/0.99/0.98
SETR	0.61/0.60/0.60	0.75/0.73/0.74	0.84/0.83/0.83	0.96/0.95/0.95	0.67/0.66/0.66	0.73/0.74/0.73	0.88/0.88/0.88	0.98/0.99/0.98
SETR-PKD (k=4)	0.63/0.62/0.62	0.78/0.79/0.78	0.89/0.90/0.89	0.96/0.95/0.95	0.74/0.73/0.73	0.86/0.85/0.85	0.96/0.95/0.95	0.98/0.99/0.98
SETR-PKD (k=8)	0.70/0.69/0.69	0.86/0.84/0.85	0.92/0.93/0.92	0.96/0.95/0.95	0.73/0.74/0.73	0.85/0.85/0.85	0.95/0.96/0.95	0.98/0.99/0.98
SETR-PKD (k=16)	0.69/0.69/0.69	0.85/0.84/0.84	0.92/0.92/0.92	0.96/0.95/0.95	0.72/0.73/0.72	0.85/0.84/0.84	0.96/0.95/0.95	0.98/0.99/0.98

input sample to a SETR block trained on a minimal fraction of the input sample (1/8, 1/4, .. 1/2) compared to when the knowledge gap is small (5/8, .. 7/8). On the other hand, our SETR-PKD approach significantly improves performance for minimal fractions of input samples on both datasets.

4 Conclusion

In this work, we show that it is possible to detect epileptic seizures from optical flow modality in a privacy-preserving manner. Moreover, to achieve real-time seizure detection, we specifically develop a novel approach using progressive knowledge distillation which proves to detect seizures more accurately during their progression itself. We believe that our proposed privacy-preserving early detection of seizures will inspire the research community to pursue real-time seizure detection in videos as well as facilitate inter-cohort studies.

References

1. Ahmedt-Aristizabal, D., et al.: A hierarchical multimodal system for motion analysis in patients with epilepsy. Epilepsy Behav. **87**, 46–58 (2018)
2. Ahmedt-Aristizabal, D., Nguyen, K., Denman, S., Sridharan, S., Dionisio, S., Fookes, C.: Deep motion analysis for epileptic seizure classification. In: 2018 40th Annual International Conference of the IEEE Engineering in Medicine and Biology Society (EMBC), pp. 3578–3581. IEEE (2018)
3. Cascino, G.D.: Video-EEG monitoring in adults. Epilepsia **43**, 80–93 (2002)
4. Cunha, J.P.S., et al.: NeuroKinect: a novel low-cost 3Dvideo-EEG system for epileptic seizure motion quantification. PloS one **11**(1), e0145669 (2016)
5. Devinsky, O., Hesdorffer, D.C., Thurman, D.J., Lhatoo, S., Richerson, G.: Sudden unexpected death in epilepsy: epidemiology, mechanisms, and prevention. Lancet Neurol. **15**(10), 1075–1088 (2016)
6. Dosovitskiy, A., et al.: An image is worth 16 × 16 words: Transformers for image recognition at scale. arXiv preprint arXiv:2010.11929 (2020)
7. Fan, M., Chou, C.A.: Detecting abnormal pattern of epileptic seizures via temporal synchronization of EEG signals. IEEE Trans. Biomed. Eng. **66**(3), 601–608 (2018)
8. Fisher, R.S., et al.: Operational classification of seizure types by the international league against epilepsy: position paper of the ILAE commission for classification and terminology. Epilepsia **58**(4), 522–530 (2017)
9. Furnari, A., Farinella, G.M.: Rolling-unrolling LSTMs for action anticipation from first-person video. IEEE Trans. Pattern Anal. Mach. Intell. (PAMI) **43**, 4021–4036 (2020)
10. Guan, W., et al.: Egocentric early action prediction via multimodal transformer-based dual action prediction. IEEE Trans. Circ. Syst. Video Technol. **33**(9), 4472–4483 (2023)
11. Horn, B.K., Schunck, B.G.: Determining optical flow. Artif. Intell. **17**(1–3), 185–203 (1981)
12. Hou, J.C., Thonnat, M., Bartolomei, F., McGonigal, A.: Automated video analysis of emotion and dystonia in epileptic seizures. Epilepsy Res. **184**, 106953 (2022)
13. Huberfeld, G., et al.: Glutamatergic pre-ictal discharges emerge at the transition to seizure in human epilepsy. Nat. Neurosci. **14**(5), 627–634 (2011)

14. Kalitzin, S., Petkov, G., Velis, D., Vledder, B., da Silva, F.L.: Automatic segmentation of episodes containing epileptic clonic seizures in video sequences. IEEE Trans. Biomed. Eng. **59**(12), 3379–3385 (2012)
15. Karayiannis, N.B., Tao, G., Frost, J.D., Jr., Wise, M.S., Hrachovy, R.A., Mizrahi, E.M.: Automated detection of videotaped neonatal seizures based on motion segmentation methods. Clin. Neurophys. **117**(7), 1585–1594 (2006)
16. Kusmakar, S., Karmakar, C.K., Yan, B., O'Brien, T.J., Muthuganapathy, R., Palaniswami, M.: Automated detection of convulsive seizures using a wearable accelerometer device. IEEE Trans. Biomed. Eng. **66**(2), 421–432 (2018)
17. Moshé, S.L., Perucca, E., Ryvlin, P., Tomson, T.: Epilepsy: new advances. Lancet **385**(9971), 884–898 (2015)
18. Nashef, L., So, E.L., Ryvlin, P., Tomson, T.: Unifying the definitions of sudden unexpected death in epilepsy. Epilepsia **53**(2), 227–233 (2012)
19. Osman, N., Camporese, G., Coscia, P., Ballan, L.: SlowFast rolling-unrolling LSTMs for action anticipation in egocentric videos. In: Proceedings of the IEEE/CVF International Conference on Computer Vision, pp. 3437–3445 (2021)
20. Pérez, J.S., Meinhardt-Llopis, E., Facciolo, G.: Tv-l1 optical flow estimation. Image Process. On Line **2013**, 137–150 (2013)
21. Pérez-García, F., Scott, C., Sparks, R., Diehl, B., Ourselin, S.: Transfer learning of deep spatiotemporal networks to model arbitrarily long videos of seizures. In: de Bruijne, M., et al. (eds.) MICCAI 2021. LNCS, vol. 12905, pp. 334–344. Springer, Cham (2021). https://doi.org/10.1007/978-3-030-87240-3_32
22. Rashid, M., Singh, H., Goyal, V.: The use of machine learning and deep learning algorithms in functional magnetic resonance imaging-a systematic review. Expert Syst. **37**(6), e12644 (2020)
23. Shih, J.J., et al.: Indications and methodology for video-electroencephalographic studies in the epilepsy monitoring unit. Epilepsia **59**(1), 27–36 (2018)
24. Siddiqui, M.K., Islam, M.Z., Kabir, M.A.: A novel quick seizure detection and localization through brain data mining on ECoG dataset. Neural Comput. Appl. **31**, 5595–5608 (2019)
25. Sivathamboo, S., et al.: Cardiorespiratory and autonomic function in epileptic seizures: a video-EEG monitoring study. Epilepsy Behav. **111**, 107271 (2020)
26. Vaswani, A., et al.: Attention is all you need. In: Advances in Neural Information Processing Systems. vol. 30 (2017)
27. Wang, L., et al.: Temporal segment networks for action recognition in videos. IEEE Trans. Pattern Anal. Mach. Intell. **41**(11), 2740–2755 (2018)
28. Wang, X., et al.: OadTR: online action detection with transformers. In: Proceedings of the IEEE/CVF International Conference on Computer Vision, pp. 7565–7575 (2021)
29. Yang, Y., Sarkis, R.A., El Atrache, R., Loddenkemper, T., Meisel, C.: Video-based detection of generalized tonic-clonic seizures using deep learning. IEEE J. Biomed. Health Inf. **25**(8), 2997–3008 (2021)
30. Yuan, Y., Xun, G., Jia, K., Zhang, A.: A multi-context learning approach for EEG epileptic seizure detection. BMC syst. Biol. **12**(6), 47–57 (2018)
31. Zheng, N., Song, X., Su, T., Liu, W., Yan, Y., Nie, L.: Egocentric early action prediction via adversarial knowledge distillation. ACM Trans. Multimedia Comput. Commun. Appl. **19**(2), 1–21 (2023)

Punctate White Matter Lesion Segmentation in Preterm Infants Powered by Counterfactually Generative Learning

Zehua Ren[1], Yongheng Sun[2], Miaomiao Wang[3], Yuying Feng[3], Xianjun Li[3], Chao Jin[3], Jian Yang[3](✉), Chunfeng Lian[2](✉), and Fan Wang[1,3](✉)

[1] The Key Laboratory of Biomedical Information Engineering of Ministry of Education, School of Life Science and Technology, Xi'an Jiaotong University, Xi'an, China
fan.wang@mail.xjtu.edu.cn

[2] School of Mathematics and Statistics, Xi'an Jiaotong University, Xi'an, China
chunfeng.lian@mail.xjtu.edu.cn

[3] Department of Radiology, The First Affiliated Hospital of Xi'an Jiaotong University, Xi'an, China
yj1118@mail.xjtu.edu.cn

Abstract. Accurate segmentation of punctate white matter lesions (PWMLs) are fundamental for the timely diagnosis and treatment of related developmental disorders. Automated PWMLs segmentation from infant brain MR images is challenging, considering that the lesions are typically small and low-contrast, and the number of lesions may dramatically change across subjects. Existing learning-based methods directly apply general network architectures to this challenging task, which may fail to capture detailed positional information of PWMLs, potentially leading to severe under-segmentations. In this paper, we propose to leverage the idea of counterfactual reasoning coupled with the auxiliary task of brain tissue segmentation to learn fine-grained positional and morphological representations of PWMLs for accurate localization and segmentation. A simple and easy-to-implement deep-learning framework (i.e., DeepPWML) is accordingly designed. It combines the lesion counterfactual map with the tissue probability map to train a lightweight PWML segmentation network, demonstrating state-of-the-art performance on a real-clinical dataset of infant T1w MR images. The code is available at https://github.com/ladderlab-xjtu/DeepPWML.

1 Introduction

Punctate white matter lesion (PWML) is a typical type of cerebral white matter injury in preterm infants, potentially leading to psychomotor developmental delay, motor delay, and cerebral palsy without timely treatment [2]. The early detection and quantitative analysis of PWMLs are critical for diagnosis and treatment, especially considering that some PWML subtypes are only detectable by magnetic resonance imaging (MRI) shortly after birth (e.g., around the third week) and will become invisible thereafter [6]. The PWMLs are small targets

H. Greenspan et al. (Eds.): MICCAI 2023, LNCS 14224, pp. 220–229, 2023.
https://doi.org/10.1007/978-3-031-43904-9_22

that typically locate anterior or adjacent to the ventricles [8]. Manually annotating them in MR images is very time-consuming and relies on expertise. Thus there is an urgent need, from neuroradiologists, to develop reliable and fully automatic methods for 3D PWML segmentation.

Automated localization and delineation of PWML are practically challenging. This is mainly because that PWMLs are isolated small objects, with typically only dozens of voxels for a lesion and varying numbers of lesions across different subjects. Also, due to underlying immature myelination of infant brains [1], the tissue-to-tissue and lesion-to-tissue contrasts are both very low, especially in T1w MR images commonly used in clinical practice. In addition to conventional methods based on thresholding [3] or stochastic likelihood estimation [4], recent works attempted to apply advanced deep neural networks in the specific task of PWML segmentation [9,10,12]. For example, Liu *et al.* [10] extended Mask R-CNN [7] to detect and segment PWMLs in 2D image slices. Li *et al.* [9] implemented a 3D ResU-Net to segment diffuse white matter abnormality from T2w images. Overall, these existing learning-based methods usually use general network architectures. They may fail to completely capture fine-grained positional information to localize small and low-contrast PWMLs, potentially resulting in high under-segmentations.

Counterfactual reasoning, explained by our task, studies how a real clinical brain image appearance (factual) changes in a hypothetical scenario (whether lesion exist or not). This idea has been applied as structural causal models (SCMs) in a deep learning way in recent years. At the theoretical level, Monteiro *et al.* [11] have presented a theoretically grounded framework to evaluate counterfactual inference models. Due to the advantage of being verifiable, this idea appeared in many medical scenarios. Pawlowski *et al.* [14] proposed a general framework for building SCMs and validated on a MNIST-like dataset and a brain MRI dataset. Reinhold *et al.* [15] developed a SCM that generates images to show what an MR image would look like if demographic or disease covariates are changed. In this paper, we propose a fully automatic deep-learning framework (DeepPWML) that leverages counterfactual reasoning coupled with location information from the brain tissue segmentation to capture fine-grained positional information for PWML localization and segmentation. Specifically, based on patch-level weak-supervision, we design a counterfactual reasoning strategy to learn voxel-wise residual maps to manipulate the classification labels of input patches (i.e., containing PWMLs or not). In turn, such fine-grained residual maps could initially capture the spatial locations and morphological patterns of potential lesions. In this article, we define this residual map as counterfactual map which may be different from the meaning of counterfactual map in other articles, hereby declare. And to refine the information learned by the counterfactual part, we further include brain tissue segmentation as an auxiliary task. Given the fact that PWMLs have specific spatial correlations with different brain tissues, the segmentation probability maps (and inherent location information) could provide a certain level anatomical contextual information to assist lesion identification. Finally, by using the counterfactual maps and segmentation proba-

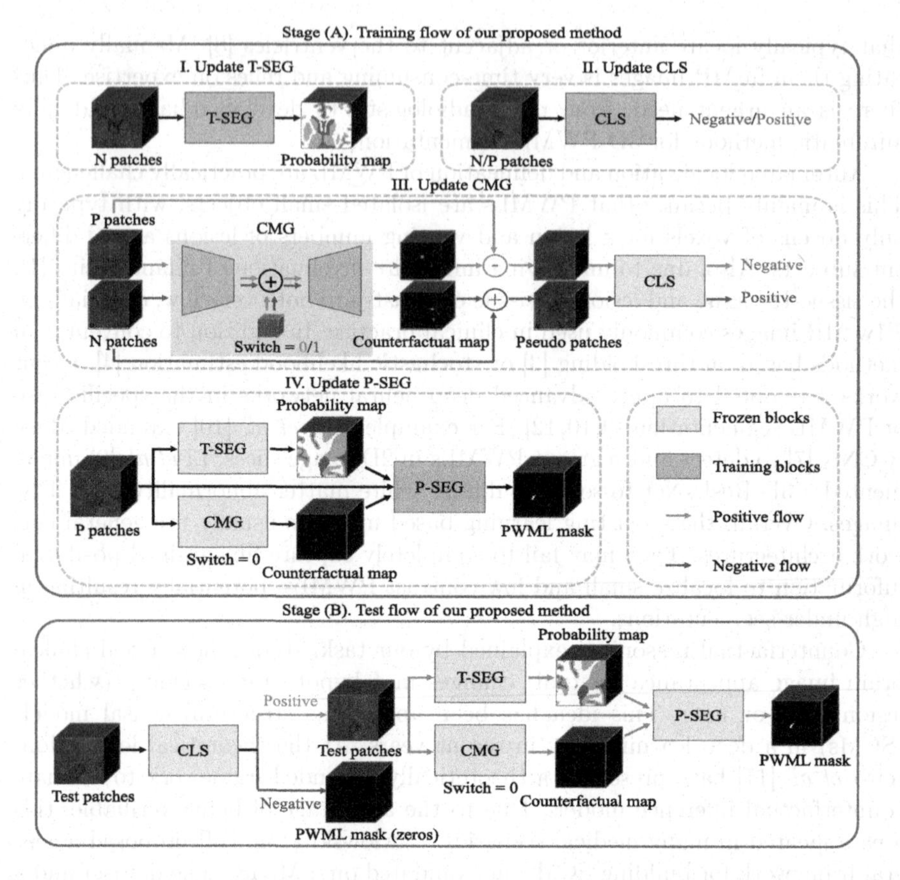

Fig. 1. Overview of the training and test steps of our DeepPWML framework that consists of four components, i.e., T-SEG, CLS, CMG, and P-SEG modules.

bility maps as the auxiliary input, we learn a lightweight sub-network for PMWL segmentation.

Overall, our DeepPWML is practically easy to implement, as the counterfactual part learns simple but effective linear manipulations, the tissue segmentation part can adopt any off-the-shelf networks, and the PWML segmentation part only needs a lightweight design. On a real-clinical dataset, our method led to a state-of-the-art performance in the infant PWML segmentation task.

2 Method

As shown in Fig. 1, our DeepPMWL consists of four parts, i.e., the tissue segmentation module (T-SEG), the classification module (CLS), the counterfactual map generator (CMG), and the PWML segmentation module (P-SEG). Specifically, in the training stage, T-SEG is learned on control data, while other modules are learned on PWML data. Given an image patch as the input, CLS is trained to distinguish positive (containing PWML) or negative (no PWML) cases, based on which CMG is further trained to produce a counterfactual map to linearly manipulate the input to change the CLS result.

The high-resolution counterfactual maps (CF maps) and segmentation probability maps (SP maps) are further combined with the input patch to train a lightweight P-SEG for PWML segmentation. In the test stage, an input patch is first determined by the CLS module whether it is positive. Positive inputs will pass through the T-SEG, CMG, and P-SEG modules to get the PWML segmentation results. It is worth noting that the test patches are generated by sliding windows, and the overlapping results are averaged to get the final segmentation results for the image, which reduces the impact of incorrect classification of the CLS module. In our experiments, T-SEG used the voxel-wise Cross-Entropy Loss. CLS used the Categorical Cross-Entropy Loss, CMG combined the sparsity loss (L1 and L2 norms) with the classification loss. Finally, P-SEG used the Dice Loss. In the following subsections, we will introduce every module in our design.

2.1 Tissue Segmentation Module

The task is to mark every pixel of the brain as cerebrospinal fluid (CSF), gray matter (GM), or white matter (WM). The choice of this module can be flexible, and there are many off-the-shelf architecture designs available. We adopt a simple Dense-Unet architecture [16] for the T-SEG module. It is trained on control premature infants' images. This module will output the SP map in which segmentation result can be obtained. Therefore, this SP map naturally contains some level anatomy information. Moreover, when an input with PWML goes through a network that has only been trained on control data, the segmentation mistake is partly due to the existence of PWML. Therebefore, this module can output a SP map carrying both potential location and anatomy guidance for PWML localization and segmentation.

2.2 Classification Module and Counterfactual Map Generator

The CLS and the CMG are trained sequentially. The CLS is trained to determine whether the current patches have lesions. The CMG is a counterfactual reasoning step for the CLS. Based on the characteristic of PWML, CMG learns a simple linear sparse transform shown as the CF map. This map aims to offset the bright PWML pixels of the image patches, which are classified as positive, or seed PWML on the patches judged as negative. In other words, CMG is learning a residual activation map for conversion between control and PWML. We adopt the T-SEG module's encoder with two fully connected layers as the CLS module. Furthermore, the architecture of CMG is a simple U-net adding a "switch" state in its skip-connection parts according to the method of Oh *et al.* [13]. Based on the nature of PWMLs, the last layer of CMG is Relu activation to ensure that the generated CF map is a positive activation.

The state of the "switch" is determined by the classifier's result on the current patch. If the judgement is positive, correspondingly, the "switch" status is 0. In this condition, the activated areas in the CF map should be where PWMLs exist. Then the pseudo patches in Fig. 1, obtained by subtracting the CF map from the input patches, should be judged as negative by the fixed CLS. Another state of the "switch" is used to generate PWMLs. When the CLS judges are negative, the "switch" status is 1. in this situation, the input patches combining the CF map should be classified as positive. This training strategy is to make CMG learn PWML features better. When it comes to the test phase, the switch status will be fixed to 0. Because in the test phase, the CF map only needs to capture PWML.

The CMG module is summarised as follows: Firstly, PWML patches C_P and control patches C_N are fed to the encoder to obtain encoded representations F_P and F_N:

$$F_P = \text{Encoder}(C_P), \tag{1}$$

$$F_N = \text{Encoder}(C_N). \tag{2}$$

Secondly, "switch" filled with zeros/ones with the same size as PWML/normal representations F_P/F_N are added to these representations and then pass through the decoder to obtain the CF maps M_P/M_N:

$$M_P = \text{Decoder}(F_P + Zeros), \tag{3}$$

$$M_N = \text{Decoder}(F_N + Ones). \tag{4}$$

Finally, the original patches C_P/C_N are added/subtracted to the CF maps M_P/M_N to yield the transformed patches $\widetilde{C}_P/\widetilde{C}_P$, which are classified by the CLS module as the opposite classes:

$$\widetilde{C}_P = C_P + M_P, \tag{5}$$

$$\widetilde{C}_N = C_N - M_N. \tag{6}$$

2.3 PWML Segmentation Module

The SP map includes the potential PWML existence, but also a lot of tissue segmentation uncertainty. The CF map directly shows the PWML location, but due to the accuracy of the CLS module, the CF map itself will also carry some false positives fault. If we synthesize the CF map, the SP map and the original input patches for appearance information, the best segmentation result can be achieved by allowing the network to verify and filter out each information in a learnable way.

The P-SEG module is implemented as a lightweight variant of the Dense-Unet. Different simplified versions have been tested, with the results summarized in Sect. 3.2. After getting the PWML segmentation result, we use the tissue segmentation result to filter out PWMLs mis-segmented at the background and CSF.

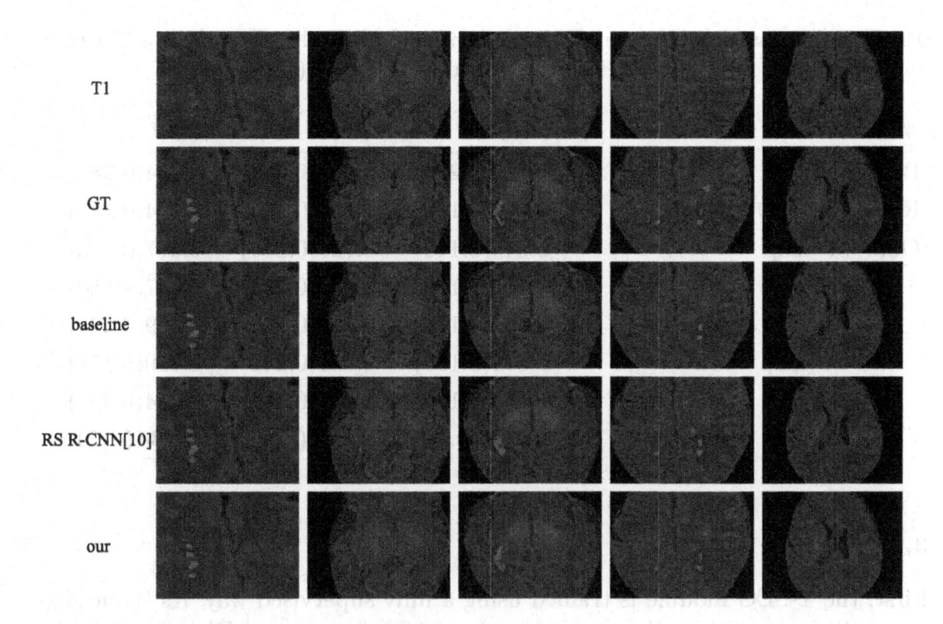

Fig. 2. Visual comparisons of the representative PWML segmentation results.

3 Experiments and Results

3.1 Dataset and Experimental Setting

Dataset: Experiments were performed on a dataset with two groups (control and PWML), where control included 52 subjects without PWML observed, and PWML included 47 subjects with PWMLs. All infants in this study were born with gestational age (GA) between 28 to 40 weeks and scanned at post-menstrual age (PMA) between 37 to 42 weeks. Two neuroscientists manually labeled PWML areas and corrected tissue labels generated by iBeat [5]. Written Consent was obtained from all parents under the institutional review board, and T1-weighted MR images were collected using a 3T MRI scanner, resampling the resolution of all images into $0.9375 \times 0.9375 \times 1\,\text{mm}^3$. All images are cropped to $130 \times 130 \times 170$.

Experimental Setting: Our method was implemented using Tensorflow. All modules were trained and tested on an NVIDIA GeForce RTX 3060 GPU. We adopted Adam as the optimizer, with the learning rate varying from 0.001 to 0.00001 according to modules. The inputs were fixed-size patches ($32 \times 32 \times 32$) cut from the T1w images. The train/validation/test ratio was 0.7/0.15/0.15 and divided on subject-level. We didn't use any data augmentation during training. We used Dice, True Positive Rate (TPR), and Positive Predictive Value (PPV) to quantitatively evaluate the segmentation performance.

Table 1. Comparison of our method (and its variants) with the state-of-the-art method and the baseline model. All metrics are presented as "mean (std)".

Methods		Dice	TPR	PPV
Baseline [16]		0.649(0.239)	0.655(0.244)	0.704(0.281)
RS R-CNN [10]		0.667(0.172)	0.754(0.250)	0.704(0.187)
Ours	SP map	0.649(0.142)	0.726(0.210)	0.677(0.213)
	CF map	0.507(0.169)	0.543(0.286)	0.647(0.180)
	SP map + T1	0.680(0.178)	0.794(0.211)	0.699(0.237)
	CF map + T1	0.672(0.198)	0.741(0.249)	0.719(0.178)
	SP map + CF map	0.670(0.184)	0.781(0.181)	0.684(0.251)
	SP map + CF map + T1	**0.721(0.177)**	**0.797(0.185)**	**0.734(0.211)**

3.2 Results

First, the T-SEG module is trained using a fully supervised way. Its tissue segmentation accuracy on the test set is about 93% in terms of Dice. Second, the CLS and other modules are trained with PWML group data. We defined the input training patches' class labels by whether they contain PWMLs or not. In other words, if any patch has at least one lesion voxel, it is positive. The accuracy of the test set can reach around 90%. Third, we train the CMG module based on the well-trained and fixed CLS module. Finally, based on T-SEG and CMG, we train P-SEG. We combine the SP map, CF map, and T1w image in a channel-wise way as the input of the module without any additional processing of these features.

Comparison Results: We compared our method with the state-of-the-art method [10]. As is shown in Table 1, our method outperforms the state-of-the-art method and the baseline model in all three indexes. The visualization results are shown in Fig. 2, from which it can be seen that our method can segment small-size PWMLs more accurately and segment PWMLs with different severities more completely.

Ablation Studies: We further evaluated the effectiveness of our design by comparing the results of the pipelines with and without SP maps and CF maps. The ablation results are shown in the last six rows of Table 1. The baseline model using the same dense-Unet is trained to segment the PWML from T1w images. Other settings are consistent with our final module. Then we will add our designed modules step by step to verify the effectiveness of two kinds of auxiliary information.

By comparing "baseline", "SP map", and "CF map", we can find that the two kinds of information individually are not good for segmenting PWMLs. The reason is as follows. The SP map mainly focuses on tissue segmentation task. The CF map has some false activation due to the offset of the highlighted areas for PWML. Fusing these two kinds of information has reduced their respective

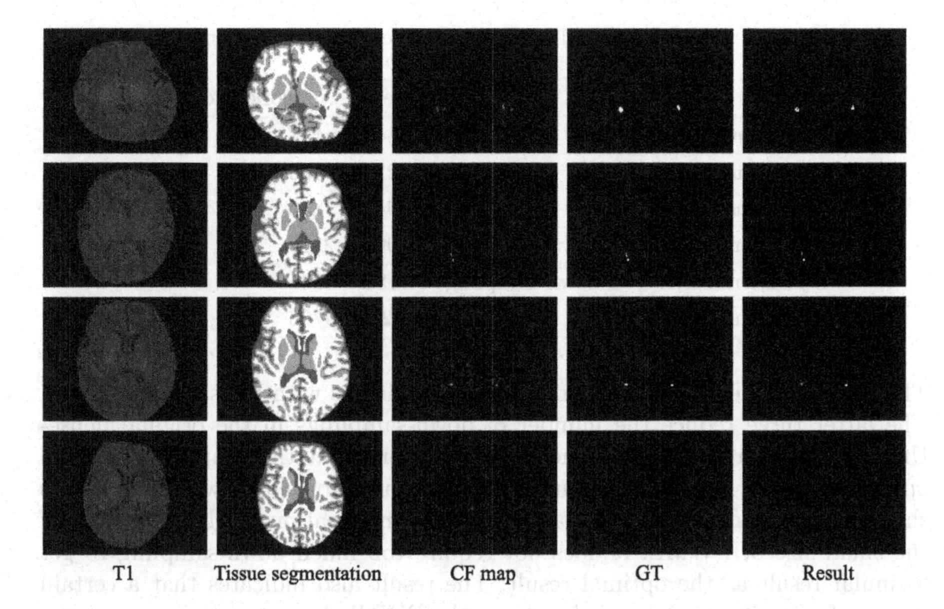

Fig. 3. Example visualization of T1w images, tissue segmentation maps, CF maps, labels, and segmentation results. Tissue segmentation maps are the final segmentation output of the T-SEG module. CF maps are the output of the CMG module.

defects ("SP map + CF map"). The icing on the cake is that when the appearance features of T1w are used again, the accuracy will be significantly improved ("SP map + T1" and "CF map + T1"). This means "SP map" and "CF map" each can be an auxiliary information but not sufficient resource for this task. Finally, after combining the three together, all indicators have been significantly improved ("SP map + CF map + T1").

Visual Analysis: Figure 3 shows the T1w images, tissue segmentation maps, CF maps, labels, and segmentation results. By selecting the most likely category from the SP map as the label, the tissue segmentation map can be obtained. As shown in the tissue segmentation maps, PWML voxels tend to be classified as gray matter surrounded by white matter which obviously does not conform to the general anatomy knowledge. The reason of this phenomenon may be that the intensity of gray matter and PWML are higher than white matter in T1w image at this age. It also can be seen from the CF maps that these maps have a preliminary localization of PWML. The last row shows the situation without PWML. It can be seen that the tissue segmentation is reasonable. The CF map has a small amount of activation and the intensity is significantly lower than the first three rows. In conclusion, these two maps complementarily provide the anatomical and morphological information for the segmentation of the PWML.

Comparison of Different Backbones of the P-SEG Module: We test from simple several layers to the whole dense-Unet to determine the required complexity in Table 2. We compared six designs with different network sizes in Table 2.

Table 2. Comparison of different backbones of the P-SEG module.

Methods	Dice	Network size (KB)
Four convolutional layers	0.5843	107
One Dense-Block	0.6220	420
Two Dense-Blocks	0.6644	836
Dense-Unet with one down-sampling	0.7099	2127
Dense-Unet with two down-sampling	0.7180	5345
Dense-Unet	**0.7214**	9736

The first three methods are several convolution layers with the same resolution. The latter three reduce the number of down-samplings in the original dense-Unet. By comparing the Dice index, it is obvious that the simple convolution operation cannot integrate the three kinds of input information well. The results show that the encoder-decoder can better fuse information. Perhaps because of the small size of PWML, it does not require too much down-sampling to get a similar result as the optimal result. The result also indicates that a certain degree of network size is needed to learn the PWMl characteristics.

4 Conclusion

In this study, we designed a simple and easy-to-implement deep learning framework (i.e. DeepPWML) to segment PWMLs. Leveraging the idea of generative counterfactual inference combined with an auxiliary task of brain tissue segmentation, we learn fine-grained positional and morphological representations of PWMLs to achieve accurate localization and segmentation. Our lightweight PWML segmentation network combines lesion counterfactual maps with tissue segmentation probability maps, achieving state-of-the-art performance on a real clinical dataset of infant T1w MR images. Moreover, our method provides a new perspective for the small-size segmentation task.

Funding. This work was supported in part by STI 2030-Major Projects (No. 2022ZD0209000), NSFC Grants (Nos. 62101430, 62101431, 82271517, 81771810), and Natural Science Basic Research Program of Shaanxi (No. 2022JM-464).

References

1. Back, S.A.: White matter injury in the preterm infant: pathology and mechanisms. Acta Neuropathol. **134**(3), 331–349 (2017)
2. de Bruïne, F.T., et al.: Clinical implications of MR imaging findings in the white matter in very preterm infants: a 2-year follow-up study. Radiology **261**(3), 899–906 (2011)
3. Cheng, I., et al.: White matter injury detection in neonatal MRI. In: Medical Imaging 2013: Computer-Aided Diagnosis, vol. 8670, pp. 664–669. SPIE (2013)

4. Cheng, I., et al.: Stochastic process for white matter injury detection in preterm neonates. NeuroImage Clin. **7**, 622–630 (2015)
5. Dai, Y., Shi, F., Wang, L., Wu, G., Shen, D.: iBEAT: a toolbox for infant brain magnetic resonance image processing. Neuroinformatics **11**, 211–225 (2013)
6. Debillon, T., N'Guyen, S., Muet, A., Quere, M., Moussaly, F., Roze, J.: Limitations of ultrasonography for diagnosing white matter damage in preterm infants. Arch. Dis. Child. Fetal Neonatal. Ed. **88**(4), F275–F279 (2003)
7. He, K., Gkioxari, G., Dollár, P., Girshick, R.: Mask R-CNN. In: Proceedings of the IEEE International Conference on Computer Vision, pp. 2961–2969 (2017)
8. Kersbergen, K.J., et al.: Different patterns of punctate white matter lesions in serially scanned preterm infants. PLoS ONE **9**(10), e108904 (2014)
9. Li, H., Chen, M., Wang, J., Illapani, V.S.P., Parikh, N.A., He, L.: Automatic segmentation of diffuse white matter abnormality on T2-weighted brain MR images using deep learning in very preterm infants. Radiol. Artif. Intell. **3**(3), e200166 (2021)
10. Liu, Y., et al.: Refined segmentation R-CNN: a two-stage convolutional neural network for punctate white matter lesion segmentation in preterm infants. In: Shen, D., et al. (eds.) MICCAI 2019. LNCS, vol. 11766, pp. 193–201. Springer, Cham (2019). https://doi.org/10.1007/978-3-030-32248-9_22
11. Monteiro, M., Ribeiro, F.D.S., Pawlowski, N., Castro, D.C., Glocker, B.: Measuring axiomatic soundness of counterfactual image models. arXiv preprint arXiv:2303.01274 (2023)
12. Mukherjee, S., Cheng, I., Miller, S., Guo, T., Chau, V., Basu, A.: A fast segmentation-free fully automated approach to white matter injury detection in preterm infants. Med. Biol. Eng. Comput. **57**, 71–87 (2019)
13. Oh, K., Yoon, J.S., Suk, H.I.: Learn-explain-reinforce: counterfactual reasoning and its guidance to reinforce an Alzheimer's disease diagnosis model. IEEE Trans. Pattern Anal. Mach. Intell. **45**(4), 4843–4857 (2022)
14. Pawlowski, N., Coelho de Castro, D., Glocker, B.: Deep structural causal models for tractable counterfactual inference. In: Advances in Neural Information Processing Systems, vol. 33, pp. 857–869 (2020)
15. Reinhold, J.C., Carass, A., Prince, J.L.: A structural causal model for MR images of multiple sclerosis. In: de Bruijne, M., et al. (eds.) MICCAI 2021. LNCS, vol. 12905, pp. 782–792. Springer, Cham (2021). https://doi.org/10.1007/978-3-030-87240-3_75
16. Zeng, Z., et al.: 3D-MASNet: 3D mixed-scale asymmetric convolutional segmentation network for 6-month-old infant brain MR images. Hum. Brain Mapp. **44**(4), 1779–1792 (2022)

Discovering Brain Network Dysfunction in Alzheimer's Disease Using Brain Hypergraph Neural Network

Hongmin Cai[1], Zhixuan Zhou[1], Defu Yang[2], Guorong Wu[2], and Jiazhou Chen[1](✉)

[1] School of Computer Science and Engineering, South China University of Technology, Guangzhou, China
csjzchen@scut.edu.cn

[2] Department of Psychiatry, University of North Carolina at Chapel Hill, Chapel Hill, USA

Abstract. Previous studies have shown that neurodegenerative diseases, specifically Alzheimer's disease (AD), primarily affect brain network function due to neuropathological burdens that spread throughout the network, similar to prion-like propagation. Therefore, identifying brain network alterations is crucial in understanding the pathophysiological mechanism of AD progression. Although recent graph neural network (GNN) analyses have provided promising results for early AD diagnosis, current methods do not account for the unique topological properties and high-order information in complex brain networks. To address this, we propose a brain network-tailored hypergraph neural network (BrainHGNN) to identify the propagation patterns of neuropathological events in AD. Our BrainHGNN approach constructs a hypergraph using region of interest (ROI) identity encoding and random-walk-based sampling strategy, preserving the unique identities of brain regions and characterizing the intrinsic properties of the brain-network organization. We then propose a self-learned weighted hypergraph convolution to iteratively update node and hyperedge messages and identify AD-related propagation patterns. We conducted extensive experiments on ADNI data, demonstrating that our BrainHGNN outperforms other state-of-the-art methods in classification performance and identifies significant propagation patterns with discriminative differences in group comparisons.

Keywords: Hypergraph Neural Network · Alzheimer's Disease · Brain Network · Propagation Patterns

1 Introduction

Alzheimer's disease (AD) is an irreversible and progressive neurodegenerative disorder, and it is the most common cause of dementia. Early diagnosis and appropriate interventions play a vital role in managing AD [22]. Existing studies suggest that AD is a disconnection syndrome manifesting the brain

H. Greenspan et al. (Eds.): MICCAI 2023, LNCS 14224, pp. 230–240, 2023.
https://doi.org/10.1007/978-3-031-43904-9_23

network alterations caused by neuropathological processes before the onset of clinical symptoms [14,15]. Therefore, identifying the propagation patterns of neuropathological burdens provides a new window to comprehend the pathophysiological mechanism of AD and predict the early stage of AD. Recent advances in neuroimaging techniques, such as diffusion-weighted imaging (DWI), allow us to observe the fiber bundles between two anatomical regions *in-vivo*, thereby encoding the structural brain network into a graph data structure [13,17].

In recent years, brain network analyses via graph neural networks have been widely used in brain disease diagnosis [10,12,21]. The advantage of graph convolution is that it takes the topological information of the graph into account, so that the information of neighbor nodes can be integrated into the target nodes via a message-passing scheme to upgrade their discriminative performance. However, these GNN-based approaches assume that the relationship between nodes is pairwise, which ignores the high-order relationships widely existing in brain networks (e.g., functional interactions between multiple brain regions) [8]. To address this limitation, hypergraph, a special graph structure where one hyperedge can contain multiple nodes, is proposed to capture the high-order information in graph [24]. Moreover, Feng et al. [2] extended the hypergraph learning method to hypergraph neural network (HGNN) to learn an optimal data representation. In neuroscience, Ji et al. [7] proposed a hypergraph attention network (FC-HAT) consisting of a dynamic hypergraph generation phase and a hypergraph attention aggregation phase to classify the functional brain networks.

Although HGNN methods have achieved promising progress in many fields, such approaches have three major limitations for brain network analyses. (1)*Lack of consideration of the unique topological properties in the brain network.* The brain networks are cost-efficient small-world networks, which contain a series of organization patterns such as hub nodes and hierarchical modularity. However, current hypergraph constructed using K-nearest neighbors (KNN) exhibit a lack of power to extract specific high-order topological information from the brain network. (2)*Lack of a robust ROI-aware encoding to overcome the anonymity of nodes in HGNN.* Current GNN or HGNN methods are permutation invariant, indicating that the node order in a graph is insensitive to the performance of the graph neural network models. However, every brain region in the brain network has its specific brain function and location; such permutation invariance and anonymity are problematic for brain network analyses. (3)*Lack of an appropriate mechanism to identify brain network dysfunction.* Compared to identifying the disease-related brain regions, less attention has been paid to detecting the propagation patterns of neuropathological burdens using the HGNN methods, despite the brain network alterations with strong interpretation being more attractive to neuroimaging researchers.

To overcome these limitations, we propose a hypergraph neural network explicitly tailored for brain networks to perform (1) identification of propagation patterns of neuropathology and (2) early diagnosis of AD. In this study, we adopt a second-order random walk on a brain network reference to generate two groups of hyperedges to depict the topology of the brain network. Additionally, an ROI identity encoding module is incorporated into our model to avoid the influence of anonymity of nodes during HGNN convolution on brain network analysis. Furthermore, we design a self-learned weighted hypergraph convolution to extract the discriminative hyperedges, which can be used to characterize the spreading pathways of neuropathological events in AD. The framework of our BrainHGNN is shown in Fig. 1. We have evaluated the effectiveness and robustness of our proposed BrainHGNN on neuroimaging data from the ADNI database. Compared to other methods, the BrainHGNN achieves enhanced discriminative capacity for CN, EMCI, and LMCI classification, as well as illustrates its potential for discovering the putative propagation patterns of neuropathology in AD.

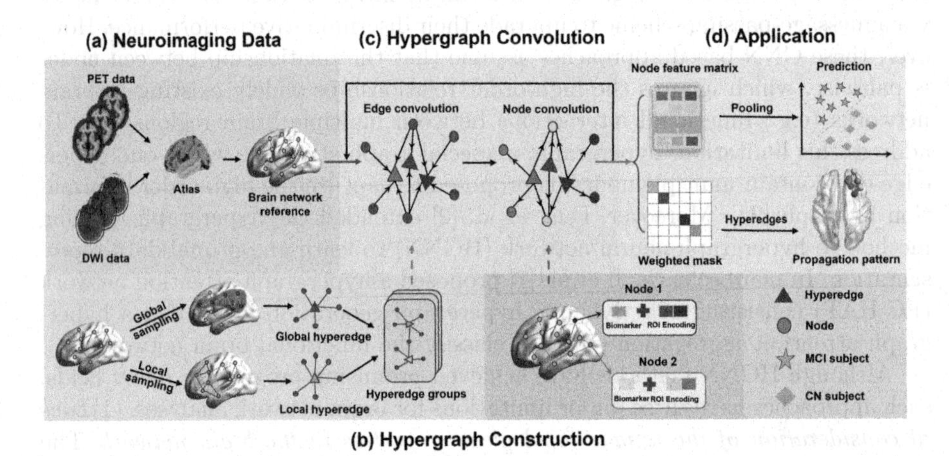

Fig. 1. Framework of our BrainHGNN for brain network analyses. (a) Utilizing DWI image and PET image data to construct the common brain network reference and its initial node feature respectively. (b) For hypergraph generation, applying biased random walk to form two hyperedge groups and then combining them together. For node feature, ROI encoding is concatenated with initial feature to obtain augmented feature. (c) Updating node feature and edge weight through two-stage edge-node convolution. (d) Classification and interpretation according to output of convolution.

2 Methodology

2.1 Hypergraph Construction

A brain network is often encapsulated into an adjacency matrix $\mathbf{A}_s \in \mathbb{R}^{N \times N} (s = 1, \cdots, S)$, where N stands for the number of brain regions in the brain network and S denotes the number of samples. To characterize the spreading pathway of neuropathological events in AD, we first calculate a group-mean adjacency matrix $\bar{\mathbf{A}} = \frac{1}{S}\sum_{s=1}^{S} \mathbf{A}_s$ as a brain network reference from a population of brain networks. Since complex brain functions are usually carried out by the interaction of multiple brain regions, a hypergraph neural network is introduced as the backbone of our proposed model to capture high-order relationships. A hyperedge in hypergraph connects more than two nodes, characterizing the high-order relationships. Let $G = \{\mathcal{V}, \mathcal{E}, \mathbf{W}\}$ denote a hypergraph with node set $\mathcal{V}$ and hyperedge set $\mathcal{E}$, where each hyperedge is actually a subset of $\mathcal{V}$ randomly sampled based on $\bar{\mathbf{A}}$. Each diagonal element in the diagonal matrix $\mathbf{W}$ represents a weight of hyperedge. The hypergraph structure is represented by an incident matrix $\mathbf{H} \in \mathbb{R}^{|\mathcal{V}| \times |\mathcal{E}|}$, where $\mathbf{H}(v, e) = 1$ means v node is in the e hyperedge. Moreover, the diagonal matrices of the hyperedge degrees $\mathbf{D}_e$ and the vertex degrees $\mathbf{D}_v$ can be calculated by $\delta(e) = \sum_{v \in \mathcal{V}} \mathbf{H}(v, e)$ and $d(v) = \sum_{e \in \mathcal{E}} w(e)\mathbf{H}(v, e)$, respectively.

Node Representation. Empirical biomarkers (such as deposition of amyloid plaques) calculated on PET neuroimaging data are the hallmarks of AD. We obtain the standard uptake value ratio for each brain region and represent them as a column vector $\mathbf{f}_s \in \mathbb{R}^N$, with each element corresponding to the feature of the node. However, the structural brain network has specific regions with distinct functions and messages, making the permutation invariance of hypergraph neural networks less suitable. We address this by generating a region-of-interest (ROI) identity encoding $\mathbf{p} \in \mathbb{R}^{N \times m}$ for each brain region and concatenating it with the empirical biomarkers, thereby reducing permutation invariance and anonymity of nodes in graph convolution. The parameter m denotes the dimension of identity encoding and is set to 2 in our experiments. The final node representation is formulated as,

$$\mathbf{x}_s^0 = Concat(\mathbf{f}_s^0, \mathbf{p}) \tag{1}$$

where $\mathbf{f}_s^0$ is a vector of the initial input biomarkers and $\mathbf{p}$ is a learnable matrix to encode the regional information.

Hyperedge Generation. The brain, a complex network, exhibits information processing with maximum efficiency and minimum costs. Such brain organization requires the brain network to be equipped with both high local clustering (hierarchical modularity) and global efficiency (shortest path length) [18]. In this context, inspired by the random walk sampling in [6], we propose a second-order biased random walk sampling strategy to each node of the weighted graph $\bar{\mathbf{A}}$ to generate a group of local hyperedges $\mathbf{H}_{local} \in \mathbb{R}^{N \times N}$ and a group of global hyperedges $\mathbf{H}_{global} \in \mathbb{R}^{N \times N}$, respectively. Specifically, the transition probability π_{ux} on edge (u, x) with source node u is decided by edge weight $\bar{\mathbf{A}}(u, x)$ and a bias term $\alpha_p(v, x)$ related to source node v of previous step, that is,

$$\pi_{ux} = \alpha_p(v, x) \cdot \bar{\mathbf{A}}(u, x) \tag{2}$$

where $\alpha_p(v, x) = p^{d_{vx}-1}, d_{vx} \in \{0, 1, 2\}$ and d_{vx} is the shortest path distance between node v and candidate target x. In Eq. (2), we can observe that when p sets to a small value $(0 < p < 1)$, node u tends to transfer to the neighbors of node v, thus forming a local network community. However, when p takes a large value $(p > 1)$, node u has a high probability of moving to a node far away from node v, thereby generating a global network community. Therefore, the hyperedge generation can be described mainly in three steps. (1) Calculate the transition probability matrix with the original edge weight matrix and parameter p according to Eq. (2). (2) For each node, set it as the beginning of the walk, and calculate which neighbor to be traversed according to transition probability. Then we sample this neighbor to the hyperedge and set this neighbor as the current node. Repeat this procedure l times to generate a hyperedge. Iterate all nodes to construct a hypergraph. (3) Execute step 1–2 for local sampling and global sampling to generate local and global hypergraphs respectively. Fuse two hypergraphs by concatenation to form the final hypergraph. In the following experiments, we set walk length $l_{local} = 14$ which is 1/4 average node degree and parameter $p = 1/4$ to generate local hyperedge group, as well as apply walk length $l_{global} = 23$ which is 1/4 of the average of shortest path length and parameter $p = 4$ to construct global hyperedge group. Finally, the incidence matrix $\mathbf{H} \in \mathbb{R}^{N \times 2N}$ can be obtained by concatenating these two hyperedge groups, i.e., $\mathbf{H} = Concat(\mathbf{H}_{local}, \mathbf{H}_{global})$.

2.2 Hypergraph Convolution

As stated in [4], the hypergraph convolution can be decoupled as a two-stage message passing procedure, which can be formulated as:

$$\begin{aligned} \mathbf{Z}^t &= \mathbf{W}\mathbf{H}^\top \mathbf{D}_e^{-1} \mathbf{X}^t \\ \mathbf{X}^{t+1} &= \sigma(\mathbf{D}_v^{-1} \mathbf{H} \mathbf{Z}^t \mathbf{\Theta}^{t+1}) \end{aligned} \tag{3}$$

where $\mathbf{X}^t \in \mathbb{R}^{|\mathcal{V}| \times M_t}$ is the node feature matrix in t^{th} layer and $\mathbf{Z}^t \in \mathbb{R}^{|\mathcal{E}| \times M_t}$ is the feature matrix of hyperedges in t^{th} layer. $\mathbf{\Theta}^{t+1} \in \mathbb{R}^{M_t \times M_{t+1}}$ is the learning parameter in $(t+1)^{th}$ layer. The intuition behind Eq. (3) is that we first use the incidence matrix $\mathbf{H}$ to guide each node to aggregate and generate the hyperedge feature matrix $\mathbf{Z}^t$, then the updated node feature matrix $\mathbf{X}^{t+1}$ can be obtained by aggregating node-relevant hyperedge features with a learning parameter $\mathbf{\Theta}^{t+1}$ and a nonlinear activation function $\sigma(\cdot)$.

In most cases, the hyperedge weight matrix $\mathbf{W}$ is pre-defined as either an identical matrix $\mathbf{I}$ or a diagonal matrix with specific elements determined by prior knowledge, for ease of use. Nevertheless, these two types of hyperedge weight matrices pose difficulties in identifying the discriminant hyperedges associated with AD-related brain network dysfunction. To address this issue, we propose using a learnable weight matrix mask to guide the learning of hyperedge weights, enabling us to identify propagation patterns of neuropathological burden in AD. The hyperedge convolution can be rewritten as,

$$\mathbf{Z}^t = \tilde{\mathbf{W}}\mathbf{H}^\top \mathbf{D}_e^{-1} \mathbf{X}^t \tag{4}$$

where $\tilde{\mathbf{W}} = sigmoid(\tilde{\mathbf{M}} \circ \mathbf{W})$ is the learned hyperedge weight matrix, $\tilde{\mathbf{M}}$ is the learnable weight mask and $\circ$ means Hadamard product.

2.3 Optimization

In this study, we use the cross entropy loss to train our model and predict disease status. To avoid overfitting and discover AD-related propagation patterns, we first impose an l_1-norm constraint to the learnable weight mask $\tilde{\mathbf{M}}$ to overshadow the irrelevant hyperedges and preserve the discriminative hyperedges that are highly related to the prediction of labels. Then, we require $\mathbf{\Theta}$ to be smooth with an l_2-norm constraint to void its divergence. Finally, the loss function can be formulated as follows:

$$loss_{ce} = \frac{1}{N} \sum_{i=1}^{N} -[y_i \cdot \log(z_i) + (1 - y_i) \cdot \log(1 - z_i)] + \lambda_1 ||\tilde{\mathbf{M}}||_1 + \lambda_2 ||\mathbf{\Theta}||_2 \quad (5)$$

where z_i is the predictive label of ith sample output from the fully connected layer by our BrainHGNN and y_i is the ground truth of ith sample.

3 Experiments

3.1 Dataset Description and Experimental Settings

Data Processing. We collect 94 structural brain networks from the ADNI database (https://adni.loni.usc.edu) to calculate a group-mean adjacency matrix $\bar{\mathbf{A}}$ to construct the incident matrix $\mathbf{H}$ of the common hypergraph. For each subject, we construct the structural brain network in the following steps. First, we apply Destrieux atlas [1] to the T1-weighted MR to obtain 148 ROIs. Then we apply surface seed-based probabilistic fiber tractography [3] to the DWI image to generate a adjacency matrix $\mathbf{A}_s$. Regarding the feature representation of each ROI, we adopt a similar image processing pipeline to pracellate the ROIs and calculate the standard update value ratio (SUVR) score of each ROI for the amyloid-PET and FDG-PET data and represent them as a column vector $\mathbf{f}_s$ of whole-brain pathological events. The detailed information on multi-modal neuroimaging data is shown in Table 1.

Parameter Settings. In our experiments, we set parameters $num_{layer} = 1$, $num_{epoch} = 250$, $batchsize = 256$, $learning\,rate = 0.05$. LeakyReLu activation function $\sigma(\cdot)$ is used for node convolution. The l_1-norm is adopted on the learnable weight mask $\tilde{\mathbf{M}}$ with a hyper-parameter $\lambda_1 = 1e - 3$ to extract the discriminative hyperedges. The l_2-norm is applied on the learning parameter matrix $\mathbf{\Theta}$ with a hyper-parameter $\lambda_2 = 1e - 5$.

Baselines. The comparison baselines incorporate two traditional methods, including support vector machines (SVM) and random forest (RF), as well as five graph-based methods, including graph convolution neural network (GCN) [9], graph attention network (GAT) [19], hypergraph neural network (HGNN) [4], Dynamic Graph CNN (DGCNN) [20], and BrainGNN [11].

Table 1. Demographic Information of Amyloid-PET and FDG-PET Data.

PET	Gender	Number	Range of Age	Average Age	CN	EMCI	LMCI
Amyloid	Male	450	55.0–91.4	73.4	136	184	130
	Female	389	55.0–89.6	71.7	148	145	96
	Total	**839**	**55.0–91.4**	**72.6**	**284**	**329**	**226**
FDG	Male	592	55.0–91.4	73.9	169	182	241
	Female	472	55.0–89.6	72.2	166	148	158
	Total	**1064**	**55.0–91.4**	**73.1**	**335**	**330**	**399**

3.2 Evaluating Diagnostic Capability on Amyloid-PET and FDG-PET Data

In this section, our objective is to evaluate the diagnostic performance of our BrainHGNN model. We first construct three group comparisons, including CN/EMCI, EMCI/LMCI, and CN/LMCI, for amyloid-PET or FDG-PET data. For each comparison, we apply our BrainHGNN on empirical biomarkers to evaluate the classification metrics, including accuracy, F1-score, sensitivity, and specificity, according to 10-fold cross validation. From Table 2, it is obvious that (1) our proposed method achieves the highest classification performance in most of the metrics on both amyloid-PET and FDG-PET data, compared to the other methods; (2) the graph-based deep learning methods (GCN, GAT, DGCNN, BrainGNN, HGNN, and BrainHGNN) are generally superior to traditional machine learning methods (SVM and RF), mainly because graph-based methods fully incorporate the topology of the brain network, thus enhancing the classification performance; (3) the BrainHGNN constructs the hypergraph customized to the brain network's unique topology, which is more suitable for brain network analyses than the traditional hypergraph neural networks.

3.3 Evaluating the Statistical Power of Identifying Brain Network Dysfunction in AD

In this section, we investigate the capability of identifying AD-related discriminative hyperedges by our BrainHGNN. Benefiting from the sparse weight mask, we can extract four discriminative hyperedges that most frequently occurred in the top-10 highest weight hyperedge list. Here, we take CN/LMCI group comparison as an example and map the discriminative hyperedges on the cortical surface for amyloid-PET data (first row in Fig. 2(a)) and FDG-PET data (second row in Fig. 2(a)). We can observe that most of nodes in these discriminative hyperedges are located in the Default Mode network, Cingulo-Opercular network, FrontoParietal network, DorsalAttention network, and VentralAttention network. Numerous studies have demonstrated a significant association between these subnetworks and the progress of AD [5,16,23].

Table 2. The Classification Results on Amyloid-PET and FDG-PET Data

Data	GROUP	MODEL	ACCURACY	F1-SCORE	SENSITIVITY	SPECIFICITY
Amyloid	CN/LMCI	SVM	0.691 ± 0.057	0.572 ± 0.086	0.476 ± 0.092	0.861 ± 0.064
		RF	0.690 ± 0.064	0.572 ± 0.086	0.555 ± 0.090	0.861 ± 0.064
		GCN	0.709 ± 0.052	0.587 ± 0.119	0.498 ± 0.154	0.863 ± 0.132
		GAT	0.718 ± 0.060	0.620 ± 0.113	0.553 ± 0.167	0.828 ± 0.161
		DGCNN	0.735 ± 0.039	0.578 ± 0.058	0.491 ± 0.077	0.841 ± 0.065
		BrainGNN	0.745 ± 0.079	0.615 ± 0.115	0.577 ± 0.122	**0.898 ± 0.088**
		HGNN	0.706 ± 0.056	0.567 ± 0.114	0.460 ± 0.143	0.891 ± 0.112
		BrainHGNN	**0.751 ± 0.040**	**0.651 ± 0.089**	**0.583 ± 0.070**	0.873 ± 0.075
	EMCI/LMCI	SVM	0.628 ± 0.067	0.353 ± 0.109	0.257 ± 0.094	0.886 ± 0.059
		RF	0.607 ± 0.059	0.396 ± 0.105	0.333 ± 0.060	0.803 ± 0.070
		GCN	0.692 ± 0.059	0.498 ± 0.160	0.422 ± 0.194	0.859 ± 0.134
		GAT	0.692 ± 0.048	0.417 ± 0.083	0.332 ± 0.023	**0.912 ± 0.089**
		DGCNN	0.672 ± 0.060	0.451 ± 0.070	0.341 ± 0.074	0.893 ± 0.054
		BrainGNN	0.673 ± 0.072	0.486 ± 0.060	0.412 ± 0.092	0.828 ± 0.162
		HGNN	0.693 ± 0.057	0.499 ± 0.161	0.423 ± 0.098	0.862 ± 0.110
		BrainHGNN	**0.695 ± 0.050**	**0.507 ± 0.139**	**0.455 ± 0.095**	0.865 ± 0.070
	CN/EMCI	SVM	0.550 ± 0.061	0.629 ± 0.060	0.731 ± 0.145	0.373 ± 0.156
		RF	0.581 ± 0.059	0.619 ± 0.066	0.642 ± 0.085	**0.512 ± 0.091**
		GCN	0.621 ± 0.046	0.655 ± 0.126	0.732 ± 0.218	0.461 ± 0.263
		GAT	0.632 ± 0.036	0.648 ± 0.127	0.701 ± 0.023	0.510 ± 0.027
		DGCNN	0.613 ± 0.025	**0.669 ± 0.033**	0.732 ± 0.072	0.470 ± 0.08
		BrainGNN	0.639 ± 0.080	0.629 ± 0.113	0.706 ± 0.166	0.473 ± 0.025
		HGNN	0.626 ± 0.045	0.648 ± 0.133	0.715 ± 0.024	0.485 ± 0.028
		BrainHGNN	**0.647 ± 0.053**	0.668 ± 0.067	**0.733 ± 0.024**	0.504 ± 0.031
FDG	CN/LMCI	SVM	0.553 ± 0.062	0.688 ± 0.047	**0.913 ± 0.061**	0.136 ± 0.083
		RF	0.617 ± 0.049	0.647 ± 0.053	0.655 ± 0.080	0.575 ± 0.086
		GCN	0.658 ± 0.050	0.673 ± 0.097	0.695 ± 0.193	0.585 ± 0.025
		GAT	0.655 ± 0.049	0.657 ± 0.100	0.651 ± 0.197	**0.638 ± 0.259**
		DGCNN	0.639 ± 0.013	0.678 ± 0.050	0.715 ± 0.115	0.539 ± 0.131
		BrainGNN	0.635 ± 0.064	0.624 ± 0.041	0.624 ± 0.133	0.593 ± 0.021
		HGNN	0.648 ± 0.046	0.673 ± 0.111	0.726 ± 0.215	0.516 ± 0.027
		BrainHGNN	**0.670 ± 0.053**	**0.707 ± 0.055**	0.741 ± 0.103	0.577 ± 0.061
	EMCI/LMCI	SVM	0.562 ± 0.064	0.691 ± 0.047	**0.902 ± 0.068**	0.166 ± 0.099
		RF	0.665 ± 0.051	0.696 ± 0.054	0.712 ± 0.070	**0.614 ± 0.081**
		GCN	0.649 ± 0.047	0.678 ± 0.089	0.718 ± 0.192	0.544 ± 0.260
		GAT	0.660 ± 0.029	0.691 ± 0.065	0.734 ± 0.020	0.555 ± 0.029
		DGCNN	0.650 ± 0.033	0.683 ± 0.043	0.723 ± 0.169	0.544 ± 0.193
		BrainGNN	0.639 ± 0.057	0.683 ± 0.070	0.637 ± 0.143	0.560 ± 0.165
		HGNN	0.651 ± 0.045	0.664 ± 0.104	0.684 ± 0.021	0.576 ± 0.028
		BrainHGNN	**0.675 ± 0.037**	**0.713 ± 0.060**	0.764 ± 0.014	0.547 ± 0.081
	CN/EMCI	SVM	0.557 ± 0.071	0.595 ± 0.129	0.712 ± 0.207	0.440 ± 0.213
		RF	0.620 ± 0.058	0.608 ± 0.066	0.605 ± 0.091	0.641 ± 0.082
		GCN	0.629 ± 0.038	0.587 ± 0.149	0.613 ± 0.254	0.599 ± 0.026
		GAT	0.631 ± 0.047	0.638 ± 0.087	0.679 ± 0.186	0.564 ± 0.117
		DGCNN	0.636 ± 0.034	0.559 ± 0.118	0.560 ± 0.160	**0.655 ± 0.106**
		BrainGNN	0.621 ± 0.069	0.557 ± 0.109	0.616 ± 0.187	0.567 ± 0.161
		HGNN	0.616 ± 0.037	0.560 ± 0.017	0.585 ± 0.175	0.596 ± 0.175
		BrainHGNN	**0.639 ± 0.042**	**0.653 ± 0.067**	**0.717 ± 0.155**	0.544 ± 0.179

Discussion. Here, we design an ablation experiment to evaluate the power of each module used in BrainHGNN. We divide the models into five different counterparts, including the HGNN, model without random walk sampling, model without self-learned edge weight, model without ROI encoding, and our BrainHGNN. We apply these five methods on three group comparisons using empirical amyloid SUVR and FDG SUVR to calculate the classification accuracy, as shown

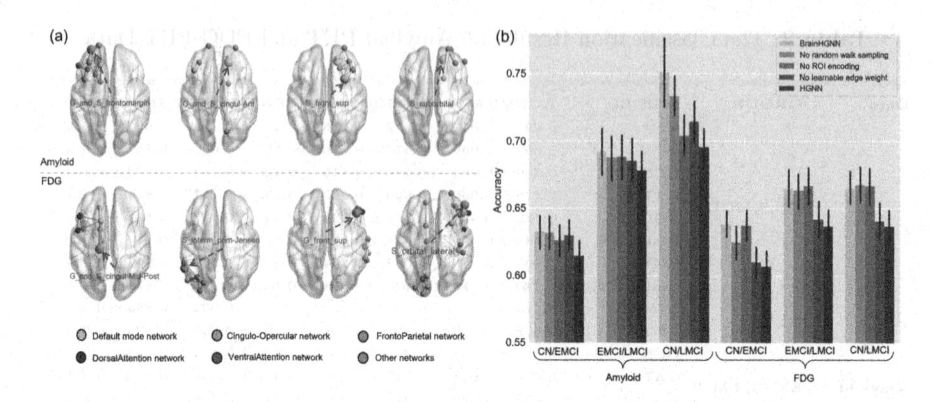

Fig. 2. (a) The brain mappings of four discriminative hyperedges on CN/LMCI group comparison. (b) The classification accuracy of ablation study.

in Fig. 2(b). It is clear that (1) compared to baseline model HGNN, all proposed modules used in HGNN show enhanced classification performance, indicating these modules are effective; (2) the BrainHGNN with full settings achieves best classification performance than other counterparts.

4 Conclusion

In this paper, we propose a random-walk-based hypergraph neural network by integrating the topological nature of the brain network to predict the early stage of AD and discover the propagation patterns of neuropathological events in AD. Compared with other methods, our proposed BrainHGNN achieve enhanced classification performance and statistical power in group comparisons on ADNI neuroimaging dataset. In the future, we plan to apply our BrainHGNN to other neurodegenerative disorders that manifest brain network dysfunction.

Acknowledgements. This work was supported in part by the National Key Research and Development Program of China (2022YFE0112200), the National Natural Science Foundation of China (U21A20520,62102153), the Science and Technology Project of Guangdong Province (2022A0505050014), the Guangdong Key Laboratory of Human Digital Twin Technology (2022B1212010004), Natural Science Foundation of Guangdong Province of China (2022A1515011162), Key-Area Research and Development Program of Guangzhou City (202206030009), and the China Postdoctoral Science Foundation (2021M691062, 2023T160226). The neuroimaging datasets used in this study were supported by the Alzheimer's Disease Neuroimaging Initiative (ADNI).

References

1. Destrieux, C., Fischl, B., Dale, A., Halgren, E.: Automatic parcellation of human cortical gyri and sulci using standard anatomical nomenclature. Neuroimage **53**(1), 1–15 (2010)
2. Feng, Y., You, H., Zhang, Z., Ji, R., Gao, Y.: Hypergraph neural networks. In: Proceedings of the AAAI Conference on Artificial Intelligence, vol. 33, pp. 3558–3565 (2019)
3. Fillard, P., et al.: Quantitative evaluation of 10 tractography algorithms on a realistic diffusion MR phantom. Neuroimage **56**(1), 220–234 (2011)
4. Gao, Y., Feng, Y., Ji, S., Ji, R.: HGNN$^+$: general hypergraph neural networks. IEEE Trans. Pattern Anal. Mach. Intell. (2022). https://doi.org/10.1109/tpami.2022.3182052
5. Greicius, M.D., Srivastava, G., Reiss, A.L., Menon, V.: Default-mode network activity distinguishes Alzheimer's disease from healthy aging: evidence from functional MRI. Proc. Natl. Acad. Sci. **101**(13), 4637–4642 (2004)
6. Grover, A., Leskovec, J.: node2vec: scalable feature learning for networks. In: Proceedings of the 22nd ACM SIGKDD International Conference on Knowledge Discovery and Data Mining, pp. 855–864 (2016)
7. Ji, J., Ren, Y., Lei, M.: Fc-hat: Hypergraph attention network for functional brain network classification. Inf. Sci. **608**, 1301–1316 (2022)
8. Jie, B., Wee, C.Y., Shen, D., Zhang, D.: Hyper-connectivity of functional networks for brain disease diagnosis. Med. Image Anal. **32**, 84–100 (2016)
9. Kipf, T.N., Welling, M.: Semi-supervised classification with graph convolutional networks. arXiv preprint arXiv:1609.02907 (2016)
10. Li, X., Dvornek, N.C., Zhou, Y., Zhuang, J., Ventola, P., Duncan, J.S.: Graph neural network for interpreting task-fMRI biomarkers. In: Shen, D., et al. (eds.) MICCAI 2019. LNCS, vol. 11768, pp. 485–493. Springer, Cham (2019). https://doi.org/10.1007/978-3-030-32254-0_54
11. Li, X., et al.: BrainGNN: Interpretable brain graph neural network for fMRI analysis. Med. Image Anal. **74**, 102233 (2021)
12. Li, X., et al.: Pooling regularized graph neural network for fMRI biomarker analysis. In: Martel, A.L., et al. (eds.) MICCAI 2020. LNCS, vol. 12267, pp. 625–635. Springer, Cham (2020). https://doi.org/10.1007/978-3-030-59728-3_61
13. Liu, J., et al.: Complex brain network analysis and its applications to brain disorders: a survey. Complexity **2017** (2017)
14. Pievani, M., Filippini, N., Van Den Heuvel, M.P., Cappa, S.F., Frisoni, G.B.: Brain connectivity in neurodegenerative diseases-from phenotype to proteinopathy. Nat. Rev. Neurol. **10**(11), 620–633 (2014)
15. Sepulcre, J., et al.: Neurogenetic contributions to amyloid beta and tau spreading in the human cortex. Nat. Med. **24**(12), 1910–1918 (2018)
16. Sorg, C., et al.: Selective changes of resting-state networks in individuals at risk for Alzheimer's disease. Proc. Natl. Acad. Sci. **104**(47), 18760–18765 (2007)
17. Sporns, O.: Structure and function of complex brain networks. Dialogues Clin. Neurosci. **15**, 247–262 (2013)
18. Stam, C.J.: Modern network science of neurological disorders. Nat. Rev. Neurosci. **15**(10), 683–695 (2014)
19. Velickovic, P., Cucurull, G., Casanova, A., Romero, A., Lio, P., Bengio, Y., et al.: Graph attention networks. Stat **1050**(20), 10–48550 (2017)

20. Wang, Y., Sun, Y., Liu, Z., Sarma, S.E., Bronstein, M.M., Solomon, J.M.: Dynamic graph CNN for learning on point clouds. ACM Trans. Graph. (TOG) **38**(5), 1–12 (2019)
21. Yang, H., et al.: Interpretable multimodality embedding of cerebral cortex using attention graph network for identifying bipolar disorder. In: Shen, D., et al. (eds.) MICCAI 2019. LNCS, vol. 11766, pp. 799–807. Springer, Cham (2019). https://doi.org/10.1007/978-3-030-32248-9_89
22. Zhang, L., Wang, M., Liu, M., Zhang, D.: A survey on deep learning for neuroimaging-based brain disorder analysis. Front. Neurosci. **14**, 779 (2020)
23. Zhang, Y., et al.: Joint assessment of structural, perfusion, and diffusion MRI in Alzheimer's disease and frontotemporal dementia. Int. J. Alzheimer's Dis. **2011** (2011)
24. Zhou, D., Huang, J., Schölkopf, B.: Learning with hypergraphs: clustering, classification, and embedding. In: Advances in Neural Information Processing Systems, vol. 19 (2006)

Improved Prognostic Prediction of Pancreatic Cancer Using Multi-phase CT by Integrating Neural Distance and Texture-Aware Transformer

Hexin Dong[1,2,3], Jiawen Yao[1,3](✉), Yuxing Tang[1], Mingze Yuan[1,2,3], Yingda Xia[1], Jian Zhou[4,5], Hong Lu[6], Jingren Zhou[1,3], Bin Dong[2,7], Le Lu[1], Zaiyi Liu[8], Li Zhang[2](✉), Yu Shi[9](✉), and Ling Zhang[1]

[1] DAMO Academy, Alibaba Group, Hangzhou, China
yaojiawen.yjw@alibaba-inc.com
[2] Peking University, Beijing, China
zhangli_pku@pku.edu.cn
[3] Hupan Lab, Hangzhou 310023, China
[4] Sun Yat-sen University Cancer Center, Guangzhou, China
[5] South China Hospital, Shenzhen University, Shenzhen, China
[6] Tianjin Medical University Cancer Institute and Hospital, Tianjin, China
[7] Peking University Changsha Institute for Computing and Digital Economy, Changsha, China
[8] Guangdong Provincial People's Hospital, Guangzhou, China
[9] Shengjing Hospital, Shenyang, China
18940259980@163.com

Abstract. Pancreatic ductal adenocarcinoma (PDAC) is a highly lethal cancer in which the tumor-vascular involvement greatly affects the resectability and, thus, overall survival of patients. However, current prognostic prediction methods fail to explicitly and accurately investigate relationships between the tumor and nearby important vessels. This paper proposes a novel learnable neural distance that describes the precise relationship between the tumor and vessels in CT images of different patients, adopting it as a major feature for prognosis prediction. Besides, different from existing models that used CNNs or LSTMs to exploit tumor enhancement patterns on dynamic contrast-enhanced CT imaging, we improved the extraction of dynamic tumor-related texture features in multi-phase contrast-enhanced CT by fusing local and global features using CNN and transformer modules, further enhancing the features extracted across multi-phase CT images. We extensively evaluated and compared the proposed method with existing methods in the multi-center (n = 4) dataset with 1,070 patients with PDAC, and statistical

H. Dong—Work was done during an internship at Alibaba DAMO Academy.

Supplementary Information The online version contains supplementary material available at https://doi.org/10.1007/978-3-031-43904-9_24.

H. Greenspan et al. (Eds.): MICCAI 2023, LNCS 14224, pp. 241–251, 2023.
https://doi.org/10.1007/978-3-031-43904-9_24

analysis confirmed its clinical effectiveness in the external test set consisting of three centers. The developed risk marker was the strongest predictor of overall survival among preoperative factors and it has the potential to be combined with established clinical factors to select patients at higher risk who might benefit from neoadjuvant therapy.

Keywords: Pancreatic ductal adenocarcinoma (PDAC) · Survival prediction · Texture-aware Transformer · Cross-attention · Nerual distance

1 Introduction

Pancreatic ductal adenocarcinoma (PDAC) is one of the deadliest forms of human cancer, with a 5-year survival rate of only 9% [16]. Neoadjuvant chemotherapy can increase the likelihood of achieving a margin-negative resection and avoid unnecessary surgery in patients with aggressive tumor types [23]. Providing accurate and objective preoperative biomarkers is crucial for triaging patients who are most likely to benefit from neoadjuvant chemotherapy. However, current clinical markers such as larger tumor size and high carbohydrate antigen (CA) 19-9 level may not be sufficient to accurately tailor neoadjuvant treatment for patients [19]. Therefore, multi-phase contrast-enhanced CT has a great potential to enable personalized prognostic prediction for PDAC, leveraging its ability to provide a wealth of texture information that can aid in the development of accurate and effective prognostic models [2,10].

Previous studies have utilized image texture analysis with hand-crafted features to predict the survival of patients with PDACs [1], but the representational

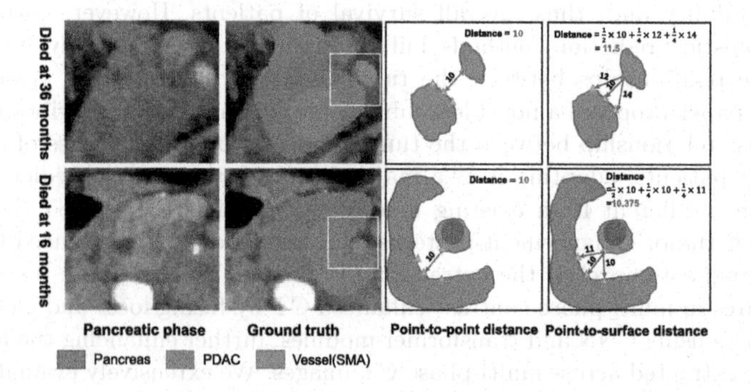

Fig. 1. Two examples of spatial information between vessel (orange region) and tumor (green region). The minimum distance, which refers to the closest distance between the Superior Mesenteric Artery (SMA) and the PDAC tumor region, is almost identical in these two cases. We define the surface-to-surface distance based on point-to-surface distance (weighted-average of red lines from $\diamondsuit$ to $\triangle$) instead of point-to-point distance (blue lines) to better capture the relationship between the tumor and the perivascular tissue. Here $\diamondsuit$ and $\triangle$ are points sampled from subset $\hat{\mathcal{V}}_c$ and $\hat{\mathcal{P}}_c$ defined in Eq. 4. The distances and weights shown in the figure is for illustration purposes only. (Color figure online)

power of these features may be limited. In recent years, deep learning-based methods have shown promising results in prognosis models [3,6,12]. However, PDACs differ significantly from the tumors in these studies. A clinical investigation based on contrast-enhanced CT has revealed a dynamic correlation between the internal stromal fractions of PDACs and their surrounding vasculature [14]. Therefore, focusing solely on the texture information of the tumor itself may not be effective for the prognostic prediction of PDAC. It is necessary to incorporate tumor-vascular involvement into the feature extraction process of the prognostic model. Although some studies have investigated tumor-vascular relationships [21,22], these methods may not be sufficiently capable of capturing the complex dynamics between the tumor and its environment.

We propose a novel approach for measuring the relative position relationship between the tumor and the vessel by explicitly using the distance between them. Typically, Chamfer distance [7], Hausdorff distance [8], or other surface-awareness metrics are used. However, as shown in Fig. 1, these point-to-point distances cannot differentiate the degree of tumor-vascular invasion [18]. To address this limitation, we propose a learnable neural distance that considers all relevant points on different surfaces and uses an attention mechanism to compute a combined distance that is more suitable for determining the degree of invasion. Furthermore, to capture the tumor enhancement patterns across multi-phase CT images, we are the first to combine convolutional neural networks (CNN) and transformer [4] modules for extracting the dynamic texture patterns of PDAC and its surroundings. This approach takes advantage of the visual transformer's adeptness in capturing long-distance information compared to the CNN-only-based framework in the original approach. By incorporating texture information between PDAC, pancreas, and peripancreatic vessels, as well as the local tumor information captured by CNN, we aim to improve the accuracy of our prognostic prediction model.

In this study, we make the following contributions: (1) We propose a novel approach for aiding survival prediction in PDAC by introducing a learnable neural distance that explicitly evaluates the degree of vascular invasion between the tumor and its surrounding vessels. (2) We introduce a texture-aware transformer block to enhance the feature extraction approach, combining local and global information for comprehensive texture information. We validate that the cross-attention is utilized to capture cross-modality information and integrate it with in-modality information, resulting in a more accurate and robust prognostic prediction model for PDAC. (3) Through extensive evaluation and statistical analysis, we demonstrate the effectiveness of our proposed method. The signature built from our model remains statistically significant in multivariable analysis after adjusting for established clinical predictors. Our proposed model has the potential to be used in combination with clinical factors for risk stratification and treatment decisions for patients with PDAC.

2 Methods

As shown in Fig. 2, the proposed method consists of two main components. The first component combines the CNN and transformer to enhance the extraction of tumor dynamic texture features. The second component proposes a neural distance metric between PDAC and important vessels to assess their involvements.

2.1 Texture-Aware Vision Transformer: Combination of CNN and Transformer

Recently, self-attention models, specifically vision transformers (ViTs [4]), have emerged as an alternative to CNNs in survival prediction [15,25]. Our proposed texture-aware transformer, inspired by MobileViT [13], aims to combine both local information (such as PDAC texture) and global information (such as the relationship between PDAC and the pancreas). This approach is different from previous methods that rely solely on either CNN-based or transformer-based backbones, focusing only on local or global information, respectively.

The texture-aware transformer (Fig. 2) comprises three blocks, each consisting of a texture-aware CNN block and a texture-aware self-attention block. These blocks encode the input feature of an image $\mathbf{F}_i \in \mathbb{R}^{H\times W\times D\times C}$ to the hidden feature $\mathbf{F}_c \in \mathbb{R}^{H\times W\times D\times C_l}$ using a $3\times 3\times 3$ convolutional layer, followed by a

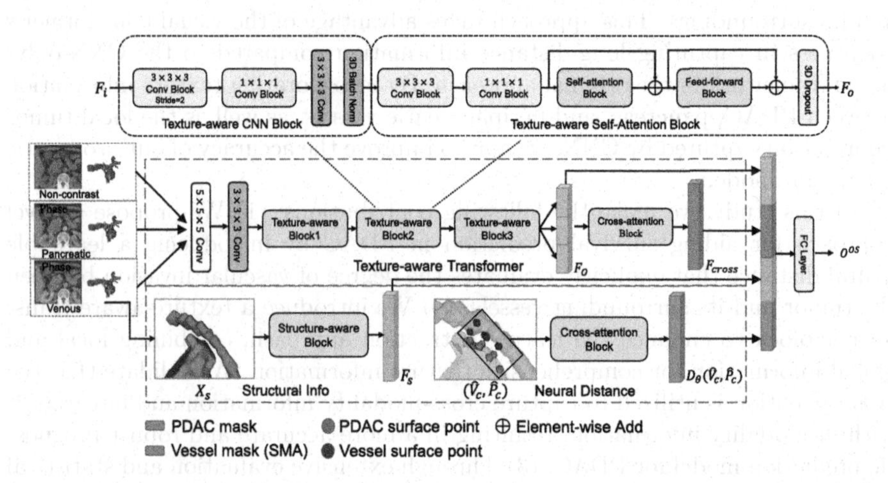

Fig. 2. An overview of the proposed method. The texture-aware transformer captures texture information among PDAC, Pancreas and vessels around Pancreas with our proposed texture-aware transformer block and a cross-attention block to fusion cross-modality features. The structure-aware block extracts the structure relationship between PDAC and four related vessels. The neural distance calculates the distances between the PDAC surface and the vessel surface with our proposed neural distance. We first select related points set from the closest sub-surface on PDAC and vessels respectively. Then we use a cross-attention block to obtain the neural distance. Finally, we concatenate features from three branches to obtain the survival outcome O_{OS}.

$1 \times 1 \times 1$ convolutional layer. The $3 \times 3 \times 3$ convolution captures local spatial information, while the $1 \times 1 \times 1$ convolution maps the input tensor to a higher-dimensional space (*i.e.*, $C_l > C$). The texture-aware CNN block downsamples the input, and the texture-aware self-attention block captures long-range non-local dependencies through a patch-wise self-attention mechanism.

In the texture-aware self-attention block, the input feature $\mathbf{F}_c$ is divided into N non-overlapping 3D patches $\mathbf{F}_u \in \mathbb{R}^{V \times N \times C_u}$, where $V = hwd$ and $N = HWD/V$ is the number of patches, and h, w, d are the height, width, and depth of a patch, respectively. For each voxel position within a patch, we apply a multi-head self-attention block and a feed-forward block following [20] to obtain the output feature $\mathbf{F}_o$. In this study, preoperative multi-phase CE-CT pancreatic imaging includes the non-contrast phase, the pancreatic phase and venous phase. Therefore, we obtain three outputs from the transformer block with the input of these phases, denoted as $\mathbf{F}_o^1, \mathbf{F}_o^2, \mathbf{F}_o^3 \in \mathbb{R}^{D \times C}$, resulting in the concatenated output $\mathbf{F}_o \in \mathbb{R}^{D \times 3C}$.

Instead of directly fusing the outputs as in previous work, we employ a 3-way cross-attention block to extract cross-modality information from these phases. The cross-attention is performed on the concatenated self-attention matrix with an extra mask $\mathbf{M} \in \{0, -\infty\}^{3C \times 3C}$, defined as:

$$\begin{aligned} \mathbf{F}_{cross} &= \text{Softmax}(\mathbf{Q}\mathbf{K}^{\mathrm{T}} + \mathbf{M})\mathbf{V}, \\ \mathbf{M}(i,j) &= \begin{cases} -\infty & kC < i, j \leq (k+1)C, \quad k = 0, 1, 2, \\ 0 & \text{otherwise,} \end{cases} \end{aligned} \tag{1}$$

Here, $\mathbf{Q}, \mathbf{K}, \mathbf{V}$ are the query, key, and value matrices, respectively, obtained by linearly projecting the input $\mathbf{F}_o^{\mathrm{T}} \in \mathbb{R}^{3C \times D}$. The cross-modality output $\mathbf{F}_{cross}$ and in-modality output $\mathbf{F}_o^{\mathrm{T}}$ are then concatenated and passed through an average pooling layer to obtain the final output feature of the texture branch, denoted as $\mathbf{F}_t \in \mathbb{R}^{C_t}$.

2.2 Neural Distance: Positional and Structural Information Between PDAC and Vessels

The vascular involvement in patients with PDAC affects the resectability and treatment planning [5]. In this study, we investigate four important vessels: portal vein and splenic vein (PVSV), superior mesenteric artery (SMA), superior mesenteric vein (SMV), and truncus coeliacus (TC). We used a semi-supervised nnUnet model to segment PDAC and the surrounding vessels, following recent work [11,21]. We define a general distance between the surface boundaries of PDAC ($\mathcal{P}$) and the aforementioned four types of vessels ($\mathcal{V}$) as $D(\mathcal{V}, \mathcal{P})$, which can be derived as follows:

$$D(\mathcal{V}, \mathcal{P}) = d_{ss}(\mathcal{V}, \mathcal{P}) + d_{ss}(\mathcal{P}, \mathcal{V}) = \frac{1}{\|\mathcal{V}\|} \int_{\mathcal{V}} d_{ps}(v, \mathcal{P}) \mathrm{d}v + \frac{1}{\|\mathcal{P}\|} \int_{\mathcal{P}} d_{ps}(p, \mathcal{V}) \mathrm{d}p, \tag{2}$$

where $v \in \mathcal{V}$ and $p \in \mathcal{P}$ are points on the surfaces of blood vessels and PDAC, respectively. The point-to-surface distance $d_{ps}(v, \mathcal{P})$ is the distance from a point v on $\mathcal{V}$ to $\mathcal{P}$, defined as $d_{ps}(v, \mathcal{P}) = \min_{p \in \mathcal{P}} \|v - p\|_2^2$, and vice versa.

To numerically calculate the integrals in the previous equation, we uniformly sample from the surfaces $\mathcal{V}$ and $\mathcal{P}$ to obtain the sets $\hat{\mathcal{V}}$ and $\hat{\mathcal{P}}$ consisting of N_v points and N_p points, respectively. The distance is then calculated between the two sets using the following equation:

$$D(\hat{\mathcal{V}}, \hat{\mathcal{P}}) = \frac{1}{N_v} \sum_{v \in \hat{\mathcal{V}}} d_{ps}(v, \hat{\mathcal{P}}) + \frac{1}{N_p} \sum_{p \in \hat{\mathcal{P}}} d_{ps}(p, \hat{\mathcal{V}}). \tag{3}$$

However, the above distance treats all points equally and may not be flexible enough to adapt to individualized prognostic predictions. Therefore, we improve the above equation in two ways. Firstly, we focus on the sub-sets $\hat{\mathcal{V}}_c$ and $\hat{\mathcal{P}}_c$ of $\hat{\mathcal{V}}$ and $\hat{\mathcal{P}}$, respectively, which only contain the K closest points to the opposite surfaces $\hat{\mathcal{P}}$ and $\hat{\mathcal{V}}$, respectively. The sub-sets are defined as:

$$\begin{aligned} \hat{\mathcal{V}}_c &= \operatorname{argmin}_{\{v_1, v_2, \cdots, v_K\} \subset \hat{\mathcal{V}}} \sum_{i=1}^{K} d_{ps}(v_i, \hat{\mathcal{P}}), \\ \hat{\mathcal{P}}_c &= \operatorname{argmin}_{\{p_1, p_2, \cdots, p_K\} \subset \hat{\mathcal{P}}} \sum_{i=1}^{K} d_{ps}(p_i, \hat{\mathcal{V}}). \end{aligned} \tag{4}$$

Secondly, we regard the entire sets $\hat{\mathcal{V}}_c$ and $\hat{\mathcal{P}}_c$ as sequences and calculate the distance using a 2-way cross-attention block (similar to Eq. 1) to build a neural distance based on the 3D spatial coordinates of each point:

$$D_\theta(\hat{\mathcal{V}}, \hat{\mathcal{P}}) = \text{CrossAttention}(\hat{\mathcal{V}}_c, \hat{\mathcal{P}}_c), \quad \hat{\mathcal{V}}_c, \hat{\mathcal{P}}_c \in \mathbb{R}^{K \times 3}. \tag{5}$$

Neural distance allows for the flexible assignment of weights to different points and is able to find positional information that is more suitable for PDAC prognosis prediction. In addition to neural distance, we use the 3D-CNN model introduced in [22] to extract the structural relationship between PDAC and the vessels. Specifically, we concatenate each PDAC-vessel pair $\mathbf{X}_s^v \in \mathbb{R}^{2 \times H \times W \times D}$, where $v \in$ {PVSV, SMV, SMA, TC} and obtain the structure feature $\mathbf{F}_s \in \mathbb{R}^{C_s}$.

Finally, we concatenate the features extracted from the two components and apply a fully-connected layer to predict the survival outcome, denoted as O^{OS}, which is a value between 0 and 1. To optimize the proposed model, we use the negative log partial likelihood as the survival loss [9].

3 Experiments

Dataset. In this study, we used data from Shengjing Hospital to train our method with 892 patients, and data from three other centers, including Guangdong Provincial People's Hospital, Tianjin Medical University and Sun Yat-sen University Cancer Center for independent testing with 178 patients. The contrast-enhanced CT protocol included non-contrast, pancreatic, and portal venous phases. PDAC masks for 340 patients were manually labeled by a radiologist from Shengjing Hospital with 18 years of experience in pancreatic cancer, while the rest were predicted using self-learning models [11,24] and checked by the same annotator. Other vessel masks were generated using the same semi-supervised segmentation models. **C-index** was used as our primary evaluation metric for survival prediction. We also reported the survival **AUC**, which estimates the cumulative area under the ROC curve for the first 36 months.

Implementation Details: We used nested 5-fold cross-validation and augmented the training data by rotating volumetric tumors in the axial direction and randomly selecting cropped regions with random shifts. We also set the output feature dimensions to $C_t = 64$ for the texture-aware transformer, $C_s = 64$ for the structure extraction and $K = 32$ for the neural distance. The batch size was 16 and the maximum iteration was set to 1000 epochs, and we selected the model with the best performance on the validation set during training for testing. We implemented our experiments using PyTorch 1.11 and trained the models on a single NVIDIA 32G-V100 GPU.

Ablation Study. We first evaluated the performance of our proposed texture-aware transformer (TAT) by comparing it with the ResNet18 CNN backbone and ViT transformer backbone, as shown in Table 1. Our model leverages the strengths of both local and global information in the pancreas and achieved the best result. Next, we compared different methods for multi-phase stages, including LSTM, early fusion (Fusion), and cross-attention (Cross) in our method. Cross-attention is more effective and lightweight than LSTM. Moreover, we separated texture features into in-phase features and cross-phase features, which is more reasonable than early fusion.

Secondly, we evaluated each component in our proposed method, as shown in Fig. 2, and presented the results in Table 1. Combining the texture-aware transformer and regular structure information improved the results from 0.630 to 0.648, as tumor invasion strongly affects the survival of PDAC patients. We also employed a simple 4-variable regression model that used only the Chamfer distance of the tumor and the four vessels for prognostic prediction. The resulting C-index of 0.611 confirmed the correlation of the distance with the survival, which is consistent with clinical findings [18]. Explicitly adding the distance measure further improved the results. Our proposed neural distance metric outperformed traditional surface distance metrics like Chamfer distance, indicating its suitability for distinguishing the severity of PDAC.

Comparisons. To further evaluate the performance of our proposed model, we compared it with recent deep prediction methods [17,21] and report the results in Table 2. We modified baseline deep learning models [12,17] and used their network architectures to take a single pancreatic phase or all three phases as inputs. DeepCT-PDAC [21] is the most recent method that considers both tumor-related and tumor-vascular relationships using 3D CNNs. Our proposed method, which uses the transformer and structure-aware blocks to capture tumor enhancement

Table 1. Ablation tests with different network backbones including ResNet18 (Res), ViT and texture-aware transformer (TAT) and methods for multi-phases including LSTM, early fusion (Fusion) and cross-attention (Cross).

Network Backbone	Structural Info	Distance	Model Size (M)	C-index
Res-LSTM	–	–	65.06	0.618 ± 0.017
Res-Cross	–	–	43.54	0.625 ± 0.016
ViT-Cross	–	–	23.18	0.628 ± 0.018
TAT-Fusion	–	–	**3.64**	0.626 ± 0.022
TAT-Cross	–	–	15.13	0.630 ± 0.019
TAT-Cross	✓	–	15.90	0.648 ± 0.021
–	–	Chamfer distance	–	0.611 ± 0.029
TAT-Cross	✓	Chamfer distance	15.93	0.652 ± 0.019
TAT-Cross	✓	Nerual distance	16.28	**0.656** ± 0.017

Table 2. Results of different methods on nested 5-fold CV and independent set.

	Nested 5-fold CV (n = 892)		Independent test (n = 178)	
	C-index	AUC	C-index	AUC
3DCNN-P [12]	0.630 ± 0.009	0.668 ± 0.019	0.674	0.740
Early Fusion [17]	0.635 ± 0.011	0.670 ± 0.024	0.696	0.779
DeepCT-PDAC [21]	0.640 ± 0.018	0.680 ± 0.036	0.697	0.773
Ours	**0.656** ± 0.017	**0.695** ± 0.023	**0.710**	**0.792**

patterns and tumor-vascular involvement, demonstrated its effectiveness with better performance in both nested 5-fold cross-validation and the multi-center independent test set.

In Table 3, we used univariate and multivariate Cox proportional-hazards models to evaluate our signature and other clinicopathologic factors in the independent test set. The proposed risk stratification was a significant prognostic factor, along with other factors like pathological TNM stages. After selecting significant variables ($p < 0.05$) in univariate analysis, our proposed staging remained strong in multivariable analysis after adjusting for important prognostic markers like pT and resection margins. Notably, our proposed marker remained the strongest among all pre-operative markers, such as tumor size and CA 19-9.

Neoadjuvant Therapy Selection. To demonstrate the added value of our signature as a tool to select patients for neoadjuvant treatment before surgery, we plotted Kaplan-Meier survival curves in Fig. 3. We further stratify patients by our signature after grouping them by tumor size and CA19-9, two clinically used preoperative criteria for selection, and also age. Our signature could significantly stratify patients in all cases and those in the high-risk group had worse outcomes and might be considered as potential neoadjuvant treatment candidates (e.g. 33 high-risk patients with larger tumor size and high CA19-9).

Table 3. Univariate and Multivariate Cox regression analysis. HR: hazard ratio.

Independent test set (n = 178)	Univariate Analysis		Multivariate Analysis	
	HR (95% CI)	p-value	HR (95% CI)	p-value
Proposed (High vs low risk)	2.42(1.64-3.58)	<0.0001	1.85(1.08-3.17)	0.027
Age (> 60 vs = 60)	1.49(1.01-2.20)	0.043	1.01(0.65-1.58)	0.888
Sex (Male vs Female)	1.28(0.86-1.90)	0.221	-	-
pT (pT3-pT4 vs pT1-pT2)	3.17(2.10-4.77)	<0.0001	2.44(1.54-3.86)	0.00015
pN (Positive ve Negative)	1.47(0.98-2.20)	0.008	1.34(0.85-2.12)	0.210
Resection margin (R1 vs R0)	2.84(1.64-4.93)	<0.0001	1.68(0.92-3.07)	0.091
CA19-9 (> 210 vs ≤ 210 U/mL)	0.94(0.64-1.39)	0.759	-	-
Tumor Size (> 25 vs ≤ 25 mm)	2.36(1.59-3.52)	<0.0001	0.99(0.52-1.85)	0.963
Tumor Location (Head vs Tail)	1.06(0.63-1.79)	0.819	-	-

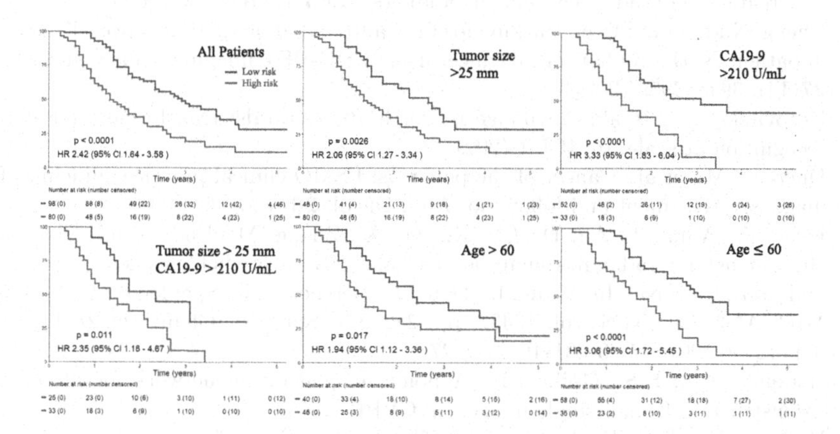

Fig. 3. Kaplan-Meier analyses of overall survival according to the proposed signature in all patients in the independent test set (n = 178) and subgroups defined by preoperative factors. High risk group indicated by the proposed method is the potential patient group that could benefit from neoadjuvant treatment before surgery.

4 Conclusion

In our paper, we propose a multi-branch transformer-based framework for predicting cancer survival. Our framework includes a texture-aware transformer that captures both local and global information about the PDAC and pancreas. We also introduce a neural distance to calculate a more reasonable distance between PDAC and vessels, which is highly correlated with PDAC survival. We have extensively evaluated and statistically analyzed our proposed method, demonstrating its effectiveness. Furthermore, our model can be combined with established high-risk features to aid in the patient selections who might benefit from neoadjuvant therapy before surgery.

Acknowledgement. This work was supported by Alibaba Group through Alibaba Research Intern Program. Bin Dong and Li Zhang was partly supported by NSFC 12090022 and 11831002, and Clinical Medicine Plus X-Young Scholars Project of Peking University PKU2023LCXQ041. Yu Shi was supported by the National Natural Science Foundation of China (No. 82071885).

References

1. Attiyeh, M.A., et al.: Survival prediction in pancreatic ductal adenocarcinoma by quantitative computed tomography image analysis. Ann. Surg. Oncol. **25**(4), 1034–1042 (2018)
2. Bian, Y., et al.: Artificial intelligence to predict lymph node metastasis at CT in pancreatic ductal adenocarcinoma. Radiology **306**(1), 160–169 (2023)
3. Cheng, N.M., et al.: Deep learning for fully automated prediction of overall survival in patients with oropharyngeal cancer using FDG-PET imaging. Clin. Cancer Res. **27**(14), 3948–3959 (2021)
4. Dosovitskiy, A., et al.: An image is worth 16x16 words: transformers for image recognition at scale. In: ICLR (2021)
5. Ducreux, M., et al.: Cancer of the pancreas: ESMO clinical practice guidelines for diagnosis, treatment and follow-up. Ann. Oncol. **26**, v56–v68 (2015)
6. Feng, Y., Wang, J., An, D., Gu, X., Xu, X., Zhang, M.: End-to-end evidential-efficient net for radiomics analysis of brain MRI to predict oncogene expression and overall survival. In: Wang, L., Dou, Q., Fletcher, P.T., Speidel, S., Li, S. (eds.) MICCAI 2022. LNCS, vol. 13433, pp. 282–291. Springer, Cham (2022). https://doi.org/10.1007/978-3-031-16437-8_27
7. Haoqiang Fan, H.S., Guibas, L.: A point set generation network for 3D object reconstruction from a single image. In: CVPR (2017)
8. Huttenlocher, D.P., Rucklidge, W.J., Klanderman, G.A.: Comparing images using the hausdorff distance under translation. In: Proceedings 1992 IEEE Computer Society Conference on Computer Vision and Pattern Recognition (2002)
9. Katzman, J.L., Shaham, U., Cloninger, A., Bates, J., Jiang, T., Kluger, Y.: Deepsurv: personalized treatment recommender system using a cox proportional hazards deep neural network. BMC Med. Res. Methodol. **18**(1), 24 (2018)
10. Koay, E.J., et al.: Computed tomography-based biomarker outcomes in a prospective trial of preoperative folfirinox and chemoradiation for borderline resectable pancreatic cancer. JCO Precis. Oncol. **3**, 1–15 (2019)
11. Koehler, G., Isensee, F., Maier-Hein, K.: A noisy nnU-Net student for semi-supervised abdominal organ segmentation. In: Ma, J., Wang, B. (eds.) MICCAI 2022. LNCS, vol. 13816, pp. 128–138. Springer, Cham (2023). https://doi.org/10.1007/978-3-031-23911-3_12
12. Lou, B., et al.: An image-based deep learning framework for individualising radiotherapy dose: a retrospective analysis of outcome prediction. Lancet Digit. Health **1**(3), e136–e147 (2019)
13. Mehta, S., Rastegari, M.: Mobilevit: light-weight, general-purpose, and mobile-friendly vision transformer. In: ICLR (2022)
14. Prokesch, R.W., Chow, L.C., Beaulieu, C.F., Bammer, R., Jeffrey, R.B., Jr.: Isoattenuating pancreatic adenocarcinoma at multi-detector row CT: secondary signs. Radiology **224**(3), 764–768 (2002)

15. Saeed, N., Sobirov, I., Al Majzoub, R., Yaqub, M.: TMSS: an end-to-end transformer-based multimodal network for segmentation and survival prediction. In: Wang, L., Dou, Q., Fletcher, P.T., Speidel, S., Li, S. (eds.) MICCAI 2022. LNCS, vol. 13437, pp. 319–329. Springer, Cham (2022). https://doi.org/10.1007/978-3-031-16449-1_31
16. Siegel, R.L., Miller, K.D., Jemal, A.: Cancer statistics, 2019. CA Cancer J. Clin. **69**(1), 7–34 (2019)
17. Tang, Z., et al.: Deep learning of imaging phenotype and genotype for predicting overall survival time of glioblastoma patients. IEEE Trans. Med. Imaging **39**(6), 2100–2109 (2020)
18. Tempero, M.A., et al.: Pancreatic adenocarcinoma, version 2.2021, NCCN clinical practice guidelines in oncology. J. Natl. Compr. Cancer Netw. **19**(4), 439–457 (2021)
19. Tsai, S., et al.: Importance of normalization of ca19-9 levels following neoadjuvant therapy in patients with localized pancreatic cancer. Ann. Surg. **271**(4), 740–747 (2020)
20. Vaswani, A., et al.: Attention is all you need. In: Guyon, I., et al. (eds.) NeurIPS, vol. 30. Curran Associates, Inc. (2017)
21. Yao, J., et al.: Deep learning for fully automated prediction of overall survival in patients undergoing resection for pancreatic cancer: a retrospective multicenter study. Ann. Surg. **278**(1), e68–e79 (2023)
22. Yao, J., et al.: Deepprognosis: Preoperative prediction of pancreatic cancer survival and surgical margin via comprehensive understanding of dynamic contrast-enhanced CT imaging and tumor-vascular contact parsing. Med. Image Anal. **73**, 102150 (2021)
23. Yuan, M., et al.: Devil is in the queries: advancing mask transformers for real-world medical image segmentation and out-of-distribution localization. In: CVPR, pp. 23879–23889 (2023)
24. Zhang, L., et al.: Robust pancreatic ductal adenocarcinoma segmentation with multi-institutional multi-phase partially-annotated CT scans. In: Martel, A.L., et al. (eds.) MICCAI 2020. LNCS, vol. 12264, pp. 491–500. Springer, Cham (2020). https://doi.org/10.1007/978-3-030-59719-1_48
25. Zheng, H., et al.: Multi-transSP: multimodal transformer for survival prediction of nasopharyngeal carcinoma patients. In: Wang, L., Dou, Q., Fletcher, P.T., Speidel, S., Li, S. (eds.) MICCAI 2022, pp. 234–243. Springer, Cham (2022). https://doi.org/10.1007/978-3-031-16449-1_23

Contrastive Feature Decoupling for Weakly-Supervised Disease Detection

Jihih-Ciang Wu[1,2], Ding-Jie Chen[1(✉)], and Chiou-Shann Fuh[2]

[1] Institute of Information Science, Academia Sinica, Taipei, Taiwan
djchen.tw@gmail.com

[2] National Taiwan University, Taipei, Taiwan

Abstract. Machine learning based Computer-Aided Diagnosis (CAD) aims to assist clinicians in the pathological diagnosis process. While dealing with video pathological diagnosis such as colonoscopy polyp detection, the recent SOTA method employs Weakly-supervised Video Anomaly Detection (WVAD) in the Multiple Instance Learning (MIL) scenarios to concern the temporal correlation within data and to formulate the concept of the interest disease simultaneously. Such a MIL-based WVAD method leverages video-level annotations to detect frame-level diseases and shows promising results. This paper casts the video pathological diagnosis as a MIL-based WVAD task and introduces Contrastive Feature Decoupling (CFD) network to decouple normal and abnormal feature ingredients per snippet. With such decoupled features, we are able to highlight the abnormal feature ingredients for accurately reasoning the disease score per snippet. The core components within our CFD model are the memory bank and contrastive loss. The former is used to learn atoms for representing normal features, and the latter is used to encourage our model to gain robust disease detection. We demonstrate that our CFD network is achieving new SOTA performance on the existing Polyp dataset and the introduced PANDA-MIL dataset. Our dataset are available at https://github.com/Jhih-Ciang/PANDA-MIL.

Keywords: Disease detection · Multiple instance learning · Weakly-supervised learning

1 Introduction

Computer-aided diagnosis utilizes machine learning techniques to conduct a pathological diagnosis concerning biomedical imaging data collected from various pathological modalities, such as computed tomography [19], magnetic resonance imaging [11], ultrasound [23], and angiography [9]. With the assistance of CAD techniques, the clinicians merely need to check the possible pathological regions narrowed down by computer-aided diagnosis method, significantly reducing the entire diagnosis time. With the recent success of deep learning, researchers are

Supplementary Information The online version contains supplementary material available at https://doi.org/10.1007/978-3-031-43904-9_25.

H. Greenspan et al. (Eds.): MICCAI 2023, LNCS 14224, pp. 252–261, 2023.
https://doi.org/10.1007/978-3-031-43904-9_25

able to raise the reliability of CAD methods and assist clinicians in diagnosing more complex clinical tasks. However, a reliable machine learning-based CAD method usually relies on the supervision of abundant annotated training data. Yet diseased pathological data are rare and diverse, and acquiring reliable pathological annotations are labor-intensive and expertise-required. As a result, the difficulty of data collection restricts the development of the supervised CAD.

Due to the difficulty of acquiring the abundant annotated training data, the current SOTA method, *i.e.*, CSM [14], proposes a MIL-based WVAD manner to specifically tackle one specific disease detection task, *i.e.*, colorectal cancer diagnosis via colonoscopy. Considering the case of colonoscopy, the CSM's anomaly detection setting is used to handle the rare and diverse diseased pathological data by commonly assuming that only video-level annotations are available for training. Furthermore, its video setting concerns the temporal correlation within data. The setting of such MIL-based weakly supervision prevents the need for abundant annotated training data by assuming that merely the video-level annotations, including normal and diseased ones, are available for training.

Similar to the previous MIL-based WVAD methods [4,13,14], our model assumes all training snippets (consecutive video frames) within a non-diseased video are all normal snippets, yet each diseased video has at least one abnormal snippet. Furthermore, the proposed contrastive feature decoupling network treats disease detection as an out-of-distribution task. Precisely, our CFD learns a memory bank to learn normal features. A snippet that failed to be well reconstructed with these normal features is considered diseased. On the other hand, the residual of a snippet and its reconstructed one reveals the snippet's abnormal ingredients. Consequently, we are able to decouple each snippet as normal ingredients (reconstructed parts) and abnormal ingredients (residual parts) by leveraging the memory bank. With the decoupled snippet-level feature ingredients, our CFD employs both the normal and abnormal feature ingredients via a contrastive learning paradigm to concurrently optimize video-level and snippet-level disease scores for pursuing more accurate detection.

To assess the proposed contrastive feature decoupling network, we conduct experiments on two datasets, *i.e.*, Polyp and PANDA-MIL. The main contributions are summarized as follows.

- Our contrastive feature decoupling network learns a memory bank to learn normal atoms for decoupling each snippet as normal and diseased feature ingredients as opposite contrastive learning samples. Such a feature decoupling intrinsically fits the contrastive learning paradigm for optimizing MIL objectives on bags and instances. The ablation study shows the decoupled diseased ingredients enable accurate disease detection, and the accompanied contrastive learning paradigm provides further improvement.
- We introduce the new biomedical imaging dataset for prostate cancer detection, *i.e.*, PANDA-MIL. The dataset is organized to fit the MIL-based WVAD task, including video-level annotations and video-wise format data.
- We demonstrate the generalization of CFD by achieving new SOTA performance on two biomedical imaging datasets concerning different pathological modalities, *i.e.*, the colonoscopy videos in Polyp and prostate tissue biopsies in PANDA-MIL.

2 Related Work

2.1 Disease Detection

With the evolution of artificial intelligence techniques in the past decades, deep learning has shown its potential for computer-aided diagnosis of various symptoms [6,8,14,17,22]. For example, Li *et al.* [8] established a large-scale attention-based database and designed a specialized model using retinal fundus images for detecting glaucoma. Windsor *et al.* [17] constructed a transformer-based model to detect spinal cancer for MRI scans. More recently, Tian *et al.* [14] formulated polyp detection in a WVAD scheme while tackling polyp detection using colonoscopy videos to search colon polyps in the temporal sequence. Unlike previous methods of handling one specific pathological modality, we simultaneously address disease detection across pathological modalities of colonoscopy videos and prostate tissue biopsies using our contrastive feature decoupling network.

2.2 Contrastive Learning

The characteristics of self-supervised learning are defining the proxy objective or addressing pretext tasks using pseudo labels for the unlabeled instances. One popular branch is contrastive learning which shows a remarkable ability to obtain the desired semantic representation from various perspectives. For example, CoLA [21] tackled action localization by proposing snippet contrast loss to refine the feature representations of hard snippets according to the easily discriminative snippets. CSM [14] borrowed the concept from CoLA and defined the hard/easy snippets for representing normal/abnormal from colonoscopy videos. They empirically selected hard snippets based on the transitional edge and missed disease snippets, such as an occlusive polyp. In this work, we employ a similar contrastive learning strategy as CSM while preventing their rule-based contrastive training samples selection by intrinsically leveraging the decoupled features derived from our feature decoupling process via memory bank.

3 Method

We aim to design a MIL-based WVAD model for tackling disease detection across different pathological modalities. Our model contains an offline trained memory bank to store feature atoms before the CFD training procedure, which associates a contrastive loss to boost the model performance using decoupled features per instance. Winin the MIL scenario, our model employs two classifiers to enable reasoning of the disease scores at instance and bag levels. Figure 1 overviews the working flow of the proposed contrastive feature decoupling network.

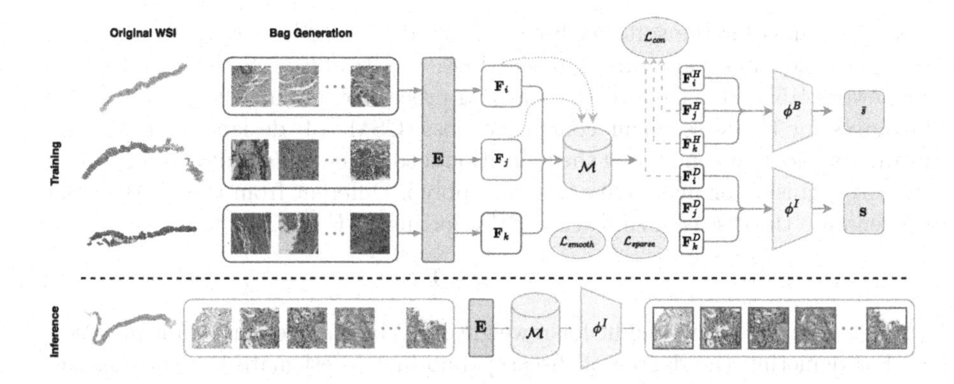

Fig. 1. Illustration of our contrastive feature decoupling (CFD) network. Given a **bag** composing multiple *instances* (also known as a **video** comprising *snippets* in Polyp or a **WSI** consisting of *patches* in PANDA-MIL), our CFD employs I3D as feature encoder $\mathbf{E}$ to generate bag-level feature $\mathbf{F}$, which subsequently decoupled as the normal proxy $\mathbf{F}^H$ and abnormal proxy $\mathbf{F}^D$ through the offline trained memory bank $\mathcal{M}$. The gray dotted arrows indicate the required normal features for constructing $\mathcal{M}$. Besides the regular MIL-based losses, *i.e.*, $\mathcal{L}_{sep}$ and $\mathcal{L}_{cls}$, we employ auxiliary losses of $\mathcal{L}_{smooth}$ and $\mathcal{L}_{sparse}$ to regularize the decoupled features within each bag and employ contrastive loss $\mathcal{L}_{con}$ to regularize the opposite decoupled features across bags. The notations ϕ^B and ϕ^I denote the bag-level and instance-level classifiers, respectively. The green box indicates a normal bag/instance, while the red box represents abnormal ones.

3.1 Memory Bank Construction

Given dataset $\mathcal{D}$ comprising normal sub-dataset $\mathcal{D}_0$ and abnormal sub-dataset $\mathcal{D}_1$, we first encode all instances per bag $\mathbf{B} \in \mathcal{D}$ into instance-level feature set $\mathbf{F} = \{f^t\}_{t=1}^{T} \in \mathbb{R}^{T\times C}$ via a pre-trained feature extractor $\mathbf{E}$. That is, $\mathbf{F} = \mathbf{E}(\mathbf{B})$, T is the number of instances, and C represents the instance-level feature dimension. We then collect all normal instance-level features $f \in \mathbb{R}^{1\times C}$ from $\mathcal{D}_0$ to learn the memory bank $\mathcal{M}$ by using the dictionary learning technique [7] via

$$\underset{\mathcal{M},\{\mathbf{w}_t\}}{\operatorname{argmin}} \sum_{B\in\mathcal{D}_0} \sum\nolimits_{t=1}^{T} (\|f^t - \mathcal{M}\mathbf{w}_t\|^2 + \lambda\|\mathbf{w}_t\|_0), \tag{1}$$

where $\mathcal{D}_0$ is the normal sub-dataset collected from the training split, $\mathbf{w}_t$ is the learned weights within the memory bank learning process, and λ is a hyperparameter to constrain the memory bank sparsity.

3.2 Contrastive Feature Decoupling

With the learned normal instance features stored in the memory bank $\mathcal{M}$, we are able to reconstruct a normal version for any given bag-level feature $\mathbf{F}$. Such a normal version is denoted as a normal-like feature $\mathbf{F}^H$. To this end, we reconstruct a given $\mathbf{F}$ concerning $\mathcal{M}$ via the following equation

$$\mathbf{F}^H = \sigma\left(\phi^q(\mathbf{F})\phi^k(\mathcal{M})^\intercal\right)\phi^v(\mathcal{M}), \tag{2}$$

where σ stands for the softmax function; ϕ^q, ϕ^k, and ϕ^v respectively represent the query, key, and value linear projections, as introduced in the self-attention framework [15]. To make a robust learning process by mining the hard and easy instances for the subsequent contrastive loss, CSM [14] designed a rule-based instances selection concerning the transitional edge and missed disease instances, such as occlusion or invisibility from the polyp. Different from the CSM model, we generate the disease-like feature $\mathbf{F}^D$ referring to $\mathbf{F}^H$ as follows:

$$\mathbf{F}^D = \omega \odot \mathbf{F}, \tag{3}$$

where $\omega \in \mathbb{R}^{1\times C}$ is the weight for reweighting $\mathbf{F}$ by channel-wise multiplication.

For depicting the degree of disease/abnormal to estimate the channel-wise weight ω, we consider attending to the distant features with respect to the normal ones, *i.e.*, $\mathbf{F}^H$. In practice, we estimate the degree of disease/abnormal based on the difference between $\mathbf{F}$ and $\mathbf{F}^H$ by

$$\omega = \text{sigmoid}(\phi^d(\mathcal{G}(\Psi(\mathbf{F}) - \Psi(\mathbf{F}^H)))), \tag{4}$$

where ϕ^d, $\mathcal{G}$, and Ψ are linear projections, global average pooling, and the multi-scale temporal network [13], respectively.

With the decoupled features $\mathbf{F}^H$ and $\mathbf{F}^D$, our MIL-based instance-level classifier aims to carry out the discriminating decision, and the bag-level classifier seeks the prediction as fitting annotation y as possible. Following recent MIL-based methods, we select top-K instances of normal and abnormal in each bag to form the separation loss as

$$\mathcal{L}_{sep} = \sum_{\substack{\mathbf{F}_i^D \in \mathcal{D}_0 \\ \mathbf{F}_j^D \in \mathcal{D}_1}} \sum_{K} (|\delta - \|\{\mathbf{F}_i^D\}_K\|_2| + \|\{\mathbf{F}_j^D\}_K\|_2), \tag{5}$$

where δ is the hyperparameter for constrained margin and $\{\cdot\}_K$ is the operator that selects top-K instances. The other common loss used in the recent MIL-based works is the classification loss building upon the binary cross entropy

$$\mathcal{L}_{cls} = BCE(\{\mathbf{S}\}_K, y) + BCE(\bar{s}, y), \tag{6}$$

where $\mathbf{S} = \phi^I(\mathbf{F}^D)$ represents the instance-level prediction inferred by an instance-level classifier ϕ^I and $\bar{s} = \phi^B(\mathcal{G}((1-\omega) \odot \mathbf{F}^H))$ means the bag-level prediction resulting from a bag-level classifier ϕ^B.

3.3 Regularization

Motivated by the recent WVAD methods [4,13,14,18], which adopt auxiliary losses to regularize the learning procedure, we consider the conventional regularization losses, such as temporal smoothness and sparsity, as follows

$$\mathcal{L}_{smooth} = \sum_{i=1}^{\|\mathcal{D}\|} \frac{1}{T-1} \sum_{t=1}^{T-1} \|f_i^{t+1} - f_i^t\|^2, \quad \mathcal{L}_{sparse} = \sum_{i=1}^{\|\mathcal{D}\|} \frac{1}{T} \sum_{t=1}^{T} \|f_i^t\|. \tag{7}$$

By using the decoupled features $\mathbf{F}^H$ and $\mathbf{F}^D$, we are ready to regularize the opposite decoupled features across bags with the aid of a contrastive loss. An expected contrastive loss aims to make our model attract features from the same category while distracting the features from distinct classes. In practice, we formulate such a contrastive loss by

$$\mathcal{L}_{con} = \sum_{\mathbf{F}_i^D, \mathbf{F}_j^H \in \mathcal{D}_0} \log \frac{\exp\left[\frac{1}{\tau}(\mathbf{F}_i^D)^\intercal \mathbf{F}_j^H\right]}{\exp\left[\frac{1}{\tau}(\mathbf{F}_i^D)^\intercal(\mathbf{F}_j^H)\right] + \sum_{\mathbf{F}_k^H \in \mathcal{D}_1} \exp\left[\frac{1}{\tau}(\mathbf{F}_i^D)^\intercal(\mathbf{F}_k^H)\right]}, \quad (8)$$

where τ denotes the temperature parameter in the normalized temperature-scaled loss. Notice that (8) is simplified for the sake of clarity. A complete objective should consider the symmetric form by switching $\mathcal{D}_1$ and $\mathcal{D}_0$ in (8).

4 Experiments

4.1 Dataset and Metric

We evaluate our model against SOTAs on the existing Polyp [14] dataset and the PANDA-MIL dataset introduced in this work. We employ the same evaluation criteria as the previous work for a fair comparison. Please refer to the supplementary material for the statistics of the two datasets.

Polyp. This dataset collects colonoscopy videos from Hyper-Kvasir [1] and LDPolypVideo [10]. Its training split contains 163 videos of video-level annotations, and the testing split includes 90 videos of frame-level annotations.

PANDA-MIL. The Prostate cANcer graDe Assessment (PANDA) challenge [2] comprises over 10K whole-slide images (WSIs) of digitized hematoxylin and eosin-stained biopsies originating from Radboud University Medical Center and Karolinska Institute. PANDA-MIL collects the eosin-stained biopsies with region-based masks indicating the benign (normal) and cancerous (abnormal) tissue, combined by stroma and epithelium. To fit the MIL-based WVAD task, we non-overlapped partition each WSI (bag) into patches (instances) and only keep those patches comprising tissue over the 50% patch size. Each kept patch gets its patch-level annotations from PANDA, and a WSI comprising any abnormal patch is treated as an abnormal WSI. In sum, PANDA-MIL's training split contains 3,925 bags of bag-level annotations, and the testing split includes 975 bags of instance-level annotations.

Metric. We follow the previous methods [4,13,14] to employ the instant-level Area Under Curve (AUC) and the Average Precision (AP) for a fair comparison. The larger values of both metrics mean better disease detection performance.

Table 1. Comparison with AUC and AP metrics on Polyp and PANDA-MIL datasets.

Method	Publication	Polyp		PANDA-MIL	
		AUC	AP	AUC	AP
MIST [4]	CVPR'21	94.53	72.85	82.84	75.45
RTFM [13]	ICCV'21	96.30	77.96	85.12	78.17
S3R [18]	ECCV'22	98.32	86.21	86.19	78.33
CSM [14]	MICCAI'22	98.41	86.63	76.52	73.12
CFD		**99.51**	**88.13**	**87.28**	**80.78**

4.2 Implementation Details

All the evaluated methods in the experiment used the same feature encoder, *i.e.*, I3D [3] pre-trained on Kinetics-400 [5], for a fair comparison. Our method is trained using Adam optimizer with the learning rate of 0.001, batch size 32, and 200 epochs. Each bag/video is encoded into $T = 32$ snippets among both datasets via linear interpolation.

4.3 Comparison Results

Table 1 shows the compared results of our CFD model against recent WVAD methods [4,13,14,18] for tackling the disease detection task. The results in Table 1 demonstrate that our CFD consistently outperforms all the other methods on two datasets. Precisely, our model achieves the new SOTA by 1.1% AUC and 1.5% AP improvements on the Polyp dataset and 1.09% AUC and 2.45% AP improvements on the PANDA-MIL dataset. Please refer to the supplementary material for the completed results, including more WVAD methods [12,16,20,24].

Figure 2 visualizes one disease detection result of our CFD model on the PANDA-MIL dataset. The disease score per instance/patch predicted by our method is close to the ground-truth annotations, in which the clear margin between cancerous and benign validates the robust prediction of the proposed CFD model.

4.4 Ablation Study

We analysis on why the CFD network performs better than other methods listed in Table 1 by ablating the contributed components in CFD. The ablation study in Table 2 is conducted on the PANDA-MIL dataset to evaluate the effectiveness of the memory bank and loss functions in our model. Row one in Table 2 indicates our CFD model without regularization, yet it has shown better AUC values than other methods besides the S3R. While employing the three loss functions for regularization, as described in Sect. 3.3, each loss function shows its improvement in our model performance. The contrastive loss contributes the most to AUC improvement, enabling our model to achieve the SOTA performance.

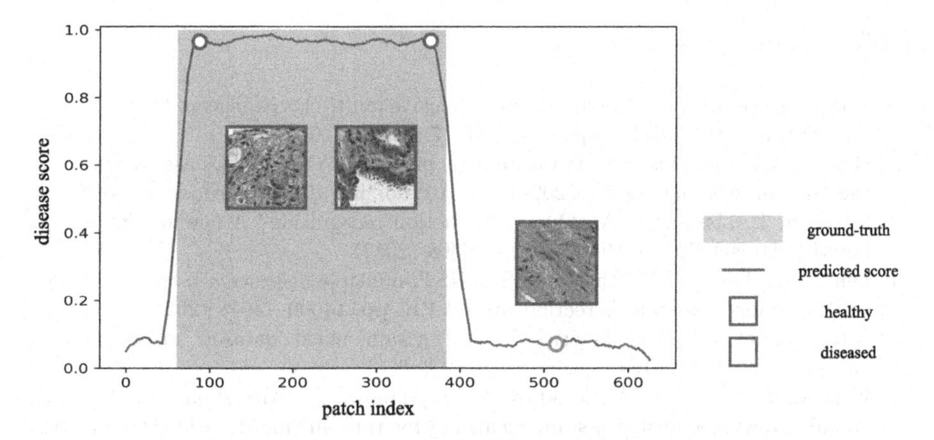

Fig. 2. Qualitative results of our CFD model on one testing prostate tissue biopsy of the PANDA-MIL dataset. Our disease detection results are close to the ground-truths.

Table 2. Ablation study on model components with the AUC metric on PANDA-MIL.

$\mathcal{M}$	$\mathcal{L}_{sparse}$	$\mathcal{L}_{smooth}$	$\mathcal{L}_{con}$	AUC
✓				85.36
✓	✓			85.79
✓	✓	✓		86.15
✓	✓	✓	✓	**87.28**

5 Conclusion

This paper casts disease detection as a MIL-based WVAD task and introduces Contrastive Feature Decoupling (CFD) network to learn a memory bank boosted with contrastive learning. With the learned feature atoms stored in the memory bank, our contrastive feature decoupling is able to decouple each snippet as normal and abnormal proxies. Further, the decoupled abnormal proxies highlight the abnormal feature ingredients for better reasoning the disease score. Our feature decoupling intrinsically fits the contrastive learning paradigm to define opposite training samples for model optimization. Besides, we introduce a new dataset of prostate cancer detection, i.e., PANDA-MIL, to provide a biomedical imaging dataset concerning a different pathological modality. Experiments demonstrate that our CFD network achieves new SOTA performance on the Polyp and PANDA-MIL datasets, indicating that our method effectively addresses the disease detection task across different pathological modalities.

Acknowledgement. This research was supported by National Science and Technology Council of Taiwan, R.O.C., under Grants NSTC 112-2221-E-002-189-MY2 and MOST 111-2221-E-002-174.

References

1. Borgli, H., et al.: Hyperkvasir, a comprehensive multi-class image and video dataset for gastrointestinal endoscopy. Sci. Data **7**(1), 283 (2020)
2. Bulten, W., et al.: The PANDA challenge: prostate cANcer graDe assessment using the Gleason grading system (2020). https://doi.org/10.5281/zenodo.3715938
3. Carreira, J., Zisserman, A.: Quo vadis, action recognition? A new model and the kinetics dataset. In: CVPR, pp. 6299–6308 (2017)
4. Feng, J.C., Hong, F.T., Zheng, W.S.: MIST: multiple instance self-training framework for video anomaly detection. In: CVPR, pp. 14009–14018 (2021)
5. Kay, W., et al.: The kinetics human action video dataset. arXiv preprint arXiv:1705.06950 (2017)
6. Kazeminia, S., Sadafi, A., Makhro, A., Bogdanova, A., Albarqouni, S., Marr, C.: Anomaly-aware multiple instance learning for rare anemia disorder classification. In: Wang, L., Dou, Q., Fletcher, P.T., Speidel, S., Li, S. (eds.) MICCAI 2022. LNCS, vol. 13438, pp. 341–350. Springer, Cham (2022). https://doi.org/10.1007/978-3-031-16452-1_33
7. Kreutz-Delgado, K., Murray, J.F., Rao, B.D., Engan, K., Lee, T.W., Sejnowski, T.J.: Dictionary learning algorithms for sparse representation. Neural Comput. **15**(2), 349–396 (2003)
8. Li, L., Xu, M., Wang, X., Jiang, L., Liu, H.: Attention based glaucoma detection: a large-scale database and CNN model. In: CVPR, pp. 10571–10580 (2019)
9. Ma, X.Y., et al.: DSP-NET: deeply-supervised pseudo-siamese network for dynamic angiographic image matching. In: Wang, L., Dou, Q., Fletcher, P.T., Speidel, S., Li, S. (eds.) MICCAI 2022. LNCS, vol. 13437, pp. 44–53. Springer, Cham (2022). https://doi.org/10.1007/978-3-031-16449-1_5
10. Ma, Y., Chen, X., Cheng, K., Li, Y., Sun, B.: LDPolypVideo benchmark: a large-scale colonoscopy video dataset of diverse polyps. In: de Bruijne, M., et al. (eds.) MICCAI 2021. LNCS, vol. 12905, pp. 387–396. Springer, Cham (2021). https://doi.org/10.1007/978-3-030-87240-3_37
11. Shao, W., et al.: Weakly supervised registration of prostate MRI and histopathology images. In: de Bruijne, M., et al. (eds.) MICCAI 2021. LNCS, vol. 12904, pp. 98–107. Springer, Cham (2021). https://doi.org/10.1007/978-3-030-87202-1_10
12. Sultani, W., Chen, C., Shah, M.: Real-world anomaly detection in surveillance videos. In: CVPR, pp. 6479–6488 (2018)
13. Tian, Y., Pang, G., Chen, Y., Singh, R., Verjans, J.W., Carneiro, G.: Weakly-supervised video anomaly detection with robust temporal feature magnitude learning. In: ICCV, pp. 4975–4986 (2021)
14. Tian, Y., et al.: Contrastive transformer-based multiple instance learning for weakly supervised polyp frame detection. In: Wang, L., Dou, Q., Fletcher, P.T., Speidel, S., Li, S. (eds.) MICCAI 2022. LNCS, vol. 13433, pp. 88–98. Springer, Cham (2022). https://doi.org/10.1007/978-3-031-16437-8_9
15. Vaswani, A., et al.: Attention is all you need. In: NIPS, pp. 5998–6008 (2017)
16. Wan, B., Fang, Y., Xia, X., Mei, J.: Weakly supervised video anomaly detection via center-guided discriminative learning. In: ICME, pp. 1–6 (2020)
17. Windsor, R., Jamaludin, A., Kadir, T., Zisserman, A.: Context-aware transformers for spinal cancer detection and radiological grading. In: Wang, L., Dou, Q., Fletcher, P.T., Speidel, S., Li, S. (eds.) MICCAI 2022. LNCS, vol. 13433, pp. 271–281. Springer, Cham (2022). https://doi.org/10.1007/978-3-031-16437-8_26

18. Wu, J.C., Hsieh, H.Y., Chen, D.J., Fuh, C.S., Liu, T.L.: Self-supervised sparse representation for video anomaly detection. In: Avidan, S., Brostow, G., Cissé, M., Farinella, G.M., Hassner, T. (eds.) ECCV 2022. LNCS, vol. 13673, pp. 729–745. Springer, Cham (2022). https://doi.org/10.1007/978-3-031-19778-9_42
19. Xia, Y., et al.: Effective pancreatic cancer screening on non-contrast CT scans via anatomy-aware transformers. In: de Bruijne, M., et al. (eds.) MICCAI 2021. LNCS, vol. 12905, pp. 259–269. Springer, Cham (2021). https://doi.org/10.1007/978-3-030-87240-3_25
20. Zaheer, M.Z., Mahmood, A., Astrid, M., Lee, S.-I.: CLAWS: clustering assisted weakly supervised learning with normalcy suppression for anomalous event detection. In: Vedaldi, A., Bischof, H., Brox, T., Frahm, J.-M. (eds.) ECCV 2020. LNCS, vol. 12367, pp. 358–376. Springer, Cham (2020). https://doi.org/10.1007/978-3-030-58542-6_22
21. Zhang, C., Cao, M., Yang, D., Chen, J., Zou, Y.: Cola: weakly-supervised temporal action localization with snippet contrastive learning. In: CVPR, pp. 16010–16019 (2021)
22. Zhang, W., et al.: A multi-task network with weight decay skip connection training for anomaly detection in retinal fundus images. In: Wang, L., Dou, Q., Fletcher, P.T., Speidel, S., Li, S. (eds.) MICCAI 2022. LNCS, vol. 13432, pp. 656–666. Springer, Cham (2022). https://doi.org/10.1007/978-3-031-16434-7_63
23. Zhao, H., et al.: Towards unsupervised ultrasound video clinical quality assessment with multi-modality data. In: Wang, L., Dou, Q., Fletcher, P.T., Speidel, S., Li, S. (eds.) MICCAI 2022. LNCS, vol. 13434, pp. 228–237. Springer, Cham (2022). https://doi.org/10.1007/978-3-031-16440-8_22
24. Zhong, J.X., Li, N., Kong, W., Liu, S., Li, T.H., Li, G.: Graph convolutional label noise cleaner: train a plug-and-play action classifier for anomaly detection. In: CVPR, pp. 1237–1246 (2019)

Uncovering Heterogeneity in Alzheimer's Disease from Graphical Modeling of the Tau Spatiotemporal Topography

Jiaxin Yue[1,2] and Yonggang Shi[1,2(✉)]

[1] Stevens Neuroimaging and Informatics Institute, Keck School of Medicine, University of Southern California (USC), Los Angeles, CA 90033, USA
yonggans@usc.edu

[2] Ming Hsieh Department of Electrical and Computer Engineering, Viterbi School of Engineering, University of Southern California (USC), Los Angeles, CA 90089, USA

Abstract. Growing evidence from post-mortem and in vivo studies have demonstrated the substantial variability of tau pathology spreading patterns in Alzheimer's disease (AD). Automated tools for characterizing the heterogeneity of tau pathology will enable a more accurate understanding of the disease and help the development of targeted treatment. In this paper, we propose a Reeb graph representation of tau pathology topography on cortical surfaces using tau PET imaging data. By comparing the spatial and temporal coherence of the Reeb graph representation across subjects, we can build a directed graph to represent the distribution of tau topography over a population, which naturally facilitates the discovery of spatiotemporal subtypes of tau pathology with graph-based clustering. In our experiments, we conducted extensive comparisons with state-of-the-art event-based model on synthetic and large-scale tau PET imaging data from ADNI3 and A4 studies. We demonstrated that our proposed method can more robustly achieve the subtyping of tau pathology with clear clinical significance and demonstrated superior generalization performance than event-based model.

Keywords: Heterogeneity · Tau pathology · Reeb graph

1 Introduction

With emerging evidence from various post-mortem and in vivo tau PET imaging studies, the existence of several well-recognized atypical patterns of neurofibrillary tangle distribution in subsets of patients and abnormal clinical presentations challenge [4] the notion that tau pathology follows a uniform Braak-like progression throughout the brain in Alzheimer's disease (AD) [2]. Disentangling tau pathology heterogeneity can thus greatly contribute to develop more accurate diagnosis, prognosis and targeted treatment for AD.

This work is supported by the National Institute of Health (NIH) under grants R01EB022744, RF1AG077578, RF1AG056573, RF1AG064584, R21AG064776, P41EB015922, U19AG078109.

H. Greenspan et al. (Eds.): MICCAI 2023, LNCS 14224, pp. 262–271, 2023.
https://doi.org/10.1007/978-3-031-43904-9_26

In contrast to many previous studies [7,14,16] utilizing clustering methods based on spatial variants for subtyping, spatiotemporal methods have recently gained popularity, combining spatial patterns and disease staging in a single framework. Event-based model [5,17] is a state-of-art spatiotemporal method that requires only cross-sectional data to automatically detect multiple spatiotemporal trajectories. However, the trajectory depends on the biomarker selection, and the predefined region-of-interest (ROI) based parcellations ignore the regional relations and the topological changes of pathology within the brain. Besides, the method focuses more on temporal transitions but neglects regional coherence. Moreover, the estimation requires large sample sizes to cover enough phenotype and intensity variants for better model fitting.

To address these limitations, this paper presents a novel Reeb graph representation that encodes the topography of tau pathology from PET imaging, and a directed-graph-based framework for uncovering spatiotemporal heterogeneity from cross-sectional tau PET data. More specifically, we first generate a pattern representation from Reeb graph analysis on cortical surfaces to encode the tau pathology pattern. Secondly, the spatial coherence and temporal consistency of tau spreading patterns across subjects are combined within a directed graph for clustering, which is robust to sample sizes and intensity deviations. With large-scale imaging data from the ADNI and A4 studies, we obtained three subtypes with systematic spatiotemporal variations in tau spreading patterns by utilizing an efficient community detection method on graphs. We also demonstrate that our method exhibits more robust generalization performance than event-based model on both synthetic and real data.

2 Method

2.1 Reeb Graph Analysis for Pathology Detection

The reeb graph encodes the topology of pathology on the cortical surface and is used to extract the salient tau pathological patterns. Given a Morse function f on the mesh $\mathcal{M}$, its Reeb graph is defined as follows [10]:

Definition 1. *Let $f : \mathcal{M} \to \mathbb{R}$. The Reeb graph $R(f)$ of f is the quotient space with its topology defined through the equivalent relation $x \simeq y$ if $f(x) = f(y)$ for $\forall x, y \in \mathcal{M}$.*

In our work, $\mathcal{M}$ is a common template surface *fsaverage* from FreeSurfer [6], and f is the tau standardized uptake value ratio (SUVR) map defined on $\mathcal{M}$. For numerical calculation of the Reeb graph of f on $\mathcal{M}$, we use the algorithm proposed in [11]. Because the topology only changes at critical points (minimum, maximum and saddle points), for a given SUVR map f on cortical surface $\mathcal{M}$, we first sort its critical points $C = \{C_1, C_2, \cdots, C_K\}$ according to their SUVR values such that $f(C_1) < f(C_2) < \cdots < f(C_K)$. The surface can be partitioned by using the level contours in the neighborhood of these critical points, and the neighboring nodes are connected in the Reeb graph through arcs by applying

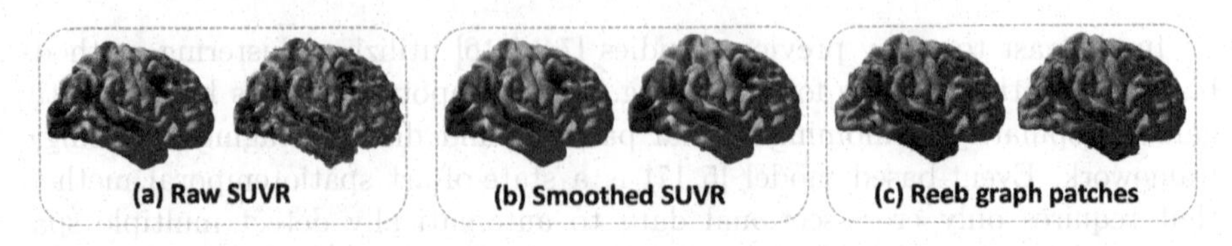

Fig. 1. Reeb graph simplification. (a) The Reeb graph of the original SUVR function is dominated by a large number of noisy peaks (purple points). (b) After graph simplification, the number of peaks (purple points) is significantly reduced. (c) Reeb graph patches can encode the salient tau pathology patterns. (color figure online)

region growing. The generated Reeb graph is represented as $R(f) = (C, E)$, where C is the nodes of the graph, and E is the edges of the graph.

We further develop a Reeb graph simplification scheme to remove the noisy peaks of the original SUVR function (Fig. 1(a)) for salient pattern detection. Firstly, we iteratively apply Laplacian smoothing on an SUVR map until the Reeb graph reaches a common level of complexity, which is measured as the number of nodes in the graph. Then we use a Reeb graph simplification algorithm [12] to merge nodes around saddle points based on their persistence. For an edge $E_k = (C_i, C_j)(f(C_j) > f(C_i))$ of the Reeb graph $R(f) = (C, E)$, its weight is defined according to its persistence:

$$w(E_k) = A_k \times f(C_j) \tag{1}$$

where A_k is area of the patch enclosed by triangles belonging to this edge, and $f(C_j)$ is the peak SUVR value of this patch. To simplify the Reeb graph, we iteratively remove saddle points and spurious edges based on the persistence threshold δ. At each iteration, we scan these saddle points sequentially and for an edge $E_k = (C_i, C_j)$ $(f(C_i) < f(C_j))$ with $w(E_k) > \delta$, we collapse this edge by removing C_i and adding all its connections to C_j. The weights of these new edges are updated according to Eq. 1. These steps can be repeated until the persistent threshold is reached. The pruned Reeb graph is illustrated in Fig. 1(c). The number of critical points in the pruned Reeb graph is determined by the complexity of SUVR function and the persistence threshold. We set $\delta = 300$ so the pruned Reeb graph typically has less than 10 patches. The edges of the simplified Reeb graph are sets of triangles with topological changes, and the patches enclosed by these triangle sets are the regions with salient tau pathology.

2.2 Directed Graph Construction for Spatiotemporal Subtyping

The Reeb graph patches of different subjects typically have distinct shapes, which makes patch matching an essential step for pattern comparison. The solution we develop here is to establish patch correspondence between subjects based on their spatial proximity by an assignment algorithm. We define the cost function for matching the patch x_i of subject x and the patch y_j of subject y as:

$$C_{x_i, y_j} = dPeak_{x_i, y_j} \times dHausdorff_{x_i, y_j} \tag{2}$$

where $dPeak$ is the distance between the peaks, and $dHausdorff$ is the Hausdorff distance between two patch sets. Both are calculated based on geodesic distances on the surface.

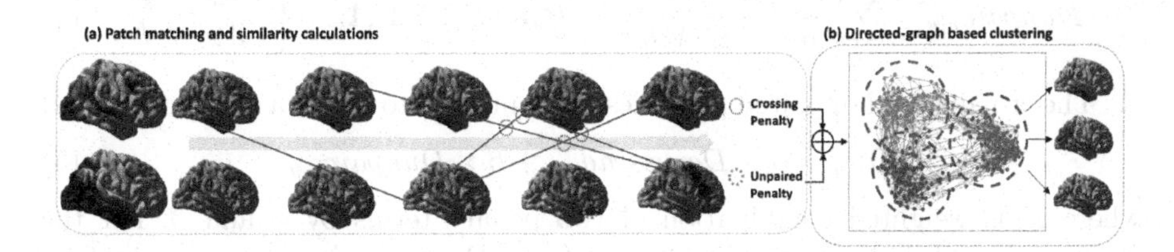

Fig. 2. Directed graph construction for clustering. (a) Patch matching and similarity calculations. The patches are arranged in a peak SUVR value decreasing order, so from left to right, the sequence shows the pathology accumulation process. The spatial approximate correspondences are linked by the red arrows, and the temporal inconsistency is highlighted by red circles. The unpaired patches are highlighted by blue circles. (b) Visualization of directed graph and graph-based clustering. (Color figure online)

Patch matching is achieved through iterative assignments to accommodate both one-to-one and many-to-one matching. The latter one is illustrated in Fig. 2(a) that multiple small patches have the same correspondences. At iteration t, a cost function C_t is calculated for the unpaired patches in x and all patches in the reference subject y. A linear assignment [3] is applied with the cost function C_t and the unpaired cost c_t to establish the one-to-one correspondences and add it into the correspondence set Φ. The iterations repeat until $min(C_t) > c_t$, and the resulting correspondences represent patches with spatial proximity across subjects.

The spatiotemporal similarity of tau pathology implies spatial coherence and temporal consistency, and the directed graph can be used to encode both the similarity as the edge weight and the disease staging as edge direction. The temporal similarity is derived from the consistency of the pathology occurring orders based on the correspondences. Based on the assumption that tau accumulation is a monotonic process, we sort the Reeb graph patches y_i in a peak SUVR descending order, and this sequence implies the patch occurring order. When comparing with subject x, patches x_i are arranged according to their correspondences y_i. For patches of x having the same correspondence in y, they are sorted based on their own peak SUVR values decreasingly. The temporal crossings between the two sequences as indicated in Fig. 2(a) imply the existence of temporal inconsistency between the two SUVR maps. A crossing penalty is then defined to quantify this inconsistency:

$$Dcrossing_{x,y} = \sum_{x_i} \sum_{x_j, \forall j > i} max(0, SUVR_{x_j} - SUVR_{x_i}) \times PatchS_{x_i}, \quad SUVR_{y_i} > SUVR_{y_j} \text{ for } i < j \tag{3}$$

where $SUVR_{x_i}$ is the peak SUVR of x_i, and $PatchS_{x_i}$ is the patch size.

The spatial coherence can be estimated from the spatial deviations caused by the unpaired patches. An unpaired penalty is defined as:

$$Dunpair_{x,y} = \sum_{x_i} \frac{SUVR_{x_i}}{\max_{x_i} SUVR_{x_i}} \times PatchS_{x_i} \times \mathbf{I}_i, \mathbf{I}_i = \begin{cases} 0 \text{ if } x_i \in \Phi \\ 1 \text{ if } x_i \notin \Phi \end{cases} \quad (4)$$

The crossing and unpaired penalties are combined to define a distance matrix:

$$D_{x,y} = \alpha \times Dcrossing_{x,y} + \beta \times Dunpair_{x,y} \quad (5)$$

where D is weighted combination of two penaty terms by α and β. For the construction of a dense graph, we transform D into a similarity matrix $S = 1/(1 + D)$ as the weights between each pair of subjects. To maintain sparsity and form clusters, we primarily focus on subjects exhibiting similar patterns and adopt a K-nearest-neighbor(KNN) approach to keep only the top K-relevant connections for each subject.

Disease staging is determined based on the assumption that later-stage subjects typically have more widespread distribution of tau pathology and/or higher peak intensities. To generate directionality between subjects in our graph representation, we calculate an unnormalized unpaired penalty as $\widetilde{Dunpair}_{x,y} = Dunpair_{x,y} \times \max_{x_i} SUVR_{x_i}$ because it is composed of both the spreading size and SUVR values. Specifically, if $\widetilde{Dunpair}_{x,y} > \widetilde{Dunpair}_{y,x}$, then subject x is considered to be in a later stage than y because of having more tau pathology and higher SUVR values, so the edge direction is $y \rightarrow x$. By applying the directions to the KNN graph, we get a directed graph for representing the spatiotemporal relationships between subjects as illustrated in Fig. 2(b). For the subtyping of tau spatiotempral topography, the Louvain community detection method [1] is applied for clustering on the directed graph by maximizing modularity. The modularity is high when the intra-subtype connections are dense while the inter-subtype connections are sparse.

3 Experiments and Results

In the current work, we choose the unpaired costs $c_1 = 300$ and $c_i = 500 (i > 1)$ in linear assignment of patches, the weights for distance matrix definition as $\alpha = 2$ and $\beta = 1$, and $K = 10$ for KNN-based graph construction. For Louvain community detection on directed graphs, we set its resolution parameter $\gamma = 0.3$. Beside clustering based on the directed graph from training data, we can apply the trained model from our method to estimate the subtype of a validation/test set via major voting by the K most similar training samples.

For the event-based SuStaIn method [17], the same experimental settings in [15] were used. The SUVR values are first normalized based on the normal control samples to create the tau z-scores. We used the z-score of 5 ROIs (parietal, frontal, occipital, temporal, and medial temporal lobe) and cut-off thresholds of z = 2, 5 and 10 for event definition. To apply the trained model to a validation/test data, we obtain subtype membership by fitting the new subject into the pre-trained SuStaIn model.

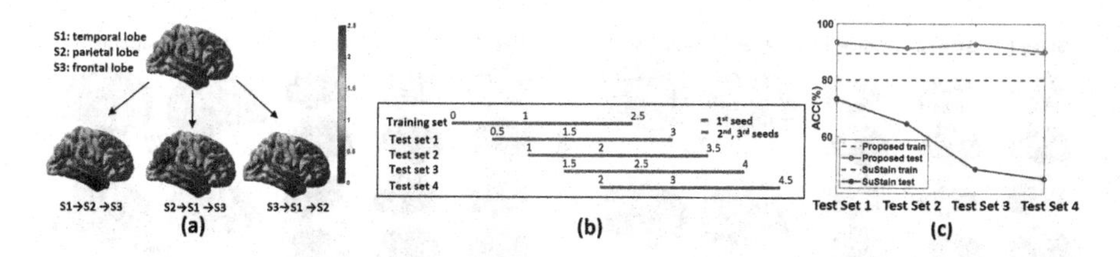

Fig. 3. Synthetic data and results. (a) Average SUVR maps of three subtypes. Each subtype has a distinct tau propagation trajectory from three seed regions (S1, S2, S3). (b) The magnitude ranges of seeds for all datasets. (c) The clustering performance on training and four test sets.

3.1 Synthetic Experiments

Synthetic Data. The synthetic tau SUVR data is generated by adding simulated values derived from a Gaussian mixture model to an SUVR template ($SUVR_{tem}$), which is calculated as the mean of normal controls, as follows:

$$\hat{SUVR} = SUVR_{tem} + \sum_{k=1,2,3} f(\mathbf{x}_k|0, \sigma) \times a_k \tag{6}$$

where f is a Gaussian distribution, $\mathbf{x}_k$ is the distance vector to seed k, and the standard deviation σ measures the spreading size. The magnitudes of the seeds a_k are sampled from the ranges shown in Fig. 3(b). A training set and four test sets with different magnitude ranges and seeds randomly sampled from temporal, parietal and frontal lobe, respectively, were generated. Each set has three subtypes with distinct tau propagation orders as shown in Fig. 3(a), and each subtype contains 50 samples.

Synthetic Results. The performances are measured by accuracy and shown in Fig. 3. First, our method (89.33%) achieved better performance over the SuStaIn method (80%) on training set. Second, we applied the trained model of both methods to the four test sets and results are shown in Fig. 3(c). With the increasing difference between the test and training set, performance of the SuStaIn method decreases quite significantly while our method maintains a stable performance. This shows the robustness of our method with respect to intensity changes from the training to the testing set and potentially better generalization ability as we will further demonstrate next in real data experiments.

3.2 Real Data Experiments

Real Dataset. 706 tau PET scans from Alzheimer's Disease Neuroimaging Initiative (ADNI) (adni.loni.usc.edu) and Anti-Amyloid Treatment in Asymptomatic Alzheimer's (A4) [13] are used. 427 amyloid-positive-tau-positive (A+T+) images are used for subtyping analysis. A+/A- labels were provided by the ADNI and A4 study, T+/T- status was defined if the image with peak

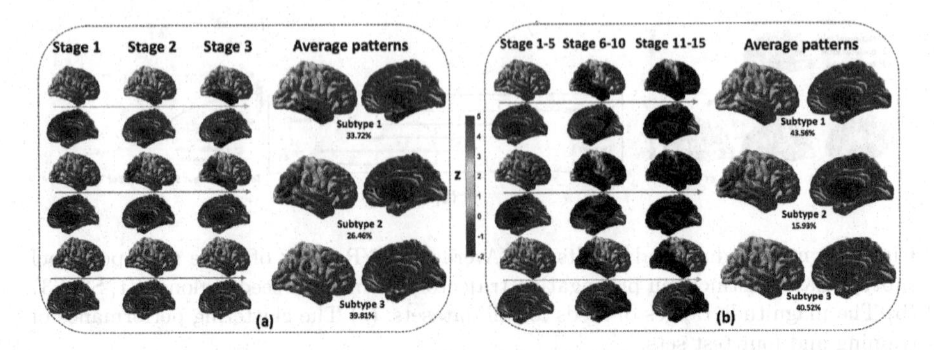

Fig. 4. Subtyping results on A+T+ subjects. (a) Patterns of three subtypes uncovered by our method. (b) Patterns of three subtypes uncovered by the SuStaIn method. 15 stages of the SuStaIn method are merged into 3 stages for comparison.

SUVR value exceeding 1.5 [9]. Besides, 279 cognitively normal A-T- subjects were selected as normal controls. The demographic information of the cohorts are listed in Table 1. All tau PET images were averaged across frames and registered to individual T1 space to obtain the skull-striped images. Besides, T1 images were preprocessed using Freesurfer [6] to get their volume parcellation using the Desikan-Killiany atlas. The registered tau PET images were intensity-normalized using an inferior cerebellar gray matter reference region, resulting in SUVR images.

Subtyping A+T+ Subjects. Using A+T+ data, we obtained three pathologically different subtypes for both methods (Fig. 4). In our method, the staging was derived from the degree of the nodes in the dense directed graph as $d_i = indegree_i - outdegree_i$. Two thresholds ($-50$ and 50) were set as the cut-offs to define three stages for each subtype. For subtype 1, more tau pathology distributes in the temporal lobe, and its propagation is consistent with the classic Braak staging [2]; for subtype 2, tau pathology primarily occurs in the occipital area and sequentially spreads to parietal and other regions; subtype 3 does not show significant differences among brain regions, and the overall SUVR is lower among all subtypes. For both methods, the average patterns are similar in three subtypes. However, comparing across stages, the intensities in early stage

Table 1. Demographic information.

Cohort	Characteristics	Control	A+T+
A4	Age	69.65 ± 4.29	72.89 ± 4.80
	Gender(M/F)	23/32	115/153
ADNI	Age	73.14 ± 6.02	78.15 ± 6.58
	Gender(M/F)	95/129	78/81

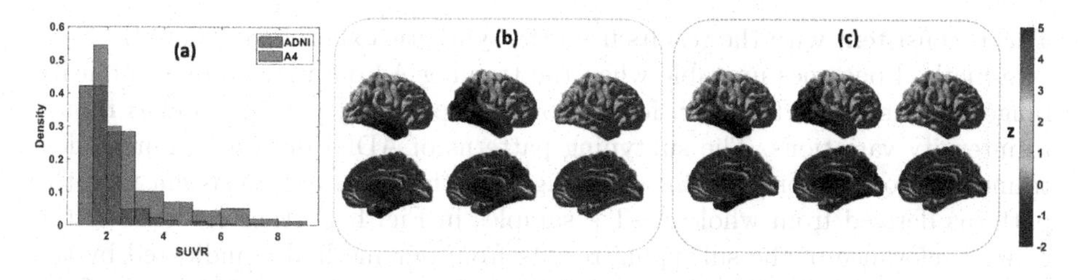

Fig. 5. Generalization results from our method. (a) Peak SUVR distributions of ADNI and A4. (b) Patterns of ADNI data based on model trained from the same cohort. (c) Patterns of ADNI data based on model trained from the A4 data.

and late stage show greater gaps for the SuStaIn method. This is caused by the unbalanced training samples across stages, as the SUVR of dominant A4 training data primarily concentrates on low intensity (Fig. 5(a)) and leads to insufficient training in late stages.

Table 2. Generalization performance measured by Rand index.

	Our method	SuStaIn
ADNI	**0.8605**	0.6648
A4	**0.8957**	0.8550

Generalization Performance Across Cohorts. The generalization performance of two methods was compared by the consistency of subtyping results of same data from models trained by different cohorts. For both methods, a model with three subtypes is first trained separately on ADNI and A4 data and then applied to both cohorts. Rand index [8] was used to measure the similarity of the subtyping results using different models and quantify the generalization.

Compared to the SuStaIn method, our method is more stable on both datasets as shown in Table 2. The inferior performance of the SuStaIn method is caused by the SUVR differences between cohorts as indicated in Fig. 5(a).

Table 3. Cognitive scores of ADNI subjects. ($*$ denotes statistical significance ($P < 0.05$) with other subtypes.)

	Control	Subtype1	Subtype2	Subtype3
MMSE	29 ± 0.98	26.21 ± 3.55	26.02 ± 3.61	27.84 ± 3.38*
ADAS11	8.69 ± 2.78	15.35 ± 7.31	14.61 ± 6.67	11.23 ± 6.28*
ADAS13	12.39 ± 4.60	22.69 ± 10.12	21.98 ± 9.53	16.70 ± 8.77*

This is consistent with the results from the synthetic experiment where the SuStaIn method becomes unstable when the training A4 data has limited intensity range. Our method, however, focuses on the spreading patterns and is robust to intensity variations. The subtyping patterns of ADNI data with our models trained on both cohorts are shown in Fig. 5. Both results are consistent with the patterns derived from whole A+T+ samples in Fig. 4. Further evidence of the clinical relevance of the subtyping results from our method is indicated by the cognitive scores of ADNI subjects in Table 3. All subtypes show deviations from the normal control, and particularly subtype 3 is different from the other two subtypes with statistical significance ($P < 0.05$).

4 Conclusions

In the current study, we proposed a novel directed-graph-based framework with a new spatiotemporal pattern representation for parsing tau pathology heterogeneity and demonstrated its improved performance over the state-of-art SuStaIn method. Application of the proposed method on large-scale tau PET imaging datasets successfully demonstrated three subtypes with clear relevance to previously well-described clinical subtypes with distinct spatiotemporal patterns.

References

1. Blondel, V.D., Guillaume, J.L., Lambiotte, R., Lefebvre, E.: Fast unfolding of communities in large networks. J. Stat. Mech. Theory Exp. **2008**(10), P10008 (2008)
2. Braak, H., Braak, E.: Neuropathological stageing of Alzheimer-related changes. Acta Neuropathol. **82**(4), 239–259 (1991)
3. Duff, I.S., Koster, J.: On algorithms for permuting large entries to the diagonal of a sparse matrix. SIAM J. Matrix Anal. Appl. **22**(4), 973–996 (2001)
4. Ferreira, D., Nordberg, A., Westman, E.: Biological subtypes of Alzheimer disease: a systematic review and meta-analysis. Neurology **94**(10), 436–448 (2020)
5. Fonteijn, H.M., et al.: An event-based model for disease progression and its application in familial Alzheimer's disease and Huntington's disease. Neuroimage **60**(3), 1880–1889 (2012)
6. Greve, D.N., et al.: Cortical surface-based analysis reduces bias and variance in kinetic modeling of brain pet data. Neuroimage **92**, 225–236 (2014)
7. Habes, M., Grothe, M.J., Tunc, B., McMillan, C., Wolk, D.A., Davatzikos, C.: Disentangling heterogeneity in Alzheimer's disease and related dementias using data-driven methods. Biol. Psychiat. **88**(1), 70–82 (2020)
8. Hubert, L., Arabie, P.: Comparing partitions. J. Classif. **2**, 193–218 (1985)
9. Jack, C.R., Jr., et al.: Defining imaging biomarker cut points for brain aging and Alzheimer's disease. Alzheimer's Dement. **13**(3), 205–216 (2017)
10. Reeb, G.: Sur les points singuliers d'une forme de pfaff completement integrable ou d'une fonction numerique [on the singular points of a completely integrable pfaff form or of a numerical function], vol. 222, pp. 847–849 (1946)
11. Shi, Y., Lai, R., Toga, A.W.: Cortical surface reconstruction via unified Reeb analysis of geometric and topological outliers in magnetic resonance images. IEEE Trans. Med. Imaging **32**(3), 511–530 (2012)

12. Shi, Y., Li, J., Toga, A.W.: Persistent Reeb graph matching for fast brain search. In: Wu, G., Zhang, D., Zhou, L. (eds.) MLMI 2014. LNCS, vol. 8679, pp. 306–313. Springer, Cham (2014). https://doi.org/10.1007/978-3-319-10581-9_38
13. Sperling, R.A., et al.: The A4 study: stopping ad before symptoms begin? Sci. Transl. Med. **6**(228), 228fs13 (2014)
14. Ten Kate, M., et al.: Atrophy subtypes in prodromal Alzheimer's disease are associated with cognitive decline. Brain **141**(12), 3443–3456 (2018)
15. Vogel, J.W., et al.: Four distinct trajectories of tau deposition identified in Alzheimer's disease. Nat. Med. **27**(5), 871–881 (2021)
16. Whitwell, J.L., et al.: [18F] AV-1451 clustering of entorhinal and cortical uptake in Alzheimer's disease. Ann. Neurol. **83**(2), 248–257 (2018)
17. Young, A.L., et al.: Uncovering the heterogeneity and temporal complexity of neurodegenerative diseases with subtype and stage inference. Nat. Commun. **9**(1), 1–16 (2018)

Text-Guided Foundation Model Adaptation for Pathological Image Classification

Yunkun Zhang[1], Jin Gao[1], Mu Zhou[2], Xiaosong Wang[3], Yu Qiao[3], Shaoting Zhang[3], and Dequan Wang[1,3](✉)

[1] Shanghai Jiao Tong University, Shanghai, China
dequanwang@sjtu.edu.cn
[2] Rutgers University, New Jersy, USA
[3] Shanghai AI Laboratory, Shanghai, China

Abstract. The recent surge of foundation models in computer vision and natural language processing opens up perspectives in utilizing multi-modal clinical data to train large models with strong generalizability. Yet pathological image datasets often lack biomedical text annotation and enrichment. Guiding data-efficient image diagnosis from the use of biomedical text knowledge becomes a substantial interest. In this paper, we propose to **C**onnect **I**mage and **T**ext **E**mbeddings (CITE) to enhance pathological image classification. CITE injects text insights gained from language models pre-trained with a broad range of biomedical texts, leading to adapt foundation models towards pathological image understanding. Through extensive experiments on the PatchGastric stomach tumor pathological image dataset, we demonstrate that CITE achieves leading performance compared with various baselines especially when training data is scarce. CITE offers insights into leveraging in-domain text knowledge to reinforce data-efficient pathological image classification. Code is available at https://github.com/Yunkun-Zhang/CITE.

Keywords: Foundation models · Multi-modality · Model Adaptation · Pathological image classification

1 Introduction

Deep learning for medical imaging has achieved remarkable progress, leading to a growing body of parameter-tuning strategies [1–3]. Those approaches are often designed to address disease-specific problems with limitations in their generalizability. In parallel, foundation models [4] have surged in computer vision [5,6] and natural language processing [7,8] with growing model capacity and data size, opening up perspectives in utilizing foundation models and large-scale clinical data for diagnostic tasks. However, pure imaging data can be insufficient to adapt foundation models with large model capacity to the medical field. Given the complex tissue characteristics of pathological whole slide images (WSI), it is crucial to develop adaptation strategies allowing (1) training data efficiency, and (2) data fusion flexibility for pathological image analysis.

H. Greenspan et al. (Eds.): MICCAI 2023, LNCS 14224, pp. 272–282, 2023.
https://doi.org/10.1007/978-3-031-43904-9_27

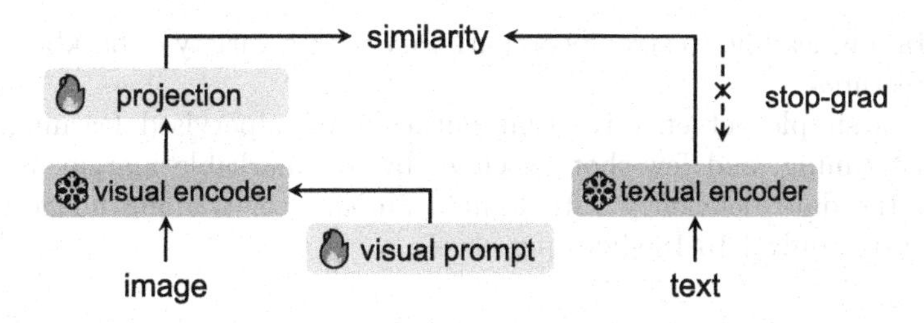

Fig. 1. Connecting Image and Text Embeddings. Our CITE emphasizes a text-guided model adaptation. An image with the visual prompt is processed through a vision encoder and a projection layer. The text knowledge is embedded by a text encoder, where a stop-gradient operation is applied. Classification prediction is made by the similarity between image and text embeddings. During adaptation, the visual prompt and the projection are tuned while the pre-trained encoders are frozen.

Although foundation models promise a strong generalization ability [4], there is an inherent domain shift between medical and natural concepts in both vision and language modalities. Pre-trained biomedical language models are increasingly applied to medical context understanding [9–11]. Language models prove to be effective in capturing semantic characteristics with a lower data acquisition and annotation cost in medical areas [12]. Such property is desired to address the dilemma of medical imaging cohorts, where well-annotated, high-quality medical imaging cohorts are expensive to collect and curate compared with text inputs [13]. In addition, vision-language models demonstrate the importance of joining multi-modal information for learning strong encoders [5,6,14]. Thus, connecting visual representations with text information from biomedical language models becomes increasingly critical to adapting foundation models for medical image classification, particularly in the challenging setting of data deficiency.

In this study, we propose CITE, a data-efficient adaptation framework that **C**onnects **I**mage and **T**ext **E**mbeddings from foundation models to perform pathological image classification with limited training samples (see Fig. 1). To enable language comprehension, CITE makes use of large language models pre-trained on biomedical text datasets [10,11] with rich and professional biomedical knowledge. Meanwhile, for visual understanding, CITE only introduces a small number of trainable parameters to a pre-trained foundation model, for example, CLIP [5] and INTERN [6], in order to capture domain-specific knowledge without modifying the backbone parameters. In this framework, we emphasize the utility of text information to play a substitutive role as traditional classification heads, guiding the adaptation of the vision encoder. A favorable contribution of our approach is to retain the completeness of both pre-trained models, enabling a low-cost adaptation given the large capacity of foundation models. Overall, our contributions are summarized as follows:

1. We demonstrate the usefulness of injecting biomedical text knowledge into foundation model adaptation for improved pathological image classification.
2. CITE introduces only a small number of extra model parameters (∼0.6% of the vision encoder), meanwhile keeping the pre-trained models frozen during

adaptation, leading to strong compatibility with a variety of backbone model architectures.
3. CITE is simple yet effective that outperforms supervised learning, visual prompt tuning, and few-shot baselines by a remarkable margin, especially under the data deficiency with limited amounts of training image samples (*e.g.*, using only 1 to 16 slides per class).

2 Related Work

Medical Image Classification. Deep learning for medical image classification has long relied on training large models from scratch [1,15]. Also, fine-tuning or linear-probing the pre-trained models obtained from natural images [16–18] is reasonable. However, those methods are supported by sufficient high-quality data expensive to collect and curate [19]. In addition, task-specific models do not generalize well with different image modalities [2]. To tackle this issue, we emphasize the adaptation of foundation models in a data-efficient manner.

Vision-Language Pre-training. Recent work has made efforts in pre-training vision-language models. CLIP [5] collects 400 million image-text pairs from the internet and trains aligned vision and text encoders from scratch. LiT [20] trains a text encoder aligned with a fixed pre-trained vision encoder. BLIP-2 [14] trains a query transformer by bootstrapping from pre-trained encoders. REACT [21] fixes both pre-trained encoders and tunes extra gated self-attention modules. However, those methods establish vision-language alignment by pre-training on large-scale image-text pairs. Instead, we combine pre-trained unimodal models on downstream tasks and build a multi-modal classifier with only a few data.

Model Adaptation via Prompt Tuning. Prompt tuning proves to be an efficient adaptation method for both vision and language models [22,23]. Originating from natural language processing, "prompting" refers to adding (manual) text instructions to model inputs, whose goal is to help the pre-trained model better understand the current task. For instance, CoOp [22] introduces learnable prompt parameters to the text branch of vision-language models. VPT [23] demonstrates the effectiveness of prompt tuning with pre-trained vision encoders. In this study, we adopt prompt tuning for adaptation because it is lightweight and only modifies the input while keeping the whole pre-trained model unchanged. However, existing prompt tuning methods lack expert knowledge and understanding of downstream medical tasks. To address this challenge, we leverage large language models pre-trained with biomedical text to inject medical domain knowledge.

Biomedical Language Model Utilization. Biomedical text mining promises to offer the necessary knowledge base in medicine [9–11]. Leveraging language models pre-trained with biomedical text for medical language tasks is a common application. For instance, Alsentzer et al. [9] pre-train a clinical text model with BioBERT [10] initialization and show a significant improvement on five clinical

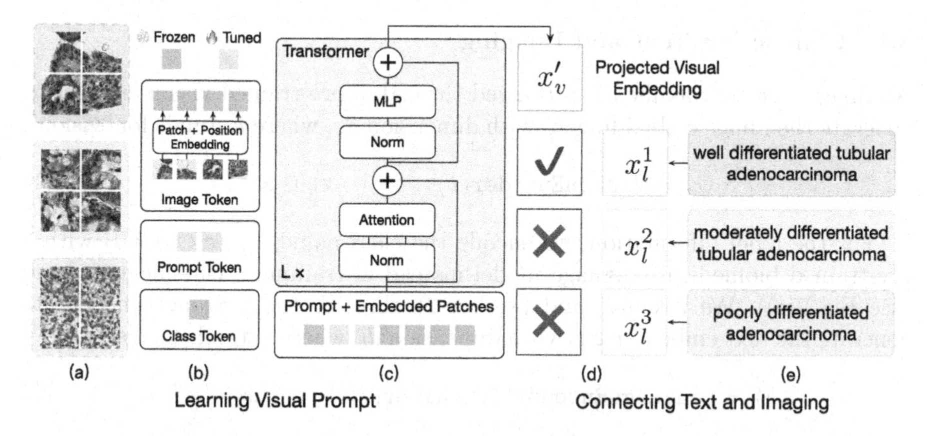

Fig. 2. An overview of CITE. (a) The pathological images are cut into patches. (b) The class token, image tokens, and learnable prompt tokens are concatenated. (c) The tokens are processed by a pre-trained vision transformer to generate image embeddings. Those 3 steps refer to *learning visual prompt* (Sect. 3.2). (d) The image is recognized as the class with maximum cosine similarity between image and text embeddings. (e) The class names are processed by a biomedical language model to generate text embeddings. Those 2 steps *connect text and imaging* (Sect. 3.1).

language tasks. However, the potential of biomedical text information in medical imaging applications has not been explicitly addressed. In our efforts, we emphasize the importance of utilizing biomedical language models for adapting foundational vision models into cancer pathological analysis.

3 Methodology

Figure 2 depicts an overview of our approach CITE for data-efficient pathological image classification. CITE jointly understands the image features extracted by vision encoders pre-trained with natural imaging, and text insights encoded in large language models pre-trained with biomedical text (*e.g.*, BioLinkBERT [11] which captures rich text insights spanning across biomedical papers via citations). We connect text and imaging by a projection and classify the images by comparing the cosine similarity between image and text embeddings.

Importantly, we introduce two low-cost sets of trainable parameters to the vision encoder in order to adapt the model with the guidance of text information. They are (1) prompt tokens in the input space to model task-specific information, and (2) a projection layer in the latent space to align image and text embeddings. During model adaptation, we freeze the pre-trained encoders and only tune the introduced parameters, which not only saves remarkable training data and computational resources but also makes our approach favorable with various foundation model architectures.

3.1 Connecting Text and Imaging

An image I to be classified is processed through a pre-trained vision encoder to generate the image embedding x_v with dimension d_v, where v stands for "vision":

$$x_v = \texttt{VisionEncoder}(I) \qquad x_v \in \mathbb{R}^{d_v}. \tag{1}$$

For the label information, we encode the class names T_c ($c \in [1, C]$) with a pre-trained biomedical language model instead of training a classification head (see Fig. 2(e)). We tokenize and process T_c through the language encoder to generate the text embedding x_l^c with dimension d_l, where l stands for "language":

$$x_l^c = \texttt{LanguageEncoder}(\texttt{Tokenizer}(T_c)) \qquad x_l^c \in \mathbb{R}^{d_l}. \tag{2}$$

Vision-language models like CLIP [5] contain both a vision encoder and a language encoder, which provide well-aligned embeddings in the same feature space. In this case, prediction $\hat{y}$ is obtained by applying softmax on scaled cosine similarities between the image and text embeddings (see Fig. 2(d)):

$$p(\hat{y} = c|I) = \frac{\exp(\mathrm{sim}(x_l^c, x_v)/\tau)}{\sum_{c'=1}^{C} \exp(\mathrm{sim}(x_l^{c'}, x_v)/\tau)}, \tag{3}$$

where $\mathrm{sim}(\cdot, \cdot)$ refers to cosine similarity and τ is the temperature parameter.

For irrelevant vision and language encoders, we introduce an extra projection layer to the end of the vision encoder to map the image embeddings to the same latent space as the text embeddings. We replace x_v in Eq. (3) with x_v':

$$x_v' = \texttt{Projection}(x_v) \qquad x_v' \in \mathbb{R}^{d_l}. \tag{4}$$

During adaptation, the extra parameters are updated by minimizing the cross-entropy of the predictions from Eq. (3) and the ground truth labels.

3.2 Learning Visual Prompt

Medical concepts exhibit a great visual distribution shift from natural images, which becomes impractical for a fixed vision encoder to capture task-specific information in few-shot scenarios. Visual prompt tuning (VPT [23]) is a lightweight adaptation method that can alleviate such an inherent difference by only tuning prompt tokens added to the visual inputs of a fixed vision transformer [24], showing impressive performance especially under data deficiency. Thus, we adopt VPT to adapt the vision encoder in our approach.

A vision transformer first cuts the image into a sequence of n patches and projects them to patch embeddings $E_0 \in \mathbb{R}^{n \times d_v}$, where d_v represents the visual embedding dimension. A CLS token $c_0 \in \mathbb{R}^{d_v}$ is prepended to the embeddings, together passing through K transformer layers $\{L_v^k\}_{k=1,2,\ldots,K}$. CLS embedding of the last layer output is the image feature x_v. Following the setting of shallow VPT, we concatenate the learnable prompt tokens $\boldsymbol{P} = [\boldsymbol{p}^1, \ldots, \boldsymbol{p}^p] \in \mathbb{R}^{p \times d_v}$, where p is the prompt length, with CLS token c_0 and patch embeddings E_0 before they are processed through the first transformer layer:

$$
\begin{aligned}
[c_1, \boldsymbol{Z}_1, E_1] &= L_v^1([c_0, \boldsymbol{P}, E_0]) \\
[c_k, \boldsymbol{Z}_k, E_k] &= L_v^k([c_{k-1}, \boldsymbol{Z}_{k-1}, E_{k-1}]) \quad k = 2, 3, \ldots, K \\
x_v &= c_K \qquad\qquad\qquad\qquad\qquad\quad x_v \in \mathbb{R}^{d_v},
\end{aligned} \tag{5}
$$

where $[\cdot, \cdot]$ refers to concatenation along the sequence length dimension, and $\boldsymbol{Z}_k \in \mathbb{R}^{p \times d_v}$ represents the output embeddings of the k-th transformer layer at the position of the prompts (see Fig. 2(a–c)). The prompt parameters are updated together with the projection layer introduced in Sect. 3.1.

4 Experimental Settings

Dataset. We adopt the PatchGastric [25] dataset, which includes histopathological image patches extracted from H&E stained whole slide images (WSI) of stomach adenocarcinoma endoscopic biopsy specimens. There are 262,777 patches of size 300 × 300 extracted from 991 WSIs at x20 magnification. The dataset contains 9 subtypes of gastric adenocarcinoma. We choose 3 major subtypes including "well differentiated tubular adenocarcinoma", "moderately differentiated tubular adenocarcinoma", and "poorly differentiated adenocarcinoma" to form a 3-class grading-like classification task with 179,285 patches from 693 WSIs. We randomly split the WSIs into *train* (20%) and *validation* (80%) subsets for measuring the model performance. To extend our evaluation into the real-world setting with insufficient data, we additionally choose 1, 2, 4, 8, or 16 WSIs with the largest numbers of patches from each class as the training set. The evaluation metric is patient-wise accuracy, where the prediction of a WSI is obtained by a soft vote over the patches, and accuracy is averaged class-wise.

Implementation. We use CLIP ViT-B/16 [5] as the visual backbone, with input image size 224 × 224, patch size 16 × 16, and embedding dimension $d_v = 512$. We adopt BioLinkBERT-large [11] as the biomedical language model, with embedding dimension $d_l = 1,024$. To show the extensibility of our approach, we additionally test on vision encoders including ImageNet-21k ViT-B/16 [24,26] and INTERN ViT-B/16 [6], and biomedical language model BioBERT-large [10]. Our implementation is based on CLIP[1], HuggingFace[2] and MMClassification[3].

Training Details. Prompt length p is set to 1. We resize the images to 224×224 to fit the model and follow the original data pipeline in PatchGastric [25]. A class-balanced sampling strategy is adopted by choosing one image from each class in turn. Training is done with 1,000 iterations of stochastic gradient descent (SGD), and the mini-batch size is 128, requiring 11.6 GB of GPU memory and 11 min on two NVIDIA GeForce RTX 2080 Ti GPUs. All our experiment results are averaged on 3 random seeds unless otherwise specified.

[1] https://github.com/openai/CLIP.
[2] https://github.com/huggingface/transformers.
[3] https://github.com/open-mmlab/mmclassification.

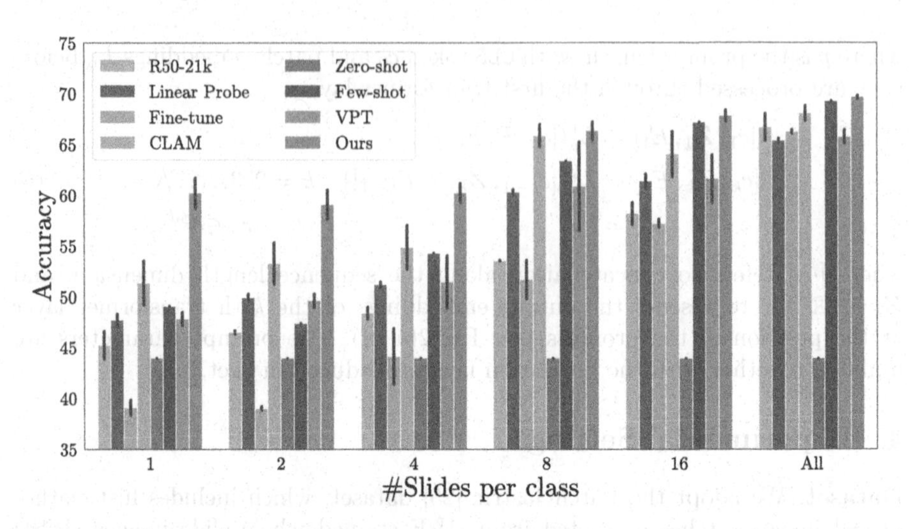

Fig. 3. Accuracy on the PatchGastric [25] 3-category classification task. R50-21k refers to ResNet50 [27] backbone pre-trained on ImageNet-21k [26]. Other methods adopt CLIP ViT-B/16 [5] backbone. Averaged results and standard deviation (error bars) of 3 runs are displayed. Our CITE consistently outperforms all baselines under all data fractions, showing a remarkable improvement under data deficiency.

5 Results

CITE Consistently Outperforms all Baselines Under all Data Scales. Figure 3 shows the classification accuracy on the PatchGastric dataset of our approach compared with baseline methods and related works, including (1) R50-21k: fine-tune the whole ResNet50 [27] backbone pre-trained on ImageNet-21k [26]. (2) Linear probe: train a classification head while freezing the backbone encoder. (3) Fine-tune: train a classification head together with the backbone encoder. (4) CLAM [18]: apply an attention network on image features to predict pseudo labels and cluster the images. (5) Zero-shot [5]: classify images to the nearest text embeddings obtained by class names, without training. (6) Few-shot [28]: cluster image features of the training data and classify images to the nearest class center. (7) VPT [23]: train a classification head together with visual prompts. Note that CLIP ViT-B/16 vision encoder is adopted as the backbone for (2)–(7). Our CITE outperforms all baselines that require training classification heads, as well as image feature clustering methods, demonstrating the key benefit of leveraging additional biomedical text information for pathological image classification.

CITE Shows a Favorable Improvement when Data is Scarce. When only one training slide per class is available, CITE achieves a remarkable performance, outperforming all baselines by a significant margin (from 51.4% to 60.2%). As data deficiency is commonly seen in medical tasks, CITE presents an appealing property to handle data-limited pathological analysis. Together, our findings

Table 1. Ablation study of CITE with and without prompt and text. We report the average accuracy and standard deviation. When prompt is not used, we fine-tune the whole vision backbone. When text is not used, we adopt the traditional classification head. Each component improves the performance.

Prompt	Text	1	2	4	8	16	All
		39.1$_{\pm0.6}$	39.0$_{\pm0.8}$	44.1$_{\pm2.2}$	51.7$_{\pm1.6}$	57.1$_{\pm0.3}$	66.0$_{\pm1.2}$
✓		47.9$_{\pm0.5}$	49.6$_{\pm0.6}$	51.5$_{\pm2.1}$	60.9$_{\pm3.6}$	61.6$_{\pm1.9}$	65.8$_{\pm0.5}$
	✓	57.6$_{\pm0.4}$	56.6$_{\pm0.5}$	57.6$_{\pm0.2}$	60.6$_{\pm0.4}$	62.2$_{\pm0.6}$	66.1$_{\pm0.9}$
✓	✓	**60.1$_{\pm0.9}$**	**59.0$_{\pm0.1}$**	**60.9$_{\pm0.9}$**	**63.2$_{\pm0.2}$**	**65.9$_{\pm0.5}$**	**68.7$_{\pm0.6}$**

Table 2. CITE fits in with various pre-trained encoders. We include CLIP ViT-B/16 [5], ImageNet-21k ViT-B/16 [26] and INTERN ViT-B/16 [6] visual encoders, combined with CLIP textual encoder [5], BioBERT (BB) [10] and BioLinkBERT (BLB) [11] language models. The highest performance of each visual encoder is bolded. For each combination, CITE consistently outperforms linear and fine-tune baselines.

Visual	Method	Textual	1	2	4	8	16	All
CLIP ViT-B/16	Linear	-	47.7$_{\pm0.1}$	49.9$_{\pm0.1}$	51.2$_{\pm0.1}$	60.3$_{\pm0.1}$	61.4$_{\pm0.1}$	65.4$_{\pm0.1}$
	Fine-tune	-	39.1$_{\pm1.2}$	39.0$_{\pm1.2}$	44.1$_{\pm1.2}$	51.7$_{\pm1.2}$	57.1$_{\pm1.2}$	66.3$_{\pm1.2}$
	CITE	CLIP	60.1$_{\pm0.9}$	59.0$_{\pm0.1}$	**60.9$_{\pm0.9}$**	63.2$_{\pm0.2}$	65.9$_{\pm0.5}$	68.7$_{\pm0.6}$
	CITE	BLB	**60.2$_{\pm1.2}$**	**59.1$_{\pm1.2}$**	60.3$_{\pm0.8}$	**66.4$_{\pm0.7}$**	**67.9$_{\pm0.4}$**	**69.7$_{\pm0.1}$**
IN-21k ViT-B/16	Linear	-	46.7$_{\pm0.7}$	45.8$_{\pm1.6}$	53.4$_{\pm1.2}$	59.5$_{\pm0.5}$	60.6$_{\pm0.6}$	66.5$_{\pm0.8}$
	Fine-tune	-	48.0$_{\pm0.3}$	49.6$_{\pm0.1}$	50.8$_{\pm0.1}$	59.3$_{\pm0.3}$	62.2$_{\pm0.4}$	66.3$_{\pm0.2}$
	CITE	BB	51.4$_{\pm1.4}$	51.8$_{\pm1.3}$	56.6$_{\pm1.9}$	62.7$_{\pm1.0}$	64.0$_{\pm0.5}$	67.2$_{\pm1.4}$
	CITE	BLB	**52.4$_{\pm1.5}$**	**52.7$_{\pm0.8}$**	**57.0$_{\pm0.9}$**	**62.8$_{\pm1.2}$**	**64.5$_{\pm1.1}$**	**67.4$_{\pm0.7}$**
INTERN ViT-B/16	Linear	-	47.3$_{\pm0.2}$	47.2$_{\pm0.2}$	52.4$_{\pm0.5}$	59.7$_{\pm0.3}$	63.1$_{\pm0.2}$	66.8$_{\pm0.7}$
	Fine-tune	-	42.0$_{\pm0.3}$	46.0$_{\pm0.3}$	51.0$_{\pm0.9}$	60.4$_{\pm0.1}$	62.7$_{\pm0.5}$	68.2$_{\pm0.4}$
	CITE	BB	**51.7$_{\pm0.1}$**	**55.4$_{\pm1.8}$**	**59.6$_{\pm0.3}$**	**66.4$_{\pm0.8}$**	**68.1$_{\pm0.8}$**	**69.7$_{\pm0.7}$**
	CITE	BLB	48.4$_{\pm5.2}$	49.1$_{\pm5.5}$	57.9$_{\pm0.8}$	65.3$_{\pm0.4}$	67.9$_{\pm0.8}$	69.4$_{\pm0.9}$

demonstrate that adding domain-specific text information provides an efficient means to guide foundation model adaptation for pathological image diagnosis.

Visual Prompt and Text Information are Both Necessary. We conduct ablation studies to show the effectiveness of visual prompt learning and text information. From the results in Table 1, we demonstrate that visual prompt learning outperforms fine-tuning as the adaptation method, and in-domain text information outperforms classification heads. Combining the two components yields the best results under all data scales. Importantly, text information is particularly effective when training data is extremely scarce (1 slide per class).

CITE Shows Model Extensibility. We evaluate our approach with additional backbones and biomedical language models to assess its potential extensibility. Table 2 displays the findings of our approach compared with linear probe and fine-tune baselines. The results demonstrate that CITE is compatible with a variety of pre-trained models, making it immune to upstream model modifications. The text information encoded in biomedical language models allows vision models pre-trained with natural imaging to bridge the domain gap without task-specific pre-training on medical imaging. Importantly, when using both the vision and language encoders of CLIP ViT-B/16, our approach still outperforms the baselines by a remarkable margin (47.7% to 60.1%), demonstrating the importance of multi-modal information. While CLIP gains such modality matching through pre-training, our CITE shows an appealing trait that irrelevant vision and language models can be combined to exhibit similar multi-modal insights on pathological tasks without a need of joint pre-training.

6 Conclusion

Adapting powerful foundation models into medical imaging constantly faces data-limited challenges. In this study, we propose CITE, a data-efficient and model-agnostic approach to adapt foundation models for pathological image classification. Our key contribution is to inject meaningful medical domain knowledge to advance pathological image embedding and classification. By tuning only a small number of parameters guided by biomedical text information, our approach effectively learns task-specific information with only limited training samples, while showing strong compatibility with various foundation models. To augment the current pipeline, the use of synthetic pathological images is promising [29]. Also, foundation training on multi-modal medical images is of substantial interest to enhance model robustness under data-limited conditions [30].

References

1. Shen, W., Zhou, M., Yang, F., Yang, C., Tian, J.: Multi-scale convolutional neural networks for lung nodule classification. In: Ourselin, S., Alexander, D.C., Westin, C.-F., Cardoso, M.J. (eds.) IPMI 2015. LNCS, vol. 9123, pp. 588–599. Springer, Cham (2015). https://doi.org/10.1007/978-3-319-19992-4_46
2. Murtaza, G., et al.: Deep learning-based breast cancer classification through medical imaging modalities: state of the art and research challenges. Artif. Intell. Rev. **53**, 1655–1720 (2020)
3. Ding, K., Zhou, M., Wang, H., Zhang, S., Metaxas, D.N.: Spatially aware graph neural networks and cross-level molecular profile prediction in colon cancer histopathology: a retrospective multi-cohort study. Lancet Digit. Health **4**(11), e787–e795 (2022)
4. Bommasani, R., et al.: On the opportunities and risks of foundation models, arXiv preprint arXiv:2108.07258 (2021)
5. Radford, A., et al.: Learning transferable visual models from natural language supervision. In: International Conference on Machine Learning, pp. 8748–8763. PMLR (2021)
6. Shao, J., et al.: Intern: a new learning paradigm towards general vision, arXiv preprint arXiv:2111.08687 (2021)
7. Devlin, J., Chang, M.-W., Lee, K., Toutanova, K.: Bert: pre-training of deep bidirectional transformers for language understanding, arXiv preprint arXiv:1810.04805 (2018)

8. Brown, T., et al.: Language models are few-shot learners. Adv. Neural. Inf. Process. Syst. **33**, 1877–1901 (2020)
9. Alsentzer, E., et al.: Publicly available clinical BERT embeddings, arXiv preprint arXiv:1904.03323 (2019)
10. Lee, J., et al.: Biobert: a pre-trained biomedical language representation model for biomedical text mining. Bioinformatics **36**(4), 1234–1240 (2020)
11. Yasunaga, M., Leskovec, J., Liang, P.: Linkbert: pretraining language models with document links. In: Association for Computational Linguistics (ACL) (2022)
12. Chen, J., Guo, H., Yi, K., Li, B., Elhoseiny, M.: VisualGPT: data-efficient adaptation of pretrained language models for image captioning. In: Proceedings of the IEEE/CVF Conference on Computer Vision and Pattern Recognition (CVPR), pp. 18030–18040 (2022)
13. Chen, C.-L., et al.: An annotation-free whole-slide training approach to pathological classification of lung cancer types using deep learning. Nat. Commun. **12**(1), 1193 (2021)
14. Li, J., Li, D., Savarese, S., Hoi, S.: Blip-2: bootstrapping language-image pretraining with frozen image encoders and large language models, arXiv preprint arXiv:2301.12597 (2023)
15. Li, Q., Cai, W., Wang, X., Zhou, Y., Feng, D.D., Chen, M.: Medical image classification with convolutional neural network. In: 2014 13th International Conference on Control Automation Robotics & Vision (ICARCV), pp. 844–848. IEEE (2014)
16. Qu, J., Hiruta, N., Terai, K., Nosato, H., Murakawa, M., Sakanashi, H.: Gastric pathology image classification using stepwise fine-tuning for deep neural networks. J. Healthc. Eng. **2018** (2018)
17. Chen, M., et al.: Classification and mutation prediction based on histopathology H&E images in liver cancer using deep learning. NPJ Precis. Oncol. **4**(1), 1–7 (2020)
18. Lu, M.Y., Williamson, D.F., Chen, T.Y., Chen, R.J., Barbieri, M., Mahmood, F.: Data-efficient and weakly supervised computational pathology on whole-slide images. Nat. Biomed. Eng. **5**(6), 555–570 (2021)
19. Tiu, E., Talius, E., Patel, P., Langlotz, C.P., Ng, A.Y., Rajpurkar, P.: Expert-level detection of pathologies from unannotated chest X-ray images via self-supervised learning. Nat. Biomed. Eng. 1–8 (2022)
20. Zhai, X., et al.: Lit: zero-shot transfer with locked-image text tuning. In: Proceedings of the IEEE/CVF Conference on Computer Vision and Pattern Recognition, pp. 18123–18133 (2022)
21. Liu, H., et al.: Learning customized visual models with retrieval-augmented knowledge. In: Proceedings of the IEEE/CVF Conference on Computer Vision and Pattern Recognition, pp. 15148–15158 (2023)
22. Zhou, K., Yang, J., Loy, C.C., Liu, Z.: Learning to prompt for vision-language models. Int. J. Comput. Vision **130**(9), 2337–2348 (2022)
23. Jia, M., et al.: Visual prompt tuning, arXiv preprint arXiv:2203.12119 (2022)
24. Dosovitskiy, A., et al.: An image is worth 16x16 words: transformers for image recognition at scale, arXiv preprint arXiv:2010.11929 (2020)
25. Tsuneki, M., Kanavati, F.: Inference of captions from histopathological patches, arXiv preprint arXiv:2202.03432 (2022)
26. Russakovsky, O., et al.: Imagenet large scale visual recognition challenge. Int. J. Comput. Vision **115**(3), 211–252 (2015)
27. He, K., Zhang, X., Ren, S., Sun, J.: Deep residual learning for image recognition. In: Proceedings of the IEEE Conference on Computer Vision and Pattern Recognition, pp. 770–778 (2016)

28. Chen, Y., Liu, Z., Xu, H., Darrell, T., Wang, X.: Meta-baseline: exploring simple meta-learning for few-shot learning. In: Proceedings of the IEEE/CVF International Conference on Computer Vision, pp. 9062–9071 (2021)
29. Ding, K., Zhou, M., Wang, H., Gevaert, O., Metaxas, D., Zhang, S.: A large-scale synthetic pathological dataset for deep learning-enabled segmentation of breast cancer. Sci. Data **10**(1), 231 (2023)
30. Gao, Y., Li, Z., Liu, D., Zhou, M., Zhang, S., Meta, D.N.: Training like a medical resident: universal medical image segmentation via context prior learning, arXiv preprint arXiv:2306.02416 (2023)

Multiple Prompt Fusion for Zero-Shot Lesion Detection Using Vision-Language Models

Miaotian Guo[1], Huahui Yi[2], Ziyuan Qin[2], Haiying Wang[1], Aidong Men[1], and Qicheng Lao[1,3(✉)]

[1] School of Artificial Intelligence, Beijing University of Posts and Telecommunications, Beijing, China
qicheng.lao@bupt.edu.cn

[2] West China Biomedical Big Data Center, West China Hospital, Sichuan University, Sichuan, China

[3] Shanghai Artificial Intelligence Laboratory, Shanghai, China

Abstract. The success of large-scale pre-trained vision-language models (VLM) has provided a promising direction of transferring natural image representations to the medical domain by providing a well-designed prompt with medical expert-level knowledge. However, one prompt has difficulty in describing the medical lesions thoroughly enough and containing all the attributes. Besides, the models pre-trained with natural images fail to detect lesions precisely. To solve this problem, fusing multiple prompts is vital to assist the VLM in learning a more comprehensive alignment between textual and visual modalities. In this paper, we propose an ensemble guided fusion approach to leverage multiple statements when tackling the phrase grounding task for zero-shot lesion detection. Extensive experiments are conducted on three public medical image datasets across different modalities and the detection accuracy improvement demonstrates the superiority of our method.

Keywords: Vision-language models · Lesion detection · Multiple prompts · Prompt fusion · Ensemble learning

1 Introduction

Medical lesion detection plays an important role in assisting doctors with the interpretation of medical images for disease diagnosing, cancer staging, etc., which can improve efficiency and reduce human errors [9,19]. Current object detection approaches are mainly based on supervised learning with abundant well-paired image-level annotations, which heavily rely on expert-level knowledge. As such, these supervised approaches may not be suitable for medical lesion detection due to the laborious labeling.

Supplementary Information The online version contains supplementary material available at https://doi.org/10.1007/978-3-031-43904-9_28.

H. Greenspan et al. (Eds.): MICCAI 2023, LNCS 14224, pp. 283–292, 2023.
https://doi.org/10.1007/978-3-031-43904-9_28

Recently, large-scale pre-trained vision-language models (VLMs), by learning the visual concepts in the images through the weak labels from text, have prevailed in natural object detection or visual grounding and shown extraordinary performance. These models, such as GLIP [11], X-VLM [10], and VinVL [24], can perform well in detection tasks without supervised annotations. Therefore, substituting conventional object detection with VLMs is possible and necessary. The VLMs are first pre-trained to learn universal representations via large-scale unlabelled data and can be effectively transferred to downstream tasks. For example, a recent study [15] has demonstrated that the pre-trained VLMs can be used for zero-shot medical lesion detection with the help of well-designed prompts.

However, current existing VLMs are mostly based on a single prompt to establish textual and visual alignment. This prompt needs refining to cover all the features of the target as much as possible. Apparently, even a well-designed prompt is not always able to combine all expressive attributes into one sentence without semantic and syntactic ambiguity, e.g., the prompt design for melanoma detection should include numerous kinds of information describing attributes complementing each other, such as shape, color, size, etc [8]. In addition, each keyword in a single lengthy prompt cannot take effect equally as we expect, where the essential information can be ignored. This problem motivates us to study alternative approaches with multiple prompt fusion.

In this work, instead of striving to design a single satisfying prompt, we aim to take advantage of pre-trained VLMs in a more flexible way with the form of multiple prompts, where each prompt can elicit respective knowledge from the model which can then be fused for better lesion detection performance. To achieve this, we propose an ensemble guided fusion approach derived from clustering ensemble learning [3], where we design a step-wise clustering mechanism to gradually screen out the implausible intermediate candidates during the grounding process, and an integration module to obtain the final results by uniting the mutually independent candidates from each prompt. In addition, we also examine the language syntax based prompt fusion approach as a comparison, and explore several fusion strategies by first grouping the prompts either with described attributes or categories and then repeating the fusion process.

We evaluate the proposed approach on a broad range of public medical datasets across different modalities including photography images for skin lesion detection ISIC 2016 [2], endoscopy images for polyp detection CVC-300 [21], and cytology images for blood cell detection BCCD. The proposed approach exhibits extraordinary superiority compared to those with single prompt and other common ensemble learning based methods for zero-shot medical lesion detection. Considering the practical need of lesion detection, we further provide significantly improved fine-tuning results with a few labeled examples.

2 Related Work

Object Detection and Vision-Language Models. In the vision-language field, phrase grounding can be regarded as another solution for object detection apart from conventional R-CNNs [5,6,18]. Recently, vision-language models have achieved exciting performance in the zero-shot and few-shot visual recognition [4,16]. GLIP [11] unifies phrase grounding and object detection tasks, demonstrating outstanding transfer capability. In addition, ViLD [7] is proposed for open-vocabulary object detection taking advantage of the rich knowledge learned from CLIP [4] and text input.

Ensemble Learning. As pointed out by a review [3], ensemble learning methods achieve better performance by producing predictions based on extracted features and fusing via various voting mechanisms. For example, a selective ensemble of classifier chains [13] is proposed to reduce the computational cost and the storage cost arose in multi-label learning [12] by decreasing the ensemble size. UNDEED [23], a semi-supervised classification method, is presented to increase the classifier accuracy on labeled data and diversity on unlabeled data simultaneously. And a hybrid semi-supervised clustering ensemble algorithm [22] is also proposed to generate basic clustering partitions with prior knowledge.

3 Method

In this section, we first briefly introduce the vision-language model for unifying object detection as phrase grounding, e.g., GLIP [11] (Sect. 3.1). Then we present a simple language syntax based prompt fusion approach in Sect. 3.2. Finally, the proposed ensemble-guided fusion approach and several fusion strategies are detailed in Sect. 3.3 to improve the zero-shot lesion detection.

3.1 Preliminaries

Phrase grounding is the task of identifying the fine-grained correspondence between phrases in a sentence and objects in an image. The GLIP model takes as input an image I and a text prompt p that describes all the M candidate categories for the target objects. Both inputs will go through specific encoders Enc_{I} and Enc_{T} to obtain unaligned representations. Then, GLIP uses a grounding module to align image boxes with corresponding phrases in the text prompt. The whole process can be formulated as follows:

$$O = Enc_{\mathrm{I}}(I),\ P = Enc_{\mathrm{T}}(p),\ S_{\mathrm{ground}} = OP^{\top},\ L_{\mathrm{cls}} = Loss(S_{\mathrm{ground}};\ T), \quad (1)$$

where $O \in \mathbb{R}^{N \times d}$, $P \in \mathbb{R}^{M \times d}$ denote the image and text features respectively for N candidate region proposals and M target objects, $S_{\mathrm{ground}} \in \mathbb{R}^{N \times M}$ represents the cross-modal alignment scores, and $T \in \{0,\ 1\}^{N \times M}$ is the target matrix.

3.2 Language Syntax Based Prompt Fusion

As mentioned above, it is difficult for a single prompt input structure such as GLIP to cover all necessary descriptions even through careful designation of the prompt. Therefore, we propose to use multiple prompts instead of a single

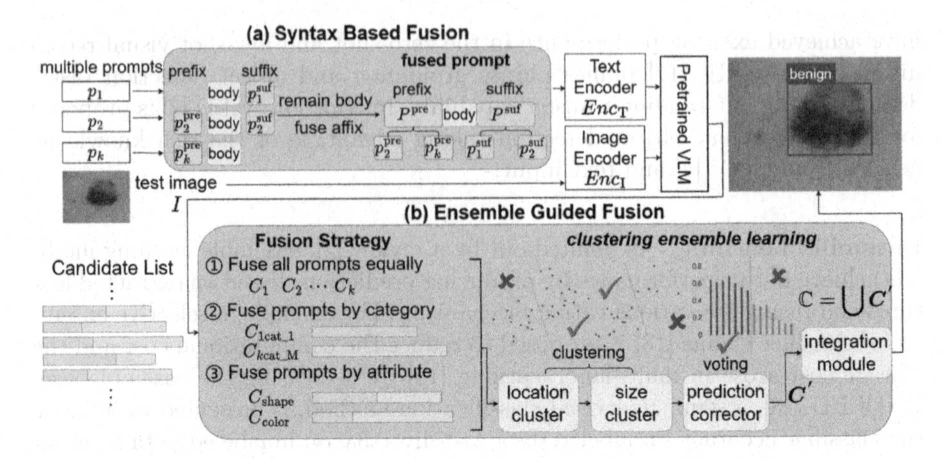

Fig. 1. Overview of the proposed approach: (a) Syntax Based Fusion, which fuses multiple prompts at the prompt level; (b) Ensemble Guided Fusion, which includes step-wise clustering mechanisms followed by voting and an integration module.

prompt for thorough and improved grounding. However, it is challenging to combine the grounding results from multiple prompts since manual integration is subjective, ineffective, and lacks uniform standards. Here, we take the first step to fuse the multiple prompts at the prompt level. We achieve this by extracting and fusing the prefixes and suffixes of each prompt based on language conventions and grammar rules. As shown in Fig. 1 (a), given serials of multiple prompts $p_1, p_2, \ldots, p_k$, the final fused prompt P_{fuse} from k single prompts is given by:

$$\begin{aligned} P_{\text{fuse}} &= p_1 + p_2 + \cdots + p_k \\ &= (p_1^{\text{pre}} + p_2^{\text{pre}} + \cdots + p_k^{\text{pre}}) + body + (p_1^{\text{suf}} + p_2^{\text{suf}} + \cdots + p_k^{\text{suf}}) \\ &= P^{\text{pre}} + body + P^{\text{suf}}. \end{aligned} \tag{2}$$

3.3 Ensemble Learning Based Fusion

Although the syntax based fusion approach is simple and sufficient, it is restricted by the form of text descriptions which may cause ambiguity in the fused prompt during processing. Moreover, the fused prompts are normally too long that the model could lose proper attention to the key information, resulting in extremely unstable performance (results shown in Sect. 4.2).

Therefore, in this subsection, we further explore fusion approaches based on ensemble learning. More specifically, the VLM outputs a set of candidate region proposals C_i for each prompt p_i, and these candidates carry more multi-dimensional information than prompts. We find in our preliminary experiments that direct concatenation of the candidates is not satisfactory and effective, since simply integration hardly screens out the bad predictions. In addition, the candidate, e.g., $c_{ij} \in C_i$, carries richer information that can be further utilized, such as central coordinate x_j and y_j, region size w_j and h_j, category label, and prediction confidence score. Therefore, we consider step-wise clustering mechanisms using the above information to screen out the implausible candidates based on clustering ensemble learning [3].

Another observation in our preliminary experiments is that most of the candidates distribute near the target if the prompt description matches better with the object. Moreover, the candidate regions of inappropriate size containing too much background or only part of the object should be abandoned directly. As such, we consider clustering the center coordinate (x_j, y_j) and region size (w_j, h_j) respectively to filter out those candidates with the wrong location and size.

This step-wise clustering with the aid of different features embodies a typical ensemble learning idea. Therefore, we propose a method called Ensemble Guided Fusion based on semi-clustering ensemble, as detailed in Fig. 1 (b). There are four sub-modules in our approach, where the location cluster f_{loc} and size cluster f_{size} discard the candidates with large deviations and abnormal sizes. Then, in the prediction corrector f_{correct}, we utilize the voting mechanism to select the remaining candidates with appropriate category tags and relatively high prediction confidence. After the first three steps of processing, the remaining candidates $C^{'}$ originated from each prompt can be written as:

$$C^{'} = C \cdot f_{\mathrm{loc}}(C) \cdot f_{\mathrm{size}}(C) \cdot f_{\mathrm{correct}}(C). \tag{3}$$

The remaining candidates are then transferred to the integration module for being integrated into the final fused result C_{fuse} that is mutually independent:

$$C_{\mathrm{fuse}} = \bigcup C^{'}. \tag{4}$$

Besides, we also propose three fusion strategies to recluster candidates in different ways before executing ensemble guided fusion, i.e., fusing the multiple prompts equally, by category, and by attribute. Compared to the first strategy, fusing by category and by attribute both have an additional step of reorgnization. Candidates whose prompts belong to the same category or have identical attributes will share the similar distribution. Accordingly, we rearrange these candidates C_i into a new set C for the subsequent fusion process.

4 Experiments and Results

4.1 Experimental Settings

We collect three public medical image datasets across various modalities including skin lesion detection dataset ISIC 2016 [2], polyp detection dataset CVC-300 [21], and blood cell detection dataset BCCD to validate our proposed approach for zero-shot medical lesion detection. For the experiments, we use the GLIP-T variant [11] as our base pre-trained model and adopt two metrics for the grounding evaluation, including Average Precision (AP) and AP50. More details on the dataset and implementation are described in the appendix.

Table 1. Our approaches v.s. single prompts and other fusion methods.

Prompt	Method	ISIC 2016		CVC-300		BCCD	
		AP	AP50	AP	AP50	AP	AP50
Single	GLIP [11]	10.5	20.0	29.8	37.9	8.9	18.4
		11.3	22.7	16.8	21.7	12.2	23.1
		13.6	25.5	20.2	30.3	9.6	18.2
Multiple	NMS [14]	12.0	20.6	27.9	37.9	11.9	21.4
	Soft-NMS [1]	18.8	30.3	24.1	31.2	11.9	21.4
	WBF [20]	1.16	5.37	3.27	9.40	1.22	4.75
	Concatenation	16.9	27.4	21.5	27.8	11.6	20.6
	Syntax based fusion	13.9	24.1	10.0	16.4	12.8	25.4
	Ours	**19.8**	**30.9**	**36.1**	**47.9**	**15.8**	**32.6**

4.2 Results

This section demonstrates that our proposed ensemble guided fusion approach can effectively benefit the model's performance.

The Proposed Approach Achieves the Best Performance in Zero-Shot Lesion Detection Compared to Baselines. To confirm the validity of our method, we conduct extensive experiments under the zero-shot setting and include a series of fusion baselines: Concatenation, Non-Maximum Suppression (NMS) [14], Soft-NMS [1] and Weighted Boxes Fusion (WBF) [20] for comparisons. As illustrated in Table 1, our ensemble guided fusion rivals the GLIP [11] with single prompt and other fusion baselines across all datasets. The first three rows in Table 1 represent the results of single prompt by only providing shape, color, and location information, respectively. Furthermore, we conduct a comparison between YOLOv5 [17] and our method on CVC-300 under 10-shot settings. Table 2 shows that our method outperforms YOLOv5, which indicates fully-supervised models such as YOLO may not be suitable for medical scenarios where a large labeled dataset is often not available. In addition, we utilize the Automatic Prompt Engineering (APE) [25] method to generate prompts. These prompts give comparable performance to our single prompt and can be still be improved by our fusion method. And the details are described in the appendix.

Table 2. Comparison with YOLOv5.

Method	Training mAP	Test mAP
YOLOv5	16.2	6.4
Ours	**60.8**	**50.2**

Table 3. Zero-shot v.s. 10-shot results.

Dataset	ISIC 2016	CVC-300
Zero-shot	19.8	36.1
10-shot	**38.2**	**50.2**

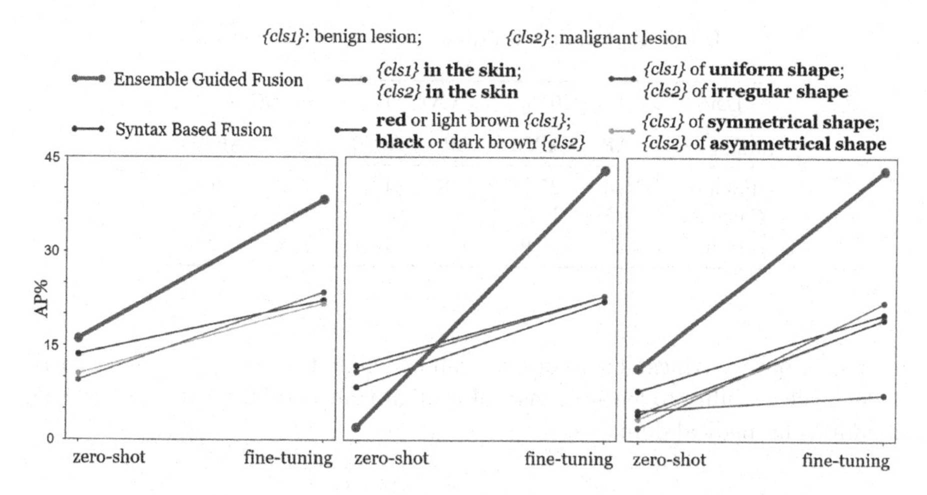

Fig. 2. Fine-tuning v.s. zero-shot results on the ISIC 2016 dataset.

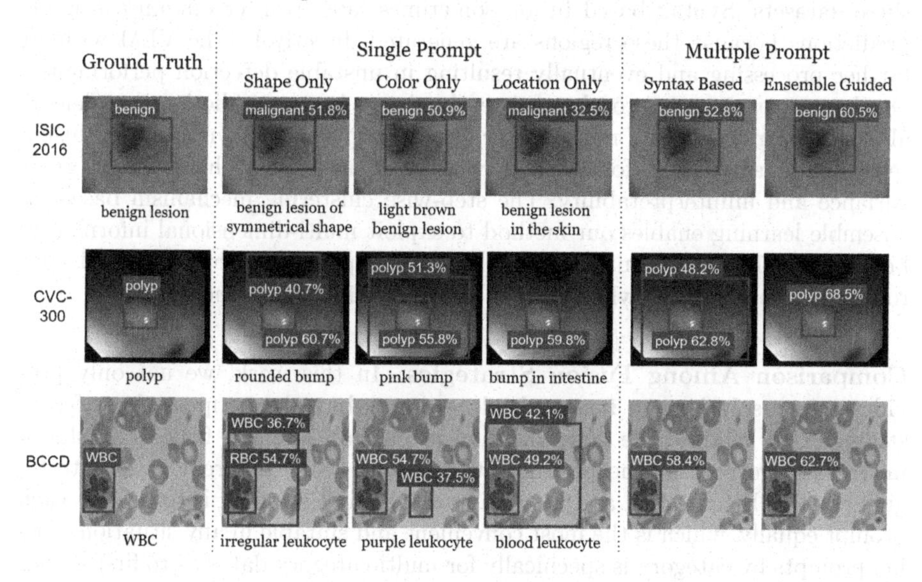

Fig. 3. Comparisons of test results between before and after multi-prompt fusion under zero-shot settings. Here we present part of the single prompts used in the experiments for illustration. The misclassification problem in some of the single prompts is corrected (i.e., malignant to benign) on the first dataset. For all datasets, the candidate boxes are more precise and associated with higher confidence scores.

Fine-Tuned Models Can Further Improve the Detection Performance. We conduct 10-shot fine-tuning experiments as a complement, and find the performance greatly improved. As shown in Table 3 and Fig. 2, with the same group of multiple prompts, the accuracy of fine-tuned model has increased almost twice as much as that of zero-shot, further demonstrating the effectiveness of our

Table 4. Results with different fusion strategies.

Strategy \ Dataset	ISIC 2016		CVC-300		BCCD	
	AP	AP50	AP	AP50	AP	AP50
Equally	16.8	25.2	30.8	40.4	12.5	21.6
Category	13.2	20.4	30.8	40.4	15.3	24.9
Attribute	19.8	30.9	36.1	47.9	15.8	32.6

method in both settings. Therefore, we can conclude that the pre-trained GLIP model has the ability to learn a reasonable alignment between textual and visual modalities in medical domains.

Visualizations. Figure 3 shows the visualization of the zero-shot results across three datasets. Syntax based fusion sometimes fails to filter out unreasonable predictions because these regions are generated directly by the VLM without further processing and eventually resulting in unstable detection performance. On the contrary, our approach consistently gives a better prediction that defeats all single prompts with a relatively proper description, yet syntax based fusion relies too much on the format and content of inputs, which results in great variance and uninterpretability. The step-wise clustering mechanism based on ensemble learning enables our method to exploit multi-dimensional information besides visual features. In addition, the key components in our proposed approach are unsupervised, which also enhances stability and generalization.

Comparison Among Fusion Strategies. In this work, we not only provide various solutions to fuse multiple prompts but also propose three fusion strategies to validate the generalization of ensemble-guided fusion. As shown in Table 4, we present the results obtained with three different fusion strategies: equally, by category, and by attribute. The first strategy is to process each prompt equally, which is the most convenient and suitable in any situation. Fusing prompts by category is specifically for multi-category datasets to first gather the prompts belonging to the same category and make further fusion. Similarly, Fusing by attribute is to fuse the candidates, whose prompts are describing the same attribute. The strategy of fusing by attribute, outperforms the other ones, due to the fact that candidates with the same attribute share a similar distribution, which is prone to obtain a more reasonable cluster. On the contrary, it is possible to neglect this distribution when fusing each prompt equally.

Ablation Study. As shown in Table 5, we perform ablation studies on three datasets. Our approach has three key components, i.e., location cluster, size cluster and prediction corrector. The location cluster filters out the candidates with severe deviation from the target. The size cluster removes those abnormal ones.

Table 5. Ablation for key components in our proposed approach.

Components			ISIC 2016		CVC-300		BCCD	
Location Cluster	Size Cluster	Prediction Corrector	AP (%)	AP50 (%)	AP (%)	AP50 (%)	AP (%)	AP50 (%)
✓			13.9	26.4	24.2	30.1	10.9	20.2
	✓		10.4	19.5	23.9	29.5	11.9	21.4
		✓	13.8	24.8	29.5	37.3	9.3	17.5
✓	✓		16.9	27.4	30.6	41.7	15.1	31.3
✓	✓	✓	**19.8**	**30.9**	**34.1**	**45.7**	**15.8**	**32.6**

Finally, the prediction corrector further eliminates the candidates that cause low accuracy. The results show that when combining the above three components, the proposed approach gives the best lesion detection performance, suggesting that all components are necessary and effective in the proposed approach.

5 Conclusion

In this paper, we propose an ensemble guided fusion approach to leverage multiple text descriptions when tackling the zero-shot medical lesion detection based on vision-language models and conduct extensive experiments to demonstrate the effectiveness of our approach. Compared to a single prompt that typically requires exhaustive engineering and designation, the multiple medical prompts provide a flexible way of covering all key information that help with lesion detection. We also present several fusion strategies for better exploiting the relationship among multiple prompts. One limitation of our method is that it requires diverse prompts for effective clustering of the candidates. However, with the help of other prompt engineering methods, the limitation can be relatively alleviated.

References

1. Bodla, N., Singh, B., Chellappa, R., Davis, L.S.: Soft-nms-improving object detection with one line of code. In: Proceedings of the IEEE International Conference on Computer Vision, pp. 5561–5569 (2017)
2. Codella, N.C., et al.: Skin lesion analysis toward melanoma detection: a challenge at the 2017 international symposium on biomedical imaging (ISBI), hosted by the international skin imaging collaboration (ISIC). In: 2018 IEEE 15th international symposium on biomedical imaging (ISBI 2018), pp. 168–172. IEEE (2018)
3. Dong, X., Yu, Z., Cao, W., Shi, Y., Ma, Q.: A survey on ensemble learning. Front. Comput. Sci. **14**, 241–258 (2020)
4. Gao, P., et al.: Clip-adapter: better vision-language models with feature adapters. arXiv preprint arXiv:2110.04544 (2021)
5. Girshick, R.: Fast R-CNN. In: Proceedings of the IEEE International Conference on Computer Vision, pp. 1440–1448 (2015)
6. Girshick, R., Donahue, J., Darrell, T., Malik, J.: Rich feature hierarchies for accurate object detection and semantic segmentation. In: Proceedings of the IEEE Conference on Computer Vision and Pattern Recognition, pp. 580–587 (2014)

7. Gu, X., Lin, T.Y., Kuo, W., Cui, Y.: Open-vocabulary object detection via vision and language knowledge distillation. arXiv preprint arXiv:2104.13921 (2021)
8. Jensen, J.D., Elewski, B.E.: The ABCDEF rule: combining the "ABCDE Rule" and the "Ugly Duckling Sign" in an effort to improve patient self-screening examinations. J. Clin. Aesthetic Dermatol. **8**(2), 15 (2015)
9. Jiang, C., Wang, S., Liang, X., Xu, H., Xiao, N.: ElixirNet: relation-aware network architecture adaptation for medical lesion detection. In: Proceedings of the AAAI Conference on Artificial Intelligence, vol. 34, pp. 11093–11100 (2020)
10. Li, J., Li, D., Xiong, C., Hoi, S.: Blip: Bootstrapping language-image pre-training for unified vision-language understanding and generation. In: International Conference on Machine Learning, pp. 12888–12900. PMLR (2022)
11. Li, L.H., et al.: Grounded language-image pre-training. In: Proceedings of the IEEE/CVF Conference on Computer Vision and Pattern Recognition, pp. 10965–10975 (2022)
12. Li, N., Jiang, Y., Zhou, Z.-H.: Multi-label selective ensemble. In: Schwenker, F., Roli, F., Kittler, J. (eds.) MCS 2015. LNCS, vol. 9132, pp. 76–88. Springer, Cham (2015). https://doi.org/10.1007/978-3-319-20248-8_7
13. Li, N., Zhou, Z.-H.: Selective ensemble of classifier chains. In: Zhou, Z.-H., Roli, F., Kittler, J. (eds.) MCS 2013. LNCS, vol. 7872, pp. 146–156. Springer, Heidelberg (2013). https://doi.org/10.1007/978-3-642-38067-9_13
14. Neubeck, A., Van Gool, L.: Efficient non-maximum suppression. In: 18th international conference on pattern recognition (ICPR 2006), vol. 3, pp. 850–855. IEEE (2006)
15. Qin, Z., Yi, H., Lao, Q., Li, K.: Medical image understanding with pretrained vision language models: a comprehensive study. arXiv preprint arXiv:2209.15517 (2022)
16. Radford, A., et al.: Learning transferable visual models from natural language supervision. In: International Conference on Machine Learning, pp. 8748–8763. PMLR (2021)
17. Redmon, J., Divvala, S., Girshick, R., Farhadi, A.: You only look once: Unified, real-time object detection. In: Proceedings of the IEEE Conference on Computer Vision and Pattern Recognition, pp. 779–788 (2016)
18. Ren, S., He, K., Girshick, R., Sun, J.: Faster R-CNN: towards real-time object detection with region proposal networks. In: Advances in Neural Information Processing Systems, vol. 28 (2015)
19. Sarcar, M., Rao, K., Narayan, K.: Computer aided design and manufacturing. PHI Learning (2008). https://books.google.co.jp/books?id=zXdivq93WIUC
20. Solovyev, R., Wang, W., Gabruseva, T.: Weighted boxes fusion: Ensembling boxes from different object detection models. Image Vis. Comput. **107**, 104117 (2021)
21. Vázquez, D., et al.: A benchmark for endoluminal scene segmentation of colonoscopy images. J. Healthcare Eng. **2017** (2017)
22. Wei, S., Li, Z., Zhang, C.: Combined constraint-based with metric-based in semi-supervised clustering ensemble. Int. J. Mach. Learn. Cybern. **9**, 1085–1100 (2018)
23. Zhang, M.L., Zhou, Z.H.: Exploiting unlabeled data to enhance ensemble diversity. Data Min. Knowl. Disc. **26**, 98–129 (2013)
24. Zhang, P., et al.: VinVL: revisiting visual representations in vision-language models. In: Proceedings of the IEEE/CVF Conference on Computer Vision and Pattern Recognition, pp. 5579–5588 (2021)
25. Zhou, Y., et al.: Large language models are human-level prompt engineers (2023)

Reversing the Abnormal: Pseudo-Healthy Generative Networks for Anomaly Detection

Cosmin I. Bercea[1,2(✉)], Benedikt Wiestler[3], Daniel Rueckert[1,3,4], and Julia A. Schnabel[1,2,5]

[1] Technical University of Munich, Munich, Germany
cosmin.bercea@tum.de
[2] Helmholtz AI and Helmholtz Center Munich, Munich, Germany
[3] Klinikum Rechts der Isar, Munich, Germany
[4] Imperial College London, London, UK
[5] King's College London, London, UK

Abstract. Early and accurate disease detection is crucial for patient management and successful treatment outcomes. However, the automatic identification of anomalies in medical images can be challenging. Conventional methods rely on large labeled datasets which are difficult to obtain. To overcome these limitations, we introduce a novel unsupervised approach, called *PHANES* (Pseudo Healthy generative networks for ANomaly Segmentation). Our method has the capability of reversing anomalies, i.e., preserving healthy tissue and replacing anomalous regions with pseudo-healthy (PH) reconstructions. Unlike recent diffusion models, our method does not rely on a learned noise distribution nor does it introduce random alterations to the entire image. Instead, we use latent generative networks to create masks around possible anomalies, which are refined using inpainting generative networks. We demonstrate the effectiveness of *PHANES* in detecting stroke lesions in T1w brain MRI datasets and show significant improvements over state-of-the-art (SOTA) methods. We believe that our proposed framework will open new avenues for interpretable, fast, and accurate anomaly segmentation with the potential to support various clinical-oriented downstream tasks. Code: https://github.com/ci-ber/PHANES

Keywords: Unsupervised Anomaly Detection · Generative Networks

1 Introduction

The early detection of lesions in medical images is critical for the diagnosis and treatment of various conditions, including neurological disorders. Stroke is a leading cause of death and disability, where early detection and treatment can significantly improve patient outcomes. However, the quantification of lesion burden is challenging and can be time-consuming and subjective when performed manually by medical professionals [14]. While supervised learning methods [10, 11] have proven to be effective in lesion segmentation, they rely heavily on large

H. Greenspan et al. (Eds.): MICCAI 2023, LNCS 14224, pp. 293–303, 2023.
https://doi.org/10.1007/978-3-031-43904-9_29

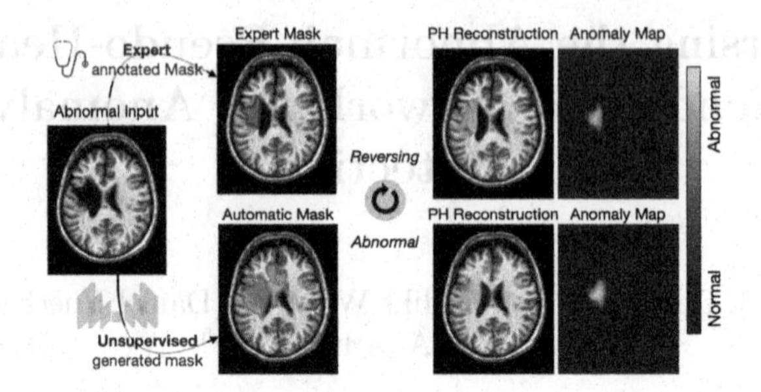

Fig. 1. Overview of PHANES (see Fig. 2). Our method can use both expert annotated- or unsupervised generated masks to reverse and segment anomalies

annotated datasets for training and tend to generalize poorly beyond the learned labels [21]. On the other hand, unsupervised methods focus on detecting patterns that significantly deviate from the norm by training only on normal data.

One widely used category of unsupervised methods is latent restoration methods. They involve autoencoders (AEs) that learn low-dimensional representations of data and detect anomalies through inaccurate reconstructions of abnormal samples [17]. However, developing compact and comprehensive representations of the healthy distribution is challenging [1], as recent studies suggest AEs perform better reconstructions on out-of-distribution (OOD) samples than on training samples [23]. Various techniques have been introduced to enhance representation learning, including discretizing the latent space [15], disentangling compounding factors [2], and variational autoencoders (VAEs) that introduce a prior into the latent distribution [26,29]. However, methods that can enforce the reconstruction of healthy generally tend to produce blurry reconstructions.

In contrast, generative adversarial networks (GANs) [8,18,24] are capable of producing high-resolution images. New adversarial AEs combine VAEs' latent representations with GANs' generative abilities, achieving SOTA results in image generation and outlier detection [1,5,6,19]. Nevertheless, latent methods still face difficulties in accurately reconstructing data from their low-dimensional representations, causing false positive detections on healthy tissues.

Several techniques have been proposed that make use of the inherent spatial information in the data rather than relying on constrained latent representations [12,25,30]. These methods are often trained on a pretext task, such as recovering masked input content [30]. De-noising AEs [12] are trained to eliminate synthetic noise patterns, utilizing skip connections to preserve the spatial information and achieve SOTA brain tumor segmentation. However, they heavily rely on a learned noise model and may miss anomalies that deviate from the noise distribution [1]. More recently, diffusion models [9] apply a more complex de-noising process to detect anomalies [25]. However, the choice and granularity of the applied noise is crucial for breaking the structure of anomalies [25]. Adapting the noise distribution to the diversity and heterogeneity of pathology is inherently difficult, and even if achieved, the noising process disrupts the structure of both healthy and anomalous regions throughout the entire image.

In related computer vision areas, such as industrial inspection [3], the top-performing methods do not focus on reversing anomalies, but rather on detecting them by using large nominal banks [7,20], or pre-trained features from large natural imaging datasets like ImageNet [4,22]. Salehi et al. [22] have employed multi-scale knowledge distillation to detect anomalies in industrial and medical imaging. However, the application of these networks in medical anomaly segmentation, particularly in brain MRI, is limited by various challenges specific to the medical imaging domain. They include the variability and complexity of normal data, subtlety of anomalies, limited size of datasets, and domain shifts.

This work aims to combine the advantages of constrained latent restoration for understanding healthy data distribution with generative in-painting networks. Unlike previous methods, our approach does not rely on a learned noise model, but instead creates masks of probable anomalies using latent restoration. These guide generative in-painting networks to reverse anomalies, i.e., preserve healthy tissues and produce pseudo-healthy in-painting in anomalous regions. We believe that our proposed method will open new avenues for interpretable, fast, and accurate anomaly segmentation and support various clinical-oriented downstream tasks, such as investigating progression of disease, patient stratification and treatment planning. In summary our main contributions are:

- We investigate and measure the ability of SOTA methods to reverse synthetic anomalies on real brain T1w MRI data.
- We propose a novel unsupervised segmentation framework, that we call *PHANES*, that is able to preserve healthy regions and utilize them to generate pseudo-healthy reconstructions on anomalous regions.
- We demonstrate a significant advancement in the challenging task of unsupervised ischemic stroke lesion segmentation.

2 Background

Latent restoration methods use neural networks to estimate the parameters θ, ϕ of an encoder E_θ and a decoder D_ϕ. The aim is to restore the input from its lower-dimensional latent representation with minimal loss. The standard objective is to minimize the residual, e.g., using mean squared error (MSE) loss: $\min_{\theta,\phi} \sum_{i=1}^{N} \|x_i - D_\phi(E_\theta(x_i))\|^2$. In the context of variational inference [13], the goal is to optimize the parameters θ of a latent variable model $p_\theta(\mathrm{x})$ by maximizing the log-likelihood of the observed samples x: $\log p_\theta(x)$. The term is intractable, but the true posterior $p_\theta(z|x)$ can be approximated by $q_\phi(z|x)$:

$$\log p_\theta(x) \geq \mathbb{E}_{q(z|x)}[\log p_\theta(x|z)] - KL[q_\phi(z|x)||p(z)] = ELBO(x). \quad (1)$$

KL is the Kullback-Leibler divergence; $q_\phi(z|x)$ and $p_\theta(x|z)$ are usually known as the encoder E_ϕ and decoder D_θ; the prior $p(z)$ is usually the normal distribution $\mathcal{N}(\mu_0, \sigma_0)$; and the ELBO denotes the Evidence Lower Bound. In unsupervised

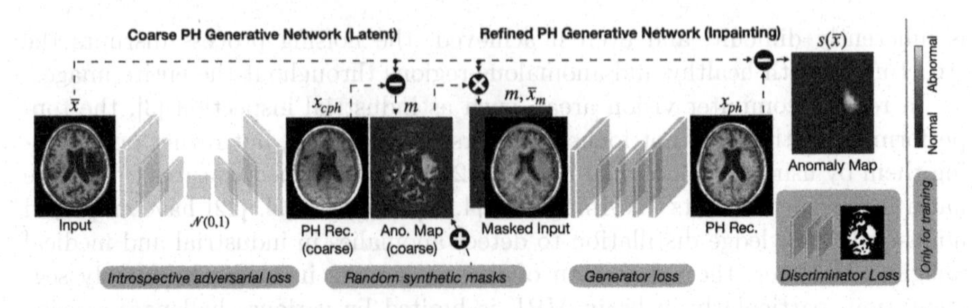

Fig. 2. *PHANES* overview. Our framework offers modularity, enabling the choice of preferred generative networks, such as adversarial or diffusion-based models. First, we use latent generative networks to learn the healthy data distribution and provide approximate pseudo-healthy reconstructions x_{cph}. Anomaly maps m obtained from this step are then used to mask out possible anomalous regions in the input. The remaining healthy tissues are used to condition the refined generative networks, which complete the image and replace anomalous regions with pseudo-healthy tissues. This results in accurate PH reconstructions x_{ph}, which enables the precise localization of diseases, as shown on the right.

anomaly detection, the networks are trained only on normal samples $x \in \mathcal{X} \subset \mathbb{R}^N$. Given an anomalous input $\overline{x} \notin \mathcal{X}$, it is assumed that the reconstruction $x_{ph} = (D_\phi(E_\theta(\overline{x}))) \in \mathcal{X}$ represents its pseudo-healthy version. The aim of the pseudo-healthy reconstructions is to accurately reverse abnormalities present in the input images. This is achieved by preserving the healthy regions while generating healthy-like tissues in anomalous regions. Thus, anomalies can ideally be directly localized by computing the difference between the anomalous input and the pseudo-healthy reconstructions: $s(\overline{x}) = |\overline{x} - x_{ph}|$.

3 Method

Figure 2 shows an overview of our proposed method. We introduce an innovative approach by utilizing masks produced by latent generative networks to condition generative inpainting networks only on healthy tissues. Our framework is modular, which allows for the flexibility of choosing a preferred generative network, such as adversarial, or diffusion-based models for predicting the pseudo-healthy reconstructions. In the following we describe each component in detail.

Latent Generative Network. The first step involves generating masks to cover potential anomalous regions in the input image. The goal of this step is to achieve unbiased detection of various pathologies and minimize false positives. It is therefore important to use a method that is restricted to in-distribution samples, particularly healthy samples, while also accurately reconstructing inputs. Here, we have adopted our previous work [1] that augments a soft introspective variational auto-encoder with a reversed embedding similarity loss with the aim to enforcing more accurate pseudo-healthy reconstructions. The training process encourages the encoder to distinguish between real and generated samples by minimizing the Kullback-Leibler (KL) divergence of the latent distribution of real samples and the prior, and maximizing the KL divergence of generated samples. On the other hand, the decoder is trained to deceive the encoder by reconstructing real data samples using the standard ELBO and minimizing the KL divergence of generated samples compressed by the encoder:

$$\mathcal{L}_{E_\phi}(x,z) = ELBO(x) - \frac{1}{\alpha}(exp(\alpha ELBO(D_\theta(z)) + \lambda \mathcal{L}_{Reversed}(x), \tag{2}$$

$$\mathcal{L}_{Reversed}(x) = \sum_{l=0}^{L}(1 - \mathcal{L}_{Sim}(E^l_\phi(x), E^l_\phi(x_{cph})) + \frac{1}{2}MSE(E^l_\phi(x), E^l_\phi(x_{cph})),$$

$$\mathcal{L}_{D_\theta}(x,z) = ELBO(x) + \gamma ELBO(D_\theta(z)),$$

where E^l_ϕ is the l-th embedding of the L encoder layers, $x_{cph} = D_\theta(E_\phi(x))$, and $\mathcal{L}_{Sim}$ is the cosine similarity.

Mask Generation. Simple residual errors have a strong dependence on the underlying intensities [16]. As it is important to assign higher values to (subtle) pathological structures, we compute anomaly masks as proposed in [1] by applying adaptive histogram equalization (eq), normalizing with the $95th$ percentile, and augmenting the errors with perceptual differences for robustness:

$$m(\overline{x}) = norm_{95}(|(eq(x_{cph}) - eq(\overline{x})|) * \mathcal{S}_{lpips}(eq(x_{cph}), eq(\overline{x})), \tag{3}$$

with $\mathcal{S}_{lpips}$ being the learned perceptual image patch similarity [28]. Finally, we binarize the masks using the 99th percentile value on the healthy validation set.

Inpainting Generative Network. The objective of the refined PH generative network is to complete the masked image by utilizing the remaining healthy tissues to generate a full PH version of the input. Considering computational efficiency, we have employed the recent in-painting AOT-GAN [27]. The method uses a generator (G) and discriminator neural network to optimize losses based on residual and perceptual differences, resulting in accurate and visually precise inpainted images. Additionally, the discriminator predicts the input mask from the inpainted image to improve the synthesis of fine textures.

Anomaly Maps. The PH reconstruction is computed as follows: $x_{ph} = \overline{x} \odot (1-m) + G(\overline{x} \odot (1-m), m) \odot m$, with $\odot$ being the pixel-wise multiplication. We compute the final anomaly maps based on residual and perceptual differences:

$$s(\overline{x}) = |x_{ph} - \overline{x}| * \mathcal{S}_{lpips}(x_{ph}, \overline{x}) \tag{4}$$

4 Experiments

Datasets. We trained our model using two publicly available brain T1w MRI datasets, including FastMRI+ (131 train, 15 val, 30 test) and IXI (581 train samples), to capture the healthy distribution. Performance evaluation was done on a large stroke T1-weighted MRI dataset, ATLAS v2.0 [14], containing 655 images

Table 1. Reversing synthetic anomalies. We evaluate the pseudo-healthy (PH) reconstruction on healthy and anomalous regions using the learned perceptual image patch similarity (LPIPS) [28] and the anomaly segmentation performance. *PHANES*^{GT} represents an upper bound and uses the ground truth anomalies to mask the image for inpainting. x% shows improvement over best baseline (RA) and x% shows the decrease in performance compared to *PHANES*.

Method	PH Reconstruction (LPIPS)		Anomaly Segmentation	
	Healthy ↓	Anomaly ↓	AUPRC ↑	⌈*DICE*⌉ ↑
PHANES^{GT} (ours)	**0.09** N/A	**0.94** ▼ 94%	**100** ▲ 37%	**100** ▲ 46%
PHANES (ours)	**2.25** ▼ 77%	**8.10** ▼ 47%	**77.93** ▲ 7%	**75.47** ▲ 10%
RA [1]	9.74 ▲ 333%	15.27 ▲ 89%	73.01 ▼ 6%	68.52 ▼ 9%
SI-VAE [6]	13.16 ▲ 485%	19.01 ▲ 135%	17.91 ▼ 77%	31.30 ▼ 59%
AnoDDPM [25]	6.64 ▲ 195%	19.46 ▲ 140%	14.85 ▼ 81%	19.89 ▼ 74%
DAE [12]	3.94 ▲ 75%	20.05 ▲ 148%	35.73 ▼ 54%	37.76 ▼ 50%
VAE [29]	33.22 ▲ 1376%	44.00 ▲ 443%	22.86 ▼ 71%	28.46 ▼ 62%

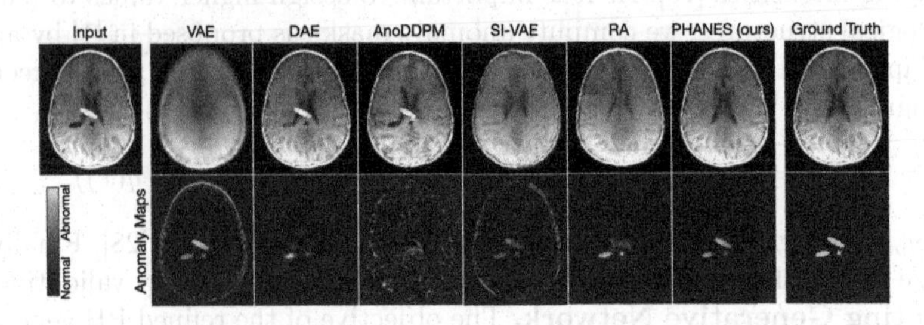

Fig. 3. Reversing synthetic anomalies. *PHANES* successfully removes synthetic anomalies and produces the most accurate pseudo-healthy reconstructions.

with manually segmented lesion masks for training and 355 test images with hidden lesion masks. We evaluated the model using the 655 training images with public annotations. The mid axial slices were normalized to the 98th percentile, padded, and resized to 128×128 resolution. During training, we performed random rotations up to 10°, translations up to 0.1, scaling from 0.9 to 1.1, and horizontal flips with a 0.5 probability. We trained for 1500 epochs, with a batch size of 8, lr of $5e^{-5}$, and early stopping.

4.1 Reversing Synthetic Anomalies

In this section, we test whether reconstruction-based methods can generate pseudo-healthy images and reverse synthetic anomalies. Results are evaluated in Table 1 and Fig. 3 using 30 test images and synthetic masks as reference. VAEs produce blurry results that lead to poor reconstruction of both healthy and anomalous regions (LPIPS) and thus poor segmentation performance. While DAEs preserve the healthy tissues well with an LPIPS of 3.94, they do not gen-

Table 2. Ischemic stroke lesion segmentation on real T1w brain MRIs. ▲ x% shows improvement over AnoDDPM, and ▼ x% shows the decrease in performance compared to *PHANES*. * marks statistical significance ($p < 0.05$).

Method	AUPRC ↑	⌈*DICE*⌉ ↑
PHANES (ours)	**19.96 ± 2.3*** ▲ 22%	**32.17 ± 2.0*** ▲ 16%
AnoDDPM [25]	16.33 ± 1.7 ▼ 18%	27.64 ± 1.4 ▼ 14%
RA [1]	12.30 ± 1.0 ▼ 38%	22.20 ± 1.5 ▼ 31%
PatchCore [20]	12.24 ± 0.7 ▼ 39%	24.79 ± 1.2 ▼ 23%
DAE [12]	9.22 ± 1.3 ▼ 54%	15.62 ± 2.1 ▼ 53%
SI-VAE [6]	6.86 ± 0.6 ▼ 66%	13.57 ± 0.9 ▼ 58%
MKD [22]	2.93 ± 0.3 ▼ 85%	5.91 ± 0.6 ▼ 82%
VAE [29]	2.76 ± 0.2 ▼ 86%	5.96 ± 0.3 ▼ 81%

erate pseudo-healthy reconstructions in anomalous regions (LPIPS ≈ 20). However, they change the intensity of some structures, e.g., hypo-intensities, allowing for improved detection accuracy (see AUPRC and Dice). Simplex noise in [25] is designed to detect large hypo-intense lesions, leaving small anomalies undetected by AnoDDPM. SI-VAEs and RA produce pseudo healthy versions of the abnormal inputs, with the latter achieving the best results among the baselines. Our proposed method, *PHANES*, successfully reverses the synthetic anomalies, with its reconstructions being the most similar to ground truth healthy samples, as can be seen in Fig. 3. It achieved an improvement of 77% and 47% in generating pseudo-healthy samples in healthy and anomalous regions, respectfully. This enables the precised localization of anomalies (see bottom row in Fig. 3).

4.2 Ischemic Stroke Lesion Segmentation on T1w Brain MRI

In this section, we evaluate the performance of our approach in segmenting stroke lesions and show the results in Table 2 and Fig. 4. For completeness, we compare our approach to teacher-student methods that use multi-scale knowledge distillation (MKD) for anomaly segmentation. The unsupervised detection of (subtle) stroke lesions is challenging. The lack of healthy data from the same scanner and the limited size of the healthy datasets limit the successful application of such methods, with a maximum achievable Dice score of just under 6%. On the other hand, PatchCore, which is currently the SOTA method in industrial anomaly detection, has demonstrated comparable performance to the top-performing baselines. VAEs yield many false positive detections due to the blurry reconstructions and achieve poor localization results. DAEs can identify anomalies that resemble the learned noise distribution and improve segmentation results (AUPRC of 9.22), despite not producing pseudo-healthy reconstructions of abnormal samples (see Subsect. 4.1). The best performing latent restoration method is RA, achieving a remarkable 79% improvement over SI-VAE. Unlike

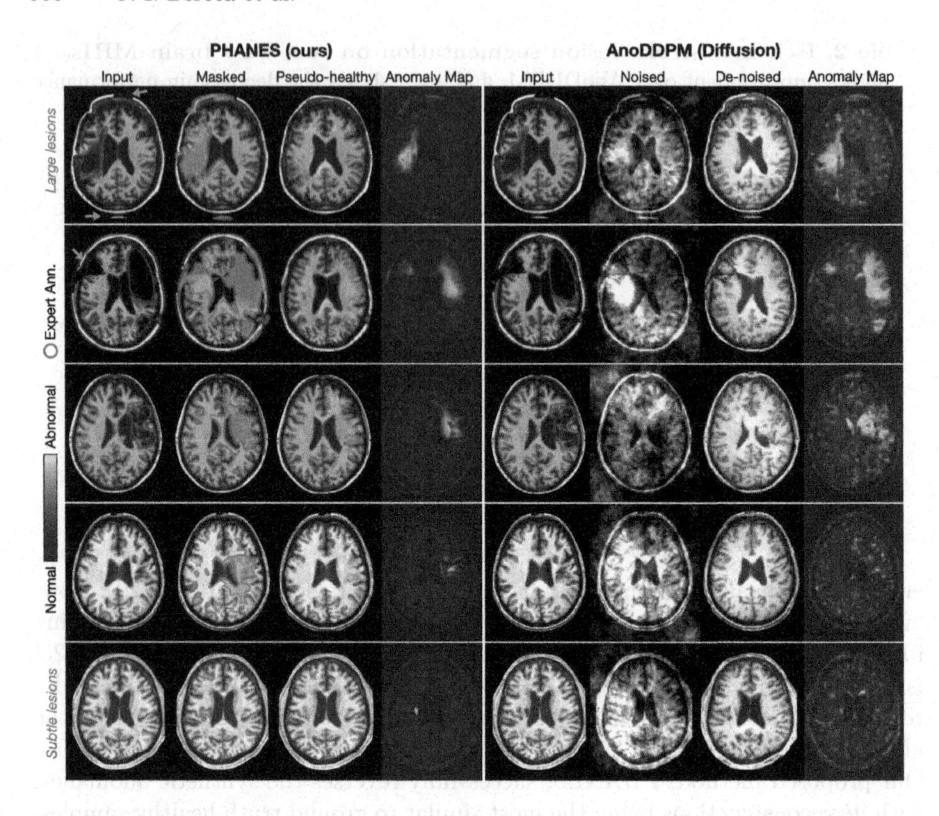

Fig. 4. Stroke lesion segmentation. We show input images with expert annotations in red along with masked images generated by the latent generative networks in Fig. 2, pseudo-healthy reconstructions, and anomaly maps. On the right, we show the performance of diffusion models on the same inputs. Different rows show cases ranging from large anomalies at the top to more subtle ones at the bottom. Green arrows mark unlabeled anomalies. *PHANES* successfully reverses the anomalies and accurately localizes even very subtle anomalies.

experiments in Subsect. 4.1, the Simplex noise aligns more closely with the hypo-intense pathology distribution of stroke in T1w brain MRI. As a result, AnoDDPM achieves the highest detection accuracy among the baselines. Compared to AnoDDPM, *PHANES* increases the detection results by 22% AUPRC. Figure 4 shows a visual comparison between the two approaches. Diffusion models tend to be more susceptible to domain shifts (first three rows) and yield more false positives. In contrast, *PHANES* demonstrates more precise localization, especially for subtle anomalies (bottom rows). Generally, unsupervised methods tend to have lower Dice scores partly due to unlabeled artefacts in the dataset. These include non-pathological (rows 1,2) as well as other pathological effects, such as changes in ventricle structure (rows 3,4). *PHANES* correctly identifies these as anomalous, but their lack of annotation limits numerical evaluations.

5 Discussion

This paper presents a novel unsupervised anomaly segmentation framework, called *PHANES*. It possesses the ability to reverse anomalies in medical images by preserving healthy tissues and substituting anomalous regions with pseudo-healthy reconstructions. By generating pseudo-healthy versions of images containing anomalies, *PHANES* can be a useful tool in supporting clinical studies.

While we are encouraged by these achievements, we also recognize certain limitations and areas for improvement. For example, the current binarization of anomaly maps does not account for the inherent uncertainty in the maps, which we aim to explore in future research. Additionally, our method relies on accurate initial estimates of the latent restoration and anomaly maps. Nevertheless, our proposed concept is independent of specific approaches and can leverage advancements in both domains. Our method is not optimized for detecting a certain anomaly distribution but rather demonstrates robustness in handling various small synthetic anomalies and diverse stroke lesions. We look forward to generalizing our method to other anatomies and imaging modalities, paving the way for exciting future research in the field of anomaly detection.

In conclusion, we demonstrated exceptional performance in reversing synthetic anomalies and segmenting stroke lesions on brain T1w MRIs. We believe that deliberate masking of (possible) abnormal regions will pave new ways for novel anomaly segmentation methods and empower further clinical applications.

Acknowledgement. C.I.B. is in part supported by the Helmholtz Association under the joint research school "Munich School for Data Science - MUDS".

References

1. Bercea, C.I., Wiestler, B., Rueckert, D., Schnabel, J.A.: Generalizing unsupervised anomaly detection: towards unbiased pathology screening. In: International Conference on Medical Imaging with Deep Learning (2023)
2. Bercea, C.I., Wiestler, B., Rueckert, D., Albarqouni, S.: Federated disentangled representation learning for unsupervised brain anomaly detection. Nat. Mach. Intell. **4**(8), 685–695 (2022)
3. Bergmann, P., Fauser, M., Sattlegger, D., Steger, C.: MVTec AD - a comprehensive real-world dataset for unsupervised anomaly detection. In: Proceedings of the IEEE/CVF Conference on Computer Vision and Pattern Recognition, pp. 9584–9592 (2019)
4. Bergmann, P., Fauser, M., Sattlegger, D., Steger, C.: Uninformed students: student-teacher anomaly detection with discriminative latent embeddings. In: Proceedings of the IEEE/CVF Conference on Computer Vision and Pattern Recognition, pp. 4183–4192 (2020)
5. Chen, X., Konukoglu, E.: Unsupervised detection of lesions in brain MRI using constrained adversarial auto-encoders. In: International Conference on Medical Imaging with Deep Learning (2018)
6. Daniel, T., Tamar, A.: Soft-IntroVAE: analyzing and improving the introspective variational autoencoder. In: Proceedings of the IEEE/CVF Conference on Computer Vision and Pattern Recognition, pp. 4391–4400 (2021)

7. Defard, T., Setkov, A., Loesch, A., Audigier, R.: PaDiM: a patch distribution modeling framework for anomaly detection and localization. In: Del Bimbo, A., et al. (eds.) ICPR 2021. LNCS, vol. 12664, pp. 475–489. Springer, Cham (2021). https://doi.org/10.1007/978-3-030-68799-1_35
8. Goodfellow, I., et al.: Generative adversarial nets. In: Advances in Neural Information Processing Systems, vol. 27 (2014)
9. Ho, J., Jain, A., Abbeel, P.: Denoising diffusion probabilistic models. Adv. Neural Inf. Process. Syst. **33**, 6840–6851 (2020)
10. Kamnitsas, K., et al.: DeepMedic for brain tumor segmentation. In: Medical Image Computing and Computer Assisted Intervention BrainLes Workshop, pp. 138–149 (2016)
11. Kamnitsas, K., et al.: Efficient multi-scale 3D CNN with fully connected CRF for accurate brain lesion segmentation. Med. Image Anal. **36**, 61–78 (2017)
12. Kascenas, A., Pugeault, N., O'Neil, A.Q.: Denoising autoencoders for unsupervised anomaly detection in brain MRI. In: International Conference on Medical Imaging with Deep Learning (2022)
13. Kingma, D.P., Welling, M.: Auto-encoding variational bayes. arXiv preprint arXiv:1312.6114 (2013)
14. Liew, S.L., Lo, B.P., Miarnda R. Donnelly, et al.: A large, curated, open-source stroke neuroimaging dataset to improve lesion segmentation algorithms. Sci. Data **9**, 230 (2022)
15. Mao, Y., Xue, F.-F., Wang, R., Zhang, J., Zheng, W.-S., Liu, H.: Abnormality detection in chest x-ray images using uncertainty prediction autoencoders. In: Martel, A.L., et al. (eds.) MICCAI 2020. LNCS, vol. 12266, pp. 529–538. Springer, Cham (2020). https://doi.org/10.1007/978-3-030-59725-2_51
16. Meissen, F., Wiestler, B., Kaissis, G., Rueckert, D.: On the pitfalls of using the residual error as anomaly score. arXiv preprint arXiv:2202.03826 (2022)
17. Pawlowski, N., et al.: Unsupervised lesion detection in brain CT using Bayesian convolutional autoencoders. In: International Conference on Medical Imaging with Deep Learning (2018)
18. Perera, P., Nallapati, R., Xiang, B.: OCGAN: one-class novelty detection using GANs with constrained latent representations. In: Proceedings of the IEEE/CVF Conference on Computer Vision and Pattern Recognition, pp. 2898–2906 (2019)
19. Pinaya, W.H., et al.: Unsupervised brain imaging 3d anomaly detection and segmentation with transformers. Med. Image Anal. **79**, 102475 (2022)
20. Roth, K., Pemula, L., Zepeda, J., Schölkopf, B., Brox, T., Gehler, P.: Towards total recall in industrial anomaly detection. In: Proceedings of the IEEE/CVF Conference on Computer Vision and Pattern Recognition, pp. 14318–14328 (2022)
21. Ruff, L., et al.: A unifying review of deep and shallow anomaly detection. In: Proceedings of the IEEE (2021)
22. Salehi, M., Sadjadi, N., Baselizadeh, S., Rohban, M.H., Rabiee, H.R.: Multiresolution knowledge distillation for anomaly detection. In: Proceedings of the IEEE/CVF Conference on Computer Vision and Pattern Recognition, pp. 14902–14912 (2021)
23. Schirrmeister, R., Zhou, Y., Ball, T., Zhang, D.: Understanding anomaly detection with deep invertible networks through hierarchies of distributions and features. Adv. Neural Inf. Proc. Syst. **33**, 21038–21049 (2020)
24. Schlegl, T., Seeböck, P., Waldstein, S.M., Langs, G., Schmidt-Erfurth, U.: f-AnoGAN: fast unsupervised anomaly detection with generative adversarial networks. Med. Image Anal. **54**, 30–44 (2019)

25. Wyatt, J., Leach, A., Schmon, S.M., Willcocks, C.G.: Anoddpm: anomaly detection with denoising diffusion probabilistic models using simplex noise. In: Proceedings of the IEEE/CVF Conference on Computer Vision and Pattern Recognition Workshops, pp. 650–656, June 2022
26. You, S., Tezcan, K.C., Chen, X., Konukoglu, E.: Unsupervised lesion detection via image restoration with a normative prior. In: International Conference on Medical Imaging with Deep Learning, pp. 540–556. PMLR (2019)
27. Zeng, Y., Fu, J., Chao, H., Guo, B.: Aggregated contextual transformations for high-resolution image inpainting. IEEE Trans. Vis. Comput. Graph. **29**, 3266–3280 (2022)
28. Zhang, R., Isola, P., Efros, A.A., Shechtman, E., Wang, O.: The unreasonable effectiveness of deep features as a perceptual metric. In: Proceedings of the IEEE/CVF Conference on Computer Vision and Pattern Recognition, pp. 586–595 (2018)
29. Zimmerer, D., Isensee, F., Petersen, J., Kohl, S., Maier-Hein, K.: Unsupervised anomaly localization using variational auto-encoders. In: Shen, D., et al. (eds.) Medical Image Computing and Computer Assisted Intervention - MICCAI 2019, LNCS, vol. 11767, pp. 289–297. Springer, Cham (2019). https://doi.org/10.1007/978-3-030-32251-9_32
30. Zimmerer, D., Kohl, S.A., Petersen, J., Isensee, F., Maier-Hein, K.H.: Context-encoding variational autoencoder for unsupervised anomaly detection. arXiv preprint arXiv:1812.05941 (2018)

What Do AEs Learn? Challenging Common Assumptions in Unsupervised Anomaly Detection

Cosmin I. Bercea[1,2(✉)], Daniel Rueckert[1,3], and Julia A. Schnabel[1,2,4]

[1] Technical University of Munich, Munich, Germany
cosmin.bercea@tum.de
[2] Helmholtz AI and Helmholtz Center Munich, Munich, Germany
[3] Imperial College London, London, UK
[4] King's College London, London, UK

Abstract. Detecting abnormal findings in medical images is a critical task that enables timely diagnoses, effective screening, and urgent case prioritization. Autoencoders (AEs) have emerged as a popular choice for anomaly detection and have achieved state-of-the-art (SOTA) performance in detecting pathology. However, their effectiveness is often hindered by the assumption that the learned manifold only contains information that is important for describing samples within the training distribution. In this work, we challenge this assumption and investigate what AEs actually learn when they are posed to solve anomaly detection tasks. We have found that standard, variational, and recent adversarial AEs are generally not well-suited for pathology detection tasks where the distributions of normal and abnormal strongly overlap. In this work, we propose *MorphAEus*, novel deformable AEs to produce pseudo-healthy reconstructions refined by estimated dense deformation fields. Our approach improves the learned representations, leading to more accurate reconstructions, reduced false positives and precise localization of pathology. We extensively validate our method on two public datasets and demonstrate SOTA performance in detecting pneumonia and COVID-19. Code: https://github.com/ci-ber/MorphAEus.

Keywords: Representation Learning · Anomaly Detection

1 Introduction

Identifying unusual patterns in data is of great interest in many applications such as medical diagnosis, industrial defect inspection, or financial fraud detection. Finding anomalies in medical images is especially hard due to large inter-patient variance of normality, the irregular appearance-, and often rare occurrence of diseases. Therefore, it is difficult and expensive to collect large amounts of annotated samples that cover the full abnormality spectrum, with supervised [7,15]

Supplementary Information The online version contains supplementary material available at https://doi.org/10.1007/978-3-031-43904-9_30.

H. Greenspan et al. (Eds.): MICCAI 2023, LNCS 14224, pp. 304–314, 2023.
https://doi.org/10.1007/978-3-031-43904-9_30

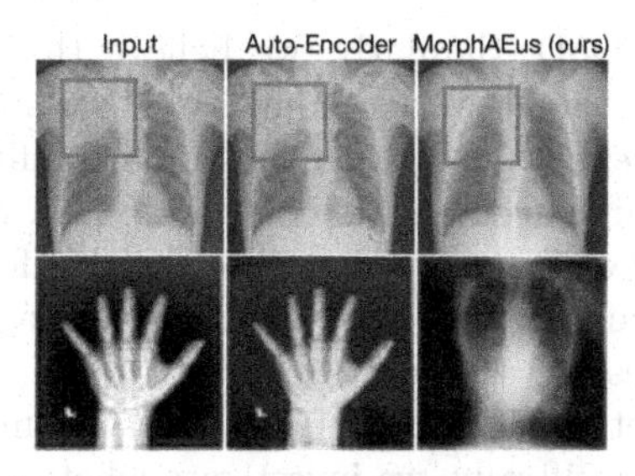

(a) Reconstruction of OoD samples.

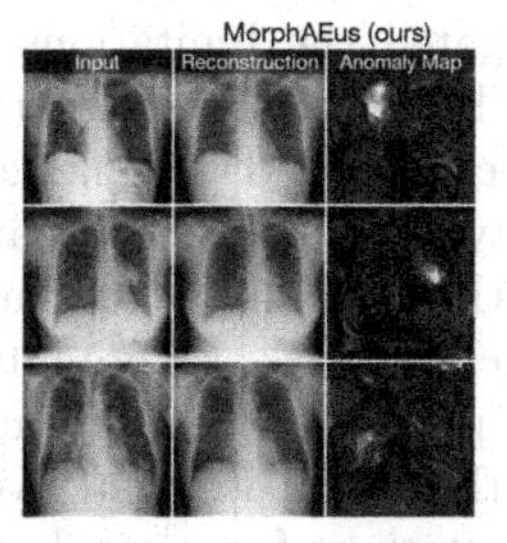

(b) Pathology Detection.

Fig. 1. *MorphAEus* outperforms AEs by generating ID reconstructions even for far OoD cases (Fig. 1a), enabling accurate pathology localization (Figure 1b).

and self-supervised [10,16] methods only capturing limited facets of the abnormal distribution [31]. However, since it is more feasible to obtain large data sets with normal samples, it is common to detect outliers by detecting patterns that deviate from the expected normative distribution.

Reconstruction-based AEs have emerged as a very popular framework for unsupervised anomaly detection and are widely adopted in medical imaging [2]. They provide straight-forward residual error maps, which are essential for safety-critical domains such as medical image analysis. However, recent work suggests that AEs might reconstruct out-of-distribution (OoD) samples even better than in-distribution (ID) samples [28], with the learned likelihood being dominated by common low-level features [32]. While this can be useful for some tasks such as reconstruction [34], or restoration [21], it often fails for pathology detection as anomalies can be missed due to small residual errors. In Fig. 1 we show that AEs that have only been trained on healthy chest X-rays are also able to reconstruct OoD samples like pathologies or hands. Similarly, Perera et al. [24] showed that AEs trained on the digit 8 can also reconstruct digits 1,5,6 and 9.

Much effort has been made in the medical imaging community to improve the limitations of traditional anomaly detection methods, particularly in the context of brain MRI. Apart from the reduced dimensionality of the bottleneck, several other techniques have been introduced to regularize AEs [11,20,29,38]. Recently, self-supervised denoising AEs [16] achieved SOTA results on brain pathology segmentation. They explicitly feed noise-corrupted inputs $\tilde{x} = x + \epsilon$ to the network with the aim at reconstructing the original input x. However, this is especially beneficial when the anomaly distribution is known *a priori*. Variational AEs (VAEs) [5,12,17,40] estimate the distribution over the latent space that is regularized to be similar to a prior distribution, usually a standard isotropic Gaussian. Generative adversarial networks have also been applied to anomaly detection [24,33]. Pidhorskyi et al. [26] trained AEs with an adversarial loss to detect OoD samples. More recently, introspective variational AEs [8] use the VAE encoder to differentiate between real and reconstructed samples, achieving SOTA image generations and outlier detection performance. Recently, Zhou et al. [39]

investigated the limitations of AEs for OoD. Similarly, we believe that AEs should have two properties: i) *minimality*: the networks should be constrained to only reconstruct ID samples and ii) *sufficiency*: the decoder should have sufficient capacity to reconstruct ID samples with high accuracy. In contrast to [39], where the authors aim at reconstructing only low-dimensional features needed for the classification task, we are interested in reconstructing pseudo-healthy images to enable pixel-wise localization of anomalies.

In this work, we first investigate whether SOTA AEs can learn meaningful representations for anomaly detection. Specifically, we investigate whether AEs can learn the healthy anatomy, i.e., absence of pathology, and generate pseudo-healthy reconstructions of abnormal samples on challenging medical anomaly detection tasks. Our findings are that SOTA AEs either do not efficiently constrain the latent space and allow the reconstruction of anomalous patterns, or that the decoder cannot accurately restore images from their latent representation. The imperfect reconstructions yield high residual errors on normal regions (false positives) that can easily overshadow residuals of interest, i.e., pathology [23]. We then propose *MorphAEus*, novel deformable AEs to learn minimal and sufficient features for anomaly detection and drastically reduce false positives. Figure 1a shows that *MorphAEus* learns the training distribution of healthy chest X-rays and yields pseudo-healthy ID reconstructions even for far OoD samples. This allows to localize pathologies, as seen in Fig. 1b.

Our manuscript advances the understanding of anomaly detection by providing insights into what AEs learn. In summary, our contributions are:

- We broaden the understanding of AEs and highlight their limitations.
- We test whether SOTA AEs can learn the training distribution of the healthy population, accurately reconstruct inputs from their latent representation and reliably detect anomalies.
- As a solution, we propose *MorphAEus*, novel deformable AEs that provide pseudo-healthy reconstructions of abnormal samples and drastically reduce false positives, achieving SOTA unsupervised pathology detection.

2 Background

The widely held popular belief is that AEs can learn the distribution of the training data and identify outliers from inaccurate reconstructions of abnormal samples [31]. This section aims to discuss the common assumptions of unsupervised anomaly detection, specifically for AEs, while also outlining the challenges and desired properties associated with these techniques.

2.1 Unsupervised Anomaly Detection: Assumptions

Let $\mathcal{X} \subset \mathbb{R}^N$ be the data space that describes normal instances for a given task. The manifold hypothesis implies that there exists a low-dimensional manifold $\mathcal{M} \subset \mathbb{R}^D \subset \mathcal{X}$ where all the points $x \in \mathcal{X}$ lie, with $D \ll N$ [9]. For example, a

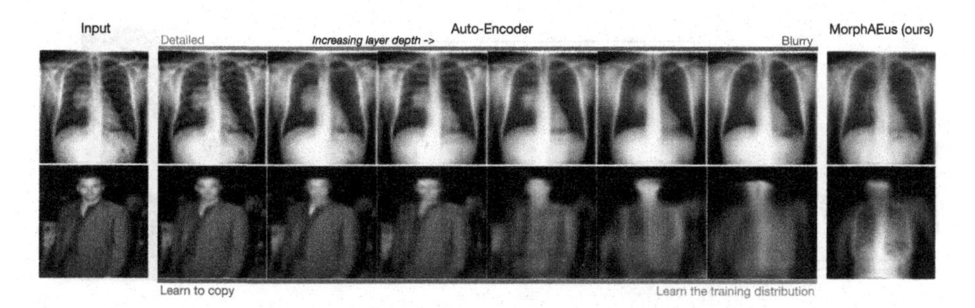

Fig. 2. What do AEs learn? Reconstruction of pathology (top row) and far OoD celebrities (bottom row) with AEs [2] (different depths) and *MorphAEus*. While shallow AEs learn to copy, deep AEs yield blurry reconstructions, rendering AEs unsuitable for anomaly detection. In contrast, *MorphAEus* generates pseudo-healthy ID reconstructions for near (pathology) and far (celebrity) OoD cases.

set of images in pixel space $\mathcal{X}$ could have a compact representation describing features like structure, shape, or orientation in $\mathcal{M}$.

Given a set of unlabeled data $x_1, .., x_n \in \mathcal{X}$ the objective of unsupervised representation learning is to find a function $f : \mathbb{R}^N \rightarrow \mathbb{R}^D$ and its inverse $g : \mathbb{R}^D \rightarrow \mathbb{R}^N$, such that $x \approx g(f(x))$, with the mapping f defining the low-dimensional manifold $\mathcal{M}$. The core assumption of unsupervised anomaly detection is that once such functions f and g are found, the learned manifold $\mathcal{M}$ would best describe the normal data samples in $\mathcal{X}$ and results in high reconstruction errors for data-points $\overline{x} \notin \mathcal{X}$, that we call anomalous. An anomaly score is therefore usually derived directly from the pixel-wise difference: $s(\overline{x}) = |\overline{x} - g(f(\overline{x}))|$.

The nominal and abnormal distributions are considerably separated from each other when $\overline{x}$ is from a different domain. However, anomalies are often defects in otherwise normal images. In medical imaging, the set $\mathcal{X}$ describes the healthy anatomy and the data set $\overline{\mathcal{X}}$ usually contains images with both healthy and pathological regions. The two distributions usually come from the same domain and might overlap considerably. The core assumption is that only the normal structures can be reconstructed from their latent representation very well, with the pathological regions ideally replaced by healthy structures. Therefore $x \approx g(f(\overline{x})) \in \mathcal{X}$ would represent the healthy synthesis of the abnormal sample $\overline{x}$ and the residual $|\overline{x} - g(f(\overline{x}))|$ would highlight only the abnormal regions.

2.2 Auto-Encoders: Challenges

AEs aim to extract meaningful representations from data, by learning to compress inputs to a lower-dimensional manifold and reconstruct them with minimal error. They use neural networks to learn the functions f and g, often denoted as encoder E_θ with parameters θ and decoder D_ϕ parameterized by a set of parameters ϕ. The embedding $z = E(x|\theta)$ is a projection of the input to a lower-dimensional manifold $\mathcal{Z}$, also referred to as the bottleneck or latent representation of x. The standard objective of AEs is finding the set of parameters θ

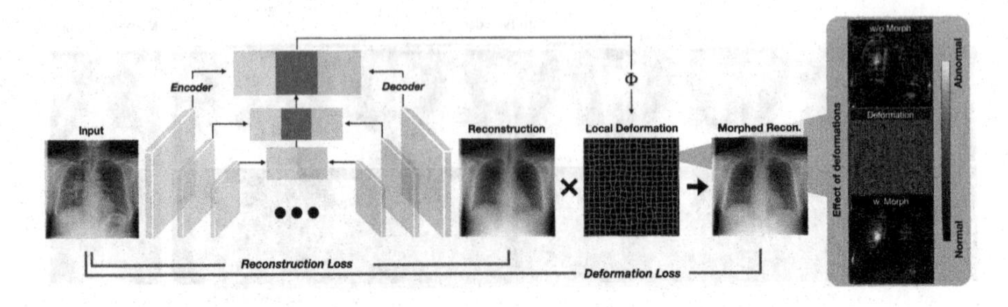

Fig. 3. MorphAEus. Deep perceptual AEs yield reconstructions matching the training distribution. We leverage the top encoding/decoding layers containing spatial information to estimate deep deformation fields. This allows the local morphometric adaptation of the predicted reconstruction to drastically reduce residual errors on normal regions (false positives), as shown on the right.

and ϕ that minimize the residual, with the mean squared error (MSE) being a popular choice for the reconstruction error: $\min_{\theta,\phi} \sum_{i=1}^{N} \|x_i - D_\phi(E_\theta(x_i))\|^2$.

In the introduction, we presented the desired properties of AEs for outlier detection, namely i) reconstructions should match the training distribution and ii) decoders have sufficient capacity to accurately restore inputs. Figure 2 shows reconstructions of spatial AEs [2] for near OoD samples, i.e., containing pathologies, and far OoD samples using real images of celebrities [19]. AEs with few encoding layers learn to copy and can reconstruct both near- and far OoD. Interestingly, with increasing layer depth, AEs can learn the prior over the training distribution, avoid the reconstruction of pathologies, and project OoD celebrities to the closest chest X-ray counterparts. However, this comes at the cost of losing spatial information and not reconstructing small details, e.g., ribs. The resulting high residual error on healthy tissues (false positives) overshadows the error on pathology [23], rendering standard AEs unsuitable for anomaly detection.

3 MorphAEus: Deformable Auto-encoders

We propose *MorphAEus*, deformable AEs that learn minimal and sufficient features for anomaly detection, see Fig. 3. We use deep perceptual AE to provide pseudo-healthy ID reconstructions and leverage estimated deep deformation fields to drastically reduce the false positives.

Pseudo-healthy Reconstructions. Given a dataset $\mathcal{X} = \{x_1, .., x_n\}$ we optimize the encoder and decoder with parameters θ, ϕ to minimize the MSE loss between the input and its reconstruction. For *minimality*, we propose to use deep AEs constrained to reconstruct only ID samples (see Fig. 2), but add a perceptual loss (PL) [14,37] to encourage reconstructions that are perceptually similar to the training distribution. The reconstruction loss is given by:

$$\mathcal{L}_{Rec}(x|\theta;\phi) = MSE(x, x_{rec}) + \alpha PL(x, x_{rec}), \quad (1)$$

with $x_{rec} = D_\phi(E_\theta(x))$, $PL(x, x_{rec}) = \sum_l (VGG_l(x) - VGG_l(x_{rec}))^2$ with VGG_l being the output of the $l \in \{1, 6, 11, 20\}$-th layer of a pre-trained VGG-19 encoder. We have empirically found $\alpha = 0.05$ to be a good weight to predict perceptually similar reconstructions without compromising pixel-wise accuracy.

Local Deformation. Imperfect reconstructions yield high residuals on normal regions which might overshadow the residuals errors associated with anomalous regions. Skip connections [30] would allow the network to bypass the learned manifold and copy anomalies at inference. Instead, we propose to align the reconstruction with the input using corresponding encoder and decoder features. We denote these shared parameters as $\theta_S \subset \theta$ and $\phi_S \subset \phi$. Inspired by the advances in image registration [1] and computer vision [4,36] we estimate dense deformation fields Φ to allow local morphometric adaptations:

$$\mathcal{L}_{Morph}(x|\psi; \theta_S; \phi_S) = 1 - LNCC(x_{morph}, x) + \beta \|\Phi\|^2, \tag{2}$$

where $x_{morph} = x_{rec} \circ \Phi$, ψ are the deformation parameters, LNCC is the local normalized cross correlation, $\circ$ is a spatial transformer and β weights the smoothness constraint on the deformation fields. We opted for the LNCC instead of the MSE to emphasize shape registration and enhance robustness to intensity variations in the inputs and reconstructions. By sharing encoder/decoder parameters, the deformation maps are not only beneficial at inference time, but also guide the training process to learn more accurate features. The full objective is given by the two losses: $\mathcal{L}(x|\theta; \phi; \psi) = \mathcal{L}_{Rec}(x|\theta; \phi) + \mathcal{L}_{Morph}(x|\psi; \theta_S; \phi_S)$. To ensure a good initialization for the deformation estimation, we introduce the deformation loss after 10 epochs. Deformable registration between normal and pathological samples is in itself an active area of research [18,25]. In particular, the deformation could mask structural abnormalities at inference if not constrained. In our experiments, we linearly increase β from $1e^{-3}$ to 3 to constrain the deformation more as the reconstruction improves (see Appendix for details). Nevertheless, recent advances allow the estimation of the optimal registration parameters automatically [13]. If not specified otherwise, we employ x_{morph} for inference.

4 Pathology Detection on Chest X-rays

In this section, we investigate whether AEs can learn the healthy anatomy, i.e., absence of pathology, and generate pseudo-healthy reconstructions of abnormal chest X-ray images. Pathology detection algorithms are often applied to finding hyper-intense lesions, such as tumors or multiple sclerosis on brain scans. However, it has been shown that thresholding techniques can outperform learning-based methods [22]. In contrast, the detection of pathology on chest radiographs is much more difficult due to the high variability and complexity of nominal features and the diversity and irregularity of abnormalities.

Datasets. We use the Covid-19 radiography database on Kaggle [6,27]. We used the RSNA dataset [35], which contains 10K CXR images of normal subjects and 6K lung opacity cases. For the detection of Covid-19, we used the Padchest

Table 1. CXR Pathology Detection. Methods marked with '*' achieve improved reconstruction accuracy, but can also reconstruct pathologies which renders them not useful for detecting anomalies, as shown in Fig. 4. In contrast, *MorphAEus* generates only ID reconstructions which enables the localization of pathologies.

Method	Healthy		Pneumonia	Covid-19	Avg.
	SSIM ↑	LPIPS ↓	AUROC ↑		
AE-S [2]	**95.9***	**1.3***	58.9 ± 0.5	59.8 ± 1.6	59.3
AE-D [2]	75.6	38.3	45.4 ± 0.5	56.1 ± 1.1	50.7
VAE [40]	72.9	38.4	33.5 ± 0.7	43.7 ± 0.7	38.6
β-VAE [12]	63.1	43.9	70.8 ± 0.9	54.5 ± 0.8	62.6
AAE [26]	66.3	15.5	58.4 ± 0.6	59.8 ± 1.2	57.8
SI-VAE [8]	71.3	17.6	49.7 ± 0.7	52.0 ± 0.7	50.9
DAE [16]	**95.8***	**2.8***	78.0 ± 0.5	76.1 ± 0.7	77.0
MorphAEus (ours)	85.7	9.5	$\mathbf{83.6 \pm 0.8}$	$\mathbf{86.0 \pm 0.7}$	**84.8**

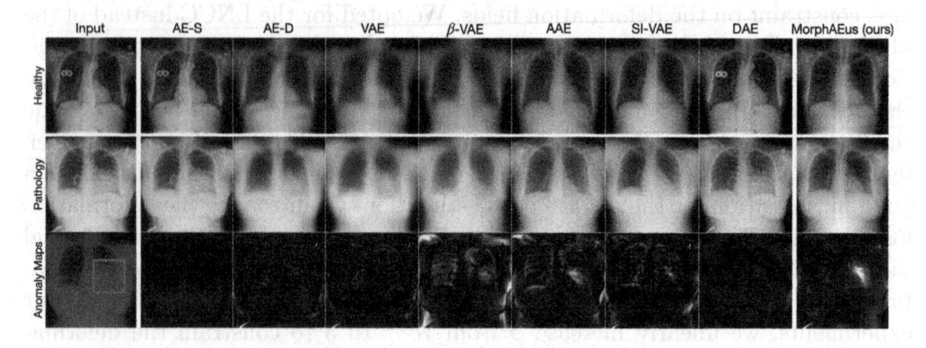

Fig. 4. The first row shows reconstructions of a healthy sample. The next rows show reconstructions of a pathological case and the corresponding anomaly maps. AE-S and DAE fully reconstruct the pathology and thereby fail the task, while the other baselines yield high residual errors on both healthy and pathological regions due to blurry (AE-D, VAE, and β-VAE) or inaccurate (AAE, SI-VAE) reconstructions. Only *MorphAEus* yields accurate pseudo-healthy reconstructions which allows to localize the pathology on the left lung.

dataset [3] containing CXR images manually annotated by trained radiologists. We used 1.3K healthy control images and 2.5K cases of Covid-19.

Results. Of the baselines, only adversarially-trained AEs can reconstruct pseudo-healthy images from abnormal samples, as shown in Fig. 4. However, their imperfect reconstructions overshadow the error on pathology leading to poor anomaly detection results, as reflected in the SSIM and AUROC in Table 1. Spatial AEs and DAEs have a tendency to reproduce the input and produce reconstructions of structures that are not included in the training distribution,

(a) Morphometric adaptations (M) considerably increase the reconstruction accuracy on healthy (H) and the pathology detection scores (AUROC).

	w/o PL		w. PL	
	w/o M	w. M	w/o M	w. M
H. (SSIM) ↑	80.2	**87.9**	72.8	85.7
H. (LPIPS) ↓	22.6	13.0	**8.3**	9.5
Pathology ↑	62.1	82.1	66.8	**84.8**

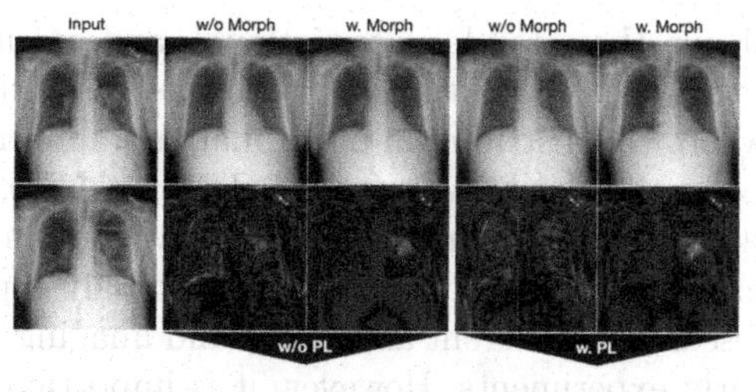

(b) Morphometric adaptations (Morph) reduce false positives and highlight pathological areas.

Fig. 5. Ablation study: Morphometric adaptations considerably improve the pathology detection accuracy with or without perceptual loss (PL).

such as medical devices and pathologies, despite not being trained on OoD data. This can lead to false negatives. DAEs can avoid the reconstruction of some pathologies and achieve good anomaly detection results. However, pathological regions that do not conform to the learned noise model are completely missed, as shown in Fig. 4. The next three methods (AE-D, VAE, and β-VAE) produce blurry reconstructions, as it can be best seen in the LPIPS score. *MorphAEus* is the only method to yield accurate pseudo-healthy reconstructions and effectively remove anomalies, such as pathology or implanted medical devices. This enables to precisely localize pathologies, considerably outperforming the baselines.

Ablation Study: Importance of Morphological Adaptations. We evaluate the effectiveness of individual components of *MorphAEus* in Fig. 5. AEs without perceptual loss tend to not reconstruct small but important features such as ribs and yield false positives on healthy tissue. Interestingly, AEs with perceptual loss achieve more visually appealing reconstructions, but fail at detecting anomalies because the pathological region is overshadowed by false positive residuals on edges and misaligned ribs. Morphometric adaptations guide the networks to learn better representations, reduce the number of false positives, and enable the localization of pathologies. This considerably improves the detection results to 84.8 and 82.1 for AEs with and without perceptual loss, respectively. It is important to note that the deformations are not only beneficial at the time of inference, but also drive the learning process towards better representations. Thereby, the average pathology detection increases from 66.8 to 80.2, even if no adaptations are made during inference, i.e., using x_{rec} instead of x_{morph} for inference (see Appendix for details).

5 Discussion

In this work, we have investigated whether AEs learn meaningful representations to solve pathology detection tasks. We stipulate that AEs should have the desired property of learning the normative distribution (*minimality*) and producing highly accurate reconstructions of ID samples (*sufficiency*). We have shown that standard, variational, and recent adversarial AEs generally do not satisfy both conditions, nor are they very suitable for pathology detection tasks where the distribution of normal and abnormal instances highly overlap.

In this paper, we introduced *MorphAEus*, a novel deformable AE that demonstrated notable performance improvement in detecting pathology. We believe our method is adaptable to various anomaly types, and we are eager to extend our research to different anatomies and imaging modalities, building upon promising early experiments. However, it is important to address false positive detection, which could be influenced by unlabelled artifacts like medical devices. Our future work aims to conduct a thorough analysis of false positives and explore strategies to mitigate their impact, ultimately enhancing the accuracy.

Although there are obstacles to overcome, AEs remain a viable option for producing easily understandable and interpretable outcomes. Nevertheless, it is crucial to continue improving the quality of the representations to advance unsupervised anomaly detection. Our findings demonstrate that *MorphAEus* is capable of learning superior representations, and can leverage the predicted dense displacement fields to refine its predictions and minimize the occurrence of false positives. This allows for accurate identification and localization of diseases, resulting in SOTA unsupervised pathology detection on chest X-rays.

Acknowledgement. C.I.B. is in part supported by the Helmholtz Association under the joint research school "Munich School for Data Science - MUDS".

References

1. Balakrishnan, G., Zhao, A., Sabuncu, M.R., Guttag, J., Dalca, A.V.: VoxelMorph: a learning framework for deformable medical image registration. IEEE Trans. Med. Imaging **38**(8), 1788–1800 (2019)
2. Baur, C., Denner, S., Wiestler, B., Navab, N., Albarqouni, S.: Autoencoders for unsupervised anomaly segmentation in brain MR images: a comparative study. Med. Image Anal. **69**, 101952 (2021)
3. Bustos, A., Pertusa, A., Salinas, J.M., de la Iglesia-Vayá, M.: Padchest: a large chest x-ray image dataset with multi-label annotated reports. Med. Image Anal. **66**, 101797 (2020)
4. Chen, R., Cong, Y., Dong, J.: Unsupervised dense deformation embedding network for template-free shape correspondence. In: ICCV, pp. 8361–8370 (2021)
5. Chen, X., You, S., Tezcan, K.C., Konukoglu, E.: Unsupervised lesion detection via image restoration with a normative prior. Med. Image Anal. **64**, 101713 (2020)
6. Chowdhury, M.E.H., et al.: Can AI help in screening viral and COVID-19 pneumonia? IEEE Access **8**, 132665–132676 (2020)
7. Ciresan, D., Giusti, A., Gambardella, L., Schmidhuber, J.: Deep neural networks segment neuronal membranes in electron microscopy images. In: NeurIPS 25 (2012)
8. Daniel, T., Tamar, A.: Soft-introvae: analyzing and improving the introspective variational autoencoder. In: CVPR, pp. 4391–4400 (2021)
9. Fefferman, C., Mitter, S., Narayanan, H.: Testing the manifold hypothesis. J. Am. Math. Soc. **29**(4), 983–1049 (2016)
10. Golan, I., El-Yaniv, R.: Deep anomaly detection using geometric transformations. In: NeurIPS vol. 31 (2018)

11. Gong, D., et al.: Memorizing normality to detect anomaly: memory-augmented deep autoencoder for unsupervised anomaly detection. In: ICCV, pp. 1705–1714 (2019)
12. Higgins, I., et al.: beta-VAE: learning basic visual concepts with a constrained variational framework. In: ICLR (2017)
13. Hoopes, A., Hoffmann, M., Fischl, B., Guttag, J., Dalca, A.V.: Hypermorph: amortized hyperparameter learning for image registration. In: IPMI, pp. 3–17 (2021)
14. Johnson, J., Alahi, A., Fei-Fei, L.: Perceptual losses for real-time style transfer and super-resolution. In: ECCV, pp. 694–711 (2016)
15. Kamnitsas, K., et al.: Efficient multi-scale 3D CNN with fully connected CRF for accurate brain lesion segmentation. Med. Image Anal. **36**, 61–78 (2017)
16. Kascenas, A., Pugeault, N., O'Neil, A.Q.: Denoising autoencoders for unsupervised anomaly detection in brain MRI. In: MIDL (2022)
17. Kingma, D.P., Welling, M.: Auto-encoding variational Bayes. arXiv preprint arXiv:1312.6114 (2013)
18. Krüger, J., Schultz, S., Handels, H., Ehrhardt, J.: Registration with probabilistic correspondences-accurate and robust registration for pathological and inhomogeneous medical data. CVIU **190**, 102839 (2020)
19. Liu, Z., Luo, P., Wang, X., Tang, X.: Deep learning face attributes in the wild. In: ICCV (2015)
20. Makhzani, A., Frey, B.: k-sparse autoencoders. In: ICLR (2014)
21. Mao, X., Shen, C., Yang, Y.B.: Image restoration using very deep convolutional encoder-decoder networks with symmetric skip connections. In: NeurIPS, vol. 29 (2016)
22. Meissen, F., Kaissis, G., Rueckert, D.: Challenging current semi-supervised anomaly segmentation methods for brain MRI. In: MICCAI brainlesion workshop, pp. 63–74 (2022)
23. Meissen, F., Wiestler, B., Kaissis, G., Rueckert, D.: On the pitfalls of using the residual as anomaly score. In: MIDL (2022)
24. Perera, P., Nallapati, R., Xiang, B.: Ocgan: one-class novelty detection using GANs with constrained latent representations. In: CVPR, pp. 2898–2906 (2019)
25. Periaswamy, S., Farid, H.: Medical image registration with partial data. Med. Image Anal. **10**(3), 452–464 (2006)
26. Pidhorskyi, S., Almohsen, R., Doretto, G.: Generative probabilistic novelty detection with adversarial autoencoders. In: NeurIPS, vol. 31 (2018)
27. Rahman, T., et al.: Exploring the effect of image enhancement techniques on COVID-19 detection using chest X-rays images (2020)
28. Ren, J., et al.: Likelihood ratios for out-of-distribution detection. In: NeurIPS, vol. 32 (2019)
29. Rifai, S., Vincent, P., Muller, X., Glorot, X., Bengio, Y.: Contractive auto-encoders: explicit invariance during feature extraction. In: ICML, pp. 833–840 (2011)
30. Ronneberger, O., Fischer, P., Brox, T.: U-net: convolutional networks for biomedical image segmentation. In: MICCAI, pp. 234–241 (2015)
31. Ruff, L., Kauffmann, J.R., et al.: A unifying review of deep and shallow anomaly detection. In: Proceeding of IEEE (2021)
32. Schirrmeister, R., Zhou, Y., Ball, T., Zhang, D.: Understanding anomaly detection with deep invertible networks through hierarchies of distributions and features. NeurIPS **33**, 21038–21049 (2020)
33. Schlegl, T., Seeböck, P., Waldstein, S.M., Langs, G., Schmidt-Erfurth, U.: f-AnoGAN: fast unsupervised anomaly detection with generative adversarial networks. Med. Image Anal. **54**, 30–44 (2019)

34. Schlemper, J., Caballero, J., Hajnal, J.V., Price, A.N., Rueckert, D.: A deep cascade of convolutional neural networks for dynamic MR image reconstruction. IEEE Trans. Med. Imaging **37**(2), 491–503 (2017)
35. Shih, G., et al.: Augmenting the national institutes of health chest radiograph dataset with expert annotations of possible pneumonia. Radiol. Artif. Intell. **1**(1), e180041 (2019)
36. Shu, Z., Sahasrabudhe, M., Guler, R.A., Samaras, D., Paragios, N., Kokkinos, I.: Deforming autoencoders: unsupervised disentangling of shape and appearance. In: ECCV, pp. 650–665 (2018)
37. Tuluptceva, N., Bakker, B., Fedulova, I., Konushin, A.: Perceptual image anomaly detection. In: ASPR, pp. 164–178 (2019)
38. Yoon, S., Noh, Y.K., Park, F.: Autoencoding under normalization constraints. In: ICML, pp. 12087–12097 (2021)
39. Zhou, Y.: Rethinking reconstruction autoencoder-based out-of-distribution detection. arXiv preprint arXiv:2203.02194 (2022)
40. Zimmerer, D., Isensee, F., Petersen, J., Kohl, S., Maier-Hein, K.: Unsupervised anomaly localization using variational auto-encoders. In: MICCAI (2019)

SwIPE: Efficient and Robust Medical Image Segmentation with Implicit Patch Embeddings

Yejia Zhang(✉), Pengfei Gu, Nishchal Sapkota, and Danny Z. Chen

University of Notre Dame, Notre Dame, IN 46556, USA
{yzhang46,pgu,nsapkota,dchen}@nd.edu

Abstract. Modern medical image segmentation methods primarily use *discrete* representations in the form of rasterized masks to learn features and generate predictions. Although effective, this paradigm is spatially inflexible, scales poorly to higher-resolution images, and lacks direct understanding of object shapes. To address these limitations, some recent works utilized implicit neural representations (INRs) to learn *continuous* representations for segmentation. However, these methods often directly adopted components designed for 3D shape reconstruction. More importantly, these formulations were also constrained to either point-based or global contexts, lacking contextual understanding or local fine-grained details, respectively—both critical for accurate segmentation. To remedy this, we propose a novel approach, **SwIPE** (Segmentation with Implicit Patch Embeddings), that leverages the advantages of INRs and predicts shapes at the patch level—rather than at the point level or image level—to enable both accurate local boundary delineation and global shape coherence. Extensive evaluations on two tasks (2D polyp segmentation and 3D abdominal organ segmentation) show that SwIPE significantly improves over recent implicit approaches and outperforms state-of-the-art discrete methods with over 10x fewer parameters. Our method also demonstrates superior data efficiency and improved robustness to data shifts across image resolutions and datasets. Code is available on Github.

Keywords: Medical Image Segmentation · Deep Implicit Shape Representations · Patch Embeddings · Implicit Shape Regularization

1 Introduction

Segmentation is a critical task in medical image analysis. Known approaches mainly utilize *discrete* data representations (e.g., rasterized label masks) with convolutional neural networks (CNNs) [6,8,12,26] or Transformers [9,10] to classify image entities in a bottom-up manner. While undeniably effective, this paradigm suffers from *two primary limitations.* (1) These approaches have limited spatial flexibility and poor computational scaling. Retrieving predictions at

Supplementary Information The online version contains supplementary material available at https://doi.org/10.1007/978-3-031-43904-9_31.

H. Greenspan et al. (Eds.): MICCAI 2023, LNCS 14224, pp. 315–326, 2023.
https://doi.org/10.1007/978-3-031-43904-9_31

higher resolutions would require either increasing the input size, which decreases performance and incurs quadratic or cubic memory increases, or interpolating output predictions, which introduces discretization artifacts. (2) Per-pixel or voxel learning inadequately models object shapes/boundaries, which are central to both robust computer vision methods and our own visual cortical pathways [23]. This often results in predictions with unrealistic object shapes and locations [24], especially in settings with limited annotations and out-of-distribution data.

Instead of segmenting structures with *discrete* grids, we explore the use of Implicit Neural Representations (INRs) which employ *continuous* representations to compactly capture coordinate-based signals (e.g., objects in images). INRs represent object shapes with a parameterized function $f_\theta : (\mathbf{p}, \mathbf{z}) \rightarrow [0, 1]$ that maps continuous spatial coordinates $\mathbf{p} = (x, y, z)$, $x, y, z \in [-1, 1]$ and a shape embedding vector $\mathbf{z}$ to occupancy scores. This formulation enables direct modeling of object contours as the decision boundary of f_θ, superior memory efficiency [5], and smooth predictions at arbitrary resolutions that are invariant to input size. INRs have been adopted in the vision community for shape reconstruction [3,4,19,22], texture synthesis [21], novel view synthesis [20], and segmentation [11]. Medical imaging studies have also used INRs to learn organ templates [31], synthesize cell shapes [29], and reconstruct radiology images [27].

The adoption of INRs for medical image segmentation, however, has been limited where most existing approaches directly apply pipelines designed for 3D reconstruction to images. These works emphasize either global embeddings $\mathbf{z}$ or point-wise ones. OSSNet [25] encodes a global embedding from an entire volume and an auxiliary local image patch to guide voxel-wise occupancy prediction. Although global shape embeddings facilitate overall shape coherence, they neglect the fine-grained details needed to delineate local boundaries. The local patches partially address this issue but lack contextual understanding beyond the patches and neglect mid-scale information. In an effort to enhance local acuity and contextual modeling, IFA [11], IOSNet [16], and NUDF [28] each extract a separate embedding for every input coordinate by concatenating point-wise features from multi-scale CNN feature maps. Although more expressive, point-wise features still lack sufficient global contextual understanding and suffer from the same unconstrained prediction issues observed in discrete segmentation methods. Moreover, these methods use components designed for shape reconstruction—a domain where synthetic data is abundant and the modeling of texture, multi-class discrimination, and multi-scale contexts are less crucial.

To address these limitations, we propose **SwIPE (Segmentation with Implicit Patch Embeddings)** which learns continuous representations of foreground shapes at the patch level. By decomposing objects into parts (i.e., patches), we aim to enable both accurate local boundary delineation and global shape coherence. This also improves model generalizability and training efficiency since local curvatures often reoccur across classes or images. SwIPE first encodes an image into descriptive patch embeddings and then decodes the point-wise occupancies using these embeddings.

To avoid polarization of patch embeddings toward either local or global features in the encoding step, we introduce a context aggregation mechanism that fuses multi-scale feature maps and propose a **Multi-stage Embedding Attention (MEA)** module to dynamically extract relevant features from all scales. This is driven by the insight that different object parts necessitate variable focus on either global/abstract (important for object interiors) or local/fine-grained information (essential around object boundaries). To enhance global shape coherence across patches in the decoding step, we augment local embeddings with global information and propose **Stochastic Patch Overreach (SPO)** to improve continuity around patch boundaries. Comprehensive evaluations are conducted on two tasks (2D polyp and 3D abdominal organ segmentation) across four datasets. SwIPE outperforms the best-known implicit methods (+6.7% & +4.5% F1 on polyp and abdominal, resp.) and beats task-specific discrete approaches (+2.5% F1 on polyp) with 10x fewer parameters. We also demonstrate SwIPE's superior model & data efficiency in terms of network size & annotation budgets, and greater robustness to data shifts across image resolutions and datasets. Our main **contributions** are as follows.

1. Away from discrete representations, we are the first to showcase the merits of patch-based implicit neural representations for medical image segmentation.
2. We propose a new efficient attention mechanism, Multi-stage Embedding Attention (MEA), to improve contextual understanding during the encoding step, and Stochastic Patch Overreach (SPO) to address boundary continuities during occupancy decoding.
3. We perform detailed evaluations of SwIPE and its components on two tasks (2D polyp segmentation and 3D abdominal organ segmentation). We not only outperform state-of-the-art implicit and discrete methods, but also yield improved data & model efficiency and better robustness to data shifts.

2 Methodology

The core idea of SwIPE (overviewed in Fig. 1) is to use patch-wise INRs for semantic segmentation. To formulate this, we first discuss the shift from discrete to implicit segmentation, then delineate the intermediate representations needed for such segmentation, and overview the major components involved in obtaining these representations. Note that for the remainder of the paper, we present formulations for 2D data but the descriptions are conceptually congruous in 3D.

In a typical discrete segmentation setting with C classes, an input image $\mathbf{X}$ is mapped to class probabilities with the same resolution $f : \mathbf{X} \in \mathbb{R}^{H \times W \times 3} \rightarrow \hat{\mathbf{Y}} \in \mathbb{R}^{H \times W \times C}$. Segmentation with INRs, on the other hand, maps an image $\mathbf{X}$ and a continuous image coordinate $\mathbf{p}_i = (x, y)$, $x, y \in [-1, 1]$, to the coordinate's class-wise occupancy probability $\hat{\mathbf{o}}_i \in \mathbb{R}^C$, yielding $f_\theta : (\mathbf{p}_i, \mathbf{X}) \rightarrow \hat{\mathbf{o}}_i$, where f_θ is parameterized by a neural network with weights θ. As a result, predictions of arbitrary resolutions can be obtained by modulating the spatial granularity of the input coordinates. This formulation also enables the direct use of discrete

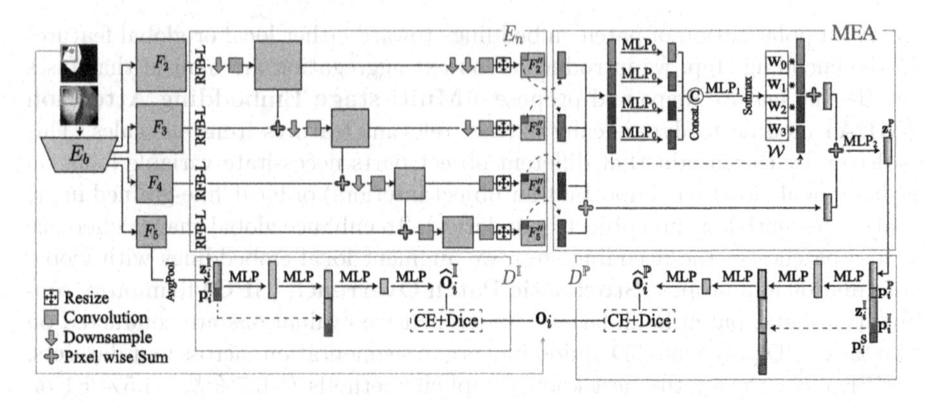

Fig. 1. At a high level, SwIPE first encodes an input image into patch $\mathbf{z}^{\mathbb{P}}$ and image $\mathbf{z}^{\mathbb{I}}$ shape embeddings, and then employs these embeddings along with coordinate information $\mathbf{p}$ to predict class occupancy scores via the patch $\mathbf{D}^{\mathbb{P}}$ and image $\mathbf{D}^{\mathbb{I}}$ decoders.

pixel-wise losses like Cross Entropy or Dice with the added benefit of boundary modeling. Object boundaries are represented as the zero-isosurface in f_θ's prediction space or, more elegantly, f_θ's decision boundary.

SwIPE builds on the INR segmentation setting (e.g., in [16]), but operates on patches rather than on points or global embeddings (see Table 1 & left of Table 3 for empirical justifications) to better enable both local boundary details and global shape coherence. This involves two main steps: (1) encode shape embeddings from an image, and (2) decode occupancies for each point while conditioning on its corresponding embedding(s). In our case, f_θ includes an encoder E_b (or backbone) that extracts multi-scale feature maps from an input image, a context aggregation module E_n (or neck) that aggregates the feature maps into vector embeddings for each patch, and MLP decoders $D^{\mathbb{P}}$ (decoder for local patches where $\mathbb{P}$ is for patch) & $D^{\mathbb{I}}$ (decoder for entire images where $\mathbb{I}$ is for image) that output smoothly-varying occupancy predictions given embedding & coordinate pairs. To encode patch embeddings in step (1), E_b and E_n map an input image $\mathbf{X}$ to a global image embedding $\mathbf{z}^{\mathbb{I}}$ and a matrix $\mathbf{Z}^{\mathbb{P}}$ containing a local patch embedding $\mathbf{z}^{\mathbb{P}}$ at each planar position. For occupancy decoding in step (2), $D^{\mathbb{P}}$ decodes the patch-wise class occupancies $\mathbf{o}_i^{\mathbb{P}}$ using relevant local and global inputs while $D^{\mathbb{I}}$ predicts occupancies $\mathbf{o}_i^{\mathbb{I}}$ for the entire image using only image coordinates $\mathbf{p}_i^{\mathbb{I}}$ and the image embedding $\mathbf{z}^{\mathbb{I}}$. Below, we detail the encoding of image & patch embeddings (Sect. §2.1), the point-wise decoding process (Sec. §2.2), and the training procedure for SwIPE (Sect. §2.3).

2.1 Image Encoding and Patch Embeddings

The encoding process utilizes the backbone E_b and neck E_n to obtain a global image embedding $\mathbf{z}^{\mathbb{I}}$ and a matrix $\mathbf{Z}^{\mathbb{P}}$ of patch embeddings. We define an image patch as an isotropic grid cell (i.e., a square in 2D or a cube with identical spacing in 3D) of length S from non-overlapping grid cells over an image. Thus, an image $\mathbf{X} \in \mathbb{R}^{H \times W \times 3}$ with a patch size S will produce $\lceil \frac{H}{S} \rceil \cdot \lceil \frac{W}{S} \rceil$ patches. For simplicity, we assume that the image dimensions are evenly divisible by S.

A fully convolutional **encoder backbone** E_b (e.g., Res2Net-50 [7]) is employed to generate multi-scale features from image $\mathbf{X}$. The entire image is processed as opposed to individual crops [3,15,25] to leverage larger receptive fields and integrate intra-patch information. Transformers [9] also model cross-patch relations and naturally operate on patch embeddings, but are data-hungry and lack helpful spatial inductive biases (we affirm this in Sec. §3.5). E_b outputs four multi-scale feature maps from the last four stages, $\{\mathbf{F}_n\}_{n=2}^{5}$ ($\mathbf{F}_n \in \mathbb{R}^{C_n \times H_n \times W_n}$, $H_n = \frac{H}{2^n}$, $W_n = \frac{W}{2^n}$).

The **encoder neck** E_n aggregates E_b's multi-scale outputs $\{\mathbf{F}_n\}_{n=2}^{5}$ to produce $\mathbf{z}^{\mathbb{I}}$ (the shape embedding for the entire image) and $\mathbf{Z}^{\mathbb{P}}$ (the grid of shape embeddings for patches). The feature maps are initially fed into a modified Receptive Field Block [17] (dubbed RFB-L or RFB-Lite) that replaces symmetric convolutions with a series of efficient asymmetric convolutions (e.g., $(3 \times 3) \rightarrow (3 \times 1) + (1 \times 3)$). The context-enriched feature maps are then fed through multiple cascaded aggregation and downsampling operations (see E_n in Fig. 1) to obtain four multi-stage intermediate embeddings with identical shapes, $\{\mathbf{F}'_n\}_{n=2}^{5} \in \mathbb{R}^{\frac{H}{32} \times \frac{W}{32} \times d}$.

To convert the intermediate embeddings $\{\mathbf{F}'_n\}_{n=2}^{5}$ to patch embeddings $\mathbf{Z}^{\mathbb{P}}$, we first resize them to $\mathbf{Z}^{\mathbb{P}}$'s final shape via linear interpolation to produce $\{\mathbf{F}''_n\}_{n=2}^{5}$, which contain low-level ($\mathbf{F}''_2$) to high-level ($\mathbf{F}''_5$) information. Resizing enables flexibility in designing appropriate patch coverage, which may differ across tasks due to varying structure sizes and shape complexities. Note that this is different from the interpolative sampling in [16] and more similar to [11], except the embeddings' spatial coverage in SwIPE are larger and adjustable. To prevent the polarization of embeddings toward either local or global scopes, we propose a **multi-stage embedding attention** (MEA) module to enhance representational power and enable dynamic focus on the most relevant abstraction level for each patch. Given four intermediate embedding vectors $\{\mathbf{e}_n\}_{n=2}^{5}$ from corresponding positions in $\{\mathbf{F}''_n\}_{n=2}^{5}$, we compute the attention weights via $\mathcal{W} = Softmax(MLP_1(cat(MLP_0(\mathbf{e}_2), MLP_0(\mathbf{e}_3), MLP_0(\mathbf{e}_4), MLP_0(\mathbf{e}_5))))$, where $\mathcal{W} \in \mathbb{R}^4$ is a weight vector, *cat* indicates concatenation, and MLP_0 is followed by a ReLU activation. The final patch embedding is obtained by $\mathbf{z}^{\mathbb{P}} = MLP_2(\sum_{n=2}^{5} \mathbf{e}_n + \sum_{n=2}^{5} w_{n-2} \cdot \mathbf{e}_n)$, where w_i is the ith weight of $\mathcal{W}$. Compared to other spatial attention mechanisms like CBAM [30], our module separately aggregates features at each position across multiple inputs and predicts a proper probability distribution in $\mathcal{W}$ instead of an unconstrained score. The output patch embedding matrix $\mathbf{Z}^{\mathbb{P}}$ is populated with $\mathbf{z}^{\mathbb{P}}$ at each position and models shape information centered at the corresponding patch in the input image (e.g., if $S = 32$, $\mathbf{Z}^{\mathbb{P}}[0, 0]$ encodes shape information of the top left patch of size 32×32 in $\mathbf{X}$). Finally, $\mathbf{z}^{\mathbb{I}}$ is obtained by average-pooling $\mathbf{F}'_5$ into a vector.

2.2 Implicit Patch Decoding

Given an image coordinate $\mathbf{p}_i^{\mathbb{I}}$ and its corresponding patch embedding $\mathbf{z}_i^{\mathbb{P}}$, the patch-wise occupancy can be decoded with decoder $D^{\mathbb{P}} : (\mathbf{p}_i^{\mathbb{P}}, \mathbf{z}_i^{\mathbb{P}}) \rightarrow \hat{\mathbf{o}}_i^{\mathbb{P}}$, where $D^{\mathbb{P}}$ is a small MLP and $\mathbf{p}_i^{\mathbb{P}}$ is the patch coordinate with respect to the patch center $\mathbf{c}_i$ associated with $\mathbf{z}_i^{\mathbb{P}}$ and is obtained by $\mathbf{p}_i^{\mathbb{P}} = \mathbf{p}_i^{\mathbb{I}} - \mathbf{c}_i$. But, this design leads to poor global shape predictions and discontinuities around patch borders.

To encourage better **global shape coherence**, we also incorporate a global image-level decoder $D^{\mathbb{I}}$. This image decoder, $D^{\mathbb{I}} : (\mathbf{p}_i^{\mathbb{I}}, \mathbf{z}^{\mathbb{I}}) \rightarrow \hat{\mathbf{o}}_i^{\mathbb{I}}$, predicts occupancies for the entire input image. To distill higher-level shape information into patch-based predictions, we also condition $D^{\mathbb{P}}$'s predictions on $\mathbf{p}_i^{\mathbb{I}}$ and $\mathbf{z}^{\mathbb{I}}$. Furthermore, we find that providing the **source coordinate** gives additional spatial context for making location-coherent predictions. In a typical segmentation pipeline, the input image $\mathbf{X}$ is a resized crop from a source image and we find that giving the coordinate $\mathbf{p}_i^{\mathbb{S}}$ ($\mathbb{S}$ for source) from the original uncropped image improves performance on 3D tasks since the additional positional information may be useful for predicting recurring structures. Our enhanced formulation for patch decoding can be described as $D^{\mathbb{P}} : (\mathbf{p}_i^{\mathbb{P}}, \mathbf{z}_i^{\mathbb{P}}, \mathbf{p}_i^{\mathbb{I}}, \mathbf{z}^{\mathbb{I}}, \mathbf{p}_i^{\mathbb{S}}) \rightarrow \hat{\mathbf{o}}_i^{\mathbb{P}}$.

To address discontinuities at patch boundaries, we propose a training technique called **Stochastic Patch Overreach** (SPO) which forces patch embeddings to make predictions for coordinates in neighboring patches. For each patch point and embedding pair $(\mathbf{p}_i^{\mathbb{P}}, \mathbf{z}_i^{\mathbb{P}})$, we create a new pair $(\mathbf{p}_i^{\mathbb{P}\prime}, \mathbf{z}_i^{\mathbb{P}\prime})$ by randomly selecting a neighboring patch embedding and updating the local point to be relative to the new patch center. This regularization is modulated by the set of valid choices to select a neighboring patch (*connectivity*, or *con*) and the number of perturbed points to sample per batch point (*occurrence*, or N_{SPO}). $con = 4$ means all adjoining patches are neighbors while $con = 8$ includes corner patches as well. Note that SPO differs from the regularization in [3] since no construction of a KD-Tree is required and we introduce a tunable stochastic component which further helps with regularization under limited-data settings.

2.3 Training SwIPE

To optimize the parameters of f_θ, we first sample a set of point and occupancy pairs $\{\mathbf{p}_i^{\mathbb{S}}, \mathbf{o}_i\}_{i \in \mathcal{I}}$ for each source image, where $\mathcal{I}$ is the index set for the selected points. We obtain an equal number of points for each foreground class using Latin Hypercube sampling with 50% of each class's points sampled within 10 pixels of the class object boundaries. The **point-wise occupancy loss**, written as $\mathcal{L}_{\text{occ}}(\mathbf{o}_i, \hat{\mathbf{o}}_i) = 0.5 \cdot \mathcal{L}_{\text{ce}}(\mathbf{o}_i, \hat{\mathbf{o}}_i) + 0.5 \cdot \mathcal{L}_{\text{dc}}(\mathbf{o}_i, \hat{\mathbf{o}}_i)$, is an equally weighted sum of Cross Entropy loss $\mathcal{L}_{\text{ce}}(\mathbf{o}_i, \hat{\mathbf{o}}_i) = -\log \hat{o}_i^c$ and Dice loss $\mathcal{L}_{\text{dc}}(\mathbf{o}_i, \hat{\mathbf{o}}_i) = 1 - \frac{1}{C}\sum_c \frac{2 \cdot o_i^c \cdot \hat{o}_i^c + 1}{(o_i^c)^2 + (\hat{o}_i^c)^2 + 1}$, where $\hat{o}_i^c$ is the predicted probability for the target occupancy with class label c. Note that in practice, these losses are computed in their vectorized forms; for example, the Dice loss is applied with multiple points per image instead of an individual point (similar to computing the Dice loss between a flattened image and its flattened mask). The **loss for patch and image decoder** predictions is $\mathcal{L}_{\mathbb{PI}}(\mathbf{o}_i, \hat{\mathbf{o}}_i^{\mathbb{P}}, \hat{\mathbf{o}}_i^{\mathbb{I}}) = \alpha \mathcal{L}_{\text{occ}}(\mathbf{o}_i, \hat{\mathbf{o}}_i^{\mathbb{P}}) + (1-\alpha)\mathcal{L}_{\text{occ}}(\mathbf{o}_i, \hat{\mathbf{o}}_i^{\mathbb{I}})$, where

α is a local-global balancing coefficient. Similarly, the **loss for the SPO** occupancy prediction $\hat{\mathbf{o}}_i'$ is $\mathcal{L}_{\text{SPO}}(\mathbf{o}_i, \hat{\mathbf{o}}_i') = \mathcal{L}_{\text{occ}}(\mathbf{o}_i, \hat{\mathbf{o}}_i')$. Finally, the **overall loss** for a coordinate is formulated as $\mathcal{L} = \mathcal{L}_{\mathbb{PI}} + \beta\mathcal{L}_{\text{SPO}} + \lambda(||\mathbf{z}_i^{\mathbb{P}}||_2^2 + ||\mathbf{z}_i^{\mathbb{I}}||_2^2)$, where β scales SPO and the last term (scaled by λ) regularizes the patch & image embeddings, respectively.

3 Experiments and Results

This section presents quantitative results from **four main studies**, analyzing overall performance, robustness to data shifts, model & data efficiency, and ablation & component studies. For more implementation details, experimental settings, and qualitative results, we refer readers to the Supplementary Material.

Table 1. Overall results versus the state-of-the-art. Starred* items indicate a state-of-the-art discrete method for each task. The Dice columns report foreground-averaged scores and standard deviations (±) across 6 runs (6 different seeds were used while train/val/test splits were kept consistent).

	2D Polyp Sessile				3D CT BCV		
Method	Params (M)	FLOPs (G)	Dice (%)	Method	Params (M)	FLOPs (G)	Dice (%)
Discrete Approaches							
U-Net$_{15}$ [26]	7.9	83.3	63.89±1.30	U-Net$_{15}$ [26]	16.3	800.9	74.47±1.57
PraNet$^*_{20}$ [6]	30.5	15.7	82.56 ± 1.08	UNETR$^*_{21}$ [10]	92.6	72.6	81.14 ± 0.85
Res2UNet$_{21}$ [7]	25.4	17.8	81.62 ± 0.97	Res2UNet$_{21}$ [7]	38.3	**44.2**	79.23 ± 0.66
Implicit Approaches							
OSSNet$_{21}$ [25]	5.2	6.4	76.11 ± 1.14	OSSNet$_{21}$ [25]	7.6	55.1	73.38 ± 1.65
IOSNet$_{22}$ [16]	4.1	**5.9**	78.37 ± 0.76	IOSNet$_{22}$ [16]	6.2	46.2	76.75 ± 1.37
SwIPE (*ours*)	**2.7**	10.2	**85.05** ± 0.82	SwIPE (*ours*)	**4.4**	71.6	**81.21** ± 0.94

3.1 Datasets, Implementations, and Baselines

We evaluate performance on two tasks: 2D binary polyp segmentation and 3D multi-class abdominal organ segmentation. For polyp segmentation, we train on the challenging **Kvasir-Sessile** dataset [13] (196 colored images of small sessile polyps), and use **CVC-ClinicDB** [2] to test model robustness. For 3D organ segmentation, we train on **BCV** [1] (30 CT scans, 13 annotated organs), and use the diverse CT images in **AMOS** [14] (200 training CTs, the same setting used in [32]) to evaluate model robustness. All the datasets are split with a 60:20:20 train:validation:test ratio. For each image in Sessile [in BCV, resp.], we obtain 4000 [20,000] background points and sample 2000 [4000] foreground points for each class with half of every class' foreground points lying within 10 pixels [voxels] of the boundary.

2D Sessile Polyp training uses a modified Res2Net [7] backbone with 28 layers, [256, 256, 256] latent MLP dimensions for $D^{\mathbb{P}}$, [256, 128] latent dimensions for $D^{\mathbb{I}}$, $d = 128$, $S = 32$, and $con = 8$. 3D BCV training uses a Res2Net-50 backbone,

[256, 256, 256, 256] latent MLP dimensions for $D^{\mathbb{P}}$, [256, 256, 128] latent MLP dimensions for $D^{\mathbb{I}}$, $d = 512$, $S = 8$, and $con = 6$ (all adjoining patches in 3D). The losses for both tasks are optimized with AdamW [18] and use $\alpha = 0.5$, $\beta = 0.1$, and $\lambda = 0.0001$. For inference, we adopt MISE like prior works [16,19,25] and evaluate on a reconstructed prediction mask equal in size to the input image. $D^{\mathbb{P}}$ segments boundaries better than $D^{\mathbb{I}}$, and is used to produce final predictions.

For fair comparisons, all the methods are trained using the same equally-weighted Dice and Cross Entropy loss for 30,000 and 50,000 iterations on 2D Sessile and 3D BCV, resp. The test score at the best validation epoch is reported. Image input sizes were 384 × 384 for Sessile and 96 × 96 × 96 for BCV. All the implicit methods utilize the same pre-sampled points for each image. For IOSNet [16], both 2D and 3D backbones were upgraded from three downsampling stages to five for fair comparisons and empirically confirmed to outperform the original. We omit comparisons against IFA [11] to focus on medical imaging approaches; plus, IFA did not outperform IOSNet [16] on either task.

3.2 Study 1: Performance Comparisons

The results for 2D Polyp Sessile and 3D CT BCV organ segmentation are presented in Table 1. FLOPs are reported from the forward pass on a single image during training.

On the smaller polyp dataset, we observe notable improvements over the best-known implicit approaches (+6.7% Dice) and discrete methods (+2.5% Dice) with much fewer parameters (9% of PraNet [6] and 66% of IOSNet [16]). For

Table 2. Left and **Middle**: Robustness to data shifts. **Right**: Efficiency studies.

Across Resolutions

Method	Size	Dice
Varying Output Size		
1 PraNet [6]	128↓	72.64
2 IOSNet [16]	128↓	76.18
3 SwIPE	128↓	81.26
4 PraNet [6]	896↑	74.95
5 IOSNet [16]	896↑	78.01
6 SwIPE	896↑	84.33
Varying Input Size		
7 PraNet [6]	128↓	68.79
8 PraNet [6]	896↑	43.92

Across Datasets

Method	Dice
Polyp Sessile → CVC	
1 PraNet [6]	68.37
2 IOSNet [16]	59.42
3 SwIPE	70.10
CT BCV → CT AMOS (liver class only)	
4 UNETR [10]	81.75
5 IOSNet [16]	79.48
6 SwIPE	82.81

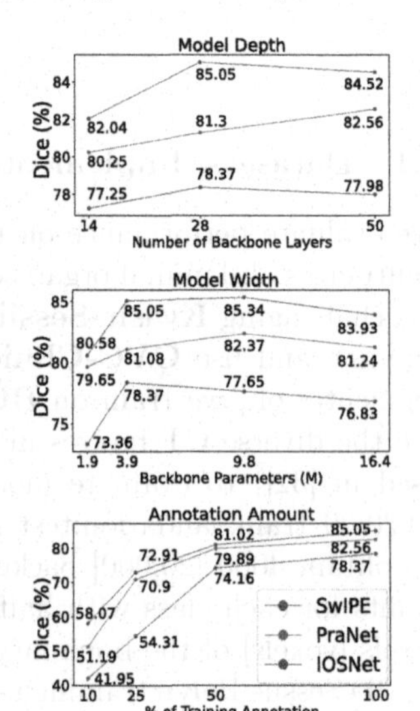

BCV, the performance gains are more muted; however, we still marginally outperform UNETR [10] with over 20x fewer parameters and comparable FLOPs.

3.3 Study 2: Robustness to Data Shifts

In this study, we explore the robustness of various methods to specified target resolutions and dataset shifts. The left-most table in Table 2 contains results for the former study conducted on 2D Sessile, where we first analyze the effect of directly resizing outputs (Table 2 left, rows 1 to 6) when given an input image that is standard during training (384 × 384). The discrete method, PraNet, outputs 384 × 384 predictions which are interpolated to the target size (Table 2 left, rows 1 & 4). This causes more performance drop-offs than the implicit methods which can naturally vary the output size by changing the resolution of the input coordinates. We also vary the input size so that no manipulations of predictions are required (Table 2 left, rows 7 & 8), which results in steep accuracy drops.

The results for the dataset shift study are given in the middle of Table 2, where CVC is another binary poly segmentation task and the liver class is evaluated on all CT scans in AMOS. Both discrete methods outperform IOSNet, which may indicate that point-based features are more prone to overfitting due to a lack of contextual regularization. Also, we highlight our method's consistent outperformance over both discrete methods and IOSNet in all of the settings.

3.4 Study 3: Model Efficiency and Data Efficiency

To analyze the model efficiency (the right-most column of charts in Table 2), we report on 2D Sessile and vary the backbone size in terms of depth and width. For data efficiency, we train using 10%, 25%, 50%, and 100% of annotations. Not only do we observe outperformance across the board in model sizes & annotation amounts, but the performance drop-off is more tapered with our method.

Table 3. Left: Ablation studies. **Right**: Design choice experiments.

Ablation Studies on 2D Sessile

Component		Incorporation							
E_n	RFB-Lite		✓	✓	✓	✓	✓	✓	✓
E_n	Cascade			✓	✓	✓	✓	✓	✓
E_n	MEA				✓	✓	✓	✓	✓
$D^{\mathbb{I}}$	$\mathbf{z}^{\mathbb{I}}, \mathbf{p}^{\mathbb{I}}$					✓	✓	✓	✓
$D^{\mathbb{P}}$	$\mathbf{z}^{\mathbb{I}}, \mathbf{p}^{\mathbb{I}}$						✓	✓	✓
$D^{\mathbb{P}}$	$\mathbf{p}^{\mathbb{S}}$							✓	✓
$D^{\mathbb{P}}$	SPO								✓
Dice (%)		76.44	76.57	78.19	80.33	80.92	82.28	83.75	**85.05**

Alternative Designs

Description	Dice (%)
Backbone	
1 CCT [9]	78.30
2 U-Net [26]	79.94
MEA Replacements for Feature Fusion	
3 Addition	84.19
4 Concat. + 1x1 Conv	83.58
5 Self-Attention	68.23
SPO	
6 $N_o = 0$	83.75
7 $N_o = 4, Con = 4$	83.71
8 $N_o = 4, Con = 8$	83.94
9 $N_o = 8, Con = 4$	84.43

3.5 Study 4: Component Studies and Ablations

The left side of Table 3 presents our ablation studies, showing the benefits enabled by context aggregation within E_n, global information conditioning, and adoption of MEA & SPO. We also explore alternative designs on the right side of Table 3 for our three key components, and affirm their contributions on achieving superior performance.

4 Conclusions

SwIPE represents a notable departure from conventional discrete segmentation approaches and directly models object shapes instead of pixels and utilizes continuous rather than discrete representations. By adopting both patch and image embeddings, our approach enables accurate local geometric descriptions and improved shape coherence. Experimental results show the superiority of SwIPE over state-of-the-art approaches in terms of segmentation accuracy, efficiency, and robustness. The use of local INRs represents a new direction for medical image segmentation, and we hope to inspire further research in this direction.

References

1. Multi-atlas labeling beyond the cranial vault (2015). https://www.synapse.org/#!Synapse:syn3193805/wiki/89480. Accessed Jan 2021
2. Bernal, J., Sánchez, F.J., Fernández-Esparrach, G., Gil, D., Rodríguez, C., Vilariño, F.: WM-DOVA maps for accurate polyp highlighting in colonoscopy: Validation vs. saliency maps from physicians. Computerized Med. Imaging Graph. **43**, 99–111 (2015)
3. Chabra, R., et al.: Deep local shapes: learning local SDF priors for detailed 3D reconstruction. In: Vedaldi, A., Bischof, H., Brox, T., Frahm, J.-M. (eds.) ECCV 2020. LNCS, vol. 12374, pp. 608–625. Springer, Cham (2020). https://doi.org/10.1007/978-3-030-58526-6_36
4. Chibane, J., Alldieck, T., Pons-Moll, G.: Implicit functions in feature space for 3D shape reconstruction and completion. CVPR, pp. 6968–6979 (2020)
5. Dupont, E., Goliński, A., Alizadeh, M., Teh, Y.W., Doucet, A.: COIN: COmpression with Implicit Neural representations. arXiv preprint arXiv:2103.03123 (2021)
6. Fan, D.-P., et al.: PraNet: parallel reverse attention network for polyp segmentation. In: Martel, A.L., et al. (eds.) MICCAI 2020. LNCS, vol. 12266, pp. 263–273. Springer, Cham (2020). https://doi.org/10.1007/978-3-030-59725-2_26
7. Gao, S.H., et al.: Res2Net: a new multi-scale backbone architecture. IEEE TPAMI **43**(2), 652–662 (2019)
8. Gu, P., Zheng, H., Zhang, Y., Wang, C., Chen, D.Z.: *k*CBAC-Net: Deeply Supervised Complete Bipartite Networks with Asymmetric Convolutions for Medical Image Segmentation. In: de Bruijne, M., et al. (eds.) MICCAI 2021. LNCS, vol. 12901, pp. 337–347. Springer, Cham (2021). https://doi.org/10.1007/978-3-030-87193-2_32
9. Hassani, A., Walton, S., Shah, N., Abuduweili, A., Li, J., Shi, H.: Escaping the big data paradigm with compact Transformers. ArXiV:2104.05704 (2021)

10. Hatamizadeh, A., et al.: UNETR: transformers for 3D medical image segmentation. In: IEEE/CVF Winter Conference on Applications of Computer Vision (WACV), pp. 574–584 (2022)
11. Hu, H., et al.: Learning implicit feature alignment function for semantic segmentation. In: ECCV, pp. 487–505. Springer, Cham (2022). https://doi.org/10.1007/978-3-031-19818-2_28
12. Isensee, F., Jaeger, P., Kohl, S., Petersen, J., Maier-Hein, K.H.: nnU-Net: a self-configuring method for deep learning-based biomedical image segmentation. Nature Methods **18**(2), 203–211 (2021)
13. Jha, D., et al.: A comprehensive study on colorectal polyp segmentation with ResUNet++, conditional random field and test-time augmentation. IEEE J. Biomed. Health Inform. **25**(6), 2029–2040 (2021)
14. Ji, Y., et al.: AMOS: a large-scale abdominal multi-organ benchmark for versatile medical image segmentation. ArXiv:2206.08023 (2022)
15. Jiang, C., Sud, A., Makadia, A., Huang, J., Nießner, M., Funkhouser, T., et al.: Local implicit grid representations for 3D scenes. In: CVPR, pp. 6001–6010 (2020)
16. Khan, M., Fang, Y.: Implicit neural representations for medical imaging segmentation. In: MICCAI (2022)
17. Liu, S., Huang, D., et al.: Receptive field block net for accurate and fast object detection. In: ECCV, pp. 385–400 (2018)
18. Loshchilov, I., Hutter, F.: Decoupled weight decay regularization. In: ICLR (2017)
19. Mescheder, L., Oechsle, M., Niemeyer, M., Nowozin, S., Geiger, A.: Occupancy networks: learning 3D reconstruction in function space. In: CVPR, pp. 4455–4465 (2018)
20. Mildenhall, B., Srinivasan, P.P., Tancik, M., Barron, J.T., Ramamoorthi, R., Ng, R.: NeRF: representing scenes as neural radiance fields for view synthesis. In: Vedaldi, A., Bischof, H., Brox, T., Frahm, J.-M. (eds.) ECCV 2020. LNCS, vol. 12346, pp. 405–421. Springer, Cham (2020). https://doi.org/10.1007/978-3-030-58452-8_24
21. Oechsle, M., Mescheder, L., Niemeyer, M., Strauss, T., Geiger, A.: Texture fields: learning texture representations in function space. In: ICCV, pp. 4531–4540 (2019)
22. Park, J., Florence, P., Straub, J., Newcombe, R., Lovegrove, S.: DeepSDF: learning continuous signed distance functions for shape representation. In: CVPR, pp. 165–174 (2019)
23. Pasupathy, A.: The neural basis of image segmentation in the primate brain. Neuroscience **296**, 101–109 (2015)
24. Raju, A., Miao, S., Jin, D., Lu, L., Huang, J., Harrison, A.P.: Deep implicit statistical shape models for 3D medical image delineation. In: Proceedings of the AAAI Conference on Artificial Intelligence, vol. 36, pp. 2135–2143 (2022)
25. Reich, C., Prangemeier, T., Cetin, O., Koeppl, H.: OSS-Net: memory efficient high resolution semantic segmentation of 3D medical data. In: British Machine Vision Conference (2021)
26. Ronneberger, O., Fischer, P., Brox, T.: U-Net: convolutional networks for biomedical image segmentation. In: Navab, N., Hornegger, J., Wells, W.M., Frangi, A.F. (eds.) MICCAI 2015. LNCS, vol. 9351, pp. 234–241. Springer, Cham (2015). https://doi.org/10.1007/978-3-319-24574-4_28
27. Shen, L., Pauly, J., Xing, L.: NeRP: implicit neural representation learning with prior embedding for sparsely sampled image reconstruction. IEEE Trans. Neural Networks Learn. Syst. (2022)

28. Sørensen, K., Camara, O., Backer, O., Kofoed, K., Paulsen, R.: NUDF: neural unsigned distance fields for high resolution 3D medical image segmentation. ISBI, pp. 1–5 (2022)
29. Wiesner, D., Suk, J., Dummer, S., Svoboda, D., Wolterink, J.M.: Implicit neural representations for generative modeling of living cell shapes. In: MICCAI, pp. 58–67. Springer, Cham (2022). https://doi.org/10.1007/978-3-031-16440-8_6
30. Woo, S., Park, J., Lee, J.Y., Kweon, I.S.: CBAM: convolutional block attention module. In: ECCV, pp. 3–19 (2018)
31. Yang, J., Wickramasinghe, U., Ni, B., Fua, P.: ImplicitAtlas: learning deformable shape templates in medical imaging. In: CVPR, pp. 15861–15871 (2022)
32. Zhang, Y., Sapkota, N., Gu, P., Peng, Y., Zheng, H., Chen, D.Z.: Keep your friends close & enemies farther: Debiasing contrastive learning with spatial priors in 3D radiology images. In: 2022 IEEE International Conference on Bioinformatics and Biomedicine (BIBM), pp. 1824–1829. IEEE (2022)

Smooth Attention for Deep Multiple Instance Learning: Application to CT Intracranial Hemorrhage Detection

Yunan Wu[1(✉)], Francisco M. Castro-Macías[2,4], Pablo Morales-Álvarez[3,4], Rafael Molina[2], and Aggelos K. Katsaggelos[1]

[1] Image and Video Processing Laboratory, Department of Electrical and Computer Engineering, Northwestern University, Evanston, USA
yunanwu2020@u.northwestern.edu

[2] Department of Computer Science and Artificial Intelligence, University of Granada, Granada, Spain

[3] Department of Statistics and Operations Research, University of Granada, Granada, Spain

[4] Research Centre for Information and Communication Technologies (CITIC-UGR), University of Granada, Granada, Spain

Abstract. Multiple Instance Learning (MIL) has been widely applied to medical imaging diagnosis, where bag labels are known and instance labels inside bags are unknown. Traditional MIL assumes that instances in each bag are independent samples from a given distribution. However, instances are often spatially or sequentially ordered, and one would expect similar diagnostic importance for neighboring instances. To address this, in this study, we propose a smooth attention deep MIL (SA-DMIL) model. Smoothness is achieved by the introduction of first and second order constraints on the latent function encoding the attention paid to each instance in a bag. The method is applied to the detection of intracranial hemorrhage (ICH) on head CT scans. The results show that this novel SA-DMIL: (a) achieves better performance than the non-smooth attention MIL at both scan (bag) and slice (instance) levels; (b) learns spatial dependencies between slices; and (c) outperforms current state-of-the-art MIL methods on the same ICH test set.

Keywords: Smooth attention · Multiple instance learning · CT hemorrhage diagnosis

1 Introduction

Multiple Instance Learning (MIL) [6,21] is a type of weakly supervised learning that has become very popular in biomedical imaging diagnostics due to the reduced annotation effort it requires [8,13]. In the case of MIL binary classification, the training set is partitioned into bags of instances. Both bags and

Supplementary Information The online version contains supplementary material available at https://doi.org/10.1007/978-3-031-43904-9_32.

H. Greenspan et al. (Eds.): MICCAI 2023, LNCS 14224, pp. 327–337, 2023.
https://doi.org/10.1007/978-3-031-43904-9_32

instances have labels, but only bag labels are observed while instance labels remain unknown. It is assumed that a bag label is positive if and only if the bag contains at least one positive instance [10]. The goal is to produce a method that, trained on bag labels only, is capable of predicting both bag and instance labels.

Among the proposed approaches for learning in the MIL scenario [21], deep learning (DL) methods stand out when dealing with highly structured data (such as medical images and videos) [17]. The most successful deep MIL approaches combine an instance-level processing mechanism (i.e., a feature extractor) with a pooling mechanism to aggregate information from instances in a bag [8,13]. Among the pooling operators, the attention-based weight pooling proposed in [15] is frequently used as a way to discover *key instances*, i.e., those responsible for the label of a bag. However, this pooling operator was formulated under strong assumptions of independence between the instances in a bag. This is a drawback in biomedical imaging problems, where instances in a bag are often spatially or sequentially ordered and their diagnostic importance is expected to be similar for neighboring instances [18,24].

In this work, we are particularly interested in the detection of intracranial hemorrhage (ICH), a serious life-threatening emergency caused by blood leakage inside the brain [5,22]. Radiologists confirm the presence of ICH by using computed tomography (CT) scans [9], which consist of a significant number of slices, each representing a section of the head at a given height. Unfortunately, the shortage of specialized radiologists and their increasing workload sometimes lead to delayed and erroneous diagnoses [3,12,20,25], which may result in potentially preventable cerebral injury or morbidity [9,11]. For this reason, there is a growing interest in the development of automated systems to assist radiologists in making rapid and reliable diagnoses.

State-of-the-art ICH detection methods rely on DL models, specifically convolutional neural networks (CNNs), to extract meaningful ICH features [31]. However, 2D CNNs need to be coupled with other mechanisms such as recurrent neural networks (RNNs) [14,30] or 3D CNNs [2,7,16,27] to account for interslice dependencies. Although these approaches are quite successful in terms of performance, their use is limited by the large amount of labeled data they require [31]. To address this issue, the ICH detection task has been formulated as an MIL problem, achieving comparable performance to fully supervised models while reducing the workload of radiologists [26,29]. Note that the MIL framework is naturally suited for the ICH detection problem since a CT scan (i.e., a bag) is considered positive if it contains at least one slice (i.e., an instance) with evidence of hemorrhage (i.e., positive instance).

In this work, we improve upon the state-of-the-art deep MIL methods by introducing dependencies between instances in a sound probabilistic manner. These dependencies are formulated over a neighborhood graph to impose smoothness on the latent function that encodes the attention given to each instance. Smoothness is achieved by introducing specific first- and second-order constraints on the latent function. Our model, called SA-DMIL, is applied to the ICH detection problem, obtaining (a) significant improvements upon the performance of non-smooth models at both scan and slice levels, (b) smoother attention weights across slices by benefiting from the inter-slice dependencies, and (c) a superior performance against other popular MIL methods on the same test set.

2 Methods

2.1 Problem Formulation

We start by formulating ICH detection as a Multiple Instance Learning (MIL) problem. To do so, we map slices to instances and CT scans to bags. The slices (instances) will be denoted by $\mathbf{x}_i^b \in \mathbb{R}^{3HW}$, where H and W are the height and width of the image, 3 is the number of color channels, b is the index of the scan to which the slice belongs to and i is the index of the slice inside the bag. We will denote the label of a slice by $y_i^b \in \{0,1\}$. If the slice contains hemorrhage, then $y_i^b = 1$, otherwise $y_i^b = 0$. Note that the slice labels remain unknown since only scan labels are given. As we know, slices are grouped to form the CT scans. Each scan (bag) will be denoted by $\mathbf{X}^b = \left[\mathbf{x}_1^b, \ldots, \mathbf{x}_{N_b}^b\right]^\top \in \mathbb{R}^{N_b \times 3HW}$. Here, N_b is the number of slices in bag b. We will assume that B CT scans are given, so $b \in \{1, \ldots, B\}$. Given a CT scan b, we will denote its label by $T^b \in \{0,1\}$. Notice that $T^b = 1$ if and only if some of ${y_i}^b = 1$, i.e., the following relationship between scan and slice labels holds,

$$T^b = \max\left\{y_1^b, \ldots, y_{N_b}^b\right\}. \tag{1}$$

2.2 Attention-Based Multiple Instance Learning Pooling

The attention-based MIL pooling was proposed in [15] as a way to discover *key instances*, i.e., those responsible for the diagnosis of a scan. It consists of a weighted average of instances (low-dimensional embeddings) where the weights are parameterized by a neural network. Formally, given a bag of N_b embeddings $\mathbf{Z}^b = \left[\mathbf{z}_1^b, \ldots, \mathbf{z}_{N_b}^b\right]^\top$, where $\mathbf{z}_i^b \in \mathbb{R}^D$, the attention-based MIL pooling computes

$$\Phi_{\text{Att}}\left(\mathbf{Z}^b\right) = \textstyle\sum_{i=1}^{N_b} s(\mathbf{z}_i^b)\mathbf{z}_i^b, \tag{2}$$

where

$$s\left(\mathbf{z}_i^b\right) = \frac{\exp\left(f\left(\mathbf{z}_i^b\right)\right)}{\sum_j^{N_b} \exp\left(f\left(\mathbf{z}_j^b\right)\right)}, \quad f\left(\mathbf{z}_i^b\right) = \mathbf{w}^\top \tanh\left(\mathbf{V}\mathbf{z}_i^b\right). \tag{3}$$

Notice that $\mathbf{w} \in \mathbb{R}^L$ and $\mathbf{V} \in \mathbb{R}^{L \times D}$ are trainable parameters, where D denotes the size of feature vectors. We refer to $s\left(\mathbf{z}_i^b\right)$ as *attention weights* and to $f\left(\mathbf{z}_i^b\right)$ as *attention values.*

This operator was proposed under the assumption that the instances in a bag show neither dependency nor order among each other. Although this may be the case in simple problems, it does not occur in problems such as ICH

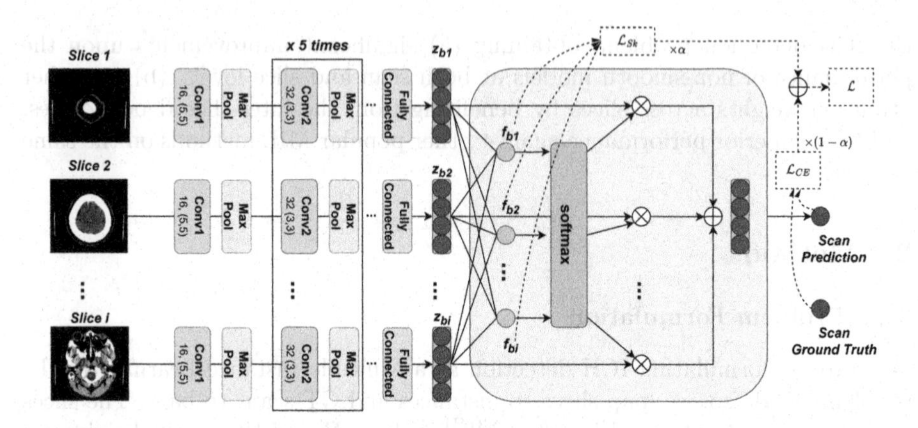

Fig. 1. SA-DMIL architecture. It consists of CNNs that extract slice level features and an attention block to aggregate slice features. The loss function is a weighted average of the binary cross entropy and a novel smooth attention loss.

detection. Note that the attention weights of slices in a bag are correlated: given a slice containing ICH, we expect that the adjacent slices will also contain ICH with high probabilities. This is essential in finding slices with ICH. In the next subsection, we show how to introduce this correlation between attention weights.

2.3 Modeling Correlation Through the Attention Mechanism

Ideally, in the case of a positive scan ($T^b = 1$), high attention weights should be assigned to slices that are likely to have a positive label ($y_i^b = 1$). Given the dependency between slices, contiguous slices should have similar attention values. In other words, the differences between the attention values of contiguous slices should be *small.* Thus, for each bag b, these quantities should be small

$$\mathcal{L}_{S1}^b = 2^{-1} \sum\nolimits_{i,j \in \text{Bag}(b)} A_{ij}^b \left(f\left(\mathbf{z}_i^b\right) - f\left(\mathbf{z}_j^b\right)\right)^2, \tag{4}$$

$$\mathcal{L}_{S2}^b = 4^{-1} \sum\nolimits_{i \in \text{Bag}(b)} \left(\sum\nolimits_{j \in \text{Bag}(b)} A_{ij}^b \left(f\left(\mathbf{z}_i^b\right) - f\left(\mathbf{z}_j^b\right)\right)\right)^2, \tag{5}$$

where $A_{ij}^b = 1$ if the slices i, j are related in bag b, and 0 otherwise. We smooth $f\left(\mathbf{z}_i^b\right)$ instead of $s\left(\mathbf{z}_i^b\right)$ because a non-constrained parameter f ensures consistent smoothing while s requires a normalization across instances in a bag.

Equations (4) and (5) correspond, respectively, to the energies of the, so called, conditional and simultaneous autoregressive models in the statistics literature [4,23]. For our problem, they model the value of f at a given location (instance) given the values at neighboring instances. From the regularization viewpoint, these terms constrain the first and second derivatives of the function f, respectively, which favors smoother functions (examine the zero of the derivative of f). That is, a *priori* all attention weights are expected to be the same because f is expected to be constant. As observations arrive, they change to reflect the importance of each instance. Note that (4) and (5) impose smoothness but they can be modified to model, for example, competition between the attention weights by simply replacing the minus sign with a plus sign.

To compute $\mathcal{L}_{S1}^b$ and $\mathcal{L}_{S2}^b$ efficiently we consider the simple graph defined by the dependency between slices. For a bag b, its adjacency matrix is $\mathbf{A}^b = \left[A_{ij}^b\right]$. The degree matrix $\mathbf{D}^b = \left[D_{ij}^b\right]$ is a diagonal matrix that contains the degree of each slice (the degree of the slice i is the number of slices j such that $A_{ij}^b = 1$). This is, $D_{ii}^b = \text{degree}(i)$ and $D_{ij}^b = 0$ if $i \neq j$. Using these, one can compute the graph Laplacian matrix of a bag as $\mathbf{L}^b = \mathbf{D}^b - \mathbf{A}^b$. It is easy to show that

$$\mathcal{L}_{S1}^b = \mathbf{f}^{b\top} \mathbf{L}^b \mathbf{f}^b, \quad \mathcal{L}_{S2}^b = \mathbf{f}^{b\top} \mathbf{L}^b \mathbf{L}^b \mathbf{f}^b, \tag{6}$$

where $\mathbf{f}^b = \left[f(\mathbf{z}_1^b), \ldots, f(\mathbf{z}_{N_b}^b)\right]^\top$. The sum of $\mathcal{L}_{Sk}^b$ over bags, where $k \in \{1, 2\}$, can be added to the loss function of a network to be minimized along the task-specific loss. Note that these two terms provide two different approaches to exploiting the correlations between instances through the loss function. We will refer to this approach as smooth attention (SA) loss. In the following subsection, we propose a model that can use either $\mathcal{L}_{S1}$ or $\mathcal{L}_{S2}$. The effect of each term will be discussed in Sect. 4.

2.4 SA-DMIL Model Description

We propose to couple the attention-based MIL pooling with the SA loss terms introduced in Subsect. 2.3. The proposed model, named Smooth Attention Deep Multiple Instance Learning (SA-DMIL), is depicted in Fig. 1. We use a Convolutional Neural Network (CNN), denoted by Φ_{CNN}, as a feature extractor to obtain a vector of low dimensional embeddings for each instance. That is, given a bag $\mathbf{X}^b = \left[\mathbf{x}_1^b, \ldots, \mathbf{x}_{N_b}^b\right]$, where $\mathbf{x}_n^b \in \mathbb{R}^{3 \times HW}$, we compute

$$\mathbf{z}_n^b = \Phi_{\text{CNN}}\left(\mathbf{x}_n^b\right) \in \mathbb{R}^D, \quad \mathbf{Z}^b = \left[\mathbf{z}_1^b, \ldots, \mathbf{z}_{N_b}^b\right]. \tag{7}$$

The CNN module in Fig. 1 is implemented with six convolutional blocks, followed by a flatten layer. $\mathbf{Z}^b$ is then fed into the attention layer Φ_{Att} described in Subsect. 2.3 to obtain a scan representation. After that, the scan representation passes through a classifier Φ_{c} (i.e., one fully connected layer with a sigmoid activation) to predict the scan labels,

$$\text{p}\left(T^b \mid \mathbf{X}^b\right) \approx \Phi\left(\mathbf{X}^b\right) = \Phi_{\text{c}}\left(\Phi_{\text{Att}}\left(\Phi_{\text{CNN}}\left(\mathbf{X}^b\right)\right)\right), \tag{8}$$

where we have written $\Phi_{\text{CNN}}\left(\mathbf{X}^b\right) = \left[\Phi_{\text{CNN}}\left(\mathbf{x}_1^b\right), \ldots, \Phi_{\text{CNN}}\left(\mathbf{x}_{N_b}^b\right)\right]$. Our model, that corresponds to the composition $\Phi = \Phi_{\text{c}} \circ \Phi_{\text{Att}} \circ \Phi_{\text{CNN}}$, is trained using the following loss function until convergence,

$$\mathcal{L} = (1 - \alpha)\, \mathcal{L}_{\text{CE}} + \alpha \mathcal{L}_{Sk}, \tag{9}$$

where $\alpha \in [0, 1]$ is an hyperparameter and $\mathcal{L}_{\text{CE}}$ the common cross-entropy loss,

$$\mathcal{L}_{\text{CE}} = \sum_b \left[T^b \log\left(\Phi\left(\mathbf{X}^b\right)\right) + \left(1 - T^b\right) \log\left(1 - \Phi\left(\mathbf{X}^b\right)\right)\right], \tag{10}$$

where $k \in \{1, 2\}$, and $\mathcal{L}_{Sk} = \sum_b \mathcal{L}_{Sk}^b$ (see Eqs. (4) and (5)). Depending on the value of k, we obtain two variations of SA-DMIL, which will be referred to as SA-DMIL-$S1$ and SA-DMIL-$S2$. The baseline model, Att-MIL (non-smooth attention), is recovered when $\alpha = 0.0$ [15]. Following the approach of previous studies [19,29], attention weights will be used to obtain predictions at the slice level (although they are not specifically designed for it). If a scan is predicted to be negative, all slices are also predicted to be negative, while if a scan is predicted to correspond to an ICH, slices whose attention weight is above a threshold (i.e., $1/N_b$, with N_b being the number of slices in that scan) are predicted as ICH.

3 Experimental Design

3.1 Data and Data Preprocessing

The dataset used in this work was obtained from the 2019 Radiological Society of North America (RSNA) challenge [1], which included 39650 CT slices from 1150 subjects. The data were split among subjects, with 1000 scans (ICH: Normal scans = 411: 589; ICH: Normal slices = 4976: 29520) used for training and validation, and the remaining 150 scans (ICH: Normal scans = 72: 78; ICH: Normal slices = 806: 4448) used for held-out testing. The number of slices in the scans varied from 24 to 57. All CT slices underwent the same preprocessing procedure as described in [29]. Each CT slice had three windows applied to its original Hounsfield Units by changing the window Width (W) and Center (C) to manipulate the display of specific tissues, as radiologists typically do when diagnosing brain CTs. Here, we selected the brain (W: 80, C:40), subdural (W:200, C:80) and soft tissue (W:380, C: 40) windows. All images were then resized to the same size of 512×512 and normalized to the range $[0, 1]$. CTs were annotated at both the scan and slice levels, but slice labels were used for evaluation only, while scan labels were used for training and evaluation.

3.2 Experimental Settings

We fix $D = 128$ and $L = 50$ in Eq. (3). We use the Adam optimizer with the learning rate starting at 10^{-4}. The batch size is set to 4, the maximum number of epochs is set to 200 and the patience for early stopping is set to 8. We test different values of the α hyperparameter, between 0 and 1 with a jump of 0.1. All experiments were run 5 independent times and the mean and standard deviation were reported in the held-old testing set at both scan and slice levels. The average training time is 10.3 h for SA-DMIL-$S1$ and 10.5 h for SA-DMIL-$S2$. The prediction time is approximately 15.8 s for each scan. All experiments were conducted using Tensorflow 2.11 in Python 3.8 on a single GPU (NVIDIA Quadro RTX 8000). The code will be available via GitHub.

4 Results and Discussion

4.1 Hyperparameters Tuning

In this subsection, we study the effect of SA loss in terms of performance. Table 1 compares the performance of models for different values of α. The standard deviation and other values of α can be found in the appendix, Tables S1 and S2. The results show that at both scan and slice levels, adding a smoothness term to the loss function ($\alpha > 0.0$) achieves better performance than Att-MIL ($\alpha = 0.0$). These improvements are significant, with increases in accuracy, F1 and AUC scores of approximately 7%, 9% and 5% respectively, at scan level, and increases in accuracy and F1 score of 8% and 11% respectively, at slice level. The recall is the only metric in which our model does not excel, where the baseline Att-MIL obtains the best value. However, this is associated with very low precision values. Note that, as α increases, the performance of the model first improves and then drops, which is consistent with the role played by the SA loss as a regularization term. The difference between $\mathcal{L}_{S1}$ and $\mathcal{L}_{S2}$ is not significant although $\mathcal{L}_{S1}$ performs slightly better. In fact, when using $\mathcal{L}_{S1}$, $\alpha = 0.5$ gives the best diagnostic performance with an AUC of 0.879 ($\pm$ 0.003) at scan level and an accuracy of 0.834 ($\pm$ 0.010) at slice level.

Table 1. Performance of SA-DMIL and other MIL methods at slice and scan levels on the RSNA dataset. The average of 5 independent runs is reported. For space constraints, the standard deviation is reported in the appendix.

Model		Scan level					Slice level			
		Acc	Pre	Rec	F1	AUC	Acc	Pre	Rec	F1
SA-DMIL-$S1$	$\alpha = 0.9$	0.753	0.803	0.681	0.735	0.839	0.789	0.670	0.541	0.598
	$\alpha = 0.7$	0.806	0.763	0.784	0.775	0.860	0.828	0.679	0.576	0.639
	$\alpha = 0.5$	**0.813**	0.805	0.806	**0.806**	**0.879**	**0.834**	0.732	**0.608**	**0.686**
	$\alpha = 0.3$	0.767	0.734	0.806	0.768	0.859	0.775	0.702	0.551	0.624
	$\alpha = 0.1$	0.747	0.783	0.652	0.712	0.841	0.766	0.649	0.540	0.584
SA-DMIL-$S2$	$\alpha = 0.9$	0.753	0.817	0.613	0.714	0.816	0.768	0.733	0.551	0.598
	$\alpha = 0.7$	0.767	0.776	0.722	0.748	0.843	0.807	0.734	0.591	0.638
	$\alpha = 0.5$	0.800	0.828	0.736	0.780	0.867	0.823	**0.748**	0.596	0.659
	$\alpha = 0.3$	0.763	0.797	0.686	0.721	0.853	0.790	0.738	0.561	0.622
	$\alpha = 0.1$	0.747	0.736	0.740	0.736	0.833	0.767	0.683	0.547	0.593
Att-MIL ($\alpha = 0.0$) [15]		0.740	0.674	0.832	0.719	0.829	0.751	0.623	0.543	0.579
MIL + Max agg. [28]		0.617	**0.856**	0.447	0.575	0.743	0.732	0.441	0.373	0.406
MIL + Mean agg. [28]		0.677	0.670	0.734	0.693	0.801	0.741	0.502	0.386	0.447
Att-CNN + VGPMIL [29]		0.765	0.724	**0.851**	0.773	0.868	0.807	0.714	0.538	0.597

4.2 Smooth Attention MIL vs. Other MIL Methods

The performance of other popular MIL methods is also included in Table 1. All method share the same CNN architecture to extract slice features, but they differ in the pooling operator they use: Max [28], Mean [28], Attention [15] or Gaussian Process (GP) [29]. These results show that the performance of SA-DMIL is consistently better than other methods across different metrics and at both scan and slice levels. Only the precision of MIL+Max agg. and the recall of AttCNN+VGPMIL at scan level are higher than those obtained by SA-DMIL. However, considering the trade-off between precision and recall given by F1, our method achieves a superior performance. In tasks like ICH detection, where neighbouring instances are expected to have similar diagnostic importance. Unlike other MIL methods that assume each instance to be independently distributed, SA-DMIL stands out by considering the spatial correlation between instances, which compels it to learn more meaningful features for making accurate bag predictions. Notably, this is achieved by simply adding a smoothing term to the loss function without increasing the number of model parameters. This can potentially be applied to existing architectures to further improve performance without adding complexity.

4.3 Visualizing Smooth Regularizing Effects at Slice Level

So far we have observed enhanced performance through the SA term. In this subsection, we visually illustrate how this novel term imposes smoothness between attention scores of consecutive slices, leading to more accurate predictions. Figure 2 shows plots of the attention scores assigned by SA-DMIL-$S1$ and Att-MIL to the slices of three different scans (Fig. S1 in the appendix contains an analogous plot for SA-DMIL-$S2$). As expected, introducing the SA loss results in smoother attention weights. Note that the smoothness constraint of SA-DMIL effectively penalizes the appearance of isolated non-smooth attention weights that incorrectly jump over or below the threshold.

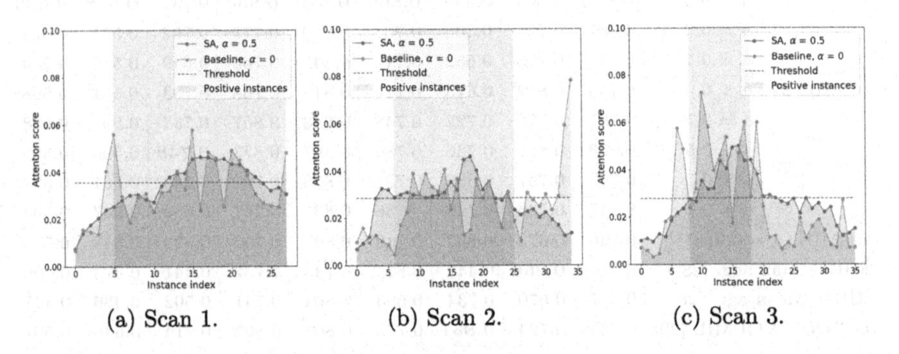

(a) Scan 1. (b) Scan 2. (c) Scan 3.

Fig. 2. Attention weights of SA-DMIL-$S1$ (blue lines, $\alpha = 0.5$) and Att-MIL [15] (orange lines, $\alpha = 0.0$). Slices with values above the threshold ($1/N_b$) are predicted as ICH, while those below are predicted as Normal. The green area highlights those slices whose ground truth label is ICH. (Color figure online)

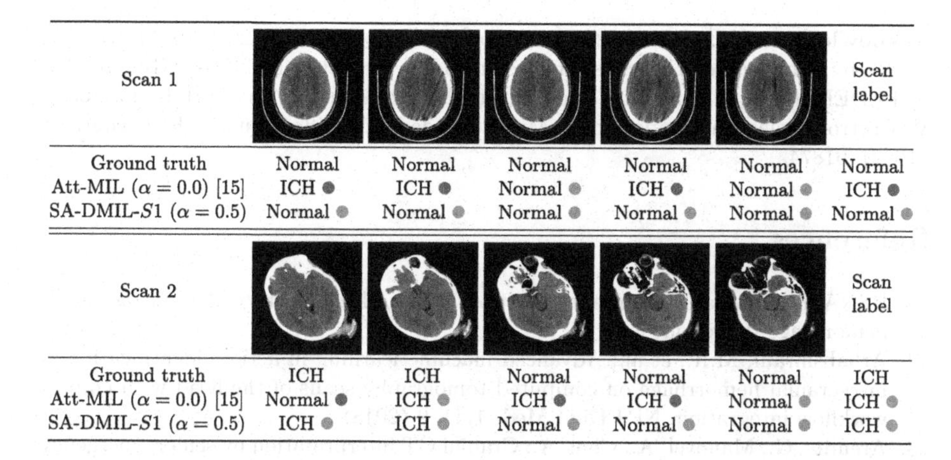

Fig. 3. Predictions of Att-MIL [15] and SA-DMIL-$S1$ at CT slice level in two different scans. SA improves predictions at both scan and slice level. Red color: incorrect prediction, green color: correct prediction.

We also include visual examples of consecutive CT slices in Fig. 3. In Scan 1, the baseline Att-MIL produces a wrong prediction at scan level. When using SA, the prediction is correct since dependencies between adjacent slices have been learned. In Scan 2, both models produce correct predictions at scan level, but SA-DMIL is more accurate at slice level. This occurs thanks to the SA loss, that turns the attention scores into smoother values and, therefore, avoids random *jumps* up and down the decision threshold.

5 Conclusion

In this study we have proposed SA-DMIL, a new model that obtains significant improvements in ICH classification compared to state-of-the-art MIL methods. This is done by adding a smoothing regularizing term to the loss function. This term imposes a smoothness constraint on the latent function that encodes the attention weights, which forces our model to learn dependencies between instances rather than training each instance independently in a bag. This flexible approach does not introduce any additional complexity, so similar ideas can be applied to other methods to model dependencies between neighboring instances.

Data Use Declaration

The dataset used in this study is from the 2019 RSNA Intracranial Hemorrhage Detection Challenge and is publicly available in this link.

Acknowledgement. This work was supported by project PID2019-105142RB-C22 funded by Ministerio de Ciencia e Innovación and by project B-TIC-324-UGR20 funded by FEDER/Junta de Andalucía and Universidad de Granada. The work by Francisco M. Castro-Macías was supported by Ministerio de Universidades under FPU contract FPU21/01874.

References

1. RSNA intracranial hemorrhage detection. https://kaggle.com/c/rsna-intracranial-hemorrhage-detection
2. Arbabshirani, M.R., et al.: Advanced machine learning in action: identification of intracranial hemorrhage on computed tomography scans of the head with clinical workflow integration. NPJ Digit.Med. **1**(1), 9 (2018)
3. Arendts, G., Manovel, A., Chai, A.: Cranial CT interpretation by senior emergency department staff. Australas. Radiol. **47**(4), 368–374 (2003)
4. Belkin, M., Niyogi, P., Sindhwani, V.: Manifold regularization: a geometric framework for learning from labeled and unlabeled examples. J. Mach. Learn. Res. **7**(11), 2399–2434 (2006)
5. Caceres, J.A., Goldstein, J.N.: Intracranial hemorrhage. Emerg. Med. Clin. North Am. **30**(3), 771–794 (2012)
6. Carbonneau, M.A., Cheplygina, V., Granger, E., Gagnon, G.: Multiple instance learning: a survey of problem characteristics and applications. Pattern Recogn. **77**, 329–353 (2018)
7. Chang, P.D., et al.: Hybrid 3d/2d convolutional neural network for hemorrhage evaluation on head CT. Am. J. Neuroradiol. **39**(9), 1609–1616 (2018)
8. Cheplygina, V., de Bruijne, M., Pluim, J.P.: Not-so-supervised: a survey of semi-supervised, multi-instance, and transfer learning in medical image analysis. Med. Image Anal. **54**, 280–296 (2019)
9. Cordonnier, C., Demchuk, A., Ziai, W., Anderson, C.S.: Intracerebral hemorrhage: current approaches to acute management. Lancet **392**(10154), 1257–1268 (2018)
10. Dietterich, T.G., Lathrop, R.H., Lozano-Pérez, T.: Solving the multiple instance problem with axis-parallel rectangles. Artif. Intell. **89**(1–2), 31–71 (1997)
11. Elliott, J., Smith, M.: The acute management of intracerebral hemorrhage: a clinical review. Anesth. Analg. **110**(5), 1419–1427 (2010)
12. Erly, W.K., Berger, W.G., Krupinski, E., Seeger, J.F., Guisto, J.A.: Radiology resident evaluation of head CT scan orders in the emergency department. Am. J. Neuroradiol. **23**(1), 103–107 (2002)
13. Gadermayr, M., Tschuchnig, M.: Multiple instance learning for digital pathology: a review on the state-of-the-art, limitations & future potential. arXiv preprint arXiv:2206.04425 (2022)
14. Grewal, M., Srivastava, M.M., Kumar, P., Varadarajan, S.: RadNet: radiologist level accuracy using deep learning for hemorrhage detection in CT scans. In: 2018 IEEE 15th International Symposium on Biomedical Imaging (ISBI 2018), pp. 281–284. IEEE (2018)
15. Ilse, M., Tomczak, J., Welling, M.: Attention-based deep multiple instance learning. In: International Conference on Machine Learning, pp. 2127–2136. PMLR (2018)
16. Ker, J., Singh, S.P., Bai, Y., Rao, J., Lim, T., Wang, L.: Image thresholding improves 3-dimensional convolutional neural network diagnosis of different acute brain hemorrhages on computed tomography scans. Sensors **19**(9), 2167 (2019)

17. LeCun, Y., Bengio, Y., Hinton, G.: Deep learning. Nature **521**(7553), 436–444 (2015)
18. Li, H., et al.: Multi-modal multi-instance learning using weakly correlated histopathological images and tabular clinical information. In: de Bruijne, M., et al. (eds.) MICCAI 2021. LNCS, vol. 12908, pp. 529–539. Springer, Cham (2021). https://doi.org/10.1007/978-3-030-87237-3_51
19. López-Pérez, M., Schmidt, A., Wu, Y., Molina, R., Katsaggelos, A.K.: Deep gaussian processes for multiple instance learning: application to CT intracranial hemorrhage detection. Comput. Methods Program. Biomed. **219**, 106783 (2022)
20. McDonald, R.J., et al.: The effects of changes in utilization and technological advancements of cross-sectional imaging on radiologist workload. Acad. Radiol. **22**(9), 1191–1198 (2015)
21. Quellec, G., Cazuguel, G., Cochener, B., Lamard, M.: Multiple-instance learning for medical image and video analysis. IEEE Rev. Biomed. Eng. **10**, 213–234 (2017)
22. Qureshi, A.I., Tuhrim, S., Broderick, J.P., Batjer, H.H., Hondo, H., Hanley, D.F.: Spontaneous intracerebral hemorrhage. New England J. Med. **344**(19), 1450–1460 (2001)
23. Ripley, B.: Spatial Statistics. Wiley, New York (1981)
24. Shao, Z., Bian, H., Chen, Y., Wang, Y., Zhang, J., Ji, X., et al.: Transmil: transformer based correlated multiple instance learning for whole slide image classification. Adv. Neural Inf. Process. Syst. **34**, 2136–2147 (2021)
25. Strub, W., Leach, J., Tomsick, T., Vagal, A.: Overnight preliminary head CT interpretations provided by residents: locations of misidentified intracranial hemorrhage. Am. J. Neuroradiol. **28**(9), 1679–1682 (2007)
26. Teneggi, J., Yi, P.H., Sulam, J.: Weakly supervised learning significantly reduces the number of labels required for intracranial hemorrhage detection on head ct. arXiv preprint arXiv:2211.15924 (2022)
27. Titano, J.J., et al.: Automated deep-neural-network surveillance of cranial images for acute neurologic events. Nat. Med. **24**(9), 1337–1341 (2018)
28. Wang, Y., Li, J., Metze, F.: A comparison of five multiple instance learning pooling functions for sound event detection with weak labeling. In: ICASSP 2019–2019 IEEE International Conference on Acoustics, Speech and Signal Processing (ICASSP), pp. 31–35. IEEE (2019)
29. Wu, Y., Schmidt, A., Hernández-Sánchez, E., Molina, R., Katsaggelos, A.K.: Combining attention-based multiple instance learning and gaussian processes for CT hemorrhage detection. In: de Bruijne, M., et al. (eds.) MICCAI 2021. LNCS, vol. 12902, pp. 582–591. Springer, Cham (2021). https://doi.org/10.1007/978-3-030-87196-3_54
30. Ye, H., et al.: Precise diagnosis of intracranial hemorrhage and subtypes using a three-dimensional joint convolutional and recurrent neural network. Eur. Radiol. **29**(11), 6191–6201 (2019). https://doi.org/10.1007/s00330-019-06163-2
31. Yeo, M., et al.: Review of deep learning algorithms for the automatic detection of intracranial hemorrhages on computed tomography head imaging. J. Neurointerventional Surg. **13**(4), 369–378 (2021)

DCAug: Domain-Aware and Content-Consistent Cross-Cycle Framework for Tumor Augmentation

Qikui Zhu[1], Lei Yin[2], Qian Tang[2], Yanqing Wang[3], Yanxiang Cheng[3], and Shuo Li[1](✉)

[1] Department of Biomedical Engineering, Case Western Reserve University, Cleveland, OH, USA
slishuo@gmail.com
[2] School of Computer Science, Wuhan University, Wuhan, China
[3] Department of Gynecology, Renmin Hospital of Wuhan University, Wuhan, China

Abstract. Existing tumor augmentation methods cannot deal with both domain and content information at the same time, causing a content distortion or domain gap (distortion problem) in the generated tumor. To address this challenge, we propose a Domain-aware and Content-consistent Cross-cycle Framework, named DCAug, for tumor augmentation to eliminate the distortion problem and improve the diversity and quality of synthetic tumors. Specifically, DCAug consists of one novel Cross-cycle Framework and two novel contrastive learning strategies: 1) Domain-aware Contrastive Learning (DaCL) and 2) Cross-domain Consistency Learning (CdCL), which disentangles the image information into two solely independent parts: 1) Domain-invariant content information; 2) Individual-specific domain information. During new sample generation, DCAug maintains the consistency of domain-invariant content information while adaptively adjusting individual-specific domain information through the advancement of DaCL and CdCL. We analyze and evaluate DCAug on two challenging tumor segmentation tasks. Experimental results (10.48% improvement in KiTS, 5.25% improvement in ATLAS) demonstrate that DCAug outperforms current state-of-the-art tumor augmentation methods and significantly improves the quality of the synthetic tumors.

Keywords: tumor aware · content-consistent · tumor augmentation

1 Introduction

Existing tumor augmentation methods, including "Copy-Paste" strategy based methods [15–17,19] and style-transfer based methods [5], only considered content or style information when synthesizing new samples, which leads to a distortion gap in content or domain space between the true image and synthetic image, and further causes a distortion problem [14] as shown in Fig. 1(1). The distortion problem damages the effectiveness of DCNNs in feature representation learning as proven in many studies [1,5,18]. Therefore, *a domain and content simultaneously aware data augmentation method is urgently needed to eliminate and avoid*

H. Greenspan et al. (Eds.): MICCAI 2023, LNCS 14224, pp. 338–347, 2023.
https://doi.org/10.1007/978-3-031-43904-9_33

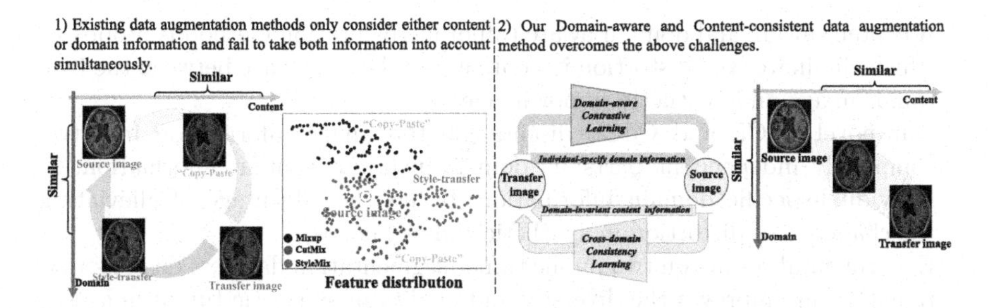

Fig. 1. 1) The t-SNE visualization of the feature distribution of synthesized tumor images from one image by various methods demonstrates that the distortion problem exists. 2) Our DCAug can solve the above challenges through two novel contrastive learning strategies and one newly designed cross-cycle framework.

the distortion challenges during tumor generation. It remains, however, a very challenging task because the content and domain space lack of clear border, and the domain information always influences the distribution of content. This is also the main reason that style transfer [7,8,10] still suffers from spurious artifacts such as disharmonious colors and repetitive patterns, and a large gap is still left between real artwork and synthetic style [2,3]. Therefore, *it's necessary to reduce the influence of the domain on content and keep the content consistent during image generation.*

To overcome the above challenges, a Domain-aware and Content-consistent tumor Augmentation method, named DCAug, is developed (Fig. 1(2)). DCAug consists of two novel contrastive learning strategies, 1) Domain-aware Contrastive Learning (DaCL) and 2) Cross-domain Consistency Learning (CdCL) and one newly designed Cross-cycle Framework, focus on both domain and content information during sample generation, which reduces the content and domain distortion challenge present in existing tumor augmentation methods. Specifically, the core idea of DaCL is to associate the transferred tumor image with target domain examples while disassociating them from the source domain examples that are regarded as "negatives". The CdCL learning strategy is designed to preserve the domain-invariant content information in the synthetic tumor images for avoiding content distortion. When generating synthetic tumor images, CdCL and DaCL disentangle the tumor information into two solely independent parts: 1) Domain-invariant content information; 2) Individual-specific domain information. The domain-invariant content information is preserved for avoiding tumor content distortion through CdCL, and the individual-specific domain information is adaptively transferred by DaCL for eliminating the domain gap between true tumor image and synthetic tumor image. The above goal is achieved via our novel designed Cross-cycle Framework.

Experimental results on two public tumor segmentation datasets show that DCAug improves the tumor segmentation accuracy compared with state-of-the-art tumor augmentation methods. In summary, our contributions are as follows:

- A content-aware and domain-aware tumor augmentation method is proposed, which eliminates the distortion in content and domain space between the true tumor image and synthetic tumor image.
- Our novel DaCL and CdCL disentangle the image information into two completely independent parts: 1) domain-invariant content information; 2) individual-specific domain information. It has the advantage of alleviating the challenge of distortion in synthetic tumor images.
- Experimental results on two public tumor segmentation datasets demonstrate that DCAug improves the diversity and quality of synthetic tumor images.

2 Method

2.1 Problem Definition

Formulation: Given two images and the corresponding tumor labels $\{X_A, Y_A\}$, $\{X_B, Y_B\}$, tumor composition process can be formulated as:

$$X_A^b = X_B \cdot Y_B + X_A \cdot (1 - Y_B), Y_A^b = Y_B + Y_A \cdot (1 - Y_B) \tag{1}$$

$$X_B^a = X_A \cdot Y_A + X_B \cdot (1 - Y_A), Y_B^a = Y_A + Y_B \cdot (1 - Y_A) \tag{2}$$

where $\cdot$ is element-wise multiplication, X_A^b represents the tumor in image X_B is copied to image X_A, X_B^a represents the tumor in image X_A is copied to image X_B, Y_A^b and Y_b^A is the corresponding new tumor labels of X_A^b, X_B^a, respectively. There are two challenges need to be solved: 1) $X_A^{b \to A}$, $X_B^{a \to B}$, by adjusting the domain information of the copied tumor, making the copied tumor have the same domain space as the target image to avoid domain distortion; 2) $X_A^{b \to A} \rightleftarrows X_A^b$, $X_B^{a \to B} \rightleftarrows X_B^a$, maintaining the domain-invariant content information consistency during tumor copy to avoid content distortion.

To achieve the above goals, a novel Cross-cycle Framework (Fig. 2) is designed, which consists of two generators and can disentangle the tumor information into two solely independent parts: 1) Domain-invariant content information, 2) Individual-specific domain information, through two new learning strategies: 1) Domain-aware contrastive learning (DaCL); 2) Cross-domain consistency learning (CdCL). When generating new sample, the domain-invariant content information is preserved by CdCL, while the individual-specific domain information is adjusted by DaCL based on the domain space of target tumor image. The details are described as follows.

2.2 Domain-Aware Contrastive Learning for Domain Adaptation

Our domain-aware contrastive learning (DaCL) strategy can adaptively adjust the domain space of the transferred tumor and makes the domain space consistent for domain adaptation. Specifically, the input of DCAug is two combined images X_A^b, X_B^a that consist of source images and tumor regions copied from another image. The synthetic tumors $X_A^{b \to A}$ generated by the generator, the

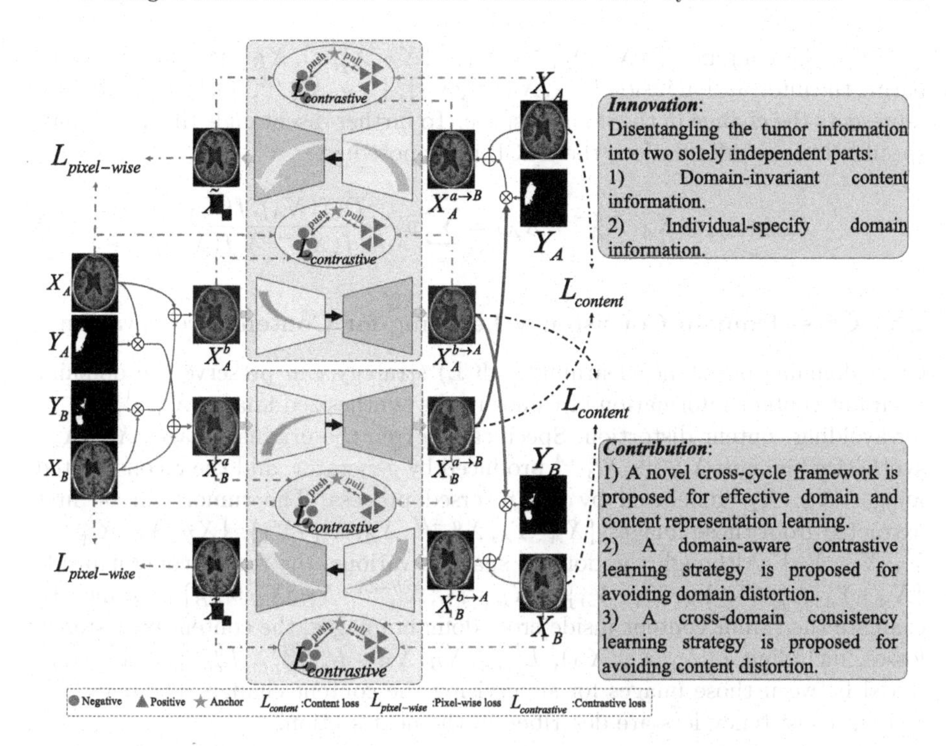

Fig. 2. Our cross-cycle framework disentangles the tumor information into two solely independent parts by two newly learning strategies for avoiding content distortion and eliminating the domain gap between true tumor and synthetic tumor.

source image X_A, and the combined image X_A^b as the anchor, the positive and the negative sample, respectively. To find the domain space of these samples for contrast, a fixed pre-trained style representation extractor f is used to obtain domain representations for different images. Thus, DaCL between the anchor, the positive, and the negative sample can be formulated as:

$$L_{contrastive}(X_A^{b\to A}, X_A, X_A^b) = \sum_{i=1}^{n} w_i \frac{D(f(X_A^{b\to A}), f(X_A))}{D(f(X_A^{b\to A}), f(X_A^b))} \tag{3}$$

where $D(x, y)$ is the L_2 distance between x and y, w_i is weighting factor.

Additionally, to further disentangle the individual-specific domain information, a reversed process is designed. By utilizing the synthetic tumors $X_B^{a\to B}$, $X_A^{b\to A}$, the reversed images $X_A^{a\to B}$,$X_B^{b\to A}$ can be construed as:

$$X_A^{a\to B} = X_B^{a\to B} \cdot Y_A + X_A \cdot (1 - Y_A), X_B^{b\to A} = X_A^{b\to A} \cdot Y_B + X_B \cdot (1 - Y_B) \tag{4}$$

The whole reversed process receives the reversed images $X_A^{a\to B}$, $X_B^{b\to A}$ as inputs and tries to restore the original domain information of the synthetic tumor $\tilde{X}_A$, $\tilde{X}_B$.

Since the content in $\{X_A, X_A^{a\to B}, \tilde{X}_A\}$, $\{X_B, X_B^{b\to A}, \tilde{X}_B\}$ is same, by comparing the information inside $\{X_A, X_A^{a\to B}, \tilde{X}_A\}$, $\{X_B, X_B^{b\to A}, \tilde{X}_B\}$, the difference represents the change in the domain space. To further disentangle the individual-specific domain information, the DaCL is proposed:

$$L_{contrastive}(X_A, X_A^{a\to B}, \tilde{X}_A) = \sum_{i=1}^{n} w_i \frac{D(f(X_A), f(\tilde{X}_A))}{D(f(X_A^{a\to B}), f(X_A))} \tag{5}$$

2.3 Cross-Domain Consistency Learning for Content Preservation

Cross-domain consistency learning (CdCL) strategy can preserve the domain-invariant content information of tumor in the synthesized images $X_A^{b\to A}$, $X_B^{a\to B}$ for avoiding content distortion. Specifically, given the original images X_A, X_B, synthesized images $X_B^{a\to B}$, $X_A^{b\to A}$ produced by generator, and the reconstructed images $\tilde{X}_A$, $\tilde{X}_B$ generated by the reversed process. The tumor can be first extracted from those images $\{X_A \cdot Y_A, X_B^{a\to B} \cdot Y_A, \tilde{X}_A \cdot Y_A\}$, $\{X_B \cdot Y_B, X_A^{b\to A} \cdot Y_B, \tilde{X}_B \cdot Y_B\}$. Although the domain space is various, the tumor content inside $\{X_A \cdot Y_A, X_B^{a\to B} \cdot Y_A, \tilde{X}_A \cdot Y_A\}$, $\{X_B \cdot Y_B, X_A^{b\to A} \cdot Y_B, \tilde{X}_B \cdot Y_B\}$ is same. To evaluate the tumor content inside cross-domain images, the content consistency losses, including $L_{pixel}^{A}(X_A, \tilde{X}_A)$, $L_{pixel}^{B}(X_B, \tilde{X}_B)$, $L_{content}^{a\to B}$, $L_{content}^{b\to A}$, are computed between those images for supervising the content change. The details of content consistency loss are described in the next section.

2.4 Loss Function

In summary, three types of losses are used to supervise the cross-cycle framework. Specifically, given the original images X_A, X_B and the combined images X_A^b, X_a^B, the synthesized images $X_B^{a\to B}$, $X_A^{b\to A}$ are produced by the generator, and the reconstructed images $\tilde{X}_A$, $\tilde{X}_B$ are generated by the reversed process.

The pixel-wise loss (L_{pixel}) computes the difference between original images and reconstructed images at the pixel level.

$$L_{pixel}^{A}(X_A, \tilde{X}_A) = \left\|X_A - \tilde{X}_A\right\|_1, L_{pixel}^{B}(X_B, \tilde{X}_B) = \left\|X_B - \tilde{X}_B\right\|_1 \tag{6}$$

To disentangle the individual-specific domain information, the higher feature representations extracted from pre-trained networks combined with CL are used:

$$\begin{aligned} L_{contrastive}^{b\to A} &= L_{contrastive}(X_A, X_A^{b\to A}, \tilde{X}_A) + L_{contrastive}(X_A, X_A^{a\to B}, \tilde{X}_A) \\ L_{contrastive}^{a\to B} &= L_{contrastive}(X_B, X_B^{a\to B}, \tilde{X}_B) + L_{contrastive}(X_B, X_A^{b\to A}, \tilde{X}_B) \end{aligned} \tag{7}$$

And two content loss $L_{content}^{b\to A}$, $L_{content}^{a\to B}$ are employed to maintain tumor content information during the domain adaptation:

$$\begin{aligned} L_{content}^{a\to B} = \sum_{i=1}^{L} (&\left\|\mu(\phi_i(X_B^{a\to B} \cdot Y_A)) - \mu(\phi_i(X_A \cdot Y_A))\right\|_2 \\ &+ \left\|\sigma(\phi_i(X_B^{a\to B} \cdot Y_A)) - \sigma(\phi_i(X_A \cdot Y_A))\right\|_2) \end{aligned} \tag{8}$$

$$L_{content}^{b\to A} = \sum_{i=1}^{L}(\left\|\mu(\phi_i(X_A^{b\to A}\cdot Y_B)) - \mu(\phi_i(X_B\cdot Y_B))\right\|_2 + \left\|\sigma(\phi_i(X_A^{b\to A}\cdot Y_B)) - \sigma(\phi_i(X_B\cdot Y_B))\right\|_2) \tag{9}$$

where ϕ denotes the ith layer of the VGG-19 network, μ and σ represent the mean and standard deviation of feature maps extracted by ϕ, respectively.

In summary, the total loss for the cross-cycle framework is

$$L_{total} = \alpha(L_{pixel}^{A} + L_{pixel}^{B}) + \beta(L_{contrastive}^{b\to A} + L_{contrastive}^{a\to B}) + \gamma(L_{content}^{b\to A} + L_{content}^{a\to B}) \tag{10}$$

where α, β, and γ represent the weight coefficients.

3 Experiments

3.1 Datasets and Implementation Details

ATLAS Dataset [11]**:** The ATLAS dataset consists of 229 T1-weighted MR images from 220 subjects with chronic stroke lesions. These images were acquired from different cohorts and different scanners. The chronic stroke lesions are annotated by a group of 11 experts. The dimension of the pre-processed images is $197 \times 233 \times 189$ with an isotropic $1mm^3$ resolution. Identical with the study in [17], We selected 50 images as the test set and the rest of the cases as the training set.

KiTS19 Dataset [4]**:** The KiTS19 consists of 210 3D abdominal CT images with kidney tumor subtypes and segmentation of kidney and kidney tumors. These CT images are from more than 50 institutions and scanned with different CT scanners and acquisition protocols. In our experiment, we randomly split the published 210 images into a training set with 168 images and a testing set with 42 images.

Training Details: The generator in DCAug is built on the RAIN [12] backbone, all of the weights in generators are shared. Our DCAug is implemented using PyTorch [13] and trained end-to-end with Adam [9] optimization method. In the training phase, the learning rate is initially set to 0.0001 and decreased by a weight decay of 1.0×10^{-6} after each epoch. The experiments were carried out on one NVIDIA RTX A4000 GPU with 16 GB memory. The weight valule of α, β, and γ is 1.0,1.0,1.0, separately.

Baseline: nnUNet [6] is selected as the baseline model. The default hyperparameters and default traditional data augmentation (TDA) including rotation, scaling, mirroring, elastic deformation, intensity perturbation are used when model training. The maximum number of training epochs was set to 500 for the two datasets. Parts of tumors generated are shown in Fig. 3. And the dice coefficients of the segmentation results on the same test set are computed to evaluate the effectiveness of methods.

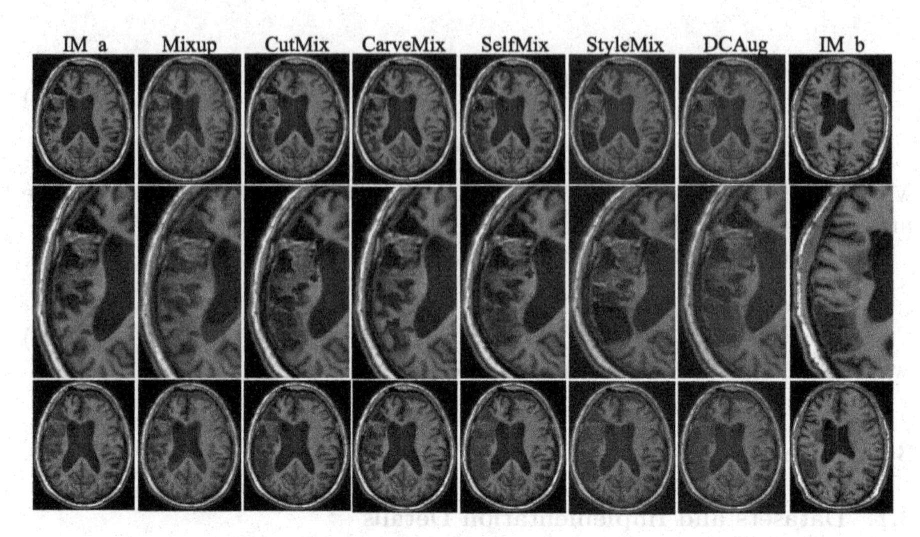

Fig. 3. Part of augmented samples produced by various data augmentation methods.

Table 1. Means and standard deviations of the Dice coefficients (%) of the segmentation results on ATLAS/KiTS19 dataset.

Dataset	Num	Means and Standard deviations of the Dice coefficients (%)						
		TDA	Mixup	CutMix	CarveMix	SelfMix	StyleMix	DCAug
ATLAS	25%	49.87 ± 32.19	49.18 ± 32.72	41.19 ± 33.98	55.16 ± 32.16	**57.89 ± 31.05**	52.84 ± 34.36	56.43 ± 32.33
	50%	56.72 ± 30.74	58.40 ± 29.35	54.25 ± 30.24	58.34 ± 31.32	58.81 ± 31.75	58.04 ± 30.39	**59.75 ± 31.41**
	100%	59.39 ± 32.45	59.33 ± 33.06	56.11 ± 32.44	62.32 ± 31.10	63.5 ± 31.06	64.00 ± 28.89	**64.64 ± 29.91**
KiTS19	25%	65.41 ± 31.93	62.82 ± 27.84	61.59 ± 30.36	64.20 ± 35.08	65.91 ± 29.54	65.97 ± 29.54	**72.29 ± 29.08**
	50%	68.25 ± 24.41	68.29 ± 22.38	62.23 ± 31.07	75.31 ± 23.38	73.95 ± 27.81	75.04 ± 27.05	**77.33 ± 27.48**
	100%	72.63 ± 24.40	73.94 ± 22.68	73.77 ± 29.68	79.99 ± 22.98	79.74 ± 20.43	79.04 ± 18.90	**83.11 ± 14.15**

3.2 Comparison with State-of-the-Art Methods

Experimental results in Table 1 and Fig. 4 show that compared with other state-of-the-art methods, including Mixup [16], CutMix [15], CarveMix [17], SelfMix [19], StyleMix [5], nnUnet combined with DCAug achieves the highest improvement on the two datasets, which convincingly demonstrates the innovations and contribution of DCAug in generating higher quality tumor. And it is worth noting that CutMix ("Copy-Paste"method that only considers content information) even degrades the segmentation performance, which indicates that both content and domain information has a significant influence on the tumor segmentation. The representative segmentation scans are shown in Fig. 4. Our DCAug produced better segmentation results than the competing methods, which further proves the effectiveness of DCAug in tumor generation.

What's more, the potential of DCAug in an extremely low-data regime is also demonstrated. We randomly select 25% and 50% of data from the training set same as training data. DCAug also assists the baseline model to achieve higher

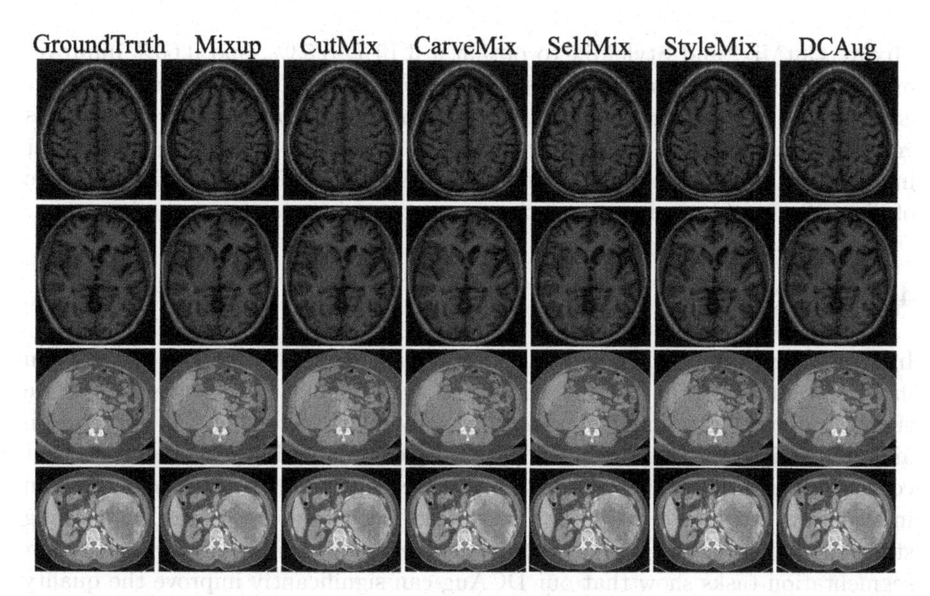

Fig. 4. Qualitative comparison with same samples segmented by nnUNet trained by various data augmentation methods.

Table 2. Means and standard deviations of the Dice coefficients (%) of the segmentation results on the test set for the ATLAS.

Method	Dice coefficients (%)	Method	Dice coefficients (%)
Mixup	59.33 ± 33.06	Mixup → DCAug	**62.48 ± 31.10** (3.15↑)
CutMix	56.11 ± 32.44	CutMix → DCAug	**64.64 ± 29.91** (8.53↑)
StyleMix	64.00 ± 28.89	StyleMix → DCAug	**64.60 ± 29.93** (0.60↑)

Dice coefficients, which convincingly demonstrates the effectiveness of DCAug in generating new tumor samples.

3.3 Significant in Improving Existing Tumor Augmentation Methods

The necessity of considering both content and domain information in the tumor generation is also demonstrated, three representative methods, Mixup ("Copy-Paste"), CutMix ("Copy-Paste"), and StyleMix (style-transfer), are selected. The DCAug optimizes generated samples from above methods from content and domain aspects to further improve the quality of generated samples. And the nnUet are trained by optimized samples. From the segmentation performances (Table 2), we can notice that DCAug can further boost the quality of generated samples produced by existing methods. Specifically, the DCAug assists the

Mixup, CutMix, and StyleMix to obtain a 3.15%, 8.53%, and 0.60% improvement in segmentation performance, respectively, which demonstrates that 1) it is necessary to consider both content and domain information during samples generation; 2) avoiding the content and domain distortion challenge can further improve the quality of generated samples; 3) DCAug can alleviate the challenge of distortion problem present in existing tumor augmentation methods.

4 Conclusion

In this paper, our domain-aware and content-consistent tumor augmentation method eliminated the content distortion and domain gap between the true tumor and synthetic tumor by simultaneously focusing the content information and domain information. Specifically, DCAug can maintain the domain-invariant content information consistency and adaptive adjust individual-specific domain information by a new cross-cycle framework and two novel contrastive learning strategies when generating synthetic tumor. Experimental results on two tumor segmentation tasks show that our DCAug can significantly improve the quality of the synthetic tumors, eliminate the gaps, and has practical value in medical imaging applications.

Acknowledgments. This work was supported by cross-innovation talent project in Renmin Hospital of Wuhan University (grant number JCRCZN-2022-016); Natural Science Foundation of Hubei Province (grant number 2022CFB252); Undergraduate education quality construction comprehensive reform project (grant number 2022ZG282) and the National Natural Science Foundation of China (grant number 81860276).

References

1. Chen, C., Li, J., Han, X., Liu, X., Yu, Y.: Compound domain generalization via meta-knowledge encoding. In: Proceedings of the IEEE/CVF Conference on Computer Vision and Pattern Recognition, pp. 7119–7129 (2022)
2. Chen, H., et al.: Artistic style transfer with internal-external learning and contrastive learning. Adv. Neural Inf. Process. Syst. **34**, 26561–26573 (2021)
3. Geirhos, R., Rubisch, P., Michaelis, C., Bethge, M., Wichmann, F.A., Brendel, W.: ImageNet-trained CNNs are biased towards texture; increasing shape bias improves accuracy and robustness. In: International Conference on Learning Representations (2018)
4. Heller, N., et al.: The kits19 challenge data: 300 kidney tumor cases with clinical context, CT semantic segmentations, and surgical outcomes. arXiv preprint arXiv:1904.00445 (2019)
5. Hong, M., Choi, J., Kim, G.: Stylemix: separating content and style for enhanced data augmentation. In: Proceedings of the IEEE/CVF Conference on Computer Vision and Pattern Recognition, pp. 14862–14870 (2021)
6. Isensee, F., Jaeger, P.F., Kohl, S.A., Petersen, J., Maier-Hein, K.H.: NNU-net: a self-configuring method for deep learning-based biomedical image segmentation. Nat. Methods **18**(2), 203–211 (2021)

7. Isola, P., Zhu, J.Y., Zhou, T., Efros, A.A.: Image-to-image translation with conditional adversarial networks. In: Proceedings of the IEEE Conference on Computer Vision and Pattern Recognition, pp. 1125–1134 (2017)
8. Jeong, S., Kim, Y., Lee, E., Sohn, K.: Memory-guided unsupervised image-to-image translation. In: Proceedings of the IEEE/CVF Conference on Computer Vision and Pattern Recognition, pp. 6558–6567 (2021)
9. Kingma, D.P., Ba, J.: Adam: a method for stochastic optimization. arXiv preprint arXiv:1412.6980 (2014)
10. Kotovenko, D., Sanakoyeu, A., Ma, P., Lang, S., Ommer, B.: A content transformation block for image style transfer. In: Proceedings of the IEEE/CVF Conference on Computer Vision and Pattern Recognition, pp. 10032–10041 (2019)
11. Liew, S.L., et al.: A large, open source dataset of stroke anatomical brain images and manual lesion segmentations. Sci. Data **5**(1), 1–11 (2018)
12. Ling, J., Xue, H., Song, L., Xie, R., Gu, X.: Region-aware adaptive instance normalization for image harmonization. In: Proceedings of the IEEE/CVF Conference on Computer Vision and Pattern Recognition, pp. 9361–9370 (2021)
13. Paszke, A., et al.: PyTorch: an imperative style, high-performance deep learning library. In: Advances in Neural Information Processing Systems, pp. 8024–8035 (2019)
14. Wang, R., Zheng, G.: CyCMIS: cycle-consistent cross-domain medical image segmentation via diverse image augmentation. Med. Image Anal. **76**, 102328 (2022)
15. Yun, S., Han, D., Oh, S.J., Chun, S., Choe, J., Yoo, Y.: CutMix: regularization strategy to train strong classifiers with localizable features. In: Proceedings of the IEEE/CVF International Conference on Computer Vision, pp. 6023–6032 (2019)
16. Zhang, H., Cisse, M., Dauphin, Y.N., Lopez-Paz, D.: mixup: beyond empirical risk minimization. arXiv preprint arXiv:1710.09412 (2017)
17. Zhang, X., et al.: CarveMix: a simple data augmentation method for brain lesion segmentation. In: de Bruijne, M., et al. (eds.) MICCAI 2021. LNCS, vol. 12901, pp. 196–205. Springer, Carvemix: A simple data augmentation method for brain lesion segmentation (2021). https://doi.org/10.1007/978-3-030-87193-2_19
18. Zhu, Q., Du, B., Yan, P.: Boundary-weighted domain adaptive neural network for prostate MR image segmentation. IEEE Trans. Med. Imaging **39**(3), 753–763 (2019)
19. Zhu, Q., Wang, Y., Yin, L., Yang, J., Liao, F., Li, S.: SelfMix: a self-adaptive data augmentation method for lesion segmentation. In: Wang, L., Dou, Q., Fletcher, P.T., Speidel, S., Li, S. (eds.) Medical Image Computing and Computer Assisted Intervention - MICCAI 2022. MICCAI 2022. LNCS, vol. 13434, pp 683–692. Springer, Cham (2022). https://doi.org/10.1007/978-3-031-16440-8_65

Beyond the Snapshot: Brain Tokenized Graph Transformer for Longitudinal Brain Functional Connectome Embedding

Zijian Dong[1,2], Yilei Wu[1], Yu Xiao[1], Joanna Su Xian Chong[1], Yueming Jin[2,4], and Juan Helen Zhou[1,2,3(✉)]

[1] Centre for Sleep and Cognition and Centre for Translational Magnetic Resonance Research, Yong Loo Lin School of Medicine, National University of Singapore, Singapore, Singapore

helen.zhou@nus.edu.sg

[2] Department of Electrical and Computer Engineering, National University of Singapore, Singapore, Singapore

[3] Integrative Sciences and Engineering Programme (ISEP), NUS Graduate School, National University of Singapore, Singapore, Singapore

[4] Department of Biomedical Engineering, National University of Singapore, Singapore, Singapore

Abstract. Under the framework of network-based neurodegeneration, brain functional connectome (FC)-based Graph Neural Networks (GNN) have emerged as a valuable tool for the diagnosis and prognosis of neurodegenerative diseases such as Alzheimer's disease (AD). However, these models are tailored for brain FC at a single time point instead of characterizing FC trajectory. Discerning how FC evolves with disease progression, particularly at the predementia stages such as cognitively normal individuals with amyloid deposition or individuals with mild cognitive impairment (MCI), is crucial for delineating disease spreading patterns and developing effective strategies to slow down or even halt disease advancement. In this work, we proposed the *first* interpretable framework for brain FC trajectory embedding with application to neurodegenerative disease diagnosis and prognosis, namely *Brain Tokenized Graph Transformer* (Brain TokenGT). It consists of two modules: 1) *Graph Invariant and Variant Embedding* (GIVE) for generation of node and spatio-temporal edge embeddings, which were tokenized for downstream processing; 2) *Brain Informed Graph Transformer Readout* (BIGTR) which augments previous tokens with trainable type identifiers and non-trainable node identifiers and feeds them into a standard transformer encoder to readout. We conducted extensive experiments on two public longitudinal fMRI datasets of the AD continuum for three tasks, including differentiating MCI from controls, predicting dementia conversion in MCI, and classification of amyloid positive or negative cognitively normal

Supplementary Information The online version contains supplementary material available at https://doi.org/10.1007/978-3-031-43904-9_34.

H. Greenspan et al. (Eds.): MICCAI 2023, LNCS 14224, pp. 348–357, 2023.
https://doi.org/10.1007/978-3-031-43904-9_34

individuals. Based on brain FC trajectory, the proposed Brain TokenGT approach outperformed all the other benchmark models and at the same time provided excellent interpretability.

Keywords: Functional connectome · Graph neural network · Tokenization · Longitudinal analysis · Neurodegenerative disease

1 Introduction

The brain functional connectome (FC) is a graph with brain regions of interest (ROIs) represented as nodes and pairwise correlations of fMRI time series between the ROIs represented as edges. FC has been shown to be a promising biomarker for the early diagnosis and tracking of neurodegenerative disease progression (*e.g.*, Alzheimer's Disease (AD)) because of its ability to capture disease-related alternations in brain functional organization [25,26]. Recently, the graph neural networks (GNN) has become the model of choice for processing graph structured data with state-of-the-art performance in different tasks [2,11,20]. With regards to FC, GNN has also shown promising results in disease diagnosis [3,4,8,15,23]. However, such studies have only focused on FC at a single time point. For neurodegenerative diseases like AD, it is crucial to investigate longitudinal FC changes [5], including graph topology *and* attributes, in order to slow down or even halt disease advancement.

Node features are commonly utilized in FC to extract important information. It is also essential to recognize the significance of edge features in FC, which are highly informative in characterizing the interdependencies between ROIs. Furthermore, node embeddings obtained from GNN manipulation contain essential information that should be effectively leveraged. Current GNNs feasible to graphs with multiple time points [16,22,24] are suboptimal to FC trajectory, as they fail to incorporate brain edge feature embeddings and/or they rely on conventional operation (*e.g.*, global pooling for readout) which introduces inductive bias and is incapable of extracting sufficient information from the node embeddings [21]. Moreover, these models lack built-in interpretability, which is crucial for clinical applications. And they are unsuitable for small-scale datasets which are common in fMRI research. The longitudinal data with multiple time points of the AD continuum is even more scarce due to the difficulty in data acquisition.

In this work, we proposed *Brain Tokenized Graph Transformer* (Brain TokenGT), the *first* framework to achieve FC trajectory embeddings with built-in interpretability, shown in Fig. 1. Our contributions are as follows: 1) Drawing on the distinctive characteristics of FC trajectories, we developed *Graph Invariant and Variant Embedding* (GIVE), which is capable of generating embeddings for both nodes and spatio-temporal edges; 2) Treating embeddings from GIVE as tokens, *Brain Informed Graph Transformer Readout* (BIGTR) augments tokens with trainable type identifiers and non-trainable node identifiers and feeds them into a standard transformer encoder to readout instead of global pooling, further extracting information from tokens and alleviating over-fitting issue by token-level task; 3) We conducted extensive experiments on two public resting state fMRI datasets (ADNI, OASIS) with three different tasks (Healthy Control (HC) vs. Mild Cognition Impairment (MCI) classification, AD conversion prediction and Amyloid positive vs. negative classification). Our model showed superior results with FC trajectory as input, accompanied by node and edge level interpretations.

2 Method

2.1 Problem Definition

The input of one subject to the proposed framework is a sequence of brain networks $\mathcal{G} = [G_1, G_2, ..., G_t, ..., G_T]$ with T time points. Each network is a graph $G = (V, E, \boldsymbol{A})$, with the node set $V = \{v_i\}_{i=1}^{M}$, the edge set $E = V \times V$, and the weighted adjacency matrix $\boldsymbol{A} \in \mathbb{R}^{M \times M}$ describing the degrees of FC between ROIs. The output of the model is an individual-level categorical diagnosis $\hat{y}_s$ for each subject s.

2.2 Graph Invariant and Variant Embedding (GIVE)

Regarding graph topology, one of the unique characteristics of FC across a trajectory is that it has invariant number and sequence of nodes (ROIs), with variant connections between different ROIs. Here, we designed GIVE, which consists of Invariant Node Embedding (INE) and Variant Edge Embedding (VEE).

Invariant Node Embedding (INE). To obtain node embeddings that capture the spatial and temporal information of the FC trajectory, we utilized evolving graph convolution [16] for the K-hop neighbourhood around each node which could be seen as a fully dynamic graph, providing a novel "zoom in" perspective to see FC. As suggested in [16], with informative node features, we chose to treat parameters in graph convolutional layers as hidden states of the dynamic system and used a gated recurrent unit (GRU) to update the hidden states.

Formally, for each node v_i in V, we define a dynamic neighbourhood graph as $\mathscr{G}_i = [g_{i1}, g_{i2}, .., g_{it}, ..., g_{iT}]$ (Fig. 1), in which g_{it} is the K-hop neighbourhood of node v_i at time point t, with adjacency matrix $\boldsymbol{A}_{it}$. At time t, for dynamic neighbourhood graph $\mathscr{G}_i$, l-th layer of evolving graph convolution first updates parameter matrix $\boldsymbol{W}_{i(t-1)}^{l}$ from the last time point to $\boldsymbol{W}_{it}^{l}$ with GRU, then the node embeddings $\boldsymbol{H}_{it}^{l}$ are updated to $\boldsymbol{H}_{it}^{l+1}$ for next layer using graph convolution network (GCN) [11]:

$$\boldsymbol{W}_{it}^{l} = \mathrm{GRU}(\boldsymbol{H}_{it}^{l}, \boldsymbol{W}_{i(t-1)}^{l}); \quad \boldsymbol{H}_{it}^{l+1} = \mathrm{GCN}(\boldsymbol{A}_{it}, \boldsymbol{H}_{it}^{l}, \boldsymbol{W}_{it}^{l}) \tag{1}$$

Variant Edge Embedding (VEE). For tasks such as graph classification, an appropriate representation of edges also plays a key role in the successful graph representation learning. To achieve edge embeddings, we first integrated graphs from multiple time points by defining *Spatial Edge* and *Temporal Edge*,

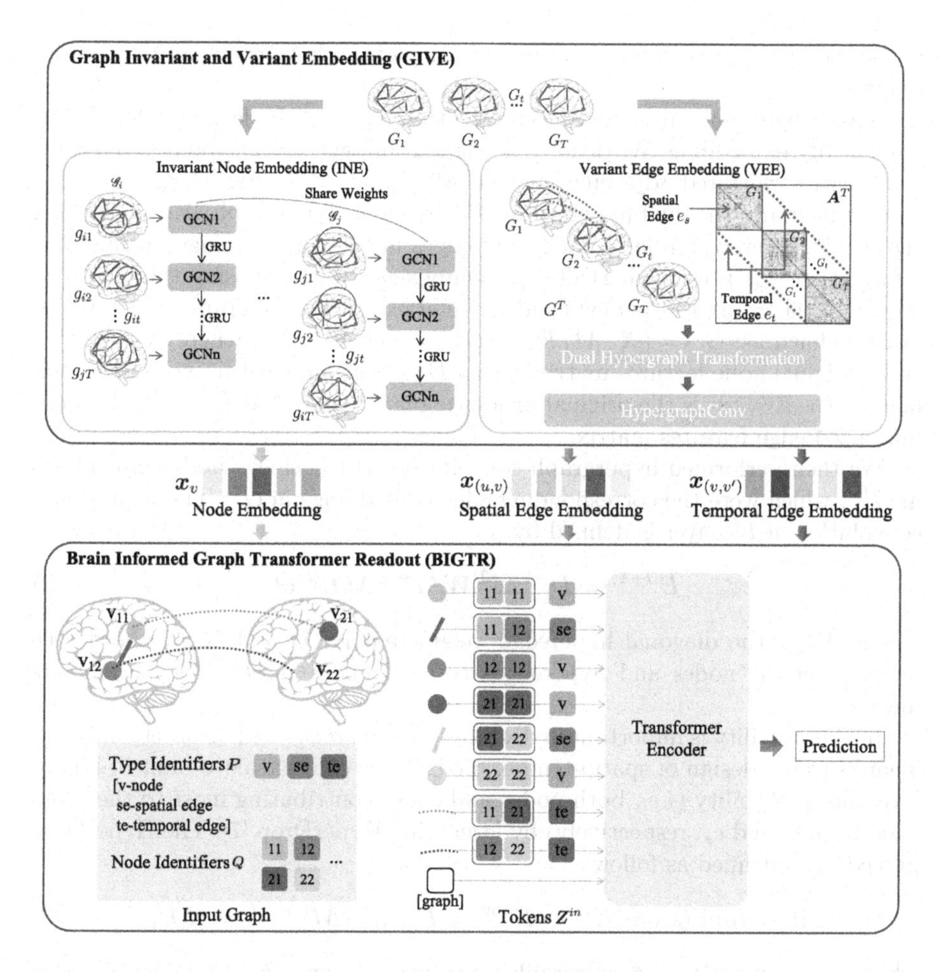

Fig. 1. An overview of Brain TokenGT. In GIVE, INE generates node embedding by performing evolving convolution on dynamic neighbourhood graph, and VEE combines different time points by defining spatio-temporal edge, and then transforms the whole trajectory into a dual hypergraph and produces spatial and temporal edge embedding. These embeddings, augmented by trainable type identifiers and non-trainable node identifiers, are used as input to a standard transformer encoder for readout within BIGTR.

and then obtained spatial and temporal edge embeddings by transforming an FC trajectory to the dual hypergraph.

For each FC trajectory, we should not only investigate the edges between different ROIs in one static FC (*i.e.*, spatial domain) but also capture the longitudinal change across different time points (*i.e.*, time domain). Instead of focusing only on intrinsic connections (*i.e.*, *spatial edges* (e_s)) between different ROIs in each FC, for each of the two consecutive graphs G_t and G_{t+1}, we added M *temporal edges* (e_t) to connect corresponding nodes in G_t and G_{t+1}, with weights initialized as 1. The attached features to spatial and temporal edges were both initialized by the concatenation of node features from both ends and their initial weights.

Accordingly, one trajectory would be treated as a single graph for downstream edge embedding. We denote the giant graph with T time points contained as G^T, with weighted adjacency matrix $\boldsymbol{A}^T \in \mathbb{R}^{TM\times TM}$ (Fig. 1). G^T was first transformed into the dual hypergraph G^{T*} by Dual Hypergraph Transformation (DHT) [7], where the role of nodes and edges in G^T was exchanged while their information was preserved. DHT is accomplished by transposing the incidence matrix of the graph to the new incidence matrix of the dual graph, which is formally defined as: $G^T = (\boldsymbol{X}, \boldsymbol{M}, \boldsymbol{E}) \mapsto G^{T*} = (\boldsymbol{E}, \boldsymbol{M}^T, \boldsymbol{X})$, where $\boldsymbol{X} \in \mathbb{R}^{M\times D}$ is the original node features matrix with a D dimensional feature vector for each node, $\boldsymbol{M} \in \mathbb{R}^{|E|\times M}$ is the original incidence matrix, and $\boldsymbol{E} \in \mathbb{R}^{|E|\times(2D+1)}$ is the initialized edge features matrix.

We then performed hypergraph convolution [1] to achieve node embeddings in G^{T*}, which were the corresponding edge embeddings in G^T. The hypergraph convolution at l^{th} layer is defined by:

$$\boldsymbol{E}^{(l+1)} = \boldsymbol{D}^{-1}\boldsymbol{M}^T\boldsymbol{W}^*\boldsymbol{B}^{-1}\boldsymbol{M}\boldsymbol{E}^{(l)}\boldsymbol{\Theta} \tag{2}$$

where $\boldsymbol{W}^*$ is the diagonal hyperedge weight matrix, $\boldsymbol{D}$ and $\boldsymbol{B}$ are the degree matrices of the nodes and hyperedges respectively, and $\boldsymbol{\Theta}$ is the parameters matrix.

Interpretability is important in decision-critical areas (*e.g.*, disorder analysis). Thanks to the design of spatio-temporal edges, we could achieve built-in binary level interpretability (*i.e.*, both nodes and edges contributing most to the given task, from e_t and e_s, respectively) by leveraging HyperDrop [7]. The HyperDrop procedure is defined as follows:

$$\text{idx} = \text{TopE}(\text{score}(\boldsymbol{E})); \quad \boldsymbol{E}^{\text{pool}} = \boldsymbol{E}_{\text{idx}}; \quad (\boldsymbol{M}^{\text{pool}})^T = (\boldsymbol{M}^T_{\text{idx}}) \tag{3}$$

where 'score' function is hypergraph convolution layers used to compute scores for each hypergraph node (e_s or e_t in the original graph). 'TopE' selects the nodes with the highest E scores (note: ranking was performed for nodes from e_s and e_t separately, and HyperDrop was only applied to nodes from e_s with hyperparameter E), and idx is the node-wise indexing vector. Finally, the salient nodes (from e_t) and edges (from e_s) were determined by ranking the scores averaged across the trajectory.

2.3 Brain Informed Graph Transformer Readout (BIGTR)

Proper readout for the embeddings from GNN manipulation is essential to produce meaningful prediction outcome for assisting diagnosis and prognosis. The vanilla ways are feeding the Node Embeddings, and Spatial and Temporal Edge Embeddings generated from the GIVE module into pooling and fully connected layers. However, this would result in a substantial loss of spatial and temporal information [21], especially under the complex settings of three types of spatial/temporal embeddings. Recently, it has been shown, both in theory and practice, that a standard transformer with appropriate token embeddings yields a powerful graph learner [10]. Here, treating embeddings output from GIVE as tokens, we leveraged graph transformer as a trainable readout function, named as Brain Informed Graph Transformer Readout (BIGTR) (Fig. 1).

We first define the Type Identifier (TI) and Node Identifier (NI) under the setting of FC trajectory. *Trainable TI* encodes whether a token is a node, spatial edge or temporal edge. They are defined as a parameter matrix $[\boldsymbol{P}_v; \boldsymbol{P}_{e_s}; \boldsymbol{P}_{e_t}] \in \mathbb{R}^{3 \times d_p}$, where $\boldsymbol{P}_v$, $\boldsymbol{P}_{e_s}$ and $\boldsymbol{P}_{e_t}$ are node, spatial edge and temporal edge identifier respectively. Specifically, we maintained a dictionary, in which the keys are types of the tokens, the values are learnable embeddings that encodes the corresponding token types. It facilitates the model's learning of type-specific attributes in tokens, compelling attention heads to focus on disease-related token disparities, thereby alleviating overfitting caused by non-disease-related attributes. Besides, it inflates 1 G^T for an individual-level task to thousands of tokens, which could also alleviate overfitting in the perspective of small-scale datasets. *Non-trainable NI* are MT node-wise orthonormal vectors $\boldsymbol{Q} \in \mathbb{R}^{MT \times d_q}$ for an FC trajectory with T time points and M nodes at each time. Then, the augmented token features become:

$$\begin{aligned} \boldsymbol{z}_v &= [\boldsymbol{x}_v, \boldsymbol{P}_v, \boldsymbol{Q}_v, \boldsymbol{Q}_v] \\ \boldsymbol{z}_{(u,v)} &= [\boldsymbol{x}_{(u,v)}, \boldsymbol{P}_{e_s}, \boldsymbol{Q}_u, \boldsymbol{Q}_v] \\ \boldsymbol{z}_{(v,v')} &= [\boldsymbol{x}_{(v,v')}, \boldsymbol{P}_{e_t}, \boldsymbol{Q}_v, \boldsymbol{Q}_{v'}] \end{aligned} \tag{4}$$

for v, e_s and e_t respectively, where node u is a neighbour to node v in the spatial domain and node v' is a neighbour to node v in the temporal domain, and $\boldsymbol{x}$ is the original token from GIVE. Thus, the augmented token features matrix is $\boldsymbol{Z} \in \mathbb{R}^{(MT+|E|T+M(T-1)) \times (h+d_p+2d_q)}$, where h is the hidden dimension of embeddings from GIVE. $\boldsymbol{Z}$ would be further projected by a trainable matrix $\omega \in \mathbb{R}^{(h+d_p+2d_q) \times h'}$. As we targeted individual-level (*i.e.*, G^T) diagnosis/prognosis, a graph token $\boldsymbol{X}_{[\text{graph}]} \in \mathbb{R}^{h'}$ was appended as well. Thus, the input to transformer is formally defined as:

$$\boldsymbol{Z}^{in} = [\boldsymbol{X}_{[\text{graph}]}; \boldsymbol{Z}\omega] \in \mathbb{R}^{(1+MT+|E|T+M(T-1)) \times h'} \tag{5}$$

3 Experiments

Datasets and Experimental Settings. We used brain FC metrics derived from ADNI [6] and OASIS-3 [13] resting state fMRI datasets, following preprocessing pipelines [12,14]. Our framework was evaluated on three classification tasks related to diagnosis or prognosis: 1) HC vs. MCI classification (ADNI: 65 HC & 60 MCI), 2) AD conversion prediction (OASIS-3: 31 MCI non-converters & 29 MCI converters), and 3) differentiating cognitively normal individuals with amyloid positivity vs. those with amyloid negativity (OASIS-3: 41 HC aβ+ve & 50 HC aβ-ve). All subjects have 2–3 time points of fMRI data and those with two time points were zero-padded to three time points. FC was built based on the AAL brain atlas with 90 ROIs [19]. The model was trained using Binary Cross-Entropy Loss in an end-to-end fashion. Implementation details could be found in supplementary materials. The code is available at https://github.com/ZijianD/Brain-TokenGT.git.

Table 1. Experimental Results reported based on five-fold cross-validation repeated five times (%, mean(standard deviation)). Our approach outperformed shallow learning (in blue), one time point feasible deep learning (in yellow), multi-time point feasible deep learning (in green) and our ablations (in pink) significantly. [* denotes significant improvement ($p < 0.05$). HC: healthy control. MCI: mild cognitive impairment. AD: Alzheimer's disease. GP: global pooling. I: identifiers. TI: type identifiers. NI: node identifiers.]

Model	HC vs. MCI		AD Conversion		Amyloid Positive vs. Negative	
	AUC	Accuracy	AUC	Accuracy	AUC	Accuracy
MK-SVM	55.00(15.31)	47.20(06.40)	56.69(14.53)	58.19(08.52)	61.31(11.16)	56.02(07.91)
RF	55.26(13.83)	57.60(07.42)	62.00(03.46)	56.44(02.50)	65.25(09.41)	60.35(08.48)
MLP	59.87(12.89)	51.20(09.26)	59.19(13.76)	58.13(13.67)	60.33(13.60)	59.30(13.52)
GCN	62.86(00.79)	58.33(01.03)	62.22(04.75)	63.33(03.78)	66.67(00.67)	66.67(00.87)
GAT	61.11(00.38)	61.54(01.47)	63.83(00.95)	46.67(07.11)	68.00(00.56)	64.44(00.69)
PNA	65.25(09.41)	60.35(08.48)	67.11(16.22)	62.57(08.38)	70.08(12.41)	62.63(05.42)
BrainNetCNN	48.06(05.72)	55.73(06.82)	60.79(09.12)	58.33(11.65)	65.80(01.60)	67.84(02.71)
BrainGNN	60.94(07.85)	51.80(07.49)	69.38(15.97)	59.93(12.30)	62.40(09.18)	63.90(10.75)
IBGNN+	67.95(07.95)	66.79(08.10)	75.98(06.36)	70.00(05.70)	73.45(05.61)	65.28(04.73)
BrainNetTF	65.03(07.11)	63.08(07.28)	73.33(10.07)	73.33(08.16)	74.32(06.46)	76.84(06.67)
OnionNet	61.52(06.16)	60.20(06.87)	65.08(03.69)	63.89(06.33)	60.74(08.47)	64.38(10.06)
STGCN	76.62(06.77)	74.77(09.59)	76.92(06.89)	75.00(04.05)	78.69(03.59)	75.03(06.00)
EvolveGCN	81.52(02.48)	80.45(02.52)	77.63(02.78)	76.99(06.53)	79.33(03.34)	75.14(09.59)
GIVE w/o e_t + GP	82.83(09.50)	83.97(04.76)	78.14(02.37)	80.02(06.62)	80.35(06.09)	75.34(09.76)
GIVE + GP	83.11(05.66)	85.50(01.47)	79.21(05.50)	81.45(04.58)	83.57(08.07)	80.85(03.67)
BIGTR itself	76.83(00.67)	71.33(00.87)	71.17(00.38)	63.33(01.47)	77.61(01.17)	74.00(00.22)
Ours w/o I	85.17(06.37)	84.11(03.20)	83.80(04.03)	83.20(02.40)	84.24(07.01)	83.20(02.40)
Ours w/o TI	86.71(08.01)	89.02(04.37)	85.17(06.37)	84.11(03.20)	88.40(06.95)	84.67(05.64)
Ours w/o NI	88.15(08.07)	**91.04(06.15)**	84.17(03.17)	85.80(07.19)	88.70(03.98)	84.36(04.32)
Ours [GIVE + BIGTR]	**90.48*(04.99)**	84.62(08.43)	**87.14*(07.16)**	**89.23*(07.84)**	**94.60*(04.96)**	**87.11*(07.88)**

Results. AUC and accuracy are presented in Table 1. (Recall and Precision could be found in supplementary materials). Brain TokenGT and its ablations were compared with three types of baseline models, including 1) shallow machine learning: MK-SVM, RF and MLP; 2) one time point feasible deep learning: three representative deep graph models GCN [11], GAT [20] and PNA [2], and four state-of-the-art deep models specifically designed for FC: BrainnetCNN [9], BrainGNN [15], IBGNN+ [4] and BrainnetTF [8]; 3) multiple time points feasible deep learning: Onionnet [24], STGCN [22] and EvolveGCN [16]. To ensure a fair comparison between models, the one-dimensional vectors flattened from FC in all time points were concatenated and used as input for the shallow learning model. For the one time point feasible deep learning models, a prediction value was generated at each time point and subsequently averaged to obtain an individual-level prediction.

The experimental results (Table 1) demonstrate that the Brain TokenGT significantly outperformed all three types of baseline by a large margin. The ablation study further revealed that GIVE w/o e_t w/ GP outperformed EvolveGCN by adding VEE without e_t, which empirically validates the importance of edge feature embeddings in FC. The performance could be further improved by incorporating e_t, suggesting the efficiency of our GIVE design with spatio-temporal edges. Interestingly, BIGTR itself (*i.e.*, the original features were directly input to BIGTR without GIVE) showed competitive performance with STGCN. Replacing GP with transformer (Ours w/o I) led to improved performance even without identifiers, indicating that the embeddings from GIVE may already capture some spatial and temporal information from the FC trajectory. The addition of identifiers further improved performance, possibly because the token-level self-supervised learning could alleviate the over-fitting issue and node identifiers could maintain the localized information effectively.

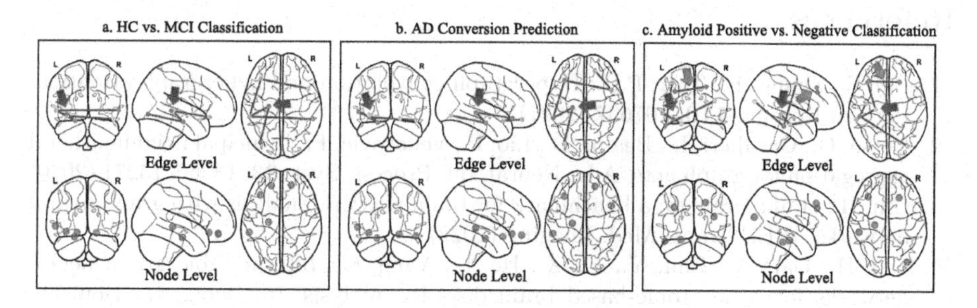

Fig. 2. HyperDrop Results. Blue arrows point to left temporal and parahippocampal regions, green arrows point to superior frontal regions. We refer readers of interest to supplementary materials for the full list of brain regions and connections involved. (Color figure online)

Interpretation. Figure 2 shows the top 5 salient edges and nodes retained by HyperDrop for each of the three tasks. Consistent with previous literature on brain network breakdown in the early stage of AD [17], parahippocampal, orbitofrontal and temporal regions and their connections contributed highly to all three tasks, underscoring their critical roles in AD-specific network dysfunction relevant to disease progression. On the other hand, superior frontal region additionally contributed to the amyloid positive vs. negative classification, which is in line with previous studies in amyloid deposition [18].

4 Conclusion

This study proposes the first interpretable framework for the embedding of FC trajectories, which can be applied to the diagnosis and prognosis of neurodegenerative diseases with small scale datasets, namely *Brain Tokenized Graph Transformer* (Brain TokenGT). Based on longitudinal brain FC, experimental results showed superior performance of our framework with excellent built-in interpretability supporting the AD-specific brain network neurodegeneration. A potential avenue for future research stemming from this study involves enhancing the "temporal resolution" of the model. This may entail, for example, incorporating an estimation of uncertainty in both diagnosis and prognosis, accounting for disease progression, and offering time-specific node and edge level interpretation.

Acknowledgement. This work was supported by National Medical Research Council, Singapore (NMRC/OFLCG19May-0035 to J-H Zhou) and Yong Loo Lin School of Medicine Research Core Funding (to J-H Zhou), National University of Singapore, Singapore. Yueming Jin was supported by MoE Tier 1 Start up grant (WBS: A-8001267-00-00).

References

1. Bai, S., Zhang, F., Torr, P.H.: Hypergraph convolution and hypergraph attention. Pattern Recogn. **110**, 107637 (2021)
2. Corso, G., Cavalleri, L., Beaini, D., Liò, P., Veličković, P.: Principal neighbourhood aggregation for graph nets. Adv. Neural. Inf. Process. Syst. **33**, 13260–13271 (2020)
3. Cui, H., et al.: Braingb: a benchmark for brain network analysis with graph neural networks. IEEE Trans. Med. Imaging (2022)
4. Cui, H., Dai, W., Zhu, Y., Li, X., He, L., Yang, C.: Interpretable graph neural networks for connectome-based brain disorder analysis. In: Wang, L., Dou, Q., Fletcher, P.T., Speidel, S., Li, S. (eds.) MICCAI 2022, Part VIII. LNCS, vol. 13438, pp. 375–385. Springer, Cham (2022). https://doi.org/10.1007/978-3-031-16452-1_36
5. Filippi, M., et al.: Changes in functional and structural brain connectome along the Alzheimer's disease continuum. Mol. Psychiatry **25**(1), 230–239 (2020)
6. Jack Jr., C.R., et al.: The Alzheimer's disease neuroimaging initiative (adni): Mri methods. J. Magnet. Resonance Imaging Official J. Int. Soc. Magnet. Resonance Med. **27**(4), 685–691 (2008)
7. Jo, J., Baek, J., Lee, S., Kim, D., Kang, M., Hwang, S.J.: Edge representation learning with hypergraphs. Adv. Neural. Inf. Process. Syst. **34**, 7534–7546 (2021)
8. Kan, X., Dai, W., Cui, H., Zhang, Z., Guo, Y., Yang, C.: Brain network transformer. In: Advances in Neural Information Processing Systems
9. Kawahara, J., et al.: Brainnetcnn: convolutional neural networks for brain networks; towards predicting neurodevelopment. Neuroimage **146**, 1038–1049 (2017)
10. Kim, J., et al.: Pure transformers are powerful graph learners. In: Advances in Neural Information Processing Systems
11. Kipf, T.N., Welling, M.: Semi-supervised classification with graph convolutional networks. In: International Conference on Learning Representations
12. Kong, R., et al.: Spatial topography of individual-specific cortical networks predicts human cognition, personality, and emotion. Cereb. Cortex **29**(6), 2533–2551 (2019)
13. LaMontagne, P.J., et al.: Oasis-3: longitudinal neuroimaging, clinical, and cognitive dataset for normal aging and Alzheimer disease. MedRxiv, pp. 2012–2019 (2019)
14. Li, J., et al.: Global signal regression strengthens association between resting-state functional connectivity and behavior. Neuroimage **196**, 126–141 (2019)
15. Li, X., et al.: Braingnn: interpretable brain graph neural network for FMRI analysis. Med. Image Anal. **74**, 102233 (2021)

16. Pareja, A., et al.: Evolvegcn: evolving graph convolutional networks for dynamic graphs. In: Proceedings of the AAAI Conference on Artificial Intelligence, vol. 34, pp. 5363–5370 (2020)
17. Sheline, Y.I., Raichle, M.E.: Resting state functional connectivity in preclinical Azheimer's disease. Biol. Psychiat. **74**(5), 340–347 (2013)
18. Thal, D.R., Rüb, U., Orantes, M., Braak, H.: Phases of aβ-deposition in the human brain and its relevance for the development of ad. Neurology **58**(12), 1791–1800 (2002)
19. Tzourio-Mazoyer, N., et al.: Automated anatomical labeling of activations in SPM using a macroscopic anatomical parcellation of the MNI MRI single-subject brain. Neuroimage **15**(1), 273–289 (2002)
20. Veličković, P., Cucurull, G., Casanova, A., Romero, A., Liò, P., Bengio, Y.: Graph attention networks. In: International Conference on Learning Representations
21. Ying, Z., You, J., Morris, C., Ren, X., Hamilton, W., Leskovec, J.: Hierarchical graph representation learning with differentiable pooling. In: Advances in Neural Information Processing Systems 31 (2018)
22. Yu, B., Yin, H., Zhu, Z.: Spatio-temporal graph convolutional networks: a deep learning framework for traffic forecasting. In: Proceedings of the 27th International Joint Conference on Artificial Intelligence, pp. 3634–3640 (2018)
23. Zhang, L., et al.: Deep fusion of brain structure-function in mild cognitive impairment. Med. Image Anal. **72**, 102082 (2021)
24. Zheng, L., Fan, J., Mu, Y.: Onionnet: a multiple-layer intermolecular-contact-based convolutional neural network for protein-ligand binding affinity prediction. ACS Omega **4**(14), 15956–15965 (2019)
25. Zhou, J., Gennatas, E.D., Kramer, J.H., Miller, B.L., Seeley, W.W.: Predicting regional neurodegeneration from the healthy brain functional connectome. Neuron **73**(6), 1216–1227 (2012)
26. Zhou, J., Liu, S., Ng, K.K., Wang, J.: Applications of resting-state functional connectivity to neurodegenerative disease. Neuroimaging Clinics **27**(4), 663–683 (2017)

CARL: Cross-Aligned Representation Learning for Multi-view Lung Cancer Histology Classification

Yin Luo[1], Wei Liu[1], Tao Fang[1], Qilong Song[2], Xuhong Min[2], Minghui Wang[1](✉), and Ao Li[1](✉)

[1] School of Information Science and Technology, University of Science and Technology of China, Hefei 230026, China
{mhwang,aoli}@ustc.edu.cn

[2] Department of Radiology, Anhui Chest Hospital, Hefei 230039, Anhui, China

Abstract. Accurately classifying the histological subtype of non-small cell lung cancer (NSCLC) using computed tomography (CT) images is critical for clinicians in determining the best treatment options for patients. Although recent advances in multi-view approaches have shown promising results, discrepancies between CT images from different views introduce various representations in the feature space, hindering the effective integration of multiple views and thus impeding classification performance. To solve this problem, we propose a novel method called cross-aligned representation learning (CARL) to learn both view-invariant and view-specific representations for more accurate NSCLC histological subtype classification. Specifically, we introduce a cross-view representation alignment learning network which learns effective view-invariant representations in a common subspace to reduce multi-view discrepancies in a discriminability-enforcing way. Additionally, CARL learns view-specific representations as a complement to provide a holistic and disentangled perspective of the multi-view CT images. Experimental results demonstrate that CARL can effectively reduce the multi-view discrepancies and outperform other state-of-the-art NSCLC histological subtype classification methods.

Keywords: Cross-view Alignment · Representation Learning · Multi-view · Histologic Subtype Classification · Non-small Cell Lung Cancer

1 Introduction

Lung cancer is currently the foremost cause of cancer-related mortalities globally, with non-small cell lung cancer (NSCLC) being responsible for 85% of reported cases [25]. Within NSCLC, squamous cell carcinoma (SCC) and adenocarcinoma (ADC) are recognized as the two principal histological subtypes. Since SCC and ADC differ in the effectiveness of chemotherapy and the risk of complications, accurate identification of different subtypes is crucial for clinical treatment options [15]. Although pathological diagnosis via lung biopsy can provide a reliable result of subtype identification, it is

H. Greenspan et al. (Eds.): MICCAI 2023, LNCS 14224, pp. 358–367, 2023.
https://doi.org/10.1007/978-3-031-43904-9_35

highly invasive with potential clinical implications [19]. Therefore, non-invasive methods utilizing computed tomography (CT) images have garnered significant attention over the last decade [15, 16].

Recently, several deep-learning methods have been put forward to differentiate between the NSCLC histological subtypes using CT images [4, 11, 13, 22]. Chaunzwa et al. [4] and Marentakis et al. [13] both employ a convolutional neural network (CNN) model with axial view CT images to classify the tumor histology into SCC and ADC. Albeit the good performance, the above 2D CNN-based models only take CT images from a single view as the input, limiting their ability to describe rich spatial properties of CT volumes [20]. Multi-view deep learning, a 2.5D method, represents a promising solution to this issue, as it focuses on obtaining a unified joint representation from different views of lung nodules to capture abundant spatial information [16, 20]. For example, Wu et al. [22] aggregate features from axial, coronal, and sagittal view CT images via a multi-view fusion model. Similarly, Li et al. [11] also extract patches from three orthogonal views of a lung nodule and present a multi-view ResNet for feature fusion and classification. By integrating multi-view representations, these methods efficiently preserve the spatial information of CT volumes while significantly reducing the required computational resource compared to 3D CNNs [9, 20, 23].

Despite the promising results of previous multi-view methods, they still confront a severe challenge for accurate NSCLC histological subtype prediction. In fact, due to the limitation of scan time and hardware capacity in clinical practice, different views of CT volumes are anisotropic in terms of in-plane and inter-plane resolution [21]. Additionally, images from certain views may inevitably contain some unique background information, e.g., the spine in the sagittal view [17]. Such anisotropy and background dissimilarity both reveal the existence of significant variations between different views, which lead to markedly various representations in feature space. Consequently, the discrepancies of distinct views will hamper the fusion of multi-view information, limiting further improvements in the classification performance.

To overcome the challenge mentioned above, we propose a novel cross-aligned representation learning (CARL) method for the multi-view histologic subtype classification of NSCLC. CARL offers a holistic and disentangled perspective of multi-view CT images by generating both view-invariant and -specific representations. Specifically, CARL incorporates a cross-view representation alignment learning network which targets the reduction of multi-view discrepancies by obtaining discriminative view-invariant representations. A shared encoder with a novel discriminability-enforcing similarity constraint is utilized to map all representations learned from multi-view CT images to a common subspace, enabling cross-view representation alignment. Such aligned projections help to capture view-invariant features of cross-view CT images and meanwhile make full use of the discriminative information obtained from each view. Additionally, CARL learns view-specific representations as well which complement the view-invariant ones, providing a comprehensive picture of the CT volume data for histological subtype prediction. We validate our approach by using a publicly available NSCLC dataset from The Cancer Imaging Archive (TCIA). Detailed experimental results demonstrate the effectiveness of CARL in reducing multi-view discrepancies and improving NSCLC histological subtype classification performance. Our contributions can be summarized as follows:

- A novel cross-aligned representation learning method called CARL is proposed for NSCLC histological subtype classification. To reduce the discrepancies of multi-view CT images, CARL incorporates a cross-view representation alignment learning network for discriminative view-invariant representations.
- We employ a view-specific representation learning network to learn view-specific representations as a complement to the view-invariant representations.
- We conduct experiments on a publicly available dataset and achieve superior performance compared to the most advanced methods currently available.

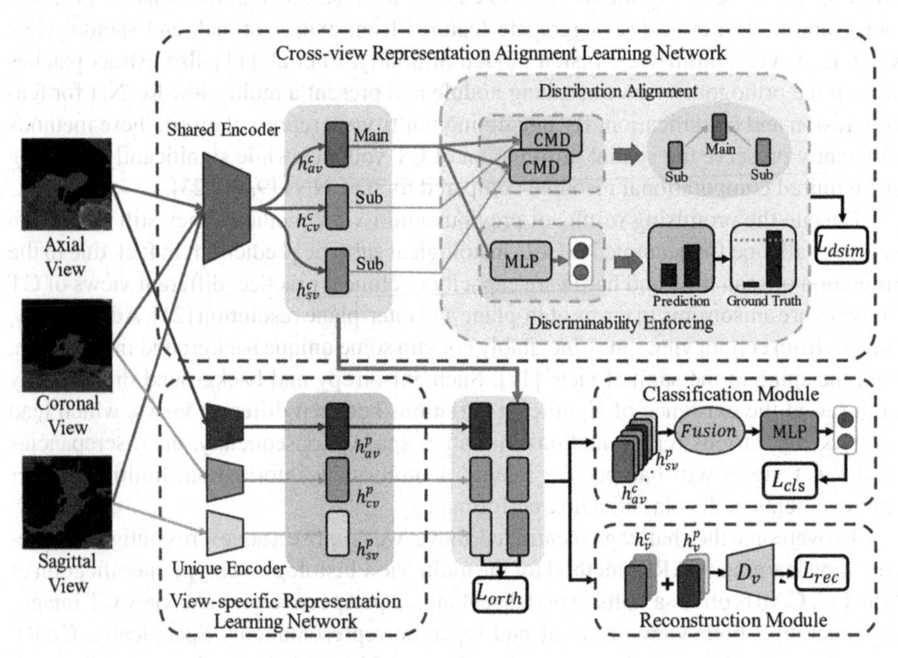

Fig. 1. Illustration of the proposed CARL.

2 Methodology

2.1 Architecture Overview

Figure 1 shows the overall architecture of CARL. The cross-view representation alignment learning network includes a shared encoder which projects patches of axial, coronal, and sagittal views into a common subspace with a discriminability-enforcing similarity constraint to obtain discriminative view-invariant representations for multi-view discrepancy reduction. In addition, CARL introduces a view-specific representation learning network consisting of three unique encoders which focus on learning view-specific representations in respective private subspaces to yield complementary information to view-invariant representations. Finally, we introduce a histological subtype classification module to fuse the view-invariant and -specific representations and make accurate NSCLC histological subtype classification.

2.2 Cross-View Representation Alignment Learning

Since the discrepancies of different views may result in divergent statistical properties in feature space, e.g., huge distributional disparities, aligning representations of different views is essential for multi-view fusion. With the aim to reduce multi-view discrepancies, CARL introduces a cross-view representation alignment learning network for mapping the representations from distinct views into a common subspace, where view-invariant representations can be obtained by cross-view alignment. Specifically, inspired by [12, 14, 24], we exert a discriminability-enforcing similarity constraint to align all sub-view representations with those of the main view, significantly mitigating the distributional disparities of multi-view representations.

Technically speaking, given the axial view image I_{av}, coronal view image I_{cv}, and sagittal view image I_{sv}, the cross-view representation alignment learning network tries to generate view-invariant representations h_v^c, $v \in \{av, cv, sv\}$ via a shared encoder based on a residual neural network [10]. This can be formulated as below:

$$h_v^c = E_c(I_v), \;\; v \in \{av, \; cv, \; sv\} \tag{1}$$

where $E_c(\cdot)$ indicates the shared encoder, and av, cv, sv represent axial, coronal, and sagittal views, respectively. In the common subspace, we hope that through optimizing the shared encoder $E_c(\cdot)$, the view-invariant representations can be matched to some extent. However, the distributions of h_{av}^c, h_{cv}^c and h_{sv}^c are very complex due to the significant variations between different views, which puts a burden on obtaining well-aligned view-invariant representations with merely an encoder.

To address this issue, we design a discriminability-enforcing similarity loss L_{dsim} to further enhance the alignment of cross-view representations in the common subspace. Importantly, considering that the axial view has a higher resolution than other views and are commonly used in clinical diagnosis, we choose axial view as the main view and force the sub-views (e.g., the coronal and sagittal views) to seek distributional similarity with the main view. Mathematically, we introduce a cross-view similarity loss L_{sim} which calculates the central moment discrepancy (CMD) metric [24] between all sub-views and the main view as shown below:

$$L_{sim} = \frac{1}{N} \sum_{i=1}^{N} CMD\left(h_i^{sub}, \, h^{main}\right) \tag{2}$$

where $CMD(\cdot)$ denotes the distance metric which measures the distribution disparities between the representations of i-th sub-view h_i^{sub} and the main view h^{main}. N is the number of sub-views. Despite the fact that minimizing the L_{sim} can efficiently mitigate the issue of distributional disparities, it may not guarantee that the alignment network will learn informative and discriminative representations. Inspired by recent work on multimodal feature extraction [12, 14], we impose a direct supervision by inputting h^{main} into a classifier $f(\cdot)$ to obtain the prediction of histological subtype, and use a cross-entropy loss to enforce the discriminability of the main-view representations. Finally, the discriminability-enforcing similarity loss L_{dsim} is as follows:

$$L_{dsim} = L_{sim} + \lambda \cdot L_{CE}\left(f\left(h^{main}\right), y\right) \tag{3}$$

where y denotes the ground-truth subtype labels, λ controls the weight of L_{CE}. We observed that L_{CE} is hundred times smaller than L_{sim}, so this study uses an empirical value of $\lambda = 110$ to balance the magnitude of two terms. By minimizing L_{dsim}, the cross-view representation alignment learning network pushes the representations of each sub-view to align with those of the main view in a discriminability-enforcing manner. Notably, the benefits of such cross-alignment are twofold. Firstly, it greatly reduces the discrepancies between the sub-views and the main view, leading to consistent view-invariant representations. Secondly, since the alignment between distinct views compels the representation distribution of the sub-views to match that of the discriminative main view, it can also enhance the discriminative power of the sub-view representations. In other words, the cross-alignment procedure spontaneously promotes the transfer of discriminative information learned by the representations of the main view to those of the sub-views. As a result, the introduced cross-view representation alignment learning network is able to generate consistent and discriminative view-invariant representations cross all views to effectively narrow the multi-view discrepancies.

2.3 View-Specific Representation Learning

On the basis of learning view-invariant representations, CARL additionally learns view-specific representations in respective private subspaces, which provides supplementary information for the view-invariant representations and contribute to subtype classification as well. To be specific, a view-specific representation learning network containing three unique encoders is proposed to learn view-specific representations $h_v^p, v \in \{av, cv, sv\}$, enabling effective exploitation of the specific information from each view. We formulate the unique encoders as follows:

$$h_v^p = E_v^p(I_v), \ v \in \{av, cv, sv\} \tag{4}$$

where $E_v^p(\cdot)$ is the encoder function dedicated to capture single-view characteristics.

To induce the view-invariant and -specific representations to learn unique characteristics of each view, we draw inspiration from [14] and adopt an orthogonality loss L_{orth} with the squared Frobenius norm between the representations in the common and private subspaces of each view, which is denoted by $L_{orth} = \|h_v^c h_v^p\|_F^2, v \in \{av, cv, sv\}$. A reconstruction module is also employed to calculate a reconstruction loss L_{rec} between original image I_v and reconstructed image I_v^r using the L_1-norm, which ensures the hidden representations to capture details of the respective view.

2.4 Histologic Subtype Classification

After obtaining view-invariant and -specific representations from each view, we integrate them together to perform NSCLC subtype classification. Specifically, we apply a residual block [10] to fuse view-invariant and -specific representations into a unified multi-view representation h. Then, h is sent to a multilayer perceptron neural network (MLP) to make the precise NSCLC subtype prediction. The NSCLC histological subtype classification loss L_{cls} can be calculated by using cross-entropy loss.

2.5 Network Optimization

The optimization of CARL is achieved through a linear combination of several loss terms, including discriminability-enforcing similarity loss L_{dsim}, orthogonality loss L_{orth}, reconstruction loss L_{rec} and the classification loss L_{cls}. Accordingly, the total loss function can be formulated as a weighted sum of these separate loss terms:

$$L_{total} = L_{cls} + \alpha L_{dsim} + \beta L_{orth} + \gamma L_{rec} \quad (5)$$

where α, β and γ denote the weights of L_{dsim}, L_{rec} and L_{orth}. To normalize the scale of L_{dsim} which is much larger than the other terms, we introduce a scaling factor $S = 0.001$, and perform a grid search for α, β and γ in the range of 0.1S-S, 0.1-1, and 0.1-1, respectively. Throughout the experiments, we set the values of α, β and γ to 0.6S, 0.4 and 0.6, respectively.

3 Experiments and Results

3.1 Dataset

Our dataset NSCLC-TCIA for lung cancer histological subtype classification is sourced from two online resources of The Cancer Imaging Archive (TCIA) [5]: NSCLC Radiomics [1] and NSCLC Radiogenomics [2]. Exclusion criteria involves patients diagnosed with large cell carcinoma or not otherwise specified, along with cases that have contouring inaccuracies or lacked tumor delineation [9, 13]. Finally, a total of 325 available cases (146 ADC cases and 179 SCC cases) are used for our study. We evaluate the performance of NSCLC classification in five-fold cross validation on the NSCLC-TCIA dataset, and measure accuracy (Acc), sensitivity (Sen), specificity (Spe), and the area under the receiver operating characteristic (ROC) curve (AUC) as evaluation metrics. We also conduct analysis including standard deviations and 95% CI, and DeLong statistical test for further AUC comparison.

For preprocessing, given that the CT data from NSCLC-TCIA has an in-plane resolution of 1 mm $\times$ 1 mm and a slice thickness of 0.7–3.0 mm, we resample the CT images using trilinear interpolation to a common resolution of 1mm $\times$ 1mm $\times$ 1mm. Then one 128 $\times$ 128 pixel slice is cropped from each view as input based on the center of the tumor. Finally following [7], we clip the intensities of the input patches to the interval ($-$1000, 400 Hounsfield Unit) and normalize them to the range of [0, 1].

3.2 Implementation Details

The implementation of CARL is carried out using PyTorch and run on a workstation equipped with Nvidia GeForce RTX 2080Ti GPUs and Intel Xeon CPU 4110 @ 2.10GHz. Adam optimizer is used with an initial learning rate of 0.00002, and the batch size is set to 8.

3.3 Results

Comparison with Existing Methods. Several subtype classification methods have been employed for comparison including: two conventional methods, four single-view and 3D deep learning methods, and four representative multi-view methods. We use publicly available codes of these comparison methods and implement models for methods without code. The experimental results are reported in Table 1. The multi-view methods are generally superior to the single-view and 3D deep learning methods. It illustrates that the multi-view methods can exploit richer spatial properties of CT volumes than the single-view methods while greatly reducing the model parameters to avoid overfitting compared to the 3D methods. The floating point operations (FLOPs) comparison between CARL (0.9 GFLOPs) and the 3D method [9] (48.4 GFLOPs) also proves the computational efficiency of our multi-view method. Among all multi-view methods, our proposed CARL achieves the best results, outperforming Wu et al. by 3.2%, 3.2%, 1.5% and 4.1% in terms of AUC, Acc, Sen and Spe, respectively. Not surprisingly, the ROC curve of CARL in Fig. 2(a) is also closer to the upper-left corner, further indicating its superior performance. These results demonstrate that CARL can effectively narrow the discrepancies of different views by obtaining view-invariant representations in a discriminative way, thus leading to excellent classification accuracy compared to other methods.

Table 1. Results on NSCLC-TCIA dataset when CARL was compared with other SOTA methods using five-fold cross validation. * indicates the p-value is less than 0.05 in DeLong test between the AUC of compared method and CARL.

Category	Methods	AUC	95% CI	Acc	Sen	Spe
Conventional	RF [3]	0.742 ± 0.061*	0.684–0.791	0.667	0.623	0.703
	SVM [6]	0.756 ± 0.069*	0.714–0.817	0.699	0.670	0.725
Deep learning	Chaunzwa et al. [4]	0.774 ± 0.051*	0.710–0.816	0.713	0.692	0.729
	Marentakis et al. [13]	0.770 ± 0.076*	0.707–0.813	0.715	0.663	0.757
	Yanagawa et al. [23]	0.777 ± 0.062*	0.718–0.822	0.719	0.650	0.776
	Guo et al. [9]	0.772 ± 0.072*	0.676–0.783	0.676	0.635	0.712
Multi-view	MVCNN [18]	0.784 ± 0.052*	0.707–0.811	0.691	0.637	0.733
	GVCNN [8]	0.767 ± 0.054*	0.704–0.809	0.685	0.581	0.770
	Wu et al. [22]	0.785 ± 0.080	0.744–0.844	0.736	0.717	0.756
	Li et al. [11]	0.782 ± 0.069*	0.719–0.822	0.729	0.705	0.748
	CARL	$\mathbf{0.817 \pm 0.055}$	**0.770–0.862**	**0.768**	**0.732**	**0.797**

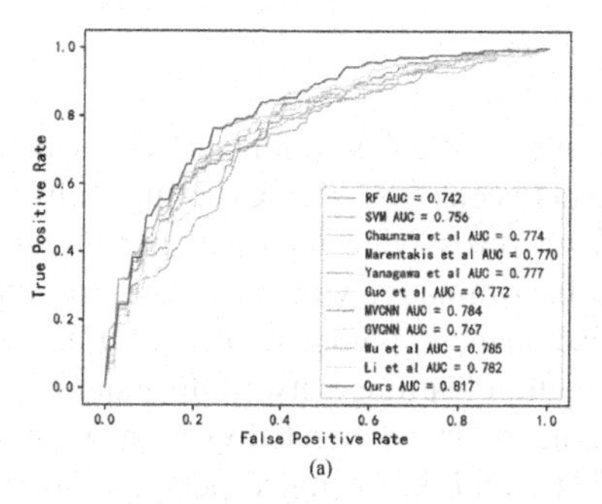

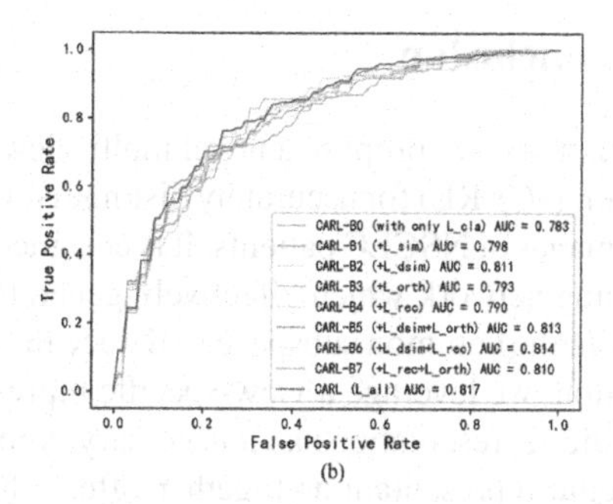

Fig. 2. ROC plots of (a) compared methods and (b) ablation analysis on NSCLC-TCIA dataset.

Table 2. Results of ablation analysis on the NSCLC-TCIA dataset.

Methods	AUC	Acc	Sen	Spe
CARL-B0 (with only L_{cla})	0.783	0.713	0.670	0.748
CARL-B1 (+L_{sim})	0.798	0.734	0.726	0.741
CARL-B2 (+L_{dsim})	0.811	0.738	0.705	0.764
CARL-B3 (+L_{orth})	0.793	0.722	0.705	0.736
CARL-B4 (+L_{rec})	0.790	0.701	0.697	0.705
CARL-B5 (+L_{dsim}+L_{orth})	0.813	0.756	0.725	0.782
CARL-B6 (+L_{dsim}+L_{rec})	0.814	0.747	0.732	0.758
CARL-B7 (+L_{orth}+L_{rec})	0.810	0.752	0.712	0.786
CARL (L_{all})	**0.817**	**0.768**	**0.732**	**0.797**

Ablation Analysis. We evaluate the efficacy of different losses in our method. The results are reported in Table 2, where CARL-B0 refers to CARL only using the classification loss, $+L_*$ indicates the loss superimposed on CARL-B0 and L_{all} denotes that we utilize all the losses in Eq. 5. We can observe that CARL-B2 performs better than CARL-B0 by employing the discriminability-enforcing similarity loss to align cross-view representations. Besides, CARL-B3 and CARL-B4 show better performance than CARL-B0, illustrating view-specific representations as a complement which can also contribute to subtype classification. Though single loss already contributes to performance improvement, CARL-B5 to CARL-B7 demonstrate that the combinations of different losses can further enhance classification results. More importantly, CARL with all losses achieves the best performance among all methods, demonstrating that our proposed method effectively reduces multi-view discrepancies and significantly improves the performance of histological subtype classification by providing a holistic and disentangled perspective of the multi-view CT images. The ROC curve of CARL in Fig. 2(b) is generally above its variants, which is also consistent with the quantitative results.

4 Conclusion

In summary, we propose a novel multi-view method called cross-aligned representation learning (CARL) for accurately distinguishing between ADC and SCC using multi-view CT images of NSCLC patients. It is designed with a cross-view representation alignment learning network which effectively generates discriminative view-invariant representations in the common subspace to reduce the discrepancies among multi-view images. In addition, we leverage a view-specific representation learning network to acquire view-specific representations as a necessary complement. The generated view-invariant and -specific representations together offer a holistic and disentangled perspective of the multi-view CT images for histological subtype classification of NSCLC. The experimental result on NSCLC-TCIA demonstrates that CARL reaches 0.817 AUC, 76.8% Acc, 73.2% Sen, and 79.7% Spe and surpasses other relative approaches, confirming the effectiveness of the proposed CARL method.

Acknowledgement. This work was supported in part by the National Natural Science Foundation of China under Grants 61971393, 62272325, 61871361 and 61571414.

References

1. Aerts, H.J.W.L. et al.: Decoding tumour phenotype by noninvasive imaging using a quantitative Radiomics approach. Nat. Commun. **5**(1), 4006 (2014). https://doi.org/10.1038/ncomms5006
2. Bakr, S. et al.: A radiogenomic dataset of non-small cell lung cancer. Sci. Data **5**(1), 180202 (2018). https://doi.org/10.1038/sdata.2018.202
3. Breiman, L.: Random forests. Mach. Learn. **45**(1), 5–32 (2001). https://doi.org/10.1023/A:1010933404324
4. Chaunzwa, T.L. et al.: Deep learning classification of lung cancer histology using CT images. Sci. Rep. **11**(1), 5471 (2021). https://doi.org/10.1038/s41598-021-84630-x
5. Clark, K., et al.: The cancer imaging archive (TCIA): Maintaining and operating a public information repository. J. Digit Imaging. **26**(6), 1045–1057 (2013). https://doi.org/10.1007/s10278-013-9622-7
6. Cortes, C., Vapnik, V.: Support-vector networks. Mach. Learn. **20**(3), 273–297 (1995). https://doi.org/10.1007/BF00994018
7. Dou, Q., et al.: Multilevel contextual 3-D CNNs for false positive reduction in pulmonary nodule detection. IEEE Trans. Biomed. Eng. **64**(7), 1558–1567 (2017). https://doi.org/10.1109/TBME.2016.2613502
8. Feng, Y., et al.: GVCNN: Group-view convolutional neural networks for 3D shape recognition. In: 2018 IEEE/CVF Conference on Computer Vision and Pattern Recognition, pp. 264–272 (2018). https://doi.org/10.1109/CVPR.2018.00035
9. Guo, Y., et al.: Histological subtypes classification of lung cancers on CT images using 3D deep learning and radiomics. Acad. Radiol. **28**(9), e258–e266 (2021). https://doi.org/10.1016/j.acra.2020.06.010
10. He, K., et al.: Deep Residual Learning for Image Recognition. http://arxiv.org/abs/1512.03385 (2015). https://doi.org/10.48550/arXiv.1512.03385
11. Li, C., et al.: Multi-view mammographic density classification by dilated and attention-guided residual learning. IEEE/ACM Trans. Comput. Biol. Bioinf. **18**(3), 1003–1013 (2021). https://doi.org/10.1109/TCBB.2020.2970713

12. Li, S., et al.: Adaptive multimodal fusion with attention guided deep supervision net for grading hepatocellular carcinoma. IEEE J. Biomed. Health Inform. **26**(8), 4123–4131 (2022). https://doi.org/10.1109/JBHI.2022.3161466
13. Marentakis, P., et al.: Lung cancer histology classification from CT images based on radiomics and deep learning models. Med. Biol. Eng. Comput. **59**(1), 215–226 (2021). https://doi.org/10.1007/s11517-020-02302-w
14. Meng, Z., et al.: MSMFN: an ultrasound based multi-step modality fusion network for identifying the histologic subtypes of metastatic cervical lymphadenopathy. In: IEEE Transactions on Medical Imaging, pp. 1–1 (2022). https://doi.org/10.1109/TMI.2022.3222541
15. Pereira, T. et al.: Comprehensive perspective for lung cancer characterisation based on AI solutions using CT images. J. Clin. Med. **10**(1), 118 (2021). https://doi.org/10.3390/jcm10010118
16. Sahu, P., et al.: A lightweight multi-section CNN for lung nodule classification and malignancy estimation. IEEE J. Biomed. Health Inform. **23**(3), 960–968 (2019). https://doi.org/10.1109/JBHI.2018.2879834
17. Sedrez, J.A., et al.: Non-invasive postural assessment of the spine in the sagittal plane: a systematic review. Motricidade **12**(2), 140–154 (2016). https://doi.org/10.6063/motricidade.6470
18. Su, H., et al.: Multi-view convolutional neural networks for 3D shape recognition. In: 2015 IEEE International Conference on Computer Vision (ICCV), pp. 945–953 (2015). https://doi.org/10.1109/ICCV.2015.114
19. Su, R., et al.: Identification of expression signatures for non-small-cell lung carcinoma subtype classification. Bioinformatics **36**(2), 339–346 (2019). https://doi.org/10.1093/bioinformatics/btz557
20. Tomassini, S., et al.: Lung nodule diagnosis and cancer histology classification from computed tomography data by convolutional neural networks: a survey. Comput. Biol. Med. **146**, 105691 (2022). https://doi.org/10.1016/j.compbiomed.2022.105691
21. Wang, J., et al.: UASSR: unsupervised arbitrary scale super-resolution reconstruction of single anisotropic 3D images via disentangled representation learning. In: Wang, L. et al. (eds.) Medical Image Computing and Computer Assisted Intervention – MICCAI 2022, pp. 453–462 Springer Nature Switzerland, Cham (2022). https://doi.org/10.1007/978-3-031-16446-0_43
22. Wu, X., et al.: Deep learning-based multi-view fusion model for screening 2019 novel coronavirus pneumonia: a multicentre study. Eur. J. Radiol. **128**, 109041 (2020). https://doi.org/10.1016/j.ejrad.2020.109041
23. Yanagawa, M., et al.: Diagnostic performance for pulmonary adenocarcinoma on CT: comparison of radiologists with and without three-dimensional convolutional neural network. Eur. Radiol. **31**(4), 1978–1986 (2021). https://doi.org/10.1007/s00330-020-07339-x
24. Zellinger, W., et al.: Central Moment Discrepancy (CMD) for Domain-Invariant Representation Learning. http://arxiv.org/abs/1702.08811 (2019)
25. Zhang, N., et al.: Circular RNA circSATB2 promotes progression of non-small cell lung cancer cells. Mol. Cancer. **19**(1), 101 (2020). https://doi.org/10.1186/s12943-020-01221-6

Style-Based Manifold for Weakly-Supervised Disease Characteristic Discovery

Siyu Liu[1(✉)], Linfeng Liu[2], Craig Engstrom[1], Xuan Vinh To[2], Zongyuan Ge[3,4,5], Stuart Crozier[1], Fatima Nasrallah[2], and Shekhar S. Chandra[1]

[1] School of Electrical Engineering and Computer Science, The University of Queensland, Brisbane, Australia
siyu.liu1@uq.edu.au
[2] The Queensland Brain Institute, The University of Queensland, Brisbane, Australia
[3] Department of Data Science and AI, Monash University, Melbourne, Australia
[4] Monash-Airdoc Research Centre, Monash University, Melbourne, Australia
[5] AIM for Health Lab, Monash University, Melbourne, Australia

Abstract. In Alzheimer's Disease (AD), interpreting tissue changes is key to discovering disease characteristics. However, AD-induced brain atrophy can be difficult to observe without Cognitively Normal (CN) reference images, and collecting co-registered AD and CN images at scale is not practical. We propose Disease Discovery GAN (DiDiGAN), a style-based network that can create representative reference images for disease characteristic discovery. DiDiGAN learns a manifold of disease-specific style codes. In the generator, these style codes are used to "stylize" an anatomical constraint into synthetic reference images (for various disease states). The constraint in this case underpins the high-level anatomical structure upon which disease features are synthesized. Additionally, DiDiGAN's manifold is smooth such that seamless disease state transitions are possible via style interpolation. Finally, to ensure the generated reference images are anatomically correlated across disease states, we incorporate anti-aliasing inspired by StyleGAN3 to enforce anatomical correspondence. We test DiDiGAN on the ADNI dataset involving CN and AD magnetic resonance images (MRIs), and the generated reference AD and CN images reveal key AD characteristics (hippocampus shrinkage, ventricular enlargement, cortex atrophies). Moreover, by interpolating DiDiGAN's manifold, smooth CN-AD transitions were acquired further enhancing disease visualization. In contrast, other methods in the literature lack such dedicated disease manifolds and fail to synthesize usable reference images for disease characteristic discovery.

Keywords: Disease learning · Weak supervision · Alzheimer's Disease

---

Supplementary Information The online version contains supplementary material available at https://doi.org/10.1007/978-3-031-43904-9_36.

H. Greenspan et al. (Eds.): MICCAI 2023, LNCS 14224, pp. 368–378, 2023.
https://doi.org/10.1007/978-3-031-43904-9_36

1 Introduction

Discovering disease characteristics from medical images can yield important insights into pathological changes. However, subtle disease characteristics are difficult to discover from imaging data. The reduction in brain volume due to Alzheimer's Disease (AD) is one such example. In this case, comparing co-registered AD and Cognitively Normal (CN) reference magnetic resonance images (MRIs) is the most effective way to reveal subtle AD characteristics. Unfortunately, collecting such co-registered reference images at scale is not practical.

Generative Adversarial Networks (GANs) has the potential to synthesize the reference images required for disease discovery. Specifically, style-based generative frameworks [4] could be suitable as they employ mechanisms to "stylize" a share anatomical structure into different disease states. However, as we will show, many GAN frameworks (style and non-style based) are unable to produce satisfactory reference AD and CN images for AD characteristic discovery.

In this work, we propose Disease Discovery GAN (DiDiGAN), a specialized style-based network for disease characteristic discovery on medical image data. On the Alzheimer's Disease Neuroimaging Initiative (ADNI) dataset [7], DiDiGAN can create not only reference AD and CN MRIs, but also smooth animated transitions between AD and CN. The highlights of DiDiGAN are:

- A learnt disease manifold that captures and encodes AD and CN disease state distributions. Style codes sampled from this manifold control the disease expression in the output reference images.
- The disease manifold is naturally smooth such that seamless transitions of AD to CN are possible via style interpolation.
- The generator uses a low-resolution input image as the source anatomical constraint (to provide coarse structural guidance), it is low-resolution to leave sufficient room for the generator to synthesize disease features.
- Anti-aliasing as a key mechanism to maintain anatomical correspondence across generated reference images. Without anti-aliasing, reference images exhibit inconsistent anatomical structures invalidating visual comparisons.
- DiDiGAN is a weakly-supervised framework requiring only class labels for images rather than pixel (voxel) labels.

DiDiGAN learns to generate representative reference AD and CN images by training on 2D coronal brain slices labelled with AD and CN. The generated reference images and animations clearly show systematic changes in the hippocampus, ventricle, and cortex areas while maintaining anatomical correspondence. The discovered characteristics are corroborated by independent tests which solidify the strengths of the findings. There has not been a previous work that can i) produce visualizations on par with DiDiGAN's and ii) learn a dedicated manifold for disease characteristic discovery.

2 Related Work

Various generative methods have attempted to synthesize pathological changes for disease modelling. HerstonNet [18] and Bernal et al. [2] synthesize brain MRIs with atrophy, but they rely on induced atrophy by altering the Partial Volume (PV)-maps and segmentation maps. Hence they do not learn disease characteristics on their own. Ravi et al. [17] and ADESyn [8] learn AD characteristics from data, and they can synthesize brain images at different stages of AD. However, ADESyn fails to capture ventricle enlargement which is a key AD while Ravi et al. failed to capture hippocampus shrinkage (which are key AD characteristics). DBCE [16] learns to deform a healthy brain image into an AD image. However, the resulting atrophy is not sufficiently distinct for AD characteristic visualization. Saliency maps must be further computed to reveal AD regions. Xia et al. [23] proposed a GAN for ageing brain synthesis. While the results clearly depict ventricle enlargement for advanced ages, other AD characteristics remain unclear as the images show significant anatomical inconsistencies.

Another limitation of the above methods is that disease features are intertwined with anatomical features. For disease modelling, a dedicated mechanism (such as a latent space) to store disease features in an explorable manner would be desirable. The objective to learn a dedicated disease latent space naturally aligns with style-based [4] frameworks, which have mostly been used for medical image translation [1,11]. Style-based frameworks have the potential to enhance generative disease modelling. Fetty et al. [5] have shown that StyleGANs can learn a dedicated latent space (manifold) of medical image features. This latent space can be smoothly interpolated to manipulate synthesis. Other works like StarGAN [4] have shown that this latent can be disentangled from the content. Similarly, in medical image analysis, Chartsias et al. [3] have shown modality information can be disentangled from anatomical content for multi-modal image translation. These findings motivate DiDiGAN to use style to learn a manifold of disease features and then apply the manifold to "stylized" an input anatomical constraint into AD and CN images. However, StyleGAN3 [9] discovered that common style-based methods suffer from aliasing [24]. When interpolating the manifold, aliasing causes undesirable texture disruptions, and it threatens to disrupt anatomical structures if applied to disease studies. Thus, applying anti-aliasing (as per Shannon-Nyquist theorem) is critical for DiDiGAN.

3 Methods

DiDiGAN projects different disease states onto a common anatomical structure to form disease reference images, and the output images must maintain a consistent anatomical structure. A detailed architecture diagram is shown in Fig. 1.

Disease Style w_c and Anatomical Constraint x_{AC}: Let x_c denote the images in a dataset and $c \in \{c_1, c_2 ... c_n\}$ are the disease classifications for the images, the generator G's aim is to synthesise reference disease images x'_c that share a consistent anatomical structure. This is done by injecting disease styles w_c into

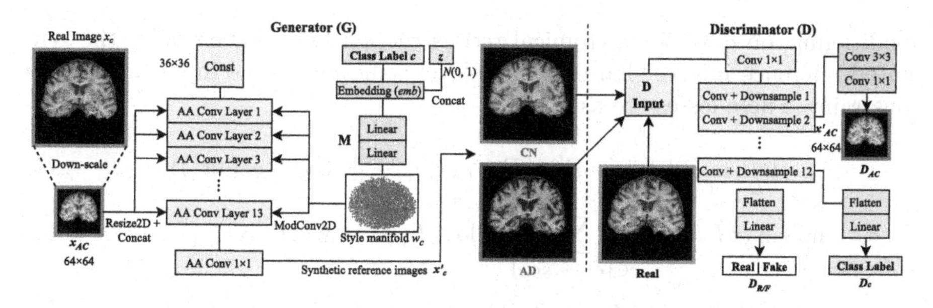

Fig. 1. Generator and discriminator architectures of DiDiGAN. Source image x_c is first down-sampled into anatomical constraint x_{AC}, then disease style codes w_c ($c \in \{c_1, c_2 ... c_n\}$) are injected into G to generate reference images x'_c for different disease states. All x'_cs maintain a consistent anatomical structure.

a shared anatomical constraint x_{AC} to produce the output $x'_c = G(w_c, x_{AC})$. w_c is created following $w_c = M(emb(c), z)$ where emb is a learned 512-d embedding specific to each class, z is a 512-d $N(0, 1)$ noise vector and M is a 2-layer mapping network made up of dense layers. Like StyleGAN, the style code space w_c is a learned manifold (of disease states), and it can be interpolated for smooth disease states transitions. The reason to include z is to introduce variability such that w_c forms a surface instead of a fixed point. x_{AC} is a $4\times$ down-sampled version of x_c which only enforces coarse anatomical structure similarities between the reference images, and it lacks resolution on purpose allowing room for disease features from w_c to manifest. As already discussed, full anatomical correspondence requires anti-aliasing to enforce.

Alias-Free Style Generator G: Generator G is alias-free as the feature maps are processed carefully following the Shannon-Nyquist sampling theorem. Like StyleGAN 3, G's convolutional (AA Conv) layers apply anti-aliasing to the leaky-relu activation function. This process involves first interpolating discrete feature maps to continuous domains. Then, low-pass filtering is used to remove offending frequencies. Finally, the activation function is applied, and the signal is re-sampled back to the discrete domain [9]. Architecture wise G is a progressively up-sampling decoder Convolutional Neural Network (CNN) with a starting resolution of 4×4 and an output resolution of 256×256. There are 13 convolutional blocks that gradually perform upsampling and synthesis. All the convolutional layers are modulated convolutions [4] to inject disease features w_c. The anatomical constraint x_{AC} is introduced by concatenating it to the input of every block, and we observed more stable training using this method.

Multi-head Discriminator D: The discriminator D is an encoder architecture with consecutive convolution and down-sampling. There are three prediction heads $D_{R/F}$, D_c, and D_{AC} which predict a realness logit, disease classification, and reconstruction of constraint x_{AC}, respectively. The main loss functions are these outputs are 1) the standard non-saturating GAN loss $L_{R/F}$ for generating realistic images, 2) the adversarial classification loss L_c which supervises the conditioning on c 3) the anatomical reconstruction loss L_{AC} ensuring the generated reference images share the high-level structure of the input anatomical constraint. These losses are as follows:

$$\min_{D}\max_{G} L_{R/F} = \mathbb{E}[\log D_{R/F}(x_c)] \; - \; \mathbb{E}[\log(D_{R/F}(x'_c))] \tag{1}$$

$$\min_{D}\max_{G} L_c = - \sum_{c \in \{c_1, c_2 \dots c_n\}} y_c \log\left(D_c(X)\right), \text{where } X \in \{x'_c, x_c\} \tag{2}$$

$$L_{AC} = ||x_{AC} - D_{AC}(G(w_c, x_{AC}))||^2 + ||x_{AC} - D_{AC}(x_c))||^2 \tag{3}$$

where x_c is a real image of class c, $x'c$ is a fake image of the same class, $w_c = M(emb(c), z)$, and x_{AC} is a down-sampled version of x_c.

There are also other standard regularization loss functions including the path-length regularization L_{pl} [9], $R1$ gradient penalty L_{R1} [14] and an explicit diversification loss $L_{div} = -||G(w, c, z_1) - G(w, c, z_2)||$ to prevent the manifold from converging to fixed points. The total loss L_{total} is then defined as

$$L_{total} = L_{R/F} + L_c + L_{AC} + L_{div} + L_{pl} + L_{R1} \tag{4}$$

4 Experiment, Results and Discussion

Data Preparation and Training. DiDiGAN is applied to the ADNI dataset to visualize MRIs pathological changes from CN to AD. The dataset consists of T1-weighted 3D MRIs from the ADNI dataset totalling 1565 scans (680 patients) including 459 CN patients and 221 AD patients. The scans are divided into training, validation, and test sets at the patient level with a ratio of 0.7, 0.2, and 0.1. Standard prepossessing procedures (N4 bias-field correction [21], skull stripping using SPM [22]) were applied to the entire dataset. For each 3D scan, the center 40 coronal slices were extracted and zero-padded to 256 × 256 pixels. Down-sampled versions of these slices (64 × 64) were used for X_{AC}. The network was implemented in Pytorch and trained for 60,000 steps (32 images per step) using Adam optimizer. The training hardware is a 32 GB Nvidia V100 graphics card to support a batch size of 32. The baseline methods include a range of style and non-style-based image translation methods. These methods can readily treat reference image generation as unpaired image translation where, for example, a CN brain is translated to an AD version of the same brain. All the methods were trained and evaluated on the exact same data and splits.

Quantitative Image Quality Evaluation Using Perceptual Metrics. The quality of the generated reference images (based on the test set, ignoring classes) from all the methods is assessed using Frechet Inception Distance (FID), Learned Perceptual Image Patch Similarity (LPIPS), and Kernel Inception Distance (KID) as they do not require paired data. As Table 1 shows, DiDiGAN generated the most realistic reference images compared to the baselines. MUNIT [6] and CUT [15] failed to reach convergence hence the metrics are omitted.

Table 1. Quality of reference images generated by DiDiGAN and the baselines measured using three popular perceptual metrics. All the baselines were run on the exact same data as DiDiGAN. DiDiGAN achieved the best image quality while CUT and MUNIT failed to converge despite all the training instructions being followed.

Method	FID↓	LPIPS↑	KID↓
CUT [15]	N/A	N/A	N/A
MUNIT [6]	N/A	N/A	N/A
CycleGAN [25]	32.52	0.04	0.041
UNIT [12]	34.93	0.01	0.039
StyleGAN-SD [10]	17.23	0.14	0.034
StarGANv2 [4]	17.82	0.07	0.037
DiDiGAN	**14.48**	**0.29**	**0.031**

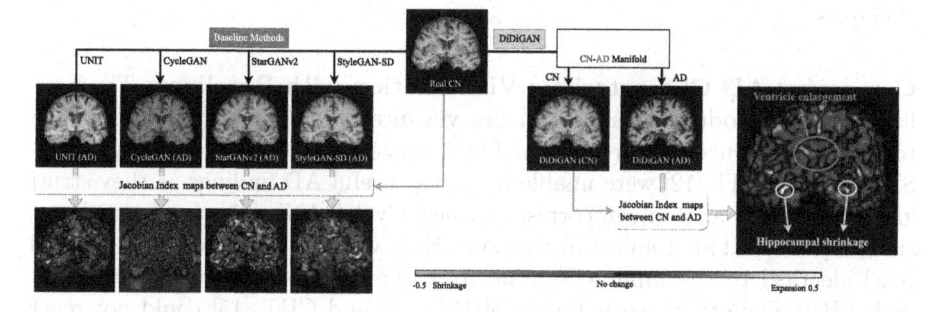

Fig. 2. AD feature discovery results for DiDiGAN and the baselines. The baselines perform image translation from an input CN to a reference AD image while DiDiGAN uses its manifold to generate a pair of reference AD and CN images. Jacobian Index heat maps were computed to show shrinkage (cool) or expansion (hot) areas between the AD and CN images. DiDiGAN's reference images reveal key AD characteristics including hippocampus shrinkage and ventricle enlargement. The baselines failed to highlight these characteristics due to anatomical disruptions.

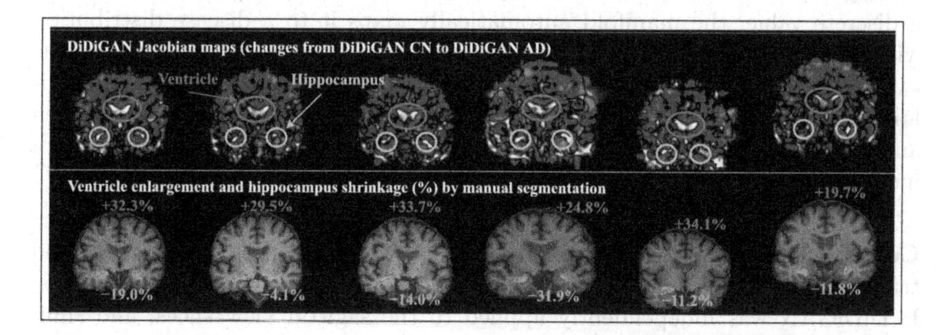

Fig. 3. More example Jacobian maps for DiDiGAN's reference AD and CN pairs. The hippocampus and ventricle areas are manually segmented with size changes labelled.

AD Characteristic Extraction and Visualization Using DiDiGAN. Figure 2 provides AD visualizations by generating reference images with DiDiGAN and the baseline methods. Since DiDiGAN's uses stylization, it can freely generate correlated AD and CN pairs using the manifold. Comparison between the reference pairs reveals AD characteristics [19,20] such as hippocampus shrinkage, ventricle enlargement, and the thinning of cortical structures. The consistent anatomical structure also facilitates the computation of Jacobian (shrinkage and expansion) maps. For DiDiGAN, the bright regions in DiDiGAN's Jacobian map more clearly highlight significant hippocampus shrinkage and ventricle enlargement from CN to AD. More examples are shown in Fig. 3. In these examples, we also manually segmented the ventricle and hippocampus, and the notable size changes marked in the figure. Anti-aliasing is crucial for maintaining anatomical correspondence across DiDiGAN's reference images. As an ablation study, a DiDiGAN without anti-aliasing was trained. The generated AD and CN pairs exhibit significant anatomical disturbances (see S1 for examples).

Comparing AD Characteristic Visualization with Baselines. The baseline methods produce reference images via direct image translation from CN to an AD reference. The results in Fig. 2 suggest StarGANv2 [4], StyleGAN-SD [10] and UNIT [12] were unable to convey useful AD findings as they struggled to maintain anatomical correspondence. CycleGAN's [25] simpler architecture helped avoid anatomical disruptions. However, the generated AD is almost pixel-identical to the input CN aside from the contrast difference (hence the high FID). The other two methods MUNIT [6] and CUT [15] could not reach convergence on the ADNI dataset despite the documentation being carefully followed.

Disease Manifold Formation and Interpolation. DiDiGAN's disease manifold formation is visualized using UMAP [13] where style codes sampled from the manifold are reduced to 2-d from 512-d. Figure 4 shows a 2D cluster containing 10,000 AD and 10,000 CN style codes. While the input disease class is a discrete value, the manifold automatically maps it to a disease distribution with AD distributed on the left and CN on the right. The overlap between the two classes suggests AD is a progressive disease. The smoothness of the manifold facilitates smooth CN-AD transition animations by interpolating between an AD style code and a CN style code. Practically, this animation provides a more intuitive visualization of AD pathological changes.

Corroborating Tests: Due to the lack of paired medical images as ground truths, we perform the following tests (I, II, III) to verify DiDiGAN's findings. **I**: DiDiGAN was independently applied to the sagittal view slices, and similar hippocampus shrinkage and ventricle enlargement were clearly observed (see examples S2). These independent findings corroborate the coronal view visualizations produced by DiDiGAN's reference images.

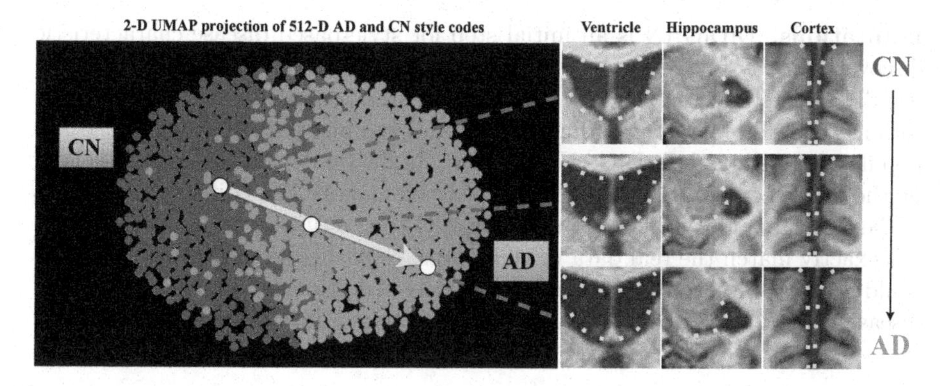

Fig. 4. 2D UMAP of 10000 CN and 10000 AD style codes sampled from the disease manifold. A linear path from CN to AD is traversed to generate an CN-AD animation. Changes in key affected regions (with reference points) are shown on the right. To help with visualization, we provide a video of several example animations.

II: SPM segmentation was applied to the test set and DiDiGAN's reference images (generated based on the test set) to help compare the magnitudes of brain volume loss. For each image, the grey matter and white matter pixels were treated as brain mass. As Fig. 5 shows, DiDiGAN's AD reference images consistently show systematic brain mass reduction across all 40 slice positions. This trend is consistent with that of the real data. The box plot suggests an average brain mass reduction of 17.3% compared to the 12.1% of the real data.

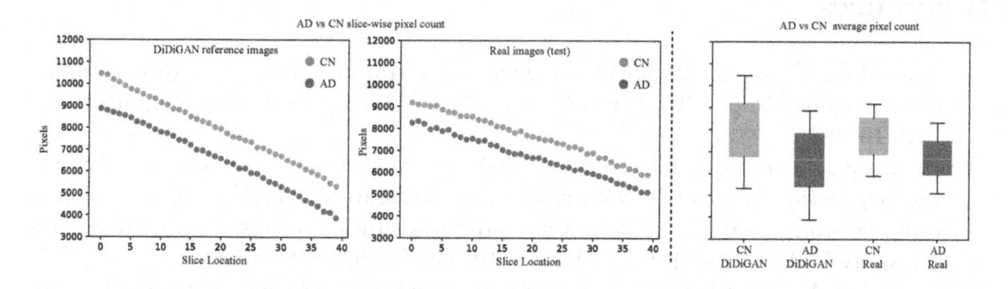

Fig. 5. Slice-wise brain mass difference (in pixels) between AD and CN as represented by DiDiGAN and real data. AD shows significant brain mass reductions in both cases.

III: After training DiDiGAN on ADNI for disease discovery, we fine-tuned D's classification head D_c for binary AD-CN classification and three-way AD-Mild Cognitive Impairment (MCI)-CN classification on the same dataset and splits. The test accuracies were 93.2% and 90.4% which are competitive with comparable methods in the literature (S3 shows a full comparison). Since D_c back-propagates disease class signals into G. This classification result suggests DiDiGAN successfully learned useful AD features.

Limitations. DiDiGAN is an initial step for style-based disease characteristic discovery. In the future, a more rigours examination of the manifold is needed to fully understand the features and trends learned especially for clinical applications. For example, longitudinal trends and atrophy patterns among the AD population. Although DiDiGAN discovered brain tissue shrinkage as indicated by the SPM segmentation analysis, the learned magnitude is different from that of the real data. This is likely a limitation of GANs as the learnt distribution may not exactly match the real data in all aspects. Nonetheless, DiDiGAN's findings could serve as disease characteristic proposals. Additionally, more datasets and diseases should be tested to more thoroughly assess DiDiGAN's generalizability.

5 Conclusion

DiDiGAN demonstrated disease characteristic discovery by generating reference images that clearly depict relevant pathological features. The main technical novelties of DiDiGAN are i) the use of a learned disease manifold to manipulate disease states AD ii) the ability to interpolate the manifold to enhance visualization and iii) mechanisms including the structural constraint and anti-aliasing to maintain anatomical correspondence without direct registration. In the experiments involving the ADNI dataset, DiDiGAN discovered key AD features such as hippocampus shrinkage, ventricular enlargement, and cortex atrophy where other frameworks failed. DiDiGAN shows potential to aid disease characteristic discovery across time of other chronic diseases such as osteoarthritis.

References

1. Armanious, K., et al.: MedGAN: medical image translation using GANs. Comput. Med. Imaging Graph. **79**, 101684 (2020). https://doi.org/10.1016/j.compmedimag.2019.101684
2. Bernal, J., Valverde, S., Kushibar, K., Cabezas, M., Oliver, A., Lladó, X.: Generating longitudinal atrophy evaluation datasets on brain magnetic resonance images using convolutional neural networks and segmentation priors. Neuroinformatics **19**(3), 477–492 (2021). https://doi.org/10.1007/s12021-020-09499-z
3. Chartsias, A., et al.: Disentangled representation learning in cardiac image analysis. Med. Image Anal. **58**, 101535 (2019). https://doi.org/10.1016/j.media.2019.101535
4. Choi, Y., Uh, Y., Yoo, J., Ha, J.W.: StarGAN v2: diverse image synthesis for multiple domains. In: 2020 IEEE/CVF Conference on Computer Vision and Pattern Recognition (CVPR). IEEE, June 2020. https://doi.org/10.1109/cvpr42600.2020.00821
5. Fetty, L., et al.: Latent space manipulation for high-resolution medical image synthesis via the StyleGAN. Z. Med. Phys. **30**(4), 305–314 (2020). https://doi.org/10.1016/j.zemedi.2020.05.001
6. Huang, X., Liu, M.-Y., Belongie, S., Kautz, J.: Multimodal unsupervised image-to-image translation. In: Ferrari, V., Hebert, M., Sminchisescu, C., Weiss, Y. (eds.) ECCV 2018. LNCS, vol. 11207, pp. 179–196. Springer, Cham (2018). https://doi.org/10.1007/978-3-030-01219-9_11

7. Jack, C.R., et al.: The Alzheimer's disease neuroimaging initiative (ADNI): MRI methods. J. Magn. Reson. Imaging **27**(4), 685–691 (2008). https://doi.org/10.1002/jmri.21049
8. Jung, E., Luna, M., Park, S.H.: Conditional GAN with an attention-based generator and a 3D discriminator for 3D medical image generation. In: de Bruijne, M., et al. (eds.) MICCAI 2021. LNCS, vol. 12906, pp. 318–328. Springer, Cham (2021). https://doi.org/10.1007/978-3-030-87231-1_31
9. Karras, T., et al.: Alias-free generative adversarial networks. In: Proceedings NeurIPS (2021)
10. Kim, K.L., Park, S.Y., Jeon, E., Kim, T.H., Kim, D.: A style-aware discriminator for controllable image translation. In: 2022 IEEE/CVF Conference on Computer Vision and Pattern Recognition (CVPR), pp. 18218–18227 (2022)
11. Liu, M., et al.: Style transfer using generative adversarial networks for multi-site MRI harmonization. In: de Bruijne, M., et al. (eds.) MICCAI 2021. LNCS, vol. 12903, pp. 313–322. Springer, Cham (2021). https://doi.org/10.1007/978-3-030-87199-4_30
12. Liu, M.Y., Breuel, T., Kautz, J.: Unsupervised image-to-image translation networks. In: Guyon, I., et al. (eds.) Advances in Neural Information Processing Systems, vol. 30. Curran Associates, Inc. (2017)
13. McInnes, L., Healy, J., Saul, N., Grossberger, L.: UMAP: uniform manifold approximation and projection. J. Open Source Softw. **3**(29), 861 (2018)
14. Mescheder, L.M., Geiger, A., Nowozin, S.: Which training methods for GANs do actually converge? In: ICML (2018)
15. Park, T., Efros, A.A., Zhang, R., Zhu, J.-Y.: Contrastive learning for unpaired image-to-image translation. In: Vedaldi, A., Bischof, H., Brox, T., Frahm, J.-M. (eds.) ECCV 2020. LNCS, vol. 12354, pp. 319–345. Springer, Cham (2020). https://doi.org/10.1007/978-3-030-58545-7_19
16. Peters, J., et al.: DBCE: a saliency method for medical deep learning through anatomically-consistent free-form deformations. In: Proceedings of the IEEE/CVF Winter Conference on Applications of Computer Vision (WACV), pp. 1959–1969, January 2023. https://doi.org/10.1109/WACV56688.2023.00200
17. Ravi, D., Blumberg, S.B., Ingala, S., Barkhof, F., Alexander, D.C., Oxtoby, N.P.: Degenerative adversarial neuroimage nets for brain scan simulations: application in ageing and dementia, vol. 75, p. 102257. Elsevier BV, January 2022. https://doi.org/10.1016/j.media.2021.102257
18. Rusak, F., et al.: Quantifiable brain atrophy synthesis for benchmarking of cortical thickness estimation methods. Med. Image Anal. **82**, 102576 (2022). https://doi.org/10.1016/j.media.2022.102576
19. Sabuncu, M.R., et al.: The dynamics of cortical and hippocampal atrophy in Alzheimer disease. Arch. Neurol. **68**(8), 1040–1048 (2011). https://doi.org/10.1001/archneurol.2011.167
20. Schuff, N., et al.: MRI of hippocampal volume loss in early Alzheimer's disease in relation to ApoE genotype and biomarkers. Brain **132**(4), 1067–1077 (2008). https://doi.org/10.1093/brain/awp007
21. Tustison, N.J., et al.: N4ITK: improved N3 bias correction **29**(6), 1310–1320 (2010). https://doi.org/10.1109/tmi.2010.2046908
22. Tzourio-Mazoyer, N., et al.: Automated anatomical labeling of activations in SPM using a macroscopic anatomical parcellation of the MNI MRI single-subject brain. Neuroimage **15**(1), 273–289 (2002). https://doi.org/10.1006/nimg.2001.0978

23. Xia, T., Chartsias, A., Tsaftaris, S.A.: Consistent brain ageing synthesis. In: Shen, D., et al. (eds.) MICCAI 2019. LNCS, vol. 11767, pp. 750–758. Springer, Cham (2019). https://doi.org/10.1007/978-3-030-32251-9_82
24. Zhang, R.: Making convolutional networks shift-invariant again. In: ICML (2019)
25. Zhu, J.Y., Park, T., Isola, P., Efros, A.A.: Unpaired image-to-image translation using cycle-consistent adversarial networks. In: 2017 IEEE International Conference on Computer Vision (ICCV). IEEE, October 2017. https://doi.org/10.1109/iccv.2017.244

COVID-19 Pneumonia Classification with Transformer from Incomplete Modalities

Eduard Lloret Carbonell[1], Yiqing Shen[2], Xin Yang[1], and Jing Ke[1,3](✉)

[1] School of Electronic Information and Electrical Engineering, Shanghai Jiao Tong University, Shanghai, China
{edu.lloret,yang_xin,kejing}@sjtu.edu.cn

[2] Department of Computer Science, Johns Hopkins University, Baltimore, MD, USA
yshen92@jhu.edu

[3] School of Computer Science and Engineering, University of New South Wales, Sydney, Australia

Abstract. COVID-19 is a viral disease that causes severe acute respiratory inflammation. Although with less death rate, its increasing infectivity rate, together with its acute symptoms and high number of infections, is still attracting growing interests in the image analysis of COVID-19 pneumonia. Current accurate diagnosis by radiologists requires two modalities of X-Ray and Computed Tomography (CT) images from one patient. However, one modality might miss in clinical practice. In this study, we propose a novel multi-modality model to integrate X-Ray and CT data to further increase the versatility and robustness of the AI-assisted COVID-19 pneumonia diagnosis that can tackle incomplete modalities. We develop a Convolutional Neural Networks (CNN) and Transformers hybrid architecture, which extracts extensive features from the distinct data modalities. This classifier is designed to be able to predict COVID-19 images with X-Ray image, or CT image, or both, while at the same time preserving the robustness when missing modalities are found. Conjointly, a new method is proposed to fuse three-dimensional and two-dimensional images, which further increase the feature extraction and feature correlation of the input data. Thus, verified with a real-world public dataset of BIMCV-COVID19, the model outperform state-of-the-arts with the AUC score of 79.93%. Clinically, the model has important medical significance for COVID-19 examination when some image modalities are missing, offering relevant flexibility to medical teams. Besides, the structure may be extended to other chest abnormalities to be detected by X-ray or CT examinations. Code is available at https://github.com/edurbi/MICCAI2023.

Keywords: COVID-19 Pneumonia Classification · Hybrid CNN-Transformer · Multi-modality

---

E. L. Carbonell and Y. Shen—Equal contributions.

H. Greenspan et al. (Eds.): MICCAI 2023, LNCS 14224, pp. 379–388, 2023.
https://doi.org/10.1007/978-3-031-43904-9_37

1 Introduction

Coronavirus disease 2019 (COVID-19) has been a highly infectious viral disease, that can affect people of all ages, with a persistently high incidence after the outbreak in 2019 [1,2]. Early detection of the lung inflammatory reaction is crucial to initiate prompt treatment decisions, where the clinical assessment typically depends on two imaging techniques in conjunction, namely X-ray and CT scans [3,4]. Specifically, CT scans can provide a three-dimensional volumetric characterization of the patient's lung; while X-rays offer a two-dimensional landscape [5]. Recently, the use of multimodality data in COVID-19 diagnosis has received increasing interest as it can significantly improve prediction accuracy with complementary information [6].

During the inference stage, modalities can be incomplete amongst some test samples, which is known as incomplete multimodal learning [7,8]. To address this, it is necessary to implement a strategy that reduces the impact of missing modalities during the inference stage while also offering clinicians the flexibility to use any possible combination of data. Various strategies have been proposed to address this issue, such as generating the missing data [9,10]. However, this approach requires training a generative model for each possible missing modality and a larger training set, making it computationally expensive. Therefore, more compact models have emerged that reduces the number of generators [7]. However, these models are prone to be biased when dealing with multiple modalities. Consequently, a modality-invariant embedding model that makes use of Transformers has been introduced [11]. Despite the excellent performance, all of the models discussed above have been applied to the datasets, where the modalities used were restricted to three-dimensional, and the structures are similar. However, in the case of COVID-19 pneumonia detection, the clinicians practically employ different dimensional data to interpret the results. One of the major problems during the analysis of medical data is that, sometimes, some data modalities are missing, e.g., the case reported negative, or the examination device was unavailable. [12] Therefore being able to have a model that is able to adapt to all possible conditions will be greatly beneficial for the detection of COVID-19 pneumonia or any other disease. To this end, a novel method to diagnose COVID-19 pneumonia which can take incomplete CT and X-Ray multimodal data is proposed. The main contributions in this model are three-fold: (1) We propose a dual feature fusion across different dimensionality data. Instead of sole pre-fusion or post-fusion, both are performed in the multi-modality prediction model. (2) We design a feature matrix dropout regularization method to improve the reliability and generalization of the model. (3) A feature correlation block is proposed between two of the given modalities to extract latent dependencies. This attention layer further improves the understanding of the patients' evolution when multi-modality images are acquired at different stages of the disease.

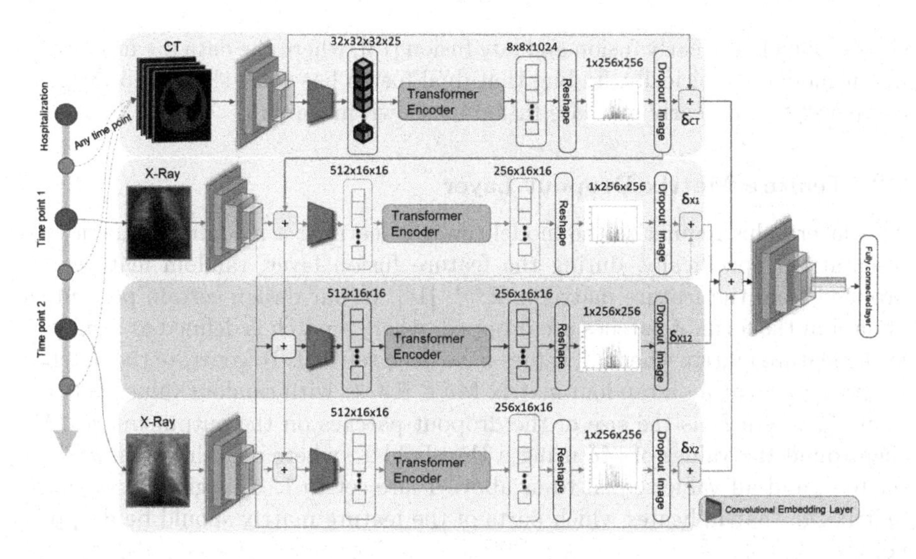

Fig. 1. Overview of the proposed model. This model is composed by three image modalities, on the first modality we use a convolutional-based feature extractor, followed by a Transformer encoder and a random features dropout. The other two modalities consist of an early fusion with the first modality features followed by a transformer encoder. Finally, all of the features have a features dropout layer followed by a Convolutional layer and a fully connected layer.

2 Methodology

2.1 Architecture Overview

We propose a model that can detect the COVID-19 pneumonia status from incomplete multi-modalities of CT scans and X-ray images. Specifically, CT scans and X-ray images are first embedded using convolutional layers and then processed by Transformer layers to obtain the global feature correlations. Then, a novel feature fusion layer can simulate incomplete modalities in the latent space while learning to fuse the features. Finally, the predictions are made using a ResNet based classification model followed by a learnable MLP layer.

2.2 Feature Fusion Layer

The objective of this layer is to combine features from different modalities, which have varying dimensions. To achieve this, we need to reduce the three-dimensional CT data to two-dimensional by convolutional layers before fusing them with two-dimensional X-ray data. We then reshape the data into smaller patches and apply a Transformer encoder to it, resulting in a two-dimensional matrix of the desired size. In this study, we investigate the data fusion at two stages, namely the early fusion and late fusion [13], where the data are fused twice in our model. Empirically, finding that dual fusion has some slight improvement compared to only early and late fusion as shown in Table 3.

2.3 Feature Matrix Dropout Layer

This layer helps regularize the model towards learning a modality agnostic representation. Specifically, during the feature fusion layer, random features are dropped from the feature matrix $\in \mathbb{R}^{X,Y}$ [14]. In our design certain percentage of data in the form of patches are dropped, where a patch is defined as a subpart of the feature matrix where all values have been set to 0. To generate the patches, we start by creating a random matrix $\mathbb{M}\delta \in \mathbb{R}^{\frac{X}{N},\frac{Y}{N}}$ with random values between 0 and 1, where N is the size of the dropout patches on the output images. We then round the values of $\mathbb{M}\delta$ using a threshold T, where all values below T are set to 0, and all values equal to or above T are set to 1. This gives us a binary matrix $\mathbb{M}_\delta$ that indicates which parts of the feature matrix should be dropped, i.e.

$$\mathbb{M}_D = R_T(\mathbb{M}_\delta) \tag{1}$$

where $R_T(\cdot)$ is the round function that converts all values of M_δ into 0 if they are under the threshold T and 1 otherwise. Finally the obtained matrix is interpolated or upsampled by a given scale N to match the size of the feature matrices $\mathbb{F}_M \in \mathbb{R}^{X,Y}$.

$$\mathbb{F}_D = I_M(\mathbb{M}_D) \odot \mathbb{F}_M \tag{2}$$

where $\mathbb{F}_D$ represents the final feature map after the patch dropout, F_M is the initial feature map, $I_M(\cdot)$ represents a nearest interpolation and $(\cdot) \odot (\cdot)$ represents an element-wise matrix multiplication. This gives us the final feature map $\mathbb{F}_D$, which has a similar structure to $\mathbb{F}_M$ but with some parts of size $N \times N$ converted to 0.

2.4 Transformer Layer

Transformers helps finding the different dependencies between the different modalities making use of their attention-based nature. Therefore, in this paper we will make use of the benefits that the transformers offers by implementing a ViT based transformer layer [15]. In this case we will use the transformer layer as a feature extractor and find the dependencies between the different embedded features. To extract the features, we will first split the image into patches of a fixed size, to afterwards be processed by an embedding layer, in this case the embedding layer will be a convolutional layer. Then each one of the embedded feature will pass through a different Self Attention Layer (SA) formulated as

$$\begin{aligned} SA_L &= \sigma\left(\frac{Q_L K^T}{\sqrt{d_L}}\right) V_L, \\ MSA_L &= \text{Concat}(SA_L^1, \cdots, SA_L^n) \cdot \mathbb{W}_0, \end{aligned} \tag{3}$$

where $\sigma(\cdot)$ represents the softmax function, d represents the size of each one of the heads, and $\mathbb{W}_0$ denotes a value embedding. Meanwhile, Q, K, V $\in \mathbb{R}^{X,d}$ represent the Query, key and embedding respectively. To constitute the ViT the values of each of the SA are concatenated forming the Multi-Headed Self Attention Layer (MSA) to afterward be normalized by a layer normalization. Finally, a Multi-Layer Perceptron (MLP) is applied to give non-linearity to the data, obtaining

$$\begin{aligned} A'_k &= MSA(LN(A_{k-1})) + A_{k-1} \\ A_k &= LN(MLP(A'_k)) + A'_k \end{aligned} \tag{4}$$

where A^{k-1} is the previous transformer layer, LN$(\cdot)$ is a Layer Normalization. In the vanilla ViT a final MLP is introduced to obtain the probabilities of each class however, in this paper the transformer layer is only used for feature extraction, therefore, the final MLP is deleted.

2.5 Dual X-Ray Attention

The Dual X-Ray attention block main idea is to find the different dependencies between the two input X-Ray images. Transformers have showed great results when looking for dependencies [11,15]. Therefore, this paper introduced a new attention layer that extracts the dependencies between two X-Ray images. The dependencies are extracted using the Transformer mentioned above Layer having as input both X-Ray images in the multimodality. This layer can also be seen as a fusion layer between two of the input modalities.

3 Experiments

3.1 Dataset

We leverage images from the BIMCV-COVID19 dataset [16], a public dataset with thousands of different positive and negative cases, for performance evaluation. This dataset is composed of 1200 unique cases from a combination of 1 CT Scan, 1 CT Scan and 1 X-Ray, 1 CT Scan and 2 X-Ray, 1 X-Ray or 2 X-Ray. One patient may have more than one image modality of CT or X-Ray. Regarding the image size of the dataset, the dimension of the CT scans is non-fixed per image, with the average image size around $500 \times 500 \times 400$. To facilitate the training on GPUs, the images perform dimension reduction with factor 2 and resulted in a final image size of $250 \times 250 \times 200$. The dataset is composed of approximately 2900 cases. From which around 1700 are negative, and 1200 are positive. The distribution is unbalanced because we wanted to extract as much data as possible with a balanced number of miss-modalities.

3.2 Implementation Details

This framework is implemented with Pytorch 1.13.0 using an NVIDIA A10 GPU with 24 GB of VRAM. The input size is $250 \times 250 \times 200$ for the CT scan and

2048×2048 with a batch size of 4 for the X-Ray. For the proposed model, we applied Adam optimizer with an cosine annealing scheduler of an initial learning rate 1e-5, and 100 epochs for training. The data partition for training, validation, and testing is 80%, 10%, and 10% on patient level.

3.3 Results

We use the following metrics to evaluate the performance of the binary classification model: the AUC score, Recall and Precision. To obtain the values on the Table 1, we made a split form the original dataset without miss-modalities. This table shows the difference between the possible combinations of models we can build to interpret the data. Taking in account the AUC, the worst model is the one using the CT as its only input with only a 65.74% in the AUC score, followed by the model with two X-Ray images as its input, obtaining a 67.98%. The result obtained in the Dual X-Ray multimodality performs around 2% worse than just having one X-Ray as an input. Finally, when having all the modalities performs 3% better in the AUC score than when only having one CT and one X-Ray which has around 71.26% in the AUC metric.

Table 1. The performance of AUC-score, Precision and Recall of the different modalities is reported. The baseline does not include any additional model proposed by us.

CT	X-Ray 1	X-Ray 2	AUC	Recall	Precision
X	✓	X	70.15	66.66	64.28
X	✓	✓	67.98	65.9	70.58
✓	X	X	65.74	61.11	75.0
✓	✓	X	71.26	74.28	69.76
✓	✓	✓	74.49	66.66	62.22

A comparison is made between our model and others with different possible variations, shown in the Table 2. The first model is used as a comparison is a fully convolutional model where all of its transformer components are converted into convolutions. This variant shows a significant reduction in all of its parameters with a sharp decrease in the performance of 70.18%, 59.9% and 68.93% in the

Table 2. The result comparisons between different variations of the proposed model.

Model	AUC	Recall	Precision
Convolutional Model	70.18	59.9	68.93
ViT Classificator	71.38	76.80	59.84
Final model	79.93	86.62	67.40

AUC, Recall, and Precision metrics, respectively. Another model is tested, where the final convolutional block is changed into a ViT base block with a final MLP layer. The AUC score remains similar to the one found in the previously tested model, obtaining 71.38%, 76.80% and 59.84% in AUC, Recall and Precision respectively. Due to the multi-modality and missing modality nature of this model and this dataset, the comparison between this model and other existing models was not feasible.

Table 3. Ablation study between the three different proposed methods by adding Dual Fusion block, Dual X-Ray attention block, and Feature matrix dropout in our method.

Dual Fusion	Dual X-Ray attention	Feature matrix dropout	AUC	Recall	Precision
X	X	X	75.28	71.14	69.87
✓	X	X	76.49	61.43	83.55
✓	✓	X	77.27	72.72	72.84
✓	✓	✓	79.93	86.62	67.40

3.4 Ablation Study

We show the classification performance on Table 3, of the different model components that introduced in the Fig. 1. We make use of the original dataset with missing modalities. From the Table 3, the addition of the Fusion Layer gives an slight increase of 0.19% in the AUC-score metric compared with the baseline. Meanwhile, we see an observable increase in the recall metric of 2.93%. The addition of the Dual X-Ray layer further increases the performance of the model increasing of 1.8% and 3.43% in the AUC-score and Precision respectively, yet a slight reduction in the recall metric by around 1.35%. Finally, the Regularization layer is topped up to further increases the AUC by a 2.66% and recall by a 13.90%, yet lower Precision score of 5.44%. Thus, compared to the baseline, the final model shows an increase of 4.65% in the AUC metric, a boost of 15.58% in the recall score and only a 2.47% decrease in the precision metric.

The saliency maps are visualized to pinpoint the diagnostic areas in a CT or X-ray image. Clinically, the saliency maps are helpful to assist radiologists. In the Fig. 2, (a) shows an example of a COVID-19 positive slice extracted from a complete CT scan. Its saliency map is situated next to the image, where a big opacity is found in the left patient's lung. The top-right figure gives an example of COVID-19 positive X-Ray which its correspondent saliency map, which mainly focus on the bottom contour of the lung. Moving to the bottom-left figure we can see a COVID-19 negative CT scan using a different angle compared to the first introduced figure. In this case, similar to what has been seen in the positive case, some opacities have been found however, in this case, the image is negative. Finally, the bottom-right figure is a negative X-Ray case, in this case, similar

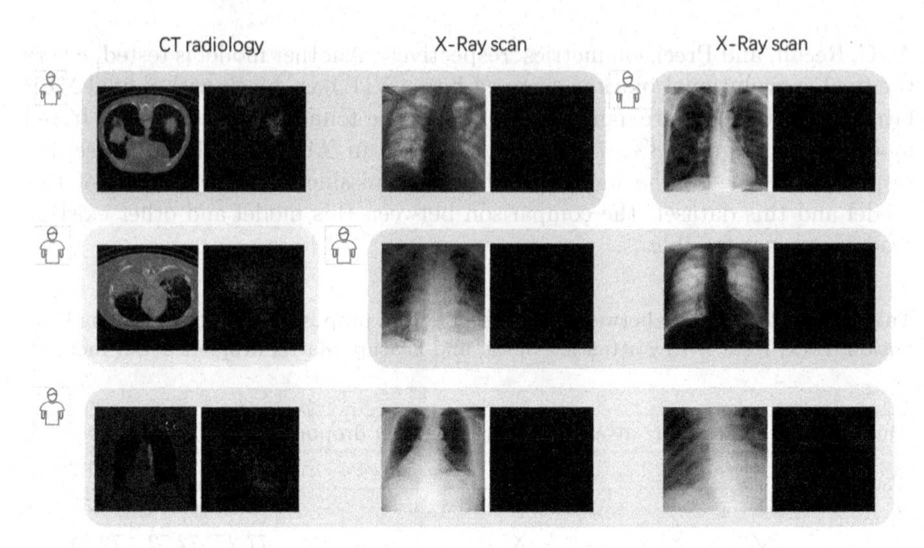

Fig. 2. X-ray and CT images with their saliency visualization.

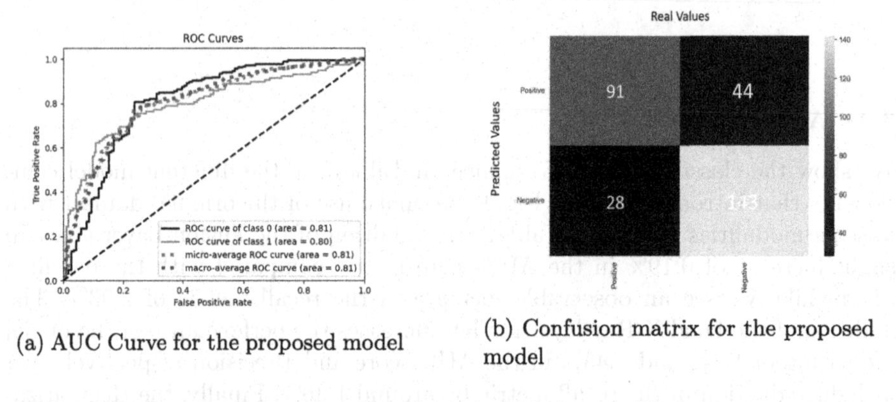

(a) AUC Curve for the proposed model

(b) Confusion matrix for the proposed model

Fig. 3. Performance of proposed model. (a) AUC Curve, (b) Confusion matrix

to what has been found in the positive X-Ray figure the model mainly looks for the contours of the lungs. Therefore, through this images, a difference can be seen between the extracted features using the CT scans and the ones extracted using the X-Ray images. Finally, in the Fig. 3 the AUC Curve obtained with the proposed model together with the confusion matrix, obtained as a result of classifying the test dataset with the model, are shown as Sub-figure a and b respectively.

4 Conclusion

In this paper, we propose a novel multi-modality framework for COVID-19 pneumonia image diagnosis. A Dual fusion layer is introduced to help establish the dependencies between the different input modalities, as well as to take in different dimensionality inputs. The proposed Dual X-Ray attention layer makes it possible to effectively extract dependencies from the two X-Ray images focusing on the salient features between multi-modality images, whereas irrelevant features are ignored. The feature deletion layer helps to regularize the model dropping random features and improving the generalization of the model. Consequently, we provide the possibility to use one modality of CT or X-Ray for COVID-19 pneumonia diagnosis. Moreover, this model has the potential to be applied to other chest abnormalities in clinical practice.

Acknowledgments. This work was supported by National Natural Science Foundation of China (Grant No. 62102247) and Natural Science Foundation of Shanghai (No. 23ZR1430700).

References

1. Percentage of Visits for COVID-19-Like Illness: Covid data page. https://www.cdc.gov/coronavirus/2019-ncov/covid-data/covidview/index.html
2. Griffin, D.O., et al.: The importance of understanding the stages of covid-19 in treatment and trials. AIDS Rev. **23**(1), 40–47 (2021). https://doi.org/10.24875/aidsrev.200001261
3. Luo, N., et al.: Utility of chest CT in diagnosis of covid-19 pneumonia. Diagn. Interv. Radiol. **26**(5), 437–442 (2020). https://doi.org/10.5152/dir.2020.20144. https://www.ncbi.nlm.nih.gov/pmc/articles/PMC7490028/. pMID: 32490829; PMCID: PMC7490028
4. Alyasseri, Z.A.A., et al.: Review on covid-19 diagnosis models based on machine learning and deep learning approaches. Exp. Syst. **39**(3), e12759 (2022). https://doi.org/10.1111/exsy.12759. https://onlinelibrary.wiley.com/doi/abs/10.1111/exsy.12759
5. Li, B., et al.: Diagnostic value and key features of computed tomography in coronavirus disease 2019. Emerg. Microbes Infect. **9**(1), 787–793 (2020). https://doi.org/10.1080/22221751.2020.1750307. pMID: 32241244
6. Abdelaziz, M., Wang, T., Elazab, A.: Alzheimer's disease diagnosis framework from incomplete multimodal data using convolutional neural networks. J. Biomed. Informatics **121**, 103863 (2021). https://doi.org/10.1016/j.jbi.2021.103863. https://www.sciencedirect.com/science/article/pii/S1532046421001921
7. Azad, R., Khosravi, N., Dehghanmanshadi, M., Cohen-Adad, J., Merhof, D.: Medical image segmentation on MRI images with missing modalities: a review (2022). https://doi.org/10.48550/ARXIV.2203.06217. https://arxiv.org/abs/2203.06217
8. Ma, M., Ren, J., Zhao, L., Tulyakov, S., Wu, C., Peng, X.: SMIL: multimodal learning with severely missing modality. Proc. AAAI Conf. Artif. Intell. **35**(3), 2302–2310 (2021). https://doi.org/10.1609/aaai.v35i3.16330. https://ojs.aaai.org/index.php/AAAI/article/view/16330
9. Jin, L., Zhao, K., Zhao, Y., Che, T., Li, S.: A hybrid deep learning method for early and late mild cognitive impairment diagnosis with incomplete multimodal data. Frontiers Neuroinf. (2022). https://doi.org/10.3389/fninf.2022.843566. https://www.ncbi.nlm.nih.gov/pmc/articles/PMC8965366/

10. Gao, X., Shi, F., Shen, D., Liu, M.: Task-induced pyramid and attention Gan for multimodal brain image imputation and classification in Alzheimer's disease. IEEE J. Biomed. Health Inform. **26**(1), 36–43 (2022). https://doi.org/10.1109/JBHI.2021.3097721
11. Zhang, Y., et al.: mmFormer: multimodal medical transformer for incomplete multimodal learning of brain tumor segmentation (2022). https://doi.org/10.48550/ARXIV.2206.02425. https://arxiv.org/abs/2206.02425
12. Altman, D.G., Bland, J.M.: Missing data. BMJ **334**(7590), 424 (2007). https://doi.org/10.1136/bmj.38977.682025.2C. https://www.bmj.com/content/334/7590/424
13. Gadzicki, K., Khamsehashari, R., Zetzsche, C.: Early vs late fusion in multimodal convolutional neural networks. In: 2020 IEEE 23rd International Conference on Information Fusion (FUSION), pp. 1–6 (2020). https://doi.org/10.23919/FUSION45008.2020.9190246
14. Choi, J.H., Lee, J.S.: EmbraceNet: a robust deep learning architecture for multimodal classification. Inf. Fusion **51**, 259–270 (2019). https://doi.org/10.1016/j.inffus.2019.02.010. https://www.sciencedirect.com/science/article/pii/S1566253517308242
15. Dosovitskiy, A., et al.: An image is worth 16x16 words: transformers for image recognition at scale (2020). https://doi.org/10.48550/ARXIV.2010.11929. https://arxiv.org/abs/2010.11929
16. de la Iglesia Vayá, M., et al.: BIMCV covid-19+: a large annotated dataset of RX and CT images from covid-19 patients with extension Part II (2023). https://doi.org/10.21227/mpqg-j236

Enhancing Breast Cancer Risk Prediction by Incorporating Prior Images

Hyeonsoo Lee(✉), Junha Kim, Eunkyung Park, Minjeong Kim, Taesoo Kim, and Thijs Kooi

Lunit Inc., Seoul, Republic of Korea
{hslee,junha.kim,ekpark,mjkim0918,taesoo.kim,tkooi}@lunit.io

Abstract. Recently, deep learning models have shown the potential to predict breast cancer risk and enable targeted screening strategies, but current models do not consider the change in the breast over time. In this paper, we present a new method, PRIME+, for breast cancer risk prediction that leverages prior mammograms using a transformer decoder, outperforming a state-of-the-art risk prediction method that only uses mammograms from a single time point. We validate our approach on a dataset with 16,113 exams and further demonstrate that it effectively captures patterns of changes from prior mammograms, such as changes in breast density, resulting in improved short-term and long-term breast cancer risk prediction. Experimental results show that our model achieves a statistically significant improvement in performance over the state-of-the-art based model, with a C-index increase from 0.68 to 0.73 ($p < 0.05$) on held-out test sets.

Keywords: Breast Cancer · Mammogram · Risk Prediction

1 Introduction

Breast cancer impacts women globally [15] and mammographic screening for women over a certain age has been shown to reduce mortality [7,10,23]. However, studies suggest that mammography alone has limited sensitivity [22]. To mitigate this, supplemental screening like MRI or a tailored screening interval have been explored to add to the screening protocol [1,13]. However, these imaging techniques are expensive and add additional burdens for the patient. Recently, several studies [8,32,33] revealed the potential of artificial intelligence (AI) to develop a better risk assessment model to identify women who may benefit from supplemental screening or a personalized screening interval and these may lead to improved screening outcomes.

In clinical practice, breast density and traditional statistical methods for predicting breast cancer risks such as the Gail [14] and the Tyrer-Cuzick models [27] have been used to estimate an individual's risk of developing breast cancer. However, these models do not perform well enough to be utilized in practical screening settings [3] and require the collection of data that is not always available. Recently, deep neural network based models that predict a patient's risk score

H. Greenspan et al. (Eds.): MICCAI 2023, LNCS 14224, pp. 389–398, 2023.
https://doi.org/10.1007/978-3-031-43904-9_38

directly from mammograms have shown promising results [3,8,9,20,33]. These models do not require additional patient information and have been shown to outperform traditional statistical models.

When prior mammograms are available, radiologists compare prior exams to the current mammogram to aid in the detection of breast cancer. Several studies have shown that utilizing past mammograms can improve the classification performance of radiologists in the classification of benign and malignant masses [11,25,26,29], especially for the detection of subtle abnormalities [25]. More recently, deep learning models trained on both prior and current mammograms have shown improved performance in breast cancer classification tasks [24]. Integrating prior mammograms into deep learning models for breast cancer risk prediction can provide a more comprehensive evaluation of a patient's breast health.

In this paper, we introduce a deep neural network that makes use of prior mammograms, to assess a patient's risk of developing breast cancer, dubbed PRIME+ (PRIor Mammogram Enabled risk prediction). We hypothesize that mammographic parenchymal pattern changes between current and prior allow the model to better assess a patient's risk. Our method is based on a transformer model that uses attention [30], similar to how radiologists would compare current and prior mammograms.

The method is trained and evaluated on a large and diverse dataset of over 9,000 patients and shown to outperform a model based on state-of-the art risk prediction techniques for mammography [33]. Although previous models such as LRP-NET and RADIFUSION [5,34] have utilized prior mammograms, PRIME+ sets itself apart by employing an attention mechanism to extract information about the prior scan.

2 Method

2.1 Risk Prediction

Survival analysis is done to predict whether events will occur sometime in the future. The data comprises three main elements: features x, time of the event t, and the occurrence of the event e [18]. For medical applications, x typically represents patient information like age, family history, genetic makeup, and diagnostic test results (e.g., a mammogram). If the event has not yet occurred by the end of the study or observation period, the data is referred to as right-censored (Fig. 1).

We typically want to estimate the hazard function $h(t)$, which measures the rate at which patients experience the event of interest at time t, given that they have survived up to that point. The hazard function can be expressed as the limit of the conditional probability of an event T occurring within a small time interval $[t, t + \Delta t)$, given that the event has not yet occurred by time t:

$$h(t) = \lim_{\Delta t \to 0} \frac{P(T \in [t, t + \Delta t) \mid T \geq t)}{\Delta t} \tag{1}$$

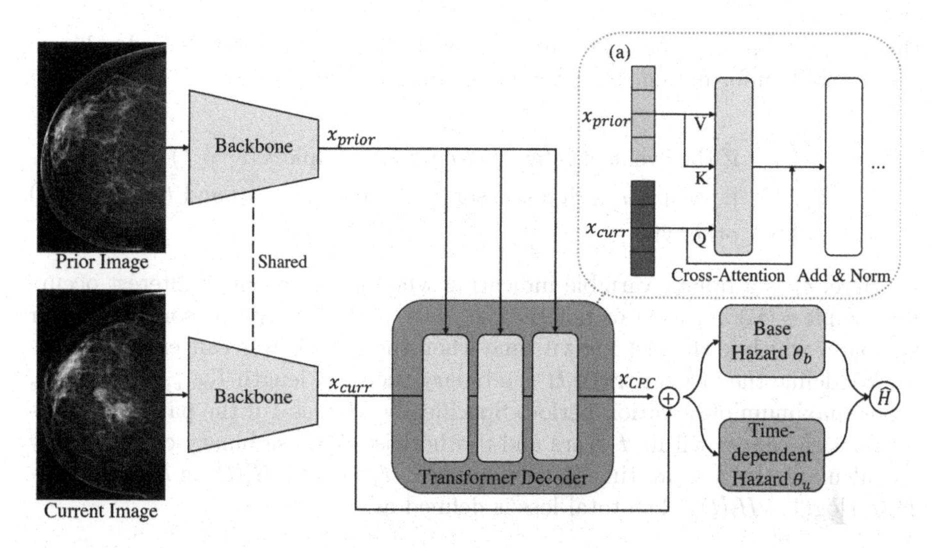

Fig. 1. We present an improved method for breast cancer risk prediction (PRIME+) by leveraging prior mammograms. A common backbone network extracts features from the prior and current images, resulting in x_{prior} and x_{curr}. We find that the transformer decoder effectively fuses relevant information from x_{prior} and x_{curr} to produce x_{CPC}. The base hazard θ_b and time-dependent hazard prediction heads θ_u use the concatenated feature to predict the cumulative hazard function $\hat{H}$. (a) illustrates the interaction between x_{prior} and x_{curr} in the cross-attention module of the transformer decoder.

The cumulative hazard function $H(t)$ is another commonly used function in survival analysis, which gives the accumulated probability of experiencing the event of interest up to time t. This function is obtained by integrating the hazard function over time from 0 to t: $H(t) = \int_0^t h(s)ds$.

2.2 Architecture Overview

We build on the current state-of-the art MIRAI [33] architecture, which is trained to predict the cumulative hazard function. We use an ImageNet pre-trained ResNet-34 [12] as the image feature backbone. The backbone network extracts features from the mammograms, and the fully connected layer produces the final feature vector x. We make use of two additional fully connected layers to calculate base hazard θ_b and time-dependent hazard θ_u, respectively.

The predicted cumulative hazard is obtained by adding the base hazard and time-dependent hazard, according to:

$$\hat{H}(t|x) = \theta_b(x) + \sum_{\tau=1}^{t} \theta_{u_\tau}(x) \tag{2}$$

When dealing with right-censored data, we use an indicator function $\delta_i(t)$ to determine whether the information for sample i at time t should be included in the loss calculation or not. This helps us exclude unknown periods and only use the available information. It is defined as follows:

$$\delta_i(t) = \begin{cases} 1, & \text{if the event of interest occurs for sample } i \ (e_i = 1) \\ 1, & \text{if sample } i \text{ is right-censored at time } t \ (e_i = 0 \text{ and } t < C_i) \\ 0, & \text{otherwise} \end{cases} \quad (3)$$

Here, e_i is a binary variable indicating whether the event of interest occurs for sample i (i.e., $e_i = 1$) or not (i.e., $e_i = 0$), and C_i is the censoring time for sample i, which is the last known time when the sample was cancer-free.

We define the ground-truth H is a binary vector of length T_{max}, where T_{max} is the maximum observation period. Specifically, $H(t)$ is 1 if the patient is diagnosed with cancer within t years and 0 otherwise. We use binary cross entropy to calculate the loss at time t for sample i: $\ell_i(t) = -H_i(t) \log \hat{H}_i(t) - (1 - H_i(t)) \log(1 - \hat{H}_i(t))$. The total loss is defined as:

$$L = \sum_{i=1}^{N} \sum_{t=1}^{T_{max}} \delta_i(t) \ell_i(t) \quad (4)$$

Here, N is the number of exams in the training set. The goal of training the model is to minimize this loss function, which encourages the model to make accurate predictions of the risk of developing breast cancer over time.

2.3 Incorporating Prior Mammograms

To improve the performance of the breast cancer risk prediction model, we incorporate information from prior mammograms taken with the same view, using a transformer decoder structure [30]. This structure allows the current and prior mammogram features to interact with each other, similar to how radiologists check for changes between current and prior mammograms.

During training, we randomly select one prior mammogram, regardless of when they were taken. This allows the model to generalize to varying time intervals. To pair each current mammogram during inference with the most relevant prior mammogram, we first select the prior mammogram taken at the time closest to the current time. This approach is based on research showing that radiologists often use the closest prior mammogram to aid in the detection of breast cancer [26].

Next, a shared backbone network is used to output the current feature x_{curr} and the prior feature x_{prior}. These features are then flattened and fed as input to the transformer decoder, where multi-head attention is used to find information related to the current feature in the prior feature. The resulting output is concatenated and passed through a linear layer to produce the current-prior comparison feature x_{CPC}. The current-prior comparison feature and current feature are concatenated to produce the final feature $x^* = x_{CPC} \oplus x_{curr}$, which is then used by the base hazard network and time-dependent hazard network to predict the cumulative hazard function $\hat{H}$.

3 Experiments

3.1 Dataset

We compiled an in-house mammography dataset comprising 16,113 exams (64,452 images) from 9,113 patients across institutions from the United States, gathered between 2010 and 2021. Each mammogram includes at least one prior mammogram. The dataset has 3,625 biopsy-proven cancer exams, 5,394 biopsy-proven benign exams, and 7,094 normal exams. Mammograms were captured using Hologic (72.3%) and Siemens (27.7%) devices. We partitioned the dataset by patient to create training, validation, and test sets. The validation set contains 800 exams (198 cancer, 210 benign, 392 normal) from 400 patients, and the test set contains 1,200 exams (302 cancer, 290 benign, 608 normal) from 600 patients. All data was de-identified according to the U.S HHS Safe Harbor Method. Therefore, the data has no PHI (Protected Health Information) and IRB (Institutional Review Board) approval is not required.

3.2 Evaluation

We make use of Uno's C-index [28] and the time-dependent AUC [16]. The C-index measures the performance of a model by evaluating how well it correctly predicts the relative order of survival times for pairs of individuals in the dataset. The C-index ranges from 0 to 1, with a value of 0.5 indicating random predictions and a value of 1 indicating that the model is perfect. Time-dependent ROC analysis generates an ROC curve and the area under the curve (AUC) for each specific time point in the follow-up period, enabling evaluation of the model's performance over time. To compare the C-index of two models, we employ the compareC [17] test, and make use of the DeLong test [6] to compare the time-dependent AUC values. Confidence bounds are generated using bootstrapping with 1,000 bootstrap samples.

We evaluate the effectiveness of PRIME+ by comparing it with two other models: (1) baseline based on MIRAI, a state-of-the art risk prediction method from [33], and (2) PRIME, a model that uses prior images by simply summing x_{curr} and x_{prior} without the use of the transformer decoder.

3.3 Implementation Details

Our model is implemented in Pytorch and trained on four V100 GPUs. We trained the model using stochastic gradient descent (SGD) for 20K iterations with a learning rate of 0.005, weight decay of 0.0001, and momentum of 0.9. We use a cosine annealing learning rate scheduling strategy [21].

We resize the images to 960 $\times$ 640 pixels and use a batch size of 96. To augment the training data, we apply geometric transformations such as vertical flipping, rotation and photometric transformations such as brightness/contrast adjustment, Gaussian noise, sharpen, CLAHE, and solarize. Empirically, we find that strong photometric augmentations improved the risk prediction model's

Table 1. Ablation analysis on the effectiveness of prior information and transformer decoder. Additional result in bottom row aims to predict unseen risks beyond visible cancer patterns by excluding early diagnosed cancer cases. The ± refers to the 95% confidence bound.

All Cases						
Prior	Decoder	C-index	Time-dependent AUC			
			1-year	2-year	3-year	4-year
✗	✗	$0.68_{\pm 0.03}$	$0.70_{\pm 0.05}$	$0.71_{\pm 0.04}$	$0.70_{\pm 0.04}$	$0.71_{\pm 0.09}$
✓	✗	$0.70_{\pm 0.03}$	$0.72_{\pm 0.05}$	$0.73_{\pm 0.05}$	$0.74_{\pm 0.04}$	$0.75_{\pm 0.07}$
✓	✓	$0.73_{\pm 0.03}$	$0.75_{\pm 0.05}$	$0.75_{\pm 0.04}$	$0.77_{\pm 0.04}$	$0.76_{\pm 0.08}$
Excluding Cancer Cases with Event Time < 180 Days						
Prior	Decoder	C-index	Time-dependent AUC			
			1-year	2-year	3-year	4-year
✗	✗	$0.63_{\pm 0.04}$	$0.64_{\pm 0.10}$	$0.66_{\pm 0.08}$	$0.64_{\pm 0.06}$	$0.64_{\pm 0.11}$
✓	✗	$0.68_{\pm 0.05}$	$0.64_{\pm 0.14}$	$0.73_{\pm 0.08}$	$0.70_{\pm 0.05}$	$0.71_{\pm 0.09}$
✓	✓	$0.70_{\pm 0.04}$	$0.68_{\pm 0.13}$	$0.76_{\pm 0.07}$	$0.73_{\pm 0.05}$	$0.71_{\pm 0.10}$

performance, while strong geometric transformations had a negative impact. This is consistent with prior work [20] showing that risk prediction models focus on overall parenchymal pattern.

3.4 Results

Ablation Study. To better understand the merit of the transformer decoder, we first performed an ablation study on the architecture. Our findings, summarized in Table 1, include two sets of results: one for all exams in the test set and the other by excluding cancer exams within 180 days of cancer diagnosis which are likely to have visible symptoms of cancer, by following a previous study [33]. This latter set of results is particularly relevant as risk prediction aims to predict unseen risks beyond visible cancer patterns. We also compare our method to two other models, the state-of-the-art baseline and PRIME models.

As shown in the top rows in Table 1, the baseline obtained a C-index of 0.68 (0.65 to 0.71). By using the transformer decoder to jointly model prior images, we observed improved C-index from 0.70 (0.67 to 0.73) to 0.73 (0.70 to 0.76). The C-index as well as all AUC differences between the baseline and the PRIME+ are all statistically significant ($p < 0.05$) except the 4-year AUC where we had a limited number of test cases.

We observe similar performance improvements when evaluating cases with at least 180 days to cancer diagnosis. Interestingly, the C-index as well as time-dependent AUCs of all three methods decreased compared to when evaluating using all cases. The intuition behind this result is that mammograms taken near the cancer diagnosis (<180 days) likely contain visible signs of cancer and thus the task of risk prediction is easier. The model must learn patterns of risk, not

Table 2. To better understand why the addition of prior images works, we split our test set into two groups based on the mammographic density: change and no change. The first and second row corresponds to performance of the baseline and PRIME+ model, respectively. Empty cell indicates an insufficient number of cases available for evaluation.

Density chg	C-index	Time-dependent AUC			
		1-year	2-year	3-year	4-year
Change	$0.63_{\pm 0.14}$	$0.74_{\pm 0.17}$	$0.66_{\pm 0.18}$	$0.56_{\pm 0.16}$	-
	$0.75_{\pm 0.10}$	$0.82_{\pm 0.13}$	$0.76_{\pm 0.14}$	$0.74_{\pm 0.14}$	-
No change	$0.69_{\pm 0.03}$	$0.70_{\pm 0.05}$	$0.72_{\pm 0.05}$	$0.72_{\pm 0.04}$	$0.71_{\pm 0.09}$
	$0.73_{\pm 0.03}$	$0.74_{\pm 0.05}$	$0.75_{\pm 0.05}$	$0.77_{\pm 0.04}$	$0.76_{\pm 0.08}$

visible signs of cancer, in order to perform well under this evaluation setting. Our results support this intuition as the performance improvements over the baseline are much more pronounced for longer term risk (3, 4-year AUC) than short term risk (1 year). The PRIME and PRIME+ models, which incorporate prior mammograms, show high performance for long-term risk prediction (3, 4-year AUC), indicating that considering changes in breast over time contain useful information for breast cancer risk prediction.

Lastly, we empirically confirm that a transformer decoder effectively models spatial relations between prior and current mammograms by demonstrating consistent performance improvements of PRIME+ across both short-term and long-term risk prediction settings. Our results suggest that incorporating changes in patients using prior mammograms and a transformer decoder improves the performance of breast cancer risk prediction models.

Analysis Based on Density. To better understand why adding prior images improves performance, we divided our test set into subgroups to examine the performance of the baseline model and the PRIME+ model on each of these groups. Mammographic breast density is one of the most important risk factor to predict breast cancer [19,31]. Women with dense breasts have a four-to six-fold higher risk of breast cancer [2]. The addition of mammographic breast density has improved the performance traditional breast cancer risk models [4] and can therefore help us understand why the addition of prior images works.

Mammographic breast density was determined using the Breast Imaging Reporting and Data System (BI-RADS) composition classification. BI-RADS category A, B are defined as fatty breasts and BI-RADS category C, D are classified as dense breasts. To determine the density category, we employed an internally developed density prediction model, as most exams lack BI-RADS ground truth. This model achieved an accuracy of 0.81 on the internal density validation set.

We categorized the exams into two groups based on changes in density: "change" and "no change". Density change was defined according to whether the BI-RADS category changed in the current image as compared to the prior

Table 3. In order to assess the performance of the models on varying levels of breast density, a critical risk factor, we divided our test set into two groups based on mammographic density: fatty and dense.

Density	C-index	Time-dependent AUC			
		1-year	2-year	3-year	4-year
Fatty	$0.70_{\pm 0.04}$	$0.73_{\pm 0.06}$	$0.74_{\pm 0.05}$	$0.70_{\pm 0.05}$	$0.70_{\pm 0.10}$
	$0.74_{\pm 0.04}$	$0.76_{\pm 0.06}$	$0.76_{\pm 0.05}$	$0.78_{\pm 0.05}$	$0.76_{\pm 0.08}$
Dense	$0.68_{\pm 0.06}$	$0.66_{\pm 0.09}$	$0.68_{\pm 0.09}$	$0.71_{\pm 0.08}$	$0.65_{\pm 0.21}$
	$0.71_{\pm 0.05}$	$0.72_{\pm 0.08}$	$0.73_{\pm 0.08}$	$0.72_{\pm 0.08}$	$0.72_{\pm 0.25}$

image. As shown in Table 2, the baseline model performs poorly for "change", with a C-index of 0.63 (0.49 to 0.77), especially for long-term risk prediction, with 3-year AUC of 0.56 (0.40 to 0.72). This suggests that the baseline model has limitations in accurately predicting long-term risk when there is a density change from the prior exam. However, PRIME+ is able to predict long-term risk accurately even when a density change has occurred (3-year AUC = 0.74 (0.60 to 0.88)), by learning to refer previous exams properly. This demonstrates the potential usefulness of incorporating past mammogram information into breast cancer risk prediction models. Thus, we believe that incorporating prior exams is important to identify changes in texture which are important for long term risk prediction (Table 3).

Lastly, we divided the exams based on the level of breast density, with a fatty group consisting of density A and B, and a dense group consisting of density C and D. Both the baseline and PRIME+ performs better in fatty group than dense group. We suspect this is because deep neural networks generally work better on low density images given that visual cues of cancer in images with lower breast density are more clearly visible.

4 Conclusion

In this paper, we introduce a novel breast cancer risk prediction method, PRIME+, which incorporates prior mammograms with a transformer decoder to capture changes in breast tissue over time. By doing so, we achieve high performance for both short-term and long-term risk prediction. Our extensive experiments on a dataset of 16,113 exams show that PRIME+ outperformed a model based on the state-of-the-art for breast cancer risk prediction [33]. Our method performed particularly well in cases where there was a change in breast density from the previous exam. We believe that our method has the potential to improve breast cancer risk prediction and ultimately contribute to earlier detection of the disease.

References

1. Bakker, M.F., et al.: Supplemental MRI screening for women with extremely dense breast tissue. N. Engl. J. Med. **381**(22), 2091–2102 (2019)
2. Boyd, N.F.: Mammographic density and risk of breast cancer. Am. Soc. Clin. Oncol. Educ. Book **33**(1), e57–e62 (2013)
3. Brentnall, A.R., Cuzick, J.: Risk models for breast cancer and their validation. Stat. Sci. Rev. J. Inst. Math. Stat. **35**(1), 14 (2020)
4. Brentnall, A.R., et al.: Mammographic density adds accuracy to both the Tyrer-Cuzick and Gail breast cancer risk models in a prospective UK screening cohort. Breast Cancer Res. **17**, 1–10 (2015)
5. Dadsetan, S., Arefan, D., Berg, W.A., Zuley, M.L., Sumkin, J.H., Wu, S.: Deep learning of longitudinal mammogram examinations for breast cancer risk prediction. Pattern Recogn. **132**, 108919 (2022)
6. DeLong, E.R., DeLong, D.M., Clarke-Pearson, D.L.: Comparing the areas under two or more correlated receiver operating characteristic curves: a nonparametric approach. Biometrics, 837–845 (1988)
7. Duffy, S.W., et al.: Effect of mammographic screening from age 40 years on breast cancer mortality (UK age trial): final results of a randomised, controlled trial. Lancet Oncol. **21**(9), 1165–1172 (2020)
8. Eriksson, M., et al.: A risk model for digital breast tomosynthesis to predict breast cancer and guide clinical care. Sci. Transl. Med. **14**(644), eabn3971 (2022)
9. Gastounioti, A., Desai, S., Ahluwalia, V.S., Conant, E.F., Kontos, D.: Artificial intelligence in mammographic phenotyping of breast cancer risk: a narrative review. Breast Cancer Res. **24**(1), 1–12 (2022)
10. Hakama, M., Coleman, M.P., Alexe, D.M., Auvinen, A.: Cancer screening: evidence and practice in Europe 2008. Eur. J. Cancer **44**(10), 1404–1413 (2008)
11. Hayward, J.H., et al.: Improving screening mammography outcomes through comparison with multiple prior mammograms. AJR Am. J. Roentgenol. **207**(4), 918 (2016)
12. He, K., Zhang, X., Ren, S., Sun, J.: Deep residual learning for image recognition. In: Proceedings of the IEEE Conference on Computer Vision and Pattern Recognition, pp. 770–778 (2016)
13. Hussein, H., et al.: Supplemental breast cancer screening in women with dense breasts and negative mammography: a systematic review and meta-analysis. Radiology **306**(3), e221785 (2023)
14. National Cancer Institute: Breast cancer risk assessment tool (2011). https://www.cancer.gov/bcrisktool/. Accessed 13 Aug 2017
15. World Cancer Research Fund International: Breast cancer statistics. https://www.wcrf.org/cancer-trends/breast-cancer-statistics/
16. Kamarudin, A.N., Cox, T., Kolamunnage-Dona, R.: Time-dependent ROC curve analysis in medical research: current methods and applications. BMC Med. Res. Methodol. **17**(1), 1–19 (2017)
17. Kang, L., Chen, W., Petrick, N.A., Gallas, B.D.: Comparing two correlated C indices with right-censored survival outcome: a one-shot nonparametric approach. Stat. Med. **34**(4), 685–703 (2015)
18. Katzman, J.L., Shaham, U., Cloninger, A., Bates, J., Jiang, T., Kluger, Y.: DeepSurv: personalized treatment recommender system using a Cox proportional hazards deep neural network. BMC Med. Res. Methodol. **18**(1), 1–12 (2018)

19. Lee, C.I., Chen, L.E., Elmore, J.G.: Risk-based breast cancer screening: implications of breast density. Med. Clin. **101**(4), 725–741 (2017)
20. Liu, Y., Azizpour, H., Strand, F., Smith, K.: Decoupling inherent risk and early cancer signs in image-based breast cancer risk models. In: Martel, A.L., et al. (eds.) MICCAI 2020. LNCS, vol. 12266, pp. 230–240. Springer, Cham (2020). https://doi.org/10.1007/978-3-030-59725-2_23
21. Loshchilov, I., Hutter, F.: SGDR: stochastic gradient descent with warm restarts. arXiv preprint arXiv:1608.03983 (2016)
22. Ontario, H.Q., et al.: Screening mammography for women aged 40 to 49 years at average risk for breast cancer: an evidence-based analysis. Ont. Health Technol. Assess. Ser. **7**(1), 1–32 (2007)
23. Paci, E.: Summary of the evidence of breast cancer service screening outcomes in Europe and first estimate of the benefit and harm balance sheet. J. Med. Screen. **19**(1_suppl), 5–13 (2012)
24. Park, J., et al.: Screening mammogram classification with prior exams. arXiv preprint arXiv:1907.13057 (2019)
25. Roelofs, A.A., et al.: Importance of comparison of current and prior mammograms in breast cancer screening. Radiology **242**(1), 70–77 (2007)
26. Sumkin, J.H., et al.: Optimal reference mammography: a comparison of mammograms obtained 1 and 2 years before the present examination. Am. J. Roentgenol. **180**(2), 343–346 (2003)
27. Tyrer, J., Duffy, S.W., Cuzick, J.: A breast cancer prediction model incorporating familial and personal risk factors. Stat. Med. **23**(7), 1111–1130 (2004)
28. Uno, H., Cai, T., Pencina, M.J., D'Agostino, R.B., Wei, L.J.: On the C-statistics for evaluating overall adequacy of risk prediction procedures with censored survival data. Stat. Med. **30**(10), 1105–1117 (2011)
29. Varela, C., Karssemeijer, N., Hendriks, J.H., Holland, R.: Use of prior mammograms in the classification of benign and malignant masses. Eur. J. Radiol. **56**(2), 248–255 (2005)
30. Vaswani, A., et al.: Attention is all you need. In: Advances in Neural Information Processing Systems, vol. 30 (2017)
31. Veronesi, U., Boyle, P., Goldhirsch, A., Orecchia, R., Viale, G.: Breast cancer. Lancet **365**, 1727–1741 (2005)
32. Yala, A., et al.: Multi-institutional validation of a mammography-based breast cancer risk model. J. Clin. Oncol. **40**(16), 1732–1740 (2022)
33. Yala, A., et al.: Toward robust mammography-based models for breast cancer risk. Sci. Transl. Med. **13**(578), eaba4373 (2021)
34. Yeoh, H.H., et al.: RADIFUSION: a multi-radiomics deep learning based breast cancer risk prediction model using sequential mammographic images with image attention and bilateral asymmetry refinement. arXiv preprint arXiv:2304.00257 (2023)

Uncertainty Inspired Autism Spectrum Disorder Screening

Ying Zhang, Yaping Huang(✉), Jiansong Qi, Sihui Zhang, Mei Tian, and Yi Tian

Beijing Key Laboratory of Traffic Data Analysis and Mining, Beijing Jiaotong University, Beijing, China
yphuang@bjtu.edu.cn

Abstract. People with autism spectrum disorder (ASD) show distinguishing preferences for specific visual stimuli compared to typically developed (TD) individuals, opening the door for objective and quantitative screening by eye-tracking data analysis. However, existing eye-tracking-based ASD screening approaches often assume that there are no individual differences and that all stimuli contribute equally to the prediction of an ASD. Consequently, a fixed number of images are usually selected by a pre-defined strategy for further training and testing, ignoring the distinct characteristics of various subjects viewing the same image. To address the aforementioned difficulties, we propose a novel Uncertainty-inspired ASD Screening Network (UASN) that dynamically modifies the contribution of each stimulus viewed by different subjects. Specifically, we estimate the uncertainty of each stimulus by considering the variation between the subject's fixation map and the ones of the two clinical groups (i.e., ASD and TD) and further utilize it for weighting the training loss. Besides, to reduce the diagnosis time, instead of the shuffle-appeared mode of image viewing, we propose an uncertainty-based personalized diagnosis method to dynamically rank the viewing images according to the preferences of different subjects, which can achieve high prediction accuracy with only a small set of images. Experiments demonstrate the superior performance of our proposed UASN.

Keywords: ASD Screening · Eye-tracking · Data Uncertainty

1 Introduction

Autism Spectrum Disorder (ASD) has been a prevalent neurodevelopmental disorder worldwide that one in 44 kids aged 8 years in the United States suffers from it as reported in 2021 [15] and there is still a steady and substantial growth in the population.

However, diagnosing ASD relies on subjective evaluations that are expensive and clinically demanding.

Supplementary Information The online version contains supplementary material available at https://doi.org/10.1007/978-3-031-43904-9_39.

H. Greenspan et al. (Eds.): MICCAI 2023, LNCS 14224, pp. 399–408, 2023.
https://doi.org/10.1007/978-3-031-43904-9_39

Seminal works [3,6,12,16,17,19] have pointed out that eye movement patterns of people with ASD play an irreplaceable vital role in identifying ASD.

Early efforts [7,13,18,20,21] usually focus on low-level behavior features combined with machine learning algorithms to identify autism, while in recent years, the eye-tracking data driven method [2,10,14] boosts the performance of ASD screening by utilizing deep neural networks (DNNs) which extract high-level semantic information of eye movement.

However, existing deep-learning-based methods usually define an image-ranking strategy as a pre-processing step to select a certain number of images. During training, each visual stimulus is treated equally, ignoring the distinct contributions of different stimuli. Besides, during the diagnosis procedure, a fixed number of images are shown to a subject which takes a relatively long time, thereby leading to poor cooperation of subjects, especially little kids.

To tackle the above issues, in this paper, we propose a novel Uncertainty-inspired ASD Screening Network, named UASN, to distinguish the importance of each visual stimulus for different individuals. Despite the success of uncertainty in computer vision [1,5,22,23], to our best knowledge, this is the first attempt to introduce uncertainty estimation into ASD screening. Our uncertainty-inspired UASN can enforce the model learning from more distinctive gaze patterns during training. Meanwhile, when the model is deployed in the real clinical scenario, we further design an efficient personalized diagnosis strategy, which can dramatically reduce the diagnosis time without a performance drop.

Specifically, the uncertainty in UASN works in two ways to ensure both higher accuracy and lower time consumption. On the one hand, given an input gaze pattern, we estimate the uncertainty by comparing the difference between the fixation map and the ones of ASD and TD groups. The uncertainty will be assigned a lower value for a larger disparity, suggesting the importance of the given gaze pattern for identifying a certain individual. Subsequently, guided by the estimated uncertainty, we design a truncated weighting loss to select the most distinctive gaze patterns and further dynamically adjust the contributions made by different stimuli, resulting in a more efficient classification. On the other hand, how to reduce the diagnosis time is also a key factor in real clinical applications, especially for preschool children. To achieve this goal, we propose a personalized diagnosis method that ranks the stimuli according to the estimated uncertainty. Instead of the random shuffle mode for image viewing, we recommend a top similar or dissimilar stimulus for the next viewing according to the decision of the previous gaze patterns. Following the proposed protocol, our method achieves state-of-the-art performance while spending much less diagnosis time.

In general, our contributions can be summarized as follows: 1) we propose the first usage of the Uncertainty-inspired ASD Screening Network, named UASN, for identifying ASD people; 2) we estimate the uncertainty of each gaze pattern and further design a truncated weighting loss, which can enforce the model to dynamically adjust the contributions of different gaze patterns during training; 3) we design a personalized online diagnosis protocol that can dramatically reduce the diagnosis time without losing accuracy; 4) we conduct comprehensive experiments on the Saliency4ASD benchmark and achieve state-of-the-art performance only using 1/5 visual stimuli compared with other leading approaches.

2 Uncertainty Inspired ASD Screening

Our UASN is built upon traditional DNNs and consists of two novel stages: 1) uncertainty-guided training, and 2) uncertainty-guided personalized diagnosis. During training, we estimate an uncertainty value for each gaze pattern and further apply it for weighting the training loss. Besides, for a more simplified diagnosis procedure, we design an uncertainty-based strategy that adaptively selects the most discriminative images based on the subject's gaze behaviors.

2.1 Uncertainty Guided Training

Figure 1 illustrates the detailed training process of UASN. Firstly, we extract the features of gaze patterns by taking the temporal eye tracking information as input and resulting in the classification prediction. Then, we design an uncertainty estimation module to compute the uncertainty values of each subject on all the visual stimuli. Moreover, we further apply the estimated uncertainty to weight the training samples by a truncated loss in a reasonable manner.

Gaze Pattern Feature Extraction. Formally, by collecting a group of ASD and TD subjects $S = \{s_i\}_{i=1}^{M}$'s eye movement data on a set of images $X = \{x_j\}_{j=1}^{N}$, we get the corresponding scanpaths which comprise each fixation point's position and duration in the temporal order. The labels of the two clinical groups are denoted as $Y = \{y_i\}_{i=1}^{M} \in \{0, 1\}$.

To generate the discriminative features of the given gaze pattern (i-th subject watching j-th image), we first feed the image x_j into a ResNet-50 [9] with the last max pooling layer removed to learn a 2048-dimension visual feature. Then the visual feature sequence taken from the scanpath of i-th subject is fed into a Long Short Term Memory (LSTM) network [8]. A final hidden state is obtained at the end of the sequence which is then fed into a fully connected (FC) layer followed by a sigmoid function to get the prediction result for the i-th subject viewing j-th image, which is denoted as $\hat{y}_i^j$. The network can be optimized by the binary cross-entropy (BCE) loss:

$$\mathcal{L}_{\text{bce}} = -\frac{1}{NM}\sum_{i}\sum_{j}(y_i^j \log \hat{y}_i^j + (1 - y_i^j)\log(1 - \hat{y}_i^j)), \tag{1}$$

where $\hat{y}_i^j$ and y_i^j denote the predicted and ground truth labels of i-th subject respectively. If belonging to ASD, $y_i^j = 1$ for all images $\{x_j\}_{j=1}^{N}$, otherwise 0.

Uncertainty Estimation. We believe that different images contribute unequally to a subject's final classification due to the subject's unique preferences for viewing images. As a result, we estimate an uncertainty value for each gaze pattern based on the variance between its fixation map and the ones

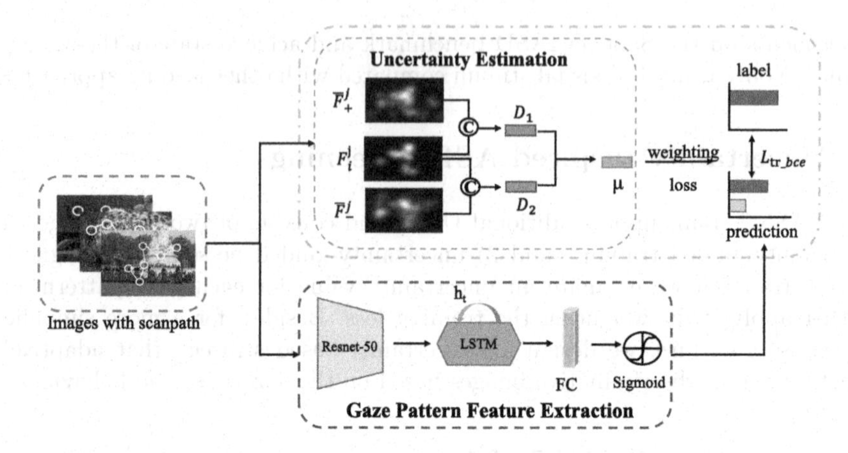

Fig. 1. Illustration of the uncertainty-guided training module of UASN. Images with corresponding gaze patterns are fed into the Gaze Pattern Feature Extraction module to get the prediction result. By applying the Uncertainty Estimation module, we compute an uncertainty for each gaze pattern and further use it to weight the training BCE loss to get a more reasonable as well as efficient result.

of the two clinical groups (*i.e.*, ASD and TD). For instance, an ASD's fixation map on a discriminative image should appear more similar to the ASD group's averaged one than the TD group's so the variance is supposed to be large.

We first generate the fixation map for each gaze pattern according to the fixation data. Then, given two groups' fixation maps on each image in the dataset, for each subject, we apply cosine similarity to compute an uncertainty measurement on each image. Let F_i^j denote the fixation map of the i-th subject's fixation map on the j-th image, and $\bar{F}_+^j$, $\bar{F}_-^j$ denote the fixation maps of ASD and TD group for j-th image respectively. The uncertainty can be written as:

$$D_i^j = \left| C(F_i^j, \bar{F}_+^j) - C(F_i^j, \bar{F}_-^j) \right|, \tag{2}$$

$$\mu_i^j = 1 - D_i^j, \tag{3}$$

where C is the cosine similarity function, D_i^j is the distinguishability of the j-th image when viewed by the i-th subject and μ_i^j denotes the uncertainty. Specifically, for images that do not contain certain subjects' eye-tracking data, we reasonably set the μ_i^j to be large because we assume that the absence of eye-tracking data is due to a lack of interest, and further signals ineffectiveness.

Truncated Weighting Loss. Upon obtaining the uncertainty value μ_i^j, we can utilize the uncertainty to re-weight the training loss by teaching the model which images to trust and which to discredit. We hope that the larger the μ is, the less the image contributes to the final classification, so the corresponding loss needs to be reduced correspondingly.

However, for some gaze patterns that are confusing and much more difficult to distinguish between ASD and healthy people, it is more suitable to discard those unreliable patterns. Considering this, we finally propose a truncated weighting BCE loss for training. Specifically, when the estimated uncertainty value is larger than a pre-defined threshold t, we set the corresponding loss to zero. In summary, The final loss function is denoted as:

$$\mathcal{L}_{\text{tr\_bce}} = \sum_i \sum_j \mathbb{I}_{[\mu \leq t]}(1 - \mu_i^j)\mathcal{L}_{\text{bce}}, \tag{4}$$

where $\mathbb{I}$ is the indicator function and t is the threshold. Only when the condition of $[\mu \leq t]$ is met, is the value of the indicator function set to 1, and the $\mathcal{L}_{\text{tr\_bce}}$ remains. Specifically, for a more reasonable computation, we do not simply set the t to be a fixed value. Instead, we determine t with an adaptive technique. For each subject, we sort the uncertainty values from low to high and choose the 1/3 of the images with the lowest uncertainty and retain the contributions they make to the prediction while discarding the remainder.

2.2 Uncertainty Guided Personalized Diagnosis

On the basis of training our model in an uncertainty-guided way, we are encouraged to go deeper to simplify the diagnostic procedure. To this end, we incorporate personalized diagnosis into our work, taking into account the gaze behavior features of each subject. Figure 2 presents the workflow of our personalized diagnosis protocol after completing the training process in Fig. 1. Specifically, by extracting features of images and forming a feature bank, we selectively choose the most suitable images to update the viewing list according to the subject's viewing pattern.

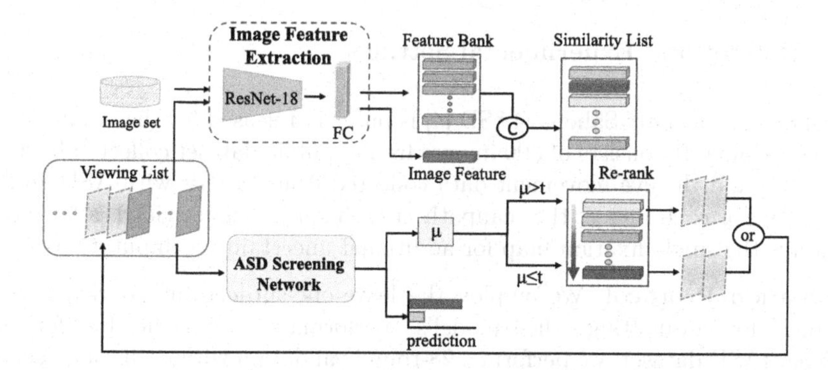

Fig. 2. Illustration of our proposed uncertainty-guided personalized diagnosis module. We feed the gaze patterns in the current viewing list into the ASD screening network to get a prediction as well as the uncertainty value μ. The μ is further used to refresh the viewing list with images based on the similarity-based ranking result.

Image Feature Extraction. First, we assume that the visual similarity between images brings the potential of personalized diagnosis that similar images may contribute similarly in distinguishing an individual. From this point, we extract the feature of all the 300 images in the dataset (the Image set in Fig. 2) using a ResNet-18 [9] followed by an FC to get a 128-dimension feature, thereby forming a feature bank getting prepared for the further dynamical ranking procedure.

Similarity-Based Image Ranking. Then, we build a viewing list simulating the diagnosis procedure where images are shown to a subject one by one and the list is updated in real-time. In the beginning, we select an image from the image set randomly to initialize the viewing list. When a trial is completed, the viewing list is subsequently updated. In each trial, we generate an average feature of all images in the viewing list. We then compute the cosine similarity between the feature of the current image and images in the feature bank to obtain a similarity list. We sort the list in similarity-ascending order.

Uncertainty-Based Viewing List Updating. To determine which images should be included in the viewing list, a method based on uncertainty is developed. First, we feed the image in the list and the corresponding eye-tracking data into the ASD screening network which is composed of the Uncertainty Estimation module and the Gaze Pattern Feature Extraction module in Fig. 1 to get the uncertainty and the prediction result. By pre-defining a threshold p, we separate the scenario into a positive case and a negative one. When the average uncertainty value is larger than p, we consider it negative so we select the top K dissimilar images to join the viewing list and vice versa. After T trials, we achieve a relatively confident prediction.

3 Experiments

3.1 Dataset and Experimental Settings

Dataset. So far, only Saliency4ASD [4] is publicly released for the evaluation of ASD screening. It consists of 300 images from a public dataset collected by Judd *et al.* [11] and the eye movement data collected from 14 kids with ASD and 14 with TD. For each image, the scanpath of each subject is provided, allowing us to derive the single fixation map for additional uncertainty computation.

Evaluation Protocol. We employ the leave-one-subject-out cross-validation method for evaluating the model's performance. Specifically, for the Saliency4ASD dataset, we perform a 28-round validation with each round selecting only one subject for testing while the remaining 27 subjects work as the training data.

Metrics. We follow the previous works [2,10] to adopt the accuracy, sensitivity, specificity, and AUC to evaluate the performance of the prediction for each subject. Besides, we assess the performance of the prediction for each scanpath to generate a more strict measurement. We still adopt accuracy, sensitivity, and specificity, called Acc_I, Sen_I, and Spe_I. Accordingly, for the previously used three metrics, we denote them as Acc_S, Sen_S, and Spe_S.

Implementation Details. The experimental setting mostly follows the [2] during training. Besides, we manually set the uncertainty values of those images without some certain subjects' eye movement data to be 1–10^{-5} which is a relatively large margin that rarely contributes to the classification. For personalized diagnosis, we initialize the viewing list by randomly selecting an image. By considering the gaze pattern with an uncertainty level lower than $p = 0.9$ a positive case, we choose the top K most similar images to update the viewing list. We set K to be 1 and update the viewing list $T = 20$ trials in total.

3.2 Comparison with State-of-the-Art

We conduct extensive experiments to explore whether our proposed model outperforms the state-of-the-art. We compare our UASN with [2] which selects a fixed 100 images subset out of a total of 300 images according to a Fisher-Score-based image selection strategy for training and testing. We design the following three UASN variants: 1) **UASN-noUnLoss** only uses uncertainty to choose the top most discriminative images and removes the weighting loss procedure for training; 2) **UASN-Fixed** selects a 100-image subset with the lowest uncertainty for each subject during training and testing without dynamically adjusting the diagnosis process; 3) **UASN-Dynamic** uses 100 images with low uncertainty for training. In the diagnosis process, we adopt the proposed uncertainty-guided personalized diagnosis to recommend a small number of images, which significantly reduces the diagnosis time while maintaining high accuracy.

Table 1. Comparison on both the subject and scanpath level with state-of-the-art [2].

Method	Subject level metrics				Scanpath level metrics		
	Acc_S	Sen_S	Spe_S	AUC	Acc_I	Sen_I	Spe_I
[2]	0.93	0.93	0.93	0.98	0.59	0.58	0.59
UASN-noUnLoss	1	1	1	1	0.76	0.75	0.72
UASN-Fixed	1	1	1	1	0.74	0.71	0.72
UASN-Dynamic	1	1	1	1	**0.79**	**0.78**	**0.78**

Table 1 shows the results. We can see that our model's three variants all outperform the baseline model by all evaluation metrics. The result demonstrates that introducing uncertainty during training can reach 100% accuracy. When removing the uncertainty weighting loss part, the UASN-noUnLoss model's performance drops slightly in scanpath level metrics than the best UASN-Dynamic, *i.e.*, 3% (Acc_I), 3% (Sen_I) and 6% (Spe_I). Besides, the personalized diagnosis strategy achieves the same accuracy but largely reduces the number of images to 1/5 (20 images), which decreases the diagnosis time dramatically. In terms of scanpath level metrics, our UASN-Dynamic outperforms the previous leading model [2] by 20% in Acc_I and Sen_I, and 19% in Spe_I.

Table 2. Comparison of using images of different uncertainty levels. The "top 100", "middle 100" and "bottom 100" denote the three subsets of 100-image with the lowest, the middle, and the highest uncertainty level.

Method	Selected ImageSet	Acc_S	Sen_S	Spe_S	AUC	Acc_I	Sen_I	Spe_I
UASN-Fixed	top-100	1	1	1	1	0.74	0.71	0.72
	middle-100	0.78	0.54	0.93	0.86	0.56	0.50	0.56
	bottom-100	0.54	0.25	0.83	0.59	0.50	0.46	0.51
UASN-Dynamic	**top-100**	1	1	1	1	**0.79**	**0.78**	**0.78**
	middle-100	0.64	0.64	0.64	0.68	0.52	0.50	0.58
	bottom-100	0.38	0.33	0.742	0.39	0.47	0.45	0.50

The results suggest introducing uncertainty both in the training and testing stage achieves the best performance with a quite small image subset at classifying the two clinical groups.

3.3 Ablation Study

Effect of the Uncertainty Estimation. To verify the effect of uncertainty estimation, we divide the 300 images into three non-overlapping subsets based on the ascending order of uncertainty level, denoted as top-100, middle-100, and bottom-100 subsets. Table 2 shows the results. It is not surprising that UASN-Dynamic with top-100 achieves the best performance and the large performance drop on both UASN-Fixed and UASN-Dynamic approves our model's strong capability of selecting the most discriminative images for classifying ASD and TD. We further visualize the relation between Acc_I and uncertainty, as well as give some samples of ASD and TD's fixation maps with different uncertainty values to support the effectiveness of our method. Details are given in Fig. I and Fig. II of the supplementary material.

Effect of Different Inference Strategies. During the diagnosis process, we need to define the number of trials (T) and the number of images (K) selected to append to the viewing list. The results are given in Table 3. We tried various permutations of T and K and finally found that appending one image each trial

Table 3. Comparison of how different inference strategies influence our model's performance. T denotes the number of inference trials during the diagnosis and K denotes the number of images inferred in each trial.

T	K	Acc_S	Sen_S	Spe_S	AUC	Acc_I	Sen_I	Spe_I
20	1	**1**	**1**	**1**	**1**	**0.79**	**0.78**	**0.78**
10	2	0.96	0.93	**1**	0.99	0.75	0.74	0.77
10	1	0.96	0.93	1	1	0.73	0.71	0.74

and performing 20 trials for one subject obtained the best performance while cost the least diagnosing time.

Effect of Different Similarity Measurements and Backbones. We conduct experiments on replacing the similarity measurement method and the model's backbone. Results can be referred to in the Table I and Table II of supplementary.

Computational Cost. Another crucial metric for assessing our model's efficiency is the time interval between every two viewing list update rounds during the personalized diagnosis process. Upon conducting experiments, we get an average of 0.038 s for each interval, demonstrating that our dynamic update strategy causes no delay in the clinical diagnosis.

4 Conclusion

In this paper, we present UASN, a novel ASD screening approach, inspired by uncertainty. The uncertainty benefits the ASD diagnosis in two ways: a weighted truncated training loss that enables the model to learn the most discriminative and effective features of gaze patterns and a personalized procedure that dynamically ranks the stimuli according to the subject's gaze behaviors. Comprehensive results show superior performance in classifying ASD people.

Acknowledgements. This work was supported by the Beijing Natural Science Foundation (M22022, L211015) and the National Natural Science Foundation of China (62271042, 61906013).

References

1. Chang, J., Lan, Z., Cheng, C., Wei, Y.: Data uncertainty learning in face recognition. In: Proceedings of the IEEE/CVF Conference on Computer Vision and Pattern Recognition (CVPR), June 2020
2. Chen, S., Zhao, Q.: Attention-based autism spectrum disorder screening with privileged modality. In: Proceedings of the IEEE/CVF International Conference on Computer Vision, pp. 1181–1190 (2019)
3. Corbetta, M., Shulman, G.L.: Control of goal-directed and stimulus-driven attention in the brain. Nat. Rev. Neurosci. **3**(3), 201–215 (2002)
4. Duan, H., et al.: A dataset of eye movements for the children with autism spectrum disorder. In: Proceedings of the 10th ACM Multimedia Systems Conference, pp. 255–260 (2019)
5. El Ghaoui, L., Lanckriet, G.R.G., Natsoulis, G., et al.: Robust classification with interval data (2003)
6. Frazier, T.W., et al.: A meta-analysis of gaze differences to social and nonsocial information between individuals with and without autism. J. Am. Acad. Child Adolesc. Psychiatry **56**(7), 546–555 (2017)
7. Freeth, M., Chapman, P., Ropar, D., Mitchell, P.: Do gaze cues in complex scenes capture and direct the attention of high functioning adolescents with ASD? Evidence from eye-tracking. J. Autism Dev. Disord. **40**(5), 534–547 (2010)

8. Graves, A.: Generating sequences with recurrent neural networks. arXiv preprint arXiv:1308.0850 (2013)
9. He, K., Zhang, X., Ren, S., Sun, J.: Deep residual learning for image recognition. In: Proceedings of the IEEE Conference on Computer Vision and Pattern Recognition, pp. 770–778 (2016)
10. Jiang, M., Zhao, Q.: Learning visual attention to identify people with autism spectrum disorder. In: Proceedings of the IEEE International Conference on Computer Vision, pp. 3267–3276 (2017)
11. Judd, T., Ehinger, K., Durand, F., Torralba, A.: Learning to predict where humans look. In: 2009 IEEE 12th International Conference on Computer Vision, pp. 2106–2113 (2009). https://doi.org/10.1109/ICCV.2009.5459462
12. Klin, A., Lin, D.J., Gorrindo, P., Ramsay, G., Jones, W.: Two-year-olds with autism orient to non-social contingencies rather than biological motion. Nature **459**(7244), 257–261 (2009)
13. Li, B., Sharma, A., Meng, J., Purushwalkam, S., Gowen, E.: Applying machine learning to identify autistic adults using imitation: an exploratory study. PLoS ONE **12**(8), e0182652 (2017)
14. Liu, W., Li, M., Yi, L.: Identifying children with autism spectrum disorder based on their face processing abnormality: a machine learning framework. Autism Res. **9**(8), 888–898 (2016)
15. Maenner, M.J., et al.: Prevalence and characteristics of autism spectrum disorder among children aged 8 years-autism and developmental disabilities monitoring network, 11 sites, United States, 2018. MMWR Surveill. Summ. **70**(11), 1 (2021)
16. McPartland, J.C., Webb, S.J., Keehn, B., Dawson, G.: Patterns of visual attention to faces and objects in autism spectrum disorder. J. Autism Dev. Disord. **41**(2), 148–157 (2011)
17. Pelphrey, K.A., Sasson, N.J., Reznick, J.S., Paul, G., Goldman, B.D., Piven, J.: Visual scanning of faces in autism. J. Autism Dev. Disord. **32**(4), 249–261 (2002)
18. Pierce, K., Marinero, S., Hazin, R., McKenna, B., Barnes, C.C., Malige, A.: Eye tracking reveals abnormal visual preference for geometric images as an early biomarker of an autism spectrum disorder subtype associated with increased symptom severity. Biol. Psychiat. **79**(8), 657–666 (2016)
19. Sasson, N.J., Turner-Brown, L.M., Holtzclaw, T.N., Lam, K.S., Bodfish, J.W.: Children with autism demonstrate circumscribed attention during passive viewing of complex social and nonsocial picture arrays. Autism Res. **1**(1), 31–42 (2008)
20. Thabtah, F.: Autism spectrum disorder screening: machine learning adaptation and DSM-5 fulfillment. In: Proceedings of the 1st International Conference on Medical and Health Informatics 2017, pp. 1–6 (2017)
21. Wang, S., et al.: Atypical visual saliency in autism spectrum disorder quantified through model-based eye tracking. Neuron **88**(3), 604–616 (2015)
22. Wang, Z., Li, Y., Guo, Y., Fang, L., Wang, S.: Data-uncertainty guided multi-phase learning for semi-supervised object detection. In: Proceedings of the IEEE/CVF Conference on Computer Vision and Pattern Recognition, pp. 4568–4577 (2021)
23. Xu, Y., et al.: Data uncertainty in face recognition. IEEE Trans. Cybern. **44**(10), 1950–1961 (2014)

Rad-ReStruct: A Novel VQA Benchmark and Method for Structured Radiology Reporting

Chantal Pellegrini(✉), Matthias Keicher, Ege Özsoy, and Nassir Navab

Computer Aided Medical Procedures, Technical University Munich, Munich, Germany
chantal.pellegrini@tum.de

Abstract. Radiology reporting is a crucial part of the communication between radiologists and other medical professionals, but it can be time-consuming and error-prone. One approach to alleviate this is structured reporting, which saves time and enables a more accurate evaluation than free-text reports. However, there is limited research on automating structured reporting, and no public benchmark is available for evaluating and comparing different methods. To close this gap, we introduce Rad-ReStruct, a new benchmark dataset that provides fine-grained, hierarchically ordered annotations in the form of structured reports for X-Ray images. We model the structured reporting task as hierarchical visual question answering (VQA) and propose hi-VQA, a novel method that considers prior context in the form of previously asked questions and answers for populating a structured radiology report. Our experiments show that hi-VQA achieves competitive performance to the state-of-the-art on the medical VQA benchmark VQARad while performing best among methods without domain-specific vision-language pretraining and provides a strong baseline on Rad-ReStruct. Our work represents a significant step towards the automated population of structured radiology reports and provides a valuable first benchmark for future research in this area. Our dataset and code is available at https://github.com/ChantalMP/Rad-ReStruct.

Keywords: Structured Report Population · VQA · X-ray diagnosis

1 Introduction

Radiology is a critical medical field that relies on accurate and efficient communication between radiologists and other healthcare professionals enabled through

C. Pellegrini and M. Keicher—Contributed equally.

Supplementary Information The online version contains supplementary material available at https://doi.org/10.1007/978-3-031-43904-9_40.

H. Greenspan et al. (Eds.): MICCAI 2023, LNCS 14224, pp. 409–419, 2023.
https://doi.org/10.1007/978-3-031-43904-9_40

radiology reports. However, generating these reports takes a lot of time and is prone to errors, as it often relies on ambiguous natural language. One alternative to free-text reports is to use structured reporting, which is endorsed by radiology societies, saves time, and offers standardized content and terminology [8,18].

Automated report generation can reduce radiologists' workload and support quick diagnostic decisions. Most current research focuses on generating free-text reports, which lack standardization, and still face challenges of ambiguity and difficulties in clinical correctness evaluation [9,14,19,23,26]. In comparison, automating structured reporting allows an accurate evaluation of clinical correctness and can enforce the prediction of detailed findings. However, for automating structured reporting, the research is limited. Some studies predict high-level abnormalities using pre-defined template sentences [10,19], or predict location and attributes for a single disease [2]. Syeda-Mahmood et al. [21] predict fine-grained but unstructured labels to retrieve and adapt free-text reports from a database. However, none of these works predict highly-detailed and structured annotations as needed to populate an entire structured report. A significant challenge towards this goal is the lack of public benchmarks with highly detailed structured annotations. To facilitate future research, we introduce Rad-ReStruct, the first dataset of publicly available, fine-grained, and structured annotations for Chest X-Rays. To create Rad-ReStruct, we define a detailed, multi-level structured reporting template and automatically populate it by parsing and analyzing unstructured finding summaries from the IU X-ray dataset [5].

Structured reports with high standardization have a structured layout and content, e.g., organized in expendable trees with drop-down menus to select answers [18]. A user interface for structured reporting would pose a series of questions that, dependent on the answer, lead to expandable follow-up questions. In this setup, structured reporting can be considered several classification tasks on different levels. We model this as a hierarchical visual question answering (VQA) task and propose hi-VQA, a hierarchical, autoregressive VQA method for populating structured reports by successively filling out all fields in the report while preserving consistency. Hi-VQA considers the prior context of previously asked questions and answers, allowing to exploit inter-dependencies between questions about the same image. For structured report population, this is essential, as lower-level questions directly depend on higher levels. Further, our autoregressive formulation enhances explainability, showing at which level and for which question type the model made mistakes. As backbone, we propose a simple yet effective VQA architecture relying on large pretrained image and text encoders and a transformer-based fusion module. Using VQA [1] allows to exploit the knowledge encoded in large language models. It has recently received attention in the medical field, mainly on small datasets, where every question is answered independently [4,11,17,20,23]. One previous work explicitly models question consistency for medical VQA in the loss [24]. Another work had promising results using an unstructured question history in a visual dialog setting [12]. They ask for high-level abnormalities, such as "Pneumonia?" and use a randomly

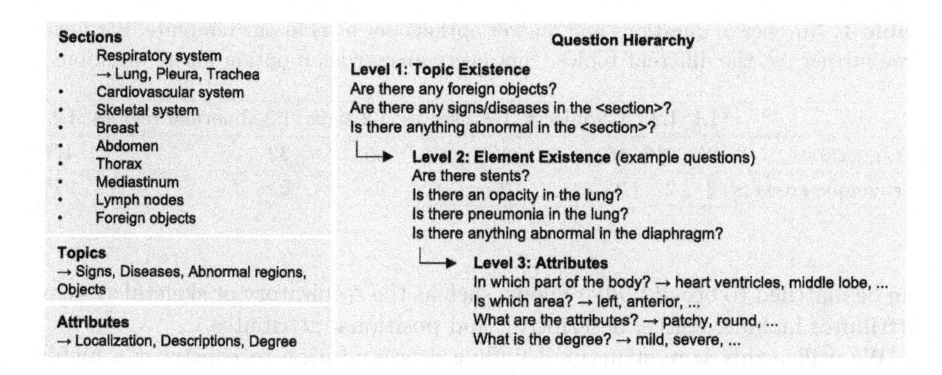

Fig. 1. Overview of report and question structure.

sampled, fixed history of other abnormality questions. In contrast, we define a hierarchical history with detailed questions and an autoregressive model.

We demonstrate the effectiveness of our streamlined design and hierarchical framework in our experimental results, reaching competitive results to the SOTA on the medical VQA benchmark VQARad and setting a baseline for Rad-ReStruct. Overall, our work is a significant step towards automating the population of structured radiology reports and provides a valuable benchmark for future research in this area.

2 Methodology

2.1 Rad-ReStruct Dataset

We propose the first benchmark dataset to enable the development and comparison of methods for the population of structured reports entailing hierarchical and fine-grained classifications for radiological images. Rad-ReStruct is based upon the IU-Xray dataset [5] and consists of X-Ray images paired with fine-grained radiological findings organized as a structured report. To create the dataset, we first define a detailed report template and then populate it automatically by parsing and analyzing the unstructured expert annotations of the reports in IU-Xray.

Creation of Structured Report Template. We build upon the semi-structured encoded findings provided for the IU-XRay data collection [5]. The encoded findings were provided by medical experts, who labeled the IU-XRay free-text reports using MeSH (Medical Subject Headings) [7] and RadLex (Radiology Lexicon) [13] codes. They accurately summarize all findings in the radiological images together with a detailed attribute description. They are an unstructured collection of findings, with a sequence of annotation terms representing each finding (e.g., "infiltrate/lung/upper lobe/left/patchy/mild"...). The codes use a controlled vocabulary containing 178 terms, which include anatomies, diseases, pathological signs, foreign objects, and attributes. Anatomies and diseases

Table 1. Number of questions and answer options per level in our template. For Level 2 we further list the different topics. *not every answer is an option for all questions

	L1	L2	L2-objects	L2-diseases	L2-signs	L2-abnormal regions	L3
Nr. questions	25	216	16	103	65	32	477
Nr. unique answers	2	2	2	2	2	2	94*

can be matched to broad body regions, such as the respiratory or skeletal system. Attributes include degree, descriptive, and positional attributes.

We utilize this semi-structured finding representation to construct a highly detailed report template as shown in Fig. 1. Our report template is structured into multiple sections and employs a multi-level hierarchy of questions that delve deeper into the findings at each level. The template can be considered a large decision tree with questions at every level. The highest level asks for the general existence of findings (signs, diseases, abnormal regions, or objects), the second level asks for specific elements, such as a certain object or disease, and the lowest-level questions ask for specific attributes. Table 1 shows how often which question type occurs. To create the template, we parse the codes of all patients and identify all occurrences of term combinations at all levels of the defined hierarchy. We then remove unseen options to produce a streamlined report template that includes only possible options for all findings. Further, we mark all questions as either single- or multi-choice and add a "no selection" option.

Overall, our structured report template provides a rigorous and comprehensive framework for classifying radiological images and mimics the style of a structured report in a clinical setting. This enables the development and comparison of methods for the population of structured reports and the prediction of fine-grained radiological findings.

Dataset and Evaluation Metrics. Our dataset consists of structured reports for each patient in the IU-XRay data collection, for which finding codes and a frontal X-Ray were available. The new dataset includes 3720 images matched to 3597 structured patient reports entailing more than 180k questions. If multiple images belong to one patient, each image is considered an independent sample. We use a 80-10-10 split to create train, validation and test set. To avoid data leakage, we ensure that different images of the same patient are in the same split.

The goal of our task is to produce fine-grained finding classifications for populating a structured report given an X-Ray image of a patient. This task involves answering a series of questions about the image, gradually adding more detail. We define several evaluation metrics for the proposed benchmark. As the distribution of questions and answers is very imbalanced, we evaluate with the macro precision, recall, and F1 score over all possible paths in the question tree to encourage methods that also perform well in under-represented question-answer combinations. One path is a unique position in the report combined with a specific answer option. Further, we employ report-level accuracy to measure how many predicted reports are entirely correct. During the evaluation, we enforce consistency within the question hierarchy. For example, if the answer to a higher-level question is "no", we prohibit to answer a lower-level question positively. This ensures the generated reports are consistent and coherent, as in a real medical report. Lastly, as multiple instances of an object, sign or pathology can occur for one patient, we iteratively ask for further occurrences, when the model predicts a positive answer. (e.g., "Are there other opacities in the lung?"). We restrict the number of follow-up questions by the maximum of per-patient occurrences in the data. As the order of occurrences is ambiguous, we apply instance matching during the metric computation. We order all predicted instances such that the highest F1 score for this finding is achieved.

2.2 Hierarchical Visual Question Answering

With Rad-ReStruct, we propose a hierarchical VQA task, where lower level questions are dependent on context information. For instance, to answer the questions "What is the degree?" it is essential to know what the question is referring to. This information is given through the previous question, which could be "Is there Pneumonia in the lung? Yes". To integrate this context, we propose a hierarchical VQA framework that can effectively answer questions about medical images by considering previously asked questions. We extend the input to the model by pre-pending the current question with the history of previously asked questions and the model's answers. This extension enables interpretable and consistent results.

We leverage a pretrained image encoder, EfficientNet-b5 [22], and a domain-specific pretrained text encoder, RadBERT [27], to extract features from the image, history, and question. The extracted features are fused by a transformer [25] layer, adapted to handle multi-modal input. The fused features are then used to perform a multi-label classification over all answer options. However, we only consider outputs that are valid answers to the current question. For single-choice questions, we predict a single label applying a softmax over all valid answers, while for multi-choice questions, we predict multiple labels using a sigmoid function. Figure 2 shows an overview of the proposed framework.

Feature Encoding. For fusing the image and text features, we construct a token sequence (Nx458x768) of the form *<image_ tokens>* *<history_ tokens>* *<question_ tokens>*. The image tokens (Nx196x768) consist of the flattened embedding representation of the image encoder, while the history and question text (Nx259x768) is encoded jointly using RadBERT. The different parts are separated by a <SEP> token and fused by a single transformer layer. We encode the type of input in the token-type IDs with different IDs for the image tokens, history questions, history answers, and the current question. Further, we use modified positional encodings to preserve the 2D spatial information of the image as well as the sequential order of the text. We create a joint positional encoding (Nx768x458) by concatenating a 1D positional encoding [25] for the text and a 2D positional encoding [3] for the image (each of shape Nx384x458). We set the non-used part of the encoding per modality to zero. The token-type IDs and positional encodings are added to the feature vector.

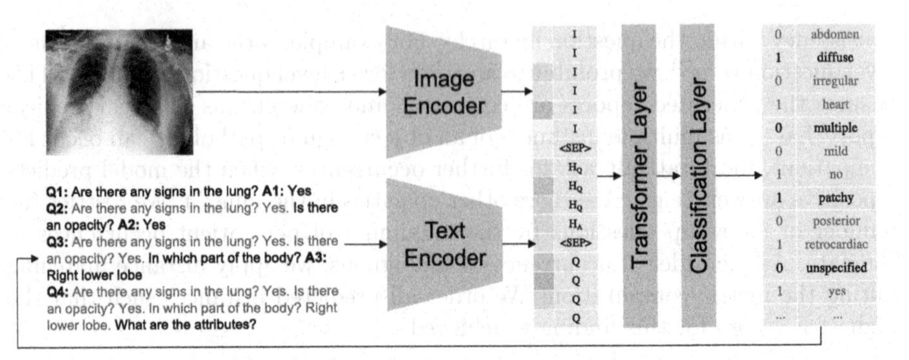

Fig. 2. Overview of our hierarchical VQA framework. The tokens representing the image (I), the history questions (H_Q) and answers (H_A), and the current question (Q) are encoded, concatenated and fused with a transformer layer. The current prediction is computed over the relevant answers and added to the history before asking the next question.

Training and Evaluation. During training, we use teacher forcing, relying on ground truth answers for prior questions in the history, allowing for efficient batch-wise processing. At inference all answers are predicted, and no ground truth is used. The model is trained end-to-end, using a weighted masked cross-entropy loss to optimize the classification performance. For every sample in a training batch, we only consider the loss for the labels corresponding to the asked question to avoid optimizing the model on currently irrelevant outputs. Further, we apply positive weighting per class. The evaluation is autoregressive, thus, the model utilizes the previously asked questions and their predicted answers as history. In a hierarchical VQA task such as Rad-ReStruct, the inference is interrupted if the model predicts a negative answer and sub-questions lower in the hierarchy are automatically answered as negative (set to No/No selection), enforcing consistency of the prediction. This also improves the explainability of predictions by allowing to track errors back to their source and showing at which level the model made a mistake. For non-hierarchical VQA tasks, the history is utilized solely as context information, allowing the model to exploit inter-dependencies between different questions about the same image.

3 Experiments and Results

We test our model on Rad-ReStruct, setting a baseline for this new dataset. To further validate our model design, we compare the performance of our model with previous work on the standard VQA benchmark VQARad. We train all models on a NVIDIA A40 GPU. We use pytorch-lightning 1.8.3. and the AdamW optimizer with a learning rate of 5e-5 for VQARad and 1e-5 for Rad-ReStruct. For all models, we set the number of epochs by maximizing validation set performance.
Rad-ReStruct. For Rad-ReStruct, the history includes all higher-level questions on the same question path. Additionally, attribute questions asked pre-

Table 2. Results of hi-VQA with and without history and our visual baseline

	domain-specific pretraining data	report acc	F1	prec	recall
Visual baseline	none (only general images)	31.3	30.7	65.6	31.2
hi-VQA - no history	radiologic reports	26.2	**31.9**	59.9	**34.1**
hi-VQA - RoBERTa$_{BASE}$	none (only general text/images)	26.2	31.6	67.9	32.4
hi-VQA	radiologic reports	**32.6**	31.7	**70.7**	32.1

Table 3. Detailed performance analysis of our model on Rad-ReStruct. F1, precision and recall are computed as macro average over all paths, where a path is a unique position in the structured report combined with an answer option. The number of answers is the mean count of answer options for all questions belonging to a level.

	report acc	F1	prec	recall	#paths	avg #answers
Level 1 - Topic Existence	36.6	63.8	79.0	63.5	50	2
Level 2 - Element Existence (all)	33.7	72.2	86.0	72.3	432	2
-Diseases	52.4	74.6	83.7	74.9	206	2
- Signs	74.3	74.1	90.1	74.1	130	2
- Abnormal body regions	58.6	69.1	86.4	69.3	64	2
- Objects	90.4	67.8	87.6	67.1	32	2
Level 3 - Attributes	32.6	3.7	60.3	4.4	1988	4.2

viously about an element, are included in the history, enabling the model to provide consistent predictions. Lastly, the history includes previously predicted instances of the same element. Table 2 and Table 3 show the overall and question-level results of our model.

We compare hi-VQA to a visual baseline, consisting of our image encoder and a classification layer. Hi-VQA achieves better performance than the visual baseline in all metrics, indicating the benefits of targeted information retrieval using a large language model. When comparing the RadBERT text encoder, a RoBERTa model [16] pretrained with radiology reports, to RoBERTa$_{BASE}$, which was pre-trained on general text, the RadBERT encoder is superior. This indicates that our method can benefit from better domain-specific language encoders. Using history information improves report accuracy and precision with a slightly decreased recall and a similar F1 score, showing the benefit of history integration. We emphasize, that the history is especially important for the low-level attribute questions, as these are only meaningful with context. Therefore, it will be even more impactful with improved performance for these questions.

Our labels' hierarchical, structured formulation enables a performance analysis on different topics and levels. Hi-VQA performs well in detecting the existence of sub-topics like objects, diseases, signs, and abnormalities. However, attribute prediction performance is much lower, likely due to the rarity and complexity of these questions and error propagation from higher levels. Such an analysis is precious to understand what a model learned and when it should be trusted.

Table 4. Results of our proposed model hi-VQA on the VQARad benchmark compared to previous work. *RepsNet used an adapted validation in their paper, where unseen answers in the test set are ignored, as they can not be predicted. We calculate their performance when keeping the unseen samples to enable a fair comparison.

	domain-specific pretraining data	acc
MEVF [17]	radiologic images	66.1
MMQ [6]	none	67.0
MM-BERT [11]	radiologic images and reports (joined PT)	72.0
CRPD [15]	radiologic images	72.7
RepsNet [23]	radiologic reports	73.5*
M3AE [4]	radiologic images and reports (joined PT)	77.0
hi-VQA - no history	radiologic reports	74.5
hi-VQA - RoBERTa$_{BASE}$	none (only general text/images)	72.5
hi-VQA	radiologic reports	76.3

VQARad is a medical VQA benchmark with 315 radiological images and 3515 questions. The task is to make a classification over 2248 possible answers. In VQARad multiple questions are asked about one image, but in previous work they are always answered separately. To make use of possible inter-dependencies between questions, we define five question levels based on question topics in VQARad, ranging from general to specific: *Modality* → *Plane* → *Organ* → *Presence, Count, Abnormality* → *Color, Position, Size, Attributes, Other.* For a certain question, previously asked questions from lower or the same level are included in the history. During training, we augment the history by randomly dropping and reordering questions within a level to prevent overfitting on this small dataset.
Table 4 shows the performance of hi-VQA compared to previous methods. Amongst the methods without domain-specific joined image-text pretraining, we reach SOTA results, even without history context. When integrating history information, our model achieves competitive results with the current SOTA method, M3AE [4]. This result demonstrates the promise of jointly answering questions about the same image in medical VQA tasks. Lastly, we again compare using the RadBERT encoder to RoBERTa$_{BASE}$ [16]. We can observe, also on VQARad, using RadBERT improves the performance notably, again indicating that VQA tasks benefit from domain-specific text encoders.

4 Discussion and Conclusion

By introducing Rad-ReStruct, the first structured radiology reporting benchmark, we create a framework to develop, evaluate, and compare structured reporting methods. The structured formulation enables an accurate evaluation of clinical correctness at different levels of granularity, focusing on levels with greater clinical importance. Moreover, such a structured finding representation could then again, rule-based, be converted to a text report while maintaining clinical accuracy. To model structured reporting, we present hi-VQA, a novel, hierarchical VQA framework with a streamlined architecture that leverages history context for multi-question and multi-level tasks. The autoregressive formulation and consistent evaluation increase interpretability and mimic the workflow of structured reporting. Moreover, as each prediction takes previous answers into account, it would allow for an interactive workflow, where the model can make predictions and react to corrections while a radiologist fills out a report.

We set a first baseline for Rad-ReStruct, with particularly good performance on higher-level questions. Although our model has limited performance on the low-level attribute questions, it performed competitive to state-of-the-art on VQARad, indicating the difficulty of our new task. We see this as an opportunity to develop methods for fine-grained understanding of radiology images, rather than solely focusing on higher-level diagnoses. Further, we show the positive effect of history integration, which is crucial for hierarchical and context-dependent tasks such as structured report population. Our work represents a significant step forward in the development of automated structured radiology report population methods, while allowing an accurate and multi-level evaluation of clinical correctness and fostering fine-grained, in-depth radiological image understanding.

Acknowledgements. The authors gratefully acknowledge the financial support by the Federal Ministry of Education and Research of Germany (BMBF) under project DIVA (FKZ 13GW0469C) and the Bavarian Research Foundation (BFS) under project PandeMIC (grant AZ-1429-20C).

References

1. Antol, S., et al.: VQA: visual question answering. In: Proceedings of the IEEE International Conference on Computer Vision, pp. 2425–2433 (2015)
2. Bhalodia, R., et al.: Improving pneumonia localization via cross-attention on medical images and reports. In: de Bruijne, M., et al. (eds.) MICCAI 2021. LNCS, vol. 12902, pp. 571–581. Springer, Cham (2021). https://doi.org/10.1007/978-3-030-87196-3_53
3. Chen, T., Saxena, S., Li, L., Fleet, D.J., Hinton, G.: Pix2Seq: a language modeling framework for object detection. arXiv preprint arXiv:2109.10852 (2021)
4. Chen, Z., et al.: Multi-modal masked autoencoders for medical vision-and-language pre-training. In: Wang, L., Dou, Q., Fletcher, P.T., Speidel, S., Li, S. (eds.) Medical Image Computing and Computer Assisted Intervention-MICCAI 2022: 25th International Conference, Singapore, 18–22 September 2022, Proceedings, Part V, pp. 679–689. Springer, Cham (2022)
5. Demner-Fushman, D., et al.: Preparing a collection of radiology examinations for distribution and retrieval. J. Am. Med. Inform. Assoc. **23**(2), 304–310 (2016)
6. Do, T., Nguyen, B.X., Tjiputra, E., Tran, M., Tran, Q.D., Nguyen, A.: Multiple meta-model quantifying for medical visual question answering. In: de Bruijne, M., et al. (eds.) MICCAI 2021. LNCS, vol. 12905, pp. 64–74. Springer, Cham (2021). https://doi.org/10.1007/978-3-030-87240-3_7

7. Rogers, F.B.: Medical subject headings. Bull. Med. Libr. Assoc. **51**, 114–116 (1963)
8. Hong, Y., Kahn, C.E.: Content analysis of reporting templates and free-text radiology reports. J. Digit. Imaging **26**, 843–849 (2013)
9. Hou, B., Kaissis, G., Summers, R.M., Kainz, B.: RATCHET: medical transformer for chest X-ray diagnosis and reporting. In: de Bruijne, M., et al. (eds.) MICCAI 2021. LNCS, vol. 12907, pp. 293–303. Springer, Cham (2021). https://doi.org/10.1007/978-3-030-87234-2_28
10. Keicher, M., Mullakaeva, K., Czempiel, T., Mach, K., Khakzar, A., Navab, N.: Few-shot structured radiology report generation using natural language prompts. arXiv preprint arXiv:2203.15723 (2022)
11. Khare, Y., Bagal, V., Mathew, M., Devi, A., Priyakumar, U.D., Jawahar, C.: MMBert: multimodal BERT pretraining for improved medical VQA. In: 2021 IEEE 18th International Symposium on Biomedical Imaging (ISBI), pp. 1033–1036. IEEE (2021)
12. Kovaleva, O., et al.: Towards visual dialog for radiology. In: Proceedings of the 19th SIGBioMed Workshop on Biomedical Language Processing, pp. 60–69 (2020)
13. Langlotz, C.P.: RadLex: a new method for indexing online educational materials (2006)
14. Li, J., Li, S., Hu, Y., Tao, H.: A self-guided framework for radiology report generation. In: Wang, L., Dou, Q., Fletcher, P.T., Speidel, S., Li, S. (eds.) Medical Image Computing and Computer Assisted Intervention-MICCAI 2022: 25th International Conference, Singapore, 18–22 September 2022, Proceedings, Part VIII, pp. 588–598. Springer, Cham (2022). https://doi.org/10.1007/978-3-031-16452-1_56
15. Liu, B., Zhan, L.-M., Wu, X.-M.: Contrastive pre-training and representation distillation for medical visual question answering based on radiology images. In: de Bruijne, M., et al. (eds.) MICCAI 2021. LNCS, vol. 12902, pp. 210–220. Springer, Cham (2021). https://doi.org/10.1007/978-3-030-87196-3_20
16. Liu, Y., et al.: RoBERTa: a robustly optimized BERT pretraining approach. arXiv preprint arXiv:1907.11692 (2019)
17. Nguyen, B.D., Do, T.-T., Nguyen, B.X., Do, T., Tjiputra, E., Tran, Q.D.: Overcoming data limitation in medical visual question answering. In: Shen, D., et al. (eds.) MICCAI 2019. LNCS, vol. 11767, pp. 522–530. Springer, Cham (2019). https://doi.org/10.1007/978-3-030-32251-9_57
18. Nobel, J.M., van Geel, K., Robben, S.G.: Structured reporting in radiology: a systematic review to explore its potential. Eur. Radiol., 1–18 (2022)
19. Pino, P., Parra, D., Besa, C., Lagos, C.: Clinically correct report generation from chest X-rays using templates. In: Lian, C., Cao, X., Rekik, I., Xu, X., Yan, P. (eds.) MLMI 2021. LNCS, vol. 12966, pp. 654–663. Springer, Cham (2021). https://doi.org/10.1007/978-3-030-87589-3_67
20. Ren, F., Zhou, Y.: CGMVQA: a new classification and generative model for medical visual question answering. IEEE Access **8**, 50626–50636 (2020)
21. Syeda-Mahmood, T., et al.: Chest X-ray report generation through fine-grained label learning. In: Martel, A.L., et al. (eds.) MICCAI 2020. LNCS, vol. 12262, pp. 561–571. Springer, Cham (2020). https://doi.org/10.1007/978-3-030-59713-9_54
22. Tan, M., Le, Q.: EfficientNet: rethinking model scaling for convolutional neural networks. In: International Conference on Machine Learning, pp. 6105–6114. PMLR (2019)
23. Tanwani, A.K., Barral, J., Freedman, D.: RepsNet: combining vision with language for automated medical reports. In: Wang, L., Dou, Q., Fletcher, P.T., Speidel, S., Li, S. (eds.) Medical Image Computing and Computer Assisted Intervention-MICCAI 2022: 25th International Conference, Singapore, 18–22 September 2022,

Proceedings, Part V, pp. 714–724. Springer, Cham (2022). https://doi.org/10.1007/978-3-031-16443-9_68
24. Tascon-Morales, S., Márquez-Neila, P., Sznitman, R.: Consistency-preserving visual question answering in medical imaging. In: Wang, L., Dou, Q., Fletcher, P.T., Speidel, S., Li, S. (eds.) Medical Image Computing and Computer Assisted Intervention-MICCAI 2022: 25th International Conference, Singapore, 18–22 September 2022, Proceedings, Part VIII, pp. 386–395. Springer, Cham (2022). https://doi.org/10.1007/978-3-031-16452-1_37
25. Vaswani, A., et al.: Attention is all you need. In: Guyon, I., et al. (eds.) Advances in Neural Information Processing Systems, vol. 30. Curran Associates, Inc. (2017)
26. Wang, Z., Tang, M., Wang, L., Li, X., Zhou, L.: A medical semantic-assisted transformer for radiographic report generation. In: Wang, L., Dou, Q., Fletcher, P.T., Speidel, S., Li, S. (eds.) Medical Image Computing and Computer Assisted Intervention-MICCAI 2022: 25th International Conference, Singapore, 18–22 September 2022, Proceedings, Part III, pp. 655–664. Springer, Cham (2022). https://doi.org/10.1007/978-3-031-16437-8_63
27. Yan, A., et al.: RadBERT: adapting transformer-based language models to radiology. Radiol. Artif. Intell. **4**(4), e210258 (2022)

Xplainer: From X-Ray Observations to Explainable Zero-Shot Diagnosis

Chantal Pellegrini[1](✉), Matthias Keicher[1], Ege Özsoy[1], Petra Jiraskova[2], Rickmer Braren[2], and Nassir Navab[1]

[1] Computer Aided Medical Procedures, Technical University Munich, Munich, Germany
chantal.pellegrini@tum.de

[2] Department of Diagnostic and Interventional Radiology, School of Medicine, Technical University of Munich, Munich, Germany

Abstract. Automated diagnosis prediction from medical images is a valuable resource to support clinical decision-making. However, such systems usually need to be trained on large amounts of annotated data, which often is scarce in the medical domain. Zero-shot methods address this challenge by allowing a flexible adaption to new settings with different clinical findings without relying on labeled data. Further, to integrate automated diagnosis in the clinical workflow, methods should be transparent and explainable, increasing medical professionals' trust and facilitating correctness verification. In this work, we introduce Xplainer, a novel framework for explainable zero-shot diagnosis in the clinical setting. Xplainer adapts the classification-by-description approach of contrastive vision-language models to the multi-label medical diagnosis task. Specifically, instead of directly predicting a diagnosis, we prompt the model to classify the existence of descriptive observations, which a radiologist would look for on an X-Ray scan, and use the descriptor probabilities to estimate the likelihood of a diagnosis. Our model is explainable by design, as the final diagnosis prediction is directly based on the prediction of the underlying descriptors. We evaluate Xplainer on two chest X-ray datasets, CheXpert and ChestX-ray14, and demonstrate its effectiveness in improving the performance and explainability of zero-shot diagnosis. Our results suggest that Xplainer provides a more detailed understanding of the decision-making process and can be a valuable tool for clinical diagnosis. Our code is available on github: https://github.com/ChantalMP/Xplainer

Keywords: Zero-Shot Diagnosis · Explainability · Contrastive Learning

C. Pellegrini, M. Keicher and E. Özsoy—These authors contributed equally.

Supplementary Information The online version contains supplementary material available at https://doi.org/10.1007/978-3-031-43904-9_41.

H. Greenspan et al. (Eds.): MICCAI 2023, LNCS 14224, pp. 420–429, 2023.
https://doi.org/10.1007/978-3-031-43904-9_41

1 Introduction

Computer-aided diagnosis systems have become a prominent tool in medical diagnosis. Yet, their adoption is limited by the need for large amounts of annotated data for training, which hinders their scalability and adaptability to new clinical findings [3,12]. Moreover, adapting to a new reporting template or clinical protocol necessitates new annotations, further reducing their feasibility in clinical settings. Recently, zero-shot [1,4,14,15,17] and few-shot [1,4,8] learning methods have been proposed as a potential solution, utilizing contrastive pretraining [13,19] on pairs of radiology reports and images, and achieving performance on par with radiologists [15]. However, these methods lack the level of detail of radiology reports and inherent explainability, impeding their adoption in clinical settings [7]. Particularly, explaining the diagnosis with image descriptors is crucial to increase trust in the system and allow radiologists to verify the results [9].

Inspired by the success of using large language models to predict image descriptors in natural images [10], we introduce Xplainer, a novel framework that enhances the explainability of zero-shot diagnosis in the clinical setting. Xplainer leverages the classification-by-description approach [10] of vision-language models and adapts it to the multi-label medical diagnosis task. Specifically, we task the model to classify the existence of descriptive observations, which a radiologist would examine on an X-Ray scan, instead of directly predicting a diagnosis. This model design imbues our framework with intrinsic explainability, as the final diagnosis prediction is predicated on the underlying descriptor predictions.

We evaluate Xplainer on two chest X-ray datasets, CheXpert [5] and ChestX-ray14 [16], and demonstrate its efficacy in enhancing the performance and explainability of zero-shot diagnosis in the clinical setting. Our results highlight that Xplainer provides a more comprehensive understanding of the diagnosis prediction process, thereby serving as a valuable tool for clinical decision-making. In summary, Xplainer presents a novel framework for zero-shot diagnosis that not only improves explainability and accuracy but also provides an invaluable tool for computer-aided diagnosis.

2 Methodology

2.1 Model Overview

We propose Xplainer, an explainable zero-shot classification-by-description approach for diagnosing pathologies from X-Ray scans. Given an image i and a list of clinical observations $o_{p_{1-n}}$ per pathology p, the goal is to make a multi-label prediction indicating the diagnosis for the patient.

Our zero-shot approach leverages the alignment of image and text embeddings provided by contrastive language-image pretraining (CLIP) [13] and therefore does not require any labeled data. We built upon BioVil [1], a CLIP model pretrained on pairs of radiology reports and images. Employing the text and image encoders from BioVil, we calculate the cosine similarity between an X-ray image and each of N pre-defined clinical observations $o_{p_{1-N}}$ describing a

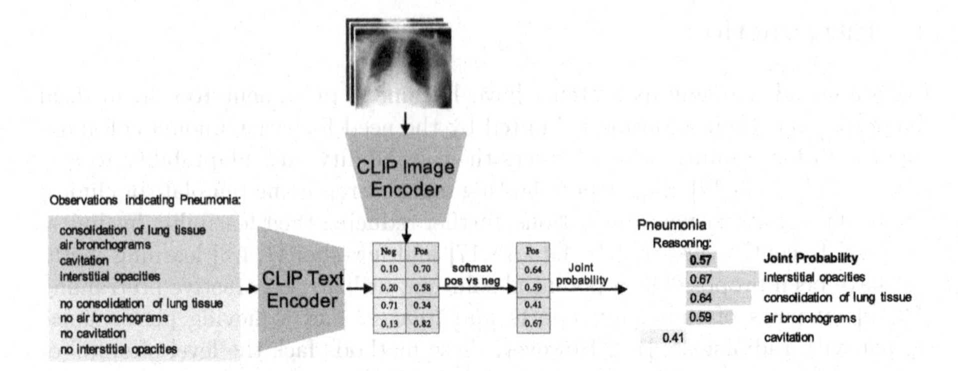

Fig. 1. Overview of Xplainer: In the first step, observation probabilities are calculated based on contrastive CLIP prompting. These are then used to make an explainable diagnosis prediction. The figure depicts an example for Pneumonia.

pathology. Then we calculate observation probabilities $P_{pos}(o_{p_i})$ for every observation. Analogously, we calculate probabilities for the absence of all observations $P_{neg}(o_{p_i})$ by defining negated prompts for all observations. Using the softmax over the positive and negative probability, we calculate the final probability of the presence of an observation $P(o_{p_i})$. Given these observation probabilities $P(o_{p_i}), i \in 1, ..., N$, we estimate a joined probability to determine the likelihood of the presence of a pathology $P(p)$:

$$P(p) = \sum_{i=1}^{N} log(P(o_{p_i})) \div N \quad (1)$$

We repeat this process for all pathologies we want to diagnose in the image. As the prediction of a pathology diagnosis is directly extracted from the observation probabilities, our method is explainable by design, producing a diagnosis prediction and the detected X-ray observations leading to that prediction. Moreover, the observation probabilities show which observations the model mainly considers for its diagnosis. Figure 1 shows an overview of our framework.

To integrate multiple images of one patient, we calculate positive and negative observation probabilities for each image and average them before calculating the pathology probability.

2.2 Prompt Engineering

Successful zero-shot inference relies on a good alignment between the contrastive pretraining and the downstream task [13]. As BioVil [1] was trained on pairs of radiological images and reports, we need to keep our observation prompts close to the style of medical reports. To initialize our prompts, we employ ChatGPT [11] and query it to describe observations in X-ray images that would occur in a radiology report indicating specific pathologies. We further refined the prompts with the help of an experienced radiologist, who manually verified and adapted the descriptors. Human refinement cost was low, taking the radiologist only a few hours. We provide a complete list of the descriptors in the supplementary.

Radiology reports often include both presence and absence of particular observations. When comparing a prompt with an image embedding, it is hard for the model to differentiate between an observation's positive and negative occurrence, as their formulation can be very similar. Previous work [14,15] has shown that introducing negative prompts can circumvent this problem. Therefore, instead of thresholding the similarity between a positive prompt and an image, we prompt the model with both a positive and a negated version of each observation prompt and compare their probabilities. We adapt our prompts in two additional steps to align them with the text in radiology reports. First, we add a disease indication, as radiology reports usually contain observations paired with conclusions. Further, this reduces the ambiguity of our prompts, as in radiology, one sign (e.g., Lung Opacity) can indicate multiple pathologies (e.g., Pneumonia, Atelectasis, or Edema). Additionally, we frame all our observations in a sentence structure sounding more like an actual report by adding "There is/are" before every observation. Putting all of this together, we define the following prompt structure: "There is/are (no) <observation> indicating <pathology>." Lastly, we define contrastive pathology-based prompts to compare to our observation-based prompting. In this setting, only two prompts, one positive and one negative prompt, are used per pathology. Overall, we compare the following styles of prompting to show the benefit of observation-based, contrastive prompting with disease indication and report style:

- **Pathology-based:** (No) <pathology>
- **Basic:** Only positive prompt per pathology: <observation>
- **Contrastive:** (No) <observation>
- **Pathology Indication:** (No) <observation> indicating <pathology>
- **Report Style:** There is/are (no) <observation> indicating <pathology>

3 Experiments and Results

We evaluate Xplainer in a zero-shot setting on the commonly used chest X-ray datasets, CheXpert [5], and ChestX-ray14 [16]. The CheXpert dataset provides a manually labeled validation and test set with 200 and 500 patients, respectively,

Table 1. AUC for zero-shot pathology classification on CheXpert and ChestX-ray14 datasets. *in-domain, as the underlying CLIP model was trained the ChestX-ray14

	CLIP pretraining data	CheXpert		ChestX-ray14
		val	test	test
CheXzero [15]	MIMIC	–	74.73	–
Seibold et al. [14]	MIMIC	78.86	–	71.23
Seibold et al. [14]	MIMIC, PadChest, ChestX-ray14	83.24	–	78.33*
Xplainer	MIMIC	**84.92**	**80.58**	**71.73**

Table 2. AUC per disease on both datasets

	CheXpert Val	CheXpert Test	ChestX-ray14
No Finding	88.82	89.94	–
Enlarged Cardiomediastinum	79.23	80.60	–
Cardiomegaly	78.62	83.32	79.71
Lung Opacity	88.18	91.76	–
Lung Lesion	91.46	69.33	–
Edema	84.84	84.55	81.46
Consolidation	91.56	85.89	71.87
Pneumonia	85.68	83.73	70.83
Atelectasis	84.64	85.46	66.86
Pneumothorax	78.09	83.75	72.18
Pleural Effusion	88.72	89.30	79.11
Pleural Other	83.92	58.67	–
Fracture	–	60.47	–
Infiltration	–	–	68.81
Mass	–	–	70.28
Nodule	–	–	64.74
Emphysema	–	–	74.02
Fibrosis	–	–	62.25
Pleural Thickening	–	–	67.44
Hernia	–	–	74.60
Support Devices/Foreign Objects	80.25	81.15	–

and 14 classes, including "No Finding", "Support Devices/Foreign Objects", and 12 pathology labels. ChestX-ray14 is evaluated on 14 pathology labels on a test set of 25.596 images. For both datasets, we use the official validation and test splits. We perform a multi-label classification for both datasets and evaluate the performance via the Area Under the ROC-curve (AUC) between the positive pathology probabilities and the labels.

Table 1 shows our results compared to previously proposed zero-shot pathology prediction approaches. On CheXpert, we compare with Seibold et al. [14] on the validation set, as they only reported validation performance. For the comparison with CheXzero [15], as well as the ChestX-ray14 dataset, we compare test set results. We outperform both previous works in an out-of-domain setting, where the zero-shot inference is performed on a different dataset than CLIP was trained on. The state-of-the-art results on both datasets show the effectiveness of our observation-based modeling. Further, in Table 2, we provide a detailed breakdown of our results per pathology and dataset.

Table 3. Comparison of different prompting styles on the validation set of CheXpert

	AUC
Contrastive pathology-based Prompting	76.14
Observation-based Prompting:	
Basic Prompt	58.65
Contrastive Prompt	77.00
+ pathology Indication	84.35
+ Report Style	84.92

Table 4. Comparison of ChatGPT prompts vs. refinement with the help of a radiologist

	CheXpert Val	CheXpert Test	ChestX-ray14
ChatGPT prompts	83.61	79.94	71.40
Refined Prompts	**84.92**	**80.58**	**71.73**

Ablation Studies. In our ablation studies, we investigate the impact of our prompt design and the effect of using multiple images. Table 3 shows the results on the CheXpert validation set using different prompting styles. We observe that pathology-based prompting, which reaches an AUC of 76.14%, is significantly worse than observation-based prompting, which reaches an AUC of 84.92%, again highlighting the benefit of observation-based prompting. Comparing the basic observation-based prompting, using only positive prompts per observation, to contrastive prompting, we see a substantial performance gap, showing the importance of using negative prompts to differentiate between positive and negative occurrences. We also show the effect of formulating our prompts unambiguously and in the style of an actual radiology report by adding pathology indication and report style. Adding pathology indication to the contrastive observation-based prompting significantly improves performance, achieving an AUC of 84.35%. Finally, incorporating report style in the prompts leads to the highest AUC of 84.92%, indicating that a contrastive observation-based prompt with pathology indication and report style is the most effective for zero-shot X-ray pathology classification.

Additionally, we compare the initial ChatGPT output to our refined prompts (Table 4). Refinement was performed by deleting irrelevant, redundant, or incorrect descriptors. We observe an improvement through the refinement, indicating that including domain knowledge further improves our method. Nevertheless, the original ChatGPT prompts already perform quite well, showing the impressive potential of combining large generic language models with large domain-specific contrastive models.

For the "No Finding" class, we compare to either define specific prompts such as "Clear lung fields" or "Normal heart size and shape" to classify "No Finding" or model it as the absence of all of the other 13 labels (Rule-based). Table 5

Table 5. Modeling of "No Finding" label with explicit prompts or rule-based definition as lack of other findings

	AUC - No Finding
Explicit Prompting	79.64
Rule-based	**88.82**

Table 6. Comparison of single-view inference to different methods for multi-image processing

	AUC
Only single Frontal View	84.19
All - Max Aggregation	84.77
All - Mean Aggregation	**84.92**

shows that a rule-based modeling of this class leads to better results. A reason for this could be that there is no clearly defined list of observations that indicate a healthy X-ray scan, which a radiologist would mention in his report.

Lastly, we investigate different image aggregation methods for pathology prediction. We compare only using a single frontal view X-ray to using all images available for a patient. For aggregation, we compute positive and negative observation probabilities for every image. In Max aggregation, we then use the highest observation probability. The intuition behind this approach is that an observation might be seen much better from one perspective than another, and then only the perspective where the model is most confident should be used. On the other hand, different views give different insights about which kind of observation a visual cue on the image indicates. To leverage this multi-view information, we test Mean aggregation, where all observation probabilities are averaged over multiple images. The results shown in Table 6 indicate Mean aggregation to be superior, while both aggregation methods outperform using just a single image.

Qualitative Results. Figure 2 shows qualitative examples of our model's predictions. For the true positive prediction, it can be seen that most of the descriptors are detected, and the model recognizes the descriptor "Mass in the mediastinum" as the main indication for the Enlarged Cardiomediastinum. For the True Negative case, the model, correctly, detected none of the descriptors. For the false positive example, one can clearly see that the model made a mistake because it detected an air bronchogram with relatively high certainty and no consolidation. Therefore, this false positive finding is easily falsified by the radiologist since an air bronchogram is a finding that co-occurs with consolidation (i.e., air-filled bronchi in consolidated areas). Thus, knowing which combination of descriptors leads to such a decision substantially improves explainability. In the false positive case, the model misses the pacemaker but detects some implant, showing the model understands there is some foreign object, but can not identify it, which is easily detected by the radiologist. Overall the classification-by-description may facilitate a plausibility check of a specific inference result and an understanding of the source of errors.

Discussion. One downside of modeling a joint probability is that it assumes that all descriptors appear simultaneously and gives all descriptors the same

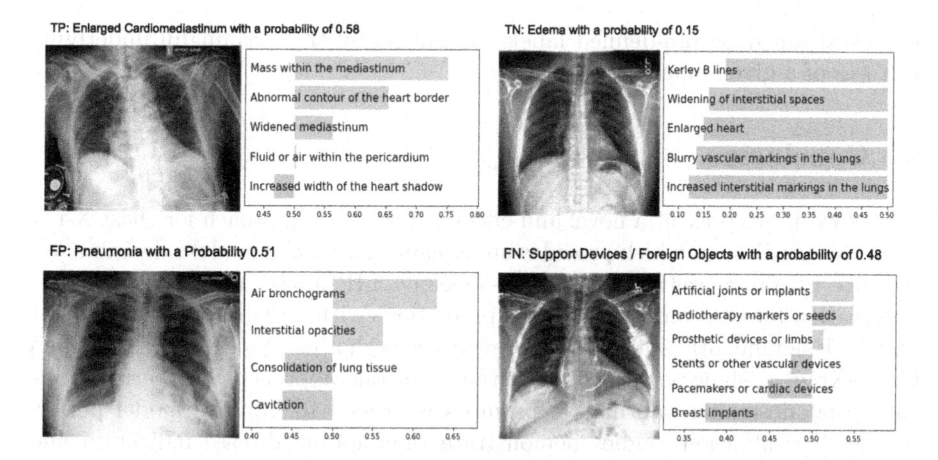

Fig. 2. Qualitative results of Xplainer

importance. While this estimation leads to good results, the assumption does not always hold, as a pathology does not always present with the same signs. Further, there might be inter-dependencies between the descriptors, e.g., there can be descriptors that strongly correlate with the presence of a disease when combined with one descriptor but much less when combined with another. As a first try to model the importance of descriptors, we look into a supervised, out-of-domain approach to model these inter-dependencies. For this, we train a Naive Bayes [2,18] CheXpert classifier on MIMIC-CXR [6], predicting a diagnosis given the descriptor probabilities, allowing the model to focus more on more relevant descriptors. While this approach relies on labels for MIMIC, these labels can be automatically generated by the CheXpert labeler [5], still not requiring human effort for labeling. We observe a slight performance increase on the test set from 80.58% to 81.37% AUC. This shows that the descriptor importance learned on MIMIC can partially be transferred to an out-of-domain dataset. We believe investigating methods to consider varying importance and complex relations between the descriptors is an essential and exciting direction to investigate in future work. Moreover, as Xplainer is not tied to specific image and text encoders, orthogonal works that lead to better encoders can be used to improve our results further.

The use of descriptors in Xplainer provides a flexible and adaptive approach to automated diagnosis prediction. By identifying and classifying the presence of descriptive observations, our model can capture the underlying characteristics of a disease without relying on labeled data. This means that our system can easily adapt to new settings with different clinical findings, including new conditions where the symptoms are known, but there is no training data available yet. Additionally, using descriptors allows for adapting the system to specific populations, where the essential descriptors can differ. This is because the model is not constrained by pre-defined labels but rather by the meaningful underlying features of a given diagnosis.

4 Conclusion

In this work, we present a novel and effective zero-shot approach for chest X-ray diagnosis prediction, which provides an explanation for the model's decision. We leverage BioVil, a pretrained, domain-specific CLIP model, and use contrastive observation-based prompting to make predictions without label supervision. Our approach significantly outperforms previous zero-shot methods on CheXpert and Chest-Xray14, showcasing the effectiveness of our approach. Furthermore, we show that designing informative prompts is crucial to improve model performance. Our ablation studies demonstrate that adding disease indication and report style formulation to observation-based prompts notably enhances performance, underscoring the importance of aligning prompts with the domain-specific language used in medical reports. Additionally, contrastive prompts significantly boost performance, suggesting that the model can benefit from explicitly contrasting positive and negative examples.

Our work highlights the potential of contrastive pretraining combined with observation-based prompting as a promising avenue for zero-shot medical image classification, where labeled data is scarce or expensive to obtain, and explainability is vital. We envision that our approach can be extended to other medical imaging domains and have practical applications in real-world scenarios. Our findings contribute to the growing body of research to improve the accuracy and interpretability of medical image diagnosis.

Acknowledgements. The authors gratefully acknowledge the financial support by the Federal Ministry of Education and Research of Germany (BMBF) under project DIVA (FKZ 13GW0469C) and the Bavarian Research Foundation (BFS) under project PandeMIC (grant AZ-1429-20C).

References

1. Boecking, B., et al.: Making the most of text semantics to improve biomedical vision-language processing. In: Avidan, S., Brostow, G., Cissé, M., Farinella, G.M., Hassner, T. (eds.) Computer Vision-ECCV 2022: 17th European Conference, Tel Aviv, Israel, 23–27 October 2022, Proceedings, Part XXXVI, pp. 1–21. Springer, Cham (2022). https://doi.org/10.1007/978-3-031-20059-5_1
2. Chan, T.F., Golub, G.H., LeVeque, R.J.: Updating formulae and a pairwise algorithm for computing sample variances. In: Caussinus, H., Ettinger, P., Tomassone, R. (eds.) COMPSTAT 1982 5th Symposium held at Toulouse 1982: Part I: Proceedings in Computational Statistics, pp. 30–41. Springer, Cham (1982). https://doi.org/10.1007/978-3-642-51461-6_3
3. Fink, O., Wang, Q., Svensen, M., Dersin, P., Lee, W.J., Ducoffe, M.: Potential, challenges and future directions for deep learning in prognostics and health management applications. Eng. Appl. Artif. Intell. **92**, 103678 (2020)

4. Huang, S.C., Shen, L., Lungren, M.P., Yeung, S.: GLoRIA: a multimodal global-local representation learning framework for label-efficient medical image recognition. In: Proceedings of the IEEE/CVF International Conference on Computer Vision, pp. 3942–3951 (2021)
5. Irvin, J., et al.: CheXpert: a large chest radiograph dataset with uncertainty labels and expert comparison. In: Proceedings of the AAAI Conference on Artificial Intelligence, vol. 33, pp. 590–597 (2019)
6. Johnson, A.E., et al.: MIMIC-CXR, a de-identified publicly available database of chest radiographs with free-text reports. Sci. Data **6**(1), 317 (2019)
7. Kayser, M., Emde, C., Camburu, O.M., Parsons, G., Papiez, B., Lukasiewicz, T.: Explaining chest X-ray pathologies in natural language. In: Wang, L., Dou, Q., Fletcher, P.T., Speidel, S., Li, S. (eds.) Medical Image Computing and Computer Assisted Intervention-MICCAI 2022: 25th International Conference, Singapore, 18–22 September 2022, Proceedings, Part V, pp. 701–713. Springer, Cham (2022). https://doi.org/10.1007/978-3-031-16443-9_67
8. Keicher, M., Mullakaeva, K., Czempiel, T., Mach, K., Khakzar, A., Navab, N.: Few-shot structured radiology report generation using natural language prompts. arXiv preprint arXiv:2203.15723 (2022)
9. McInerney, D.J., Young, G., van de Meent, J.W., Wallace, B.C.: CHiLL: zero-shot custom interpretable feature extraction from clinical notes with large language models. arXiv preprint arXiv:2302.12343 (2023)
10. Menon, S., Vondrick, C.: Visual classification via description from large language models. arXiv preprint arXiv:2210.07183 (2022)
11. OpenAI: Chatgpt. chat.openai.com. Accessed 8 Mar 2023
12. Qin, C., Yao, D., Shi, Y., Song, Z.: Computer-aided detection in chest radiography based on artificial intelligence: a survey. Biomed. Eng. Online **17**(1), 1–23 (2018)
13. Radford, A., et al.: Learning transferable visual models from natural language supervision. In: International Conference on Machine Learning, pp. 8748–8763. PMLR (2021)
14. Seibold, C., Reiß, S., Sarfraz, M.S., Stiefelhagen, R., Kleesiek, J.: Breaking with fixed set pathology recognition through report-guided contrastive training. In: Wang, L., Dou, Q., Fletcher, P.T., Speidel, S., Li, S. (eds.) Medical Image Computing and Computer Assisted Intervention – MICCAI 2022: 25th International Conference, Singapore, 18–22 September 2022, Proceedings, Part V, pp. 690–700. Springer, Heidelberg (2022). https://doi.org/10.1007/978-3-031-16443-9_66
15. Tiu, E., Talius, E., Patel, P., Langlotz, C.P., Ng, A.Y., Rajpurkar, P.: Expert-level detection of pathologies from unannotated chest X-ray images via self-supervised learning. Nat. Biomed. Eng., 1–8 (2022)
16. Wang, X., Peng, Y., Lu, L., Lu, Z., Bagheri, M., Summers, R.M.: ChestX-ray8: hospital-scale chest X-ray database and benchmarks on weakly-supervised classification and localization of common thorax diseases. In: Proceedings of the IEEE Conference on Computer Vision and Pattern Recognition, pp. 2097–2106 (2017)
17. Wang, Z., Wu, Z., Agarwal, D., Sun, J.: MedCLIP: contrastive learning from unpaired medical images and text. arXiv preprint arXiv:2210.10163 (2022)
18. Zhang, H.: The optimality of Naive Bayes. In: Barr, V., Markov, Z. (eds.) Proceedings of the Seventeenth International Florida Artificial Intelligence Research Society Conference (FLAIRS 2004). AAAI Press (2004)
19. Zhang, Y., Jiang, H., Miura, Y., Manning, C.D., Langlotz, C.P.: Contrastive learning of medical visual representations from paired images and text. In: Machine Learning for Healthcare Conference, pp. 2–25. PMLR (2022)

Towards Generalizable Diabetic Retinopathy Grading in Unseen Domains

Haoxuan Che[1], Yuhan Cheng[1], Haibo Jin[1], and Hao Chen[1,2(✉)]

[1] Department of Computer Science and Engineering, The Hong Kong University of Science and Technology, Kowloon, Hong Kong
{hche,ychengbj,hjinag,jhc}@cse.ust.hk

[2] Department of Chemical and Biological Engineering, The Hong Kong University of Science and Technology, Kowloon, Hong Kong

Abstract. Diabetic Retinopathy (DR) is a common complication of diabetes and a leading cause of blindness worldwide. Early and accurate grading of its severity is crucial for disease management. Although deep learning has shown great potential for automated DR grading, its real-world deployment is still challenging due to distribution shifts among source and target domains, known as the domain generalization problem. Existing works have mainly attributed the performance degradation to limited domain shifts caused by simple visual discrepancies, which cannot handle complex real-world scenarios. Instead, we present preliminary evidence suggesting the existence of three-fold generalization issues: visual and degradation style shifts, diagnostic pattern diversity, and data imbalance. To tackle these issues, we propose a novel unified framework named Generalizable Diabetic Retinopathy Grading Network (GDRNet). GDRNet consists of three vital components: fundus visual-artifact augmentation (FundusAug), dynamic hybrid-supervised loss (DahLoss), and domain-class-aware re-balancing (DCR). FundusAug generates realistic augmented images via visual transformation and image degradation, while DahLoss jointly leverages pixel-level consistency and image-level semantics to capture the diverse diagnostic patterns and build generalizable feature representations. Moreover, DCR mitigates the data imbalance from a domain-class view and avoids undesired over-emphasis on rare domain-class pairs. Finally, we design a publicly available benchmark for fair evaluations. Extensive comparison experiments against advanced methods and exhaustive ablation studies demonstrate the effectiveness and generalization ability of GDRNet. The source code is released at https://github.com/chehx/DGDR.

1 Introduction

Diabetic Retinopathy (DR) is a leading cause of blindness, affecting millions of people worldwide, and early severity grading is vital for disease management [1].

Supplementary Information The online version contains supplementary material available at https://doi.org/10.1007/978-3-031-43904-9_42.

H. Greenspan et al. (Eds.): MICCAI 2023, LNCS 14224, pp. 430–440, 2023.
https://doi.org/10.1007/978-3-031-43904-9_42

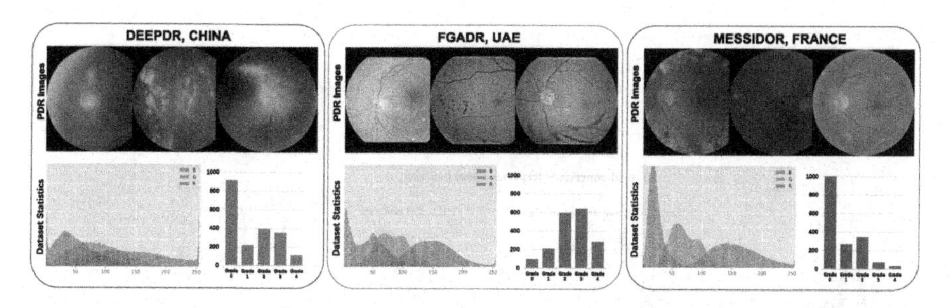

Fig. 1. The RGB statistics, category histograms and proliferative DR (PDR) samples from different datasets/domains. It can be observed that the existence of visual differences, image degradations and diverse diagnostic patterns from PDR samples and RGB statistics. Besides, the divergence among label histograms shows the data imbalance problem across domains and categories.

Although deep learning (DL) has shown promising results in automatic DR grading [2–4], its real-world deployment is still challenging. For instance, Google's DR grading system performed ideally in controlled lab settings [4], but failed to generalize well to complex scenarios which suffer from data shifts [5]. It is a common problem known as domain generalization (DG) [6], where the model performance significantly drops when applied to unseen domains different from the training data. Such an issue hinders the wide adoption and success of DL-based diagnostic tools in clinical practice [7].

Recently, several studies have explored the DG problem and reported significant performance drops in the retinal vessel segmentation [8,9]. Similarly, in DR grading, previous works showed a significant decrease in performance when presented with unseen domains and attempted to solve this problem through the perspective of feature disentanglement [10] and domain-invariant feature learning [11]. Although these methods have improved performance towards unseen domains, they may not be effective in more complex real-world scenarios because they attribute the generalization issue only to limited domain shifts, such as simple visual discrepancies. However, the generalization issues across domains cannot be solely attributed to visual discrepancies [12].

In contrast to previous works, we argue that three factors contribute to poor generalization in DGDR: visual and degradation style shifts, diagnostic pattern diversity, and data imbalance. Specifically, as shown in Fig. 1, we first conduct a preliminary analysis of three public datasets/domains. First, style shifts arise due to various factors, not only limited to visual style discrepancies, but also factors such as variations in lighting conditions [13], image resolution [14], or the presence of artifacts or noise [15]. These factors have been neglected in previous works yet are essential for building a generalizable model. Second, domains may contain diverse diagnostic patterns, such as variations in lesion types, distribution, combination and severity in certain categories [16,17]. This diversity makes learning a generalizable model towards unseen domains challenging because they may contain partially-overlapped or even unknown diagnostic patterns. Finally, data

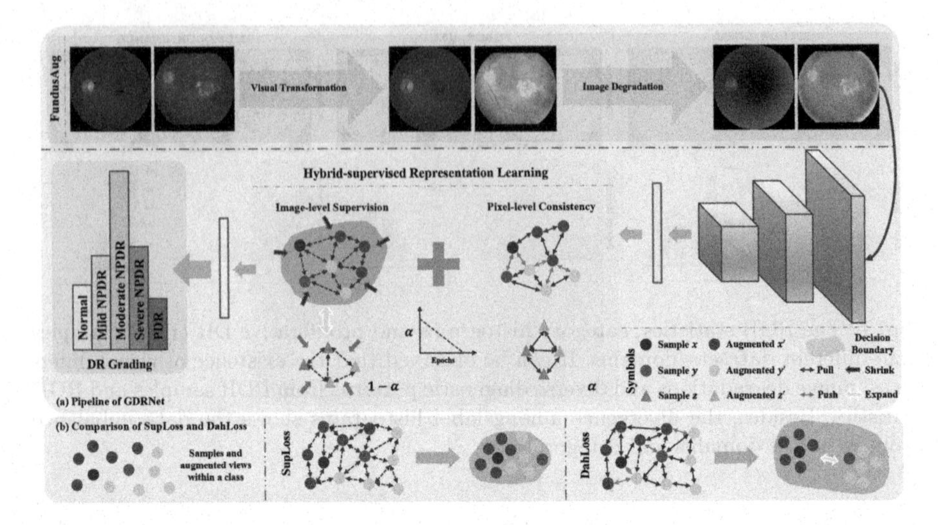

Fig. 2. The pipeline of GDRNet and a high-level visual understanding of DahLoss. FundusAug generates diverse, realistic augmented views, then leverages DahLoss to preserve pixel-level diagnostic patterns and learn generalizable features with sufficient intra-class variations. Moreover, DCR is applied to prevent minority classes from being underrepresented.

imbalance across categories and domains causes samples from specific datasets and minority classes to be underrepresented. Moreover, this imbalance can exacerbate the issue of omitting rare diagnostic patterns, leading to shortcuts in learning and poor generalization [18].

In this paper, we propose a novel framework, Generalizable Diabetic Retinopathy Grading Network (GDRNet) to address the DGDR problem. Our framework consists of three critical components: fundus visual-artifact augmentation (FundusAug), dynamic hybrid-supervised loss (DahLoss), and domain-class-aware re-balancing (DCR). By simulating visual transformations and image degradations, FundusAug enables the model to learn robust features that are less sensitive to style shifts caused by factors such as lighting conditions or artifacts and noise. DahLoss employs a hybrid-supervised learning paradigm to handle diagnostic pattern diversity and dynamically balances the influence of supervised and unsupervised learning. Jointly functioning with FundusAug, it enables the model to preserve pixel-level diagnostic information and learn generalizable features with sufficient intra-class variations. Furthermore, DCR assigns soft-balanced weights to each domain-class pair to prevent underrepresentation caused by data imbalance while avoiding undesired over-emphasis introduced by hard weighting. Finally, to evaluate generalization ability, we design a publicly available benchmark named Generalizable Diabetic Retinopathy Grading Benchmark (GDRBench), comprising eight popular datasets and two evaluation settings.

2 Methodology

An overview of GDRNet is shown in Fig. 2. It addresses the mentioned generalization issues, including style shifts, diagnostic pattern diversity, and domain-class data imbalance, by the proposed FundusAug, DahLoss, and DCR, respectively. Overall, GDRNet provides a unified solution to improve the generalization ability in unseen domains. This section will introduce each component in detail.

Fundus Visual-Artifact Augmentation. The external machine and internal retinal illumination conditions can cause differences in visual attributes [13], such as contrast and brightness, while image degradations, like spot artifacts and blurring, are also very common in fundus imaging [15,19], as depicted in Fig. 1. To bridge the style shifts caused by visual discrepancies and image degradations, we developed FundusAug, which is parameter-free and plug-and-use, designed to generate diverse and realistic augmented views. It employs five basic image transformations, including brightness, contrast, saturation, hue, and sharpness adjustments, to fill visual discrepancy gaps. Moreover, it uses extra four degradation-based image transformations, including halo simulation, hole generation, spot addition, and image blur, to address image degradation gaps. Specifically, given training data x with label y, the augmented view $\hat{x}$ of original data can be derived through the following equation:

$$\hat{x} := \text{FundusAug}(x) = \pi^{n}_{p_n,m_n}(\pi^{n-1}_{p_{n-1},m_{n-1}}(...\pi^{1}_{p_1,m_1}(x))), \tag{1}$$

where $\pi^{n}_{p_n,m_n}$ denotes transformation n with probability p_n and random intensity m_n. To reduce the parameter space of FundusAug while still ensuring image diversity, we implemented FundusAug by applying each operation with a parameter-free procedure that uniformly selects a process with a probability of 0.5. FundusAug can generate realistic augmented views while preserving their diagnostic semantics, as well as providing a robust foundation for subsequent generalizable feature learning by increasing image diversity. A detailed description and visualization of operations can be found in the appendix.

Dynamic Hybrid-Supervised Loss. While the supervised loss (SupLoss), e.g., the cross-entropy loss (CE), effectively guides the model to learn effective feature representations [20], it has two disadvantages in DGDR. First, SupLoss leads to dense features within categories while sufficient intra-class variations are crucial for effectively generalizing to unseen domains [21]. Second, the potential variety of diagnostic patterns in unseen domains requires the models to learn pixel-level lesion semantics as much as possible, while SupLoss lacks such functionality [22]. To tackle these issues, we proposed DahLoss to encourage models to learn features with sufficient intra-class variations and preserve diagnostic patterns, by introducing a hybrid-supervised paradigm to jointly leverage image-level severity supervision and pixel-level semantics consistency. A basic form of DahLoss is as

$$\mathcal{L}_{dhl} = (1-\alpha)\mathcal{L}_{sup} + \alpha\mathcal{L}_{scon}, \tag{2}$$

where α decreasing within range $[0, 1]$ is a hyper-parameter to dynamically control the task focus, and $\mathcal{L}_{sup}$ and $\mathcal{L}_{scon}$ could be any supervised and self-supervised contrastive loss functions. In this paper, we adopt CE and instance discrimination loss [23] as $\mathcal{L}_{sup}$ and $\mathcal{L}_{scon}$. Specifically, within a multiviewed batch, let $i \in I \equiv \{1...N\}$ be the index of arbitrary augmented samples by FundusAug and $j(i) \in J \equiv \{1'...N'\}$ be the index of weakly-augmented samples originating from the same source sample, we denote $\mathcal{L}^i_{dhs}$ for sample i as

$$\mathcal{L}^i_{dhl} = -(1-\alpha)\log p^i_t - \alpha \log \frac{\exp(f_i \cdot f_{j(i)}/\tau)}{\sum_{a \in A(i)} \exp(f_i \cdot f_a)/\tau)}, \quad (3)$$

where p^i_t denotes the predicted probability of the true class under one-hot encoding label, $f.$ denotes the l_2 normalized feature, the $\cdot$ symbol denotes the inner product, τ is the temperature parameter and $A(i) \equiv (J \cup I)/i$.

As illustrated in Fig. 2 (b), the use of $\mathcal{L}_{sup}$ alone tends to force samples within the same class to cluster tightly together, resulting in a highly concentrated feature representation that may not be beneficial for generalization to unseen domains. However, by incorporating $\mathcal{L}_{scon}$, DahLoss can achieve a balance between the intra-class variation and inter-class distance in learned feature representations. Specifically, while $\mathcal{L}_{sup}$ encourages samples within the same class to cluster together, $\mathcal{L}_{scon}$ simultaneously pulls the augmented views of these samples closer to each other and pushes them far away from those of other samples. It results in a feature representation with both sufficient intra-class variation and clear inter-class separation. Moreover, DahLoss also leverages pixel-level consistency to preserve crucial diagnostic patterns. It is achieved by enforcing the model to maintain feature representations with semantics similarity between strong-weak augmented views originating from the same sample, which ensures that critical details and structures are learned. By jointly considering both image-level supervision and pixel-level consistency, DahLoss encourages the model to learn features with sufficient intra-class variations and preserve crucial diagnostic information, improving generalization performance in unseen domains. Finally, gradually decreasing the value of α during training would guide the model to focus more on the grading task, leading to a balance between representation learning and grading performance. In this paper, we simply set α decay from 1 to 0 linearly across training epochs to verify the effectiveness of DahLoss.

Domain-Class-Aware Re-balancing. The domain-class data imbalance can result in certain categories and diagnostic patterns in specific datasets being underrepresented, leading to biased and inaccurate model predictions [12]. To complicate the situation, adopting a hard balancing style, such as weighting domain-class pairs based on the reciprocal of occurrence frequency, could potentially cause performance degradation [24]. Such decay is owing to the fact that underrepresented domain-class pairs often have a small ratio, resulting in excessively high weights to dominate gradients. Inspired by [25], we designed the DCR method to address this issue. It assigns weights to each sample based on

Table 1. Comparison with state-of-the-art approaches under the DG test.

Target	APTOS			DeepDR			FGADR			IDRID			Messidor			RLDR			Average		
Metrics	AUC	ACC	F1	AUC	ACC	F1	AUC	ACC	F1	AUC	ACC	F1	AUC	ACC	F1	AUC	ACC	F1	AUC	ACC	F1
ERM	75.0	44.4	38.9	77.0	39.5	34.3	66.2	32.0	27.1	82.3	50.0	**44.1**	79.1	60.7	43.4	75.9	36.5	35.7	75.9	43.8	37.3
DRGen [11]	79.9	58.1	40.2	83.0	38.7	34.1	69.4	41.7	24.7	**84.7**	44.6	37.4	79.0	60.1	40.5	79.5	43.1	37.0	79.3	47.7	37.3
Mixup [26]	75.3	62.6	43.2	75.3	29.0	25.2	66.7	42.3	32.3	78.8	39.0	27.6	76.7	54.7	32.6	76.9	43.6	37.7	75.0	45.2	33.1
MixStyle [27]	79.0	65.8	39.9	76.9	32.9	27.9	71.2	35.8	22.7	83.0	**51.4**	39.2	75.2	62.2	36.5	75.5	41.1	31.4	76.8	48.2	32.9
GREEN [3]	75.1	53.8	38.9	76.4	28.1	24.9	69.5	41.3	31.5	79.9	41.3	32.2	75.8	52.0	36.8	74.8	34.0	34.4	75.3	41.8	33.1
CABNet [2]	75.8	55.5	39.4	75.2	42.7	31.8	73.2	43.7	34.8	79.2	44.8	37.3	74.2	56.1	34.1	75.8	37.0	35.6	75.6	46.6	35.5
DDAIG [28]	78.0	**67.1**	41.0	75.6	37.6	32.2	73.6	42.0	33.8	82.1	37.4	27.0	76.6	58.4	35.3	75.6	36.1	27.7	76.9	46.4	32.8
ATS [29]	77.1	56.9	38.3	79.4	36.1	31.6	74.7	**46.7**	33.4	83.0	41.5	34.9	77.2	64.7	35.8	76.5	37.4	34.9	78.0	47.2	34.8
Fishr [30]	79.2	66.6	43.4	81.1	48.1	34.4	73.3	44.4	34.4	82.7	40.3	27.6	76.4	**65.1**	41.1	77.4	36.8	34.7	78.4	50.2	35.9
MDLT [12]	77.3	57.2	41.5	80.0	39.5	36.2	74.1	45.7	29.0	81.5	44.2	35.4	75.4	58.9	36.9	75.7	37.6	35.0	77.3	47.2	35.7
GDRNet	**79.9**	66.8	**46.0**	**84.7**	**53.1**	**45.3**	**80.8**	45.3	**39.4**	84.0	40.3	35.9	**83.2**	63.4	**50.9**	**82.9**	**45.8**	**43.5**	**82.6**	**52.5**	**43.5**

the occurrence probability q_c^d of its category c and domain d. Specifically, DCR calculates the weight w_c^d for category c in domain d as

$$w_c^d = \frac{\sum_{d' \in D} \sum_{j=1}^{N} (q_j^{d'})^\beta}{(q_c^d)^\beta}, \tag{4}$$

where D denotes the set of domains, N is the amount of classes, and β with a range of $[0, 1]$ is a hyperparameter that adjusts the balancing intensity. When β approaches to 0, DCR assigns weight more equally, and when β closes to 1, it acts more like the naive hard balancing method. By introducing β, DCR enables more nuanced weighting of samples based on their domain and class, reducing the risk of over-emphasizing underrepresented samples in the loss function. By considering the occurrence probabilities of all categories across all domains, DCR mitigates the data imbalance problem of underrepresented class-domain pairs. These two make DCR an effective solution for handling domain-class data imbalance and improving the generalizability of DR grading models.

3 Experiments

Experimental Settings, Implementation Details and Evaluation Metrics. To comprehensively analyze and evaluate our framework, we designed the GDRBench involving two generalization ability evaluation settings and eight popular public datasets. First, GDRBench preserves the classic leave-one-domain-out protocol (DG test), which requires leaving one domain for evaluation and training models on the rest. It involves six datasets, including DeepDR [16], Messidor [17], IDRID [31], APTOS [32], FGADR [33], and RLDR [34]. Further, we designed an extreme single-domain generalization setting (ESDG test), which follows the train-on-single-domain protocol with datasets mentioned above but adds two extra large-scale datasets, DDR [35] and EyePACS [36] for evaluation. It simulates real-world generalization issues, in which models are trained only on thousands of samples but are required to generalize well on one hundred thousand images from multiple hospitals and areas. We used ResNet50 pre-trained on ImageNet as the backbone and a fully connected layer as the linear classifier.

Table 2. Ablation studies on proposed components under the DG test.

Method	VT	ID	DCR	$\mathcal{L}_{dhl}$	APTOS	DeepDR	FGADR	IDRiD	Messidor	RLDR	Average
ERM	–	–	–	–	75.04	77.02	66.19	82.31	79.10	75.86	75.92
Model A	✓	–	–	–	77.16	80.22	74.87	82.83	80.48	80.23	79.30
Model B	–	✓	–	–	75.38	79.13	68.35	82.72	77.94	78.93	77.08
Model C	–	–	✓	–	75.58	78.76	67.34	82.59	78.86	78.39	76.92
Model D	–	–	–	✓	77.67	81.22	75.34	81.98	79.09	80.57	79.31
Model E	✓	✓	–	–	77.28	83.41	77.44	83.63	80.22	80.82	80.46
Model F	✓	✓	✓	–	77.37	83.90	77.91	83.44	82.43	83.33	81.39
Model G	✓	✓	–	✓	80.40	85.07	78.65	83.68	84.91	81.97	82.45
GDRNet	✓	✓	✓	✓	77.67	84.69	80.78	84.01	83.16	82.91	82.57

For evaluation, we report three critical metrics, namely accuracy (ACC), the area under the ROC curve (AUC), and macro F1-score (F1). We used **bold** and underline to indicate the first- and second-highest scores. A detailed illustration of the datasets and implementation settings can be found in the appendix.

Comparison with Other Methods. We conducted a comprehensive experiment to evaluate our framework, comparing it with a vanilla baseline (ERM) and other state-of-the-art methods from various categories, including ophthalmic disease diagnosis (OSD) [2,3], domain generalization techniques (DGT) [11,26–29], and feature representation learning (FRL) [12,30]. These chosen methods can be adopted in DGDR with minimal or no modifications, and their brief descriptions are in the appendix. We employed a standard DG augmentation pipeline [12] for all methods except DRGen with a default augmentation strategy [11]. Table 1 shows the quantitative results of the DG test, where the row of "target" indicates the test domain. GDRNet performs better than other methods and significantly improves at least two of AUC, ACC, and F1 on all sub-tests except the test on IDRiD, whose small scale leads to unobvious diagnostic pattern diversity. Typically, ODS methods do not consider domain shifts, and they thus fail to generalize as well as other methods. As expected, DGT and FRL methods consistently improve the performance compared with ERM due to specific designs towards limited domain shifts. However, GDRNet still outperforms these methods markedly due to it handles three-fold generalization issues via increasing training diversity via realistic augmentation, learning generalizable features with sufficient intra-class variations, preserving diagnostic patterns, and rebalancing the minority of samples. Overall, the result shows the effectiveness of GDRNet.

Ablation Study of Proposed Components. To evaluate the effectiveness of our proposed components, we conducted an extensive ablation study under the DG test and presented the AUC score achieved by different models in Table 2. We first examined the individual effects of FundusAug with only visual transformation (VT), FundusAug with only image degradation (ID), DCR and DahLoss

Table 3. Comparison with state-of-the-art approaches under the ESDG test.

Source	APTOS			DeepDR			FGADR			IDRID			Messidor			RLDR			Average		
Metrics	AUC	ACC	F1	AUC	ACC	F1	AUC	ACC	F1	AUC	ACC	F1	AUC	ACC	F1	AUC	ACC	F1	AUC	ACC	F1
ERM	66.4	53.2	31.6	70.7	47.3	31.2	55.3	5.6	7.1	69.6	56.5	33.9	70.6	51.3	33.7	70.1	27.3	26.4	67.1	40.2	27.3
DRGen [11]	69.4	60.7	**35.7**	**78.5**	39.4	31.6	59.8	6.8	8.4	70.8	67.7	30.6	77.0	64.5	37.4	78.9	19.0	21.2	72.4	43.0	27.5
Mixup [26]	65.5	49.4	30.2	70.7	49.7	33.3	58.8	5.8	7.4	70.2	64.0	32.6	71.5	63.0	32.6	72.9	27.7	27.0	68.3	43.3	27.2
MixStyle [27]	62.0	48.8	25.0	53.3	32.0	14.6	51.0	7.0	7.9	53.0	53.5	19.4	51.4	57.6	16.8	53.5	18.3	6.4	54.0	36.2	15.0
GREEN [3]	67.5	52.6	33.3	71.2	44.6	31.1	58.1	5.7	6.9	68.5	60.7	33.0	71.3	54.5	33.1	71.0	31.9	27.8	67.9	41.7	27.5
CABNet [2]	67.3	52.2	30.8	70.0	55.4	32.0	57.1	6.1	7.5	67.4	62.7	31.7	72.3	63.8	35.3	75.2	23.0	25.4	68.2	43.8	27.2
DDAIG [28]	67.4	48.7	31.6	73.2	38.5	29.7	59.9	5.0	5.5	70.2	60.2	33.4	73.5	**69.1**	35.6	74.4	25.4	23.5	69.8	41.2	26.7
ATS [29]	68.8	51.7	32.4	72.7	52.4	33.5	60.3	5.3	5.7	69.1	66.6	30.6	73.4	64.8	32.4	75.0	24.2	23.9	69.9	44.2	26.4
Fishr [30]	64.5	**61.7**	31.0	72.1	**61.0**	30.1	56.3	6.0	7.2	71.8	48.0	30.6	74.3	52.0	33.8	78.6	19.3	21.3	69.6	41.3	25.7
MDLT [12]	67.6	53.3	32.4	73.1	50.2	33.7	57.1	7.1	7.8	71.9	61.7	32.4	73.4	58.9	34.1	76.6	29.0	30.0	70.0	43.4	28.4
GDRNet	**69.8**	52.8	35.2	76.1	40.0	**35.0**	**63.7**	**7.5**	**9.2**	**72.9**	**70.0**	**35.1**	**78.1**	65.7	**40.5**	**79.7**	**44.3**	**37.9**	**73.4**	**46.7**	**32.2**

(Models A-D, respectively). Our results demonstrate that all proposed components effectively improve model generalization performance and address the three-fold generalization issues discussed in this paper. We then combined these components to analyze their joint effects (Models E-G). Notably, DahLoss and FundusAug contribute significantly to the improvement, demonstrating their capability to increase the training diversity, handle diagnostic pattern diversity, and dynamically balance the influence of supervised and unsupervised learning. Finally, we achieved the best performance by combining all the components in GDRNet, showing that our components play complementary roles. It is worth mentioning that we can observe different performance trends across the various datasets in the ablation study, which indicates the complexity of generalization issues in DGDR. It also suggests that different domains may have varying grades of severity in the three-fold generalization issues identified in this paper.

Generalization from a Single Source Domain. Further, to comprehensively investigate the generalization performance, we introduced the ESDG test. It is a more demanding evaluation setting than the DG test because it requires models trained on a single source dataset to generalize to new domains with significantly larger scales and different data distributions. In contrast, the DG test involves training on multiple domains to fill domain gaps naturally. The quantitative results are presented in Table 3, where the row of "source" indicates the training dataset. As expected, the average performance of all methods decreases significantly, indicating the difficulty of ESDG. Although some methods outperform others in the DG test, such as MixStyle, due to their specific design to leverage the discrepancy of domains, they fail to generalize in the strict ESDG test. Despite more strict requirements, GDRNet still outperforms other methods in average performance and at least one metric in all sub-tests, owing to its effective designs towards three-fold generalization issues.

4 Conclusion

In this paper, we tackled the three-fold generalization issues that hinder the generalizability of DR grading, including style shifts, diagnostic pattern diversity, and data imbalance. To overcome these challenges, we proposed a novel and unified framework called GDRNet, incorporating three effective components: FundusAug, DahLoss, and DCR. FundusAug enables the generation of realistic augmented views, while DahLoss leverages supervised and unsupervised learning to preserve diagnostic patterns and increase intra-class variations of features. Finally, DCR softly handles the data imbalance across categories and domains to avoid potential performance decay. Together, these three components work synergistically to improve the generalization performance of the model. GDRNet achieved superior performance on both DG and ESDG tests of the proposed publicly available GDRBench, demonstrating its effectiveness and robustness in addressing the three-fold generalization issues in DR grading. Overall, our work provides valuable insights and practical solutions for improving the generalization capability of deep learning in medical image analysis, and has the potential to benefit real-world clinical applications.

Acknowledgement. This work was supported by the Hong Kong Innovation and Technology Fund (Project No. ITS/028/21FP), Shenzhen Science and Technology Innovation Committee Fund (Project No. SGDX20210823103201011) and HKUST 30 for 30 Research Initiative Scheme.

References

1. Cho, N.H., Shaw, J., Karuranga, S., Huang, Y., da Rocha Fernandes, J., et al.: IDF diabetes atlas: global estimates of diabetes prevalence for 2017 and projections for 2045. Diabetes Res. Clin. Pract. **138**, 271–281 (2018)
2. He, A., Li, T., Li, N., Wang, K., Fu, H.: CabNet: category attention block for imbalanced diabetic retinopathy grading. IEEE Trans. Med. Imaging **40**(1), 143–153 (2020)
3. Liu, S., Gong, L., Ma, K., Zheng, Y.: GREEN: a graph REsidual rE-ranking network for grading diabetic retinopathy. In: Martel, A.L., et al. (eds.) MICCAI 2020. LNCS, vol. 12265, pp. 585–594. Springer, Cham (2020). https://doi.org/10.1007/978-3-030-59722-1_56
4. Beede, E., et al.: A human-centered evaluation of a deep learning system deployed in clinics for the detection of diabetic retinopathy. In: Proceedings of the 2020 CHI Conference on Human Factors in Computing Systems, pp. 1–12 (2020)
5. Heaven, W.D.: Google's medical AI was super accurate in a lab. Real life was a different story. MIT Technol. Rev. **4**, 27 (2020)
6. Wang, J., et al.: Generalizing to unseen domains: a survey on domain generalization. IEEE Trans. Knowl. Data Eng. (2022)
7. Li, T., et al.: Applications of deep learning in fundus images: a review. Med. Image Anal. **69**, 101971 (2021)
8. Wang, S., Yu, L., Li, K., Yang, X., Fu, C.W., Heng, P.A.: DoFE: domain-oriented feature embedding for generalizable fundus image segmentation on unseen datasets. IEEE Trans. Med. Imaging **39**(12), 4237–4248 (2020)
9. Lyu, J., Zhang, Y., Huang, Y., Lin, L., Cheng, P., Tang, X.: AADG: automatic augmentation for domain generalization on retinal image segmentation. IEEE Trans. Med. Imaging **41**(12), 3699–3711 (2022)

10. Che, H., Jin, H., Chen, H.: Learning robust representation for joint grading of ophthalmic diseases via adaptive curriculum and feature disentanglement. In: Wang, L., Dou, Q., Fletcher, P.T., Speidel, S., Li, S. (eds.) International Conference on Medical Image Computing and Computer-Assisted Intervention. pp. 523–533. Springer, Cham (2022). https://doi.org/10.1007/978-3-031-16437-8_50
11. Atwany, M., Yaqub, M.: DRGen: domain generalization in diabetic retinopathy classification. In: Wang, L., Dou, Q., Fletcher, P.T., Speidel, S., Li, S. (eds.) International Conference on Medical Image Computing and Computer-Assisted Intervention, pp. 635–644. Springer, Cham (2022). https://doi.org/10.1007/978-3-031-16434-7_61
12. Yang, Y., Wang, H., Katabi, D.: On multi-domain long-tailed recognition, imbalanced domain generalization and beyond. In: Avidan, S., Brostow, G., Cissé, M., Farinella, G.M., Hassner, T. (eds.) European Conference on Computer Vision. pp. 57–75. Springer, Cham (2022). https://doi.org/10.1007/978-3-031-20044-1_4
13. Shen, Z., Fu, H., Shen, J., Shao, L.: Modeling and enhancing low-quality retinal fundus images. IEEE Trans. Med. Imaging **40**(3), 996–1006 (2020)
14. Wang, X., Xu, M., Zhang, J., Jiang, L., Li, L.: Deep multi-task learning for diabetic retinopathy grading in fundus images. In: Proceedings of the AAAI Conference on Artificial Intelligence, vol. 35, pp. 2826–2834 (2021)
15. Liu, H., Li, H., Wang, X., Li, H., et al.: Understanding how fundus image quality degradation affects CNN-based diagnosis. In: 44th Annual International Conference of the IEEE Engineering in Medicine & Biology Society, pp. 438–442. IEEE (2022)
16. Liu, R., Wang, X., Wu, Q., Dai, L., Fang, X., et al.: DeepDRiD: diabetic retinopathy-grading and image quality estimation challenge. Patterns **3**(6), 100512 (2022)
17. Abràmoff, M.D., Lou, Y., Erginay, A., et al.: Improved automated detection of diabetic retinopathy on a publicly available dataset through integration of deep learning. Invest. Ophthalmol. Vis. Sci. **57**(13), 5200–5206 (2016)
18. Geirhos, R., Jacobsen, J.H., Michaelis, C., Zemel, R., et al.: Shortcut learning in deep neural networks. Nat. Mach. Intell. **2**(11), 665–673 (2020)
19. Che, H., Chen, S., Chen, H.: Image quality-aware diagnosis via meta-knowledge co-embedding. In: Proceedings of the IEEE/CVF Conference on Computer Vision and Pattern Recognition, pp. 19819–19829 (2023)
20. Zhang, Y., Hooi, B., Hu, D., Liang, J., Feng, J.: Unleashing the power of contrastive self-supervised visual models via contrast-regularized fine-tuning. Adv. Neural. Inf. Process. Syst. **34**, 29848–29860 (2021)
21. Duboudin, T., Dellandréa, E., et al.: Encouraging intra-class diversity through a reverse contrastive loss for single-source domain generalization. In: Proceedings of the IEEE/CVF International Conference on Computer Vision, pp. 51–60 (2021)
22. Islam, A., Chen, C.F.R., Panda, R., et al.: A broad study on the transferability of visual representations with contrastive learning. In: Proceedings of the IEEE/CVF International Conference on Computer Vision, pp. 8845–8855 (2021)
23. Chen, T., Kornblith, S., Norouzi, M., Hinton, G.: A simple framework for contrastive learning of visual representations. In: International Conference on Machine Learning, pp. 1597–1607. PMLR (2020)
24. Ren, M., Zeng, W., Yang, B., Urtasun, R.: Learning to reweight examples for robust deep learning. In: International Conference on Machine Learning, pp. 4334–4343. PMLR (2018)
25. Conneau, A., Lample, G.: Cross-lingual language model pretraining. In: Advances in Neural Information Processing Systems, vol. 32 (2019)

26. Zhang, H., Cisse, M., Dauphin, Y.N., Lopez-Paz, D.: mixup: beyond empirical risk minimization. In: International Conference on Learning Representations (2017)
27. Zhou, K., Yang, Y., Qiao, Y., Xiang, T.: Domain generalization with MixStyle. In: International Conference on Learning Representations
28. Zhou, K., Yang, Y., Hospedales, T., Xiang, T.: Deep domain-adversarial image generation for domain generalisation. In: Proceedings of the AAAI Conference on Artificial Intelligence, vol. 34, pp. 13025–13032 (2020)
29. Yang, F.E., Cheng, Y.C., Shiau, Z.Y., Wang, Y.C.F.: Adversarial teacher-student representation learning for domain generalization. Adv. Neural. Inf. Process. Syst. **34**, 19448–19460 (2021)
30. Rame, A., Dancette, C., Cord, M.: Fishr: invariant gradient variances for out-of-distribution generalization. In: International Conference on Machine Learning, pp. 18347–18377. PMLR (2022)
31. Porwal, P., Pachade, S., Kamble, R., Kokare, M., Deshmukh, G., et al.: Indian diabetic retinopathy image dataset (IDRiD): a database for diabetic retinopathy screening research. Data **3**(3), 25 (2018)
32. APTOS: Aptos 2019 blindness detection website. https://www.kaggle.com/c/aptos2019-blindness-detection. Accessed 20 Feb 2022
33. Zhou, Y., Wang, B., Huang, L., Cui, S., Shao, L.: A benchmark for studying diabetic retinopathy: segmentation, grading, and transferability. IEEE Trans. Med. Imaging **40**(3), 818–828 (2020)
34. Wei, Q., et al.: Learn to segment retinal lesions and beyond. In: International Conference on Pattern Recognition, pp. 7403–7410. IEEE (2021)
35. Li, T., Gao, Y., Wang, K., Guo, S., Liu, H., Kang, H.: Diagnostic assessment of deep learning algorithms for diabetic retinopathy screening. Inf. Sci. **501**, 511–522 (2019)
36. EYEPACS: Kaggle eyepacs dataset. https://paperswithcode.com/dataset/kaggle-eyepacs. Accessed 20 Feb. 2023

Boosting Breast Ultrasound Video Classification by the Guidance of Keyframe Feature Centers

Anlan Sun[1], Zhao Zhang[2], Meng Lei[2], Yuting Dai[1], Dong Wang[3], and Liwei Wang[3,4](✉)

[1] Yizhun Medical AI Co., Ltd., Beijing, China
{anlan.sun,yuting.dai}@yizhun-ai.com
[2] Center for Data Science, Peking University, Beijing, China
{zhangzh,leimeng}@stu.pku.edu.cn
[3] National Key Laboratory of General Artificial Intelligence, School of Intelligence Science and Technology, Peking University, Beijing, China
{wangdongcis,wanglw}@pku.edu.cn
[4] Center for Machine Learning Research, Peking University, Beijing, China

Abstract. Breast ultrasound videos contain richer information than ultrasound images, therefore it is more meaningful to develop video models for this diagnosis task. However, the collection of ultrasound video datasets is much harder. In this paper, we explore the feasibility of enhancing the performance of ultrasound video classification using the static image dataset. To this end, we propose KGA-Net and coherence loss. The KGA-Net adopts both video clips and static images to train the network. The coherence loss uses the feature centers generated by the static images to guide the frame attention in the video model. Our KGA-Net boosts the performance on the public BUSV dataset by a large margin. The visualization results of frame attention prove the explainability of our method. We release the code and model weights in https://github.com/PlayerSAL/KGA-Net.

Keywords: Breast ultrasound classification · Ultrasound video · Coherence loss

1 Introduction

Breast cancer is a life-threatening disease that has surpassed lung cancer as leading cancer in some countries and regions [20]. Breast ultrasound is the primary screening method for diagnosing breast cancer, and accurately distinguishing between malignant and benign breast lesions is crucial. This task is also an essential component of computer-aided diagnosis. Since each frame in an ultrasound video can only capture a specific view of a lesion, it is essential to aggregate information from the entire video to perform accurate automatic lesion diagnosis. Therefore, in this study, we focus on the classification of breast ultrasound videos for detecting malignant and benign breast lesions.

A. Sun and Z. Zhang—Equal contribution.

H. Greenspan et al. (Eds.): MICCAI 2023, LNCS 14224, pp. 441–451, 2023.
https://doi.org/10.1007/978-3-031-43904-9_43

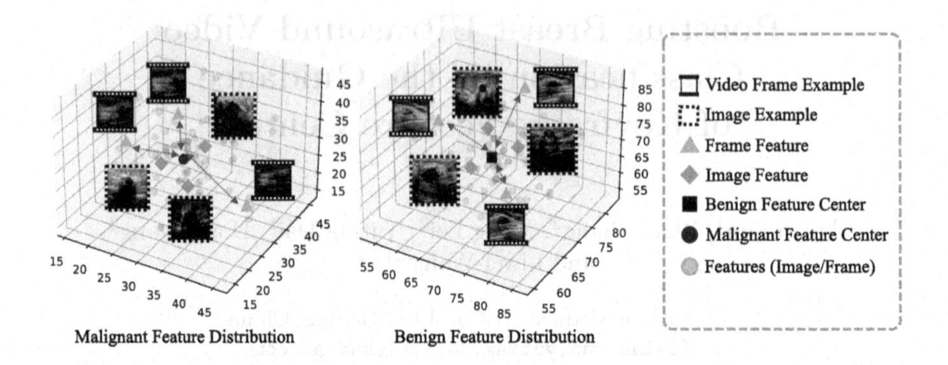

Fig. 1. Feature distribution of video frames from BUSV [15] and static images from BUSI [1]. We use a 2D ResNet trained on ultrasound images to get the features.

While ultrasound videos offer more information, prior studies have primarily focused on static image classification [2,11,27]. Obtaining ultrasound video data with pathology gold standard results poses a major challenge. Sonographers typically record keyframe images during general ultrasound examinations, not entire videos. Prospective collection requires additional efforts to track corresponding pathological results. Consequently, while there are many breast ultrasound image datasets [1,28], breast ultrasound video datasets remain scarce, with only one relatively small dataset [15] containing 188 videos available currently.

Given the difficulties in collecting ultrasound video data, we investigate the feasibility of enhancing the performance of ultrasound video classification using a static image dataset. To achieve this, we first analyze the relationship between ultrasound videos and images. The images in the ultrasound dataset are keyframes of a lesion that exhibit the clearest appearance and most typical symptoms, making them more discriminative for diagnosis. Although ultrasound videos provide more information, the abundance of frames may introduce redundancy or vagueness that could disrupt classification. From the aspect of feature distribution, as shown in Fig. 1, the feature points of static images are more concentrated, while the feature of video frames sometimes are away from the class centers. Frames far from the centers are harder to classify. Therefore, it is a promising approach to guide the video model to pay more attention to important frames close to the class center with the assistance of static keyframe images. Meanwhile, our approach aligns with the diagnosis of ultrasound physicians, automatically evaluates the importance of frames, and diagnoses based on the information of key frames. Additionally, our method provides interpretability through key frames.

In this paper, we propose a novel Keyframe Guided Attention Network (KGA-Net) to boost ultrasound video classification. Our approach leverages both image (keyframes) and video datasets to train the network. To classify videos, we use frame attention to predict feature weights for all frames and aggregate them to make the final classification. The feature weights determine the contribution of each frame for the final diagnosis. During training, we construct category feature centers for malignant and benign examples respectively using center loss [26] on static image inputs and use the centers to guide the training of video frame attention. Specifically, we propose coherence loss, which promotes the frames close to the centers to have high attention weights and decreases the weights for frames far from the centers. Due to the feature centers being generated by the larger scale image dataset, it provides more accurate and discriminative feature centers which can guide the video frame attention to focus on important frames, and finally leads to better video classification.

Our experimental results on the public BUSV dataset [15] show that our KGA-Net significantly outperforms other video classification models by using an external ultrasound image dataset. Additionally, we visualized attention values guided by the coherence loss. The frames with clear diagnostic characteristics are given higher attention values. This phenomenon makes our method more explainable and provides a new perspective for selecting keyframes from video.

In conclusion, our contributions are as follows:

1. We analyze the relationship between ultrasound video data and image data, and propose the coherence loss to use image feature centers to guide the training of frame attention.
2. We propose KGA-Net, which adopts a static image dataset to boost the performance of ultrasound video classification. KGA-Net significantly outperforms other video baselines on the BUSV dataset.
3. The qualitative analysis of the frame attention verifies the explainability of our method and provides a new perspective for selecting keyframes.

2 Related Works

Breast Ultrasound Classification. Breast ultrasound (BUS) plays an important supporting role in the diagnosis of breast-related diseases. Recent research demonstrated the potential of deep learning for breast lesion classification tasks [6,18,19,23,27]. [6,18] design ensemble methods to integrate the features of multiple models to obtain higher accuracy. [19,23,27] utilize multi-task learning to improve the model performance. However, all of them are based on image datasets, such as BUSI [1], while few works focus on the video modality. [14] design a pre-training model based on contrastive learning for ultrasound video classification. [13,25] develop a keyframe extraction model for ultrasound videos and utilized the extracted keyframes to perform various classification tasks. However, these methods rely on keyframe supervision, which limits their applicability. Fortunately, the recent publicly available dataset BUSV [15] has made the research on the task of BUS video-based classification possible. In this paper, we build our model based on this dataset.

Video Recognition Based on Neural Networks. Traditional methods are based on Two-stream networks [9,10,24]. Since I3D [3] was proposed, 3D CNNs have dominated video understanding for a long time. [21,22] decompose 3D convolution in different ways to reduce computation complexity without losing

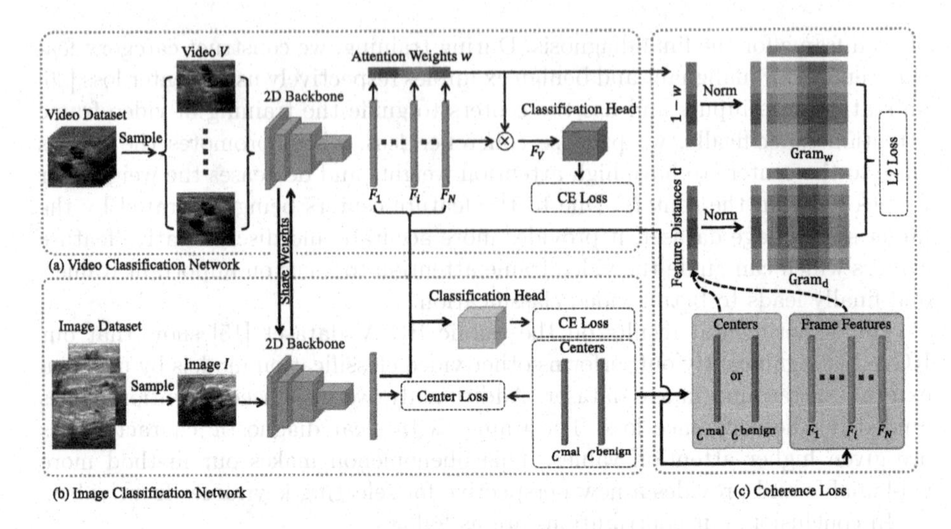

Fig. 2. Overview of our proposed keyframe-guided attention network.

performance. [8] designed two branches to focus on temporal information and spatial features, respectively. However, 3D CNNs have a limited receptive field, and thus struggle to capture long-range dependency. Vision Transformers [5,16] have become popular for their ability to aggregate spatial-temporal information. To address computational complexity, MViT [7] employed a hierarchical structure and Video Swin [17] introduced 3D shifted window attention. Our proposed KGA-Net is a simple framework that leverages the frame attention module to aggregate multi-frame features efficiently.

3 Methodology

As shown in Fig. 2, our KGA-Net takes the video inputs and static image inputs simultaneously to train the network. The coherence loss is proposed to guide the frame attention by using the feature centers generated by the images. We will then elaborate on each component in the following sections.

3.1 Video and Image Classification Network

The video classification network is illustrated in Fig. 2 (a). The model is composed of a 2D CNN backbone, a frame attention module, and a classification head. For an input video clip V composed of N frames, it is first processed by the backbone network and the feature vectors of the frames $\{F_i\}_{i=1}^{N}$ are obtained. Then, the frame attention module predicts the attention weight for each frame using a FC and sigmoid layer, and then the features are aggregated by the weights to form an integrated feature vector. Formally,

$$w_i = \text{Sigmoid}(\text{FC}(F_i)) \tag{1}$$

where w_i denotes the weight for the i_{th} frame and FC is the fully-connected layer. Then, the features are aggregated by $F_V = \sum_{i=1}^{N} w_i \cdot F_i$. Finally, the classification head is applied to the final result of lesion classification. To train the model, the cross-entropy loss (CE Loss) is applied to the classification prediction of the video.

The image classification network is used to assist in training the video model. We use the same 2D CNN as the backbone network in the video classification network. The model weights are shared for the two backbones for better generalization. To promote the formation of feature centers, we apply the center loss [26] to the image model besides the cross-entropy loss. In addition, the frame-level cross-entropy loss is also applied to the video frames to facilitate training.

3.2 Training with Coherence Loss

In this section, we introduce the coherence loss to guide the frame attention with the assistance of the category feature centers. We use the same method as center loss [26] to obtain the feature centers for the malignancy and benign lesions, which are denoted as $\mathcal{C}^{\text{mal}}$ and $\mathcal{C}^{\text{benign}}$, respectively.

The distances of frame features and the feature centers can measure the quality of the frames. The frame features close to the centers are more discriminative for the classification task. Therefore, we use these distances to guide the generation of frame attention. Specifically, we push the frames close to the centers to have higher attention weights and decrease the weights far from the centers. To do this, for each video frame with feature F_i, we first calculate the feature distance from its corresponding class center. Formally,

$$d_i = \|F_i - \mathcal{C}^{Y}\|_2, \tag{2}$$

where $Y \in \{\text{mal}, \text{benign}\}$ is the label of the video V and d_i is the computed distance of frame i.

Afterward, we apply coherence loss to the attention weights $\mathbf{w} = [w_1, w_2, ..., w_N]^\intercal$ to make them have a similar distribution with the feature distances $\mathbf{d} = [d_1, d_2, ..., d_N]^\intercal$. To supervise the distribution, the coherence loss is defined as the L2 loss of the gram matrix of these two vectors

$$\mathrm{L}_{\mathrm{Coh}} = \|\mathrm{Gram}_{\mathbf{w}} - \mathrm{Gram}_{\mathbf{d}}\|_2, \tag{3}$$

where $\mathrm{Gram}_{\mathbf{w}} = \frac{(\mathbf{1}-\mathbf{w})\cdot(\mathbf{1}-\mathbf{w})^\intercal}{\|\mathbf{1}-\mathbf{w}\|_2^2}$ is the gram matrix of normalized attention weights, and $\mathrm{Gram}_{\mathbf{d}} = \frac{\mathbf{d}\cdot\mathbf{d}^\intercal}{\|\mathbf{d}\|_2^2}$ is the gram matrix of normalized feature distances. Note that lower distances correspond to stronger attention, hence we use the opposite of $\mathbf{w}$ to get $\mathrm{Gram}_{\mathbf{w}}$.

3.3 Total Training Loss

To summarize, the total training loss of our KGA-Net

$$L_{\text{total}} = L_{\text{CE}}^{V} + L_{\text{CE}}^{I} + L_{\text{Center}} + \lambda \cdot L_{\text{Coh}}. \tag{4}$$

L_{CE}^V and L_{CE}^I denote the cross-entropy for video classification and image and frame classification. L_{Center} means the center loss. λ is the weight for coherence loss. Empirically, we set $\lambda = 1$ in our experiments.

During inference, to perform classification on video data, the video classification network can be utilized individually for prediction.

4 Experiments

4.1 Implementation Details

Datasets. We use the public BUSV dataset [15] for video classification and the BUSI dataset [1] as the image dataset. BUSV consists of 113 malignant videos and 75 benign videos. BUSI contains 445 images of benign lesions and 210 images of malignant lesions. For the BUSV dataset, we use the official data split in [15]. All images of the BUSI dataset are adopted to train our KGA-Net.

Model Details. ResNet-50 [12] pretrained on ImageNet [4] is used as backbone. We use SGD optimizer with an initial learning rate of 0.005, which is reduced by 10× at the 4,000th and 6,000th iteration. The total learning iteration number is 8,000. The learning rate warmup is used in the first 1,000 iterations. For each batch, the video clips and static images are both sampled and sent to the network. We use a total batchsize of 16 and the sample probability of video clips and images is 1:1. We implement the model based on Pytorch and train it with NVIDIA Titan RTX GPU cards.

During inference, we use the video classification network individually. In order to satisfy the fixed video length requirement of MViT [7], we sample up to 128 frames of each video to form a video clip and predict its classification result using all the models in experiments.

4.2 Comparison with Video Models

In this section, we compare our KGA-Net with other competitive video classification models. However, comparing with ultrasound-video-based work is challenging due to limited code accessibility and lack of keyframe detection model in existing methods [13,14,25]. Therefore, we compare our method with strong video baselines on natural images. We include CNN-based models, I3D [3], SlowFast [8], R(2+1)D [22], and CSN [21], along with the popular transformer-based model MViT [7]. For fairness comparison, we train these models using both video and image data, treating images as static videos. Evaluation metrics are reported on the BUSV test set for performance assessment.

As shown in Table 1, by leveraging the guidance of the image dataset, our KGA-Net significantly surpasses all other models on all of the metrics. The video classification model of our KGA-Net is composed of a standard 2D ResNet-50 and a light feature attention module, while the baseline models are with net structures carefully designed for video analysis. Therefore, the success of our KGA-Net lies in the correct usage of the image guidance. The feature centers

Table 1. Comparison with other video models. Classification thresholds are determined by Youden index.

Model	AUC(%)	ACC(%)	Sensitivity(%)	Specificity(%)
I3D [3]	88.31	81.58	84.00	76.92
SlowFast [8]	82.54	79.49	76.92	84.62
R(2+1)D [22]	86.46	81.58	84.00	76.92
CSN [21]	83.38	81.58	84.00	76.92
MViT [7]	90.53	82.05	80.77	84.62
KGA-Net (Our)	**94.67**	**89.74**	**88.46**	**92.31**

formed by the image dataset with larger data size and clear appearance effectively improve the accuracy of frame attention hence boosting the video classification performance.

4.3 Ablation Study

In this section, we ablate the contribution of each key design in our KGA-Net. We observe their importance by removing these key components from the whole network. The results are shown in Table 2. The results of KGA-Net are shown in the last row in Table 2, while the components are ablated in the first three rows. We use the same training schedule for all of the experiments.
Image guidance is the main purpose of our method. To portray the effect of using the image dataset, we train the KGA-Net using BUSV dataset alone in the first row of Table 2. Without the image dataset, we generate the feature centers from the video frames. As a result, the performance significantly drops due to the decrease in dataset scale. It also shows that the feature centers generated by the image dataset are more discriminative than that of the video dataset. It is not only because the lesion number of BUSI is larger than BUSV, but also because the images in BUSI are all the keyframes that contain typical characteristics of lesions.
Frame attention and coherence loss are two essential modules of our KGA-Net. We train a KGA-Net without the coherence loss in the third row of Table 2. In the second row, we further replace the feature attention module with feature averaging of video frames. It can be seen that both of these two modules contribute to the overall performance according to AUC and ACC. It is worth noting that these two models without coherence loss obtain very low sensitivity and high specificity, which means the model predictions are imbalanced and intend to make benign predictions. It is because that clear malignant appearances usually only exist in limited frames in a malignant video. Without our coherence loss or frame attention, it is difficult for the model to focus on typical frames that possess malignant features. This phenomenon certifies the effectiveness of our KGA-Net to prevent false negatives in diagnosis.

Table 2. Ablation studies. Model components are removed in the first three lines to analyze their contributions in KGA-Net. Classification thresholds are determined by Youden index.

Model	AUC(%)	ACC(%)	Sensitivity(%)	Specificity(%)
w/o image guidance	85.21	76.92	73.08	84.62
w/o coherence loss & attention	88.17	74.36	61.54	**100.0**
w/o coherence loss	92.90	87.18	80.77	**100.0**
KGA-Net	**94.67**	**89.74**	**88.46**	92.31

4.4 Visual Analysis

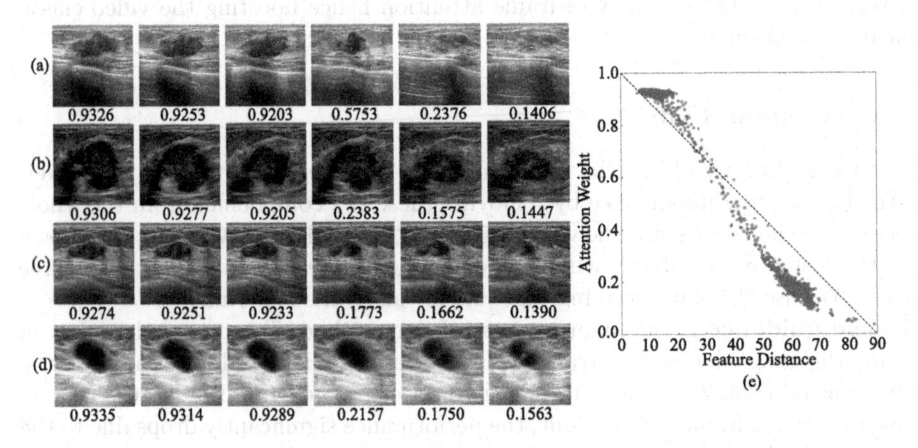

Fig. 3. Visual Analysis. (a–d) Visualization of video frames and corresponding frame attention weights. (e) Relationship between attention weight and feature distance.

In Fig. 3, we illustrate video frames with their corresponding frame attention weights predicted by KGA-Net. Overall speaking, the frames with high attention weights do have clear image appearances for diagnosis. For example, the first three frames in Fig. 3(b) clearly demonstrate the edge micro-lobulation and irregular shapes, which lead to malignant judgment. Furthermore, we plot the relationships between the predicted attention values and the feature distances to the centers. As shown in Fig. 3(e), these two variables are linearly related, which indicates that KGA-Net the attention weights are effectively guided by the feature distances.

The qualitative analysis proves the interpretability of our method, which will benefit clinical usage. Moreover, the attention weights reveal the importance of each frame for lesion diagnosis. Therefore, it can provide a new perspective for the keyframe extraction task of ultrasound videos.

5 Conclusion

We propose KGA-Net, a novel video classification model for breast ultrasound diagnosis. Our KGA-Net takes as input both the video data and image data to train the network. We propose the coherence loss to guide the training of the video model by the guidance of feature centers of the images. Our method significantly exceeds the performance of other competitive video baselines. The visualization of the attention weights validates the effectiveness and interpretability of our KGA-Net.

Acknowledgements. This work is supported by National Key R&D Program of China (2022ZD0114900) and National Science Foundation of China (NSFC62276005).

References

1. Al-Dhabyani, W., Gomaa, M., Khaled, H., Fahmy, A.: Dataset of breast ultrasound images. Data Brief **28**, 104863 (2020)
2. Byra, M.: Breast mass classification with transfer learning based on scaling of deep representations. Biomed. Signal Process. Control **69**, 102828 (2021)
3. Carreira, J., Zisserman, A.: Quo vadis, action recognition? A new model and the kinetics dataset. In: proceedings of the IEEE Conference on Computer Vision and Pattern Recognition, pp. 6299–6308 (2017)
4. Deng, J., Dong, W., Socher, R., Li, L.J., Li, K., Fei-Fei, L.: ImageNet: a large-scale hierarchical image database. In: 2009 IEEE Conference on Computer Vision and Pattern Recognition, pp. 248–255. IEEE (2009)
5. Dosovitskiy, A., et al.: An image is worth 16x16 words: transformers for image recognition at scale. arXiv preprint arXiv:2010.11929 (2020)
6. Eroğlu, Y., Yildirim, M., Çinar, A.: Convolutional neural networks based classification of breast ultrasonography images by hybrid method with respect to benign, malignant, and normal using mRMR. Comput. Biol. Med. **133**, 104407 (2021)
7. Fan, H., et al.: Multiscale vision transformers. In: Proceedings of the IEEE/CVF International Conference on Computer Vision, pp. 6824–6835 (2021)
8. Feichtenhofer, C., Fan, H., Malik, J., He, K.: SlowFast networks for video recognition. In: Proceedings of the IEEE/CVF International Conference on Computer Vision, pp. 6202–6211 (2019)
9. Feichtenhofer, C., Pinz, A., Wildes, R.: Spatiotemporal residual networks for video action recognition. In: Advances in Neural Information Processing Systems (NIPS), pp. 3468–3476 (2016)
10. Feichtenhofer, C., Pinz, A., Zisserman, A.: Convolutional two-stream network fusion for video action recognition. In: Conference on Computer Vision and Pattern Recognition (CVPR) (2016)
11. Gheflati, B., Rivaz, H.: Vision transformers for classification of breast ultrasound images. In: 2022 44th Annual International Conference of the IEEE Engineering in Medicine & Biology Society (EMBC), pp. 480–483. IEEE (2022)

12. He, K., Zhang, X., Ren, S., Sun, J.: Deep residual learning for image recognition. In: Proceedings of the IEEE Conference on Computer Vision and Pattern Recognition, pp. 770–778 (2016)
13. Huang, R., et al.: Extracting keyframes of breast ultrasound video using deep reinforcement learning. Med. Image Anal. **80**, 102490 (2022)
14. Lin, Z., Huang, R., Ni, D., Wu, J., Luo, B.: Masked video modeling with correlation-aware contrastive learning for breast cancer diagnosis in ultrasound. In: Xu, X., Li, X., Mahapatra, D., Cheng, L., Petitjean, C., Fu, H. (eds.) Resource-Efficient Medical Image Analysis: First MICCAI Workshop, REMIA 2022, Singapore, 22 September 2022, Proceedings, pp. 105–114. Springer, Cham (2022). https://doi.org/10.1007/978-3-031-16876-5_11
15. Lin, Z., Lin, J., Zhu, L., Fu, H., Qin, J., Wang, L.: A new dataset and a baseline model for breast lesion detection in ultrasound videos. In: Wang, L., Dou, Q., Fletcher, P.T., Speidel, S., Li, S. (eds.) Medical Image Computing and Computer Assisted Intervention-MICCAI 2022: 25th International Conference, Singapore, 18–22 September 2022, Proceedings, Part III, pp. 614–623. Springer, Cham (2022). https://doi.org/10.1007/978-3-031-16437-8_59
16. Liu, Z., et al.: Swin transformer: hierarchical vision transformer using shifted windows. In: Proceedings of the IEEE/CVF International Conference on Computer Vision, pp. 10012–10022 (2021)
17. Liu, Z., et al.: Video swin transformer. In: Proceedings of the IEEE/CVF Conference on Computer Vision and Pattern Recognition, pp. 3202–3211 (2022)
18. Moon, W.K., Lee, Y.W., Ke, H.H., Lee, S.H., Huang, C.S., Chang, R.F.: Computer-aided diagnosis of breast ultrasound images using ensemble learning from convolutional neural networks. Comput. Methods Programs Biomed. **190**, 105361 (2020)
19. Podda, A.S., Balia, R., Barra, S., Carta, S., Fenu, G., Piano, L.: Fully-automated deep learning pipeline for segmentation and classification of breast ultrasound images. J. Comput. Sci. **63**, 101816 (2022)
20. Siegel, R.L., et al.: Colorectal cancer statistics, 2017. CA: Cancer J. Clin. **67**(3), 177–193 (2017)
21. Tran, D., Wang, H., Torresani, L., Feiszli, M.: Video classification with channel-separated convolutional networks. In: Proceedings of the IEEE/CVF International Conference on Computer Vision, pp. 5552–5561 (2019)
22. Tran, D., Wang, H., Torresani, L., Ray, J., LeCun, Y., Paluri, M.: A closer look at spatiotemporal convolutions for action recognition. In: Proceedings of the IEEE conference on Computer Vision and Pattern Recognition, pp. 6450–6459 (2018)
23. Wang, J., et al.: Information bottleneck-based interpretable multitask network for breast cancer classification and segmentation. Med. Image Anal. **83**, 102687 (2023)
24. Wang, L., et al.: Temporal segment networks: towards good practices for deep action recognition. In: Leibe, B., Matas, J., Sebe, N., Welling, M. (eds.) ECCV 2016. LNCS, vol. 9912, pp. 20–36. Springer, Cham (2016). https://doi.org/10.1007/978-3-319-46484-8_2
25. Wang, Y., et al.: Key-frame guided network for thyroid nodule recognition using ultrasound videos. In: Wang, L., Dou, Q., Fletcher, P.T., Speidel, S., Li, S. (eds.) Medical Image Computing and Computer Assisted Intervention-MICCAI 2022: 25th International Conference, Singapore, 18–22 September 2022, Proceedings, Part IV, pp. 238–247. Springer, Cham (2022). https://doi.org/10.1007/978-3-031-16440-8_23

26. Wen, Y., Zhang, K., Li, Z., Qiao, Yu.: A discriminative feature learning approach for deep face recognition. In: Leibe, B., Matas, J., Sebe, N., Welling, M. (eds.) ECCV 2016. LNCS, vol. 9911, pp. 499–515. Springer, Cham (2016). https://doi.org/10.1007/978-3-319-46478-7_31
27. Zhang, G., Zhao, K., Hong, Y., Qiu, X., Zhang, K., Wei, B.: SHA-MTL: soft and hard attention multi-task learning for automated breast cancer ultrasound image segmentation and classification. Int. J. Comput. Assist. Radiol. Surg. **16**, 1719–1725 (2021)
28. Zhang, Y., et al.: BUSIS: a benchmark for breast ultrasound image segmentation. In: Healthcare, vol. 10, p. 729. MDPI (2022)

Learning with Domain-Knowledge for Generalizable Prediction of Alzheimer's Disease from Multi-site Structural MRI

Yanjie Zhou, Youhao Li, Feng Zhou, Yong Liu, and Liyun Tu(✉)

School of Artificial Intelligence, Beijing University of Posts and Telecommunications, Beijing 100876, China
tuliyun@bupt.edu.cn

Abstract. Construct a generalizable model for the diagnosis of Alzheimer's disease (AD) is an important task in medical imaging. While deep neural networks have recently advanced classification performance for various diseases using structural magnetic resonance imaging (sMRI), existing methods often provide suboptimal and untrustworthy results because they do not incorporate domain-knowledge and global context information. Additionally, most state-of-the-art deep learning methods rely on multi-stage preprocessing pipelines, which are inefficient and prone to errors. In this paper, we propose a novel domain-knowledge-constrained neural network for automatic diagnosis of AD using multi-center sMRI. Specifically, we incorporate domain-knowledge into a ResNet-like architecture. We explicitly enforce the network to learn domain invariant and domain specific features by jointly training multiple weighted classifiers, so that pixel-wise predictive performance generalizes to unseen images. In addition, the network directly takes segmentation-free and patch-free images in original resolution as input, which offers accurate inference with global context information and accurate individualized abnormalities to further refines reproducible predictions. The framework was evaluated on a set of sMRI collected from 7 independent centers. The proposed approach identifies important discriminative brain abnormalities associated with AD. Experimental results demonstrate superior performance of our method compared to state-of-the-art methods.

Keywords: Domain-knowledge encoding · Patch-free · Structural magnetic resonance imaging (sMRI) · Alzheimer's disease

1 Introduction

Alzheimer's disease (AD) is one of the most pervasive neurodegenerative disorders, causing an increasing morbidity burden that may outstrip diagnosis and management capacity with the population ages. The assessment of AD usually involves the acquisition of structural magnetic resonance imaging (sMRI) images, since it offers accurate visualization of the anatomy and pathology of the brain. Brain abnormalities (e.g., atrophy, enlargement, malformation) are known to be

H. Greenspan et al. (Eds.): MICCAI 2023, LNCS 14224, pp. 452–461, 2023.
https://doi.org/10.1007/978-3-031-43904-9_44

the most discriminative and reliable biomarkers [1] of AD that can be observed and analyzed through sMRI. However, automatic and reproducible identification of AD remains challenging due to heterogeneous of sMRI collected from different centers.

Recently, convolutional neural networks (CNN) have been used for automatic classification of AD from sMRI. Many methods [2,3] use a bag of patches selected from the skull-stripped brain region, which ignores global context information that can play a significant role in identifying lesions for accurate inference [4]. Many studies [5–8] proposed to characterize AD using segmented anatomies (e.g., gray matter or hippcampus). These methods rely on the accurate segmentation of the anatomies which is usually performed in a multi-stage data processing pipeline with the help of third-party softwares (e.g., FreeSurfer [9]) driven by a prior template. However, template-driven methods depend on variable image registration accuracy and highly affected by the anatomical variability between subjects, introducing errors to the characterization of individualized abnormalities. Similarly, methods (e.g., [10]) use detected landmarks also depend on a template-driven pipeline. Taking advantage of attention mechanism, some methods [5] proposed to diagnose AD using sMRI images from multiple centers. However, the classification performance is either hardly reproducible or difficult to compare across studies. One of the major reasons is that existing methods are often trained with samples from the same training (source) domain, while testing samples come from an independent new (target) domain with a different feature distribution. In the literature, this situation relates to domain adaptation [11–16] or domain generalization [17–19]. A widely used solution for the problem is to learn a domain-invariant latent feature space [20]. Unfortunately, there is no guarantee that the target samples' features will fall into the shared source domain-invariant representation, and in practice it is that new domains typically do not.

In this paper, we propose a novel domain-knowledge-constrained neural network for the diagnosis of AD using sMRI from multiple source domains. We designed a new domain-knowledge encoding module into a ResNet-like architecture for feature learning that yields a latent feature space with domain specific and domain shared information. In addition, we propose to use segmentation-free, resampling-free, patch-free 3D sub-images, which offers global context information and subject-level abnormalities to further refines generalizable and reproducible predictions.

2 Methods

We propose to design and implement an end-to-end neural network (Fig. 1) for automatic, robust, and reproducible diagnosis of AD using sMRI images, with the hope to identify and understand the most discriminative anatomical regions associate with AD. The model operates in 3 major steps: a) crop the input sMRI image to keep a sub-region (red rectangle), containing relevant anatomy structures (e.g., hippocampus, caudate, ventricles) associate with AD; b) extract features shared by all training sources based on ResNet [21]; c) design a domain-knowledge encoding module and a set of label predictors to constrain the feature learning process for better generalization.

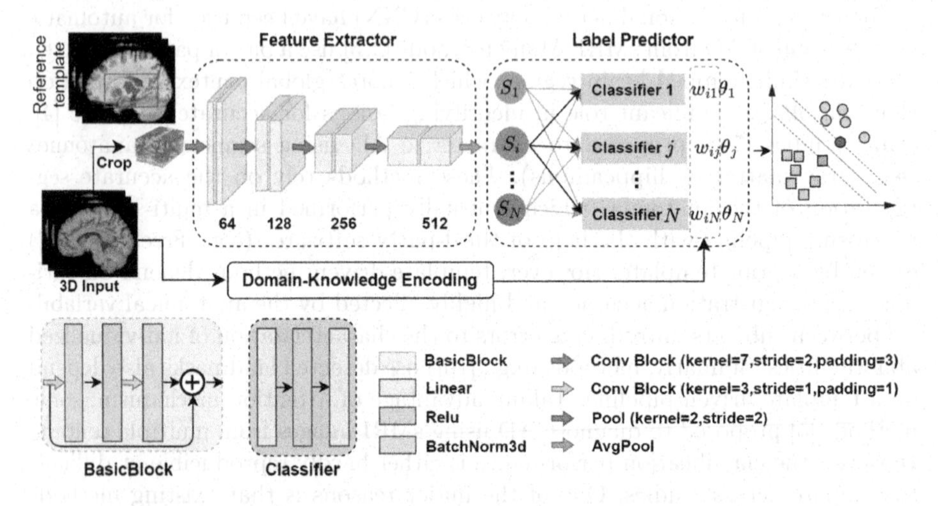

Fig. 1. Schematic of the proposed generalizable classification model. **Feature extractor** is a ResNet18-like 3D network that extracts high-dimensional features from MRI images for classification using 3D convolution and residual connection. **Basic block** is the basic component of the feature extractor and consists of two 3D convolutional layers, two BatchNorm layers, a ReLu layer and residual connection. **Classifier** is a multilayer perceptron (MLP), consisting of two linear layers and a ReLu layer. **Domain-Knowledge Encoding** captures domain invariant features and domain-specific features and generates weights for classifiers based on domain similarity. **Label Predictor** specifies that our model has multiple mutually independent classifiers, and the predictions of all classifiers are weighted and summed to obtain the final output. (Color figure online)

2.1 Patch-Free 3D Feature Extractor

We first estimate a bounding box around relevant anatomical objects in the input sMRI. The objects are automatically identified by affine registration, which transforms the reference template to each image in the dataset to estimate label for the image. We note that, the estimated labels are only used to locate the bounding box, it has no effect on the individual's atrophy since we pad extra space to ensure the cropped image contain all interested objects with respect to registration errors. Then, we crop the input image using the located bounding box to obtain the sub-image as input to our network. It need to be clarified that the cropping size is a fixed tuple determined by the maximum bounding box containing informative anatomical objects associated with AD.

To encode global context information, we propose a patch-free 3D feature extractor for different source domains, which is expected to learn domain-invariant features while not eliminating domain-specific features. Each domain has a unique label classifier, allowing adjustments for domain differences. Based on ResNet, we design our feature extractor as shown in Fig. 1. Each basic block consists of two convolutional layers. Each convolutional layer is followed by a batch normalization and a nonlinear activation function LeakyReLU. The basic block can be wrote as:

$$X_{l+1} = F(W_i, X_l) + W_s X_l, \tag{1}$$

where X_l and X_{l+1} are the input and output of the basic block and $F(W_i, X_l)$ denotes the nonlinear mapping in the basic block. Since the dimensions of X and $F(W_i, X)$ must be the same for summation, we use the linear mapping W_s to adjust the dimensions of X in the shortcut connection.

In the proposed method, we use global average pooling function which is more suitable for disease classification, because the global average pooling operation reflects the information of gray matter volume in brain regions and preserves the relative position relationship between different channels of the feature map.

In the output layer, we use a softmax classifier based on cross-entropy loss to calculate the loss between the predicted and true labels.

$$\mathcal{L} = \textit{cross-entropy}(\widehat{Y}_i(X_i \in D_s; \omega), Y_i) \tag{2}$$

2.2 Global Average Pooling

Global average pooling solves the problem of excessive image feature dimensions. If the feature maps of 3D images are directly expanded for classification, it will significantly increase the number of classifier parameters and increase the time and space complexity of training. Global average pooling averages the 3D feature maps in the channel dimension, preserving the relative position relationship between channels and reducing the resources required for model training.

The dimension change in the global average pooling is $[B, C, D, H, W] \rightarrow [B, C, 1, 1, 1]$, where B denotes the batch-size and C denotes the channel number.

$$GAP(\delta) = \frac{1}{D \times H \times W} \sum_{i=1}^{D} \sum_{j=1}^{H} \sum_{k=1}^{W} \delta_{i,j,k} \tag{3}$$

where δ denotes the image feature extracted by ResNet, and D, H, W denote the three dimensions of the feature.

Since global average pooling has fewer parameters, it can prevent over-fitting to some extent, further more, global average pooling sums out the spatial information, thus it is more robust to spatial translation of the input.

2.3 Domain-Knowledge Encoding

The domain-knowledge encoding module is designed to give relative similarity weights to source domains from a new sample. The weights reflect the similarity between the testing sample and source domains, allowing the module to share strength only between similar domains.

Our model uses multiple classifiers for prediction from the features extracted by the feature extractor. The classifiers are independent from each other. We feed the image features to different classifiers and generate weights to each classifier, summing the predictions of each classifier according to the weights as the final output.

$$\widehat{Y} = \sum_{j=1}^{c\_num} \omega_{ij} \cdot classifier_j(\delta(X \in D_i), \theta_j) \tag{4}$$

where $\widehat{Y}$ denotes the prediction result of X, c_num denotes the number of classifiers, D_i denotes the center which X belongs, δ denotes the extracted feature from X, $classifier_j$ denotes one classifier and θ_j are the parameters in $classifier_j$.

Multiple classifiers can capture the invariant and specific feature distributions between different domains, comparing the similarity of feature distributions between training source and unseen target domains by a joint training of the admixture classifiers, generating weights to integrate the feature distributions of known domains to fit the unknown domain feature distributions.

3 Experiments and Results

3.1 Data Description

Structural T1-weighted brain MRI data of 809 subjects (468 male, 341 female, age 68.16 ± 8.12 years, range 42–89 year) were acquired from 7 in-house independent multiple centers as detailed in [5,22]. In total, 552 subjects (295 of normal control (NC), 257 of AD) were used for leave-center-out training. The rest 257 subjects with mild cognitive impairment (MCI) were used as an independent dataset for evaluation and compared with clinical diagnosis metrics.

3.2 Implementation Details

We first evaluated the model using leave-center-out cross-validation, where one center was selected for testing at a time and all remaining centers were used for training. Then, we applied the trained model on an independent validation set of unseen images for subjects with MCI. All images were cropped to have the same size of $[80, 128, 72]$. Image features were extracted with $3 \times 3 \times 3$ convolution in the network and $2 \times 2 \times 2$ convolution with a stride of 2 replacing the maximum pooling. The extracted features were passed through a global average pooling layer (Sect. 2.1). $N = 6$ independent classifiers were used.

During training, we sorted all training centers and feed the image features from $site_i$ to all classifiers, and set the weight of $classifier_{j(j=i)}$ to 1 and the weight of the rest classifiers to 0. We used cross-entropy to calculate the prediction error and update the parameters of the feature extractor and $classifier_j$ by backpropagation. In testing stage, we feed the image features from the test center to all classifiers, and the final prediction was used the weighted average of predicted probability over all classifiers as the final prediction.

We used SGD algorithm to optimize the model coefficients, and set the initial learning rate to 0.001 and reduce the learning rate to one-tenth of the previous value every 50 epochs. The method was implemented using PyTorch 1.1 with Python 3.7. The experiments were run on an Intel Xeon CPU with 16 cores, 43 GB. RAM and a NVIDIA A5000 GPU with 24 GB memory. The code and model are available at https://github.com/Yanjie-Z/DomainKnowledge4AD.

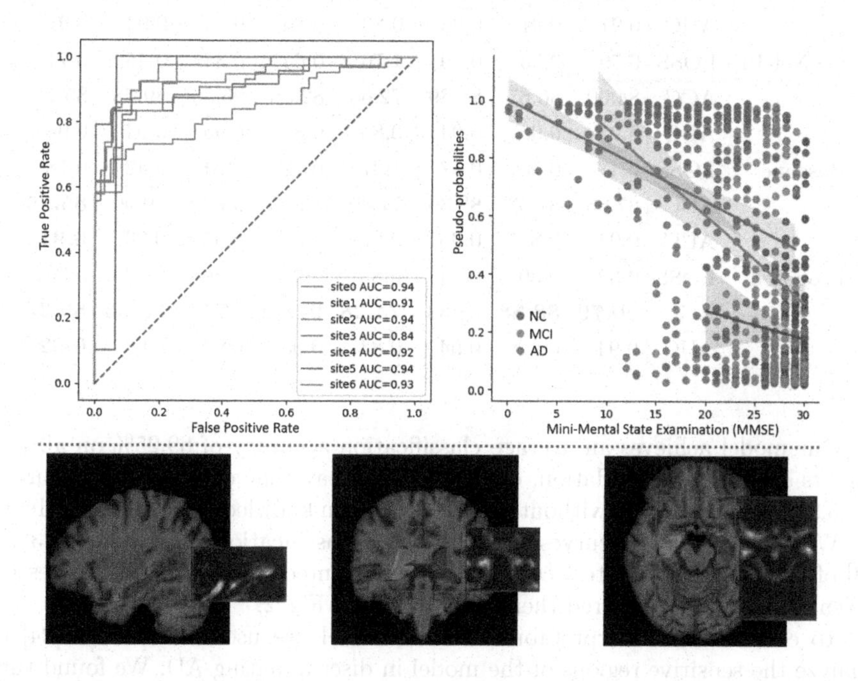

Fig. 2. First row: the left panel evaluates the AUC-ROC curve for each domain through leave-center-out cross validation, and the right panel investigates the association between the predicted probabilities and clinical measure (MMSE) in subjects with Alzheimer's disease (AD), mild cognitive impairment (MCI), and healthy controls (NC). Second row: attention map for an arbitrary example sMRI of a subject with AD, illustrating the most discriminative features learnt from the proposed approach.

3.3 Performance Evaluation

To evaluate the proposed approach, we feed 2 different types of input to the conventional 3D-ResNet [21] and each obtains a models: 1) ResNet, which use the original image as input, and 2) Baseline, which use the bounding box cropping strategy as proposed in Sect. 2.1. In addition, we incorporated the patch-free cropping strategy inspired by [4] to crop the middle-half sub-region of the original input sMRI image of the brain, and feed to ResNet, which we denote as ResNet-PF. The prediction performance are compared in Table 1.

Table 1. Comparisons among different methods with leave-center-out cross-validation. Abbreviations: ACC = accuracy, AUC = area under the curve of the receiver operating characteristic, AVG = average performance over centers. ACC in percentage.

		S0	S1	S2	S3	S4	S5	S6	AVG
ResNet	LOSS	0.90	0.53	0.35	1.34	0.53	0.38	0.56	0.66
	ACC	87.05	88.31	**90.83**	72.14	79.39	87.57	87.71	84.71
	AUC	0.91	0.91	0.96	0.83	0.86	0.94	0.94	0.91
ResNet-PF	LOSS	0.79	0.53	0.39	1.76	0.64	0.38	0.45	0.71
	ACC	84.00	86.67	88.89	72.63	82.86	95.56	89.78	85.77
	AUC	0.91	0.90	0.94	0.83	0.86	0.95	0.94	0.90
Baseline	LOSS	0.47	0.50	0.37	1.37	0.72	0.24	0.42	0.58
	ACC	87.66	87.03	88.36	71.40	82.85	95.40	89.00	**85.95**
	AUC	0.93	0.86	0.95	0.83	0.92	0.95	0.93	**0.91**
Proposed	LOSS	0.32	0.39	0.34	0.85	0.34	0.20	0.33	0.39
	ACC	**90.79**	**88.88**	88.33	**74.28**	**91.42**	**97.77**	**93.33**	**89.25**
	AUC	0.94	0.92	0.94	0.84	0.93	0.94	0.93	**0.92**

Our model achieves an average classification accuracy of 89.25% on all test centers during cross-validation, compared to the average classification accuracy of 85.95% with baseline (without the use of domain knowledge encoding module).

We used AUC-ROC curves to evaluate the classification effectiveness [13,17, 23] of the model on the test centers, and we counted the AUC-ROC curves for seven centers and compared them accordingly in Fig. 2.

To evaluate the interpretability of the model, we used Grad-CAM [24] to analyze the sensitive regions of the model in discriminating AD. We found that the model focused on the hippocampus in the images during prediction, which confirms that AD and the hippocampus have a significant correlation. We also find that the model pays more attention to the hippocampus in discriminating AD than healthy controls. Figure 3 compares the 3D attention map between a subject with AD and a healthy subject who never has AD, demonstrating obvious higher values in hippocampus region.

4 Discussion

We proposed a novel reproducible and generalizable neural network to assist the automatically diagnosis of AD that benefits from domain knowledge and

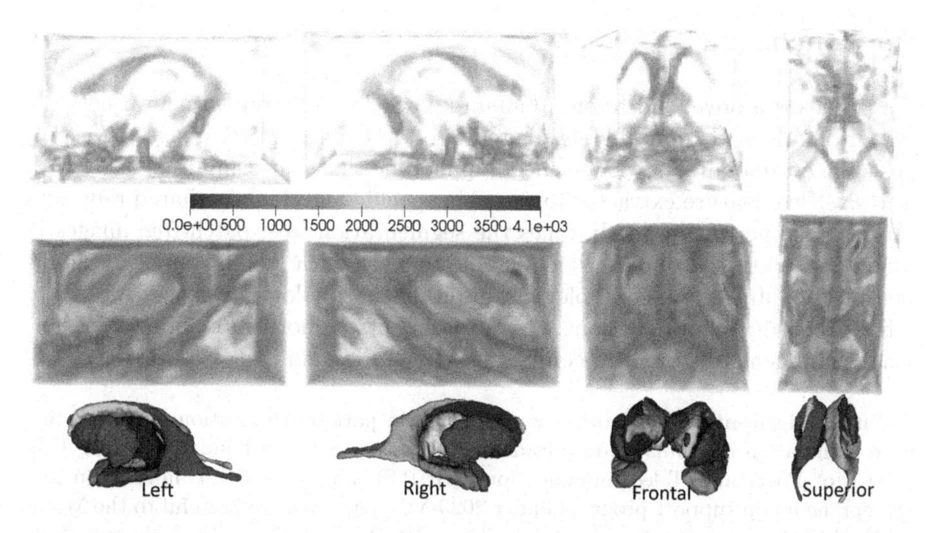

Fig. 3. 3D attention maps for a healthy subject (first row) and a subject with AD (second row) in 4 different views (column). The bottom row shows a visual navigator.

global contextual information with the help of segmentation-free, resampling-free, patch-free sub-image. The model was evaluated with leave-center-out cross-validation and with an independent set of unseen images for subjects with MCI (Fig. 2). It obtains an average accuracy of 89.25%, loss of 0.39 and AUC of 0.92 comparing with 85.95%, 0.58 and 0.91 using ResNet. We apply the proposed model to images from a new domain (never used during training), demonstrating promising results.

We did ablation studies to evaluate the proposed method (Table 1), unsurprisingly, the cropped images obtain the best performance. Figures 2 and 3 evaluated the explainability of the proposed neural network. The results suggest that the hippocampus and ventricles regions suffer the most in AD, which are consistent with multi-stage segmentation-based methods [5], and clinical measures (in terms of MMSE) on an independent dataset (Fig. 2).

Our results and all comparative frameworks tend to predict worse for center 3, probably because it has some subjects with AD who have higher MMSE (Fig. 2) making the diagnosis challenging. As opposite, all models provide the best accuracy for center 5. We will further explore possible reasons of this center imbalance in future work. Another limitation of the presented study is the empirical estimation of early stop strategy during leave-center-out cross validation based on the observed loss ranges. In future work, we will also explore a more automated mechanism to increase model robustness for images from more center.

5 Conclusion

We proposed a novel end-to-end domain-knowledge constrained neural network for automatic and reproducible diagnosis of AD using sMRI images. We proposed a new domain-knowledge encoding module that learn simultaneously with a ResNet-like feature extractor for domain specific and domain shared representations. The network directly takes the segmentation-free, patch-free images in original resolution as input, which is able to learn with global contextual information for subject-level pathological brain dysmorphologies features to further refines reproducible predictions. Our experiments demonstrate superior performance and generalize well to completely unseen domain.

Acknowledgments. This work was supported in part by the National Natural Science Foundation of China under Grant 62201091, the Startup Funds at Beijing University of Posts and Telecommunications (BUPT), and the BUPT innovation and entrepreneurship support program under 2023-YC-A208. We are grateful to the Multicenter Alzheimer Disease Imaging Consortium (PI: Prof. Xi Zhang, Prof. Yuying Zhou, Prof. Ying Han, and Prof. Qing Wang). The content is solely the responsibility of the authors and does not necessarily represent the official views of any of the funding agencies or sponsors.

References

1. Guptha, S.H., Holroyd, E., Campbell, G.: Progressive lateral ventricular enlargement as a clue to Alzheimer's disease. Lancet **359**(9322), 2040 (2002). https://doi.org/10.1016/S0140-6736(02)08806-2
2. Zhu, W., Sun, L., Huang, J., Han, L., Zhang, D.: Dual attention multi-instance deep learning for Alzheimer's disease diagnosis with structural MRI. IEEE Trans. Med. Imaging **40**(9), 2354–2366 (2021)
3. Wen, J., et al.: Convolutional neural networks for classification of Alzheimer's disease: overview and reproducible evaluation. Med. Image Anal. **63**, 101694 (2020). https://www.sciencedirect.com/science/article/pii/S1361841520300591
4. Wang, H., et al.: Super-resolution based patch-free 3D medical image segmentation with self-supervised guidance (2022). https://arxiv.org/abs/2210.14645
5. Jin, D., et al.: Generalizable, reproducible, and neuroscientifically interpretable imaging biomarkers for Alzheimer's disease. Adv. Sci. **7**(14), 2000675 (2020)
6. Goenka, N., Tiwari, S.: Deep learning for Alzheimer prediction using brain biomarkers. Artif. Intell. Rev. **54**(7), 4827–4871 (2021)
7. Gutiérrez-Becker, B., Wachinger, C.: Deep multi-structural shape analysis: application to neuroanatomy. In: Frangi, A.F., Schnabel, J.A., Davatzikos, C., Alberola-López, C., Fichtinger, G. (eds.) MICCAI 2018. LNCS, vol. 11072, pp. 523–531. Springer, Cham (2018). https://doi.org/10.1007/978-3-030-00931-1_60
8. Nguyen, H.-D., Clément, M., Mansencal, B., Coupé, P.: Interpretable differential diagnosis for Alzheimer's disease and Frontotemporal dementia. In: Wang, L., Dou, Q., Fletcher, P.T., Speidel, S., Li, S. (eds.) Medical Image Computing and Computer Assisted Intervention – MICCAI 2022: 25th International Conference, Singapore, 18–22 September 2022, Proceedings, Part I, pp. 55–65. Springer, Heidelberg (2022). https://doi.org/10.1007/978-3-031-16431-6_6

9. Hedges, E.P., et al.: Reliability of structural MRI measurements: the effects of scan session, head tilt, inter-scan interval, acquisition sequence, freesurfer version and processing stream. NeuroImage **246**, 118751 (2022). https://www.sciencedirect.com/science/article/pii/S1053811921010235
10. Zhang, J., Gao, Y., Gao, Y., Munsell, B.C., Shen, D.: Detecting anatomical landmarks for fast Alzheimer's disease diagnosis. IEEE Trans. Med. Imaging **35**(12), 2524–2533 (2016)
11. Danig, S., Orsborn, A.L., Moorman, H.G., Carmena, J.M.: Design and analysis of closed-loop decoder adaptation algorithms for brain-machine interfaces. Technical report 7 (2013)
12. Li, Y., Murias, M., Major, S., Dawson, G., Carlson, D.E.: On target shift in adversarial domain adaptation. In: AISTATS, March 2019
13. Hoffman, J., et al.: CyCADA: cycle-consistent adversarial domain adaptation. Int. Conf. Mach. Learn. **5**(11), 3162–3174 (2018). http://arxiv.org/abs/1711.03213
14. Sun, S., Shi, H., Wu, Y.: A survey of multi-source domain adaptation. Inf. Fusion **24**, 84–92 (2015)
15. Dozat, T.: Incorporating Nesterov momentum into Adam. In: ICLR Workshop, vol. 1, pp. 2013–2016 (2016)
16. Jiang, J.: A literature survey on domain adaptation of statistical Classifiers. UIUC Technical report, pp. 1–12, March 2008
17. Balaji, Y., Sankaranarayanan, S., Chellappa, R.: MetaReg: towards domain generalization using meta-regularization. In: NeurIPS, vol. 2018-Decem, pp. 998–1008 (2018). http://papers.nips.cc/paper/7378-metareg-towards-domain-generalization-using-meta-regularization
18. Li, D., Yang, Y., Song, Y.-Z., Hospedales, T.M.: Learning to generalize: meta-learning for domain generalization. In: Thirty-Second AAAI Conference on Artificial Intelligence, vol. 4 (2018). https://www.aaai.org/ocs/index.php/AAAI/AAAI18/paper/viewPaper/16067
19. Li, D., Zhang, J., Yang, Y., Liu, C., Song, Y.-Z., Hospedales, T.M.: Episodic training for domain generalization. In: IEEE International Conference on Computer Vision (2019). https://arxiv.org/pdf/1902.00113.pdf
20. Johansson, F.D., Sontag, D., Ranganath, R.: Support and invertibility in domain-invariant representations. In: Chaudhuri, K., Sugiyama, M. (eds.) Proceedings of the Twenty-Second International Conference on Artificial Intelligence and Statistics, ser. Proceedings of Machine Learning Research, vol. 89, pp. 527–536. PMLR, 16–18 April 2019
21. He, K., Zhang, X., Ren, S., Sun, J.: Deep residual learning for image recognition. arXiv e-prints, arXiv:1512.03385, December 2015
22. Zhao, K., et al.: Independent and reproducible hippocampal radiomic biomarkers for multisite Alzheimer's disease: diagnosis, longitudinal progress and biological basis. Sci. Bull. **65**(13), 1103–1113 (2020). https://www.sciencedirect.com/science/article/pii/S2095927320302140
23. Tu, L., Talbot, A., Gallagher, N.M., Carlson, D.E.: Supervising the decoder of variational autoencoders to improve scientific utility. IEEE Trans. Signal Process. **70**, 5954–5966 (2022)
24. Selvaraju, R.R., Cogswell, M., Das, A., Vedantam, R., Parikh, D., Batra, D.: Grad-CAM: visual explanations from deep networks via gradient-based localization. In: IEEE International Conference on Computer Vision (ICCV), pp. 618–626 (2017)

GSDG: Exploring a Global Semantic-Guided Dual-Stream Graph Model for Automated Volume Differential Diagnosis and Prognosis

Shouyu Chen[1(✉)], Xin Guo[2], Jianping Zhu[2], and Yin Wang[1]

[1] Tongji University, Shanghai, China
{1910667,yinw}@tongji.edu.cn

[2] Dalian University of Technology, Dalian, China
{guoxinguo,zhujp}@mail.dlut.edu.cn

Abstract. Three-dimensional medical images are crucial for the early screening and prognosis of numerous diseases. However, constructing an accurate computer-aided prediction model is challenging when dealing with volumes of different sizes due to numerous slices (native nodes) in a single case and variable-length slice sequence. We propose a Global Semantic-guided Dual-stream Graph model to address this issue. Our approach differs from the existing solution that aligns volumes with varying numbers of slices through downsampling. Instead, we leverage global semantic vectors to guide the grouping of native nodes, construct supernodes, and build dual-stream graphs by incorporating the sequential association of each volume's unique slices and the feature association of global semantic vectors. Specifically, we propose a shared global semantic vectors-based grouping method that aligns the number and the semantic distribution of nodes among different volumes without discarding slices. Furthermore, we construct a dual-stream graph module that enables Graph Convolutional Networks (GCN) to make clinical predictions from computer tomography (CT) volumes through the natural sequence association between native nodes and, simultaneously, the latent feature association between semantic vectors. We provide interpretability by visualizing the distribution of native nodes within each group and weakly-supervised slice localization. The results demonstrate that our method outperforms previous work in diagnostic (96.74%, +2.81%) and prognostic accuracy (84.56%, +1.86%) while being more interpretable, making it a promising approach for medical image analysis scenarios with limited fine-grained annotation.

Keywords: Diagnosis · Prognosis · Semantic-guided grouping · Graph convolutional networks · Lesion localization

---

Supplementary Information The online version contains supplementary material available at https://doi.org/10.1007/978-3-031-43904-9_45.

H. Greenspan et al. (Eds.): MICCAI 2023, LNCS 14224, pp. 462–471, 2023.
https://doi.org/10.1007/978-3-031-43904-9_45

1 Introduction and Related Works

Deep learning algorithms have shown success in performing computer-aided diagnosis (CAD) tasks using high-dimensional medical images, such as classification [20], detection [18], and segmentation [19]. Physicians typically review slices sequentially in CT and diagnose based on changes in lesion morphology and knowledge within the key slices. However, the variability in the number of slices between volumes challenges the CAD model in capturing the complex associations between slices and assisting medical decision-making.

The diagnosis of COVID-19 is challenging. Convolutional Neural Networks (CNNs) and their variants, such as 3D CNNs [15,21] and 2.5D CNNs [18], have shown promise, CNNs required pre-extracted regions of interest (ROI) with aligned size, fine-grained annotation, and high computational complexity. For instance, [21] employed a two-stage process where a segmentation model is trained first using segmentation masks. Then, a fixed-size 3D tensor is cropped from the lung region, transformed into a 4D tensor, and fed into the 3D classification net. Additionally, CT slices have intrinsic non-Euclidean associations, which has led to recent interest in using Transformers and Graph Neural Networks (GNNs) to handle them. For example, ViT [3], and Swin Transformer [8] are variants of Transformers that use multi-head self-attention to learn fully-connected associations between image patches. However, this architecture had primarily been applied to medical 3D patches [12,16] rather than the complete sequence. Regarding GNNs, ViG [4] organized images into patch sequences but had yet to extend the model to variable-length sequences. Another earlier work [7] used systematic sampling to align the number of slices, introducing sampling bias. We identify a common issue that existed in CNNs [15,18], Transformers [12,16], and GNNs [7] that they cannot be directly trained end-to-end on variable-length slice sequences. Therefore, this paper proposes a graph model to break through this limitation by reconstructing node knowledge at the super-node level.

One of the major challenges is achieving consistency training in GCNs while preserving the integrity of the slice information. Existing graph model [7] down-sampled slices to align nodes, resulting in the loss of some critical information. This approach also treated slices at the same location after sampling from different volumes equally, assuming they have the same semantics, which contradicts semantic consistency and clinical meanings. Moreover, the complex associations between slices further complicate the modeling. In three-dimensional medical images, the slice sequence dynamics naturally encode critical knowledge about morphology changes, which is of great diagnostic value. A recent study [4] showed that using sparse connections can improve the efficiency of GNNs, at least for natural images, and lead to better performance than other architectures such as CNNs and Transformers. Existing research mainly utilized GNNs to extract associations among slices or patches, constructing topology connections using methods such as k-nearest neighbors [4,13] and cosine similarity [7], as well as a learnable adjacency matrix [13]. Such approaches have limitations in capturing

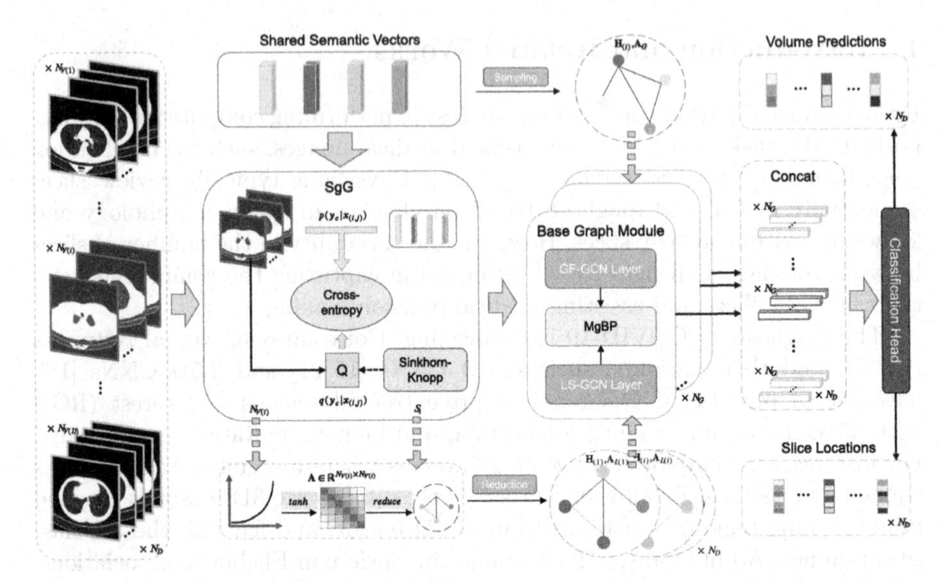

Fig. 1. Global Semantic-guided Dual-stream Graph (GSDG) model schematic diagram. Native CT nodes are first transformed into super-nodes guided by shared semantic vectors. Global feature and local sequence adjacency matrices are then generated to facilitate learning in the dual-stream base graph modules. The classification head produces diagnostic or prognostic probabilities and weakly-supervised slice localization upon concatenated multi-modal bilinear pooling features from all graph layers. *GF*: global feature, *LS*: local sequence. (Color figure online)

the task-specific local sequential associations between slices and the higher-level global feature associations.

Besides, various approaches have been developed to locate key slices in CT volumes under weak supervision. For instance, [7] proposed an end-to-end node masking method. In contrast, [11] used CNNs as feature extractors and selected a fixed number of substantial slices based on Shannon entropy under a multiple-instance learning framework. This method calculates the average prediction distribution of slices under random noise but cannot be trained end-to-end. This limitation is precisely the problem we aim to address. Our contributions are: (1) We propose a Global Semantic-guided Dual-stream Graph model for weakly-supervised graph classification tasks. It contains an unsupervised grouping algorithm called Semantic-guided Grouping, which aligns variable-length slice sequences using shared global semantic vectors to enable precise prediction by aligning both the node numbers and semantics. (2) We develop a dual-stream Base Graph Module incorporating the local slice sequence and global semantic knowledge by learning these two representations jointly. (3) We thoroughly evaluate the effectiveness of our proposed method through comparison and ablation experiments. Our method outperforms the weakly-supervised benchmark GCNs in terms of accuracy of diagnosis and prognosis on a publicly available

CT dataset while maintaining similar slice localization performance and offering more interpretability.

2 Method

2.1 Problem Statement

Given a dataset $\mathbf{D} = \{(\mathbf{V}_i, \mathbf{y}_i)\}_{n=1}^{N_D}$, which consists of N_D volumes. Each volume is represented by a set of slice nodes $\{\mathbf{v}_{i,j}\}_{j=1}^{N_{V(i)}}$, where $N_{V(i)}$ is the cardinality and varies for each volume. The volume-level label is given by $\mathbf{y}_i$. We first extract the native descriptors of the slices $\mathbf{X}_i = \{\mathbf{x}_{i,j} | \mathbf{x}_{i,j} = \mathrm{F}_{\mathrm{ext}}(\mathbf{v}_{i,j})\}_{j=1}^{N_{V(i)}} \in \mathcal{R}^{m_0 \times N_{V(i)}}$ using a spatial feature extractor, where the output of $\mathrm{F}_{\mathrm{ext}}$ is an m_0-dimensional vector. To perform volume-level prediction under the guidance of shared semantic vectors, we introduce the **G**lobal **S**emantic-guided **D**ual-stream **G**raph (GSDG) model: $\hat{\mathbf{y}} = \mathrm{F}_{\mathrm{GSDG}}(\mathbf{X}, \mathbf{C})$. $\mathbf{C}$ indicate semantic vectors and will be introduced in the following. Figure 1 provides a schematic diagram of our method.

2.2 Constructing Super-Nodes

We introduce a method for grouping native nodes $\mathbf{X}$ into super-nodes $\mathbf{H}$ which we denote as $\mathbf{H} = \mathrm{F}_{\mathrm{Gro}}(\mathbf{X}, \mathbf{C})$. To accomplish this, we propose semantic vectors $\mathbf{C} = \{\mathbf{c}_1, \cdots, \mathbf{c}_K\} \in \mathcal{R}^{m_0 \times K}$ that correspond to K groups and are shared across all volumes, end-to-end updated with the model. In an unsupervised setting, previous work [17] has extended cross-entropy minimization to optimal transportation. We build on this approach and draw inspiration from [1] to propose our unsupervised grouping algorithm, **S**emantic-**g**uided **G**rouping (SgG). SgG utilizes semantic vectors to guide the grouping process, ensuring that the resulting super-nodes align both the semantic and the number of nodes simultaneously on the variable-length volumes. However, minimizing cross-entropy in unsupervised classification can result in degeneration, where all slices are assigned to a single label. To address this issue, we encode the grouping label of a slice as the posterior distribution $q(\mathbf{y}_c|\mathbf{x}_{i,j})$ and re-express cross-entropy as:

$$E(p, q) = -\frac{1}{N_D}\sum_{i=1}^{N_D} \frac{1}{N_{V(i)}} \sum_{j=1}^{N_{V(i)}} \sum_{y_c=1}^{K} q(\mathbf{y}_c|\mathbf{x}_{i,j}) \log p(\mathbf{y}_c|\mathbf{x}_{i,j}) \tag{1}$$

To achieve semantic uniformity, we reformulate $p(\mathbf{y}_c|\mathbf{x}_{i,j})$ as $p(\mathbf{y}_c|\mathbf{x}_{i,j}, \mathbf{c}_{y_c})$:

$$p(\mathbf{y}_c|\mathbf{x}_{i,j}, \mathbf{c}_{y_c}) = \frac{\exp(\mathbf{x}_{i,j}^\top \mathbf{c}_{y_c}/\tau)}{\sum_{y'_c} \exp(\mathbf{x}_{i,j}^\top \mathbf{c}_{y'_c}/\tau)} \tag{2}$$

where τ is a temperature hyper-parameter. The mapping from native nodes to semantic vectors is described by Eq. 2. We represent this mapping using $\mathrm{Q} \in \mathcal{R}^{K \times N_{V(i)}}$, and optimize it to maximize the similarity between the native node features and the semantic vectors of their corresponding groups. Therefore, the optimization objective of F_{Gro} can be formulated as follows:

$$\min_{p,q} E(p,q) \text{ s.t. } \forall \mathbf{y}_c : q(\mathbf{y}_c|\mathbf{x}_{i,j}) \in \{0,1\} \text{ and } \sum_{j=1}^{N_{V(i)}} q(\mathbf{y}_c|\mathbf{x}_{i,j}) = \frac{N_{V(i)}}{K} \quad (3)$$

We utilize the Sinkhorn-Knopp (SK) algorithm [2] to handle the constraint term, which aims to distribute $N_{V(i)}$ native nodes uniformly into K groups. The SK algorithm produces an assignment matrix $\mathbf{S} \in \mathcal{R}^{N_V \times K}$, where each row is a one-hot vector indicating the group index to which a native node belongs. Consequently, we compute $\mathbf{H} = FC(\mathbf{X}\frac{\mathbf{S}}{\|\mathbf{S}\|_0})$, where FC is a linear layer and $\|\cdot\|_0 : \mathcal{R}^{N_{V(i)} \times K} \to \mathcal{R}^{1 \times K}$ computes the column-wise 0-norm of a matrix.

2.3 Bi-level Adjacency Matrices

We aim to train GCN on a dataset of variable-length volumes; and have already grouped the native-nodes into super-nodes, $\mathbf{H}$. Then, we could depict the adjacency relationships between nodes in $\mathbf{H}$ by semantic vectors or native nodes. Inspired by the multi-resolution model design in CNNs, we explicitly model the global and local adjacency relations: $\mathbf{A}_G, \mathbf{A}_L = \mathrm{F}_{Adj}(\mathbf{X}, \mathbf{S}, \mathbf{C})$. $\mathbf{A}_G$ represents the global semantic adjacency matrix, while $\mathbf{A}_L$ the local sequence adjacency matrix of the learned super-nodes from $\mathbf{X}$.

Global Adjacency Matrix Based on Grouping Semantic Vectors. The existing study [13] utilized the Gumbel reparameterization trick [6,10] to allow gradient flow through the adjacency matrix. Building upon the global semantic vectors $\mathbf{C}$ constructed in Sect. 2.2, we learn representations of the commonality between super-nodes over different volumes. Exactly, a link predictor, constructed by a 2-layer MLP, takes the concatenation of semantic vectors $\mathbf{c}_i$ and $\mathbf{c}_j$ as its input and produces the output $\theta_{i,j}$. We calculate the corresponding value, $\mathbf{A}_{G(i,j)} = sigmoid\left(\left(\log\left(\theta_{i,j}/\left(1-\theta_{i,j}\right)\right) + \left(g_{i,j}^1 - g_{i,j}^2\right)\right)/s\right)$, in the global adjacency matrix, $g_{i,j}^1, g_{i,j}^2 \sim \mathrm{Gumbel}(0,1)$ and s is a hyper-parameter.

Local Adjacency Matrix Based on Native Sequence Association. To account for the associations varying with relative distance between slices within a volume, we utilize exponential smoothing to create a sequence adjacency matrix $\mathbf{A} \in \mathcal{R}^{N_{V(i)} \times N_{V(i)}}$ for each volume. The adjacency value between native nodes i and j is calculated by: $\mathbf{A}_{i,j} = \tanh(\sum_{d=1}^{D}(1-\mathbf{s}_d)\mathbf{s}_d^{|i-j|})$. Here, $\mathbf{s}$ represents the output of the *sigmoid* function applied to a learnable vector $\mathbf{w}_L \in \mathcal{R}^D$, and D is a hyper-parameter. We combine the connectivities of native nodes belonging to the same group using the allocation matrix $\mathbf{S}$ introduced in Sect. 2.2 and obtain the reduced local adjacency matrix appliable to super-nodes: $\mathbf{A}_L = \mathbf{S}^{\mathrm{T}}\mathbf{A}\mathbf{S}$.

2.4 Dual-Stream Graph Classifier

We introduce a graph classification module, denoted as $\hat{\mathbf{y}} = \mathrm{F}_{\mathrm{Cls}}(\mathbf{H}, \mathbf{A}_G, \mathbf{A}_L)$, consisting of stacked **B**ase **G**raph **M**odules (BGM) and a classifier. The BGM comprises two parallel isomorphic graph convolutions, a global feature GCN layer and a local sequence GCN layer, and a **M**ulti-graph **B**ilinear **P**ooling (MgBP) module.

Base Graph Module. The two GCN layers pass messages between super-nodes from distinct perspectives and output $\mathbf{H}_G, \mathbf{H}_L \in \mathcal{R}^{m \times K}$. Then the MgBP module extracts fine-grained graph-level representation, $\mathbf{F}$, using a low-rank multi-modal bilinear module: $\mathbf{F} = \mathbf{P}\sigma\left(\mathbf{U}\mathbf{H}_G \circ \mathbf{V}\mathbf{H}_L\right) + b$, where trainable weights $\mathbf{U}, \mathbf{V} \in \mathcal{R}^{d \times m}$, $\mathbf{P} \in \mathcal{R}^{\frac{m}{N_G} \times d}$, $b \in \mathcal{R}$, $d < m$, and N_G is the number of BGM layers. The resulting matrix, $\mathbf{F} \in \mathcal{R}^{\frac{m}{N_G} \times K}$, represents the output of one BGM layer.

Classification Head. To obtain hierarchical features, we concatenate $\mathbf{F}_i$ from each BGM layer along the feature dimension, resulting in $\mathbf{F}_{final} = \|_{i=1}^{N_G} \mathbf{F}_i \in \mathcal{R}^{m \times K}$. We then compute the mean and max along the node dimension separately, resulting in two length-m vectors. These vectors are concatenated and passed through a 2-layer MLP and softmax activation for classification.

Weakly-Supervised Informative Slice Localization. Firstly, we obtain the predicted probability p_{base} for the target class from $\mathbf{F}_{final}$ using the Classification Head. Then, we mask each super-node in turn to create K sub-matrices of size $m \times (K-1)$. The Head is utilized again to calculate the new probabilities, which results in a vector $\mathbf{p} \in \mathcal{R}^K$. The groups are ranked by $\mathbf{d}_{sn} = p_{base} - \mathbf{p}$. Within a group, the distances between the native nodes and the super-node are measured using the dot product and normalized to the interval [0,1], which results in $\mathbf{d}_{rn}$. Slices' global importance within the group i is $\mathbf{d}_{sn(i)}/\mathbf{d}_{rn}$. We repeat this procedure for all groups and select the top k slices globally.

3 Experiments and Discussion

3.1 Dataset and Pre-processing

Our experiment used a public CT volume dataset 2019nCoVR [21], which contains complete chest CT scans from 929 COVID-19 (NCP) patients, 964 patients with common pneumonia (CP), and 849 healthy individuals (Normal). Among them, 408 patients are annotated with prognosis labels and some pneumonia patients with slice-level lesion annotations. To make a fair comparison with [7], we divided the dataset into training, validation, and testing sets using the same method with 20 random seeds. Each slice was resized to 224 × 224 pixels and normalized with $mean = 0.449$ and $std = 0.226$.

We chosen ResNet-50 [5] as the F_{ext} corresponding to $m^0 = 2048$. Only the frozen F_{ext} module was pre-trained on ImageNet, while all other modules were trained end-to-end on the 2019nCoVR dataset. The model hyper-parameters were set to $m = 256$, $K = 6$, $N_G = 2$, $d = 64$, $k = 10$, $s = 0.5$, $\tau = 1$, and $D = 8$. AdamW [9] served as the optimizer with a learning rate of $3e^{-3}$ and a batch size of 64. The model was trained with the cross-entropy loss for 20 epochs. After each feature aggregation, a Dropout [14] layer with a rate of 0.1 was added. The selection of hyper-parameter values is mainly based on experience and constraints in the formula above. The diagnostic model, which has one output node activated by the sigmoid function, was trained from scratch. In contrast, the prognosis model with three output nodes activated by softmax was initialized with the weights of the trained diagnostic model, except for the classification head. We used the same evaluation metric, precision and recall, for weakly-supervised localization as [7].

3.2 Differential Diagnosis, Prognosis and Weakly-Supervised Localization

Table 1 presents the diagnostic and prognostic performance of two state-of-the-art architectures and our proposed method. The clinical AI system based on 3D CNNs [21] was trained with volume-level and additional pixel-level labels. Our proposed method outperforms the state-of-the-art weakly-supervised graph model GCN-DAP [7] and 3D CNNs [21] in terms of diagnostic accuracy and AUC scores. Furthermore, GSDG also surpassed [7] in the prognostic task. These results demonstrate the superiority and effectiveness of our method in modeling full-size variable-length volumes. The left panel of Fig. 2 compares the performance of GSDG and experienced radiologists [21] in diagnostic tasks. For the NCP, GSDG outperformed the radiologists. Moreover, when identifying Normal and CP cases, GSDG achieved similarly high levels of AUCs. These results suggest that our method is advantageous over radiologists in NCP diagnosis. It is worth noting that our model required fewer training epochs than [21], as shown in the right panel of Fig. 2, which highlights the faster convergence of GSDG compared to GCN-DAP [7]. GSDG located the most informative CT

Table 1. Diagnostic and prognostic performance comparison. *: Results come from the original paper rather than 20 runs. *ACC*: macro accuracy, *AUC*: macro area under the receiver operating characteristic curve, *SD*: standard deviation, *CI*: confidence interval. Prefix *D* denotes the diagnosis task, and *P* stands for the prognosis task. All scores are multiplied by 100 to simplify the table.

Method	D-ACC (SD)	D-AUC (95% CI)	P-ACC (SD)	P-AUC (95% CI)
3D-CNN [21]	92.49 (N/A)*	98.13 (96.91–99.02)*	N/A	N/A
GCN-DAP [7]	93.93 (0.41)	99.00 (N/A)	82.70 (3.90)	N/A (N/A)
GSDG (Ours)	96.74 (0.64)	99.65 (99.61–99.69)	84.56 (2.35)	90.89 (88.77–89.88)

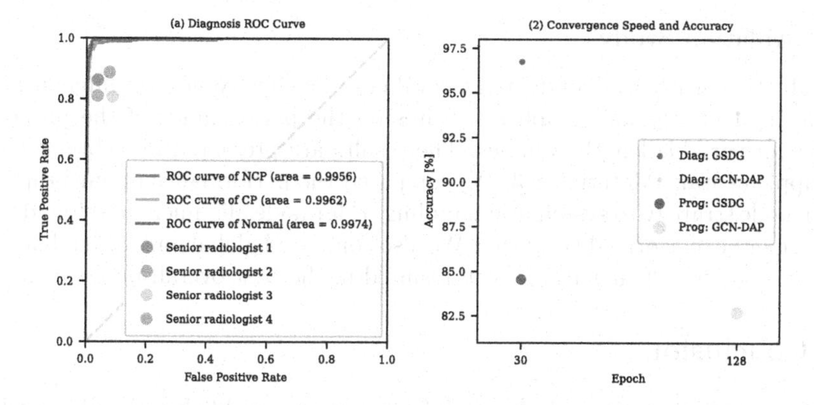

Fig. 2. The left panel shows the diagnostic ROC curves of GSDG and the NCP diagnostic scores of four senior radiologists with 15 to 25 years of clinical experience [21]. The right panel demonstrates convergence speed and accuracy of GSDG and [7] on diagnostic and prognostic tasks. The diameter of each circle represents its standard deviation, multiplied by 10 for ease of observation. *NCP*, *CP*, and *Normal* represent COVID-19, common pneumonia, and healthy individuals, respectively (Color figure online)

slices, achieved 51.60% (2.75%) and 89.27% (8.95%) for precision and recall, respectively, slightly worse than [7], 57.39% (3.32%), in precision, but outperformed [7], 79.89% (3.94%), in recall. During our experiments, we frequently observed that the model became unstable as the precision score increased further and the predictions degraded to a single category. This may be due to the loss of some information that only exists in the Hounsfield Unit, as the 2019nCoVR dataset uses JPG instead of DCM format to save the slices. With the emergence of possible technologies that can recover the JPGs losslessly to DCMs, training the model on lung-masked slices is expected to produce better results.

3.3 Visualization of Grouping and Slice Localization

We visualized the grouping of native slices in Supplementary Material S.1 and the weakly-supervised slice localization for two patients in S.2. It can be observed that slices belonging to the same group display a remarkable degree of visual resemblance. Besides, the group importance distribution of NCP and CP cases are more similar to each other than to the Normal case, which reflects patterns differences between positive and negative cases. Within the positive cases, our model exhibits a tendency to concentrate more on the lung base in the CP case, as evidenced by columns 4 and 5 of Fig. 2 (S.1), in comparison to the NCP case, where the attention is around the middle lobe, as demonstrated in columns 4, 5 and 6 of Fig. 1 (S.1). These observations may reveal group semantic differences among positive cases and provide a new perspective for clinical diagnosis. Regarding localization, our method's ability to identify lesion slices is superior to its precision performance, as indicated by Figs. 4 and 5 (S.2).

3.4 Ablation Study

The ablation study was conducted to evaluate the effectiveness of different node alignment methods and graph structures on the performance of the proposed model for variable-length volumes. The results are presented in Tables 1 and 2 in Supplementary Material, S.3. We compared the performance of our proposed super-node strategy to systematic sampling [7] for node alignment and found that the former outperformed the latter. We also compared global and local adjacency matrices and found that using both resulted in the best overall performance.

4 Conclusion

This paper proposes a novel approach for handling variable-length volume while preserving the integrity of the data by not discarding any slices, which is a departure from the previous method. Our approach first introduces a shared global semantic vectors-guided native node grouping scheme. Then we present an efficient and effective dual-stream graph module for simultaneously learning representations from global semantic vectors and sequence associations specific to each volume. Additionally, our approach offers informative slice localization and visually-consistent grouping outcomes, which enhances interpretability for clinical purposes. Moreover, the current dataset format prevents us from using existing semantic segmentation techniques to remove non-pulmonary noise. We will delve deeper into this direction to enhance localization accuracy.

Acknowledgments. I would like to thank my wife, Yang Feng, for her support during my doctoral studies.

References

1. Caron, M., Misra, I., Mairal, J., Goyal, P., Bojanowski, P., Joulin, A.: Unsupervised learning of visual features by contrasting cluster assignments. Adv. Neural Inf. Process. Syst. **33**, 9912–9924 (2020)
2. Cuturi, M.: Sinkhorn distances: lightspeed computation of optimal transport. In: Advances in Neural Information Processing Systems, vol. 26 (2013)
3. Dosovitskiy, A., et al.: An image is worth 16 × 16 words: transformers for image recognition at scale. In: International Conference on Learning Representations (2021). https://openreview.net/forum?id=YicbFdNTTy
4. Han, K., Wang, Y., Guo, J., Tang, Y., Wu, E.: Vision GNN: an image is worth graph of nodes. arXiv preprint arXiv:2206.00272 (2022)
5. He, K., Zhang, X., Ren, S., Sun, J.: Deep residual learning for image recognition. In: Proceedings of the IEEE Conference on Computer Vision and Pattern Recognition, pp. 770–778 (2016)
6. Jang, E., Gu, S., Poole, B.: Categorical reparameterization with Gumbel-Softmax. arXiv preprint arXiv:1611.01144 (2016)
7. Liu, C., Cui, J., Gan, D., Yin, G.: Beyond COVID-19 diagnosis: prognosis with hierarchical graph representation learning. In: de Bruijne, M., et al. (eds.) MICCAI 2021. LNCS, vol. 12907, pp. 283–292. Springer, Cham (2021). https://doi.org/10.1007/978-3-030-87234-2_27

8. Liu, Z., et al.: Swin transformer: hierarchical vision transformer using shifted windows. In: Proceedings of the IEEE/CVF International Conference on Computer Vision, pp. 10012–10022 (2021)
9. Loshchilov, I., Hutter, F.: Decoupled weight decay regularization. In: International Conference on Learning Representations (2019). https://openreview.net/forum?id=Bkg6RiCqY7
10. Maddison, C.J., Mnih, A., Teh, Y.W.: The concrete distribution: a continuous relaxation of discrete random variables. arXiv preprint arXiv:1611.00712 (2016)
11. Meng, Y., et al.: Bilateral adaptive graph convolutional network on CT based COVID-19 diagnosis with uncertainty-aware consensus-assisted multiple instance learning. Med. Image Anal. **84**, 102722 (2023)
12. Niu, C., Wang, G.: Unsupervised contrastive learning based transformer for lung nodule detection. Phys. Med. Biol. **67**(20), 204001 (2022)
13. Shang, C., Chen, J., Bi, J.: Discrete graph structure learning for forecasting multiple time series. In: International Conference on Learning Representations (2021). https://openreview.net/forum?id=WEHSlH5mOk
14. Srivastava, N., Hinton, G., Krizhevsky, A., Sutskever, I., Salakhutdinov, R.: Dropout: a simple way to prevent neural networks from overfitting. J. Mach. Learn. Res. **15**(1), 1929–1958 (2014)
15. Taleb, A., et al.: 3D self-supervised methods for medical imaging. Adv. Neural Inf. Process. Syst. **33**, 18158–18172 (2020)
16. Tang, Y., et al.: Self-supervised pre-training of Swin transformers for 3D medical image analysis. In: Proceedings of the IEEE/CVF Conference on Computer Vision and Pattern Recognition (CVPR), pp. 20730–20740, June 2022
17. Vedaldi, A., Asano, Y., Rupprecht, C.: Self-labelling via simultaneous clustering and representation learning (2020)
18. Wang, X., Han, S., Chen, Y., Gao, D., Vasconcelos, N.: Volumetric attention for 3D medical image segmentation and detection. In: Shen, D., et al. (eds.) MICCAI 2019. LNCS, vol. 11769, pp. 175–184. Springer, Cham (2019). https://doi.org/10.1007/978-3-030-32226-7_20
19. Yeung, P.-H., Namburete, A.I.L., Xie, W.: Sli2Vol: annotate a 3D volume from a single slice with self-supervised learning. In: de Bruijne, M., et al. (eds.) MICCAI 2021. LNCS, vol. 12902, pp. 69–79. Springer, Cham (2021). https://doi.org/10.1007/978-3-030-87196-3_7
20. Yuan, Z., Yan, Y., Sonka, M., Yang, T.: Large-scale robust deep AUC maximization: a new surrogate loss and empirical studies on medical image classification. In: Proceedings of the IEEE/CVF International Conference on Computer Vision, pp. 3040–3049 (2021)
21. Zhang, K., et al.: Clinically applicable AI system for accurate diagnosis, quantitative measurements, and prognosis of COVID-19 pneumonia using computed tomography. Cell **181**(6), 1423–1433 (2020)

Improving Image-Based Precision Medicine with Uncertainty-Aware Causal Models

Joshua Durso-Finley[1,4(✉)], Jean-Pierre Falet[1,3,4], Raghav Mehta[1,4], Douglas L. Arnold[1,3], Nick Pawlowski[2], and Tal Arbel[1,4]

[1] Center for Intelligent Machines, McGill University, Quebec, Canada
jddursof@cim.mcgill.ca
[2] Microsoft Research, Cambridge, UK
[3] Montreal Neurological Institute, McGill University, Montreal, Canada
[4] MILA (Quebec AI institute), Quebec, Canada

Abstract. Image-based precision medicine aims to personalize treatment decisions based on an individual's unique imaging features so as to improve their clinical outcome. Machine learning frameworks that integrate uncertainty estimation as part of their treatment recommendations would be safer and more reliable. However, little work has been done in adapting uncertainty estimation techniques and validation metrics for precision medicine. In this paper, we use Bayesian deep learning for estimating the posterior distribution over factual and counterfactual outcomes on several treatments. This allows for estimating the uncertainty for each treatment option and for the individual treatment effects (ITE) between any two treatments. We train and evaluate this model to predict future new and enlarging T2 lesion counts on a large, multi-center dataset of MR brain images of patients with multiple sclerosis, exposed to several treatments during randomized controlled trials. We evaluate the correlation of the uncertainty estimate with the factual error, and, given the lack of ground truth counterfactual outcomes, demonstrate how uncertainty for the ITE prediction relates to bounds on the ITE error. Lastly, we demonstrate how knowledge of uncertainty could modify clinical decision-making to improve individual patient and clinical trial outcomes.

1 Introduction

Precision medicine permits more informed treatment decisions to be made based on individual patient characteristics (e.g. age, sex), with the goal of improving patient outcomes. Deep causal models based on medical images can significantly improve personalization by learning individual, data-driven features to predict the effect of treatments.[1] As a result, they could significantly improve patient

[1] See [24] for a review on causality in medical imaging.

Supplementary Information The online version contains supplementary material available at https://doi.org/10.1007/978-3-031-43904-9_46.

H. Greenspan et al. (Eds.): MICCAI 2023, LNCS 14224, pp. 472–481, 2023.
https://doi.org/10.1007/978-3-031-43904-9_46

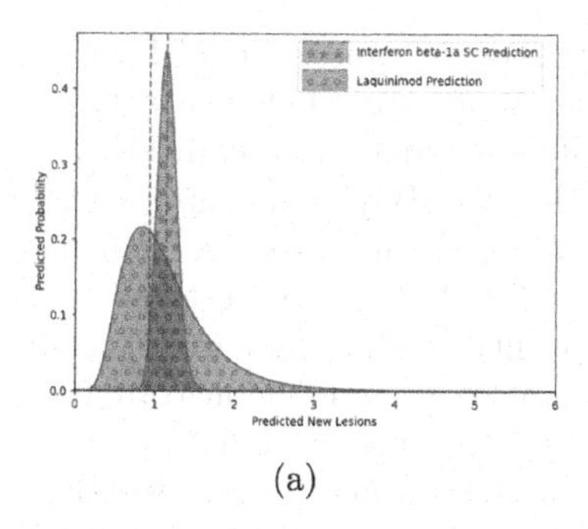

(a)

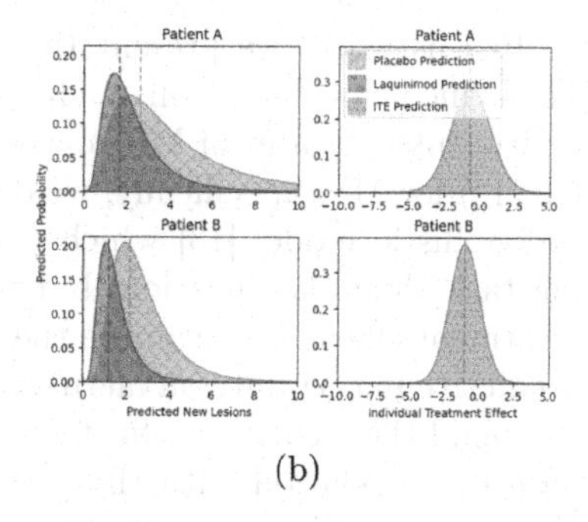

(b)

Fig. 1. (a) Probability distributions for a MS patient's predicted future new lesions on two different drugs (laquinimod) and INFB-SC). A patient might prefer INFB-SC if they are willing to make the tradeoff between slightly larger mean (dashed line) and lower variance (spread) in potential outcomes. (b) Predicted future outcomes for two patients for laquinimod and placebo drugs. Patients have similar expected (dashed line) ITE (difference between drugs), laquinimod, and placebo outcomes, but with different levels of confidence. Here, patient B is a better candidate for trial enrichment. (Color figure online)

outcomes, particularly in the context of chronic, heterogeneous diseases [18], potentially non-invasively.

However, despite significant advances, predictive deep learning models for medical image analysis are not immune to error, and severe consequences for the patient can occur if a clinician trusts erroneous predictions. A provided measure of uncertainty for each prediction is therefore essential to trust the model [26]. Although uncertainty is now commonly embedded in predictive medical image analysis (e.g. [1,15,22]), it is not well-studied for precision medicine.

Image-based precision-medicine is highly relevant in multiple sclerosis (MS), a chronic disease characterized by the appearance over time of new or enlarging T2 lesions (NE-T2) on MRI [9,17]. Several treatment options exist to suppress future NE-T2 lesions, but their level of efficacy and side effects are heterogeneous across the population [12]. Although one other model has been proposed for estimating the individual treatment effect (ITE) based on MR images [2], it does not incorporate uncertainty. Figure 1 illustrates how knowledge of the model's uncertainty could improve treatment recommendations.

To integrate uncertainty into clinical decision making, new validation measures must be defined. The usual strategy for validating uncertainty estimates, discarding uncertain predictions [8,14] and examining performance on the remaining predictions, is not always appropriate when predicting treatment effects. For example, discarding uncertain predictions could result in discarding predictions for the most responsive of individuals. A better strategy for this individual would be to consider the level of response and uncertainty jointly when making a treatment decision.

In this work, we present the first uncertainty-aware causal model for precision medicine based on medical images. We validate our model on a large, multi-center dataset of MR images from four different randomized clinical trials (RCTs) for MS. Specifically, we develop a multi-headed, Bayesian deep learning probabilistic model [13] which regresses future lesion counts, a more challenging task than classification, but which provides more fine-grained estimates of treatment effect. We evaluate the model's uncertainty by showing correlation of predictive uncertainty on factual and counterfactual error, and demonstrate how to bound the treatment effect error using group-level ground truth data to evaluate its correlation with the predicted personalized treatment effect. We then show the use of incorporating predictive uncertainty to improve disease outcomes by better treatment recommendations. Lastly, we demonstrate how uncertainty can be used to enrich clinical trials and increase their statistical power [21].

2 Methods

2.1 Background on Individual Treatment Effect Estimation

We frame precision medicine as a causal inference problem. Specifically, we wish to predict *factual* outcomes (on the treatment a patient received), *counterfactual* outcomes (on treatments a patient did not receive), as well as the individual treatment effect (ITE, the difference between the outcomes on two treatments). Let $X \in \mathbb{R}^d$ be the input features, $Y \in \mathbb{R}$ be the outcome of interest, and $T \in \{0, 1, ..., m\}$ be the treatment allocation with $t = 0$ as a control (e.g. placebo) and the remaining are m treatment options. Given a dataset containing triples $\mathcal{D} = \{(x^i, y^i, t^i)\}_{i=1}^n$, the ITE for patient i and a drug $T = t$ can be defined using the Neyman/Rubin Potential Outcome Framework [16] as $\text{ITE}_t = y_t - y_0$, where y_t and y_0 represents *potential* outcomes on treatment and control, respectively. The ITE_t is an unobservable causal quantity because only one of the two potential outcomes is observed. The average treatment effect (ATE_t) is defined as $\mathbb{E}[\text{ITE}_t] = \mathbb{E}[y_t] - \mathbb{E}[y_0]$ and is an observable quantity. Treatment effect estimation in machine learning therefore relies on a related causal estimand, τ_t:

$$\tau_t(x) = \mathbb{E}[\text{ITE}_t|x] = \mathbb{E}[y_t - y_0|x] = \mathbb{E}[y_t|x] - \mathbb{E}[y_0|x]. \tag{1}$$

$\tau_t(x)$[2] can be identified from RCT data (as in our case), where $(y_0, y_t) \perp\!\!\!\perp T|X$ [5]. Individual treatment outcomes y_t and y_0, and ITE_t, can therefore be estimated using machine learning models such that $\widehat{\text{ITE}}_t(x) = \hat{y}_t(x) - \hat{y}_0(x)$ [11].

2.2 Probabilistic Model of Individual Treatment Effects

In this work, we seek to learn the probability distribution of individual potential outcome predictions $\hat{y}_t(x)$ and the effect estimates $\widehat{\text{ITE}}_t(x)$. Let $\hat{y}_t(x) \sim \mathcal{N}(\hat{\mu}_t(x), \hat{\sigma}_t{}^2(x))$ be a normal distribution for potential outcome predictions

[2] Also known as conditional average treatment effect (CATE).

whose parameters are outputs of a neural network. This probabilistic framework conveniently allows for propagating the uncertainty estimates for each potential outcome to an uncertainty estimate for personalized treatment effects. Assuming independence between the two Gaussian distributions, $\widehat{\mathrm{ITE}}_t(x) \sim \mathcal{N}(\hat{\mu_t}(x) - \hat{\mu_0}(x), \hat{\sigma_t}(x)^2 + \hat{\sigma_0}(x)^2)$.

For our specific context, the input x to our model consists of multi-sequence patient MRI, lesion maps, and clinical and demographic features at baseline. The model is based on a multi-headed network for treatment response estimation [2,19]. Each head predicts $\hat{\mu_t}(x)$ and $\hat{\sigma_t}^2(x)$ for a particular treatment. For the case of MS, the model maximizes the log likelihood of the observed number of log NE-T2 lesions formed between 1 year and 2 years in the future (Fig. 2).

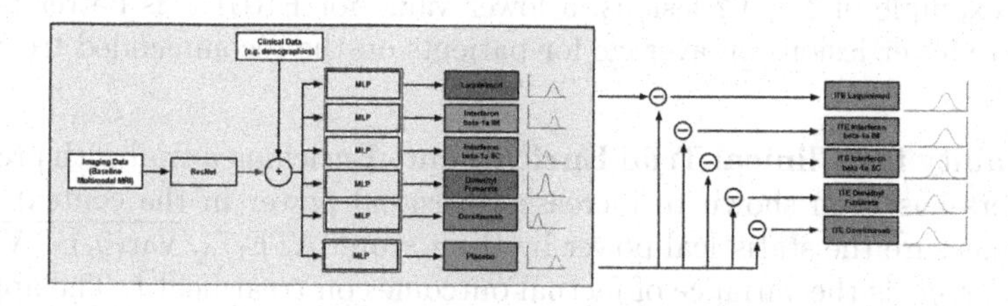

Fig. 2. Multi-head ResNet architecture for treatment effect prediction (based on [2]). It is modified to generate probabilistic estimates of individual outcomes. Specific architecture details can be found in the Appendix.

2.3 Evaluating Probabilistic Predictions

Bounds for the ITE Error We can validate the quality of the estimated uncertainty for factual outcome predictions through the correlation between predictive uncertainty and Mean Squared Error (MSE) error. However, given that ground truth for the individual treatment effects are not available, we cannot compute MSE between ITE_t and $\widehat{\mathrm{ITE}}_t(x)$. In this work, we choose to compute the upper and lower bounds for this MSE. We validate our uncertainty estimates by showing that selecting patients with the highest confidence in their predictions reduces the bounds on the ITE error. The bounds serve as an approximation to the true ITE error, and can validate models even if the ground truth ITE is not available. We use the upper bound for the MSE as in [19]. Jensen's inequality can be used to obtain a lower bound on the MSE as follows:

$$\mathbb{E}[(\mathrm{ITE}_t - \widehat{\mathrm{ITE}}_t(x))^2] \geq (\mathbb{E}[\mathrm{ITE}_t] - \mathbb{E}[\widehat{\mathrm{ITE}}_t(x)])^2 = (\mathrm{ATE}_t - \mathbb{E}[\widehat{\mathrm{ITE}}_t(x)])^2 \quad (2)$$

Evaluating Individual Treatment Recommendations. Predictive uncertainty can be used to improve treatment recommendations for the individual. Let $\pi(x^i, t^i) \in \{0, 1\}$ be a treatment recommendation policy taking as input a patient's features x_i and their factual treatment assignment t^i. The binary output of $\pi(x^i, t^i)$ denotes whether t^i is recommended under π. In this work, we set π to be a function of the model's predictions, $\hat{\mu_t}(x)$ and $\hat{\sigma_t}(x)$. For example, π can be defined such that a treatment is recommended if the number of predicted NE-T2 lesions on a particular treatment are less than 2 [4]. An uncertainty aware policy could instead recommend a drug according to $P(\hat{y_t}(x) < 2)$. The expected response under proposed treatments (ERUPT) [27] can then be used to quantify the effectiveness of that policy:

$$\text{ERUPT}_{\pi} = \sum_{i=1}^{n} y^i * \pi(x^i, t^i) / \sum_{i=1}^{n} \pi(x^i, t^i) \tag{3}$$

For the example of NE-T2 lesions, a lower value for ERUPT is better because there were fewer lesions on average for patients on the recommended treatment.

Uncertainty for Clinical Trial Enrichment. Enriching a trial with predicted responders has been shown to increase statistical power in the context of MS [3]. We measure the statistical power by the z-score: $\text{ATE}_t / \sqrt{\text{Var}(y_t) + \text{Var}(y_0)}$. Where $\text{Var}(y_t)$ is the variance of factual outcomes on treatment t. The approach taken by [3] achieves higher statistical power by selecting a subset of the population with larger ATE_t. Our proposed uncertainty-based enrichment selects patients with lower ITE uncertainty $(\hat{\sigma_t}(x)^2 + \hat{\sigma_0}(x)^2)$, with the goal of reducing the population variance $(\text{Var}(y_t) + \text{Var}(y_0))$. The benefit of this approach is most apparent if we inspect a specific population (defined by a particular value for ATE in the numerator).

3 Experiments and Results

3.1 Dataset

The dataset is composed of patients from four randomized clinical trials: BRAVO [25], OPERA 1 [6], OPERA 2 [6], and DEFINE [7]. Each trial enrolled patients with relapsing-remitting MS. Each patient sample consists of multi-sequence patient MRI (T1 weighted pre-contrast, T1 weighted post-contrast, FLAIR, T2-weighted, and proton density weighted), lesion maps (T2 hyperintense and gadolinium-enhancing lesions), as well as relevant clinical and demographic features (age, sex, expanded disability status scale scores [10]) at baseline. The number of NE-T2 lesions between 1 and 2 years after trial initiation were provided for each patient. Excluding patients with incomplete data resulted in a dataset with $n = 2389$ patients. In total the dataset contains the following treatment arms: placebo ($n = 406$), laquinimod ($n = 273$), interferon beta-1a intramuscular (INFB-IM) ($n = 304$), interferon beta-1a subcutaneous (INFB-SC) ($n = 564$), dimethyl fumarate (DMF) ($n = 225$), and ocrelizumab ($n = 627$). We perform 4 fold nested cross validation on this dataset. [23]

3.2 Evaluation of Factual Predictions and Uncertainty Estimation

Each patient is given a single treatment. The MSE for the future log-NE-T2 lesion count on the observed (factual) treatment and $\hat{\mu}_t(x)$ is used as a measure of the model's predictive accuracy. Taking all treatments in aggregate, the model achieves an MSE of 0.59 ± 0.03. Separating each treatment, it achieves an error of 0.84 ± 0.10 for placebo, 0.95 ± 0.07 for laquinimod, 0.70 ± 0.05 for INFB-IM, 0.76 ± 0.08 for INFB-SC, 0.62 ± 0.08 for DMF, and 0.04 ± 0.02 for ocrelizumab. Next we evaluate the correlation between the model error and the predicted variance. An accurate uncertainty estimate should be positively correlated with prediction accuracy [14]. This relationship is shown in Fig. 3a, where the MSE for the factual predictions decreases as we select a sub-group of patients with lower predictive uncertainty.

Next, we examine the results for the ITE error. Figure 3b and Fig. 3c show the upper and lower bounds (Eq. 2). Similarly to the factual error, the lower bound and upper bound on the ITE error decrease with decreasing ITE uncertainty.

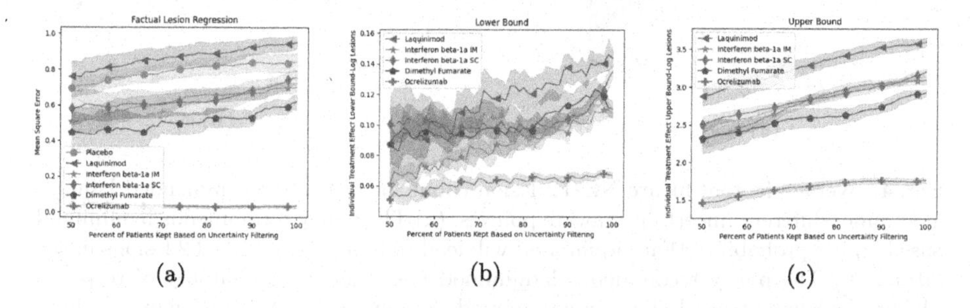

Fig. 3. (a) MSE for the log-lesion outcome as a function of predictive uncertainty. MSE is plotted separately for each treatment, using only patients who factually received the particular treatment. Uncertainty is computed according to the variance of the normal distribution predicted by the model, and the x-axis refers to the percent of kept patients based on uncertainty filtering (i.e. At 100, all patients are kept when computing MSE). (b) Lower bound for the ITE error as a function of predicted ITE uncertainty. (c) Upper bound for the ITE error as a function of predicted ITE uncertainty.

3.3 Uncertainty for Individual Treatment Recommendations

The effect of integrating uncertainty into treatment recommendations is evaluated by defining a policy using this uncertainty (Eq. 3). Here, we report outcomes on the lesion values (as opposed to log-lesions) for interpretability. In Fig. 4a, a treatment, laquinimod[3], is recommended if the predicted probability of having fewer than 2 NE-T2 lesions is greater than a threshold k: $P(\hat{y}_t(x) < 2) > k$. A policy requiring greater confidence indeed selects patients who more often

[3] Results on other treatments can be found in the Appendix.

have fewer than 2 lesions. It is worth noting that laquinimod was not found to be efficacious at the whole group level in clinical trials [25] and is therefore not approved, but this analysis shows that using personalized recommendations based on uncertainty can identify a sub-group of individuals that can benefit.

In Fig. 4b, a treatment effect-based policy is used such that laquinimod is recommended if the probability of any treatment response is greater than a threshold k: $P(\widehat{\text{ITE}}_t(x) \leq 0) > k$. As certainty in response grows, the difference between the treated and placebo groups grows suggesting an uncertainty aware policy better identifies patients for which the drug will have an effect.

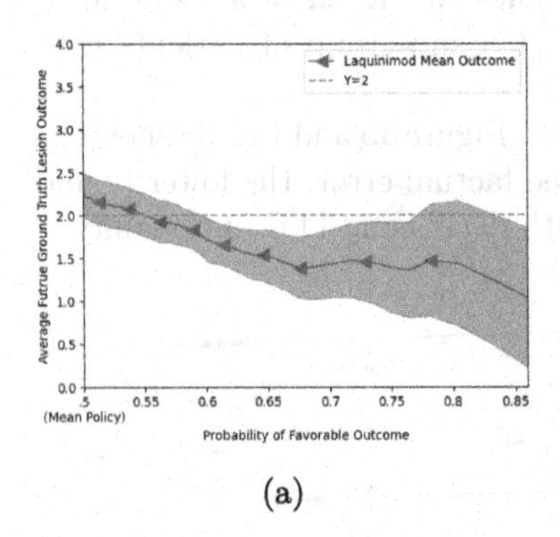

(a)

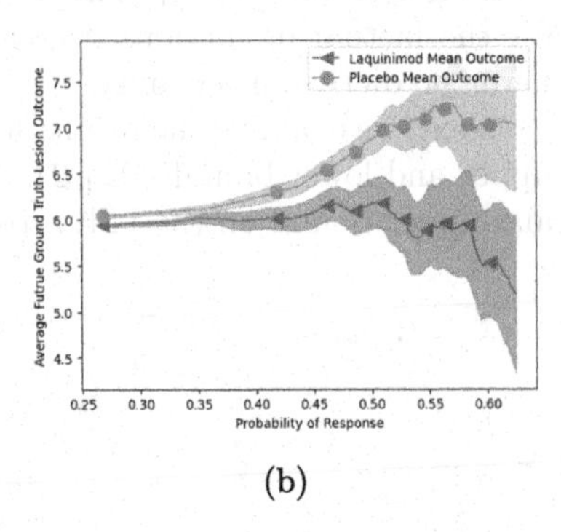

(b)

Fig. 4. Average factual future NE-T2 lesion count for patients recommended laquinimod under different uncertainty-aware policies. (a) The policy recommends laquinimod based on the probability that laquinimod will lead to fewer than 2 NE-T2 lesions in the future. (b) The policy recommends laquinimod based on the probability of response (defined as having fewer lesions on laquinimod than on placebo). (Color figure online)

Uncertainty can be useful when we wish to attribute a cost, or risk, to a certain range of outcomes. In our case, we assume a hypothetical non-linear cost c for having more NE-T2 lesions, where $c = (\text{NE-T2 lesions} + 1)^2$. Figure 5a describes a case where the recommended treatment (in terms of the mean) changes if this cost transformation is applied. In this case, the shape of the distribution over possible outcomes (which informs our uncertainty about this outcome) affects how much the mean of the distribution shifts under this transformation. This analysis is extended to the entire laquinimod cohort in Fig. 5b. We compute the average cost (Eq. 3) rather than the number of future NE-T2 lesions (as in Fig. 4a) for two types of policies. In the uncertainty-aware policy, the predicted distribution is used to make the treatment decision, whereas for the mean policy, the decision is based on only on $\hat{\mu_t}(x)$. As expected, uncertainty-aware recommendations incur a lower expected cost across the entire cohort compared to the mean policy. The advantage is most visible for intermediate values on the x-axis, because at the far right all patients are recommended laquinimod, and at the far left patients have closer to 0 NE-T2 lesions on average and the magnitude of the improvement due to the uncertainty-aware policy lessens.

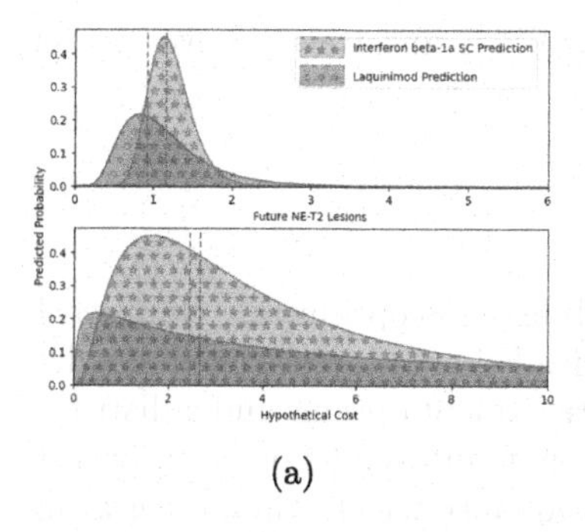

(a)

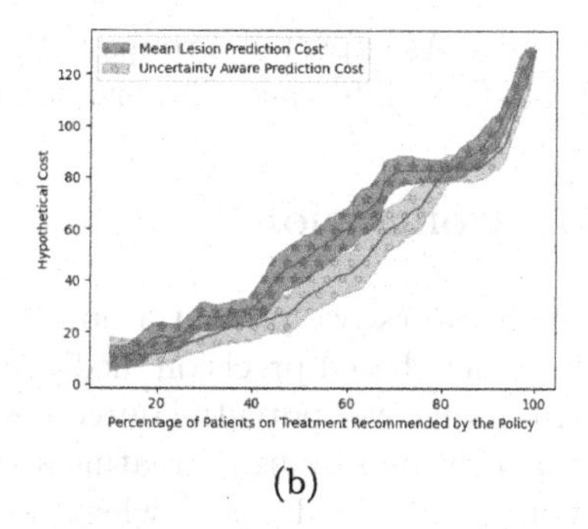

(b)

Fig. 5. (a) Example of the predicted outcomes for a single patient on two drugs (laquinimod and INFB-SC) before [top] and after [bottom] a hypothetical cost transformation. Note that the transformation causes the recommended treatment (as defined by the mean of the distribution, see dashed line) to switch from laquinimod to INFB-SC. (b) The expected cost under the mean and uncertainty-aware policies at the level of the entire laquinimod cohort. (Color figure online)

3.4 Uncertainty for Clinical Trial Enrichment

Uncertainty estimation can also be useful when selecting a sub-population of to enroll in a clinical trial, in a technique called predictive enrichment [20]. Figure 1b, shows an example where two patients have similar estimated future lesions but different ITE uncertainties. For trial enrichment, the second patient is more likely to experience a significant effect from this drug, and therefore enriching the trial with such patients could increase it's statistical power to detect an effect if done appropriately (see Sect. 2.3). In Fig. 6 we show the effect of uncertainty-aware trial enrichment. For a population with a particular effect size, we remove patients (right to left) with high ITE uncertainty and compute the z-score between the untreated and treated populations for the remaining

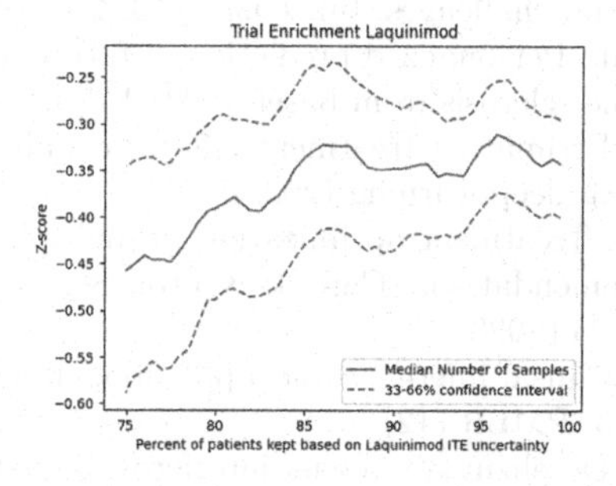

Fig. 6. To isolate the effect of uncertainty on enrichment, we fixed the ATE to be equal to 0 to -2 NE-T2 lesions by including patients with fixed placebo ($2 < \hat{y_0}(x) < 3$) and treatment ($1 < \hat{y_t}(x) < 2$) outcomes. The z-score then decreases for patient groups with smaller predicted ITE uncertainty.

groups. As expected, groups with smaller average ITE uncertainty have greater statistical differences (lower z scores).

4 Conclusion

In this work, we present a novel, causal, probabilistic, deep learning framework for image-based precision medicine. Our multi-headed architecture produces distributions over potential outcomes on multiple treatment options and a distribution over personalized treatment effects. We evaluate our model on a real-world, multi-trial MS dataset, where we demonstrate quantitatively that integrating the uncertainties associated with each prediction can improve treatment-related outcomes in several real clinical scenarios compared to a simple mean prediction. The evaluation methods used in this work are agnostic to the method of uncertainty quantification which permits flexibility in the choice of measure. Overall, this work has the potential to greatly increase trust in the predictions of causal models for image-based precision medicine in the clinic.

Acknowledgement. This investigation was supported by the International Progressive Multiple Sclerosis Alliance (PA-1412-02420), the companies who generously provided the data: Biogen, BioMS, MedDay, Novartis, Roche/Genentech, and Teva, the Canada Institute for Advanced Research (CIFAR) AI Chairs program, the Natural Sciences and Engineering Research Council of Canada, the Multiple Sclerosis Society of Canada, Calcul Quebec, and the Digital Research Alliance of Canada (alliance.can.ca). The authors would like to thank Louis Collins and Mahsa Dadar for preprocessing the MRI data, Zografos Caramanos, Alfredo Morales Pinzon, Charles Guttmann and István Mórocz for collating the clinical data, and Sridar Narayanan. Maria-Pia Sormani for their MS expertise.

References

1. Abdar, M., et al.: A review of uncertainty quantification in deep learning: techniques, applications and challenges. Inf. Fusion **76**, 243–297 (2021)
2. Durso-Finley, J., et al.: Personalized prediction of future lesion activity and treatment effect in multiple sclerosis from baseline MRI (2022)
3. Falet, J.P.R., et al.: Estimating treatment effect for individuals with progressive multiple sclerosis using deep learning (2021)
4. Freedman, M., et al.: Treatment optimization in multiple sclerosis: Canadian ms working group recommendations. Can. J. Neurol. Sci./J. Canadien des Sciences Neurologiques **47**, 1–76 (2020)
5. Gutierrez, P., et al.: Causal inference and uplift modelling: a review of the literature, vol. 67, pp. 1–13. PMLR (12)
6. Hauser, S.L., et al.: Ocrelizumab versus interferon beta-1a in relapsing multiple sclerosis. N. Engl. J. Med. **376**(3), 221–234 (2017)
7. Havrdova, E., et al.: Oral BG-12 (dimethyl fumarate) for relapsing-remitting multiple sclerosis: a review of DEFINE and CONFIRM. Evaluation of: Gold R, Kappos L, Arnold D, and others Placebo-controlled phase 3 study of oral BG-12 for relapsing multiple sclerosis. Expert Opin. Pharmacother. **14**(15), 2145–2156 (2013)

8. Jesson, A., et al.: Identifying causal effect inference failure with uncertainty-aware models (2020)
9. Kappos, L., et al.: Predictive value of gadolinium-enhanced magnetic resonance imaging for relapse rate and changes in disability or impairment in multiple sclerosis: a meta-analysis. Lancet (London, England) **353**, 964–969 (1999)
10. Kurtzke, J.F.: Rating neurologic impairment in multiple sclerosis. Neurology. **33**, 1444 (1983)
11. Künzel, S.R., Sekhon, J.S., Bickel, P.J., Yu, B.: Metalearners for estimating heterogeneous treatment effects using machine learning. Proc. Natl. Acad. Sci. **116**(10), 4156–4165 (2019)
12. Lucchinetti, C., et al.: Heterogeneity of multiple sclerosis lesions: implications for the pathogenesis of demyelination. Ann. Neurol. **47**, 707–17 (2000)
13. MacKay, D.J.C.: A practical Bayesian framework for backpropagation networks. Neural Comput. **4**(3), 448–472 (1992)
14. Nadeem, M.S.A., et al.: Accuracy-rejection curves (arcs) for comparing classification methods with a reject option. In: Proceedings of the third International Workshop on Machine Learning in Systems Biology. Proceedings of Machine Learning Research, vol. 8, pp. 65–81. PMLR, Ljubljana, Slovenia, 05–06 September 2009
15. Nair, T., et al.: Exploring uncertainty measures in deep networks for multiple sclerosis lesion detection and segmentation. Med. Image Anal. **59**, 101557 (2020)
16. Rubin, D.B.: Estimating causal effects of treatments in randomized and nonrandomized studies. J. Educ. Psychol. **66**, 688–701 (1974)
17. Rudick, R., et al.: Significance of t2 lesions in multiple sclerosis: a 13-year longitudinal study. Ann. Neurol. **60**, 236–242 (2006)
18. Sanchez, P., et al.: Causal machine learning for healthcare and precision medicine. R. Soc. Open Sci. **9**(8), 1–14 (2022)
19. Shalit, U., et al.: Estimating individual treatment effect: generalization bounds and algorithms (2017)
20. Simon, R., Maitournam, A.: Evaluating the efficiency of targeted designs for randomized clinical trials. Clin. Cancer Res. Off. J. Am. Assoc. Cancer. Res. **10**, 6759–6763 (2004)
21. Temple, R.: Enrichment of clinical study populations. Clin. Pharmacol. Therap. **88**(6), 774–778 (2010)
22. Tousignant, A., et al.: Prediction of disease progression in multiple sclerosis patients using deep learning analysis of MRI data. In: International Conference on Medical Imaging with Deep Learning, pp. 483–492. PMLR (2019)
23. Vabalas, A., et al.: Machine learning algorithm validation with a limited sample size. PLoS ONE **14**(11), 1–17 (2019)
24. Vlontzos, A., et al.: A review of causality for learning algorithms in medical image analysis (2022)
25. Vollmer, T.L., et al.: A randomized placebo-controlled phase III trial of oral Laquinimod for multiple sclerosis. J. Neurol. **261**(4), 773–783 (2014)
26. Zhang, Y., et al.: Effect of confidence and explanation on accuracy and trust calibration in AI-assisted decision making. CoRR (2020)
27. Zhao, Y., et al.: Uplift modeling with multiple treatments and general response types (2017)

Diversity-Preserving Chest Radiographs Generation from Reports in One Stage

Zeyi Hou[1], Ruixin Yan[2], Qizheng Wang[2], Ning Lang[2], and Xiuzhuang Zhou[1](✉)

[1] School of Artificial Intelligence, Beijing University of Posts and Telecommunications, Beijing, China
xiuzhuang.zhou@bupt.edu.cn
[2] Department of Radiology, Peking University Third Hospital, Beijing, China

Abstract. Automating the analysis of chest radiographs based on deep learning algorithms has the potential to improve various steps of the radiology workflow. Such algorithms require large, labeled and domain-specific datasets, which are difficult to obtain due to privacy concerns and laborious annotations. Recent advances in generating X-rays from radiology reports provide a possible remedy for this problem. However, due to the complexity of medical images, existing methods synthesize low-fidelity X-rays and cannot guarantee image diversity. In this paper, we propose a diversity-preserving report-to-X-ray generation method with one-stage architecture, named DivXGAN. Specifically, we design a domain-specific hierarchical text encoder to extract medical concepts inherent in reports. This information is incorporated into a one-stage generator, along with the latent vectors, to generate diverse yet relevant X-ray images. Extensive experiments on two widely used datasets, namely Open-i and MIMIC-CXR, demonstrate the high fidelity and diversity of our synthesized chest radiographs. Furthermore, we demonstrate the efficacy of the generated X-rays in facilitating supervised downstream applications via a multi-label classification task.

Keywords: Chest X-ray generation · Radiology report · Generative adversarial networks · One-stage architecture

1 Introduction

Chest radiography is currently the most common imaging examination, playing a crucial role in epidemiological studies [3] and clinical diagnosis [10]. Nowadays, the automated analysis of chest X-rays using deep learning algorithms has attracted increasing attention due to its capability of significantly reducing the workload of radiologists and expediting clinical practice. However, training deep learning models to achieve expert-level performance on various medical imaging tasks requires large, labeled datasets, which are typically difficult to obtain due to data privacy and workload concerns. Developing generative models for high-fidelity X-rays that faithfully represent various medical concepts in radiology reports presents a possible remedy for the lack of datasets in the medical

H. Greenspan et al. (Eds.): MICCAI 2023, LNCS 14224, pp. 482–492, 2023.
https://doi.org/10.1007/978-3-031-43904-9_47

domain [5,15]. This approach may substantially improve traditional supervised downstream tasks such as disease diagnosis [10] and medical image retrieval [8].

Generating chest radiographs based on radiology reports can be thought of as transforming textual input into visual output, while current methods typically rely on text-to-image generation in computer vision. The fidelity and diversity of synthesized images are two major qualities of generative models [1], where fidelity means the generated images should be close to the underlying real data distribution, while diversity means the output images should ideally cover a large variability of real-world images. Recently, several works have been extensively proposed to generate high-fidelity images using GANs according to text prompts. StackGAN [25] stacked multiple generators and discriminators to gradually increase the resolution of the generated images. AttnGAN [23] synthesized images with fine-grained details by introducing a cross-modal attention mechanism between subregions of images and relevant words. DM-GAN [27] introduced a memory module to refine fuzzy image contents caused by inferior initial images in stacked architecture. MirrorGAN [16] reconstructed the text from the synthesized images to preserve cross-domain semantic consistency. Despite the progress of text-to-image generation in the general domain, generating X-rays from radiology reports remains challenging in terms of word embedding, handling the linguistic structure in reports, cross-modal feature fusion, etc.

The first work to explore generating chest X-rays conditioned on clinical text prompts is XrayGAN [24], which synthesized high-fidelity images in a progressive way. Additionally, XrayGAN [24] proposed a hierarchical attentional text encoder to handle the linguistic structure of radiology reports, as well as a pre-trained view-consistency network to constrain the generators. Although impressive results have been presented, three problems still exist: 1) The progressive generation stacks multiple generators of different scales trained separately in an adversarial manner (see Fig. 1), where visual features of different scales are difficult to be fused uniformly and smoothly, making the final refined X-rays look like a simple combination of blurred contours and some details (see Fig. 2). 2) A high proportion of reconstruction loss and a pre-trained view-consistency network are used at each layer for the convergence of training stacked generators, which severely limits the diversity of generated X-rays (only one chest radiograph can be generated from one report). 3) The word vectors in [24] are based on the order in which words appear in the vocabulary, ignoring the information presented in medical-specific structures and the internal structure of words. More recently, another line of works [1,2] investigated adapting pre-trained visual-language foundational models to generate chest X-rays. However, transferring the diffusion models [18] trained with large multi-modal datasets to the medical imaging domain typically has high computational requirements.

In this paper, we propose a new report-to-X-ray generation method called DivXGAN to address the above issues. As illustrated in Fig. 1, DivXGAN allows for the synthesis of various X-rays containing relevant clinical concepts from a single report. The following contributions are made: 1) Inspired by the one-stage architecture [21], we propose to directly synthesize high-fidelity X-rays without

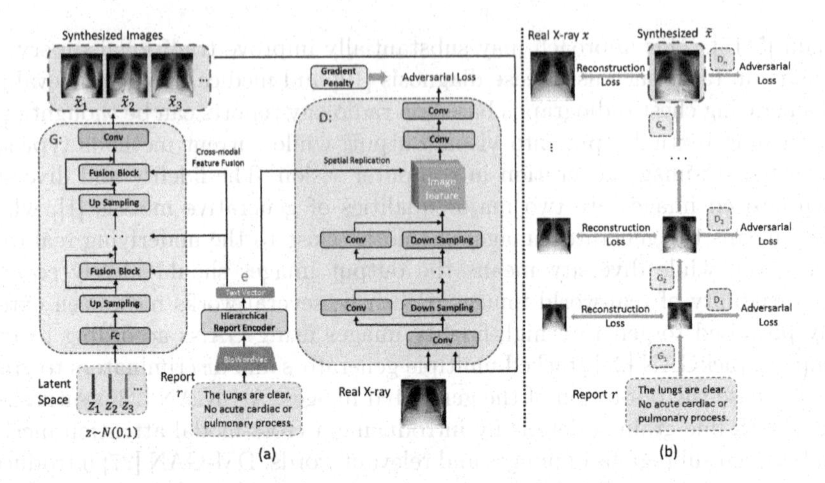

Fig. 1. (a) Overview of the proposed diversity-preserving report-to-X-ray generation with one-stage architecture. (b) Existing report-to-X-ray generative model [24]. DivXGAN discards the stacked structure with strong constraints and incorporates necessary variability via latent vectors, thus synthesizing X-rays with high fidelity and diversity.

entangling different generators. 2) We discard the pixel-level reconstruction losses and introduce noise vectors in the latent space of the generator to provide the variability, thus allowing the diversity of generated chest radiographs. 3) Lastly, we design a domain-specific hierarchical text encoder to represent semantic information in reports and perform multiple cross-modal feature fusions during X-ray generation. We demonstrate the superiority of our method on two benchmark datasets and a downstream task of multi-label classification.

2 Method

Let $\mathcal{X}$ and $\mathcal{Z}$ denote the image space and the low-dimensional latent space, respectively. Given a training set $\{x_i, r_i\}_{i=1}^{N}$ of N X-ray images, each of which x_i is associated with a radiology report r_i. The task of report-to-X-ray generation aims to synthesize multiple high-fidelity chest radiographs $\left\{\tilde{x}_i^{(j)} \in \mathcal{X}\right\}_{j=1,2,...,}$ from the corresponding report r_i and latent noises $\{z_j \in \mathcal{Z}\}_{j=1,2,...,}$. The generative models are expected to produce X-rays with high fidelity and diversity, so as to be used for data augmentation of downstream applications.

2.1 Fidelity of Generated X-Rays

One-Stage Generation. Existing generative method [24] uses a stacked structure to progressively synthesize high-fidelity X-rays. The stacked structure stabilizes the training of GANs but induces entanglements between generators trained separately in an adversarial way at different scales, resulting in fuzzy or discontinuous images. We draw inspiration from the one-stage architecture [21] and propose to directly generate high-fidelity X-rays using a single pair of generator G and discriminator D. The network architecture of our method is illustrated in Fig. 1. The generator contains many up-sampling layers to increase the resolution of the synthesized X-ray $\tilde{x}_i$, while the corresponding discriminator also requires many down-sampling operations to compute the adversarial loss. To stabilize the training of deep networks in this design, we introduce residual connections [6] in both the generator and the discriminator.

Distill and Incorporate Report Knowledge. Semantic information and medical concepts in radiology reports should be fully interpreted and incorporated into visual features to reduce the distance between the generated data distribution and the real data distribution, thereby improving fidelity. We design a medical domain-specific text encoder with hierarchical structure to extract the embeddings of the free-text reports. At the word level, each sentence is represented as a sequence of T word tokens, plus a special token $[SENT]$. We embed each word token w_t with an embedding matric W_e, i.e., $e_t = W_e w_t$. Unlike previous work [24] that use one-hot embedding, we initialize our word embeddings with the pre-trained biomedical embeddings BioWordVec [26], which can capture the semantic information in the medical domain and the internal structure of words. Then, we use a Transformer encoder with positional embedding to capture the contextual information of each word and aggregate the holistic representations of the sentence into the embedding of the special token $e_{[SENT]}$:

$$e_{[SENT]} = TrsEncoder\left(\{e_{[SENT]}, e_1, e_2, ..., e_T\}\right) \tag{1}$$

At the sentence level, a report consists of a sequence of S sentences, each of which is represented as $e^{(i)}_{[SENT]}$ using the word-level encoder described above. We also utilize a Transformer to learn the contextual importance of each sentence and encode them into a special token embedding $e_{[REPO]}$, which serves as the holistic representation of the report:

$$e_{[REPO]} = TrsEncoder\left(\left\{e_{[REPO]}, e^{(1)}_{[SENT]}, e^{(2)}_{[SENT]}, ..., e^{(S)}_{[SENT]}\right\}\right) \tag{2}$$

Moreover, we perform cross-modal feature fusion after each up-sampling module of the generator (see Fig. 1), to make the synthesized X-rays more faithful to the report. The fusion block contains two channel-wise Affine transformations and two ReLU layers. The scaling and shifting parameters of each Affine transformation are predicted by two MLPs (Multilayer Perceptron), using the vector $e_{[REPO]}$ as input. The Affine transformation expands the representation space of the generator G, allowing for better fusion of features from different modalities.

2.2 Diversity of Generated X-Rays

A radiology report is a medical interpretation of the corresponding chest radiograph, describing the clinical information included and assessing the patient's physical condition. Reports that describe chest radiographs of different patients with similar physical conditions are often consistent. Ideally, multiple X-ray images with the same health conditions could be generated from a single report, only with some differences in irrelevant factors such as body size, etc.

To this end, we omit the pixel-wise reconstruction loss and introduce noise vectors z in the latent space $\mathcal{Z}$ as one of the inputs to our one-stage generator, thereby providing the model with the necessary variability to ensure the diversity of synthesized X-rays. In this case, the generator G maps the low-dimensional latent space $\mathcal{Z}$ into a specific X-ray image space $\mathcal{X}_r$, conditioned on the report vector $e^i_{[REPO]}$:

$$\tilde{x}_i^{(j)} \leftarrow G\left(z_j, e^i_{[REPO]}\right), \ j = 1, 2, ... \tag{3}$$

where $\tilde{x}_i^{(j)}$ denotes the j-th synthesized X-ray from the i-th report r_i. The noise vector $z_j \in \mathcal{Z}$ follows a standard multivariate normal distribution $\mathcal{N}(0, I)$. In this way, given a radiology report, noise vectors can be sampled to generate various chest X-rays matching the medical description in the report.

2.3 Learning Objectives and Training Process

Since DivXGAN uses a one-stage generator to directly generate high-fidelity chest radiographs, only one level of generator and discriminator needs to be alternately trained. The discriminator D outputs a scalar representing the probability that the input X-ray came from the real dataset and is faithful to the input report. There are three kinds of inputs that the discriminator can observe: real X-ray with matching report, synthesized X-ray with matching report, and real X-ray with mismatched report. The discriminator $D\left(x, e_{[REPO]}; \theta_d\right)$ is trained to maximize the probability of assigning the report vector $e_{[REPO]}$ to the corresponding real X-ray x_i, while minimizing the probability of the other two kinds of inputs. Due to multiple down-sampling blocks and residual connections, we employ the hinge loss [13] to stabilize the training of D:

$$\begin{aligned} L_D = & \mathbb{E}_{x \sim p_{data}} \left[max\left(0, 1 - D\left(x, e_{[REPO]}\right)\right)\right] \\ & + \mathbb{E}_{G\left(z, e_{[REPO]}\right) \sim p_g} \left[max\left(0, 1 + D\left(G\left(z, e_{[REPO]}\right), e_{[REPO]}\right)\right)\right] \\ & + \mathbb{E}_{x \sim p_{mis}} \left[max\left(0, 1 + D\left(x, e_{[REPO]}\right)\right)\right] \end{aligned} \tag{4}$$

where p_{data}, p_g and p_{mis} denote the data distribution, implicit generative distribution (represented by G) and mismatched data distribution, respectively.

The generator $G\left(z, e_{[REPO]}; \theta_g\right)$ builds a mapping from the latent noise distribution to the X-ray image distribution based on the correlated reports, fooling the discriminator to obtain high scores:

$$L_G = -\mathbb{E}_{G\left(z, e_{[REPO]}\right) \sim p_g} \left[D\left(G\left(z, e_{[REPO]}\right), e_{[REPO]}\right)\right] \tag{5}$$

It is worth noting that the parameters θ_t of the text encoder in Eqs. (1) and (2) are learned simultaneously during the training of the generator G.

3 Experiments and Results

3.1 Datasets and Experimental Settings

We use two public datasets, namely Open-i [4] and MIMIC-CXR [9], to evaluate our generative model. The public subset of Open-i [4] consists of 7,470 chest X-rays with 3,955 associated reports. Following previous works [14,24], we select studies with two-view X-rays and a report, then end up with 2,585 such studies. As for MIMIC-CXR [9], which contains 377,110 chest X-ray images and 227,827 radiology reports, for a fair comparison with alternative methods, we also conduct experiments on the p10 subset with 6,654 cases to verify the effectiveness of our approach. Moreover, we adopt the same data split protocol as used in XRayGAN [24] for these two datasets, where the ratio of the train, validation, and test sets are 70%, 10%, and 20%. For consistency, we follow the set-up of XRayGAN [24] to focus on two major sections in each free-text radiology report, namely the "findings" section and the "impression" section.

Our network is trained from scratch using the Adam [11] optimizer with β_1=0.0 and β_2=0.9. The learning rates for G and D are set to 0.0001 and 0.0004, respectively, according to the Two Timescale Update Rule [7]. The hidden dimension of the Transformer in the text encoder is 512. The noise vector z in the latent space is sampled from a standard multivariate normal distribution with a dimension of 100. The resolution of synthesized X-rays is 512×512. We implemented our method using PyTorch 1.7 and two GeForce RTX 3090 GPUs. We use Inception Score (IS) [19] and Fréchet Inception Distance (FID) [7] to assess the fidelity and diversity of the synthesized X-rays. Typically, IS and FID are calculated using an Inception-V3 model [20] pre-trained on ImageNet, which might fail in capturing relevant features of the chest X-ray modality [12]. Therefore, we calculate these metrics from the intermediate layer of a pre-trained CheXNet [17]. Higher IS and lower FID indicate that the generated X-rays are more similar to the original X-rays. In addition, we also calculate the pairwise Structural Similarity Index Metric (SSIM) [22] to evaluate the diversity of X-rays generated by different methods. A lower SSIM indicates a smaller structural similarity between images, which combined with a low FID, can be interpreted as higher generative diversity [1].

3.2 Results and Analysis

We compare our approach with several state-of-the-art methods based on generative adversarial networks, including text-to-image generation: StackGAN [25], AttnGAN [23], and report-to-X-ray generation: XRayGAN [24]. The performance of different approaches on the test sets of Open-i and MIMIC-CXR is shown in Table 1. We can observe that XRayGAN [24] achieves better IS and FID than other text-to-image generation baselines. This is because the pixel-wise reconstruction loss imposes strong constraints on the stacked generators at different scales to help avoid generating insensible images. However, this strong constraint severely reduces the diversity of generated X-rays, resulting in the

Table 1. Performance of different methods on Open-i and MIMIC-CXR.

Method	Open-i			MIMIC-CXR		
	IS↑	FID↓	SSIM↓	IS↑	FID↓	SSIM↓
StackGAN [25]	1.043	243.4	0.138	1.063	245.5	0.212
AttnGAN [23]	1.055	226.6	0.171	1.067	232.7	0.231
XRayGAN [24]	1.081	141.5	0.343	1.112	86.15	0.379
DivXGAN	**1.105**	**93.42**	**0.114**	**1.119**	**62.06**	**0.143**

worst SSIM. Our method consistently outperforms other alternatives by achieving both the lowest FID and the lowest SSIM, which means the generated X-rays have better fidelity and diversity. The reason lies in that the latent noise vectors impose the necessary variation factor, and the one-stage generation process and multiple cross-modal feature fusions improve the image quality.

We conduct ablation experiments to quantify the impact of different components in DivXGAN. The results in Table 2 show that our one-stage architecture definitely improves performance due to better cross-modal feature fusion. The domain-specific encoder outperforms the hierarchical attentional encoder [24], regardless of the backbone structure, indicating the advantage of the domain-specific embedding matrix, especially for medical reports with very rare and specific vocabulary. The comparison of SSIM for different components demonstrates that the latent vector input preserves the diversity of synthesized X-rays.

Visualization of chest X-rays synthesized from a report using different methods is shown in Fig. 2. As we can see, the X-rays generated by text-to-image baselines are very coarse, and even the outline of the hearts can be barely recognized. XRayGAN [24] alleviates the blur and generates X-rays with relatively obvious chest contours, because of the strong constraints that prevent the generative model from producing abnormal samples. However, this strongly constrained approach still fails to preserve a clear outline of the heart and ribs, due to the entanglements between generators introduced by the stacked architecture. In particular, the lack of variability in the strong constraints results in only one X-ray being generated per report. Our method prevails over other alternatives as the generated X-rays are obviously clearer and more realistic, and even generate annotation information in the top right corner (seen in almost all samples). This phenomenon indicates the efficacy of the one-stage generation and multiple cross-modal feature fusion in generating high-fidelity X-rays. Furthermore, our method can generate various X-rays from one report, each of which manifests the relevant clinical findings. For example, the regions marked by red arrows in Fig. 2 show that our method can synthesize various different X-rays matching the clinical finding "Cardiomegaly" described in the report. Although our model is capable of generating X-rays based on radiology reports, due to the complexity of medical images, the synthesized X-rays have a limited range of gray-scale values and may not effectively capture high-frequency information such as subtle lung markings.

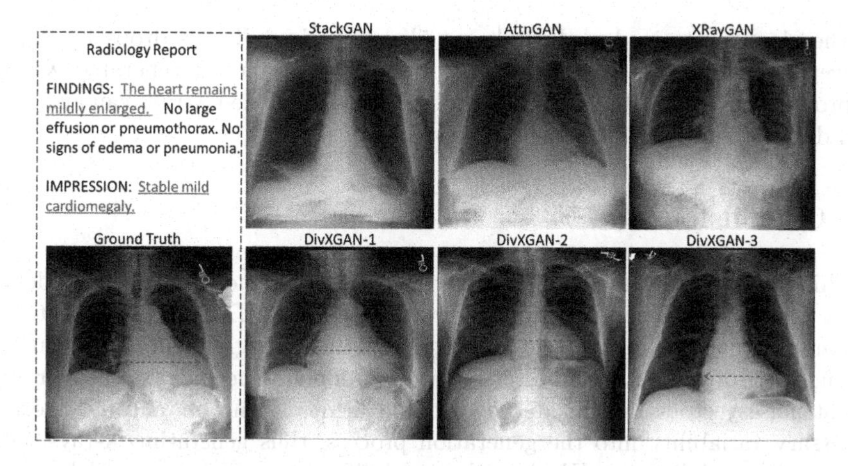

Fig. 2. X-ray images generated by different methods from a radiology report, where the red arrows mark the clinical finding "Cardiomegaly" described by the underlined sentences in the report. (Color figure online)

Table 2. Ablation study on Open-i (One-stage structure has latent vector input).

Methods	FID ↓	SSIM ↓
Stack w/ Hia-encoder	141.5	0.343
Stack w/ Med-encoder	139.3	0.358
One-stage w/ Hia-encoder	98.91	0.151
One-stage w/ Med-encoder	**93.42**	**0.114**

Table 3. Classification performance of a DenseNet-121 trained with various splits.

Experiment	Training Data		AUROC
	Real	Synthetic	
Real	5k	–	0.683
Synth	–	5k	0.664(↓0.019)
Real+Synth	5k	5k	0.714(↑0.031)

Furthermore, the ethical implications associated with the misuse of generated X-rays are significant. They should not be solely relied upon for clinical decision-making or used to train inexperienced medical students as radiologists. While the direct use of generated X-rays in clinical studies or medical training programs may not be appropriate, they can still serve valuable purposes in research, data augmentation, and other potential applications within the medical field. Here, we train a DenseNet-121 from scratch using various splits of real data (from MIMIC-CXR) and synthesized data (from DivXGAN) to demonstrate that the generated chest radiographs can be used for data augmentation when training downstream tasks. The task is a multi-label classification of four findings ("Cardiomegaly", "Consolidation", "Pleural Effusion" and "No Findings"). We randomly sample 5k real images with corresponding reports from the test set. These 5k real reports are input into our generative model, generating one image per report using a single latent vector, resulting in 5k generated images. When training the multi-label classifier, both real and generated images undergo the same general data augmentation techniques, such as rotation and scaling. As shown in Table 3, compared to the baseline trained exclusively on 5k real X-rays, the AUROC of the classifier trained exclusively on 5k synthesized X-rays drops by 0.019. However, training the classifier with 5k real X-rays and 5k generated X-rays improves AUROC by 0.031, suggesting that the synthesized X-rays can augment real data for supervised downstream applications.

4 Conclusion

In this paper, we have devised a diversity-preserving method for high-fidelity chest radiographs generation from the radiology report. Different from state-of-the-art alternatives, we propose to directly synthesize high-fidelity X-rays using a single pair of generator and discriminator. A domain-specific text encoder and latent noise vectors are introduced to distill medical concepts and incorporate necessary variability into the generation process, thus generating X-rays with high fidelity and diversity. We show the capability of our generative model in data augmentation for supervised downstream applications. Investigation of capturing high-frequency information of X-rays in generative models can be an interesting and challenging direction of future work.

Acknowledgements. This work is partially supported by the National Natural Science Foundation of China under grants 61972046, the Beijing Natural Science Foundation under grant Z190020, and the Proof of Concept Program of Zhongguancun Science City and Peking University Third Hospital under grant HDCXZHKC2022202.

References

1. Chambon, P., et al.: RoentGen: vision-language foundation model for chest X-ray generation. arXiv preprint arXiv:2211.12737 (2022)
2. Chambon, P., Bluethgen, C., Langlotz, C.P., Chaudhari, A.: Adapting pretrained vision-language foundational models to medical imaging domains. arXiv preprint arXiv:2210.04133 (2022)
3. Cherian, T., et al.: Standardized interpretation of paediatric chest radiographs for the diagnosis of pneumonia in epidemiological studies. Bull. World Health Organ. **83**, 353–359 (2005)
4. Demner-Fushman, D., et al.: Preparing a collection of radiology examinations for distribution and retrieval. J. Am. Med. Inform. Assoc. **23**(2), 304–310 (2016)
5. Ganesan, P., Rajaraman, S., Long, R., Ghoraani, B., Antani, S.: Assessment of data augmentation strategies toward performance improvement of abnormality classification in chest radiographs. In: 2019 41st Annual International Conference of the IEEE Engineering in Medicine and Biology Society (EMBC), pp. 841–844. IEEE (2019)
6. He, K., Zhang, X., Ren, S., Sun, J.: Deep residual learning for image recognition. In: Proceedings of the IEEE Conference on Computer Vision and Pattern Recognition, pp. 770–778 (2016)
7. Heusel, M., Ramsauer, H., Unterthiner, T., Nessler, B., Hochreiter, S.: GANs trained by a two time-scale update rule converge to a local nash equilibrium. In: Advances in Neural Information Processing Systems, vol. 30 (2017)

8. Huang, P., Zhou, X., Wei, Z., Guo, G.: Energy-based supervised hashing for multimorbidity image retrieval. In: de Bruijne, M., et al. (eds.) MICCAI 2021. LNCS, vol. 12905, pp. 205–214. Springer, Cham (2021). https://doi.org/10.1007/978-3-030-87240-3_20
9. Johnson, A.E., et al.: MIMIC-CXR, a de-identified publicly available database of chest radiographs with free-text reports. Sci. Data **6**(1), 317 (2019)
10. Khan, A.I., Shah, J.L., Bhat, M.M.: Coronet: a deep neural network for detection and diagnosis of Covid-19 from chest X-ray images. Comput. Methods Programs Biomed. **196**, 105581 (2020)
11. Kingma, D., Ba, J.: Adam: a method for stochastic optimization. In: Proceedings of the International Conference on Learning Representations (ICLR) (2014)
12. Kynkäänniemi, T., Karras, T., Aittala, M., Aila, T., Lehtinen, J.: The role of imagenet classes in frechet inception distance. arXiv preprint arXiv:2203.06026 (2022)
13. Lim, J.H., Ye, J.C.: Geometric GAN. arXiv preprint arXiv:1705.02894 (2017)
14. Liu, F., Wu, X., Ge, S., Fan, W., Zou, Y.: Exploring and distilling posterior and prior knowledge for radiology report generation. In: Proceedings of the IEEE/CVF Conference on Computer Vision and Pattern Recognition, pp. 13753–13762 (2021)
15. Madani, A., Moradi, M., Karargyris, A., Syeda-Mahmood, T.: Chest X-ray generation and data augmentation for cardiovascular abnormality classification. In: Medical Imaging 2018: Image Processing, vol. 10574, pp. 415–420. SPIE (2018)
16. Qiao, T., Zhang, J., Xu, D., Tao, D.: MirrorGAN: learning text-to-image generation by redescription. In: Proceedings of the IEEE/CVF Conference on Computer Vision and Pattern Recognition, pp. 1505–1514 (2019)
17. Rajpurkar, P., et al.: CheXNet: radiologist-level pneumonia detection on chest X-rays with deep learning. arXiv preprint arXiv:1711.05225 (2017)
18. Rombach, R., Blattmann, A., Lorenz, D., Esser, P., Ommer, B.: High-resolution image synthesis with latent diffusion models. In: Proceedings of the IEEE/CVF Conference on Computer Vision and Pattern Recognition, pp. 10684–10695 (2022)
19. Salimans, T., Goodfellow, I., Zaremba, W., Cheung, V., Radford, A., Chen, X.: Improved techniques for training GANs. In: Advances in Neural Information Processing Systems, vol. 29 (2016)
20. Szegedy, C., Vanhoucke, V., Ioffe, S., Shlens, J., Wojna, Z.: Rethinking the inception architecture for computer vision. In: Proceedings of the IEEE Conference on Computer Vision and Pattern Recognition, pp. 2818–2826 (2016)
21. Tao, M., Tang, H., Wu, F., Jing, X.Y., Bao, B.K., Xu, C.: DF-GAN: a simple and effective baseline for text-to-image synthesis. In: Proceedings of the IEEE/CVF Conference on Computer Vision and Pattern Recognition, pp. 16515–16525 (2022)
22. Wang, Z., Bovik, A.C., Sheikh, H.R., Simoncelli, E.P.: Image quality assessment: from error visibility to structural similarity. IEEE Trans. Image Process. **13**(4), 600–612 (2004)
23. Xu, T., et al.: AttnGAN: fine-grained text to image generation with attentional generative adversarial networks. In: Proceedings of the IEEE Conference on Computer Vision and Pattern Recognition, pp. 1316–1324 (2018)
24. Yang, X., Gireesh, N., Xing, E., Xie, P.: XrayGAN: consistency-preserving generation of X-ray images from radiology reports. arXiv preprint arXiv:2006.10552 (2020)
25. Zhang, H., et al.: StackGAN: text to photo-realistic image synthesis with stacked generative adversarial networks. In: Proceedings of the IEEE International Conference on Computer Vision, pp. 5907–5915 (2017)

26. Zhang, Y., Chen, Q., Yang, Z., Lin, H., Lu, Z.: BioWordVec, improving biomedical word embeddings with subword information and MeSH. Scientific data **6**(1), 52 (2019)
27. Zhu, M., Pan, P., Chen, W., Yang, Y.: DM-GAN: Dynamic memory generative adversarial networks for text-to-image synthesis. In: Proceedings of the IEEE/CVF Conference on Computer Vision and Pattern Recognition, pp. 5802–5810 (2019)

Contrastive Masked Image-Text Modeling for Medical Visual Representation Learning

Cheng Chen[1], Aoxiao Zhong[2], Dufan Wu[1], Jie Luo[1], and Quanzheng Li[1,3](✉)

[1] Center for Advanced Medical Computing and Analysis, Massachusetts General Hospital and Harvard Medical School, Boston, MA, USA
[2] School of Engineering and Applied Sciences, Harvard University, Boston, MA, USA
[3] Data Science Office, Massachusetts General Brigham, Boston, MA, USA
`li.quanzheng@mgh.harvard.edu`

Abstract. Self-supervised learning (SSL) of visual representations from paired medical images and text reports has recently shown great promise for various downstream tasks. However, previous work has focused on investigating the effectiveness of two major SSL techniques separately, i.e., contrastive learning and masked autoencoding, without exploring their potential synergies. In this paper, we aim to integrate the strengths of these two techniques by proposing a contrastive masked image-text modeling framework for medical visual representation learning. On one hand, our framework conducts cross-modal contrastive learning between masked medical images and text reports, with a representation decoder being incorporated to recover the misaligned information in the masked images. On the other hand, to further leverage masked autoencoding, a masked image is also required to be able to reconstruct the original image itself and the masked information in the text reports. With pre-training on a large-scale medical image and report dataset, our framework shows complementary benefits of integrating the two SSL techniques on four downstream classification datasets. Extensive evaluations demonstrate consistent improvements of our method over state-of-the-art approaches, especially when very scarce labeled data are available. code is available at https://github.com/cchen-cc/CMITM.

Keywords: Image-text representation learning · masked autoencoding · contrastive learning

1 Introduction

Deep learning models have demonstrated undoubted potential in achieving expert-level medical image interpretation when powered by large-scale labeled datasets [4,7,20]. However, medical image annotations require expert knowledge thus are extremely costly and difficult to obtain at scale. Such an issue has even become a bottleneck for advancing deep learning models in medical applications. To tackle this issue, recent efforts have resorted to the text reports that

H. Greenspan et al. (Eds.): MICCAI 2023, LNCS 14224, pp. 493–503, 2023.
https://doi.org/10.1007/978-3-031-43904-9_48

are paired with the medical images, aiming to leverage the detailed text interpretation provided by radiologists to assist the representation learning of medical images without relying on any manual labels [3,10,18,25–27]. The learned image representations have proven to be generalizable to other downstream tasks of medical image analysis, which can significantly reduce the amount of labeled data required for fine-tuning. This topic has been actively studied and seen rapid progress with evaluation on chest X-ray datasets, because of both the clinical importance of radiograph screening and the availability of large-scale datasets of public chest X-ray images with paired radiology reports [14].

The current mainstream approaches for medical image-text pre-training are based on the popular self-supervised learning (SSL) technique known as contrastive learning [2,9], which maximizes agreement between global and local representations of paired images and reports with a contrastive loss [1,10,15, 22,26,27]. These methods have demonstrated the effectiveness of using medical reports as a form of free supervision to enhance the learning of general image representations. Another well-demonstrated SSL method is masked autoencoding, which achieves representation learning via solving the pretext task of recovering masked image patches with unmasked ones [8]. Until very recently, the potential of masked autoencoding had only begun to be explored for medical image-text pre-training. In a latest work, Zhou et al. [28] propose to learn radiograph representations with a unified framework that requires the unmasked image patches to recover masked images and complete text reports. While both contrastive learning and masked autoencoding have demonstrated their ability to learn effective image representations, the two major SSL techniques have only been separately explored. The latest research has started to combine the two SSL techniques for joint benefits, but their focus is on the image domain rather than cross-modal learning in medical images and paired text reports [11,13,17]. It remains an interesting and unexplored question whether contrastive learning and masked autoencoding can benefit each other and how to jointly exploit their strengths for medical image-text pre-training.

In fact, the learning principles of contrastive learning and masked autoencoding suggest that they could be complementary to each other. Contrastive image-text learning explicitly discriminates the positive and negative pairs of images and text reports, making it good at promoting strong discriminative capabilities of image representations. Instead, masked autoencoding aims to reconstruct masked image/text tokens, which emphasizes learning local image structures, but may be less effective in capturing discriminative representations. This motivates us to propose a novel contrastive masked image-text modeling (CMITM) method for medical visual representation learning. Our framework is designed to accomplish three self-supervised learning tasks: First, aligning the representations of masked images with text reports. Second, reconstructing the masked images themselves. Third, reconstructing the masked text reports using the learned image representations. To reduce the information misalignment between the masked images and text reports, we incorporate a representation decoder to recover the missed information in images, which benefits the cross-modal learning. Moreover, the synergy of contrastive learning and masked autoencoding is

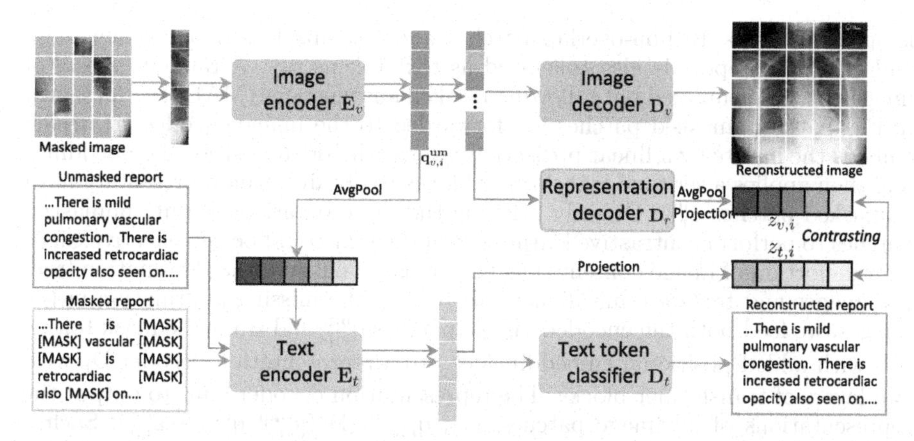

Fig. 1. Overview of proposed *contrastive masked image-text modeling* (CMITM) framework for medical visual representation learning. The masked image patches are required to align with the text reports, reconstructing original images, and reconstructing original text reports. The data flow in gray and orange line corresponds to first- and second-stage of pre-training respectively.

unleashed via a cascaded training strategy. Our framework is pre-trained on a large-scale medical dataset MIMIC-CXR with paired chest X-ray images and reports, and is extensively validated on four downstream classification datasets with improved fine-tuning performance. Combining the two techniques yields consistent performance increase and the improvement of our method even surpasses the benefits of adding data from 1% to 100% labels on CheXpert dataset.

2 Method

Figure 1 illustrates our contrastive masked image-text modeling (CMITM) framework. In this section, we first present how the cross-modal contrastive learning and masked autoencoding are realized in the framework respectively. Then we introduce the training procedures and implementation details of the framework.

2.1 Cross-Modal Contrastive Learning with Masked Images

Cross-modal contrastive learning has demonstrated to be an effective tool to align the representations of a medical image with that of its paired text report. In this way, the network is guided to interpret the image contents with the knowledge provided by medical reports. Different from previous methods, the cross-modal contrastive learning in our framework is between the representations of masked images and unmasked reports, aiming to integrate the benefits of both contrastive learning and masked image modeling.

Specifically, denote $\mathcal{D} = \{x_{v,i}, x_{t,i}\}_{i=1}^{N}$ as the multi-modal dataset consisting of N pairs of medical images $x_{v,i}$ and medical reports $x_{t,i}$. Each input image is split into 16×16 non-overlap patches and tokenized as image tokens $a_{v,i}$, and each text report is also tokenized as text tokens $a_{t,i}$. A random subset of image patches is masked out following masked autoencoder (MAE) [8]. As shown in Fig. 1, the unmasked patches are forwarded to the image encoder E_v, which embeds the inputs by a linear projection layer with added positional embeddings and then applies a series of transformer blocks to obtain token representations of unmasked patches $\mathrm{q}^{\mathrm{um}}_{v,i}$. Directly utilizing the representations of only unmasked patches to perform contrastive learning with the text could be less effective, since a large portion of image contents has been masked out and the information from the images and texts are misaligned. To recover the missing information in the images, we feed both the encoded visible patches $\mathrm{q}^{\mathrm{um}}_{v,i}$ and trainable mask tokens $\mathrm{q}^{\mathrm{m}}_{v,i}$ with added positional embeddings $\mathrm{e}^{p}_{v,i}$ to a representation decoder D_r with two layers of transformer blocks. The representation decoder aims to output the representations of all image patches, i.e., $\hat{\mathrm{q}}_{v,i} = \mathrm{D}_r([\mathrm{q}^{\mathrm{um}}_{v,i}, \mathrm{q}^{\mathrm{m}}_{v,i}] + \mathrm{e}^{p}_{v,i})$. Such a design helps to avoid that the contrastive learning is confused by the misaligned information between masked images and text reports. Finally we apply a global average pooling operation and a project layer h_v to obtain the image embeddings $z_{v,i}$, i.e., $z_{v,i} = \mathrm{h}_v(\mathrm{AvgPool}(\hat{\mathrm{q}}_{v,i}))$. For text branch, we consider the full reports could give more meaningful guidance to image understanding than masked ones. So we forward the full text tokens without masking to the text encoder E_t and the text project layer h_t to obtain the global text embeddings $z_{t,i}$, i.e., $z_{t,i} = \mathrm{h}_t(\mathrm{E}_t(a_{t,i}))$. To ensure that the embeddings of images are aligned with those of paired texts while remaining distant from unpaired texts, we employ cross-modal contrastive learning with the following symmetric InfoNCE loss [19].

$$\mathcal{L}_c = -\frac{1}{B}\sum_{i=1}^{B}\left[\log\frac{\exp(s^{vt}_{i,i}/\tau)}{\sum\exp(s^{vt}_{i,k}/\tau)} + \log\frac{\exp(s^{tv}_{i,i}/\tau)}{\sum\exp(s^{tv}_{i,k}/\tau)}\right], \tag{1}$$

where $s^{vt}_{i,j} = z^{T}_{v,i} z_{t,j}$, $s^{tv}_{i,j} = z^{T}_{t,i} z_{v,j}$, τ denotes the temperature which is set to be 0.07 following common practice, and B is the number of image-report pairs in a batch. The cross-modal contrastive loss is used to supervise the network training associated with the data flow in orange line in Fig. 1.

2.2 Masked Image-Text Modeling

The masked image-text modeling component in our framework consists of two parallel tasks, i.e., masked image reconstruction with image only information and masked text reconstruction with cross-modal information. We follow the design in [8,28] for the masked image-text modeling since our main focus is whether masked autoencoding and contrastive learning can have joint benefits.

Masked Image Reconstruction. As aforementioned, the input images are masked and processed by the image encoder E_v to obtain $\mathrm{q}^{\mathrm{um}}_{v,i}$. As shown in Fig. 1, besides the representation decoder to reconstruct image representations, we also connect both the encoded visible patches and learnable unmasked tokens with added positional embeddings to an image decoder D_v to reconstruct the pixel values of masked patches, i.e., $\hat{x}_{v,i} = \mathrm{D}_v([\mathrm{q}^{\mathrm{um}}_{v,i}, \mathrm{q}^{\mathrm{m}}_{v,i}] + \mathrm{e}^{p}_{v,i}))$. The image decoder consists of a series of transformer blocks and a linear projection layer to predict the values for each pixel in a patch. To enhance the learning of local details, the image decoder is required to reconstruct a high-resolution patch, which is twice the input resolution [28]. The training of image reconstruction is supervised by a mean squared error loss $\mathcal{L}_{vr}$ which computes the difference between the reconstructed and original pixel values for the masked patches:

$$\mathcal{L}_{vr} = \frac{1}{B}\sum_{i=1}^{B}\frac{\sum(\hat{x}_{v,i}^{\mathrm{m}} - \mathrm{norm}(\tilde{x}_{v,i}^{\mathrm{m}}))^2}{|\tilde{x}_{v,i}^{\mathrm{m}}|}, \tag{2}$$

where $\hat{x}_{v,i}^{\mathrm{m}}$, $\tilde{x}_{v,i}^{\mathrm{m}}$ denote the predicted and the original high-resolution masked patches, $|\cdot|$ calculates the number of masked patches, and norm denotes the pixel normalization with the mean and standard deviation of all pixels in a patch suggested in MAE [8]. The loss is only computed on the masked patches.

Cross-Modal Masked Text Reconstruction. To make the most of the text reports paired with imaging data for learning visual representations, the task of cross-modal masked text modeling aims to encourage the encoded visible image tokens $\mathrm{q}_{v,i}^{\mathrm{um}}$ to participate in completing the masked text reports. Specifically, besides the full texts, we also forward a masked text report with a masking ratio of 50% to the text encoder E_t. Following [28], this value is deliberately set to be higher than the masking ratio of 15% in BERT [5] in order to enforce the image encoder to better understand the image contents by trying to reconstruct a large portion of masked texts. Then the global embedding of corresponding unmasked image patches $\mathrm{AvgPool}(\mathrm{q}_{v,i}^{\mathrm{um}})$ is added to the text token embeddings $\mathrm{q}_{t,i}^{\mathrm{um}}$ to form a multi-modal embeddings. To reconstruct the masked text tokens, the multi-modal embeddings are processed by the text encoder E_t and a text token classifier D_t, i.e., $\hat{a}_{t,i} = \mathrm{D}_t(\mathrm{E}_t(\mathrm{q}_{t,i}^{\mathrm{um}} + \mathrm{AvgPool}(\mathrm{q}_{v,i}^{\mathrm{um}})))$. The training of text reconstruction is supervised by the cross entropy loss between the predictions and original text tokens as follows:

$$\mathcal{L}_{tr} = \frac{1}{B}\sum_{i=1}^{B}\mathcal{H}(a_{t,i}^{\mathrm{m}}, \hat{a}_{t,i}^{\mathrm{m}}), \tag{3}$$

where $a_{t,i}^{\mathrm{m}}$, $\hat{a}_{t,i}^{\mathrm{m}}$ denote the original and recovered masked text tokens respectively, $\mathcal{H}$ denotes the cross entropy loss. Similar to masked image reconstruction, the loss is also only computed on the masked text tokens. The image and text reconstruction losses are used to supervise the network training associated with the data flow in gray line in Fig. 1.

2.3 Training Procedures and Implementation Details

Training a framework that combines cross-modal contrastive learning and masked autoencoding is non-trivial. As observed in prior work [13], forming the training as a parallel multi-task learning task can lead to decreased performance, which might be caused by the conflicting gradients of the contrastive and reconstruction losses. Similar phenomenon has also been observed in our experiments. We therefore adopt a cascaded training strategy, that is the framework is first trained with the reconstruction loss $\mathcal{L}_r = \mathcal{L}_{vr} + \mathcal{L}_{tr}$ and is further trained with the contrastive loss $\mathcal{L}_c$. Such a training order is considered based on the insights that masked autoencoding focuses more on the lower layers with local details while contrastive learning is effective in learning semantic information for higher layers. Specifically, the first stage of pre-training follows [28] to employ the loss $\mathcal{L}_r$ for training 200 epochs. The model is trained using AdamW [16] optimizer with a learning rate of 1.5e-4 and a weight decay of 0.05. In the second stage of pre-training, the image encoder, text encoder, and representation decoder are further trained with the loss $\mathcal{L}_c$ for 50 epochs. Similarly a AdamW optimizer with a learning rate of 2e-5 and a weight decay of 0.05 is adopted. The framework is implemented on 4 pieces of Tesla V100 GPU with a batch size of 256. For network configurations, we use ViT-B/16 [6] as the image encoder and BERT [5] as the text encoder. The image decoder and representation decoder consists of four transformer blocks and two transformer blocks respectively.

3 Experiments

To validate our framework, the image encoder of the pre-trained model is used to initialize a classification network with a ViT-B/16 backbone and a linear classification head. We adopt the fine-tuning strategy as used in [26,28], where both the encoder and classification head are fine-tuned. This fine-tuning setting reflects how the pre-trained weights can be applied in practical applications. For each dataset, the model is fine-tuned with 1%, 10%, and 100% labeled training data to extensively evaluate the data efficiency of different pre-trained models. The dataset split remains consistent across all approaches.

Pre-training Dataset. To pre-train our framework, we utilize **MIMIC-CXR** dataset [14], which is a large public chest X-ray collection. The dataset contains 377,110 images extracted from 227,835 radiographic studies and each radiograph is associated with one radiology report. For pre-training, images are resized and randomly cropped into the size of 448×448 and 224×224 as the high-resolution image reconstruction ground truth and low-resolution inputs respectively.

Fine-Tuning Datasets. We transfer the learned image representations to four datasets for chest X-ray classification. **NIH ChestX-ray** [24] includes 112,120 chest X-ray images with 14 disease labels. Each chest radiograph can associate with multiple diseases. We follow [28] to split the dataset into 70%/10%/20% for training, validation, and testing. **CheXpert** [12] comprises 191,229 chest X-ray images for multi-label classification, i.e., atelectasis, cardiomegaly, consolidation, edema, and pleural effusion. We follow previous work to use the official validation set as the test images and randomly split 5,000 images from training data as the validation set. **RSNA** Pneumonia [21] dataset contains 29,684 chest X-rays for a binary classification task of distinguishing between normal and pneumonia. Following [28], the dataset is split as training/validation/testing with 25,184/1,500/3,000 images respectively. **COVIDx** [23] is a three-class classification dataset with 29,986 chest radiographs from 16,648 patients. The task

Table 1. Quantitative comparison with different pre-training methods on four chest X-ray datasets when fine-tuning with 1%, 10%, and 100% training data.

Methods	NIH X-ray (AUC)			CheXpert (AUC)			RSNA (AUC)			COVIDx (ACC)		
	1%	10%	100%	1%	10%	100%	1%	10%	100%	1%	10%	100%
Random init	60.0	65.2	72.8	70.4	81.1	85.8	71.9	82.2	88.5	64.2	75.4	87.7
ImageNet init	69.8	74.4	80.0	80.1	84.8	87.6	83.1	87.3	90.8	72.0	84.4	90.3
MAE [8]	74.7	81.3	85.1	80.7	86.0	86.7	84.2	89.6	91.3	69.8	82.3	90.7
MRM [28]	79.4	84.0	85.9	88.5	88.5	88.7	91.3	**92.7**	93.3	78.0	90.2	92.5
GLoRIA [10]	77.7	82.8	85.0	86.5	87.5	87.8	89.7	91.2	92.1	76.7	**91.7**	94.8
MGCA [22]	78.2	82.7	85.0	87.0	88.4	88.5	90.7	92.6	**93.4**	75.2	91.5	94.3
CMITM (ours)	**80.4**	**84.1**	**86.0**	**89.0**	**89.0**	**89.2**	**91.6**	92.6	**93.4**	**79.5**	90.2	**95.3**

is to classify each image into normal, COVID-19 and non-COVID pneumonia. Same as [22], we use the official validation data as the test data and split 10% images from the training data for validation. For fine-tuning, images are also resized and randomly cropped into the size of 224 × 224.

Comparison with State-of-the-Art Methods. We compare our method with four state-of-the-art SSL methods including two masked autoencoding-based methods and two contrastive learning-based methods. **MAE (CVPR 2022)** [8] is the representative work on masked image autoencoding. **MRM (ICLR 2023)** [28] is the latest work on medical image-text pre-training by using both the self- and report-completion objectives based on the masked record modeling. **GLoRIA (ICCV 2021)** [10] and **MGCA (NeurIPS 2022)** [22] are two cross-modal medical visual representation learning methods based on multi-scale image-text contrastive learning. For a fair comparison, the results of MRM model on the datasets NIH ChestX-ray, CheXpert, and RSNA are directly obtained from original paper since we use the same data split and fine-tuning strategy as theirs. The other comparison results are obtained by re-implementing corresponding methods with their released code with the same network backbone and fine-tuning strategy as ours. We also compare with models fine-tuned with random initialization and with weights pre-trained on ImageNet data, denoted as "Random init" and "ImageNet init" respectively. We use the area under the ROC curve (AUC) on NIH ChestX-ray, CheXpert, and RSNA datasets and accuracy (ACC) on COVIDx dataset as the evaluation metric following [22,28].

Table 1 shows the results on four downstream datasets for chest X-ray classification. We can see that, compared to "Random init" and "ImageNet init" models, pre-training on medical datasets significantly improve the fine-tuning performance on all the datasets. This shows the importance of medical visual representation learning. Compared to MAE model that only uses image data for pre-training, the other methods leveraging cross-modal image-text pre-training obtain higher performance, demonstrating the great benefits of detailed description in text reports. Our CMITM model generally outperforms methods that use

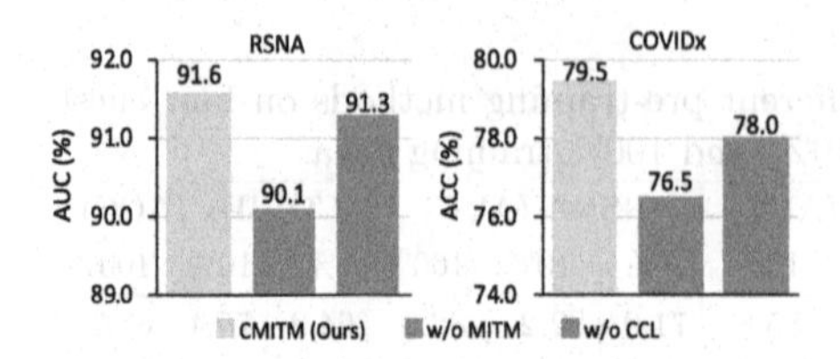

Fig. 2. Effects of two SSL components in our method (MITM and CCL).

Table 2. Ablation on removing either design in our framework.

Methods	RSNA	COVIDx
CMITM (ours)	**91.6**	**79.5**
- cascaded training	90.8	76.2
- image masking	90.9	77.0
- representation decoder	91.2	79.0

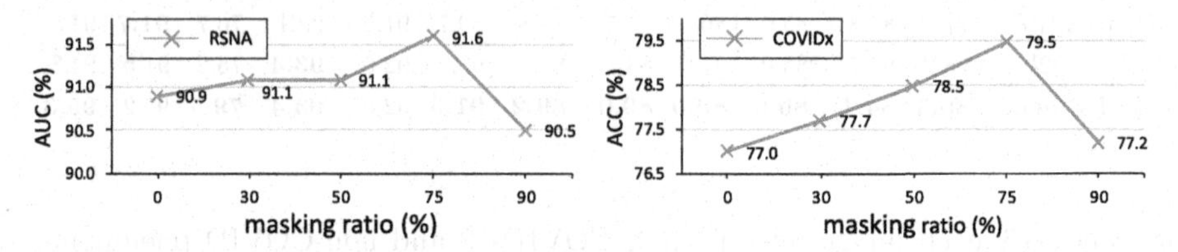

Fig. 3. Effect of masking ratio for cross-modal contrastive learning in our method.

either masked autoencoding or contrastive learning alone, showing the effectiveness of combining the two SSL techniques. It can be observed that our method shows the most obvious improvements in scenarios with limited data. When fine-tuning with 1% labeled data, our method outperforms the current best-performing method MRM by 1.0% on NIH ChestX-ray dataset and 1.5% on COVIDx dataset. On CheXpert dataset, MRM model shows 0.2% performance gain when increasing labeled data from 1% to 100%, but with our method, fine-tuning on 1% labeled data already outperforms MRM model fine-tuned on 100% labeled data. These results may indicate that masked autoencoding and contrastive learning benefit each other more in data-scare scenarios.

Ablation Study. We perform ablation analysis on RSNA and COVIDx datasets with 1% labeled data to investigate the effect of each component in the proposed method. In Fig. 2, we can see that removing either the masked image-text modeling (MITM) or cross-modal contrastive learning (CCL) in our method lead to a decrease in fine-tuning results. This again reflects the complementary role of the two SSL components. In Table 2, we show that the designs of cascaded training strategy, image masking in the contrastive learning, and representation decoder play important role to the performance of our method. Notably, results significantly decrease if not using cascaded training, that means directly using reconstruction loss and contrastive loss for joint training bring negative effects. These ablation results show that combing the two SSL approaches is non-trivial and requires careful designs to make it to be effective. For masking ratio, previous work [28] has shown that the masking ratio of 75% works well in masked medical image-text reconstruction. So we directly adopt the masking ratio of 75% for the masked image-text modeling in the first pre-training stage, but we analyze how the performance changes with the masking ratio for the contrastive learning in the second pre-training stage. As shown in Fig. 3, it is interesting to see that the optimal masking ratio is also 75%. This might indicate that the masking ratio should keep as the same during the cascaded training.

4 Conclusion

We present a novel framework for medical visual representation learning by integrating the strengths of both cross-modal contrastive learning and masked image-text modeling. With careful designs, the effectiveness of our method is demonstrated on four downstream classification datasets, consistently improving data efficiency under data-scarce scenarios. This shows the complementary benefits of the two SSL techniques in medical visual representation learning. One limitation of the work is that the pre-training model is evaluated solely on classification tasks. A compelling extension of this work would be to conduct further evaluation on a broader spectrum of downstream tasks, including organ segmentation, lesion detection, and image retrieval, thereby providing a more comprehensive evaluation of our model's capabilities.

Acknowledgements. This work was supported by NIH R01HL159183.

References

1. Boecking, B., et al.: Making the most of text semantics to improve biomedical vision–language processing. In: Avidan, S., Brostow, G., Cissé, M., Farinella, G.M., Hassner, T. (eds.) Computer Vision – ECCV 2022. ECCV 2022. Lecture Notes in Computer Science, vol. 13696, pp. 1–21. Springer, Cham (2022). https://doi.org/10.1007/978-3-031-20059-5_1
2. Chen, T., Kornblith, S., Norouzi, M., Hinton, G.: A simple framework for contrastive learning of visual representations. In: International conference on machine learning, pp. 1597–1607. PMLR (2020)
3. Chen, Z., Du, Y., Hu, J., Liu, Y., Li, G., Wan, X., Chang, T.: Multi-modal masked autoencoders for medical vision-and-language pre-training. In: Wang, L., Dou, Q., Fletcher, P.T., Speidel, S., Li, S. (eds.) Medical Image Computing and Computer Assisted Intervention - MICCAI, Singapore, September 18–22, 2022, Proceedings, Part V, vol. 13435, pp. 679–689. Springer (2022), https://doi.org/10.1007/978-3-031-16443-9_65
4. De Fauw, J., et al.: Clinically applicable deep learning for diagnosis and referral in retinal disease. Nat. Med. **24**(9), 1342–1350 (2018)
5. Devlin, J., Chang, M., Lee, K., Toutanova, K.: BERT: pre-training of deep bidirectional transformers for language understanding. In: Burstein, J., Doran, C., Solorio, T. (eds.) Proceedings of the 2019 Conference of the North American Chapter of the Association for Computational Linguistics: Human Language Technologies, NAACL-HLT, Minneapolis, MN, USA, June 2–7, 2019, vol. 1, pp. 4171–4186 (2019), https://doi.org/10.18653/v1/n19-1423
6. Dosovitskiy, A., et al.: An image is worth 16×16 words: transformers for image recognition at scale. In: 9th International Conference on Learning Representations, ICLR (2021)

7. Esteva, A., et al.: Dermatologist-level classification of skin cancer with deep neural networks. Nature **542**(7639), 115–118 (2017)
8. He, K., Chen, X., Xie, S., Li, Y., Dollár, P., Girshick, R.: Masked autoencoders are scalable vision learners. In: Proceedings of the IEEE/CVF Conference on Computer Vision and Pattern Recognition, pp. 16000–16009 (2022)
9. He, K., Fan, H., Wu, Y., Xie, S., Girshick, R.: Momentum contrast for unsupervised visual representation learning. In: Proceedings of the IEEE/CVF Conference on Computer Vision and Pattern Recognition, pp. 9729–9738 (2020)
10. Huang, S.C., Shen, L., Lungren, M.P., Yeung, S.: Gloria: a multimodal global-local representation learning framework for label-efficient medical image recognition. In: Proceedings of the IEEE/CVF International Conference on Computer Vision, pp. 3942–3951 (2021)
11. Huang, Z., et al.: Contrastive masked autoencoders are stronger vision learners. arXiv preprint arXiv:2207.13532 (2022)
12. Irvin, J., et al.: CheXpert: a large chest radiograph dataset with uncertainty labels and expert comparison. In: Proceedings of the AAAI Conference on Artificial Intelligence, vol. 33, pp. 590–597 (2019)
13. Jiang, Z., et al.: Layer grafted pre-training: bridging contrastive learning and masked image modeling for label-efficient representations. arXiv preprint arXiv:2302.14138 (2023)
14. Johnson, A.E., et al.: MIMIC-CXR, a de-identified publicly available database of chest radiographs with free-text reports. Sci. Data **6**(1), 317 (2019)
15. Liao, R., et al.: Multimodal representation learning via maximization of local mutual information. In: de Bruijne, M., Cattin, P.C., Cotin, S., Padoy, N., Speidel, S., Zheng, Y., Essert, C. (eds.) Medical Image Computing and Computer Assisted Intervention - MICCAI, Strasbourg, France, September 27 - October 1, 2021, Proceedings, Part II, vol. 12902, pp. 273–283. Springer (2021), https://doi.org/10.1007/978-3-030-87196-3_26
16. Loshchilov, I., Hutter, F.: Decoupled weight decay regularization. arXiv preprint arXiv:1711.05101 (2017)
17. Mishra, S., et al.: A simple, efficient and scalable contrastive masked autoencoder for learning visual representations. arXiv preprint arXiv:2210.16870 (2022)
18. Müller, P., Kaissis, G., Zou, C., Rueckert, D.: Radiological reports improve pre-training for localized imaging tasks on chest X-rays. In: Wang, L., Dou, Q., Fletcher, P.T., Speidel, S., Li, S. (eds.) Medical Image Computing and Computer Assisted Intervention - MICCAI, Singapore, September 18–22, 2022, Proceedings, Part V, vol. 13435, pp. 647–657. Springer (2022), https://doi.org/10.1007/978-3-031-16443-9_62
19. Oord, A.v.d., Li, Y., Vinyals, O.: Representation learning with contrastive predictive coding. arXiv preprint arXiv:1807.03748 (2018)
20. Rajpurkar, P., et al.: CheXNet: radiologist-level pneumonia detection on chest X-rays with deep learning. arXiv preprint arXiv:1711.05225 (2017)
21. Shih, G., et al.: Augmenting the national institutes of health chest radiograph dataset with expert annotations of possible pneumonia. Radiol. Artif. Intell. **1**(1), e180041 (2019)
22. Wang, F., Zhou, Y., Wang, S., Vardhanabhuti, V., Yu, L.: Multi-granularity cross-modal alignment for generalized medical visual representation learning. Adv. Neural. Inf. Process. Syst. **35**, 33536–33549 (2022)
23. Wang, L., Lin, Z.Q., Wong, A.: Covid-Net: a tailored deep convolutional neural network design for detection of Covid-19 cases from chest X-ray images. Sci. Rep. **10**(1), 1–12 (2020)

24. Wang, X., Peng, Y., Lu, L., Lu, Z., Bagheri, M., Summers, R.: ChestX-ray8: hospital-scale chest X-ray database and benchmarks on weakly-supervised classification and localization of common thorax diseases. In: 2017 IEEE Conference on Computer Vision and Pattern Recognition (CVPR), pp. 3462–3471 (2017)
25. Wu, C., Zhang, X., Zhang, Y., Wang, Y., Xie, W.: Medklip: medical knowledge enhanced language-image pre-training. medRxiv pp. 2023–01 (2023)
26. Zhang, Y., Jiang, H., Miura, Y., Manning, C.D., Langlotz, C.P.: Contrastive learning of medical visual representations from paired images and text. In: Machine Learning for Healthcare Conference, pp. 2–25. PMLR (2022)
27. Zhou, H.Y., Chen, X., Zhang, Y., Luo, R., Wang, L., Yu, Y.: Generalized radiograph representation learning via cross-supervision between images and free-text radiology reports. Nature Mach. Intell. **4**(1), 32–40 (2022)
28. Zhou, H., Lian, C., Wang, L., Yu, Y.: Advancing radiograph representation learning with masked record modeling. In: The Eleventh International Conference on Learning Representations, ICLR 2023, Kigali, Rwanda, May 1–5, 2023 (2023)

Adjustable Robust Transformer for High Myopia Screening in Optical Coherence Tomography

Xiao Ma[1], Zetian Zhang[1], Zexuan Ji[1], Kun Huang[1], Na Su[2], Songtao Yuan[2], and Qiang Chen[1(✉)]

[1] School of Computer Science and Engineering, Nanjing University of Science and Technology, Nanjing, China
chen2qiang@njust.edu.cn
[2] Department of Ophthalmology, The First Affiliated Hospital of Nanjing Medical University, Nanjing, China

Abstract. Myopia is a manifestation of visual impairment caused by an excessively elongated eyeball. Image data is critical material for studying high myopia and pathological myopia. Measurements of spherical equivalent and axial length are the gold standards for identifying high myopia, but the available image data for matching them is scarce. In addition, the criteria for defining high myopia vary from study to study, and therefore the inclusion of samples in automated screening efforts requires an appropriate assessment of interpretability. In this work, we propose a model called adjustable robust transformer (ARTran) for high myopia screening of optical coherence tomography (OCT) data. Based on vision transformer, we propose anisotropic patch embedding (APE) to capture more discriminative features of high myopia. To make the model effective under variable screening conditions, we propose an adjustable class embedding (ACE) to replace the fixed class token, which changes the output to adapt to different conditions. Considering the confusion of the data at high myopia and low myopia threshold, we introduce the label noise learning strategy and propose a shifted subspace transition matrix (SST) to enhance the robustness of the model. Besides, combining the two structures proposed above, the model can provide evidence for uncertainty evaluation. The experimental results demonstrate the effectiveness and reliability of the proposed method. Code is available at: https://github.com/maxiao0234/ARTran.

Keywords: High myopia screening · Optical coherence tomography · Adjustable model · Label noise learning

1 Introduction

Myopia, resulting in blurred distance vision, is one of the most common eye diseases, with a rising prevalence around the world, particularly among schoolchildren and young adults [11,19,22]. Common wisdom attributes the causes of myopia to excessive elongation of the eyeball, the development of which can

H. Greenspan et al. (Eds.): MICCAI 2023, LNCS 14224, pp. 504–514, 2023.
https://doi.org/10.1007/978-3-031-43904-9_49

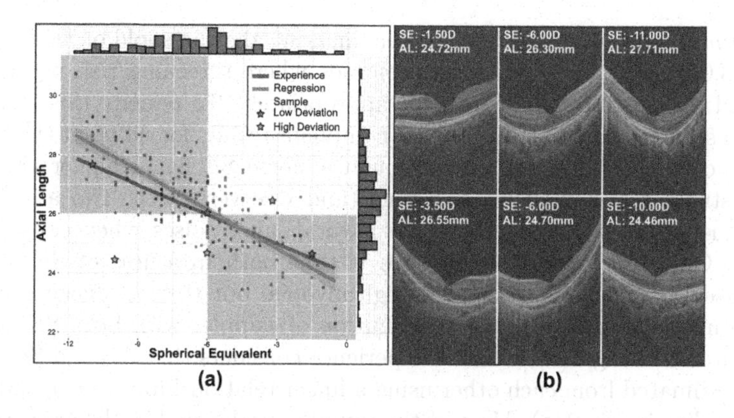

Fig. 1. Illustration of spherical equivalent (SE) and axial length (AL) for some samples in the dataset. (a) Scatterplot on two measured values. The red line indicates the rough estimation relationship from clinical experience, the blue line indicates the linear representation of regression. (b) OCT examples for low deviation of the experience trend (the top row) and with high deviation (the bottom row). (Color figure online)

continue throughout childhood, and especially in patients with high myopia, throughout adulthood [1,5]. In the coming decades, the prognosis for patients with high myopia will continue to deteriorate, with some developing pathological myopia, leading to irreversible vision damage involving posterior scleral staphyloma, macular retinoschisis, macular hole, retinal detachment, choroidal neovascularization, dome-shaped macula, etc. [3,7,12,23].

As a crucial tool in the study of high myopia, fundus images demonstrate the retinal changes affected by myopia. Myopic maculopathy in color fundus photographs (CFP) can be important evidence in the evaluation of high myopia and myopic fundus diseases [21]. However, some myopic macular lesions such as myopic traction maculopathy and domeshaped macula are usually not observed in CFP. Optical coherence tomography (OCT), characterized by non-invasive and high-resolution three-dimensional retinal structures, has more advantages in the examination of high myopia [9,15]. Several studies have used convolutional neural networks to automatically diagnose high myopia and pathological myopia [6,16]. Choi *et al.* employed two ResNet-50 [10] networks to inference vertical and horizontal OCT images simultaneously. Li *et al.* introduced focal loss into Inception-Resnet-V2 [24] to enhance its identification ability. However, the different classes of images in these tasks have significant differences, and the performance of the model, when used for more complex screening scenarios, is not discussed. Hence, this work aims to design a highly generalizable screening model for high myopia on OCT images.

There exist challenges to developing an automatic model that meets certain clinical conditions. For the inclusion criteria for high myopia, different studies will expect different outputs under different thresholds. High myopia is defined by a spherical equivalent (SE) ≤ -6.0 dioptres (D) or an axial length (AL)

$\geq 26.0mm$ in most cases, but some researchers set the threshold of SE to $-5.0D$ [4] or $-8.0D$ [20]. Moreover, some scenarios where screening risk needs to be controlled may modify the thresholds appropriately. To remedy this issue, the screening scheme should ideally output compliant results for different thresholds. For the accuracy of supervision, although a worse SE has a higher chance of causing structural changes in the retina (and vice versa), the two are not the single cause and effect, i.e., there are natural label noises when constructing datasets. One direct piece of evidence is that both measures of SE and AL can be used as inclusion criteria for high myopia, but there is disagreement in some samples. We illustrate the distribution of samples with both SE and AL in our dataset in Fig. 1(a). Clinical experience considers that AL and SE can be roughly estimated from each other using a linear relationship, which is indicated by the red line in Fig. 1(a). Most of the samples are located in the area (purple) that satisfies both conditions, but the remaining samples can only satisfy one. In more detail, Fig. 1(b) shows three samples with a low deviation of the experience trend (the top row) and three samples with a high deviation (the bottom row), where the worse SE does not always correspond to longer AL or more retinal structural changes. To mitigate degenerate representation caused by noisy data, the screening scheme should avoid over-fitting of extremely confusing samples. Besides, rich interpretable decisions support enhancing confidence in screening results. To this end, the screening scheme should evaluate the uncertainty of the results.

The contributions of our work are summarized as: (1) we propose a novel adjustable robust transformer (ARTran) for high myopia screening to adapt variable inclusion criteria. We design an anisotropic patch embedding (APE) to encode more myopia-related information in OCT images, and an adjustable class embedding (ACE) to obtain adjustable inferences. (2) We propose shifted subspace transition matrix (SST) to mitigate the negative impacts of label noise, which maps the class-posteriors to the corresponding distribution range according to the variable inclusion criteria. (3) We implement our ARTran on a high myopia dataset and verify the effectiveness of screening, and jointly use the proposed modules to generate multi-perspective uncertainty evaluation results.

2 Method

In this section, we propose a novel framework for high myopia screening in OCT called adjustable robust transformer (ARTran). This model can obtain the corresponding decisions based on the given threshold of inclusion criteria for high myopia and is trained end-to-end for all thresholds at once. During the testing phase, the screening results can be predicted interactively for a given condition.

2.1 Modified Vision Transformer for Screening

Transformers have shown promising performance in visual tasks attributed to long-range dependencies. Inspired by this, we use ViT [8] as the backbone of our

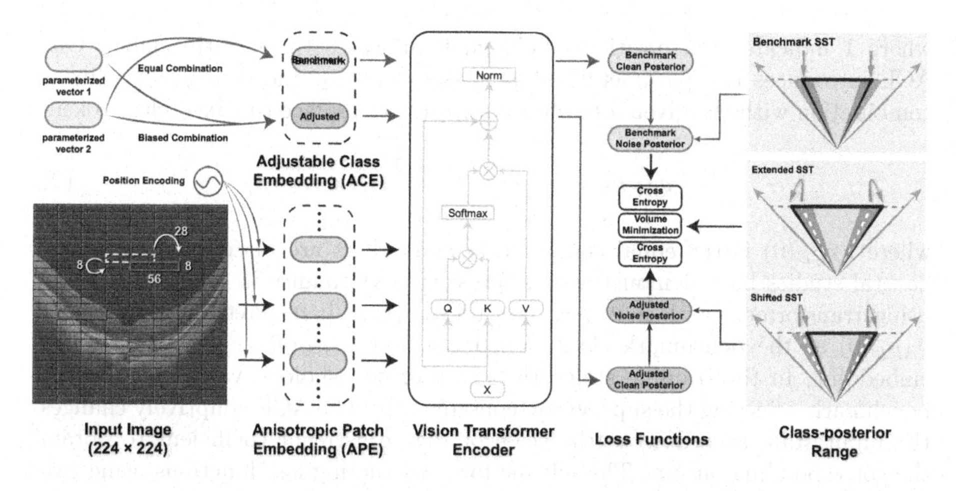

Fig. 2. The framework of our proposed adjustable robust transformer (ARTran). The proposed APE and ACE encode the image information and adjustment information as the input to transformer. The proposed SST learns the label noise and establishes the connection between the class-posteriors. The range of class-posteriors are also shown.

framework and make improvements for the task of screening high myopia OCT images.

Patients with high myopia are usually accompanied by directional structural changes, such as increased retinal curvature and choroidal thinning. On the other hand, due to the direction of light incidence perpendicular to the macula, OCT images have unbalanced information in the horizontal and vertical directions. Therefore, we propose a novel non-square strategy called anisotropic patch embedding (APE) to replace vanilla patch embedding for perceiving finer structural information, where an example is shown in Fig. 2. First, we reduced the height size of the patches. This strategy increases the number of patches per column, i.e., the number of patches within the retinal height range, which enhances the information perception of overall and individual layer thickness. In order not to introduce extra operations, we reduce the number of patches per row. We also use an overlapping sampling strategy for sliding windows in the horizontal direction, where a wider patch captures more information about the peripheral structure of the layer in which it is located. Specifically, based on the 16×16 size of ViT, our APE changed the height to 8 pixels and the width to 56 pixels with an overlap of 28 pixels, which keeps the number of embedded patches.

Considering that different researchers may have different requirements for inclusion criteria or risk control, we propose adjustable class embedding (ACE) to adapt to variable conditions. Take the $-6.0D$ as the benchmark of inclusion criteria, we define the relationship of the biased label, SE, and the adjustment coefficient Δ:

$$Y(SE = s, \Delta = \delta) := \begin{cases} 0 & , -0.25D \cdot \delta + s > -6.0D \\ 1 & , -0.25D \cdot \delta + s \leq -6.0D \end{cases}, -1 \leq \delta \leq 1 \quad (1)$$

where 1 indicates the positive label and 0 indicates the negative label. Our ACE introduces two parameterized vectors v_1 and v_2, and constructs a linear combination with the given δ to obtain $v_{APE}(\delta)$ to replace the fixed class token:

$$v_{APE}(\delta) = \frac{1+\delta}{2} \cdot v_1 + \frac{1-\delta}{2} \cdot v_2 \tag{2}$$

where $v_{APE}(0)$ is the equal combination and others are biased combinations. Several studies have demonstrated impressive performance in multi-label tasks using transformers with multiple class tokens [13,27]. Inspired by this, we set $v_{APE}(0)$ as the benchmark class embedding and $v_{APE}(\delta)$ as the biased class embedding. In the training stage, the inclusion threshold is varied around the benchmark, affecting the supervision consequently. The ACE adaptively changes the input state according to the scale of the adjustment coefficient to obtain the corresponding output. The scheme for constructing loss functions using two outputs is described in Sect. 2.2. In the testing stage, the ACE interactively makes the model output different results depending on the given conditions. Furthermore, we did not add the position encoding to ACE, so both tokens are position-independently involved to multi-head self-attention and are only distinguished by adjustment coefficient.

2.2 Shifted Subspace Transition Matrix

To enhance the robustness of the model, we follow the common assumptions of some impressive label noise learning methods [26,28]. Conventional wisdom is that the clean class-posterior $P(Y|X=x)$ can be inferred by utilizing the noisy class-posterior $P(\tilde{Y}|X=x)$ and the transition matrix $\boldsymbol{T}(x)$:

$$P(\tilde{\boldsymbol{Y}}|X=x) = \boldsymbol{T}(x)P(\boldsymbol{Y}|X=x) \tag{3}$$

where the transition matrix guarantees statistical consistency. Li *et al.* [14] have proved that identifying the transition matrix can be treated as the problem of recovering simplex for any x, i.e., $\boldsymbol{T}(x) = \boldsymbol{T}$. Based on this theory, in this work, we further propose a class-dependent and adjustment-dependent transition matrix $\boldsymbol{T}(x,\delta) = \boldsymbol{T}(\delta)$ called shifted subspace transition matrix (SST) to adapt to the variable class-posteriors distribution space.

Simplistically, this work only discusses the binary classification approach applicable to screening. For each specific inclusion threshold $\delta \in [-1,1]$, the range of the noisy class-posterior is determined jointly by the threshold and the SST:

$$\begin{bmatrix} P(\tilde{Y}=0|x,\delta) \\ P(\tilde{Y}=1|x,\delta) \end{bmatrix} = \begin{bmatrix} \boldsymbol{T}_{1,1}(\delta) & 1-\boldsymbol{T}_{2,2}(\delta) \\ 1-\boldsymbol{T}_{1,1}(\delta) & \boldsymbol{T}_{2,2}(\delta) \end{bmatrix} \begin{bmatrix} P(Y=0|x,\delta) \\ P(Y=1|x,\delta) \end{bmatrix} \tag{4}$$

where $T_{i,i}(\delta) > 0.5$ is the i_{th} diagonal element of SST, and the sum of each column of the matrix is 1. Thus any class-posterior of x is inside the simplex form columns of $\boldsymbol{T}(\delta)$ [2]. As shown in Fig. 2, the benchmark simplex is formed by $\boldsymbol{T}(0)$, where the orange arrows indicate the two sides of the simplex. When adjusting the inclusion criteria for high myopia, we expect the adjusted class-posterior to prefer same the adjustment direction compared to the benchmark. One solution is to offset both the upper and lower bounds of the class-posterior according to the scales of the adjustment coefficient:

$$\boldsymbol{T}_{i,i}(\delta) = \frac{1 + S(\theta_0) \cdot S(\theta_i)}{2} + \frac{1 - S(\theta_0)}{4} \cdot \begin{cases} 1 - \delta & ,i = 1 \\ 1 + \delta & ,i = 2 \end{cases} \tag{5}$$

where the $S(\cdot)$ is the *Sigmoid* function, θ_i is the parameter used only for column, and θ_0 is the parameter shared by both columns. Adjusting the δ is equivalent to shifting the geometric space of the simplex, where the spaces with a closer adjustment coefficient share a more common area. This ensures that the distribution of the noise class-posterior has a strong consistency with the adjustment coefficient. Furthermore, the range distribution of any $\boldsymbol{T}(\delta)$ is one subspace of an extended transition matrix $\boldsymbol{T}^{\Sigma}$, whose edges is defined as the edges of $\boldsymbol{T}(-1)$ and $\boldsymbol{T}(1)$:

$$\boldsymbol{T}^{\Sigma}_{i,i} = 1 + \frac{S(\theta_0) \cdot S(\theta_i) - S(\theta_0)}{2} \tag{6}$$

To train the proposed ARTran, we organize a group of loss functions to jointly optimize the proposed modules. The benchmark noise posteriors and the adjusted noise posteriors are optimized with classification loss with benchmark and adjusted labels respectively. Following the instance-independent scheme, we optimize the SST by minimizing the volume of the extended SST [14]. The total loss function $\mathcal{L}$ we give is as follows:

$$\mathcal{L} = \mathcal{L}_{cls}(P(\tilde{\boldsymbol{Y}}|x,0), Y(s,0)) + \mathcal{L}_{cls}(P(\tilde{\boldsymbol{Y}}|x,\delta), Y(s,\delta)) + \mathcal{L}_{vol}(\boldsymbol{T}^{\Sigma}) \tag{7}$$

where $\mathcal{L}_{cls}(\cdot)$ indicates the cross entropy function, and $\mathcal{L}_{vol}(\cdot)$ indicates the volume minimization function.

3 Experiments

3.1 Dataset

We conduct experiments on an OCT dataset including 509 volumes of 290 subjects from the same OCT system (RTVue-XR, Optovue, CA) with high diversity in SE range and retinal shape. Each OCT volume has a size of 400 (frames) $\times$ 400 (width) $\times$ 640 (height) corresponding to a $6mm \times 6mm \times 2mm$ volume centered at the retinal macular region. The exclusion criteria were as follows: eyes with the opacity of refractive media that interfered with the retinal image quality, and eyes that have undergone myopia correction surgery. Our dataset contains 234 low (or non) myopia volumes, and 275 high myopia volumes, where labels are determined according to a threshold spherical equivalent $-6.0D$. We divide the dataset evenly into 5 folds for cross-validation according to the principle of subject independence for all experiments. For data selection, we select the center 100 frames of each volume for training and testing, so a total of 50,900 images were added to the experiment. And the final decision outcome of one model for each volume is determined by the classification results of the majority of frames. For data augmentation, due to the special appearance and location characteristics of high myopia in OCT, we only adopt random horizontal flipping and random vertical translation with a range of $[0, 0.1]$.

Table 1. Comparison of classification methods with benchmark inclusion criteria.

Model	Parameters(M)	Accuracy(%)	Precision(%)	Recall(%)
ResNet-50 [10]	23.5	82.97±9.0	85.4±27.4	83.3±4.0
EfficientNet-B5 [25]	28.3	81.1±7.0	83.1±14.2	80.4±5.6
ViT-Small [8]	21.7	82.7±4.1	85.0±4.6	82.5±4.7
Swin-Tiny [18]	27.5	83.3±13.3	83.9±19.4	85.5±26.1
Swin-V2-Tiny [17]	27.5	83.3±7.4	83.2±23.2	86.2±34.7
Choi's Method [6]	47.0	83.5±5.7	84.3±12.3	86.2±10.4
Li's Method [16]	54.3	84.8±10.7	85.6±14.5	86.5±30.2
ARTran(Ours)	21.9	**86.2**±1.4	**87.5**±2.4	**86.9** ±2.1

Table 2. The ablation study of our anisotropic patch embedding (APE), adjustable class embedding (ACE), and shifted subspace transition matrix (SST).

APE	ACE	SST	Accuracy(%)	Precision(%)	Recall(%)
✓	✓	✓	86.2	87.5	86.9
✗	✓	✓	84.1	86.6	84.7
✓	✓	✗	84.7	86.5	86.2
✗	✓	✗	83.9	84.8	85.6
✓	✗	✗	83.3	86.2	84.4
✗	✗	✗	82.7	85.0	82.5

3.2 Comparison Experiments and Ablations

To evaluate the proposed ARTran in predicting under a benchmark inclusion criteria ($-6.0D$), we compare it with two convolution-based baselines: ResNet-50 [10] and EfficientNet-B5 [25]; three transformer-based baselines: ViT-Small [8], Swin-Tiny [18] and Swin-V2-Tiny [17]; and two state-of-the-art high myopia screening methods: Choi's method and Li's method [16].

Table 1 presents the classification accuracy, precision, and recall by our ARTran and comparison methods. The proposed ARTran already outperforms baseline methods remarkably with a 2.9% to 5.1% higher accuracy and a lower variance. This is because we design modules to capture the features of high myopia, which brings effectiveness and robustness. Although consisting of fewer parameters, our ARTran obtains higher accuracy than two state-of-the-art screening methods. Moreover, clinical practice requirements generally advocate the model that predicts smaller false negative samples, i.e., a higher recall. It is observed that the recall of our approach is the best performance, which means that our model has minimal risk of missing highly myopic samples.

We further perform ablations in order to better understand the effectiveness of the proposed modules. Table 2 presents quantitative comparisons between different combinations of our APE, ACE, and SST. As can be seen, ablation of APE leads to a rapid decrease in recall, which means that this patch embedding approach captures better features about the propensity for high myopia. The ablation of ACE represents the ablation of the proposed adjustment scheme, which leads to lower accuracy. The ACE drives the network to learn more discriminative features for images in the adjustment range during the training process. The ablation of SST leads to a rapid decrease in precision. The possible reason is that the label noise may be more from negative samples, leading to increased false positive samples.

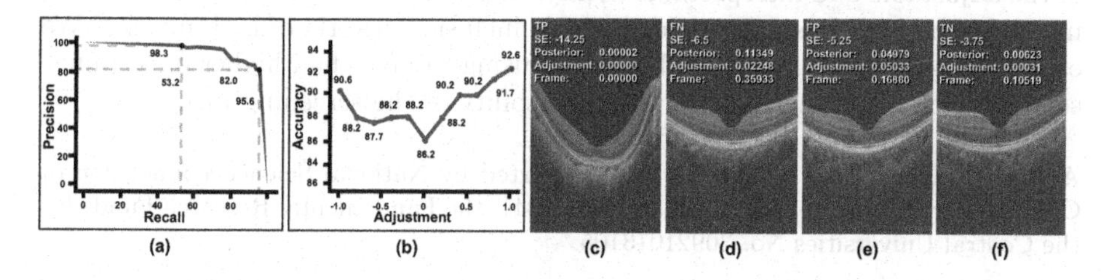

Fig. 3. Adjustable and uncertainty analyses. (a) PR curve of the adjustment coefficient. (b) Classification accuracy using biased labels with adjustment coefficients. (c) A true positive example. (d) A false negative example. (e) A false positive example. (f) A true negative example.

3.3 Adjustable Evaluation and Uncertainty Evaluation

To evaluate the effectiveness of the adjustment module, we change the adjustment coefficient several times during the testing phase to obtain screening results at different thresholds. Figure 3(a) depicts the PR curve when adjusting the adjustment coefficient. The performance of the two endpoints ($\Delta = -1$ and $\Delta = 1$) is marked. Even with a high recall rate, the precision is not low. Figure 3(b) shows the performance of the biased labels for different adjustment coefficients. As can be seen, the accuracy improves when offsetting the inclusion criteria, which on the one hand may be due to the difficulty of classification near the benchmark criteria, and on the other hand proves that the proposed model is effective for the adjustment problem.

To evaluate the interpretability of our ARTran, we propose novel uncertainty scores based on the proposed adjustable scheme: (1) The closer the posteriors closer to 0.5 indicates larger uncertainty. (2) More frames in a volume with disagreement indicate larger uncertainty. (3) Based on a set of adjustment coefficients, the greater the variance, the greater the difficulty or uncertainty. Some examples are shown in Fig. 3. The two correctly classified (TP&TN) examples are less difficult and therefore smaller uncertainty score. Large uncertainty scores are more likely to occur around inclusion criteria. Each of the two error examples (FP&FN) contains at least one uncertainty score higher than the other examples.

4 Conclusion

In this work, we proposed ARTran to screen high myopia using OCT images. Experimental results demonstrated that our approach outperforms baseline classification methods and other screening methods. The ablation results also demonstrated that our modules helps the network to capture the features associated with high myopia and to mitigate the noise of labels. We organized the evaluation of the adjustable and interpretable ability. Experimental results showed that our method exhibits robustness under variable inclusion criteria of high myopia. We evaluated uncertainty and found that confusing samples had higher uncertainty scores, which could increase the interpretability of the screening task.

Acknowledgements. This work was supported by National Science Foundation of China under Grants No. 62172223 and 62072241, the Fundamental Research Funds for the Central Universities No. 30921013105.

References

1. Baird, P.N., et al.: Myopia. Nature Rev. Dis. Primers **6**(1), 99 (2020)
2. Boyd, S., Boyd, S.P., Vandenberghe, L.: Convex Optimization. Cambridge University Press (2004)
3. Buch, H., Vinding, T., La Cour, M., Appleyard, M., Jensen, G.B., Nielsen, N.V.: Prevalence and causes of visual impairment and blindness among 9980 scandinavian adults: the Copenhagen city eye study. Ophthalmology **111**(1), 53–61 (2004)
4. Bullimore, M.A., Brennan, N.A.: Myopia control: why each diopter matters. Optom. Vis. Sci. **96**(6), 463–465 (2019)
5. Bullimore, M.A., Ritchey, E.R., Shah, S., Leveziel, N., Bourne, R.R., Flitcroft, D.I.: The risks and benefits of myopia control. Ophthalmology **128**(11), 1561–1579 (2021)
6. Choi, K.J., et al.: Deep learning models for screening of high myopia using optical coherence tomography. Sci. Rep. **11**(1), 21663 (2021)
7. Choudhury, F., et al.: Prevalence and characteristics of myopic degeneration in an adult Chinese American population: the chinese American eye study. Am. J. Ophthalmol. **187**, 34–42 (2018)

8. Dosovitskiy, A., et al.: An image is worth 16×16 words: transformers for image recognition at scale. arXiv preprint arXiv:2010.11929 (2020)
9. Fang, Y., et al.: Oct-based diagnostic criteria for different stages of myopic maculopathy. Ophthalmology **126**(7), 1018–1032 (2019)
10. He, K., Zhang, X., Ren, S., Sun, J.: Deep residual learning for image recognition. In: Proceedings of the IEEE Conference on Computer Vision and Pattern Recognition, pp. 770–778 (2016)
11. Holden, B.A., et al.: Global prevalence of myopia and high myopia and temporal trends from 2000 through 2050. Ophthalmology **123**(5), 1036–1042 (2016)
12. Iwase, A., et al.: Prevalence and causes of low vision and blindness in a Japanese adult population: the tajimi study. Ophthalmology **113**(8), 1354–1362 (2006)
13. Lanchantin, J., Wang, T., Ordonez, V., Qi, Y.: General multi-label image classification with transformers. In: Proceedings of the IEEE/CVF Conference on Computer Vision and Pattern Recognition, pp. 16478–16488 (2021)
14. Li, X., Liu, T., Han, B., Niu, G., Sugiyama, M.: Provably end-to-end label-noise learning without anchor points. In: International Conference on Machine Learning, pp. 6403–6413. PMLR (2021)
15. Li, Y., et al.: Advances in oct imaging in myopia and pathologic myopia. Diagnostics **12**(6), 1418 (2022)
16. Li, Y., et al.: Development and validation of a deep learning system to screen vision-threatening conditions in high myopia using optical coherence tomography images. Br. J. Ophthalmol. **106**(5), 633–639 (2022)
17. Liu, Z., et al.: Swin transformer V2: scaling up capacity and resolution. In: Proceedings of the IEEE/CVF Conference on Computer Vision and Pattern Recognition, pp. 12009–12019 (2022)
18. Liu, Z., et al.: Swin transformer: Hierarchical vision transformer using shifted windows. In: Proceedings of the IEEE/CVF International Conference On Computer Vision, pp. 10012–10022 (2021)
19. Morgan, I.G., Ohno-Matsui, K., Saw, S.M.: Myopia. The Lancet **379**(9827), 1739–1748 (2012)
20. Nakao, N., Igarashi-Yokoi, T., Takahashi, H., Xie, S., Shinohara, K., Ohno-Matsui, K.: Quantitative evaluations of posterior staphylomas in highly myopic eyes by ultra-widefield optical coherence tomography. Invest. Ophthalmol. Vis. Sci. **63**(8), 20–20 (2022)
21. Ohno-Matsui, K., et al.: International photographic classification and grading system for myopic maculopathy. Am. J. Ophthalmol. **159**(5), 877–883 (2015)
22. Resnikoff, S., et al.: Myopia-a 21st century public health issue. Invest. Ophthalmol. Vis. Sci. **60**(3), 1–12 (2019)
23. Ruiz-Medrano, J., Montero, J.A., Flores-Moreno, I., Arias, L., García-Layana, A., Ruiz-Moreno, J.M.: Myopic maculopathy: current status and proposal for a new classification and grading system (ATN). Prog. Retin. Eye Res. **69**, 80–115 (2019)
24. Szegedy, C., Ioffe, S., Vanhoucke, V., Alemi, A.: Inception-v4, inception-ResNet and the impact of residual connections on learning. In: Proceedings of the AAAI Conference on Artificial Intelligence, vol. 31 (2017)
25. Tan, M., Le, Q.: EfficientnNet: rethinking model scaling for convolutional neural networks. In: International Conference on Machine Learning, pp. 6105–6114. PMLR (2019)
26. Xia, X., et al.: Are anchor points really indispensable in label-noise learning? In: Advances in Neural Information Processing Systems, vol. 32 (2019)

27. Xu, L., Ouyang, W., Bennamoun, M., Boussaid, F., Xu, D.: Multi-class token transformer for weakly supervised semantic segmentation. In: Proceedings of the IEEE/CVF Conference on Computer Vision and Pattern Recognition, pp. 4310–4319 (2022)
28. Yao, Y., et al.: Dual t: reducing estimation error for transition matrix in label-noise learning. Adv. Neural. Inf. Process. Syst. **33**, 7260–7271 (2020)

Improving Outcome Prediction of Pulmonary Embolism by De-biased Multi-modality Model

Zhusi Zhong[1,2,3], Jie Li[1], Shreyas Kulkarni[2,3], Yang Li[4], Fayez H. Fayad[2,3], Helen Zhang[2,3], Sun Ho Ahn[2,3], Harrison Bai[5], Xinbo Gao[1], Michael K. Atalay[2,3], and Zhicheng Jiao[2,3](✉)

[1] School of Electronic Engineering, Xidian University, Xi'an, China
[2] Warren Alpert Medical School, Brown University, Providence, USA
zhicheng_jiao@brown.edu
[3] Department of Diagnostic Imaging, Rhode Island Hospital, Providence, USA
[4] School of Computer Science and Engineering, Central South University, Changsha, China
[5] Department of Radiology and Radiological Sciences, Johns Hopkins University School of Medicine, Baltimore, USA

Abstract. Bias in healthcare negatively impacts marginalized populations with lower socioeconomic status and contributes to healthcare inequalities. Eliminating bias in AI models is crucial for fair and precise medical implementation. The development of a holistic approach to reducing bias aggregation in multimodal medical data and promoting equity in healthcare is highly demanded. Racial disparities exist in the presentation and development of algorithms for pulmonary embolism (PE), and deep survival prediction model can be de-biased with multimodal data. In this paper, we present a novel survival prediction (SP) framework with demographic bias disentanglement for PE. The CTPA images and clinical reports are encoded by the state-of-the-art backbones pretrained with large-scale medical-related tasks. The proposed de-biased SP modules effectively disentangle latent race-intrinsic attributes from the survival features, which provides a fair survival outcome through the survival prediction head. We evaluate our method using a multimodal PE dataset with time-to-event labels and race identifications. The comprehensive results show an effective de-biased performance of our framework on outcome predictions.

Keywords: Pulmonary Embolism · Deep Survival Prediction · De-Bias learning · Multi-modal learning

1 Introduction

Bias in medicine has demonstrated a notable challenge for providing comprehensive and equitable care. Implicit biases can negatively affect patient care, particularly for marginalized populations with lower socioeconomic status [30]. Evidence

Supplementary Information The online version contains supplementary material available at https://doi.org/10.1007/978-3-031-43904-9_50.

H. Greenspan et al. (Eds.): MICCAI 2023, LNCS 14224, pp. 515–525, 2023.
https://doi.org/10.1007/978-3-031-43904-9_50

has demonstrated that implicit biases in healthcare providers could contribute to exacerbating these healthcare inequalities and create a more unfair system for people of lower socioeconomic status [30]. Based on the data with racial bias, the unfairness presents in developing evaluative algorithms. In an algorithm used to predict healthcare costs, black patients who received the same health risk scores as white patients were consistently sicker [21]. Using biased data for AI models reinforces racial inequities, worsening disparities among minorities in healthcare decision-making [22].

Within the radiology arm of AI research, there have been significant advances in diagnostics and decision making [19]. Along these advancements, bias in healthcare and AI are exposing poignant gaps in the field's understanding of model implementation and their utility [25,26]. AI model quality relies on input data and addressing bias is a crucial research area. Systemic bias poses a greater threat to AI model's applications, as these biases can be baked right into the model's decision process [22].

Pulmonary embolism (PE) is an example of health disparities related to race. Black patients exhibit a 50% higher age-standardized PE fatality rate and a twofold risk for PE hospitalization than White patients [18,24]. Hospitalized Black patients with PE were younger than Whites. In terms of PE severity, Blacks received fewer surgical interventions for intermediate PE but more for high-severity PE [24]. Racial disparities exist in PE and demonstrate the inequities that affect Black patients. The Pulmonary Embolism Severity Index (PESI) is a well-validated clinical tool based on 11 clinical variables and used for outcome prediction measurement [2]. Survival analysis is often used in PE to assess how survival is affected by different variables, using a statistical method like Kaplan-Meier method and Cox proportional-hazards regression model [7,12,14].

However, one issue with traditional survival analysis is bias from single modal data that gets compounded when curating multimodal datasets, as different combinations of modes and datasets create with a unified structure. Multimodal data sets are useful for fair AI model development as the bias complementary from different sources can make de-biased decisions and assessments. In that process, the biases of each individual data set will get pooled together, creating a multimodal data set that inherits multiple biases, such as racial bias [1,15,23]. In addition, it has been found that creating multimodal datasets without any de-biasing techniques does not improve performance significantly and does increase bias and reduce fairness [5]. Overall, a holistic approach to model development would be beneficial in reducing bias aggregation in multimodal datasets. In recent years, Disentangled Representation Learning (DRL) [4] for bias disentanglement improves model generalization for fairness [3,6,27].

We developed a PE outcome model that predicted mortality and detected bias in the output. We then implemented methods to remove racial bias in our dataset and model and output unbiased PE outcomes as a result. Our contributions are as follows: (1) We identified bias diversity in multimodal information using a survival prediction fusion framework. (2) We proposed a de-biased survival prediction framework with demographic bias disentanglement. (3) The multimodal CPH learning models improve fairness with unbiased features.

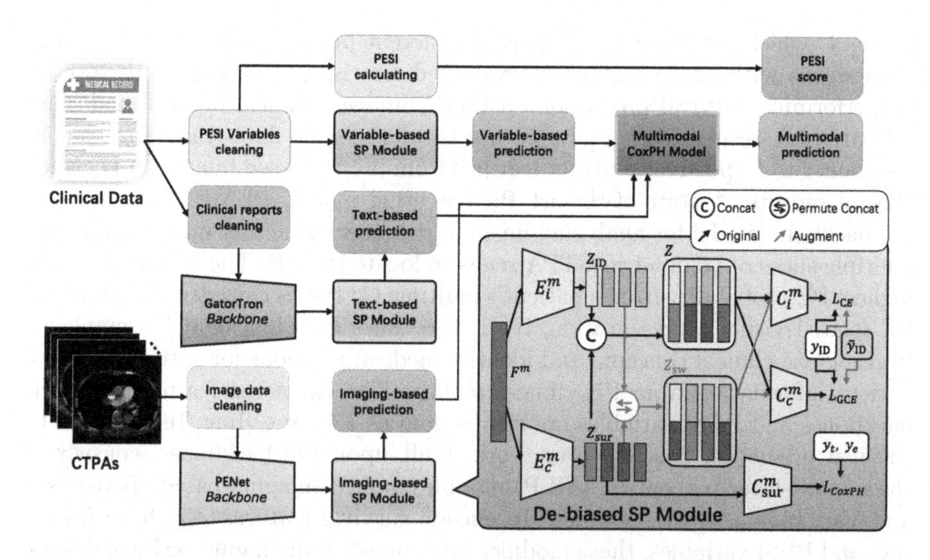

Fig. 1. Overview of the Survival Prediction (SP) framework and the proposed de-biased SP module (lower right). ID branch (E_i;C_i) and Survival branch (E_c;C_c) are trained to disentangle race-intrinsic attributes and survival attributes with the feature swapping augmentation, respectively. The survival head predicts the outcomes based on the de-biased survival attributes.

2 Bias in Survival Prediction

This section describes the detail of how we identify the varying degrees of bias in multimodal information and illustrates bias using the relative difference in survival outcomes. We will first introduce our pulmonary embolism multimodal datasets, including survival and race labels. Then, we evaluate the baseline survival learning framework without de-biasing in the various racial groups.

Dataset. The Pulmonary Embolism dataset used in this study from 918 patients (163 deceased, median age 64 years, range 13–99 years, 52% female), including 3978 CTPA images and 918 clinical reports, which were identified via retrospective review across three institutions. The clinical reports from physicians that provided crucial information are anonymized and divided into four parts: medical history, clinical diagnosis, observations and radiologist's opinion. For each patient, the race labels, survival time-to-event labels and PESI variables are collected from clinical data, and the 11 PESI variables are used to calculate the PESI scores, which include age, sex, comorbid illnesses (cancer, heart failure, chronic lung disease), pulse, systolic blood pressure, respiratory rate, temperature, altered mental status, and arterial oxygen saturation at the time of diagnosis [2].

Diverse Bias of Multimodal Survival Prediction Model. We designed a deep survival prediction (SP) baseline framework for multimodal data as shown in Fig. 1, which compares the impact of different population distributions. The frameworks without de-basing are evaluated for risk prediction in the test set by performing survival prediction on CTPA images, clinical reports, and clinical variables, respectively. First, we use two large-scale data-trained models as backbones to respectively extract features from preprocessed images and cleaned clinical reports. A state-of-the-art PE detecting model, PENet [11] is used as the backbone model for analyzing imaging risk and extracting information from multiple slices of volumetric CTPA scans to locate the PE. The feature with the highest PE probability from a patient's multiple CTPAs is considered as the most PE-related visual representation. Next, the GatorTron [29] model is employed to recognize clinical concepts and identify medical relations for getting accurate patient information from PE clinical reports. The extracted features from the backbones and PESI variables are represented as $F^m, m \in$ [Img, Text, Var]. The survival prediction baseline framework, built upon the backbones, consists of three multi-layer perceptron (MLP) modules named Imaging-based, Text-based and Variable-based SP modules. To encode survival features Z^m_{sur} from image, text and PESI variables, these modules are trained to distinguish critical disease from non-critical disease with Cox partial log-likelihood loss (CoxPHloss) [13]. The framework also consists of a Cox proportional hazard (CoxPH) model [7] that is trained to predict patient ranking using a multimodal combination of risk predictions from the above three SP modules. These CoxPH models calculate the corresponding time-to-event evaluation and predict the fusion of patients' risk as the survival outcome. We evaluate the performance of each module with concordance probability (C-index), which measures the accuracy of prediction in terms of ranking the order of survival times [8]. For reference, the C-index of PESI scores is additionally provided for comparative analysis.

In Table 1 (Baseline), we computed the C-index between the predicted risk of each model and time-to-event labels. When debiasing is not performed, significant differences exist among the different modalities, with the image modality exhibiting the most pronounced deviation, followed by text and PESI variables. The biased performance of the imaging-based module is likely caused by the richness of redundant information in images, which includes implicit features such as body structure and posture that reflect the distribution of different races. This redundancy leads to model overfitting on race, compromising the fairness of risk prediction across different races. Besides, clinical data in the form of text reports and PESI variables objectively reflect the patient's physiological information and the physician's diagnosis, exhibiting smaller race biases in correlation with survival across different races. Moreover, the multimodal fusion strategy is found to be effective, yielding more relevant survival outcomes than the clinical gold standard PESI scores.

3 De-biased Survival Prediction Model

Based on our SP baseline framework and multimodal findings from Sect. 2, we present a feature-level de-biased SP module that enhances fairness in survival outcomes by decoupling race attributes, as shown in the lower right of Fig. 1. In the de-biased SP module, firstly, two separate encoders E_i^m and E_c^m are formulated to embed features F^m into disentangled latent vectors for race-intrinsic attributes z_{ID} or race-conflicting attributes z_{sur} implied survival information [16]. Then, the linear classifiers C_i^m and C_c^m constructed to predict the race label y_{ID} with concatenated vector $z = [z_{\mathrm{ID}}; z_{\mathrm{sur}}]$. To disentangle survival features from the race identification, we use the generalized cross-entropy (GCE) loss [31] to train E_c^m and C_c^m to overfit to race label while training E_i^m and C_i^m with cross-entropy (CE) loss. The relative difficulty scores W as defined in Eq. 1 reweight and enhance the learning of the race-intrinsic attributes [20]. The objective function for disentanglement shown in Eq. 2, but the parameters of ID or survival branch are only updated by their respective losses:

$$W(z) = \frac{CE\left(C_c(z), y_{\mathrm{ID}}\right)}{CE\left(C_c(z), y_{\mathrm{ID}}\right) + CE\left(C_i(z), y_{\mathrm{ID}}\right)} \tag{1}$$

$$L_{\mathrm{dis}} = W(z) CE\left(C_i(z), y_{\mathrm{ID}}\right) + GCE\left(C_c(z), y_{\mathrm{ID}}\right) \tag{2}$$

To promote race-intrinsic learning in E_i^m and C_i^m, we apply diversify with latent vectors swapping. The randomly permuted $\tilde{z}_{\mathrm{sur}}$ in each mini-batch concatenate with z_{ID} to obtain $z_{\mathrm{sw}} = [z_{\mathrm{ID}}; \tilde{z}_{\mathrm{sur}}]$. The two neural networks are trained to predict y_{ID} or $\tilde{y}_{\mathrm{ID}}$ with CE loss or GCE loss. As the random combination are generated from different samples, the swapping decreases the correlation of these feature vectors, thereby enhancing the race-intrinsic attributes. The loss functions of swapping augmentation added to train two neural networks is defined as:

$$L_{\mathrm{sw}} = W(z) CE\left(C_i(z_{\mathrm{sw}}), y_{\mathrm{ID}}\right) + GCE\left(C_c(z_{\mathrm{sw}}), \tilde{y}_{\mathrm{ID}}\right) \tag{3}$$

The survival prediction head C_{sur}^m predicts the risk on the survival feature z_{sur}. CoxPH loss function [13], which optimizes the Cox partial likelihood, is used to maximize concordance differentiable and update model weights of the survival branch. Thus, CoxPH loss and overall loss function are formulated as:

$$L_{\mathrm{CoxPH}} := -\sum_{i:y_e^i=1} \left(C_{\mathrm{sur}}(z_{\mathrm{sur}}^i) - \log \sum_{j:y_t^j > y_t^i} e^{C_{\mathrm{sur}}(z_{\mathrm{sur}}^j)} \right) \tag{4}$$

$$L_{\mathrm{overall}} = L_{\mathrm{dis}} + \lambda_{\mathrm{sw}} L_{\mathrm{sw}} + \lambda_{\mathrm{sur}} L_{\mathrm{CoxPH}} \tag{5}$$

where Y_t and Y_e are survival labels including the survival time and the event, respectively. The weights λ_{sw} and λ_{sur} are assigned as 0.5 and 0.8, respectively, to balance the feature disentanglement and survival prediction.

4 Experiment

We validate the proposed de-biased survival prediction frameworks on the collected multi-modality PE data. The data from 3 institutions are randomly split

Table 1. Performance comparison of the proposed de-biased SP framework and baseline using C-index values on multiple modal outcomes. The larger C-index value is better and the lower bias is fairer.

Method	Baseline				De-biased SP model			
Dataset	Overall	White	Color	Bias	Overall	White	Color	Bias
Imaging	**0.662**	0.736	0.422	0.314	0.646	0.656	0.622	<u>0.035</u>
Text	0.657	0.642	0.714	0.071	**0.719**	0.689	0.746	<u>0.057</u>
Variable	0.668	0.669	0.741	<u>0.072</u>	**0.698**	0.683	0.778	0.095
Multimodal	0.709	0.692	0.816	0.124	**0.759**	0.756	0.768	<u>0.012</u>

into training, validation, and testing sets, with a ratio of 7:1:2, the same ratio of survival events is maintained in each institution. We apply race-balanced resampling to the training and validation sets to eliminate training bias caused by minority groups.

The lung region of CPTA images is extracted with a slice thickness of 1.25 mm and scaled to $N \times 512 \times 512$ pixels [10]. Hounsfield units (HU) of all slices are clipped to the range of $[-1000, 900]$ and applied with zero-centered normalization. The PENet-based imaging backbone consists of a 77-layer 3D convolutional neural network and linear regression layers. It takes in a sliding window of 24 slices at a time, resulting in a window-level prediction that represents the probability of PE for the current slices [11]. The PENet is pre-trained on large-scale CTPA studies and shows excellent PE detection performance with an AUROC of 0.85 on our entire dataset. The 2048 dimensional features from the last convolution with the highest probability of PE, are designated as the imaging features.

The GatorTron [29] uses a transformer-based architecture to extract features from the clinical text, which was pre-trained on over 82 billion words of de-identified clinical text. We used the Huggingface library [28] to deploy the 345m-parameter cased model as the clinical report feature extractor. The outputs from each patient's medical history, clinical diagnosis, observations, and radiologist impression are separately generated and concatenated to form the 1024×4 features.

We build the encoders of the baseline SP modules and de-biased SP modules with multi-layer perceptron (MLP) neural networks and ReLu activation. The MLPs with 3 hidden layers are used to encode image and text features, and another MLPs with 2 layers encodes the features of PESI variables. A fully connected layer with sigmoid activation acts as a risk classifier $C^m_{\text{sur}}(z^m_{\text{sur}})$ for survival prediction, where z^m_{sur} is the feature encoded from single modal data. For training the biased and de-biased SP modules, we collect data from one modality as a batch with synchronized batch normalization. The SP modules are optimized using the *AdamW* [17] optimizer with a momentum of 0.9, a weight decay of 0.0005, and a learning rate of 0.001. We apply early stopping when validation loss doesn't decrease for 600 epochs. Experiments are conducted on an Nvidia GV100 GPU.

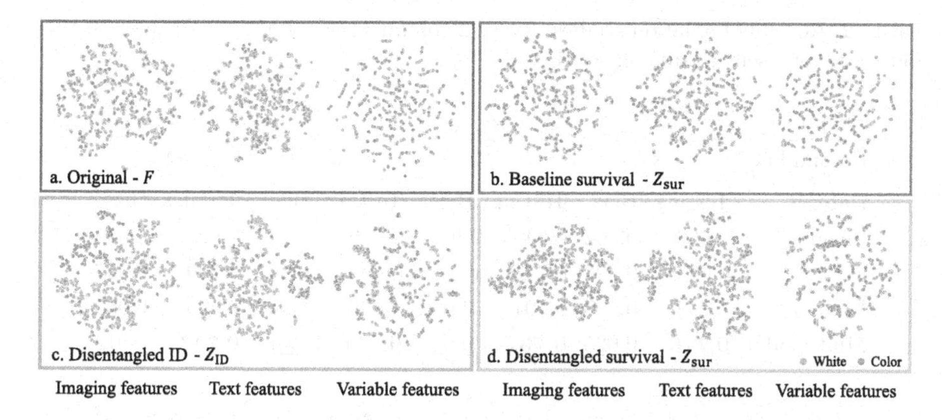

Fig. 2. tSNE visualizations of the features from multimodal data. Based on the comparison between the ID features and others, it is observed that the clusters containing race obtained from the same class are more compact.

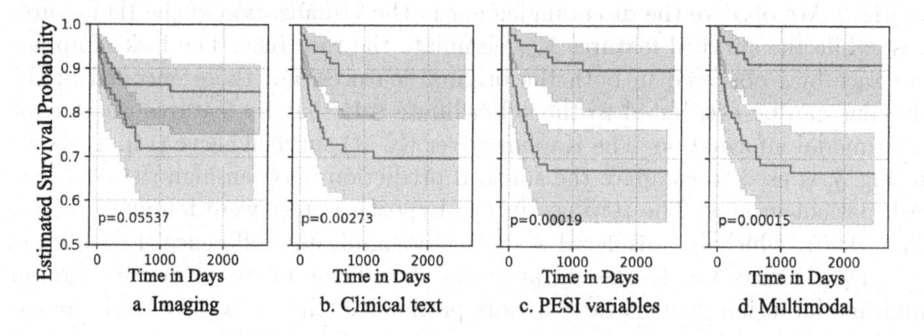

Fig. 3. Kaplan-Meier survival curves of our 3 de-biased SP modules and the multimodal CoxPH model. High-risk and low-risk groups are plotted as red and blue lines, respectively. The x-axis shows the time in days, and y-axis presents the estimated survival probability. Log-rank p value is shown on each figure. (Color figure online)

4.1 Results

Table 1 shows the quantitative comparisons of the baseline and de-biased frameworks with the C-indexes of the multimodal survival predictions. In general, our framework including de-biased SP modules shows significantly better predictions in testing set than the PESI-based outcome estimation with C-indexes of 0.669, 0.654, 0.697, 0.043 for the overall testset, White testset, Color testset and race bias. The de-biased results outperform the baseline in overall survival C-index and show a lower race bias, especially in imaging- and fusion-based predictions. The results indicate the effectiveness of the proposed de-biasing in mitigating race inequity. The results also prove the observations for the different biases present in different modalities, especially in the CTPA images containing more abundant race-related information. It also explains the limited effectiveness of de-biasing the clinical results, which contain less racial identification. The pre-

Table 2. Results of ablation studies. Every 2 columns (overall performance of Testing and Bias) represent a training setting.

Swapping	×		✓		×		✓	
Resampling	×		×		✓		✓	
Dataset	Testing	Bias	Testing	Bias	Testing	Bias	Testing	Bias
Imaging	0.666	0.062	0.641	0.014	0.649	0.050	0.622	0.035
Text	0.684	0.090	0.711	0.123	0.698	0.102	0.709	0.057
Variable	0.702	0.095	0.701	0.052	0.697	0.082	0.699	0.095
Multimodal	**0.716**	0.025	**0.737**	0.041	**0.741**	0.011	**0.743**	0.012

diction performance based on multiply modalities is significantly better than the PESI-based outcome estimation. The disentangled representations, transformed from latent space to a 2D plane via tSNE and color-coded by race [9], are shown in Fig. 2. We observe the disentanglement in the visualization of the ID features z_{ID}, while the survival features z_{sur} eliminate the race bias. The lack of apparent race bias observed in both the original features and those encoded in the baseline can be attributed to the subordinate role that ID features play in the multimodal information. The Kaplan-Meier (K-M) survival curve [14], as shown in Fig. 3, is used to compare the survival prediction between high-risk and low-risk patient groups. The p-values in the hypothesis test were found to be less than 0.001, which is considered statistically significant difference. In addition, the predictions of the de-biased framework show favorable performance, and our multimodal fusion demonstrates a more pronounced discriminative ability in the K-M survival analysis compared to the single-modal results.

We conducted ablation studies to examine the effect of the two key components, including swapping feature augmentation and race-balance resampling. As shown in Table 2, the different training settings show significant differences in survival prediction performance across modalities. The swapping augmentation provides a strong bias correction effect for image data with obvious bias. For clinical data, the resampling generally improves performance in most cases. Overall, multimodal fusion approaches are effective in all training settings, and the CoxPH model can actively learn the optimal combination of multimodal features to predict survival outcomes.

5 Discussions and Conclusions

In this work, we developed a de-biased survival prediction framework based on the race-disentangled representation. The proposed de-biased SP framework, based on the SOTA PE detection backbone and large-scale clinical language model, can predict the PE outcome with a higher survival correlation ahead of the clinical evaluation index. We detected indications of racial bias in our dataset and conducted an analysis of the multimodal diversity. Experimental results illustrate that our approach is effective for eliminating racial bias while resulting in an overall improved model performance. The proposed technique is clinically relevant as it can address the pervasive presence of racial bias in healthcare systems and offer a solution for minimizing or eliminating bias without pausing to evaluate their affection for the models and tools. Our study is significant as it highlights and evaluates the negative impact of racial bias on deep learning models. The proposed de-biased method has already shown the capacity to relieve them, which is vital when serving patients with an accurate analysis. The research in our paper demonstrates and proves that eliminating racial biases from data improves performance, and yields a more precise and robust survival prediction tool. In the future, these de-biased SP modules can be plugged into other models, offering a fairer method to predict survival outcomes.

References

1. Acosta, J.N., Falcone, G.J., Rajpurkar, P., Topol, E.J.: Multimodal biomedical AI. Nat. Med. **28**, 1773–1784 (2022). https://doi.org/10.1038/s41591-022-01981-2
2. Aujesky, D., et al.: Derivation and validation of a prognostic model for pulmonary embolism. Am. J. Respir. Crit. Care Med. **172**, 1041–1046 (2005). https://doi.org/10.1164/rccm.200506-862OC
3. Bahng, H., Chun, S., Yun, S., Choo, J., Oh, S.J.: Learning de-biased representations with biased representations. In: International Conference on Machine Learning, pp. 528–539. PMLR (2020)
4. Bengio, Y., Courville, A., Vincent, P.: Representation learning: a review and new perspectives. IEEE Trans. Pattern Anal. Mach. Intell. **35**(8), 1798–1828 (2013)
5. Booth, B.M., Hickman, L., Subburaj, S.K., Tay, L., Woo, S.E., D'Mello, S.K.: Bias and fairness in multimodal machine learning: a case study of automated video interviews. In: Proceedings of the 2021 International Conference on Multimodal Interaction, pp. 268–277 (2021). https://doi.org/10.1145/3462244.3479897
6. Creager, E., et al.: Flexibly fair representation learning by disentanglement. In: International Conference on Machine Learning, pp. 1436–1445. PMLR (2019)
7. Fox, J., Weisberg, S.: Cox proportional-hazards regression for survival data. An R and S-PLUS companion to applied regression 2002 (2002)
8. Harrell, F.E., Jr., Lee, K.L., Califf, R.M., Pryor, D.B., Rosati, R.A.: Regression modelling strategies for improved prognostic prediction. Stat. Med. **3**(2), 143–152 (1984)
9. Hinton, G., van der Maaten, L.: Visualizing data using t-SNE journal of machine learning research (2008)
10. Hofmanninger, J., Prayer, F., Pan, J., Röhrich, S., Prosch, H., Langs, G.: Automatic lung segmentation in routine imaging is primarily a data diversity problem, not a methodology problem. Eur. Radiol. Exp. **4**(1), 1–13 (2020). https://doi.org/10.1186/s41747-020-00173-2
11. Huang, S.C., et al.: PENet-a scalable deep-learning model for automated diagnosis of pulmonary embolism using volumetric CT imaging. NPJ Digit. Med. **3**(1), 61 (2020)
12. Kaplan, E.L., Meier, P.: Nonparametric estimation from incomplete observations. J. Am. Stat. Assoc. **53**(282), 457–481 (1958)

13. Katzman, J.L., Shaham, U., Cloninger, A., Bates, J., Jiang, T., Kluger, Y.: Deepsurv: personalized treatment recommender system using a cox proportional hazards deep neural network. BMC Med. Res. Methodol. **18**(1), 1–12 (2018)
14. Klok, F.A.: Patient outcomes after acute pulmonary embolism a pooled survival analysis of different adverse events. Am. J. Respir. Crit. Care Med. **181**, 501–506 (2009)
15. Lahat, D., Adali, T., Jutten, C.: Multimodal data fusion: an overview of methods, challenges, and prospects. Proc. IEEE **103**, 144–1477 (2015). https://doi.org/10.1109/JPROC.2015.2460697
16. Lee, J., Kim, E., Lee, J., Lee, J., Choo, J.: Learning debiased representation via disentangled feature augmentation. Adv. Neural. Inf. Process. Syst. **34**, 25123–25133 (2021)
17. Loshchilov, I., Hutter, F.: Decoupled weight decay regularization. In: International Conference on Learning Representations
18. Martin, K.A., McCabe, M.E., Feinglass, J., Khan, S.S.: Racial disparities exist across age groups in illinois for pulmonary embolism hospitalizations. Arterioscler. Thromb. Vasc. Biol. **40**, 2338–2340 (2020). https://doi.org/10.1161/ATVBAHA.120.314573
19. Perera, N, Perchik, J.D., Perchik, M.C., Tridandapani, S.: Trends in medical artificial intelligence publications from 2000–2020: where does radiology stand? Open J. Clin. Med. Images **2** (2022)
20. Nam, J., Cha, H., Ahn, S., Lee, J., Shin, J.: Learning from failure: de-biasing classifier from biased classifier. In: Larochelle, H., Ranzato, M., Hadsell, R., Balcan, M., Lin, H. (eds.) Advances in Neural Information Processing Systems, vol. 33, pp. 20673–20684. Curran Associates, Inc. (2020), https://proceedings.neurips.cc/paper/2020/file/eddc3427c5d77843c2253f1e799fe933-Paper.pdf
21. Obermeyer, Z., Powers, B., Vogeli, C., Mullainathan, S.: Dissecting racial bias in an algorithm used to manage the health of populations. Science **366**, 447–453 (2019). https://doi.org/10.1126/science.aax234
22. Parikh, R.B., Teeple, S., Navathe, A.S.: Addressing bias in artificial intelligence in health care. JAMA **322**, 2377–2378 (2019). https://doi.org/10.1001/jama.2019.18058
23. Pena, A., Serna, I., Morales, A., Fierrez, J.: Bias in multimodal AI: testbed for fair automatic recruitment. In: 2020 IEEE/CVF Conference on Computer Vision and Pattern Recognition Workshops (CVPRW), pp. 129–137 (2022). https://doi.org/10.1109/CVPRW50498.2020.00022
24. Phillips, A.R., et al.: Association between black race, clinical severity, and management of acute pulmonary embolism: a retrospective cohort study. J. Am. Heart Assoc. **10** (2021). https://doi.org/10.1161/JAHA.121.021818
25. Rajpurkar, P., Chen, E., Banerjee, O., Topol, E.J.: AI in health and medicine. Nat. Med. **28**, 31–38 (2022). https://doi.org/10.1038/s41591-021-01614-0
26. Rouzrokh, P., et al.: Mitigating bias in radiology machine learning: 1. data handling. Radiol. Artif. Intell. **4**, 1–10 (2022). https://doi.org/10.1148/ryai.210290
27. Song, J., Kalluri, P., Grover, A., Zhao, S., Ermon, S.: Learning controllable fair representations. In: The 22nd International Conference on Artificial Intelligence and Statistics, pp. 2164–2173. PMLR (2019)
28. Wolf, T., Debut, L., Sanh, V., Chaumond, J., Delangue, C., Moi, A., Cistac, P., Rault, T., Louf, R., Funtowicz, M., et al.: Huggingface's transformers: State-of-the-art natural language processing. arXiv preprint arXiv:1910.03771 (2019)
29. Yang, X., et al.: A large language model for electronic health records. NPJ Digit. Med. **5**(1), 194 (2022)

30. Zestcott, C.A., Blair, I.V., Stone, J.: Examining the presence, consequences, and reduction of implicit bias in health care: a narrative review. Group Process. Intergroup Relat. **19**, 528–542 (2016). https://doi.org/10.1177/1368430216642029
31. Zhang, Z., Sabuncu, M.: Generalized cross entropy loss for training deep neural networks with noisy labels. In: Advances in Neural Information Processing Systems, vol. 31 (2018)

Recruiting the Best Teacher Modality: A Customized Knowledge Distillation Method for if Based Nephropathy Diagnosis

Ning Dai[1], Lai Jiang[1](✉), Yibing Fu[1], Sai Pan[2], Mai Xu[1], Xin Deng[3], Pu Chen[2], and Xiangmei Chen[2]

[1] School of Electronic and Information Engineering, Beihang University, Beijing, China
jianglai.china@buaa.edu.cn

[2] National Clinical Research Center for Kidney Diseases, State Key Laboratory of Kidney Diseases, Institute of Nephrology of Chinese PLA, Department of Nephrology, General Hospital of Chinese PLA, Medical School of Chinese PLA, Beijing, China

[3] School of Cyber Science and Technology, Beihang University, Beijing, China

Abstract. The joint use of multiple imaging modalities for medical image has been widely studied in recent years. The fusion of information from different modalities has demonstrated the performance improvement for some medical tasks. For nephropathy diagnosis, immunofluorescence (IF) is one of the most widely-used medical image due to its ease of acquisition with low cost, which is also an advanced multi-modality technique. However, the existing methods mainly integrate information from diverse sources by averaging or combining them, failing to exploit multi-modality knowledge in details. In this paper, we observe that the 7 modalities of IF images have different impact on different nephropathy categories. Accordingly, we propose a knowledge distillation framework to transfer knowledge from the trained single-modality teacher networks to a multi-modality student network. On top of this, given a input IF sequence, a recruitment module is developed to dynamically assign weights to teacher models and optimize the performance of student model. By applying on several different architectures, the extensive experimental results verify the effectiveness of our method for nephropathy diagnosis.

Keywords: Nephropathy diagnosis · IF image · Knowledge distillation

This work was supported by NSFC under Grant 62250001, and Alibaba Innovative Reaserch.

Supplementary Information The online version contains supplementary material available at https://doi.org/10.1007/978-3-031-43904-9_51.

H. Greenspan et al. (Eds.): MICCAI 2023, LNCS 14224, pp. 526–536, 2023.
https://doi.org/10.1007/978-3-031-43904-9_51

1 Introduction

Nephropathy is a progressive and incurable disease with high mortality, occurring commonly in the general adult population, with a world-wide prevalence of 10% [11]. Therefore, early detection and treatment is of pivotal importance, as it can prevent the death or inevitable renal failure that requires renal dialysis or replacement therapy. Due to the low cost and sensitivity for certain lesion [20], immunofluorescence (IF) images have been increasingly used in the diagnostic process of nephropathy. Most recently, benefiting from the development of deep learning, a couple of deep neural networks (DNNs) have been developed for nephropathy related tasks on IF images [7,8,13,18]. For instance, Ligabue *et al.* [8] proposed a residual convolutional neural network (CNN) for IF nephropathy reporting. Similarly, in [18], a DNN-based model with a pre-segmentation module and a classification module was introduced to automatically detect the different glomerulus in IF images. Kitamura *et al.* [7] designed a CNN structure for diabetic nephropathy (DN) diagnosis on IF images, and further visualized where the CNN focused on for diagnosis.

There exist only a few IF image based DNN methods, probably because the properties of IF images are complicated and have not been fully exploited. Different from the natural and other medical images, a IF sequence usually include multiple modalities from different types of fluorescent [2]. On the other hand, the collected modalities of a IF sequence are usually incomplete, due to the medical reasons and acquiring processes. This leads to various modality combination in a IF dataset, therefore significantly reducing the learning efficiency of a DNN. More importantly, the correlation between different IF modalities and nephropathy categories is complicated. For instance, as shown in Fig. 1, the experiments in this paper find that anti-neutrophil cytoplasmic antibodies (ANCA) disease is strongly related to the modalities of Immunoglobulin G (IgG) and Immunoglobulin A (IgA). Meanwhile, other modalities like Complement 3

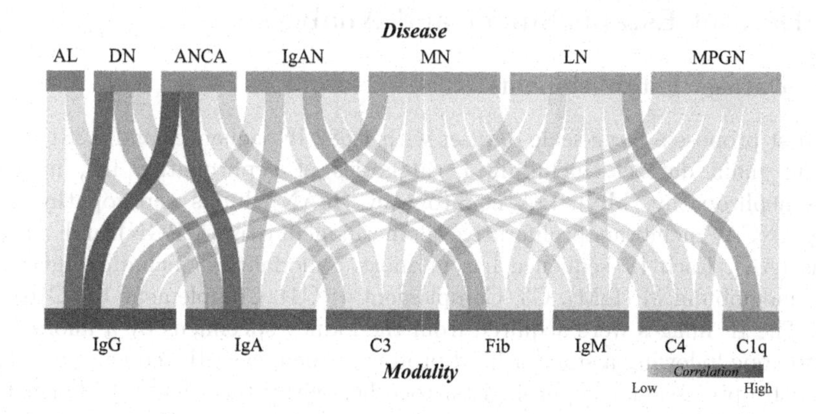

Fig. 1. The importance of a individual modality contributing to the diagnosis on each nephropathy. The detailed calculation is introduced in the data analysis section.

(C3) and Fibronectin (Fib) sometimes mislead the DNN and thus degrade the ANCA diagnosis performance. Unfortunately, all above DNN methods assume different modalities have the equal effect on the diagnosis task, neglecting the medical prior of the importance of individual modality.

To address above issue, this paper proposes a novel customized multi-teacher knowledge distillation framework for nephropathy diagnosis on IF images. Specifically, we establish a large-scale IF image dataset including 7 types of nephropathy from 1,582 patients. By mining our dataset, we conduct experiments to explore the importance of a individual modality contributing to each nephropathy, as the empirical medical prior (see Fig. 1). Then, we develop a multi-teacher knowledge distillation framework, in which the knowledge is transferred from the teacher networks trained by individual modalities. Different from the traditional knowledge distillation [5,9], we propose a customized framework with a recruitment module, which learns to select the "best" teacher networks based on the medical priors. Benefiting from this, the student network can effectively learns from the individual modalities, thus achieving better overall performance for the clinical IF sequence with incomplete modalities. We show the effectiveness of the proposed method over our dataset and another external dataset for nephropathy diagnosis. In summary, the main contributions of this paper are three-fold.

- We establish a large-scale IF dataset containing 7 nephropathy categories and 1,582 IF sequences, with the data analysis for the importance of a individual modality contributing to the diagnosis on each nephropathy category.
- We propose a new customized knowledge distillation framework for nephropathy diagnosis, which transfers the knowledge from individual modalities for improving the performance over IF sequence with multiple modalities.
- We develop a recruitment module in our knowledge distillation method, which learns to select the "best" teacher modalities based on the medical priors of input IF sequence and its nephropathy.

2 Dataset Establishment and Analysis

2.1 Dataset Establishment

We first propose a large-scale dataset with 1,582 IF sequences and 6,381 images as our main dataset, including 7 categories of nephropathy, i.e., membranous nephropathy (MN), IgA nephropathy (IgAN), lupus nephropathy (LN), DN, ANCA, membranoproliferative glomerulonephritis (MPGN) and Amyloidosis (AL). Each IF sequence has at most 7 modalities, including IgG, IgA, Immunoglobulin M (IgM), C3, Complement 4 (C4), Complement 1q (C1q) and Fib. The IF images were acquired from the kidney specimens by a fluorescence microscope following a standardized protocol. Then, each IF sequence was diagnosed by professional nephrologists, together with other medical information, such as medical history and examination results of patient. Besides, with similar processes, we further establish an external dataset collected from another

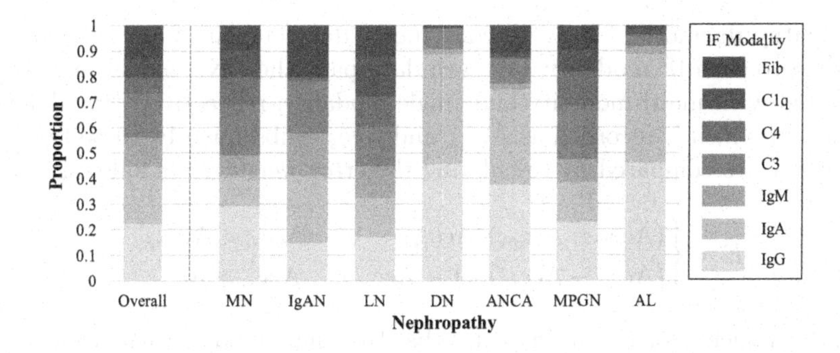

Fig. 2. The proportions of each IF modality for the 7 categories of nephropathy.

hospital including 69 IF sequences with 348 images. Note that our work is conducted according to the Declaration of Helsinki. Compared with the existing IF image datasets [7,8], our dataset collects most patients including most categories of nephropathy. In the experiments, our main dataset is randomly divided into training and test sets at a ratio of 4 to 1, while the external set is only used for test.

2.2 Dataset Analysis

Based on our main dataset, we further conduct data analysis to obtain the following findings about the relationship between different modalities and nephropathy.

Finding 1: *The proportions of each IF modality vary greatly in different nephropathy.*

Analysis: As introduced above, the collected modalities of a IF sequence are usually incomplete. This is partially due to the importance of each IF modality for the specific nephropathy, since the patient may not have the anti-body of the useless fluorescent. Therefore, we count the proportions of 7 IF modalities on each nephropathy over 1,582 IF sequences in our main set. As shown in Fig. 2, there exists a significant inconsistency of the modality proportions between different nephropathy. For instance, IgAN does not have the modality of C1q, while LN has almost equal proportions for all IF modalities. Besides, the proportions of a certain IF modality vary a lot in different nephropathy. For example, the proportions of IgM are 10.9%, 1.4%, 11.9%, 0, 0.7%, 13.5%, 0.7% in 7 modalities, respectively. The above analysis completes the analysis of *Finding 1*.

Finding 2: *For certain combinations, the single-modality IF image achieves better diagnosis accuracy than multi-modality IF sequences over DNN models.*

Analysis: To explore the impact of each modality in DNN models, we conduct experiments to evaluate the effectiveness of single IF modality for diagnosing each nephropathy. Specifically, we compare the nephropathy diagnosis accuracy of the same DNN model, when trained and tested over multi-modality IF sequences versus single-modality IF images. First, two widely-used classification models (ResNet-18 [4] and ECANet [15]) are implemented for 7-class nephropathy diagnosis. Then, we construct 7 dataset pairs from our main set, according to each IF modality. For each data pair, the DNN models are trained and tested with multi-modality and single-modality, respectively, the diagnosis accuracy of which is recorded as Acc_m and Acc_s. Subsequently, these two kinds of accuracy are compared by calculating the error weight E as follows,

$$E = \begin{cases} (\text{Acc}_s - \text{Acc}_m)/\text{Acc}_m, & \text{Acc}_s < \text{Acc}_m \\ (\text{Acc}_s - \text{Acc}_m)/(1 - \text{Acc}_m), & \text{Acc}_s > \text{Acc}_m \end{cases} \tag{1}$$

Thus, the higher error weight indicates that the single-modality can achieve more accurate diagnosis compared with using all modalities. Table 1 tabulates the error weights between each pair of IF modality and nephropathy, over ResNet-18 and ECANet. As shown, for certain combinations, such as IgG to ANCA, IgA to DN, and Fib to IgAN, the single-modality can even achieve better performance than using multi-modality IF sequences. Besides, the phenomenon is consistent over multiple DNN models. This implies that there exists a correlation for each single modality for contributing to the diagnosis on each nephropathy.

Table 1. The error weight results over ResNet-18/ECANet. Note that the error weights are marked in **bold** when consistently positive over two DNN models.

		IF Modality						
		IgG	IgA	IgM	C3	C4	C1q	Fib
Nephropathy	MN	**0.42/0.33**	-0.42/-0.31	-0.47/-0.53	-0.27/-0.34	-0.30/-0.37	-0.08/-0.26	-0.85/-0.80
	IgAN	-0.03/-0.03	**0.29/0.17**	-0.20/-0.11	-0.08/-0.09	-0.76/-0.81	0/0	**0.50/0.25**
	LN	-0.26/-0.22	-0.35/-0.28	-0.18/-0.15	**0.08/0.17**	-0.18/-0.14	**0.27/0.38**	-0.61/-0.11
	DN	**0.10/0.50**	**0.40/0.38**	0/0	-0.25/0	0/0	0/0	0/0
	ANCA	**0.86/0.80**	**0.63/0.88**	0/0	-0.50/-0.88	0/0	0/0	-0.80/-0.44
	MPGN	-0.50/-0.73	-0.29/-0.50	-0.51/-0.51	0/-0.50	**0.17**/0	-0.50/-0.89	-0.50/-0.60
	AL	0.22/0.27	**0.24/0.16**	0/0	0/0	0/0	0/0	0/0

3 Proposed Method

In this section, we introduce a customized knowledge distillation method for nephropathy diagnosis over IF sequence. Figure 3 illustrates the overall framework of the proposed method. As shown in the figure, a student network N_s is constructed to diagnose the nephropathy from the input IF sequence with multiple modalities, which is same as the piratical scenario. Besides, we further develop M teacher networks $\{\text{N}_\text{t}^\text{i}\}_{i=1}^{M}$ with the similar structure as the student network, but were trained over each single IF modality. During the training on the student network, given a input IF sequence, a learnable customized recruitment module is developed to adaptively select and fuse the knowledge (i.e., the intermediate features and diagnosis logits) from teacher networks, based on the

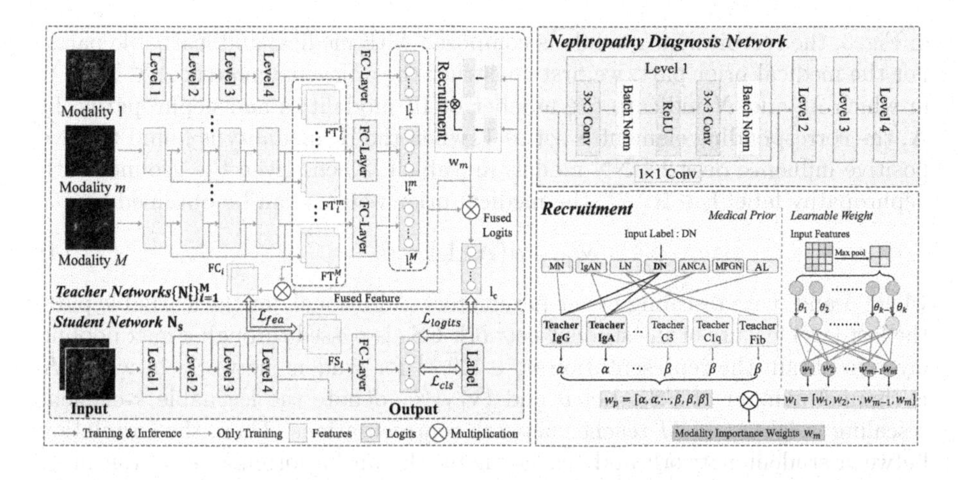

Fig. 3. The overall architecture of the proposed framework.

medical prior from our findings. Then, we transfer the fused knowledge to optimize the student network, via the developed multi-level distillation losses. This way, the student network can dynamically learn from the individual modalities, and finally achieve much better performance overs multi-modality IF sequence.

3.1 Nephropathy Diagnosis Network

Here, we introduce the detailed structure of the nephropathy diagnosis network, as the backbone structure for both student and teacher networks. Note that the backbone structure is flexible, and we implement 4 advanced DNNs in the experimental section. Taking ResNet-18 for the student network as an example, as illustrated in Fig. 3, ResNet-18 is implemented by 4 residual blocks, to extract the features with multiple levels. Each residual block consists of two 3×3 convolutional layers, two batch normalization layers and a ReLU activation layer. Given the input IF sequence $\mathbf{X}$, the multi-level features $\{\mathbf{FS}_i\}_{i=1}^4$ and prediction logits $\mathbf{l_s}$ of the student network $\mathrm{N_s}(\cdot)$ can be obtained as

$$\{\mathbf{FS}_i\}_{i=1}^4, \mathbf{l_s} = \mathrm{N_s}(\mathbf{X}), \mathbf{FS}_{i+1}{=}\mathrm{Res}(\mathbf{FS}_i) + \mathrm{Conv}_{1\times1}(\mathbf{FS}_i), \mathbf{l_s} = \mathrm{MLP}(\mathbf{FS}_4). \quad (2)$$

In (7), $\mathrm{Res}(\cdot)$, $\mathrm{Conv}_{1\times1}(\cdot)$, and $\mathrm{MLP}(\cdot)$ indicate the residual block, 1 × 1 convolutional layer, and multilayer perceptron, respectively. As the supervision of N-class nephropathy diagnosis, the cross-entropy loss is calculated upon the one-hot ground-truth diagnosis label $\hat{\mathbf{l}}_\mathrm{s}$ and the output predicted logits $\mathbf{l_s}$:

$$\mathcal{L}_{\mathrm{cls}} = -\sum\nolimits_{n=1}^{N} \hat{l}_s^n \log l_s^n. \quad (3)$$

3.2 Customized Recruitment Module

A customized recruitment module is developed to adaptively select the effective teacher networks, on the top of the medical priors from our findings. As shown in Fig. 3, the recruitment module is composed with medical and learnable parts. For the medical prior part, we first construct the adjacency matrix $\mathbf{A} \in \mathbb{R}^{M \times N}$, in which M and N indicate the number of IF modalities and nephropathy. In $\mathbf{A}$, the corresponding element is set as 1, when the IF modality is found to have positive influence over 2 DNN models in Table 1. Then, given the ground-truth nephropathy label $\hat{\mathbf{l}}_{\mathrm{s}} \in \mathbb{R}^{N \times 1}$, the medical prior weights can be obtained as:

$$\mathbf{w_p} = \alpha(\mathbf{A} \cdot \hat{\mathbf{l}}_{\mathrm{s}}) + \beta, \tag{4}$$

where α and β are rescaling hyper-parameters. Additional, for the learnable part, the last level feature of the student network $\mathbf{FS}_4$ is passed through a max pooling layer to obtain the representation $\mathbf{v}_{\mathrm{s}} \in \mathbb{R}^{K \times 1}$ for student network, in which K is channel number of $\mathbf{FS}_4$. Let $\boldsymbol{\theta}$ and $\{\mathbf{v}_{\mathrm{t,i}}\}_{i=1}^{M}$ denote the learnable k-element rescaling weights and M teacher network representations. Then, the correlation between student network and teacher network can be formulated as the inner product between vector representations $\langle \mathbf{v}_{\mathrm{t,i}}, \boldsymbol{\theta} \odot \mathbf{v}_{\mathrm{s}} \rangle$. In summary, the learnable importance weights $\mathbf{w_l}$ can be presented as

$$\mathbf{w_l} = \mathbf{V}(\boldsymbol{\theta} \odot (\text{Maxpooling}(\mathbf{FS_4}))), \tag{5}$$

where $\mathbf{V}$ is the matrix of $\{\mathbf{v}_{\mathrm{t,i}}\}_{i=1}^{M}$, while $\odot$ denotes element-wise multiplication. Note that $\mathbf{V}$ and $\boldsymbol{\theta}$ are initialized with 1, and optimized during training. Finally, a overall modality importance weight $\mathbf{w_t}$ can be obtained as

$$\mathbf{w_m} = \mathbf{w_p} \odot \mathbf{w_l}. \tag{6}$$

3.3 Multi-level Knowledge Distillation

Here, a multi-level knowledge distillation is developed between teacher and student networks. Based on the modality importance weight $\mathbf{w_t}$ from our recruitment module, the multi-level features $\{\mathbf{FC}_i\}_{i=1}^{M}$ and predicted logits $\mathbf{l_c}$ of are fused from multiple teacher networks:

$$\mathbf{l_c} = (\sum\nolimits_{j=1}^{M} w_t^j \mathbf{l}_{\mathbf{t}}^j)/M, \qquad \mathbf{FC}_i = \sum\nolimits_{j=1}^{M} w_t^j \mathbf{FT}_i^j / M \quad (i = 1, 2, 3, 4). \tag{7}$$

In above equation, w_t^j, $\mathbf{l}_{\mathbf{t}}^j$, $\mathbf{FT}_i^j$ are the importance weight, predicted logits, and the i-th level features for the j-th teacher networks, respectively. After that, to transfer the learned knowledge from teacher to student network, the logit loss is introduced by calculating the Kullback-Leibler (KL) divergence between the fused $\mathbf{l_c}$ and predicted logits $\mathbf{l_s}$ from the student network:

$$\mathcal{L}_{\text{logits}} = \sum\nolimits_{n=1}^{N} \mathbf{l}_{\mathbf{c}}^n \log(\frac{\mathbf{l}_{\mathbf{c}}^n}{\mathbf{l}_{\mathbf{s}}^n}). \tag{8}$$

Meanwhile, in order to make the student network fully learn from diagnosis processes of teacher networks, mean square error (MSE) losses are conducted on each level of fused $\{\mathbf{FC}_i\}_{i=1}^{4}$ and student network features $\{\mathbf{FS}_i\}_{i=1}^{4}$:

$$\mathcal{L}_{\text{fea}} = \sum\nolimits_{i=1}^{4} \|\mathbf{FC}_i - \mathbf{FS}_i\|_2^2. \tag{9}$$

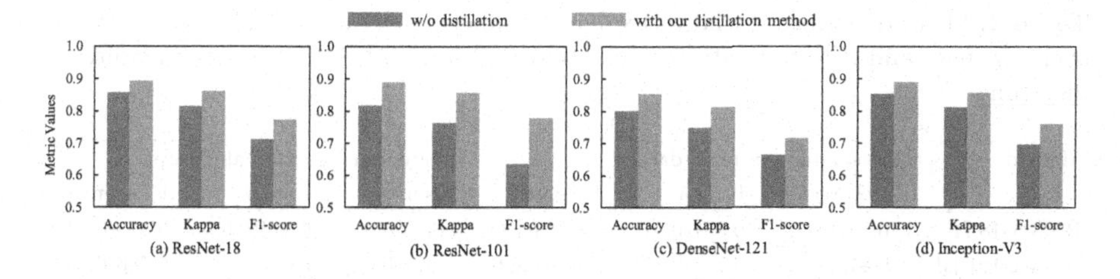

Fig. 4. Diagnosis results of 4 backbone models with our distillation method.

Finally, the overall loss function for the student network can be written as

$$\mathcal{L} = \lambda_{fea}\mathcal{L}_{\text{fea}} + \lambda_{logits}\mathcal{L}_{\text{logits}} + \lambda_{cls}\mathcal{L}_{\text{cls}}, \tag{10}$$

where λ_{fea}, λ_{logits} and λ_{cls} are the hyper-parameters for balancing single losses.

4 Experiment

4.1 Experimental Settings

In our experiments, all IF images are resized to 512×512 for consistency. During training, the parameters are updated by Adam optimizer with an initial learning rate of 0.0001. Then, each teacher network is pre-trained for 70 epochs, and then the student network with our distillation method is trained for 460 epochs. Finally, our and 8 other compared methods are trained over the training set of main set, and evaluated over the main test set and the external set, by adopting 3 evaluation metrics of accuracy, kappa and F1-score.

4.2 Evaluation on Knowledge Distillation

To evaluate the effectiveness of the proposed customized knowledge distillation method, we implement it over 4 different backbone models of ResNet-18 [4], ResNet-101 [4], DenseNet-121 [6] and Inception-V3 [12]. Figure 4 compares the nephropathy diagnosis results of 4 backbone models after conducting our distillation method. As shown, all backbone models obtain significant improvements when applying the proposed distillation methods. For instance, benefiting the transferred knowledge from single-modality, DenseNet-121 improves 0.053, 0.065, and 0.051 in accuracy, kappa and F1-score over our main dataset. This validates the effectiveness of the proposed method. Note that, we select ResNet-18 as the final model, due to its best performance among 4 backbone structures.

4.3 Comparisons with the State-of-the-Art Models

We evaluate the diagnosis performance of our method over our main dataset, compared with 10 DNN based methods, i.e., ShuffleNet [19], EfficientNet [14], GCNet [1], ECANet [15], KNet [7], MANet [18], Hao *et al.* [3], Ada-CCFNet

Table 2. Mean (standard deviation) values of 3 metrics of our and compared methods over our main and external datasets. The best and second best results are in **bold** and underline.

Method	Evaluation on main dataset			Evaluation on external dataset		
	Accuracy	Kappa	F1-score	Accuracy	Kappa	F1-score
ResNet-18 [4]	0.857(0.03)	0.831(0.01)	0.721(0.01)	0.525(0.02)	0.424(0.03)	0.353(0.03)
ResNet-101 [4]	0.813(0.01)	0.760(0.01)	0.620(0.02)	0.673(0.02)	0.600(0.02)	0.586(0.05)
ShuffleNet [19]	0.833(0.01)	0.785(0.01)	0.691(0.02)	0.714(0.04)	0.645(0.05)	0.649(0.04)
EfficientNet [14]	0.841(0.01)	0.795(0.01)	0.676(0.01)	0.437(0.04)	0.306(0.03)	0.264(0.05)
GCNet [1]	0.839(0.01)	0.794(0.01)	0.683(0.02)	0.666(0.03)	0.590(0.03)	0.535(0.03)
DenseNet-121 [6]	0.842(0.01)	0.797(0.01)	0.720(0.01)	0.471(0.04)	0.413(0.03)	0.424(0.03)
ECANet [15]	0.833(0.01)	0.786(0.01)	0.637(0.01)	0.634(0.03)	0.544(0.04)	0.506(0.05)
Inception-V3 [12]	0.855(0.01)	0.817(0.01)	0.709(0.01)	0.709(0.02)	0.646(0.03)	0.605(0.02)
KNet [7]	0.850(0.01)	0.807(0.01)	0.750(0.01)	0.668(0.02)	0.595(0.03)	0.623(0.04)
MANet [18]	0.861(0.01)	0.820(0.01)	0.678(0.02)	0.702(0.03)	0.634(0.04)	0.624(0.02)
Hao *et al.* [3]	0.842(0.01)	0.802(0.01)	0.739(0.02)	0.666(0.03)	0.586(0.04)	0.569(0.02)
Ada-CCFNet [16]	0.851(0.01)	0.820(0.01)	0.669(0.02)	0.678(0.02)	0.521(0.03)	0.520(0.02)
MCL [17]	0.842(0.01)	0.803(0.01)	0.661(0.01)	0.528(0.02)	0.496(0.03)	0.436(0.02)
ITRD [10]	0.806(0.01)	0.773(0.01)	0.646(0.02)	0.618(0.02)	0.593(0.03)	0.507(0.02)
Ours	**0.892(0.01)**	**0.861(0.01)**	**0.773(0.01)**	**0.757(0.02)**	**0.706(0.01)**	**0.655(0.01)**

[16], MCL [17] and ITRD [10]. Among them, KNet [7], MANet [18], Hao *et al.* [3] and Ada-CCFNet [16] are nephropathy diagnosis methods, while MCL [17] and ITRD [10] are knowledge distillation methods. All compared methods are re-trained with the same settings as ours. As shown in Table 2, our proposed method achieves the best performance on IF based nephropathy diagnosis over 3 metrics and 2 datasets. For example, compared with the second best method, our method can improve 0.035/0.043, 0.030/0.060 and 0.023/0.006 in accuracy, kappa and F1-score over our main/external dataset, respectively. This validates the superior performance and generalization ability of the proposed method.

4.4 Ablation Study

We ablate different components of our method to thoroughly analyze their effects on nephropathy diagnosis. Specifically, the accuracy degrades 0.028, 0.023, and 0.031, when ablating medical prior, learnable weights, and recruitment module (equal distillation), respectively. Besides, other ablations, such as the number of teacher networks and distillation loss, are also analyzed. The ablation experiments are reported in Fig. reffinding1 of the supplementary.

5 Conclusion

In this paper, we propose a customized knowledge distillation method for IF based nephropathy diagnosis. Different from the existing methods that averagely integrate information of different IF modalities, we propose a knowledge distillation framework to transfer knowledge from the trained single-modality teacher networks to a multi-modality student network. In particular, a recruitment module and multi-level knowledge distillation are developed to dynamically select and fuse the knowledge from teacher networks. The extensive experiments on several backbone networks verify the effectiveness of our proposed framework.

References

1. Cao, Y., Xu, J., Lin, S., Wei, F., Hu, H.: Gcnet: non-local networks meet squeeze-excitation networks and beyond. In: Proceedings of the IEEE/CVF International Conference on Computer Vision Workshops (2019)
2. Gomariz, A., et al.: Modality attention and sampling enables deep learning with heterogeneous marker combinations in fluorescence microscopy. Nature Mach. Intell. **3**(9), 799–811 (2021)
3. Hao, F., Liu, X., Li, M., Han, W.: Accurate kidney pathological image classification method based on deep learning and multi-modal fusion method with application to membranous nephropathy. Life **13**(2), 399 (2023)
4. He, K., Zhang, X., Ren, S., Sun, J.: Deep residual learning for image recognition. In: Proceedings of the IEEE Conference on Computer Vision and Pattern Recognition, pp. 770–778 (2016)
5. Hinton, G., Vinyals, O., Dean, J., et al.: Distilling the knowledge in a neural network. arXiv preprint arXiv:1503.02531 2(7) (2015)
6. Huang, G., Liu, Z., Van Der Maaten, L., Weinberger, K.Q.: Densely connected convolutional networks. In: Proceedings of the IEEE Conference on Computer Vision and Pattern Recognition, pp. 4700–4708 (2017)
7. Kitamura, S., Takahashi, K., Sang, Y., Fukushima, K., Tsuji, K., Wada, J.: Deep learning could diagnose diabetic nephropathy with renal pathological immunofluorescent images. Diagnostics **10**(7), 466 (2020)
8. Ligabue, G., et al.: Evaluation of the classification accuracy of the kidney biopsy direct immunofluorescence through convolutional neural networks. Clin. J. Am. Soc. Nephrol. **15**(10), 1445–1454 (2020)
9. Liu, Y., Cao, J., Li, B., Hu, W., Maybank, S.: Learning to explore distillability and sparsability: a joint framework for model compression. IEEE Trans. Pattern Anal. Mach. Intell. (2022)
10. Miles, R., Lopez-Rodriguez, A., Mikolajczyk, K.: Information theoretic representation distillation
11. Romagnani, P., et al.: Chronic kidney disease. Nat. Rev. Dis. Primers. **3**(1), 1–24 (2017)
12. Szegedy, C., Vanhoucke, V., Ioffe, S., Shlens, J., Wojna, Z.: Rethinking the inception architecture for computer vision. In: Proceedings of the IEEE Conference on Computer Vision and Pattern Recognition, pp. 2818–2826 (2016)
13. Takahashi, K., Kitamura, S., Fukushima, K., Sang, Y., Tsuji, K., Wada, J.: The resolution of immunofluorescent pathological images affects diagnosis for not only artificial intelligence but also human. J. Nephropathol. **10**(3), e26 (2021)
14. Tan, M., Le, Q.: Efficientnet: rethinking model scaling for convolutional neural networks. In: International Conference on Machine Learning, pp. 6105–6114. PMLR (2019)
15. Wang, Q., Wu, B., Zhu, P.F., Li, P., Zuo, W., Hu, Q.: Eca-net: efficient channel attention for deep convolutional neural networks. 2020 IEEE/CVF Conference on Computer Vision and Pattern Recognition (CVPR), pp. 11531–11539 (2019)

16. Wang, R., et al.: Ada-ccfnet: classification of multimodal direct immunofluorescence images for membranous nephropathy via adaptive weighted confidence calibration fusion network. Eng. Appl. Artif. Intell. **117**, 105637 (2023)
17. Yang, C., An, Z., Cai, L., Xu, Y.: Mutual contrastive learning for visual representation learning. In: Proceedings of the AAAI Conference on Artificial Intelligence, vol. 36, pp. 3045–3053 (2022)
18. Zhang, L., et al.: Classification of renal biopsy direct immunofluorescence image using multiple attention convolutional neural network. Comput. Methods Programs Biomed. **214**, 106532 (2022)
19. Zhang, X., Zhou, X., Lin, M., Sun, J.: Shufflenet: an extremely efficient convolutional neural network for mobile devices. In: Proceedings of the IEEE Conference on Computer Vision and Pattern Recognition, pp. 6848–6856 (2018)
20. Zhou, X., Laszik, Z., Silva, F.: Algorithmic approach to the interpretation of renal biopsy. In: Zhou, X.J., Laszik Z., Nadasdy, T., D'Agati, V.D., Silva, F.G. (eds.) Silva's Diagnostic Reanal Pathology, pp. pp. 55–57. Cambridge University Press, New York (2009)

Text-Guided Cross-Position Attention for Segmentation: Case of Medical Image

Go-Eun Lee[1], Seon Ho Kim[2], Jungchan Cho[3], Sang Tae Choi[4(✉)], and Sang-Il Choi[1(✉)]

[1] Dankook University, Yongin, Gyeonggi-do, Korea
choisi@dankook.ac.kr
[2] University of Southern California, Los Angeles, USA
[3] Gachon University, Seongnam, Gyeonggi-do, Korea
[4] Chung-Ang University College of Medicine, Seoul, Korea
beconst@cau.ac.kr

Abstract. We propose a novel text-guided cross-position attention module which aims at applying a multi-modality of text and image to position attention in medical image segmentation. To match the dimension of the text feature to that of the image feature map, we multiply learnable parameters by text features and combine the multi-modal semantics via cross-attention. It allows a model to learn the dependency between various characteristics of text and image. Our proposed model demonstrates superior performance compared to other medical models using image-only data or image-text data. Furthermore, we utilize our module as a region of interest (RoI) generator to classify the inflammation of the sacroiliac joints. The RoIs obtained from the model contribute to improve the performance of classification models.

Keywords: Image Segmentation · Multi Modal Learning · Cross Position Attention · Text-Guided Attention · Medical Image

1 Introduction

Advances in deep learning have been witnessed in many research areas over the past decade. In medical field, automatic analysis of medical image data has actively been studied. In particular, segmentation which identify region of interest (RoI) in an automatic way is an essential medical imaging process. Thus, deep learning-based segmentation has been utilized in various medical domains such as brain, breast cancers, and colon polyps. Among the popular architectures, variants of U-Net have been widely adopted due to their effective encoder-decoder structure, proficient at capturing the characteristics of cells in images. Recently, it has been demonstrated that the attention modules [4,17,20] enable deep learning networks to better extract robust features, which can be applied in medical image segmentation to learn subtle medical features and achieve higher performance [14,16,18,21].

H. Greenspan et al. (Eds.): MICCAI 2023, LNCS 14224, pp. 537–546, 2023.
https://doi.org/10.1007/978-3-031-43904-9_52

However, as image-only training trains a model with pixels that constitute an image, there is a limit in extracting fine-grained information about a target object even if transfer learning is applied through a pre-trained model. Recently, to overcome this limitation, multi-modality studies have been conducted, aiming to enhance the expressive power of both text and image features. For instance, CLIP [12] used contrastive learning based on image-text pairs to learn the similarity between the image of an object and the text describing it, achieving significant performance gains in a variety of computer vision problems.

The trend of text-image multi-modality-based research on image processing has extended to the medical field. [19] proposed a semantic matching loss that learns medical knowledge to supplement the disadvantages of CLIP that cannot capture uncertain medical semantic meaning. In [2], they trained to increase the similarity between the image and text by calculating their influence on each other as a weighted feature. For the segmentation task, LViT [10] generated the positional characteristics of lesions or target objects as text labels. Furthermore, it proposed a Double U-Shaped structure consisting of a U-Shaped ViT that combines image and text information and a U-Shaped CNN that produces a segmentation mask. However, when combining medical images with non-fine-grained text information, noise can affect the outcome.

In this paper, we propose a new text-guided cross-position attention module ($CPAM^{TG}$) that combines text and image. In a medical image, a position attention module (PAM) effectively learns subtle differences among pixels. We utilized PAM which calculates the influence among pixels of an image to capture the association between text and image. To this end, we converted the global text representation generated from the text encoder into a form, such as an image feature map, to create keys and values. The image feature map generated from an image encoder was used as a query. Learning the association between text and image enables us to learn positional information of targets in an image more effectively than existing models that learned multi-modality from medical images. $CPAM^{TG}$ showed an excellent segmentation performance in our comprehensive experiments on various medical images, such as cell, chest X-ray, and magnetic resonance image (MRI). In addition, by applying the proposed technique to the automatic RoI setting module for the deep learning-based diagnosis of sacroiliac arthritis, we confirmed that the proposed method could be effective when it is used in a practical application of computer-aided diagnosis.

Our main contributions are as follows:

- We devised a text-guided cross-position attention module ($CPAM^{TG}$) that efficiently combines text information with image feature maps.
- We demonstrated the effect of $CPAM^{TG}$ on segmentation for various types of medical images.
- For a practical computer-aided diagnosis system, we confirm the effectiveness of the proposed method in a deep learning-based sacroiliac arthritis diagnosis system.

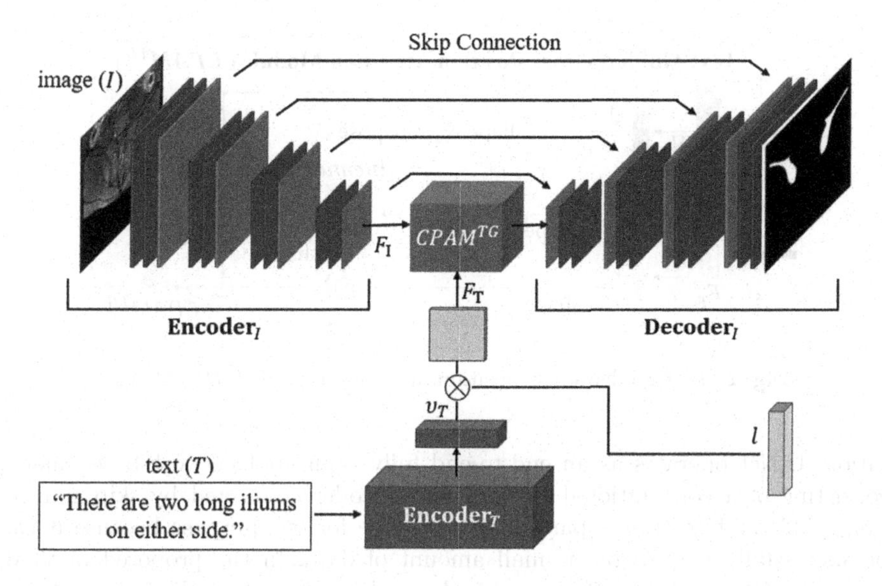

Fig. 1. Overview of our proposed segmentation model.

2 Methods

In this section, we propose text-guided segmentation model that can effectively learn the multi-modality of text and images. Figure 1 shows the overall architecture of the proposed model, which consists of an image encoder for generating a feature map from an input image, a text encoder for embedding a text describing the image, and a cross-attention module. The cross-attention module allows the text to serve as a guide for image segmentation by using the correlation between the global text representation and the image feature map. To achieve robust text encoding, we adopt a transformer [17] structure which performs well in Natural Language Processing (NLP). For image encoding and decoding, we employed U-Net, widely used as a backbone in medical image segmentation. To train our proposed model, we utilize a dataset consisting of image and text pairs.

2.1 Configuration of Text-Image Encoder and Decoder

As Transformer has demonstrated its effectiveness in handling the long-range dependency in sequential data through self-attention [1], it performs well in various fields requiring NLP or contextual information analysis of data. We used a Transformer ($\mathbf{Encoder}_T$) to encode the semantic information of the text describing a medical image into a global text representation $v_T \in \mathbb{R}^{1\times 2C}$ as $v_T = \mathbf{Encoder}_T(T)$. Here, the text semantics (T) can be a sentence indicating the location or characteristics of an interested region in an image such as a lesion shown in Fig. 1.

To create a segmentation mask from medical images (I), we used U-Net [13] which has a relatively simple yet effective structure for biomedical image segmen-

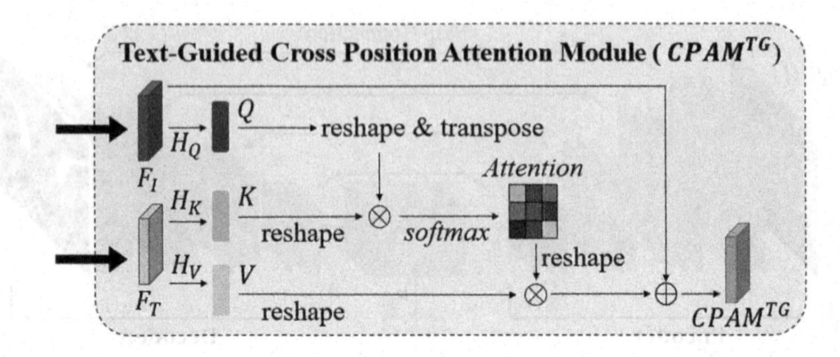

Fig. 2. Text-guided cross-position attention module ($CPAM^{TG}$).

tation. U-Net operates as an end-to-end fully connected network-based model consisting of a convolutional encoder and decoder connected by skip connections. This architecture is particularly suitable for our purpose because it can be successfully trained on a small amount of data. In the proposed method, we used VGG-16 [15] as the encoder ($\mathbf{Encoder}_I$) to obtain the image feature $F_I \in \mathbb{R}^{C\times H\times W}$ as $F_I = \mathbf{Encoder}_I(I)$ and the decoder ($\mathbf{Decoder}_I$) that will generate the segmented image from the enhanced encoding vector obtained by the cross-position attention which will be described in the following subsection.

The weights of text and image encoders were initialized by the weights of CLIP's pre-trained transformer and VGG16 pre-trained on ImageNet, respectively, and fine-tuned by a loss function for segmentation which will be described in Sect. 3.

2.2 Text-Guided Cross Position Attention Module

We introduce a text-guided cross-position attention module ($CPAM^{TG}$) that integrates cross-attention [3] with the position attention module (PAM) [5] to combine the semantic information of text and image. This module utilizes not only the image feature map from the image encoder but also the global text representation from the text encoder to learn the dependency between various characteristics of text and image. PAM models rich contextual relationships for local features generated from FCNs. It effectively captures spatial dependencies among pixels by generating keys, queries, and values from feature maps. By encoding broad contextual information into local features, and then adaptively gathering spatial contexts, PAM improves representation capability. In particular, this correlation analysis among pixels can effectively analyze medical images in which objects are relatively ambiguous compared to other types of natural images.

In Fig. 2, we multiply the learnable parameter ($l \in \mathbb{R}^{1\times(HW)}$) by the global text representation (v_T) to match the dimension of the text feature with that of the image feature map as $F_T = \mathcal{R}(G(v_T)^\intercal \times l)$, where $G(\cdot)$ is a fully connected layer that adjusts the $2C$ channel of the global text representation v_T to the image feature map channel C. $\mathcal{R}(\cdot)$ is a reshape operator to $C \times H \times W$.

The text feature map F_T is used as key and value, and the image feature map F_I is used as a query to perform self-attention as

$$Q = H_Q(F_I), \quad K = H_K(F_T), \quad V = H_V(F_T), \tag{1}$$

where H_Q, H_K, and H_V are convolution layers with a kernel size of 1, and Q, K, and V are queries, keys, and values for self-attention.

$$Attention = softmax(Q^{\intercal}K) \tag{2}$$

$$CPAM^{TG} = Attention^{\intercal}V + F_I \tag{3}$$

Finally, by upsampling the low-dimensional $CPAM^{TG}$ obtained through cross-attention of text and image together with skip-connection, more accurate segmentation prediction can express the detailed information of an object.

3 Experiments

3.1 Setup

Medical Datasets. We evaluated $CPAM^{TG}$ using three datasets: MoNuSeg [8] dataset, QaTa-COV19 [6] dataset, and sacroiliac joint (SIJ) dataset. The first two datasets are the same benchmark datasets used in [10]. MoNuSeg [8] contains 30 digital microscopic tissue images of several patients and QaTa-COV19 are COVID-19 chest X-ray images. The ratio of training, validation, and test sets was the same as in [10]. SIJ is the dataset privately prepared for this study which consists of 804 MRI slices of nineteen healthy subjects and sixty patients diagnosed with axial spondyloarthritis. Among all MRI slices, we selected the gadolinium-enhanced fat-suppressed T1-weighted oblique coronal images, excluding the first and last several slices in which the pelvic bones did not appear, and added the text annotations for the slices.

Training and Metrics. For a better training, data augmentation was used. We randomly rotated images by $-20° \sim +20°$ and conducted a horizontal flip with 0.5 probability for only the MoNuSeg and QaTa-COV19 datasets. The batch size and learning rate were set to 2 and 0.001, respectively. The loss function ($\mathcal{L}_T$) for training is the sum of the binary cross-entropy loss ($\mathcal{L}_{BCE}$) and the dice loss ($\mathcal{L}_{DICE}$): $\mathcal{L}_T = \mathcal{L}_{BCE} + \mathcal{L}_{DICE}$. The mDice and mIoU metrics, widely used to measure the performance of segmentation models, were used to evaluate the performance of object segmentation. For experiments, PyTorch (v1.7.0) were used on a computer with NVIDIA-V100 32 GB GPU.

3.2 Segmentation Performance

Table 1 presents the comparison of image segmentation performance among the proposed model and the U-Net [13], U-Net++ [22], Attention U-Net [11],

MedT [16], and LViT [10] methods. Analyzing the results in Table 1, unlike natural image segmentation, the attention module-based method (Attention U-Net) and transformer-based method (MEdT) did not achieve significant performance gains compared to U-Net based methods (U-Net and U-Net++).

By contrast, LViT and $CPAM^{TG}$, which utilize both text and image information, significantly improved image segmentation performance because of multimodal complementarity, even for medical images with complex and ambiguous object boundaries. Furthermore, $CPAM^{TG}$ achieves a better performance by 1 to 3% than LViT [10] on all datasets. This means that the proposed $CPAM^{TG}$ helps to improve segmentation performance by allowing text information to serve as a guide for feature extraction for segmentation.

Figure 3 shows the examples of segmentation masks obtained using each method. In Fig. 3, we marked the boundary of the target object with a red box and showed the ground truth masks for these objects in the last column. Similar to the analysis that can be derived from Table 1, Fig. 3 shows that $CPAM^{TG}$ and LViT, which use text information together for image segmentation, create a segmentation mask with more distinctive borders than other methods. In particular, with SIJ, $CPAM^{TG}$ accurately predicted the boundaries of even thin bone parts compared to LViT. Figure 3 also shows that even on the QaTa-COV19 and MoNuSeg datasets, $CPAM^{TG}$ predicted the most accurate segmentation masks (see the red box areas). From these results, we conjecture that the reasons for the performance improvement of $CPAM^{TG}$ are as follows. $CPAM^{TG}$ independently encodes the input text and image and then combines semantic information via a cross-attention module. Consequently, the two types of information (text and image) do not act as noise from each other, and $CPAM^{TG}$ achieves an improved performance compared to LViT.

3.3 Ablation Study

To validate the design of our proposed model, we perform an ablation study on position attention and $CPAM^{TG}$. Specifically, for the SIJ dataset, we examined the effect of attention in extracting feature maps through comparison with backbone networks (U-Net) and PAM. In addition, we investigated whether text information about images serves as a guide in the position attention process for image segmentation by comparing it with $CPAM^{TG}$. Table 2 summarizes the result of each case. As can be observed in Table 2, the performance of PAM was higher than that of the backbone. This indicates that PAM improves performance by learning associations between pixels for ambiguous targets, as in medical images. In addition, the best performance results of $CPAM^{TG}$ show that text information provided helpful information in an image segmentation process using the proposed model.

Table 1. Performance comparison of medical segmentation models with three datasets

	Qata-COV19		MoNuSeg		SIJ	
	mDice	mIoU	mDice	mIoU	mDice	mIoU
U-Net	0.7902	0.6946	0.7645	0.6286	0.7395	0.6082
U-Net++	0.7962	0.7025	0.7701	0.6304	0.7481	0.6124
AttUNet	0.7931	0.7004	0.7667	0.6347	0.7770	0.6487
MedT	0.7747	0.6751	0.7746	0.6337	0.7914	0.6600
LViT	0.8366	0.7511	0.8036	0.6731	0.8572	0.7572
Proposed	**0.8425**	**0.7598**	**0.8105**	**0.6775**	**0.8800**	**0.7887**

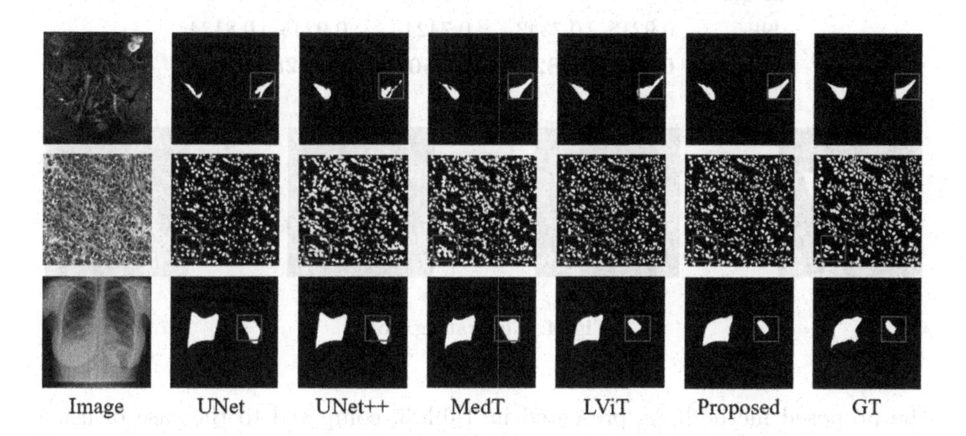

Fig. 3. Qualitative results of segmentation models.

3.4 Application: Deep-Learning Based Disease Diagnosis

In this section, we confirm the effectiveness of the proposed segmentation method through a practical bio-medical application as a deep learning-based active sacroiliitis diagnosis system.

MRI is a representative means for early diagnosis of "active sacroiliitis in axSpA". As active sacroiliitis is a disease that occurs between the pelvic bone and sacral bone, when a MR slice is input, the diagnostic system first separates the area around the pelvic bone into an RoI patch and uses it as an input for the active sacroiliitis classification network [7]. However, even in the same pelvis, the shape of the bone shown in MR slices varies depending on the slice position of the MRI and the texture of the tissue around the bone is complex. This makes finding an accurate RoI a challenge.

We segmented the pelvic bones in MRI slices using the proposed method to construct a fully automatic deep learning-based active sacroiliitis diagnosis system, including RoI settings from MRI input images. Figure 4 shows the results of generating RoI patches by dividing the pelvic bone from MRI slices using

Table 2. Ablation study of the effectiveness of our proposed module. "PAM" means we used it instead of $CPAM^{TG}$ for image-only training.

Backbone	PAM	$CPAM^{TG}$	mDice	mIoU
✓			0.7395	0.6082
✓	✓		0.7886	0.6591
✓		✓	**0.8800**	**0.7887**

Table 3. The results of classification models.

	Recall	Precision	Specificity	NPV	F1
Origin	0.8500	0.6640	0.7346	0.8880	0.7456
[9]	0.9215	0.7343	0.7424	0.9245	0.8174
Proposed	**0.9217**	**0.8281**	**0.8503**	**0.9328**	**0.8724**

Fig. 4. Generating RoI.

the proposed method. As presented in Table 3, compared to the case of using the original MRI image without the RoI setting, using the hand-crafted RoI patch [9] showed an average of 7% higher performance in recall, precision, and f1. It is noticeable that the automatically set RoI patch showed similar or better performance than the manual RoI patch for each measurement. This indicates that the proposed method can be effectively utilized in practical applications of computer-aided diagnosis.

4 Conclusion

In this study, we developed a new text-guided cross-attention module ($CPAM^{TG}$) that learns text and image information together. The proposed model has a composite structure of position attention and cross-attention in that the key and value are from text data, and the query is created from the image. We use a learnable parameter to convert text features into a tensor of the same dimension as the image feature map to combine text and image information effectively. By calculating the association between the reshaped global text representation and each component of the image feature map, the proposed method outperformed image segmentation performance compared to previous studies using both text and image or image-only training method. We also confirmed that it could be utilized for a deep-learning-based sacroiliac arthritis diagnosis system, one of the use cases for practical medical applications. The proposed method can be further used in various medical applications.

Acknowledgements. This work was supported by the MSIT (Ministry of Science, ICT), Korea, under the High-Potential Individuals Global Training Program (RS-2022-00155227) supervised by the IITP (Institute for Information & Communications Technology Planning & Evaluation), the National Research Foundation of Korea (NRF) grant funded by the Korea government (MSIT) (2021R1A2B5B01001412), and the Bio & Medical Technology Development Program of the National Research Foundation (NRF) funded by the Korean government (MSIT) (RS-2023-00220408).

References

1. Bahdanau, D., Cho, K., Bengio, Y.: Neural machine translation by jointly learning to align and translate. arXiv preprint arXiv:1409.0473 (2014)
2. Bhalodia, R., et al.: Improving pneumonia localization via cross-attention on medical images and reports. In: de Bruijne, M., et al. (eds.) MICCAI 2021. LNCS, vol. 12902, pp. 571–581. Springer, Cham (2021). https://doi.org/10.1007/978-3-030-87196-3_53
3. Chen, C.F.R., Fan, Q., Panda, R.: Crossvit: cross-attention multi-scale vision transformer for image classification. In: Proceedings of the IEEE/CVF International Conference on Computer Vision, pp. 357–366 (2021)
4. Dosovitskiy, A., et al.: An image is worth 16×16 words: transformers for image recognition at scale. arXiv preprint arXiv:2010.11929 (2020)
5. Fu, J., et al.: Dual attention network for scene segmentation. In: Proceedings of the IEEE/CVF Conference on Computer Vision and Pattern Recognition, pp. 3146–3154 (2019)
6. Haghanifar, A., Majdabadi, M.M., Choi, Y., Deivalakshmi, S., Ko, S.: Covid-cxnet: detecting covid-19 in frontal chest x-ray images using deep learning. Multimedia Tools Appl. **81**(21), 30615–30645 (2022)
7. He, K., Zhang, X., Ren, S., Sun, J.: Deep residual learning for image recognition. In: Proceedings of the IEEE Conference on Computer Vision and Pattern Recognition, pp. 770–778 (2016)
8. Kumar, N., Verma, R., Sharma, S., Bhargava, S., Vahadane, A., Sethi, A.: A dataset and a technique for generalized nuclear segmentation for computational pathology. IEEE Trans. Med. Imaging **36**(7), 1550–1560 (2017)
9. Lee, K.H., Choi, S.T., Lee, G.Y., Ha, Y.J., Choi, S.I.: Method for diagnosing the bone marrow edema of sacroiliac joint in patients with axial spondyloarthritis using magnetic resonance image analysis based on deep learning. Diagnostics **11**(7), 1156 (2021)
10. Li, Z., et al.: Lvit: language meets vision transformer in medical image segmentation. arXiv preprint arXiv:2206.14718 (2022)
11. Oktay, O., et al.: Attention u-net: learning where to look for the pancreas. arXiv preprint arXiv:1804.03999 (2018)
12. Radford, A., et al.: Learning transferable visual models from natural language supervision. In: International Conference on Machine Learning, pp. 8748–8763 (2021)

13. Ronneberger, O., Fischer, P., Brox, T.: U-Net: convolutional networks for biomedical image segmentation. In: Navab, N., Hornegger, J., Wells, W.M., Frangi, A.F. (eds.) MICCAI 2015. LNCS, vol. 9351, pp. 234–241. Springer, Cham (2015). https://doi.org/10.1007/978-3-319-24574-4_28
14. Shome, D., et al.: Covid-transformer: interpretable covid-19 detection using vision transformer for healthcare. Int. J. Environ. Res. Public Health **18**(21), 11086 (2021)
15. Simonyan, K., Zisserman, A.: Very deep convolutional networks for large-scale image recognition. arXiv preprint arXiv:1409.1556 (2014)
16. Valanarasu, J.M.J., Oza, P., Hacihaliloglu, I., Patel, V.M.: Medical transformer: gated axial-attention for medical image segmentation. In: de Bruijne, M., Cattin, P.C., Cotin, S., Padoy, N., Speidel, S., Zheng, Y., Essert, C. (eds.) MICCAI 2021. LNCS, vol. 12901, pp. 36–46. Springer, Cham (2021). https://doi.org/10.1007/978-3-030-87193-2_4
17. Vaswani, A., et al.: Attention is all you need. Adv. Neural Inf. Process. Syst. **30**, 1–11 (2017)
18. Wang, W., Chen, C., Ding, M., Yu, H., Zha, S., Li, J.: TransBTS: multimodal brain tumor segmentation using transformer. In: de Bruijne, M., et al. (eds.) MICCAI 2021. LNCS, vol. 12901, pp. 109–119. Springer, Cham (2021). https://doi.org/10.1007/978-3-030-87193-2_11
19. Wang, Z., Wu, Z., Agarwal, D., Sun, J.: Medclip: contrastive learning from unpaired medical images and text. arXiv preprint arXiv:2210.10163 (2022)
20. Woo, S., Park, J., Lee, J.Y., Kweon, I.S.: Cbam: convolutional block attention module. In: Proceedings of the European Conference on Computer Vision (ECCV), pp. 3–19 (2018)
21. Xing, Z., Yu, L., Wan, L., Han, T., Zhu, L.: Nestedformer: nested modality-aware transformer for brain tumor segmentation. In: Wang, L., Dou, Q., Fletcher, P.T., Speidel, S., Li, S. (eds.) MICCAI 2022. LNCS, vol. 13435, pp. 140–150. Springer, Heidelberg (2022)
22. Zhou, Z., Rahman Siddiquee, M.M., Tajbakhsh, N., Liang, J.: UNet++: a nested U-net architecture for medical image segmentation. In: Stoyanov, D., et al. (eds.) DLMIA/ML-CDS -2018. LNCS, vol. 11045, pp. 3–11. Springer, Cham (2018). https://doi.org/10.1007/978-3-030-00889-5_1

Visual-Attribute Prompt Learning for Progressive Mild Cognitive Impairment Prediction

Luoyao Kang[1,2], Haifan Gong[1,2], Xiang Wan[1], and Haofeng Li[1(✉)]

[1] Shenzhen Research Institute of Big Data, Shenzhen, China
lhaof@sribd.cn

[2] The Chinese University of Hong Kong, Shenzhen, China

Abstract. Deep learning (DL) has been used in the automatic diagnosis of Mild Cognitive Impairment (MCI) and Alzheimer's Disease (AD) with brain imaging data. However, previous methods have not fully exploited the relation between brain image and clinical information that is widely adopted by experts in practice. To exploit the heterogeneous features from imaging and tabular data simultaneously, we propose the Visual-Attribute Prompt Learning-based Transformer (VAP-Former), a transformer-based network that efficiently extracts and fuses the multi-modal features with prompt fine-tuning. Furthermore, we propose a Prompt fine-Tuning (PT) scheme to transfer the knowledge from AD prediction task for progressive MCI (pMCI) diagnosis. In details, we first pre-train the VAP-Former without prompts on the AD diagnosis task and then fine-tune the model on the pMCI detection task with PT, which only needs to optimize a small amount of parameters while keeping the backbone frozen. Next, we propose a novel global prompt token for the visual prompts to provide global guidance to the multi-modal representations. Extensive experiments not only show the superiority of our method compared with the state-of-the-art methods in pMCI prediction but also demonstrate that the global prompt can make the prompt learning process more effective and stable. Interestingly, the proposed prompt learning model even outperforms the fully fine-tuning baseline on transferring the knowledge from AD to pMCI.

Keywords: Alzheimer's disease · Prompt learning · Magnetic resonance imaging · Multi-modal classification · Transformer · Attention modeling

This work is supported by Chinese Key-Area Research and Development Program of Guangdong Province (2020B0101350001), and the National Natural Science Foundation of China (No.62102267), and the Guangdong Basic and Applied Basic Research Foundation (2023A1515011464), and the Shenzhen Science and Technology Program (JCYJ20220818103001002), and the Guangdong Provincial Key Laboratory of Big Data Computing, The Chinese University of Hong Kong, Shenzhen.

L. Kang and H. Gong—Contribute equally to this work.

Supplementary Information The online version contains supplementary material available at https://doi.org/10.1007/978-3-031-43904-9_53.

H. Greenspan et al. (Eds.): MICCAI 2023, LNCS 14224, pp. 547–557, 2023.
https://doi.org/10.1007/978-3-031-43904-9_53

1 Introduction

Alzheimer's disease (AD) is one of the most common neurological diseases in elderly people, accounting for 50–70% of dementia cases [31]. The progression of AD triggers memory deterioration, impairment of cognition, irreversible loss of neurons, and further genetically complex disorders as well. Mild Cognitive Impairment (MCI), the prodromal stage of AD, has been shown as the optimal stage to be treated to prevent the MCI-to-AD conversion [29]. Progressive MCI (pMCI) group denotes those MCI patients who progress to AD within 3 years, while stable MCI (sMCI) patients remained stable over the same time period. pMCI is an important group to study longitudinal changes associated with the development of AD [26]. By predicting the pMCI progress of the patients, we can intervene and delay the progress of AD. Thus, it is valuable to distinguish patients with pMCI from those with sMCI in the early stage with a computer-aided diagnosis system.

In recent years, deep learning (DL) based methods [2,8,11,13] have been widely used to identify pMCI or AD based on brain MRI data. Several works [9,24,29] take both brain MRI and clinical tabular data into account, using convolutional neural networks (CNNs) and multi-layer perceptions as the feature encoder. Due to the limited data in pMCI diagnosis, some works [19,29] resort to transfer learning to fine-tune the model on the pMCI-related task, by pre-training weights on the AD detection task. Previous CNN-based approaches [9,24,29] may fail on the lack of modeling long-range relationship, while the models based on two-stage fine-tuning [19,29] are inefficient and may cause networks to forget learned knowledge [32].

Inspired by the advance in transformers [6,17,28] and prompt learning [4,15, 25], we tailor-design an effective multi-modal transformer-based framework based on prompt learning for pMCI detection. The proposed framework is composed of a transformer-based visual encoder, a transformer-based attribute encoder, a multi-modal fusion module, and a prompt learning scheme. Clinical attributes and brain MR images are sent to the attribute encoder and visual encoder with the prompt tokens, respectively. Then the high-level visual and tabular features are aggregated and sent to fully-connected layers for final classification. For the proposed prompt tuning scheme, we first pre-train the neural network on the AD classification task. After that, we fine-tune the neural network by introducing and updating only a few trainable parameters. Importantly, we observe that the number of image patches and input tokens have a gap between 2D natural images and 3D MR images, due to the different dimensions. To complement local interactions between prompt and other tokens, we develop a global prompt token to strengthen the global guidance and make the visual feature extraction process more efficient and stable.

Our contributions have three folds: (1) We propose a visual-attribute prompt learning framework based on transformers (VAP-Former) for pMCI detection. (2) We design a global prompt to adapt to high-dimension MRI data and build a prompt learning framework for transferring knowledge from AD diagnosis to pMCI detection. (3) Experiments not only show that our VAP-Former obtains state-of-the-art results on pMCI prediction task by exceeding the full fine-tuning methods, but also verify the global prompt can make the training more efficient and stable.

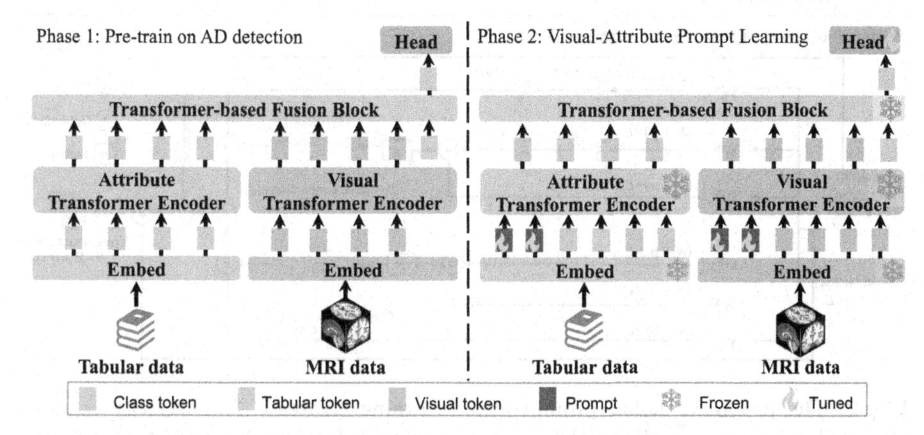

Fig. 1. The proposed Visual-Attribute Prompt learning transformer framework. The left part shows the architecture of VA-Former without using prompts. The right part shows the proposed framework with the prompt tuning strategy (VAP-Former) that only updates the learnable prompts.

2 Methodology

We aim to predict if an MCI patient will remain stable or develop Alzheimer's Disease, and formulate the problem as a binary classification task based on the brain MR image and the corresponding tabular attribute information of an MCI patient from baseline examination. As Fig. 1 shows, we adopt the model weights learned from AD identification to initialize the prediction model of MCI conversion. In the prompt fine-tuning stage, we keep all encoders frozen, only optimizing the prompts concatenated with feature tokens.

2.1 Network Architecture

We propose a transformer-based framework for pMCI prediction (VAP-Former) based on visual and attribute data, which is shown in Fig. 1. VAP-Former is mainly composed of three parts: a visual encoder for processing MRI data, an attribute encoder for processing attribute data, and a transformer-based fusion block for combining the multi-modal feature. Considering that capturing the long-range relationship of MRI is important, we employ the encoder of 3D UNETR++ [28] as our visual encoder. For the attribute encoder, since the clinical variables have no order or position, we embed the tabular data with the transformer blocks [30]. Followed by [6,7], we prepend a class token for dual-modality before the transformer-based fusion block. A class token is a learnable vector concatenated with dual-modal feature vectors. The class token of dual-modal is further processed by fully-connected layers for the final classification.

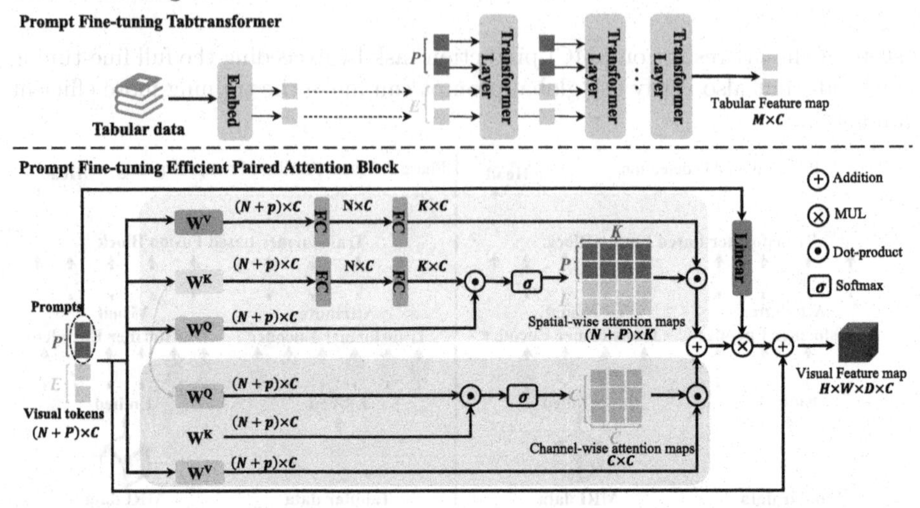

Fig. 2. The proposed global prompt-based learning strategy. The strategy replaces the attribute encoder with the Prompt Fine-tuning Tab-transformer (upper part) and replaces the Efficient Paired Attention (EPA) block with the prompt fine-tuning EPA block (lower part).

2.2 Knowledge Transfer with Multi-modal Prompt Learning

To effectively transfer the knowledge learned from AD prediction task to the pMCI prediction task, we propose a multi-modal prompt fine-tuning strategy that adds a small number of learnable parameters (i.e., prompt tokens) to the input of the transformer layer and keeps the backbone frozen. The overall pipeline is shown in Fig. 2, where the upper part indicates adding the prompt tokens to the attribute transformer, while the lower part denotes sending the prompt tokens to the visual transformer.

Tabular Context Prompt Learning. In the attribute encoder, we insert prompts into every transformer layer's input [15]. For the (i)-th Layer L_i of SA transformer block, we denote collection of p prompts as $P = \{p^k \in \mathbb{R}^C \| k \in \mathbb{N}, 1 \leq k \leq p\}$, where k is the number of the tabular prompt. The prompt fine-tuning Tabtransformer can be formulated as:

$$\begin{aligned} [\_\_, X_i] &= L_i([P_{i-1}, X_{i-1}]), \\ [\_\_, X_{i+1}] &= L_{i+1}([P_i, X_i]), \end{aligned} \tag{1}$$

where $X_i \in \mathbb{R}^{M \times C}$ denotes the attribute embedding at L_i's output space and P_i denotes the attribute prompt at L_{i+1}'s input space concatenated with X_i.

Visual Context Prompt Learning with Paired Attention. For the visual prompt learning part, we concatenate a small number of prompts with visual embedding to take part in the spatial-wise and channel-wise attention block [12,16,18] denoted as the prompt fine-tuning efficient paired attention block in Fig. 2. Within the prompt fine-tuning efficient paired attention block, we insert shared prompts into the spatial-wise attention module (SWA) and the channel-wise attention module (CWA), respectively. With a shared keys-queries scheme, queries, keys, and values are noted as Q^p_{shared}, K^p_{shared}, $V^p_{spatial}$, and $V^p_{channel}$. Let $[\cdot, \cdot]$ be the concatenation operation, the SWA and CWA can be formulated as:

$$\begin{aligned} [P_{spatial}, I_S] &= SWA(Q^p_{shared}, K^p_{shared}, V^p_{spatial}), \\ [P_{channel}, I_C] &= CWA(Q^p_{shared}, K^p_{shared}, V^p_{channel}), \end{aligned} \tag{2}$$

where $I_S \in \mathbb{R}^{N \times C}$ and $P_{spatial} \in \mathbb{R}^{\frac{P}{2} \times C}$ are spatial-wise visual embedding and prompts, and $I_C \in \mathbb{R}^{N \times C}$ and $P_{channel} \in \mathbb{R}^{\frac{P}{2} \times C}$ are channel-wise visual embedding and prompts, and P is the number of visual prompt. After that, the initial feature map I is added to the attention feature map using a skip connection. Let $+$ be the element-wise addition, this process is formulated as $I = I + (I_S + I_C)$.

2.3 Global Prompt for Better Visual Prompt Learning

Compared to natural images with relatively low dimensionality (e.g., shaped 224 × 224 × 3) and the salient region usually locates in a small part of the image, brain MRIs for diagnosis of Alzheimer's are usually high dimensional (e.g., shaped 144 × 144 × 144) and the cues to diagnosis disease (e.g., cortical atrophy and beta protein deposition) can occupy a large area of the image. Therefore, vanilla prompt learning [32] methods that are designed for natural images may not be directly and effectively applied to MRI's Recognition of Alzheimer's disease. We consider that the above differences lead to the following two problems: (1) vanilla prompt token often focuses on local visual information features; (2) the interaction between vanilla prompt token and visual feature is insufficient. Therefore, we consider that a sophisticated prompt learning module should be able to address the above-mentioned issues with the following feature: (1) the prompt token can influence the global feature; (2) the prompt token can effectively interact with the visual input features. A simple approach to achieve the second goal is to increase the number of prompt tokens so that they can better interact with other input features. However, the experiment proves that this is not feasible. We think it is because too many randomly initialized prompt tokens (i.e., unknown information) will make the model hard to train.

Thus, we tailor-design a global prompt token g to achieve the above two goals. Specifically, we apply a linear transformation T to the input prompt tokens P to obtain the global prompt token g (i.e., a vector), which further multiplies the global feature map. Since the vector is directly multiplied with the global feature, we can better find the global feature responses in each layer of the visual network, which enables the model to better focus on some important global features for pMCI diagnosis, such as cortical atrophy. Since this linear transformation, T operation is learnable in the prompt training stage, the original prompt token can better interact with other features through the global token. To embed the global prompt token into our framework, we rewrite Eq. 3 as:

$$I = I + (I_S + I_C) \times T([P_{spatial}, P_{channel}]), \quad (3)$$

where T denotes the linear transformation layer with $P \times C$ elements as input and $1 \times C$ element as output. Experiments not only demonstrate the effectiveness of the above method but also make the training stage more stable.

3 Experiments and Results

3.1 Datasets and Implementation Details

The datasets are from Alzheimer's Disease Neuroimaging Initiative (ADNI) [14], including ADNI-1 and ADNI-2. Following [19,22], we adopt ADNI-1/ADNI-2 as the train/test set. The subjects that exist in both datasets are excluded from ADNI-2. There are 1340 baseline T1-weighted structure MRI scans of 4 categories including Alzheimer's disease (AD), normal control (NC), stable mild cognitive impairment (sMCI), and progressive mild cognitive impairment (pMCI). 707 subjects (158 AD, 193 NC, 130 pMCI, 226 sMCI) are from ADNI-1, while 633 subjects (137 AD, 159 NC, 78 pMCI, 259 sMCI) are from ADNI-2. We pre-process the MR image as [22]. All MRI scans are pre-processed via 4 steps: (1) motion correction, (2) intensity normalization, (3) skull stripping, and (4) spatial normalization to a template. We use the Computational Anatomy Toolbox (CAT12)[1] via Statistical Parametric Mapping software (SPM12)[2] to perform the above procedures. Then all images are re-sampled as the size of 113×137×113 and the resolution of $1 \times 1 \times 1\,\text{mm}^3$. For tabular clinical data, we select 7 attributes including age, gender, education, ApoE4, P-tau181, T-tau, and a summary measure (FDG) derived from 18F-fluorodeoxyglucose PET imaging. For tabular clinical data, we apply one-hot encoding to the categorical variables and min-max normalization to the numerical variables. We adopt the model weights learned from AD identification to initialize the prediction model of MCI conversion. Our model is trained using 1 NVIDIA V100 GPU of 32GB via AdamW optimizer [20] with 30 epochs for AD identification and 20 epochs for pMCI detection. The batch size is set to 4. We adopt the ReduceLROnPlateau [1] learning rate decay strategy with an initial learning rate of 1e-5. The loss function is binary cross-entropy loss. The number of visual and tabular prompts is 10 and 5, respectively. We take F1-score [23], class balanced accuracy (BACC) [3], and the area under the receiver operating characteristic curve (AUC) [23] as evaluation metrics.

3.2 Comparison with the State-of-the-Art Methods

To validate the proposed VAP-Former and prompt fine-tuning (PT) strategy, we compare our model with three unimodal baselines: 1) UNETR++, which denotes

[1] https://neuro-jena.github.io/cat/.
[2] https://www.fil.ion.ucl.ac.uk/spm/software/spm12/.

Table 1. Comparison with existing pMCI classification methods. 'FT' represents full fine-tuning without prompts and 'PT' represents prompt fine-tuning. '# params' denotes the number of parameters being tuned in the fine-tuning stage.

Method	Modal		Fine-tuning	#params (M)	BACC	F1	AUC
	Vis	Tab					
UNETR++ [28]	✓	✗	FT	64.59	$65.94_{\pm 1.19}$	$47.42_{\pm 1.20}$	$70.46_{\pm 1.67}$
Tabformer [21]	✗	✓	FT	27.13	$78.05_{\pm 0.52}$	$60.29_{\pm 0.27}$	$84.26_{\pm 0.17}$
4-Way Classifier [27]	✓	✗	FT	12.27	$71.86_{\pm 2.07}$	$53.49_{\pm 3.44}$	$75.81_{\pm 1.01}$
DFAF [10]	✓	✓	FT	69.32	$77.14_{\pm 0.35}$	$59.66_{\pm 0.85}$	$84.41_{\pm 0.49}$
HAMT [5]	✓	✓	FT	71.83	$75.07_{\pm 0.91}$	$55.58_{\pm 1.11}$	$84.23_{\pm 0.31}$
DAFT [24]	✓	✓	FT	67.89	$78.06_{\pm 0.55}$	$62.92_{\pm 0.71}$	$85.33_{\pm 0.41}$
VA-Former	✓	✓	FT	70.19	$78.29_{\pm 0.52}$	$62.93_{\pm 0.29}$	$84.77_{\pm 0.35}$
VAP-Former	✓	✓	PT	0.59	$\mathbf{79.22}_{\pm 0.58}$	$\mathbf{63.13}_{\pm 0.11}$	$\mathbf{86.31}_{\pm 0.25}$

the encoder of UNETR++ [28] only using MRI data as input, 2) Tabformer [21], which only uses tabular data and is similar to the attribute encoder work in our model. 3) 4-Way Classifier [27] which used 3D DenseNet as the backbone to construct Alzheimer's disease diagnosis model only using MRI data as input. To further evaluate the efficiency of our model and the proposed PT strategy, we integrate the DAFT [24], HAMT [5], and DFAF [10] into the same attribute and visual encoder for a fair comparison. And we fine-tune the proposed model with two strategies including full fine-tuning (FT) and prompt tuning (PT).

In Table 1, the proposed model with PT strategy achieves 79.22% BACC, 63.13% F1, and 86.31% AUC. VAP-Former and VA-Former outperform all uni-modal baselines in all metrics, indicating that our model can effectively exploit the relation between MRI data and tabular data to improve the prediction of pMCI. By using the PT strategy, the VAP-Former outperforms the second-best model, DAFT, by 1.16% BACC, 0.21% F1, and 0.98% AUC, demonstrating that the proposed PT strategy can efficiently adapt learned knowledge to pMCI detection and even achieves better classification results. Besides that, VAP-Former achieves the best results by tuning the minimum number of parameters.

3.3 Ablation Study and Investigation of Hyper-parameters

In this section, we study the internal settings and mechanism of the proposed PT strategy. First, we investigate how the prompts affect the VAP-Former performance. So we remove the visual prompts from the VAP-Former (denoted as TabPrompt in Table 2) and tabular prompts (denoted as VisPrompt), respectively. Compared with VA-Former, VisPrompt outperforms it by 0.07% AUC and TabPrompt degrades the performance. However, VAP-Former, which combines tabular prompts and visual prompts, significantly outperforms VA-Former by 1.54% AUC, indicating that introducing both types of prompts into the model simultaneously results in a more robust pMCI classification model. We further validate the importance of the global prompt module in the VAP-Former by

Table 2. Ablation study for prompt fine-tuning. 'Vis-GP' denotes the visual global prompt. The best results are in **bold**.

Method	Prompt		Vis-GP	BACC	F1	AUC
	Visual	Tabular				
VA-Former	✗	✗	✗	78.29$\pm$0.52	62.93$\pm$0.29	84.77$\pm$0.35
VisPrompt	✓	✗	✗	77.95$\pm$0.66	60.15$\pm$2.51	84.84$\pm$1.28
TabPrompt	✗	✓	✗	77.39$\pm$0.80	60.49$\pm$0.60	84.70$\pm$0.81
Vis-TabPrompt	✓	✓	✗	78.19$\pm$0.77	63.11$\pm$1.45	85.23$\pm$0.53
VAP-Former	✓	✓	✓	**79.22$\pm$0.58**	**63.13$\pm$0.11**	**86.31$\pm$0.25**

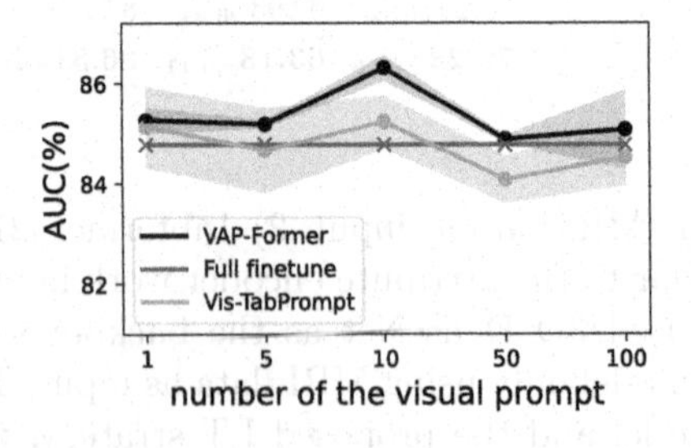

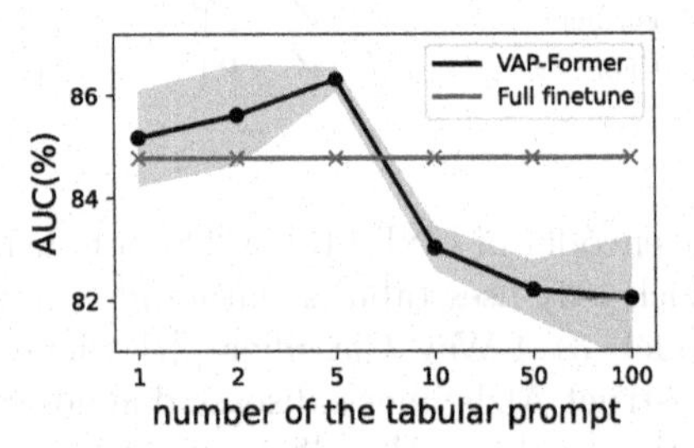

Fig. 3. Investigation of the number of prompt tokens. The line in red, black, and orange denote the performance of fully-finetuned VA-Former, VAP-Former, and Vis-TabPrompt, respectively. The light blue and orange area represents the error interval obtained by 3 different running seeds. The left figure shows that our global prompt improves the AUC and alleviates the performance fluctuations [15] in prompt learning. (Color figure online)

removing it from the model (denoted as Vis-TabPrompt). As shown in Table 2, removing the global prompt results in degraded performance by 1.08% AUC.

To investigate the impact of the number of prompts on performance, we evaluate VAP-Former with varying numbers of prompts. Given that the number of visual tokens exceeds that of tabular tokens. We fix the number of tabular prompts at 5 (left plot in Fig. 3) and fix the number of visual prompts at 10 (right plot), respectively while changing the other one. We hypothesize that the interaction between prompts and feature tokens is insufficient, so we gradually increase the number of tokens. As shown in Fig. 3, we observe that the model's performance ceases to increase after a certain number of prompts, confirming our assumption in Sect. 3.2 that too many randomly initialized prompt tokens make the model difficult to train. Conversely, when the number of prompts is small, the prompts can not learn enough information for the task and the interaction between prompt and feature tokens is not sufficient. Furthermore, as depicted in Fig. 3, the light blue area is smaller than the orange area, suggesting that the global prompt module makes VAP-Former more robust by helping the model to focus on important global features for pMCI detection.

4 Conclusion

To detect pMCI with visual and attribute data, we propose a simple but effective transformer-based model, VA-Former, to learn multi-modal representations. Besides, we propose a global prompt-based tuning strategy, which is integrated with the VA-Former to obtain our overall framework VAP-Former. The proposed framework can efficiently transfer the learned knowledge from AD classification to the pMCI prediction task. The experimental results not only show that the VAP-Former performs better than uni-modal models, but also suggest that the VAP-Former with the proposed prompt tuning strategy even surpasses the full fine-tuning while dramatically reducing the tuned parameters.

References

1. Al-Kababji, A., Bensaali, F., Dakua, S.P.: Scheduling techniques for liver segmentation: Reducelronplateau vs OneCycleLR. In: Bennour, A., Ensari, T., Kessentini, Y., Eom, S. (eds.) ISPR 2022, vol. 1589, pp. 204–212. Springer, Heidelberg (2022). https://doi.org/10.1007/978-3-031-08277-1_17
2. Arbabshirani, M.R., Plis, S., Sui, J., Calhoun, V.D.: Single subject prediction of brain disorders in neuroimaging: promises and pitfalls. Neuroimage **145**, 137–165 (2017)
3. Brodersen, K.H., Ong, C.S., Stephan, K.E., Buhmann, J.M.: The balanced accuracy and its posterior distribution. In: 20th International Conference on Pattern Recognition, pp. 3121–3124. IEEE (2010)
4. Brown, T., et al.: Language models are few-shot learners. Adv. Neural Inf. Process. Syst. **33**, 1877–1901 (2020)
5. Chen, S., Guhur, P.L., Schmid, C., Laptev, I.: History aware multimodal transformer for vision-and-language navigation. Adv. Neural Inf. Process. Syst. **34**, 5834–5847 (2021)
6. Devlin, J., Chang, M.W., Lee, K., Toutanova, K.: Bert: pre-training of deep bidirectional transformers for language understanding. In: Proceedings of NAACL-HLT, pp. 4171–4186 (2019)
7. Dosovitskiy, A., et al.: An image is worth 16×16 words: transformers for image recognition at scale. In: International Conference on Learning Representations (2021). https://openreview.net/forum?id=YicbFdNTTy
8. Ebrahimighahnavieh, M.A., Luo, S., Chiong, R.: Deep learning to detect Alzheimer's disease from neuroimaging: a systematic literature review. Comput. Methods Prog. Biomed. **187**, 105242 (2020)
9. El-Sappagh, S., Abuhmed, T., Islam, S.R., Kwak, K.S.: Multimodal multitask deep learning model for Alzheimer's disease progression detection based on time series data. Neurocomputing **412**, 197–215 (2020)
10. Gao, P., et al.: Dynamic fusion with intra-and inter-modality attention flow for visual question answering. In: Proceedings of the IEEE/CVF Conference on Computer Vision and Pattern Recognition, pp. 6639–6648 (2019)
11. Gong, H., Chen, G., Mao, M., Li, Z., Li, G.: Vqamix: conditional triplet mixup for medical visual question answering. IEEE Trans. Med. Imaging **41**(11), 3332–3343 (2022)

12. He, X., Yang, S., Li, G., Li, H., Chang, H., Yu, Y.: Non-local context encoder: robust biomedical image segmentation against adversarial attacks. In: Proceedings of the AAAI Conference on Artificial Intelligence, vol. 33, pp. 8417–8424 (2019)
13. Huang, J., Li, H., Li, G., Wan, X.: Attentive symmetric autoencoder for brain MRI segmentation. In: Wang, L., Dou, Q., Fletcher, P.T., Speidel, S., Li, S. (eds.) MICCAI 2022. LNCS, pp. 203–213. Springer, Heidelberg (2022). https://doi.org/10.1007/978-3-031-16443-9_20
14. Jack, C.R., Jr., et al.: The Alzheimer's disease neuroimaging initiative (ADNI): MRI methods. J. Magn. Reson. Imaging **27**(4), 685–691 (2008)
15. Jia, M., et al.: Visual prompt tuning. In: Avidan, S., Brostow, G., Cisse, M., Farinella, G.M., Hassner, T. (eds.) ECCV 2022. LNCS, pp. 709–727. Springer, Heidelberg (2022). https://doi.org/10.1007/978-3-031-19827-4_41
16. Li, H., Chen, G., Li, G., Yu, Y.: Motion guided attention for video salient object detection. In: Proceedings of the IEEE/CVF International Conference on Computer Vision, pp. 7274–7283 (2019)
17. Li, H., et al.: View-disentangled transformer for brain lesion detection. In: IEEE 19th International Symposium on Biomedical Imaging (ISBI), pp. 1–5 (2022)
18. Li, H., Li, G., Yang, B., Chen, G., Lin, L., Yu, Y.: Depthwise nonlocal module for fast salient object detection using a single thread. IEEE Trans. Cybern. **51**(12), 6188–6199 (2020)
19. Lian, C., Liu, M., Zhang, J., Shen, D.: Hierarchical fully convolutional network for joint atrophy localization and alzheimer's disease diagnosis using structural MRI. IEEE Trans. Pattern Anal. Mach. Intell. **42**(4), 880–893 (2020)
20. Loshchilov, I., Hutter, F.: Fixing weight decay regularization in adam (2018). https://openreview.net/forum?id=rk6qdGgCZ
21. Padhi, I., et al.: Tabular transformers for modeling multivariate time series. In: IEEE International Conference on Acoustics, Speech and Signal Processing (ICASSP), pp. 3565–3569 (2021)
22. Pan, Y., Chen, Y., Shen, D., Xia, Y.: Collaborative image synthesis and disease diagnosis for classification of neurodegenerative disorders with incomplete multi-modal neuroimages. In: de Bruijne, M., et al. (eds.) MICCAI 2021. LNCS, vol. 12905, pp. 480–489. Springer, Cham (2021). https://doi.org/10.1007/978-3-030-87240-3_46
23. Pan, Y., Liu, M., Lian, C., Xia, Y., Shen, D.: Spatially-constrained fisher representation for brain disease identification with incomplete multi-modal neuroimages. IEEE Trans. Med. Imaging **39**(9), 2965–2975 (2020)
24. Pölsterl, S., Wolf, T.N., Wachinger, C.: Combining 3D image and tabular data via the dynamic affine feature map transform. In: de Bruijne, M., et al. (eds.) MICCAI 2021. LNCS, vol. 12905, pp. 688–698. Springer, Cham (2021). https://doi.org/10.1007/978-3-030-87240-3_66
25. Radford, A., et al.: Learning transferable visual models from natural language supervision. In: International Conference on Machine Learning, pp. 8748–8763. PMLR (2021)
26. Risacher, S.L., Saykin, A.J., Wes, J.D., Shen, L., Firpi, H.A., McDonald, B.C.: Baseline MRI predictors of conversion from MCI to probable AD in the ADNI cohort. Curr. Alzheimer Res. **6**(4), 347–361 (2009)
27. Ruiz, J., Mahmud, M., Modasshir, Md., Shamim Kaiser, M.: 3D DenseNet ensemble in 4-way classification of alzheimer's disease. In: Mahmud, M., Vassanelli, S., Kaiser, M.S., Zhong, N. (eds.) BI 2020. LNCS (LNAI), vol. 12241, pp. 85–96. Springer, Cham (2020). https://doi.org/10.1007/978-3-030-59277-6_8

28. Shaker, A., Maaz, M., Rasheed, H., Khan, S., Yang, M.H., Khan, F.S.: Unetr++: delving into efficient and accurate 3d medical image segmentation. arXiv preprint arXiv:2212.04497 (2022)
29. Spasov, S., Passamonti, L., Duggento, A., Lio, P., Toschi, N., Initiative, A.D.N., et al.: A parameter-efficient deep learning approach to predict conversion from mild cognitive impairment to Alzheimer's disease. Neuroimage **189**, 276–287 (2019)
30. Vaswani, A., et al.: Attention is all you need. Adv. Neural Inf. Process. Syst. **30** (2017)
31. Winblad, B., et al.: Defeating Alzheimer's disease and other dementias: a priority for European science and society. Lancet Neurol. **15**(5), 455–532 (2016)
32. Zhou, K., Yang, J., Loy, C.C., Liu, Z.: Learning to prompt for vision-language models. Int. J. Comput. Vision **130**(9), 2337–2348 (2022)

Acute Ischemic Stroke Onset Time Classification with Dynamic Convolution and Perfusion Maps Fusion

Peng Yang[1], Yuchen Zhang[2], Haijun Lei[2], Yueyan Bian[3], Qi Yang[3,4](✉), and Baiying Lei[1](✉)

[1] Guangdong Key Laboratory for Biomedical Measurements and Ultrasound Imaging, National-Regional Key Technology Engineering Laboratory for Medical Ultrasound, School of Biomedical Engineering, Shenzhen University Medical School, Shenzhen 518060, China
leiby@szu.edu.cn

[2] Key Laboratory of Service Computing and Applications, Guangdong Province Key Laboratory of Popular High Performance Computers, College of Computer Science and Software Engineering, Shenzhen University, Shenzhen 518060, China

[3] Department of Radiology, Beijing Chaoyang Hospital, Capital Medical University, Beijing, China
yangyangqiqi@gmail.com

[4] Laboratory for Clinical Medicine, Capital Medical University, Beijing, China

Abstract. In treating acute ischemic stroke (AIS), determining the time since stroke onset (TSS) is crucial. Computed tomography perfusion (CTP) is vital for determining TSS by providing sufficient cerebral blood flow information. However, the CTP has small samples and high dimensions. In addition, the CTP is multi-map data, which has heterogeneity and complementarity. To address these issues, this paper demonstrates a classification model using CTP to classify the TSS of AIS patients. Firstly, we use dynamic convolution to improve model representation without increasing network complexity. Secondly, we use multi-scale feature fusion to fuse the local correlation of low-order features and use a transformer to fuse the global correlation of higher-order features. Finally, multi-head pooling attention is used to learn the feature information further and obtain as much important information as possible. We use a five-fold cross-validation strategy to verify the effectiveness of our method on the private dataset from a local hospital. The experimental results show that our proposed method achieves at least 5% higher accuracy than other methods in TTS classification task.

Keywords: Computed tomography perfusion · Dynamic convolution · Multi-map

P. Yang and Y. Zhang—These authors contributed equally to this work.

H. Greenspan et al. (Eds.): MICCAI 2023, LNCS 14224, pp. 558–568, 2023.
https://doi.org/10.1007/978-3-031-43904-9_54

1 Introduction

Acute ischemic stroke (AIS) is a disease of ischemic necrosis or softening of localized brain tissue caused by cerebral blood circulation disturbance, ischemia, and hypoxia [1, 2]. Intravenous thrombolysis can be used to open blood vessels within 4.5 h. For patients with large vessel occlusion, the internal blood vessels can be opened by removing the thrombus within 6 h. Therefore, determining the time since stroke onset (TSS) is crucial for creating a treatment plan for AIS patients, with a TSS of less than 6 h being critical. However, approximately 30% of AIS occurs at unknown time points due to wake-up stroke (WUS) and unknown onset stroke (UOS) [3]. For such patients, determining the TSS accurately is challenging, and they may be excluded from appropriate treatment. Computed tomography perfusion (CTP) is processed with special software to generate perfusion maps: cerebral blood volume (CBF), cerebral blood volume (CBV), mean transit time (MTT), and peak response time (Tmax) [4]. These perfusion maps can provide sufficient information on cerebral blood flow, ischemic penumbra, and infarction core area.

There are some machine learning methods to determine the TSS of AIS by automatic discrimination [5–8]. For example, Ho *et al.* [5] first used auto-encoder (AE) to learn magnetic resonance imaging (MRI) and then put the learned and original features into the classifier for TSS classification. Lee *et al.* [7] analyzed diffusion-weighted imaging (DWI) and fluid-attenuated inversion recovery (FLAIR) using automatic image processing methods to obtain appropriate dimensional features and used machine learning to classify the TSS. The above machine learning-based TSS classification methods often use regions of interest (ROI) while ignoring the spatial correlation between neural images. Several researchers try to employ deep learning techniques to classify AIS TSS, considering the spatial correlation among neural images. Zhang *et al.* [9] designed a new intra-domain task adaptive migration learning method to classify the TSS of AIS. Polson *et al.* [10] designed the neighborhood and attention network with segmented weight sharing to learn DWI, apparent diffusion coefficient (ADC), and FLAIR, then used weighted softmax to aggregate sub-features and achieve TSS classification.

With the small samples and the high dimension of CTP, the convolution neural network (CNN) cannot effectively extract features, resulting in the problem of network non-convergence. Additionally, some existing TSS classification methods leverage multi-map mainly by simple linear connections, which do not thoroughly learn the supplementary information of CTP [6, 10]. Therefore, we design a classification model based on dynamic convolution and multi-map fusion to classify the TSS. Firstly, we replace the ordinary convolution in the feature extraction network with dynamic convolution (DConv) [11] to improve the network performance without increasing the network complexity. Secondly, low-order multi-map features are fused to enhance the acquisition of local information by multi-scale feature fusion (MFF). Then high-order features are fused to obtain the global association information by transformer fusion (Transfus). Finally, multi-head pooling attention (MPA) is used to emphasize the high-order features, and the most discriminative features are selected to classify TSS.

2 Methodology

The main framework of our proposed method is depicted in Fig. 1. Specifically, the feature extraction of each map is performed independently using four feature extraction networks, each consisting of five stages. Dconv [11] is used to improve network performance without increasing complexity. In the first three stages, MFF is used to capture and fuse the details of different scales of multi-map features. In the last two stages, Transfus is used to fuse the global correlation of the high-order features. The learned multi-map high-order features are put into the MPA to learn them further and merge the potential tensor sequence. Finally, the selected features are put into the full connection layer to achieve the classification of TSS.

2.1 Dynamic Convolution Feature Extraction Network

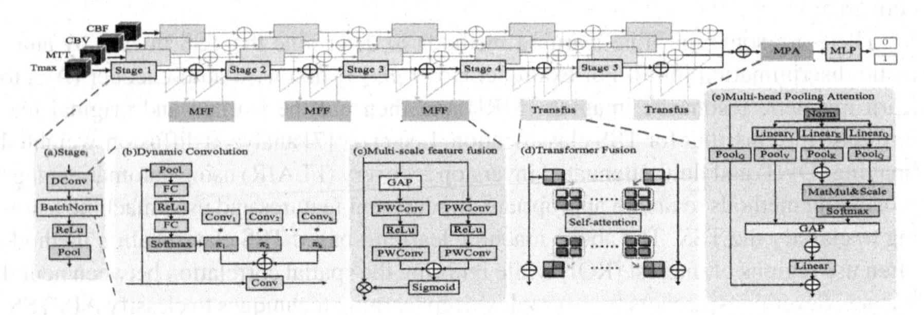

Fig. 1. The framework of our proposed method.

The feature extraction network of a single map consists of five stages. The structure of each stage is shown in Fig. 1 (a). Considering the high data dimension and the small number of samples, the network layers are too deep to cause over-fitting. We replace traditional convolution with dynamic convolution [11]. DConv improves the model expression ability by fusing multiple convolution cores, and its structure is shown in Fig. 1 (b). DConv will can not increase the network complexity and thus improve the network performance. It will not increase too many parameters and calculation amount while increasing the capacity of the model. Inspired by static perceptron [12], Dconv has k convolution cores, which share the same core size and input/output dimensions. An attention block is used [13] to generate the attention weight of k convolution cores, and finally aggregate the results through convolution. The formula is shown in Eq. (1):

$$y = \Sigma_{i=1}^{k} \pi_i(x) Conv(x), \ s.t\, 0 \leq \pi_i(x) \leq 1, \ \Sigma_{i=1}^{k} \pi_i(x) = 1 \tag{1}$$

where $\pi_i(x)$ is the weight of the i-th convolution, which varies with each input x. . The attention block compresses the features of each channel through global average pooling and then uses two fully connected layers (with ReLU activation function between them) and a Softmax function to generate the attention weight of k convolution cores.

2.2 Multi-map Fusion Module

For multi-map information fusion of low-order features, considering the small area of acute stroke focus, a multi-scale attention module is used to fuse multi-map features in the first three stages. Its structure is shown in Fig. 1 (c). Record the output feature of each stage as $x_{ij}(i = 1, 2, 3; j = 1, 2, 3, 4)$, where i represents the output feature of the i-th stage, and j represents the j-th map. The feature x_i of the input multi-scale channel attention module are denoted as:

$$x_i = x_{i1} \oplus x_{i2} \oplus x_{i3} \oplus x_{i4} \tag{2}$$

By setting different global average pooling (GAP) sizes, we can focus on the interactive feature information in channel dimensions at multiple scales, and aggregate local and global features. Through point-wise convolution (PWConv), point-wise channel interaction is used for each spatial location to realize the integration of local information. The local channel features are calculated as follows:

$$L(x_i) = PWConv(ReLu(PWConv(GAP(x_i)))) \tag{3}$$

The core size of $PWConv$ is $\frac{C}{r} \times C \times 1 \times 1 \times 1$ and $C \times \frac{C}{r} \times 1 \times 1 \times 1$. The global channel features are calculated as follows:

$$G(x_i) = PWConv(ReLu(PWConv(x_i))) \tag{4}$$

The final feature $x_i^{'}$ is calculated as follows:

$$x_i\prime = x_i \otimes \sigma(L(x_i) \oplus G(x_i)) \tag{5}$$

For the fusion of multi-map information of high-order features, considering the relevance of global information between different modes, the self-attention mechanism [14] in the transformer is used to learn multi-map information, and its structure is shown in Fig. 1 (d). Specifically, the output characteristic of each stage is $x_{ij}(i = 4, 5; j = 1, 2, 3, 4)$, where i represents the output feature of the i-th stage, and j represents the j-th mode. Similar to the previous work of some scholars [15–18], we think that the middle feature map of each mode is a set rather than a patch, and treat each element in the set as a token [19]. At this time, each token takes into account all the token information of the four branches. Finally, the fused features are superimposed on the branches for the next stage. Let the characteristic $x \in \mathbb{R}^{N \times D}$ of the input transformer block, where N is the number of tokens in the sequence and each token is represented by the feature vector of dimension D. It uses the scaling dot product between Query and Key to calculate the focus weight and aggregates the value of each Query. Finally, nonlinear transformation is used to calculate the fused output features x_{out}.

Table 1. Comparison of results of different methods (%). (p-value < 0.05)

Methods	Accuracy	Sensitivity	Precision	F1-score	Kappa	AUC
ResNet18	75.04 ± 4.39	92.42 ± 6.04	75.59 ± 3.22	83.07 ± 3.25	37.05 ± 10.17	69.57 ± 5.03
ResNet34	77.56 ± 4.57	91.74 ± 6.07	78.44 ± 4.07	84.43 ± 3.36	44.82 ± 11.65	77.07 ± 5.15
ResNet50	74.01 ± 2.63	91.62 ± 8.80	75.22 ± 3.46	82.31 ± 2.65	34.31 ± 6.70	67.60 ± 8.44
VGG11	73.52 ± 4.68	86.38 ± 7.62	76.92 ± 3.87	81.17 ± 3.90	36.71 ± 10.48	63.59 ± 5.28
AlexNet	74.02 ± 4.37	84.9 ± 7.28	78.31 ± 4.48	81.23 ± 3.51	38.93 ± 11.06	69.54 ± 6.24
CoAtNet0 [20]	76.00 ± 4.21	91.00 ± 3.25	77.38 ± 4.76	83.51 ± 2.30	40.17 ± 13.63	67.48 ± 8.29
C3D [21]	74.95 ± 2.98	87.09 ± 8.89	78.22 ± 3.54	82.11 ± 2.77	40.18 ± 7.28	67.78 ± 5.31
I3D [22]	73.09 ± 4.84	**92.48 ± 7.08**	74.07 ± 4.86	82.03 ± 3.17	30.28 ± 14.93	69.15 ± 7.35
MFNet [23]	74.55 ± 4.49	90.91 ± 6.97	75.99 ± 4.47	82.58 ± 3.23	36.11 ± 12.57	71.08 ± 4.78
SSFTTNet [24]	75.95 ± 4.16	89.32 ± 10.00	78.09 ± 3.60	83.00 ± 3.82	41.70 ± 8.58	74.74 ± 6.05
Slowfast [25]	76.05 ± 4.82	85.75 ± 4.70	79.79 ± 4.01	82.62 ± 3.73	44.16 ± 10.84	74.21 ± 5.58
Ours	**82.99 ± 4.14**	89.40 ± 5.77	**85.89 ± 4.61**	**87.45 ± 3.12**	**60.97 ± 9.65**	**81.48 ± 5.74**

2.3 Multi-head Pooling Attention

With the deepening of the network layers, the semantic information contained in the output features becomes higher and higher. After the post-fusion of the branch network, we use an MPA to learn the high-order semantic details further. Here, a smaller number of tokens is used to increase the dimension of each token to facilitate the storage of more information. Unlike the original multi-head attention (MHA) operator [14], the multi-head pooling attention module gathers the potential tensor sequence to reduce the length of the input sequence, and its structure is shown in Fig. 1 (e). Like MHA [14], Query, Key, and Value are obtained through the linear operation. Add the corresponding pooling layer to Query, Key, and Value to further sample it.

3 Experiments

3.1 Experimental Configuration

Dataset and Data Preprocessing. The dataset of 200 AIS patients in this paper is from a local hospital. The patients are divided into two categories: positive (TSS < 6 h) and negative (TSS ≥ 6 h). Finally, 133 in the positive subjects and 67 in the negative subjects are included. Each subject contains CBF, CBV, MTT, and Tmax. The size of all CTP images is set to 256 × 256 × 32.

Experimental Setup. The network structure is based on PyTorch 1.9.0 framework and CUDA 11.2 Titan × 2. We use a five-fold cross-validation method to verify the effectiveness of our method. 80% of the data is used as a training set and 20% as a test set. During the training process, the Adam optimizer optimizes the parameters, and the learning rate is set to 0.00001. The learning strategy of fixed step attenuation is adopted, in which the step size is set to 15, γ is 0.8. The number of iterations of training is 50.

3.2 Experimental Results and Analysis

Comparative Study. We evaluate the effectiveness of our method by comparing it with other approaches on the same dataset [20–25]. The results in Table 1 demonstrate that our model achieves at least a 5% higher accuracy than the other methods and outperforms them in other evaluation indicators. Table 1 also shows each method's area under the curve (AUC). Our method achieves an 81% AUC in the TSS classification task, indicating its superiority. To provide a more intuitive comparison of the model's performance under different indicators, we created radar charts to represent the evaluation results of the comparative experiment, as shown in Fig. 2. These charts indicate that our method performs well in all evaluate indicators. To verify the reliability of our method, we conduct T-test verification on the comparison methods and find the p-value to be less than 0.05. Therefore, we believe that our method is valid.

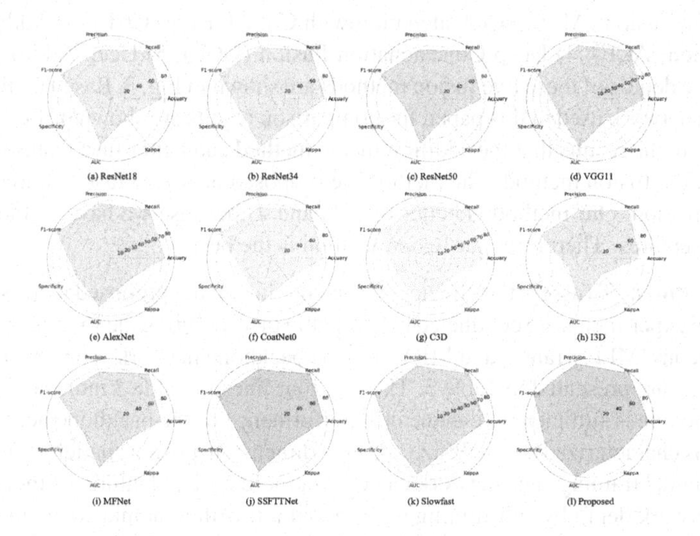

Fig. 2. Radar chart of different model performances.

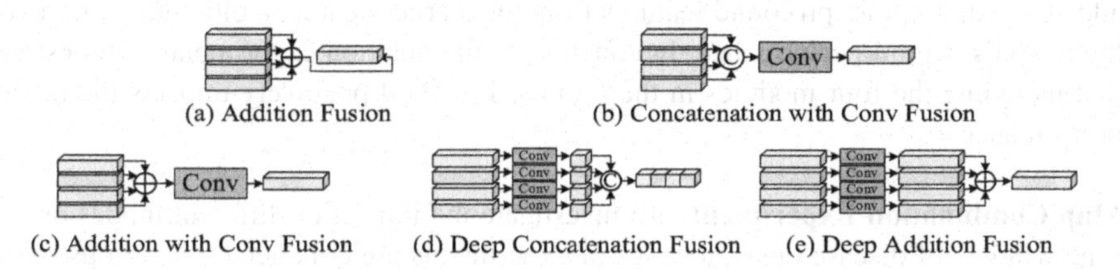

Fig. 3. The flow chart of different fusion methods.

Table 2. The results of different fusion methods (%).

Method	Accuracy	Sensitivity	Precision	F1-score	Specificity	Kappa
AF	79.44 ± 3.70	90.91 ± 5.19	80.68 ± 2.33	85.43 ± 2.80	56.59 ± 7.08	50.76 ± 8.54
CCF	79.41 ± 6.04	90.97 ± 7.77	81.05 ± 5.52	85.45 ± 4.06	56.37 ± 19.05	50.05 ± 17.31
ACF	79.45 ± 2.56	90.17 ± 4.44	81.19 ± 2.75	85.36 ± 1.91	58.02 ± 9.37	50.99 ± 6.59
DCF	77.99 ± 2.80	**95.50 ± 3.17**	77.19 ± 3.72	85.27 ± 1.32	43.19 ± 13.94	43.46 ± 10.32
DAF	80.48 ± 3.90	93.16 ± 5.08	80.62 ± 3.05	86.37 ± 2.90	55.27 ± 8.82	52.46 ± 9.26
Ours	**82.99 ± 4.14**	89.40 ± 5.77	**85.89 ± 4.61**	**87.45 ± 3.12**	**70.33 ± 11.18**	**60.97 ± 9.65**

Fusion Effectiveness. To effectively fuse the image features of CBV, CBF, MTT, and Tmax and realize the task of classifying TSS, This section verifies the fusion method in this paper. We mainly compare it with five different feature fusion methods. They are 1) Addition Fusion (AF); 2) Concatenation with Conv Fusion (CCF); 3) Addition with Conv Fusion (ACF); 4) Deep Concatenation Fusion (DCF); 5) Deep Addition Fusion (DAF). The details of these five fusion methods are shown in Fig. 3. Based on the feature extraction network used in this paper, the comparison results are shown in Table 2. It can be seen from the results that the addition fusion method can better fuse features than the concatenation fusion method. The method we proposed is also to fuse features with the addition method. Our method is better than 1) and 4) because we have further learned the fused features. Therefore, our fusion method is the best.

Ablation Study. To assess the efficacy of each module in the proposed method, a series of ablation experiments are conducted by gradually incorporating the four main modules, namely Dconv, MFF, TransF, and MPA, into the backbone network. The results of these experiments are presented in Table 3. The first four lines in Table 3 indicate that adding a single module is sufficient to enhance the performance of the backbone net-work. The CTP data is characterized by small size and high dimension, posing challenges to the deep learning model training. The network can extract more critical features without increasing the network depth by substituting the convolution with dynamic convolution on the backbone network. Map fusion on the backbone net-work improves the model accuracy by exploiting complementary information from multiple maps. Subsequently, MPA is added to extract more profound features from the learned features, ultimately improves the model's overall performance. In conclusion, the ablation experiments demonstrate that including the four modules in the proposed method positively impacts the model performance.

Map Combination Experiment. To investigate the impact of different modes on the time window of disease onset, a series of experiments are conducted on various mode combinations using the techniques proposed in this study. The results of different map combinations are presented in detail in Table 4. Comparing the results of the second to fourth rows in Table 4 shows that the CBV mode fusion exhibits a higher classification rate, which is superior to the fusion of other groups of modes. Furthermore, the experimental outcomes of the single map are slightly lower than the experimental results after fusion, indicating that multi-map fusion can assist the feature extraction network

Table 3. Ablation study (%) (D: DConv, M1: MFF, T: TransF, and M2: MPA).

Module	Accuracy	Sensitivity	Precision	F1-score	Kappa	AUC
D	77.05 ± 2.70	90.23 ± 3.32	78.54 ± 2.58	83.94 ± 2.01	44.33 ± 6.48	66.9 + 7.91
M1	77.48 ± 2.06	89.34 ± 9.20	79.72 ± 3.14	83.92 ± 2.60	46.78 ± 14.24	66.15 ± 23.87
T	76.91 ± 4.62	85.67 ± 5.76	80.9 ± 3.37	83.12 ± 3.51	43.51 ± 14.76	73.10 ± 7.71
M1 + T	78.01 ± 1.81	86.44 ± 4.39	81.98 ± 4.65	83.96 ± 0.73	54.60 ± 15.90	72.96 ± 3.60
M1 + M2	78.48 ± 2.48	90.94 ± 5.75	79.66 ± 1.50	84.83 ± 2.25	51.78 ± 14.24	76.04 ± 5.45
T + M2	78.00 ± 2.74	**91.60 ± 6.90**	78.96 ± 2.95	84.65 ± 2.40	56.54 ± 10.46	74.69 ± 3.08
M1 + T + M2	80.04 ± 5.07	90.04 ± 4.23	81.82 ± 4.57	85.74 ± 3.51	58.26 ± 6.18	77.82 ± 12.48
Ours	**82.99 ± 4.14**	89.40 ± 5.77	**85.89 ± 4.61**	**87.45 ± 3.12**	**60.97 ± 9.65**	**81.48 ± 5.74**

in obtaining more crucial disease information, thereby enhancing the effectiveness of our network. These observations demonstrate that our approach can improve the TSS classification result by learning the multi-map relationship.

Comparison with SOTA Methods. In Table 5, we have chosen relevant works for comparison. These studies aim to classify TSS based on brain images, with a time threshold of 4.5 h. The results demonstrate that our method performs relatively well in comparison. Specifically, our method achieves the best AUC and ACC among all methods, as reported in [8]. This may be attributed to the relatively large size of their dataset. Compared to studies such as [7, 26], our method achieves better results despite using the same amount of data.

Table 4. Map combination study (%) (V: CBV, F: CBF, M: MTT, and T: Tmax).

Map				Accuracy	Sensitivity	Precision	F1-score	Kappa	AUC
V	F	M	T						
√				77.01 ± 4.68	85.67 ± 7.26	81.02 ± 3.66	83.11 ± 3.77	46.90 ± 10.30	76.58 ± 4.74
√		√		78.99 ± 5.18	83.50 ± 10.65	85.22 ± 5.46	83.85 ± 5.05	53.23 ± 10.13	71.61 ± 11.49
	√	√		79.03 ± 3.94	90.14 ± 6.48	81.08 ± 5.16	85.11 ± 2.58	49.61 ± 11.83	75.46 ± 5.71
	√		√	78.03 ± 3.43	90.91 ± 6.46	79.49 ± 4.38	84.59 ± 2.41	46.54 ± 9.99	75.22 ± 7.24
√	√	√		80.99 ± 2.93	91.68 ± 5.66	81.89 ± 0.90	86.44 ± 2.53	54.77 ± 5.19	75.18 ± 3.86
√		√	√	80.50 ± 5.66	89.40 ± 5.09	83.41 ± 7.98	86.00 ± 3.71	53.90 ± 15.11	75.93 ± 5.73
	√	√	√	79.96 ± 3.79	**95.41 ± 5.02**	78.96 ± 3.04	86.34 ± 2.75	49.95 ± 8.97	76.94 ± 8.99
√	√	√	√	**82.99 ± 4.14**	89.40 ± 5.77	**85.89 ± 4.61**	**87.45 ± 3.12**	**60.97 ± 9.65**	**81.48 ± 5.74**

Table 5. Comparison of results of SOTA methods.

Ref.	Data	Subjects	AUC	sensitivity	specificity	Accuracy
[9]	DWI; FLAIR	422	0.74	0.70	0.81	0.758
[8]	DWI; FLAIR	342;245	0.896	0.823	0.827	0.878
[27]	DWI; FLAIR	404;368	–	0.777	0.802	0.791
[28]	DWI; MRI; ADC	25;26	0.754	0.952	0.500	0.788
[26]	DWI; FLAIR	173;95	–	0.769	0.840	0.805
[6]	DWI; FLAIR; MR perfusion	85;46	0.765	0.788	–	–
[7]	DWI;FLAIR;ADC	149;173	–	0.48	0.91	–
Ours	CTP	133;67	0.807	0.859	0.894	0.830

4 Conclusion

In this study, we propose a TSS classification model that integrates dynamic convolution and multi-map fusion to enable rapid and accurate diagnosis of unknown stroke cases. Our approach leverages the dynamic convolution mechanism to enhance model representation without introducing additional network complexity. We also employ a multi-map fusion strategy, consisting of MFF and TransF, to incorporate local and global correlations across low-order and high-order features, respectively. Furthermore, we introduce an MPA module to extract and incorporate as much critical feature information as possible. Through a series of rigorous experiments, our proposed method outperforms several state-of-the-art models in accuracy and robustness. Our findings suggest that our approach holds immense promise in assisting medical practitioners in making effective diagnosis decisions for TSS classification.

Acknowledgement. This work was supported National Natural Science Foundation of China (Nos. 62201360, 62101338, 61871274, and U1902209), National Natural Science Foundation of Guangdong Province (2019A1515111205), Guangdong Basic and Applied Basic Research (2021A1515110746), Shenzhen Key Basic Research Project (KCXFZ20201221173213036, JCYJ20220818095809021, SGDX202011030958020–07, JCYJ201908081556188–06, and JCYJ20190808145011259) Capital's Funds for Health Improvement and Research (No. 2022-1-2031), Beijing Hospitals Authority's Ascent Plan (No. DFL20220303), and Beijing Key Specialists in Major Epidemic Prevention and Control.

References

1. Phipps, M.S., Cronin, C.A.: Management of acute ischemic stroke. RMD Open **368**, l6983 (2020)
2. Paciaroni, M., Caso, V., Agnelli, G.: The concept of ischemic penumbra in acute stroke and therapeutic opportunities. Eur. Neurol. **61**(6), 321–330 (2009)
3. Peter-Derex, L., Derex, L.: Wake-up stroke: from pathophysiology to management. Sleep Med. Rev. **48**, 101212 (2019)

4. Konstas, A., Goldmakher, G., Lee, T.-Y., Lev, M.: Theoretic basis and technical implementations of CT perfusion in acute ischemic stroke, part 1: theoretic basis. Am. J. Neuroradiol. **30**(4), 662–668 (2009)
5. Ho, K.C., Speier, W., El-Saden, S., Arnold, C.W.: Classifying acute ischemic stroke onset time using deep imaging features. In: AMIA Annual Symposium Proceedings, pp. 892–901 (2017)
6. Ho, K.C., Speier, W., Zhang, H., Scalzo, F., El-Saden, S., Arnold, C.W.: A machine learning approach for classifying ischemic stroke onset time from imaging. IEEE Trans. Med. Imaging **38**(7), 1666–1676 (2019)
7. Lee, H., et al.: Machine learning approach to identify stroke within 4.5 hours. Stroke **51**(3), 860–866 (2020)
8. Jiang, L., et al.: Development and external validation of a stability machine learning model to identify wake-up stroke onset time from MRI. Eur. Radiol. **32**(6), 3661–3669 (2022)
9. Zhang, H., et al.: Intra-domain task-adaptive transfer learning to determine acute ischemic stroke onset time. Comput. Med. Imaging Graph. **90**, 101926 (2021)
10. Polson, J.S., et al.: Identifying acute ischemic stroke patients within the thrombolytic treatment window using deep learning. J. Neuroimaging **32**(6), 1153–1160 (2022)
11. Chen, Y., Dai, X., Liu, M., Chen, D., Yuan, L., Liu, Z.: Dynamic convolution: attention over convolution kernels. In: Proceedings of the IEEE/CVF Conference on Computer Vision and Pattern Recognition, pp. 11030–11039 (2020)
12. Rosenblatt, F.: The perceptron, a perceiving and recognizing automaton Project Para. Cornell Aeronautical Laboratory (1957)
13. Hu, J., Shen, L., Sun, G.: Squeeze-and-excitation networks. In: Proceedings of the IEEE Conference on Computer Vision and Pattern Recognition, pp. 7132–7141 (2018)
14. Vaswani, A., et al.: Attention is all you need. In: Advances in Neural Information Processing Systems. Vol. 30 (2017)
15. Dosovitskiy, A., et al.: An image is worth 16x16 words: Transformers for image recognition at scale. arXiv preprint arXiv:2010.11929 (2020)
16. Qi, D., Su, L., Song, J., Cui, E., Bharti, T., Sacheti, A.: ImageBERT: cross-modal pre-training with large-scale weak-supervised image-text data. arXiv preprint arXiv:2001.07966 (2020)
17. Chen, M., Radford, A., Child, R., Wu, J., Jun, H., Luan, D., Sutskever, I.: Generative pretraining from pixels. In: International Conference on Machine Learning, pp. 1691–1703. PMLR (2020)
18. Sun, C., Myers, A., Vondrick, C., Murphy, K., Schmid, C.: VideoBERT: a joint model for video and language representation learning. In: Proceedings of the IEEE/CVF International Conference on Computer Vision, pp. 7464–7473 (2019)
19. Chitta, K., Prakash, A., Jaeger, B., Yu, Z., Renz, K., Geiger, A.: TransFuser: Imitation with transformer-based sensor fusion for autonomous driving. IEEE Trans. Pattern Anal. Mach. Intell. (2022). https://doi.org/10.1109/TPAMI.2022.3200245
20. Dai, Z., Liu, H., Le, Q.V., Tan, M.: CoatNet: marrying convolution and attention for all data sizes. Adv. Neural. Inf. Process. Syst. **34**, 3965–3977 (2021)
21. Tran, D., Bourdev, L., Fergus, R., Torresani, L., Paluri, M.: Learning spatiotemporal features with 3D convolutional networks. In: Proceedings of the IEEE International Conference on Computer Vision, pp. 4489–4497 (2015)
22. Carreira, J., Zisserman, A.: Quo vadis, action recognition? A new model and the kinetics dataset. In: proceedings of the IEEE Conference on Computer Vision and Pattern Recognition, pp. 6299–6308 (2017)
23. Chen, Y., Kalantidis, Y., Li, J., Yan, S., Feng, J.: Multi-fiber networks for video recognition. In: Proceedings of the European Conference on Computer Vision, pp. 352–367 (2018)
24. Sun, L., Zhao, G., Zheng, Y., Wu, Z.: Spectral–spatial feature tokenization transformer for hyperspectral image classification. IEEE Trans. Geosci. Remote Sens. **60**, 1–14 (2022)

25. Feichtenhofer, C., Fan, H., Malik, J., He, K.: SlowFast networks for video recognition. In: Proceedings of the IEEE/CVF International Conference on Computer Vision, pp. 6202–6211 (2019)
26. Zhu, H., Jiang, L., Zhang, H., Luo, L., Chen, Y., Chen, Y.: An automatic machine learning approach for ischemic stroke onset time identification based on DWI and FLAIR imaging. NeuroImage: Clin. **31**, 102744 (2021)
27. Polson, J.S., et al.: Deep learning approaches to identify patients within the thrombolytic treatment window (2022).https://doi.org/10.1111/jon.13043
28. Zhang, Y.-Q., et al.: MRI radiomic features-based machine learning approach to classify ischemic stroke onset time. J. Neurol. **269**(1), 350–360 (2022)

Self-supervised Learning for Endoscopic Video Analysis

Roy Hirsch[1], Mathilde Caron[2], Regev Cohen[1](✉), Amir Livne[1], Ron Shapiro[1], Tomer Golany[1], Roman Goldenberg[1], Daniel Freedman[1], and Ehud Rivlin[1]

[1] Verily AI, Tel Aviv, Israel
regevcohen@google.com
[2] Google Research, Grenoble, France

Abstract. Self-supervised learning (SSL) has led to important breakthroughs in computer vision by allowing learning from large amounts of *unlabeled* data. As such, it might have a pivotal role to play in biomedicine where annotating data requires a highly specialized expertise. Yet, there are many healthcare domains for which SSL has not been extensively explored. One such domain is endoscopy, minimally invasive procedures which are commonly used to detect and treat infections, chronic inflammatory diseases or cancer. In this work, we study the use of a leading SSL framework, namely Masked Siamese Networks (MSNs), for endoscopic video analysis such as colonoscopy and laparoscopy. To fully exploit the power of SSL, we create sizable *unlabeled* endoscopic video datasets for training MSNs. These strong image representations serve as a foundation for secondary training with limited annotated datasets, resulting in state-of-the-art performance in endoscopic benchmarks like surgical phase recognition during laparoscopy and colonoscopic polyp characterization. Additionally, we achieve a 50% reduction in annotated data size without sacrificing performance. Thus, our work provides evidence that SSL can dramatically reduce the need of annotated data in endoscopy.

Keywords: Artificial intelligence · Self-Supervised Learning · Endoscopy Video Analysis

1 Introduction

Endoscopic operations are minimally invasive medical procedures which allow physicians to examine inner body organs and cavities. During an endoscopy, a thin, flexible tube with a tiny camera is inserted into the body through a small orifice or incision. It is used to diagnose and treat a variety of conditions, including ulcers, polyps, tumors, and inflammation. Over 250 million endoscopic

Supplementary Information The online version contains supplementary material available at https://doi.org/10.1007/978-3-031-43904-9_55.

H. Greenspan et al. (Eds.): MICCAI 2023, LNCS 14224, pp. 569–578, 2023.
https://doi.org/10.1007/978-3-031-43904-9_55

procedures are performed each year globally and 80 million in the United States, signifying the crucial role of endoscopy in clinical research and care.

A cardinal challenge in performing endoscopy is the limited field of view which hinders navigation and proper visual assessment, potentially leading to high detection miss-rate, incorrect diagnosis or insufficient treatment. These limitations have fostered the development of computer-aided systems based on artificial intelligence (AI), resulting in unprecedented performance over a broad range of clinical applications [10,11,17,23–25]. Yet the success of such AI systems heavily relies on acquiring annotated data which requires experts of specific knowledge, leading to an expensive, prolonged process. In the last few years, Self-Supervised Learning (SSL [5–8]) has been shown to be a revolutionary strategy for unsupervised representation learning, eliminating the need to manually annotate vast quantities of data. Training large models on sizable unlabeled data via SSL leads to powerful representations which are effective for downstream tasks with few labels. However, research in endoscopic video analysis has only scratched the surface of SSL which remains largely unexplored.

This study introduces Masked Siamese Networks (MSNs [2]), a prominent SSL framework, into endoscopic video analysis where we focus on laparoscopy and colonoscopy. We first experiment solely on public datasets, Cholec80 [32] and PolypsSet [33], demonstrating performance on-par with the top results reported in the literature. Yet, the power of SSL lies in large data regimes. Therefore, to exploit MSNs to their full extent, we collect and build two sizable *unlabeled* datasets for laparoscopy and colonoscopy with 7,700 videos (>23M frames) and 14,000 videos (>2M frames) respectively. Through extensive experiments, we find that scaling the data size necessitates scaling the model architecture, leading to state-of-the-art performance in surgical phase recognition of laparoscopic procedures, as well as in polyp characterization of colonoscopic videos. Furthermore, the proposed approach exhibits robust generalization, yielding better performance with only 50% of the annotated data, compared with standard supervised learning using the complete labeled dataset. This shows the potential to reduce significantly the need for expensive annotated medical data.

2 Background and Related Work

There exist a wide variety of endoscopic applications. Here, we focus on colonoscopy and laparoscopy, which combined covers over 70% of all endoscopic procedures. Specifically, our study addresses two important common tasks, described below.

Cholecystectomy Phase Recognition. Cholecystectomy is the surgical removal of the gallbladder using small incisions and specialized instruments. It is a common procedure performed to treat gallstones, inflammation, or other conditions affecting the gallbladder. Phase recognition in surgical videos is an important task that aims to improve surgical workflow and efficiency. Apart from measuring quality and monitoring adverse event, this task also serves in facilitating education, statistical analysis, and evaluating surgical performance.

Furthermore, the ability to recognize phases allows real-time monitoring and decision-making assistance during surgery, thus improving patient safety and outcomes. AI solutions have shown remarkable performance in recognizing surgical phases of cholecystectomy procedures [17,18,32]; however, they typically require large labelled training datasets. As an alternative, SSL methods have been developed [12,28,30], however, these are early-days methods that based on heuristic, often require external information and leads to sub-optimal performance. A recent work [27] presented an extensive analysis of modern SSL techniques for surgical computer vision, yet on relatively small laparoscopic datasets.

Optical Polyp Characterization. Colorectal cancer (CRC) remains a critical health concern and significant financial burden worldwide. Optical colonoscopy is the standard of care screening procedure for preventing CRC through the identification and removal of polyps [3]. According to colonoscopy guidelines, all identified polyps must be removed and histologically evaluated regardless of their malignant nature. Optical biopsy enables practitioners to remove pre-cancerous adenoma polyps or leave distal hyperplastic polyps in situ without the need for pathology examination, by visually predicting histology. However, this technique is highly dependent on operator expertise [14]. This limitation has motivated the development of AI systems for automatic optical biopsy, allowing non-experts to also effectively perform optical biopsy during polyp management. In recent years, various AI systems have been developed to this end [1,19]. However, training such automatic optical biopsy systems relies on a large body of annotated data, while SSL has not been investigated in this context, to the best of our knowledge.

3 Self-supervised Learning for Endoscopy

SSL approaches have produced impressive results recently [5–8], relying on two key factors: (i) effective algorithms for unsupervised learning and (ii) training on large-scale datasets. Here, we first describe Masked Siamese Networks [2], our chosen SSL framework. Additionally, we present our large-scale data collection (see Fig. 2). Through extensive experiments in Sect. 4, we show that training MSNs on these substantial datasets unlocks their potential, yielding effective representations that transfer well to public laparoscopy and colonoscopy datasets.

3.1 Masked Siamese Networks

SSL has become an active research area, giving rise to efficient learning methods such as SimCLR [7], SwAV [5] and DINO [6]. Recently, Masked Siamese Networks [2] have set a new state-of-the-art among SSL methods on the ImageNet benchmark [29], with a particular focus on the low data regime. This is of great interest for us since our downstream datasets are typically of small size [32,33]. We briefly describe MSNs below and refer the reader to [2] for further details.

During pretraining, on each image $\mathbf{x}_i \in \mathbb{R}^n$ of a mini-batch of $B \geq 1$ samples (e.g. laparoscopic images) we apply two sets of random augmentations to generate anchor and target views, denoted by $\mathbf{x}_i^a$ and $\mathbf{x}_i^t$ respectively. We convert each view into a sequence of non-overlapping patches and perform an additional masking ("random" or "focal" styles) step on the anchor view by randomly discarding some of its patches. The resultant anchor and target sequences are used as inputs to their respective image encoders f_{θ^a} and f_{θ^t}. Both encoders share the same Vision Transformer (ViT [16]) architecture where the parameters θ^t of the target encoder are updated via an exponential moving average of the anchor encoder parameters θ^a. The outputs of the networks are the representation vectors $\mathbf{z}_i^a \in \mathbb{R}^d$ and $\mathbf{z}_i^t \in \mathbb{R}^d$, corresponding to the [CLS] tokens of the networks. The similarity between each view and a series of $K > 1$ learnable prototypes is then computed, and the results undergo a softmax operation to yield the following probabilities $\mathbf{p}_i^a = softmax\left(\frac{\mathbf{Q}\mathbf{z}_i^a}{\tau^a}\right)$ and $\mathbf{p}_i^t = softmax\left(\frac{\mathbf{Q}\mathbf{z}_i^t}{\tau^t}\right)$ where $0 < \tau^t < \tau^a < 1$ are temperatures and $\mathbf{Q} \in \mathbb{R}^{K \times d}$ is a matrix whose rows are the prototypes. The probabilities are promoted to be the same by minimizing the cross-entropy loss $H(p_i^t, p_i^a)$, as illustrated in Fig. 1.

In practice, a sequence of $M \geq 1$ anchor views are generated, leading to multiple probabilities $\{\mathbf{p}_{i,m}^a\}_{m=1}^M$. Furthermore, to prevent representation collapse and encourage the model to fully exploit the prototypes, a mean entropy maximization (me-max) regularizer [2,22] is added, aiming to maximize the entropy $H(\bar{\mathbf{p}}^a)$ of the average prediction across all the anchor views $\bar{\mathbf{p}}^a \triangleq \frac{1}{MB}\sum_{i=1}^B \sum_{m=1}^M \mathbf{p}_{i,m}^a$. Thus, the overall training objective to be minimized for both θ^a and $\mathbf{Q}$ is where $\lambda > 0$ is an hyperparameter and the gradients are computed only with respect to the anchor predictions $\mathbf{p}_{i,m}^a$ (not the target predictions $\mathbf{p}_i^t$). Applying MSNs on the large datasets described below, generates representations that serve as a strong basis for various downstream tasks, as shown in the next section.

3.2 Private Datasets

Laparoscopy. We compiled a dataset of laparoscopic procedures videos exclusively performed on patients aged 18 years or older. The dataset consists of 7,877 videos recorded at eight different medical centers in Israel. The dataset predominantly consists of the following procedures: cholecystectomy (35%), appendectomy (20%), herniorrhaphy (12%), colectomy (6%), and bariatric surgery (5%). The remaining 21% of the dataset encompasses various standard laparoscopic operations. The recorded procedures have an average duration of 47 min, with a median duration of 40 min. Each video recording was sampled at a rate of 1 frame per second (FPS), resulting in an extensive dataset containing 23.3 million images. Further details are given in the supplementary materials.

Colonoscopy. We have curated a dataset comprising 13,979 colonoscopy videos of patients aged 18 years or older. These videos were recorded during standard colonoscopy procedures performed at six different medical centers between the years 2019 and 2022. The average duration of the recorded procedures is 15 min, with a median duration of 13 min. To identify and extract polyps from the videos, we employed a pretrained polyp detection model [21,25,26]. Using this model, we obtained bounding boxes around the detected polyps. To ensure high-quality data, we filtered out detections with confidence scores below 0.5. For each frame, we cropped the bounding boxes to generate individual images of the polyps. This process resulted in a comprehensive collection of 2.2 million polyp images.

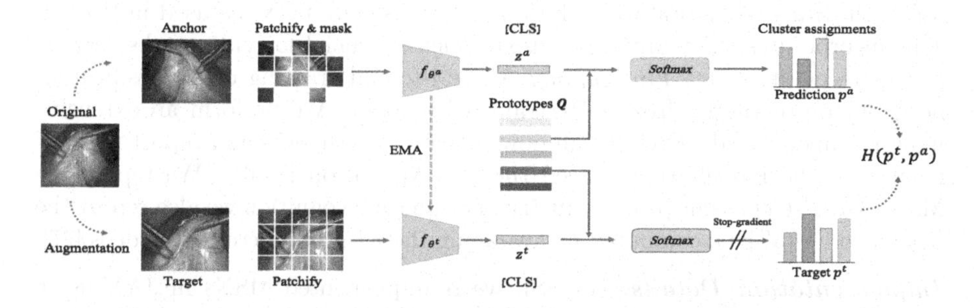

Fig. 1. Schematic of Masked Siamese Networks.

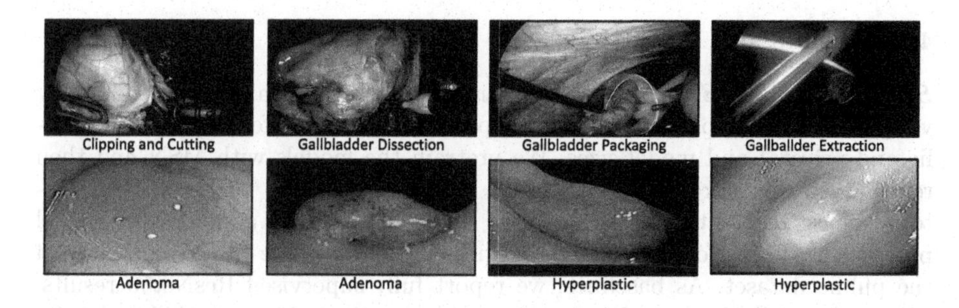

Fig. 2. Data Samples. Top: Laparoscopy. Bottom: Colonoscopy.

4 Experiments

In this section, we empirically demonstrate the power of SSL in the context of endoscopy. Our experimental protocol is the following: (i) first, we perform *SSL pretraining* with MSNs over our unlabeled private dataset to learn informative and generic representations, (ii) second we probe these representations by utilizing them for different public *downstream tasks*. Specifically, we use the following two benchmarks. (a) *Cholec80* [32]: 80 videos of cholecystectomy procedures resulting in nearly 200k frames at 1 FPS. Senior surgeons annotated each frame to one out of seven phases. (b) *PolypsSet* [33]: A unified dataset of 155 colonoscopy videos (37,899 frames) with labeled polyp classes (hyperplastic or adenoma) and bounding boxes. We use the provided detections to perform binary classification.

Downstream Task Evaluation Protocols. (a) *Linear evaluation:* A standard protocol consisting in learning a linear classifier on top of frozen SSL features [6, 20]. (b) *Temporal evaluation:* A natural extension of the linear protocol where we learn a temporal model on top of the frame-level frozen features. We specifically use Multi-Stage Temporal Convolution Networks (MS-TCN) as used in [13,27]. This incorporates the temporal context which is crucial for video tasks such as phases recognition. (c) *Fine-tuning:* An end-to-end training of a classification head on top of the (unfrozen) pretrained backbone. We perform an extensive hyperparameter grid search for all downstream experiments and report the test results for the models that exceed the best validation results. We report the Macro F1 (F-F1) as our primary metric. For phase recognition we also report the per-video F1 (V-F1), computed by averaging the F1 scores across all videos [27].

Implementation Details. For SSL we re-implemented MSNs in JAX using Scenic library [15]. As our image encoders we train Vision Transformer (ViT [16]) of different sizes, abbreviated as ViT-S/B/L, using 16 TPUs. Downstream experiments are implemented in TensorFlow where training is performed on 4 Nvidia Tesla V100 GPUs. See the supplementary for further implementation details.[1]

4.1 Results and Discussion

Scaling Laws of SSL. We explore large scale SSL pretraining for endoscopy videos. Table 1 compares the results of pretraining with different datasets (public and private) and model sizes. We pretrain the models with MSN and then report their downstream performances. We present results for the cholecystectomy phase recognition task based on fine-tuned models and for the optical polyp characterization task based on linear evaluation, due to the small size of the public dataset. As baselines, we report fully-supervised ResNet50 results, trained on public datasets. We find that replacing ResNet50 with ViT-S, despite comparable number of parameters, yields sub-optimal performance.

SSL pretraining on public datasets (without labels) provides comparable or better results than fully supervised baselines. The performance in per-frame phase recognition is comparable with the baseline. Phase recognition per-video results improve by 1.3 points when using the MSN pretraining, while polyp characterization improve by 2.2 points. Importantly, we see that the performance gap becomes prominent when using the large scale private datasets for SSL pretraining. Here, per-frame and per-video phase recognition performances improve by 6.7% and 8.2%, respectively. When using the private colonoscopy dataset the Macro F1 improves by 11.5% compared to the fully supervised baseline. Notice that the performance improves with scaling both model and private data sizes, demonstrating that both factors are crucial to achieve optimal performance.

Low-Shot Regime. Next, we examine the benefits of using MSNs to improve downstream performance in a *low-shot* regime with few annotated samples.

[1] For reproducibility purposes, code and model checkpoints are available at https://github.com/RoyHirsch/endossl.

Table 1. Comparing the downstream F1 performances of: (i) Models trained on the private (Pri) and public (Pub) datasets using SSL. (ii) Fully supervised baselines pre-trained on ImageNet-1K (IN1K). Best results are highlighted.

Method	Arch	Pretrain	Cholec80 frame	Cholec80 temporal		PolypsSet
				F-F1	V-F1	
Fully Supervised						
FS [27]	RN50	IN1K	71.5	-	80.3	72.1
TeCNO	RN50	IN1K	–	83.3	–	–
OperA	RN50	IN1K	-	84.4	–	–
Self Supervised						
DINO	ViT-S	IN1K	64.9	77.4	72.4	61.0
DINO [27]	RN50	Pub	71.1	-	81.6	72.4
MSN	ViT-S	Pub	65.0	83.4	80.9	70.6
MSN	ViT-B	Pub	71.2	82.6	82.9	74.6
MSN	ViT-L	Pub	65.6	84.0	82.0	73.6
MSN	ViT-S	Pri	70.7	87.0	84.3	78.5
MSN	ViT-B	Pri	73.5	87.3	85.8	78.2
MSN	ViT-L	Pri	76.3	89.6	86.9	80.4

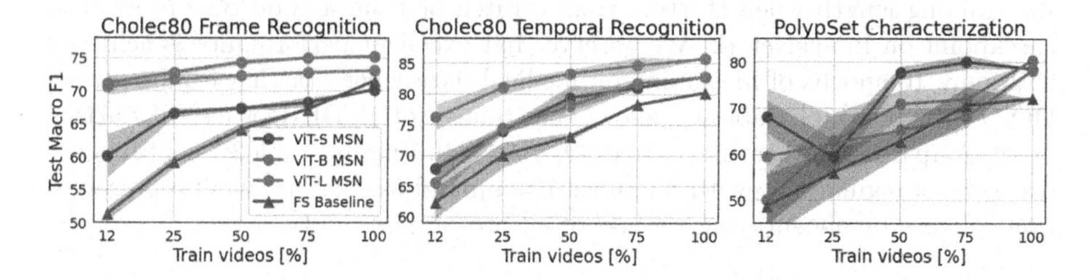

Fig. 3. Low-shot evaluation comparing MSN to fully supervised baselines.

Note that MSNs have originally been found to produce excellent features for low data regime [2]. We train a linear classifier on top of the extracted features and report the test classification results. Figure 3 shows the low-shot performance for the two endoscopic tasks. We report results using a fraction $k = \{12\%, 25\%, 50\%, 75\%, 100\%\}$ of the annotated public videos. We also report results for fully-supervised baselines trained on the same fraction of annotated samples. Each experiment is repeated three times with a random sample of train videos, and we report the mean and standard deviation (shaded area).

As seen, SSL-based models provide enhanced robustness to limited annotations. When examining the cholecystectomy phase recognition task, it is evident that we can achieve comparable frame-level performance by using only 12% of the annotated videos. Using 25% of the annotated videos yields comparable results to the fully supervised temporal models. Optical polyp characterization results show a similar trend, but with a greater degree of variability. Using small portions of PolypSet (12% and 25%) hindered the training process and increased sensitivity to the selected portions. However, when using more than 50% of PolypSet, the training process stabilized, yielding results comparable to the fully supervised baseline. This feature is crucial for medical applications, given the time and cost involved in expert-led annotation processes.

4.2 Ablation Study

Table 2 details different design choices regarding our SSL pretraining. Ablations are done on ViT-S trained over the public Cholec80. We report results on the validation set after linear evaluation. In Table 2a), we see that the method is robust to the number of prototypes, though over-clustering [4] with 1k prototypes is optimal. In Table 2b) and Table 2c), we explore the effect of random and focal masking. We see that 50% random masking (i.e. we keep 98 tokens out of 196 for the global view) and using 4 local views gives the best of performance. In Table 2d) we study the effect of data augmentation. SSL augmentation pipelines have been developed on ImageNet-1k [7], hence, it is important to re-evaluate these choices for medical images. Surprisingly, we see that augmentations primarily found to work well on ImageNet-1k are also effective on laparoscopic videos (e.g. color jiterring and horizontal flips). In Table2e), we look at the effect of the training length when starting from scratch or from a good SSL pretrained checkpoint on ImageNet-1k. We observe that excellent performance is achieved with only 10 epochs of finetuning on medical data when starting from a strong DINO checkpoint [6]. Table 2g) shows that ImageNet-1k DINO is a solid starting point compared to other alternatives [9,20,31,34]. Finally, Table2f) confirms the necessity of regularizing with Sinkhorn-Knopp and me-max to avoid representation collapse by encouraging the use of all prototypes.

Table 2. Ablation study of different design choices (default setting is highlighted).

a) Number of prototypes

K	10^1	10^2	10^3	10^4
val	65.4	67.8	69.8	69.1

b) Effect of random masking

%	0	50	70	90
val	69.1	69.8	68.4	68.2

c) Local crops (focal masking)

#	0	2	4	8
val	67.7	69.1	69.8	68.1

d) Data augmentation

color jit	flip (hor)	blur	val
✓	✓	✓	69.8
✓	✓		69.8
✓		✓	68.6
	✓	✓	67.4

e) Training length

epochs	10	100	200	500
scratch	33.8	63.3	65.5	66.5
SSL init	68.2	69.3	69.8	68.4

f) Avoiding collapse.

SK+me-max	SK	∅
69.8	67.7	34.0

g) ImNet-1k initialization

weights. (ViT-B/16)	val
MAE [20]	53.5
Supervised [31]	63.1
MoCo-v3 [9]	63.3
iBOT [34]	65.7
DINO [6]	65.9

5 Conclusion

This study showcases the use of Masked Siamese Networks to learn informative representations from large, unlabeled endoscopic datasets. The learnt representations lead to state-of-the-art results in identifying surgical phases of laparoscopic procedures and in optical characterization of colorectal polyps. Moreover, this methodology displays strong generalization, achieving comparable performance with just 50% of labeled data compared to standard supervised training on the complete labeled datasets. This dramatically reduces the need for annotated medical data, thereby facilitating the development of AI methods for healthcare.

References

1. Antonelli, G., Rizkala, T., Iacopini, F., Hassan, C.: Current and future implications of artificial intelligence in colonoscopy. Ann. Gastroenterol. **36**(2), 114–122 (2023)
2. Assran, M., et al.: Masked siamese networks for label-efficient learning. In: ECCV (2022). https://doi.org/10.1007/978-3-031-19821-2_26
3. Byrne, M.F., Shahidi, N., Rex, D.K.: Will computer-aided detection and diagnosis revolutionize colonoscopy? Gastroenterology **153**(6), 1460–1464 (2017)
4. Caron, M., Bojanowski, P., Joulin, A., Douze, M.: Deep clustering for unsupervised learning of visual features. In: Ferrari, V., Hebert, M., Sminchisescu, C., Weiss, Y. (eds.) Computer Vision – ECCV 2018. LNCS, vol. 11218, pp. 139–156. Springer, Cham (2018). https://doi.org/10.1007/978-3-030-01264-9_9
5. Caron, M., Misra, I., Mairal, J., Goyal, P., Bojanowski, P., Joulin, A.: Unsupervised learning of visual features by contrasting cluster assignments. In: NeurIPS (2020)
6. Caron, M., et al.: Emerging properties in self-supervised vision transformers. In: ICCV (2021)
7. Chen, T., Kornblith, S., Norouzi, M., Hinton, G.: A simple framework for contrastive learning of visual representations. In: ICML (2020)
8. Chen, X., He, K.: Exploring simple siamese representation learning. In: CVPR (2021)
9. Chen, X., Xie, S., He, K.: An empirical study of training self-supervised vision transformers. In: ICCV (2021)
10. Cohen, R., Blau, Y., Freedman, D., Rivlin, E.: It has potential: gradient-driven denoisers for convergent solutions to inverse problems. Adv. Neural. Inf. Process. Syst. **34**, 18152–18164 (2021)
11. Cohen, R., Elad, M., Milanfar, P.: Regularization by denoising via fixed-point projection (RED-PRO). SIAM J. Imag. Sci. **14**(3), 1374–1406 (2021)
12. da Costa Rocha, C., Padoy, N., Rosa, B.: Self-supervised surgical tool segmentation using kinematic information. In: 2019 International Conference on Robotics and Automation (ICRA), pp. 8720–8726. IEEE (2019)
13. Czempiel, T., et al.: TeCNO: surgical phase recognition with multi-stage temporal convolutional networks. In: Martel, A.L., et al. (eds.) MICCAI 2020. LNCS, vol. 12263, pp. 343–352. Springer, Cham (2020). https://doi.org/10.1007/978-3-030-59716-0_33
14. Dayyeh, B.K.A., et al.: Asge technology committee systematic review and meta-analysis assessing the asge pivi thresholds for adopting real-time endoscopic assessment of the histology of diminutive colorectal polyps. Gastrointest. Endosc. **81**(3), 502.e1–502.e16 (2015)
15. Dehghani, M., Gritsenko, A., Arnab, A., Minderer, M., Tay, Y.: Scenic: a jax library for computer vision research and beyond. In: CVPR, pp. 21393–21398 (2022)
16. Dosovitskiy, A., et al.: An image is worth 16×16 words: transformers for image recognition at scale. arXiv preprint arXiv:2010.11929 (2020)

17. Golany, T., et al.: AI for phase recognition in complex laparoscopic cholecystectomy. Surgical Endoscopy, 1–9 (2022)
18. Goldbraikh, A., Avisdris, N., Pugh, C.M., Laufer, S.: Bounded future MS-TCN++ for surgical gesture recognition. In: ECCV 2022 Workshops, October 23–27, 2022, Proceedings, Part III, pp. 406–421. Springer (2023). https://doi.org/10.1007/978-3-031-25066-8_22
19. Hassan, C., et al.: Performance of artificial intelligence in colonoscopy for adenoma and polyp detection: a systematic review and meta-analysis. Gastrointest. Endosc. **93**(1), 77–85 (2021)
20. He, K., Chen, X., Xie, S., Li, Y., Dollár, P., Girshick, R.: Masked autoencoders are scalable vision learners. In: CVPR (2022)
21. Intrator, Y., Aizenberg, N., Livne, A., Rivlin, E., Goldenberg, R.: Self-supervised polyp re-identification in colonoscopy. arXiv preprint arXiv:2306.08591 (2023)
22. Joulin, A., Bach, F.: A convex relaxation for weakly supervised classifiers. arXiv preprint arXiv:1206.6413 (2012)
23. Katzir, L., et al.: Estimating withdrawal time in colonoscopies. In: ECCV, pp. 495–512. Springer (2022). https://doi.org/10.1007/978-3-031-25066-8_28
24. Kutiel, G., Cohen, R., Elad, M., Freedman, D., Rivlin, E.: Conformal prediction masks: visualizing uncertainty in medical imaging. In: ICLR 2023 Workshop on Trustworthy Machine Learning for Healthcare (2023)
25. Livovsky, D.M., et al.: Detection of elusive polyps using a large-scale artificial intelligence system (with videos). Gastrointest. Endosc. **94**(6), 1099–1109 (2021)
26. Ou, S., Gao, Y., Zhang, Z., Shi, C.: Polyp-YOLOv5-Tiny: a lightweight model for real-time polyp detection. In: International Conference on Information Technology, Big Data and Artificial Intelligence (ICIBA), vol. 2, pp. 1106–1111 (2021
27. Ramesh, S., et al.: Dissecting self-supervised learning methods for surgical computer vision. Med. Image Anal. **88**, 102844 (2023)
28. Ross, T., et al.: Exploiting the potential of unlabeled endoscopic video data with self-supervised learning. Int. J. Comput. Assist. Radiol. Surg. **13**, 925–933 (2018)
29. Russakovsky, O., et al.: Imagenet large scale visual recognition challenge. In: IJCV (2015)
30. Sestini, L., Rosa, B., De Momi, E., Ferrigno, G., Padoy, N.: A kinematic bottleneck approach for pose regression of flexible surgical instruments directly from images. IEEE Robotics Autom. Lett. **6**(2), 2938–2945 (2021)
31. Touvron, H., Cord, M., Jégou, H.: DeIT III: Revenge of the ViT. arXiv preprint arXiv:2204.07118 (2022)
32. Twinanda, A.P., Shehata, S., Mutter, D., Marescaux, J., De Mathelin, M., Padoy, N.: Endonet: a deep architecture for recognition tasks on laparoscopic videos. IEEE Trans. Med. Imaging **36**(1), 86–97 (2016)
33. Wang, G.: Replication data for: colonoscopy polyp detection and classification: dataset creation and comparative evaluations. Harvard Dataverse (2021). https://doi.org/10.7910/DVN/FCBUOR
34. Zhou, J., et al.: ibot: image bert pre-training with online tokenizer. arXiv preprint arXiv:2111.07832 (2021)

Fast Non-Markovian Diffusion Model for Weakly Supervised Anomaly Detection in Brain MR Images

Jinpeng Li[1], Hanqun Cao[1], Jiaze Wang[1], Furui Liu[3], Qi Dou[1,2], Guangyong Chen[3(✉)], and Pheng-Ann Heng[1,2]

[1] Department of Computer Science and Engineering, The Chinese University of Hong Kong, Hong Kong, China
[2] Institute of Medical Intelligence and XR, The Chinese University of Hong Kong, Hong Kong, China
[3] Zhejiang Lab, Hangzhou, China
gychen@zhejianglab.com

Abstract. In medical image analysis, anomaly detection in weakly supervised settings has gained significant interest due to the high cost associated with expert-annotated pixel-wise labeling. Current methods primarily rely on auto-encoders and flow-based healthy image reconstruction to detect anomalies. However, these methods have limitations in terms of high-fidelity generation and suffer from complicated training processes and low-quality reconstructions. Recent studies have shown promising results with diffusion models in image generation. However, their practical value in medical scenarios is restricted due to their weak detail-retaining ability and low inference speed. To address these limitations, we propose a fast non-Markovian diffusion model (FNDM) with hybrid-condition guidance to detect high-precision anomalies in the brain MR images. A non-Markovian diffusion process is designed to enable the efficient transfer of anatomical information across diffusion steps. Additionally, we introduce new hybrid pixel-wise conditions as more substantial guidance on hidden states, which enables the model to concentrate more efficiently on the anomaly regions. Furthermore, to reduce computational burden during clinical applications, we have accelerated the encoding and sampling procedures in our FNDM using multi-step ODE solvers. As a result, our proposed FNDM method outperforms the previous state-of-the-art diffusion model, achieving a 9.56% and 19.98% improvement in Dice scores on the BRATS 2020 and ISLES datasets, respectively, while requiring only six times less computational cost.

Keywords: Anomaly Detection · Diffusion Probabilistic Model · Medical Image Analysis

---

Supplementary Information The online version contains supplementary material available at https://doi.org/10.1007/978-3-031-43904-9_56.

H. Greenspan et al. (Eds.): MICCAI 2023, LNCS 14224, pp. 579–589, 2023.
https://doi.org/10.1007/978-3-031-43904-9_56

1 Introduction

Weakly supervised anomaly detection holds significant potential in real-world clinical applications [27], particularly for new pandemic diseases where obtaining pixel-wise annotations from human experts is challenging or even impossible [16]. However, dominant anomaly detection methods based on the one-class classification paradigm [8,11] often overlook the binary labels of healthy and disease samples available in clinical centers, limiting their detection granularity. Traditional clustering techniques like z-score thresholding [15], PCA [14], and SVM [26] have limited clinical value as they mainly generate image-level anomaly results. To address this paradox, deep generative models leverage additional pixel-wise domain knowledge captured through adversarial training [1,12,24] or latent representation encoding [3,4,6,32]. While these approaches allow obtaining pixel-wise anomaly maps through Out-Of-Distribution (OOD) detection, they often fail to generate high-fidelity segmentation maps due to inadequate utilization of image conditions. Thus, the utilization of domain knowledge from weakly supervised data becomes a crucial factor in achieving high-quality anomaly detection.

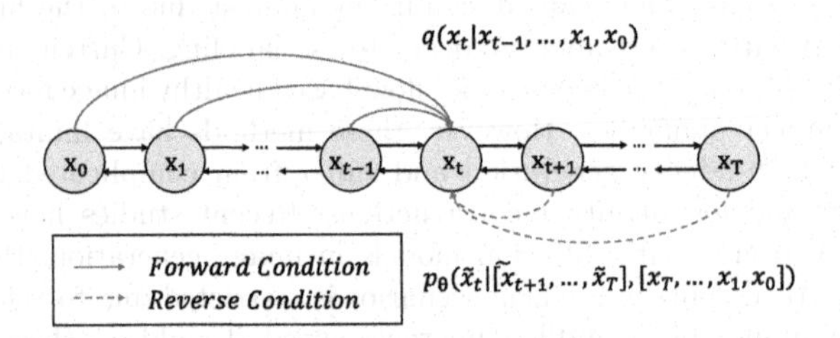

Fig. 1. Non-Markovian Diffusion Framework: To enhance information transfer, we make an assumption that the forward states and reverse states at the same time step t follow different distributions. Based on this assumption, we introduce conditions in the forward process, where all previous states are used as conditions for the current state. This enables efficient information transfer between states. And the final state x_T incorporates comprehensive information from the forward states. Similarly, the reverse process incorporates information from previous reverse states.

Moreover, traditional generative models, such as GANs [1,12] and normalizing flows [11], commonly used for pixel-wise anomaly detection, are constrained by one-step data projection in handling complex data distributions. This dilemma can be overcome by employing probabilistic diffusion models [13,25] that capture data knowledge through a series of step-by-step Markovian processes [5]. The high-fidelity generation proficiency, flexible network architecture, and stable training scheme of existing diffusion-based anomaly detection approaches have demonstrated promising performance in detecting anomaly regions [23,27–29]. However, diffusionnv-based approaches have high computational costs due to iterative evaluations. [20] explores the diffusion on the latent space with smaller sizes instead of pixel space to reduce the computations. Moreover, all these methods still face challenges including the lack of fine-grained guidance and gradual loss of anatomical information in Markovian chains.

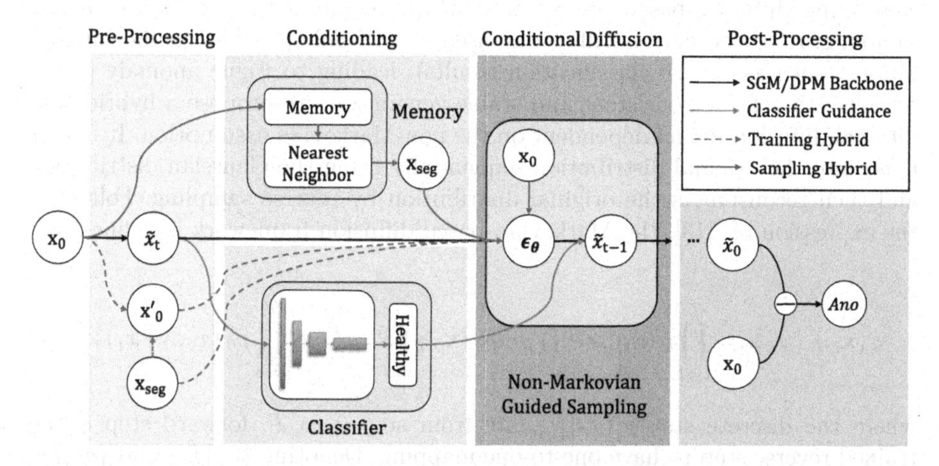

Fig. 2. Overall Framework of FNDM: The hybrid conditional noise prediction network ϵ_θ and binary classifier are separately trained. FNDM combines image state $\tilde{x}_t$, original data x_0 and coarse segmentation map x_{seg} into ϵ_θ, processing multi-step ODE sampling together with classifier gradient from current data. Finally, we obtain anomaly maps by subtraction.

To address these limitations, we propose a novel non-Markovian hybrid-conditioned diffusion model with fast samplers. Our approach utilizes strong hybrid image conditions that provide powerful sampling guidance by integrating coarse segmentation maps and original instance information based on the non-Markovian assumption (as shown in Fig. 1). Additionally, we modify the forward and reverse process as a higher-order deterministic Ordinary Differential Equation (ODE) sampler to accelerate inference. We validate our framework on two brain medical datasets, demonstrating the effectiveness of the framework components and showing more accurate detection results of anomaly regions.

2 Method

In this section, we present a fast non-Markovian diffusion model that utilizes pixel-wise strong conditions and encoding/sampling accelerator for anomaly segmentation to enhance generation fidelity and sampling speed. Section 2.1 introduces the non-Markonvian model and hybrid conditions for guided sampling. Section 2.2 proposes the acceleration approach for encoding and sampling.

2.1 Non-Markovian Diffusion Model with Hybrid Condition

Learning deterministic mappings between diseased and healthy samples sharing the same anatomical structures is essential to enhance inaccurate and time-consuming diffusion-based approaches, which require strong guidance during sampling. However, current diffusion-based models only provide insufficient conditions (such as binary classification results), leading to vague anomaly distributions. To achieve consistent and stable generation, we propose a hybrid conditional diffusion model dependent on the non-Markovian assumption. It injects noise into the original distribution sequentially using the Gaussian distribution and then reconstructs the original distribution by reverse sampling. Following the expression of [13], the Markovian-based diffusion framework is defined as:

$$q\left(\mathbf{x}_{1:T} \mid x_0\right) := \prod_{t=1}^{T} q\left(x_t \mid x_{t-1}\right), \quad p_\theta\left(\mathbf{x}_{0:T}\right) := p\left(x_T\right) \prod_{t=1}^{T} p_\theta\left(x_{t-1} \mid x_t\right), \tag{1}$$

where the discrete states $\{x_t\}_{t=0}^{T}$ are from step 0 to T, forward step q and trained reverse step p_θ have one-to-one mapping. Denoting $\{\alpha_t\}_{t=0}^{T}$ and $\{\sigma_t\}_{t=0}^{T}$ as variance scales for noise perturbation, the Gaussian transition kernels are:

$$\begin{aligned} q\left(x_t \mid x_{t-1}\right) &:= \mathcal{N}\left(x_t; \sqrt{\alpha_t} x_{t-1}, (1-\alpha_t)\mathbf{I}\right) \\ p_\theta\left(x_{t-1} \mid x_t\right) &:= \mathcal{N}\left(x_{t-1}; \boldsymbol{\mu}_\theta\left(x_t, t\right), \boldsymbol{\sigma}_t\right). \end{aligned} \tag{2}$$

To keep the anatomical information across states, the proposed non-Markovian anatomy structure mappings are built by adding previous-state information into forward and reverse states, which preserves distribution prior for high-quality reconstruction. Denoting all accumulated states from the forward process as c and the state in the backward step t as $\tilde{x}_t$, we formulate the Generalized Non-Markovian Diffusion Framework (GNDF) as:

$$q(\mathbf{x}_{0:T}) := q(x_0) \prod_{t=1}^{T} q(x_t | x_{i<t}), \quad p_\theta\left(\tilde{\mathbf{x}}_{0:T} | c\right) := p(\tilde{x}_T | c) \prod_{t=T}^{1} p_\theta(\tilde{x}_{t-1} | \tilde{x}_{i \geq t}, c) \tag{3}$$

Similar to vanilla DDPM [13], our conditional noise prediction network is trained according to the negative log-likelihood (NLL) lower bound minimization of generated distributions. It is further transformed into the L2 loss between the estimated conditional noise and the ground-truth Gaussian noise as:

$$\begin{aligned} \mathcal{L} &:= \mathbb{E}\left[-\log p_\theta\left(x_0\right)\right] \\ &\leq \mathbb{E}_{x_0,t}\left[\mathcal{KL}(p_\theta(\tilde{x}_{t-1} | \tilde{x}_t, c) | q(x_{t-1} | x_t, x_{t-2}, ..., x_0))\right] \\ &= \mathbb{E}_{x_0,\epsilon,t}\left[\omega(t) \left\|\epsilon_\theta\left(\tilde{x}_t, t, c\right) - \epsilon\right\|_2^2\right]. \end{aligned} \tag{4}$$

To enable the diffusion model to effectively differentiate between anatomical and anomaly information from previous states, we introduce a hybrid condition that includes the input state x_0, coarse segmentation maps, and classifier gradients derived from healthy labels. In order to simplify the computational complexity and leverage the rich information contained in x_0, we replace the forward state conditions c with the original state x_0.

Regarding hybrid condition implementation, we train a binary classifier with healthy and diseased images to provide further guidance on anomaly regions independently, following class-conditional methods [10,27]. A memory bank [21] is applied to store representative features of healthy samples, which enables quick generation of coarse segmentation maps x_{seg} during the testing phase, addressing the issue of knowledge forgetting in original diffusion models. To keep the active segmentation map as a condition in the diffusion model training, a health image x_0 is transformed into a diseased image x_0^n based on a random x_{seg}.

Overall, we train our non-Markovian diffusion model depending on the current states, the coarse segmentation maps, image labels, and the original data during the diffusion process with healthy and diseased images. The training objective is given by:

$$\min_{\theta} \mathbb{E}_{x_0,\epsilon,t} \left[\omega(t) \left\| \epsilon_\theta \left(\tilde{x}_t^{'}, t, y \right) - \epsilon \right\|_2^2 \right], \tag{5}$$

where $\tilde{x}_t^{'}$ is the concatenation of $\tilde{x}_t$, x_0^n, and x_{seg}. y denotes the corresponding binary label for each data. $\omega(t)$ is a weighted function for providing dynamic weights to explore density regions. In this work, we simply set $\omega(t)$ as 1. The full framework of our model is shown in Fig. 2.

2.2 Accelerated Encoding and Sampling

Diffusion acceleration, which is equivalent to reconstruction error minimization [7], is implemented by balancing the trade-off between sample quality and computation cost. Existing diffusion-based anomaly detectors [27] require a significant number of evaluation steps for encoding and sampling, which makes them impractical for clinical applications. To address this paradox, we adapt an Ordinary Differential Equation solver [17] for fast and stable encoding & sampling independent of the complex trade-off.

Following the setting of continuous-time Ito stochastic differential equation, where the drift coefficient, the diffusion coefficient, and the Wiener process are denoted as $\mathbf{f}(x,t)$, $g(t)$, $\mathbf{w}$, we have our forward process and probability flow ODE sampling [25] as:

$$\mathrm{d}x = \mathbf{f}(x,t)\mathrm{d}t + g(t)\mathrm{d}\mathbf{w}, \quad \frac{\mathrm{d}x}{\mathrm{d}t} = \mathbf{f}(x,t) - g(t)^2 \nabla_x \log p_t(x). \tag{6}$$

By decomposing the conditional sampling scheme according to Bayes Theorem, the guided diffusion process can be achieved by mixture guidance composed of conditional noise prediction model and classifier gradient:

$$p_{\theta,\phi}(\tilde{x}_{t-1}, y | \tilde{x}_t, x_0) \propto p_\theta(\tilde{x}_{t-1} | \tilde{x}_t, x_0) p_\phi(y | \tilde{x}_t). \tag{7}$$

Denote the classifier as $\mathcal{C}$, the binary label as y, the signal-to-noise coefficient as λ, and the conditional noise prediction network as $\boldsymbol{\epsilon}_\theta\left(\tilde{x}',\lambda,y\right)$, we further have the hybrid conditional diffusion network $\hat{\boldsymbol{\epsilon}}_\theta$ as:

$$\hat{\boldsymbol{\epsilon}}_\theta(\tilde{x}',\lambda,y) := \boldsymbol{\epsilon}_\theta(\tilde{x}',\lambda,y) + s\cdot\mathcal{C}(\tilde{x}_t,t,y). \tag{8}$$

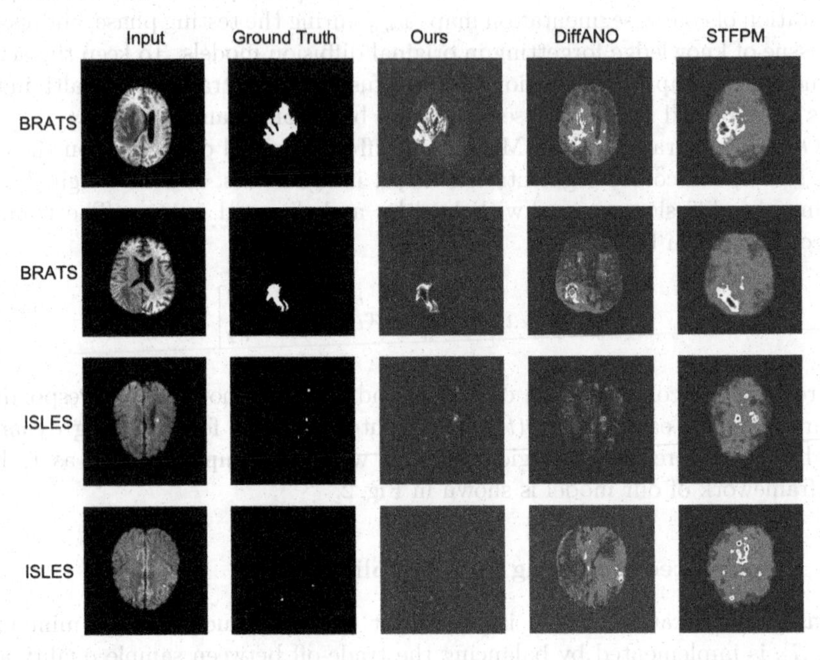

Fig. 3. Qualitative comparisons on BRATS and ISLES datasets. The lesions in ISLES are small, zooming in this figure to get better visualization. Thanks to the pixel-wise hybrid guidance, our results have few noise.

The semi-linear probability flow ODE is solved reversely with second-order multi-step numerical methods [18]. Then, we apply hybrid conditional sampling to noisy data $\tilde{x}_t$ to reproduce a healthy one with the same anatomy structure by conditional data prediction network and the binary classifier [10]. Following the symbol of [17,18], the conditional sampling is:

$$\frac{x_t}{\alpha_t} = \frac{x_s}{\alpha_s} - \int_{\lambda_s}^{\lambda_t} e^{-\lambda}\hat{\boldsymbol{\epsilon}}_\theta\left(\tilde{x}',\lambda,y\right)\mathrm{d}\lambda. \tag{9}$$

Finally, we post-process the reconstructed samples by subtracting original inputs and performing Otsu's threshold to obtain the anomaly segmentation map.

Table 1. Comparison with state-of-the-art anomaly detection methods. The best performances are bolded. The second-best performances are underlined.

	BRATS 2020			ISLES		
Method	Dice↑	HDis↓	VSim↑	Dice↑	HDis↓	VSim↑
PadimCore [8]	36.76	5.78	57.83	4.89	8.35	4.89
CFlow [11]	38.95	5.41	58.40	3.68	9.08	3.38
CSFlow [22]	22.33	7.05	44.09	7.52	6.02	12.53
FastFlow [31]	32.34	7.86	35.93	5.55	8.27	5.65
RevDis [9]	34.46	7.46	36.77	17.87	6.17	20.8
STFPM [33]	53.83	4.82	72.20	10.32	6.75	12.12
PatchCore [21]	51.78	4.29	62.83	5.83	7.91	5.83
DiffANO [27]	66.65	4.82	81.74	34.46	3.38	48.68
FNDM(Ours)	**76.21**	**3.80**	**82.28**	**54.44**	**1.99**	**75.41**

3 Experiment

3.1 Dataset and Evaluation Metric

BRATS 2020 [2] is a brain tumor segmentation dataset containing the MR sequences of T1, T1Gd, T2, and FLAIR. Following the preprocessing approach of [27], we concatenate all image modalities along the channel dimension, prune the upper and lower axial slices, and pad each slice into 256 × 256. All tumor classes are merged into a single class in the segmentation mask. The training set contains 10,410 slices with tumors and 5,809 healthy slices. The testing set includes 1,316 images with tumors. ISLES 2022 [19] is an MR image dataset for stroke lesions segmentation. It contains ischemic strokes of various sizes and from different disease stages. We extract the axial slices from DWI sequence and resize them into 256 × 256. The training set includes 2,707 healthy slices and 1,483 slices with lesions. The testing set contains 282 slices with lesions. Note that only the image-level binary labels are used in our training. The evaluation metrics include Dice, Volumetric Similarity, and Hausdorff Distance, which are calculated in slice level and volume level for BRATS and ISLES, respectively.

3.2 Implementation Details

To fairly compare with previous state-of-the-art diffusion models, we use the same network architectures as DiffANO [27]. We set the batch sizes of diffusion model and classifier as 3 and 10, respectively. The Adam optimizer with a learning rate of 0.0001 is used to train the diffusion model for 130,000 iterations and train the classifier for 150,000 iterations. The training diffusion step is 1000. The forward encoding and sampling steps are both set to 50 in the inference. We randomly generate the lesion masks as [30] and corrupt the corresponding regions on the conditioning images in the training phase.

3.3 Comparison with State-of-the-Art Methods

To compare the performance, we choose the anomaly detection methods including memory-based methods (such as PatchCore [8] and PadimCore [8]), normalizing flow based methods (such as CSFlow [22] and FastFlow [31]), distillation-based methods (such as Reverse Distillation [9] and STFPM [33]), and a diffusion-based method (DiffANO [27]) which also utilizes image-level binary labels. We train them on BRATS2020 and ISLES datasets. Table 1 shows the segmentation results. Our FNDM outperforms the existing methods in all metrics on both datasets. Diffusion methods have a more powerful generation ability than non-diffusion methods, and our method outperforms the best non-diffusion methods over 20% Dice. FNDM also outperforms the previous state-of-the-art diffusion method, DiffANO, by a large gap of +9.56% Dice, –0.98% HDis, and +0.54% VSim on BRATS dataset, and +19.98% Dice, –1.39% HDis, and +26.73% VSim on ISLES dataset, revealing that our FNDM is effective to reconstruct the healthy image from diseased to detect the anomaly regions in brain MR images. Thanks to non-Markovian procedure and pixel-wise hybrid guidance, the performance improvement of our method is larger on the ISLES dataset where stroke lesions are more challenging due to smaller sizes and irregular shapes.

Table 2. Ablation study on the hybrid guidance in our method.

Modules			BRATS		
CG	NM	MB	Dice	HDis	VSim
√			63.26	4.83	79.25
√	√		73.87	3.82	79.80
√	√	√	76.26	3.75	81.16

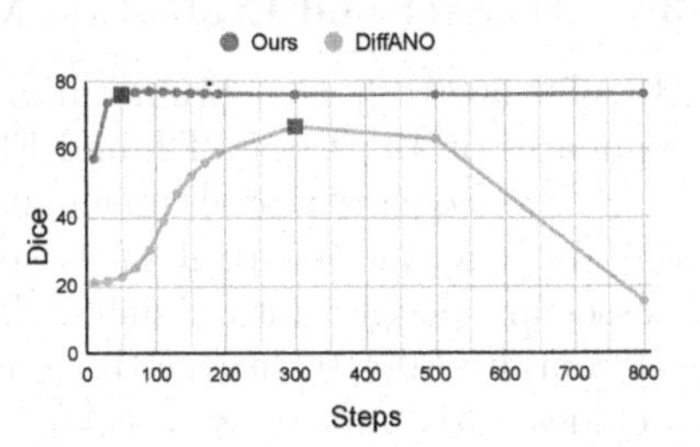

Fig. 4. Ablation on steps. We use the same step value in encoding and sampling.

3.4 Ablation Study

We conduct ablation studies for hybrid conditions and the steps of encoding and sampling procedures. From Table 2, we decompose the overall pixel-wise hybrid condition into classifier gradient (CG), Non-Markovian (NM), and Memory Bank (MB), comparing all possible combinations on BRATS2020 dataset. We observe that combinations with more components achieve better performance, and NM module achieves a higher increase than MB module. In Fig. 4, we further evaluate the segmentation performance of DiffANO [27] and ours across diverse steps on BRATS2020 dataset. DiffANO performs best at 300 steps, concluding that it is the optima where anomaly information vanishes and anatomy information preserves partly. Our method only needs 50 steps which achieve 6-time acceleration compared to DiffANO when both approaches reach the best Dice scores. Besides, our method performs stably as the step amount exceeds 50 since non-Markovain strong guidance ensures high-quality information transition along the timeline, independent of the loss of anatomy structure.

4 Conclusion and Discussion

We propose a Fast Non-Markovian Diffusion Model (FNDM) for weakly supervised anomaly detection. FNDM first encodes the images into noisy ones, then applies hybrid conditional generation to reconstructed original images without anomalies. FNDM achieves high-fidelity generation on weak labels by leveraging non-Markovian modeling and pixel-wise hybrid conditions. Besides, FNDM conducts ODE fast solver for encoding and sampling to reach 6-time acceleration. Extensive experiments on two brain datasets reveal the effectiveness and superiority of our approach for anomaly detection. The limitation of our method is that, as a diffusion-based method, it still needs more evaluation steps than GANs. In the future, we could investigate the knowledge distillation techniques to further reduce the sampling steps and apply FNDM in other modalities.

Acknowlegdement. This work described in this paper was supported in part by the Shenzhen Portion of Shenzhen-Hong Kong Science and Technology Innovation Cooperation Zone under HZQB-KCZYB-20200089. The work was also partially supported by a grant from the Research Grants Council of the Hong Kong Special Administrative Region, China (Project Number: T45-401/22-N) and by a grant from the Hong Kong Innovation and Technology Fund (Project Number: MHP/085/21). The work was also partially supported by a grant from the National Key R&D Program of China (2022YFE0200700), a grant from the National Natural Science Foundation of China (Project No. 62006219), and a grant from the Natural Science Foundation of Guangdong Province (2022A1515011579).

References

1. Alex, V., KP, M.S., Chennamsetty, S.S., Krishnamurthi, G.: Generative adversarial networks for brain lesion detection. In: Medical Imaging 2017: Image Processing (2017)
2. Bakas, S., et al.: Identifying the best machine learning algorithms for brain tumor segmentation, progression assessment, and overall survival prediction in the BRATS challenge (2018)
3. Baur, C., Denner, S., Wiestler, B., Navab, N., Albarqouni, S.: Autoencoders for unsupervised anomaly segmentation in brain mr images: a comparative study. Med. Image Anal. (2021)
4. Baur, C., Wiestler, B., Albarqouni, S., Navab, N.: Deep autoencoding models for unsupervised anomaly segmentation in brain mr images. In: Brainlesion (2019)
5. Cao, H., Tan, C., Gao, Z., Chen, G., Heng, P.A., Li, S.Z.: A survey on generative diffusion model. arXiv (2022)

6. Chen, X., Konukoglu, E.: Unsupervised detection of lesions in brain mri using constrained adversarial auto-encoders. In: Medical Imaging with Deep Learning (2018)
7. Chung, H., Sim, B., Ye, J.C.: Come-closer-diffuse-faster: accelerating conditional diffusion models for inverse problems through stochastic contraction. In: CVPR (2022)
8. Defard, T., Setkov, A., Loesch, A., Audigier, R.: Padim: A patch distribution modeling framework for anomaly detection and localization. In: Bimbo, A.D., et al. (eds.) ICPR (2020)
9. Deng, H., Li, X.: Anomaly detection via reverse distillation from one-class embedding. In: CVPR (2022)
10. Dhariwal, P., Nichol, A.: Diffusion models beat gans on image synthesis. In: NIPS (2021)
11. Gudovskiy, D.A., Ishizaka, S., Kozuka, K.: CFLOW-AD: real-time unsupervised anomaly detection with localization via conditional normalizing flows. In: WACV (2022)
12. Han, C., et al.: Madgan: unsupervised medical anomaly detection gan using multiple adjacent brain mri slice reconstruction. BMC Bioinform. (2021)
13. Ho, J., Jain, A., Abbeel, P.: Denoising diffusion probabilistic models. In: NIPS (2020)
14. Kim, C.M., Hong, E.J., Park, R.C.: Chest x-ray outlier detection model using dimension reduction and edge detection. IEEE Access (2021)
15. Li, W., et al.: Burn injury diagnostic imaging device's accuracy improved by outlier detection and removal. In: Algorithms and technologies for multispectral, hyperspectral, and ultraspectral imagery (2015)
16. Li, Y., Luo, L., Lin, H., Chen, H., Heng, P.-A.: Dual-consistency semi-supervised learning with uncertainty quantification for COVID-19 lesion segmentation from CT Images. In: de Bruijne, M., et al. (eds.) MICCAI 2021. LNCS, vol. 12902, pp. 199–209. Springer, Cham (2021). https://doi.org/10.1007/978-3-030-87196-3_19
17. Lu, C., Zhou, Y., Bao, F., Chen, J., Li, C., Zhu, J.: Dpm-solver: a fast ode solver for diffusion probabilistic model sampling in around 10 steps. In: NIPS (2022)
18. Lu, C., Zhou, Y., Bao, F., Chen, J., Li, C., Zhu, J.: Dpm-solver++: fast solver for guided sampling of diffusion probabilistic models. CoRR (2022)
19. Petzsche, M.R.H., et al.: ISLES 2022: a multi-center magnetic resonance imaging stroke lesion segmentation dataset. CoRR (2022)
20. Pinaya, W.H.L., et al.: Fast unsupervised brain anomaly detection and segmentation with diffusion models. In: Medical Image Computing and Computer Assisted Intervention - MICCAI 2022 (2022). https://doi.org/10.1007/978-3-031-16452-1_67
21. Roth, K., Pemula, L., Zepeda, J., Schölkopf, B., Brox, T., Gehler, P.: Towards total recall in industrial anomaly detection. In: CVPR (2022)
22. Rudolph, M., Wehrbein, T., Rosenhahn, B., Wandt, B.: Fully convolutional cross-scale-flows for image-based defect detection. In: WACV (2022)
23. Sanchez, P., Kascenas, A., Liu, X., O'Neil, A.Q., Tsaftaris, S.A.: What is healthy? generative counterfactual diffusion for lesion localization. In: MICCAI (2022). https://doi.org/10.1007/978-3-031-18576-2_4
24. Schlegl, T., Seeböck, P., Waldstein, S.M., Schmidt-Erfurth, U., Langs, G.: Unsupervised anomaly detection with generative adversarial networks to guide marker discovery. In: IPMI (2017)

25. Song, Y., Sohl-Dickstein, J., Kingma, D.P., Kumar, A., Ermon, S., Poole, B.: Score-based generative modeling through stochastic differential equations. In: ICML (2020)
26. Tax, D.M., Duin, R.P.: Uniform object generation for optimizing one-class classifiers. J. Mach. Learn. Res. (2001)
27. Wolleb, J., Bieder, F., Sandkühler, R., Cattin, P.C.: Diffusion models for medical anomaly detection. In: MICCAI (2022). https://doi.org/10.1007/978-3-031-16452-1_4
28. Wolleb, J., Sandkühler, R., Bieder, F., Cattin, P.C.: The swiss army knife for image-to-image translation: multi-task diffusion models. arXiv (2022)
29. Wyatt, J., Leach, A., Schmon, S.M., Willcocks, C.G.: Anoddpm: anomaly detection with denoising diffusion probabilistic models using simplex noise. In: CVPR (2022)
30. Yu, J., Lin, Z., Yang, J., Shen, X., Lu, X., Huang, T.S.: Free-form image inpainting with gated convolution. In: ICCV (2019)
31. Yu, J., et al.: Fastflow: unsupervised anomaly detection and localization via 2d normalizing flows. CoRR (2021)
32. Zimmerer, D., Isensee, F., Petersen, J., Kohl, S., Maier-Hein, K.: Unsupervised anomaly localization using variational auto-encoders. In: Shen, D., et al. (eds.) MICCAI 2019. LNCS, vol. 11767, pp. 289–297. Springer, Cham (2019). https://doi.org/10.1007/978-3-030-32251-9_32
33. Zolfaghari, M., Sajedi, H.: Unsupervised anomaly detection with an enhanced teacher for student-teacher feature pyramid matching. In: 27th CSICC (2022)

Self-supervised Polyp Re-identification in Colonoscopy

Yotam Intrator, Natalie Aizenberg, Amir Livne(✉), Ehud Rivlin, and Roman Goldenberg

Verily AI, Haifa, Israel
amirlivne@verily.com

Abstract. Computer-aided polyp detection (CADe) is becoming a standard, integral part of any modern colonoscopy system. A typical colonoscopy CADe detects a polyp in a single frame and does not track it through the video sequence. Yet, many downstream tasks including polyp characterization (CADx), quality metrics, automatic reporting, require aggregating polyp data from multiple frames. In this work we propose a robust long term polyp tracking method based on re-identification by visual appearance. Our solution uses an attention-based self-supervised ML model, specifically designed to leverage the temporal nature of video input. We quantitatively evaluate method's performance and demonstrate its value for the CADx task.

Keywords: Colonoscopy · Re-Identification · Optical Biopsy · Attention · Self Supervised

1 Introduction

Optical colonoscopy is the standard of care screening procedure for the prevention and early detection of colorectal cancer (CRC). The primary goal of a screening colonoscopy is polyp detection and preventive removal. It is well known that many polyps go unnoticed during colonoscopy [22]. To deal with this problem, computer-aided polyp detector (CADe) was introduced [13–16] and recently became commercially available [3]. The success of polyp detector sparkled the development of new CAD tools for colonoscopy, including polyp characterization (CADx, or optical biopsy), extraction of various quality metrics, and automatic reporting. Many of those new CAD applications require aggregation of all available data on a polyp into a single unified entity. For example, one would expect higher accuracy for CADx when it analyzes all frames where a polyp is observed. Clustering polyp detections into polyp entities is a prerequisite for computing such quality metrics as Polyp Detection Rate (PDR) and Polyps Per Colonoscopy (PPC), and for listing detected polyps in a report.

Supplementary Information The online version contains supplementary material available at https://doi.org/10.1007/978-3-031-43904-9_57.

H. Greenspan et al. (Eds.): MICCAI 2023, LNCS 14224, pp. 590–600, 2023.
https://doi.org/10.1007/978-3-031-43904-9_57

One may notice that the described task generally falls into the category of the well known multiple object tracking (MOT) problem [26,27]. While this is true, there are a few factors specific to the colonoscopy setup: (a) Due to abrupt endoscope camera movements, targets (polyps) often go out of the field of view, (b) Because of heavy imaging conditions (liquids, debris, low illumination) and non-rigid nature of the colon, targets may change their appearance significantly, (c) Many targets (polyps) are quite similar in appearance. Those factors limit the scope and accuracy of existing frame-by-frame spatio-temporal tracking methods, which typically yield an over-fragmented result. That is, the track is often lost, resulting in relatively short tracklets (temporal sequences of same target detections in multiple near-consecutive frames), see Supplementary Fig.1.

A recently published method [2] addresses this limitation by combining spatial target proximity and visual similarity to match a polyp detected in the current frame to "active" polyp tracklets dynamically maintained by the system. The tracklets are built incrementally, by adding a single frame detection to the matched tracklet, one-by-one. However, this approach limits itself to use of close-in-time consistent detections, and cannot handle the frequent cases where polyp gets out of the field of view and long range association is required.

In this work we propose an alternative approach that allows polyp detections grouping over an extended period of time (up to 10 min), relaxing the spatio-temporal proximity limitation. It involves two steps: (I) a short-term multi-object tracking, which forms initial, relatively short tracklets, followed by (II) a longer-term tracklets grouping by appearance-based polyp re-identification (ReID). As the first step can be done by any generic multiple object tracking algorithm (e.g. we use a tracking by detection method [27]), in this paper we focus on the second step.

To avoid manual data annotation, which is extremely ineffective in our case, we turn to self-supervision and adapt the widely used contrastive learning approach [5] to video input and object tracking scenario.

As tracklet re-identification is a sequence-to-sequence matching problem, the standard solution is comparing sequences element-wise and then aggregating the per-element comparisons, e.g. by averaging or max/min pooling [21] - the so-called late fusion technique. We, on the other hand, follow an early fusion approach by building a joint representation for the whole sequence. We use an advanced transformer network [23] to leverage the attention paradigm for non-uniform weighing and "knowledge exchange" between tracklet frames.

We extensively test the proposed method on hundreds of colonoscopy videos and evaluate the contribution of method components using an ablation study. Finally, we demonstrate the effectiveness of the proposed ReID method for improving the accuracy of polyp characterization (CADx).

To summarize, the three main contributions of the paper are:

- An adaptation of contrastive learning to video input for the purpose of appearance based object tracking.
- An early fusion, joint multi-view object representation for ReID, based on transformer networks.
- The application of polyp ReID to boost the polyp CADx performance.

2 Methods

This work assumes the availability of an automatic polyp detector. Quite a few highly accurate polyp detectors were recently reported [14–16], detecting (multiple) polyps in a single frame. Our ultimate goal is to group those detections into sets corresponding to distinct polyps.

As briefly mentioned above, the proposed approach starts with an initial grouping of polyp detections using an off-the-shelf multiple object tracking algorithm. Such a tracker is expected to track polyps through consecutive frames as long as they do not leave the camera field of view, forming disjoint, time separated polyp tracklets. In this work we use the ByteTrack [27] "tracking by detection" algorithm, but, in principle, any other tracker could be used instead.

The resulting tracklets are typically relatively short, and there are quite a few tracklets corresponding to the same polyp. To improve the result, we propose an Appearance-based Polyp Re-Identification (ReID), which groups multiple disjoint tracklets by their visual appearance into a joint tracklet, associated with a single polyp. In what follows we describe in detail the proposed ReID component.

As stated above, the objective of ReID is to ascertain whether two time-separated, disjoint tracklets belong to the same polyp. To this end we seek a tracklet representation that allows measuring visual similarity between tracklets. The two basic alternatives are either a single representation for the whole tracklet, or a sequence of single-frame representations for each tracklet frame. We will consider both options below.

2.1 Single-Frame Representation for ReID

To generate a single frame representation we train an embedding model that maps a polyp image into a latent space, s.t. the vectors of different views of the same polyp are placed closer, and of different polyps away from each other [11].

A straightforward approach to train such model is supervised learning, which requires forming a large collection of polyp image pairs, manually labeled as same/not same polyp [1]. Such annotation turned out to be inaccurate and expensive. In addition, finding hard negative pairs is especially challenging, as images of two randomly sampled polyps are usually very dissimilar. Moreover, self-supervised techniques using extensive unannotated datasets has exhibited substantial advantages within the medical domain [12].

Hence, we turn to SimCLR [5], a contrastive self-supervised learning technique, which requires no manual labeling. In SimCLR the loss is calculated over the whole batch where all input samples serve as negatives of each other and positive samples are generated via image augmentations. Combined with the temperature mechanism this allows for hard negative mining by prioritizing hard-to-distinguish pairs, resulting in a more effective loss weighting scheme.

One caveat of SimCLR is the difficulty to generate augmentations beneficial for the learning process [5]. Specifically for colonoscopy, the standard image augmentations do not capture the diversity of polyp appearances in different views (see Fig. 1(c)).

Instead of customizing the augmentations to fit the colonoscopy setup, we leverage the temporal nature of videos, and take different polyp views from the same tracklet as positive samples (see Fig. 1(b)).

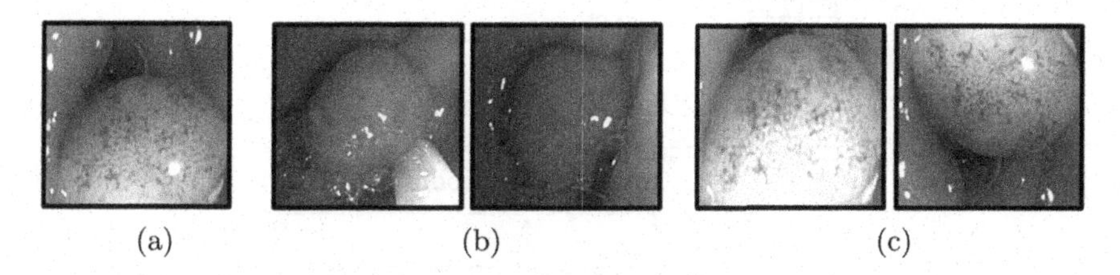

Fig. 1. (a) A polyp image, (b) two additional views of the polyp in (a) taken from the same tracklet, (c) two typical augmentations of the polyp in (a). Images in (b) offer more realistic variations, such as different texture, tools, etc.

Formally, a batch is formed by sampling one tracklet from N different procedures to ensure the tracklets belong to different polyps. Two polyp views i, j are sampled from each tracklet as positive pairs (same polyp). Let f be the embedding model. The loss function for the positive pair (i, j) is defined as:

$$\ell_{i,j} = -log \frac{exp(sim(f(i), f(j))/\tau)}{\sum_{k=1}^{2N} \mathbb{1}_{k \neq i} exp(sim(f(i), f(k))/\tau)} \tag{1}$$

where sim is the dot product and τ is the temperature parameter [24]. The final loss is computed across all positive pairs in the batch.

Tracklets represented as sequences of per-frame embeddings can be matched by computing pair-wise distances between frames, followed by an aggregation - e.g. min/max/mean distance [4,10]. An example of similarities between frames can be seen in Supplementary Fig. 2.

2.2 Multi-view Tracklet Representation for ReID

As discussed earlier, an alternative to the single frame approach, is a unified representation for the whole tracklet. A commonly used practice is to compute single frame embedding (for each view) and fuse them [8,21], e.g. by averaging. The downside of those simple techniques is that they treat every frame in the same way, including bad quality, repeating, non-informative views. We postulate that learning a joint embedding of multiple views in an end-to-end manner will produce a better representation of the visual properties of a polyp, by allowing "knowledge exchange" between the tracklet frames.

To achieve this, we employ a transformer network [23], with the addition of BERT [7] classification token (CLS). The attention mechanism enables both frame based intra attention and selective weighting of the frames thus providing a more comprehensive tracklet representation. The overview of the architecture is presented in Fig. 2. Training this multi-view encoder is done similarly to training a single-view encoder using SimCLR, but now, instead of pairs of frames, we deal with pairs of tracklet.

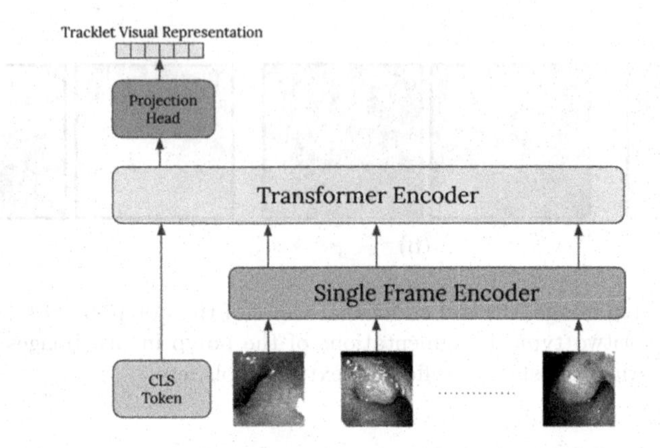

Fig. 2. Multi-view transformer encoder. Tracklet frames are passed through a single frame encoder to generate frame embedding. The embeddings then go through the transformer encoder, concatenated with the CLS token. Finally, the contextualized CLS token from the transformer encoder output goes through a projection head, resulting with the tracklet visual representation.

To generate positive tracklet pairs, we cannot apply the trick used for single frames, where positive pairs are sampled within the same tracklet. Instead we generate "pseudo positive" pairs from existing tracklets. We artificially split a tracklet into 3 disjoint segments, where the middle segment is discarded, and the first and the last segments are used as a positive pair, thus providing sufficiently different appearances of the same polyp as would happen in real procedures. In addition, this type of sampling approach, which effectively discards highly correlated samples from training, has been shown to improve model performance in [17].

3 Experiments

This section includes two parts. The first provides a stand-alone evaluation of the proposed ReID method. The second assesses the impact of ReID on polyp classification accuracy.

3.1 ReID Standalone Evaluation

Dataset. We use 22,283 colonoscopy videos, split into training (21,737) and test (546) sets. These recordings were captured from standard colonoscopy procedures conducted at six medical centers during the period of 2019 to 2022. The average length of the recorded procedures is 15 min, with a median duration of

13 min. For training, we automatically generated polyp tracklets using automatic polyp detection and tracking as described in Sect. 2.

The tracking algorithm might produce short and uninformative tracklets as well as outliers. The following clean up steps were performed on the training set: we filtered out tracklets shorter than 1 s or having less than 15 high confidence detections, as defined in [27], and took only the longest tracklet from every procedure. The thresholds were determined using analysis of the training set tracklets distribution. This yielded the training set of 15,465 tracklets (mean duration of 377 frames or 29 s). For evaluation, the test set polyp tracklets were manually annotated (timestamps and bounding boxes) by certified physicians. In addition, tracklet pairs from the same procedure were manually labeled as either belonging to the same polyp or not. This yielded 348 negative and 252 positive tracklet pairs.

Training. We utilize ResNet50V2 [9] as the single frame encoder, with an MLP head projecting the representation into a 128-dimensional embedding vector. We initialize the model using pre-trained ImageNet [6] weights. While ImageNet weights are not optimal for medical tasks [18,19], they offer training speedups [18]. The multi-view encoder consists of 3 transformer encoder blocks with an MLP projection head. We use LARS optimizer [25] with the learning rate of 0.01 and $\tau = 0.1$ as suggested in [5]. The batch size is set to 1024 for training both the single frame and the multi-view encoder.

We first train the single frame encoder and use its weights to initialize the single frame module of the multi-view encoder. Due to memory limitations, we use 8 views per tracklet during training, resulting in $1024 * 8 = 8192$ images per training step. The model was trained for 5,000 steps using cloud v3 TPUs with 16 cores. The single frame encoder has $24M$ parameters, and the multi-view encoder adds an additional $1M$ parameters.

Evaluation. We start by comparing various ReID techniques described in Sect. 2. Namely, we evaluate the accuracy of tracklet re-identification using: (a) single-frame representation with pairwise distances aggregation by Min / Max / Mean functions [4,10]; (b) multi-view representation by frame embeddings averaging; and, finally, (c) the joint embedding multi-view model. We evaluate the performance using AUC of the ROC and precision-recall curve (PRC) for tracklet similarity scores over the test set (see Table 1 and Supplementary Fig. 3). One can see that the joint embedding multi-view model outperforms all other techniques both on ROC and PRC.

In addition, we evaluate the effectiveness of ReID by measuring the average polyp fragmentation rate (FR), defined as the average number of tracklets polyps are split into. Obviously, lower fragmentation rate means better result (with the best fragmentation of 1), but it may come at the expense of wrong tracklet matching (false positive). We measure the fragmentation rate at the operating point of 5% false positive rate. The number of polyp fragments is determined by matching tracklets to manually annotated polyps and counting

Table 1. Polyp ReID accuracy for various ReID techniques.

	Single-frame			Multi-view	
	Min	Max	Mean	Averaging	Joint Embedding
AUROC	0.60	0.74	0.72	0.75	0.77
AUPRC	0.50	0.65	0.62	0.67	0.69

their number. Results presented in Table 2 demonstrate that ReID can reduce the fragmentation rate by over 50%, compared to a tracking only solution [27].

Table 2. Fragmentation Rate (FR) statistics before and after the ReID. FR STD is the FR standard deviation. Fragmented Polyps Ratio is the percentage of polyps divided into more than one tracklet.

	FR	FR STD	Fragmented Polyps Ratio
Tracking	3.3	3.3	0.64
Tracking+ReID	1.86	1.49	0.45

3.2 ReID for CADx

In this section, we investigate the potential benefits of using polyp ReID as part of a CADx system. Polyp CADx aims to assist physicians to figure out, in real time, during the procedure, whether the detected polyp is an adenoma.

Most reported CADx systems compute a classification score for each frame, and aggregate scores from multiple frames to determine the final polyp classification. Grouping polyp frames into a tracklet, to be fed into the CADx, is usually done by a spatio-temporal tracker [2]. Longer tracklets provide more information for polyp classification.

Here, we investigate if the proposed ReID model, used to group disjoint tracklets of the same polyp, can increase the accuracy of CADx.

Data. We use 3290 colonoscopy videos split into train, validation, and test sets (2666, 296, and 328 videos respectively). The videos are processed by a polyp detector and tracker to form polyp tracklets. The tracklets are then manually grouped together to build a single sequence for every polyp. Each polyp is annotated by a certified gastroenterologist as either adenoma or non-adenoma.

CADx. We trained a simple image classification CNN, composed of a MobileNet [20] backbone, followed by an MLP layer with a sigmoid activation, to predict the non-adenoma/adenoma score in $[0, 1]$, for each frame. The chosen architecture has 2.4M parameters and can run in real-time. The model was trained on Nvidia Tesla V100 GPU for 200 epochs with a learning rate of 0.001, using Adam optimizer.

For evaluation, we used the model to predict the classification score for each frame and aggregated the scores using soft voting to achieve the final prediction for each tracklet.

Evaluation. To assess the contribution of the ReID to polyp classification, we compare the CADx results on the test set, while using different grouping methods to merge multiple polyp detections into tracklets. The 3 evaluated methods are: (1) manual annotation (2) grouping by tracking, and (3) grouping by ReID. The manually annotated tracklets - the ground truth (GT) - are the longest sequences, containing all frames of each polyp in the test set. In grouping by tracking, we use tracklets generated by the spatio-temporal tracking algorithm [27]. Finally, for ReID, we merge disjoint tracklets by their appearance using the ReID model. By construction, tracklets generated by methods (2) and (3) are subsets of the corresponding manually annotated GT tracklet, and are assigned its polyp classification label. A visualization of the resulting tracklets using different grouping methods is provided in Supplementary Fig. 4. The number of resulting tracklets in the test set for each grouping method and polyp labels distribution are summarized in Table 3.

Table 3. CADx test data distribution and fragmentation rate (FR).

Grouping	Tracklets	FR	Adenoma	Adenoma FR	Non-Adenoma	Non-Adenoma FR
Annotation	608	1.0	464	1.0	144	1.0
Tracking	3161	5.20	2537	5.47	624	4.33
Tracking+ReID	1023	1.68	813	1.75	210	1.46

We ran the CADx model on tracklets generated by the 3 grouping methods. We compute the F_1 score and the AUC for the tracklet classification task. In addition, we measure the CADx sensitivity at specificity=0.9. The results are summarized in Table 4. The result on the manually annotated data is the accuracy upper-bound and is brought as a reference point. One can see that the ReID based approach significantly improves the CADx accuracy compared to the tracking-based grouping.

Table 4. Optical biopsy result per grouping method.

Grouping	AUC	F1 (Macro)	F1 (Micro)	Sensitivity @ Specificity=0.9
Annotation	0.95	0.88	0.91	0.86
Tracking	0.86	0.77	0.83	0.71
Tracking+ReID	0.90	0.82	0.88	0.79

4 Conclusions

In this study we present a novel multi-view self-supervised learning method for learning informative representations of a sequence of video frames. By jointly encoding multiple views of the same object, we get more discriminative features in comparison to traditional embedding fusion techniques. This approach can be used to group disjoint tracklets generated by a spatio-temporal tracking algorithm based on their appearance, by measuring the similarity between tracklets representations. Its applicability to medical contexts is of particular relevance, as medical data annotation often requires specific expertise and may be costly and time consuming. We use this method to train a polyp re-identification model (ReID) from large unlabeled data, and show that using the ReID model as part of a CADx system enhances the performance of polyp classification. There are some limitations however in identifying polyps based on their appearance, as it may be changed drastically during the procedure (for example, during resection). In future work we may examine the use of ReID for additional medical applications, such as listing detected polyps in an automatic report, bookmarking of specific areas of the colon during the procedure, and calculation of clinical metrics such as Polyp Detection Rate and Polyps Per Colonoscopy.

References

1. Ahmed, E., Jones, M., Marks, T.K.: An improved deep learning architecture for person re-identification. In: Proceedings of the IEEE Conference on Computer Vision and Pattern Recognition, pp. 3908–3916 (2015)
2. Biffi, C., Salvagnini, P., Dinh, N.N., Hassan, C., Sharma, P., Cherubini, A.: A novel ai device for real-time optical characterization of colorectal polyps. NPJ Digital Med. **5**(1), 84 (2022)
3. Brand, M., et al.: Frame-by-frame analysis of a commercially available artificial intelligence polyp detection system in full-length colonoscopies. Digestion **103**(5), 378–385 (2022)
4. Breckon, T.P., Alsehaim, A.: Not 3d re-id: simple single stream 2d convolution for robust video re-identification. In: 2020 25th International conference on pattern recognition (ICPR), pp. 5190–5197. IEEE (2021)
5. Chen, T., Kornblith, S., Norouzi, M., Hinton, G.: A simple framework for contrastive learning of visual representations. In: International Conference on Machine Learning, pp. 1597–1607. PMLR (2020)

6. Deng, J., Dong, W., Socher, R., Li, L.J., Li, K., Fei-Fei, L.: Imagenet: a large-scale hierarchical image database. In: 2009 IEEE Conference on Computer Vision and Pattern Recognition, pp. 248–255. IEEE (2009)
7. Devlin, J., Chang, M.W., Lee, K., Toutanova, K.: Bert: pre-training of deep bidirectional transformers for language understanding. arXiv preprint arXiv:1810.04805 (2018)
8. Gao, J., Nevatia, R.: Revisiting temporal modeling for video-based person reid. arXiv preprint arXiv:1805.02104 (2018)
9. He, K., Zhang, X., Ren, S., Sun, J.: Identity mappings in deep residual networks. In: Leibe, B., Matas, J., Sebe, N., Welling, M. (eds.) ECCV 2016. LNCS, vol. 9908, pp. 630–645. Springer, Cham (2016). https://doi.org/10.1007/978-3-319-46493-0_38
10. He, T., Jin, X., Shen, X., Huang, J., Chen, Z., Hua, X.S.: Dense interaction learning for video-based person re-identification. In: Proceedings of the IEEE/CVF International Conference on Computer Vision, pp. 1490–1501 (2021)
11. Hermans, A., Beyer, L., Leibe, B.: In defense of the triplet loss for person re-identification. arXiv preprint arXiv:1703.07737 (2017)
12. Hirsch, R., et al.: Self-supervised learning for endoscopic video analysis. In: International Conference on Medical Image Computing and Computer-Assisted Intervention (2023)
13. Lachter, J., et al.: Novel artificial intelligence-enabled deep learning system to enhance adenoma detection: a prospective randomized controlled study. iGIE (2023)
14. Livovsky, D.M., et al.: Detection of elusive polyps using a large-scale artificial intelligence system (with videos). Gastrointest. Endosc. **94**(6), 1099–1109 (2021)
15. Ou, S., Gao, Y., Zhang, Z., Shi, C.: Polyp-yolov5-tiny: a lightweight model for real-time polyp detection. In: 2021 IEEE 2nd International Conference on Information Technology, Big Data and Artificial Intelligence (ICIBA), vol. 2, pp. 1106–1111. IEEE (2021)
16. Pacal, I., Karaboga, D.: A robust real-time deep learning based automatic polyp detection system. Comput. Biol. Med. **134**, 104519 (2021)
17. Qian, R., et al.: Spatiotemporal contrastive video representation learning. In: Proceedings of the IEEE/CVF Conference on Computer Vision and Pattern Recognition, pp. 6964–6974 (2021)
18. Raghu, M., Zhang, C., Kleinberg, J., Bengio, S.: Transfusion: understanding transfer learning for medical imaging (2019)
19. Rajpurkar, P., et al.: Chexnet: radiologist-level pneumonia detection on chest x-rays with deep learning (2017)
20. Sandler, M., Howard, A., Zhu, M., Zhmoginov, A., Chen, L.C.: Mobilenetv 2: inverted residuals and linear bottlenecks. In: Proceedings of the IEEE Conference on Computer Vision and Pattern Recognition, pp. 4510–4520 (2018)
21. Seeland, M., Mäder, P.: Multi-view classification with convolutional neural networks. PLoS ONE **16**(1), e0245230 (2021)
22. Van Rijn, J.C., Reitsma, J.B., Stoker, J., Bossuyt, P.M., Van Deventer, S.J., Dekker, E.: Polyp miss rate determined by tandem colonoscopy: a systematic review. Official J. Am. College Gastroenterology| ACG **101**(2), 343–350 (2006)
23. Vaswani, A., et al.: Attention is all you need. In: Advances in Neural Information Processing Systems 30 (2017)
24. Wang, F., Liu, H.: Understanding the behaviour of contrastive loss. In: Proceedings of the IEEE/CVF Conference on Computer Vision and Pattern Recognition, pp. 2495–2504 (2021)

25. You, Y., et al.: Large batch optimization for deep learning: training bert in 76 minutes. arXiv preprint arXiv:1904.00962 (2019)
26. Yu, T., et al.: An end-to-end tracking method for polyp detectors in colonoscopy videos. Artif. Intell. Med. **131**, 102363 (2022)
27. Zhang, Y., et al.: Bytetrack: multi-object tracking by associating every detection box. In: Computer Vision-ECCV 2022: 17th European Conference, Tel Aviv, Israel, 23–27 October 2022, Proceedings, Part XXII, pp. 1–21. Springer (2022). https://doi.org/10.1007/978-3-031-20047-2_1

A Multimodal Disease Progression Model for Genetic Associations with Disease Dynamics

Nemo Fournier[(✉)] and Stanley Durrleman

Sorbonne Université, Institut du Cerveau - Paris Brain Institute - ICM, CNRS, Inria, Inserm, AP-HP, Hôpital de la Pitié Salpêtrière, 75013 Paris, France
nemo.fournier@icm-institute.org

Abstract. We introduce a disease progression model suited for neurodegenerative pathologies that allows to model associations between covariates and dynamic features of the disease course. We establish a statistical framework and implement an algorithm for its estimation. We show that the model is reliable and can provide uncertainty estimates of the discovered associations thanks to its Bayesian formulation. The model's interest is showcased by shining a new light on genetic associations.

Keywords: Multimodal Disease Progression Modelling · Alzheimer's Disease · Genetic Associations

1 Introduction

The clinical courses of neurodegenerative pathologies such as Alzheimer's or Parkinson's Diseases span multiple years and encompass intricate evolution of patients' cognitive abilities, physiological biomarkers and brain structure. Longitudinal studies are an essential tool for clinicians to uncover the diseases' mechanisms. In such studies, biomarkers and cognitive scores of patients are repeatedly measured at different times and need to be analyzed together, usually with a two-sided scope. First, to describe the general process at play across a whole cohort of patients: this is *population-level* modelling and allows to describe the average course of the disease. A second layer aims at explaining and predicting the variability observed among individuals: this is *personalized-level* modelling.

Mixed-effect frameworks are widely adopted to address these multi-layered prospects, offering to disentangle *fixed effects* (population level) from *random*

This work was funded in part by the French government under management of Agence Nationale de la Recherche as part of the Investissements d'avenir program, reference ANR-19-P3IA- 0001 (PRAIRIE 3IA Institute), ANR-19-JPW2-000 (E-DADS), and ANR10-IAIHU-06 (IHU ICM), as well as by the European Research council reference ERC-678304 and the H2020 programme via grant 826421 (TVB-Cloud).

H. Greenspan et al. (Eds.): MICCAI 2023, LNCS 14224, pp. 601–610, 2023.
https://doi.org/10.1007/978-3-031-43904-9_58

effects (individual level) to explain the variability of the disease. Linear mixed-effects models are the simplest instances of such models. Generalized linear and non-linear mixed-models are now often prefered to account for the neurodegenerative diseases' peculiarities, and most state of the art disease progression models (*e.g.* [2,14,16]) belong to these categories. They are indeed better suited to describe phenomena whose complex dynamic spans multiple years. They have been used with success to describe the natural history of diseases [8] or make individualized predictions, for instance to enrich clinical trials [11]. A general formulation is as follows, where η is a non linear mapping between timepoints and clinical markers, parametrized by fixed-effects $\boldsymbol{\alpha}$ and random effects $\boldsymbol{\beta}_i$ (methods differ by the chosen non-linearity η and how $\boldsymbol{\alpha}$ and $\boldsymbol{\beta}$ parametrize the disease course):

$$\boldsymbol{y}_i = \eta_{\boldsymbol{\alpha}}(\boldsymbol{t}_i \mid \boldsymbol{\beta}_i)$$

Inter-patient variability is thereby modelled through random perturbations $\boldsymbol{\beta}_i$ around a fixed reference $\boldsymbol{\alpha}$. However, it is known that some of this clinical variability between patients is explained by external factors (and thus hardly explained entirely by *random* perturbations). Genetic mutations or external factors such as gender, family history, education or socio-economics levels can influence the course of pathologies. Accumulating evidence suggests that the variability induced by such covariates stems from general mechanisms shared across the population, for instance in Alzheimer's and Parkison's diseases [3,6,7,10]. In the presented models, observed covariates $\boldsymbol{c}_i$ are not taken into account and only the repeated observations $\boldsymbol{y}_i$ are modelled as a function of the patient's ages $\boldsymbol{t}_i$. Thus, random effects might be such that $\mathbf{E}\left[\boldsymbol{\beta}_i \mid \boldsymbol{c}_i\right] \neq 0$. This shows that some signal present in covariates to explain the progression of the disease has not been fully exploited.

Our contribution is to propose a slight change in this mixed-effect paradigm to allow non-linear models to also be influenced by these variables. Instead of estimating a fixed effect $\boldsymbol{\alpha}$ (parametrizing the average disease course) as well as random effects $\boldsymbol{\beta}_i$, we introduce a *link function* f_φ that can predict, given a set of covariates $\boldsymbol{c}_i$ (*e.g.* sex, education level, SNP arrays, genetic risk scores), an expected trajectory of the disease conditionned by these covariates.

The main difference between the standard approach and our method is that the previously introduced fixed-effects $\boldsymbol{\alpha}$ are now estimated for each subject as a deterministic function of their covariates $f_\varphi(\boldsymbol{c}_i)$. It also differs from accounting for the heterogeneity through hierarchical progression models [13,17]) since covariates are, in our case, supervisingly used during model calibration and used to navigate through a *continuum* of disease models, instead of having defined clusters.

We demonstrate the value of this approach by adapting a general modelling framework, namely a non-linear Bayesian model: the *Disease Course Mapping* (DCM) [16]. We show that accounting for time-independent covariates in the longitudinal modelling with this approach can be done in a reasonable statistical setting. A stochastic estimation algorithm can be devised and we propose an instantiation and implementation of our model, which we validate first on synthetic. We then use clinical data from the Alzheimer's Disease Neuroimaging (ADNI) cohort and further demonstrate the clinical interest of the method by estimating new associations between genetics and disease dynamics.

2 Method

We derive here an algorithm that learns to model repeated observations while accounting for the heterogeneity explained by additional covariates.

2.1 A Generic Mixed-Effects Geometric Model

In their seminal paper [16], Schirrati *et al.* introduced a generic framework to model a dataset $(\boldsymbol{y}_{i,j})$ of multimodal longitudinal measurements. Here $\boldsymbol{y}_{i,j}$ is a vector of N biomarkers measured for the i-th subject at their j-th visit – *i.e.* at age $t_{i,j}$. Each observation $\boldsymbol{y}_{i,j}$ is assumed to lie on Riemannian submanifold $(\mathcal{M}, g)$ of $\mathbf{R}^N$. The *average course* of the disease is posited to be such that individual progressions stem from a geodesic trajectory γ_0 on the manifold surface. Geodesic equations imply that γ_0 is entirely characterized by its initial position $p_0 \in \mathcal{M}$ and speed $v_0 \in \mathrm{T}_{p_0}\mathcal{M}$ (tangent space of $\mathcal{M}$ at p_0) at time t_0. *Individual trajectories* are obtained from this reference trajectory γ_0 via a temporal reparametrization $t \mapsto \psi_i(t)$, used to derive what we name a *disease age* and enables registering patient's chronological ages onto a common disease timeline. Spatial effects w_i are applied to the reference trajectory thanks to an exp-parallelization procedure that identifiably deforms geodesics in the manifold space. We denote $\eta_{\gamma_0}^{w_i}$ the resulting geodesic. We refer to [16] for extensive details on the geometric properties of these operations (such as commutativity and identifiability of both temporal and spatial effects).

The choice of the manifold's metric shapes the geodesic trajectories and thus the disease model [4,15,16]. Clinical knowledge of Alzheimer's Disease suggests that sigmoid shapes as sound candidates to model biomarkers' evolutions (see [5] for clinical considerations, or [12] for the logistic dynamic of imaging-derived features such as brain-averaged protein loads backed by prion-like diffusion hypothesis). We therefore consider a product-metric g_p such that geodesic are sigmoids: $g_p(u, v) = u \cdot M(p) \cdot v$ with $M(p) = \frac{1}{p^2(1-p)^2}$, which gives the trajectories of Eq. (1). The resulting trajectories and the geometric interpretation of the (v_0, p_0, t_0) parameters are also presented in Fig. 1.

$$\eta_{\gamma_0}^{w_i}(\psi_i(t_{i,j}))^{(k)} = \left(1 + \left(\frac{1}{p_0^{(k)}} - 1\right) \exp\left(-\frac{v_0^{(k)} \psi_i(t_{i,j}) + w_i^{(k)}}{p_0^{(k)}(1 - p_0^{(k)})}\right)\right)^{-1} \quad (1)$$

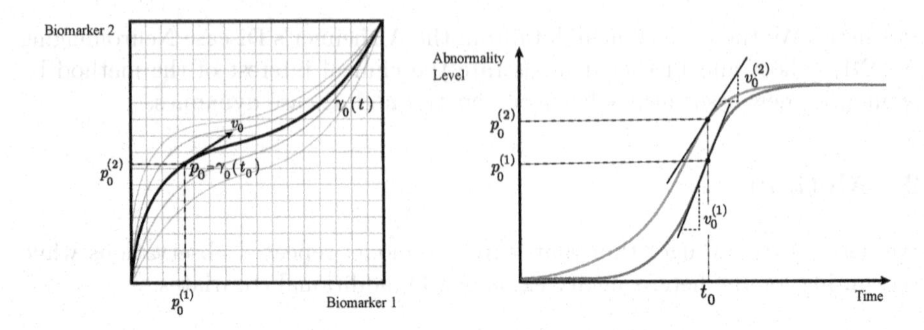

Fig. 1. A two feature model, with geodesic trajectories on the manifold (left) and the biomarkers observation space (right). This provides the intuition over the effect of the initial position p_0 and the initial velocity v_0 at time t_0.

2.2 Covariate Association and Statistical Framework

We provide here a statistical instantiation of the previous geometric model. As described, given a geodesic trajectory γ_0 (fully specified by its position p_0 and speed v_0 at initial time t_0), and a set of random effect ψ_i and w_i, individual trajectories of an individual i observed at times $(t_{i,j})_j$ are modelled by the curve $\eta_{\gamma_0}^{w_i}(\psi_i(t_{i,j}))$. We propose that γ_0 (which represents the reference disease course, as a fixed-effect of the model) is to be computed for each subject i from the measured covariates $\boldsymbol{c}_i$ as:

$$\gamma_{0,(i)} \cong \left(p_{0,(i)}, v_{0,(i)}, t_{0,(i)}\right) = f_{\varphi}(\boldsymbol{c}_i)$$

where f belongs to a parametrized family of functions and φ are its parameters treated as the new fixed-effect of the model. The individual effects to register this computed $\gamma_{0,(i)}$ onto observations are characterized by two random effects: an acceleration factor ξ_i and a time-shift τ_i such that $\psi_i(t) = e^{\xi_i}\left(t - t_{0,(i)} - \tau_i\right)$. On top of these are space-shifts $w_i \in \mathbf{R}^N$, computed thanks to an ICA: $\boldsymbol{w}_i = A\boldsymbol{s}_i$, where A is a latent matrix of independent directions (fixed effect) and $\boldsymbol{s}_i$ is the corresponding individual latent source vector (random effect).

Our hierarchical statistical model treats the fixed and random effects as a set of latent variables $\boldsymbol{z}$ which is the reunion of the population and individual variables $\boldsymbol{z}_{\text{pop}} = \{\varphi, A\}$ and $\boldsymbol{z}_{\text{indiv}} = \{(\boldsymbol{s}_i)_i, (\tau_i)_i, (\xi_i)_i\}$. We posit the following priors on these latent parameters, where $\theta_{\text{hyper}} = \{\sigma_\varphi, \sigma_A\}$ are fixed hyperparameters and $\theta_{model} = \left\{\overline{\varphi}, \overline{A}, \sigma_\tau^2, \sigma_\xi^2, \sigma^2\right\}$ are the parameters of the model to be estimated:

$$\varphi \sim \mathcal{N}\left(\overline{\varphi}, \sigma_\varphi^2\right) \quad A \sim \mathcal{N}\left(\overline{A}, \sigma_A^2\right) \quad \xi_i \sim \mathcal{N}\left(0, \sigma_\xi^2\right) \quad \tau_i \sim \mathcal{N}\left(0, \sigma_\tau^2\right) \quad s_i \sim \mathcal{N}(0, 1)$$

A non-informative prior is used over these model parameters due to the lack of a-priori knowledge. We seek to maximize a posteriori the joint-likelihood under the following additive Gaussian noise modelling $\boldsymbol{y}_{i,j} = \eta_{f_\varphi(\boldsymbol{c}_i)}^{A\boldsymbol{s}_i}(\psi_i(t_{i,j})) + \varepsilon_{i,j}$

$$q(\boldsymbol{y}, \boldsymbol{z}, \theta_{\text{model}}) = \underbrace{q(\boldsymbol{y} \mid \boldsymbol{z}, \theta_{\text{model}})}_{\text{data attachment}} \times \underbrace{q(\boldsymbol{z} \mid \theta_{\text{model}}, \theta_{\text{hyper}})}_{\substack{\text{regularization of the} \\ \text{latent variables}}} \times \underbrace{q_{\text{prior}}(\theta_{\text{model}} \mid \theta_{\text{hyper}})}_{\substack{\text{prior on model parameters} \\ \text{(taken non informative)}}}$$

It can be shown that the model's likelihood function lies in the curved exponential family. That is there exist two smooth functions Φ and Ψ functions of θ_{model} and a measurable sufficient statistics function $S(\boldsymbol{y}, \boldsymbol{z})$ of the data and the latent realizations such that $\log q(\boldsymbol{y}, \boldsymbol{z}, \theta_{\text{model}})$ factors as:

$$\log q\left(\boldsymbol{y}, \boldsymbol{z}, \theta_{\text{model}}\right) = -\Phi(\theta_{\text{model}}) + \langle S(\boldsymbol{y}, \boldsymbol{z}), \Psi(\theta_{\text{model}})\rangle$$

This allows estimating our model with a Monte-Carlo Markov-Chain Stochastic Approximation version of the Expectation Maximization algorithm (MCMC-SAEM) while enjoying theoretical guarantees of convergence [1]. The expectation phase is therefore built upon a sampling scheme to sample from the posterior distribution of the latent parameters (namely Metropolis-Hastings within Gibbs sampler). The maximization phase follows update rules established by finding critical points of $\theta \mapsto -\Phi(\theta) + \langle \tilde{S}_p, \Psi(\theta)\rangle$ ($\tilde{S}_p$ is the stochastic approximation of the sufficient statistics built at step p), which yields analytic expressions.

We choose to parametrize the link function as a linear mapping between covariates and dynamic parameters $f_{\varphi}(\boldsymbol{c}_i) = \boldsymbol{\varphi}_{\text{slope}} \cdot \boldsymbol{c}_i + \boldsymbol{\varphi}_{\text{intercept}}$. This will provide an interpretable model to explain the general processes linking covariates to dynamic features such as the base pace of the disease or average onset time. The coefficients of $\boldsymbol{\varphi}$ that correspond to the mapping between the covariates $\boldsymbol{c}_i$ and v_0 measure how much a given covariate impact the progression speed of each feature, and can be analyzed easily. Model parameters are initialized by setting their intercept to the models learned by a regular DCM model without considering covariates, while latent parameters are initialized at random.

3 Evaluation, Clinical Results and Discussion

3.1 Simulated Data

We used the generative abilities of the DCM[8] to simulate multimodal longitudinal datasets with covariates influencing the dynamic of the progression. To this end, we fixed some reference models corresponding each to a slightly different *pure form* of a fictional disease. Then, covariates were simulated either:

- as binary covariates that directly dictated which hardcoded model is used to simulate the repeated measurements (covariate thought as a mutation-status or sex for instance).
- as continuous covariates, influencing the simulated progression by using them as convex coefficients in a combination of the reference models. The covariates are seen as continuous risk factors of following one form or another.

Our simulated datasets typically included 500 subjects with an average of 5 visits and an average follow-up duration of about 5±2 years and a measurement noise of around 5%. Such an experiment is summarized in Fig. 2, where three continuous covariates were simulated on the [0, 1] range, the first one being a risk to develop a motor form of the disease, the second one a memory-form risk and the third a covariate without any influence on the disease.

In this example, our calibrated model correctly matches each covariate to its simulated effect. For instance: the coefficients of the link function related to the disease initial speed v_0 (we refer to Fig. 1 for its interpretation) associated with the memory-risk covariate show an acceleration of the decline in memory (multiplicative factor of 1.32 [1.15, 1.62] - credible interval at 95%) contrasted to the two other biomarkers (factor of 0.79 [0.73, 0.90] for the motor and 0.89 [0.80, 1.01] for the language). These intervals are represented in Fig . 2b. These two other features are slowed down relatively to the memory in order for the model to capture the change in the slope-ratio of different features on the fixed effects. If we translate the effects of these coefficients into effects on *slope-ratio*, we obtain indeed obtain that the ratio between memory-speed and motor-speed goes from 4.22 [3.54, 5.02] (no extra memory-risk) to 7.05 [5.54, 11.26] (maximum extra memory-risk). The ground truth change of slope (from the reference models) was from 4.48 (no particular risk) to 9.19 (full risk) and is therefore covered by our credible intervals.

Similarly, the coefficients associated with the irrelevant covariate did not capture any significant effect (factors of 1.03 [0.94, 1.12], 1.05 [0.98, 1.12] and 1.04 [0.97, 1.11] for the memory, motor and language features), which validates the ability of the model to discard covariates without influence on the disease dynamic.

3.2 Multimodal Clinical Data

Data used in the preparation of this article were obtained from the Alzheimer's Disease Neuroimaging Initiative (ADNI) database (adni.loni.usc.edu). ADNI was launched in 2003 as a public-private partnership led by Michael W. Weiner, MD. For up-to-date information, see https://www.adni-info.org. We selected subjects that eventually converted to an MCI or AD stage during their follow-up. This amounted to 1440 patients for a total of 9343 visits. The follow-up duration was 4.069 (± 3.190) years, with a baseline age of 73.683 (± 7.508) years old. We processed and included biomarkers relevant to monitor AD progression:

- Two cognitive scores: the Mini-Mental State Exam (MMSE) and the AD Assessment Scale-Cognitive (ADAS-Cog). We normalized and inverted them so that they both cover the [0, 1] interval, (1 being the highest abnormality).
- Hippocampus and Ventricles volumes, measured by structural T1 MRI and normalized by patient's Intracranial Volume (ICV). As for the cognitive scores, these measurements were rescaled to [0, 1] interval.
- Contrasted PET imaging derived brain-averaged amyloid $\beta 42$ and phosphorylated τ proteins loads, also rescaled to [0, 1].

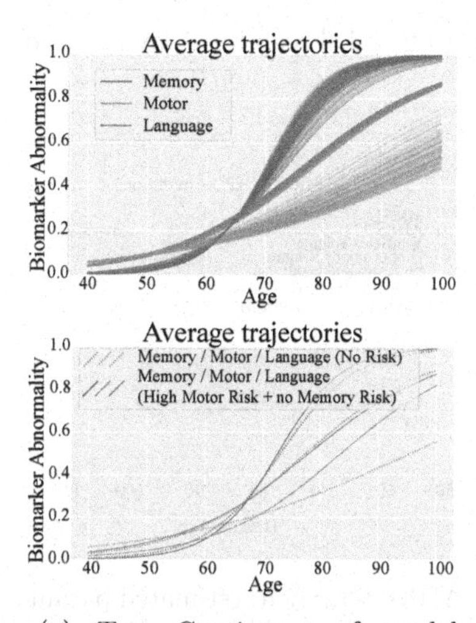

(a) *Top.* Continuum of models used to simulate data. *Bottom.* recovered disease course and effect of the motor-risk covariate.

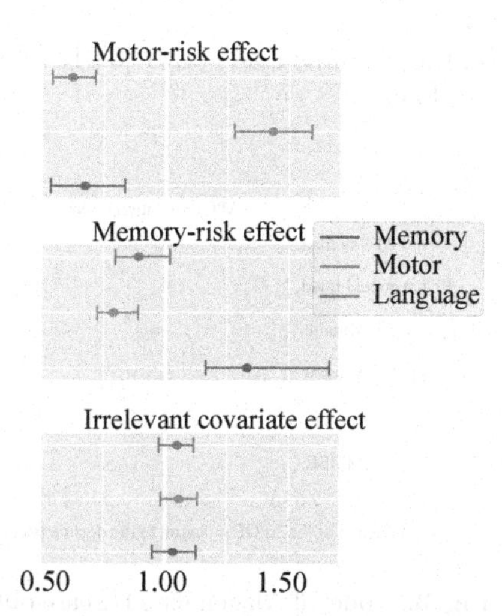

(b) Posterior mode and credible intervals on the *covariates - progression - speed* interaction coefficient

Fig. 2. Risk factors to develop an acute memory-form or acute motor-form of the disease are sampled continuously from the [0, 1] interval, and used as convex coefficients to combine three reference models (a standard form, a motor-dominant form and an acute memory form), yielding the continuum of possible trajectories presented in (a). We also sample an *irrelevant* covariate that is never used to modulate the disease course. In (a) is the resulting model, which we can visualize for any combination of covariates (two combinations are presented). We also plot the credible intervals (at 95%) for the coefficients linking covariates to the speed of progression on each feature (multiplicative effect, thus 1.0 stands for no influence while a coefficient of 1.5 stands for an expected progression speed greater by 50%).

APOE-ε4. We calibrated our model by including a covariate known to modulate Alzheimer's Disease course, namely the patient's APOE mutation status. The results of this model are showcased in Fig. 3. It shows that the APOE mutation, which is a known risk factor for AD, has a clear effect on the disease dynamic: in the obtained disease course map, the mutation is associated with earlier and faster abnormalities on most biomarkers. We also investigate the learned linked function f_φ and its coefficients dictating the interaction between the covariates and the speed of progression. This is presented in Fig. 3. This showcases how the contribution of the APOE to the speed of progression (as in the coefficient linking the covariate to the coordinates of v_0) is different among biomarkers. Features can be grouped into a *cognitive scores* group, more impacted than the other features by the mutation (1.49 [1.31, 1.66] and 1.34 [1.21, 1.46] respectively for MMSE and ADAS), a structural subgroup that also shows significative (even though slower) increase of progression speed (1.08 [1.00 1.14] and 1.07 [1.0, 1.15] resp. for Hippocampus and Ventricles volume), while exhibiting less clear effects on the proteins loads (1.01 [0.91, 1.12] and 1.03 [0.92, 1.18] resp for Amyloid and tau loads).

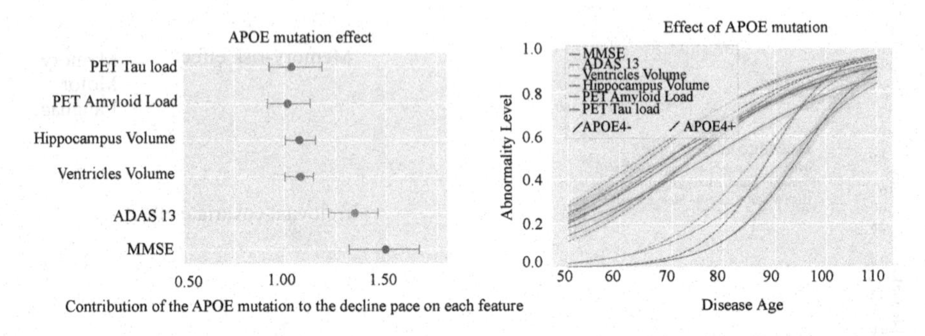

Fig. 3. Model of Alzheimer's Disease obtained from ADNI data. Left: estimated parameters of the link function φ (mode of the posterior and 95% credible interval). Right: difference in the trajectory conditioned by the mutation status (the represented trajectories are for 0 copy vs 2 copies of the APOE-ε4 allele).

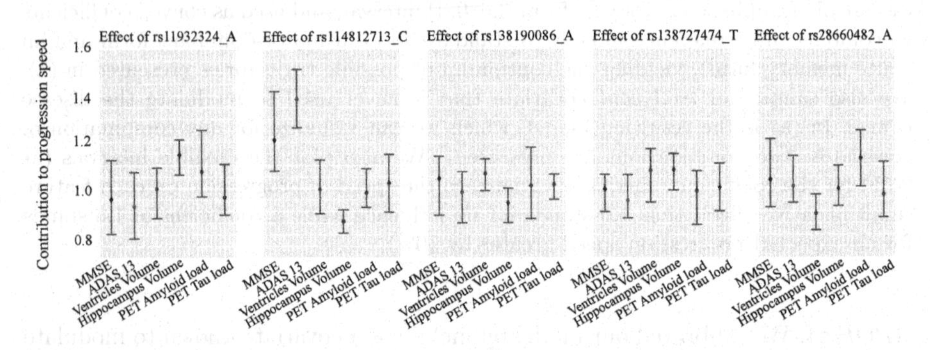

Fig. 4. Analysis of the learned interaction between some of the SNP included in the analysis and the speed of progression of each of the 6 measured features. Some SNPs present no significant interaction with the progression speed of any of the variables (*e.g.* rs138727474T) to SNPs that are associated with a group of feature (*e.g.* cognitive domain for rs114812713C) or single features (*e.g.* rs11932324A or rs286604821A, in either a protective or risk-inducing direction).

SNP Associations. We selected a subset of 69 Single Nucleotide Polymorphisms (SNP) among the top associations with AD diagnosis from a reference Genome-Wide Association Study (GWAS) [9]. We included them in our model as covariates. The results suggest that being associated with the diagnosis does not inform a priori on the influence of each SNP on the disease course. In Fig. 4 we show that, even though all these SNP were selected for being significatively associated with the diagnosis, they can exhibit differences in their association with the disease dynamic.

4 Conclusion

We proposed a framework to adapt a state of the art Bayesian non-linear mixed-effect disease progression model to capture the effects of external covariates into the disease dynamic. We implemented an estimation algorithm, and show that it reliably provides new interpretable measures of interaction between covariates and the disease course. For instance, we recover the (clinically known) association between the APOE-ε4 mutation and cognitive dysfunction. In particular, its use on genetic data (either single mutation status or SNP arrays) could help to go beyond associations with the sole diagnosis and provide complementary tools to GWAS.

References

1. Allassonnière, S., Kuhn, E., Trouvé, A.: Construction of Bayesian deformable models via a stochastic approximation algorithm: a convergence study. Bernoulli **16**(3), 641–678 (2010). https://doi.org/10.3150/09-BEJ229
2. Donohue, M.C., et al.: Estimating long-term multivariate progression from short-term data. Alzheimer's & Dementia **10**(5S), S400–S410 (2014). https://doi.org/10.1016/j.jalz.2013.10.003
3. Greenland, J.C., Williams-Gray, C.H., Barker, R.A.: The clinical heterogeneity of Parkinson's disease and its therapeutic implications. Eur. J. Neurosci. **49**(3), 328–338 (2019). https://doi.org/10.1111/ejn.14094
4. Gruffaz, S., Poulet, P.E., Maheux, E., Jedynak, B., Durrleman, S.: Learning Riemannian metric for disease progression modeling. In: Advances in Neural Information Processing Systems, vol. 34, pp. 23780–23792. Curran Associates, Inc. (2021)
5. Jack, C.R., et al.: Tracking pathophysiological processes in Alzheimer's disease: an updated hypothetical model of dynamic biomarkers. Lancet Neurol. **12**(2), 207–216 (2013). https://doi.org/10.1016/S1474-4422(12)70291-0
6. Jutten, R.J., Sikkes, S.A., Van der Flier, W.M., Scheltens, P., Visser, P.J., Tijms, B.M.: for the Alzheimer's disease neuroimaging initiative: finding treatment effects in Alzheimer trials in the face of disease progression heterogeneity. Neurology **96**(22), e2673–e2684 (2021). https://doi.org/10.1212/WNL.0000000000012022
7. Komarova, N.L., Thalhauser, C.J.: High degree of heterogeneity in Alzheimer's disease progression patterns. PLoS Comput. Biol. **7**(11), e1002251 (2011). https://doi.org/10.1371/journal.pcbi.1002251
8. Koval, I., et al.: AD course map charts Alzheimer's disease progression. Sci. Rep. **11**(1), 8020 (2021). https://doi.org/10.1038/s41598-021-87434-1
9. Kunkle, B.W., et al.: Genetic meta-analysis of diagnosed Alzheimer's disease identifies new risk loci and implicates Aβ, tau, immunity and lipid processing. Nat. Genet. **51**(3), 414–430 (2019). https://doi.org/10.1038/s41588-019-0358-2
10. Livingston, G., et al.: Dementia prevention, intervention, and care: 2020 report of the lancet commission. The Lancet **396**(10248), 413–446 (2020). https://doi.org/10.1016/S0140-6736(20)30367-6

11. Maheux, E., et al.: Forecasting individual progression trajectories in Alzheimer's disease. Nat. Commun. **14**(1), 761 (2023). https://doi.org/10.1038/s41467-022-35712-5
12. Meisl, G., et al.: In vivo rate-determining steps of tau seed accumulation in Alzheimer's disease. Sci. Adv. **7**(44), eabh1448 (2021). https://doi.org/10.1126/sciadv.abh1448
13. Poulet, P.-E., Durrleman, S.: Mixture modeling for identifying subtypes in disease course mapping. In: Feragen, A., Sommer, S., Schnabel, J., Nielsen, M. (eds.) IPMI 2021. LNCS, vol. 12729, pp. 571–582. Springer, Cham (2021). https://doi.org/10.1007/978-3-030-78191-0_44
14. Raket, L.L.: Statistical disease progression modeling in alzheimer disease. Front. Big Data **3** (2020). https://doi.org/10.3389/fdata.2020.00024
15. Sauty, B., Durrleman, S.: Riemannian metric learning for progression modeling of longitudinal datasets. In: 2022 IEEE 19th International Symposium on Biomedical Imaging (ISBI), pp. 1–5 (Mar 2022). https://doi.org/10.1109/ISBI52829.2022.9761641
16. Schiratti, J.B., Allassonnière, S., Colliot, O., Durrleman, S.: A Bayesian mixed-effects model to learn trajectories of changes from repeated manifold-valued observations. J. Mach. Learn. Res. **18**(1), 4840–4872 (2017)
17. Young, A.L., et al.: Uncovering the heterogeneity and temporal complexity of neurodegenerative diseases with subtype and stage inference. Nat. Commun. **9**(1), 4273 (2018). https://doi.org/10.1038/s41467-018-05892-0

Visual Grounding of Whole Radiology Reports for 3D CT Images

Akimichi Ichinose[1(✉)], Taro Hatsutani[1], Keigo Nakamura[1], Yoshiro Kitamura[1], Satoshi Iizuka[2], Edgar Simo-Serra[3], Shoji Kido[4], and Noriyuki Tomiyama[4]

[1] Medical Systems Research and Development Center, FUJIFILM Corporation, Tokyo, Japan
akimichi.ichinose@fujifilm.com

[2] Center for Artificial Intelligence Research, University of Tsukuba, Ibaraki, Japan

[3] Department of Computer Science and Engineering, Waseda University, Tokyo, Japan

[4] Graduate School of Medicine, Osaka University, Osaka, Japan

Abstract. Building a large-scale training dataset is an essential problem in the development of medical image recognition systems. Visual grounding techniques, which automatically associate objects in images with corresponding descriptions, can facilitate labeling of large number of images. However, visual grounding of radiology reports for CT images remains challenging, because so many kinds of anomalies are detectable via CT imaging, and resulting report descriptions are long and complex. In this paper, we present the first visual grounding framework designed for CT image and report pairs covering various body parts and diverse anomaly types. Our framework combines two components of 1) anatomical segmentation of images, and 2) report structuring. The anatomical segmentation provides multiple organ masks of given CT images, and helps the grounding model recognize detailed anatomies. The report structuring helps to accurately extract information regarding the presence, location, and type of each anomaly described in corresponding reports. Given the two additional image/report features, the grounding model can achieve better localization. In the verification process, we constructed a large-scale dataset with region-description correspondence annotations for 10,410 studies of 7,321 unique patients. We evaluated our framework using grounding accuracy, the percentage of correctly localized anomalies, as a metric and demonstrated that the combination of the anatomical segmentation and the report structuring improves the performance with a large margin over the baseline model (66.0% vs 77.8%). Comparison with the prior techniques also showed higher performance of our method.

Keywords: Deep Learning · Vision Language · Visual Grounding · Computed Tomography

Supplementary Information The online version contains supplementary material available at https://doi.org/10.1007/978-3-031-43904-9_59.

H. Greenspan et al. (Eds.): MICCAI 2023, LNCS 14224, pp. 611–621, 2023.
https://doi.org/10.1007/978-3-031-43904-9_59

1 Introduction

In recent years, a number of medical image recognition systems have been developed [6] to alleviate the increasing burden on radiologists [2,21,22]. In the development of such systems, the task of manually labeling images is a significant bottleneck. Auto-labeling, the process of automatically assigning labels to images using machine learning algorithms, has emerged as a promising solution to this problem. In cases where there are plenty of image and caption pairs, one potential approach to auto-labeling is visual grounding [12], which utilizes natural language descriptions to identify and localize objects in images.

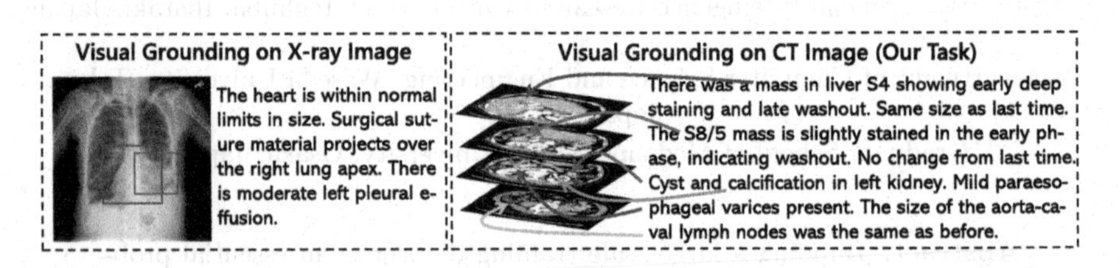

Fig. 1. Comparison of the visual grounding task on X-ray image and on CT image.

With the recent advances in cross-modal technology based on deep learning, many frameworks for visual grounding has been proposed [7,11]. Within the medical domain, several large scale datasets with radiology reports are available (e.g. OpenI [3], MIMIC-CXR [9]), and these produced researches on medical image visual grounding [1,25]. However, to the best of our knowledge, prior studies have focused on 2D X-ray images [28] or videos [15], and there has been no research applying visual grounding to 3D computed tomography (CT) images so far. Visual grounding on CT images has the following difficulties: 1) **Large number of anomaly types to detect**: Existing researches on visual grounding using X-ray images handled only chest X-ray images. The number of anomaly types to detect is at most dozen or so (e.g. 13 findings [8]). In contrast, our research handles CT images including various parts of the human body. Consequently, the number of anomaly types to be detected is larger than one hundred. 2) **Long and complex sentences**: Radiology reports on X-ray images are often simple, noting only the presence or absence of anomalies. On the other hand, in CT examinations, the qualitative diagnosis of each anomaly is often performed. In cases, multiple anomalies are simultaneously described in a sentence. Therefore, the description tend to be long and complicated with multiple sentences (Fig. 1). Visual grounding for CT images requires the extraction of information about the location and type of each anomaly from these complex sentences.

In this work, we propose a novel visual grounding framework for 3D CT images and radiology reports. The main idea is to separate the task into three parts: 1) anatomical segmentation on images, 2) report structuring, and 3) localization of described anomalies. In the anatomical segmentation, multiple organs and tissues are extracted using the deep learning based segmentation model and provided as landmarks. The report structuring model, which is based on BERT [5], is also introduced to extract information of each anomaly from a complex report. Both of these features are fed into the grounding model (3) to extrapolate medical domain knowledge, thereby enabling accurate visual grounding.

Our contributions are as follows:

- We show the first visual grounding results for 3D CT images that covers various body parts and anomalies.
- We introduce a novel grounding architecture that can leverage report structuring results of presence/type/location of described anomalies.
- We validate the efficacy of the proposed framework using a large-scale dataset with region-description correspondence annotations.

2 Related Work

Visual Grounding. Visual grounding task involves learning the correspondences between descriptions in the text and image regions from a given training set of region-description pairs [12]. There are mainly two approaches: one-stage approach and two-stage approach. Most studies follow a two-stage approach [14,17]. However, this approach usually employs a pre-trained object detector, and it leads to restrict the capability of categories and attributes in grounding. Accordingly, recent studies is shifting to employ the one-stage approach, in which visual grounding is performed by end-to-end training [4,10,27].

Vision-Language Tasks on Medical Image. The existence of public datasets with paired images and reports [3,9,26] has accelerated research on cross-modal tasks in the medical field [16,25]. Inspired by the success of visual grounding, several studies of visual grounding for medical images and radiology reports have also been reported [1,23,28]. These studies utilized a large scale dataset and an attention-based language interpretation model such as BERT [5] to ground the descriptions in the report. However, these studies have focused on X-ray images, and to the best of our knowledge, there have been no studies on CT images, which cover the entire body and have a complex report.

3 Methods

We first formulate the problem. Next, we explain three key components of anatomical segmentation, report structuring, and anomaly localization in our framework. In our framework, multiple organ labels obtained as the output of anatomical segmentation encourage the grounding model to learn detailed anatomy, and report structuring allows the grounding model to accurately extract the features of the target anomaly from complex sentences.

3.1 Problem Formulation

Our research assumes that a dataset of image-report pairs with region-description correspondence annotations is provided for training. We show the overall framework in Fig. 2. We denote an image and a paired report as I and T respectively. Let I_a be a label image in which multiple organs are extracted from I. Each report T contains descriptions of multiple (image) anomalies. We denote each anomalies as $t_i \in \{t_1, t_2, ..., t_N\}$. Given an image I and corresponding organ label images I_a encoded as $V \in \mathbb{R}^{d_z \times d_y \times d_x \times d}$ and a description about an anomaly t_i encoded as $L_{t_i} \in \mathbb{R}^d$, the goal of our framework is to generate a segmentation map M_{t_i} that represents the location of the anomaly t_i.

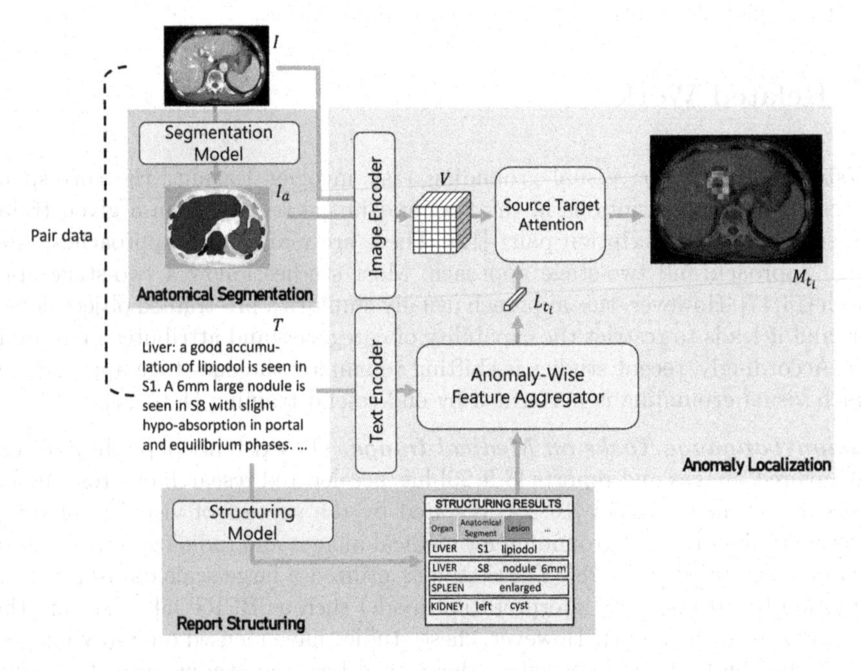

Fig. 2. The proposed framework for 3D-CT visual grounding.

3.2 Anatomical Segmentation

The task of the anatomical segmentation is to extract relevant anatomies that can be clues for visual grounding. We use the commercial version of the 3D image analysis software (Synapse 3D V6.8, FUJIFILM corporation, Japan) to extract 32 organs and tissues (See Appendix Table. A1). In this software, anatomies are extracted using U-Net based architectures [13,18]. The extracted anatomical label images are I_a.

3.3 Report Structuring

The tasks of the report structuring are as follows: 1) anatomical prediction, 2) phrase recognition, and 3) relationship estimation between phrases (See Appendix Fig. A1). The anatomical prediction is a sentence-wise prediction to determine which organ or body part is mentioned in each sentence. The organs and body parts to be recognized are shown in Appendix Table. A2. The sentences belonging to the same class are concatenated, then the phrase recognition and the relationship estimation are performed for each class.

The phrase recognition module extracts phrases and classifies each of them into 9 classes (See Appendix Table. A2). Subsequently, the relationship estimation module determines whether there is a relationship between anomaly phrases (e.g. 'nodule', 'fracture') and other phrases (e.g. '6mm', 'Liver S6'), resulting in the grouping of phrases related to the same anomaly. If multiple anatomical phrases are grouped in the same group, they are split into separate groups on a rule basis (e.g. ['right S1', 'left S6', 'nodule'] -> ['right S1', 'nodule'], ['left S6', 'nodule']). More details of implementation and training methods are reported in Nakano et al. [20] and Tagawa et al [24].

3.4 Anomaly Localization

The task of the anomaly localization is to output a localization map of the anomaly mentioned in the input report T. The CT image I and the organ label image I_a are concatenated along the channel dimension and encoded by a convolutional backbone to generate a visual embedding V. The sentences in the report T are encoded by BERT [5] to generate embeddings for each character. Let $r = \{r_1, r_2, ..., r_{N_C}\}$ be the set of character embeddings where N_C is the number of characters. Our framework next adopt the Anomaly-Wise Feature Aggregator (AFA). For each anomaly t_i, AFA generates a representative embedding L_{t_i} by aggregating the embeddings of related phrases based on report structuring results. The final grounding result M_{t_i} is obtained by the following Source-Target Attention.

$$M_{t_i} = \mathrm{sigmoid}(L_{t_i} W_Q (V W_K)^T) \tag{1}$$

where W_Q, $W_K \in \mathbb{R}^{d \times d_n}$ are trainable variables.

The overall architecture of this module is illustrated in Appendix Fig. A2.

Anomaly-Wise Feature Aggregator. The results of the report structuring $m_{t_i} \in \mathbb{R}^{N_C}$ are defined as follows:

$$m_{t_{ij}} = \begin{cases} c_j & \text{if a } j\text{-th character is related to an anomaly } t_i, \\ 0 & \text{else.} \end{cases} \tag{2}$$

$$m_{t_i} = \{m_{t_{i1}}, m_{t_{i2}}, ... m_{t_{iN_C}}\} \tag{3}$$

where c_j is the class index labeled by the phrase recognition module (Let C be the number of classes). In this module, aggregate character-wise embeddings based on the following formula.

$$e_k = \{r_j | m_{t_{ij}} = k\} \tag{4}$$

$$L_{t_i} = \text{LSTM}([v_{organ}; p_1; e_1; p_2; e_2; ..., p_C; e_C]) \tag{5}$$

where v_{organ} and p_k are trainable embeddings for each organ and each class label respectively. $[\cdot;\cdot]$ stands for concatenation operation. In this way, embeddings of characters related to the anomaly t_i are aggregated and concatenated. Subsequently, representative embeddings of the anomaly are generated by an LSTM layer. In the task of visual grounding focused on 3D CT images, the size of the dataset that can be created is relatively small. Considering this limitation, we use an LSTM layer with strong inductive bias to achieve high generalization performance.

4 Dataset and Implementation Details

4.1 Clinical Data

We retrospectively collected 10,410 CT studies (11,163 volumes/7,321 unique patients) and 671,691 radiology reports from one university hospital in Japan. We assigned a bounding box to each anomaly described in the reports as shown in Appendix Fig. A3. The total category number is about 130 in combination of anatomical regions and anomaly types (The details are in Fig. 4) For each anomaly, a correspondence annotation was made with anomaly phrases in the report. The total number of annotated regions is 17,536 (head: 713 regions, neck: 285 regions, chest: 8,598 regions, and abdomen: 7,940 regions). We divide the data into 9,163/1,000/1,000 volumes as a training/validation/test split.

4.2 Implementation Details

We use a VGG-like network as Image Encoder, with 15 3D-convolutional layers and 3 max pooling layers. For training, the voxel spacings in all three dimensions are normalized to 1.0 mm. CT values are linearly normalized to obtain a value of [0–1]. The anatomy label image, in which only one label is assigned to each voxel, is also normalized to the value [0–1], and the CT image and the label image are concatenated along the channel dimension. As our Text Encoder, we use a BERT with 12 transformer encoder layers, each with hidden dimension of 768 and 12 heads in the multi-head attention. At first, we pre-train the BERT using 6.7M sentences extracted from the reports in a Masked Language Model task. Then we train the whole architecture jointly using dice loss [19] with the first 8 transformer encoder layers of the BERT frozen. Further information about implementation are shown in Appendix Table. A3.

5 Experiments

We did two kinds of experiments for comparison and ablation studies. The comparison study was made against TransVG [4] and MDETR [10] that are one-stage visual grounding approaches and established state-of-the-art performances on photos and captions. To adapt TransVG and MDETR for the 3D modality, the backbone was changed to a VGG-like network with 3D convolution layers, the same as the proposed method. We refer one of the proposed method without anatomical segmentation and report structuring as the baseline model.

5.1 Evaluation Metrics

We report segmentation performance using Dice score, mean intersection over union (mIoU), and the grounding accuracy. The output masks are thresholded to compute mIoU and grounding accuracy score. The mIoU is defined as an average IoU over the thresholds [0.1, 0.2, 0.3, 0.4, 0.5]. The grounding accuracy is defined as the percentage of anomalies for which the IoU exceeds 0.1 under the threshold 0.1.

5.2 Results

The experimental results of the two studies are shown in Table. 1. Both of MDETR and TransVG failed to achieve stable grounding in this task. A main difference between these models and our baseline model is using a source-target attention layer instead of the transformer. It is known that a transformer-based algorithm with many parameters and no strong inductive bias is difficult to generalize with such a relatively limited number of training data. For this reason, the baseline model achieved a much higher accuracy than the comparison methods.

Table 1. Results of the comparison/ablation studies. '-' represents 'not converged'.

Method	Anatomical Seg.	Report Struct.	Dice [%]	mIoU [%]	Accuracy [%]
MDETR [10]	–	–	N/A	–	–
TransVG [4]	–	–	N/A	8.5	21.8
Baseline	✗	✗	27.4	15.6	66.0
Proposed	✓	✗	28.1	16.6	67.9
	✗	✓	33.0	20.3	75.9
	✓	✓	**34.5**	**21.5**	**77.8**

The ablation study showed that the anatomical segmentation and the report structuring can improve the performance. In Fig. 3 (upper row), we demonstrate several cases that facilitate an intuitive understanding of each effect. Longer reports often mention more than one anomaly, making it difficult to recognize the grounding target and cause localization errors. The proposed method can explicitly indicate phrases such as the location and size of the target anomaly, reducing the risk of failure. Figure 3 (lower row) shows examples of grounding results when a query that is not related to the image is inputted. In this case, the grounding results were less consistent with the anatomical phrases. The results suggest that the model performs grounding with an emphasis on anatomical information against the backdrop of abundant anatomical knowledge.

The grounding performance for each combination of organ and anomaly type is shown in Fig. 4. The performance is relatively high for organ shape abnormalities (e.g. swelling, duct dilation) and high-frequency anomalies in small organs (e.g. thyroid/prostate mass). For these anomaly types, our model is considered to be available for automatic training data generation. On the other hand, the performance tends to be low for rare anomalies (e.g. mass in small intestine) and anomalies in large body part (e.g. limb). Improving grounding performance for these targets will be an important future work.

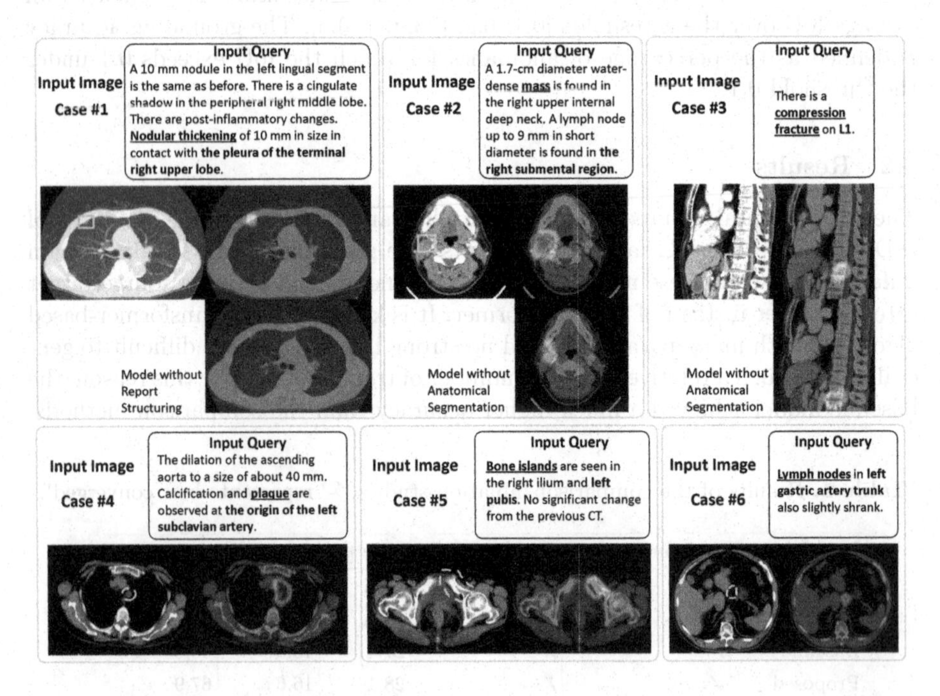

Fig. 3. The grounding results for several input queries. Underlines in the input query indicate the target anomaly phrase to be grounded. The phrases highlighted in bold blue indicate the anatomical locations of the target anomaly. The red rectangles indicate the ground truth regions. Case #4-#6 are the grounding results when an unrelated input query is inputted. The region surrounded by the red dashed line indicates the anatomical location corresponding to the input query.

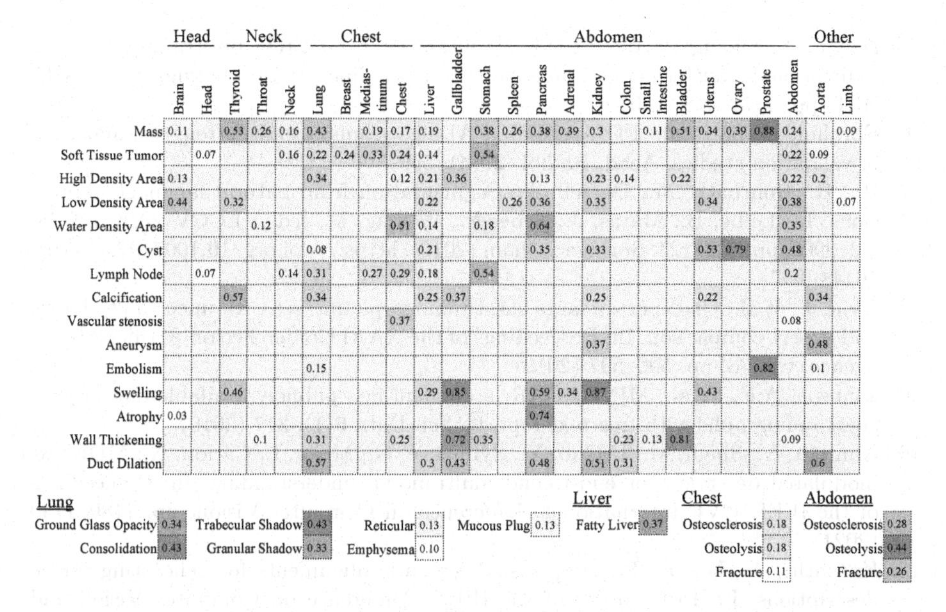

Fig. 4. Grounding performance for representative anomalies. The value in each cell is the average dice score of the proposed method.

6 Conclusion

In this paper, we proposed the first visual grounding framework for 3D CT images and reports. To deal with various type of anomalies throughout the body and complex reports, we introduced a new approach using anatomical recognition results and report structuring results. The experiments showed the effectiveness of our approach and achieved higher performance compared to prior techniques. However, in clinical practice, radiologists write reports from comparing multiple images such as time-series images, or multi-phase scans. Realizing such sophisticated diagnose process by a visual grounding model will be a future research.

References

1. Bhalodia, R., et al.: Improving pneumonia localization via cross-attention on medical images and reports. In: de Bruijne, M., et al. (eds.) MICCAI 2021. LNCS, vol. 12902, pp. 571–581. Springer, Cham (2021). https://doi.org/10.1007/978-3-030-87196-3_53
2. Dall, T.: The Complexities of Physician Supply and Demand: Projections from 2016 to 2030. IHS Markit Limited (2018)
3. Demner-Fushman, D., et al.: Preparing a collection of radiology examinations for distribution and retrieval. J. Am. Med. Inform. Assoc. **23**(2), 304–310 (2016)
4. Deng, J., Yang, Z., Chen, T., Zhou, W., Li, H.: TransVG: end-to-end Visual Grounding With Transformers. In: Proceedings of the IEEE/CVF International Conference on Computer Vision, pp. 1769–1779 (2021)

5. Devlin, J., Chang, M.W., Lee, K., Toutanova, K.: BERT: pre-training of deep bidirectional transformers for language understanding. In: Proceedings of NAACL-HLT, pp. 4171–4186 (2019)
6. Ebrahimian, S., et al.: FDA-regulated AI algorithms: trends, strengths, and gaps of validation studies. Acad. Radiol. **29**(4), 559–566 (2022)
7. Hu, R., Rohrbach, M., Darrell, T.: Segmentation from natural language expressions. In: Leibe, B., Matas, J., Sebe, N., Welling, M. (eds.) ECCV 2016. LNCS, vol. 9905, pp. 108–124. Springer, Cham (2016). https://doi.org/10.1007/978-3-319-46448-0_7
8. Irvin, J., et al.: CheXpert: a large chest radiograph dataset with uncertainty labels and expert comparison. In: Proceedings of the AAAI Conference on Artificial Intelligence, vol. 33, pp. 590–597 (2019)
9. Johnson, A.E., et al.: MIMIC-CXR, a de-identified publicly available database of chest radiographs with free-text reports. Sci. Data **6**(1), 317 (2019)
10. Kamath, A., Singh, M., LeCun, Y., Synnaeve, G., Misra, I., Carion, N.: MDETR-modulated detection for end-to-end multi-modal understanding. In: Proceedings of the IEEE/CVF International Conference on Computer Vision, pp. 1780–1790 (2021)
11. Karpathy, A., Fei-Fei, L.: Deep visual-semantic alignments for generating image descriptions. In: Proceedings of the IEEE Conference on Computer Vision and Pattern Recognition, pp. 3128–3137 (2015)
12. Karpathy, A., Joulin, A., Fei-Fei, L.F.: Deep fragment embeddings for bidirectional image sentence mapping. In: Proceedings of Advances in Neural Information Processing System, pp. 1889–1897 (2014)
13. Keshwani, D., Kitamura, Y., Li, Y.: Computation of total kidney volume from ct images in autosomal dominant polycystic kidney disease using multi-task 3D convolutional neural networks. In: Shi, Y., Suk, H.-I., Liu, M. (eds.) MLMI 2018. LNCS, vol. 11046, pp. 380–388. Springer, Cham (2018). https://doi.org/10.1007/978-3-030-00919-9_44
14. Lee, K.H., Chen, X., Hua, G., Hu, H., He, X.: Stacked cross attention for image-text matching. In: Proceedings of the European Conference on Computer Vision, pp. 201–216 (2018)
15. Li, B., Weng, Y., Sun, B., Li, S.: Towards visual-prompt temporal answering grounding in medical instructional video. arXiv preprint arXiv:2203.06667 (2022)
16. Li, Y., Wang, H., Luo, Y.: A comparison of pre-trained vision-and-language models for multimodal representation learning across medical images and reports. In: Proceedings of the IEEE International Conference on Bioinformatics and Biomedicine, pp. 1999–2004. IEEE (2020)
17. Lu, J., Batra, D., Parikh, D., Lee, S.: ViLBERT: pretraining task-agnostic visiolinguistic representations for vision-and-language tasks. Adv. Neural. Inf. Process. Syst. **32**, 13–23 (2019)
18. Masuzawa, N., Kitamura, Y., Nakamura, K., Iizuka, S., Simo-Serra, E.: Automatic segmentation, localization, and identification of vertebrae in 3d ct images using cascaded convolutional neural networks. In: Martel, A.L., et al. (eds.) MICCAI 2020. LNCS, vol. 12266, pp. 681–690. Springer, Cham (2020). https://doi.org/10.1007/978-3-030-59725-2_66
19. Milletari, F., Navab, N., Ahmadi, S.A.: V-Net: fully convolutional neural networks for volumetric medical image segmentation. In: Proceedings of the International Conference on 3D Vision, pp. 565–571. IEEE (2016)

20. Nakano, N., et al.: Pre-training methods for creating a language model with embedded knowledge of radiology reports. In: Proceedings of the annual meeting of the Association for Natural Language Processing (2022)
21. Nishie, A., et al.: Current radiologist workload and the shortages in Japan: how many full-time radiologists are required? Jpn. J. Radiol. **33**, 266–272 (2015)
22. Rimmer, A.: Radiologist shortage leaves patient care at risk, warns royal college. BMJ: British Med. J. (Online) 359 (2017)
23. Seibold, C., et al.: Detailed Annotations of Chest X-Rays via CT Projection for Report Understanding. arXiv preprint arXiv:2210.03416 (2022)
24. Tagawa, Y., et al.: Performance improvement of named entity recognition on noisy data using teacher-student training. In: Proceedings of the annual meeting of the Association for Natural Language Processing (2022)
25. Wang, X., Peng, Y., Lu, L., Lu, Z., Summers, R.M.: TieNet: text-image embedding network for common thorax disease classification and reporting in Chest X-rays. In: Proceedings of the IEEE Conference on Computer Vision and Pattern Recognition, pp. 9049–9058 (2018)
26. Yan, K., Wang, X., Lu, L., Summers, R.M.: DeepLesion: automated mining of large-scale lesion annotations and universal lesion detection with deep learning. J. Med. Imaging **5**(3), 036501–036501 (2018)
27. Yang, Z., Gong, B., Wang, L., Huang, W., Yu, D., Luo, J.: A fast and accurate one-stage approach to visual grounding. In: Proceedings of the IEEE/CVF International Conference on Computer Vision, pp. 4683–4693 (2019)
28. You, D., Liu, F., Ge, S., Xie, X., Zhang, J., Wu, X.: AlignTransformer: hierarchical alignment of visual regions and disease tags for medical report generation. In: de Bruijne, M., et al. (eds.) MICCAI 2021. LNCS, vol. 12903, pp. 72–82. Springer, Cham (2021). https://doi.org/10.1007/978-3-030-87199-4_7

Identification of Disease-Sensitive Brain Imaging Phenotypes and Genetic Factors Using GWAS Summary Statistics

Duo Xi, Dingnan Cui, Jin Zhang, Muheng Shang, Minjianan Zhang, Lei Guo, Junwei Han(✉), and Lei Du(✉)

School of Automation, Northwestern Polytechnical University, Xi'an 710072, China
{jhan,dulei}@nwpu.edu.cn

Abstract. Brain imaging genetics is a rapidly growing neuroscience area that integrates genetic variations and brain imaging phenotypes to investigate the genetic underpinnings of brain disorders. In this field, using multi-modal imaging data can leverage complementary information and thus stands a chance of identifying comprehensive genetic risk factors. Due to privacy and copyright issues, many imaging and genetic data are unavailable, and thus existing imaging genetic methods cannot work. In this paper, we proposed a novel multi-modal brain imaging genetic learning method that can study the associations between imaging phenotypes and genetic variations using genome-wide association study (GWAS) summary statistics. Our method leverages the powerful multi-modal of brain imaging phenotypes and GWAS. More importantly, it does not need to access the imaging and genetic data of each individual. Experimental results on both Alzheimer's Disease Neuroimaging Initiative (ADNI) database and GWAS summary statistics suggested that our method has the same learning ability, including identifying associations between genetic biomarkers and imaging phenotypes and selecting relevant biomarkers, as those counterparts depending on the individual data. Therefore, our learning method provides a novel methodology for brain imaging genetics without individual data.

Keywords: Brain imaging genetics · GWAS summary statistics · Multi-modal brain image analysis

L. Du—This work was supported in part by National Natural Science Foundation of China [61973255, 62136004, 61936007] at Northwestern Polytechnical University.
Alzheimer's Disease Neuroimaging Initiative: Data used in preparation of this article were obtained from the Alzheimer's Disease Neuroimaging Initiative (ADNI) database (adni.loni.usc.edu). As such, the investigators within the ADNI contributed to the design and implementation of ADNI and/or provided data but did not participate in analysis or writing of this report. A complete listing of ADNI investigators can be found at: http://adni.loni.usc.edu/wp-content/uploads/how_to_apply/ADNI_Acknowledgement_List.pdf.

H. Greenspan et al. (Eds.): MICCAI 2023, LNCS 14224, pp. 622–631, 2023.
https://doi.org/10.1007/978-3-031-43904-9_60

1 Introduction

Nowadays, brain imaging genetics has gained increasing attention in the neuroscience area. This interdisciplinary field refers to integrates genetic variations (single nucleotide polymorphisms, SNPs) and structural or functional neuroimaging quantitative traits (QTs). Different imaging technologies can capture different knowledge of the brain and thus are a better choice in brain imaging genetics [12,17]. Over the past decade, genome-wide association studies (GWAS) have proven to be a powerful tool in finding the genetic effects on imaging phenotypes in single SNP level [7,10,14]. However, GWAS can only investigate the single-SNP-single-QT relationship, and thus may lose the information among multiple SNPs and/or multiple QTs due to the polygenic inheritance of brain disorders [9].

To leverage the multi-modal brain imaging QTs and identify the joint effect of multiple SNPs, many learning methods were proposed for multi-modal brain imaging genetics [5,6,15]. The dirty multi-task sparse canonical correlation analysis (DMTSCCA) is a bi-multivariate learning method for multi-modal brain imaging genetics [4]. DMTSCCA can disentangle the specific patterns of multi-modal imaging QTs from shared ones and thus is state-of-the-art. However, DMTSCCA depends on individual-level imaging and genetic data, and cannot work when the original imaging and genetic data are unavailable.

Since GWAS studies usually release their summary statistics results for academic use, we here developed a novel DMTSCCA method using GWAS summary statistics rather than individual data. The method, named S-DMTSCCA, has the same ability as DMTSCCA in modeling the association between multi-modal imaging QTs and SNPs and does not require raw imaging and genetic data. We investigated the performance of S-DMTSCCA based on two kinds of experiments. Firstly, we applied S-DMTSCCA to GWAS summary statistics from Alzheimer's Disease Neuroimaging Initiative (ADNI) and compared it to DMTSCCA which directly ran on the original imaging genetic data of ADNI. Results suggested that S-DMTSCCA and DMTSCCA obtained equivalent results. Secondly, we applied S-DMTSCCA to a GWAS summary statistics from the UK Biobank. The experiment results showed that S-DMTSCCA can identify meaningful genetic markers for brain imaging QTs. More importantly, the structure information of SNPs was also captured which was usually missed by GWAS. It is worth noting that all these results were obtained without assessing the original neuroimaging genetic data. This demonstrates that our method is a powerful tool and provides a novel method for brain imaging genetics.

2 Method

In this article, we represent scalars with italicized letters, column vectors with boldface lowercase letters, and matrices with boldface capitals. For $\mathbf{X} = (x_{ij})$, the ith row is denoted as $\mathbf{x}^i$, jth column as $\mathbf{x}_j$, and the ith matrix as $\mathbf{X}_i$. $\|\mathbf{X}\|_2$ denotes the Euclidean norm, $\|\mathbf{X}\|_{2,1}$ denotes the sum of the Euclidean norms of the rows of $\mathbf{X}$. Suppose $\mathbf{X} \in \mathbb{R}^{n \times p}$ load the genetic data with n subjects and p biomarkers, and $\mathbf{Y}_c \in \mathbb{R}^{n \times q} (c = 1, ..., C)$ load the cth modality of phenotype data, where q and C is the number of imaging QTs and imaging modalities (tasks) respectively.

2.1 DMTSCCA

DMTSCCA identifies the genotype-phenotype associations between SNPs and multi-modal imaging QTs using the following model [4],

$$\begin{aligned} \min_{\mathbf{S},\mathbf{W},\mathbf{v}_c} \sum_{c=1}^{C} \left[\kappa_c \|\mathbf{X}(\mathbf{s}_c + \mathbf{w}_c) - \mathbf{Y}_c \mathbf{v}_c\|_2^2 + \lambda_v \|\mathbf{v}_c\|_1\right] + \lambda_s \|\mathbf{S}\|_{2,1} + \lambda_w \|\mathbf{W}\|_{1,1} \\ s.t.\ \|\mathbf{X}(\mathbf{s}_c + \mathbf{w}_c)\|_2^2 = 1, \|\mathbf{Y}_c \mathbf{v}_c\|_2^2 = 1, \forall c. \end{aligned} \tag{1}$$

In this model, $\kappa \in \mathbb{R}^{1 \times C}$ $(0 \leq \kappa_c \leq 1, \sum_c \kappa_c = 1)$ is a weight vector to balance among multiple sub-tasks. In this paper, κ ensures an equal optimization for each imaging modality. $\mathbf{v}_c$ is the cth vector in $\mathbf{V}$, where $\mathbf{V} = [\mathbf{v}_1, ..., \mathbf{v}_C] \in \mathbb{R}^{q \times C}$ denotes the canonical weight for phenotypic data. $\mathbf{S} \in \mathbb{R}^{p \times C}$ and $\mathbf{W} \in \mathbb{R}^{p \times C}$ are the canonical weights for genotypes, where $\mathbf{S}$ is the task-consistent component being shared by all tasks and $\mathbf{W}$ is the task-dependent component being associated with a single task. λ_v, λ_s and λ_w are nonnegative tuning parameters.

2.2 Summary-DMTSCCA (S-DMTSCCA)

Now we propose S-DMTSCCA only using summary statistics from GWAS. It does not need individual-level imaging and genetic data.

For ease of presentation, we derive our method by first introducing GWAS. GWAS uses linear regression to study the effect of a single SNP on a single imaging QT. Let $\mathbf{x}_d$ denotes the genotype data with p SNPs and $\mathbf{y}_l$ denotes the phenotype data with q imaging QTs, a typical GWAS model can be defined as

$$\mathbf{y}_l = \alpha + \mathbf{x}_d b_{dl} + \epsilon, \tag{2}$$

where b_{dl} is the effect size of the d-th SNP on the l-th imaging QT. α is the y-intercept, and ϵ is the error term which is independent of $\mathbf{x}_d$. When the SNPs and imaging QTs were normalized to have zero mean and unit variance, b_{dl} will equal to the covariance between $\mathbf{x}_d$ and $\mathbf{y}_l$, i.e., $b_{dl} = \frac{\mathbf{x}_d^T \mathbf{y}_l}{n-1}$. On this account, we can construct $\mathbf{B} \in \mathbb{R}^{p \times q}$ by loading $p \times q$ summary statistics of GWAS. Obviously, $\mathbf{B}$ will be the covariance between multiple SNPs and multiple imaging QTs since its element is the covariance of a single SNP and a single imaging QT. Let $\mathbf{B}_c$ denotes covariance of the c-th modality from GWAS, we have

$$\mathbf{X}^T \mathbf{Y}_c = (n-1)\,\mathbf{B}_c. \tag{3}$$

Further, we use $\hat{\Sigma}_{XX}$ denote an estimated covariance of genetic data, i.e., $\hat{\Sigma}_{XX} = \frac{\overline{\mathbf{X}}^T \overline{\mathbf{X}}}{n-1}$. $\overline{\mathbf{X}}$ can be obtained from n subjects of the same or similar population since we do not have the original data.

According the phenotype correlation [8], the covariance of phenotype data of the c-th modality can be calculated by

$$\hat{\Sigma}_{YYc} = \frac{\mathbf{Y}_c^T \mathbf{Y}_c}{n-1} = corr\left(\mathbf{B}_c\right). \tag{4}$$

Let $\mathbf{s}_c^*$, $\mathbf{w}_c^*$ and $\mathbf{v}_c^*$ denote the final results, we will present how to solve them only using GWAS results ($\mathbf{B}$) and several subjects of a public reference database.

Solving S and W: Since our method is bi-convex, we can solve one variable by fixing the remaining variables as constants. The model of S-DMTSCCA and DMTSCCA are the same [4], and thus we can solve each $\hat{\mathbf{s}}_c$ and $\hat{\mathbf{w}}_c$ by substituting Σ_{XX}, Σ_{YY}, Σ_{XY} and Σ_{YX}. Specifically, we have the following closed-form solution,

$$\hat{\mathbf{s}}_c = \frac{(n-1)\mathbf{B}_c\mathbf{v}_c}{(n-1)\hat{\Sigma}_{XX}+\frac{\lambda_s}{\kappa_c}\mathbf{D}} = \frac{\mathbf{B}_c\mathbf{v}_c}{\hat{\Sigma}_{XX}+\frac{\lambda_s}{\kappa_c}\tilde{\mathbf{D}}}, \tag{5}$$

$$\hat{\mathbf{w}}_c = \frac{(n-1)\mathbf{B}_c\mathbf{v}_c}{(n-1)\hat{\Sigma}_{XX}+\frac{\lambda_w}{\kappa_c}\breve{\mathbf{D}}_c} = \frac{\mathbf{B}_c\mathbf{v}_c}{\hat{\Sigma}_{XX}+\frac{\lambda_w}{\kappa_c}\tilde{\mathbf{D}}_c}. \tag{6}$$

In both equations, $\mathbf{D}$ and $\tilde{\mathbf{D}}$ are diagonal matrices, and their i-th diagonal element are $\frac{1}{2\|\mathbf{s}^i\|_2}$ and $\frac{1}{2(n-1)\|\mathbf{s}^i\|_2}$ for $i = 1, ..., p$ respectively. $\breve{\mathbf{D}}_c$ and $\tilde{\mathbf{D}}_c$ are also diagonal matrices, and their i-th diagonal element are $\frac{1}{2|w_{ic}|}$ and $\frac{1}{2(n-1)|w_{jc}|}$ for $i = 1, ..., p$.

Finally, to satisfy the equality constraints, $\mathbf{S}$ and $\mathbf{W}$ are respectively scaled by

$$\mathbf{s}_c^* = \frac{\hat{\mathbf{s}}_c}{\sqrt{(n-1)(\hat{\mathbf{s}}_c+\hat{\mathbf{w}}_c)^T\hat{\Sigma}_{XX}(\hat{\mathbf{s}}_c+\hat{\mathbf{w}}_c)}}, \; \mathbf{w}_c^* = \frac{\hat{\mathbf{w}}_c}{\sqrt{(n-1)(\hat{\mathbf{s}}_c+\hat{\mathbf{w}}_c)^T\hat{\Sigma}_{XX}(\hat{\mathbf{s}}_c+\hat{\mathbf{w}}_c)}}. \tag{7}$$

Solving V: If $\mathbf{S}$ and $\mathbf{W}$ are solved, we can fix them to solve for $\mathbf{V}$. In line with Eqs. (5–6), substituting Eq. (3) and Eq. (4) into the equation of $\hat{\mathbf{v}}_c$ in [4], we can get the solution formulas for $\mathbf{V}$, i.e.,

$$\hat{\mathbf{v}}_c = \frac{\mathbf{B}_c^T(\mathbf{s}_c+\mathbf{w}_c)}{\hat{\Sigma}_{YYc}+\frac{\lambda_v}{\kappa_c}\mathbf{Q}_c}, \; \mathbf{v}_c^* = \frac{\hat{\mathbf{v}}_c}{\sqrt{(n-1)\hat{\mathbf{v}}_c^T\hat{\Sigma}_{YYc}\mathbf{v}_c}}. \tag{8}$$

$\mathbf{Q}$ is a diagonal matrix with the j-th element being $\frac{1}{2(n-1)|v_{jc}|}$ $(j = 1, ..., q)$.

Now we have obtained all the solutions to DMTSCCA without using original imaging and genetic data. In contrast, we use the GWAS summary statistics to obtain the covariance between imaging QTs and SNPs. The in-set covariance Σ_{YY} can also be calculated based on the results of GWAS. The in-set covariance Σ_{XX} can be approximated using subjects of the same population. In this paper, we used the public 1000 genome project (1kGP) database to generate $\hat{\Sigma}_{XX}$. In practice, using $\hat{\Sigma}_{XX}$ of the same population could yield acceptable results [1,13]. Therefore, our S-DMTSCCA is quite meaningful since it does not depend on raw neuroimaging genetic data.

3 Experiment and Results

We conducted two kinds of experiments to evaluate S-DMTSCCA. First, we used the ADNI data set where the original brain imaging phenotypes and genotypes are available. Specifically, our method cannot access individual-level imaging and genetic data. Instead, S-DMTSCCA can only work on the GWAS summary statistics obtained from this ADNI data set. At the same time, DMTSCCA directly ran on the original imaging and genetic data. By comparison, we can observe the performance difference between S-DMTSCCA and DMTSCCA. This can help evaluate the usefulness of our method. Second, we ran our method on a public GWAS result which studied the associations of imaging phenotypes and SNPs from the UK Biobank.

To find the best parameters for λ_s, λ_w and λ_v in DMTSCCA, we employed the grid search strategy with a moderate candidate parameter range $10^i (i = -5, -4, ..., 0, ..., 4, 5)$. Since S-DMTSCCA takes summary statistics as the input data without using the individual data, the conventional regularization parameters procedure is impracticable. Therefore, we used the grid search method with the same range based on the data set whose individual-level data was accessible to find the optimal parameters for S-DMTSCCA. Besides, to ensure equal optimization for each imaging modality, we used the same constants for the task weight parameters κ_c in different sub-tasks.

We used the 1kGP data sets as the reference samples. By conducting whole-genome sequencing on individuals from a range of ethnicity, the 1kGP institute obtains an extensive collection of human prevalent genetic variants [3]. We used individuals of British in England and Scotland (GBR) from 1kGP (release 20130502) to compute $\hat{\Sigma}_{XX}$ in S-DMTSCCA since UK Biobank GWAS results from the European ancestors. All experiments ran on the same platform, and all methods employed the same stopping condition, i.e. both $\max_c |(\mathbf{s}_c + \mathbf{w}_c)^{t+1} - (\mathbf{s}_c + \mathbf{w}_c)^t| \leq 10^{-5}$ and $\max_c |\mathbf{v}_c^{t+1} - \mathbf{v}_c^t| \leq 10^{-5}$.

3.1 Study on the ADNI Dataset

Data Source. The individual-level brain genotype and imaging data we used were downloaded from the Alzheimer's Disease Neuroimaging Initiative (ADNI) database (adni.loni.usc.edu). One goal of ADNI is to investigate the feasibility of utilizing a multi-modal approach that combines serial magnetic resonance imaging (MRI), positron emission tomography (PET), other biological markers, and clinical and neuropsychological assessment to measure the progression of Alzheimer's disease (AD). For the latest information, see https://www.adni-info.org.

We used three modalities, i.e. the 18-F florbetapir PET (AV45) scans, fluorodeoxyglucose positron emission tomography (FDG) scans, and structural MRI (sMRI) scans. These data had been aligned to the same visit of each subject. The sMRI data were analyzed with voxel-based morphometry (VBM) by SPM. All scans were aligned to a T1-weighted template image, segmented into gray matter (GM), white matter (WM), and cerebrospinal fluid (CSF) maps, normalized to the standard MNI space, and smoothed with an 8 mm FWHM kernel. Additionally, the FDG and AV45 scans were registered into the same MNI space. Then 116 regions of interest (ROIs) level measurements were extracted based on the MarsBaR automated anatomical labeling (AAL) atlas. These 116 imaging QTs were pre-adjusted to remove the effects of the baseline age, gender, education, and handedness by the regression weights generated from healthy controls. The genotype data were also from the ADNI database. Specifically, we studied 5000 SNPs of chromosome 19: 46670909 - 46167652 including the well-known AD risk genes such as *APOE* [11]. As a methodology paper, we aimed to develop a GWAS summary statistics-based imaging genetics method that could handle GWAS summary statistics rather than individual-level imaging and genetic data. Thus, although using a large number of SNPs could be more interesting, it might go beyond the key research topic of this paper. The goal of the method was to explore the relationships between the multiple modalities of QTs (GM densities for VBM scans, amyloid values for AV45 scans and glucose utilization for FDG scans) and SNPs. For clarity, we respectively denoted the canonical weights of imaging QTs for AV45-PET, FDG-PET, and VBM-MRI as $\mathbf{v}_1$, $\mathbf{v}_2$ and $\mathbf{v}_3$.

Biomarkers Identification. We presented the identified SNPs and imaging QTs for each imaging modality based on the estimated canonical weights in Fig. 1 (a, b, and c). DMTSCCA and S-DMTSCCA decomposed the canonical weight of SNPs into two components, i.e., the multi-modal shared component $\mathbf{S}$ and modality-specific component $\mathbf{W}$. Thus, we presented both of them here. Both $\mathbf{S}$ and $\mathbf{W}$ of S-DMTSCCA identified the famous AD-related SNP rs429358, and the top SNPs marked in this figure were all related to AD, demonstrating the effectiveness of our method. In addition, S-DMTSCCA presented a task-consistent pattern, indicating that SNPs such as rs12721051 (*APOC1*), rs56131196 (*APOC1*) and rs44203638 (*APOC1*) could contribute to all three imaging modalities. S-DMTSCCA also presented a task-specific pattern, including rs7256200 (*APOC1*), rs73052335 (*APOC1*) and rs10414043 (*APOC1*) which were associated with Av45 and FDG scans, rs483082 (*APOC1*) which was associated with Av45 and VBM-MRI scans, rs12721046 (*APOC1*) and rs5117 (*APOC1*) which were only associated with VBM imaging scans. Importantly, our method identified the same top SNPs as the conventional DMTSCCA, suggesting that S-DMTSCCA possesses an equivalent feature selection capacity to DMTSCCA. In the heat maps of imaging QTs, S-DMTSCCA was able to identify different biomarkers for different scans as the conventional one. For example, the *Frontal-Med-Orb-Right* and *Frontal-Med-Orb-Left* of AV45-PET scans, *Cingulum-Post-Right* of FDG-PET scans and *Hippocampus-Left* of VBM-MRI scans. All in all, S-DMTSCCA presented a good agreement with the conventional method in feature selection. These results demonstrated that our method could be a very promising and meaningful method in multi-modal brain imaging genetics since it did not use individual-level brain imaging genetic data.

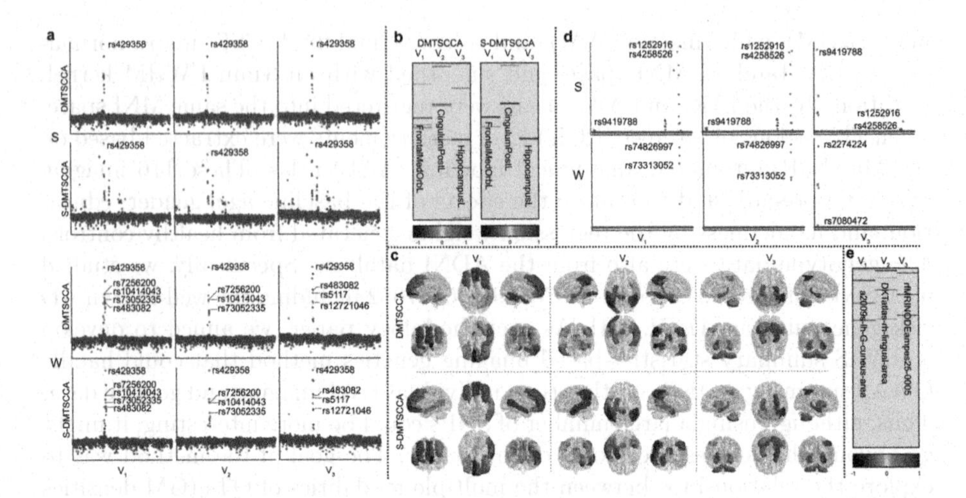

Fig. 1. Comparison of canonical weights from DMTSCCA and S-DMTSCCA when applied to ADNI data set (**a**, **b** and **c**), and Comparison of canonical weights when applied to IDPs GWAS (**d** and **e**). **a** The canonical weights ($70 + \log_2 |\mathbf{u}|$) of SNPs. **b** The canonical weights of QTs. **c** Visualization of the top 10 identified ROIs mapped on the brain, and the different ROIs marked with different colors. This sub-figure was drawn by BrainNet Viewer toolbox [16]. **d** The canonical weights of SNPs. **e** The canonical weights of QTs. Within each sub-figure, there are three columns for three modalities, i.e., $\mathbf{v}_1$ for AV45, $\mathbf{v}_2$ for FDG, and $\mathbf{v}_3$ for VBM.

Bi-multivariate Associations. We summarized the CCCs of conventional DMTSCCA and our method in Table 1. The values here represented the strength of the identified correlations. Since we have 3 imaging modalities in this section, 3 groups of CCCs were shown. Firstly, we can see that all CCCs of S-DMTSCCA were comparable to those of DMTSCCA, and all CCCs were relatively high (>0.2). These results indicated that our method can identify equivalent bi-multivariate associations between genetic variants and multi-modal phenotypes without using individual-level data.

Table 1. CCCs values between SNPs and three modalities imaging QTs.

Model	SNP-$\mathbf{v}_1$	SNP-$\mathbf{v}_2$	SNP-$\mathbf{v}_3$
DMTSCCA	0.4722	0.3306	0.2450
S-DMTSCCA	0.4559	0.3122	0.2018

3.2 Application to Summary Statistics from Brain Imaging GWAS

Summary Statistics from GWAS. The GWAS summary data were associations between brain imaging-derived phenotypes (IDPs) and SNPs. This GWAS studied a comprehensive set of imaging QTs derived from different types of brain imaging data of 8428 individuals from the UK Biobank database [7]. We used three modalities of imaging QTs, including two structural volumetric measurements obtained by FreeSurfer, i.e., the Desikan-Killiany-Tourville atlas and the Destrieux atlas, and one from resting-state functional MRI (rfMRI). We used summary statistics of these three modalities of the whole brain and denoted their canonical weights as $\mathbf{v}_1$, $\mathbf{v}_2$ and $\mathbf{v}_3$ respectively. In particular, there were 148,64 and 76 imaging QTs for three imaging modalities. Genotype data for these imaging QTs were obtained from the UKB database [2]. We used 5000 SNPs in chromosome 10: 95952196 - 96758136 and chromosome 14: 59072418 - 59830880. We aimed to evaluate our method with the expectation to gain a comprehensive understanding of the genetic basis of multi-modal brain imaging phenotypes.

Biomarkers Identification. Figure 1 (d and e) presented the heat maps of the identified SNPs and imaging QTs for all three modalities, with modality-consistent and modality-specific SNPs separately shown. In this figure, **S** identified rs9419788 (*PLCE1*) in chromosome 10, and rs1252916 (*DAAM1*) and rs4258526 (*LINC01500*) in chromosome 14. This implied that these loci are shared by all three modalities. In addition, in heat maps of **W**, rs7080472 (*PLCE1*) only can be identified by rfMRI scans, and rs74826997 (*LINC01500*) and rs73313052 only can be identified by the other two scans. All these identified biomarkers shared by three modalities or related to a specific modality were consistent with the results of GWAS. This suggested that S-DMTSCCA can select leading genetic variations contributing to related imaging QTs. From the heat maps of imaging QTs, our method simultaneously identified the specific related imaging QTs for a specific task while GWAS cannot. These specific patterns included the *a2009s-lh-G-cuneus-area* for a2009s imaging QTs, the *DKTatlas-rh-lingual-area* for DKT IDPs, and *rfMRI-NODEampes25-0005* for rfMRI scan. In summary, these results demonstrated that S-DMTSCCA can simultaneously identify the important genetic variants and imaging phenotypes of multiple modalities only depending on summary statistics data.

Bi-multivariate Associations. In addition to feature selection, we calculated CCCs of S-DMTSCCA. The values for the three modalities were 0.1455, 0.1354, and 0.1474 respectively. This indicated that S-DMTSCCA could identify substantial bi-multivariate associations for each modality which might be attributed to its good modeling capability. These results again suggested that S-DMTSCCA can work well with summary statistics.

4 Conclusion

In brain imaging genetics, DMTSCCA can identify the genetic basis of multi-modal phenotypes. However, DMTSCCA depends on individual-level genetic and imaging data and thus was infeasible when the raw data cannot be obtained. In this paper, we developed a source-free S-DMTSCCA method using GWAS summary statistics rather than the original imaging and genetic data. Our method had the same modeling ability as the conventional methods. When applied to multi-modal phenotypes from ADNI, S-DMTSCCA showed an agreement in feature selection and canonical correlation coefficient results with DMTSCCA. When applied to multiple modalities of GWAS summary statistics, our method can identify important SNPs and their related imaging QTs simultaneously. In the future, it is essential to consider the pathway and brain network information in our method to identify higher-level biomarkers of biological significance.

References

1. Barbeira, A.N., et al.: Exploring the phenotypic consequences of tissue specific gene expression variation inferred from gwas summary statistics. Nat. Commun. **9**(1), 1825 (2018)
2. Bycroft, C., et al.: The UK Biobank resource with deep phenotyping and genomic data. Nature **562**(7726), 203–209 (2018)
3. Consortium,G.P., et al.: A global reference for human genetic variation. Nature **526**(7571), 68 (2015)
4. Du, L., et al.: Associating multi-modal brain imaging phenotypes and genetic risk factors via a dirty multi-task learning method. IEEE Trans. Med. Imaging **39**(11), 3416–3428 (2020)
5. Du, L., et al.: Fast multi-task scca learning with feature selection for multi-modal brain imaging genetics. In: 2018 IEEE International Conference on Bioinformatics and Biomedicine (BIBM), pp. 356–361. IEEE (2018)
6. Du, L., et al.: Multi-task sparse canonical correlation analysis with application to multi-modal brain imaging genetics. IEEE/ACM Trans. Comput. Biol. Bioinf. **18**(1), 227–239 (2019)
7. Elliott, L.T., et al.: Genome-wide association studies of brain imaging phenotypes in UK Biobank. Nature **562**(7726), 210–216 (2018)
8. Li, T., Ning, Z., Shen, X.: Improved estimation of phenotypic correlations using summary association statistics. Front. Genet. **12**, 665252 (2021)
9. Manolio, T.A., et al.: Finding the missing heritability of complex diseases. Nature **461**(7265), 747–753 (2009)
10. Marouli, E., et al.: Rare and low-frequency coding variants alter human adult height. Nature **542**(7640), 186–190 (2017)
11. Ramanan, V.K., et al.: APOE and BCHE as modulators of cerebral amyloid deposition: a florbetapir pet genome-wide association study. Mol. Psychiatry **19**(3), 351–357 (2014)
12. Shen, L., Thompson, P.M.: Brain imaging genomics: integrated analysis and machine learning. Proc. IEEE **108**(1), 125–162 (2019)
13. Turley, P., et al.: Multi-trait analysis of genome-wide association summary statistics using MTAG. Nat. Genet. **50**(2), 229–237 (2018)

14. Uffelmann, E., et al.: Genome-wide association studies. Nat. Rev. Methods Primers **1**(1), 59 (2021)
15. Wei, K., Kong, W., Wang, S.: An improved multi-task sparse canonical correlation analysis of imaging genetics for detecting biomarkers of Alzheimer's disease. IEEE Access **9**, 30528–30538 (2021)
16. Xia, M., Wang, J., He, Y.: BrainNet Viewer: a network visualization tool for human brain connectomics. PLoS ONE **8**(7), e68910 (2013)
17. Zhuang, X., Yang, Z., Cordes, D.: A technical review of canonical correlation analysis for neuroscience applications. Hum. Brain Mapp. **41**(13), 3807–3833 (2020)

Revisiting Feature Propagation and Aggregation in Polyp Segmentation

Yanzhou Su[1], Yiqing Shen[3], Jin Ye[2], Junjun He[2], and Jian Cheng[1](✉)

[1] School of Information and Communication Engineering, University of Electronic Science and Technology of China, Chengdu 611731, China
chenjian@uestc.edu.cn

[2] Shanghai AI Laboratory, Shanghai 200232, China

[3] Johns Hopkins University, Baltimore, MD 21218, USA

Abstract. Accurate segmentation of polyps is a crucial step in the efficient diagnosis of colorectal cancer during screening procedures. The prevalent UNet-like encoder-decoder frameworks are commonly employed, due to their capability of capturing multi-scale contextual information efficiently. However, two major limitations hinder the network from achieving effective feature propagation and aggregation. Firstly, the skip connection only transmits a single scale feature to the decoder, which can result in limited feature representation. Secondly, the features are transmitted without any information filter, which is inefficient for performing feature fusion at the decoder. To address these limitations, we propose a novel feature enhancement network that leverages feature propagation enhancement and feature aggregation enhancement modules for more efficient feature fusion and multi-scale feature propagation. Specifically, the feature propagation enhancement module transmits all encoder-extracted feature maps from the encoder to the decoder, while the feature aggregation enhancement module performs feature fusion with gate mechanisms, allowing for more effective information filtering. The multi-scale feature aggregation module provides rich multi-scale semantic information to the decoder, further enhancing the network's performance. Extensive evaluations on five datasets demonstrate the effectiveness of our method, particularly on challenging datasets such as CVC-ColonDB and ETIS, where it can outperform the previous state-of-the-art models by a significant margin (3%) in terms of mIoU and mDice.

Keywords: Polyp Segmentation · Feature Propagation · Feature Aggregation

1 Introduction

Colorectal cancer is a life-threatening disease that results in the loss of millions of lives each year. In order to improve survival rates, it is essential to identify colorectal polyps early. Hence, regular bowel screenings are recommended, where

H. Greenspan et al. (Eds.): MICCAI 2023, LNCS 14224, pp. 632–641, 2023.
https://doi.org/10.1007/978-3-031-43904-9_61

endoscopy is the gold standard. However, the accuracy of endoscopic screening can heavily rely on the individual skill and expertise of the domain experts involved, which are prone to incorrect diagnoses and missed cases. To reduce the workload on physicians and enhance diagnostic accuracy, computer vision technologies, such as deep neural networks, are involved to assist in the pre-segmentation of endoscopic images.

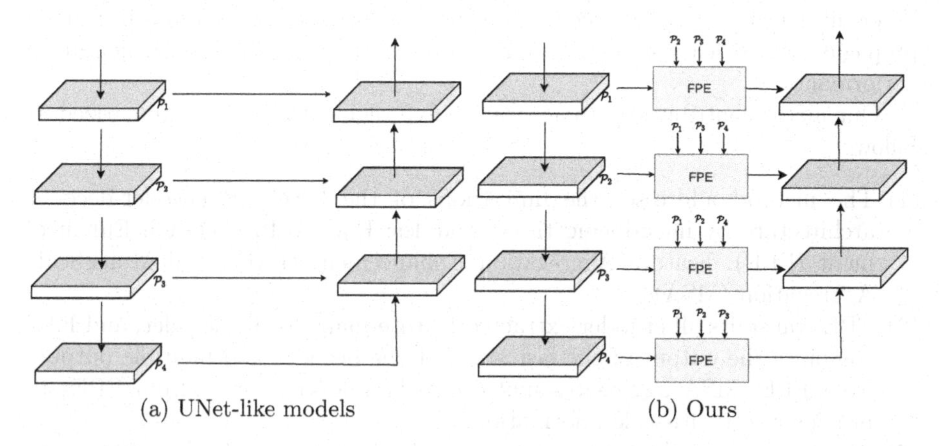

Fig. 1. Comparison of feature propagation methods. The UNet-like model uses skip connections that transmit only single-stage features. In contrast, our approach utilizes FPE to propagate features from all stages, incorporating a gate mechanism to regulate the flow of valuable information.

Deep learning-based image segmentation methods have gained popularity in recent years, dominated by UNet [11] in the field of medical image segmentation. UNet's success has led to the development of several other methods that use a similar encoder-decoder architecture to tackle polyp segmentation, including ResUNet++ [7], PraNet [3], CaraNet [10] and UACANet [8]. However, these methods are prone to inefficient feature fusion at the decoder due to the transmission of multi-stage features without filtering out irrelevant information.

To address these limitations, we propose a novel feature enhancement network for polyp segmentation that employs Feature Propagation Enhancement (FPE) modules to transmit multi-scale features from all stages to the decoder. Figure 1 illustrates a semantic comparison of our feature propagation scheme with the UNet-like model. While the existing UNet-like models use skip connections to propagate a single-scale feature, our method utilizes FPE to propagate multi-scale features from all stages in encoder. More importantly, this research highlights the usage of FPE can effectively replace skip connections by providing more comprehensive multi-scale characteristics from full stages in encoder. To further address the issue of high-level semantics being overwhelmed in the progressive feature fusion process, we also integrate a Feature Aggregation Enhancement (FAE) module that aggregates the outputs of FPE from previous stages at decoder. Moreover, we introduce gate mechanisms in both FPE and FAE to filter out redundant information, prioritizing informative features for efficient feature fusion. Finally, we propose a Multi-Scale Aggregation (MSA) module appended to the output of the encoder to capture multi-scale features and provide the decoder with rich multi-scale semantic information. The MSA incorporates a cross-stage multi-scale feature aggregation scheme to facilitate the aggregation of multi-scale features. Overall, our proposed method improves upon existing UNet-like encoder-decoder architectures by addressing the limitations in feature propagation and feature aggregation, leading to improved polyp segmentation performance.

Our major contributions to accurate polyp segmentation are summarized as follows.

(1) The method addresses the limitations of the UNet-like encoder-decoder architecture by introducing three modules: Feature Propagation Enhancement (FPE), Feature Aggregation Enhancement (FAE), and Multi-Scale Aggregation (MSA).
(2) FPE transmits all encoder-extracted feature maps to the decoder, and FAE combines the output of the last stage at the decoder and multiple outputs from FPE. MSA aggregates multi-scale high-level features from FPEs to provide rich multi-scale information.
(3) The proposed method achieves state-of-the-art results in five polyp segmentation datasets and outperforms the previous cutting-edge approach by a large margin (3%) on CVC-ColonDB and ETIS datasets.

2 Method

Overview. Our proposed feature enhancement network illustrated in Fig. 2(a), is also a standard encoder-decoder architecture. Following Polyp-PVT [2], we adopt PVT [16] pretrained on ImageNet as the encoder. The decoder consists of three feature aggregation enhancement modules (FAE) and a multi-scale aggregation module (MSA). Given an input image $\mathcal{I}$, we first extract the pyramidal features using the encoder, which is defined as follows,

$$\mathcal{P}_1, \mathcal{P}_2, \mathcal{P}_3, \mathcal{P}_4 = \mathrm{PVT}(\mathcal{I}) \tag{1}$$

where, $\{\mathcal{P}_1, \mathcal{P}_2, \mathcal{P}_3, \mathcal{P}_4\}$ is the set of pyramidal features from four stages with the spatial size of 1/4, 1/8, 1/16, 1/32 of the input respectively. Features with lower spatial resolution usually contain richer high-level semantics. Then, these features are transmitted by the feature propagation enhancement module (FPE) to yield the feature set $\{\mathcal{C}_1, \mathcal{C}_2, \mathcal{C}_3, \mathcal{C}_4\}$, which provides multi-scale information from all the stages. This is different from the skip connection which only transmits the single-scale features at the present stage. Referring to [3,10], three highest-level features generated by FPEs are subsequently fed into a MSA module for aggregating rich multi-scale information. Afterwards, feature fusion is performed by

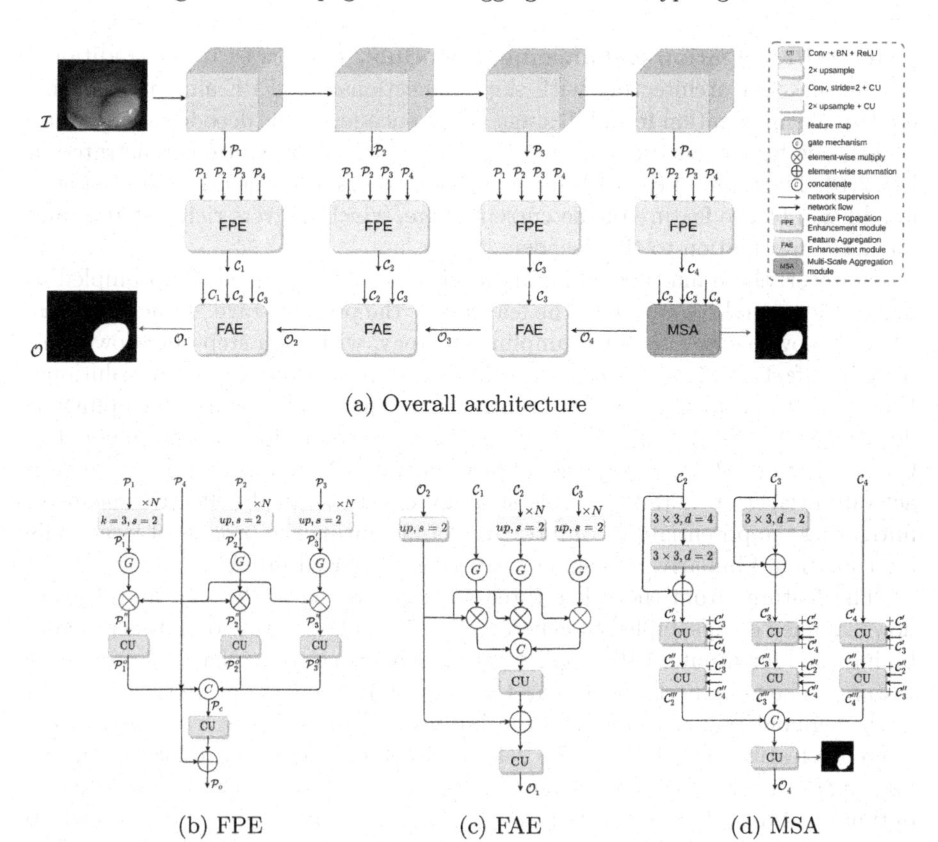

(a) Overall architecture

(b) FPE (c) FAE (d) MSA

Fig. 2. An overview of the proposed framework. (a) Overall architecture; (b) FPE: Feature propagation enhancement module; (c) FAE: Feature aggregation enhancement module; (d) MSA: Multi-scale aggregation module

FAE in the decoder, whereby it progressively integrates the outputs from FPE and previous stages. The higher-level semantic features of the FPE output are capable of effectively compensating for the semantics that may have been overwhelmed during the upsampling process. This process is formulated as,

$$\begin{aligned} \mathcal{O}_4 &= \mathrm{MSA}(\{\mathcal{C}_2, \mathcal{C}_3, \mathcal{C}_4\}) \\ \mathcal{O}_i &= \mathrm{FAE}(\{\mathcal{C}_i, \mathcal{C}_{i+1}, ..., \mathcal{C}_3\}, O_{i+1}) \quad i = 1, 2, 3. \end{aligned} \tag{2}$$

A noteworthy observation is that the gating mechanism has been widely utilized in both FPE and FAE to modulate the transmission and integration of features. By selectively controlling the flow of relevant information, this technique has shown promise in enhancing the overall quality of feature representations [4,9]. The final features $\mathcal{O}_1$ are passed through the classifier (i.e., a 1×1 convolutional layer) to get the final prediction result in $\mathcal{O}$. Further details on FPE, FAE, and MSA will be provided in the following sections.

Feature Propagation Enhancement Module. In contrast to the traditional encoder-decoder architecture with skip connections, the FPE aims to transmit multi-scale information from full stage at the encoder to the decoder, rather than single-scale features at the current stage. The FPE architecture is illustrated in Fig. 2(b). The input of the FPE includes the features from the other three stages, in addition to the feature of the current stage, which delivers richer spatial and semantic information to the decoder.

However, these multi-stage inputs need to be downsampled or upsampled to match the spatial resolution of the features at the present stage. To achieve this, FPE employs a stepwise downsampling strategy, with each step only downsampling by a factor of 2, and each downsampling step is followed by a Convolutional Unit (CU) to perform feature transformation. The number of downsamplings is denoted as $N = \log_2^T$, where T stands for the scale factor for downsampling. The CU consists of a 3×3 convolutional layer, a Batch Normalization layer, and an activation layer (i.e., ReLU). This strategy can be a highly effective means of mitigating the potential loss of intricate details during the process of large-scale interpolation. Similarly, this same strategy is employed in FAE.

The features from the other three stages, denoted as $\mathcal{P}_1$, $\mathcal{P}_2$, and $\mathcal{P}_3$, are downsampled or upsampled to generate $\mathcal{P}_1^{'}$, $\mathcal{P}_2^{'}$, and $\mathcal{P}_3^{'}$. Instead of directly combining the four inputs, FPE applies gate mechanisms to emphasize informative features. The gate mechanism takes the form of $\mathcal{Y}' = \mathrm{G}(\mathcal{X}) * \mathcal{Y}$, where G (in this work, Sigmoid is used) measures the importance of each feature vector in the reference feature $\mathcal{X} \in \mathbf{R}^{H \times W}$. By selectively enhancing useful information and filtering out irrelevant information, the reference features $\mathcal{X}$ assist in identifying optimal features $\mathcal{Y}$ at the current level. The output of G is in $[0,1]^{H \times W}$, which controls the transmission of informative features from $\mathcal{Y}$ or helps filter useless information. Notably, $\mathcal{X}$ can be $\mathcal{Y}$ itself, serving as a reference feature. FPE leverages such gate mechanism to obtain informative features in $\mathcal{P}$ and passes them through a CU respectively. After that, FPE concatenates features from the four branches to accomplish feature aggregation. A CU is followed to boost the feature fusion.

Feature Aggregation Enhancement Module. The FAE is a novel approach that integrates the outputs of the last stages at the decoder with the FPE's outputs at both the current and deeper stages to compensate for the high-level semantics that may be lost in the process of progressive feature fusion. In contrast to the traditional encoder-decoder architecture with skip connections, the FAE assimilates the output of the present and higher-stage FPEs, delivering richer spatial and semantic information to the decoder.

The FAE, depicted in Fig. 2(c), integrates the outputs of the current and deeper FPE stages with high-level semantics. As an example, the last FAE takes as inputs $\mathcal{O}_2$ (output of the penultimate FAE), $\mathcal{C}_1$ (output of the current FPE stage), and $\mathcal{C}_2$ and $\mathcal{C}_3$ (outputs of FPE from deeper stages). Multiple outputs from deeper FPE stages are introduced to compensate for high-level semantics. Furthermore, gate mechanisms are utilized to filter out valueless features for fusion, and the resulting enhanced feature is generated by a CU after concatenating the filtered features. Finally, $\mathcal{O}_2$ is merged with the output feature through element-wise summation, followed by a CU to produce the final output feature $\mathcal{O}_1$.

Multi-Scale Aggregation Module. The MSA module in our proposed framework, inspired by the Parallel Partial Decoder in PraNet [3], integrates three highest-level features $\mathcal{C}_2$, $\mathcal{C}_3$, and $\mathcal{C}_4$ from the FPE output. This provides rich multi-scale information for subsequent feature aggregation in the FAE and also helps to form a coarse localization of polyps under supervision. It benefits from an additional supervision signal, as observed in PraNet [3], CaraNet [8], and etc. by aiding in forming a coarse location of the polyp and contributing to improved accuracy and performance. As depicted in Fig. 2(d), the MSA module first processes these three features separately. $\mathcal{C}_2$, which has the highest feature resolution, is processed with multiple dilated convolutions to capture its multi-scale information while keeping its spatial resolution unchanged. $\mathcal{C}_3$ is processed with only one dilated convolution due to its higher spatial resolution, while $\mathcal{C}_4$ is not processed since it already contains the richest contextual information. The output features of the three branches are then upsampled to the size of $\mathcal{C}_2$. To better integrate these three multi-scale high-level features for subsequent fusion, additional cross-feature fusion operations (i.e., 2 CU layer) are performed in the MSA module.

3 Experiments

Datasets. We conduct extensive experiments on five polyp segmentation datasets, including Kvasir [6], CVC-ClinicDB [1], CVC-ColonDB [13], ETIS [12] and CVC-T [14]. Following the setting in [2,3,5,10,10,17,19], the model is trained using a fraction of the images from CVC-ClinicDB and Kvasir, and its performance is evaluated by the remaining images as well as those from CVC-T, CVC-ColonDB, and ETIS. In particular, there are 1450 images in the training set, of which 900 are from Kvasir and 550 from CVC-ClinicDB. The test set contains all of the images from CVC-T, CVC-ColonDB, and ETIS, which have 60, 380, and 196 images, respectively, along with the remaining 100 images from Kvasir and the remaining 62 images from CVC-ClinicDB.

Implementations. We utilize PyTorch 1.10 to run experiments on an NVIDIA RTX3090 GPU. We set an initial learning rate to 1e-4 and halve it after 80 epochs. We train the model for 120 epochs. The same multi-scale input and gradient clip strategies used in [3,8,10] are employed in the training phase. The batch size is 16 by default. AdamW is selected as the optimizer with a weight decay of 1e-4. We adopt the same data augmentation techniques as UACANet [8], including random flip, random rotation, and color jittering. In evaluation phase, we mainly focus on mDice, mIoU, the two most common metrics in medical image segmentation, to evaluate the performance of the model. Referring to [3,8,10], we use a combination loss consisting of weighted Dice loss and weighted IoU loss to supervise network optimization.

Table 1. Quantitative results on CVC-ClinicDB, Kvasir, CVC-T, CVC-ColonDB and ETIS.

Methods	Published Venue	CVC-ClinicDB		Kvasir		CVC-ColonDB		ETIS		CVC-T	
		mDice	mIoU	mDice	mIoU	mDice	mIoU	mDice	mIoU	mDice	mIoU
UNet [11]	MICCAI'15	0.823	0.755	0.818	0.746	0.512	0.444	0.398	0.335	0.710	0.627
PraNet [3]	MICCAI'19	0.899	0.849	0.898	0.840	0.709	0.640	0.628	0.567	0.871	0.797
ResUNet++ [7]	JBHI'21	0.846	0.786	0.807	0.727	0.588	0.497	0.337	0.275	0.687	0.598
SANet [17]	MICCAI'21	0.916	0.859	0.904	0.847	0.752	0.669	0.750	0.654	0.888	0.815
MSNet [19]	MICCAI'21	0.915	0.866	0.902	0.847	0.747	0.668	0.720	0.650	0.862	0.796
UACANet-S [8]	MM'21	0.916	0.870	0.905	0.852	0.783	0.704	0.694	0.615	0.902	0.837
UACANet-L [8]	MM'21	0.926	0.880	0.912	0.859	0.751	0.678	0.766	0.689	**0.910**	**0.849**
Polyp-PVT [2]	arxiv'21	**0.937**	**0.889**	0.917	0.864	0.808	0.727	0.787	0.706	0.900	0.833
CaraNet [10]	SPIE MI'22	0.921	0.876	0.913	0.859	0.775	0.700	0.740	0.660	0.902	0.836
LDNet [18]	MICCAI'22	0.923	0.872	0.912	0.855	0.794	0.715	0.778	0.707	0.893	0.826
SSFormer-S [15]	MICCAI'22	0.916	0.873	0.925	0.878	0.772	0.697	0.767	0.698	0.887	0.821
SSFormer-L [15]	MICCAI'22	0.906	0.855	0.917	0.864	0.802	0.721	0.796	0.720	0.895	0.827
Ours	-	0.931	0.885	**0.928**	**0.880**	**0.837**	**0.759**	**0.822**	**0.746**	0.905	0.839

Comparison with State-of-the-Art Methods. We compared our proposed method with previous state-of-the-art methods. According to the experimental settings, the results on CVC-ClinicDB and Kvasir demonstrate the learning ability of the proposed model, while the results on CVC-T, CVC-ColonDB, and ETIS demonstrate the model's ability for cross-dataset generalization. The experimental results are listed in Table.1. It can be seen that our model is slightly inferior to Polyp-PVT on the CVC-ColonDB, but the gap is quite small, e.g., 0.6% in mDice and 0.4% in mIoU. On Kvasir, we are ahead of the previous best model by 1.1% in mDice and 1.6% in mIoU. This shows that our model is second to none in terms of learning ability, which demonstrates the effectiveness of our model.

Furthermore, our proposed method demonstrates strong cross-dataset generalization capability on CVC-T, CVC-ColonDB, and ETIS datasets, with particularly good performance on the latter two due to their larger and more representative datasets. Our model outperforms state-of-the-art models by 2.9% mDice and 3.2% mIoU on CVC-ColonDB and 3.5% mDice and 4.0% mIoU on ETIS. These results validate the effectiveness of feature-level enhancement and highlight the superior performance of our method. We also provide visual results in Fig. 3, where our predictions are shown to be closer to the ground truth.

Ablation Study. We carried out ablation experiments to verify the effectiveness of the proposed FPE, FAE, and MSA. For our baseline, we use the simple encoder-decoder structure with skip connections for feature fusion and perform element-wise summation at the decoder. Table 2 presents the results of our ablation experiments. Following the ablation study conducted on our proposed approach, it is with confidence that we assert the significant contribution of each module to the overall performance enhancement compared to the baseline. Our results indicate that the impact of each module on the final performance is considerable, and their combination yields the optimal overall performance. Specifically, across the five datasets, our proposed model improves the mDice score by at least 1.4% and up to 3.4% on CVC-T, compared to the baseline. For mIoU, the improvements are 1.5% and 3.5% on the corresponding datasets. In summary, our ablation study underscores the crucial role played by each component of our approach, and establishes its potential as a promising framework for future research in this domain.

Table 2. Ablation study on FPE, MSA, and FAE. Each item presents mDice/mIoU.

FPE	MSA	FAE	CVC-T	CVC-ClinicDB	Kvasir	CVC-ColonDB	ETIS
			87.1/80.4	91.6/87.0	91.4/86.3	81.0/72.8	79.9/71.5
✓			90.2/83.7	92.7/87.8	91.5/86.4	81.0/73.1	80.2/71.9
	✓		89.9/83.2	91.6/86.8	91.9/87.0	82.1/73.9	81.3/73.1
		✓	90.0/83.3	91.8/87.2	91.7/86.6	81.4/73.3	81.2/73.3
✓	✓		90.1/83.4	92.6/87.7	92.2/87.2	81.6/73.4	80.9/72.9
✓		✓	90.2/83.5	93.0/88.4	92.1/87.2	81.5/73.4	80.9/72.9
	✓	✓	90.0/83.6	92.7/88.0	92.6/87.8	81.2/73.1	81.6/73.9
			90.5/83.9	**93.1/88.5**	**92.8/88.0**	**83.7/75.9**	**82.2/74.6**

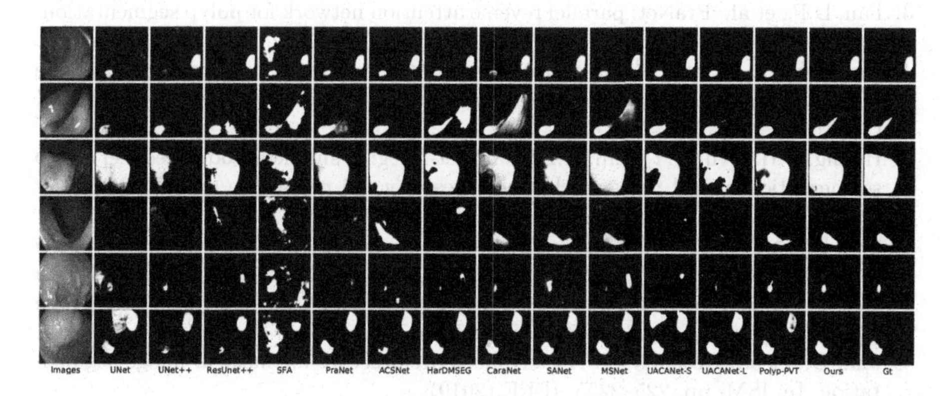

Fig. 3. Exemplary images and results that are segmented by different approaches.

4 Conclusion

We introduce a new approach to polyp segmentation that addresses inefficient feature propagation in existing UNet-like encoder-decoder networks. Specifically, a feature propagation enhancement module is introduced to propagate multi-scale information over full stages in the encoder, while a feature aggregation enhancement module is attended at the decoder side to prevent the loss of high-level semantics during progressive feature fusion. Furthermore, a multi-scale aggregation module is used to aggregate multi-scale features to provide rich information for the decoder. Experimental results on five popular polyp datasets demonstrate the effectiveness and superiority of our proposed method. Specifically, it outperforms the previous cutting-edge approach by a large margin (3%) on CVC-ColonDB and ETIS datasets. To extend our work, our future direction focuses on exploring more effective approaches to feature utilization, such that efficient feature integration and propagation can be achieved even on lightweight networks.

Acknowledgements. This research was partly supported by the National Natural Science Foundation of China (No. 62071104), partly supported by the Sichuan Science and Technology Program (No. 2021YFG0328), partly supported by the NSFC&CAAC (No. U2233209, No.U2133211).

References

1. Bernal, J., Sánchez, F.J., Fernández-Esparrach, G., Gil, D., Rodríguez, C., Vilariño, F.: Wm-dova maps for accurate polyp highlighting in colonoscopy: Validation vs. saliency maps from physicians. CMIG **43**, 99–111 (2015)
2. Bo, D., Wenhai, W., Jinpeng, L., Deng-Ping, F.: Polyp-pvt: Polyp segmentation with pyramid vision transformers. arXiv preprint arXiv:2108.06932v3 (2021)
3. Fan, D.P., et al.: PraNet: parallel reverse attention network for polyp segmentation. In: MICCAI (2020)
4. Hu, J., Shen, L., Sun, G.: Squeeze-and-excitation networks. In: Proceedings of the IEEE Conference on Computer Vision and Pattern Recognition, pp. 7132–7141 (2018)
5. Huang, C.H., Wu, H.Y., Lin, Y.L.: Hardnet-mseg: A simple encoder-decoder polyp segmentation neural network that achieves over 0.9 mean dice and 86 fps. arXiv preprint arXiv:2101.07172 (2021)
6. Jha, D., et al.: Kvasir-SEG: a segmented polyp dataset. In: Ro, Y.M., et al. (eds.) MultiMedia Modeling: 26th International Conference, MMM 2020, Daejeon, South Korea, January 5–8, 2020, Proceedings, Part II, pp. 451–462. Springer, Cham (2020). https://doi.org/10.1007/978-3-030-37734-2_37
7. Jha, D., et al.: Resunet++: An advanced architecture for medical image segmentation. In: ISM, pp. 225–2255. IEEE (2019)
8. Kim, T., Lee, H., Kim, D.: UACANet: Uncertainty augmented context attention for polyp segmentation. In: ACM MM, pp. 2167–2175 (2021)
9. Li, X., Zhao, H., Han, L., Tong, Y., Yang, K.: GFF: Gated fully fusion for semantic segmentation. arxiv 2019. arXiv preprint arXiv:1904.01803
10. Lou, A., Guan, S., Ko, H., Loew, M.H.: CaraNet: context axial reverse attention network for segmentation of small medical objects, pp. 81–92. SPIE (2022)
11. Ronneberger, O., Fischer, P., Brox, T.: U-net: Convolutional networks for biomedical image segmentation. In: MICCAI (2015)

12. Silva, J., Histace, A., Romain, O., Dray, X., Granado, B.: Toward embedded detection of polyps in WCE images for early diagnosis of colorectal cancer. IJCARS **9**(2), 283–293 (2014)
13. Tajbakhsh, N., Gurudu, S.R., Liang, J.: Automated polyp detection in colonoscopy videos using shape and context information. TMI **35**(2), 630–644 (2015)
14. Vázquez, D., et al.: A benchmark for endoluminal scene segmentation of colonoscopy images. JHE (2017)
15. Wang, J., Huang, Q., Tang, F., Meng, J., Su, J., Song, S.: Stepwise feature fusion: local guides global. In: Wang, L., Dou, Q., Fletcher, P.T., Speidel, S., Li, S. (eds.) Medical Image Computing and Computer Assisted Intervention – MICCAI 2022: 25th International Conference, Singapore, September 18–22, 2022, Proceedings, Part III, pp. 110–120. Springer, Cham (2022). https://doi.org/10.1007/978-3-031-16437-8_11
16. Wang, W., et al.: Pyramid vision transformer: a versatile backbone for dense prediction without convolutions. In: ICCV, pp. 568–578 (2021)
17. Wei, J., Hu, Y., Zhang, R., Li, Z., Zhou, S.K., Cui, S.: Shallow Attention Network for Polyp Segmentation. In: de Bruijne, M., et al. (eds.) Medical Image Computing and Computer Assisted Intervention – MICCAI 2021: 24th International Conference, Strasbourg, France, September 27–October 1, 2021, Proceedings, Part I, pp. 699–708. Springer, Cham (2021). https://doi.org/10.1007/978-3-030-87193-2_66
18. Zhang, R., et al.: Lesion-aware dynamic kernel for polyp segmentation. In: Wang, L., Dou, Q., Fletcher, P.T., Speidel, S., Li, S. (eds.) Medical Image Computing and Computer Assisted Intervention – MICCAI 2022: 25th International Conference, Singapore, September 18–22, 2022, Proceedings, Part III, pp. 99–109. Springer, Cham (2022). https://doi.org/10.1007/978-3-031-16437-8_10
19. Zhao, X., Zhang, L., Lu, H.: Automatic polyp segmentation via multi-scale subtraction network. In: de Bruijne, M., et al. (eds.) Medical Image Computing and Computer Assisted Intervention – MICCAI 2021: 24th International Conference, Strasbourg, France, September 27–October 1, 2021, Proceedings, Part I, pp. 120–130. Springer, Cham (2021). https://doi.org/10.1007/978-3-030-87193-2_12

MUVF-YOLOX: A Multi-modal Ultrasound Video Fusion Network for Renal Tumor Diagnosis

Junyu Li[1,2,3], Han Huang[1,2,3], Dong Ni[1,2,3], Wufeng Xue[1,2,3], Dongmei Zhu[4(✉)], and Jun Cheng[1,2,3(✉)]

[1] National-Regional Key Technology Engineering Laboratory for Medical Ultrasound, School of Biomedical Engineering, Shenzhen University Medical School, Shenzhen University, Shenzhen, China
chengjun583@qq.com
[2] Medical UltraSound Image Computing (MUSIC) Lab, Shenzhen University, Shenzhen, China
[3] Marshall Laboratory of Biomedical Engineering, Shenzhen University, Shenzhen, China
[4] Department of Ultrasound, The Affiliated Nanchong Central Hospital of North Sichuan Medical College, Nanchong, China
zdm596987@gmail.com

Abstract. Early diagnosis of renal cancer can greatly improve the survival rate of patients. Contrast-enhanced ultrasound (CEUS) is a cost-effective and non-invasive imaging technique and has become more and more frequently used for renal tumor diagnosis. However, the classification of benign and malignant renal tumors can still be very challenging due to the highly heterogeneous appearance of cancer and imaging artifacts. Our aim is to detect and classify renal tumors by integrating B-mode and CEUS-mode ultrasound videos. To this end, we propose a novel multi-modal ultrasound video fusion network that can effectively perform multi-modal feature fusion and video classification for renal tumor diagnosis. The attention-based multi-modal fusion module uses cross-attention and self-attention to extract modality-invariant features and modality-specific features in parallel. In addition, we design an object-level temporal aggregation (OTA) module that can automatically filter low-quality features and efficiently integrate temporal information from multiple frames to improve the accuracy of tumor diagnosis. Experimental results on a multicenter dataset show that the proposed framework outperforms the single-modal models and the competing methods. Furthermore, our OTA module achieves higher classification accuracy than the frame-level predictions. Our code is available at https://github.com/JeunyuLi/MUAF.

Keywords: Multi-modal Fusion · Ultrasound Video · Object Detection · Renal Tumor

H. Greenspan et al. (Eds.): MICCAI 2023, LNCS 14224, pp. 642–651, 2023.
https://doi.org/10.1007/978-3-031-43904-9_62

1 Introduction

Renal cancer is the most lethal malignant tumor of the urinary system, and the incidence is steadily rising [13]. Conventional B-mode ultrasound (US) is a good screening tool but can be limited in its ability to characterize complicated renal lesions. Contrast-enhanced ultrasound (CEUS) can provide information on microcirculatory perfusion. Compared with CT and MRI, CEUS is radiation-free, cost-effective, and safe in patients with renal dysfunction. Due to these benefits, CEUS is becoming increasingly popular in diagnosing renal lesions. However, recognizing important diagnostic features from CEUS videos to diagnose lesions as benign or malignant is non-trivial and requires lots of experience.

To improve diagnostic efficiency and accuracy, many computational methods were proposed to analyze renal US images and could assist radiologists in making clinical decisions [6]. However, most of these methods only focused on conventional B-mode images. In recent years, there has been increasing interest in multi-modal medical image fusion [1]. Directly concatenation and addition were the most common methods, such as [3,4,12]. These simple operations might not highlight essential information from different modalities. Weight-based fusion methods generally used an importance prediction module to learn the weight of each modality and then performed sum, replacement, or exchange based on the weights [7,16,17,19]. Although effective, these methods did not allow direct interaction between multi-modal information. To address this, attention-based methods were proposed. They utilized cross-attention to establish the feature correlation of different modalities and self-attention to focus on global feature modeling [9,18]. Nevertheless, we prove in our experiments that these attention-based methods may have the potential risks of entangling features of different modalities.

In practice, experienced radiologists usually utilize dynamic information on tumors' blood supply in CEUS videos to make diagnoses [8]. Previous researches have proved that temporal information is effective in improving the performance of deep learning models. Lin et al.[11] proposed a network for breast lesion detection in US videos by aggregating temporal features, which outperformed other image-based methods. Chen et al. [2] showed that CEUS videos can provide more detailed blood supply information of tumors allowing a more accurate breast lesion diagnosis than static US images.

In this work, we propose a novel multi-modal US video fusion network (MUVF-YOLOX) based on CEUS videos for renal tumor diagnosis. Our main contributions are fourfold. (1) To the best of our knowledge, this is the first deep learning-based multi-modal framework that integrates both B-mode and CEUS-mode information for renal tumor diagnosis using US videos. (2) We propose an attention-based multi-modal fusion (AMF) module consisting of cross-attention and self-attention blocks to capture modality-invariant and modality-specific features in parallel. (3) We design an object-level temporal aggregation (OTA) module to make video-based diagnostic decisions based on the information from multi-frames. (4) We build the first multi-modal US video datatset containing

B-mode and CEUS-mode videos for renal tumor diagnosis. Experimental results show that the proposed framework outperforms single-modal, single-frame, and other state-of-the-art methods in renal tumor diagnosis.

2 Methods

2.1 Overview of Framework

The proposed MUVF-YOLOX framework is shown in Fig. 1. It can be divided into two stages: single-frame detection stage and video-based diagnosis stage. (1) In the single-frame detection stage, the network predicts the tumor bounding box and category on each frame in the multi-modal CEUS video clips. Dual-branch backbone is adopted to extract the features from two modalities and followed by the AMF module to fuse these features. During the diagnostic process, experienced radiologists usually take the global features of US images into consideration [20]. Therefore, we modify the backbone of YOLOX from CSP-Darknet to Swin-Transformer-Tiny, which is a more suitable choice by the virtue of its global modeling capabilities [15]. (2) In the video-based diagnosis stage, the network automatically chooses high-confidence region features of each frame according to the single-frame detection results and performs temporal aggregation to output a more accurate diagnosis. The above two stages are trained successively. We first perform a strong data augmentation to train the network for tumor detection and classification on individual frames. After that, the first stage model is switched to the evaluation mode and predicts the label of each frame in the video clip. Finally, we train the OTA module to aggregate the temporal information for precise diagnosis.

2.2 Dual-Attention Strategy for Multimodal Fusion

Using complementary information between multi-modal data can greatly improve the precision of detection. Therefore, we propose a novel AMF module to fuse the features of different modalities. As shown in Fig. 1, the features of each modality will be input into cross-attention and self-attention blocks in parallel to extract modality-invariant features and modality-specific features simultaneously.

Taking the B-mode as an example, we first map the B-mode features F_B and the CEUS-mode features F_C into (Q_B, K_B, V_B) and (Q_C, K_C, V_C) using linear projection. Then cross-attention uses scaled dot-product to calculate the similarity between Q_B and K_C. The similarity is used to weight V_C. Cross-attention extracts modality-invariant features through correlation calculation but ignores modality-specific features in individual modalities. Therefore, we apply self-attention in parallel to highlight these features. The self-attention calculates the similarity between Q_B and K_B and then uses the similarity to weight V_B. Similarly, the features of the CEUS modality go through the same process in parallel. Finally, we merge the two cross-attention outputs by addition

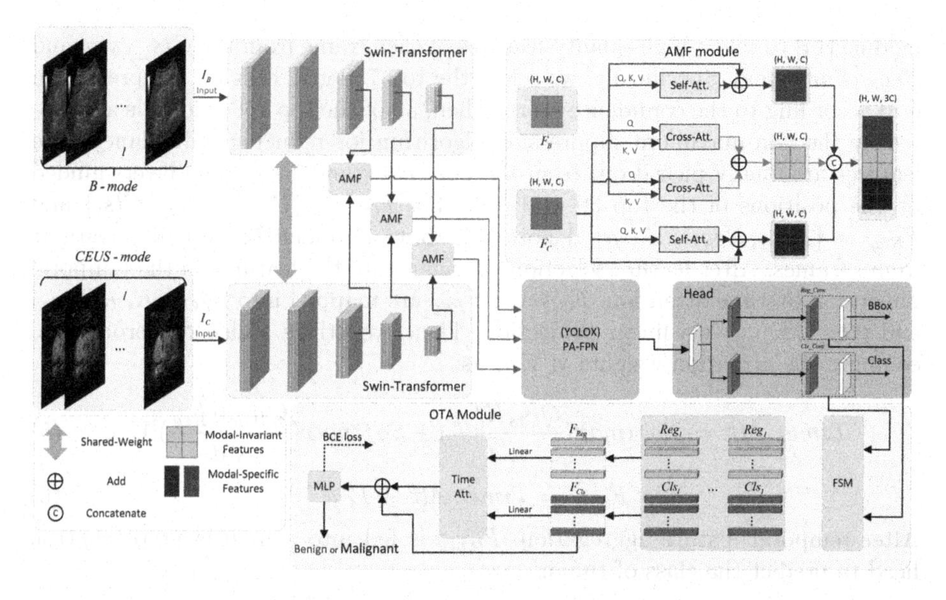

Fig. 1. Framework of MUVF-YOLOX. AMF module is used to fuse multi-modal features. OTA module is used to classify the tumor as benign or malignant based on videos. FSM means feature selection module.

since they are both invariant features of two modalities and concatenate the obtained sum and the outputs of the two self-attention blocks. The process mentioned above can be formulated as follows:

$$F_{invar} = Softmax(\frac{Q_B K_C^T}{\sqrt{d}})V_C + Softmax(\frac{Q_C K_B^T}{\sqrt{d}})V_B \tag{1}$$

$$F_{B-spec} = Softmax(\frac{Q_B K_B^T}{\sqrt{d}})V_B + F_B \tag{2}$$

$$F_{C-spec} = Softmax(\frac{Q_C K_C^T}{\sqrt{d}})V_C + F_C \tag{3}$$

$$F_{AMF} = Concat(F_{B-spec}, F_{invar}, F_{C-spec}) \tag{4}$$

where, F_{invar} represents the modality-invariant features. F_{B-spec} and F_{C-spec} represent the modal-specific features of B-mode and CEUS-mode respectively. F_{AMF} is the output of the AMF module.

2.3 Video-Level Decision Generation

In clinical practice, the dynamic changes in US videos provide useful information for radiologists to make diagnoses. Therefore, we design an OTA module that aggregates single-frame renal tumor detection results in temporal dimension for diagnosing tumors as benign and malignant. First, we utilize a feature selection module [14] to select high-quality features of each frame from the Cls_conv and Reg_conv layers. Specifically, we select the top 750 grid cells on the prediction grid according to the confidence score. Then, 30 of the top 750 grid cells are chosen by the non-maximum suppression algorithm for reducing redundancy. The features are finally picked out from the Cls_conv and Reg_conv layers guided by the positions of the top 30 grid cells. Let $F_{Cls} = \{Cls_1, Cls_2, ...Cls_l\}$ and $F_{Reg} = \{Reg_1, Reg_2, ...Reg_l\}$ denote the above obtained high-quality features from l frames. After feature selection, we aggregate the features in the temporal dimension by time attention. F_{Cls} and F_{Reg} are mapped into $(Q_{Cls}, K_{Cls}, V_{Cls})$ and (Q_{Reg}, K_{Reg}) via linear projection. Then, we utilize scaled dot-product to compute the attention weights of V_{Cls} as:

$$Time\_Att. = [Softmax(\frac{Q_{Cls}K_{Cls}^T}{\sqrt{d}}) + Softmax(\frac{Q_{Reg}K_{Reg}^T}{\sqrt{d}})]V_{Cls} \quad (5)$$

$$F_{temp} = Time\_Att. + F_{Cls} \quad (6)$$

After temporal feature aggregation, F_{temp} is fed into a multilayer perceptron head to predict the class of tumor.

3 Experimental Results

3.1 Materials and Implementations

We collect a renal tumor US dataset of 179 cases from two medical centers, which is split into the training and validation sets. We further collect 36 cases from the two medical centers mentioned above (14 benign cases) and another center (Fujian Provincial Hospital, 22 malignant cases) to form the test set. Each case has a video with simultaneous imaging of B-mode and CEUS-mode. Some examples of the images are shown in Fig. 2. There is an obvious visual difference between the images from the Fujian Provincial Hospital (last column in Fig. 2) and the other two centers, which raises the complexity of the task but can better verify our method's generalization ability. More than two radiologists with ten years of experience manually annotate the tumor bounding box and class label at the frame level using the Pair annotation software package (https://www.aipair.com.cn/en/, Version 2.7, RayShape, Shenzhen, China) [10]. Each case has 40–50 labeled frames, and these frames cover the complete contrast-enhanced imaging cycle. The number of cases and annotated frames is summarized in Table 1.

Weights pre-trained from ImageNet are used to initialize the Swin-Transformer backbone. Data augmentation strategies are applied synchronously to B-mode and CEUS-mode images for all experiments, including random rotation, mosaic, mixup, and so on. All models are trained for 150 epochs. The batch size is set to 2. We use the SGD optimizer with a learning rate of 0.0025. The weight decay is set to 0.0005 and the momentum is set to 0.9. In the test phase, we use the weights of the best model in validation to make predictions. All Experiments are implemented in PyTorch with an NVIDIA RTX A6000 GPU. AP_{50} and AP_{75} are used to assess the performance of single-frame detection. Accuracy and F1-score are used to evaluate the video-based tumor diagnosis.

Table 1. The details of our dataset. Number of cases in brackets.

Category	Training	Validation	Test
Benign	2775(63)	841(16)	875(14)
Malignant	4017(81)	894(19)	1701(22)
Total	6792(144)	1735(35)	2576(36)

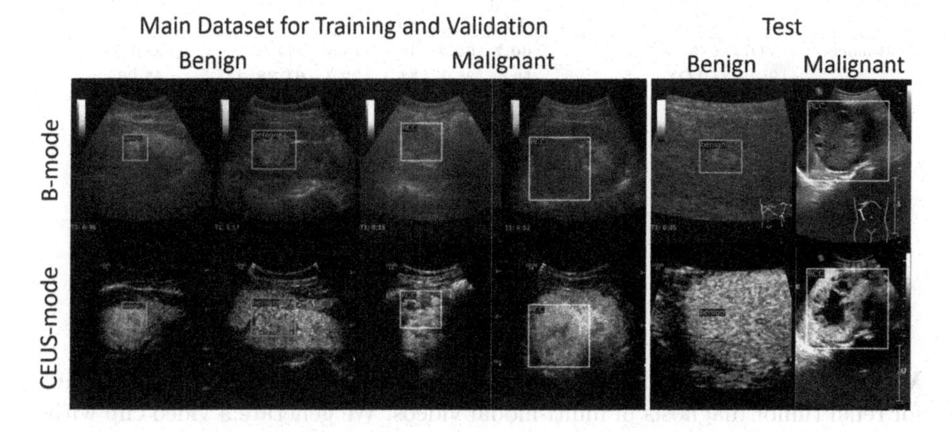

Fig. 2. Examples of the annotated B-mode and CEUS-mode US images.

3.2 Ablation Study

Single-Frame Detection. We explore the impact of different backbones in YOLOX and different ways of multi-modal fusion. As shown in Table 2, using Swin-Transformer as the backbone in YOLOX achieves better performance than the original backbone while reducing half of the parameters. The improvement may stem from the fact that Swin-Transformer has a better ability to characterize global features, which is critical in US image diagnosis. In addition, we explore the role of cross-attention and self-attention blocks in multi-modal tasks, as well as the optimal strategy for combining their outputs. Comparing row 5 with row 7 and row 8 in Table 2, the dual-attention mechanism outperforms the single cross-attention. It indicates that we need to pay attention to both modality-invariant and modality-specific features in our multi-modal task through cross-attention and self-attention blocks. However, "CA+SA" (row 6 in Table 2) obtains inferior performance than "CA" (row 5 in Table 2). We conjecture that connecting the two attention modules in series leads to the entanglement of modality-specific and modality-invariant information, which would disrupt the model training. On the contrary, the "CA//SA" method, combining two attention modules in parallel, enables the model to capture and digest modality-specific and modality-invariant features independently. For the same reason, we concatenate the outputs of the attention blocks rather than summing, which further avoids confusing modality-specific and modality-invariant information. Therefore, the proposed method achieves the best performance.

Table 2. The results of ablation study. "CA" and "SA" denote cross-attention and self-attention respectively. "//" and "+" mean parallel connection and series connection.

Modal	Network	Validation		Test		Flops (GLOPS)	Params (M)
		AP_{50}	AP_{75}	AP_{50}	AP_{75}		
B-mode	YOLOX [5]	**60.7**	**39.7**	48.5	19.6	140.76	99.00
	Swin-YOLOX	59.6	38.0	**58.1**	**22.1**	**61.28**	**45.06**
CEUS-mode	YOLOX [5]	52.3	29.7	49.1	**17.8**	140.76	99.00
	Swin-YOLOX	**60.1**	**30.6**	**52.1**	14.1	**61.28**	**45.06**
Multi-modal	CA (CMF [18])	81.4	54.2	75.2	35.2	103.53	51.26
	CA+SA (TMM [9])	80.8	52.7	74.3	37.0	109.19	57.46
	CA//SA (Ours w/o Concat)	82.0	56.9	74.6	35.0	109.19	57.46
	Ours	**82.8**	**60.6**	**79.5**	**39.2**	117.69	66.76

Video-Based Diagnosis. We investigate the performance of the OTA module for renal tumor diagnosis in multi-modal videos. We generate a video clip with l frames from annotated frames at a fixed interval forward. As shown in Table 3, gradually increasing the clip length can effectively improve the accuracy. This suggests that the multi-frame model can provide a more comprehensive characterization of the tumor and thus achieves better performance. Meanwhile, increasing the sampling interval tends to decrease the performance (row 4 and row 5 in Table 3). It indicates that continuous inter-frame information is beneficial for renal tumor diagnosis.

Table 3. The results of video-based diagnosis.

Clip Length	Sampling Interval	Validation		Test	
		Accuracy (%)	F1-score (%)	Accuracy (%)	F1-score (%)
1	1	81.6	81.6	90.3	89.2
2	1	82.1	82.1	90.5	89.3
2	2	82.9	82.9	90.0	88.7
4	1	83.7	83.7	**91.0**	**90.0**
4	2	82.6	82.6	90.8	89.7
8	1	**84.0**	**84.0**	90.9	89.9

3.3 Comparison with Other Methods

The comparison results are shown in Table 4. Compared to the single-modal models, directly concatenating multi-modal features (row 3 in Table 4) improves

AP_{50} and AP_{75} by more than 15%. This proves that complementary information exists among different modalities. For a fair comparison with other fusion methods, we embed their fusion modules into our framework so that different approaches can be validated in the same environment. CMML [19] and CEN [17] merge the multi-modal features or pick one of them by automatically generating channel-wise weights for each modality. They score higher AP in the validation set but lower one in the test set than "Concatenate". This may be because the generated weights are biased to make similar decisions to the source domain, thereby reducing model generalization in the external data. Moreover, CMF only highlights similar features between two modalities, ignoring that each modality contains some unique features. TMM focuses on both modality-specific and modality-invariant information, but the chaotic confusion of the two types of information deteriorates the model performance. Therefore, both CMF [17] and TMM [9] fail to outperform weight-based models. On the contrary, our AMF module prevents information entanglement by conducting cross-attention and self-attention blocks in parallel. It achieves $AP_{50} = 82.8$, $AP_{75} = 60.6$ in the validation set and $AP_{50} = 79.5$, $AP_{75} = 39.2$ in the test set, outperforming all competing methods while demonstrating superior generalization ability. Meanwhile, the improvement of the detection performance is beneficial to our OTA module to obtain lesion features from more precise locations, thereby improving the accuracy of benign and malignant renal tumor diagnosis.

Table 4. Diagnosis results of different methods.

Fusion Methods	Validation				External Test			
	AP_{50}	AP_{75}	Accuracy	F1-score	AP_{50}	AP_{75}	Accuracy	F1-score
B-mode	59.6	38.0	76.4	76.3	58.1	22.1	80.4	79.1
CEUS-mode	60.1	30.6	78.2	78.1	52.1	14.1	70.5	69.3
Concatenate	78.8	50.5	79.6	79.5	76.8	38.8	86.8	85.7
CMML [19]	80.7	54.4	80.1	80.1	76.0	37.2	87.4	86.2
CEN [17]	81.4	56.2	83.0	83.0	74.3	36.3	85.1	83.8
CMF [18]	81.4	54.8	79.7	79.7	75.2	35.2	87.8	86.8
TMM [9]	80.8	52.7	80.1	80.1	74.3	37.0	84.4	83.2
Ours	**82.8**	**60.6**	**84.0**	**84.0**	**79.5**	**39.2**	**90.9**	**89.9**

4 Conclusions

In this paper, we create the first multi-modal CEUS video dataset and propose a novel attention-based multi-modal video fusion framework for renal tumor diagnosis using B-mode and CEUS-mode US videos. It encourages interactions between different modalities via a weight-sharing dual-branch backbone and automatically captures the modality-invariant and modality-specific information by the AMF module. It also utilizes a portable OTA module to aggregate information in the temporal dimension of videos, making video-level decisions. The design of the AMF module and OTA module is plug-and-play and could be applied to other multi-modal video tasks. The experimental results show that the proposed method outperforms single-modal, single-frame, and other state-of-the-art multi-modal approaches.

Acknowledgment. Our dataset was collected from The Affiliated Nanchong Central Hospital of North Sichuan Medical College, Shenzhen People's Hospital, and Fujian Provincial Hospital hospitals. This study was approved by local institutional review boards. This work is supported by the Guangdong Basic and Applied Basic Research Foundation (No. 2021B1515120059), National Natural Science Foundation of China (No. 62171290), and Shenzhen Science and Technology Program (No. SGDX20201103095613036 and 20220810145705001).

References

1. Azam, M.A., et al.: A review on multimodal medical image fusion: compendious analysis of medical modalities, multimodal databases, fusion techniques and quality metrics. Comput. Biol. Med. **144**, 105253 (2022)
2. Chen, C., Wang, Y., Niu, J., Liu, X., Li, Q., Gong, X.: Domain knowledge powered deep learning for breast cancer diagnosis based on contrast-enhanced ultrasound videos. IEEE Trans. Med. Imaging **40**(9), 2439–2451 (2021)
3. Chen, H., Li, Y., Su, D.: Multi-modal fusion network with multi-scale multi-path and cross-modal interactions for RGB-D salient object detection. Pattern Recogn. **86**, 376–385 (2019)
4. Fang, J., et al.: Weighted concordance index loss-based multimodal survival modeling for radiation encephalopathy assessment in nasopharyngeal carcinoma radiotherapy. In: Wang, L., Dou, Q., Fletcher, P.T., Speidel, S., Li, S. (eds.) Medical Image Computing and Computer Assisted Intervention-MICCAI 2022: 25th International Conference, Singapore, 18–22 September 2022, Proceedings, Part VII, vol. 13437, pp. 191–201. Springer, Cham (2022). https://doi.org/10.1007/978-3-031-16449-1_19
5. Ge, Z., Liu, S., Wang, F., Li, Z., Sun, J.: YOLOx: exceeding YOLO series in 2021. arXiv preprint arXiv:2107.08430 (2021)
6. George, M., Anita, H.: Analysis of kidney ultrasound images using deep learning and machine learning techniques: a review. Pervasive Comput. Soc. Networking Proc. ICPCSN **2021**, 183–199 (2022)
7. Huang, H., et al.: Personalized diagnostic tool for thyroid cancer classification using multi-view ultrasound. In: Wang, L., Dou, Q., Fletcher, P.T., Speidel, S., Li, S. (eds.) Medical Image Computing and Computer Assisted Intervention-MICCAI 2022: 25th International Conference, Singapore, 18–22 September 2022, Proceedings, Part III, vol. 13433, pp. 665–674. Springer, Cham (2022). https://doi.org/10.1007/978-3-031-16437-8_64
8. Kapetas, P., et al.: Quantitative multiparametric breast ultrasound: application of contrast-enhanced ultrasound and elastography leads to an improved differentiation of benign and malignant lesions. Invest. Radiol. **54**(5), 257 (2019)
9. Li, X., Ma, S., Tang, J., Guo, F.: TranSiam: fusing multimodal visual features using transformer for medical image segmentation. arXiv preprint arXiv:2204.12185 (2022)

10. Liang, J., et al.: Sketch guided and progressive growing GAN for realistic and editable ultrasound image synthesis. Med. Image Anal. **79**, 102461 (2022)
11. Lin, Z., Lin, J., Zhu, L., Fu, H., Qin, J., Wang, L.: A new dataset and a baseline model for breast lesion detection in ultrasound videos. In: Wang, L., Dou, Q., Fletcher, P.T., Speidel, S., Li, S. (eds.) Medical Image Computing and Computer Assisted Intervention-MICCAI 2022: 25th International Conference, Singapore, 18–22 September 2022, Proceedings, Part III, vol. 13433, pp. 614–623. Springer, Cham (2022). https://doi.org/10.1007/978-3-031-16437-8_59
12. Liu, Y., Chen, X., Cheng, J., Peng, H.: A medical image fusion method based on convolutional neural networks. In: 2017 20th International Conference on Information Fusion (Fusion), pp. 1–7. IEEE (2017)
13. Ljungberg, B., et al.: European association of urology guidelines on renal cell carcinoma: the 2019 update. Eur. Urol. **75**(5), 799–810 (2019)
14. Shi, Y., Wang, N., Guo, X.: YOLOV: making still image object detectors great at video object detection. arXiv preprint arXiv:2208.09686 (2022)
15. Wang, W., Chen, C., Ding, M., Yu, H., Zha, S., Li, J.: TransBTS: multimodal brain tumor segmentation using transformer. In: de Bruijne, M., et al. (eds.) Medical Image Computing and Computer Assisted Intervention-MICCAI 2021: 24th International Conference, Strasbourg, France, 27 September–1 October 2021, Proceedings, Part I 24, pp. 109–119. Springer, Cham (2021). https://doi.org/10.1007/978-3-030-87193-2_11
16. Wang, Y., Huang, W., Sun, F., Xu, T., Rong, Y., Huang, J.: Deep multimodal fusion by channel exchanging. Adv. Neural. Inf. Process. Syst. **33**, 4835–4845 (2020)
17. Wang, Y., Sun, F., Huang, W., He, F., Tao, D.: Channel exchanging networks for multimodal and multitask dense image prediction. IEEE Trans. Pattern Anal. Mach. Intell. **45**(5), 5481–5496 (2022)
18. Xu, J., et al.: RemixFormer: a transformer model for precision skin tumor differential diagnosis via multi-modal imaging and non-imaging data. In: Wang, L., Dou, Q., Fletcher, P.T., Speidel, S., Li, S. (eds.) Medical Image Computing and Computer Assisted Intervention-MICCAI 2022: 25th International Conference, Singapore, 18–22 September 2022, Proceedings, Part III, vol. 13433, pp. 624–633. Springer, Cham (2022). https://doi.org/10.1007/978-3-031-16437-8_60
19. Yang, Y., Wang, K.T., Zhan, D.C., Xiong, H., Jiang, Y.: Comprehensive semi-supervised multi-modal learning. In: IJCAI, pp. 4092–4098 (2019)
20. Zhu, J., et al.: Contrast-enhanced ultrasound (CEUS) of benign and malignant renal tumors: distinguishing CEUS features differ with tumor size. Cancer Med. **12**(3), 2551–2559 (2022)

Unsupervised Classification of Congenital Inner Ear Malformations Using DeepDiffusion for Latent Space Representation

Paula López Diez[1(✉)], Jan Margeta[3,4], Khassan Diab[5], François Patou[2], and Rasmus R. Paulsen[1]

[1] DTU Compute, Technical University of Denmark, Kongens Lyngby, Denmark
plodi@dtu.dk
[2] Oticon Medical, Research and Technology, Smørum, Denmark
[3] Oticon Medical, Research and Technology, Vallauris, France
[4] KardioMe, Research and Development, Nova Dubnica, Slovakia
[5] Tashkent International Clinic, Tashkent, Uzbekistan

Abstract. The identification of congenital inner ear malformations is a challenging task even for experienced clinicians. In this study, we present the first automated method for classifying congenital inner ear malformations. We generate 3D meshes of the cochlear structure in 364 normative and 107 abnormal anatomies using a segmentation model trained exclusively with normative anatomies. Given the sparsity and natural unbalance of such datasets, we use an unsupervised method for learning a feature representation of the 3D meshes using DeepDiffusion. In this approach, we use the PointNet architecture for the network-based unsupervised feature learning and combine it with the diffusion distance on a feature manifold. This unsupervised approach captures the variability of the different cochlear shapes and generates clusters in the latent space which faithfully represent the variability observed in the data. We report a mean average precision of 0.77 over the seven main pathological subgroups diagnosed by an ENT (Ear, Nose, and Throat) surgeon specialized in congenital inner ear malformations.

Keywords: Unsupervised · Classification · DeepDiffusion · Inner Ear

1 Introduction

Inner ear malformations are found in 20–30% of children with congenital hearing loss [1]. While the prevalence of bilateral congenital hearing loss is estimated to be 1.33 per 1000 live births in North America and Europe, it is much higher in sub-Saharan Africa (19 per 1,000 newborns) and South Asia (up to 24 per 1,000) [8]. Early detection of sensorineural hearing loss is crucial for appropriate intervention, such as cochlear implant therapy, which is prescribed to approximately

H. Greenspan et al. (Eds.): MICCAI 2023, LNCS 14224, pp. 652–662, 2023.
https://doi.org/10.1007/978-3-031-43904-9_63

80,000 infants and toddlers annually worldwide [16]. Radiological examination is essential for an early diagnosis of congenital inner ear malformation, particularly when cochlear implant therapy is planned. However, detecting and classifying such malformations from standard imaging modalities is a complex task even for expert clinicians, and presents challenges during CI surgery [2]. Previous studies have proposed methods to classify congenital inner ear malformations based on explicit measurements and visual analysis of CT scans [5]. These methods are time-consuming and subject to clinician subjectivity. A suggested approach for the automated detection of inner ear malformation has relied on deep reinforcement learning trained for landmark location in normal anatomies based on an anomaly detection technique [9]. However, this method is only limited to the detection of a malformation but does not attempt to classify them.

Currently, supervised deep metric learning garners significant interest due to its exceptional efficacy in data clustering and pathology classification. Most of these approaches are fully supervised and use supervisory signals that model the training by creating tuples of labeled training data. These tuples are then used to optimize the intra-class distance of the different samples in the latent space, as has been done mostly for 2D images [15,20,21] and 2D representation of 3D images [4]. Several recent studies have demonstrated promising outcomes from unsupervised contrastive learning from natural images. However, their utility in the medical image domain is limited due to the high degree of inter-class similarity. Particularly in heterogeneous real clinical datasets in which the image quality and appearance can significantly impact the performance of such methods, rendering them less effective. In [22] an unsupervised strategy to learn medical visual representations by exploiting naturally occurring paired descriptive text in 2D images is proposed. Typically, in 3D images, an unsupervised low-dimensional representation is utilized for further clustering, as demonstrated in [14]. Nonetheless, such approaches are commonly developed using quite homogeneous datasets that are not representative of real-world applications and the diverse clinical settings in which they must operate.

Our objective is to develop a fully automated pipeline for the classification of inner ear malformations, utilizing a relatively large and unique dataset of such anomalies. The pipeline's design necessitates a profound comprehension of this data type and the congenital malformations themselves. Given the CT scans in this region are complex, and the images originate from diverse sources, we employ an unsupervised approach, uniquely based on the 3D shape of the cochlear structure. We have observed that the cochlear structure can be roughly but consistently segmented by a 3D-UNET model trained exclusively on normal cochlear anatomies. We then use these segmentations and adopt an entirely unsupervised approach, meaning the deep learning model is trained from scratch on these segmentations, and the class labels are not used for training. To map these shapes to an optimal latent space representation, we utilize DeepDiffusion, which combines the diffusion distance on a feature manifold with the feature learning of the encoder.

In this paper, we present the first automatic approach for the classification of congenital inner ear malformations. We use an unsupervised method to find the latent space representation of cochlear shapes, which allows for their further classification. We demonstrate that shapes from a segmentation model trained on normative cases, albeit imperfect, can be used to represent abnormalities. Moreover, our results indicate the potential for successfully applying this approach to other anatomies.

2 Data

Our dataset comprises a total of 485 clinical CT scans, consisting of 364 normal scans and 121 scans with various types of inner ear malformations. The distribution of inner ear scans for each type of malformation is shown in Fig. 1. We utilized the region-of-interest (ROI) extraction technique developed by [18], which involves selecting anatomical points of interest that are not part of the inner ear region to achieve a standardized and robust image orientation. To ensure consistency, all images were resampled to a spacing of 0.125 mm, and their intensities were normalized by scaling the 5th and 95th percentiles of the intensity distribution of each image to 0 and 1, respectively. Figure 1 also shows the data split used for training our model. We chose to use an approximate 50% split for abnormal cases, while the vast majority of normal cases, approximately 86%, were used for training. Other configurations were explored, including using only normal cases for training. However, it was demonstrated that while this approach may work for anomaly detection, it does not adequately categorize the different types of malformations.

3 Methods

3.1 Anatomical Representation

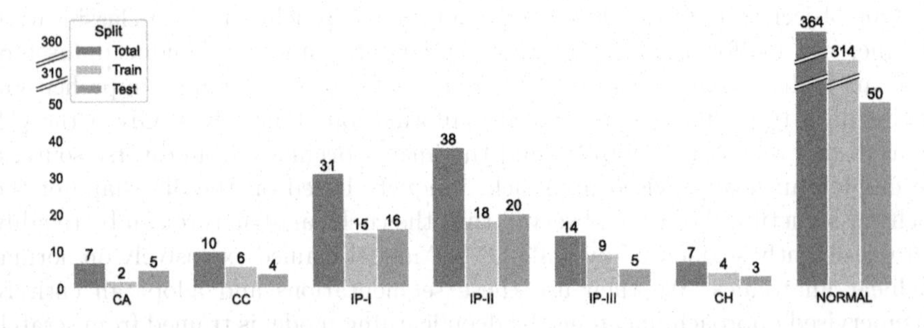

Fig. 1. Distribution of cases among the different classes and the split used for our approach. Cochlear aplasia (CA), common cavity (CC) incomplete partitioning type I, II, and III (IP-I, IP-II, IP-III), cochlear hypoplasia (CH), and normal.

Our aim is to find a parametrized shape that is representative of the anatomy of the patient. We decided to focus on the cochlear structure as it is the main structure of interest when trying to identify a malformation in the inner ear. To obtain a 3D segmentation of this structure we use the 3D-UNET [19] presented in [10] which has been trained exclusively in normal anatomies (130 images from diverse imaging equipment) and built using MONAI [12]. Even though no abnormal anatomies have been used for training, given the high contrast between the soft tissue of the cochlear structure and the bony structure that surrounds it, the model still performs quite well to segment the abnormal cases. This can be seen in Fig. 2 where an example of each of the types of malformations used in this study and an anatomically normal case are shown. The largest connected component of the segmentation has been selected to generate the final 3D meshes.

An overview of our pipeline is presented in Fig. 3. Each 3D mesh obtained from a CT image is transformed into a 1024 point cloud using the Ohbuchi method [13]. Each shape is then normalized by centering its origin in its center of gravity and enclosing the shape within a unit sphere, resulting in the point cloud representation of the shape S. Before the shape S is fed to the encoder, the shape is augmented into shape $\hat{S}$ with a probability of 0.8. This augmentation consists of a random rotation with U($-5°$,$5°$), an anisotropic scaling sampled from U(0.8,1), and a shearing and translation in each axes sampled from U(-0.2,0.2) for both actions.

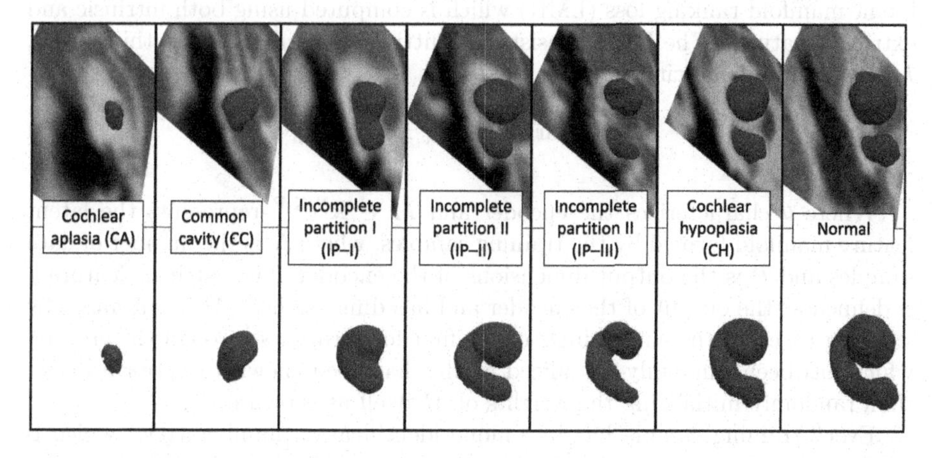

Fig. 2. Representative 3D-UNET segmentation meshes from each type of cochlear anatomy used in this study. Top row shows the 3D mesh with the original CT scan image; the bottom row shows exclusively the 3D mesh.

3.2 Deep Diffusion Algorithm

The DeepDiffusion (DD) algorithm [7] incorporates the manifold ranking [23] technique, which uses similarity diffusion on the manifold graph to learn a distance metric among the samples. The DD algorithm optimizes both the feature extraction and the embeddings produced by the encoder, which results in

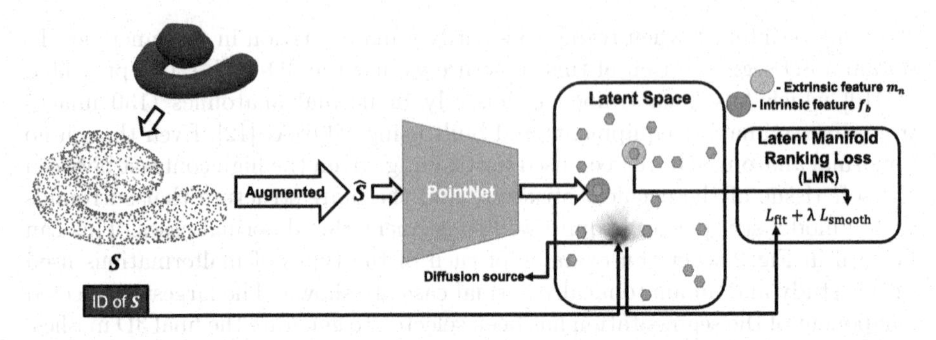

Fig. 3. Sketch of the DeepDiffusion used for latent space representation of the cochlear 3D meshes. The pointcloud extracted from the mesh is fed to the PointNet encoder which generates the corresponding latent feature which is optimized by minimizing the LMR loss so the encoder and the latent feature manifold are optimized for the comparison of data samples.

salient features in a continuous and smooth latent space. In this latent space, the Euclidean distance among the latent features approximates the diffusion distance on the latent feature manifold. The crux behind this algorithm is the latent manifold ranking loss (LMR) which is computed using both intrinsic and extrinsic features. The LMR consists of a fitting term, L_{fit}, a smoothing term, L_{smooth}, and a balancing term, λ.

$$\text{LMR} = \underset{M,\theta}{\arg\min} \; L_{\text{fit}} \pm \lambda L_{\text{smooth}} \tag{1}$$

Where θ characterizes the encoder and $M \in \mathbb{R}^{\mathrm{N \; X \; P}}$ represents the latent feature manifold formed by the training samples, where N is the number of data samples and P is the output dimensions of the encoder. The extrinsic feature f is defined as the output of the encoder and has dimension P. M is initialized by stacking together the embeddings of the first forward pass through the encoder which has been randomly initialized as this has been shown to perform better than randomly initializing the weights of M itself as shown in [7].

Every training sample has its unique identification number (ID_b) which is used to specify a diffusion source y_b that is consistent throughout the training procedure. L_{fit} constrains the ranking vector r_b to being close to the diffusion source y_b, which is defined as the vector containing one-hot encoding of ID_b. The ranking vector is defined as $r_b = \text{softmax}(f_b M^T)$ and represents the probabilistic similarities between the feature f_b and all the intrinsic features contained in M. The fitting term is therefore defined as

$$L_{\text{fit}} = \sum_b \text{CrossEntropy}(r_b, y_b)$$

its minimization results in all the extrinsic features being embedded farther away from each other as they are being pulled toward their respective and unique diffusion source vectors. The smoothing term is defined as

$$L_{\text{smooth}} = \sum_{b} \sum_{n} w_{bn} \text{Dissimilarity}(r_b, t_n) \tag{2}$$

where the dissimilarity operator is the Jensen-Shannon divergence [6] and $t_n = \text{softmax}(m_n M^T)$ being m_n the nth row of the matrix M so that t_n contains the ranking score of the intrinsic feature m_n to all the intrinsic features. w_{bn} indicates the similarity between the extrinsic feature f_b and the neighboring intrinsic feature m_n and it is defined as:

$$w_{bn} = \begin{cases} f_b m_n^T, & \text{m}_n \in \text{kNN}(f_b) \\ 0 & \text{otherwise} \end{cases} \tag{3}$$

Minimizing L_{smooth} pulls extrinsic features and their neighboring intrinsic features together which implies that an extrinsic feature is more likely to be projected onto the surface of the latent feature manifold of the intrinsic features when L_{smooth} is smaller.

3.3 Implementation

For our encoder, we use the PointNet [3] architecture which takes 1024 3D points as input, applies input and feature transformations, and then aggregates point features by max pooling to a feature of dimensionality 1024 which is then compressed into dimensionality 254 with two sets of fully connected layers. The network has been trained by using mini-batch (of size 8) gradient descent using the Adam optimizer with a learning rate of 10^{-8} and ReLU as the activation function. The DD algorithm is implemented in PyTorch [17] and the code used for this study is available at https://github.com/paulalopez10/Deep-Diffusion-Unsupervised-Classification-3D-Mesh. The models are trained on an NVIDIA GeForce RTX 3070 Laptop GPU with 8GB VRAM. The different hyper-parameters related to the approach have been explored and it has been empirically found for this specific configuration $\lambda = 0.6$ and $k = 10$ produce the best results that will be analyzed in the following section.

4 Results

We evaluate the classification performance of our pipeline by analyzing the embeddings generated by the trained encoder. To visualize the projection of the features of the test in 2D we use the U-MAP [11], as illustrated in Fig. 4. The U-MAP visualization demonstrates the clustering of different classes in the latent space. Furthermore, it is very interesting to notice how the latent space representation displays the anatomical changes of the anatomy where the more extreme types of malformations (CA and CC) are the most distant to the normative cochlear structures. The transition between the different classes shown in the latent space properly represents the pathological variations in this anatomy.

We have also included, in Fig. 4, the projection of the features projected in the 2D-PCA space defined by the training set, where both, training and testing, sets are included to show not only the clustering in this space but also the similar distribution of the different classes in both sets within the PCA projection.

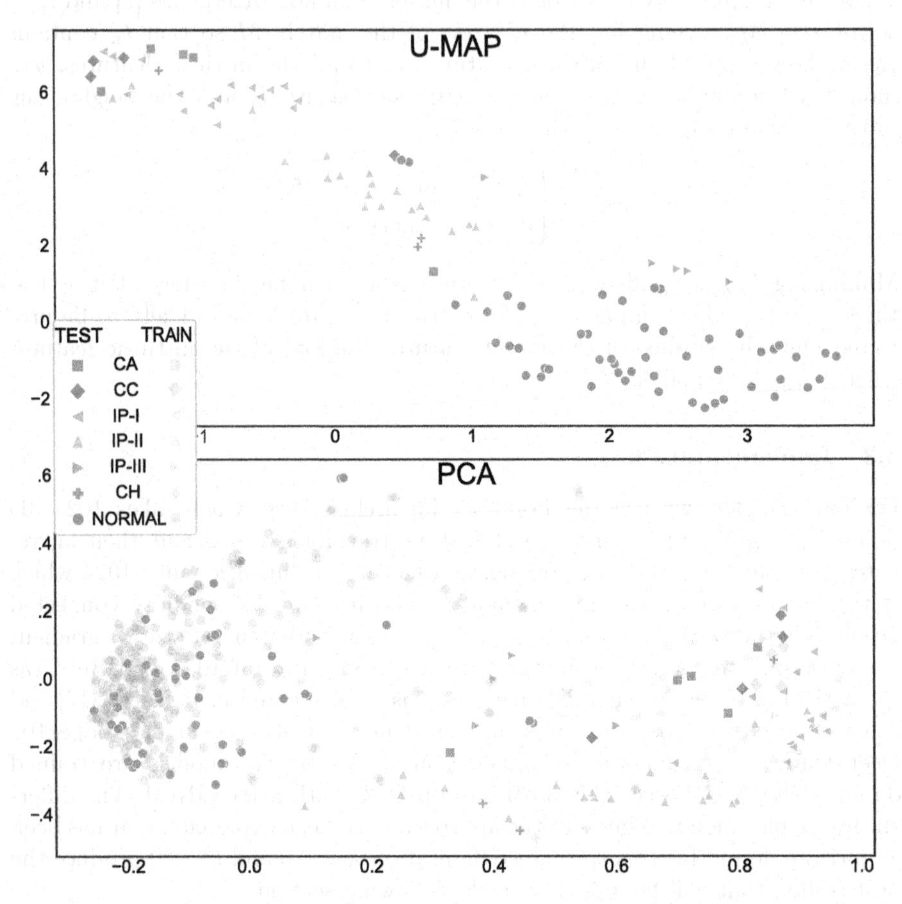

Fig. 4. Top: U-MAP representation of the test features where it can be observed how the different classes group together and how the anatomical variation is represented as there is a progression from the most abnormal cases towards fully normal cases. **Bottom**: Test and train features projected into the 2D PCA space defined by the training samples, where the classes are separated and show consistency between training and testing samples.

For a further analysis of the performance, we compute some evaluation metrics based on the pairwise cosine distance between samples that can be seen in Fig. 5 c). The average ROC and precision-recall curves for each of the classes can be seen in Fig. 5 a) and b). To calculate those, each test feature vector f_b is considered to be the centroid of a kNN(f_b) which consists in the k nearest

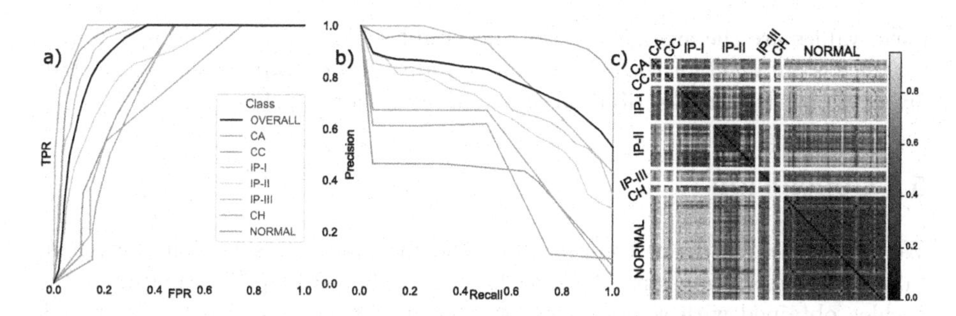

Fig. 5. Evaluation plots. **a)**Mean ROC curves for each class **b)**Mean Recall-Precision curve for each class **c)**Pairwise cosine distance between test embeddings used to evaluate the performance.

Table 1. Evaluation metrics reported in our experiment. ROC-Receiver operating characteristic, AUC- Area under the curve, AP-Average precision, PR - Precision-recall

	CA	CC	IP-I	IP-II	IP-III	CH	NORMAL	Overall
Max Accuracy	0.98	0.96	0.92	0.96	0.98	0.99	0.91	0.93
Mean Accuracy	0.73 ±0.25	0.77 ±0.26	0.87 ±0.03	0.77 ±0.08	0.96 ±0.03	0.69 ±0.01	0.75 ±0.05	0.78 ±0.12
Max ROC-AUC	0.99	0.99	0.98	0.96	0.99	0.98	0.99	0.99
Mean ROC-AUC	0.75 ±0.25	0.79 ±0.25	0.94 ±0.04	0.84 ±0.10	0.98 ±0.02	0.71 ±0.08	0.95 ±0.09	0.91 ±0.13
Max AP	0.57	0.70	0.87	0.88	0.92	0.51	0.99	0.91
Mean AP	0.41 ±0.42	0.38 ±0.33	0.70 ±0.23	0.65 ±0.32	0.82 ±0.25	0.50 ±0.42	0.94 ±0.12	0.77 ±0.29
Max f1-score	0.67	0.57	0.67	0.88	0.86	0.67	0.91	0.75
Max PR-AUC	0.63	0.80	0.84	0.85	0.91	0.71	0.95	0.88
Mean PR-AUC	0.40 ±0.26	0.37 ±0.25	0.67 ±0.04	0.62 ±0.11	0.78 ±0.02	0.48 ±0.08	0.90 ±0.10	0.74 ±0.24

features from other samples using the cosine pairwise distance shown in Fig. 5 c). We vary k until all the features from the corresponding class are within the cluster and compute the precision and false positive rate per the different recall steps, the shown results are the average among each class and overall. With the same procedure, different evaluation metrics have been obtained and are shown in Table 1. These metrics encompass the area under the curve (AUC) for the curves shown in Fig. 5 a) and b), both for the average curve and the optimal curve for each class. Furthermore, the maximum and average accuracy has been computed together with the maximum f1-score. Considering the dataset's significant class imbalance, these metrics provide a comprehensive assessment of the performance achieved. Finally, the mean average precision is also included in the table together with the optimal one for each class. The optimal or maximum value of each metric corresponds to when the optimal sample within our test features distribution is being evaluated as the centroid of its own class and the mean values are the average over all the samples. We can observe how a bigger variance is obtained for the classes that contain a few examples as it is expected, given the nature and distribution of our dataset shown in Fig. 1.

5 Conclusion

We have presented the first approach for the automatic classification of congenital inner ear malformations. We show how using the 3D shape information of the cochlea obtained with a model only trained in normative anatomies is enough to classify the malformations and reduces the influence of the image's source, which is crucial in a clinical application setting.

Our method shows a mean average precision of 0.77 with a mean ROC-AUC of 0.91, indicating its effectiveness in classifying inner ear malformations. Furthermore, the representation of the different cases in the latent space shows spatial relation between classes, which is correlated with the anatomical appearance of the different malformations. These promising results pave the way towards assisting clinicians in the challenging assessment of congenital inner ear malformations potentially leading to improved patient outcome of cochlear surgery.

References

1. Brotto, D., et al.: Genetics of inner ear malformations: a review. Audiol. Res. **11**(4), 524–536 (2021). https://doi.org/10.3390/audiolres11040047
2. Chakravorti, S., et al.: Further evidence of the relationship between cochlear implant electrode positioning and hearing outcomes. Otol. Neurotol. **40**(5), 617–624 (2019). https://doi.org/10.1097/MAO.0000000000002204
3. Charles, R.Q., Su, H., Kaichun, M., Guibas, L.J.: Pointnet: Deep learning on point sets for 3D classification and segmentation. In: 2017 IEEE Conference on Computer Vision and Pattern Recognition (CVPR), pp. 77–85 (2017). https://doi.org/10.1109/CVPR.2017.16
4. Chen, X., Wang, W., Jiang, Y., Qian, X.: A dual-transformation with contrastive learning framework for lymph node metastasis prediction in pancreatic cancer. Med. Image Anal. **85**, 102753 (2023). https://doi.org/10.1016/j.media.2023.102753, https://www.sciencedirect.com/science/article/pii/S1361841523000142
5. Dhanasingh, A.E., et al.: A novel three-step process for the identification of inner ear malformation types. Laryngoscope Investigative Otolaryngology (2022). https://doi.org/10.1002/lio2.936, https://onlinelibrary.wiley.com/doi/10.1002/lio2.936
6. Fuglede, B., Topsoe, F.: Jensen-shannon divergence and hilbert space embedding. In: International Symposium onInformation Theory, 2004. ISIT 2004. Proceedings, pp. 31- (2004). https://doi.org/10.1109/ISIT.2004.1365067
7. Furuya, T., Ohbuchi, R.: Deepdiffusion: unsupervised learning of retrieval-adapted representations via diffusion-based ranking on latent feature manifold. IEEE Access **10**, 116287–116301 (2022). https://doi.org/10.1109/ACCESS.2022.3218909

8. Korver, A.M., et al.: Congenital hearing loss. Nature Rev. Disease Primers **3**(1), 1–17 (2017)
9. López Diez, P., et al.: Deep reinforcement learning for detection of inner ear abnormal anatomy in computed tomography. In: Wang, L., Dou, Q., Fletcher, P.T., Speidel, S., Li, S. (eds.) Medical Image Computing and Computer Assisted Intervention - MICCAI 2022, pp. 697–706. Springer Nature Switzerland, Cham (2022). https://doi.org/10.1007/978-3-031-16437-8_67
10. Margeta, J., et al.: A web-based automated image processing research platform for cochlear implantation-related studies. J. Clin. Med. **11**(22) (2022). https://doi.org/10.3390/jcm11226640, https://www.mdpi.com/2077-0383/11/22/6640
11. McInnes, L., Healy, J., Saul, N., Großberger, L.: Umap: Uniform manifold approximation and projection. J. Open Source Softw. **3**(29), 861 (2018). https://doi.org/10.21105/joss.00861
12. MONAI-Consortium: Monai: Medical open network for AI (2022). https://doi.org/10.5281/zenodo.7459814
13. Ohbuchi, R., Minamitani, T., Takei, T.: Shape-similarity search of 3D models by using enhanced shape functions. Int. J. Comput. Appl. Technol. **23**(2–4), 70–85 (2005)
14. Onga, Y., Fujiyama, S., Arai, H., Chayama, Y., Iyatomi, H., Oishi, K.: Efficient feature embedding of 3d brain mri images for content-based image retrieval with deep metric learning. In: 2019 IEEE International Conference on Big Data (Big Data), pp. 3764–3769. IEEE (2019)
15. Pal, A., et al.: Deep metric learning for cervical image classification. IEEE Access **9**, 53266–53275 (2021)
16. Paludetti, G., et al.: Infant hearing loss: from diagnosis to therapy official report of xxi conference of Italian society of pediatric otorhinolaryngology. Acta Otorhinolaryngol. Italica **32**(6), 347 (2012)
17. Paszke, A., et al.: Pytorch: An imperative style, high-performance deep learning library. In: Advances in Neural Information Processing Systems 32, pp. 8024–8035. Curran Associates, Inc. (2019). http://papers.neurips.cc/paper/9015-pytorch-an-imperative-style-high-performance-deep-learning-library.pdf
18. Radutoiu, A.T., Patou, F., Margeta, J., Paulsen, R.R., López Diez, P.: Accurate localization of inner ear regions of interests using deep reinforcement learning. In: Lian, C., Cao, X., Rekik, I., Xu, X., Cui, Z. (eds.) Machine Learning in Medical Imaging. pp. 416–424. Springer Nature Switzerland, Cham (2022). https://doi.org/10.1007/978-3-031-21014-3_43
19. Ronneberger, O., Fischer, P., Brox, T.: U-net: convolutional networks for biomedical image segmentation. In: Navab, N., Hornegger, J., Wells, W.M., Frangi, A.F. (eds.) Medical Image Computing and Computer-Assisted Intervention - MICCAI 2015, pp. 234–241. Springer International Publishing, Cham (2015)
20. Sundgaard, J.V., et al.: Deep metric learning for otitis media classification. Med. Image Anal. **71**, 102034 (2021). https://doi.org/10.1016/j.media.2021.102034, https://www.sciencedirect.com/science/article/pii/S1361841521000803
21. Zhang, Y., Luo, L., Dou, Q., Heng, P.A.: Triplet attention and dual-pool contrastive learning for clinic-driven multi-label medical image classification. Med. Image Anal. **86**, 102772 (2023). https://doi.org/10.1016/j.media.2023.102772, https://www.sciencedirect.com/science/article/pii/S1361841523000336

22. Zhang, Y., Jiang, H., Miura, Y., Manning, C.D., Langlotz, C.P.: Contrastive learning of medical visual representations from paired images and text. In: Proceedings of the 7th Machine Learning for Healthcare Conference. Proceedings of Machine Learning Research, vol. 182, pp. 2–25. PMLR (2022). https://proceedings.mlr.press/v182/zhang22a.html
23. Zhou, D., Weston, J., Gretton, A., Bousquet, O., Schölkopf, B.: Ranking on data manifolds. In: Thrun, S., Saul, L., Schölkopf, B. (eds.) Advances in Neural Information Processing Systems. vol. 16. MIT Press (2003). https://proceedings.neurips.cc/paper/2003/file/2c3ddf4bf13852db711dd1901fb517fa-Paper.pdf

How Does Pruning Impact Long-Tailed Multi-label Medical Image Classifiers?

Gregory Holste[1], Ziyu Jiang[2], Ajay Jaiswal[1], Maria Hanna[3], Shlomo Minkowitz[3], Alan C. Legasto[3], Joanna G. Escalon[3], Sharon Steinberger[3], Mark Bittman[3], Thomas C. Shen[4], Ying Ding[1], Ronald M. Summers[4], George Shih[3], Yifan Peng[3(✉)], and Zhangyang Wang[1(✉)]

[1] The University of Texas at Austin, Austin, TX, USA
atlaswang@utexas.edu
[2] Texas A&M University, College Station, TX, USA
[3] Weill Cornell Medicine, New York, NY, USA
yip4002@med.cornell.edu
[4] Clinical Center, National Institutes of Health, Bethesda, MD, USA

Abstract. Pruning has emerged as a powerful technique for compressing deep neural networks, reducing memory usage and inference time without significantly affecting overall performance. However, the nuanced ways in which pruning impacts model behavior are not well understood, particularly for *long-tailed*, *multi-label* datasets commonly found in clinical settings. This knowledge gap could have dangerous implications when deploying a pruned model for diagnosis, where unexpected model behavior could impact patient well-being. To fill this gap, we perform the first analysis of pruning's effect on neural networks trained to diagnose thorax diseases from chest X-rays (CXRs). On two large CXR datasets, we examine which diseases are most affected by pruning and characterize class "forgettability" based on disease frequency and co-occurrence behavior. Further, we identify individual CXRs where uncompressed and heavily pruned models disagree, known as pruning-identified exemplars (PIEs), and conduct a human reader study to evaluate their unifying qualities. We find that radiologists perceive PIEs as having more label noise, lower image quality, and higher diagnosis difficulty. This work represents a first step toward understanding the impact of pruning on model behavior in deep long-tailed, multi-label medical image classification. All code, model weights, and data access instructions can be found at https://github.com/VITA-Group/PruneCXR.

Keywords: Pruning · Chest X-Ray · Imbalance · Long-Tailed Learning

Supplementary Information The online version contains supplementary material available at https://doi.org/10.1007/978-3-031-43904-9_64.

H. Greenspan et al. (Eds.): MICCAI 2023, LNCS 14224, pp. 663–673, 2023.
https://doi.org/10.1007/978-3-031-43904-9_64

1 Introduction

Deep learning has enabled significant progress in image-based computer-aided diagnosis [8,10,23,26,33]. However, the increasing memory requirements of deep neural networks limit their practical deployment in hardware-constrained environments. One promising approach to reducing memory usage and inference latency is model ***pruning***, which aims to remove redundant or unimportant model weights [21]. Since modern deep neural networks are often overparameterized, they can be heavily pruned with minimal impact on overall performance [6,20,22,34]. This being said, the impact of pruning on model behavior *beyond* high-level performance metrics like top-1 accuracy remain unclear. This gap in understanding has major implications for real-world deployment of neural networks for high-risk tasks like disease diagnosis, where pruning may cause unexpected consequences that could potentially threaten patient well-being.

To bridge this gap, this study aims to answer the following guiding questions by conducting experiments to dissect the differential impact of pruning:

Q1. What is the impact of pruning on overall performance in long-tailed multi-label medical image classification?
Q2. Which disease classes are most affected by pruning and why?
Q3. How does disease co-occurrence influence the impact of pruning?
Q4. Which individual images are most vulnerable to pruning?

We focus our experiments on thorax disease classification on chest X-rays (CXRs), a challenging ***long-tailed*** and ***multi-label*** computer-aided diagnosis problem, where patients may present with multiple abnormal findings in one exam and most findings are rare relative to the few most common diseases [12].

This study draws inspiration from Hooker *et al.* [13], who found that pruning disparately impacts a small subset of classes in order to maintain overall performance. The authors also introduced ***pruning-identified exemplars (PIEs)***, images where an uncompressed and heavily pruned model disagree. They discovered that PIEs share common characteristics such as multiple salient objects and noisy, fine-grained labels. While these findings uncover what neural networks "forget" upon pruning, the insights are limited to highly curated natural image datasets where each image belongs to one class. Previous studies have shown that pruning can enhance fairness [31], robustness [1], and efficiency for medical image classification [4,7,32] and segmentation [3,16,24,25,29] tasks. However, these efforts also either focused solely on high-level performance or did not consider settings with severe class imbalance or co-occurrence.

Unlike existing work, we explicitly connect class "forgettability" to the unique aspects of our problem setting: disease frequency (long-tailedness) and disease co-occurrence (multi-label behavior). Since many diagnostic exams, like CXR, are long-tailed and multi-label, this work fills a critical knowledge gap enabling more informed deployment of pruned disease classifiers. We hope that our findings can provide a foundation for future research on pruning in clinically realistic settings.

2 Methods

2.1 Preliminaries

Datasets. For this study, we use expanded versions of NIH ChestXRay14 [30] and MIMIC-CXR [18], two large-scale CXR datasets for multi-label disease classification.[1] As described in Holste *et al.* [12], we augmented the set of possible labels for each image by adding five new rare disease findings parsed from radiology reports. This creates a challenging long-tailed classification problem, with training class prevalence ranging from under 100 to over 70,000 (Supplement). ***NIH-CXR-LT*** contains 112,120 CXRs, each labeled with at least one of 20 classes, while ***MIMIC-CXR-LT*** contains 257,018 frontal CXRs labeled with at least one of 19 classes. Each dataset was split into training (70%), validation (10%), and test (20%) sets at the patient level.

Model Pruning & Evaluation. Following Hooker *et al.* [13], we focus on global unstructured ***L1 pruning*** [34]. After training a disease classifier, a fraction k of weights with the smallest magnitude are "pruned" (set to zero); for instance, $k = 0.9$ means 90% of weights have been pruned. While area under the receiver operating characteristic curve is a standard metric on related datasets [26,28,30], it can become heavily inflated in the presence of class imbalance [2,5]. Since we seek a metric that is both resistant to imbalance and captures performance across thresholds (as choosing a threshold is non-trivial in the multi-label setting [27]), we use **average precision (AP)** as our primary metric.

2.2 Assessing the Impact of Pruning

Experimental Setup. We first train a baseline model to classify thorax diseases on both NIH-CXR-LT and MIMIC-CXR-LT. The architecture used was a ResNet50 [9] with ImageNet-pretrained weights and a sigmoid cross-entropy loss. For full training details, please see the Supplemental Materials and code repository. Following Hooker *et al.* [13], we then repeat this process with 30 unique random initializations, performing L1 pruning at a range of sparsity ratios $\boldsymbol{k \in \{0, 0.05, \ldots, 0.9, 0.95\}}$ on each model and dataset. Using a "population" of 30 models allows for reliable estimation of model performance at each sparsity ratio. We then analyze how pruning impacts overall, disease-level, and image-level model behavior with increasing sparsity as described below.

Overall and Class-Level Analysis. To evaluate the overall impact of pruning, we compute the mean AP across classes for each sparsity ratio and dataset. We use Welch's t-test to assess performance differences between the 30 uncompressed models and 30 k-sparse models. We then characterize the class-level impact of pruning by considering the relative change in AP from an uncompressed model

[1] NIH ChestXRay14 can be found here, and MIMIC-CXR can be found here.

to its k-sparse counterpart for all k. Using relative change in AP allows for comparison of the *impact* of pruning regardless of class difficulty. We then define the ***forgettability curve*** of a class c as follows:

$$\left\{ \text{med} \left\{ \frac{\text{AP}_{i,k,c} - \text{AP}_{i,0,c}}{\text{AP}_{i,0,c}} \right\}_{i \in \{1,\ldots,30\}} \right\}_{k \in \{0,0.05,\ldots,0.9,0.95\}} \tag{1}$$

where $\text{AP}_{i,k,c} :=$ AP of the ith model with sparsity k on class c, and $\text{med}(\cdot) :=$ median across all 30 runs. We analyze how these curves relate to class frequency and co-occurrence using Pearson (r) and Spearman (ρ) correlation tests.

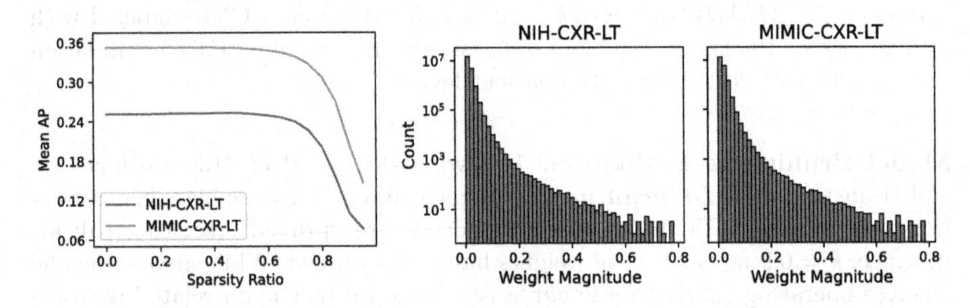

Fig. 1. Overall effect of pruning on disease classification performance. Presented is the mean AP (median across 30 runs) for sparsity ratios $k \in \{0, \ldots, 0.95\}$ (left) and log-scale histogram of model weight magnitudes (right).

Incorporating Disease Co-occurrence Behavior. For each *unique pair* of NIH-CXR-LT classes, we compute the ***Forgettability Curve Dissimilarity (FCD)***, the mean squared error (MSE) between the forgettability curves of each disease. FCD quantifies how similar two classes are with respect to their forgetting behavior over all sparsity ratios. Ordinary least squares (OLS) linear regression is employed to understand the interaction between difference in class frequency and class co-occurrence with respect to FCD for a given disease pair.

2.3 Pruning-Identified Exemplars (PIEs)

Definition. After evaluating the overall and class-level impact of pruning on CXR classification, we investigate which individual images are most vulnerable to pruning. Like Hooker *et al.* [13], we consider PIEs to be images where an uncompressed and pruned model disagree. Letting C be the number of classes, we compute the average prediction $\frac{1}{30}\sum_i \hat{\boldsymbol{y}}_{\mathbf{0}} \in \mathbb{R}^C$ of the uncompressed models and average prediction $\frac{1}{30}\sum_i \hat{\boldsymbol{y}}_{\mathbf{0.9}} \in \mathbb{R}^C$ of the L1-pruned models at 90% sparsity for all NIH-CXR-LT test set images. Then the Spearman rank correlation $\sigma(\frac{1}{30}\sum_i \hat{\boldsymbol{y}}_{\mathbf{0}}, \frac{1}{30}\sum_i \hat{\boldsymbol{y}}_{\mathbf{0.9}})$ represents the agreement between the uncompressed and heavily pruned models for each image; we define PIEs as images whose correlation falls in the bottom 5th percentile of test images.

Analysis and Human Study. To understand the common characteristics of PIEs, we compare how frequently (i) each class appears and (ii) images with $d = 0, \ldots, 3, 4+$ simultaneous diseases appear in PIEs relative to non-PIEs. To further analyze qualities of CXRs that require domain expertise, we conducted a human study to assess radiologist perceptions of PIEs. Six board-certified attending radiologists were each presented with a unique set of 40 CXRs (half PIE, half non-PIE). Each image was presented along with its ground-truth labels and the following three questions:

1. **Do you fully agree with the given label?** [Yes/No]
2. **How would you rate the image quality?** [1–5 Likert]
3. **How difficult is it to properly diagnose this image?** [1–5 Likert]

We use the Kruskal-Wallis test to evaluate differential perception of PIEs.

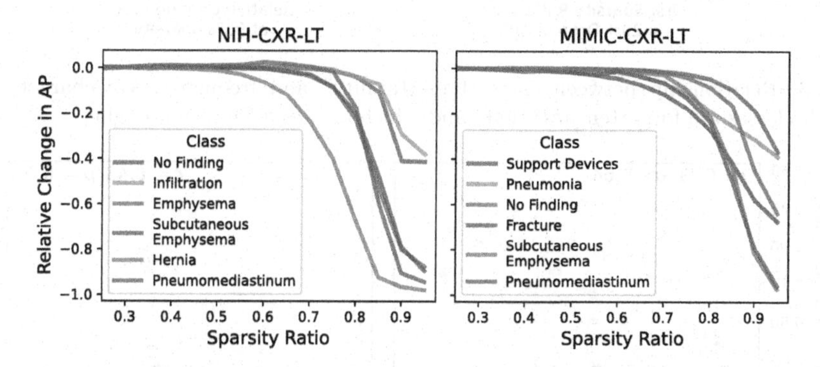

Fig. 2. "Forgettability curves" depicting relative change in AP (median across 30 runs at each sparsity ratio) upon L1 pruning for a subset of classes.

3 Results

3.1 What is the Overall Effect of Pruning?

We find that under L1 pruning, the first sparsity ratio causing a significant drop in mean AP is 65% for NIH-CXR-LT ($P < 0.001$) and 60% for MIMIC-CXR-LT ($P < 0.001$) (Fig. 1, left). This observation may be explained by the fact that ResNet50 is highly overparameterized for this task. Since only a subset of weights are required to adequately model the data, the trained classifiers have naturally sparse activations (Fig. 1, right). For example, over half of all learned weights have magnitude under 0.01. However, beyond a sparsity ratio of 60%, we observe a steep decline in performance with increasing sparsity for both datasets.

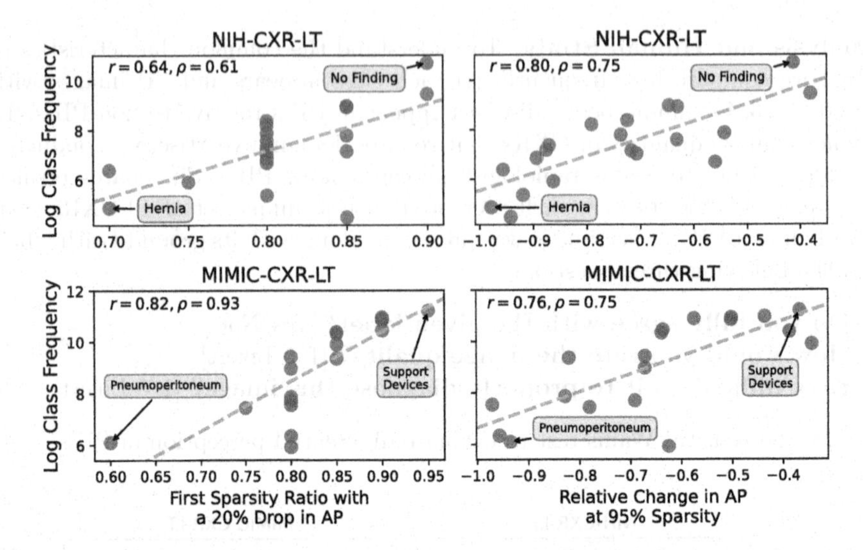

Fig. 3. Relationship between class "forgettability" and frequency. We characterize which classes are forgotten *first* (left) and which are *most* forgotten (right).

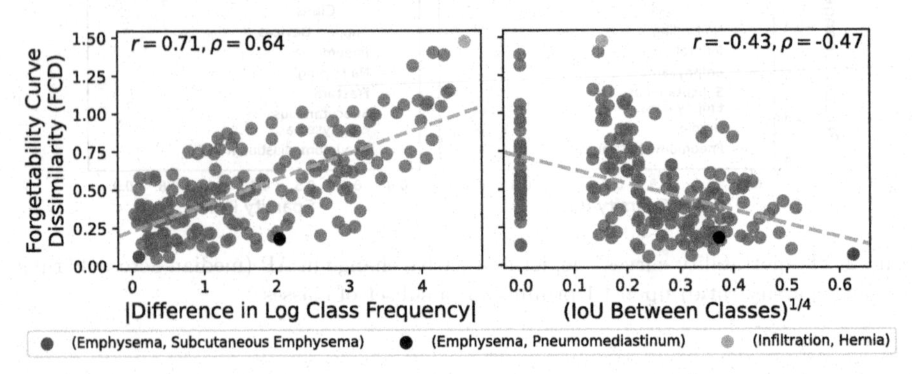

Fig. 4. Mutual relationship between pairs of diseases and their forgettability curves. For each pair of NIH-CXR-LT classes, FCD is plotted against the absolute difference in log frequency (left) and the IoU between the two classes (right).

3.2 Which Diseases are Most Vulnerable to Pruning and Why?

Class forgettability curves in Fig. 2 depict the relative change in AP by sparsity ratio for a representative subset of classes. Although these curves follow a similar general trend to Fig. 1, some curves (i) drop earlier and (ii) drop more considerably at high sparsity. Notably, we find a strong positive relationship between training class frequency and (i) the first sparsity ratio at which a class experienced a median 20% relative drop in AP ($\rho = 0.61, P = 0.005$ for NIH-CXR-LT; $\rho = 0.93, P \ll 0.001$ for MIMIC-CXR-LT) and (ii) the median relative change in AP at 95% sparsity ($\rho = 0.75, P < 0.001$ for NIH-CXR-LT; $\rho = 0.75, P < 0.001$ for MIMI-CXR-LT). These findings indicate that, in general, **rare diseases are**

forgotten earlier (Fig. 3, left) **and are more severely impacted at high sparsity** (Fig. 3, right).

3.3 How Does Disease Co-occurrence Influence Class Forgettability?

Our analysis reveals that for NIH-CXR-LT, the absolute difference in log test frequency between two diseases is a strong predictor of the pair's FCD ($\rho = 0.64, P \ll 0.001$). This finding suggests that **diseases with larger differences in prevalence exhibit more distinct forgettability behavior** upon L1 pruning (Fig. 4, left). To account for the multi-label nature of thorax disease classification, we also explore the relationship between intersection over union (IoU) – a measure of co-occurrence between two diseases – and FCD. Our analysis indicates that the IoU between two diseases is negatively associated with FCD ($\rho = -0.47, P \ll 0.001$). This suggests that **the more two diseases co-occur, the more similar their forgetting trajectories are across all sparsity ratios** (Fig. 4, right). For example, the disease pair (Infiltration, Hernia) has a dramatic difference in prevalence ($|\text{LogFreqDiff}| = 4.58$) *and* rare co-occurrence ($\text{IoU}^{1/4} = 0.15$), resulting in an extremely high FCD for the pair of diseases.

We also find, however, that there is a push and pull between differences in individual class frequency and class co-occurrence with respect to FCD. To illustrate, consider the disease pair (Emphysema, Pneumomediastinum) marked in black in Fig. 4. These classes have an absolute difference in log frequency of 2.04, which would suggest an FCD of around 0.58. However, because Emphysema and Pneumomediastinum co-occur relatively often ($\text{IoU}^{1/4} = 0.37$), their forgettability curves are *more similar than prevalence alone would dictate*, resulting in a lower FCD of 0.18. To quantify this effect, we obtain an OLS model that fitted FCD as a function of $|\text{LogFreqDiff}|$, $\text{IoU}^{1/4}$, and their interaction:

$$\text{FCD} = 0.27 + 0.21|\text{LogFreqDiff}| - 0.05(\text{IoU})^{1/4} - 0.31|\text{LogFreqDiff}| * (\text{IoU})^{1/4} \quad (2)$$

We observe a statistically significant interaction effect between the difference in individual class frequency and class co-occurrence on FCD ($\beta_3 = -0.31, P = 0.005$). Thus, for disease pairs with a very large difference in prevalence, the effect of co-occurrence on FCD is even more pronounced (Supplement).

3.4 What Do Pruning-Identified CXRs have in Common?

For NIH-CXR-LT, we find that **PIEs are more likely to contain rare diseases** and **more likely to contain 3+ simultaneous diseases** when compared to non-PIEs (Fig. 5). The five rarest classes appear 3–15x more often in PIEs than non-PIEs, and images with 4+ diseases appear 3.2x more often in PIEs.

In a human reader study involving 240 CXRs from the NIH-CXR-LT test set (120 PIEs and 120 non-PIEs), **radiologists perceived that PIEs had more label noise, lower image quality, and higher diagnosis difficulty** (Fig. 6). However, due to small sample size and large variability, these differences are not statistically significant. Respondents fully agreed with the label 55% of the time

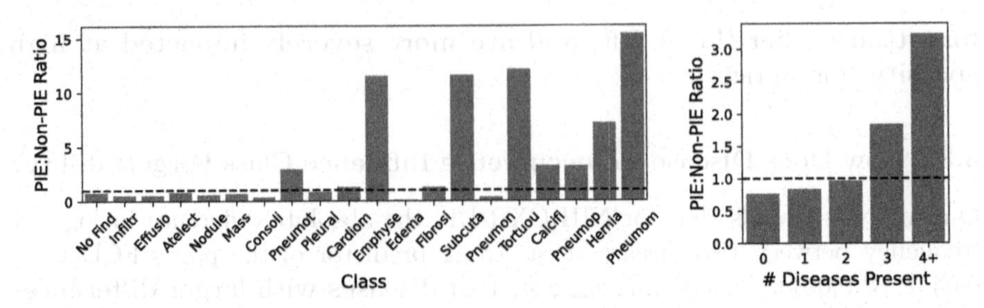

Fig. 5. Unique characteristics of PIEs. Presented is the *ratio* of class prevalence (left) and number of diseases per image (right) in PIEs relative to non-PIEs. The dotted line represents the 1:1 ratio (equally frequent in PIEs vs. non-PIEs).

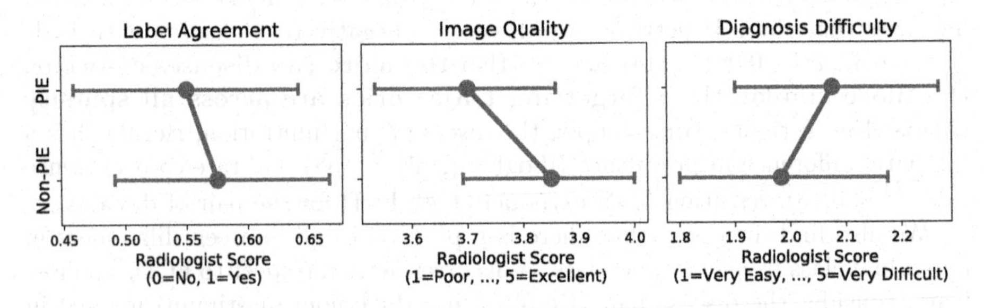

Fig. 6. Human study results describing radiologist perception of PIEs vs. non-PIEs. Mean ± standard deviation (error bar) radiologist scores are presented.

for PIEs and 57.5% of the time for non-PIEs ($P = 0.35$), gave an average image quality of 3.6 for PIEs and 3.8 for non-PIEs ($P = 0.09$), and gave an average diagnosis difficulty of 2.5 for PIEs and 2.05 for non-PIEs ($P = 0.25$).

Overall, these findings suggest that pruning identifies CXRs with many potential sources of difficulty, such as containing underrepresented diseases, (partially) incorrect labels, low image quality, and complex disease presentation.

4 Discussion and Conclusion

In conclusion, we conducted the first study of the effect of pruning on multi-label, long-tailed medical image classification, focusing on thorax disease diagnosis in CXRs. Our findings are summarized as follows:

1. As observed in standard image classification, CXR classifiers can be heavily pruned (up to 60% sparsity) before dropping in overall performance.
2. Class frequency is a strong predictor of both *when* and *how severely* a class is impacted by pruning. Rare classes suffer the most.
3. Large differences in class frequency lead to dissimilar "forgettability" behavior and stronger co-occurrence leads to more similar forgettability behavior.
   - Further, we discover a significant interaction effect between these two factors with respect to how similarly pruning impacts two classes.

4. We adapt PIEs to the multi-label setting, observing that PIEs are far more likely to contain rare diseases and multiple concurrent diseases.
   - A radiologist study further suggests that PIEs have more label noise, lower image quality, and higher diagnosis difficulty.

It should be noted that this study is limited to the analysis of global unstructured L1 (magnitude-based) pruning, a simple heuristic for post-training network pruning. Meanwhile, other state-of-the-art pruning approaches [6,20,22] and model compression techniques beyond pruning (e.g., weight quantization [14] and knowledge distillation [11]) could be employed to strengthen this work. Additionally, since our experiments only consider the ResNet50 architecture, it remains unclear whether other training approaches, architectures, or compression methods could mitigate the adverse effects of pruning on rare classes. In line with recent work [15,17,19], future research may leverage the insights gained from this study to develop an algorithm for improved long-tailed learning on medical image analysis tasks. For example, PIEs could be interpreted as salient, difficult examples that warrant greater weight during training. Conversely, PIEs may just as well be regarded as noisy examples to be ignored, using pruning as a tool for data cleaning.

Acknowledgments. This project was supported by the Intramural Research Programs of the National Institutes of Health, Clinical Center. It also was supported by the National Library of Medicine under Award No. 4R00LM013001, NSF CAREER Award No. 2145640, Cornell Multi-Investigator Seed Grant (Peng and Shih), and Amazon Research Award.

References

1. Chen, L., Zhao, L., Chen, C.Y.C.: Enhancing adversarial defense for medical image analysis systems with pruning and attention mechanism. Med. Phys. **48**(10), 6198–6212 (2021)
2. Davis, J., Goadrich, M.: The relationship between precision-recall and roc curves. In: ICML, pp. 233–240 (2006)
3. Dinsdale, N.K., Jenkinson, M., Namburete, A.I.: Stamp: simultaneous training and model pruning for low data regimes in medical image segmentation. Med. Image Anal. **81**, 102583 (2022)
4. Fernandes, F.E., Yen, G.G.: Automatic searching and pruning of deep neural networks for medical imaging diagnostic. IEEE Trans. Neural Netw. Learn. Syst. **32**(12), 5664–5674 (2020)
5. Fernández, A., García, S., Galar, M., Prati, R.C., Krawczyk, B., Herrera, F.: Learning from Imbalanced Data Sets. Springer, Cham (2018). https://doi.org/10.1007/978-3-319-98074-4
6. Frankle, J., Carbin, M.: The lottery ticket hypothesis: finding sparse, trainable neural networks. arXiv preprint arXiv:1803.03635 (2018)
7. Hajabdollahi, M., Esfandiarpoor, R., Najarian, K., Karimi, N., Samavi, S., Soroushmehr, S.R.: Hierarchical pruning for simplification of convolutional neural networks in diabetic retinopathy classification. In: IEEE Engineering in Medicine and Biology Society (EMBC), pp. 970–973. IEEE (2019)

8. Han, Y., Holste, G., Ding, Y., Tewfik, A., Peng, Y., Wang, Z.: Radiomics-Guided Global-Local Transformer For Weakly Supervised Pathology Localization in Chest X-rays. IEEE Trans. Med, Imaging PP (Oct (2022)
9. He, K., Zhang, X., Ren, S., Sun, J.: Deep residual learning for image recognition. In: Proceedings of the IEEE Conference on Computer Vision and Pattern Recognition, CVPR, pp. 770–778 (2016)
10. Hesamian, M.H., Jia, W., He, X., Kennedy, P.: Deep learning techniques for medical image segmentation: achievements and challenges. J. Digit. Imaging **32**, 582–596 (2019)
11. Hinton, G., Vinyals, O., Dean, J.: Distilling the knowledge in a neural network. arXiv preprint arXiv:1503.02531 (2015)
12. Holste, G., et al.: Long-tailed classification of thorax diseases on chest x-ray: A new benchmark study. In: Data Augmentation, Labelling, and Imperfections: Second MICCAI Workshop, pp. 22–32. Springer (2022). https://doi.org/10.1007/978-3-031-17027-0_3
13. Hooker, S., Courville, A., Clark, G., Dauphin, Y., Frome, A.: What do compressed deep neural networks forget? arXiv preprint arXiv:1911.05248 (2019)
14. Jacob, B., et al.: Quantization and training of neural networks for efficient integer-arithmetic-only inference. In: Proceedings of the IEEE Conference on Computer Vision and Pattern Recognition, pp. 2704–2713 (2018)
15. Jaiswal, A., Chen, T., Rousseau, J.F., Peng, Y., Ding, Y., Wang, Z.: Attend who is weak: Pruning-assisted medical image localization under sophisticated and implicit imbalances. In: WACV, pp. 4987–4996 (2023)
16. Jeong, T., Bollavaram, M., Delaye, E., Sirasao, A.: Neural network pruning for biomedical image segmentation. In: Medical Imaging 2021: Image-Guided Procedures, Robotic Interventions, and Modeling. vol. 11598, pp. 415–425. SPIE (2021)
17. Jiang, Z., Chen, T., Mortazavi, B.J., Wang, Z.: Self-damaging contrastive learning. In: International Conference on Machine Learning, pp. 4927–4939. PMLR (2021)
18. Johnson, A.E., et al.: MIMIC-CXR, a de-identified publicly available database of chest radiographs with free-text reports. Scientific Data (2019)
19. Kong, H., Lee, G.H., Kim, S., Lee, S.W.: Pruning-guided curriculum learning for semi-supervised semantic segmentation. In: Proceedings of the IEEE/CVF Winter Conference on Applications of Computer Vision, pp. 5914–5923 (2023)
20. Kurtic, E., Alistarh, D.: Gmp*: Well-tuned global magnitude pruning can outperform most bert-pruning methods. arXiv preprint arXiv:2210.06384 (2022)
21. LeCun, Y., Denker, J., Solla, S.: Optimal brain damage. Adv. Neural Inform. Process. Syst. **2** (1989)
22. Lee, N., Ajanthan, T., Torr, P.H.: Snip: single-shot network pruning based on connection sensitivity. arXiv preprint arXiv:1810.02340 (2018)
23. Lin, M., et al.: Automated diagnosing primary open-angle glaucoma from fundus image by simulating human's grading with deep learning. Sci. Rep. **12**(1), 14080 (2022)
24. Lin, X., Yu, L., Cheng, K.T., Yan, Z.: The lighter the better: rethinking transformers in medical image segmentation through adaptive pruning. arXiv preprint arXiv:2206.14413 (2022)
25. Mahbod, A., Entezari, R., Ellinger, I., Saukh, O.: Deep neural network pruning for nuclei instance segmentation in hematoxylin and eosin-stained histological images. In: Wu, S., Shabestari, B., Xing, L. (eds.) Applications of Medical Artificial Intelligence: First International Workshop, AMAI 2022, Held in Conjunction with MICCAI 2022, Singapore, September 18, 2022, Proceedings, pp. 108–117. Springer, Cham (2022). https://doi.org/10.1007/978-3-031-17721-7_12

26. Rajpurkar, P., et al.: Chexnet: radiologist-level pneumonia detection on chest -X-rays with deep learning. arXiv preprint arXiv:1711.05225 (2017)
27. Rethmeier, N., Augenstein, I.: Long-tail zero and few-shot learning via contrastive pretraining on and for small data. In: Computer Sciences & Mathematics Forum. vol. 3, p. 10. MDPI (2022)
28. Seyyed-Kalantari, L., Liu, G., McDermott, M., Chen, I.Y., Ghassemi, M.: Chex-clusion: fairness gaps in deep chest x-ray classifiers. In: BIOCOMPUTING 2021: proceedings of the Pacific symposium, pp. 232–243. World Scientific (2020)
29. Valverde, J.M., Shatillo, A., Tohka, J.: Sauron u-net: Simple automated redundancy elimination in medical image segmentation via filter pruning. arXiv preprint arXiv:2209.13590 (2022)
30. Wang, X., Peng, Y., Lu, L., Lu, Z., Bagheri, M., Summers, R.M.: ChestX-Ray8: Hospital-scale chest x-ray database and benchmarks on weakly-supervised classification and localization of common thorax diseases. In: IEEE Conference on Computer Vision and Pattern Recognition, CVPR, pp. 3462–3471 (2017)
31. Wu, Y., Zeng, D., Xu, X., Shi, Y., Hu, J.: FairPrune: achieving fairness through pruning for dermatological disease diagnosis. In: Wang, L., Dou, Q., Fletcher, P.T., Speidel, S., Li, S. (eds.) Medical Image Computing and Computer Assisted Intervention – MICCAI 2022: 25th International Conference, Singapore, September 18–22, 2022, Proceedings, Part I, pp. 743–753. Springer, Cham (2022). https://doi.org/10.1007/978-3-031-16431-6_70
32. Yang, B., et al.: Network pruning for OCT image classification. In: Fu, H., Garvin, M.K., MacGillivray, T., Xu, Y., Zheng, Y. (eds.) Ophthalmic Medical Image Analysis: 6th International Workshop, OMIA 2019, Held in Conjunction with MICCAI 2019, Shenzhen, China, October 17, Proceedings, pp. 121–129. Springer, Cham (2019). https://doi.org/10.1007/978-3-030-32956-3_15
33. Zhou, S.K., et al.: A review of deep learning in medical imaging: imaging traits, technology trends, case studies with progress highlights, and future promises. Proc. IEEE **109**(5), 820–838 (2021)
34. Zhu, M., Gupta, S.: To prune, or not to prune: exploring the efficacy of pruning for model compression. arXiv preprint arXiv:1710.01878 (2017)

Multimodal Deep Fusion in Hyperbolic Space for Mild Cognitive Impairment Study

Lu Zhang[1], Saiyang Na[1], Tianming Liu[2], Dajiang Zhu[1], and Junzhou Huang[1(✉)]

[1] Department of Computer Science and Engineering, The University of Texas at Arlington, Arlington, TX 76019, USA
jzhuang@exchange.uta.edu

[2] Department of Computer Science, University of Georgia, Athens, GA 30602, USA

Abstract. Multimodal fusion of different types of neural image data offers an invaluable opportunity to leverage complementary cross-modal information and has greatly advanced our understanding of mild cognitive impairment (MCI), a precursor to Alzheimer's disease (AD). Current multi-modal fusion methods assume that both brain's natural geometry and the related feature embeddings are in Euclidean space. However, recent studies have suggested that non-Euclidean hyperbolic space may provide a more accurate interpretation of brain connectomes than Euclidean space. In light of these findings, we propose a novel graph-based hyperbolic deep model with a learnable topology to integrate the individual structural network with functional information in hyperbolic space for the MCI/NC (normal control) classification task. We comprehensively compared the classification performance of the proposed model with state-of-the-art methods and analyzed the feature representation in hyperbolic space and its Euclidean counterparts. The results demonstrate the superiority of the proposed model in both feature representation and classification performance, highlighting the advantages of using hyperbolic space for multimodal fusion in the study of brain diseases. (Code is available here (https://github.com/nasyxx/MDF-HS).)

Keywords: Hyperbolic Space · Multimodal Fusion · MCI

1 Introduction

Alzheimer's disease (AD) is an irreversible progressive neurodegenerative brain disorder that ranks as the sixth leading cause of death in the United States [2]. While AD cannot currently be prevented or cured once established, early

Supplementary Information The online version contains supplementary material available at https://doi.org/10.1007/978-3-031-43904-9_65.

H. Greenspan et al. (Eds.): MICCAI 2023, LNCS 14224, pp. 674–684, 2023.
https://doi.org/10.1007/978-3-031-43904-9_65

diagnosis and intervention during the mild cognitive impairment (MCI - precursor of AD) stage offer a feasible way for patients to plan for the future. Although the neuropathological mechanism of MCI is not fully understood, accumulating evidence suggests that both structural and functional brain alterations have been found in MCI patients [8]. Consequently, numerous multimodal fusion approaches have been published [10,14,20,22,24,27,29], significantly enhancing our comprehension of MCI and AD. Combining various modalities of brain data through multimodal fusion provides an invaluable chance to exploit complementary cross-modal information. The outstanding outcomes obtained by these multimodal fusion techniques in MCI/AD research highlight the essential importance of integrating multimodal brain networks and comprehending their intricate connections in the investigation of brain disorders.

In the most of existing multimodal fusion methods, Euclidean space is typically assumed as the natural geometry of the brain. As such, both feature embedding and model establishment are conducted in the Euclidean space. However, recent studies have suggested that non-Euclidean hyperbolic space may offer a more accurate interpretation of brain connectomes than Euclidean space [1,28]. For example, [1] studied the navigability properties of structural brain networks across species and found that the Euclidean distance cannot fully explain the structural organization of brain connectomes, while hyperbolic space provides almost perfectly navigable maps of connectomes for all species. Similarly, the research in [28] found that the structure of the human brain remains self-similar (a non-Euclidean property) when the resolution of the network is progressively decreased by hierarchical coarse graining of the anatomical regions. Additionally, in the general AI domain, consistent results have been reported, demonstrating that learning hierarchical representations of graphs is easier in the hyperbolic space due to its curvature and geometrical properties [3,7,15]. These compelling findings raise the question of whether hyperbolic space is a better choice for multimodal fusion in the study of brain diseases.

To answer this question, we propose a novel graph-based hyperbolic deep model to conduct multimodal fusion in hyperbolic space for MCI study. Specifically, we embedded brain functional activities into hyperbolic space, where a hyperbolic graph convolution neural network (HGCN) [7] is developed to integrate structural substrate and functional hyperbolic embeddings. The HGCN takes into two inputs: the topology of the network and the associated node features. The input topology is initialized by the individual structural network and iteratively updated by incorporating the corresponding individual functional hyperbolic features to maximize the classification power between elder normal control (NC) and MCI patients. This results in a learned topology that becomes a deeply hybrid connectome embedded in hyperbolic space, combining both brain structural substrate and functional influences. The associated node features are the functional similarities derived from the functional hyperbolic features. By graph convolution in hyperbolic space, node features are aggregated along the topology and used to conduct MCI/NC classification. In the experiments, we comprehensively evaluate the proposed model from three perspectives: Firstly, we compared the classification performance of our proposed model with state-of-the-art methods. Secondly, we conducted an ablation study to evaluate the effectiveness of both the hyperbolic feature embedding and the hyperbolic graph convolutional neural network. Finally, we visualized, analyzed, and compared the feature representation in hyperbolic space and its Euclidean counterpart. The proposed model achieves the outstanding classification performance, and the results demonstrate the advantages of using hyperbolic space for multimodal fusion in the study of brain diseases.

2 Related Work

Gromov has demonstrated that hyperbolic spaces are particularly well-suited for representing tree-like structures [13], as they allow for the representation of objects that would require an exponential number of dimensions in Euclidean space in only a polynomial number of dimensions in hyperbolic space, with low distortion. To make full use of the geometric properties of hyperbolic space, novel hyperbolic deep neural networks have been proposed and shown to have superior model compactness and predictive performance when compared to their counterparts in Euclidean space. Pioneering works, such as [6,17], have introduced the concept of graph embeddings in hyperbolic space. Recently, HGNN [15] and HGCN [7] were proposed to build graph neural networks in hyperbolic space and achieved remarkable success in graph-related tasks, including node classification [7], graph classification [15], link prediction [7], and graph embedding [3]. Inspired by these works, we attempted to apply the HGCN to brain multimodal fusion studies. In our work, the topology of the HGCN is learnable and incorporates multimodal information from different modalities.

3 Methods

3.1 Data Description and Preprocessing

We used 209 subjects, comprising 116 individuals from the NC group (60 females, 56 males; 74.26 ± 8.42 years) and 93 subjects from the MCI group (53 females, 40 males; 74.24 ± 8.63 years) from ADNI dataset. Each subject has structural MRI (T1-weighted), resting-state fMRI (rs-fMRI), and DTI. We performed standard preprocessing procedures as described in [25,26]. In summary, these steps involved skull removal for all modalities. For rs-fMRI images, we conducted spatial smoothing, slice time correction, temporal pre-whitening, global drift removal, and band-pass filtering (0.01–0.1 Hz) using FMRIB Software Library (FSL) FEAT. As for DTI images, we performed eddy current correction using FSL and fiber tracking using MedINRIA. T1-weighted images were registered to DTI space using FSL FLIRT, followed by segmentation using the FreeSurfer package. Subsequently, the Destrieux Atlas [9] was utilized for ROI labeling, and the brain cortex was partitioned into 148 regions. We calculated the average fMRI signal for each brain region as node features and generated the individual structural network using fiber count.

3.2 Preliminary

The hyperbolic space $\mathbb{H}^n_K$ is a specific type of n-dimension *Riemannian manifold* $(\mathcal{M}, g)$ with a constant negative sectional curvature K [4]. It has some unique properties, such as its exponential growth rate, which makes it useful in areas like geometry, topology, and computer science [18]. In hyperbolic space, isometric hyperbolic models are used to maintain the relationship between points without distorting distances. The hyperbolic space is usually defined via five isometric hyperbolic models [19], among which the n-dimensional Poincaré ball model is frequently used because it provides a more intuitive visualization of hyperbolic space, which can be easier to understand than other models. Therefore, in this work, we align with [7] and build our model upon the Poincaré ball model $\mathbb{B}^n_K$. Existing studies [11,16,21] adopt the Möbius transformations as the non-associative algebraic formalism for hyperbolic geometry. Three basic operations of two elements $\mathbf{x}, \mathbf{y}$ on Poincaré ball model, including addition, scalar multiplication, and matrix-vector multiplication, are defined as:

$$\mathbf{x} \oplus_K \mathbf{y} := \frac{(1 - 2K < \mathbf{x}, \mathbf{y} > -K||\mathbf{y}||^2)\mathbf{x} + (1 + K||\mathbf{x}||^2)\mathbf{y}}{1 - 2K < \mathbf{x}, \mathbf{y} > +K^2||\mathbf{x}||^2||\mathbf{y}||^2} \tag{1}$$

$$r \otimes_K \mathbf{x} = \mathrm{Exp}^K_{\mathbf{0}}(r\mathrm{Log}^K_{\mathbf{0}}(\mathbf{x})) \tag{2}$$

$$M \otimes_K \mathbf{x} = \mathrm{Exp}^K_{\mathbf{0}}(M\mathrm{Log}^K_{\mathbf{0}}(\mathbf{x})) \tag{3}$$

A *manifold* $\mathcal{M}$ of n-dimension is a topological space that is locally Euclidean. For all $\mathbf{x} \in \mathcal{M}$, the *tangent space* $\mathcal{T}_{\mathbf{x}}\mathcal{M}$ at a point $\mathbf{x}$ is the vector space of the same n-dimension as $\mathcal{M}$, containing all tangent vector tangentially pass through $\mathbf{x}$. The *metric tensor* $g_{\mathbf{x}}$ [12] on point $\mathbf{x}$ defines an inner product on the associated tangent space. Accordingly, two mappings are defined to achieve transformation between the Euclidean *tangent space* $\mathcal{T}_{\mathbf{x}}\mathcal{M}$ and hyperbolic space $\mathcal{M}$: **Exponential map** $\mathrm{Exp}^K_{\mathbf{x}} : \mathcal{T}_{\mathbf{x}}\mathcal{M} \to \mathcal{M}$ that maps an arbitrary tangent vector $\mathbf{v} \in \mathcal{T}_{\mathbf{x}}\mathcal{M}$ to $\mathcal{M}$ and **logarithmic map** $\mathrm{Log}^K_{\mathbf{x}} : \mathcal{M} \to \mathcal{T}_{\mathbf{x}}\mathcal{M}$ that maps an arbitrary point $\mathbf{y} \in \mathcal{M}$ to $\mathcal{T}_{\mathbf{x}}\mathcal{M}$:

$$\mathrm{Exp}^K_{\mathbf{x}}(\mathbf{v}) := \mathbf{x} \oplus_K (\tanh(\sqrt{|K|}\lambda^K_{\mathbf{x}}\frac{||\mathbf{v}||}{2}\frac{\mathbf{v}}{\sqrt{K}||\mathbf{v}})) \tag{4}$$

$$\mathrm{Log}^K_{\mathbf{x}}(\mathbf{y}) := \frac{2}{\sqrt{|K|}\lambda^K_{\mathbf{x}}} \tanh^{-1}(\sqrt{|K|}|| - \mathbf{x} \oplus_K \mathbf{y}||)\frac{-\mathbf{x} \oplus_K \mathbf{y}}{|| - \mathbf{x} \oplus_K \mathbf{y}||} \tag{5}$$

where $\lambda^K_{\mathbf{x}} = \frac{2}{1+K||\mathbf{x}||^2}$ is the conformal factor at point $\mathbf{x}$.

3.3 Functional Profile Learning in Hyperbolic Space

In this work, we aim to learn a disease-related functional profile in the hyperbolic space. To this end, we firstly mapped the averaged functional signal of region i in Euclidean – f^E_i, to hyperbolic space via (4): $f^H_i = \mathrm{Exp}^K_{\mathbf{0}}(f^E_i)$, here we choose point 0, the origin in hyperbolic space to conduct transformation. Then upon

f_i^H, we defined the parameterized functional-pairwise distance between region i and region j in hyperbolic space by a learnable mapping matrix M:

$$\phi(M, f_i^H, f_j^H) = ||M \otimes_K (f_i^H \oplus_K -f_j^H)||_2^2, \forall i, j \in 1..N \tag{6}$$

It is worth noting that (6) is a linear projection. It is inadequate for modeling the distance/similarity of the complicated fMRI signals. To alleviate this issue, we introduced nonlinear projection by applying Gaussian kernel:

$$A_{i,j}^{f^H} = \mathrm{Exp}_0^K(\exp(-\frac{\mathrm{Log}_0^K(\phi(M, f_i^H, f_j^H))}{2\sigma^2})) \tag{7}$$

Here, $A_{i,j}^{f^H} \in \mathbb{B}_K^n$ represents the pairwise functional profile between brain regions i and j in hyperbolic space. σ is the bandwidth parameter of Gaussian kernel and is treated as a hyper-parameter. In the proposed model, M is initialized as identity matrix to avoid introducing any bias and iteratively updated during the training process based on classification results.

3.4 Multimodal Fusion by HGCN

A major goal of this work is to conduct effective multimodal fusion of brain structural and functional data in hyperbolic space for MCI study. To achieve this aim, we combined the learned functional profile with the brain structural network in the hyperbolic space by:

$$\hat{A}^H = I \oplus_K (\theta \otimes_K A^{S^H}) \oplus_K ((1-\theta) \otimes_K A^{f^H}) \tag{8}$$

where I is the identity matrix, A^{S^E} is the original individual structural network calculated by fiber count, and $A^{S^H} = \mathrm{Exp}_0^K(A^{S^E})$ is the hyperbolic counterpart of A^{S^E}. $\theta \in (0, 1)$ is a learnable parameter to control the contributions of structural and functional components in the combined new brain connectome $\hat{A}^H$. In the training process, A^{f^H} and A^{S^H} will be iteratively updated, and disease-related knowledge (from classification) is extracted and passed to functional profile (A^{f^H}) and structural network (A^{S^H}) and then transferred to the new brain connectome $\hat{A}^H$. Next, $\hat{A}^H$ will be used as the new topology along with node features in graph convolution conducted by HGCN.

HGCN performs graph convolution within the hyperbolic space in two steps:

$$\tilde{\mathbf{h}}_i^{(l)} = (\mathbf{W} \otimes_K \mathbf{h}_i^{(l)}) \oplus_K \mathbf{b} \tag{9}$$

$$\mathbf{h}_i^{(l+1)} = \mathrm{Exp}_{\tilde{\mathbf{h}}_i^{(l)}}^K(\sum_j^n A\mathrm{Log}_{\tilde{\mathbf{h}}_i^{(l)}}^K(\tilde{\mathbf{h}}_j^{(l)})) \tag{10}$$

where $\mathbf{h}_i^{(l)}$ is the input of i-th node in the l-th layer, $\mathbf{W} \in \mathbb{R}^{d \times d}$ and $\mathbf{b} \in \mathbb{R}^d$ are the weight and bias of the l-th layer. (9) is the operation of input features and model parameters. After that, for each node i, the feature vector of its neighbors

– $\tilde{\mathbf{h}}_j^{(l)}$, will firstly be mapped to i's local tangent space by $\mathrm{Log}^K_{\tilde{\mathbf{h}}_i^{(l)}}$ to conduct graph convolution with adjacency matrix A, then the results will be mapped back to the hyperbolic space by $\mathrm{Exp}^K_{\tilde{\mathbf{h}}_i^{(l)}}$ and used as the input in next layer in HGCN. In a previous study [26], seven different methods for creating functional connectivity were compared, and it was found that the Pearson correlation coefficient (PCC) outperformed other approaches. Therefore, in our work, the PCC of f_i^H is used as node features ($\mathbf{h}_i^{(1)}$) and $\hat{A}^H$ is used as the adjacency matrix A in Eq. 10.

4 Results

4.1 Experimental Setting

Data Settings. In this study, the entire brain was partitioned into 148 distinct regions using the well-established Destrieux Atlas. Averaged fMRI signals were then calculated for each brain region, and the brain structural network (A^S) was generated for each individual. 5-fold cross-validation was performed using a cohort of 209 individuals, consisting of 116 elder normal control (NC) and 93 MCI patients.

Model Settings. The HGCN model in this work has two layers, wherein the output of each HGCN layer was set to 148 and 296, respectively. These outputs were subsequently combined using a hyperbolic fully connected layer [11] with an output size denoted by C, where C corresponds to the number of classes for classification purposes (in our work, $C = 2$). To optimize the model's parameters, an end-to-end training approach was adopted, and the Xavier initialization scheme was utilized. During the training process, the Riemannian SGD [5] optimizer was applied with a standard learning rate of 1.

Table 1. Classification performance comparison with other state-of-the-art methods. ACC: Accuracy; SPE: Specificity; SEN: Sensitivity; AUC: Area under the ROC Curve.

Study	Modality	NC/MCI	ACC(%)	SEN(%)	SPE(%)	AUC
Fang et al. (2020) [10]	MRI+PET	101/204	88.25	79.74	91.58	0.79
Li et al. (2020) [14]	DTI+rsfMRI	37/36	87.70	88.90	86.50	0.89
Zhou et al. (2021) [29]	MRI+PET	142/82	90.35	88.36	92.56	-
Zhang et al. (2021) [22]	Multi-level MRI	275/162	87.82	87.56	88.84	0.93
Zhang et al. (2021) [23]	DTI+rsfMRI	116/98	92.70	-	89.90	-
Shi et al. (2022) [20]	MRI+PET	52/99	80.73	85.98	70.90	0.79
[23] re-implementation	DTI+rsfMRI	116/93	86.23	78.11	87.74	0.83
Proposed	DTI+rsfMRI	116/93	92.30	83.22	89.99	0.87

4.2 Classification Performance Comparison

We compared the classification performance of our proposed model with other state-of-the-art methods on MCI/NC classification task using multi-modal data and presented the results in Table 1. The feature embedding and model development of all the other studies were conducted in Euclidean space. In addition, some of them employed PET data, which tends to yield higher classification accuracy than MRI-based modalities. However, it is noteworthy that our approach utilized noninvasive DTI/fMRI data and achieved outstanding performance with a classification accuracy of 92.3% for MCI/NC classification. Although study [23] reported a slightly higher accuracy than ours, we sought to eliminate biases arising from factors other than the methods by implementing the methods used in [23] with the same data settings as ours and showed results in the " [23] re-implement" row. Notably, our model outperforming [23] in MCI/NC classification when using the same data setting. It is worth mentioning that the only difference between " [23] re-implement" and our method is the space, suggesting that hyperbolic space is a superior choice for multi-modal fusion in MCI/NC classification.

Table 2. Ablation Study. Acc_b: Best Accuracy; Acc_m: Average accuracy. The best and worst results are highlighted by blue and red, respectively.

Model	Manifold	-K	Acc_b(%)↑	Acc_m(%)↑	SPE(%)↑	SEN(%)↑	AUC↑
GCN	Euclidean	0	86.23	83.47	87.74	78.11	0.8293
GCN	Poincaré	1.4	88.02	85.17	89.44	79.86	0.8465
GCN	Poincaré	1.2	87.43	84.93	89.23	79.59	0.8441
GCN	Poincaré	1.0	87.43	84.93	88.80	80.13	0.8446
GCN	Poincaré	0.8	88.02	85.05	89.23	79.85	0.8454
GCN	Poincaré	0.6	87.43	85.05	89.44	79.60	0.8452
GCN	Poincaré	0.4	87.43	85.05	89.22	79.86	0.8454
HGCN	Poincaré	1.4	89.29	83.96	88.83	78.24	0.8353
HGCN	Poincaré	1.2	89.88	83.60	88.22	78.23	0.8323
HGCN	Poincaré	1.0	89.88	85.49	86.81	79.37	0.8309
HGCN	Poincaré	0.8	88.69	83.61	87.71	78.63	0.8317
HGCN	Poincaré	0.6	92.26	86.96	89.99	83.22	0.8660
HGCN	Poincaré	0.4	89.22	86.13	88.38	83.28	0.8583

4.3 Ablation Study

The proposed model implements both feature embedding and graph neural network establishment in hyperbolic space. We conducted ablation studies to evaluate the impact of the two factors on the classification performance of MCI/NC classification. In contrast to Euclidean space, where the curvature is a constant 0, hyperbolic space has negative curvature, and we evaluated a range of curvature values from -0.4 to -1.4. The results were summarized in Table 2, with the best and worst performances highlighted in blue and red colors, respectively. The best results were obtained when both feature embedding and graph neural network establishment were implemented in hyperbolic space, while the worst results were observed in the pure Euclidean setting. We found that the hybrid combination of GCN established in Euclidean space with the feature embedding in hyperbolic space achieved better performance compared to the pure Euclidean setting, but in most cases, it performed worse than the pure hyperbolic setting. These findings indicate the superiority of hyperbolic feature embedding and graph neural network over their Euclidean counterparts.

4.4 Feature Representation

To gain further insight into the impact of space choice on feature representation, we visualized the distribution of two distinct feature representations in Euclidean space and hyperbolic space. Firstly, we employed principal component analysis (PCA) to project the high-dimensional input feature vectors from both the Euclidean space and the hyperbolic space onto a two-dimensional space. The resulting visualizations are presented in Fig. 1 (a). Notably, the feature vectors in hyperbolic space exhibited a more uniform distribution, with samples distributed more evenly across the space. In contrast, the feature distribution in Euclidean space was non-uniform, with some samples scattered far away from others (highlighted by purple arrows). These differences highlight the advantages of hyperbolic space for representing diverse and complex relationships of input data. Secondly, we projected the latent feature vectors before the classification layer onto a two-dimensional space and presented the results in Fig. 1 (b). These feature vectors are directly related to the classification performance, and their distribution can have a significant impact on classification accuracy. In the hyperbolic space, it is evident that the feature vectors of the two classes have no overlapping areas, while in the Euclidean space, there is a considerable amount

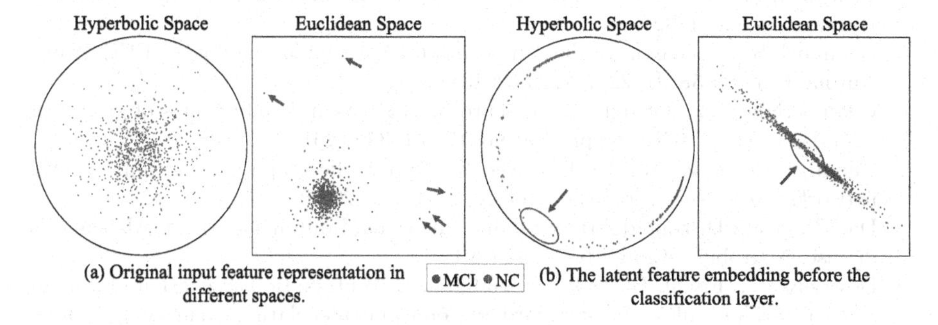

Fig. 1. Distributions of two distinct feature representations in the Euclidean space and the hyperbolic space.

of mixed data where the two types are handed over (highlighted by oval circles). This difference highlights the superiority of hyperbolic space in achieving a clear and distinguishable feature representation for classification tasks.

5 Conclusion

It is widely believed that the AD/MCI related brain alterations involve both brain structure and function. However, effectively modeling the complex relationships between structural and functional data and integrating them at the network level remains a challenge. Recent advances in graph modeling in hyperbolic space have inspired us to integrate multimodal brain networks via graph convolution conducted in the hyperbolic space. To this end, we mapped brain structural and functional features into hyperbolic space and conducted graph convolution by a hyperbolic graph convolution neural network, which enabled us to obtain a new brain connectome that incorporates multimodal information. Our developed model has demonstrated exceptional performance when compared to state-of-the-art methods. Furthermore, the results suggest that feature embedding and graph convolution neural network establishment in hyperbolic space are both crucial for enhancing performance.

Acknowledgement. This work was supported by National Institutes of Health (R01AG075582 and RF1NS128534).

References

1. Allard, A., Serrano, M.Á.: Navigable maps of structural brain networks across species. PLoS Comput. Biol. **16**(2), e1007584 (2020)
2. Association, A.: 2019 Alzheimer's disease facts and figures. Alzheimer's Dementia **15**(3), 321–387 (2019)
3. Bachmann, G., Bécigneul, G., Ganea, O.: Constant curvature graph convolutional networks. In: International Conference on Machine Learning, pp. 486–496. PMLR (2020)
4. Benedetti, R., Petronio, C.: Lectures on hyperbolic geometry. Springer Science & Business Media (1992)
5. Bonnabel, S.: Stochastic gradient descent on Riemannian manifolds. IEEE Trans. Autom. Control **58**(9), 2217–2229 (2013)
6. Chamberlain, B.P., Clough, J., Deisenroth, M.P.: Neural embeddings of graphs in hyperbolic space. arXiv preprint arXiv:1705.10359 (2017)
7. Chami, I., Ying, Z., Ré, C., Leskovec, J.: Hyperbolic graph convolutional neural networks. Adv. Neural Inform. Process. Syst. **32** (2019)
8. Dai, Z., et al.: Disrupted structural and functional brain net-works in Alzheimer's disease. Neurobiol. Aging **75**, 71–82 (2019)
9. Destrieux, C., Fischl, B., Dale, A., Halgren, E.: Automatic parcellation of human cortical gyri and sulci using standard anatomical nomenclature. Neuroimage **53**(1), 1–15 (2010)

10. Fang, X., Liu, Z., Xu, M.: Ensemble of deep convolutional neural networks based multi-modality images for Alzheimer's disease diagnosis. IET Image Proc. **14**(2), 318–326 (2020)
11. Ganea, O., Bécigneul, G., Hofmann, T.: Hyperbolic neural networks. Adv. Neural Inform. Process. Syst. **31** (2018)
12. Greene, R.E.: S. gallot, d. hulin and j. lafontaine, riemannian geometry (1989)
13. Gromov, M.: Hyperbolic groups. In: Gersten, S.M., et al. (eds.) Essays in Group Theory, pp. 75–263. Springer, New York, NY (1987). https://doi.org/10.1007/978-1-4613-9586-7_3
14. Li, Y., Liu, J., Tang, Z., Lei, B.: Deep spatial-temporal feature fusion from adaptive dynamic functional connectivity for mci identification. IEEE Trans. Med. Imaging **39**(9), 2818–2830 (2020)
15. Liu, Q., Nickel, M., Kiela, D.: Hyperbolic graph neural networks. Adv. Neural Inform. Process. Syst. **32** (2019)
16. Mathieu, E., Le Lan, C., Maddison, C.J., Tomioka, R., Teh, Y.W.: Continuous hierarchical representations with poincaré variational auto-encoders. Adv. Neural Informa. Process. Syst. **32** (2019)
17. Muscoloni, A., Thomas, J.M., Ciucci, S., Bianconi, G., Cannistraci, C.V.: Machine learning meets complex networks via coalescent embedding in the hyperbolic space. Nat. Commun. **8**(1), 1615 (2017)
18. Newman, M.E.: Power laws, pareto distributions and Zipf's law. Contemp. Phys. **46**(5), 323–351 (2005)
19. Peng, W., Varanka, T., Mostafa, A., Shi, H., Zhao, G.: Hyperbolic deep neural networks: a survey. IEEE Trans. Pattern Anal. Mach. Intell. **44**(12), 10023–10044 (2021)
20. Shi, Y., et al.: Asmfs: Adaptive-similarity-based multi-modality feature selection for classification of Alzheimer's disease. Pattern Recogn. **126**, 108566 (2022)
21. Shimizu, R., Mukuta, Y., Harada, T.: Hyperbolic neural networks++. arXiv preprint arXiv:2006.08210 (2020)
22. Zhang, J., Zheng, B., Gao, A., Feng, X., Liang, D., Long, X.: A 3D densely connected convolution neural net-work with connection-wise attention mechanism for Alzheimer's disease classification. Magn. Reson. Imaging **78**, 119–126 (2021)
23. Zhang, L., et al.: Roimaging Initiative, A.D.N., et al.: Deep fusion of brain structure-function in mild cognitive impairment. Medical Image Anal. **72**, 102082 (2021)
24. Zhang, L., Wang, L., Zhu, D.: Jointly analyzing Alzheimer's disease related structure-function using deep cross-model attention net-work. In: 2020 IEEE 17th International Symposium on Biomedical Imaging (ISBI), pp. 563–567. IEEE (2020)
25. Zhang, L., Wang, L., Zhu, D.: Recovering brain structural connectivity from functional connectivity via multi-GCN based generative adversarial network. In: Martel, A.L., et al. (eds.) Medical Image Computing and Computer Assisted Intervention – MICCAI 2020: 23rd International Conference, Lima, Peru, October 4–8, 2020, Proceedings, Part VII, pp. 53–61. Springer, Cham (2020). https://doi.org/10.1007/978-3-030-59728-3_6
26. Zhang, L., Wang, L., Zhu, D., Initiative, A.D.N., et al.: Predicting brain structural network using functional connectivity. Med. Image Anal. **79**, 102463 (2022)
27. Zhang, L., Zaman, A., Wang, L., Yan, J., Zhu, D.: A cascaded multi-modality analysis in mild cognitive impairment. In: Suk, H.-I., Liu, M., Yan, P., Lian, C. (eds.) MLMI 2019. LNCS, vol. 11861, pp. 557–565. Springer, Cham (2019). https://doi.org/10.1007/978-3-030-32692-0_64

28. Zheng, M., Allard, A., Hagmann, P., Alemán-Gómez, Y., Serrano, M.Á.: Geometric renormalization unravels self-similarity of the multiscale human connectome. Proc. Natl. Acad. Sci. **117**(33), 20244–20253 (2020)
29. Zhou, P., et al.: Use of a sparse-response deep belief network and extreme learning machine to discriminate Alzheimer's dis-ease, mild cognitive impairment, and normal controls based on amyloid pet/mri images. Front. Med. **7**, 621204 (2021)

Hierarchical Vision Transformers for Disease Progression Detection in Chest X-Ray Images

Amarachi B. Mbakwe[1], Lyuyang Wang[2], Mehdi Moradi[2], and Ismini Lourentzou[1](✉)

[1] Virginia Tech, Blacksburg, VA, USA
{bmamarachi,ilourentzou}@vt.edu
[2] McMaster University, Hamilton, ON, Canada
{wangl307,moradm4}@mcmaster.ca

Abstract. Chest radiography is a commonly used diagnostic imaging exam for monitoring disease progression and treatment effectiveness. While machine learning has made significant strides in tasks such as image segmentation, disease diagnosis, and automatic report generation, more intricate tasks such as disease progression monitoring remain fairly underexplored. This task presents a formidable challenge because of the complex and intricate nature of disease appearances on chest X-ray images, which makes distinguishing significant changes from irrelevant variations between images challenging. Motivated by these challenges, this work proposes `CheXRelFormer`, an end-to-end siamese Transformer disease progression model that takes a pair of images as input and detects whether the patient's condition has improved, worsened, or remained unchanged. The model comprises two hierarchical Transformer encoders, a difference module that compares feature differences across images, and a final classification layer that predicts the change in the patient's condition. Experimental results demonstrate that `CheXRelFormer` outperforms previous counterparts. Code is available at https://github.com/PLAN-Lab/CheXRelFormer.

Keywords: Vision Transformers · Disease Progression · Chest X-Ray Comparison Relations · Longitudinal CXR Relationships

1 Introduction

Chest X-rays (CXRs) are frequently used for disease detection and disease progression monitoring. However, interpreting CXRs can be challenging and time-consuming, particularly in regions with a shortage of radiologists. This can lead to delayed or inaccurate diagnoses and management, potentially harming

---

Supplementary Information The online version contains supplementary material available at https://doi.org/10.1007/978-3-031-43904-9_66.

H. Greenspan et al. (Eds.): MICCAI 2023, LNCS 14224, pp. 685–695, 2023.
https://doi.org/10.1007/978-3-031-43904-9_66

patients. Automating the CXR interpretation process can lead to faster and more accurate diagnoses. Advances in Artificial Intelligence (AI), particularly in the field of computer vision for medical imaging, have significantly alleviated the challenges faced in radiology. The availability of large labeled collections of CXRs has been instrumental in driving progress in this area [8,10]. Both CXR disease detection and automatic report generation have witnessed substantial improvements [15,16,23]. Remarkably, AI-based methods for finding detection are now approaching the performance level of experienced radiologists [21,27]. Moreover, just as vision transformers have revolutionized various areas of computer vision [12], they have also become an integral part of automatic CXR analysis [18].

Although significant strides have been made in AI-assisted medical image segmentation and disease detection, tasks requiring intricate reasoning have received less attention. One such complex task is monitoring disease progression in a sequence of images, which is particularly critical in assessing patients with pneumonia and other CXR findings. For example, temporal lung changes serve as vital indicators of patient outcomes and are routinely mentioned in radiology reports for determining the course of treatment [20]. Prior work has investigated tracking the progression of COVID-19 pulmonary diseases and predicting outcomes [13]. Recently, CheXRelNet was proposed, which utilizes graph attention networks to capture anatomical correlations and detect changes in CXRs using both local and global anatomical information [11]. Change detection between longitudinal patient visits has also been studied in modalities beyond CXR, such as osteoarthritis in knee radiographs and retinopathy in retinal photographs [14]. Nonetheless, prior works have faced limitations in effectively attending to fine-grained relevant changes while disregarding irrelevant variations. Additionally, it is important to capture long-range spatial and temporal information to identify pertinent changes in medical images effectively.

Inspired by the success of Transformer models in remote sensing change detection tasks [3,28], we introduce `CheXRelFormer`, an end-to-end siamese disease progression model. `CheXRelFormer` takes a pair of CXR images as input, extracts visual features with a hierarchical vision Transformer module, and subsequently computes multi-level feature differences with a difference module. A self-attention mechanism allows the model to identify the most informative regions of the input images. By attending to fine-grained relevant changes, our model can accurately detect whether the patient's condition has improved, worsened, or remained unchanged. We evaluate the performance of our model on a large dataset of paired CXR images with corresponding disease progression labels. Experimental results demonstrate that our model outperforms existing state-of-the-art methods in detecting disease progression in CXR images. Our model has the potential to improve the efficiency and accuracy of CXR interpretation and thereby lead to more personalized treatment plans for patients. The contributions of our work can be summarized as follows:

(1) We propose `CheXRelFormer`, an end-to-end siamese disease progression model that can accurately detect changes in CXR image pairs by attend-

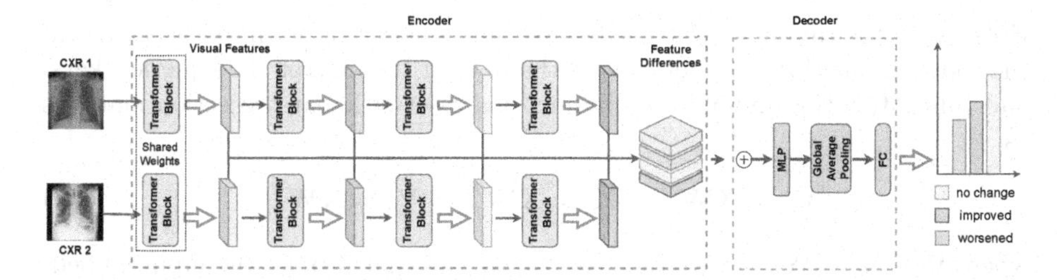

Fig. 1. CheXRelFormer overview. Given two CXR images $(\mathbf{X}, \mathbf{X}')$ a shared Transformer encoder extracts visual features at multiple resolutions. These feature maps are then fed into a difference module that computes visual differences across multiple scales. The difference module enhances the ability of the model to capture disease progression by effectively attending to relevant changes in the image pair. The decoder fuses information and performs the final disease progression classification task.

ing to informative regions and identifying fine-grained relevant visual differences.

(2) CheXRelFormer leverages hierarchical vision Transformers and a difference module to compute multi-level feature differences across CXR images, allowing the model to capture long-range spatial and temporal information.

(3) We experimentally demonstrate that CheXRelFormer outperforms existing state-of-the-art baselines in detecting disease progression in CXR images.

2 Methodology

Let $\mathcal{C} = \{(\mathbf{X}, \mathbf{X}')_i\}_{i=1}^{N}$ be a set of CXR image pairs, where $\mathbf{X}, \mathbf{X}' \in \mathbb{R}^{H \times W \times C}$, and H, W, and C are the height, width, and number of channels, respectively. Each image pair $(\mathbf{X}, \mathbf{X}')_i$ is associated with a set of labels $\mathcal{Y}_i = \{y_{i,m}\}_{m=1}^{M}$, where $y_{i,m} \in \{0, 1, 2\}$ indicates whether the pathology m appearing in the image pair has improved, worsened, or remained the same. The goal is to design a model that accurately predicts the disease progression labels for an unseen image pair $(\mathbf{X}, \mathbf{X}')$ and a wide range of pathologies.

To this end, we use a hierarchical Transformer [12] encoder to process each image pair. Specifically, let $\mathbf{X}, \mathbf{X}' \in \mathbb{R}^{H \times W \times C}$ be the input image pair. The encoder consists of L identical Transformer layers, each with a multi-head self-attention block followed by a position-wise feedforward network. The multi-head self-attention block contains a series of self-attention heads and is defined as

$$\text{MultiHead}(\mathbf{Q}, \mathbf{K}, \mathbf{V}) = \text{Concat}(\text{head}_1, \text{head}_2, \ldots, \text{head}_J)\mathbf{W}^O, \tag{1}$$

where $\mathbf{Q}, \mathbf{K}, \mathbf{V} \in \mathbb{R}^{N \times C}$ are the queries, keys, and values, respectively; $\mathbf{W}^O$ is a learned weight matrix, J is the number of heads, and head_j is the j-th attention head, computed as

$$\text{head}_j = \text{Attention}(\mathbf{Q}_j, \mathbf{K}_j, \mathbf{V}_j) = \text{softmax}\left(\frac{\mathbf{Q}_j \mathbf{K}_j^T}{\sqrt{d_k}}\right)\mathbf{V}_j. \tag{2}$$

Here, d_k is the dimensionality of the key and query vectors in each head, and the softmax function is applied along the rows of the matrix. The queries, keys, and values $\mathbf{Q}_j$, $\mathbf{K}_j$, and $\mathbf{V}_j$ are obtained via a set of linear projection matrices as

$$\mathbf{Q}_j = \mathbf{X}\mathbf{W}_j^Q, \quad \mathbf{K}_j = \mathbf{X}\mathbf{W}_j^K, \quad \mathbf{V}_j = \mathbf{X}\mathbf{W}_j^V, \tag{3}$$

where $\mathbf{W}^Q, \mathbf{W}^K, \mathbf{W}^V$ are learned weight tensors that project the input embeddings onto a lower-dimensional space. Similarly, the query, key, and value matrices for the second image in the pair are computed as

$$\mathbf{Q}_j = \mathbf{X}'\mathbf{W}_j^Q, \quad \mathbf{K}_j = \mathbf{X}'\mathbf{W}_j^K, \quad \mathbf{V}_j = \mathbf{X}'\mathbf{W}_j^V, \tag{4}$$

where the weight tensors $\mathbf{W}^Q, \mathbf{W}^K, \mathbf{W}^V$ are shared across the two images in the pair. The output of each multi-head self-attention block for each image pair, denoted by $(\mathbf{F}, \mathbf{F}')$, is then fed into a position-wise feedforward network which consists of two linear transformations and a depth-wise convolution [4] that captures local spatial information:

$$\mathbf{F}_c = f_1(\text{Conv2D}(f_2(\mathbf{F}), \mathbf{W}_d)) \tag{5}$$

$$\mathbf{F}_c' = f_1(\text{Conv2D}(f_2(\mathbf{F}'), \mathbf{W}_d)). \tag{6}$$

Here, $\mathbf{W_d}$ is the shared depth-wise convolution weight matrix and each feed-forward layer f_1, f_2 consists of a linear transformation followed by a non-linear activation. The difference module then processes the visual features from each Transformer layer to compute multi-level feature differences as follows:

$$\mathbf{C}^l = \phi\Big(\text{Conv2D}\left([\mathbf{F}_l, \mathbf{F}_l'], \mathbf{W}_l\right)\Big). \tag{7}$$

Here, $l = 1, \ldots, L$ denotes the l-th Transformer layer, with initial inputs $(\mathbf{F}_1, \mathbf{F}_1') = (\mathbf{F}_c, \mathbf{F}_c')$. Furthermore, ϕ is a non-linear activation, $\mathbf{W}_l$ is a learned weight parameter that essentially represents a multi-scale trainable distance metric, and $[\cdot, \cdot]$ is the concatenation operation. By computing differences between features at different scales, the proposed model can capture local and global structures that are relevant to the disease progression task.

Each multi-scale feature difference map is then passed through a feed-forward layer that maps the input features to a common feature space

$$\mathbf{C}_{out}^l = f_{\theta_l}(\mathbf{C}^l), \ \forall l \in [1, L], \tag{8}$$

where θ_l represents the set of learnable parameters for the l-th feed-forward network. The concatenated feature tensor combines information from multiple scales and is denoted as

$$\mathbf{C}_{out} = \Big[\mathbf{C}_{out}^1, \ldots, \mathbf{C}_{out}^l, \ldots, \mathbf{C}_{out}^L\Big], \tag{9}$$

where $[\cdot]$ denotes concatenation along the channel dimension. A feed-forward network with a global average pooling step, denoted by g, creates a fused feature representation with fixed dimensionality, which is finally passed through the final classification layer, denoted by h, to obtain the label predictions

$$\hat{y} = h\big(g(\mathbf{C}_{out})\big). \tag{10}$$

The network is trained end-to-end with a multi-label cross-entropy classification loss

$$\mathcal{L} = \frac{1}{N}\frac{1}{M}\sum_{i=1}^{N}\sum_{m=1}^{M} y_{i,m}\log(\sigma(\hat{y}_{i,m})) + (1 - y_{i,m})\log(1 - \sigma(\hat{y}_{i,m})), \tag{11}$$

where σ represents the sigmoid function and $\hat{y}_{i,m}, y_{i,m}$ are the model prediction and the ground truth for example $(\mathbf{X}, \mathbf{X}')_i$. An overview of the model architecture is represented in Fig. 1.

3 Experiments

3.1 Implementation Details

`CheXRelFormer` is implemented in Pytorch [19]. The encoder comprises four Transformer blocks with embedding dimensions 32, 64, 128, and 256, respectively. The number of heads on each multi-head attention block is 2, 2, 4, and 8. We train the encoder with a stochastic depth decay rule [6], with depths 3, 3, 6, and 18, for each Transformer block. To decrease the spatial dimension and reduce complexity, we perform spatial-reduction operations [22]. The position-wise feedforward network uses Gaussian Error Linear Unit (GELU) [5] activation functions. The multi-level image features extracted from the Transformer encoder are passed to the difference module. The difference module is composed of 2D convolutions, ReLU activations [2], and Batch Normalization [7]. The feed-forward layers consist of 64 neurons. The outputs from the difference module are upsampled, concatenated, and passed through a linear fusion layer followed by global average pooling. The model is trained using AdamW optimizer [17] with a learning rate of 6×10^{-5} and 16 batch size.

3.2 Dataset

We make use of the Chest ImaGenome dataset [25], which comprises 242,072 frontal MIMIC-CXRs [9] that were locally labeled using a combination of rule-based natural language processing (NLP) and CXR atlas-based bounding box detection techniques [24,26] to generate the annotations. Chest ImaGenome is represented as an anatomy-centered scene graph with 1,256 combinations of relation annotations between 29 CXR anatomical locations and their attributes. Each image is structured as one scene graph, resulting in approximately 670,000 localized comparison relations between the anatomical locations across sequential exams. In this work, we focus on the localized comparison relations data

Table 1. Dataset Statistics: pathology ID and label, number of training, validation and test CXR image pairs, and total number of CXR pairs

ID	Pathology Label	Train	Val	Test	Total Pairs
LO	Lung Opacity	5,516	912	1,579	8,007
PE	Pleural Effusion	7,450	1,089	2,165	10,704
AT	Atelectasis	77	11	26	114
EC	Enlarged Cardiac Silhouette	5,251	715	1,555	7,521
HO	Hazy Opacity/Pulmonary Edema	2,939	454	884	4,277
PX	Pneumothorax	979	132	261	1,372
CO	Consolidation	142	15	46	203
HF	Heart Failure/Fluid Overload	791	113	223	1,127
PN	Pneumonia	1,757	283	543	2,583
	Total	24,902	3,724	7,282	35,908

within Chest ImaGenome that pertains to cross-image relations for nine diseases of interest. Each comparison relation in the Chest ImaGenome dataset includes the DICOM identifiers of the two CXRs being compared, the comparison label, and the disease label name. The comparison is labeled as "no change", "improved" or "worsened", which indicates whether the patient's condition w.r.t. the disease has remained stable, improved, or worsened, respectively. The dataset contains 122,444 unique comparisons. We use 35,908 CXR pairs in total that pertain to the nine diseases of interest. The distribution of the data is improved (12,396), worsened (12,287) and no change (11,205). Table 1 presents high-level dataset statistics and training/validation/test splits employed in our experiments.

3.3 Baselines

To assess the performance of the proposed `CheXRelFormer` model, we conduct a comparative analysis with several baselines.

Local: This model employs a previously proposed siamese network [11] that only focuses on specific regions of the image, without considering inter-region dependencies or global information. The Local model is essentially a siamese network with a pretrained ResNet101 autoencoder trained on cropped Regions-of-Interest (RoIs), which are available in the Chest ImaGenome dataset.

Global: The Global model is a siamese network similar to the Local model but encodes global image-level information.

CheXRelNet: CheXRelNet combines global image-level information with local intra-image and inter-image information [11]. This model consists of a 2-layer graph neural network with a ResNet101 autoencoder for feature extraction.

Table 2. Quantitative comparison against the baselines

Method	LO	PE	AT	EC	HO	PX	CO	HF	PN	**All**
Local	0.41	0.37	0.41	0.29	0.37	**0.37**	0.49	0.29	0.42	0.43
Global	0.45	0.47	0.44	0.48	0.48	0.36	0.47	0.50	0.43	0.45
CheXRelNet	**0.49**	0.47	0.44	**0.49**	0.49	0.36	0.47	0.44	0.47	0.47
`CheXRelFormer`	0.48	**0.51**	**0.54**	0.40	**0.58**	0.35	**0.59**	**0.53**	**0.51**	**0.49**

Table 3. Ablation analysis of `CheXRelFormer` variants.

Method	LO	PE	AT	EC	HO	PX	CO	HF	PN	**All**
`CheXRelFormer_AbsDiff`	0.35	0.37	0.27	0.29	0.35	**0.37**	0.24	0.34	0.38	0.34
`CheXRelFormer_Local`	0.33	0.39	0.12	0.26	0.38	0.27	0.24	0.46	0.49	0.35
`CheXRelFormer`	**0.48**	**0.51**	**0.54**	0.40	**0.58**	0.35	**0.59**	**0.53**	**0.51**	**0.49**

3.4 Experimental Results

Table 2 lists the CXR change detection accuracy of all models across the nine diseases. We also report the mean weighted overall accuracy. `CheXRelFormer` outperforms baselines with a mean accuracy of 0.493 ± 0.0012 in this three-way classification task. The closest baseline is CheXRelNet with an accuracy of 0.468 ± 0.0041. Additionally, we perform a one-tailed t-test between `CheXRelFormer` and CheXRelNet, with $p = 0.00027$ indicating that `CheXRelFormer` significantly outperforms CheXRelNet in seven of the nine diseases (pleural effusion, atelectasis, pulmonary edema/hazy opacity, heart failure, pneumonia, and consolidation). Most importantly, we observe up to 12% performance gains for pathology labels with limited amounts of data, such as atelectasis and consolidation. These findings suggest that `CheXRelFormer` has the potential to be a valuable tool for detecting changes in CXR images associated with various common diseases.

3.5 Ablations on `CheXRelFormer` Architecture Components

We perform an ablation study to understand the impact of four factors, the difference module, the use of global vs. localized visual information, and the impact of multi-level features. Specifically, in Table 3, we present a comparison of in `CheXRelFormer` against `CheXRelFormer_AbsDiff`, where we replace the proposed difference module with an absolute difference component that subtracts two visual features from the last Transformer block. We also consider `CheXRelFormer_Local`, a variant that is trained on cropped anatomical RoIs instead of the entire image. Additional model variants are presented in the supplementary material. Table 4 presents the number of trainable parameters (*i.e.*, model capacity) and training time per epoch for each model variant.

Table 4. Model Capacity Comparison

Measures	CheXRelFormer_AbsDiff	CheXRelFormer_Local	CheXRelFormer
Number of Parameters (M)	28.9	41.0	41.0
Time per epoch (minutes)	148.8	56.4	171

An interesting observation is that CheXRelFormer_Local underperforms as the focus on specific anatomies limits the visual information available to the model. Given the highly fine-grained nature of this task, this result suggests that the relationship between an area and its surroundings is critical to a radiologist's perception of change, and the local Transformer cannot provide the necessary second-order information to the model. Therefore, our results show that global image-level information is crucial for accurately predicting disease change.

In addition, the absolute difference model, CheXRelFormer_AbsDiff, failed to perform the disease change classification task, indicating the importance of the proposed difference module. By incorporating the difference module and computing multi-level feature differences at multiple resolutions, CheXRelFormer learns to focus on the changes between two CXRs and to ignore irrelevant information. Our results demonstrate that multi-level feature differences are critical for improving performance in predicting disease change.

3.6 Qualitative Analysis

In Fig. 2, we visualize the model predictions from CheXRelFormer using attention rollout [1]. The produced attention maps clearly show the model's focus regions, which confirm that the model concentrated on the correct region in each image pair. CheXRelFormer can better differentiate between important and extraneous visual signals, allowing it to more accurately predict the 'no change' label. The model's ability to learn the optimal distance metric for each scale allows differentiating between relevant and irrelevant differences. In addition, analyzing multiple scales of visual features enables capturing subtle changes in pairs of CXR images, resulting in a better predictive performance for the 'improved' label. In contrast, the CheXRelFormer_AbsDiff model has difficulty in predicting both 'no change' and 'improved' labels (as shown in Fig. 3) due to the fact that images are not co-registered and exhibit several differences in their spatial or spectral characteristics – even though there was no actual change in the observed pathology.

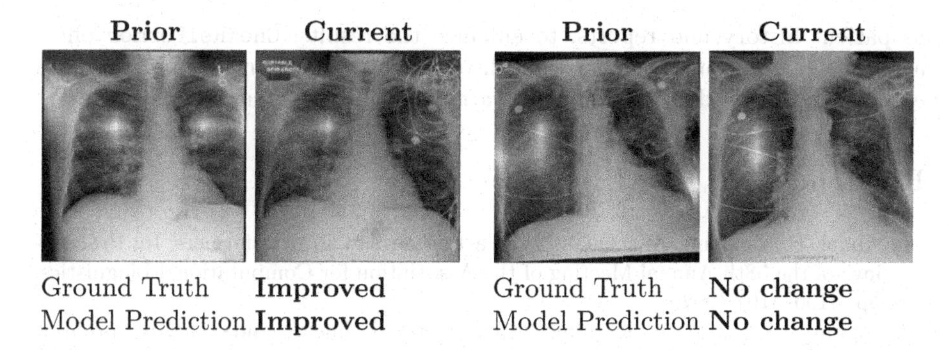

Fig. 2. Examples of model predictions obtained by `CheXRelFormer` compared against the ground-truth labels. Pathology labels LO: Lung Opacity (left) and PN: Pneumothorax (right). `CheXRelFormer` predicts the correct labels.

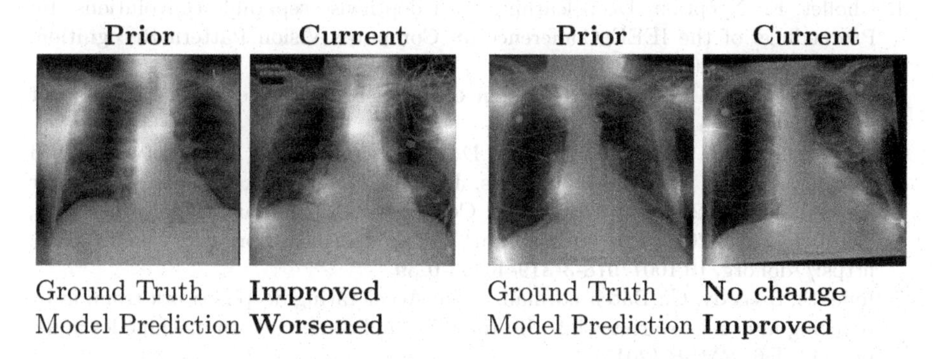

Fig. 3. Examples of model predictions obtained by `CheXRelFormer_AbsDiff` model compared against the ground-truth labels. Pathology labels LO: Lung Opacity (left) and PN: Pneumothorax (right). `CheXRelFormer_AbsDiff` struggles with differentiating between relevant and extraneous visual changes.

4 Conclusion

Monitoring disease progression is a critical aspect of patient management. This task requires skilled clinicians to carefully reason and evaluate changes in a patient's condition. In this paper, we propose `CheXRelFormer`, a hierarchical Transformer with a multi-scale difference module, trained on global image pair information to detect disease changes. Our model is inspired by the way clinicians monitor changes between CXRs, and improves the state of the art in this challenging medical imaging task. Our ablation studies show that global attention and the proposed difference module are critical components, and both help detect fine-grained changes between images. While our work shows significant progress, given the fine-grained nature of visual features that characterize findings in CXRs, disease progression remains a challenging task. In future work, we intend to include multimodal contextual information beyond the images, such as patient history and reports, to enhance the results. CheXRelFormer offers a promising solution for monitoring disease progression, and future work can extend the proposed methodology to various medical imaging modalities.

References

1. Abnar, S., Zuidema, W.: Quantifying attention flow in transformers. In: Proceedings of the 58th Annual Meeting of the Association for Computational Linguistics, pp. 4190–4197 (2020)
2. Agarap, A.F.: Deep learning using rectified linear units. arXiv preprint arXiv:1803.08375 (2018)
3. Bandara, W.G.C., Patel, V.M.: A transformer-based siamese network for change detection. In: IGARSS 2022–2022 IEEE International Geoscience and Remote Sensing Symposium, pp. 207–210. IEEE (2022)
4. Chollet, F.: Xception: Deep learning with depthwise separable convolutions. In: Proceedings of the IEEE Conference on Computer Vision Pattern Recognition, pp. 1251–1258 (2017)
5. Hendrycks, D., Gimpel, K.: Gaussian error linear units (gelus). arXiv preprint arXiv:1606.08415 (2016)
6. Huang, G., Sun, Yu., Liu, Z., Sedra, D., Weinberger, K.Q.: Deep networks with stochastic depth. In: Leibe, B., Matas, J., Sebe, N., Welling, M. (eds.) Computer Vision – ECCV 2016: 14th European Conference, Amsterdam, The Netherlands, October 11–14, 2016, Proceedings, Part IV, pp. 646–661. Springer, Cham (2016). https://doi.org/10.1007/978-3-319-46493-0_39
7. Ioffe, S., Szegedy, C.: Batch normalization: Accelerating deep network training by reducing internal covariate shift. In: International conference on machine learning, pp. 448–456. PMLR (2015)
8. Irvin, J., et al.: Chexpert: a large chest radiograph dataset with uncertainty labels and expert comparison. In: Proceedings of the AAAI Conference on Artificial Intelligence. vol. 33, pp. 590–597 (2019)
9. Johnson, A.E., Pollard, T.J., Berkowitz, S.J., et al.: Mimic-cxr, a de-identified publicly available database of chest radiographs with free-text reports. Scientific data, pp. 1–8 (2019)
10. Johnson, A.E., et al.: Mimic-iii, a freely accessible critical care database. Sci. Data **3**(1), 1–9 (2016)
11. Karwande, G., Mbakwe, A.B., Wu, J.T., Celi, L.A., Moradi, M., Lourentzou, I.: CheXRelNet: an anatomy-aware model for tracking longitudinal relationships between chest X-rays. In: Wang, L., Dou, Q., Fletcher, P.T., Speidel, S., Li, S. (eds.) Medical Image Computing and Computer Assisted Intervention – MICCAI 2022: 25th International Conference, Singapore, September 18–22, 2022, Proceedings, Part I, pp. 581–591. Springer, Cham (2022). https://doi.org/10.1007/978-3-031-16431-6_55
12. Kolesnikov, A., et al.: An image is worth 16x16 words: Transformers for image recognition at scale. In: ICLR (2021)
13. Li, M.D., et al.: Automated assessment and tracking of Covid-19 pulmonary disease severity on chest radiographs using convolutional siamese neural networks. Radiology: Artif. Intell. **2**(4) (2020)
14. Li, M.D., et al.: Siamese neural networks for continuous disease severity evaluation and change detection in medical imaging. NPJ digital medicine **3**(1), 1–9 (2020)

15. Liu, F., Yin, C., Wu, X., Ge, S., Zhang, P., Sun, X.: Contrastive attention for automatic chest x-ray report generation. In: Findings of the Association for Computational Linguistics: ACL-IJCNLP 2021, pp. 269–280 (2021)
16. Liu, G., et al.: Clinically accurate chest x-ray report generation. In: Doshi-Velez, F., Fackler, J., Jung, K., Kale, D., Ranganath, R., Wallace, B., Wiens, J. (eds.) Proceedings of the 4th Machine Learning for Healthcare Conference. Proceedings of Machine Learning Research, vol. 106, pp. 249–269 (2019)
17. Loshchilov, I., Hutter, F.: Decoupled weight decay regularization. In: International Conference on Learning Representations (ICLR) (2018)
18. Park, S., et al.: Multi-task vision transformer using low-level chest x-ray feature corpus for Covid-19 diagnosis and severity quantification. Med. Image Anal. **75**, 102299 (2022)
19. Paszke, A., et al.: Pytorch: An imperative style, high-performance deep learning library. In: Proceedings of the 32nd Annual Conference on Neural Information Processing Systems (NeurIPs), pp. 8024–8035 (2019)
20. Rousan, L.A., Elobeid, E., Karrar, M., Khader, Y.: Chest x-ray findings and temporal lung changes in patients with Covid-19 pneumonia. BMC Pulm. Med. **20**(1), 1–9 (2020)
21. Tang, Y.X., et al.: Automated abnormality classification of chest radiographs using deep convolutional neural networks. NPJ Digital Med. **3**(1), 70 (2020)
22. Wang, W., et al.: Pyramid vision transformer: A versatile backbone for dense prediction without convolutions. In: Proceedings of the IEEE/CVF International Conference on Computer Vision, pp. 568–578 (2021)
23. Wang, X., Peng, Y., Lu, L., Lu, Z., Summers, R.M.: Tienet: Text-image embedding network for common thorax disease classification and reporting in chest x-rays. In: Proceedings of the IEEE Conference On Computer Vision and Pattern Recognition, pp. 9049–9058 (2018)
24. Wu, J., et al.: Automatic bounding box annotation of chest x-ray data for localization of abnormalities. In: Proceedings of the 17th International Symposium on Biomedical Imaging (ISBI), pp. 799–803. IEEE (2020)
25. Wu, J.T., et al.: Chest imagenome dataset for clinical reasoning. In: Thirty-fifth Conference on Neural Information Processing Systems Datasets and Benchmarks Track (Round 2) (2021)
26. Wu, J.T., Syed, A., Ahmad, H., et al.: Ai accelerated human-in-the-loop structuring of radiology reports. In: Proceedings of the Americal Medical Informatics Association (AMIA) Annual Symposium (2020)
27. Wu, J.T., et al.: Comparison of chest radiograph interpretations by artificial intelligence algorithm vs radiology residents. JAMA Netw. Open **3**(10) (2020)
28. Zhang, M., Liu, Z., Feng, J., Liu, L., Jiao, L.: Remote sensing image change detection based on deep multi-scale multi-attention siamese transformer network. Remote Sens. **15**(3), 842 (2023)

Improved Flexibility and Interpretability of Large Vessel Stroke Prognostication Using Image Synthesis and Multi-task Learning

Minyan Zeng[1,2](✉), Yutong Xie[1], Minh-Son To[1,3], Lauren Oakden-Rayner[1,4], Luke Whitbread[1,5], Stephen Bacchi[3,4], Alix Bird[1,2], Luke Smith[1,2], Rebecca Scroop[4], Timothy Kleinig[4], Jim Jannes[4], Lyle J Palmer[1,2], and Mark Jenkinson[1,5,6]

[1] Australian Institute for Machine Learning, University of Adelaide, Adelaide, Australia
minyan.zeng@adelaide.edu.au
[2] School of Public Health, University of Adelaide, Adelaide, Australia
[3] Flinders Health and Medical Research Institute, Flinders University, Bedford Park, Australia
[4] Royal Adelaide Hospital, Adelaide, Australia
[5] School of Computer and Mathematical Sciences, University of Adelaide, Adelaide, Australia
[6] South Australian Health and Medical Research Institute, Adelaide, Australia

Abstract. While acute ischemic stroke due to large vessel occlusion (LVO) may be life-threatening or permanently disabling, timely intervention with endovascular thrombectomy (EVT) can prove life-saving for affected patients. Appropriate patient selection based on prognostic prediction is vital for this costly and invasive procedure, as not all patients will benefit from EVT. Accurate prognostic prediction for LVO presents a significant challenge. Computed Tomography Perfusion (CTP) maps can provide additional information for clinicians to make decisions. However, CTP maps are not always available due to variations in available equipment, funding, expertise and image quality. To address these gaps, we test (i) the utility of acquired CTP maps in a deep learning prediction model, (ii) the ability to improve flexibility of this model through image synthesis, and (iii) the added benefits of including multi-task learning with a simple clinical task to focus the synthesis on key clinical features. Our results demonstrate that network architectures utilising a full set of images can still be flexibly deployed if CTP maps are unavailable as their benefits can be effectively synthesized from more widely available images (NCCT and CTA). Additionally, such synthesized images may help with interpretability and building a clinically trusted model.

Supplementary Information The online version contains supplementary material available at https://doi.org/10.1007/978-3-031-43904-9_67.

H. Greenspan et al. (Eds.): MICCAI 2023, LNCS 14224, pp. 696–705, 2023.
https://doi.org/10.1007/978-3-031-43904-9_67

1 Introduction

Ischemic stroke caused by large vessel occlusion (LVO) is one of the leading causes of death and disability worldwide [1]. These occlusions restrict cerebral blood supply, resulting in rapid and irreversible damage to brain tissues. As such, prompt reperfusion treatment is critical to restore blood flow and salvage brain tissues from permanent damage. Endovascular thrombectomy (EVT) is the standard treatment recommended for LVO patients [2]. However, this treatment is costly and invasive and does not improve prognoses for all LVO patients [3]. Therefore, it is vital to identify those most likely to benefit from EVT.

There have been a number of models proposed for the prognostication of LVO in recent years [4–6]. The majority are traditional statistical models using non-imaging data, with the identification of clinical predictors for prognostication having shown limited capability [7]. A few models incorporate raw image data using deep-learning techniques, while most of these [4,5] utilize a single image modality, such as Non-Contrast Computed Tomography (NCCT).

CT Perfusion (CTP) maps are increasingly utilized for the prediction of LVO outcomes because they offer quantitative information on blood flow and volume in the brain, as well as the arrival time of blood bolus to brain tissues [8]. This information can show the brain perfusion status and identify the brain tissues that are irreversibly damaged (ischemic core), and those that are potentially salvageable (penumbra) [8]. In contrast, NCCT images can highlight the ischemic core as these tissues normally appear hypodense and CT angiography (CTA) can illustrate the collateral supply via highlighting vessel structure using a low dose contrast injection [9]. These two modalities are routinely collected at patients' admission because of their rapid acquisition and high tolerance [2]. Although the information estimated by NCCT and CTA can reflect the status of blood perfusion to some extent [9], these images are not as clinically useful as CTP maps since they do not directly show the most prominent neuropathological changes of ischemic stroke—ischemic core and salvageable penumbra. An example of ischemic changes shown by NCCT, CTA and CTP maps is presented in Fig. 1, where it can be seen that CTP maps provide clearer visual information, and hence are likely to lead to more accurate and clinically trusted prediction models. As such, it is important to investigate the benefits of incorporating CTP maps into deep learning prediction models.

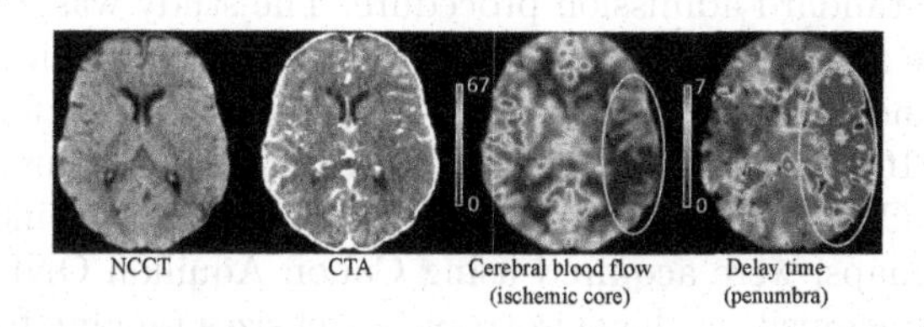

Fig. 1. Examples of NCCT, CTA, and two CTP maps

Despite the advantages of CTP maps, several factors have limited their utilization: (i) it is time-consuming to acquire, process and interpret the images [2]; (ii) poor quality maps are more likely due to motion artefacts and inadequate contrast agent, potentially making them uninterpretable [10]; and (iii) they require substantial investment to purchase advanced scanners, hire professional staff to run and maintain the scanner, which limits usage in hospitals with limited funding, such as in many rural and remote areas [11]. Therefore, it will be helpful if the flexibility of models incorporating CTP maps can be improved through image synthesis from commonly available image modalities (*e.g.,* NCCT and CTA). In this way, patients without access to a CTP scanner, or with uninterpretable CTP maps, can still benefit from the prognostic prediction that uses clinically-relevant features of CTP maps to inform treatment selection.

In recent years, techniques of image synthesis have shown promising potential in medicine [12–14]. However, most of these models have focused on image-to-image conversion tasks. In this paper, we propose a two-stage model that incorporates clinical knowledge for image synthesis and multimodal prognostic prediction. Specifically, in the image synthesis stage, the model was assigned to optimize a joint task, including a generative task, a discriminative task and a clinically-focused task. In the multimodal prognostic prediction stage, the model utilized imaging and non-imaging clinical information for predicting the dichotomised modified Rankin Scale (mRS) score 3 months post-stroke — the main outcome in stroke prognosis.

2 Method

2.1 Dataset and Pre-processing

Data utilized in this research was collected from the Royal Adelaide Hospital, which provides the sole EVT service to all stroke patients in South Australia and Northern Territory. There were 460 LVO patients included in the study, admitted between 01 Dec 2016 and 01 Dec 2021, and treated with EVT with full image modalities (NCCT, CTA and CTP maps). Of these, 256 achieved functional independency (mRS ≤ 2) 3 months post-stroke. The non-imaging data (*i.e.,* age, stroke severity, blood glucose, pre-admission functional status, use of intravenous thrombolysis, onset-to-groin puncture time, stroke hemisphere, and 3-month mRS score) was collected by experienced neurologists and nurses adhering to the standard admission procedure. The study was approved by The Central Adelaide Local Health Network Human Research Ethics Committee.

The NCCT and CTA images were skull-stripped with the attenuation clipped between 0 and 100 Hounsfield Units (HU) for the NCCT images and 0 and 750 HU for the CTA images. Multimodal CT imaging data, including NCCT, CTA and CTP maps, were acquired using Canon Aquilion ONE scanners. The NCCT and CTA acquisitions have isotropic voxel sizes ranging from 0.4-0.6 mm and 0.4-0.7 mm, respectively. The acquisition voxel size of the CTP maps is $0.4 \times 0.4 \times 4.9\ mm^3$. Four CTP maps, including cerebral blood volume (CBV), cerebral blood flow (CBF), mean transit time (MTT), and relative arrival time of contrast (Delay), were selected for their clinical utility, based on consultations with two senior neurologists. To rule out the impact of different brain sizes, affine registration to a CT template was performed for each modality [15,16]. The CTP maps were linearly scaled versions of the "true" (quantitative) maps using the within-scan relative values for prognostic prediction, rather than absolute values that may be influenced by a range of nuisance factors (e.g. head size, blood pressure) [17]. The image intensities for each modality were normalised to the interval of [0,1] by rescaling the min-max range for prognostication. Images were resampled to a 1mm isotropic resolution.

2.2 Models

Problem Statement. We take $\{\mathbf{X}^i_{\text{NCCT}}, \mathbf{X}^i_{\text{CTA}}, \mathbf{X}^i_{\text{CTP}}, \mathbf{X}^i_{\text{Cli\_var}}, y_{\text{mRS}}{}^i\}_{i=1}^N$ to be a set of data for N patients, where $\mathbf{X}^i_{\text{NCCT}}$, $\mathbf{X}^i_{\text{CTA}}$, and $\mathbf{X}^i_{\text{Cli\_var}}$ are the NCCT, CTA and clinical non-imaging data (*i.e.*, age, stroke severity, blood glucose, pre-admission functional status, use of intravenous thrombolysis, and onset-to-groin puncture time) for the ith patient. Four CTP maps for the ith patient are denoted by the set $\mathbf{X}^i_{\text{CTP}}$ which is defined as $\left\{\mathbf{X}^i_{\text{CBF}}, \mathbf{X}^i_{\text{CBV}}, \mathbf{X}^i_{\text{MTT}}, \mathbf{X}^i_{\text{Delay}}\right\}$. The dichotomized prognostic outcome for the ith patient is denoted by $y_{\text{mRS}}{}^i$ defined as $\{0 \text{ if } mRS \leq 2, \text{ or } 1 \text{ if } mRS > 2\}$.

We aim to: (i) evaluate the performance of CTP maps in predicting the dichotomized mRS score; and (ii) synthesize the CTP maps using two commonly used image modalities (NCCT and CTA) at admission for prognostic prediction. For the first aim, the model can be written as:

$$\hat{y}^{\,i}_{acq} = F_{\text{acq}}\left(\mathbf{X}^i_{\text{CBF}}, \mathbf{X}^i_{\text{CBV}}, \mathbf{X}^i_{\text{MTT}}, \mathbf{X}^i_{\text{Delay}}, \mathbf{X}^i_{\text{Cli\_var}}\right) \tag{1}$$

where $\hat{y}^{\,i}_{acq}$ is the predicted outcome and F_{acq} is the mapping function from the acquired CTP maps and clinical information to the dichotomized mRS score.

For the second aim, there are two tasks, including (i) learning a mapping function G for CTP map generation, and (ii) learning a function F_{syn} to map synthetic CTP maps and clinical information to the dichotomized mRS score. That is:

$$\left(\hat{\mathbf{X}}^i_{\text{CBF}}, \hat{\mathbf{X}}^i_{\text{CBV}}, \hat{\mathbf{X}}^i_{\text{MTT}}, \hat{\mathbf{X}}^i_{\text{Delay}}\right) = G\left(\mathbf{X}^i_{\text{NCCT}}, \mathbf{X}^i_{\text{CTA}}\right), \tag{2}$$

$$\hat{y}^{\,i}_{syn} = F_{\text{syn}}\left(\hat{\mathbf{X}}^i_{\text{CBF}}, \hat{\mathbf{X}}^i_{\text{CBV}}, \hat{\mathbf{X}}^i_{\text{MTT}}, \hat{\mathbf{X}}^i_{\text{Delay}}, \mathbf{X}^i_{\text{Cli\_var}}\right) \tag{3}$$

where $\hat{\mathbf{X}}^i_{\text{CBF}}, \hat{\mathbf{X}}^i_{\text{CBV}}, \hat{\mathbf{X}}^i_{\text{MTT}}, \hat{\mathbf{X}}^i_{\text{Delay}}$ are the predicted CBF, CBV, MTT, and Delay maps and $\hat{y}^{\,i}_{syn}$ is the predicted outcome from synthetic CTP maps for the ith patient. To fulfil the second aim, we propose a two-stage deep learning framework, including a clinical-guided synthesis and a multimodal prognostic prediction. The network architecture and loss function are detailed below.

Stage 1: Clinical-Guided Synthesis. The method for synthesizing CTP maps utilizes a 3D generative adversarial network (GAN) model, which is illustrated in Fig. 2 - stage 1. The encoder-decoder is a U-net architecture [18]. The encoder contains one convolutional layer (32 filters) and five 3D-residual blocks (32, 64, 128, 256, 256 filters) with $3 \times 3 \times 3$ kernels. Each convolutional layer in the residual block is followed by instance normalization and LeakyReLU activation. The decoder contains five 3D-residual blocks (128, 64, 32, 32, 4 filters) with $3 \times 3 \times 3$ kernels. After each residual block, features were upsampled and combined with encoder outputs, as usual. NCCT and CTA inputs were concatenated at the channel level. The discriminator contains one convolutional layer (16 filters) and four 3D-residual blocks (32, 64, 128, 256 filters). Real CTP maps and synthetic CTP maps (*e.g.*, $\mathbf{X}^i_{\text{CBF}}$ and $\hat{\mathbf{X}}^i_{\text{CBF}}$) were input to the discriminator, where two classification heads were designed for: a discriminative task with four fully connected (FC) layers (filters from 128 to 2) to distinguish real from synthetic maps; and a clinical task with a FC layer (filters from 128 to 2) to identify the cerebral hemisphere of the occlusion. The loss function is:

$$\begin{aligned} Loss_{\text{Stage1}} &= Loss_{\text{mse}}\left(G\left(\mathbf{X}^i_{\text{NCCT}}, \mathbf{X}^i_{\text{CTA}}\right), \mathbf{X}^i_{\text{CTP}}\right) \\ &- \left[Loss_{\text{bce}}\left(D\left(G\left(\mathbf{X}^i_{\text{NCCT}}, \mathbf{X}^i_{\text{CTA}}\right)\right), 0\right) + Loss_{\text{bce}}\left(D\left(\mathbf{X}^i_{\text{CTP}}\right), 1\right)\right] \\ &- \left[Loss_{\text{bce}}\left(D\left(G\left(\mathbf{X}^i_{\text{NCCT}}, \mathbf{X}^i_{\text{CTA}}\right)\right), y^i_{\text{hemi}}\right) + Loss_{\text{bce}}\left(D\left(\mathbf{X}^i_{\text{CTP}}\right), y^i_{\text{hemi}}\right)\right] \end{aligned} \tag{4}$$

where $Loss_{\text{mse}}$ and $Loss_{\text{bce}}$ calculate the mean square error and the binary cross entropy, respectively. D is a mapping function for discriminative and clinical tasks. y^i_{hemi} is the label of the clinical task $\{0$: occlusion in the left hemishpere; 1 : occlusion in the right hemishpere$\}$. We used the total loss for the set of synthetic CTP maps for backpropagation.

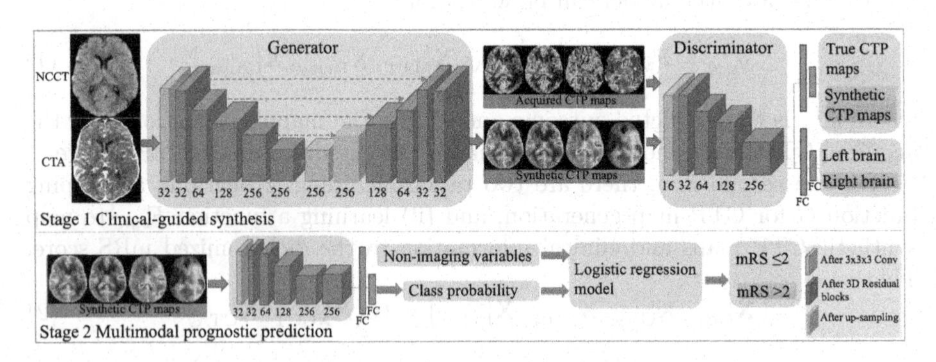

Fig. 2. Overview of the proposed method for image synthesis and prognostic prediction

Stage 2: Multimodal Prognostic Prediction. In the second stage, we used synthetic CTP images $\hat{\mathbf{X}}^i_{\text{CBF}}, \hat{\mathbf{X}}^i_{\text{CBV}}, \hat{\mathbf{X}}^i_{\text{MTT}}, \hat{\mathbf{X}}^i_{\text{Delay}}$ from stage 1 incorporating non-imaging variables to predict the dichotomized mRS score. The architecture is illustrated in Fig. 2 - stage 2. Specifically, the first step is to train a model with synthetic image inputs. The model contains one convolutional layer (32 filters), five 3D-residual blocks (32, 64, 128, 256, 256 filters) with $3 \times 3 \times 3$ kernels and a FC layer (filters from 128 to 2). The second step is to train a logistic regression model (single node with sigmoid activation) with the inputs including the predicted class probability derived from the first step and non-imaging variables. The loss functions of stage 2 can be formulated as:

$$Step\ 1: Loss_{\mathrm{img}} = Loss_{\mathrm{bce}}\left(F_{\mathrm{img}}\left(\hat{\mathbf{X}}^{i}_{\mathrm{CTP}}\right), y_{\mathrm{mRS}}{}^{i}\right) \tag{5}$$

$$Step\ 2: Loss_{\mathrm{logistic}} = Loss_{\mathrm{bce}}\left(F_{\mathrm{logistic}}\left(F_{\mathrm{img}}\left(\hat{\mathbf{X}}^{i}_{\mathrm{CTP}}\right), \mathbf{X}^{i}_{\mathrm{Cli\_var}}\right), y_{\mathrm{mRS}}{}^{i}\right) \tag{6}$$

where F_{img} is a mapping function of $\hat{\mathbf{X}}^{i}_{\mathrm{CTP}}$ to a binary mRS score and F_{logistic} is a mapping function of step 1 outputs and $\mathbf{X}^{i}_{\mathrm{Cli\_var}}$ to a binary mRS score. $\hat{\mathbf{X}}^{i}_{\mathrm{CTP}}$ is composed of $\hat{\mathbf{X}}^{i}_{\mathrm{CBF}}, \hat{\mathbf{X}}^{i}_{\mathrm{CBF}}, \hat{\mathbf{X}}^{i}_{\mathrm{MTT}}, \hat{\mathbf{X}}^{i}_{\mathrm{Delay}}$, concatenated as channels.

3 Experiments and Results

We performed two sets of experiments in the current study. In the first set of experiments, we compared prognostic prediction performance between models using different modalities, including (i) NCCT and CTA, (ii) CTP maps, (iii) NCCT, CTA and CTP maps, (iv) non-imaging data, (v) NCCT, CTA and non-imaging data, (vi) CTP maps and non-imaging data, and (vii) NCCT, CTA, CTP maps and non-imaging data. Images were input into the models with the architecture described in Sect. 2.2 stage 2, where inputs were replaced with corresponding imaging modalities concatenated at the channel level. In the second set of experiments, we evaluated the quality of the synthetic images and the performance when using them for prognostic prediction. We initially compared our model to four synthesis models: UNET, WGAN, CycleGAN and L2GAN. The L2GAN has the same architecture as our model but is not assigned the additional clinical task in the discriminator. To evaluate the quality of the synthetic images, we compared the structural similarity index measure (SSIM) and peak signal-to-noise ratio (PSNR) between synthesis models. Area under the ROC curve (AUC), accuracy (ACC), and F1-Score were used to assess the performance of prognostic prediction. We also compared our model to three state-of-the-art models that used raw images and clinical non-imaging data [4,5].

We randomly split the data into a training and a testing dataset. We used 4-fold cross-validation for training. For image synthesis, the models were trained for 200 epochs with a batch size of 4 using the Adam optimizer with learning rates of 2×10^{-4} and 2×10^{-5} for the generator and discriminator, respectively. For prognostic prediction, models with image inputs were trained for 100 epochs with a batch size of 4 using the Adam optimizer with a learning rate of 1×10^{-5}. All of the models were trained independently. The experiments above were implemented using PyTorch on the NVIDIA 3090 24GB GPU. Logistic regression models were trained with hyperparameters using grid search (Supplementary Table S1) based on Scikit-learn.

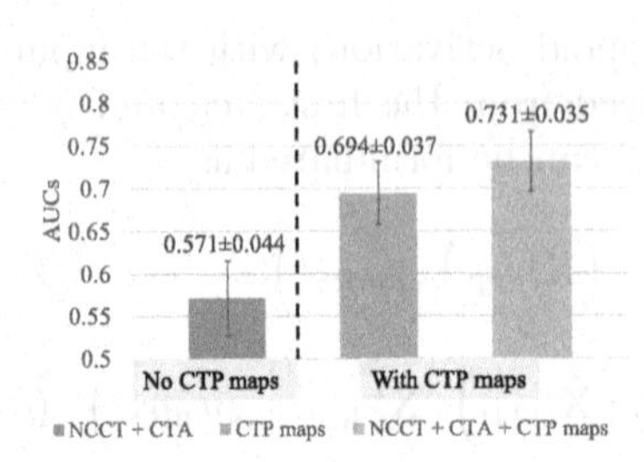

Fig. 3. AUCs of models using imaging data only

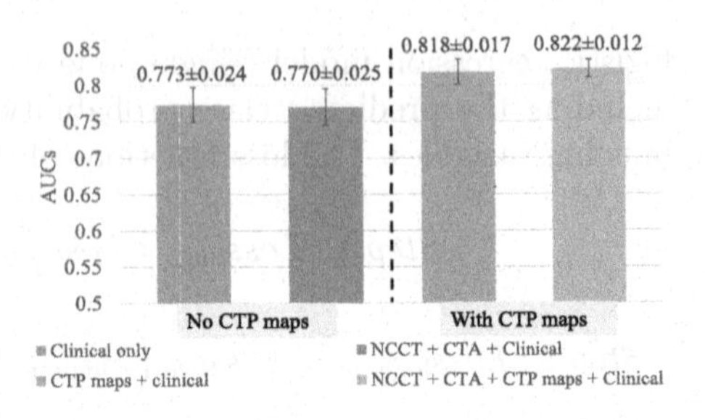

Fig. 4. AUCs of models incorporating imaging and clinical non-imaging data

Table 1. Image quality of five generative models

Models	SSIM (%)				PSNR			
	CBF	CBV	MTT	Delay	CBF	CBV	MTT	Delay
UNET	75.2 ± 0.5	79.3 ± 0.6	54.4 ± 1.3	61.7 ± 0.5	31.0 ± 0.2	34.7 ± 0.4	**20.0 ± 0.1**	**20.3 ± 0.1**
WGAN	79.0 ± 1.2	83.2 ± 0.8	46.4 ± 4.0	55.2 ± 2.1	31.0 ± 1.0	34.2 ± 0.5	19.5 ± 0.4	19.1 ± 0.3
CycleGAN	80.9 ± 1.2	84.5 ± 1.0	57.7 ± 0.5	66.6 ± 0.5	31.0 ± 0.3	34.1 ± 0.6	19.3 ± 0.1	19.3 ± 0.1
L2GAN	82.2 ± 0.3	85.8 ± 0.4	59.8 ± 0.5	67.7 ± 0.7	**31.7 ± 0.1**	35.1 ± 0.2	19.9 ± 0.3	20.2 ± 0.3
Ours	**82.5 ± 0.4**	**86.6 ± 0.2**	**60.7 ± 0.4**	**69.2 ± 0.5**	31.6 ± 0.3	**35.2 ± 0.1**	**20.0 ± 0.1**	20.2 ± 0.1

Table 2. Performance of models using acquired NCCT and CTA, and synthetic CTP maps, compared to the ideal case with acquired CTP maps (bottom row), for prediction of 3-month mRS.

	AUC(%)	ACC(%)	F1(%)
NCCT+CTA	77.0 ± 2.5	69.6 ± 3.9	68.0 ± 4.3
UNET	78.3 ± 1.0	69.3 ± 2.4	67.5 ± 3.0
WGAN	78.9 ± 3.3	70.9 ± 5.1	69.5 ± 6.0
CycleGAN	77.8 ± 1.9	70.1 ± 3.4	68.1 ± 3.4
L2GAN	79.5 ± 1.1	71.8 ± 2.0	69.5 ± 1.1
Our method	**80.7 ± 1.0**	**72.0 ± 2.4**	**70.3 ± 2.0**
Acquired CTP	81.8 ± 1.7	73.4 ± 4.2	71.4 ± 3.1

3.1 Results of Image Synthesis and Prognostic Prediction

Data Modalities for Prognostic Prediction. The performance of models using different combinations of data modalities is shown in Figs. 3 and 4 and supplementary Table S2. Models that included CTP maps clearly had better performance (Fig. 3). When non-imaging data were incorporated into the models (Fig. 4), those with CTP maps outperformed those without. This demonstrates that the inclusion of CTP maps can increase the performance of prognostic prediction. The validation results are similar to these test data results (Table S2).

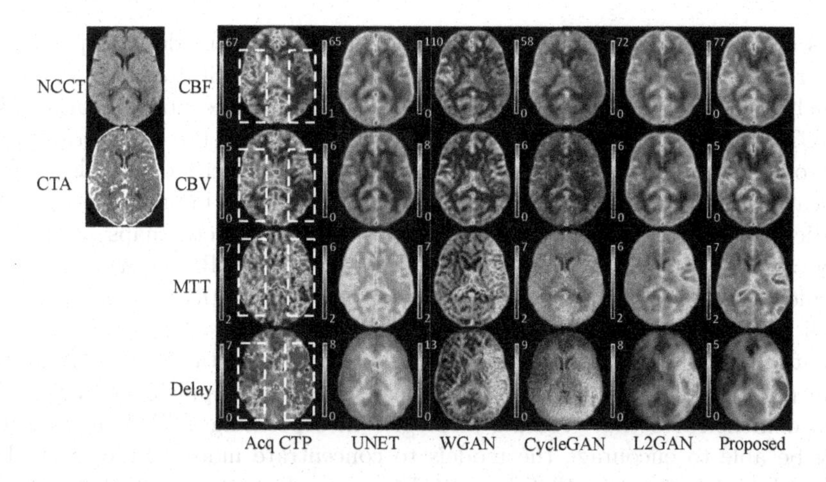

Fig. 5. Illustration of synthetic CTP maps generated by models (rightmost 5 columns). The dashed squares indicate the location of the lesion in the acquired images (Acq CTP, which is the ground truth for image generation). The solid square indicates the non-ischemic regions for comparison. The colour bars show the corresponding CTP values of the colours.

Table 3. Performance of models for prediction of 3-month mRS. Bacchi S [4] and Samak Z [5] are state-of-the-art models. The bottom row shows the ideal case for comparison.

	AUC(%)	ACC(%)	F1(%)
Bacchi $S_{NCCTonly}$ [4]	74.6 ± 3.0	69.6 ± 3.7	63.2 ± 5.9
Bacchi $S_{NCCT+CTA}$ [4]	76.8 ± 0.8	67.9 ± 0.6	63.4 ± 2.6
Samak Z [5]	76.6 ± 0.7	70.4 ± 1.4	63.7 ± 3.1
Our method	$\mathbf{80.7 \pm 1.0}$	$\mathbf{72.0 \pm 2.4}$	$\mathbf{70.3 \pm 2.0}$
Acquired CTP	81.8 ± 1.7	73.4 ± 4.2	71.4 ± 3.1

Image Quality of Synthetic CTP Maps. The synthetic CTP maps generated by different methods are shown in Fig. 5 and Supplementary Fig. S1. The acquired CBF and CBV maps show the ischemic core with hypointense signals on the right side, while MTT and Delay maps show the penumbra with hyperintense signals when compared to the contralateral side. Despite some visible differences from the acquired CTP maps, our synthesis method successfully learned the key clinical features related to the ischemic changes. The CTP maps generated by L2GAN are visually similar to our synthetic maps, except for the hypointense signals in the CBF and hyper-intense signals in Delay being weaker in the L2GAN. This shows that adding the clinical task enables the model to better learn key clinical features. Each of the other synthetic maps (UNET, WGAN and CycleGAN) missed even more of the important hyper- and hypointense features across the synthesized CTP images. The proposed model also performed better than the other models in image quality metrics (SSIM and PSNR).

Synthetic CTP Maps for Prognostic Prediction. The model incorporating the CTP maps generated by our proposed method shows the best performance of prognostic prediction compared to other synthesis methods (Table 2). This indicates that the inclusion of another clinical task can improve the outcome prediction. Also, the predictive performance of our synthetic maps is considerably closer to that of the acquired maps (bottom row) (ROC curves shown in Supplementary Fig. S2), indicating that the proposed method can recover most of its predictive ability when CTP maps are unavailable. Moreover, our model with the synthetic CTP maps also outperformed three state-of-the-art models trained on our dataset (Table 3) (ROC curves shown in Supplementary Fig. S2). This demonstrates that training strategies incorporating CTP map synthesis may be able to encourage the models to concentrate more on the most clinically relevant features in NCCT and CTA images for outcome prediction. Such training strategies may help build models that not only have better performance but are also clinically trusted, given their ability to demonstrate the replication of key clinical imaging features. Validation set results are similar to these (Supplementary Tables S3–S5).

4 Conclusion

This study demonstrates that CTP maps, which are known to provide critical information for clinicians, also benefit prognostic prediction using deep learning techniques. When CTP maps are not available at hospital admission, their benefits can still be largely retained through image synthesis. Using multi-task learning with a simple clinical task, our model outperformed other synthesis methods in both image quality and the performance of prognostic prediction. Our synthetic CTP maps show key clinical features that are able to be readily discerned upon visual inspection. These findings verify the advantages of including additional CTP maps in LVO prognostication and establish the ability to effectively synthesize such maps to retain their benefits. While we acknowledge that our network architectures are not novel, we highlight the novelty of our architectures for stroke prognostication. The proposed framework can provide significant utility in the future to aid in the selection of patients for high-stakes time-critical EVT, particularly for those who have limited access to advanced imaging. Furthermore, by demonstrating the key clinical imaging features, our framework may improve confidence in building a clinically trusted model.

References

1. Malhotra, K., Gornbein, J., Saver, J.L.: Ischemic strokes due to large-vessel occlusions contribute disproportionately to stroke-related dependence and death: a review. Front. Neurol. **8**, 651 (2017)

2. Powers, W.J., et al.: 2018 guidelines for the early management of patients with acute ischemic stroke: a guideline for healthcare professionals from the American Heart Association/American Stroke Association. Stroke **49**(3), e46–e99 (2018)
3. Wolman, D.N., et al.: Endovascular versus medical therapy for large-vessel anterior occlusive stroke presenting with mild symptoms. Int. J. Stroke **15**(3), 324–331 (2020)
4. Bacchi, S., Zerner, T., Oakden-Rayner, L., Kleinig, T., Patel, S., Jannes, J.: Deep learning in the prediction of ischaemic stroke thrombolysis functional outcomes: a pilot study. Acad. Radiol. **27**(2), e19–e23 (2020)
5. Samak, Z.A., Clatworthy, P., Mirmehdi, M.: Transop: transformer-based multimodal classification for stroke treatment outcome prediction. arXiv preprint arXiv:2301.10829 (2023)
6. Zeng, M., et al.: Pre-thrombectomy prognostic prediction of large-vessel ischemic stroke using machine learning: a systematic review and meta-analysis. Front. Neurol. **13**, 945813 (2022)
7. Saleem, Y., et al.: Acute neurological deterioration in large vessel occlusions and mild symptoms managed medically. Stroke **51**(5), 1428–1434 (2020)
8. Nogueira, R.G., et al.: Thrombectomy 6 to 24 hours after stroke with a mismatch between deficit and infarct. New Engl. J. Med. **378**(1), 11–21 (2018)
9. Bal, S., et al.: Time dependence of reliability of noncontrast computed tomography in comparison to computed tomography angiography source image in acute ischemic stroke. Int. J. Stroke **10**(1), 55–60 (2015)
10. Chu, Y., et al.: Comparison of time consumption and success rate between CT angiography-and CT perfusion-based imaging assessment strategy for the patients with acute ischemic stroke. BMC Med. Imaging **22**(1), 1–8 (2022)
11. Chung, C.Y., Hu, R., Peterson, R.B., Allen, J.W.: Automated processing of head CT perfusion imaging for ischemic stroke triage: a practical guide to quality assurance and interpretation. Am. J. Roentgenol. **217**(6), 1401–1416 (2021)
12. Goodfellow, I., et al.: Generative adversarial networks. Commun. ACM **63**(11), 139–144 (2020)
13. Arjovsky, M., Chintala, S., Bottou, L.: Wasserstein generative adversarial networks. In: International Conference on Machine Learning, pp. 214–223. PMLR (2017)
14. Zhu, J.-Y., Park, T., Isola, P., Efros, A.A.: Unpaired image-to-image translation using cycle-consistent adversarial networks. In: Proceedings of the IEEE International Conference on Computer Vision, pp. 2223–2232 (2017)
15. Jenkinson, M., Smith, S.: A global optimisation method for robust affine registration of brain images. Med. Image Anal. **5**(2), 143–156 (2001)
16. Muschelli, J.: A publicly available, high resolution, unbiased CT brain template. In: Lesot, M.-J., et al. (eds.) IPMU 2020. CCIS, vol. 1239, pp. 358–366. Springer, Cham (2020). https://doi.org/10.1007/978-3-030-50153-2_27
17. Waaijer, A., et al.: Reproducibility of quantitative CT brain perfusion measurements in patients with symptomatic unilateral carotid artery stenosis. Am. J. Neuroradiol. **28**(5), 927–932 (2007)
18. Ronneberger, O., Fischer, P., Brox, T.: U-net: convolutional networks for biomedical image segmentation. In: Navab, N., Hornegger, J., Wells, W.M., Frangi, A.F. (eds.) MICCAI 2015. LNCS, vol. 9351, pp. 234–241. Springer, Cham (2015). https://doi.org/10.1007/978-3-319-24574-4_28

Transformer-Based Tooth Segmentation, Identification and Pulp Calcification Recognition in CBCT

Shangxuan Li[1,2], Chichi Li[2], Yu Du[3,4,5], Li Ye[3,4,5], Yanshu Fang[6], Cheng Wang[2(✉)], and Wu Zhou[1(✉)]

[1] School of Medical Information Engineering, Guangzhou University of Chinese Medicine, Guangzhou, China
zhouwu@gzucm.edu.cn
[2] Hanglok-Tech Co., Ltd., Zhuhai, China
cheng.wang@hanglok-tech.cn
[3] Department of Operative Dentistry and Endodontics, Sun Yat-sen University, Guangzhou, China
[4] Affiliated Stomatological Hospital, Guangzhou, China
[5] Guangdong Provincial Key Laboratory of Stomatology, Guangzhou, China
[6] First Clinical Medical College, Guangzhou University of Chinese Medicine, Guangzhou, China

Abstract. The recognition of dental pulp calcification has important value for oral clinic, which determines the subsequent treatment decision. However, the recognition of dental pulp calcification is remarkably difficult in clinical practice due to its atypical morphological characteristics. In addition, pulp calcification is also difficult to be visualized in high-resolution CBCT due to its small area and weak contrast. In this work, we proposed a new method of tooth segmentation, identification and pulp calcification recognition based on Transformer to achieve accurate recognition of pulp calcification in high-resolution CBCT images. First, in order to realize that the network can handle extremely high-resolution CBCT, we proposed a coarse-to-fine method to segment the tooth instance in the down-scaled low-resolution CBCT image, and then back to the high-resolution CBCT image to intercept the region of the tooth as the input for the fine segmentation, identification and pulp calcification recognition. Then, in order to enhance the weak distinction between normal teeth and calcified teeth, we proposed tooth instance correlation and triple loss to improve the recognition performance of calcification. Finally, we built a multi-task learning architecture based on Transformer to realize the tooth segmentation, identification and calcification recognition for mutual promotion between tasks. The clinical data verified the effectiveness of the proposed method for the recognition of pulp calcification in high-resolution CBCT for digital dentistry.

Supplementary Information The online version contains supplementary material available at https://doi.org/10.1007/978-3-031-43904-9_68.

H. Greenspan et al. (Eds.): MICCAI 2023, LNCS 14224, pp. 706–714, 2023.
https://doi.org/10.1007/978-3-031-43904-9_68

Keywords: Pulp calcification recognition · Transformer · Instance correlation · Tooth Segmentation · CBCT

1 Introduction

Pulp calcification is a type of pulp degeneration, characterized by the deposition of calcified tissue in the root canal. Clinically, pulp calcification usually occurs with pulp periapical diseases, which brings great challenges to root canal therapy. Finding pulp calcification before root canal treatment is very important for dentists to decide treatment strategies [1]. However, teeth with pulp calcification usually show few clinical symptoms, which are mainly found and diagnosed by radiographic examination. Compared with X-ray, cone beam computed tomography (CBCT) performs better in displaying root canal structure and pulp disease, so it is widely used for pulp calcification detection [2]. On CBCT images, pulp calcification showed partial or complete high attenuation root canal occlusion as shown in Fig. 1. Currently, the diagnosis of pulp calcification mainly depends on the image recognition of root canal occlusion by dentists. On the one hand, although high-resolution CBCT is used, the image contrast of calcified area is rather low, and human cannot easily find all calcified tubes. On the other hand, human recognition is time-consuming and laborious, and the agreement between observers is far from satisfactory. Therefore, an intelligent recognition method of pulp calcification is urgently needed for digital dentistry in clinical practice.

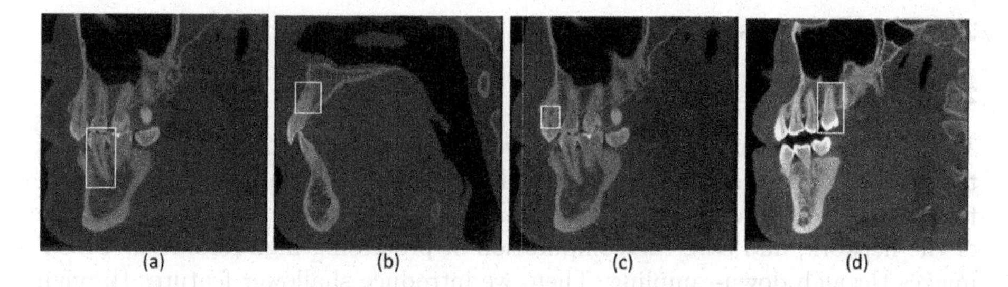

Fig. 1. Representative tooth images in CBCT. (a) Normal teeth; (b) Diffuse calcified teeth with a few root canal residues; (c) Calcified teeth with pulpal stones; (d) Calcified teeth with difficulty in recognition. The red box corresponds to the pulp area. (Color figure online)

With the introduction of artificial intelligence into various fields of medical images, research has proposed tooth and root canal segmentation models based on deep learning networks and CBCT images [3–7], which have achieved good performance in segmentation. However, how to intelligently recognize pulp calcification in small root canals is still very difficult and has not yet been explored. First, the calcified area has no clear morphological characteristics and is difficult for feature representation. Then, the resolution of CBCT image has a high resolution ($672 \times 688 \times 688$ in this work), while the tooth calcification areas are relatively small and in low contrast. It is very difficult to directly process such large volume data based on deep learning networks. Reducing the high-resolution of CBCT image will weaken the local tooth calcification area information, which brings certain challenges to the intelligent recognition of pulp calcification in CBCT. In addition, the current digital dentistry needs the whole process from 3D volume input, tooth segmentation, identification, and lesion recognition. However, the current relevant research [3–7] separately focuses on functions such as tooth segmentation, root canal segmentation or lesion detection, and has not yet built an integrated intelligent diagnosis process.

To this end, we propose a new method of tooth segmentation, identification and pulp calcification recognition based on Transformer to achieve accurate recognition of pulp calcification in high-resolution CBCT images. Specifically, we propose a coarse-to-fine method to segment the tooth instance in the low-resolution CBCT image, and back to the high-resolution CBCT image to intercept the region of the tooth as the input for the fine segmentation, identification and calcification recognition of the tooth instance. In order to enhance the weak distinction between normal teeth and calcified teeth, we put forward tooth instance correlation and triple loss to further improve the recognition performance of calcification. Finally, we introduce transformer to realize the above three tasks in an integrated way, and achieve mutual promotion of task performance. The clinical oral CBCT image data is used to verify the effectiveness of the proposed method.

2 The Proposed Method

2.1 The Proposed Framework

The network structure designed in this work is shown in Fig. 2. It mainly includes two modules: tooth segmentation and identification module, and pulp calcification recognition module. First, we stack the swin-transformer as the backbone of the network, and save the computation of processing high-resolution CBCT images through down-sampling. Then, we introduce shallower features through skip connection. Those features with higher resolution and shallower layers will contain relatively rich low-level information, which is more conducive to accurate tooth segmentation and identification recognition. In addition, through the multi-task learning mechanism based on Transformer, the performance of tooth segmentation, identification and calcification recognition can be mutually improved.

In the pulp calcification recognition module, we extract the features of each tooth from the deep feature through the results of tooth segmentation, and input them into the pulp calcification recognition module. Specifically, we design an instance correlation transformer (ICT) block. This block allows teeth to learn information from other teeth, so that different teeth can interact, which enables the network itself to explore the relationship between instances, thus improving the recognition performance of calcified teeth. In addition, we introduce a discriminator in the ICT block, which uses triple loss to learn the spatial distribution between categories, so as to learn better classification embedding.

2.2 Tooth Segmentation and Identification Module

In order to obtain the features of each tooth from the high-resolution CBCT image, we first recognize the segmented and identificated teeth,and combine the result of them for tooth instance segmentation. We use swin-transformer as the network backbone. Swin-transformer [8] is based on the idea of ViT model, which innovatively introduces the sliding window mechanism, so that the model can learn cross-window information. Meanwhile, through the down-sampling layer, the model can process super-resolution images, save computation and focus on global and local information. In addition, it has been proved to have advantages in the process of modeling an end-to-end multi-task system [9]. Unlike the typical swin-transformer, which only uses the Patch Merging layer in the deep layer, we first reduce the resolution of the input feature map through the Patch Merging layer in each stage, and then conduct the next sampling operation. This is conducive to our processing of high-resolution CBCT images. In addition, we have adopted the reverse skip connection (RSC). The information from the deep layer can reversely guide the learning of the shallow details to achieve better segmentation effect, which has been proved effective in segmentation task [10]. In this way, we can distinguish the edges of adjacent teeth more clearly, thus promoting better segmentation recognition.

The segmentation loss L_{seg} of the model can be defined as $L_{seg} = L_{cs} + \gamma_1 L_{dice} + \gamma_2 L_{bs}$, where γ_1,γ_2 are the balance parameters, where L_{cs} is the 33 categories of pixels (32 teeth classes + background) cross entropy loss. L_{dice} is the dice loss of each tooth instance segmentation. L_{bs} is binary segmentation. We calculate the cross entropy loss between teeth and background. The purpose of this step is to assist the model to distinguish foreground and background

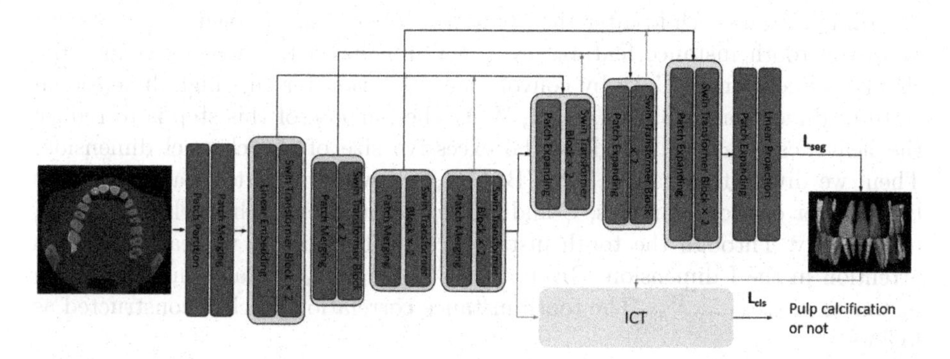

Fig. 2. The proposed framework. The orange arrow indicates the RSC process. (Color figure online)

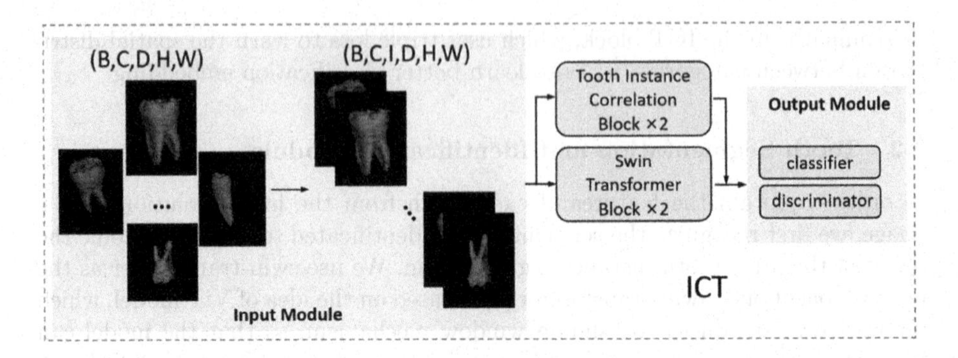

Fig. 3. The instance relevance transformer module

2.3 Calcification Recognition Module and Tooth Instance Correlation Block

Calcified root canals, especially small root canal calcification, are shown on CBCT images as the shadow in the center of the cross section faded and disappeared, and the density of the shadow is close to or the same as that of the surrounding dentin, which is significantly different from the root canal images of other root canals of the same affected tooth and normal adjacent teeth. Based on the above clinical observation, we design the tooth instance correlation block for better calcification recognition. The multi-head attention layer is widely used to build image relationship models in channel and spatial dimensions [11,12], so we believe it can be extended to explore the relationship between tooth instances. Specifically, we propose an ICT related to tooth instances to better identify calcified teeth. The ICT module is shown in Fig. 3.

First, in the input module of ICT, we extract the tooth deep feature from the output feature of the encoder through the prediction of instance segmentation. The specific method is to extract the tooth deep feature one by one according to the tooth id after the prediction label is de-sampled. We splice the extracted tooth instance features in a new dimension I. Then, we reduce the channel dimension to C/16 by convolution operation for this high-dimensional feature (dimensions: (B, C, I, D, H, W)). The purpose of this step is to reduce the heavy calculation caused by the excessive size of the channel dimension. Then, we divide the reshape into (B, C, I, N), where B is the batch size, I is the number of tooth samples, C and N are the number of channels and pixels, respectively. Through the tooth instance correlation block, we learn the cross attention in the I dimension. Given a CBCT image X_i contains multiple tooth instances $x_{i,1}$,$x_{i,2}$,...,$x_{i,n}$. The tooth instance correlation block is constructed as follows:

$$X_i^0 = (x_{i,class}; f(x_{i,1}); f(x_{i,2}); ...; f(x_{i,n})), X_i^0 \in R(n+1) \times d \tag{1}$$

$$X_i^l = MSA(LN(X_i^{l-1})) + X_i^{l-1}, l = 1....L \tag{2}$$

$$X_i^l = LN(MLP(X_i^l)) + X_i^l \tag{3}$$

where MSA is multi-head self attention, L is the layer of MSA, and LN is the standardization layer.

2.4 Triple Loss and Total Loss Function

In order to make the model more discriminative for the recognition of calcified teeth, a discriminator is designed in this work, which uses triplet loss to make the embedding of the input classifier more discriminative. This process is to make the features of the same category as close as possible, and the features of different categories as far away as possible. Meanwhile, in order to prevent the features of the instance from converging into a very small space, it is required that for two positive cases and one negative case, the negative case should be at least margin away from the positive case [13].

Specifically, we randomly selected an instance in the I dimension: Anchor (a), and randomly selected an instance that belongs to the same class as Anchor: Positive (p), and randomly selected an instance that belongs to a different class from Anchor: Negative (n). The goal of model learning is to make the distance $D(a, p)$ between Positive and Anchor as small as possible, and the distance $D(a, n)$ between Negative and Anchor as large as possible L_t. It is defined as follows:

$$L_t = max(D(a,p) - D(a,n) + \alpha), 0) \tag{4}$$

where D is the European distance, α is a margin between positive and negative pairs.

The classification loss is defined as $L_{cls} = L_{pc} + \gamma_3 L_t$, where γ_3 is the balance parameter, L_{pc} is the cross entropy loss as the loss of calcified tooth classification. Finally, the total loss function of the network is defined as $L_{total} = L_{seg} + L_{cls}$.

2.5 Implementation

The initialization setting of the learning rate is 1e-3, with 60000 iterations. The Adam algorithm is used to minimize the objective function. Two RTX3090 GPUs are used, each with 24G memory. The attenuation setting of the learning rate is 0.99 for every 500 iterations. All parameters, including weights and deviations, are initialized using a truncated normal distribution with a standard deviation of 0.1. In the tooth instance segmentation task, we use connected component analysis to extract the maximum area of predicted voxels and remove some small false-positive voxels. Code is available at: https://github.com/Lsx0802/ToothICT.

3 Experimental Results

3.1 Clinical Data, Experimental Setup and Evaluation Metric

This study was performed in line with the principles of the Declaration of Helsinki. In this work, 151 CBCT imaging data from the Imaging department of the local institute were acquired. The image resolution of the CBCT equipment used was $0.2 \sim 1.0$ mm, and the size of the CBCT volume is $672 \times 688 \times 688$. The bulb voltage was $60 \sim 90$ kV, and the bulb current was $1 \sim 10$ mA. 151 cases of dental symptoms were identified as CBCT oral indications by two dentists with 10 years of clinical experience. Among them, 60 patients had dental pulp calcification. In addition, each tooth was also marked for calcification. One dentist is responsible for the data label, and two doctors review it. When they disagree, they will reach an agreement through negotiation.

The CBCT data is preprocessed as follows. First, considering the balance between computational efficiency and instance segmentation accuracy, all CBCT images are normalized to $0.4 \times 0.4 \times 0.4\,\text{mm}^3$. Then, in order to reduce the impact of extreme values, especially in the metal artifact area, we cut the voxel-level intensity value of each CBCT scan to [0, 2500], and finally normalized the pixel value to the interval [0,1]. For Pulp calcification recognitionin, the adopted evaluation metrics include: Accuracy, Precision, Recall, F1, Dice. The measurement results were conducted with 10 times of four-fold cross validation. In addition, we have compared the performance of the proposed method with the relevant tooth segmentation methods, we used typical segmentation metrics for performance evaluation: Dice, Jaccard similarity coefficient (Jaccard), 95% Hausdorff distance (HD95), Average surface distance (ASD) and have conducted the ablation study of the proposed method.

3.2 Performance Evaluation

As shown in Table 1, Dice is based on the instance segmentation performance of each tooth. Backbone1 uses the swin transformer with skip connection to segment the teeth. w/o RSC is a model for eliminating reverse skip connection design. The segmentation results of tooth instance show that the proposed method is superior to other relevant segmentation methods. The main reason is that the proposed method uses the transformer structure. Its multi-head attention mechanism can capture global information, which is superior to the U-net structure based on CNN local features in the relevant methods. In addition, the ablation study for instance segmentation also shows the effectiveness of the proposed module.

Our model adopts a network based on swin transformer. Through its powerful global and local modeling ability, while retaining the jumping connection in UNet to retain the shallow features, the performance of Backbone1 is better than the previous segmentation model. In particular, we use reverse skip connection and use deep features to guide shallow feature learning, which has achieved obvious improvement in segmentation performance. After combining the task of calcification classification, the segmentation network has been improved a little, which benefits from the ICT module we adopted, because it not only learns the correlation characteristics between calcified teeth and normal teeth, but also learns the morphological correlation between teeth, which is beneficial to tooth segmentation.

Table 2 shows the ablation experimental results of calcified tooth recognition. Backbone 2 is the calcified tooth recognition of the whole module of tooth instance segmentation+classifier. w/o ICT is to remove the tooth instance correlation block, and w/o L_t is to remove the discriminator. We can find that the proposed two modules can effectively improve the performance of calcification recognition.

The accuracy of the model is only 74.62% when only swin transformer is used to classify tooth samples, while our proposed model can improve the performance of pulp calcification recognition by 3.85%. Especially, in the ablation experiment, when the ICT module is removed, the model performance drops obviously, which proves that our proposed ICT module can effectively learn the relationship between dental examples. In addition, after the loss Lt of the discriminant module is removed, the accuracy of the model decreases by about 1.16%, which proves that this method can effectively reduce the distance between similar samples and increase the distance between different samples. (See the supplementary materials for more visualization results)

Table 1. Performance comparison of instance segmentatio)

Method	Dice (%) ↑	Jaccard (%) ↑	HD95 ↓	ASD ↓
ToothNet [4]	90.84 ± 1.10	82.67 ± 1.59	3.13 ± 1.31	0.36 ± 0.17
CGDNet [5]	92.07 ± 0.91	84.93 ± 1.34	2.61 ± 1.011	0.31 ± 0.13
ToothSeg [6]	92.68 ± 0.86	87.39 ± 0.88	2.13 ± 0.47	0.27 ± 0.11
Backbone1	93.90 ± 0.51	88.67 ± 0.82	1.66 ± 0.26	0.25 ± 0.09
w/o RSC	93.99 ± 0.46	88.77 ± 0.93	1.72 ± 0.31	0.25 ± 0.05
Proposed	94.23 ± 0.61	89.64 ± 0.86	1.50 ± 0.27	0.23 ± 0.06

Table 2. Performance of pulp calcification recognition

Method	Accuracy (%) ↑	Precision (%) ↑	Recall (%) ↑	F1 (%) ↑
Backbone2	74.62 ± 0.41	72.67 ± 0.34	77.31 ± 0.43	76.79 ± 0.37
w/o ICT	75.31 ± 0.29	74.48 ± 0.36	78.13 ± 0.31	78.31 ± 0.33
w/o L_t	77.31 ± 0.36	$75,81 \pm 0.27$	82.06 ± 0.27	81.15 ± 0.41
Proposed	78.47 ± 0.25	77.31 ± 0.37	82.08 ± 0.23	82.41 ± 0.32

4 Conclusion

In this study, we proposed a calcified tooth recognition method based on transformer, which can detect calcified teeth in high-resolution CBCT images while achieving tooth instance segmentation and identification. Specifically, we proposed a coarse-to-fine processing method to make it possible to process high-resolution CBCT with deep network for calcification recognition. In addition, the design of instance correlation and triple loss further improved the accuracy of calcification detection. The validation of clinical data showed the effectiveness and advantages of the proposed method. We believe that this research will bring help to the intellectualization of oral imaging diagnosis and the navigation of oral surgery.

Acknowledgement. This research is supported by the school-enterprise cooperation project (No.6401-222-127-001).

References

1. Yang, Y.M., et al.: CBCT-aided microscopic and ultrasonic treatment for upper or middle thirds calcified root canals. BioMed Res. Int. **1–9**, 2016 (2016)
2. Patel, S., Brown, J., Pimental, T., Kelly, R., Abella, F., Durack, C.: Cone beam computed tomography in endodontics - a review of the literature. Int. Endodont. J. (2019)
3. Duan, W., Chen, Y., Zhang, Q., Lin, X., Yang, X.: Refined tooth and pulp segmentation using u-net in CBCT image. Dentomaxillofacial Radiol. 20200251 (2021)
4. Cui, Z., Li, C., Wang, W.: Toothnet: automatic tooth instance segmentation and identification from cone beam CT images. In: IEEE Conference on Computer Vision and Pattern Recognition, CVPR 2019, Long Beach, 16–20 June 2019, pp. 6368–6377. Computer Vision Foundation/IEEE (2019)
5. Wu, X., Chen, H., Huang, Y., Guo, H., Qiu, T., Wang, L.: Center-sensitive and boundary-aware tooth instance segmentation and classification from cone-beam CT. In: 2020 IEEE 17th International Symposium on Biomedical Imaging (ISBI), pp. 939–942 (2020)
6. Cui, Z., et al.: Hierarchical morphology-guided tooth instance segmentation from CBCT images. In: Feragen, A., Sommer, S., Schnabel, J., Nielsen, M. (eds.) IPMI 2021. LNCS, vol. 12729, pp. 150–162. Springer, Cham (2021). https://doi.org/10.1007/978-3-030-78191-0_12
7. Cui, Z., et al.: A fully automatic AI system for tooth and alveolar bone segmentation from cone-beam CT images. Nat. Commun. **13**, 2096 (2022)
8. Liu, Z., et al.: Swin transformer: hierarchical vision transformer using shifted windows. In: Proceedings of the IEEE/CVF International Conference on Computer Vision, pp. 10012–10022 (2021)
9. Bhattacharjee, D., Zhang, T., Süsstrunk, S., Salzmann, M.: Mult: an end-to-end multitask learning transformer. In: Proceedings of the IEEE/CVF Conference on Computer Vision and Pattern Recognition, pp. 12031–12041 (2022)
10. Xia, L., et al.: 3d vessel-like structure segmentation in medical images by an edge-reinforced network. Med. Image Anal. **82**, 102581 (2022)
11. Shao, Z., et al.: Transmil: transformer based correlated multiple instance learning for whole slide image classification. Adv. Neural Inf. Process. Syst. **34**, 2136–2147 (2021)
12. Hou, Z., Yu, B., Tao, D.: Batchformer: learning to explore sample relationships for robust representation learning. In: Proceedings of the IEEE/CVF Conference on Computer Vision and Pattern Recognition, pp. 7256–7266 (2022)
13. Hermans, A., Beyer, L., Leibe, B.: In defense of the triplet loss for person re-identification. arXiv preprint arXiv:1703.07737 (2017)

Treatment Outcome Prediction for Intracerebral Hemorrhage via Generative Prognostic Model with Imaging and Tabular Data

Wenao Ma[1], Cheng Chen[2], Jill Abrigo[3], Calvin Hoi-Kwan Mak[4], Yuqi Gong[1], Nga Yan Chan[3], Chu Han[5,6], Zaiyi Liu[5,6], and Qi Dou[1](✉)

[1] Department of Computer Science and Engineering, The Chinese University of Hong Kong, Hong Kong, China
qidou@cuhk.edu.hk

[2] Center for Advanced Medical Computing and Analysis, Harvard Medical School, Boston, USA

[3] Department of Imaging and Interventional Radiology, The Chinese University of Hong Kong, Hong Kong, China

[4] Department of Neurosurgery, Queen Elizabeth Hospital, Hong Kong, China

[5] Department of Radiology, Guangdong Provincial People's Hospital (Guangdong Academy of Medical Sciences), Southern Medical University, Guangzhou, China

[6] Guangdong Provincial Key Laboratory of Artificial Intelligence in Medical Image Analysis and Application, Guangzhou, China

Abstract. Intracerebral hemorrhage (ICH) is the second most common and deadliest form of stroke. Despite medical advances, predicting treatment outcomes for ICH remains a challenge. This paper proposes a novel prognostic model that utilizes both imaging and tabular data to predict treatment outcome for ICH. Our model is trained on observational data collected from non-randomized controlled trials, providing reliable predictions of treatment success. Specifically, we propose to employ a variational autoencoder model to generate a low-dimensional prognostic score, which can effectively address the selection bias resulting from the non-randomized controlled trials. Importantly, we develop a variational distributions combination module that combines the information from imaging data, non-imaging clinical data, and treatment assignment to accurately generate the prognostic score. We conducted extensive experiments on a real-world clinical dataset of intracerebral hemorrhage. Our proposed method demonstrates a substantial improvement in treatment outcome prediction compared to existing state-of-the-art approaches. Code is available at https://github.com/med-air/TOP-GPM.

Supplementary Information The online version contains supplementary material available at https://doi.org/10.1007/978-3-031-43904-9_69.

H. Greenspan et al. (Eds.): MICCAI 2023, LNCS 14224, pp. 715–725, 2023.
https://doi.org/10.1007/978-3-031-43904-9_69

Keywords: Prognostic Model · Intracerebral Hemorrhage · Mutli-modaltiy

1 Introduction

Intracerebral Hemorrhage (ICH) is a bleeding into the brain parenchyma, which has the second-highest incidence of stroke (accounts for more than 10% of strokes) and remains the deadliest type of stroke with mortality more than 40% [4–6]. Timely and proper treatments are crucial in reducing mortality [15], as well as improving functional outcomes, which is clinically deemed more valuable for prognostic model [2]. However, the treatment decision-making of ICH still remains problematic despite progression of clinical practice [20]. It is widely accepted that there is currently no effective approach in clinical practice to aid in decision-making regarding the evaluation of risks and benefits of a treatment [7,9,30]. Thus, there is an urgent need for reliable treatment recommendation model in clinical practice. Unfortunately, existing works of ICH treatment outcome prediction can either predict the outcome under a certain type of treatment [11,14,29], or consider treatment assignment as an input variable but ignore potential differences in outcomes due to varying treatment assignments [7,10,18], making it still challenging to determine from data which treatment would yield better outcomes. For this reason, we seek to provide a treatment recommendation model that outputs the reliable outcomes of all potential treatment assignments and focuses on the effect of different treatments.

One of the major challenges in treatment effect estimation is missing *counterfactual* outcome [19,24]. This means that we can only observe the outcomes of the actual treatment decision made for an individual. As a consequence, the counterfactuals, that are, the outcomes that would have resulted from treatment decisions not given to the patient are missing. Another challenge is *selection bias* brought by non-randomized controlled trials [1], that the treatment assignments may highly depend on patients' characteristics. For instance, for the ICH patients with a Glasgow Coma Scale (GCS) score 9–12 [31], early surgery is generally preferred over conservative treatment [30]. This selection bias can thus make the model unreliable in predicting the outcome of conservative treatment for patients with GCS 9–12 due to lack of observational data. These factors lead to inaccurate comparisons of treatment effects, as we can only get reliable outcome on one side (i.e., early surgery or conservative treatment).

To handle these challenges, some related works were based on the concept of balanced representation learning, which proposes to use additional loss to mitigate the aforementioned selection bias in the representation space [19,26,27,35]. Other approaches attempted to tackle this issue by utilizing generative models, such as variational autoencoder (VAE) [23,34] and generative adversarial network (GAN) [36], which utilize the favorable characteristics of generative models to generate either hidden unobserved variables, balanced latent variable, or uncertainties of counterfactual outcomes. These mentioned works have only shown encouraging results in estimating treatment effects from single-modality data. In practical scenarios, however, doctors routinely integrate both imaging and non-imaging data when making prognoses, and the interpretation of imaging data is substantially impacted by clinical information [17]. In this regard, we consider two key ingredients. Firstly, the selection bias commonly exists in clinical scenarios, and the sysematic imbalance brought by this bias is amplified in high-dimensional data, as the higher number of covariates makes it more challenging to establish and verify overlap [3]. Therefore, it would be significant if we could map imbalanced high-dimensional data into a balanced low-dimensional representation. We thus seek to generate the distribution of low-dimensional prognostic score [12], which we will explain in Sect. 2 later. Secondly, motivated by the existing multi-modality VAE models [21,22,28,33], we can fuse multi-modality distributions into a joint distribution with reasonable feasibility, which can be leveraged to construct a multi-modality model for prognosis.

In this paper, we propose a novel prognostic model that leverages both imaging and tabular data to achieve accurate treatment outcome prediction. This model is intended to be trained on observational data obtained from the non-randomized controlled trials. Specifically, to increase the reliability of the model, we employ a variational autoencoder model to generate a low-dimensional prognostic score that alleviates the problem of selection bias. Moreover, we introduce a variational distributions combination module that integrate information from imaging data and non-imaging clinical data to generate the aforementioned prognostic score. We evaluate our proposed model on a clinical dataset of intracerebral hemorrhage and demonstrate a significant improvement in treatment outcome prediction compared to existing treatment effect estimation techniques.

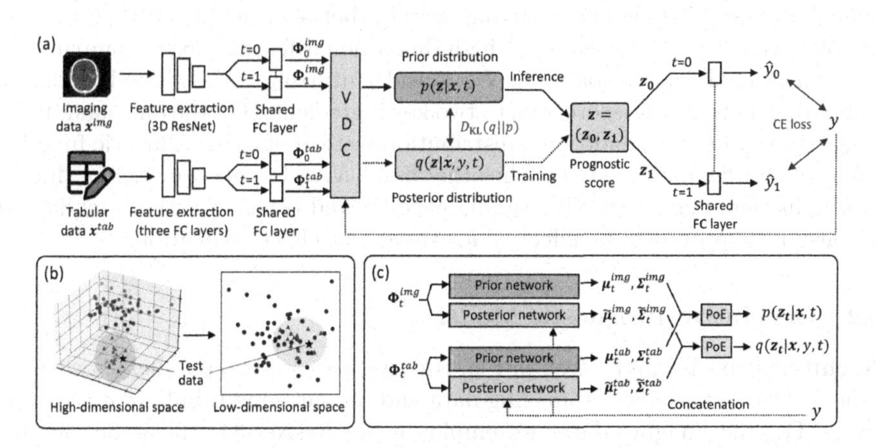

Fig. 1. (a) Overview of proposed generative prognostic model. (b) Illustrative example: dimension reduction can increase the overlap between representation spaces of training samples with $T = 0$ (blue points) and $T = 1$ (red points), so that improves outcome prediction reliability for all treatment assignments (e.g., test data fall within the overlap). (c) Variational distribution combination (VDC) module; PoE denotes using the product-of-experts to generate joint distributions from different means and covariances. (Color figure online)

2 Method

2.1 Formulation and Motivation

We aim to predict the individualized treatment outcome based on a set of observations that include the actual treatment T, observed covariates X, and factual outcome Y. In this paper, we study the one-year functional outcome of patient who underwent either conservative treatment ($T = 0$) or surgery ($T = 1$). For each individual, let $t \in \{0, 1\}$ denote the treatment assignment, $\boldsymbol{x} = (\boldsymbol{x}^{img}, \boldsymbol{x}^{tab})$ represent the observed covariates comprising imaging data $\boldsymbol{x}^{img}$ and non-imaging tabular data $\boldsymbol{x}^{tab}$, and y indicate the factual outcome. In this study, the treatment outcome was assessed using 1-year modified Rankin Scale (mRS) [32]. Our objective is to estimate $\mathbb{E}[Y \mid X = \boldsymbol{x}, T = t]$.

The non-randomized controlled trials impacted by treatment preference can lead to selection bias, rendering the model unreliable due to potential encounters with unobserved scenarios during training. To address this issue, our model is inspired by the approach commonly used by doctors in clinical practice: using a combination of imaging data and non-imaging biomarkers to generate a prognostic score (e.g., GCS score) that predicts the likelihood of good or poor condition after treatment. In this study, a prognostic score is defined as any function $f_T(X)$ of X and T that Markov separates Y and X, such that $Y \perp\!\!\!\perp X | f_T(X)$. The insight is that a patient's health status can be effectively captured by a low-dimensional score $Z = f_T(X)$, which is a form of dimension reduction that is sufficient for causal inference and can naturally mitigate the problem brought by non-randomized controlled trials. This is because, as illustrated in Fig. 1(b), the difficulty of establishing and verifying overlap (between samples with $T = 0$ and samples with $T = 1$) increases in high-dimensional feature space compared to low-dimensional feature space [3]. We consider utilizing a VAE-based model for generating a prognostic score, due to two key ingredients: On the one hand, modeling score through a conditional distribution instead of a deterministic function offers greater flexibility [34]. On the other hand, VAE is a good model for dimension reduction, compared with vanilla encoder and other generative models. It has also been proved to be effective for treatment effect estimation.

2.2 Generative Prognostic Model

Architecture. As can be seen in Fig. 1(a), we first use two parallel networks to generate latent variables of imaging data and non-imaging tabular data, respectively. For the imaging data, we employ a 3D ResNet-34 [13] as our feature extraction network and modify the final fully connected layers. We then generate the features conditioned on different treatments by concatenating the extracted features with their respective treatment assignments t and forwarding them to a shared fully connected layer (FC layer), yielding $\boldsymbol{\Phi}_0^{img}$ and $\boldsymbol{\Phi}_1^{img}$ respectively. This allows us to incorporate treatment assignment information and generate the prognostic score more effectively. For the non-imaging tabular data, we employ three blocks of a FC layer, followed by a Batch Normalization layer, and a ReLU activation function to generate the features $\boldsymbol{\Phi}_0^{tab}$ and $\boldsymbol{\Phi}_1^{tab}$. These features are then forwarded to a variational distribution combination (VDC) module, which we will describe in detail later. Through the VDC module, we can estimate the prior distribution $p(\boldsymbol{z} \mid \boldsymbol{x}, t)$ of the prognostic score $\boldsymbol{z} = (\boldsymbol{z_0}, \boldsymbol{z_1})$. In addition, during the training phase, the true posterior distribution $q(\boldsymbol{z} \mid \boldsymbol{x}, y, t)$ can be approximated, which is additionally conditioned on y and can help the model learn how to estimate an accurate prior distribution. The prognostic score $\boldsymbol{z_0}, \boldsymbol{z_1}$ are then concatenated with treatment $t = 1$ and $t = 0$ respectively and are passed through a decoder consisting of a shared FC layer, to output the predicted potential outcomes with different treatment assignments, i.e., $\hat{y}_0$ and $\hat{y}_1$. Notably, during the inference phase, we use the prior distribution $p(\boldsymbol{z} \mid \boldsymbol{x}, t)$ to generate $\boldsymbol{z}$. In contrast, during the training phase, we use the posterior distribution $q(\boldsymbol{z} \mid \boldsymbol{x}, y, t)$ to generate $\boldsymbol{z}$ and predict the outcomes.

Training Scheme. The evidence lower bound (ELBO) of our model is given by:

$$ELBO = \mathbb{E}_{z\sim q} \log p(y \mid \boldsymbol{z}, t) - \beta D_{\mathrm{KL}} \left(q(\boldsymbol{z} \mid \boldsymbol{x}, y, t) || p(\boldsymbol{z} \mid \boldsymbol{x}, t)\right), \quad (1)$$

where $D_{\mathrm{KL}} \left(q(\boldsymbol{z} \mid \boldsymbol{x}, y, t) || p(\boldsymbol{z} \mid \boldsymbol{x}, t)\right)$ is the Kullback-Leibler (KL) divergence between distributions $q(\boldsymbol{z} \mid \boldsymbol{x}, y, t)$ and $p(\boldsymbol{z} \mid \boldsymbol{x}, t)$, and β is the weight balancing the terms in the ELBO. Note that the first term of Eq. 1 corresponds the classification error, which is minimized by the cross entropy loss. The second term of Eq. 1 uses KL divergence to encourage convergence of the prior distribution towards the posterior distribution. The training objective is to maximize the ELBO given the observational data, so that the model can be optimized.

2.3 Variational Distributions Combination

Once the features of imaging and non-imaging tabular data have been extracted, the primary challenge is to effectively integrate the multi-modal information. One approach that is immediately apparent is to train a single encoder network that takes all modalities as input, which can explicitly parameterize the joint distribution. Another commonly used method called Mixture-of-Experts (MoE) proposes to fuse the distributions from different modalities by weighting [28]. However, for this study, we generate the distribution of each modality separately and then use the Product-of-Experts (PoE) method to combine the two distributions into a single one [22,33]. As can be seen in Fig. 1(c), assuming that we have generated distributions of two modalities with the means $\boldsymbol{\mu}_t^{img}$ and $\boldsymbol{\mu}_t^{tab}$, and covariances $\boldsymbol{\Sigma}_t^{img}$ and $\boldsymbol{\Sigma}_t^{tab}$ respectively from the prior network, we can use PoE to generate the joint distributions $p(\boldsymbol{z}_t \mid \boldsymbol{x}, t) = \mathcal{N}(\boldsymbol{\mu}_t, \boldsymbol{\Sigma}_t)$ by:

$$\boldsymbol{\mu}_t = \left(\boldsymbol{\mu}_t^{pri} / \boldsymbol{\Sigma}_t^{pri} + \boldsymbol{\mu}_t^{img} / \boldsymbol{\Sigma}_t^{img} + \boldsymbol{\mu}_t^{tab} / \boldsymbol{\Sigma}_t^{tab}\right) \boldsymbol{\Sigma}_t, \quad (2)$$

$$\boldsymbol{\Sigma}_t = \left(1 / \boldsymbol{\Sigma}_t^{pri} + 1 / \boldsymbol{\Sigma}_t^{img} + 1 / \boldsymbol{\Sigma}_t^{tab}\right)^{-1}, \quad (3)$$

where $\boldsymbol{\mu}_t^{pri}$ and $\boldsymbol{\Sigma}_t^{pri}$ are mean and covariance of universal prior expert, which is typically a spherical Gaussian ($\mathcal{N}(0, 1)$). For posterior distribution $q(\boldsymbol{z}_t \mid \boldsymbol{x}, y, t)$, we first additionally concatenate the features and y together, and then generate the joint distribution by the same way. The PoE for generating joint distributions offers several advantages over the aforementioned approaches. Compared with the approaches that simply combing the features and then generating the joint distributions, PoE not only can effectively address the potential issue of prediction outcomes being overly influenced by the modality with a more abundant feature [17], but also is more flexible and has the potential to handle missing modalities. Compared to that using a mixture-of-experts, it can produce sharper distributions [16,22], which is desirable for our multi-modality data with complementary or unevenly distributed information.

3 Experiment

3.1 Dataset and Experimental Setup

Datasets. We utilized an in-house dataset of intracerebral hemorrhage cases obtained from the Hong Kong Hospital Authority. The dataset comprises 504 cases who underwent head CT scans and were diagnosed with ICH. Among them, 364 cases received conservative treatment, and 140 cases underwent surgery treatment. For each case, we collected both CT imaging and non-imaging clinical data. The non-imaging data have 17 clinical characteristics which have been proved to be potentially associated with the treatment outcome in clinical practice [18], including gender, age, admission type, GCS, the history of smoking and drinking, hypertension, diabetes mellitus, hyperlipidemia, history of atrial fibrillation, coronary heart disease, history of stroke, pre-admission anticoagulation, pre-admission antiplatelet, pre-admission statin, small-vessel vascular disease and lower cholesterol. To address the selection bias resulting from the non-randomized controlled trials, we intentionally increased the imbalance of the dataset. Specifically, we selected 50 out of 68 cases who had IVH (another subtype of brain hemorrhage which can be infered from the CT image) and were treated conservatively, and 50 out of 61 cases who had a GCS score below 9 and underwent surgery. We used these samples for testing and reserved the remaining cases for training our model. As a result, the dataset is systematically imbalanced, which presents challenges for the model in producing reliable outcomes on test set. We also conducted additional experiments with different setting, as shown in the supplementary. A favorable outcome was defined as an mRS score of 0 to 3 (247 cases in total), while an unfavorable outcome was defined as an mRS score of 4 to 6 (257 cases in total) [8,11,29].

Evaluation Metrics. We employed three evaluation metrics that are commonly used in treatment effect estimation and outcome prediction in our experiments, including the policy risk (P_{ROL}), the accuracy (Acc) and the area under the ROC curve (AUC). P_{ROL} measures the average loss incurred when utilizing the treatment predicted by the treatment outcome estimator [26], which is a lower-is-better metric. Besides, we calculate Acc_0/AUC_0 for samples factually treated with $T = 0$ and Acc_1/AUC_1 for samples factually treated with $T = 1$.

Implementation Details. In preprocessing the imaging data, raw image intensity values were truncated to $[-20, 100]$, normalized to have zero mean and unit variance, and slices were uniformly resized to 224×224 in the axial plane. We implemented our model using PyTorch and executed it on an NVIDIA A100 SXM4 card. For training, we used the Adam optimizer, a weight decay of 5×10^{-3}, and an initial learning rate of 5×10^{-3}. The training process lasted for a total of 2 h, consisting of 1000 epochs with a batch size of 128. Our reported results are the average and standard deviation obtained from three independent runs.

Table 1. Comparison with state-of-the-art methods on the ICH dataset.

Method	Evaluation metrics				
	R_{POL} ↓	AUC_0 ↑	AUC_1 ↑	Acc_0 ↑	Acc_1 ↑
BNN [19]	.581 ± .028	.770 ± .032	.724 ± .018	.720 ± .035	.720 ± .053
CFR-WASS [26]	.568 ± .020	.792 ± .013	.743 ± .013	.747 ± .012	.727 ± .046
SITE [35]	.536 ± .040	.789 ± .022	.741 ± .022	.733 ± .012	.740 ± .020
β-Intact-VAE [34]	.533 ± .044	.797 ± .020	.774 ± .012	.773 ± .011	.753 ± .023
DAFT [25]	.575 ± .023	.782 ± .021	.732 ± .024	.740 ± .020	.733 ± .031
Ours	**.502 ± .023**	**.820 ± .019**	**.801 ± .018**	**.793 ± .012**	**.780 ± .040**

3.2 Experiment Results

Comparison with State-of-the-Art Methods. We benchmarked our method against state-of-the-art approaches for treatment effect estimation, which are recognized as strong competitors in this field. These approaches include **BNN** [19], which is a representative work that balances the distribution of different treatment groups by discrepancy distance minimization, **CFR-WASS** [26], which use separate heads to estimate the treatment effect and use Wasserstein Distance to balance the distribution, **SITE** [35], which prioritizes hard samples to preserve local similarity while also balancing the distribution of data, and **β-Intact-VAE** [34], which proposes to use a novel VAE model to generate low-dimensional representation conditioned on both covariates and treatment assignments to handle the selection bias problem. These methods were primarily developed for estimating treatment effects on single-modality data. To apply these methods to our multi-modality data, we utilized the same feature extraction architectures as our approach. The features extracted from these networks are concatenated. The fused features are then forwarded to the specific networks described in these papers. Moreover, we compared our method with **DAFT** [25], which designs a block to suppress high-level concepts from 3D images while considering both image and tabular data, making it great for processing multi-modal data.

Table 1 shows a comparison of results from different methods for estimating treatment outcomes in ICH. Our proposed method demonstrated a significant improvement in model performance compared to other methods, as evaluated using all five metrics. Compared with the classic methods based on balanced presentation learning, β-Intact-VAE seeks to generate balanced latent variable so that is more effective and can achieve more accurate performance in our experiment setting. However, this strategy lacks the ability to extract complementary information as it does not include a specially designed module for combining the distributions extracted from different modalities. Instead of simply concatenating features and generating a low-dimensional representation, our model utilizes the PoE technique to combine two low-dimensional distributions generated from two distinct modalities, which can effectively mitigate the risk of prediction outcomes being disproportionately influenced by the more feature-rich modality [17]. For these reasons, our proposed method improved performances by 3.1% on R_{POL}, 2.3%/2.7% on AUC_0/AUC_1, and 2.0%/2.7% on Acc_0/Acc_1, respectively.

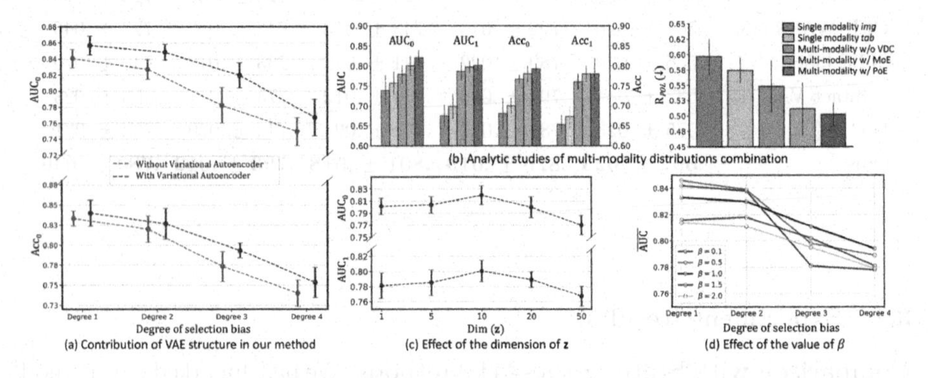

Fig. 2. Ablation analysis of each component of our method.

Ablation Analysis. We then conducted comprehensive ablation studies. Initially, we studied the necessity of using a VAE structure for obtaining a low-dimensional prognostic score instead of a vanilla encoder. As shown in Fig. 2(a), we systematically changed the degree of selection bias by varying the number of cases with IVH who underwent conservative treatment (68 in total) in the training set and test set. The ratios of training set/test set were: 68/0 (Degree 1), 48/20 (Degree 2), 18/50 (Degree 3) and 0/68 (Degree 4). When there are fewer cases in the training set, the selection bias increases, leading to a reduced ability of the model to predict such cases in the test set. The experiment showed that as the selection bias increased in the training set (ranging from Degree 1 to Degree 3), the difference between using a VAE structure and a vanilla encoder became more prominent. This is due to the VAE's effective dimension reduction. When there were no cases related to the outcome of interest (Degree 4) in the training set, further increases were stopped. This is expected since in such situations, there are no related cases that the model can learn from, thus rendering the advantages of dimension reduction ineffective.

Next, we studied the contributions of multi-modality distribution combination. As can be seen in Fig. 2(b), despite using the proposed generative prognostic model, satisfactory performance cannot be achieved by simply using a single modality. Furthermore, compared to other commonly used approaches for integrating two generated distributions, such as simply combining the feature maps before generating the prognostic score (Multi-modality w/o VDC) and Mixture-of-Experts (Multi-modality w/ MoE), our proposed model (Multi-modality w/ PoE) achieved better performance. This highlights the effectiveness of distribution combination via PoE. Additionally, in Fig. 2(c) and (d), we compared the performance of the model trained with different dimensions of generated prognostic score and the values of β in Eq. 1. The dimension of the generated prognostic score is a trade-off between the degree of eliminating selection bias and the amount of accessible information. The results demonstrate that the optimal choice of dimension is 10. Moreover, note that β controls the trade-off between outcome reconstruction and prognostic score recovery. Figure 2(d) suggests β should be 0.5 for low imbalance degree and 1.0 for high imbalance degree.

4 Conclusion

This paper introduces a novel generative prognostic model for predicting ICH treatment outcomes using imaging and non-imaging data. The model is designed to be trained on data collected from non-randomized controlled trials, addressing the imbalance problem with a VAE model and integrating multi-modality information using a variational distribution combination module. The model was evaluated on a large-scale dataset, confirming its effectiveness.

Acknowledgement. This work was supported in part by Shenzhen Portion of Shenzhen-Hong Kong Science and Technology Innovation Cooperation Zone under HZQB-KCZYB-20200089, in part by Hong Kong Innovation and Technology Commission Project No. ITS/238/21, in part by Hong Kong Research Grants Council Project No. T45-401/22-N, in part by Science, Technology and Innovation Commission of Shenzhen Municipality Project No. SGDX20220530111201008. We also thank the Hong Kong Hospital Authority Data Collaboration Laboratory (HADCL) for their support of this study.

References

1. Bica, I., Alaa, A.M., Lambert, C., Van Der Schaar, M.: From real-world patient data to individualized treatment effects using machine learning: current and future methods to address underlying challenges. Clin. Pharmacol. Therap. **109**(1), 87–100 (2021)
2. Cheung, R.T.F., Zou, L.Y.: Use of the original, modified, or new intracerebral hemorrhage score to predict mortality and morbidity after intracerebral hemorrhage. Stroke **34**(7), 1717–1722 (2003)
3. D'Amour, A., Ding, P., Feller, A., Lei, L., Sekhon, J.: Overlap in observational studies with high-dimensional covariates. J. Econometrics **221**(2), 644–654 (2021)

4. Ducruet, A.F., et al.: The complement cascade as a therapeutic target in intracerebral hemorrhage. Exp. Neurol. **219**(2), 398–403 (2009)
5. Feigin, V.L., Lawes, C.M., Bennett, D.A., Barker-Collo, S.L., Parag, V.: Worldwide stroke incidence and early case fatality reported in 56 population-based studies: a systematic review. Lancet Neurol. **8**(4), 355–369 (2009)
6. Flaherty, M., et al.: Long-term mortality after intracerebral hemorrhage. Neurology **66**(8), 1182–1186 (2006)
7. Godoy, D.A., Pinero, G., Di Napoli, M.: Predicting mortality in spontaneous intracerebral hemorrhage: can modification to original score improve the prediction? Stroke **37**(4), 1038–1044 (2006)
8. Gregório, T., et al.: Prognostic models for intracerebral hemorrhage: systematic review and meta-analysis. BMC Med. Res. Methodol. **18**, 1–17 (2018)
9. Gregson, B.A., Mitchell, P., Mendelow, A.D.: Surgical decision making in brain hemorrhage: new analysis of the stich, stich ii, and stitch (trauma) randomized trials. Stroke **50**(5), 1108–1115 (2019)
10. Guo, R., et al.: Machine learning-based approaches for prediction of patients' functional outcome and mortality after spontaneous intracerebral hemorrhage. J. Personal. Med. **12**(1), 112 (2022)
11. Hall, A.N., et al.: Identifying modifiable predictors of patient outcomes after intracerebral hemorrhage with machine learning. Neurocrit. Care **34**, 73–84 (2021)
12. Hansen, B.B.: The prognostic analogue of the propensity score. Biometrika **95**(2), 481–488 (2008)
13. He, K., Zhang, X., Ren, S., Sun, J.: Deep residual learning for image recognition. In: CVPR, pp. 770–778 (2016)
14. Hemphill, J.C., III, Bonovich, D.C., Besmertis, L., Manley, G.T., Johnston, S.C.: The ICH score: a simple, reliable grading scale for intracerebral hemorrhage. Stroke **32**(4), 891–897 (2001)
15. Hemphill, J.C., III, et al.: Guidelines for the management of spontaneous intracerebral hemorrhage: a guideline for healthcare professionals from the American Heart Association/American Stroke Association. Stroke **46**(7), 2032–2060 (2015)
16. Hinton, G.E.: Training products of experts by minimizing contrastive divergence. Neural Comput. **14**(8), 1771–1800 (2002)
17. Huang, S.C., Pareek, A., Seyyedi, S., Banerjee, I., Lungren, M.P.: Fusion of medical imaging and electronic health records using deep learning: a systematic review and implementation guidelines. NPJ Digit. Med. **3**(1), 1–9 (2020)
18. Ji, R., et al.: A novel risk score to predict 1-year functional outcome after intracerebral hemorrhage and comparison with existing scores. Critical Care **17**, 1–10 (2013)
19. Johansson, F., Shalit, U., Sontag, D.: Learning representations for counterfactual inference. In: ICML, pp. 3020–3029. PMLR (2016)
20. Kim, J.Y., Bae, H.J.: Spontaneous intracerebral hemorrhage: management. J. Stroke **19**(1), 28 (2017)
21. Kingma, D.P., Welling, M.: Auto-encoding variational Bayes. In: ICLR (2014)
22. Lee, C., van der Schaar, M.: A variational information bottleneck approach to multi-omics data integration. In: International Conference on Artificial Intelligence and Statistics, pp. 1513–1521. PMLR (2021)
23. Louizos, C., Shalit, U., Mooij, J.M., Sontag, D., Zemel, R., Welling, M.: Causal effect inference with deep latent-variable models. NeurIPS **30** (2017)
24. Peters, J., Janzing, D., Schölkopf, B.: Elements of Causal Inference: Foundations and Learning Algorithms. The MIT Press (2017)

25. Pölsterl, S., Wolf, T.N., Wachinger, C.: Combining 3D image and tabular data via the dynamic affine feature map transform. In: de Bruijne, M., et al. (eds.) MICCAI 2021. LNCS, vol. 12905, pp. 688–698. Springer, Cham (2021). https://doi.org/10.1007/978-3-030-87240-3_66
26. Shalit, U., Johansson, F.D., Sontag, D.: Estimating individual treatment effect: generalization bounds and algorithms. In: ICML, pp. 3076–3085. PMLR (2017)
27. Shi, C., Blei, D., Veitch, V.: Adapting neural networks for the estimation of treatment effects. NeurIPS **32** (2019)
28. Shi, Y., Paige, B., Torr, P., et al.: Variational mixture-of-experts autoencoders for multi-modal deep generative models. NeurIPS **32** (2019)
29. Stein, M., Luecke, M., Preuss, M., Boeker, D.K., Joedicke, A., Oertel, M.F.: Spontaneous intracerebral hemorrhage with ventricular extension and the grading of obstructive hydrocephalus: the prediction of outcome of a special life-threatening entity. Neurosurgery **67**(5), 1243–1252 (2010)
30. Steiner, T., et al.: European Stroke Organisation (ESO) guidelines for the management of spontaneous intracerebral hemorrhage. Int. J. Stroke **9**(7), 840–855 (2014)
31. Teasdale, G., Jennett, B.: Assessment of coma and impaired consciousness: a practical scale. The Lancet **304**(7872), 81–84 (1974)
32. Van Swieten, J., Koudstaal, P., Visser, M., Schouten, H., Van Gijn, J.: Interobserver agreement for the assessment of handicap in stroke patients. Stroke **19**(5), 604–607 (1988)
33. Wu, M., Goodman, N.: Multimodal generative models for scalable weakly-supervised learning. NeurIPS **31** (2018)
34. Wu, P., Fukumizu, K.: \beta-intact-VAE: identifying and estimating causal effects under limited overlap. In: ICLR (2022)
35. Yao, L., Li, S., Li, Y., Huai, M., Gao, J., Zhang, A.: Representation learning for treatment effect estimation from observational data. NeurIPS **31** (2018)
36. Yoon, J., Jordon, J., Van Der Schaar, M.: Ganite: estimation of individualized treatment effects using generative adversarial nets. In: ICLR (2018)

Open-Ended Medical Visual Question Answering Through Prefix Tuning of Language Models

Tom van Sonsbeek(✉), Mohammad Mahdi Derakhshani, Ivona Najdenkoska, Cees G. M. Snoek, and Marcel Worring

University of Amsterdam, Amsterdam, The Netherlands
{t.j.vansonsbeek,m.m.derakhshani,i.najdenkoska,c.g.m.snoek, m.worring}@uva.nl

Abstract. Medical Visual Question Answering (VQA) is an important challenge, as it would lead to faster and more accurate diagnoses and treatment decisions. Most existing methods approach it as a multi-class classification problem, which restricts the outcome to a predefined closed-set of curated answers. We focus on open-ended VQA and motivated by the recent advances in language models consider it as a generative task. Leveraging pre-trained language models, we introduce a novel method particularly suited for small, domain-specific, medical datasets. To properly communicate the medical images to the language model, we develop a network that maps the extracted visual features to a set of learnable tokens. Then, alongside the question, these learnable tokens directly prompt the language model. We explore recent parameter-efficient fine-tuning strategies for language models, which allow for resource- and data-efficient fine-tuning. We evaluate our approach on the prime medical VQA benchmarks, namely, Slake, OVQA and PathVQA. The results demonstrate that our approach outperforms existing methods across various training settings while also being computationally efficient.

Keywords: Visual Question Answering · Language Models · Prompting · Prefix Tuning

1 Introduction

Images and text are inherently intertwined in clinical diagnosis and treatment. Having an automated approach that is able to answer questions based on images, giving insight to clinicians and patients, can be a valuable asset. In such a medical Visual Question Answering (VQA) setting the common approach is to treat VQA as a multi-class classification problem solved by neural networks. Given a joint encoded representation of the image and question, the model classifies it into a predefined set of answers. Although these approaches yield good performance [5, 18,24,33], they deal with closed-set predictions, which is not an ideal solution

T. van Sonsbeek, M. M. Derakhshani and I. Najdenkoska—Equal contribution.

H. Greenspan et al. (Eds.): MICCAI 2023, LNCS 14224, pp. 726–736, 2023.
https://doi.org/10.1007/978-3-031-43904-9_70

for VQA. For instance, medical VQA datasets commonly contain hundreds to thousands of free-form answers [11], which is suboptimal to be treated as a classification task. Moreover, the severe class imbalance and out-of-vocabulary answers further hinder the generalizability of these classification methods.

We believe that a possible solution can be found in the generative capability of language models, since they are able to produce free text, instead of being limited to closed-set predictions. However, leveraging language models for solving open-ended medical VQA is limited due to several challenges, such as finding ways to properly communicate the visual features and letting such large-scale models be employed on small-sized medical VQA datasets.

Inspired by recent image captioning models [22], we propose to use the medical images by converting them into a set of learnable tokens through a small-scale mapping network. These tokens can then be interpreted as a visual prefix for the language model [1,4,23]. Afterward, the visual prefix is used together with the question as input to the language model, which generates the answer token by token [25].

Furthermore, large-scale language models can generalize across domains while keeping their weights frozen [30]. This makes them very appealing for the medical domain, which inherently does not possess large quantities of labeled data required to train these models from scratch [29]. Models like BioGPT [21] and BioMedLM [31] are based on the generic GPT2 language model [26] and are trained on biomedical text corpora. They perform quite well compared to their general counterparts on specific biomedical language tasks, like question answering or relation extraction. We design our model in a flexible manner, which allows us to incorporate any of these pre-trained language models.

In summary, we contribute in three major aspects: (i) We propose the first large-scale language model-based method for open-ended medical VQA. (ii) We adopt parameter-efficient tuning strategies for the language backbone, which gives us the ability to fine-tune a large model with a small dataset without the danger of overfitting. (iii) We demonstrate through extensive experiments on relevant benchmarks that our model yields strong open-ended VQA performance without the need for extensive computational resources.

2 Related Works

To describe existing medical VQA methods, we make a distinction between classification methods and generative methods. The majority of methods are classification-based and make use of different types of encoders, such as CNNs or Transformers [3,9,11,17,32] followed by a classification layer.

Classification-Based VQA. We highlight a number of methods that showed good performance on current competitive medical VQA datasets. The Mixture Enhanced Visual Features (MEVF) [24] is initialized based on pre-trained weights from the Model-Agnostic Meta-Learning (MAML) model [7] in combination with image feature extraction from Conditional Denoising Auto-Encoders

(CDAE) to generate a joint question-answer representation using Bilinear (BAN) or Stacked (SAN) Attention Networks. Do *et al.* [5] create a similar embedding space by extracting annotations from multiple pre-trained meta-models, and learning meta-annotations by training each meta-model. Linear combinations [10] or question-conditioned selections [34] from this multi-modal embedding space can further enhance performance. The use of Transformer [14] and CLIP [6,25] encoders also results in strong VQA classification performance.

Open-Ended VQA. MedFuseNet [28] is one of the few methods performing and reporting open-ended visual question answering on recent public datasets. They do so by creating a BERT-based multi-modal representation of image and question and subsequently passing it through an LSTM decoder. Ren *et al.* [27] create open-ended answers by using the masked token prediction functionality of BERT. We aim to show that generative language models are more versatile and better suited for this task.

3 Methodology

3.1 Problem Statement

Given an input image $\mathbf{I}$ and an input question in natural language $\mathbf{Q}$, our method aims to sequentially generate an answer $\mathbf{A} = \{A_0, A_1, ..., A_N\}$ composed of N tokens, by conditioning on both inputs. From a model definition perspective, we aim to find the optimal parameters θ^* for a model by maximizing the conditional log-likelihood as follows:

$$\theta^* = \arg\max_{\theta} \sum_{i=1}^{N} \log p_\theta(\mathbf{A}_i|\mathbf{Q}, \mathbf{I}, \mathbf{A}_{i-1}). \quad (1)$$

3.2 Model Architecture

Our VQA model is designed as an encoder-decoder architecture, with a two-stream encoder and a language model (LM) as a decoder, as illustrated in Fig. 1. Specifically, the two streams encode the two input modalities, namely the image $\mathbf{I}$ and the question $\mathbf{Q}$. The language model is defined as a causal language Transformer [26], and it generates the answer $\mathbf{A}$ in an autoregressive manner. It closely follows the prefix tuning technique for prompting a language model to produce an output of a particular style [16], such as in our case an answer given a question and an image[1].

Vision Encoding Stream. For encoding the image, we employ a pre-trained vision encoder to extract visual features $\{x_1, x_2 ... x_{\ell_x}\}$. To use these features as input to the decoder, they should be mapped into the latent space of the

[1] Code available at: github.com/tjvsonsbeek/open-ended-medical-vqa.

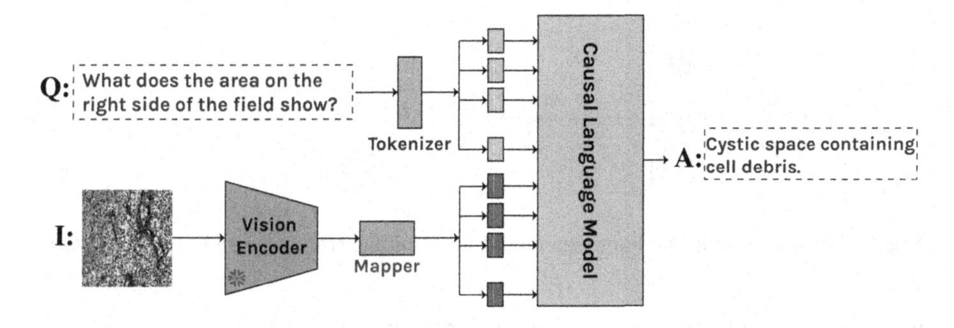

Fig. 1. Model architecture of our proposed open-ended generative VQA method.

language decoder. Following [22], we define a mapping network $\mathbf{f_M}$, implemented as a three-layer MLP. This network maps the visual features into a visual prefix $\{v_1, v_2, \dots v_x\} \in \mathbb{R}^{\ell_x \times e}$ for the language model, where e is the embedding size.

Language Encoding Stream. Regarding the encoding of the textual part, firstly we utilize a standard tokenization process to obtain a sequence of tokens, both for the question $\mathbf{Q} = \{q_1, q_2 ... q_{\ell_q}\} \in \mathbb{R}^{\ell_q \times e}$ and answer $\mathbf{A} = \{a_1, a_2 ... a_{\ell_a}\} \in \mathbb{R}^{\ell_a \times e}$. This is followed by embedding the tokens using the embedding function of a pre-trained language model.

Prompt Structure. To create a structured prompt, following existing QA methods using language models [2,26], we prepend the question, image, and answer tokens with tokenized descriptive strings, namely `question:`, `context:` and `answer:`. By placing the embeddings of the question before the visual tokens we mitigate the problem of fixation of the language model on the question [21,22]. As an example this would yield the following prompt template: $p =$`[question: What does the right side of the field show? context:` $v_1, v_2, \dots v_x$ `answer: ]` which is fed as input to the language model.

Language Model. Following standard language modeling systems, we treat VQA as a conditional generation of text, and we optimize the standard maximum likelihood objective during training. The language model receives the prompt sequence p as input and outputs the answer $\mathbf{A}$, token by token. Specifically, at each time step i, the output of the model are the logits parametrizing a categorical distribution $p_\theta(\mathbf{A})$ over the vocabulary tokens. This distribution is represented as follows:

$$\log p_\theta(\mathbf{A}) = \sum_{l_a} \log\ p_\theta(a_i | q_1, ... q_{\ell_q}, v_1, ... v_x, a_1, ... a_{i-1}). \tag{2}$$

The parameters of the language model are initialized from a pre-trained model, which has been previously pre-trained on huge web-collected datasets.

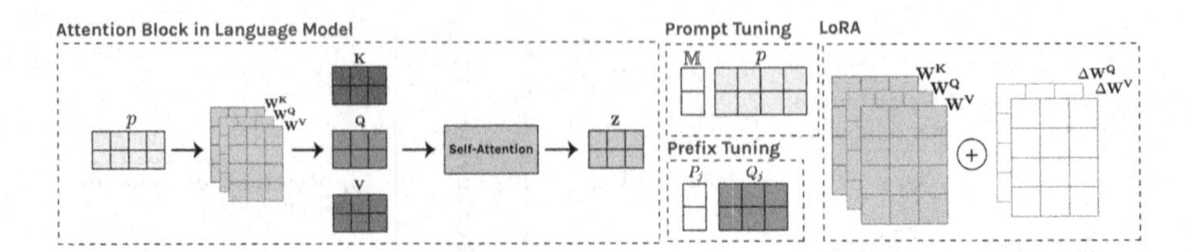

Fig. 2. Parameter-efficient language model fine-tuning strategies used in our method.

3.3 Parameter-Efficient Strategies for Fine-Tuning the Language Model

Standard fine-tuning of language models can hurt the generalization capabilities of the model, especially if small, domain-specific datasets are used as in our case. Therefore, we consider four different parameter-efficient strategies that adapt the attention blocks of language models, as illustrated in Fig. 2 and outlined below:

Frozen Method: the parameters of the language model are kept entirely frozen during training, following [30]. In this setting, only the mapping network is updated through backpropagation. **Prompt Tuning**: we prepend a set of m learnable tokens $\mathbb{M} \in \mathbb{R}^{m \times e}$ to the input prompt sequence, which yields $[\mathbb{M}, p]$ [15] as input to the frozen language model. Besides updating the mapping network, this approach also involves updating these learnable tokens through backpropagation. **Prefix Tuning**: we prepend a learnable prefix P_j to the query Q_j of each attention block j in the Transformer, such that $Q_j^{ft} = [P_j, Q_j]$ [16]. Similar as in prompt tuning, we update both the mapping function and the learnable prefixes of the queries. **Low-Rank Adaptation (LoRA)**: We add learnable weight matrices to the query Q and value V of the attention blocks in each layer of the frozen language model as $\mathbf{W} + \Delta\mathbf{W}$ following [12]. Again, the mapping function is trained together with the learnable weight matrices.

4 Experimental Setup

Datasets. The three datasets used for the evaluation of our method are Slake [20], PathVQA [11], and OVQA [13]. These three datasets are the current most suitable VQA datasets given their large variety in answers and the manual curation of answers by domain experts. Each dataset is split 50/50 between 'yes/no' and open-set answers. See the datatset details in Table 1. We use the official train/validation/test splits across all three datasets.

Evaluation Protocol. We evaluate our approach using the conventional metrics BLEU-1 and F1 Score. Additionally, we measure the contextual capturing of information with BERTScore [35] this method can handle synonyms. Lastly to allow for comparison against existing classification-based methods we also report accuracy and F1 score.

Table 1. Statistics of the medical VQA datasets used in this paper.

	Slake	OVQA	PathVQA
Number of images	642	2,001	4,998
Number of questions	14,028	19,020	32,799
Mean length of questions	4.52	8.98	6.36
Mean length of answers	1.21	3.31	1.80
Number of unique answers	461	641	3,182

Implementation Details. We extract the visual features using a pre-trained CLIP model with ViT backbone [25], having a dimensionality of 512. The MLP layers of the mapping network $\mathbf{f_M}$ have sizes $\{512, (\ell_x \cdot e)/2, \ell_x \cdot e\}$. The length of ℓ_x is set at 8. The lengths ℓ_q and ℓ_a are dataset dependent and defined by the mean number of tokens in the train set plus three times its standard deviation. Zero padding is added to the right side of the sequence for batch-wise learning.

We use the following language models: GPT2-XL [26], a causal language model with 1.5B parameters trained on WebText [26]. BioMedLM [31] and BioGPT [21] are both GPT2-based models, pre-trained on PubMed and biomedical data from The Pile [8], with a size of 1.5B and 2.7B parameters, respectively. All models are able to train on a single NVIDIA RTX 2080ti GPU (average training time $\approx$ 3 h). We use the AdamW optimizer with 600 warmup steps and a learning rate of 5e-3 and apply early stopping with a tolerance of 3.

5 Results

Benefits of Parameter-Efficient Fine-Tuning. The evaluation of our method across various language models and fine-tuning settings in Table 2 shows

Table 2. Performance across different language models and fine-tuning strategies, measured in BLEU1 (BL1), BERTScore (BS), F1 and accuracy. Params% is the amount of trainable parameters in the language model. Our method using GPT2 in combination with LoRA yields the best performance across all datasets.

	LM fine-tuning	LM size	Params%	Slake				OVQA				PathVQA			
				BL1	BS	F1	Acc.	BL1	BS	F1	Acc.	BL1	BS	F1	Acc.
MedFuseNet [28]				-	-	-	-	-	-	-	-	60.5	-	38.1	-
Ours w/ BioGPT	Frozen	1.5B	0%	64.5	69.9	57.7	66.5	32.4	71.9	52.5	53.5	36.9	57.6	31.0	45.3
	Prefix [16]		0.487%	58.1	74.1	54.1	67.4	37.9	65.0	46.1	53.2	53.6	61.8	34.8	46.7
	Prompt [15]		0.001%	44.2	**75.6**	47.6	53.7	47.5	62.9	34.6	50.3	28.0	58.7	**43.8**	33.2
	LoRA [12]		0.311%	**59.2**	72.2	**63.1**	**71.9**	**41.0**	**68.5**	**57.7**	**57.3**	**57.8**	**62.9**	40.4	**47.9**
Ours w/ BioMedLM	Frozen	2.7B	0%	70.2	77.8	47.8	66.0	55.2	72.9	54.2	61.1	61.2	66.1	52.4	53.0
	Prefix [16]		0.753%	64.3	79.4	60.9	63.3	49.1	76.9	51.5	60.1	59.7	60.7	48.9	52.3
	Prompt [15]		0.009%	44.6	73.5	38.8	41.6	48.9	72.8	44.3	59.5	51.9	59.8	38.9	49.3
	LoRA [12]		0.101%	**72.3**	**80.6**	**62.4**	**71.7**	**59.0**	76.2	**62.6**	**67.8**	**67.9**	**76.0**	**54.4**	**57.2**
Ours w/ GPT2	Frozen	1.5B	0%	65.1	83.3	57.7	71.2	60.2	79.8	59.4	66.1	64.2	74.6	47.5	58.1
	Prefix [16]		0.492%	70.0	86.5	66.3	74.1	61.2	83.9	65.5	68.9	67.5	76.2	52.5	60.5
	Prompt [15]		0.003%	57.8	80.3	49.9	60.0	57.8	78.3	55.2	63.1	54.4	72.0	38.1	46.6
	LoRA [12]		0.157%	**78.6**	**91.2**	**78.1**	**83.3**	**61.8**	**85.4**	**69.1**	**71.0**	**70.3**	**78.5**	**58.4**	**63.6**

Table 3. Comparison of the accuracy between open-ended VQA against classification-based VQA methods, split between yes/no and open-set answers. Our method performs particularly well on both types of answers compared to the state-of-the-art methods.

	Slake			OVQA			PathVQA		
	Open-set	Yes/no	All	Open-set	Yes/no	All	Open-set	Yes/no	All
MEVF-SAN [24]	75.3	78.4	76.5	36.9	72.8	58.5	6.0	81.0	43.6
MEVF-BAN [24]	77.8	79.8	78.6	36.3	76.3	60.4	8.1	81.4	44.8
MEVF-SAN+VQAMix [10]	-	-	-	-	-	-	12.1	84.4	48.4
MEVF-BAN+VQAMix [10]	-	-	-	-	-	-	13.4	83.5	48.6
MMQ-SAN [5]	-	-	-	56.9	76.2	68.5	9.6	83.7	46.8
MMQ-BAN [5]	-	-	-	48.2	76.2	65.0	11.8	82.1	47.1
QCR-BAN [34]	78.8	82.0	80.0	52.6	77.7	67.7	-	-	-
CRPD-BAN [19]	81.2	84.4	82.1	-	-	-	-	-	-
MMBERT [14]	-	-	-	37.9	80.2	63.3	-	-	-
QCR-CLIP [6]	78.4	82.5	80.1	-	-	-	-	-	-
Ours w/ BioGPT (LoRA)	71.1	72.7	71.9	48.3	66.5	57.3	30.2	65.5	47.9
Ours w/ BioMedLM (LoRA)	72.1	71.4	71.7	55.3	80.3	67.8	34.1	80.4	57.2
Ours w/ GPT2 (LoRA)	**84.3**	82.1	**83.3**	**62.6**	**84.7**	**71.0**	**40.0**	**87.0**	**63.6**

that language models can perform open-ended medical VQA. Specifically, we outperform the only existing method MedFuseNet [28] that does open-ended VQA, due to the capability of pre-trained language models to capture long-term dependencies when generating free-form answers. Additionally, prefix [16] and prompt tuning [15] do not improve the performance of the model as much as using LoRA [12] which directly adapts the Q and V weight matrices of the attention blocks. Moreover, larger datasets show the most consistent performance gain of parameter-efficient fine-tuning across all metrics.

Comparison Between Standard and Medical LMs. Using a language model pre-trained on a general text corpus, such as GPT2 [26], improves the overall performance compared to its medically-trained models (e.g. BioGPT or BioMedLM), as can be observed in Table 2. BioGPT and BioMedLM could be overoptimized to their medical text corpora, which leads to lack of generalization to different downstream domains.

As mentioned in [21,31], these models require full fine-tuning on the respective downstream tasks, to achieve the desired performance. On the other hand, GPT2 benefits from observing diverse data during pre-training which also encompasses medically oriented text. This enables GPT2 models to generalize easily to other domains, which is relevant for our different VQA datasets.

Benefit of Open-Ended Answer Generation. Our method is performing significantly better on the open-set answering, in comparison to classification-based methods, as shown in Table 3. We also confirm that CLIP based image embeddings perform well in the medical domain [6] compared to the conventional use of CNNs. Since our approach is generative, it is not bounded by the class imbalance issue, which is considered a bottleneck of classification-based VQA

Table 4. Effect of using different prompt structures. Note that **Q** and **I** denote the question and image respectively. The regular setting with the question embeddings followed by the visual prefix (Fig. 1) leads to the best overall performance.

Setting	Slake				OVQA				PathVQA			
	B1	BS	F1	Acc.	B1	BS	F1	Acc.	B1	BS	F1	Acc.
w/o **Q**	29.4	48.4	14.3	22.1	33.2	41.9	18.1	27.6	42.9	43.8	18.3	24.6
w/o **I**	54.8	79.3	49.5	50.9	45.5	77.1	49.5	54.4	65.9	72.8	47.2	46.3
Swap **Q** and **I**	73.3	88.7	73.2	74.9	60.0	84.2	67.3	66.9	70.2	78.0	57.2	58.7
Regular	**78.6**	**91.2**	**78.1**	**83.3**	**61.8**	**85.4**	**69.1**	**71.0**	**70.3**	**78.5**	**58.4**	**63.6**

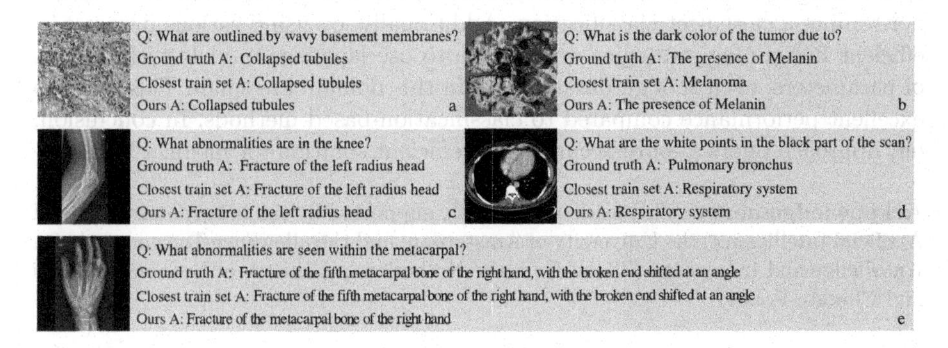

Fig. 3. Outputs of our open-ended VQA method, with GPT2 and LoRA fine-tuning, using data samples from PathVQA (a, b, d) and OVQA (c, e).

models. Our method performs especially well compared to other method on PathVQA, which relatively has the largest class imbalance, accentuating this effect. Even on the simple 'yes/no' questions, the performance is better, showing that this simple yet effective method provides a more natural way of doing VQA.

It worth noting that the comparison of accuracy as a metric for exact matches, between classification and generation methods is not in favor of generative methods. Despite that, we outperform existing methods on all datasets and metrics, which is a testament to the benefit of phrasing VQA as an open-ended generation problem.

In Fig. 3(a–c), we show qualitative examples of capability of the language model to successfully predict the correct answer. However, in Fig. 3 (d, e) we show cases where our method predicts a factually correct answer which is not specific enough.

Effect of Using Different Prompt Structures. We also investigate the influence of the prompt structure on the overall performance, demonstrated in Table 4. It can be observed that the performance largely decreases when the question is removed, compared to when the visual information is removed. This suggests that the question plays a more important role in answer generation.

Interestingly, the model is sensitive the order of the elements in the prompt, as the swapping of the question embeddings and the visual prefix yields decreases the performance. The reason for this is that the language model conveys lower to no importance the visual information if it is located in front of the question. In this situation the language model basically generates blind answers. This highlights the importance of prompt structure.

6 Conclusion

In this paper, we propose a new perspective on medical VQA. We are using generative language models to generate answers in an open-ended manner, instead of performing a closed-set classification. Additionally, by using various parameter-efficient fine-tuning strategies we are able to use language models with billions of parameters, even though dataset sizes in this domain are small. This leads to excellent performance compared to classification-based methods. In conclusion, our approach offers a more accurate and efficient solution for medical VQA.

Acknowledgements. This work is financially supported by the Inception Institute of Artificial Intelligence, the University of Amsterdam and the allowance Top consortia for Knowledge and Innovation (TKIs) from the Netherlands Ministry of Economic Affairs and Climate Policy.

References

1. Barraco, M., Cornia, M., Cascianelli, S., Baraldi, L., Cucchiara, R.: The unreasonable effectiveness of CLIP features for image captioning: an experimental analysis. In: CVPR, pp. 4662–4670 (2022)
2. Brown, T., et al.: Language models are few-shot learners. NeurIPS **33**, 1877–1901 (2020)
3. Cong, F., Xu, S., Guo, L., Tian, Y.: Caption-aware medical VQA via semantic focusing and progressive cross-modality comprehension. In: ACM Multimedia, pp. 3569–3577 (2022)
4. Derakhshani, M.M., et al.: Variational prompt tuning improves generalization of vision-language models. arXiv:2210.02390 (2022)
5. Do, T., Nguyen, B.X., Tjiputra, E., Tran, M., Tran, Q.D., Nguyen, A.: Multiple meta-model quantifying for medical visual question answering. In: de Bruijne, M., et al. (eds.) MICCAI 2021. LNCS, vol. 12905, pp. 64–74. Springer, Cham (2021). https://doi.org/10.1007/978-3-030-87240-3_7
6. Eslami, S., de Melo, G., Meinel, C.: Does CLIP benefit visual question answering in the medical domain as much as it does in the general domain? arXiv:2112.13906 (2021)
7. Finn, C., Abbeel, P., Levine, S.: Model-agnostic meta-learning for fast adaptation of deep networks. In: ICLR, pp. 1126–1135 (2017)
8. Gao, L., et al.: The pile: an 800 GB dataset of diverse text for language modeling. arXiv:2101.00027 (2020)
9. Gong, H., Chen, G., Liu, S., Yu, Y., Li, G.: Cross-modal self-attention with multi-task pre-training for medical visual question answering. In: ICMR, pp. 456–460 (2021)

10. Gong, H., Chen, G., Mao, M., Li, Z., Li, G.: Vqamix: conditional triplet mixup for medical visual question answering. IEEE Trans. Med. Imaging (2022)
11. He, X., Zhang, Y., Mou, L., Xing, E., Xie, P.: Pathvqa: 30000+ questions for medical visual question answering. arXiv:2003.10286 (2020)
12. Hu, E.J., et al.: Lora: low-rank adaptation of large language models. arXiv:2106.09685 (2021)
13. Huang, Y., Wang, X., Liu, F., Huang, G.: OVQA: A clinically generated visual question answering dataset. In: ACM SIGIR, pp. 2924–2938 (2022)
14. Khare, Y., Bagal, V., Mathew, M., Devi, A., Priyakumar, U.D., Jawahar, C.: MMBERT: multimodal BERT pretraining for improved medical VQA. In: ISBI, pp. 1033–1036. IEEE (2021)
15. Lester, B., Al-Rfou, R., Constant, N.: The power of scale for parameter-efficient prompt tuning. In: EMNLP, pp. 3045–3059 (2021)
16. Li, X.L., Liang, P.: Prefix-tuning: optimizing continuous prompts for generation. In: ACL, pp. 4582–4597 (2021)
17. Li, Y., et al.: A bi-level representation learning model for medical visual question answering. J. Biomed. Inf. **134**, 104183 (2022)
18. Lin, Z., et al.: Medical visual question answering: a survey. arXiv:2111.10056 (2021)
19. Liu, B., Zhan, L.-M., Wu, X.-M.: Contrastive pre-training and representation distillation for medical visual question answering based on radiology images. In: de Bruijne, M., et al. (eds.) MICCAI 2021. LNCS, vol. 12902, pp. 210–220. Springer, Cham (2021). https://doi.org/10.1007/978-3-030-87196-3_20
20. Liu, B., Zhan, L.M., Xu, L., Ma, L., Yang, Y., Wu, X.M.: Slake: a semantically-labeled knowledge-enhanced dataset for medical visual question answering. In: ISBI, pp. 1650–1654. IEEE (2021)
21. Luo, R., et al.: BioGPT: generative pre-trained transformer for biomedical text generation and mining. Briefings Bioinformat. **23**(6) (2022)
22. Mokady, R., Hertz, A., Bermano, A.H.: Clipcap: clip prefix for image captioning. arXiv:2111.09734 (2021)
23. Najdenkoska, I., Zhen, X., Worring, M.: Meta learning to bridge vision and language models for multimodal few-shot learning. In: ICLR (2023)
24. Nguyen, B.D., Do, T.-T., Nguyen, B.X., Do, T., Tjiputra, E., Tran, Q.D.: Overcoming data limitation in medical visual question aswering. In: Shen, D., et al. (eds.) MICCAI 2019. LNCS, vol. 11767, pp. 522–530. Springer, Cham (2019). https://doi.org/10.1007/978-3-030-32251-9_57
25. Radford, A., et al.: Learning transferable visual models from natural language supervision. In: ICML, pp. 8748–8763. PMLR (2021)
26. Radford, A., et al.: Language models are unsupervised multitask learners. OpenAI blog **1**(8), 9 (2019)
27. Ren, F., Zhou, Y.: Cgmvqa: a new classification and generative model for medical visual question answering. IEEE Access **8**, 50626–50636 (2020)
28. Sharma, D., Purushotham, S., Reddy, C.K.: MedFuseNet: an attention-based multimodal deep learning model for visual question answering in the medical domain. Sci. Rep. **11**(1), 19826 (2021)
29. Taylor, N., Zhang, Y., Joyce, D., Nevado-Holgado, A., Kormilitzin, A.: Clinical prompt learning with frozen language models. arXiv:2205.05535 (2022)
30. Tsimpoukelli, M., Menick, J.L., Cabi, S., Eslami, S., Vinyals, O., Hill, F.: Multimodal few-shot learning with frozen language models. NeurIPS **34**, 200–212 (2021)
31. Venigalla, A., Frankle, J., Carbin, M.: BioMedLM: a domain-specific large language model for biomedicine. www.mosaicml.com/blog/introducing-pubmed-gpt (2022). Accessed 06 Mar 2022

32. Wang, J., Huang, S., Du, H., Qin, Y., Wang, H., Zhang, W.: MHKD-MVQA: multimodal hierarchical knowledge distillation for medical visual question answering. In: 2022 IEEE International Conference on Bioinformatics and Biomedicine (BIBM), pp. 567–574. IEEE (2022)
33. Wu, Q., Wang, P., Wang, X., He, X., Zhu, W.: Medical VQA. In: Visual Question Answering: From Theory to Application, pp. 165–176. Springer, Singapore (2022). https://doi.org/10.1007/978-981-19-0964-1_11
34. Zhan, L.M., Liu, B., Fan, L., Chen, J., Wu, X.M.: Medical visual question answering via conditional reasoning. In: ACM Multimedia, pp. 2345–2354 (2020)
35. Zhang, T., Kishore, V., Wu, F., Weinberger, K.Q., Artzi, Y.: Bertscore: evaluating text generation with bert. In: ICLR (2020)

Flexible Unfolding of Circular Structures for Rendering Textbook-Style Cerebrovascular Maps

Leonhard Rist[1,2(✉)], Oliver Taubmann[2], Hendrik Ditt[2], Michael Sühling[2], and Andreas Maier[1]

[1] Friedrich-Alexander-Universität Erlangen-Nürnberg, Erlangen, Germany
leonhard.rist@fau.de

[2] CT R&D Image Analytics, Siemens Healthineers, Forchheim, Germany

Abstract. Comprehensive, contiguous visualizations of the main cerebral arteries and the surrounding parenchyma offer considerable potential for improving diagnostic workflows in cerebrovascular disease, e.g., for fast assessment of vascular topology and lumen in stroke patients. Unfolding the brain vasculature into a 2D overview is, however, infeasible using common Curved Planar Reformation (CPR) due to the circular structure of the Circle of Willis (CoW) and the spatial configuration of the vessels typically rendering them unsuitable for mapping onto simple geometric primitives. We propose CeVasMap, extending the As-Rigid-As-Possible (ARAP) deformation by a smart initialization of the required mesh to map the CoW as well as a merging of neighboring vessels depending on the resulting degree of distortion. Otherwise, vessels are unfolded and attached individually, creating a textbook-style overview image. We provide an extensive distortion analysis, comparing the vector fields of individual and merged unfoldings of each vessel to their CPR results. In addition to enabling unfolding of circular structures, our method is on par in terms of incurred distortions to optimally oriented CPRs for individual vessels and comparable to unfavorable CPR orientations when merging the complete CoW with a median distortion of 65 µm/mm.

Keywords: Vessel unfolding · Distortion · Cerebrovascular disease

1 Introduction

Assessing vascular lumen and topology is among the most critical tasks for timely diagnosis of acute cerebrovascular disease, including the detection of vessel occlusions (e.g., by left-right hemisphere comparisons) and the assessment of redundant blood flow paths via the communicating arteries [2] in stroke diagnosis. Identifying impairments such as thrombi for stroke analysis in the nested cerebral artery system from a Computed Tomography Angiography (CTA) scan

Supplementary Information The online version contains supplementary material available at https://doi.org/10.1007/978-3-031-43904-9_71.

H. Greenspan et al. (Eds.): MICCAI 2023, LNCS 14224, pp. 737–746, 2023.
https://doi.org/10.1007/978-3-031-43904-9_71

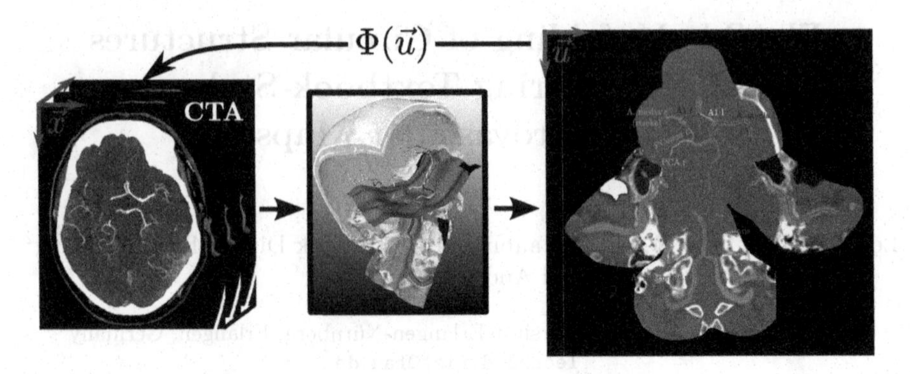

Fig. 1. Unfolding of a labeled cerebrovascular vessel graph from a CTA volume to a 2D overview map with centerline, label and area overlay.

can be inefficient due to the need to interact with the visualization to properly review all relevant vessels. Instead of manually tracking individual vessels across multiple slices, it is also possible to project their complete lumen into a single image plane, called unfolding. Yet, analyzing the global appearance of all cerebral vessels at once, primarily the main arteries forming and surrounding a ring structure called Circle of Willis (CoW; exists in many common norm variants) can be equally important.

However, this circular structure cannot be properly unfolded with the common Curved Planar Reformation (CPR) technique [3], its untangled extension or through conformal mappings [15] due to imperfect circle symmetry leading to different lengths when unfolding the opposing sides. Hence, unfolding is done individually per vessel [12] or by simply showing the configuration in bullet-maps [9]. For many other anatomical structures exist already flattening techniques [4], e.g., using ray-casting for rip unfolding [8]. Relevant unfolding techniques in the brain include aneurysm maps [10] or cortex flattening using mesh deformation [1].

In computer graphics, disk-like mesh parameterization [6] or angle-preserving conformal maps [7] can be used to bring objects in a planar representation. For the involved task of mesh deformation, algorithms such as the As-Rigid-As-Possible (ARAP) method [13] aim to preserve shape during the process, which was already adapted for medical purposes such as pelvis [5] or heart (vessel) unfolding. However, as shown in the vessel tree in Fig. 1, cerebral vessels often bifurcate in perpendicular directions or run in parallel, eliminating the possibility to use simple geometric primitives for the whole CoW. Consequently, comprehensive, contiguous unfoldings of the cerebrovascular system have not been demonstrated before.

This work aims to generate a complete textbook-style vessel map of the cerebral vasculature along the CoW to generate a standardized overview image. Building on the ARAP algorithm, we propose CeVasMap (Ce*rebral* Vas*culature* Map*ping*): a method to unfold circular structures which can be flexibly extended with peripheral vessels, either by including them in the unfolding directly or attaching them individually. This results in locally restricted distortions, retaining curvature information and keeping most of the image distortion-free. Given labeled centerlines, we create a smooth initial mesh with optimal viewing direction dependent on the vessels' principal components and deform it as rigidly as possible to jointly display all vessels of interest. We provide a comprehensive vessel-level evaluation by calculating and visualizing the distortions resulting from the underlying 2D-3D vector field.

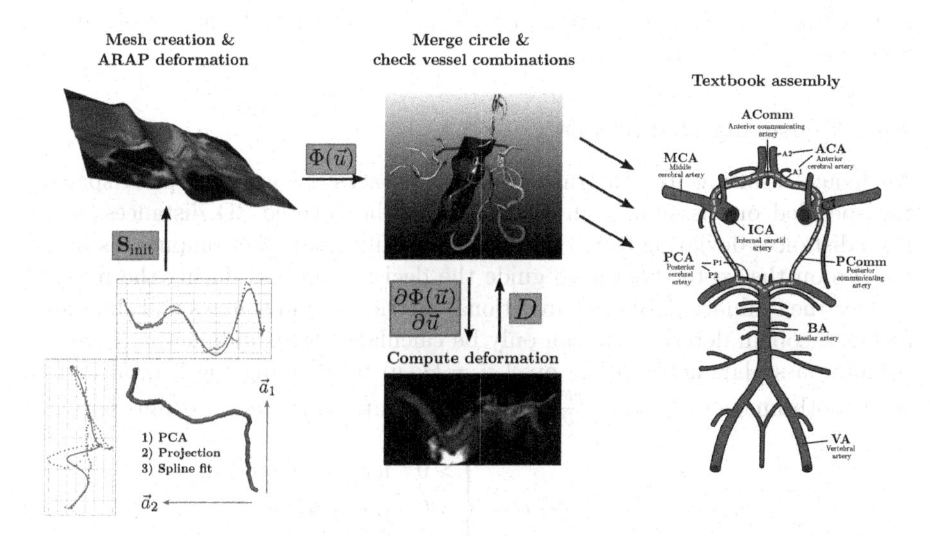

Fig. 2. Unfolding pipeline for an individual vessel (ICA), starting from centerline (left). CoW textbook scheme (right), circle highlighted in yellow. (Color figure online)

2 Methods

2.1 Data

We use a data set consisting of 30 CTA scans (Siemens Somatom Definition AS+) from stroke patients (63.3% males, $74,5 \pm 12,5$ years, 33.3% MCA stroke) with an average voxel spacing of 0.634 mm (in-plane) and 0.57 mm (axial). The brain vasculature is segmented and labeled [11,14]. An example is displayed in the CoW scheme in Fig. 2.

2.2 Rationale

The goal of this work is to generate a single 2D image which contains all vessels of interest jointly with the surrounding parenchyma. Hence, the task at hand is to find an unfolding transformation $\Phi(\vec{u}) = \vec{x} : \mathbb{R}^2 \rightarrow \mathbb{R}^3$ mapping from a 2D position $\vec{u}$ in the unfolded target image to its 3D CTA volume location $\vec{x}$. The corresponding coordinate systems are illustrated in Fig. 1.

Clinically useful images should be merged at the bifurcations to allow consistent path tracing. Additionally, anatomical properties such as vessel curvature should be preserved whereas strong distortions are to be avoided. The main concept of the proposed method is the generation of a joint mapping for the CoW. Further attached vessels can then be merged to it to form a complete overview at the cost of a higher distortion, see Sect. 2.3. For configurations that lead to strong distortions, outer vessels are unfolded individually and attached to the main component. Their arrangement is inspired by the common textbook CoW schematic for better orientation, see Fig. 2. Unfolding of individual as well as combined vessels is described in Sect. 2.4, merging and image assembly in Sect. 2.5.

2.3 Measuring Distortions

An image resulting from a transformation Φ lacks a consistent pixel spacing, the specified one is simply an average from the covered 3D distances, hence pixel distances deviating from that average are distorted. We compute distortion metrics on the vector field Φ to guide the decision process during the merging and for the evaluation. Since conventional metrics to find sources and sinks such as the Jacobian determinant can only be calculated for equidimensional Φs, we compute a scalar metric d per pixel $\vec{u} = (u, v)$ by deriving the transformation w.r.t. both image directions $\frac{\partial \Phi(\vec{u})}{\partial \vec{u}} \in \mathbb{R}^{3 \times 2}$ and applying the Frobenius norm,

$$d_{uv} = ||\frac{\partial \Phi(\vec{u})}{\partial \vec{u}}||_F - \alpha = \begin{cases} > 0 & \text{locally stretched,} \\ = 0 & \text{no distortion,} \\ < 0 & \text{locally contracted.} \end{cases} \tag{1}$$

When sampling a 2D image of size $N \times M$ with isotropic sampling μ, a pixel gradient of 1 in one image direction implies a sampling in 3D with μ, meaning no distortion. To normalize distortion-free values of d_{uv} to 0, the constant $\alpha = \sqrt{2}$ is subtracted. A global image metric is then calculated as $D = \frac{1}{N \cdot M} \sum_{u,v}^{N,M} |d_{uv}|$.

2.4 ARAP Vessel Unfolding

CPR unfolding can lead to displeasing results, e.g., at highly curved segments perpendicular to image read-out direction, impairing the quality of the displayed lumen and parenchyma in the whole image row, see bottom right in Fig. 4. Instead of sampling read-out points $\vec{p} \in \mathbb{R}^3$ line by line starting from the centerline as done by Kanitsar et al. [3], one could also assume a continuous (triangular) mesh surface $\mathbf{S}$ with $\vec{p}_i$ being the vertex position i and downstream image read-out points. One would then find and deform an initial mesh representation, fitting it to the vessels of interest. When deforming such a mesh into $\mathbf{S}'$ with $\vec{p}'_i$, one has direct control of the rigidity between neighboring vertices by measuring the local rigidity energy [13]:

$$E(\mathbf{S}') = \sum_{i=1..N} w_i \sum_{j \in \mathcal{N}(i)} w_{ij} ||(\vec{p}_i{}' - \vec{p}_j{}') - \mathbf{R}_i(\vec{p}_i - \vec{p}_j)||^2 \tag{2}$$

with $\mathcal{N}(i)$ being the 1-step neighborhood of i. By minimizing the norm of the rigid error on the right-hand side, one can approximate a rigid transformation, since actual rigidness is infeasible. For the derivation of the rotation $\mathbf{R}_i$ and the weights w_{ij} and w_i, we refer to Sorkine et al. [13].

On one hand, we lack a fully defined object surface due to our sparse vessel structure. On the other hand, staying close to a meshed plane would mimic multiplanar reformation images and help with the orientation. Since this structure would lead to strong distortion due to its simplicity compared to the centerlines, we define our read-out mesh in a two-step approach. First, we use the principal component vectors stored column-wise in $\mathbf{A} \in \mathbb{R}^{3\times3}$, using column $\vec{a}_3$ as the optimal viewing direction by projecting the points $\vec{c}_j$ from the set of centerline points $j \in \mathcal{C}$ along the two maximum principal components to acquire two 1D distributions $k_{1,2}$, see Fig. 2. To create a smooth surface independent of point density in certain areas, a B-spline b_k is fitted for each of the distributions with the number of knots dependent on the vessel length. Sampling from those functions creates an initial smooth shape $\mathbf{S}_{\text{init}}$ with

$$\vec{p}_{u,v}^{\,(\text{init})} = \mathbf{A}\vec{q}_{u,v} + \bar{\vec{c}} \quad \text{with} \quad \vec{q}_{u,v} = (\mu u, \mu v, b_1(u) + b_2(v))^\top \tag{3}$$

as vertex positions where u and v are sampled uniformly from $[-\frac{N}{2}, \frac{N}{2}]$, $[-\frac{M}{2}, \frac{M}{2}]$ and $\bar{\vec{c}}$ is the centerline mean position. To avoid escalating border regions, the last value of $\vec{q}_{u,v}^{\,(\text{init})}$ is set to 0 outside of the centerline bounding box. Next, $\mathbf{S}_{\text{init}}$ is deformed to contain all vessel points following the ARAP minimization in Eq. 2 by restraining a subset of $\vec{p}_i{}'$ to $\vec{c}_j$. Volume intensities can be read out at the vertex positions of the resulting mesh $\mathbf{S}$, forming the target vector field $\Phi_{\text{CeVasMap}}(\vec{u}) = \vec{p}$ beginning from the unfolded image.

2.5 Merging and Image Assembly

In principle, the described approach can be applied to all vessels of interest at once, introducing high distortion at nearly perpendicular bifurcations. Due to a varying amount of contrast agent; scan, segmentation or labeling quality; or occlusions in stroke, the vessels can differ greatly in length. Long, curved structures influence the unfolding more severely since $\mathbf{S}_{\text{init}}$ has to adapt to more points $\vec{c}_j$, increasing the initial fitting error. In such cases, individual vessels are unfolded separately to maintain sufficient image quality and included in the cerebral vessel map at the position and rotation angle according to the textbook schematic, see Fig. 2. To obtain a standardized image, the CoW (AComm, A1, PComm, P1, ICA C7) is always unfolded jointly to form the center of the image. Afterwards, vessels can be merged based on their importance to the task at hand and by setting an upper threshold $D < \beta$. An exemplary strategy in stroke scenarios would encourage a merging of the often-affected MCAs, together with the anterior part, followed by the PCAs.

2.6 Evaluation

The distortion by the proposed transformation and the baseline is evaluated quantitatively and qualitatively using the metrics from 2.3. The baseline method is the CPR transformation $\Phi_{\mathrm{CPR}}(\vec{u})$ following Kanitsar et al. [3], linearly sampling image rows starting from equidistant points on the centerline. We present the results of both the best and worst CPR viewing. All vessels are unfolded individually with approx. the same isotropic spacing (0.256 mm) and zoom level. Metrics are computed only for pixels that are sampled from a conservatively chosen radius of 10 mm around the centerline in the volume, since larger images favor our approach due to its distortion-free property in areas distant to the vessel.

The pixel-wise metric d_{uv} is used to visually inspect the results. Quantitatively, distortion is assessed showing the mean-like metric D as a distribution over all 30 patients. Additionally, the median of $|d_{uv}|$ per vessel unfolding averaged over all patients is shown to also investigate the maximal distortion of the lesser affected half of the pixels. To assess the expected distortion increase caused by merging, we split the cerebral vessels into two groups: For the CoW vessel segments (merged by default), D and median metric are calculated over the image patches of each vessel, both for individual unfoldings and extracted from the merging. Similarly, the D-distribution and frequency of successfully merging outer vessel pairs to the unfolded circle structure is reported when setting an upper distortion threshold of $D < 25\%$ for the vessel of interest. To avoid that simultaneously merged vessel pairs influence each other or the circle, we compute $\mathbf{S}_{\mathrm{init}}$ only from the circle and then merge vessel pairs using ARAP. This also eliminates the need to evaluate distortions for all possible vessel combinations.

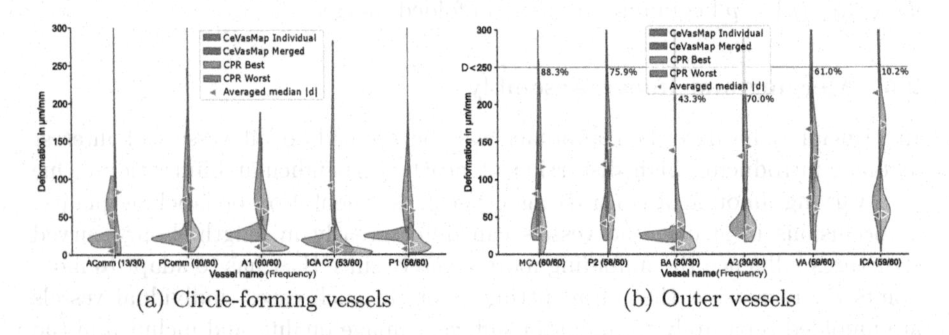

(a) Circle-forming vessels (b) Outer vessels

Fig. 3. Quantitative distortion metrics per individual vessel unfolding over all 30 patients. D is shown as violin distributions comparing our method with best and worst CPR angle. Averaged medians are computed from $|d_{uv}|$. (a) Only the circle is merged. (b) Vessel pairs are merged to circle. Fraction of merged vessels and median distortions provided for merged vessels with $D < 25\%$ (red threshold).

3 Results

Quantitative metrics per vessel (pair) are presented in Fig. 3 and a qualitative example of a cerebral vessel map with distortion heatmaps is shown in Fig. 4. Even using the narrow 10 mm distance threshold, CeVasMap consistently has lower median values than the best CPR angles, except for the short ICA C7 and the VA. Within the CoW, all median values are below 10 μm/mm distortion, the outer vessels are all below 50 μm/mm, except for VA. However, the distributions generally show a slightly lower mean for CPR (best), together with a lower average standard deviation for all vessels. Especially for long vessels (MCA, VA, P2, ICA), we observe higher maximum values in the D-distribution and standard deviation. Qualitative analysis (cf. the example in Fig. 4) confirms the theoretical properties of the two compared approaches. While our approach has locally restricted distortions around the centerline (ARAP target points) with a higher amplitude, CPR affects the distortion along an image row, potentially resulting in inflated vessels due to oversampling inside the lumen at curvature points, see Fig. 4 on the bottom right. For CeVasMap, high distortion in the VA occurs a the end of the labeled vessel, caused by absent neighboring target points. Slight distortions for our approach outside of the vessel (orange shade in ICA r) are caused by the sampling of $\mathbf{S}_{\text{init}}$. For more qualitative examples, see Suppl. Fig. 1.

When merging the CoW (Fig. 3a), the median distortions (65 μm/mm for the complete circle) naturally increase but are still almost on par with (P1, A1) or even outperform (AComm, PComm) the individually unfolded worst CPR angle, except for the ICA C7 segment which is often oriented nearly perpendicular to the remaining circle segments. Using our threshold for the outer vessels, it is possible to merge the MCA in 88.3% of the cases. Vessels perpendicular to the CoW plane could however only be merged in 10.2% (ICA) or 43.3% (BA).

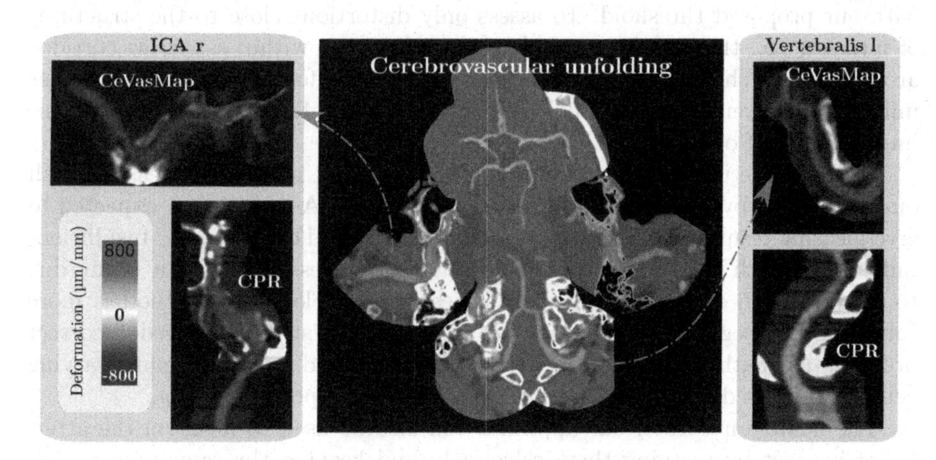

Fig. 4. Exemplary qualitative result of cerebrovascular unfolding: Merged circle with A2, MCA and PCA. ICAs unfolded individually. BA and VA are merged together independent of CoW. Comparison to CPR with distortion heatmap (left/right) using a global color scale (bottom left).

4 Discussion and Conclusion

We present CeVasMap, a flexible method to generate textbook-style vessel maps of the main cerebral arteries, including an unfolding technique capable of displaying circular structures infeasible for the widely used CPR approach. To unfold individual vessels or vessel groups, a read-out mesh is generated by fitting splines to centerline projections along the principal components. This initial mesh is deformed using the ARAP algorithm to display the full vessel lumen. In our approach, we suggest jointly mapping the inner circular structure of the CoW by default and extending it with outer vessels depending on the task at hand and the introduced distortion. For this purpose, a gradient-based distortion metric is calculated on the 2D-3D unfolding vector field, which also facilitates quantitative and qualitative assessment of the mapping quality.

The method results in higher standard deviations and slightly higher mean values of the proposed gradient-based distortion metric D compared to the best possible CPR angle. However, CeVasMap generally achieves lower median values ($\leq$10 µm/mm for the CoW vessels, and mostly $\leq$50 µm/mm for the longer outer segments), meaning larger parts of the image are distortion-free. Even when merging the complete CoW (average distortion median of 65 µm/mm), our method can still compete with the individual CPR unfolding using less favorable angles. When merged with the CoW, distortion for outer vessels, especially those nearly perpendicular to the CoW plane, is quite high, and only the MCAs, PCAs and A2 arteries achieve satisfying results due to their orientation. For instance, the ICA could only be merged in 10.2% while the MCA has a 88.3% success rate with our proposed threshold. To assess only distortions close to the structures of interest, i.e., the vessels, we evaluated our metrics within a narrow corridor around them. When generating full cerebrovascular unfoldings, one would rather unfold more parenchyma for a hole-free image (cf. Fig. 4), leading to a significant improvement of our metrics.

Rotating the unfolded views as often done in CPR is in theory feasible with our approach—by rotating the principal components **A**—but is not expected to give pleasing results as ARAP target points are rotated out-of-plane. Insufficient unfolding length or asymmetric manifestations of vessel pairs can occur due to segmentation or labeling inconsistencies. Nevertheless, our method is more robust against segmentation or labeling errors causing strongly curved, incorrect or fistula-like pathways. In contrast to CPR, our method is robust against varying centerline point distances and inconsistent centerline point ordering.

The flexible nature of our approach can also be used to focus on the arteries of interest by merging them selectively and keeping the remaining vessels separate but arranged intuitively inspired by textbook presentations, effectively reducing overall distortion. Such a selective merging could also be done interactively as, compared to higher computation times for segmentation and labeling, most unfolding steps are parallelizable and can be computed in few seconds.

The clinical applications of a cerebrovascular overview map are manifold. The configuration of the CoW and surrounding major vessels is visualized, giving insight into the collateral blood supply or supporting contralateral comparisons for time-critical stroke detection while being able to view the vessel lumen and its pathologies at a glance. To evaluate challenging real-world scenarios, this work is specifically tested using stroke data as it one of the main reasons for topological impairments of the CoW. This unique representation (compared to volume renderings or slice images) of the cerebrovascular system can be useful for diagnostic reports, comparing patients, disease tracking or to simply to mark locations of possible findings. The maps are also helpful for navigation within the volume as the transformation allows us to readily display multiplanar reformations of the original volume centered at 3D positions corresponding to (manually selected) positions in the unfolded overview image ("picking"). Finally, this condensed lower-dimensional representation could also be beneficial for training downstream deep learning models more efficiently.

Acknowledgments. We would like to acknowledge our collaborators at the University Hospital Schleswig-Holstein in Lübeck for providing the data used in this work. It was collected in a retrospective study which received Institutional Review Board approval prior to starting. The need for informed consent was waived.

References

1. Balasubramanian, M., Polimeni, J.R., Schwartz, E.L.: Near-isometric flattening of brain surfaces. NeuroImage **51**(2), 694–703 (2010). https://doi.org/10.1016/j.neuroimage.2010.02.008
2. Gunnal, S.A., Farooqui, M.S., Wabale, R.N.: Anatomical variability of the posterior communicating artery. Asian J. Neurosurg. **13**, 363 (2018). https://doi.org/10.4103/AJNS.AJNS_152_16
3. Kanitsar, A., Fleischmann, D., Wegenkittl, R., Felkel, P., Groller, E.: CPR - curved planar reformation. In: IEEE Visualization, 2002 (VIS 2002), pp. 37–44 (2002). https://doi.org/10.1109/VISUAL.2002.1183754
4. Kreiser, J., Meuschke, M., Mistelbauer, G., Preim, B., Ropinski, T.: A survey of flattening-based medical visualization techniques. Comput. Graph. Forum **37**, 597–624 (2018). https://doi.org/10.1111/cgf.13445
5. Kretschmer, J., Soza, G., Tietjen, C., Suehling, M., Preim, B., Stamminger, M.: ADR - anatomy-driven reformation. IEEE Trans. Visual. Comput. Graph. **20**(12), 2496–2505 (2014). https://doi.org/10.1109/TVCG.2014.2346405
6. Liu, L., Zhang, L., Xu, Y., Gotsman, C., Gortler, S.J.: A local/global approach to mesh parameterization. Comput. Graph. Forum **27**, 1495–1504 (2008). https://doi.org/10.1111/j.1467-8659.2008.01290.x
7. Lévy, B., Petitjean, S., Ray, N., Maillot, J.: Least squares conformal maps for automatic texture atlas generation. ACM Trans. Graph. **21**, 362–371 (2002). https://doi.org/10.1145/566654.566590
8. Martinke, H., et al.: Bone fracture and lesion assessment using shape-adaptive unfolding. In: Eurographics Workshop on Visual Computing for Biology and Medicine. The Eurographics Association (2017). https://doi.org/10.2312/vcbm.20171249
9. Miao, H., Mistelbauer, G., Nasel, C., Gröller, E.: Visual quantification of the circle of Willis: an automated identification and standardized representation. Comput. Graph. Forum **36**(6), 393–404 (2017). https://doi.org/10.1111/cgf.12988

10. Neugebauer, M., Gasteiger, R., Beuing, O., Diehl, V., Skalej, M., Preim, B.: Map displays for the analysis of scalar data on cerebral aneurysm surfaces. Comput. Graph. Forum **28**(3), 895–902 (2009). https://doi.org/10.1111/j.1467-8659.2009.01459.x
11. Rist, L., Taubmann, O., Thamm, F., Ditt, H., Sühling, M., Maier, A.: Bifurcation matching for consistent cerebral vessel labeling in CTA of stroke patients. Int. J. Comput. Assist. Radiol. Surg. **18**(3), 509–516 (2022). https://doi.org/10.1007/s11548-022-02750-9
12. Shen, M., et al.: Automatic cerebral artery system labeling using registration and key points tracking. In: Knowledge Science, Engineering and Management, pp. 355–367 (2020). https://doi.org/10.1007/978-3-030-55130-8_31
13. Sorkine, O., Alexa, M.: As-rigid-as-possible surface modeling. In: Proceedings of the Fifth Eurographics Symposium on Geometry Processing (SGP 2007), pp. 109–116. Eurographics Association, Goslar, DEU (2007). https://doi.org/10.2312/SGP/SGP07/109-116
14. Thamm, F., et al.: An algorithm for the labeling and interactive visualization of the cerebrovascular system of ischemic strokes. Biomed. Phys. Eng. Exp. **8**(6) (2022). https://doi.org/10.1088/2057-1976/ac9415
15. Zhu, L., Haker, S., Tannenbaum, A.: Flattening maps for the visualization of multibranched vessels. IEEE Trans. Med. Imaging **24**(2), 191–198 (2005). https://doi.org/10.1109/TMI.2004.839368

Dynamic Curriculum Learning via In-Domain Uncertainty for Medical Image Classification

Chaoyi Li[1], Meng Li[1(✉)], Can Peng[2], and Brian C. Lovell[1]

[1] The University of Queensland, School of EECS, St. Lucia, QLD 4072, Australia
chaoyi.li@uq.net.au, meng.li@uq.edu.au, lovell@eecs.uq.edu.au
[2] CSIRO DATA 61, Robotics and Autonomous Systems Group, Pullenvale, Australia
Can.peng@Data61.csiro.au

Abstract. This paper presents an innovative approach to curriculum learning, which is a technique used to train learning models. Curriculum learning is inspired by the way humans learn, starting with simple examples and gradually progressing to more challenging ones. There are currently two main types of curriculum learning: fixed curriculum generated by transfer learning, and self-paced learning based on loss functions. However, these methods have limitations that can hinder their effectiveness. To overcome these limitations, this article proposes a new approach called Dynamic Curriculum Learning via In-Domain Uncertainty (DCLU), which is derived from uncertainty estimation. The proposed approach utilizes a Dirichlet distribution classifier to obtain prediction and uncertainty estimates from the network, which can be used as a metric to quantify the difficulty level of the data. An uncertainty-aware sampling pacing function is also introduced to adapt the curriculum according to the difficulty metric. This new approach has been evaluated on two medical image datasets, and the results show that it outperforms other curriculum learning methods. The source code for this approach will be released at https://github.com/Joey2117/DCLU.

Keywords: Medical image classification · Curriculum learning · Uncertainty estimation

1 Introduction

Curriculum learning methods in deep learning are inspired by human education and involve structuring the training data from easy to hard to teach networks progressively. However, developing a good difficulty metric or measurer for curriculum learning is a challenge. Recently, there are two main categories of approaches to designing the difficulty metrics. The first is that human experts

Supplementary Information The online version contains supplementary material available at https://doi.org/10.1007/978-3-031-43904-9_72.

H. Greenspan et al. (Eds.): MICCAI 2023, LNCS 14224, pp. 747–757, 2023.
https://doi.org/10.1007/978-3-031-43904-9_72

quantify data difficulty based on data characteristics such as complexity [19]. These metrics not only require sufficient domain knowledge, but they also run the risk that metrics of difficulty from a human perspective may not be applicable to a learning model due to different decision boundaries between the model and the human [25].

Another popular approach is difficulty measurers based on the network, including transfer learning [7,26] and loss function [14]. Transfer learning is the scoring of samples using predictions of a reference model on the same training data. These measurers are fixed and do not take feedback from the progress of the current model into account during the training process. Difficulty measurers based on loss function tend to select samples with small training losses as a priority to train the model. However, this type of approach suffers from the uncertainty of quantifying the difficulty of data due to insufficient training in the early stages. In addition, while deep learning models have achieved impressive performance in the medical image analysis field, there remain challenges in measuring and developing a model with low in-domain uncertainty. Recently, uncertainty estimation has emerged as an effective tool for measuring the in-domain uncertainty of models. However, reducing the in-domain uncertainty of models is an active research direction [6].

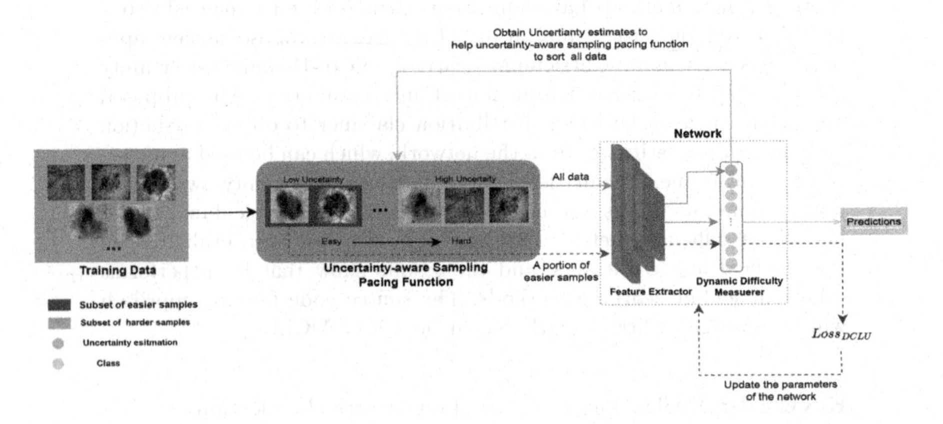

Fig. 1. The pipeline of our DCLU with the dynamic difficulty measurer (DDM) and uncertainty-aware sampling pacing function (UAS). First, DDM provides both uncertainty estimates and predictions for all training data simultaneously in every iteration. Then, based on uncertainty estimates, we use UAS to sort all data from easy to hard and select easier samples to update the parameters of the network. This process ensures that the network is trained on easier samples at the beginning when it is less mature and gradually moves towards harder samples as the network improves.

We propose a new approach to address the challenges of curriculum learning, which we call Dynamic Curriculum Learning via In-Domain Uncertainty (DCLU). Our approach is motivated by two key observations: 1) sample difficulty is influenced by both the complexity of the data and the model's inability to explain data, related to in-domain uncertainty, and 2) reducing in-domain uncertainty by improving the learning process can boost model performance. To estimate in-domain uncertainty, we use a Dirichlet distribution classifier, which provides uncertainty estimates and predictions simultaneously. DCLU then sorts the training data from easy to hard based on in-domain uncertainty estimation, allowing the model to focus on easier samples first. Our approach does not require additional networks to capture uncertainty and is end-to-end.

In particular, Our dynamic difficulty measurer (DDM) generates uncertainties and predictions for each image simultaneously. Uncertainties reflect the difficulty and we use these as the criteria for data rearrangement. Uncertainty estimation runs at each iteration to ingest the feedback of the current network. We also propose an effective uncertainty-aware sampling pacing function (UAS) to sort all training data according to the latest results of DDM and gradually introduce progressively harder samples to learn new parameter vectors and update the network until the entire dataset is covered. The full process of the proposed method is shown in Fig. 1. We evaluate our method on two medical image datasets ISIC 2018 task 3 and Chest-Xray8 (COVID-19). Results indicate that our method outperforms other curriculum learning works. In addition, with our proposed approach, the uncertainty of the model can be mitigated effectively.

2 Related Work

In-Domain Uncertainty. In-domain uncertainty represents the uncertainty associated with inputs extracted from a data distribution equivalent to the training data distribution [6]. Deep learning models can experience in-domain uncertainty as they lack the necessary in-domain knowledge to interpret in-domain samples. In-domain uncertainty is caused by two types of uncertainty: data uncertainty and model uncertainty. Data uncertainty represents the complexity of data, which is related to noise and variations in observations [6]. On the other hand, model uncertainty arises due to shortcomings of the model such as a poor fit to the training dataset or lack of knowledge [6].

In-Domain Uncertainty Estimate. MC dropout [4] and deep ensemble [15] have emerged as two widely adopted techniques for estimating uncertainty in recent years. These methods need significant computing sources and extra metrics to quantify in-domain uncertainty. As our proposal exploits in-domain uncertainties from the current network at every iteration, we want to avoid using additional modules and metrics to obtain similar uncertainty estimates from the above methods and thereby reduce computational resource requirements. Therefore, we apply a classifier with Dirichlet distribution to gain direct both the predictions and uncertainty estimation simultaneously.

Curriculum Learning. Bengio *et al.* [2] first brought curriculum learning into the field of machine learning, which has prompted quite a bit of interest in the field of computer vision [1,5,16,20,21]. Hacohen *et al.* [7] used the confidence score obtained from transfer learning or bootstrapping to determine a fixed curriculum. Self-paced learning(SPL) was proposed by Kumar *et al.* [14], which applied example-wise training loss at each iteration as the difficulty measurer. The drawback of SPL is that some easier data appear all the time since SPL tended to select data with lower current losses. Jiang *et al.* [11] indicated a novel curriculum learning method named self-paced curriculum learning (SPCL) that combined predefined curriculum learning and self-paced learning so that prior knowledge before training and information during training are used effectively. In addition, Kong *et al.* [12] presented a linear combination of the current model loss and prior knowledge to adapt the difficulty measurer of the current model. In our work, we want to build a dynamic difficulty measurer through the uncertainty estimates generated by each iteration of the network. This means our approach does not require the prior knowledge provided by the pre-trained model and is able to update the curricula during the training process.

3 Method

3.1 Overview

In DCLU, we randomly input data into the dynamic difficulty measurer (DDM) at the first epoch to obtain the uncertainty for each data point. Next, we apply the uncertainty-aware sampling pacing function (UAS) to present the training data to the DDM in order of the ascending uncertainties to synchronously produce predictions and new uncertainty estimates. New uncertainties can be used as a criterion for sorting data in the next iteration. Additionally, our pacing function selects a fraction of easier samples and learns a parameter vector to update the network. With the training process progressing, the proportion of selected samples increases until it eventually comprises the entire dataset. The pseudo-code for our method is given in the Appendix.

3.2 Dynamic Difficulty Measurer

The difficulty measurer is used to measure the difficulty of the data to decide on the order of the training data, which is a crucial component of curriculum learning. Our dynamic difficulty measurer (DDM) is a multi-class classifier with Dirichlet distribution, based on evidential deep learning [22]. The setting of our work is focusing on the K-classes classification task. In detail, first of all, we assume $e_k \geq 0$ to be evidence of k-th output of the activation function (e.g. softplus [17]) of our DDM. Then, DDM assigns a belief mass b_k for each category and an overall uncertainty mass u, which can be defined as (1):

$$u + \sum_{k=1}^{K} b_k = 1, \tag{1}$$

where K is the total number of classes. We allocate b_k in correspondence to Dirichlet distribution with parameters $\alpha_k = e_k + 1$. So, the belief mass and uncertainty can be formulated as $b_k = \frac{\alpha_k - 1}{S}$ and $u = \frac{K}{S}$, where $S = \sum_{k=1}^{K}(\alpha_k)$.

Additionally, the Dirichlet distribution of DDM can be defined with parameters $\alpha = [\alpha_1, ..., \alpha_K]$ as (2):

$$D(p_k|\alpha_k) = \frac{1}{\mathrm{B}(\boldsymbol{\alpha})} \prod_{k=1}^{K} p_k^{\alpha_k - 1}, \tag{2}$$

where $\sum_{k=1}^{K} p_k = 1$ and $0 \leq p_1, ..., p_K \leq 1$. $\mathrm{B}(\boldsymbol{\alpha})$ is a K-dimensional multinomial beta function [13]. It represents the density of each probability distribution, which is based on parameters derived from the evidence vector [10]. The expected probability of the k-th output of the classifier can be formulated as (3):

$$\hat{p_k} = \frac{\alpha_k}{S}. \tag{3}$$

Thus, DDM can generate both predictions and in-domain uncertainty estimates for each sample simultaneously. To be specific, in-domain uncertainty estimates include data and model uncertainty. Data uncertainty cannot be eliminated with training and model uncertainty can be reduced by improving the learning process [6]. Inspired by this phenomenon, we employ in-domain uncertainty to measure the difficulty of data at each iteration. This not only allows the data uncertainty as a prior for the criterion of difficulty measure but also allows the measure of data difficulty to be updated efficiently based on model uncertainty estimated from the current state of the model. The method can achieve a dynamic curriculum.

3.3 Uncertainty-Aware Sampling Pacing Function

The pacing function is based on using the difficulty measurer to determine how training examples (data) are fed into the network during the training process. Our dynamic difficulty measurer (DDM) can provide difficulty scores for all data at each iteration. However, some existing pacing functions attempt to partition the dataset into multiple subsets and gradually feed them into the network during training - this fails to satisfy the requirement of our difficulty measurer. To address this problem, we propose the uncertainty-aware sampling pacing function (UAS) consisting of two modules: the reorder and sampling modules. Within the reorder module, UAS sorts all training data from easy to hard according to ascending uncertainties from DDM at the last epoch and sends them into the network to yield predictions and new uncertainty estimates. The sampling module specifies a fraction of easier samples to update to the network. The weight assigned to the selected examples is set to 1, while the weight for other data is set to 0. These weights α will be applied to the objective function, which allows the parameters learned from the specified examples to update the network. We increase the fraction exponentially in each epoch until it eventually comprises the entire dataset.

In our work, we implement UAS through two approaches. Firstly, UAS (exponential) incorporates both the reorder and sampling modules, prompting the network to prioritize learning easier examples during the initial stages of training and then gradually learn more difficult examples. In each epoch, UAS (exponential) first utilizes the reorder module to sort all training data from easy to hard. Then, it applies the sampling module to select a fraction of easier examples to update the network. Additionally, UAS (full) only contains the reorder module to enable the network to learn all sorted data, ranging from easy to hard examples.

3.4 Loss Function

Assume $D(p_k|\alpha_k)$ is the prior on the cross-entropy loss. The classification loss function for each sample can be formulated by the Bayes risk by (4):

$$\mathcal{L}_i^{cls}(\Theta) = \int \left[\sum_{k=1}^{K} -y_{ik} \log(p_{ik}) \right] D(p_{ik}|\alpha_{ik}) d\mathbf{p}_i = \sum_{k=1}^{K} y_{ik} \left(\psi(S_i) - \psi(\alpha_{ik}) \right), \tag{4}$$

where ψ is the digamma function, p_{ik} is the estimated probability of the k-th class of the i-th sample. Additionally, Kullback-Leibler (KL) divergence is used to reduce the total evidence to zero under the condition of incorrect classification, which can be denoted by (5):

$$\mathcal{L}_i^{KL} = \log \left(\frac{\Gamma\left(\sum_{k=1}^{K} \tilde{\alpha}_{ik}\right)}{\Gamma(K) \prod_{k=1}^{K} \Gamma(\tilde{\alpha}_{ik})} \right) + \sum_{k=1}^{K} (\tilde{\alpha}_{ik} - 1) \left[\psi(\tilde{\alpha}_{ik}) - \psi\left(\sum_{k=1}^{K} \tilde{\alpha}_{ik} \right) \right], \tag{5}$$

where Γ denotes the gamma function. To achieve pace control and reduce the effects of overfitting due to easier data that may always appear at the beginning of training, the final objective function is:

$$\mathcal{L}_{DCLU} = \sum_{i=1}^{N} \alpha_i * (\mathcal{L}_i^{cls} + \lambda_t \mathcal{L}_i^{KL}) + \frac{1}{2} ||w||_2^2, \tag{6}$$

where α_i is the weight from the UAS pacing function to control the pace, $\lambda_t = min(1, t/50)$ is the annealing parameter, t is the current training epoch and w is weights of the network.

4 Experiment

4.1 Dataset and Experimental Setup

Dataset. We evaluated our method on two public medical image datasets including ISIC 2018 Task 3 [8,23] and Chest-Xray8 (COVID-19) [18]. ISIC 2018 Task 3 dataset has 10,015 training images and 194 images for validation. It aims to classify 7 skin cancer types including Melanoma, Melanocytic nevus, Basal cell carcinoma, Actinic keratosis, Benign keratosis Dermatofibroma and Vascular lesion. Chest-Xray 8 (COVID-19) contains 1125 X-ray images of the chest of the individuals studied, including 125 images labeled COVID-19 taken from [3], 500 images labeled pneumonia and 500 images labeled no findings were randomly taken from the ChestX-ray8 [24]. We employ training and validation data from ISIC 2018 Task 3 dataset as our training and test sets. Chest-Xray 8 (COVID-19) dataset is randomly divided into two parts, with 80% of the images being the training set and 20% of the images being the test set.

Evaluation Metrics. For both datasets, the performance of diagnosis is evaluated with both accuracy and F1 score. Moreover, for the assessment of uncertainty estimation, we apply expected calibration error (ECE) as the metric.

Implementation Details. We employ ResNet-18 [9] as the backbone for both tasks. In our experiments, we used the TensorFlow framework and trained on an NVIDIA 3090 GPU with 32G of RAM. An Adam optimizer with $\beta = 0$ and $\alpha = 0.99$ and a linear learning rate scheduler are used to tune the network. The initial learning rate is 0.0001 and the learning rate is decreased to 0.00001 after 20 epochs with batch size 16. The total epoch is 50.

4.2 Experimental Results

Comparison with State-of-the-Art. Our method is compared with various curriculum learning methods: 1) Vanilla samples batches randomly on the entire dataset without any curriculum learning techniques; 2) Fixed curriculum learning (FCL) [7] sorts data with the confidence score derived from transfer learning; 3) Self-pace learning (SPL) [14] chooses the data with minimum losses to train the network in advance; 4) Self-pace curriculum learning (SPCL) [11] is the combination of self-paced learning and predefined curriculum learning; 5. Adaptive Curriculum learning (Adaptive CL) [12] adjusts the difficulty measurer by incorporating the prior knowledge and feedback from the current model. The results in Table 1 show that our method (DDM) with UAS (full) outperforms both datasets. Our method with UAS (exponential) performs better than other methods on Chest-Xray 8 (COVID-19) and has the second best performance on ISIC 2018 Task 3. The performance gap between our two methods may be

Table 1. Comparison with state-of-the-art curriculum learning methods on ISIC 2018 Task 3 and Chest-Xray 8 (COVID-19)

Method	ISIC 2018 Task 3		Chest-Xray8 (COVID-19)	
	Accuracy	F1 Score	Accuracy	F1 Score
Vanilla	81.73	38.8	71.88	57.51
FCL [7]	82.85	39.87	72.32	58.86
SPL [14]	82.69	39.49	74.55	68.77
SPCL [11]	82.21	39.19	73.21	63.08
Adaptive CL [12]	83.17	37.64	72.77	64.13
Ours (Exponential)	82.93	39.26	81.7	73.69
Ours (Full)	**85.1**	**40.58**	**83.04**	**82.03**

due to the fact that our objective function is constructed based on the Dirichlet distribution and the presence of class imbalance within both datasets. When the samples are selected by UAS (exponential) at the early training stage, samples from smaller numbered classes may not be chosen which leads to the optimization may tend to fall into local extrema. Our approach clearly outperforms FCL, demonstrating the necessity to design a dynamic curriculum. Compared with loss function-based curriculum learning (SPL, SPCL and Adaptive CL), our methods obtain better performance, showing the importance of using uncertainty estimates as the criterion for difficulty measurement of data. Furthermore, our method is more effective than SPL at an early training stage. Details are referred to in the Appendix.

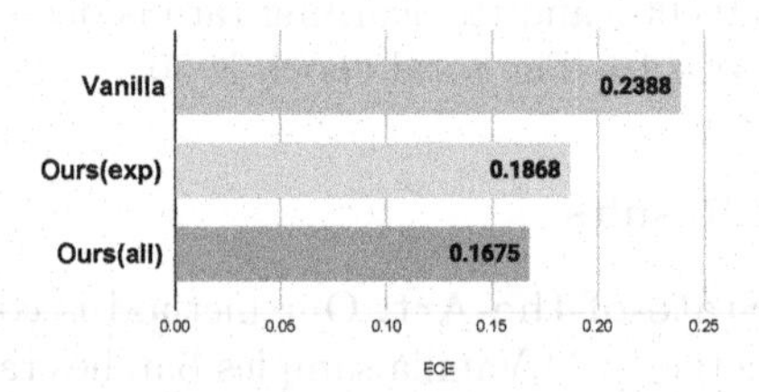

Fig. 2. Compare our method and vanilla under ECE metric on Chest-Xray 8 (COVID-19)

Evaluation In-Domain Uncertainty. To verify that our method is effective in reducing in-domain uncertainty, our method is compared with vanilla. Figure 2 shows that our method is more robust than vanilla. Details of the comparison with other curriculum learning methods are referred to in the Appendix.

Table 2. Ablation analysis of our method on Chest-Xray 8 (COVID-19) dataset.

Method	DDM	Exp	Single-step	Linear	UAS	Accuracy
Vanilla						71.88
FCL		✓				72.32
Ours	✓	✓				81.25
Ours	✓		✓			80.36
Ours	✓			✓		77.23
Ours with full UAS(w/o L2)	✓				✓	82.14
Ours with full UAS(w L2)	✓				✓	83.04

Ablation Study. We present the results of ablation studies on Chest-Xray 8 (COVID-19) in Table 2 to show the effectiveness of key components, *i.e.*, DDM, UAS pacing function and the loss function with L2 regular term. First of all, our method performs 8.93% better than FCL using the same exponential pacing functions. This indicates that DDM performs quite well compared to the fixed curriculum obtained by transfer learning. Moreover, In order to demonstrate that the order generated by DDM is robust, we have conducted experiments on different backbones, details of which can be found in the Appendix. Next, we compare our proposed pacing function to three pacing functions, which are fixed exponential, single-step and linear pacing functions [7]. Both single-step and linear pacing functions divide the entire data set into subsets and put them into the model progressively. The difference between the two is in the number of subsets. Single-step pacing function divides into two subsets and the linear pacing function splits into several subsets (*e.g.* 3 subsets). The results in Table 2 show that UAS is more suited to DDM than the other pacing functions. When the other three pacing functions are used, new samples added to the current subset will disrupt the existing order of data when the size of the subset is increased. It causes the network to have to relearn and reorder the new subset. Finally, We compare the effect of using the loss function with and without L2 regular term on the model performance, finding that using the loss function with L2 can eliminate the effect of overfitting.

5 Conclusion

Our approach, called Dynamic Curriculum Learning via In-Domain Uncertainty (DCLU), introduces a new perspective on curriculum design by utilizing the current stage of the network to estimate the difficulty of data based on its in-domain uncertainty. To support this, we also introduce an uncertainty-aware sampling pacing function that is compatible with our dynamic difficulty measurer. Our experimental results confirm that DCLU is successful in reducing uncertainty, and we demonstrate the effectiveness of our approach on two medical image datasets through extensive experimentation.

References

1. Appalaraju, S., Chaoji, V.: Image similarity using deep CNN and curriculum learning. arXiv preprint arXiv:1709.08761 (2017)
2. Bengio, Y., Louradour, J., Collobert, R., Weston, J.: Curriculum learning. In: Proceedings of the 26th Annual International Conference on Machine Learning, pp. 41–48 (2009)
3. Cohen, J.P., Morrison, P., Dao, L.: Covid-19 image data collection. arXiv preprint arXiv:2003.11597 (2020)
4. Gal, Y., Ghahramani, Z.: Dropout as a bayesian approximation: representing model uncertainty in deep learning. In: International Conference on Machine Learning, pp. 1050–1059. PMLR (2016)
5. Ganesh, M.R., Corso, J.J.: Rethinking curriculum learning with incremental labels and adaptive compensation. arXiv preprint arXiv:2001.04529 (2020)

6. Gawlikowski, J., et al.: A survey of uncertainty in deep neural networks. arXiv preprint arXiv:2107.03342 (2021)
7. Hacohen, G., Weinshall, D.: On the power of curriculum learning in training deep networks. In: International Conference on Machine Learning, pp. 2535–2544. PMLR (2019)
8. Hardie, R.C., Ali, R., De Silva, M.S., Kebede, T.M.: Skin lesion segmentation and classification for ISIC 2018 using traditional classifiers with hand-crafted features. arXiv preprint arXiv:1807.07001 (2018)
9. He, K., Zhang, X., Ren, S., Sun, J.: Deep residual learning for image recognition. In: Proceedings of the IEEE Conference on Computer Vision and Pattern Recognition, pp. 770–778 (2016)
10. Hemmer, P., Kühl, N., Schöffer, J.: Deal: Deep evidential active learning for image classification. Deep Learn. Appl. **3**, 171–192 (2022)
11. Jiang, L., Meng, D., Zhao, Q., Shan, S., Hauptmann, A.G.: Self-paced curriculum learning. In: Twenty-Ninth AAAI Conference on Artificial Intelligence (2015)
12. Kong, Y., Liu, L., Wang, J., Tao, D.: Adaptive curriculum learning. In: Proceedings of the IEEE/CVF International Conference on Computer Vision, pp. 5067–5076 (2021)
13. Kotz, S., Balakrishnan, N., Johnson, N.L.: Continuous multivariate distributions. In: Models and applications, vol. 1. John Wiley & Sons (2004)
14. Kumar, M., Packer, B., Koller, D.: Self-paced learning for latent variable models. Adv. Neural Inf. Process. Syst. **23** (2010)
15. Lakshminarayanan, B., Pritzel, A., Blundell, C.: Simple and scalable predictive uncertainty estimation using deep ensembles. Adv. Neural Inf. Process. Syst. **30** (2017)
16. Morerio, P., Cavazza, J., Volpi, R., Vidal, R., Murino, V.: Curriculum dropout. In: Proceedings of the IEEE International Conference on Computer Vision, pp. 3544–3552 (2017)
17. Nair, V., Hinton, G.E.: Rectified linear units improve restricted Boltzmann machines. In: ICML (2010)
18. Ozturk, T., Talo, M., Yildirim, E.A., Baloglu, U.B., Yildirim, O., Acharya, U.R.: Automated detection of covid-19 cases using deep neural networks with x-ray images. Comput. Biol. Med. **121**, 103792 (2020)
19. Platanios, E.A., Stretcu, O., Neubig, G., Poczos, B., Mitchell, T.M.: Competence-based curriculum learning for neural machine translation. arXiv preprint arXiv:1903.09848 (2019)
20. Qu, M., Tang, J., Han, J.: Curriculum learning for heterogeneous star network embedding via deep reinforcement learning. In: Proceedings of the Eleventh ACM International Conference on Web Search and Data Mining, pp. 468–476 (2018)
21. Ruiter, D., España-Bonet, C., van Genabith, J.: Self-induced curriculum learning in neural machine translation (2019)
22. Sensoy, M., Kaplan, L., Kandemir, M.: Evidential deep learning to quantify classification uncertainty. Adv. Neural Inf. Process. Syst. **31** (2018)
23. Tschandl, P., Rosendahl, C., Kittler, H.: The ham10000 dataset, a large collection of multi-source dermatoscopic images of common pigmented skin lesions. Scientific Data **5**(1), 1–9 (2018)
24. Wang, X., Peng, Y., Lu, L., Lu, Z., Bagheri, M., Summers, R.M.: Chestx-ray8: hospital-scale chest x-ray database and benchmarks on weakly-supervised classification and localization of common thorax diseases. In: Proceedings of the IEEE Conference on Computer Vision and Pattern Recognition, pp. 2097–2106 (2017)

25. Wang, X., Chen, Y., Zhu, W.: A survey on curriculum learning. IEEE Trans. Pattern Anal. Mach. Intell. (2021)
26. Weinshall, D., Cohen, G., Amir, D.: Curriculum learning by transfer learning: theory and experiments with deep networks. In: International Conference on Machine Learning, pp. 5238–5246. PMLR (2018)

Joint Prediction of Response to Therapy, Molecular Traits, and Spatial Organisation in Colorectal Cancer Biopsies

Ruby Wood[1,3], Enric Domingo[4], Korsuk Sirinukunwattana[1,3], Maxime W. Lafarge[5], Viktor H. Koelzer[5], Timothy S. Maughan[4], and Jens Rittscher[1,2,3](✉)

[1] Department of Engineering Science, University of Oxford, Oxford, UK
{ruby.wood,jens.rittscher}@eng.ox.ac.uk
[2] Nuffield Department of Medicine, University of Oxford, Oxford, UK
[3] Big Data Institute, University of Oxford, Li Ka Shing Centre for Health Information and Discovery, Oxford, UK
[4] Department of Oncology, University of Oxford, Oxford, UK
[5] Computational and Translational Pathology Group, Department of Pathology and Molecular Pathology, University Hospital and University of Zürich, Zürich 8091, Switzerland

Abstract. Existing methods for interpretability of model predictions are largely based on technical insights and are not linked to clinical context. We use the question of predicting response to radiotherapy in colorectal cancer patients as an exemplar for developing prediction models that do provide such contextual information and therefore can effectively support clinical decision making. There is a growing body of evidence that about 30% of colorectal cancer patients do not respond to radiotherapy and will need alternative treatment. The consensus molecular subtypes for colorectal cancer (CMS) provide one such approach to categorising patients based on their disease biology. Here we select the CMS4 subtype as a proxy for stromal infiltration. By jointly predicting a patient's response to radiotherapy, the presence of CMS4, and the epithelial tissue map from morphological features extracted from standard H&E slides we provide a comprehensive clinically relevant assessment of a biopsy. A graph neural network is trained to achieve this joint prediction task, which subsequently provides novel interpretability maps to aid clinicians in their cancer treatment decision making process. Our model is trained and validated on two private rectal cancer datasets.

Keywords: Histology · Graphs · Colorectal Cancer · Response to Therapy · Interpretability

---

Supported by Cancer Research UK.

Supplementary Information The online version contains supplementary material available at https://doi.org/10.1007/978-3-031-43904-9_73.

H. Greenspan et al. (Eds.): MICCAI 2023, LNCS 14224, pp. 758–767, 2023.
https://doi.org/10.1007/978-3-031-43904-9_73

1 Introduction

In the UK, approximately 11,500 patients are diagnosed with rectal cancer each year [19]. A common form of treatment for such patients is neoadjuvant therapy, including chemotherapy and radiotherapy, which can be given to patients with locally advanced rectal cancer to shrink the tumour prior to surgery. Recent evidence suggests that 10–20% of patients will have a complete pathological response to neoadjuvant therapy and can therefore avoid surgery altogether [2,5]. However, one third of patients do not benefit from radiotherapy treatment prior to surgery [8], hence it is important to determine how a patient will respond to radiotherapy with a personalized approach in order to avoid overtreatment.

Histology-based digital biomarkers enable the possibility to predict a patient's response to therapy. The consensus molecular subtypes (CMS) classification system derived from gene expressions [9] has been developed to provide biological insight into metastatic colorectal cancer. It has been shown that these four CMS classes can be predicted directly from the standard haematoxylin and eosin (H&E) stained slide images using deep learning [18]. Various studies have investigated the link between CMS and patient outcomes, suggesting that patients with tumour classified as CMS4, which features stromal invasion [9] and shows significantly higher stroma content [15], have worse survival rates compared to the other CMS classes [5]. Increased stromal content has independently been shown to be a predictor for increased risk of recurrence in early rectal cancer [10], and tumour immune infiltrate evaluated with Immunoscore is a useful prognostic marker [3]. The spatial organisation of the cancerous tissue has been identified as a biomarker for aggressiveness or recurrence [12], and Qi et al. [15] found that the features they developed representing spatial organisation reflected characteristics of the four CMS classes. Interactions between the epithelial tissue (cellular tissue lining) and other prevalent tissue types in the tumour microenvironment are also indicators of prognosis [15], since progression of colorectal cancer is dependent on both the epithelial and stromal tissues [20]. Other work has looked at predicting chemoradiotherapy response in rectal cancer patients from H&E images using different approaches, but without providing contextual interpretations [19,22].

As opposed to predicting response to radiotherapy alone, we aim to analyse this prediction in the context of the overall tissue architecture and the tumour biology as captured by CMS. Input to our model is a standard H&E Whole Slide Image (WSI) which is split into smaller patches to overcome the memory limitations of existing GPUs. To achieve our goal we need to capture the heterogeneity at the slide level, which is why applying full or semi-supervised approaches on individual tiles followed by a slide aggregation method is not suitable. Instead, we build on recent graph neural network (GNN) approaches that allow us to model the entire WSI as a graph. As local cell communities form the nodes of such a graph it can effectively model the micro-anatomy of the tissue. At the same time it is possible to make predictions at the node-, graph-, and slide-level.

Related Work. To predict the grading of colorectal cancer (CRC), both cell-based and patch-based graphs have been used in separate works [16,23], setting the nodes of the graph as either cell nuclei or square patches, defining the node features as either handcrafted or learned features, and then applying a GNN for outcome prediction. Another patch-based GNN approach to predicting genetic mutations in CRC from H&E slides found their model trained on colon cancer generalised well to rectal cancer. For other cancers, the SlideGraph pipeline clusters nuclei for the graph nodes, and provides node-level predictions to make their model more interpretable [13]. Other approaches to setting the graph nodes include using subgraphs to represent regions [14], and creating superpatches by combining patches [11]. Edges between the nodes are usually defined by a spatial distance metric, which helps model the spatial organisation of the tissue. Common choices for GNNs include a Graph Isomorphism Network (GIN) with jumping connectivity [7,13,14], as we use in this research.

Our methodology proposes a novel and disease relevant approach to a more interpretable model that effectively supports a diagnostic task. Pathologists and oncologists can use this information to inspect the validity of the prediction result and interrogate key aspects of the spatial biology that is critical for patient management. Ultimately, this type of information that is not available today will help to characterise interactions between the tumour and the host tissue and therefore help to support choice of therapy. The developed framework combines self-supervised training of a Vision Transformer (ViT) to extract morphological features, a superpixel algorithm for determining nodes of a graph, and a GNN for predictions. We achieve 0.82 AUC predicting complete response to radiotherapy using deep learning on WSIs for CRC patients, whilst providing novel interpretability of the results.

2 Methods

In this section we present the patch-level feature extraction, provide the detail of the superpixel segmentation of the WSI, and illustrate the resulting graph presentation. A GNN with three branches for our output predictions is used to simultaneously make the three different predictions as shown in Fig. 1.

Pipeline. For computational reasons, all images are split into patches of size 256×256 pixels. In order to have a common feature set all the way up to the last layer of the GNN, individual patches should be represented by morphological features that are label-agnostic. This last layer of the GNN then splits into three branches to predict response to radiotherapy, the CMS4 subtype classification for CRC, and epithelial tissue regions. This way we can guarantee the common latent features and derivation across branches, maintaining the contextual importance of each branch. The DINO framework [4] uses a self-distillation training approach, using data augmentation to locally crop the patches and train with a local-global student-teacher approach. We use the DINO framework to train a ViT in a self-supervised manner on our H&E slides [6], representing each patch with 384 features. We use only the training set to train this model, and use the image patches at 20x magnification.

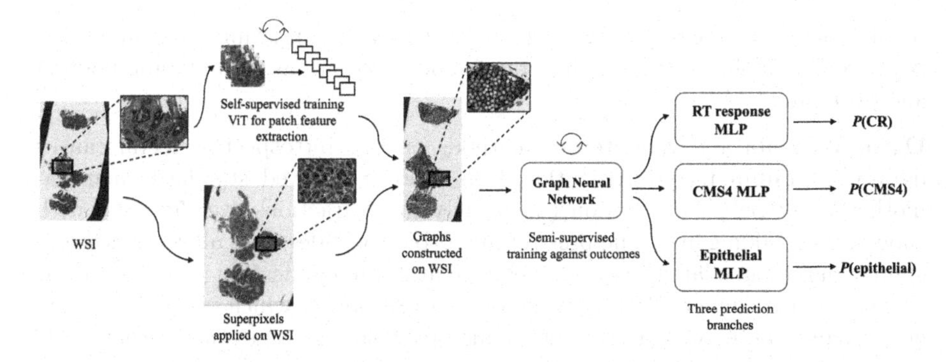

Fig. 1. Approach. We extract patch-level features from each WSI using self-supervised DINO training with a ViT model [4]. The SLIC superpixel algorithm segments the entire slide into smaller regions [1]. We calculate the mean patch features for these superpixel regions, and use the superpixel features and centers as our graph nodes, applying Delaunay triangulation to generate the edges of the graph. A GNN consisting of GINConv layers is trained on these fixed graphs, and the final layer splits into three separate MLP branches to provide predictions of three different outcomes, complete response (CR) to radiotherapy (RT), CMS4 classification, and epithelial tissue. An example output is visualized in Fig. 2.

To find the nodes of the WSI graphs, we apply the SLIC superpixel algorithm [1] on the WSIs at 5x magnification to segment the tissue to capture cellular neighbourhoods that are roughly between 80–100 μm^2/pixels in size. It can be seen that the superpixel boundaries consistently align with the boundaries of tissue compartments.

The superpixels centers are used as the nodes of the graph, and the node features are the weighted mean of the corresponding patch features which overlap with the superpixel region. The edges of the graph are determined by nearest neighbours from Delaunay triangulation, as in SlideGraph [13].

Building on the ideas introduced by SlideGraph [13] we use GINConv layers [21], adding tempering to avoid overfitting, and replace their logistic regression scaler with a simple sigmoid function. We add three branches to the final layer of the GNN, in the form of three separate multilayer perceptrons (MLPs). Two of these MLPs return a graph-level prediction, for the response to RT and CMS4 predictions, and the final branch returns node-level predictions, predicting whether each node is epithelial tissue or not. Our loss function is defined as

$$\mathcal{L} = \text{BCE}(\hat{y}_{RT}, y_{RT}) + \text{BCE}(\hat{y}_{CMS4}, y_{CMS4}) + \text{BCE}(\hat{\mathbf{y}}_{epi}, \mathbf{y}_{epi}),$$

where BCE is the binary cross entropy loss, $\hat{y}_{RT} \in \mathbb{R}$ is the slide-level prediction of response to radiotherapy, $\hat{y}_{CMS4} \in \mathbb{R}$ is the slide-level prediction of CMS4, $\hat{\mathbf{y}}_{epi} \in \mathbb{R}^{n_i}$ are the node-level predictions of epithelial tissue and n_i is the number of nodes in the ith WSI graph.

For each prediction branch, we can visualize the individual node predictions from the WSI graph, overlaid on the WSI itself, to get an idea of how the node predictions vary across the different tissue regions. Each graph-level prediction is derived from the corresponding branch node predictions, by applying pooling and dropout.

Data. We train and validate our methods on two retrospective rectal cancer datasets, Grampian and Aristotle. Both cohorts received standard chemoradiotherapy of pelvic irradiation (45–50.4Gy in 25 fractions over 5 weeks) with capecitabine 900mg/m^2. The pre-treatment biopsy slides were all sectioned and stained in the same laboratory, and scanned at 20x magnification (0.5 μm^2/pixel) on an Aperio scanner. Pathological complete response, which we use as a target outcome here, was derived from histopathological assessment from post-treatment resections.

The CMS labels for this data are derived from three different transcriptomic versions (single cohort, combined cohort correcting batch effects and combined cohort including 2036 cases run with the same platform), in order to generate robust classifications. In all cases the CMS call was calculated using the CMSclassifier random forest and single sample predictor [9]. Final CMS calls are based on matching calls between the three transcriptomic versions. Despite our efforts to minimise the noise from RNA sequencing, we still expect a certain level of noise in our ground truth data, which we discuss in the Results section.

The epithelial labels for each graph node are calculated from epithelial masks for each WSI. These epithelial segmentation masks were generated at 10x magnification (1 μm^2/pixel) with a U-Net [17] which was trained and validated on 666 full tissue sections belonging to 362 patients from the FOCUS cohort [18]. The ground truth annotations for the training of this model were generated by VK.

For consistency the tumour regions were marked up by an expert pathologist. We use these masks in our analysis to filter out background and irrelevant tissue from the images. Grampian and Aristotle are used in both training and validation, with a 70/30% training-validation split, keeping any WSIs from a single patient in the same dataset. We predict complete response to radiotherapy against all other responses, such as partial response and no response. The datasets are unbalanced, since in Grampian only 61/244 slides have complete response, and in Aristotle only 24/121 slides have complete response. They are even more unbalanced for CMS4, since only 28/244 slides in Grampian and 17/121 slides in Aristotle are labelled with CMS4. We address this imbalance in the Supplementary Materials. There are 365 slides total in our dataset, from 249 patients.

3 Experiments

Implementation. We use the default DINO parameters, but train for 20 epochs with 5 warmup epochs. We apply the SLIC algorithm [1] with compactness of 20, setting the number of segments for each WSI as half the mean size of the WSI. Prior to fitting the graph model we normalize the node features relative to the whole dataset. We train our graph model for 30 epochs using Adam optimizer with learning rate 0.001 and weight decay 0.0001. Our graph model has three GINConv layers [21] with dimensions 64, 32 and 16 respectively. We apply dropout of 0.5 in-between graph layers, use minimum aggregation for message passing between nodes and use maximum pooling for concatenating the node activations. We apply tempering to the outcome of the graph model, dividing the output by 1.5. We evaluate the best validation epoch by finding the best mean AUC across the three prediction branches. We run the whole pipeline on four folds with different random data splits for training and validation. The code for this research will be made available upon request.

Table 1. For each fold, we take the mean metrics for the three branch predictions from the best model on our validation data, with the best epoch chosen based on mean AUC for the three predictions. The standard deviation of the metrics across the four folds is provided in brackets. Each prediction uses an optimised threshold value determined from the validation set in order to round the output probabilities to a binary prediction. We use weighted metrics due to the class imbalance in our dataset.

Response branch	Response to RT	CMS4	Epithelial
Mean AUC (std)	0.819 (0.04)	0.819 (0.04)	0.760 (0.01)
Mean accuracy (std)	0.795 (0.05)	0.750 (0.04)	0.691 (0.01)
Mean balanced accuracy (std)	0.774 (0.05)	0.719 (0.05)	0.691 (0.01)
Mean weighted F1 (std)	0.810 (0.04)	0.791 (0.02)	0.700 (0.01)
Mean weighted precision (std)	0.843 (0.02)	0.870 (0.02)	0.725 (0.00)
Mean weighted recall (std)	0.795 (0.05)	0.750 (0.04)	0.691 (0.01)

Results. Despite the noise in our reference data used for training, our model achieves good performance in terms of mean AUC scores on all three prediction branches of our model, predicting complete response to radiotherapy (RT) with 0.819 AUC, CMS4 with 0.819 AUC and epithelial tissue at the node level with 0.760 AUC across folds. Further metrics are provided in Table 1. The prediction performance of the model could be improved by utilising a larger training dataset and performing more exhaustive parameter searches, however the current performance of the model is sufficient to demonstrate the impact of this approach. The predicted response to radiotherapy can now be viewed in the context of disease biology as captured by CMS4. For example, the model demonstrates that CMS4 patients are less likely to respond to radiotherapy. In addition, it is now possible to view the spatial distribution of CMS4 active regions in the tissue architecture context as shown in Fig. 2. Additional samples are presented in the Supplementary Materials.

An example of our proposed prediction maps on two slides can be seen in Fig. 2, with further slides in the Supplementary Materials. A pathologist reviewing these maps assesses that the observed patterns fit the known interplay of response to therapy, CMS4 activation, and the spatial localisation of these signals. In the top slide, we observe high CMS4 activation in stromal rich regions, and interestingly also high CMS4 activation in the bottom center, dissociating from the response to RT activation map. This could be explained by the lymphocyte content, supported by the higher epithelial map activations in the same location. Expert pathologists highlight a similar pattern in certain regions of the maps for the bottom slide. Different from the slide above, the CMS4 and response to RT maps have some overlap with moderate activations here, encouraging discovery into tumour-host interactions. Ultimately, a pathologist confirmed that these maps support an interpretable and trustworthy prediction in the context of response to radiotherapy. While we cannot present a more extensive interpretation of these results due to space limitations, these examples already indicate that the proposed approach enables a level of analysis that has not been possible before.

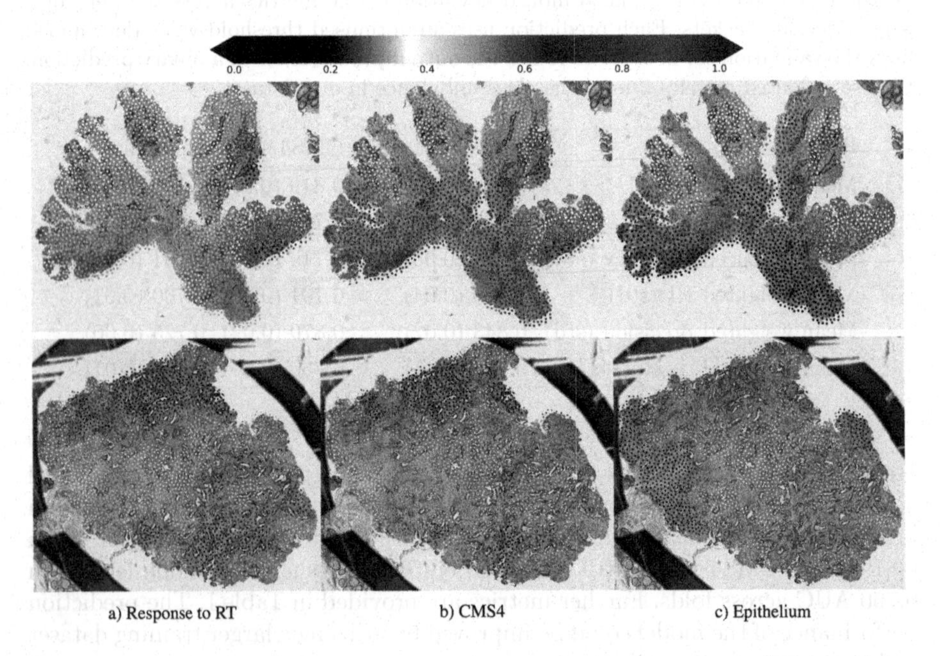

Fig. 2. Node activation maps from the three prediction branches on two different slides, top and bottom. The nodes are coloured by their predictions. Both slides are classified as CMS4 and the patients did not have a complete response to radiotherapy.

Ablation Studies. Using ablation studies, we prove our model and the prediction maps it produces are robust. Changing the dropout, loss weights, loss function, and message passing aggregation methods only changes prediction AUC scores by absolute values up to 0.03. The node activation maps are also very visually similar across ablation study models.

We find that predicting these outcomes individually in a single branch model, particularly with response to radiotherapy, can result in slightly higher AUC scores, but we consciously make this trade-off in order to provide better interpretability of the model predictions. The focus of this research is not to achieve the best possible metrics, but to develop robust methods which can add context and explanation to clinical black box deep learning model predictions, with the view to ease clinical translation of such models.

To explore the effects of the noisy CMS4 ground truth labels, we remove from our dataset any WSIs classified as 'Unmatched' for the CMS call, which for the main results of this paper we defined as 'Not CMS4'. Removing this data and rerunning our analysis improved our predictions for CMS4 by +0.06 AUC, and reduced our response to radiotherapy and epithelial predictions by −0.02 and −0.01 respectively. The results can be found in the Supplementary Materials. These small changes indicate that the noise in our data does not degrade the performance of our classifier, reinforcing it as a robust and accurate model.

4 Conclusion

By setting the prediction of response to therapy in context with disease biology and spatial organisation of the tissue we are providing a novel approach for enhancing the interpretablity of complex prediction tasks. These results do not only enhance the interpretability, they also provide new ways to utilise large retrospective clinical trial cohorts for which no additional molecular data is available. Extending the amount of training data and improving model training will improve model performance, which is already impressive.

We argue that this work also advances the state of the art in feature representation and analysis. Our prediction maps derive from the same graph model, and hence they share underlying graph features. The prediction branches only diverge at the final stage of translating these graph features into outcome predictions for our three clinically relevant outcomes. Importantly, this level of visualisation is not only accessible to pathologists, this joint prediction model also enhances the communication between pathologists and oncologists which is critical for patient management. By cross-referencing these prediction maps with our prior understanding of cancer biology, this approach can help to establish trust in the prediction model and also help to identify potential failure cases.

This work relies on access to well annotated clinical trial samples which will limit our ability to include more data for training and testing. In future, we plan to use these methods to help better characterise tumour-stromal interactions of the tissue. We also plan to use a denser graph with less connectivity to be able to better predict the heterogeneous epithelial tissue.

Data Use Declaration and Acknowledgements. The Grampian, Aristotle and FOCUS datasets used in this study were available through the S:CORT consortium. Grampian and Aristotle will be made publicly available shortly. Additional data is available on request to the S:CORT consortium. The S:CORT consortium was reviewed and approved by the South Cambs Research Ethics committee (REC ref 15/EE/0241; IRAS reference 169363). The Stratification in Colorectal Cancer Consortium (S:CORT) was funded by the Medical Research Council and Cancer Research UK (MR/M016587/1).

The Aristotle trial was funded by Cancer Research UK (CRUK/08/032). The funders played no role in the analyses performed or the results presented. Financial support: RW - EPSRC Center for Doctoral Training in Health Data Science (EP/S02428X/1), Oxford CRUK Cancer Centre; VHK - Promedica Foundation (F-87701-41-01) and Swiss National Science Foundation (P2SKP3_168322/1, P2SKP3_168322/2); TSM - S:CORT (see above); JR, KS - Oxford NIHR National Oxford Biomedical Research Centre and the PathLAKE consortium (InnovateUK). The computational aspects of this research were funded from the NIHR Oxford BRC with additional support from the Wellcome Trust Core Award Grant Number 203141/Z/16/Z. The views expressed are those of the author(s) and not necessarily those of the NHS, the NIHR or the Department of Health.

References

1. Achanta, R., Shaji, A., Smith, K., Lucchi, A., Fua, P., Süsstrunk, S.: Slic superpixels compared to state-of-the-art superpixel methods. IEEE Trans. Pattern Anal. Mach. Intell. **34**(11), 2274–2282 (2012)
2. Alkan, A., Hofving, T., Angenete, E., Yrlid, U.: Biomarkers and cell-based models to predict the outcome of neoadjuvant therapy for rectal cancer patients. Biomark. Res. **9**(1) (2021)
3. Anitei, M.G., et al.: Prognostic and predictive values of the immunoscore in patients with rectal cancer. Clin. Cancer Res. **20**(7), 1891–1899 (2014)
4. Caron, M., et al.: Emerging properties in self-supervised vision transformers. In: Proceedings of the International Conference on Computer Vision (ICCV) (2021)
5. Chatila, W., Kim, J., Walch, H.: Genomic and transcriptomic determinants of response to neoadjuvant therapy in rectal cancer. Nat. Med. **28**, 1646–1655 (2022)
6. Chen, R.J., Krishnan, R.G.: Self-supervised vision transformers learn visual concepts in histopathology (2022)
7. Ding, K., Liu, Q., Lee, E., Zhou, M., Lu, A., Zhang, S.: Feature-enhanced graph networks for genetic mutational prediction using histopathological images in colon cancer. In: Martel, A.L., et al. (eds.) MICCAI 2020. LNCS, vol. 12262, pp. 294–304. Springer, Cham (2020). https://doi.org/10.1007/978-3-030-59713-9_29
8. George, T.J.J., Allegra, C.J., Yothers, G.: Neoadjuvant rectal (NAR) score: a new surrogate endpoint in rectal cancer clinical trials. Curr. Colorect. Cancer Rep. **11**(5), 275–280 (2015)
9. Guinney, J., et al.: The consensus molecular subtypes of colorectal cancer. Nat. Med. **21**, 1350–1356 (2015)
10. Jones, H.J.S., et al.: Stromal composition predicts recurrence of early rectal cancer after local excision. Histopathology **79**, 947–956 (2021)
11. Lee, Y., Park, J., Oh, S.: Derivation of prognostic contextual histopathological features from whole-slide images of tumours via graph deep learning. Nat. Biomed. Eng. (2022)
12. Lipkova, J., et al.: Artificial intelligence for multimodal data integration in oncology. Cancer Cell **40**(10), 1095–1110 (2022)
13. Lu, W., Toss, M., Dawood, M., Rakha, E., Rajpoot, N., Minhas, F.: Slidegraph+: whole slide image level graphs to predict her2 status in breast cancer. Med. Image Anal. **80**, 102486 (2022)

14. Pina, O., Vilaplana, V.: Self-supervised graph representations of WSIS. In: Bekkers, E., Wolterink, J.M., Aviles-Rivero, A. (eds.) Proceedings of the First International Workshop on Geometric Deep Learning in Medical Image Analysis. Proceedings of Machine Learning Research, vol. 194, pp. 107–117. PMLR (2022)
15. Qi, L., et al.: Identification of prognostic spatial organization features in colorectal cancer microenvironment using deep learning on histopathology images. Med. Omics **2**, 100008 (2021)
16. Raju, A., Yao, J., Haq, M.M.H., Jonnagaddala, J., Huang, J.: Graph attention multi-instance learning for accurate colorectal cancer staging. In: Martel, A.L., et al. (eds.) MICCAI 2020. LNCS, vol. 12265, pp. 529–539. Springer, Cham (2020). https://doi.org/10.1007/978-3-030-59722-1_51
17. Ronneberger, O., Fischer, P., Brox, T.: U-net: convolutional networks for biomedical image segmentation (2015)
18. Sirinukunwattana, K., et al.: Image-based consensus molecular subtype (imcms) classification of colorectal cancer using deep learning. Gut **70**(3), 544–554 (2021)
19. Wood, R., et al.: Enhancing local context of histology features in vision transformers. In: Artificial Intelligence over Infrared Images for Medical Applications and Medical Image Assisted Biomarker Discovery. pp. 154–163. Springer, Cham (2022). https://doi.org/10.1007/978-3-031-19660-7_15
20. Xu, J., Luo, X., Wang, G., Gilmore, H., Madabhushi, A.: A deep convolutional neural network for segmenting and classifying epithelial and stromal regions in histopathological images. Neurocomputing **191**, 214–223 (2016)
21. Xu, K., Hu, W., Leskovec, J., Jegelka, S.: How powerful are graph neural networks? (2018)
22. Zhang, F., et al.: Predicting treatment response to neoadjuvant chemoradiotherapy in local advanced rectal cancer by biopsy digital pathology image features. Clin. Transl. Med. **10**(2), e110 (2020)
23. Zhou, Y., Graham, S., Koohbanani, N.A., Shaban, M., Heng, P.A., Rajpoot, N.M.: CGC-net: cell graph convolutional network for grading of colorectal cancer histology images. In: IEEE/CVF International Conference on Computer Vision Workshops, pp. 388–398. IEEE (2019)

Distributionally Robust Image Classifiers for Stroke Diagnosis in Accelerated MRI

Boran Hao[1], Guoyao Shen[2], Ruidi Chen[1], Chad W. Farris[3], Stephan W. Anderson[3], Xin Zhang[2], and Ioannis Ch. Paschalidis[1(✉)]

[1] Department of Electrical and Computer Engineering, Boston University, Boston, MA 02215, USA
yannisp@bu.edu

[2] Department of Mechanical Engineering and the Photonics Center, Boston University, Boston, MA 02215, USA

[3] Boston Medical Center and Boston University Chobanian & Avedisian School of Medicine, Boston, MA 02118, USA

Abstract. *Magnetic Resonance Imaging (MRI)* acceleration techniques using *k-space sub-sampling (KS)* can greatly improve the efficiency of MRI-based stroke diagnosis. Although *Deep Neural Networks (DNN)* have shown great potential on stroke lesion recognition tasks when the MR images are reconstructed from the full k-space, they are vulnerable to the lower quality MR images generated by KS. In this paper, we propose a *Distributionally Robust Learning (DRL)* approach to improve the performance of stroke recognition DNN models when the MR images are reconstructed from the sub-sampled k-space. For *Convolutional Neural Network (CNN)* and *Vision Transformer (ViT)*-based models, our methods improve the stroke classification AUROC and AUPRC by up to 11.91% and 9.32% on the KS-perturbed brain MR images, respectively, compared against *Empirical Risk Minimization (ERM)* and other baseline defensive methods. We further show that DRL models can successfully recognize the stroke cases from highly perturbed MR images where clinicians may fail, which provides a solution for improved diagnosis in an accelerated MRI setting.

Keywords: Stroke diagnosis · MRI acceleration · Distributionally robust optimization · Deep learning

1 Introduction

Magnetic Resonance Imaging (MRI) has been extensively applied to clinical diagnosis [16]. Compared with *Computed Tomography (CT)*, a brain MRI is more

This work was supported by the Rajen Kilachand Fund for Integrated Life Science and Engineering and by the NSF under grant CCF-2200052 and IIS-1914792.
Ruidi Chen contributed to this work while at Boston University, before moving to her current position at Amazon SCOT.

Supplementary Information The online version contains supplementary material available at https://doi.org/10.1007/978-3-031-43904-9_74.

H. Greenspan et al. (Eds.): MICCAI 2023, LNCS 14224, pp. 768–777, 2023.
https://doi.org/10.1007/978-3-031-43904-9_74

sensitive for multiple stroke types [3], therefore considered as the gold standard for stroke diagnosis. Nevertheless, the long acquisition time for a brain MRI (20 to 30 min) imposes challenges, especially in cases of acute stroke where rapid diagnosis is essential and patient movement during this distressing period of time commonly limits evaluation. As a result, MRI acceleration techniques have been developed to achieve more rapid diagnosis, increasing resource availability while reducing costs [14,18]. A *k-space sub-sampling (KS)* approach serves as a simple MRI acceleration solution [20], compared with other hardware-based acceleration methods. However, the signal loss by KS leads to blurry reconstructed MR images that are less than ideal for a reliable clinical diagnosis.

Artificial Intelligence (AI) plays an increasingly important role in MRI-based diagnosis, for both MR image reconstruction and clinical decision making. *Deep Neural Networks (DNN)* were trained to reconstruct the MR images from the sub-sampled k-space [10,13], which provides a better reconstruction than the *Inverse Fast Fourier Transform (IFFT)*. Nevertheless, detailed information in the brain may still be lost in the reconstructed MR images due to the signal sparsity in the k-space. On the other hand, traditional *Convolutional Neural Network (CNN)* [15] and the latest *Vision Transformer (ViT)*-based [6] predictive models have shown impressive prediction accuracy on stroke diagnosis tasks, such as slice classification and lesion segmentation [7,11]. However, these DNNs trained on clean images through *Empirical Risk Minimization (ERM)* are vulnerable to perturbations in the input images [2]. Whatever the reconstruction method used, even the slightest perturbation in accelerated MR images can lead to a wrong stroke prediction from the AI models. Therefore, building robust DNN models to handle the perturbed MR image input is important for MRI acceleration.

In this paper, we introduce a *Distributionally Robust Learning (DRL)*-based approach [4] into the deep MR image classifier training, in order to improve the model robustness to the image perturbation resulting from the signal sparsity in accelerated MRI. Compared with ERM, DRL is an optimization method minimizing the worst-case loss over an ambiguity set, therefore, can tolerate outliers in the data [5]. We implemented DRL to different linear layers in deep CNN/ViT classifiers, and applied a randomized training approach to improve the training efficiency. Our results show that on a real-world dataset, DRL can significantly improve the stroke classification performance of ERM and other baseline defensive training methods, when the signal sparsity and noise in accelerated MRI are generated by the *Cartesian Undersampling (CU)* method [20] and *White Gaussian Noise (WGN)*. We further show that in highly perturbed MR images where the ERM model and even clinicians cannot give a reliable diagnosis, our DRL model can still correctly recognize stroke, which establishes that our method can assist accelerated MRI diagnosis.

2 Methodology

2.1 Distributionally Robust Learning

We will use the DRL framework under a multi-class classification setting developed by [4] in a DNN-based stroke diagnosis application. We provide a brief overview of the DRL model. Assume that there are K classes, and our goal is to classify an example with an input feature $\mathbf{x} \in \mathbb{R}^d$ to one of the K classes with a one-hot class label $\mathbf{y} \in \{0,1\}^K$. Logistic regression solves this problem by minimizing the following expected true risk

$$\inf_{\mathbf{B}} \mathbb{E}^{\mathbb{P}^*}\big[h_{\mathbf{B}}(\mathbf{x},\mathbf{y})\big], \tag{1}$$

where $\mathbf{B} \triangleq [\mathbf{w}_1 \cdots \mathbf{w}_K]$ is the coefficient matrix, $\mathbb{P}^*$ is the true distribution of the data $(\mathbf{x},\mathbf{y})$, $h_{\mathbf{B}}(\mathbf{x},\mathbf{y}) \triangleq \log \mathbf{1}' e^{\mathbf{B}'\mathbf{x}} - \mathbf{y}'\mathbf{B}'\mathbf{x}$ is the loss function to be minimized, and $\mathbb{E}^{\mathbb{P}^*}$ denotes the expectation under the distribution $\mathbb{P}^*$. Since $\mathbb{P}^*$ is usually unknown, Problem (1) cannot be solved directly. The ERM approach tackles this by replacing the expected true risk by a sample averaged risk. Given N realizations of $(\mathbf{x},\mathbf{y})$, ERM minimizes the following empirical risk

$$\inf_{\mathbf{B}} \tfrac{1}{N}\textstyle\sum_{i=1}^{N} h_{\mathbf{B}}(\mathbf{x}_i,\mathbf{y}_i). \tag{2}$$

ERM can produce unreliable solutions when the samples are contaminated by noise or drawn from an outlying distribution. To obtain robust estimators that can hedge against noise in the training data and generalize well out-of-sample, [4] proposed the DRL framework under the Wasserstein metric. Specifically, it minimizes the worst-case expected loss over a set of probability distributions

$$\inf_{\mathbf{B}} \sup_{\mathbb{Q}\in\Omega} \mathbb{E}^{\mathbb{Q}}\big[h_{\mathbf{B}}(\mathbf{x},\mathbf{y})\big], \tag{3}$$

where Ω contains a set of probability distributions that are close to the empirical distribution $\hat{\mathbb{P}}_N$ measured by the Wasserstein metric, $\Omega \triangleq \{\mathbb{Q} \in \mathcal{P}(\mathcal{Z}) : W_1(\mathbb{Q},\ \hat{\mathbb{P}}_N) \leq \epsilon\}$, where $\mathcal{Z}$ is the set of possible values for $(\mathbf{x},\mathbf{y})$, $\mathcal{P}(\mathcal{Z})$ is the space of all probability distributions supported on $\mathcal{Z}$, ϵ is a pre-specified radius of the ambiguity set Ω, $\hat{\mathbb{P}}_N$ is the empirical distribution that assigns an equal probability $1/N$ to each observed sample $(\mathbf{x}_i,\mathbf{y}_i)$, and $W_1(\mathbb{Q},\ \hat{\mathbb{P}}_N)$ is the order-1 Wasserstein distance between $\mathbb{Q}$ and $\hat{\mathbb{P}}_N$ defined as

$$W_1(\mathbb{Q},\ \hat{\mathbb{P}}_N) \triangleq \min_{\Pi\in\mathcal{P}(\mathcal{Z}\times\mathcal{Z})}\Big\{\textstyle\int_{\mathcal{Z}\times\mathcal{Z}} l(\mathbf{z}_1,\mathbf{z}_2)\ \Pi(\mathrm{d}\mathbf{z}_1,\mathrm{d}\mathbf{z}_2)\Big\},$$

where Π is the joint distribution of $\mathbf{z}_1 \triangleq (\mathbf{x}_1,\mathbf{y}_1)$ and $\mathbf{z}_2 \triangleq (\mathbf{x}_2,\mathbf{y}_2)$ with marginals $\mathbb{Q}$ and $\hat{\mathbb{P}}_N$, respectively, and l is a distance metric on the data space.

An equivalent reformulation (4) of (3) was developed by [4] when $l(\mathbf{z}_1,\mathbf{z}_2) \triangleq \|\mathbf{W}^{1/2}(\mathbf{x}_1-\mathbf{x}_2)\|_2 + M\|\mathbf{y}_1-\mathbf{y}_2\|_2$, where $\mathbf{W}$ is a positive semidefinite weight matrix to account for any transformation on the input feature $\mathbf{x}$ and can be estimated from data using metric learning (see Sec. 2.2) and with $M \to \infty$:

$$\inf_{\mathbf{B}} \tfrac{1}{N}\textstyle\sum_{i=1}^{N} h_{\mathbf{B}}(\mathbf{x}_i,\mathbf{y}_i) + \epsilon 2^{1/2}\|\mathbf{W}^{-1/2}\mathbf{B}\|_2, \tag{4}$$

where $\|\mathbf{W}^{-1/2}\mathbf{B}\|_2$ is the induced ℓ_2-norm of $\mathbf{W}^{-1/2}\mathbf{B}$.

2.2 DRL for Deep Stroke Diagnosis Networks

We apply DRL to ViT and CNN-based MR image stroke classification models, in order to enhance their robustness against image perturbations in accelerated MRI. We apply the DRL reformulation (4) to the last layer B, and certain intermediate linear layers in a deep MR image classifier. For a ViT model (cf. Fig. 1), we apply DRL to the patch projection layer P and the final linear classification layer B, in order to predict if an MR image slice is normal (label 0) or depicts a stroke lesion (label 1). To speed up the training process, during each epoch, we randomly pick one layer L to train while keeping all other layers frozen. A validation set $\mathcal{V}$ is used to tune the hyperparameters (e.g., regularization coefficients). To account for the non-linear transformation of the raw image resulted from all layers before L, we solve the following *Linear Matrix Inequality (LMI)* problem [4] to estimate a weight matrix $\mathbf{W}$:

$$\begin{aligned} \min_{\mathbf{W}\succcurlyeq 0} \quad & \sum_{\mathbf{x}_i\in\mathcal{D}} \sum_{t=1}^{T} \|\mathbf{W}^{1/2}(\phi_L^{(t)}(\tilde{\mathbf{x}}_i) - \phi_L^{(t)}(\mathbf{x}_i))\|_2^2 \\ \text{s.t.} \quad & \frac{1}{|\mathcal{S}|} \sum_{(i,j)\in\mathcal{S}} \sum_{t=1}^{T} \|\mathbf{W}^{1/2}(\phi_L^{(t)}(\tilde{\mathbf{x}}_i) - \phi_L^{(t)}(\tilde{\mathbf{x}}_j))\|_2^2 \geq c, \\ & \frac{1}{|\mathcal{S}|} \sum_{(i,j)\in\mathcal{S}} \sum_{t=1}^{T} \|\mathbf{W}^{1/2}(\phi_L^{(t)}(\mathbf{x}_i) - \phi_L^{(t)}(\mathbf{x}_j))\|_2^2 \geq c, \end{aligned} \tag{5}$$

where $\tilde{\mathbf{x}}_i$ is the perturbed version of an MR image slice $\mathbf{x}_i$, $\mathcal{D}$ the training set, $\mathcal{S} \triangleq \{(i,j)|\mathbf{x}_i, \mathbf{x}_j \in \mathcal{D}, \mathbf{y}_i \neq \mathbf{y}_j\}$, $|\mathcal{S}|$ denotes the cardinality of the set $\mathcal{S}$, ϕ_L is the input to L and $\phi_L^{(t)}$ is the t-th hidden state (i.e., the vector representation for each instance in the sequence, output by and fed into different layers in ViT) in the sequence ϕ_L of length T, and c is a fixed parameter. $T = 1$ if L refers to the B layer. The intuition of (5) is that in the transformed feature space, distance between the clean and perturbed version of a slice will be minimized, while slices from different classes (normal and stroke) are sufficiently far away. For a CNN model, we only applied DRL to the final linear layer B.

We chose two approaches to generate the perturbation in accelerated MRI. *Cartesian Undersampling (CU)* perturbation [20] keeps only the central and a few randomly-sampled parts of the k-space; the corresponding reconstructed MR image only keeps the main structural information in the brain and introduces misalignments. A smaller central fraction f used in k-space indicates a larger perturbation. Noise might be introduced during the signal transmission, so we add *White Gaussian Noise (WGN)* as another type of perturbation, where the standard deviation σ is regarded as the perturbation intensity. To show the strength of DRL, in addition to ERM, we also apply *Brute-force Adversarial Training (BAT)* [2] and *Projected Gradient Descent (PGD)* [12] as baseline methods. Among all of the current defensive training methods which improve the model robustness against perturbations, BAT and PGD are representative. BAT adds noisy samples into the training set, therefore is simple and widely used. PGD is known to be robust to a wide range of image perturbations, and is considered a state-of-the-art method.

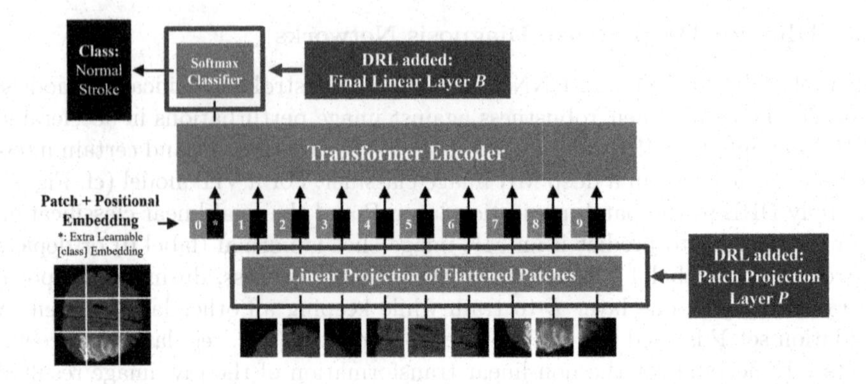

Fig. 1. Adding DRL into different layers in ViT.

3 Experiments

3.1 Experimental Materials and Settings

Our dataset included MRI brain scans from 226 patients performed at an urban tertiary referral academic medical center that is a comprehensive stroke center. Clinical scans of adult patients aged 18–89 years with recent (acute or subacute) strokes were identified between 1/1/2013 and 1/1/2021 for inclusion in this study via a search of the Philips Performance Bridge. Scans meeting this criteria were downloaded and simultaneously anonymized to preserve patient anonymity and prevent disclosure of protected health information as part of this IRB exempt study. No patient demographic information was retained for the scans, as it was considered to represent an unnecessary risk for accidental release of protected health information. The diffusion weighted images with a gradient of B=1000 were utilized for the analysis (see the Supplement[1] for information about the MRI scanner and parameter settings). Each MR image contains multiple slices, and every slice was annotated as normal or stroke by a board-certified neuroradiologist with a subspecialty certification. Annotation of the strokes was performed on the diffusion weighted images using ITK-SNAP (ver. 3.80) [19], and all included MRI examinations were reviewed by the neuroradiologist during the annotation process to ensure that the images were of diagnostic quality without significant motion degradation or other artifacts. To avoid the dependency among the slices from the same subject, we applied a 2-d acquisition during the MR imaging, and implemented a slice-level MR image preprocessing. While the whole dataset includes 4,883 (74.7%) normal slices and 1,650 (25.3%) stroke slices, we further randomly split them into training/validation/test sets using the ratio 80%/10%/10%. For the training set, we implemented data augmentation strategies by rotating or flipping each slice. Finally, the training/validation/test set contains 31,356/653/654 slices, correspondingly.

[1] Supplement and source code are available at https://github.com/noc-lab/drl_mri.

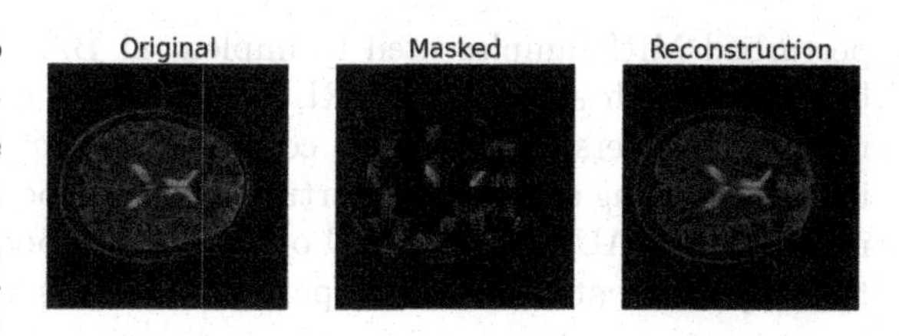

Fig. 2. MAE pre-trained models reconstruct the masked MR images. This example uses a T2-weighted image.

We implemented DRL to both CNN and ViT models. For the CNN model, we used a ResNet-18 [9] architecture, while for the ViT model, we first pre-trained a 4-layer ViT using a self-supervised pre-training method called Masked Autoencoder (MAE) [8], using the T1/T2-weighted brain MR images in the IXI dataset [1]. MAE pre-training first randomly masks 75% of the image patches in an MRI slice input, and then uses a ViT encoder-decoder architecture to reconstruct the masked MRI patches, in order to learn the dependency among different locations in the brain. After 400 pre-training epochs, an overall satisfying reconstruction result can be observed in Fig. 2.

To evaluate the binary classification performance of different models, we use the *Area Under the Receiver Operating Characteristic (AUROC)* curve as our main metric. As our dataset is unbalanced, we also considered the *Area Under Precision-Recall Curve (AUPRC)*. We ran the experiments 3 times using different random seeds. The training of our DNNs were implemented on 3 NVIDIA RTX A6000 (48GB VRAM) GPUs, and each DRL training epoch can be completed within 0.03 GPU hours. We used a learning rate of 1×10^{-5} and batchsize of 128 for DRL training, while no weight decay was applied. To solve the LMI problem in (5), we used SDPT3 v4.0 [17] as the solver. We set the CU perturbation with the acceleration factor of 4, 6, 8, 12 with the central fraction of 8%, 6%, 4% and 2% in k-space respectively, and the remaining parts were chosen randomly in the peripheral region accordingly.

3.2 Results

We show the stroke classification AUROC in Fig. 3. When the k-space subsampling fraction decreased and the signal became sparser, the performance of both ViT and CNN models trained under ERM dropped significantly, from around 95% to below 80%. DRL significantly improved the AUROC of the ERM-based ViT model from 74.5% to 83.1% when the MR images were under extreme CU perturbation, while only slightly influenced model performance on the clean test set. For WGN, the largest improvement brought to ERM-based ViT model was 16.9%. Although we only applied DRL to the last layer of the CNN model, the improvement against ERM was still remarkable, up to 11.9%/4.9% for CU/WGN. With BAT and PGD adversarial training, the corresponding ViT or CNN models were also improved, though when DRL was combined with BAT and PGD, the model robustness can be further enhanced. Table 1 shows the maximum AUROC and AUPRC improvement that DRL can bring to different baseline methods. For ViT and CNN models, the AUROC improvement w.r.t BAT/PGD defensive methods is up to 23.9%/12.2%, respectively. Note that the perturbed MRI samples used to implement BAT were the same as those used by DRL, which shows that DRL is a more effective way to exploit the information in adversarial samples, compared to simply adding the blurry images into the training set. For CU perturbation, our best combined model using DRL improved the AUROC/AUPRC of the ERM model by up to 15%/12.5%, while this improvement under WGN perturbation was up to 18.8%/36.2%.

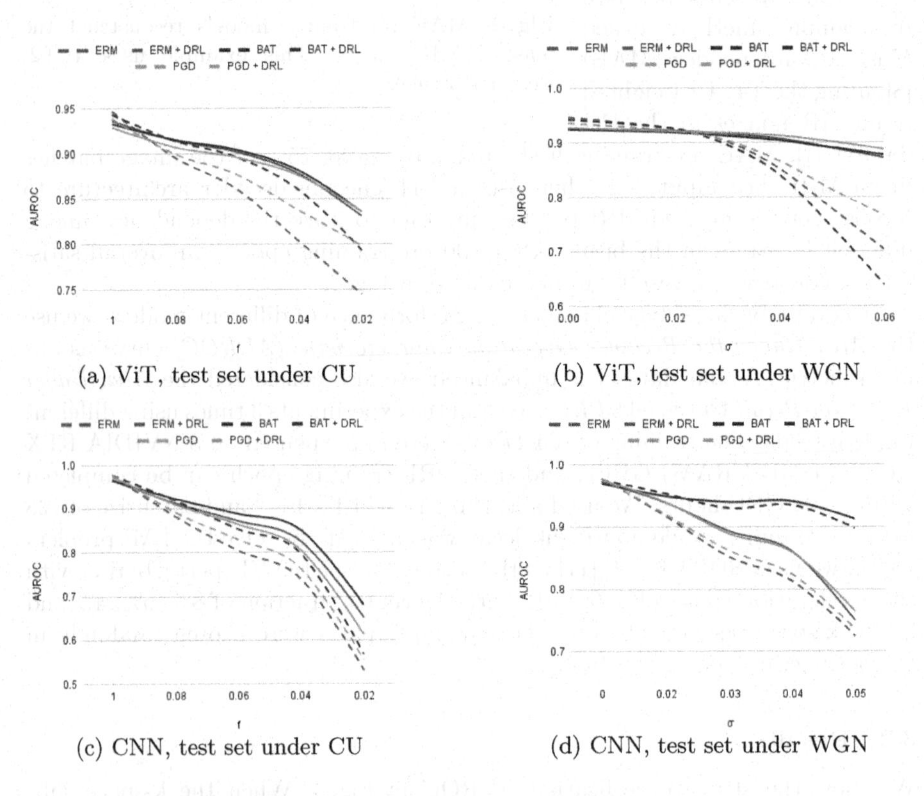

(a) ViT, test set under CU (b) ViT, test set under WGN

(c) CNN, test set under CU (d) CNN, test set under WGN

Fig. 3. AUROC of different methods using ViT and CNN models.

Under CU perturbation, we further show that our DRL model can recognize stroke while clinicians may fail to. In Fig. 4, the stroke MRI slices from the test sets are under different levels of CU perturbations. For both ERM- and DRL-based ViT models, we maximized the F1 score of the stroke class on the training set to calculate the optimal decision threshold for stroke prediction, in order to balance the precision and recall. When the k-space signal becomes more sparse, the reconstructed MRI slices get more blurry and the lesion areas become less recognizable, even for human eyes. As a result, the ERM model fails to detect stroke under high perturbation levels. Nevertheless, the DRL model can tolerate more intense CU perturbation and recognize stroke slices that may even be misclassified by clinicians, which reveals its value in improving the diagnosis in

Table 1. AUROC and AUPRC improvement of DRL over three baseline methods using ViT/CNN models under CU and WGN perturbations on the test set.

		ViT				CNN			
		CU		WGN		CU		WGN	
		Mean	Std.	Mean	Std.	Mean	Std.	Mean	Std.
ERM+DRL	AUROC	8.65%	0.00%	16.87%	0.04%	11.91%	0.18%	4.92%	0.24%
	AUPRC	8.31%	0.02%	29.77%	0.10%	8.65%	0.15%	6.81%	0.79%
BAT+DRL	AUROC	4.29%	0.00%	23.90%	0.05%	11.26%	0.07%	1.42%	0.04%
	AUPRC	4.37%	0.01%	39.75%	0.12%	9.32%	0.11%	3.59%	0.22%
PGD+DRL	AUROC	3.83%	0.01%	12.19%	0.05%	6.00%	0.14%	3.45%	0.28%
	AUPRC	2.88%	0.02%	18.68%	0.06%	3.85%	0.21%	4.13%	0.77%

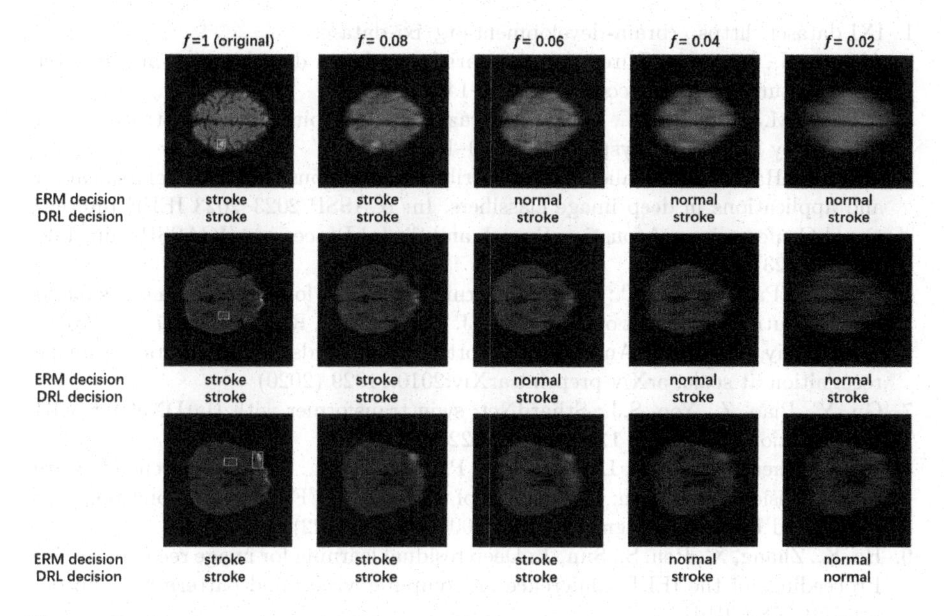

Fig. 4. Stroke slices where the ERM model and clinicians may fail when the CU perturbation is large, while the DRL model succeeds. The red boxes indicate the stroke lesion areas in the clean images. (Color figure online)

an accelerated MRI mode. We verified the effectiveness of our approach on the actual clinical scans acquired for clinical care and not just for research purposes, suggesting that the methods and findings in the current study should be generalizable to routine clinical practice conditions and potentially other types of clinical image-based diagnosis (e.g., brain tumor) as well. In addition, our DRL framework does not necessarily need to be used in isolation, rather it can also be combined with other performance boosting methods in accelerated MRI to further improve them, just like for BAT and PGD.

4 Conclusions

In this study, we implemented a DRL-based robust learning approach to improve the robustness of deep image classifiers, in order to achieve more accurate stroke classification from brain MR images reconstructed from a sub-sampled k-space. Our work can potentially be applied to accelerate and improve time-critical stroke diagnosis. Future work can apply DRL to more MRI diagnosis tasks (e.g., lesion area segmentation), justifying its effectiveness on more types of subsampling methods in MRI acceleration.

References

1. IXI dataset. https://brain-development.org/ixi-dataset/
2. Akhtar, N., Mian, A.: Threat of adversarial attacks on deep learning in computer vision: a survey. IEEE Access **6**, 14410–14430 (2018)
3. Alberts, M.J., Horner, J., Gray, L., Brazer, S.R.: Aspiration after stroke: lesion analysis by brain mri. Dysphagia **7**, 170–173 (1992)
4. Chen, R., Hao, B., Paschalidis, I.C.: Distributionally robust multiclass classification and applications in deep image classifiers. In: ICASSP 2023–2023 IEEE International Conference on Acoustics, Speech and Signal Processing (ICASSP). pp. 1–2. IEEE (2023)
5. Chen, R., Paschalidis, I.C.: A robust learning approach for regression models based on distributionally robust optimization. J. Mach. Learn. Res. **19**(1), 517–564 (2018)
6. Dosovitskiy, A., et al.: An image is worth 16x16 words: transformers for image recognition at scale. arXiv preprint arXiv:2010.11929 (2020)
7. Gu, Y., Piao, Z., Yoo, S.J.: SthardNet: swin transformer with HarDNet for MRI segmentation. Appl. Sci. **12**(1), 468 (2022)
8. He, K., Chen, X., Xie, S., Li, Y., Dollár, P., Girshick, R.: Masked autoencoders are scalable vision learners. In: Proceedings of the IEEE/CVF Conference on Computer Vision and Pattern Recognition. pp. 16000–16009 (2022)
9. He, K., Zhang, X., Ren, S., Sun, J.: Deep residual learning for image recognition. In: Proceedings of the IEEE conference on computer vision and pattern recognition. pp. 770–778 (2016)
10. Hyun, C.M., Kim, H.P., Lee, S.M., Lee, S., Seo, J.K.: Deep learning for undersampled MRI reconstruction. Phy. Med. Bio. **63**(13), 135007 (2018)
11. Liu, L., Chen, S., Zhang, F., Wu, F.X., Pan, Y., Wang, J.: Deep convolutional neural network for automatically segmenting acute ischemic stroke lesion in multimodality MRI. Neural Comput. Appl. **32**, 6545–6558 (2020)
12. Madry, A., Makelov, A., Schmidt, L., Tsipras, D., Vladu, A.: Towards deep learning models resistant to adversarial attacks. arXiv preprint arXiv:1706.06083 (2017)
13. Montalt-Tordera, J., Muthurangu, V., Hauptmann, A., Steeden, J.A.: Machine learning in magnetic resonance imaging: image reconstruction. Physica Medica **83**, 79–87 (2021)
14. Plein, S., Ryf, S., Schwitter, J., Radjenovic, A., Boesiger, P., Kozerke, S.: Dynamic contrast-enhanced myocardial perfusion MRI accelerated with k-t sense. Magnetic Resonance in Medicine. Official J. Int. Soc. Magn. Reson. Med. **58**(4), 777–785 (2007)

15. Ronneberger, O., Fischer, P., Brox, T.: U-Net: Convolutional Networks for Biomedical Image Segmentation. In: Navab, N., Hornegger, J., Wells, W.M., Frangi, A.F. (eds.) Medical Image Computing and Computer-Assisted Intervention – MICCAI 2015: 18th International Conference, Munich, Germany, October 5-9, 2015, Proceedings, Part III, pp. 234–241. Springer, Cham (2015). https://doi.org/10.1007/978-3-319-24574-4_28
16. Stahl, R., et al.: Assessment of cartilage-dedicated sequences at ultra-high-field MRI: comparison of imaging performance and diagnostic confidence between 3.0 and 7.0 t with respect to osteoarthritis-induced changes at the knee joint. Skeletal Radiol. **38**, 771–783 (2009)
17. Toh, K.C., Todd, M.J., Tütüncü, R.H.: SDPT3 - a MATLAB software package for semidefinite programming, version 1.3. Optim. Methods Softw. **11**(1–4), 545–581 (1999)
18. Wang, H., Peng, H., Chang, Y., Liang, D.: A survey of GPU-based acceleration techniques in MRI reconstructions. Quant. Imaging Med. Surg. **8**(2), 196 (2018)
19. Yushkevich, P.A., et al.: User-guided 3d active contour segmentation of anatomical structures: significantly improved efficiency and reliability. Neuroimage **31**(3), 1116–1128 (2006)
20. Zbontar, J., et al.: FastMRI: an open dataset and benchmarks for accelerated MRI. arXiv preprint arXiv:1811.08839 (2018)

M&M: Tackling False Positives in Mammography with a Multi-view and Multi-instance Learning Sparse Detector

Yen Nhi Truong Vu(✉), Dan Guo, Ahmed Taha, Jason Su, and Thomas Paul Matthews

WhiteRabbit.AI, Santa Clara, USA
nhi@whiterabbit.ai

Abstract. Deep-learning-based object detection methods show promise for improving screening mammography, but high rates of false positives can hinder their effectiveness in clinical practice. To reduce false positives, we identify three challenges: (1) unlike natural images, a malignant mammogram typically contains only one malignant finding; (2) mammography exams contain two views of each breast, and both views ought to be considered to make a correct assessment; (3) most mammograms are negative and do not contain any findings. In this work, we tackle the three aforementioned challenges by: (1) leveraging Sparse R-CNN and showing that sparse detectors are more appropriate than dense detectors for mammography; (2) including a multi-view cross-attention module to synthesize information from different views; (3) incorporating multi-instance learning (MIL) to train with unannotated images and perform breast-level classification. The resulting model, M&M, is a **M**ulti-view and **M**ulti-instance learning system that can both localize malignant findings and provide breast-level predictions. We validate M&M's detection and classification performance using five mammography datasets. In addition, we demonstrate the effectiveness of each proposed component through comprehensive ablation studies.

Keywords: Mammography · Detection · Classification · False positive

1 Introduction

Screening mammography helps detect breast cancer earlier and has reduced the breast cancer mortality rate significantly [4]. Computer-aided diagnosis (CAD) software was developed to aid radiologists, but its effectiveness has been questioned following recent large-scale clinical studies [6]. In particular, the high

Y.N.Truong Vu and D. Guo—Equal Contribution.

Supplementary Information The online version contains supplementary material available at https://doi.org/10.1007/978-3-031-43904-9_75.

H. Greenspan et al. (Eds.): MICCAI 2023, LNCS 14224, pp. 778–788, 2023.
https://doi.org/10.1007/978-3-031-43904-9_75

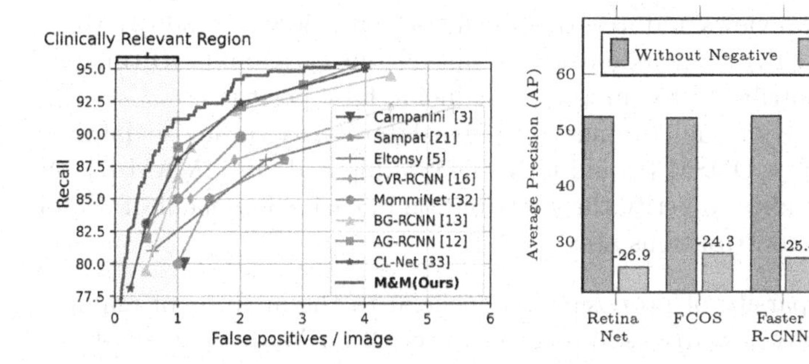

(a) Free response operating characteristic (FROC) curves on DDSM.

(b) Quantitative detection evaluation with and without negative images on OPTIMAM.

Fig. 1. Two gaps between deep learning literature and clinical applicability. **(a)** Few works report detailed performance in the clinically relevant region of less than 1 FP/image. M&M surpasses previous works by a large margin in this region. **(b)** Typical evaluation datasets are not representative: they contain from zero (CBIS-DDSM [9]) to few negative cases (DDSM [8], INBreast [17]). To illustrate the distribution shift, we train four popular dense detectors using a standard setup that includes only annotated malignant and benign cases [1,13,16]. We utilize OPTIMAM [7], a large dataset with a significant proportion of negatives (Table 1), for training and evaluation. Across all dense models, there is a large performance drop in the clinically representative setting that includes negative images. This means that the dense models are producing too many FPs on negative images. Our model, M&M, successfully tackles this performance gap.

rate of false positive (FP) predictions of CAD can cause a significant reduction in radiologists' specificity [6]. Surprisingly, recent deep learning literature [3,5,13,16,20,21,32] focuses on improving recall without considering the need to operate at low FP rates. As shown in Fig. 1a, most works focus on reporting recalls outside the clinically relevant region of less than 1 FP/image.

To tackle the high rate of false positives in mammography, we identify three challenges: (1) A malignant mammogram typically contains only one malignant finding. This is different from natural images: for example, an image in COCO contains on average 7.7 objects [11]. This calls into question the usage of dense detectors for mammography; (2) A standard screening exam consists of two views per breast. Both views are essential in making a clinical decision because a finding may appear suspicious in one view but not the other; (3) Most mammograms are negative: they do not contain any findings. However, excluding negative images from training and evaluation leads to a distribution shift since negative images are abundant in clinical practice. Concretely, the false positive rate is low for a typical evaluation data distribution but much higher for a clinically-representative data distribution, as shown in Fig. 1b.

In this work, we tackle these challenges and propose a **M**ulti-view and **M**ulti-instance learning system, **M&M**. M&M is an end-to-end system that detects malignant findings and provides breast-level classification. To achieve these goals, M&M leverages three components: (1) Sparse R-CNN to replace dense anchors with a set of sparse proposals; (2) Multi-view cross-attention to synthesize information from two views and iteratively refine the predictions, and (3) Multi-instance learning (MIL) to include negative images during training. Ultimately, each component contributes to our goal of reducing false positives.

We validate M&M through evaluation on five datasets: two in-house datasets, two public datasets — DDSM [8] and CBIS-DDSM [9], and OPTIMAM [7]. We perform ablation studies to verify the contribution of each component of M&M. To summarize, our contributions are:

1. We show that sparsity of proposals is beneficial to the analysis of mammograms, which have low disease prevalence (Sec. 2.1). With Sparse R-CNN, M&M generalizes better to clinically-representative data, where the majority of images are negative, *i.e.*, have no findings (Table 2);
2. We incorporate a simple and efficient cross-view multi-head attention module for mammography analysis (Sec. 2.2). With multi-view reasoning, M&M improves the recall at 0.1 FP/image by 8.6%, as shown in Fig. 4;
3. We leverage MIL to include images without bounding boxes during training (Sec. 2.3). Accordingly, M&M sees seven times more images during training. With MIL, M&M improves the recall at 0.1 FP/image by 12.6% (Fig. 4). Furthermore, M&M can provide breast-level classification predictions, achieving AUCs of more than 0.88 on four different datasets (Table 3).

2 M&M: A Multi-view and MIL System

2.1 Sparse R-CNN with Dual Classification Heads

The sparsity of malignant findings calls into question the use of dense detectors. As shown in Fig. 1b, dense detectors generalize poorly to negative images as they produce too many false positives. Thus, we propose to use Sparse R-CNN [24].

Sparse R-CNN utilizes a sparse set of N learnable proposals consisting of $\mathbf{b_0} \in \mathbb{R}^{N\times 4}$ coordinates and $\mathbf{h_0} \in \mathbb{R}^{N\times D}$ features. The architecture uses 6 cascading heads to iteratively refine the proposals. Within the ith head, the proposals $\mathbf{h}_{i-1}$ first interact with themselves via self-attention, and then generate DynamicConv (Fig. 4, [24]) to interact with RoI features cropped by $\mathbf{b}_{i-1}$. The resulting outputs $\mathbf{h}_i \in \mathbb{R}^{N\times D}$ are features for the $(i+1)^{\text{th}}$ head. In addition, a regression module is applied to $\mathbf{h}_i$ to generate boxes $\mathbf{b}_i \in \mathbb{R}^{N\times 4}$, and a classification module generates scores $\mathbf{p}_i \in \mathbb{R}^{N\times C}$, with C being the number of classes.

We modify Sparse R-CNN to include dual classification modules (Fig. 2). First, an objectness module produces objectness logits $\mathbf{o}_i \in \mathbb{R}^N$ to distinguish all findings — malignant and benign — from the background. By utilizing all findings, the objectness head increases the training sample size [1,13,16], but also increases FPs because it flags benign findings. To mitigate this side effect, we include a dedicated malignancy module $[\mathbf{W}_i, \mathbf{b}_i]$ to generate malignancy logits $\mathbf{m}_i \in \mathbb{R}^N$ that is trained to distinguish malignant from benign findings:

$$\mathbf{m}_i = \mathbf{o}_i - \text{SoftPlus}(\mathbf{W}_i\mathbf{h}_i + \mathbf{b}_i). \tag{1}$$

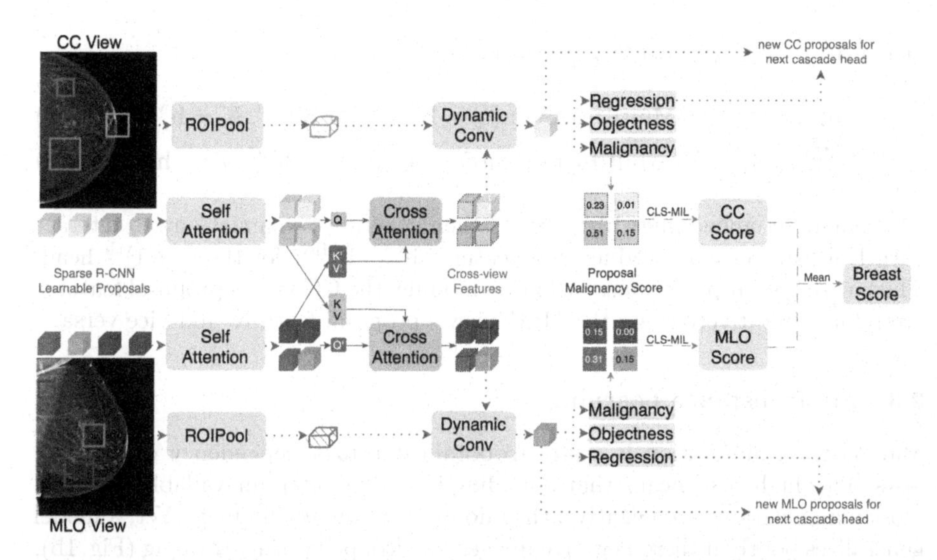

Fig. 2. M&M tackles false positives through (1, blue, dotted arrows) leveraging the Sparse R-CNN cascade architecture to iteratively refine sparse learnable proposals into predictions, (2, red, solid arrows) incorporating a cross-attention module to reason about relations between objects across two views, and (3, green, dashed arrows) utilizing image and breast MIL pooling to train with images that do not have lesion annotations. (Color figure online)

The strictly positive function $\text{SoftPlus}(x) = \log(1 + e^x)$ is chosen to enforce consistency: a high objectness logit $\mathbf{o}_i$ is required to generate a high malignancy logit $\mathbf{m}_i$. Thus, at the finding level, we obtain the following loss

$$\mathcal{L}_{\text{lesion}} = \mathcal{L}_{\text{malignant}} + \mathcal{L}_{\text{objectness}} + 2\mathcal{L}_{\text{giou}} + 5\mathcal{L}_{\text{L1}}, \tag{2}$$

where $\mathcal{L}_{\text{giou}}$ and $\mathcal{L}_{\text{L1}}$ are regression losses as in Sparse R-CNN. $\mathcal{L}_{\text{objectness}}$ and $\mathcal{L}_{\text{malignancy}}$ are focal losses applied to the predicted objectness $\mathbf{o}_i$ and the predicted malignancy $\mathbf{m}_i$ across all cascading heads $1 \leq i \leq 6$, respectively.

2.2 Multi-view Reasoning

A standard screening exam includes two standard views of each breast. The craniocaudal (CC) view is taken from the top down, while the mediolateral oblique (MLO) view is captured from the side at an oblique angle. Radiologists examine both views when making a clinical decision as a finding may look innocuous in one view but suspicious in the other.

To enable multi-view reasoning, M&M incorporates a cross-attention module [28] into every cascading head. Recall that within the ith cascading head, self-attention is first applied to proposal features $\mathbf{h}_{i-1}$ to reason about the relations between objects. After this self-attention module, we introduce a cross-attention module (Fig. 2, Appendix Algo. 1) to reason about the relations between CC view feature $\mathbf{h}_{i-1}^{\mathrm{CC}}$ and MLO view feature $\mathbf{h}_{i-1}^{\mathrm{MLO}}$:

$$\tilde{\mathbf{h}}_{i-1}^{\mathrm{CC}} = \mathbf{h}_{i-1}^{\mathrm{CC}} + \mathrm{MultiHeadAttn}(Q = \mathbf{h}_{i-1}^{\mathrm{CC}}, V = \mathbf{h}_{i-1}^{\mathrm{MLO}}, K = \mathbf{h}_{i-1}^{\mathrm{MLO}}), \quad (3)$$

$$\tilde{\mathbf{h}}_{i-1}^{\mathrm{MLO}} = \mathbf{h}_{i-1}^{\mathrm{MLO}} + \mathrm{MultiHeadAttn}(Q = \mathbf{h}_{i-1}^{\mathrm{MLO}}, V = \mathbf{h}_{i-1}^{\mathrm{CC}}, K = \mathbf{h}_{i-1}^{\mathrm{CC}}). \quad (4)$$

The enhanced embeddings $\tilde{\mathbf{h}}_{i-1}^{\mathrm{CC}}$, $\tilde{\mathbf{h}}_{i-1}^{\mathrm{MLO}}$ then generate DynamicConv to interact with RoI features and produce new features $\mathbf{h}_{i}^{\mathrm{CC}}$, $\mathbf{h}_{i}^{\mathrm{MLO}}$ for the $(i+1)^{\mathrm{th}}$ head. Thus, with the proposed cross-attention module, the CC view's proposal features are refined iteratively using the MLO view's proposal features and vice versa.

2.3 Multi-instance Learning

Mammogram annotation is costly to obtain due to a dependency on radiologists. This high cost means that bounding boxes are often unavailable. Further, most mammograms are negative: they do not contain any findings. Yet, a model generalizes poorly if these negative images are dropped during training (Fig. 1b).

Since image- and breast-level labels are available, we adopt an MIL module to include images without bounding boxes during training. To compute image- and breast-level scores, we leverage the proposal malignancy logits $\mathbf{m}_i$ (Eq. (1)). Since an image is malignant if it contains a malignant lesion, we obtain image-level scores by applying the NoisyOR function $f(\mathbf{x}) = 1 - \prod_{k=1}^{N}(1 - \mathbf{x}[k])$ to the malignancy probabilities $\mathbf{p}_i = \mathrm{Sigmoid}(\mathbf{m}_i) \in \mathbb{R}^N$. Next, as CC and MLO views offer complimentary information on a breast, we obtain breast-level malignancy score by averaging the image-level scores across these views.

We apply cross-entropy losses $\mathcal{L}_{\mathrm{image}}$ and $\mathcal{L}_{\mathrm{breast}}$ at the image and breast level for all training samples. The lesion loss $\mathcal{L}_{\mathrm{lesion}}$ (Eq. (2)) is only applied for annotated lesions. We thus obtain the following total training loss for M&M:

$$\mathcal{L} = \mathbb{1}_{\text{annotated lesion}} \mathcal{L}_{\mathrm{lesion}} + 0.5\mathcal{L}_{\mathrm{image}} + 0.5\mathcal{L}_{\mathrm{breast}}. \quad (5)$$

3 Experiments

Implementation Details. We use PyTorch 1.10. The training settings follow Sparse R-CNN [24]. We apply random horizontal flipping and random rotation. We resize the images' shorter edges to 2560 with the larger edges no longer than 3328. We utilize a COCO-pretrained PVT-B2-Li backbone [30]. We use AdamW optimizer with 5×10^{-5} learning rate and 0.0001 weight decay. The model is trained for 9000 iterations, and the learning rate is scaled by 0.1 at the 6750 and 8250 iterations. Each batch contains 16 breasts (32 images). We employ a 1:1 sampling ratio between unannotated and annotated images.

Datasets. We utilize three 2D digital mammography datasets: (1) *OPTIMAM*: a development dataset derived from the OPTIMAM database [7], which is funded by Cancer Research UK. We split the data into train/val/test with an 80:10:10 ratio at the patient level; (2) *Inhouse-A*: an evaluation dataset collected from

Table 1. Dataset statistics. We report the number of breasts in each dataset, broken down by 3 categories: malignant, benign, and negative. Malignant breasts contain findings with positive biopsy outcomes. Benign breasts contain findings that are determined to be non-malignant after additional follow-up. Negative breasts do not contain any radiologist-marked findings. In the parentheses, we report the number of breasts with bounding box annotations. "Bbox" indicates whether bounding box annotations are available.

Datasets	Bbox	Malignant	(Ann.)	Benign	(Ann.)	Negative	(Ann.)
OPTIMAM	✓	4,838	(4,245)	1,999	(567)	26,003	(2)
Inhouse-A		496	(0)	2,128	(0)	2074	(0)
Inhouse-B		243	(0)	7,797	(0)	47,929	(0)
DDSM	✓	624	(624)	555	(555)	2,877	(1)
CBIS-DDSM	✓	312	(310)	347	(336)	0	(0)

Table 2. Quantitative detection evaluation on OPTIMAM. Δ denotes the AP gap between evaluating with and without negative images.

Model	AP_{mb}	AP	Δ	R@0.1	R@0.25	R@0.5
RetinaNet [10]	52.4	25.5	-26.9	53.3	73.1	83.0
FCOS [26]	52.2	27.9	-24.3	52.0	77.4	87.0
Faster R-CNN [19]	52.5	27.1	-25.4	51.5	71.2	84.1
Cascade R-CNN [2]	52.7	29.7	-23.0	54.9	77.0	86.2
Sparse R-CNN [24]	53.2	36.2	-17.0	64.3	77.0	85.5
M&M (ours)	**57.1**	**53.6**	**-3.5**	**87.7**	**90.9**	**92.5**

a U.S. multi-site mammography operator; (3) *Inhouse-B*: an evaluation dataset collected from a U.S. academic hospital (see [18], Sec. 2.2 for more details on the inhouse datasets). We also utilize two film mammography datasets: (4) DDSM: a dataset maintained at the University of South Florida [8]. We followed the methods by [3,5,13,16] to split the test set; (5) CBIS-DDSM: a curated subset of DDSM [9]. We only include breasts that have one CC view and one MLO view. Dataset statistics are reported in Table 1.

Metrics. We report average precision with Intersection over Union from 0.25 to 0.75. AP_{mb} denotes average precision on the set of annotated malignant and benign images. AP denotes average precision when all data is included. We report free response operating characteristic (FROC) curves and recalls at various FP/image (R@t). Following [3,5,16,29], a proposal is considered true positive if its center lies within the ground truth box. For classification, we report the area under the receiver operating characteristic curve (AUC).

Detection Results. Table 2 presents quantitative detection evaluation on OPTIMAM. All dense detectors [2,10,19,26] suffer a large Δ gap of more than

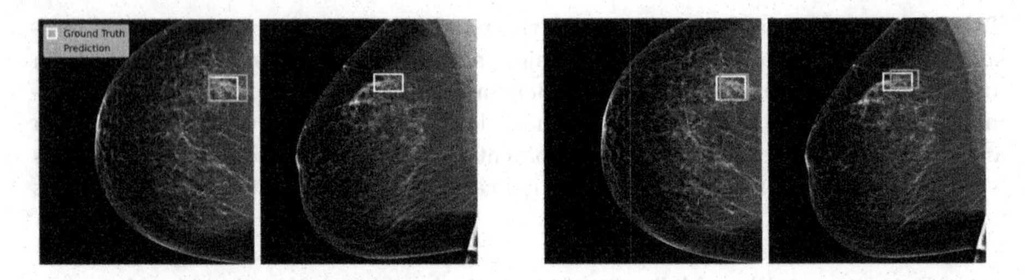

Fig. 3. Qualitative Evaluation. **Left**: Model without multi-view (row 4 of Fig. 4) produces a loose box on the CC view and misses the finding on the MLO view. **Right**: M&M produces tight boxes around the finding in both views.

Table 3. Quantitative classification evaluation. (a) On three private datasets, we use two open-sourced mammography classifiers as baselines [23,25]. All models were trained only on OPTIMAM. We report AUC at both the breast and the exam level, except for Inhouse-A, where breast-level labels are unavailable. (b) We train M&M on CBIS-DDSM and compare breast AUC with recent literature. (* Tulder *et al.* [27] report results using five-fold cross validation.)

(a) Private Datasets

Model	OPTIMAM		Inhouse-A	Inhouse-B	
	Breast AUC	Exam AUC	Exam AUC	Breast AUC	Exam AUC
GMIC [23]	0.911	0.896	0.814	0.815	0.796
HCT [25]	0.923	0.912	0.816	0.817	0.793
M&M (ours)	**0.960**	**0.942**	**0.920**	**0.910**	**0.898**

(b) CBIS-DDSM

Model	Breast AUC
ResNet50 [14]	0.724
Shared ResNet [31]	0.735
PHResNet50 [14]	0.739
Cross-view Transformer [27]	0.803*
M&M (ours)	**0.883**

23 points (pt) between excluding and including negative images. Large Δ means the models are producing too many FPs on negative images. Sparse R-CNN [24] generalizes significantly better with a gap of 17pt. This shows the importance of sparsity for reducing FP. By adding both multi-view and MIL, M&M successfully reduces the Δ gap to 3.5pt. With this performance gap closed, M&M is able to achieve a high recall of 87.7% at just 0.1 FP/image.

Figure 1a compares M&M with recent literature evaluated on DDSM. M&M adopts the same DDSM splits used by [3,12,13,16,33], while [5,21,32] use other splits. M&M (87% R@0.5) outperforms all recent SOTA with the same test split, including 2022 SOTA [33] (83% R@0.5), by at least 4%.

Classification Results. Table 3a reports M&M's breast-level and exam-level classification results on OPTIMAM and the two inhouse datasets. We use GMIC [23] and HCT [25] as baselines since they are open-sourced classifiers developed for mammography. All three models were trained only on OPTIMAM. For all models, the breast-level score is the average of the CC score and MLO score, while the exam-level score is the max of the left breast score and right breast score. Both baseline models suffer large generalization drops of approximately

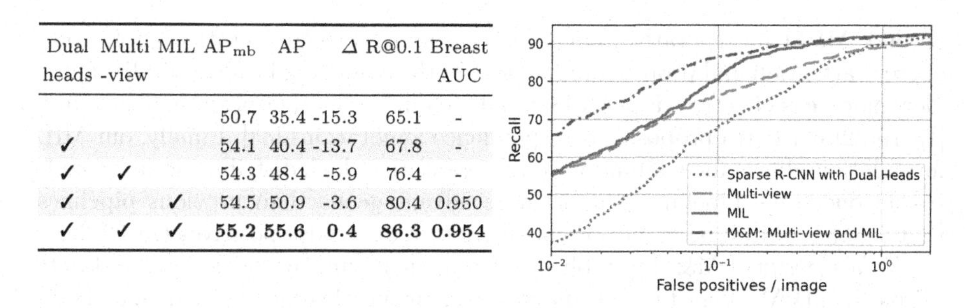

Dual heads	Multi -view	MIL	AP_{mb}	AP	Δ	R@0.1	Breast AUC
			50.7	35.4	-15.3	65.1	-
✓			54.1	40.4	-13.7	67.8	-
✓	✓		54.3	48.4	-5.9	76.4	-
✓		✓	54.5	50.9	-3.6	80.4	0.950
✓	✓	✓	**55.2**	**55.6**	**0.4**	**86.3**	**0.954**

Fig. 4. Effect of M&M's components on classification and detection performance.

0.08–0.12 exam AUC when evaluated on Inhouse-A and Inhouse-B. In comparison, M&M has smaller performance gaps of 0.02 on Inhouse-A and 0.04 on Inhouse-B. Similar observations for other classifiers, such as EfficientNet, are reported in the appendix.

Table 3b compares M&M with recent literature reporting on the public CBIS-DDSM dataset. In particular, M&M outperforms the cross-view transformer [27] and PHResNet50 [14] by 0.08 and 0.14 breast AUC, respectively.

Qualitative Evaluation. Figure 3 presents a qualitative evaluation of the multi-view module. With multi-view, M&M produces a tighter box on the CC view and recovers a missed finding on the MLO view.

Ablation Studies. Figure 4 presents ablation results using the OPTIMAM validation split. On the left, we demonstrate how each component of M&M contributes to closing the gap Δ between evaluating with and without negative images. Notably, without using any extra training samples, multi-view reasoning reduces Δ to only -5.9pt (Row 3). MIL allows the model to train with significantly more negative images, reducing Δ to -3.6pt (Row 4). On the right of Fig. 4, the FROC curves show how each component of M&M improves recall significantly at low FP/image. In particular, M&M's recall at 0.1FP/image is 86.3%, +21.2% over vanilla Sparse R-CNN.

Further studies. In the appendix, we present more qualitative evaluation as well as further ablation studies on (1) number of learnable proposals, (2) different MIL schemes, (3) backbone choices and (4) positional encoding.

4 Discussion and Conclusion

We present M&M, an end-to-end model leveraging multi-view reasoning and multi-instance learning for mammography detection and classification.

As a detector, M&M offers significant improvement in recall at low FP/image (Fig. 1a, Table 2). This success comes from three points of advancement. First, unlike previous works that do not consider the impact of sparsity [13,16,33], we show that sparsity of proposals is beneficial for false positive reduction (Table 2).

Second, M&M incorporates multi-view reasoning through iterative application of cross-attention and proposal refinement in the cascading heads. M&M's multi-view module is effective (Fig. 4) yet simple, requiring neither positional encoding [13,16,32] nor extra proposal correspondence annotations [33]. Finally, our MIL formulation allows for training with representative data distribution in an end-to-end one stage pipeline. This is more advantageous than previous pipelines that require additional stages or classifiers to reduce false positives [15,22,29].

As a classifier, M&M establishes strong performance on several datasets (Table 3). M&M offers two advantages over image classifiers: (1) Image classifiers are often pre-trained as patch classifiers with patches cropped from bounding box annotations [14,23,25]. In comparison, M&M utilizes these bounding boxes to learn localization and can be trained directly in a single stage from COCO/ImageNet weights; (2) Image classifiers offer limited explainability, while M&M's breast-level prediction is more interpretable through its localization ability.

References

1. Agarwal, R., Diaz, O., Lladó, X., Yap, M.H., Martí, R.: Automatic mass detection in mammograms using deep convolutional neural networks. J. Med. Imaging **6**(3), 031409 (2019)
2. Cai, Z., Vasconcelos, N.: Cascade R-CNN: delving into high quality object detection. In: CVPR (2018)
3. Campanini, R., et al.: A novel featureless approach to mass detection in digital mammograms based on support vector machines. Phys. Med. Bio. **49**(6), 961 (2004)
4. Duffy, S.W., et al.: Mammography screening reduces rates of advanced and fatal breast cancers: results in 549,091 women. Cancer **126**(13), 2971–2979 (2020)
5. Eltonsy, N.H., Tourassi, G.D., Elmaghraby, A.S.: A concentric morphology model for the detection of masses in mammography. IEEE Trans. Med. Imaging **26**(6), 880–889 (2007)
6. Fenton, J.J., et al.: Effectiveness of computer-aided detection in community mammography practice. J. National Cancer Inst. **103**(15), 1152–1161 (2011)
7. Halling-Brown, M.D., et al.: OPTIMAM mammography image database: a large-scale resource of mammography images and clinical data. Radiol. Artif. Intell. **3**, e200103 (2020)
8. Heath, M., Bowyer, K., Kopans, D., Moore, R., Kegelmeyer, P.: The digital database for screening mammography. In: Proceedings of the Fifth International Workshop on Digital Mammography. Medical Physics Publishing (2001)
9. Lee, R.S., Gimenez, F., Hoogi, A., Miyake, K.K., Gorovoy, M., Rubin, D.L.: A curated mammography data set for use in computer-aided detection and diagnosis research. Sci. Data **4**(1), 1–9 (2017)
10. Lin, T.Y., Goyal, P., Girshick, R., He, K., Dollár, P.: Focal loss for dense object detection. In: ICCV. pp. 2980–2988 (2017)
11. Lin, T.-Y., et al.: Microsoft COCO: common objects in context. In: Fleet, D., Pajdla, T., Schiele, B., Tuytelaars, T. (eds.) Computer Vision – ECCV 2014: 13th European Conference, Zurich, Switzerland, September 6-12, 2014, Proceedings, Part V, pp. 740–755. Springer, Cham (2014). https://doi.org/10.1007/978-3-319-10602-1_48

12. Liu, Y., Zhang, F., Chen, C., Wang, S., Wang, Y., Yu, Y.: Act like a radiologist: towards reliable multi-view correspondence reasoning for mammogram mass detection. PAMI **44**(10), 5947–5961 (2021)
13. Liu, Y., Zhang, F., Zhang, Q., Wang, S., Wang, Y., Yu, Y.: Cross-view correspondence reasoning based on bipartite graph convolutional network for mammogram mass detection. In: CVPR. pp. 3812–3822 (2020)
14. Lopez, E., Grassucci, E., Valleriani, M., Comminiello, D.: Multi-view breast cancer classification via hypercomplex neural networks. arXiv:2204.05798 (2022)
15. Lotter, W., et al.: Robust breast cancer detection in mammography and digital breast tomosynthesis using an annotation-efficient deep learning approach. Nature Med. **27**(2), 244–249 (2021)
16. Ma, J., Li, X., Li, H., Wang, R., Menze, B., Zheng, W.S.: Cross-view relation networks for mammogram mass detection. In: 2020 25th International Conference on Pattern Recognition (ICPR). pp. 8632–8638. IEEE (2021)
17. Moreira, I.C., Amaral, I., Domingues, I., Cardoso, A., Cardoso, M.J., Cardoso, J.S.: INBreast: toward a full-field digital mammographic database. Acad. Radiol. **19**(2), 236–248 (2012)
18. Pedemonte, S., et al.: A deep learning algorithm for reducing false positives in screening mammography. arXiv preprint arXiv:2204.06671 (2022)
19. Ren, S., He, K., Girshick, R., Sun, J.: Faster R-CNN: towards real-time object detection with region proposal networks. Adv. Neural Inf. Process. Syst. **28** (2015)
20. Ren, Y., et al.: Retina-Match: ipsilateral mammography lesion matching in a single shot detection pipeline. In: de Bruijne, M., et al. (eds.) Medical Image Computing and Computer Assisted Intervention – MICCAI 2021: 24th International Conference, Strasbourg, France, September 27 – October 1, 2021, Proceedings, Part V, pp. 345–354. Springer, Cham (2021). https://doi.org/10.1007/978-3-030-87240-3_33
21. Sampat, M.P., Bovik, A.C., Whitman, G.J., Markey, M.K.: A model-based framework for the detection of spiculated masses on mammography a. Med. Phys. **35**(5), 2110–2123 (2008)
22. Sarath, C.K., Chakravarty, A., Ghosh, N., Sarkar, T., Sethuraman, R., Sheet, D.: A two-stage multiple instance learning framework for the detection of breast cancer in mammograms. In: EMBC. IEEE (2020)
23. Shen, Y.: An interpretable classifier for high-resolution breast cancer screening images utilizing weakly supervised localization. Med. Image Anal. **68**, 101908 (2021)
24. Sun, P., et al.: Sparse R-CNN: end-to-end object detection with learnable proposals. In: CVPR (2021)
25. Taha, A., Truong Vu, Y.N., Mombourquette, B., Matthews, T.P., Su, J., Singh, S.: Deep is a Luxury We Don't Have. In: Wang, L., Dou, Q., Fletcher, P.T., Speidel, S., Li, S. (eds.) Medical Image Computing and Computer Assisted Intervention – MICCAI 2022: 25th International Conference, Singapore, September 18–22, 2022, Proceedings, Part III, pp. 25–35. Springer, Cham (2022). https://doi.org/10.1007/978-3-031-16437-8_3
26. Tian, Z., Shen, C., Chen, H., He, T.: Fcos: fully convolutional one-stage object detection. In: CVPR (2019)
27. van Tulder, G., Tong, Y., Marchiori, E.: Multi-view Analysis of Unregistered Medical Images Using Cross-View Transformers. In: de Bruijne, M., et al. (eds.) MICCAI 2021. LNCS, vol. 12903, pp. 104–113. Springer, Cham (2021). https://doi.org/10.1007/978-3-030-87199-4_10
28. Vaswani, A., et al.: Attention is all you need. NeurIPS **30** (2017)

29. Vu, Y.N.T., Mombourquette, B., Matthews, T.P., Su, J., Singh, S.: WRDet: a breast cancer detector for full-field digital mammograms. In: Medical Imaging 2022: Computer-Aided Diagnosis. vol. 12033, pp. 219–230. SPIE (2022)
30. Wang, W., et al.: Pvt v2: improved baselines with pyramid vision transformer. Comput. Visual Media **8**(3), 415–424 (2022)
31. Wu, N., Jastrzębski, S., Park, J., Moy, L., Cho, K., Geras, K.J.: Improving the ability of deep neural networks to use information from multiple views in breast cancer screening. In: Medical Imaging with Deep Learning. PMLR (2020)
32. Yang, Z., et al.: MommiNet-v2: mammographic multi-view mass identification networks. Med. Image Anal. **73**, 102204 (2021)
33. Zhao, Z., Wang, D., Chen, Y., Wang, Z., Wang, L.: Check and link: pairwise lesion correspondence guides mammogram mass detection. In: ECCV (2022). https://doi.org/10.1007/978-3-031-19803-8_23

Convolving Directed Graph Edges via Hodge Laplacian for Brain Network Analysis

Joonhyuk Park[1], Yechan Hwang[1], Minjeong Kim[2], Moo K. Chung[3], Guorong Wu[4], and Won Hwa Kim[1(✉)]

[1] Pohang University of Science and Technology, Pohang, South Korea
{pjh1023,yechan99,wonhwa}@postech.ac.kr

[2] University of North Carolina at Greensboro, Greensboro, USA

[3] University of Wisconsin - Madison, Madison, USA

[4] University of North Carolina at Chapel Hill, Chapel Hill, USA

Abstract. A brain network, viewed as a graph wiring different regions of interest (ROIs) in the brain, has been widely used to investigate brain dysfunction with various graph neural networks (GNNs). However, existing GNNs are built upon graph convolution that transforms measurements on the nodes, where ROI-wise features are not always guaranteed for brain networks. Therefore, the majority of existing graph analysis methods that rely on node features are inapplicable for network analysis unless a proxy such as node degree is provided. Moreover, the complex neurological interactions across different brain regions cannot be directly expressed in a simple node-to-node (i.e., 0-simplex) representation. In this paper, we propose a novel method, Hodge-Graph Neural Network (Hodge-GNN), that allows the GNN to directly derive desirable representations of graph edges and capture complex edge-wise topological features spatially via the Hodge Laplacian. Specifically, representing a graph as a simplicial complex holds a significant advantage over conventional methods that extract higher-order connectivity of a graph through hierarchical convolution in the spatial domain or graph transformation. The superiority of our method is validated in the Alzheimer's Disease Neuroimaging Initiative (ADNI) study, in comparison to benchmarking GNNs as well as state-of-the-art graph classification models.

Keywords: Hodge Laplacian · Brain Network · Alzheimer's Disease

1 Introduction

The wiring system within the human brain can be modeled as a complex graph, where the anatomical regions of interest (ROIs) are represented as nodes, and

J. Park and Y. Hwang—contributed equally to this paper.

Supplementary Information The online version contains supplementary material available at https://doi.org/10.1007/978-3-031-43904-9_76.

H. Greenspan et al. (Eds.): MICCAI 2023, LNCS 14224, pp. 789–799, 2023.
https://doi.org/10.1007/978-3-031-43904-9_76

the white matter connectomes define edges between them [9,18]. The graph-based representation explains actual connections between different nodes, and thus brain network analysis has a significant advantage over traditional spatial analysis to investigate interactions of different ROIs. As various neurodegenerative disorders such as Alzheimer's Disease (AD) are understood as a disconnection syndrome [2,9], studies on the structural connectivities in the brain are of significant interest from both machine learning and clinical perspectives.

Recently, variants of graph neural networks (GNNs) have been successful in brain network analysis with feature aggregation and message-passing mechanism on graph nodes [4,5,18]. Notice that, regardless of whether it is spatial or spectral, existing methods heavily rely on graph convolution that operates with the node signals (i.e., ROI measures). Here, the topology of graph plays an indirect role as the domain of the signal, merely selecting specific neighborhood for feature aggregation. The problem becomes even more severe when it comes to actual connectivity analysis *without* ROI-wise measurements. To utilize GNN methods, auxiliary node-wise measures such as node degree and clustering coefficients are required to perform prediction tasks on the brain networks, and the contribution of connectomic features as a biomarker is often not fully investigated. To perform convolution on edges directly, heuristics such as line graph [15] or defining orthogonal matrices from graph Laplacian [21] exists, but they have disadvantages when the graph is directed or has many components.

In order to analyze the connectivity of brain networks directly, we propose a novel graph learning framework that allows a neural network to utilize the topological features by representing the brain network as a simplicial complex. We utilize the Hodge 1-Laplacian, i.e., $\mathcal{L}_1$; in geometry, the 1-simplex denotes a line segment, and the Hodge Laplacian $\mathcal{L}_1$ includes connection between different line segments (i.e., edges) in a graph depending on their directions. Leveraging the Hodge Laplacian lets us obtain directed relationships between graph edges (i.e., a brain network) as an undirected graph (i.e., Hodge Laplacian), and the edge weights in the original graph become a signal on the nodes comprehended by the Hodge Laplacian. Spatial convolution with the $\mathcal{L}_1$ combines directed edge weights of the original graph based on the nodes that they are sharing. Together with a spatial graph convolution formulation, we construct our Hodge-Graph Neural Network (Hodge-GNN) which predicts labels of graphs purely based on the topology of the graphs without any node measures.

The **contributions** of our work are **1)** proposing a novel graph edge-learning framework on higher-order connectivity (i.e., connectivity between edges) of graphs with Hodge Laplacian, **2)** defining spatial edge convolution layer that operates on graph edges directly, and **3)** demonstrating superior performance on graph classification with brain connectivity from Alzheimer's Disease Neuroimaging Initiative (ADNI) with interpretability. Using Hodge-GNN, we depict brain connectivities that are highly associated with AD classification, which are corroborated by prior AD literature.

2 Related Work

Higher-order connectivity in Spatial domain. To capture the relation of higher-order graph structures, Morris et al. [24,25] proposed hierarchical k-GNNs, which are hierarchical GNN architectures based on the k-dimensional Weisfeiler-Lehman (WL) algorithm. By performing message passing directly between subgraph structures, k-GNNs enable the network to capture structural information that is not observable at the node-level. Despite their superiority over 1-GNN, k-GNNs require large memory and high computational cost due to their stacking of models, showing limitation in scalability and effectiveness on large graphs. The authors in [4] define range with diffusion instead of hop-distances and train on the range to obtain desirable node-embeddings.

Line Graphs. Line graph transformation interchanges the nodes and edges in the original graph respectively, allowing the node-wise graph convolution to be performed edge-wise, which makes the learning of edge-embeddings feasible [15]. However, line graphs lack the property of injectivity, which implies that different graphs can be transformed into a same line graph.

Edge Convolution. [30] performed *edge convolution* by creating edge-embeddings from neighboring pair of point clouds. However, the suggested method requires node features to generate the edge-embeddings without any graph prior, and thus it requires rich node features to construct strong embeddings.

Spectral Filtering of Graph Edges. Huang et al. [14] defined spectral filters for graph signals of nodes and edges using the k-th Hodge Laplacian (HL) operators, i.e., HL-node and HL-edge, and showed effectiveness of capturing the edge-wise relation in heterogeneous brain functional networks. The authors in [21] performed kernel filtering in the spectral domain with a specialized orthonormal graph transform.

3 Preliminaries: Simplicial Complex Representation

A simplicial complex is a collection of simplices with various dimensional representations, where simplices refer to the basic blocks to represent objects in topological space. In detail, each simplex of various dimension can be seen as nodes (0-simplex), edges (1-simplex), triangles (2-simplex), and other higher dimensional counterparts. A simplicial complex composed of only 0-simplices is called a 0-skeleton, likewise, p-skeleton is composed of 0 to p-simplices. A graph, therefore, is a 1-skeleton with 0- and 1-simplices (nodes and edges) [1,14,19].

In a simplicial complex, a p-chain is defined as a sum of p-simplicies, denoted as $c = \sum_i \alpha_i \sigma^i$, where σ^i are the p-simplices and the α_i are either 0 or 1 [8]. A chain complex is defined as the sequence of groups, each of which is made

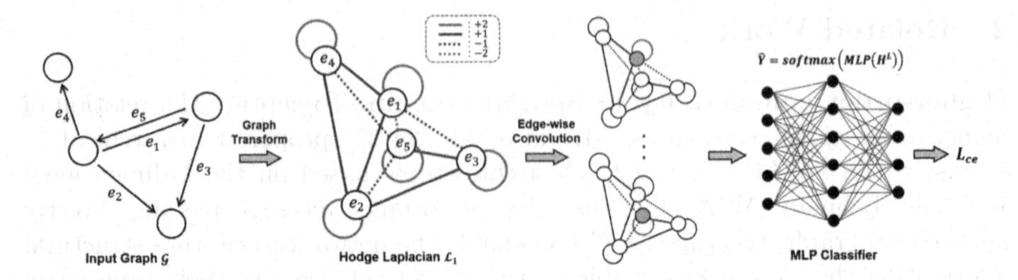

Fig. 1. The overall process of the proposed method: 1) deriving Hodge Laplacian $\mathcal{L}_1$ from input graph $\mathcal{G}$. 2) edge-wise graph convolution on e with $\mathcal{L}_1$. 3) Each edge (orange node) aggregates the information of neighboring edges in different weights given by $\mathcal{L}_1$. 4) MLP takes edge-convolved features as an input and is updated by the loss.

up of p-chains. To represent the relationship between chain groups, a boundary operator $\partial_p : C_p \to C_{p-1}$ is defined. Here, C_p denotes the p-th chain group, and the boundary operator ∂_p maps p-simplex to its boundaries ($(p-1)$-simplex). For an oriented p-simplex σ^p, the boundary operator can be defined as

$$\partial_p(\sigma^p) = \sum_{i=0}^{p}(-1)^i[v_0, v_1, \cdots, \hat{v}_i, \cdots, v_p], \tag{1}$$

where $[v_0, v_1, \cdots, \hat{v}_i, \cdots, v_p]$ is a $(p-1)$-simplex that is created by removing the vertex $\hat{v}_i$ from the p-simplex $\sigma^p = [v_0, v_1, \cdots, v_p]$ [1]. Also, the boundary operator ∂_p is represented using a boundary matrix $\mathcal{B}_p$ to facilitate efficient computation of the Hodge Laplacian. The p-th boundary matrix $\mathcal{B}_p$ can be defined as [23],

$$\mathcal{B}_p(i,j) = \begin{cases} 1, & \text{if } \sigma_i^{p-1} \subset \sigma_j^p \text{ and } \sigma_i^{p-1} \sim \sigma_j^p \\ -1, & \text{if } \sigma_i^{p-1} \subset \sigma_j^p \text{ and } \sigma_i^{p-1} \nsim \sigma_j^p \\ 0, & \text{if } \sigma_i^{p-1} \not\subset \sigma_j^p \end{cases} \tag{2}$$

where σ_i^p is the i-th p-simplex, and $\sim$ and $\nsim$ denote similar and dissimilar orientations respectively. The boundary matrix $\mathcal{B}_p$ relates the two adjacent simplices, i.e., p- and $(p-1)$-simplex, which will be used to define the Hodge Laplacian for higher-order graph representation in Sec. 4.1.

4 Proposed Method

The proposed method is composed of two components; graph transformation of adjacency matrix to Hodge Laplacian $\mathcal{L}_1$ with edge-wise features, and edge-wise graph convolution using the $\mathcal{L}_1$. With the traditional graph convolution formulation, the network can conduct transform of topological features directly instead of using them as indirect measures in previous GNNs.

4.1 Hodge Laplacian of Brain Network Data

Let $\mathcal{G} = (\mathcal{V}, \mathcal{E})$ be a directed weighted graph, where $\mathcal{V}$ is a set of nodes, and $\mathcal{E}$ is a set of directed edges consisting of ordered tuples, (u, v), s.t. $u, v \in \mathcal{V}$, which denotes an edge from u to v. $\mathcal{E}$ is indexed with $\{e_i\}_{i=1}^{E}$, $|\mathcal{E}| = E$ is the number of edges, and $|\mathcal{V}| = N$ is the number of nodes.

Hodge Laplacian $\mathcal{L}_p$, also known as the p-Laplacian is a generalization of graph Laplacian on higher simplices, i.e., nodes (0-simplices) to p-simplices. The Hodge Laplacian $\mathcal{L}_p$ is defined using the $\mathcal{B}_p(i,j)$ in Eq. (2) as:

$$\mathcal{L}_p = \mathcal{B}_p^T \mathcal{B}_p + \mathcal{B}_{p+1} \mathcal{B}_{p+1}^T. \tag{3}$$

From Eq. (3), the Hodge 0-Laplacian, $\mathcal{L}_0$, is equivalent to graph Laplacian, defined as $\mathcal{L}_0 = \mathcal{B}_1 \mathcal{B}_1^T$, where $\mathcal{B}_1 \in \mathbb{R}^{N \times E}$ is a boundary matrix for the 1-simplex, i.e., an incidence matrix relating nodes to edges.

To enable the graph representation to hold connectivity over edges, we construct Hodge Laplacian $\mathcal{L}_1$. As a 1-skeleton, i.e. topological graph, is composed of 0 and 1 simplex only (i.e., $\mathcal{B}_2 = 0$), the $\mathcal{L}_1 \in \mathbb{R}^{E \times E}$ is derived as:

$$\mathcal{L}_1 = \mathcal{B}_1^T \mathcal{B}_1. \tag{4}$$

Considering the vertices $u, v, t \in \mathcal{V}$, s.t. $u \neq v \neq t$, each element $\mathcal{L}_1(i,j)$ is defined as:

$$\mathcal{L}_1(i,j) = \begin{cases} 2 & e_i = e_j \\ -2 & e_i = (u,v), e_j = (v,u) \\ 1 & (e_i = (u,v), e_j = (u,t)) \text{ or } (e_i = (u,v), e_j = (t,v)) \\ -1 & (e_i = (u,v), e_j = (t,u)) \text{ or } (e_i = (u,v), e_j = (v,t)) \\ 0 & \text{otherwise} \end{cases}. \tag{5}$$

A directed weighted graph $\mathcal{G}$ can be represented as a binary adjacency matrix $A \in \mathbb{R}^{N \times N}$ and a weight matrix $\mathcal{W} \in \mathbb{R}^{N \times N}$ for A. Since $\mathcal{W}$ holds features for each edge, we can extract the non-zero components of $\mathcal{W}$, which serves as a signal $\mathcal{W}_{\mathcal{E}} \in \mathbb{R}^{E}$ on $\mathcal{L}_1$.

4.2 Convolving Graph Edges via Hodge Laplacian $\mathcal{L}_1$

GCN [31] performs aggregation of the neighboring node features as:

$$H^{(l+1)} = \sigma(AH^{(l)}W^{(l)}), \qquad l = 0, \cdots, L, \tag{6}$$

where $A \in \mathbb{R}^{N \times N}$ is an adjacency matrix, and $W^{(l)}$ is a parameter matrix of the l-th layer. $H^{(l)} \in \mathbb{R}^{N \times K}$ is the output of the l-th convolution layer with K features, where $H^{(0)}$ is the input of node feature vectors, and $\sigma(\cdot)$ is a non-linear activation function. From a spatial perspective, the graph convolution relates the neighboring node features to generate the node embedding utilizing the adjacency matrix as a relational matrix that provides the direct neighboring information.

From a weighted adjacency matrix $A \in \mathbb{R}^{N\times N}$, we can extract Hodge Laplacian $\mathcal{L}_1 \in \mathbb{R}^{E\times E}$ and $W_{\mathcal{E}}$ which is a vector of edge weights considered as measurements on the nodes of $\mathcal{L}_1$. With $\mathcal{L}_1$, we construct a Hodge graph neural network (Hodge-GNN) whose l-th layer is defined as:

$$H^{(l+1)} = \sigma(\mathcal{L}_1 H^{(l)} W^{(l)}), \qquad l = 0, \cdots, L, \tag{7}$$

where $W^{(l)}$ is a learnable weight parameter and $H^{(l)} \in \mathbb{R}^{E\times K}$ is the output from the l-th convolution layer, with $H^{(0)} = \mathcal{W}_{\mathcal{E}}$. The key component here is $\mathcal{L}_1 H^{(l)}$, described as Edge-wise Convolution in Fig. 1.

When it comes to graph analysis, most of existing GNN methods assume that features on the nodes exist for node-wise analysis. However, when the measurements on the nodes do not exist, and the analysis must be performed solely with the graph topology and edge information, other GNN methods must define an auxiliary node-wise measures such as node degree and clustering coefficients. Unlike the previous approaches, our framework enables the information from adjacent edges to be given different weights, either positive or negative, depending on the topology of the graph, and the edge-wise convolution can now relate the edge features and generate edge embeddings, allowing the network to utilize the hidden topological features that were not seen in the original input graph form.

Finally, the class prediction $\hat{Y}^c$ for each class c is obtained by flattening the $H^{(L)} \in \mathbb{R}^{E\times K}$ and passing it through multi-layer perceptron (MLP), and applying a softmax yields

$$\hat{Y}^c = \frac{\text{MLP}(H^{(L)})^c}{\sum_{c' \in C} \text{MLP}(H^{(L)})^{c'}}. \tag{8}$$

The objective function defined by cross-entropy over all T samples is:

$$L_{ce} = -\sum_{t=1}^{T} \sum_{c \in C} Y_t^c log \hat{Y}_t^c, \tag{9}$$

where $Y_t^c = 1$ if the class of t-th graph is c, otherwise $Y_t^c = 0$.

4.3 Interpretability of the Connectomes in Brain Dysfunction

To provide interpretability to the framework, we define gradient-based class activation map on the graph edges using the formulation in [27,33]. Specifically, when a graph (i.e., $\mathcal{L}_1$ and $\mathcal{W}_{\mathcal{E}}$) is inputted to the network, by tracking the backpropagating gradients of the score for a specific class c (i.e., $\hat{Y}^c$) with regard to each feature vectors $H^k \in \mathbb{R}^{E\times 1}$ of the final convolution layer of GNN (i.e., $H^{(L)}$) [27], the importance of each activation α_k^c and the heatmap $\mathcal{H}$ of the specific class c can be computed as:

$$\mathcal{H}^c = ReLU\left(\frac{1}{K}\sum_k \alpha_k^c H^k\right), \quad \alpha_k^c = \frac{1}{Z}\sum_i \sum_j \left(\frac{\partial \hat{Y}^c}{\partial H_{ij}^k}\right). \tag{10}$$

Performing the edge-wise convolution from Hodge Laplacian $\mathcal{L}_1$, this heatmap $\mathcal{H}^c$ holds the contribution of each connectome to the classification of developmental stages in brain dysfunction (i.e., Alzheimer's Disease).

5 Experiments

5.1 Dataset and Experimental Settings

Dataset. Our dataset contains structural brain connectivity data derived from Diffusion Tensor Images (DTI) in Alzheimer's Disease Neuroimaging Initiative (ADNI) with tractography. Each sample is given as a directed weighted graph whose weights denote the *number of white matter fiber tracts* connecting two different ROIs and its corresponding diagnostic label. The ROIs and their connectomes were defined by the Destrieux atlas [7] with 148 cortical and 12 sub-cortical ROIs. As tractography involves probabilistic calculation, the connectivity matrix becomes non-symmetric with varying weights on the same connectivity matrices. The dataset is composed of n=1824 subjects within Control (CN, n=844), Early Mild Cognitive Impairment (EMCI, n=490), Late Mild Cognitive Impairment (LMCI, n=250), and AD (n=240) groups.

Table 1. Quantitative Comparison of Hodge-GNN with Baselines.

	Methods	Accuracy	Precision	Recall	F1-score
Conventional	SVM	58.31 ± 1.40	70.67 ± 3.44	44.49 ± 1.57	46.94 ± 2.36
	SLP	71.42 ± 3.42	68.91 ± 4.46	70.27 ± 3.39	69.16 ± 3.99
	MLP	72.41 ± 1.14	73.66 ± 1.29	67.47 ± 2.59	69.45 ± 1.22
	GCN ($\mathcal{G}$) [31]	70.16 ± 1.82	67.67 ± 2.46	66.82 ± 1.30	67.00 ± 1.88
	GCN ($\mathcal{G}_L$)	80.85 ± 3.10	81.21 ± 3.34	77.55 ± 2.93	79.09 ± 3.10
Spatial domain	1-2-GNN [25]	73.58 ± 1.12	71.44 ± 0.87	71.31 ± 1.41	71.15 ± 0.43
	1-2-3-GNN [25]	73.81 ± 1.91	72.39 ± 1.82	71.04 ± 2.64	71.25 ± 1.51
Spectral domain	MENET [21]	80.47 ± 1.82	79.30 ± 0.69	77.09 ± 1.59	77.92 ± 1.12
Edge Convolution	DGCNN [30]	72.61 ± 2.01	69.69 ± 2.29	69.11 ± 1.51	69.11 ± 1.90
Ours	Hodge-GNN	**83.60±1.93**	**84.75±2.19**	**80.10±1.87**	**82.03±1.93**

Edge Preprocessing. Unlike the widely used $A \in \mathbb{R}^{N \times N}$ or $\mathcal{L}_0 \in \mathbb{R}^{N \times N}$, the adoption of $\mathcal{L}_1 \in \mathbb{R}^{E \times E}$ incurs an exponential rise in both memory and computational costs. Also, the E across different samples varies from 2138 to 11802 (6766 in average). To handle such problems, we used the intersection of edges across entire samples. This preprocessing step yields the number of edges to be $E = 530$, common across all subjects.

Baselines. Our method is validated on various approaches, including conventional classification methods, neural networks, and graph methods of both spatial and spectral domain. In detail, we used support vector machine (SVM), single layer perceptron (SLP), multi-layer perceptron (MLP), and GCN [31] with original graph $\mathcal{G}$ and line graph $\mathcal{G}_L$ as the conventional classification method, as well as the hierarchical k-GNNs (1-2-GNN, 1-2-3-GNN) [25] and MENET [21] for spatial and spectral baselines respectively. Also, DGCNN [30] which extracts edge-embeddings with edge convolution from node features is compared as well.

Evaluation. 4-way classification is performed on the AD-specific groups. All models are evaluated using 5-fold cross validation (CV) for unbiased results. On the 4-way classification task for AD-specific groups, average accuracy, Macro-precision, Macro-recall, and Macro-F1-score (Table 1) are compared. Qualitative results of our experiment reveal the important connectivities in classifying AD stages (Fig. 2) from the trained model.

5.2 Experimental Results

Our Hodge-GNN is evaluated based on the 4-way classification of CN, EMCI, LMCI, and AD. The quantitative comparisons are shown in Tab. 1. As k-GNN, MEMET, and GCN with $\mathcal{G}_L$ capture higher-order connectivity information, they performed better than the conventional GCN with $\mathcal{G}$. However, Hodge-GNN was more effective in the AD classification, achieving the highest performance in all measures by ~3%p over the second best method.

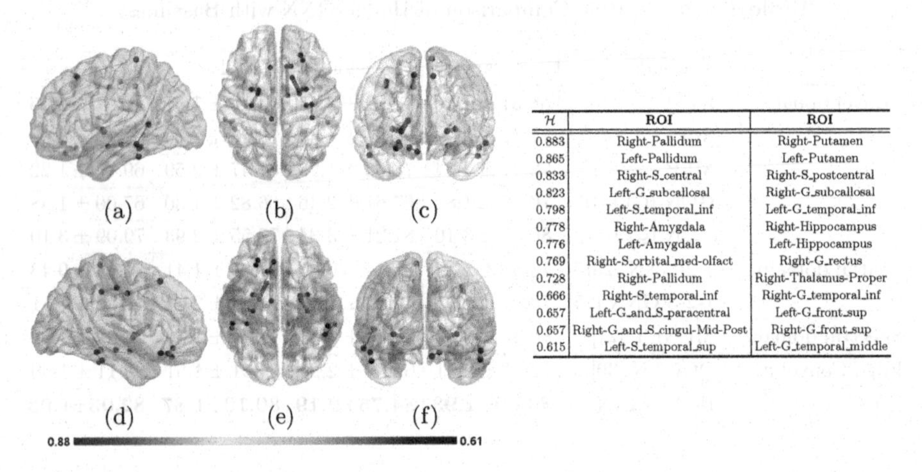

$\mathcal{H}$	ROI	ROI
0.883	Right-Pallidum	Right-Putamen
0.865	Left-Pallidum	Left-Putamen
0.833	Right-S_central	Right-S_postcentral
0.823	Left-G_subcallosal	Right-G_subcallosal
0.798	Left-S_temporal_inf	Left-G_temporal_inf
0.778	Right-Amygdala	Right-Hippocampus
0.776	Left-Amygdala	Left-Hippocampus
0.769	Right-S_orbital_med-olfact	Right-G_rectus
0.728	Right-Pallidum	Right-Thalamus-Proper
0.666	Right-S_temporal_inf	Right-G_temporal_inf
0.657	Left-G_and_S_paracentral	Left-G_front_sup
0.657	Right-G_and_S_cingul-Mid-Post	Right-G_front_sup
0.615	Left-S_temporal_sup	Left-G_temporal_middle

Fig. 2. Significant edges depicted from the AD analysis; (a),(d) outer view of left/right hemisphere, (b),(e) top/bottom, (c),(f) front/rear view. Blue node: cortical, Red node: subcortical region, Edge color/thickness denote class activation $\mathcal{H}$ from Eq. (10 (Color figure online)).

5.3 Interpretation of AD via Trained Hodge-GNN

Using the computed importance from Eq. (10), significant edges for classifying the stages of Alzheimer's Disease are depicted (Fig. 2). The significant edges are selected by taking intersection of the top-k edges for each class label. The top-k edges are the distinct edges obtained from the top-10 edges across the 5-fold cross validation. Thus, the selected edges represent common edges that shows high importance for classifying the brain dysfunction. During the progress of AD, ROIs of the brain show not only the shrinkage of volume but also weakening of connectivity [2,6]. As in Fig. 2, our classifier picked up connectome of ROIs in the subcortical regions (i.e., amygdala, hippocampus, pallidum, putamen, and thalamus) [3,10,26], temporal lobe (i.e., inferior, middle, and superior temporal cortex) [3,10,11], frontal lobe (i.e., superior frontal gyrus) [10,16], and other important regions that are highly related to AD [12,20].

In addition, the depicted edges showed several symmetry found in both left and right hemispheres, such as pallidum-putamen and amygdala-hippocampus connectomes, both of which play a crucial role in the development of AD [13,29,32]. Interestingly, right pallidum was shared on two detected connectivities (i.e., with right-putamen and right-thalamus-proper), which highlights the importance of pallidum connectomes [17]. Also, out of the 17 ROIs consisting the detected connectivities, 5 ROIs (pallidum, amygdala, putamen, hippocampus, and thalamus) were from subcortical regions, denoting that they are critical for our AD-stage classification and subcortical areas are highly implicated in AD as in prior works [22,28].

6 Conclusion

We proposed a novel framework for extracting edge-to-edge relations in graph spatial domain using Hodge 1-Laplacian, i.e., Hodge-GNN. The Hodge-GNN performs graph convolution on edges via shared nodes among edges and allows a downstream predictor to accurately classify different stages of AD. The validation experiment showed superiority of performance in prediction together with interpretable outcomes depicting specific connectomes and ROIs for effective AD analysis.

Acknowledgement. This research was supported by NRF-2022R1A2C2092336 (50%), IITP-2022-0-00290 (20%), IITP-2019-0-01906 (AI Graduate Program at POSTECH, 10%) funded by MSIT, HU22C0171 (10%) and HU22C0168 (10%) funded by MOHW in South Korea, and NIH R03AG070701 from the US, and Foundation of Hope.

References

1. Anand, D.V., Chung, M.K.: Hodge-laplacian of brain networks and its application to modeling cycles. arXiv preprint arXiv:2110.14599 (2021)
2. Catani, M., Ffytche, D.H.: The rises and falls of disconnection syndromes. Brain **128**(10), 2224–2239 (2005)
3. Cheyuo, C., et al.: Connectomic neuromodulation for Alzheimer's disease: a systematic review and meta-analysis of invasive and non-invasive techniques. Transl. Psychiatry **12**(1), 490 (2022)
4. Choi, I., Wu, G., Kim, W.H.: How much to aggregate: learning adaptive node-wise scales on graphs for brain networks. In: Wang, L., Dou, Q., Fletcher, P.T., Speidel, S., Li, S. (eds.) Medical Image Computing and Computer Assisted Intervention – MICCAI 2022: 25th International Conference, Singapore, September 18–22, 2022, Proceedings, Part I, pp. 376–385. Springer, Cham (2022). https://doi.org/10.1007/978-3-031-16431-6_36

5. Cui, H., et al.: BrainGB: a benchmark for brain network analysis with graph neural networks. IEEE Trans. Med. Imaging **42**(2), 493–506 (2023)
6. Delbeuck, X., Van der Linden, M., Collette, F.: Alzheimer'disease as a disconnection syndrome? Neuropsychol. Rev. **13**, 79–92 (2003)
7. Destrieux, C., et al.: Automatic parcellation of human cortical gyri and sulci using standard anatomical nomenclature. Neuroimage **53**(1), 1–15 (2010)
8. Edelsbrunner, H., Harer, J.L.: Computational topology: an introduction. American Mathematical Society (2022)
9. Farahani, F.V., et al.: Application of graph theory for identifying connectivity patterns in human brain networks: a systematic review. Front. Neurosci. **13**, 585 (2019)
10. Filippi, M., et al.: Changes in functional and structural brain connectome along the Alzheimer's disease continuum. Mol. Psychiatry **25**(1), 230–239 (2020)
11. Galton, C.J., Patterson, K., et al.: Differing patterns of temporal atrophy in Alzheimer's disease and semantic dementia **57**(2), 216–225 (2001)
12. Gan, C., O'Sullivan, M., Metzler-Baddeley, C., et al.: Association of imaging abnormalities of the subcallosal septal area with Alzheimer's disease and mild cognitive impairment. Clin. Radiol. **72**(11), 915–922 (2017)
13. Guo, Z., et al.: Disrupted topological organization of functional brain networks in Alzheimer's disease patients with depressive symptoms. BMC Psychiatry **22**(1), 1–10 (2022)
14. Huang, J., Chung, M.K., Qiu, A.: Heterogeneous Graph Convolutional Neural Network via Hodge-Laplacian for Brain Functional Data. In: Frangi, A., de Bruijne, M., Wassermann, D., Navab, N. (eds.) Information Processing in Medical Imaging: 28th International Conference, IPMI 2023, San Carlos de Bariloche, Argentina, June 18–23, 2023, Proceedings, pp. 278–290. Springer, Cham (2023). https://doi.org/10.1007/978-3-031-34048-2_22
15. Jiang, X., Ji, P., Li, S.: CensNet: convolution with edge-node switching in graph neural networks. In: IJCAI, pp. 2656–2662 (2019)
16. Johnson, J.K., et al.: Clinical and pathological evidence for a frontal variant of Alzheimer disease. Arch. Neurol. **56**(10), 1233–1239 (10 1999)
17. Lehéricy, S., Hirsch, E.C., Hersh, L.B., et al.: Cholinergic neuronal loss in the globus pallidus of Alzheimer disease patients. Neurosci. Lett. **123**(2), 152–155 (1991)
18. Li, X., et al.: BrainGNN: interpretable brain graph neural network for fMRI analysis. Med. Image Anal. **74**, 102233 (2021)
19. Lim, L.H.: Hodge laplacians on graphs. arXiv preprint arXiv:1507.05379 (2015)
20. Lu, J., et al.: Functional connectivity between the resting-state olfactory network and the hippocampus in Alzheimer's disease. Brain Sci. **9**(12), 338 (2019)
21. Ma, X., Wu, G., Hwang, S.J., Kim, W.H.: Learning multi-resolution graph edge embedding for discovering brain network dysfunction in neurological disorders. In: Feragen, A., Sommer, S., Schnabel, J., Nielsen, M. (eds.) Information Processing in Medical Imaging: 27th International Conference, IPMI 2021, Virtual Event, June 28–June 30, 2021, Proceedings, pp. 253–266. Springer, Cham (2021). https://doi.org/10.1007/978-3-030-78191-0_20
22. McDuff, T., Sumi, S.: Subcortical degeneration in Alzheimer's disease. Neurology **35**(1), 123–123 (1985)
23. Meng, Z., Xia, K.: Persistent spectral-based machine learning (PerSpect ML) for protein-ligand binding affinity prediction. Sci. Adv. **7**(19), eabc5329 (2021)
24. Morris, C., Lipman, Y., Maron, H., et al.: Weisfeiler and leman go machine learning: the story so far. arXiv preprint arXiv:2112.09992 (2021)

25. Morris, C., Ritzert, M., Fey, M., et al.: Weisfeiler and leman go neural: Higher-order graph neural networks. In: AAAI, vol.33, pp.4602–4609 (2019)
26. Persson, K., Bohbot, V., Bogdanovic, N., et al.: Finding of increased caudate nucleus in patients with Alzheimer's disease. Acta Neurologica Scandinavica **137**(2), 224–232 (2018)
27. Selvaraju, R.R., Das, A., Vedantam, R., Cogswell, M., Parikh, D., Batra, D.: Grad-cam: Why did you say that? visual explanations from deep networks via gradient-based localization. CoRR abs/1610.02391 (2016)
28. Tentolouris-Piperas, V., Ryan, N.S., Thomas, D.L., Kinnunen, K.M.: Brain imaging evidence of early involvement of subcortical regions in familial and sporadic Alzheimer's disease. Brain Res. **1655**, 23–32 (2017)
29. Vogt, L.K., Hyman, B., Van Hoesen, G., Damasio, A.: Pathological alterations in the amygdala in Alzheimer's disease. Neuroscience **37**(2), 377–385 (1990)
30. Wang, Y., Sun, Y., Liu, Z., Sarma, S.E., Bronstein, M.M., Solomon, J.M.: Dynamic graph CNN for learning on point clouds. ACM Trans. Graph. **38**(5), 1–12 (2019)
31. Welling, M., Kipf, T.N.: Semi-supervised classification with graph convolutional networks. In: ICLR (2016)
32. West, M., Coleman, P., Flood, D., Troncoso, J.: Differences in the pattern of hippocampal neuronal loss in normal ageing and Alzheimer's disease. The Lancet **344**(8925), 769–772 (1994), originally published as Volume 2, Issue 8925
33. Zhou, B., Khosla, A., Lapedriza, A., et al.: Learning deep features for discriminative localization. In: CVPR. pp. 2921–2929 (2016)

Author Index

H. Greenspan et al. (Eds.): MICCAI 2023, LNCS 14224, pp. 801–806, 2023.
https://doi.org/10.1007/978-3-031-43904-9

Printed in the United States
by Baker & Taylor Publisher Services

Printed in the United States
by Baker & Taylor Publisher Services